ORGANIC ELECTROCHEMISTRY

An Introduction and a Guide

ORGANIC ELECTROCHEMISTRY

An Introduction and a Guide

edited by MANUEL M. BAIZER
Corporate Research Department
Monsanto Company
St. Louis, Missouri

1973

MARCEL DEKKER, INC. New York

MARCEL DEKKER, INC.

95 Madison Avenue, New York, New York 10016

LIBRARY OF CONGRESS CATALOG CARD NUMBER: 78-179383

ISBN: 0-8247-1029-0

PRINTED IN THE UNITED STATES OF AMERICA

PREFACE

The organic chemist who wishes to employ a pressure catalytic hydrogenation in a synthetic sequence learns how to operate the necessary equipment and does not hesitate to carry out the experiment. While doing so, he is aware that formidable fundamental problems of mass transfer, adsorption, surface chemistry, kinetics, etc., are involved and that, while these are research areas in their own right, they are beyond his immediate objective. He derives his guidance probably from a book on organic chemistry. Unfortunately, however, this chemist's confrontation, if any, with electrochemistry has been first through thermodynamics or polarography or the intricacies of double-layer theory. In other words, he may well have been introduced to the discipline by way of problems which are the special areas of concern to the analytical and physical chemists for whom the notation $\mathrm{Ox} \overset{ne}{\rightleftharpoons} \mathrm{Red}$ may serve as well for ferric-ferrous as for nitrobenzene-nitrobenzene radical ion. He may well have obtained the highest academic degree without having performed a single synthetic or kinetic organic electrochemical experiment.

This book is directed primarily to the organic chemist. Its aim is to present to him the capabilities and limitations of organic electrochemistry; its principles, methods, problems, utility in synthesis and in elucidation of mechanisms of organic reactions, and fertility as a research area.

Fundamental broad problems relating to physical phenomena before and during electron transfer are presented only in essence since excellent treatments dealing extensively and in depth with the component segments are available. The major emphasis here is upon the elucidation of the chemical events which accompany electron-transfer steps. It is recognized that this leads to only a partial description and rationalization of the phenomena; in most cases no more is possible at this stage of the science.

Organic electrochemistry has expanded in many directions in the last twenty-five years. Advances in theory, in the design of equipment, and in the employment of new techniques have brought to the subject a degree of sophistication formerly lacking and have rendered obsolete the previous monographs (now out-of-print): C. J. Brockman's *Electro-organic Chemistry* (1926), S. Glasstone and A. Hickling's *Electrolytic Oxidation and Reduction* (1935), F. Fichter's *Organische Elektrochemie* (1942), and M. J.

Allen's Organic Electrode Processes (1958). Professor S. Swann, Jr.'s review (1956) is deliberately largely pragmatic.

The chapters chosen for inclusion in this book illustrate but do not exhaust the total scope of the subject. Specific anodic and cathodic reaction types are illustrated but are not usually, because of limitations of space, followed by comprehensive tables of data. The literature citations, on the other hand, are reasonably complete (at least to September 1970).

Each author represented here is actively engaged in research in one or another area of organic electrochemistry and has contributed to the common goal of replacing "art" by science. The original members of our group (Cauquis, Dietz, Eberson, and Lund) have seen all segments of the book while it was in various stages of preparation and have offered many valuable comments. I am particularly grateful to Dr. Eberson for his detailed constructive criticism.

It has become routine, at least for reviewers, to point out that in a multiauthored book there is overlap of material, nonuniformity of style, repetition, etc. This book was not hastily assembled, and there was adequate time to achieve uniformity if that had been desired. But the material was deliberately organized so that a given segment might appear in two or more contexts and, further, where unanimity of opinion does not exist, more than one current viewpoint might be expressed.

We all owe a tremendous debt to Miss Pat Merlenbach who, though only part time on this project, handled magnificently not only a superabundance of manuscripts but a flood of polyglot correspondence. The Monsanto Company, the Director of Central Research (Dr. B. L. Williams, Jr.), and the Section Director (Mr. H. W. Mohrman) were most kind in providing facilities, time, encouragement, and tolerance in the course of this undertaking.

Translations from French and German were made by the Editor.

St. Louis, Missouri
September 1972

Manuel M. Baizer

CONTRIBUTORS TO THIS VOLUME

MANUEL M. BAIZER, Corporate Research Department, Monsanto Company, St. Louis, Missouri

GEORGES CAUQUIS, Centre d'Études Nucléaires de Grenoble, Grenoble, France

D. DANLY, Monsanto Textiles Company, Pensacola, Florida

ROY DIETZ, Division of Molecular Science, National Physical Laboratory, Teddington, Middlesex, England

L. EBERSON, Division of Organic Chemistry, Chemical Center, University of Lund, Lund, Sweden

L. G. FEOKTISTOV, Institute of Electrochemistry, Academy of Sciences, Moscow, USSR

L. HORNER, Organisch-Chemisches Institut, University of Mainz, Mainz, Germany

P. IVERSEN, Department of Organic Chemistry, University of Aarhus, Aarhus, Denmark

H. LEHMKUHL, Max-Planck-Institut, Mülheim-Ruhr, Germany

HENNING LUND, Department of Organic Chemistry, University of Aarhus, Aarhus, Denmark

KLAS NYBERG, Division of Organic Chemistry, Chemical Center, University of Lund, Lund, Sweden

VERNON D. PARKER, Department of General and Organic Chemistry, University of Copenhagen, Copenhagen, Denmark

G. PARRAVANO, Department of Chemical and Metallurgical Engineering, University of Michigan, Ann Arbor, Michigan

M. R. RIFI, Union Carbide Corporation, Bound Brook, New Jersey

D. SERVE, Centre d'Études Nucléaires de Grenoble, Grenoble, France

PETR ZUMAN, Department of Chemistry, Clarkson College of Technology, Potsdam, New York

CONTENTS

SECTION B: PRACTICAL PROBLEMS

SECTION C: SYNTHETIC AND MECHANISTIC ASPECTS OF CATHODIC REACTIONS OF ORGANIC COMPOUNDS CLASSIFIED BY ELECTROPHORES

SECTION F: ELECTRODE REACTIONS CLASSIFIED BY REACTION TYPE

SECTION G: INDIRECT ELECTRO-OXIDATION AND REDUCTION

SECTION H: (HETEROGENEOUS) ORGANIC ELECTRO-CHEMISTRY VIS-À-VIS HOMOGENEOUS ORGANIC CHEMISTRY

SECTION I: PRESENT AND FUTURE APPLICATIONS OF ORGANIC ELECTROCHEMISTRY

GLOSSARY* AND ABBREVIATIONS

Cells: Containers holding the solution and electrodes and whatever auxiliary equipment is desirable in a particular case.

Divided: The region in which the cathode is located is separated by an ion-permeable membrane from the region of the anode.

H: A cell in the shape of a capital H, the vertical portions holding the electrodes and the horizontal portions the membrane(s).

Current

Concentration: Current/volume, e.g., A cm^{-3}.

Density: The current/unit area flowing into the effective surface of an electrode, i.e., the portion of the surface that is not obstructed and that faces the counterelectrode. Usually expressed in A cm^{-2}, A dm^{-2}, or A ft^{-2}.

Efficiency or current yield: The actual amount of product obtained in an electrolysis divided by the amount to be expected (by Faraday's law) from a given quantity of electricity. Usually expressed as a percentage.

Intensity: The unit is the ampere (A) = 1 C s^{-1}.

Quantity: The unit is the coulomb. One faraday (F) ≡ ~96,500 C ≡ 26.8 A hr is the quantity of electricity required for one equivalent of a substance to exchange one electron with an electrode.

EC, ECE, etc.: The sequence of discrete electrochemical and chemical steps which together comprise an electrochemical mechanism.

Electroactivity range or electrochemical domain: For a given system of solvent and supporting electrolyte the range of potential within which oxidations or reductions of the substrate can be effected without significantly oxidizing or reducing the constituents of the medium.

*The definitions are meant to be descriptive rather than rigorous.

Electrodes: The materials in contact with electrolyte through which current flows to and from a power source through the solution.

Anode: The electrode to which electrons are transferred from solution.

Cathode: The electrode at which electrons are supplied to the solution.

DME: The dropping mercury electrode used chiefly as a cathode in polarography; microdrops falling from the end of a capillary at the rate of about one every 3 seconds.

RDE: Rotating disk electrode.

Reference: An electrode, usually an electrochemical half-cell, of relatively stable and known potential. Commonly used are: SCE (saturated calomel electrode), NCE (normal calomel electrode), and NHE (normal hydrogen electrode, the conventional primary standard).

RRDE: Rotating ring-disk electrode.

Secondary, auxiliary, or counter: The electrode required in a cell in order to complete the circuit but at which the reactions occurring are of secondary interest.

Working or indicating: The electrode at which the reaction of interest is occurring.

Electrolysis: Passage of current through a solution.

Cpe: Controlled potential electrolysis. The potential vs. a reference at the working electrode is manually or automatically maintained at a chosen value.

Electrolytes: Solutions which conduct electricity.

Anolyte: The solution in the anode chamber of a divided cell.

Catholyte: The solution in the cathode chamber of a divided cell.

Counterion ("gegenion"): The anion or cation of a supporting electrolyte which balances the positive or negative charge, respectively, of a charged species produced during electrolysis.

Depolarizer: The substance which undergoes electron transfer with an electrode and thereby tends to decrease the polarization of the electrode. The term is becoming obsolete.

Supporting, inert: Ionic substances included in the electrolyte in order to promote flow of current but which do not themselves undergo electron-transfer reactions at the electrodes.

Hydrodimerization: Strictly, reductive dimerization.

ORGANIC ELECTROCHEMISTRY

An Introduction and a Guide

Chapter 0

INTRODUCTORY SURVEY

Manuel M. Baizer

Corporate Research Department
Monsanto Company
St. Louis, Missouri 63166

I. GENERAL CONSIDERATIONS

This book is concerned almost exclusively with the liquid-phase electrochemistry of organic substances capable of undergoing electron-exchange reactions with electrodes. This capability, constantly being extended, may be defined in terms of the presence of "electrophores" within the molecule. (The term "electrophore" seems to have occurred simultaneously to a number of people and is gaining widespread use; it is meant to designate a function which renders a molecule electroreducible or electrooxidizable.)

A discussion of the experimental problems is presented in Chapter IV. In simplest terms all the experimental procedures require in common: a container which may or may not have ion-permeable membranes to separate it into several chambers; a source of (usually) direct current; a means for voltage control; an ammeter; at least one positive (anode) and one negative (cathode) electrode; a liquid dielectric medium within which is usually dissolved a preferably inert ("supporting") electrolyte together with the usually poorly conducting substrate. Passage of current through an appropriate assemblage inaugurates a complex sequence of events.

Organic electrochemistry has relevance to analysis and synthesis of organic compounds, elucidation or confirmation of structure, investigations into the nature of catalytic activity, the study of transient intermediates, the generation of chemiluminescence, the nature of electron-transfer processes, structure-electroactivity relationships, initiation of polymerization, synthesis of catalysts and catalyst components, degradation, biological redox systems, and so on. It is germane to work on kinetics, reaction mechanisms, salt effects, solvation, effect of electric field upon chemical reactions, support of molecular orbital theory, and other areas of inquiry. It is, therefore, fortunate that in this as in other developing branches of science and technology [1] investigators of a variety of talents, backgrounds, and personal and professional motivations have chosen to dedicate their efforts to a furtherance of knowledge in a variety of the components of organic electrochemistry. This pluralism has made possible simultaneous advances on many fronts.[1] Furthermore, since this is an interdisciplinary subject, most participants have brought to it expertise in more than one field. The result is that the organic chemist who is active or becoming active in this area has at his disposal a spectrum of challenging problems ranging from the theoretical to the applied, one or more of which may well coincide with his own special interests.

A simple example will illustrate for <u>one particular system</u> the diverse aspects of it which are amenable to study by electrochemical means. The electrolytic reduction of nitrobenzene under specified conditions yields p-aminophenol, azoxybenzene, azobenzene, hydrazobenzene, benzidine, or aniline [2]. The detailed mechanism of the electron-transfer steps for this or for most other organic compounds is under intensive investigation. Polarography (p. 106) permits one to determine the cathode voltage and the number of electrons involved under given conditions of pH, etc. (Polarography of a series of substituted nitrobenzenes has been part of a larger project [3] concerning the relationship between structure and reducibility.) Cyclic voltammetry (p. 115) indicates that the first step of the reduction is the transfer of one electron to yield an ion radical and permits the estimation of its half-life. Electrolysis within the cavity of an electron spin resonance apparatus (p. 145) is an elegant means for preparing the intermediate and recording its spectrum. (The nitrobenzene anion radical may be used to initiate the polymerization [4] of acrylonitrile.) Controlled potential electrolysis (p. 166) of nitrobenzene makes possible the preparation of one or the other of the several products in high yield. Continuous (p. 183) as well as batch cells may be used [5]. The yield and the time for the completion of the reaction are profoundly influenced by rotating the cathode [6].

In order to orient the experimental organic chemist who is hesitantly first becoming acquainted with this "novel" area it is proposed to use an unorthodox approach in this Introductory Survey and in the first instant to illustrate the extensive scope of organic electrochemistry by the following experimental results chosen from the areas of synthesis, analytical and physical chemistry, and polymer chemistry. Review articles covering individual areas are listed in Appendix I.

II. EXPERIMENTAL RESULTS

A. Synthetic

1. Anodic Preparation of Dimethyl 1,14-Tetradecanedioate, a Decarboxylative Coupling [7]

The electrolysis cell was a multinecked round-bottom flask. One platinum anode (4 × 5 cm) spaced between two platinum cathodes (5.5 × 2.5 cm) was used. A solution in 500 ml of refluxing methanol of 282 g (1.5 moles) monomethyl suberate and ca. 0.1 mole of sodium methoxide was electrolyzed at about 1.6 A. The polarity of the electrodes was reversed periodically to remove coatings from the electrode. When the pH of the solution became greater than 7, the solution was made acid with acetic acid and heated to remove easily volatile material in vacuo. The higher-boiling material was dissolved in a mixture of ether and benzene, the solution filtered, washed with aqueous sodium bicarbonate, and dried. After removal of the solvents, the crude product was recrystallized from methanol, 225.7 g (84.7%), mp 43-45°.

2. Anodic Preparation of 3-Acetoxycyclohexene-1, an Allylic Substitution [8]

The cell contained a ceramic diaphragm (p. 195), a carbon anode, and cathode. The electrolysis solution contained 1.67 moles of acetic acid in which was dissolved 0.122 mole of cyclohexene and 0.03 mole of tetraethylammonium p-toluenesulfonate. Electrolysis for 30 h at 0.2 A yielded 54.8% of the product (volume percent analysis).

3. Cathodic Preparation of Diethyl Adipate from Ethyl Acrylate, a Reductive Coupling ("Hydrodimerization") [9]

A divided cell (p. 171) was used. The membrane was an Alundum cup. The cathode was mercury and the anode platinum. The catholyte was a solution of 76.0 g of ethyl acrylate containing a trace of polymerization inhibitor, 75.8 g of a 76.5% solution of methyltriethylammonium p-toluenesulfonate in 88.6 g of dimethylformamide (DMF). The anolyte contained 40 ml of a 38%

solution of the quaternary salt. The electrolysis was carried out at 3 A for 2 h at a temperature of 35-40°. As product formed, additional DMF was added to the catholyte to maintain homogeneity. The pH of the catholyte was maintained at 7-9 by occasional addition of drops of acetic acid. The catholyte was then separated from Hg diluted with water and extracted thoroughly with methylene chloride. The washed and dried extracts were distilled first to remove solvents and excess starting material and then to recover 23.6 g of product, bp 142-148° (30 mm). The yield was ca. 100% based on the amount of current that had been passed.

4. Cathodic Preparation of Metal Carbonyls [10]

In a special undivided cell constructed so as to permit electrolysis at 50-200 atm a solution of, e.g., 7% of chromium acetylacetonate in anhydrous pyridine containing 1-5% of tetrabutylammonium bromide was electrolyzed at 80° and 160 atm of carbon monoxide. The anode was aluminum and the cathode graphite. A 60% yield of chromium hexacarbonyl was obtained.

5. Cathodic Preparation of 1, 3-Dimethylbicyclobutane, a Reductive Cyclization [11]

An H cell (p. 175) with sintered glass divider was used. The catholyte contained a solution of 50 g 1, 3-dibromo-1, 3-dimethylcyclobutane in 230 ml of dimethylformamide-lithium bromide. The anolyte was 150 ml of the solvent-salt solution. Electrolysis at 0.5 A for 16 h yielded 9-15 g (55-94%) of the product, bp 54.5° (760 mm). The yield was poorer when a higher concentration of starting material was used.

B. Analytical

Since polarography is now said to be the most widely universally used analytical method, it is hopeless to choose one out of thousands of examples of the application of this method. The reader is referred to Zuman's book [12] and to the biennial reviews in Analytical Chemistry. Other electroanalytical techniques of great value to organic chemists are discussed in Chapter II.

C. Physical

1. Generation and Observation of the Nitrobenzene Anion Radical [13]

A special electrolysis cell was provided which permitted electrolysis within the cavity of an electron spin resonance (ESR) spectrometer (p. 145). The cathode was mercury, the solvent acetonitrile, the supporting electrolyte tetra-n-propylammonium perchlorate. The electrolysis was conducted at the first reduction wave for nitrobenzene (-1.3 V vs. the saturated calomel electrode). The ESR spectrum simultaneously obtained showed hyperfine structure with 54 components.

2. Generation and Observation of the 1, 1, 3, 3-Tetraphenylpropene Radical [14]

A two-cell flow system was used [14]. The first cell had a silver gauze anode separated by an alumina thimble from a platinum gauze cathode. Tetraphenylallene in dimethylformamide containing tetra-n-butylammonium iodide was reduced at ca. -2.1 V (vs. SCE) to its anion radical. The latter was rapidly protonated and further reduced to the monoanion. The solution containing the latter species flowed into the second cell, electrically insulated from the first, where oxidation to the radical occurred. The effluent from the second cell was passed through the cavity of an ESR spectrometer and the spectrum was obtained.

3. Evidence for ECE (Electrochemical-Chemical-Electrochemical) rather than EEC Mechanism in the Anodic Oxidation of Aromatic Hydrocarbons [15]

Platinum rotated disk electrodes (p. 96) were used. As a model, 9, 10-diphenylanthracene was oxidized in the presence of pyridine and derivatives, nucleophiles of differing electronegativities. The overall process is a two-electron transfer. The analysis of the data obtained relating the number of electrons transferred to rotation speed and presence/absence of nucleophile indicated that reaction of the first-formed cation radical with the nucleophile intervened between the two electron-transfer steps. The data, presented graphically, matched very well the curves obtained by digital computer simulation of an ECE reaction using given values for the chemical rate constants.

D. Polymerization

"Living" anionic polymers [16] were cathodically generated in an H cell (p. 175). Materials were rigorously purified and handled by vacuum-line technique. The solvent was tetrahydrofuran, the electrolytes alkali metal tetraphenylborons or alkali metal tetraalkylaluminums, and the monomers styrene, α-methylstyrene, or other vinyl compounds with electron-withdrawing groups. The concentration of "living ends," measured spectrophotometrically, was directly proportional to the amount of current passed. Further, the polarity of the cell could be reversed and a calculated fraction of the "living ends" destroyed. The surviving dianions could initiate new polymerization with the same or another monomer so that block copolymers of given composition could be made. In addition, by using an appropriate feedback device, it was possible to maintain a predetermined concentration of "living ends" in the polymerization of monomers (e.g., methyl methacrylate) which tend to terminate by intramolecular reaction.

The subject of electroinitiated polymerization is treated extensively in Chapter XXIX.

III. CONCLUDING REMARKS

In the recent past a recurrent pattern has begun to appear in the presentation of a "compleat" paper describing a research project in organic electrochemistry [17, 18]. The compound(s) is studied by polarography in anhydrous media and in the presence of proton donors. Tentative conclusions are drawn about the nature of presumed intermediates. Cyclic voltammetry is then applied in order to determine the degree of reversibility — if any — of the electrode reaction and the kinetics of the electrode process. Conclusions are drawn concerning the (usual) ECE steps. If ion radical intermediates are involved, and they usually are, an attempt is made to obtain a resolved electron spin resonance spectrogram. Sites of highest and lowest electron density are then established as the most probable sites for follow-up nucleophilic and electrophilic reactions, respectively. Finally, macroscale electrolyses are carried out, the products isolated and identified, and their structure and genesis rationalized on the basis of the analytical electrochemical data that had been obtained.

The scope of present-day organic electrochemistry may be further illustrated by examining the question: What are some of the active areas of research which presage exciting new developments? The range of useful solvents and supporting electrolytes [19] is steadily being extended so that organic compounds formerly considered inert can be made to undergo electron transfer at an electrode. Osa and co-workers [20], using acetonitrile and tetra-n-butylammonium hexafluorophosphate, were able to oxidize benzene. The first intermediate was a cation radical; after loss of a proton, phenyl radicals were formed which produced polyphenyls. Fleischmann and Pletcher [21] had earlier reported that, in a system containing acetonitrile and tetrafluoroborate or hexafluorophosphate, aliphatic hydrocarbons having allylic or tertiary C-H bonds were electrooxidized. It was suggested that a two-electron oxidation was involved and that the carbonium ion formed probably attacked solvent in a Ritter-type reaction [22].

The electrochemical generation of solvated electrons, particularly in hexamethylphosphoramide [23] has made possible the indirect electroreduction of benzene and even of an unconjugated double bond (Chapter XXV). While numerous indirect metal ion-assisted electrooxidations have been reported, the catalytic electrooxidation of cyanide ion by tri-p-anisylamine cation radical is novel [24].

We are currently witnessing spectacular progress in the elucidation of organic electrode processes [15, 18, 25, 26]: the unscrambling of the E and C sequences and the determination of the kinetics of each step. These advances have been facilitated by the application of an ensemble of sophisticated experimental techniques (Chapter II) to a given problem; they are fortified by computer simulation techniques [27, 28]. The organic chemist's specialized experience with stereochemistry, "equivalent" nonelectrochemical reactions, isotope effects, etc., has been brought to bear on the C reactions, particularly as it concerns the question of whether the reactions are occurring at the electrode surface or in the "bulk" of the solution [29-32].

Techniques are being developed which permit the product of the first electron exchange to be removed from the cell rapidly and used in subsequent chemical/electrochemical steps. For example, Sioda [33] has described a flow system in which 9, 10-diphenylanthracene is electrooxidized to its cation at a platinum grid electrode; the relatively stable oxidized species, now outside the cell, is available for further operations. Exhaustive electrolyses of substrates in thin layer cells [34-36], employed so far only for analytical and mechanistic studies, might make available for synthetic use reactive intermediates that are not, as is often the case, substantially consumed by reaction with starting material. Rapid current reversal [25] allowed Dietz and Peover [26] to reduce polycyclic aromatic hydrocarbons to dianions (protonate these intermediates to monoanions), and then electrooxidize to radicals. A two-cell system for first oxidizing and then reducing [14] has been mentioned above.

It was long ago [37] suggested that two coupled (organic) syntheses might be made to occur simultaneously in one cell at the anode and cathode, respectively. Newer efforts in this direction involve the use of bipolar electrodes [38, 39]. Many electrode matrices incorporating catalysts have been developed during the worldwide program on fuel cells [40]; applications in synthesis are awaited.

The fluidized electrode systems being developed at the University of Newcastle and at Constructors John Brown in England [5] may profoundly affect electrochemical technology.

Electroorganic synthesis is treated extensively throughout this book. The number of research papers in this area is increasing rapidly. Especially active are the categories of reductive coupling (Chapter XIX), anodic substitution (Chapter XXII), oxidative coupling (Chapter XX), the formation of small carbocyclic rings (Chapter VI), and the synthesis of heterocycles (Chapter XVII). Syntheses by means of active species generated from simple inexpensive raw materials are being explored, e.g., the reported formation of malic acid from carbon dioxide [41] and the use of electrogenerated superoxide ion as a nucleophile [42]. Isolated applications of organic electrochemistry are sometimes arresting: anodic oxidation of bicyclo [2.2.0] hexa-5-ene-2, 3-dicarboxylic acid was a more elegant final step in the preparation of "Dewar benzene" [43, 44] than had been lead tetraacetate oxidation. Perhaps Fichter would have called this also a "Spielerei."

Electropolymerization is still in its infancy (Chapter XXIX). Aside from demonstrations that it can be done, that it can be controlled, that it can lead to "tailored" block copolymers, the further unique potentialities of this method remain to be explored.

The work of Dessy and his students [45] and of Vlček and Masek and co-workers [46] has pointed the way for new syntheses using intermediates prepared by electrolysis of organometallic and coordination compounds, respectively (Chapter XVIII).

Finally, the entire area of ion radical chemistry [47] is being aided in its rapid development by the employment of electrochemical techniques.

The view has been expressed [48] that most organic reactions are in their initial stages electron-transfer reactions. Perhaps organic electrochemistry is, in addition to what is encompassed in the following pages, simply a philosophy of organic chemistry.

NOTE

[1]As Professor Cauquis has pointed out, the organic chemist first dipping into the electrochemical literature may be struck by the disadvantages rather than the advantages of this situation. He has been accustomed to a relatively homogeneous body of doctrine; he now finds papers by coteries of polarographers, electrokineticists, electroanalytical chemists, and other highly specialized scientists. Each group has developed its own doctrines and terminology and at first glance seems intent on communicating only to other members of the same group.

REFERENCES

[1]. J. O'M. Bockris, J. Electroanal. Chem., 9, 408 (1965).

[2]. S. Swann, Jr., "Electrolytic Reactions," in Technique of Organic Chemistry (A. Weissberger, ed.) 2nd ed., Vol. II, Wiley-Interscience, New York, 1956, pp. 385-523.

[3]. P. Zuman, Substituent Effects in Organic Polarography, Plenum, New York, 1967.

[4]. G. Mengoli, G. Farina, and E. Vianello, Eur. Polym. J., 5, 61 (1969).

[5]. F. Goodridge, Chem. Process Eng., 49, 93 (1968).

[6]. K. S. Udupa, G. S. Subramanian, and H. V. K. Udupa, J. Electrochem. Soc., 108, 373 (1961).

[7]. J. D. Anderson, M. M. Baizer, and J. P. Petrovich, J. Org. Chem., 31, 3890 (1966).

[8]. T. Shono and T. Kosaka, Tetrahedron Lett., 1968, 6207.

[9]. M. M. Baizer, ibid., 1963, 973.

[10]. R. Ercoli, M. Guainazzi, and G. Silvestri, Chem. Commun., 1967, 927.

[11]. M. R. Rifi, J. Amer. Chem. Soc., 89, 4442 (1967).

[12]. P. Zuman, Organic Polarographic Analysis, Macmillan, New York, 1964.

[13]. D. H. Geske and A. H. Maki, J. Amer. Chem. Soc., 82, 2671 (1960).

[14]. A. E. J. Forno, Chem. Ind., 1968, 1728.

[15]. G. Manning, V. D. Parker, and R. N. Adams, J. Amer. Chem. Soc., 91, 4584 (1969).

[16]. B. L. Funt, U.S. Pat. 3,448,020, June 3, 1969.

[17]. J. P. Petrovich, M. M. Baizer, and M. R. Ort, J. Electrochem. Soc., 116, 743, 749 (1969).

[18]. J. L. Sadler and A. J. Bard, J. Amer. Chem. Soc., 90, 1979 (1968).

[19]. C. K. Mann, in Electroanalytical Chemistry (A. J. Bard, ed.), Vol. III, Dekker, New York, 1969.

[20]. T. Osa, A. Yildiz, and T. Kuwana, J. Amer. Chem. Soc., 91, 3994 (1969).

[21]. M. Fleischmann and D. Pletcher, Tetrahedron Lett., 1968, 6255.

[22]. L. Eberson and K. Nyberg, Acta Chem. Scand., 18, 1567 (1964).

[23]. H. W. Sternberg, R. E. Markby, I. Wender, and D. M. Mohilner, J. Amer. Chem. Soc., 91, 4191 (1969).

[24]. L. Papouchado, R. N. Adams, and S. W. Feldberg, J. Electroanal. Chem. Interfacial Electrochem., 21, 408 (1969).

[25]. L. Eberson and S. Nilsson, Discuss. Faraday Soc., 45, 242 (1968).

[26]. R. Dietz and M. E. Peover, Trans. Faraday Soc., 62, 3535 (1966).

[27]. S. W. Feldberg, in Electroanalytical Chemistry (A. J. Bard, ed.), Vol. III, Dekker, New York, 1969.

[28]. F. H. Covitz, in Symposium on Synthetic and Mechanistic Aspects of Electroorganic Chemistry, U.S. Army Research Office, Durham, N.C., 14-16 Oct. 1968, p. 99.

[29]. A. J. Fry and R. H. Moore, J. Org. Chem., 33, 1283 (1968).

[30]. R. Annino, R. E. Erickson, J. Michalovic, and B. McKay, J. Amer. Chem. Soc., 88, 4424 (1966).

[31]. L. Eberson, Electrochim. Acta, 12, 1473 (1967).

[32]. A. K. Vijh and B. E. Conway, Chem. Rev., 67, 623 (1967).

[33]. R. E. Sioda, Electrochim. Acta, 13, 375 (1968).

[34]. C. R. Christiansen and F. C. Anson, Anal. Chem., 35, 205 (1963); A. T. Hubbard and F. C. Anson, in Electroanalytical Chemistry (A. Bard, ed.), Vol. IV, Dekker, New York, 1970, p. 129 ff.

[35]. L. B. Anderson and C. N. Reilley, J. Electroanal. Chem., 10, 538 (1965).

[36]. J. E. McClure and D. L. Maricle, Anal. Chem., 39, 236 (1967).

[37]. C. J. Thatcher, Trans. Amer. Electrochem. Soc., 51, 175 (1927); U.S. Pat. 1,544,357 (1923).

[38]. W. Tomassi, H. Jankowska, S. Pietrzyk, and J. Brylka, Electrochim. Acta, 10, 605 (1965).

[39]. R. G. M. Savidze, Brit. Pat. 1,136,869, Dec. 18, 1968.

[40]. R. J. Jasinski, U.S. Pat. 3,410,727, Nov. 12, 1968.

[41]. A. Bewick and G. P. Greener, Symposium, Society for Electrochemistry, Thornton Research Centre, Chester, England, 15-16 April 1969, p. 5.

[42]. R. Dietz and M. E. Peover, ibid., p. 8; R. Dietz, A. E. J. Forno, B. E. Larcombe, and M. E. Peover, J. Chem. Soc. (B), 1970, 816.

[43]. P. Radlick, R. Klem, S. Spurlock, J. J. Sims, E. E. van Tamelen, and T. Whiteside, Tetrahedron Lett., 1968, 5117.

[44]. H. H. Westberg and H. Dauben, ibid., 1968, 5123.

[45]. For example, see R. E. Dessy and R. L. Pohl, J. Amer. Chem. Soc., 90, 2005 (1968).

[46]. A. A. Vlček, Pure Appl. Chem., 10, 61 (1965); J. Masek, Talanta, 12, 1173 (1965).

[47]. E. T. Kaiser and L. Kevan (eds.), Radical Ions, Wiley-Interscience, New York, 1968.

[48]. K. A. Bilevich and O. Yu. Okhlobystin, Russ. Chem. Rev., 1968, 954.

Section A

PRINCIPLES AND METHODS

Chapter I

BASIC CONCEPTS

Georges Cauquis

Centre d'Études Nucléaires de Grenoble
Grenoble, France

I. INTRODUCTION

In accordance with the previously mentioned purposes of this book, it will be the objective of this chapter to familiarize the organic chemist with the fundamental electrochemical concepts he will require in order to understand the several aspects of organic electrochemistry. The exposition will be relatively elementary, largely qualitative,[1] and will assume no prior knowledge of the field by the reader. The illustrative examples have been chosen more for their instructional than their intrinsic scientific value. The synthetic chemist will find abundant important reactions throughout the main body of this book.

Electrochemistry encompasses a vast realm of phenomena. Even the simple experiment described below illustrates this complexity and, in addition, includes several elements which are the particular concern of this chapter.

Consider the assembly schematically presented in Fig. 1. A battery provides a direct current across two conductors and an aqueous solution in

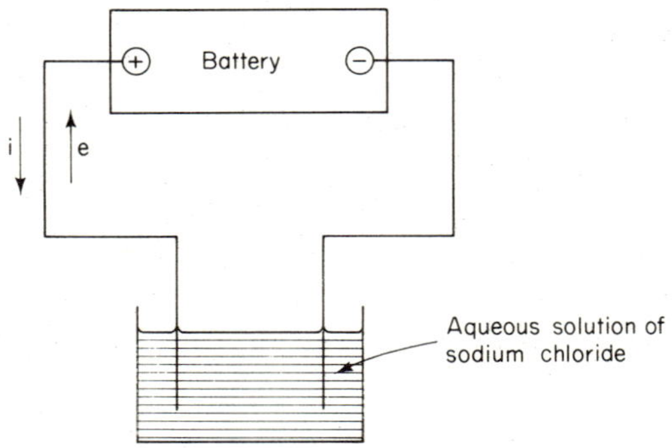

Fig. 1. Simple experiment illustrating various aspects of electrochemistry.

which sodium chloride is dissociated. The means whereby the current flows is different in different parts of the circuit:

(a) In the metallic conductors it occurs by a flux of electrons. Because of present conventions, the electrons are considered to move in a direction opposite to that of the current.

(b) In the solution it occurs by migration of the Na^+ and Cl^- ions which move in directions opposite to their respective signs and in accordance with the direction of the electric field which exists between the two metal conductors (electrodes).

(c) At the metal-solution interface it occurs by an exchange of electrons between the metal and certain species in solution. This exchange constitutes the electrochemical reaction.

Electrochemistry is limited to the study of the latter two types of phenomena. In spite of this limitation it is immediately obvious that this science must comprise two groups of phenomena: those associated with the movement of charged (or uncharged) particles within the solution (sometimes the study of this movement is called ionics), and those associated with the exchange of electrons at electrodes (electrodics).

We shall be concerned largely with the latter aspect of electrochemistry, charge transfer, directing our attention to the movements within the solution only when it is necessary to understand how the species capable of undergoing electron transfer arrive at or leave the electrode surface by the processes called in general mass transfer.

The following comments may be made concerning the simple experiment outlined above:

(a) It may appear from the above description of electrochemical phenomena that only charged particles can reach the electrode by migration and that consequently only they can, strictly speaking, participate in an electrochemical reaction. This is not so. A neutral molecule and even an ion having the same charge as the electrode — and, therefore, supposedly repelled from it — can undergo electrochemical reactions. In these cases mass transfer occurs not by migration but by diffusion and convection. Diffusion is considered to occur when disappearance of a substance at an electrode due to electron transfer sets up a concentration gradient for that substance between the electrode surface and the solution. Convection is the means of mass transfer in which the solution is by one of several means set into motion with respect to the electrode, e.g., by stirring, by rotating or vibrating the electrode, or by periodic changes in the volume of the electrode, as in the case of mercury dropping from the bottom of a capillary. In many instances the rate of mass transfer is the factor limiting the rate at which a given species can participate in an electrochemical reaction.

This intimate relationship between mass transfer and charge transfer is so fundamental to the understanding of electrochemical phenomena that it will be discussed in greater detail later (p. 30).

It should be noted that studies of these phenomena are greatly simplified if several modes of mass transfer are not simultaneously involved. Generally, at least in the laboratory, it is simplest to suppress migration. This is accomplished by adding to the solution a high concentration (e.g., 10^{-1} M) of a salt which is highly dissociated in the medium used. The substance whose electrochemical behavior is being studied is present in much lower concentration (e.g., 10^{-3} M) and under these conditions undergoes mass transfer only by diffusion or convection. The highly dissociated electrolyte present in relatively high concentrations is called the supporting or background or inert electrolyte; the latter terminology implies that it is not being made the subject of an electrochemical study.[2]

(b) The fundamental electrochemical event, i.e., the charge transfer, occurs at the electrode surface. By definition, then, electrochemistry is a heterogeneous chemistry. There is nothing in this fact to disturb the organic chemist who, for example, frequently uses hydrogenation with a heterogeneous catalyst in the course of a synthesis. But the new factors he may now be discovering in electrochemistry are that there is in the vicinity of the electrode an electric field which in a special way complicates the phenomena of adsorption and desorption at the surface and that the first layer of solution which is in contact with the electrode possesses a special structure. The study of these phenomena is so critical from the experimental and interpretative points of view that entire schools of electrochemists have been and are devoted to this field alone.

(c) Whatever may be the means by which the species to be involved in the electrochemical reaction arrive at the electrode and whatever adsorption, desorption, distortion, etc., these species may undergo at the electrode surface the fundamental phenomenon occurring at the latter is exchange of electrons. Electrochemistry is, therefore, a means before all else of carrying out redox reactions; consequently, it would be of limited interest to the organic chemist if it were not for the more or less complex reactions which precede, follow, or intervene between charge transfers. These coupled chemical reactions are responsible for the great variety of organic electrochemical reactions.

In turn, these coupled chemical reactions introduce greater complexity into organic than into inorganic electrochemistry because, of course, organic compounds are capable of participating in a greater variety of chemical reactions. For the organic chemist the problem of elucidating an organic electrochemical reaction is very frequently in the first instant the question of determining the chemical reactions accompanying the charge transfers.

There are several ways in which the principles of electrochemistry may be presented. A reasonable approach is first to discuss the special structure which a solution in contact with an electrode acquires and then to derive from this the laws of charge transfer which occur through this structure. The influence of mass transfer must likewise be taken into account. A discussion of the effect of coupled chemical reactions may be deferred until later.

The mode of exposition to be presented here differs from the above in that it will describe and explain electrochemical phenomena on the basis of current intensity-potential curves; these constitute themselves a method of studying the overall effects of the phenomena mentioned above. This "experimental" approach will, we believe, facilitate an understanding of the electrochemical laws. But, it must be remembered that in adopting this approach we appear to place special emphasis on one particular experimental method. Many other methods (Chapter II), some of which are often more capable of furnishing information on the nature of electrochemical reactions, have been devised.[3]

II. FUNDAMENTAL ELECTRODE PHENOMENA

A. Chemical and Electrochemical Oxidation and Reduction of Organic Compounds

The concepts of oxidation and reduction are much more precise in inorganic than in organic chemistry. It is for example quite plain that the oxidation $Fe^{2+} \rightarrow Fe^{3+}$ involves the loss of one electron and that the reduction $Zn^{2+} \rightarrow Zn^{0}$, the gain of two electrons. The situation is seldom so clear in the case of organic compounds in which exchange of electrons often involves making and breaking of covalent bonds.

Nevertheless, certain organic compounds undergo redox reactions which are clearly electron transfers. A now classic example is found in the reduction by an alkali metal in an anhydrous ether of an aromatic hydrocarbon to its anion radical [2]. In the same way variously substituted phenothiazines (I) may be successively oxidized to radical cations (II) and to dications (III); if the medium is properly chosen, reactions involving bond scission [→ (IV)] or bond formation [→ (V)] can be avoided.

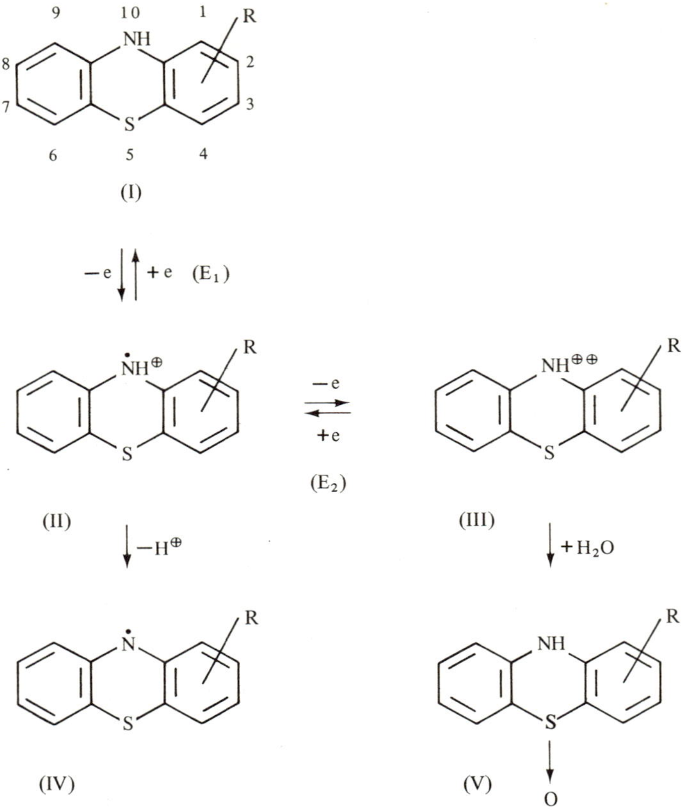

In this section we shall be concerned only with the above types of compounds which are able to undergo electron exchange without chemical reaction. It must be remarked that all compounds are apparently in this category because there are many reasons for believing that exchange of electrons occurs one electron at a time without changing molecular structure. In the case of the types of compounds discussed here only the solvation shell is modified by the electron transfer. However, in the majority of cases the formation or rupture of covalent bonds occurs too rapidly after (or before) electron transfer to permit detection by present experimental means of the intermediate formed by the charge-transfer step only.

Two stable compounds which are interconvertible by gain or loss of one electron form a <u>simple system</u>. For such systems it is well known [4] that a standard redox potential can be defined. This potential is referred to the potential of a particular system. For example, in inorganic chemistry, oxidation-reduction potentials in aqueous medium are referred to the H_2/H^+

couple which is not a simple system but which has a very well-defined potential under convenient conditions [5]. Similarly, in all organic or aqueous-organic media the standard potential of a simple system as defined above may be referred to a judiciously chosen reference (Fig. 2).

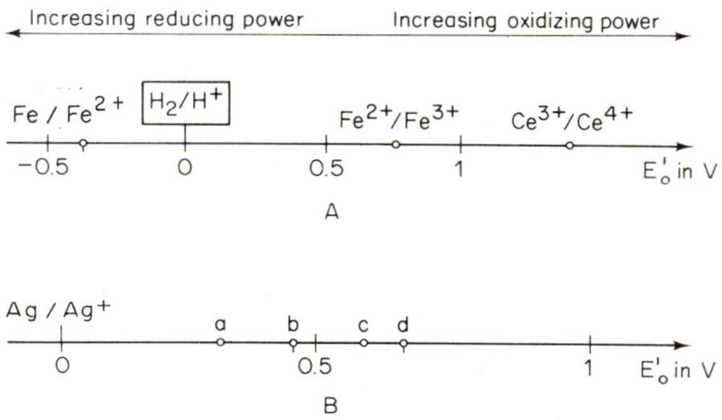

Fig. 2. Two examples of potential scales: A is the potential scale measured in aqueous solutions vs. H_2/H^+; B is a relative scale in CH_3CN. In this solvent the reference is the simple system Ag/Ag^+ 10^{-2} M; the potentials of the couples of type (I)/(II) are shown for several phenothiazine derivatives in which R is: (a) 3-C_2H_5O; (b) H; (c) 3-CHO; (d) 3-NO_2.

The ensemble of numbers representing potentials derived in this manner constitutes a potential scale. In this type of scale the power of the oxidant or the reductant in the couple Red/Ox depends on the position within the scale of the standard potential of this couple. Thus, the oxidant Ce^{4+} in the couple Ce^{3+}/Ce^{4+} is more powerful than Fe^{3+} in the couple Fe^{2+}/Fe^{3+}. It follows that Ce^{4+} is capable of oxidizing Fe^{2+} to Fe^{3+} (Fig. 2). In the same way the radical cation of the phenothiazine (I) d (R = 3-nitro) is able to oxidize (I) a (R = 3-OC_2H_5) to its cation radical.

The above discussion considers electron transfer between two chemical species. But the same redox phenomena obtain if the electron transfer occurs between a compound in solution and a metallic conductor (electrode) placed in this solution. If the potential acquired by this electrode with respect to the solution is referred to the reference which was used to construct a scale of potentials similar to that of Fig. 2, the scale again permits certain predictions to be made. Thus, it is to be expected that the phenothiazine (I) d (R = 3-nitro) will be oxidized in acetonitrile at an electrode only if its potential is higher than 0.65 V (vs. Ag/Ag^+).

Until now it has been implicitly assumed that electron transfers either between two compounds or between a compound and an electrode are very rapid. This assumption is taken for granted when it is stated that in order to oxidize the "Red" of a Red/Ox couple of standard potential E_0^1 it is sufficient to bring it into contact with the "Ox" of the Red/Ox couple E_0^2, if $E_0^2 > E_0^1$, or with an electrode (placed within the solution) having a potential E ($> E_0^1$). Actually, this is not the true situation and we must add to these thermodynamic considerations the kinetic aspect, taking into account both the rates of electron transfer and of mass transfer which likewise may be limiting.

These considerations are the basis for current intensity-potential (or voltamperometric or polarization) curves. This concept will first be presented qualitatively; a simple mathematical study will then provide the equations of the curves and at the same time a better understanding of their physical significance.

B. The Concept of Current Intensity-Potential Curves in Simple Systems. Qualitative Aspects

1. Three-Electrode Assembly

To begin the discussion it is necessary to modify the electrical diagram that was given in Fig. 2. We shall be referring to the potential of the electrode at which an electrochemical reaction is occurring. By this we do not mean the difference in potential between the two electrodes of Fig. 1 but that which exists between the electrode in question and a third electrode, the reference electrode, whose potential relative to the solution is independent of the composition of the latter and of the intensity of the current flowing between the two other electrodes.

Therefore, an assembly used for determining a current intensity-voltage[4] curve comprises three electrodes (Fig. 3). The electrode A at which the electrochemical reaction being studied occurs is called the

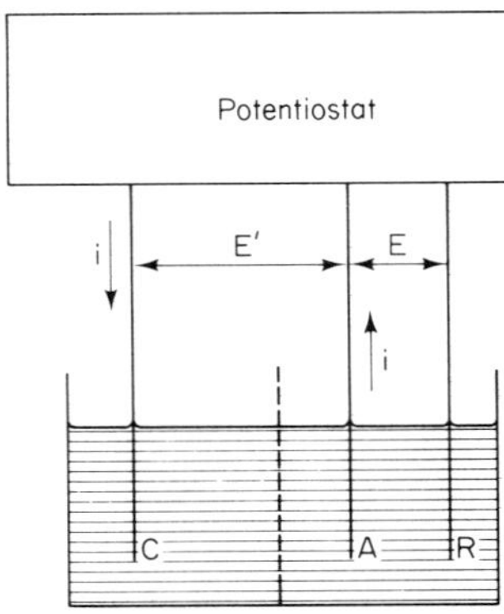

Fig. 3. Three-electrode assembly for obtaining a current intensity-potential curve for the electrochemical phenomena occurring at electrode A.

indicating or working electrode. The electrode C which completes the circuit is the counter or auxiliary electrode. A device called a potentiostat (p. 232) maintains a potential difference E between the reference electrode R and A and supplies the current i needed for affecting the changes occurring at A. The i-E curve is obtained by recording i as E is slowly varied linearly with time. The phenomena at C are usually of little concern and usually it is sufficient merely to separate A from C by a porous membrane or a salt bridge in order to avoid the possibility that the products formed at C will interfere with the phenomena being studied at A (two-compartment cell). The reference electrode R sustains very little or no current; therefore no significant electrochemical changes occur in it. It will be noted that the potential E' existing between the working electrode A and the auxiliary electrode C is of little interest and is not associated in any simple way with the occurrences at A. As will be seen, both oxidations and reductions may occur successively at one and the same electrode in the course of obtaining an i-E curve. The terms anode and cathode are therefore only relative.

2. Current-Potential Curve for One Reductant in Solution

Keeping in mind Fig. 3, consider the i-E curve of a simple system comprising phenothiazine [(I), R = H] and its radical cation [(II), R = H] or, at the moment, of a solution of (I) in a solvent like acetonitrile (Fig. 4) [3].[5] The working electrode is a 1 mm disk of polished platinum formed by the section of a wire sealed in thick-wall glass tubing. As the potential E of the indicating electrode increases (numerically) the current, which at

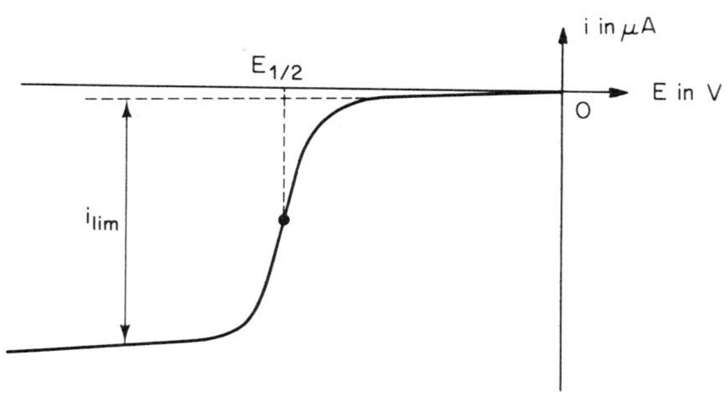

Fig. 4. Current intensity-potential curve for a 10^{-3} M solution of phenothiazine in acetonitrile. The supporting electrolyte is 10^{-1} M $LiClO_4$. The reference electrode is the Ag/Ag^+ 10^{-2} M couple and the working electrode is a 1 mm rotating disk of polished platinum. The counter electrode is a platinum wire immersed in a compartment which is separated from the electrolysis cell by a glass frit. Only the portion of the curve concerning the transformation (I) $- \bar{e} \rightarrow$ (II) is shown [3].

first is practically zero, increases rapidly after a certain value of the potential is reached. Then after reaching an appropriate value the current attains a limiting value.

This type of curve is characteristic of the reaction:

(I) - $\bar{e}$ → (II),

or, more exactly, of the overall kinetics of this reaction since the intensity of the current is proportional to the rate at which the electron exchange occurs at the metal-solution interface. The overall kinetics is the resultant of the kinetics of mass transfer and of charge transfer, two distinct factors which as already mentioned are always involved together in all electrochemical phenomena:

(a) The exchange of electrons at the surface of the electrode occurs at a rate which increases with potential (see below). The current i therefore increases as E increases.

(b) The transport of the phenothiazine molecules in solution to the surface of the electrode is itself governed by two factors: first, convection due to the rotation of the electrode is responsible for bringing the substance toward the electrode; and second, the concentration gradient existing in a thin (ca. 0.1 to 10^{-3} cm) layer around the electrode, the diffusion (or Nernst) layer, ensures the movement of the molecules to the electrode surface.

Appreciation of the fact that there are two kinetic factors involved helps to clarify the form of the i-E curve. In the first stage the increasing potential is accompanied by increasing rate of exchange of electrons and therefore by an increase in current (foot of the curve). But the local depletion of (I) due to the electrode reaction eventually causes the kinetics of the mass transfer of (I) to be the limiting factor. The current intensity becomes independent of potential (upper horizontal portion of the curve). In the case of the simple examples discussed here it has been shown that diffusion limits the value of maximum current, consequently called the diffusion current i_d. Under certain experimental conditions which will be described shortly i_d is proportional to the concentration of the substance in the solution and at the electrode surface.

Another important characteristic of the i-E curve for a simple system is the half-wave potential $E_{1/2}$, i.e., the potential at which the current is $\frac{1}{2}i_d$. It is approximately equal to the potential E_o previously defined.

The experimental conditions under which the above statements hold are the following:

(a) The phenomena of migrations have been suppressed by dissolving in the medium a high concentration of an indifferent electrolyte, the ions of which are not oxidized or reduced within the potential range under investigation.

(b) Stable conditions of convection in which the movement of solution relative to the electrode is well defined.

(c) Utilization of an indicating electrode having a very small area so that the quantity of material oxidized in the course of obtaining the i-E curve is negligible compared to the quantity in solution.

(d) The scanning of the potential is slow enough.

3. Current-Potential Curve for an Oxidant Alone and for a Mixture of Reductant and Oxidant

Consider first a solution containing only the cation radical (II) of phenothiazine. An i-E curve (Fig. 5, curve 2) is obtained which has the same

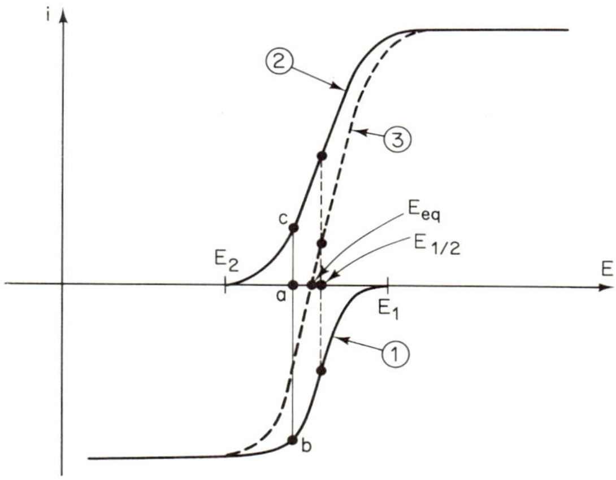

Fig. 5. Current intensity-potential curves for a solution of phenothiazine (curve 1), its cation radical (curve 2), and a mixture of the two (curve 3). The experimental conditions are identical to those of Fig. 4.

characteristics as the curve of Fig. 4 but which now corresponds to a cathodic current. The same half-wave potential is found and this time the limiting current is proportional to the concentration of the substance being reduced at the electrode following the reaction

(II) + $\bar{e}$ → (I).

If the solution contains both the oxidant and the reductant of a simple system, the phenomena are additive. The curve obtained is the summation of the curves for the individual compounds (Fig. 5, curve 3). If we call E_1

the potential above which the oxidation of (I) can be observed and E_2 the potential under which the reduction of (II) can be observed, three cases must be considered following the potential of the electrode:

(a) If the potential of the electrode is less than E_1, only reduction of the cation (II) will occur. Conversely, above E_2 only the oxidation of (I) will occur.

(b) If the potential of the electrode has a value between E_1 and E_2, there will be simultaneous oxidation and reduction whose respective rates are measured by $i_1 = ab$ and $i_2 = ac$. The current actually flowing across the electrode is the algebraic sum of i_1 and i_2.

(c) In particular, if the potential of the electrode is such that $i_1 = -i_2$, no overall change seems to be produced at the macroscopic scale in spite of the fact that both phenomena are occurring. But they are occurring with exactly the same rate measured by the two equal currents. The potential E_{eq} which is thus defined is the equilibrium potential [6]; it is given by the Nernst equation

$$E_{eq} = E_0 + \frac{RT}{F} \ln \frac{[Ox]}{[Red]}$$

in which the quantities in brackets are the concentrations of the two species in the solution.

4. Rapid (or Reversible) and Slow (or Irreversible) Systems

A simple system which provides a group of curves similar to those of Fig. 5 is called a reversible system. This term connotes that an oxidation or a reduction at a finite rate can be achieved by bringing the electrode potential slightly below or slightly above the equilibrium potential E_{eq}.

Certain redox systems do not yield i-E curves of the general appearance of Fig. 5 (again shown in Fig. 6, A). The cathodic branch and the anodic

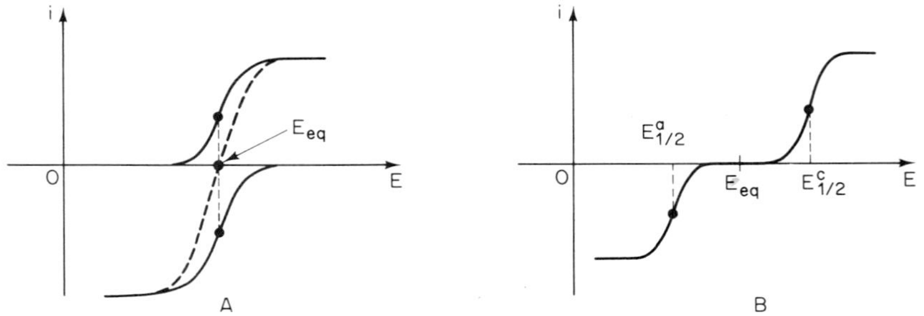

Fig. 6. Distinction between a rapid (A) and a slow (B) system.

branch (Fig. 6, B) are in such different ranges that it is impossible to realize either an oxidation or a reduction at a finite rate in the neighborhood of E_{eq}. Such a system is said to be irreversible. The E_{eq} is, as before, precisely defined but it cannot be determined by recording the i-E curve of a mixture of the forms "Ox" and "Red."

The expressions reversible and irreversible are not very satisfactory. Actually, in a system which gives curves similar to that of Fig. 6, B, it is still possible to transform Ox into Red and vice versa. It is simply necessary to increase or decrease the electrode potential sufficiently: the system is therefore not "irreversible" in the usual chemical sense. Furthermore, if the conversion of Ox to Red (and vice versa) cannot be brought about near E_{eq}, it is for kinetic reasons. The currents i_1 and i_2 which were discussed in connection with Fig. 5 are here simply too low when a potential near E_{eq} is used. The distinction between the systems represented in Fig. 6 by A and B, respectively, is therefore of kinetic order. It has therefore seemed more logical to designate the two systems, respectively, as rapid and slow.

Quantitative study of i-E curves (p. 30 ff.) confirms the kinetic bases of this distinction in the case of simple systems. It should be noted that if charge transfer is followed by a chemical step which is itself irreversible [e.g., the formation of sulfoxide (V) from the phenothiazine dication (III)], then the total electrochemical plus chemical reaction causes the system phenothiazine (I) and its sulfoxide (V) to be a truly irreversible system.

5. Current-Potential Curves with Several Waves

These may result from two different causes:

(a) First, the substance, like phenothiazine (I), may be oxidized in two successive steps. A curve with two plateaus is then obtained (Fig. 7); the

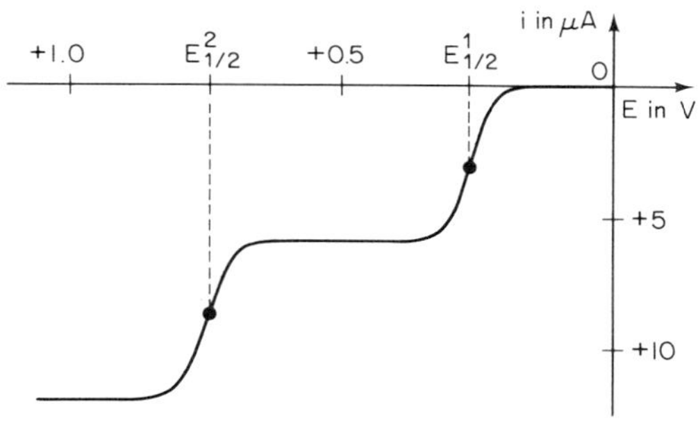

Fig. 7. Complete current intensity-potential curve for phenothiazine. The experimental conditions are identical to those of Fig. 4 [3].

limiting current of the second stage, i^2_{lim} is twice i^1_{lim}. Each "wave" of the curve has a "half-wave" potential close to the standard potential — which is characteristic of the individual process which is occurring,

$$(I) - \bar{e} \rightleftarrows (II), \quad (II) - \bar{e} \rightleftarrows III .$$

At an electrode held at potential E_1 phenothiazine (I) is converted exclusively to its radical cation (II), while at a potential E_2 it yields directly the divalent cation (III).

(b) Since current intensities are generally additive, an i-E curve may have two plateaus if the solution contains two substances, each belonging to a simple system, possessing half-waves of $E^1_{1/2}$ and $E^2_{1/2}$, respectively (Fig. 8). An electrode at potential E_1 will simultaneously act upon both substances

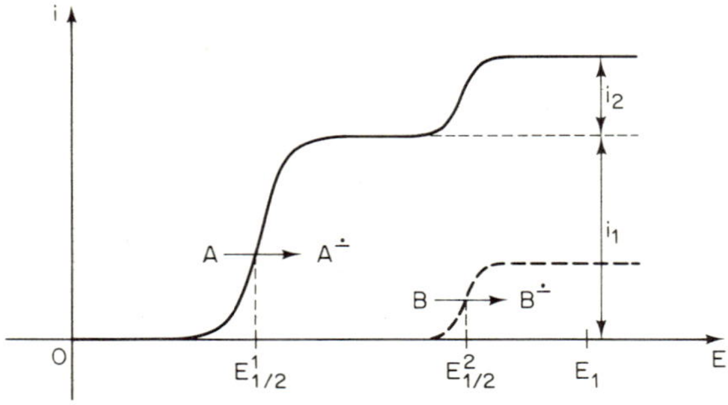

Fig. 8. Current intensity-potential curve for a mixture of two compounds A and B which reduce at respective half-wave potentials of $E^1_{1/2}$ and $E^2_{1/2}$.

with rates which are reflected, respectively, in the values of i_1 and i_2.

6. Special Case in Which the Electrode Material Dissolves. An Example of a Reference Electrode

In certain cases the electrode material itself may participate in the electrode reaction. Then the i-E curve does not exhibit a plateau because the local depletion of material discussed previously does not occur. Figure 9 (curve 1) is illustrative: a silver electrode is placed in acetonitrile containing 10^{-1} M $LiClO_4$ as supporting electrode. It is quite obvious that such an electrode cannot be used in experiments which require potentials greater than that necessary for the oxidation of silver.

Suppose now that the solution mentioned above contains in addition 10^{-3} M $AgClO_4$ (Fig. 9, curve 2). By analogy with curve 3 of Fig. 5 an i-E curve will be obtained having a cathodic branch corresponding to the reduction

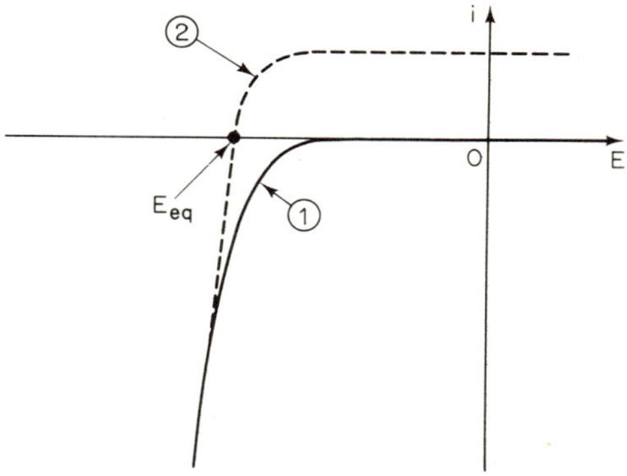

Fig. 9. Oxidation of a silver electrode in acetonitrile containing 10^{-1} M $LiClO_4$. Curve 1, no other compound in solution; curve 2, in the presence of 10^{-3} M $AgClO_4$.

$$Ag^+ + \bar{e} \rightarrow Ag.$$

Since Ag/Ag^+ is a rapid system, the i-E curve intersects the potential axis at a point corresponding to the potential E_{eq}. The latter is given by Nernst's equation which in this case [6] becomes

$$E = E_0 + \frac{RT}{F} \ln[Ag^+].$$

Since the i-E curve is very steep in the region where it intersects the potential axis, the electrode potential will be perturbed only slightly even if there is a small current flowing. This type of assembly, consisting of a wire of an oxidizable metal placed in a solution containing a defined concentration of a salt of the same metal, constitutes a reference electrode similar to those discussed on pp. 205 and 206. Obviously, many other types of reference electrodes are conceivable (cf. Chapter IV).

7. Oxidation and Reduction of the Medium. Concepts of Electrochemical Domain[6] and Residual Current

An i-E curve has no limiting current if the latter is due to the oxidation or reduction of the solvent or of the "supporting" electrolyte which is always present in high concentration.

Under these circumstances the local depletion of oxidizable or reducible species is meaningless. Such curves are obtained if the potential of the working electrode is made very negative or very positive. In Fig. 10 is shown schematically the curve which is obtained when a platinum electrode is placed in a 10^{-1} M tetraethylammonium perchlorate solution in nitrobenzene.

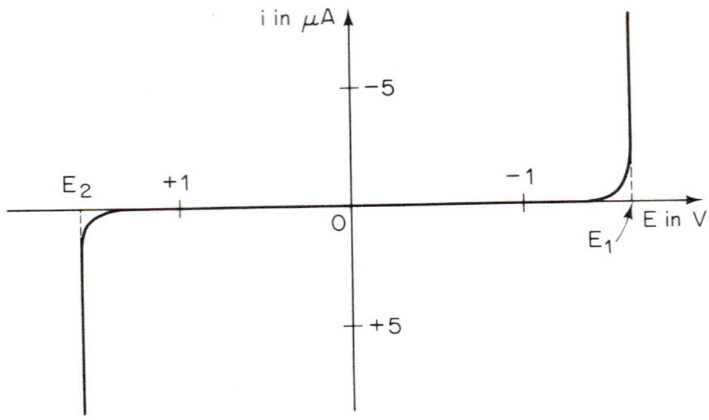

Fig. 10. Domain of electroactivity of 10^{-1} M nitrobenzene in Et_4NClO_4 using a polished platinum rotating disk electrode. The reference electrode is $Ag/AgClO_4$ 5×10^{-3} M [7].

It has been established [7] that the reaction which is responsible for the sudden increase in current at about -1.5 V (vs. $Ag/AgClO_4$, 5×10^{-3} M) is the reduction of the solvent to its radical anion. On the positive side the sharp rise in current at ca. 1.5 V apparently must be attributed to the oxidation of ClO_4^- [8].

The electrochemical domain of the medium is bounded by the sudden increases in current at both the positive and negative potential regions. It is the domain of potentials within which oxidations and reductions can be effected without oxidizing or reducing the constituents of the medium at a significant rate. The potential boundaries of this domain are not precisely defined because they depend on what is meant by "significant rate." The latter depends, among other things, on the nature of the electrode (which must always be specified) and the current density (at this electrode) beyond which the electrochemical reaction of the medium may conceal the reactions of the substances being studied. In practice the electrochemical domain is within the vertical lines of E_1 and E_2; beyond these lines, electrochemical reactions of the substance in question cannot be investigated by classical electrochemical methods.

Strictly speaking, since the electrode itself may eventually contribute to limiting the electrochemical domain (see p. 27 regarding the Ag electrode), it would be preferable to speak of the electrochemical domain of the total system electrode-medium, the latter including solvent and supporting electrolyte.

The concept of electroactivity of a substance is intimately tied to that of electrochemical domain: a substance is electroactive in a given medium and at a given electrode if it can be oxidized or reduced within the electrochemical domain of this electrode-medium system.

With the above in mind it is interesting to compare the electroactivity ranges of several electrode-medium systems among themselves. With the aid of this comparison it is possible, e. g., to determine what system is suitable for the oxidation of a difficultly oxidized material, i. e., one requiring very positive potentials. This type of comparison, however, presents a fundamental problem: it presupposes that there is a universal reference electrode having the same potential in all the media under consideration. But the potential of a reference electrode is dependent on the free energies of the solvation of the two species which comprise it (p. 202); these free energies generally vary with the medium. This dilemma has been solved largely by using a reference electrode whose constituents, by all the evidence available, are poorly solvated in any medium. For example, the couple ferrocene-ferricinium comprises a rapid simple system whose potential, measured at Pt, is practically independent of the solvent. Both species of this couple have large volumes; the free energies involved in their solvation are very small [92].

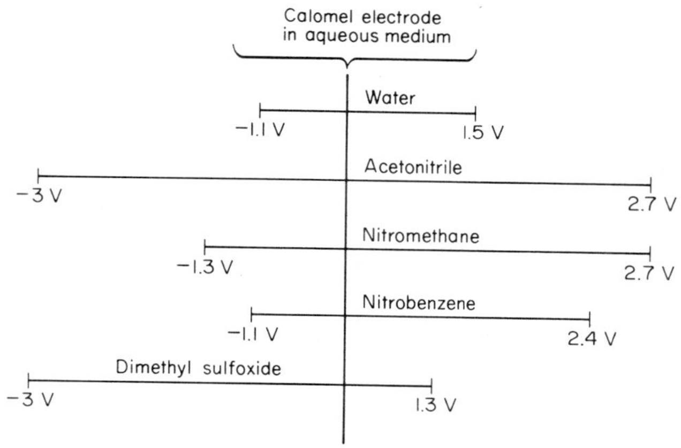

Fig. 11. Comparison of the electrochemical domain of several solvents containing 10^{-1} M Et_4NClO_4; platinum electrode. These domains are referred to a calomel electrode in aqueous medium.

Using such a system comparisons such as those in Fig. 11 can be set up in which the potentials have been referred to the classic saturated calomel electrode in aqueous medium. It will be noted that electroactivity ranges are usually broader in organic than in aqueous media.

In curves similar to that of Fig. 10, carefully drawn at high sensitivity of the recording system, it is found that there is a very small current flow within the potential range between E_1 and E_2. This small current, the residual current, has several sources. In addition to reflecting the oxidation or reduction of trace impurities inevitably present in the medium, it results from capacity phenomena (p. 58) at the electrode or from oxidation or reduction of films on the electrode surface (p. 71).

C. Equation of the Current Intensity-Potential Curve in Simple Systems

1. General Considerations

The equation of an i-E curve is the mathematical relationship between the intensity of the current flowing through the indicating electrode and the potential at which the current is transported (referred to a reference electrode); in certain special cases the numerical value for the one can be given as a function of the other.

In the following paragraphs we shall derive this equation taking into account the conditions and restrictions which have been specified on p. 22. In order to accomplish this, it will be necessary to consider the mathematics of mass transfer and charge transfer successively. Finally, we shall compare these two phenomena, keeping in mind that at the surface of the electrode there is neither creation nor accumulation of charge or of matter.

In the study of mass transfer we shall limit the discussion to the case of a planar electrode placed in a solution; we shall consider that the solution extends in an infinite direction from the electrode. Furthermore, the special phenomena occurring at the edges of the electrode will be neglected.

2. Study of Mass Transfer

It has been shown in hydrodynamic studies [9] which are quite independent of any electrochemical phenomena that the equation governing the transport of a substance in solution to a plane surface in contact with this solution is:

$$D \frac{\delta^2 C}{\delta x^2} - V(x) \frac{\delta C}{\delta x} = \frac{\delta C}{\delta t},$$

in which x is the distance from a given point in solution to the surface of the plane, C is the concentration of the substance at this point at time t, and D is the diffusion coefficient of this substance, expressed, e.g., in square centimeters per second. The function V(x) is the component, normal to the plane, of the relative velocity of the solution; V(x), expressed in centimeters per second, represents then the phenomenon of convection. In the case of pure diffusion, in which the solution does not move with respect to the plane, $V(x) = 0$.

In this section only the case of the rotating disk electrode, defined on p. 96, will be considered. As has been said, the electroactive substance arrives at an electrode of this type by a combination of convection and diffusion phenomena, sometimes called convective diffusion. Before recording an i-E curve using this electrode, it is ascertained that the potential scan is slow enough so that at any moment a state of equilibrium is attained both in the solution and at the surface of the electrode. Under these conditions $\delta C/\delta t = 0$, whatever the value of x. An electrochemical method which satisfies the latter condition is called a stationary method.[7]

Under these conditions the above equation reduces to

$$D\frac{\delta^2 C}{\delta x^2} - V(x)\frac{\delta C}{\delta x} = 0 \qquad (1)$$

with $C = C^*$ for $x = 0$ (concentration at the electrode) and $C = C_\infty$ for $x = \infty$ (concentration in the solution).

By definition, the flux of material arriving at the plane surface is given by

$$\Phi^0 = D\left(\frac{dC}{dx}\right)_{x=0} .$$

Obviously, differential equation (1) cannot be solved unless V(x) is known. This function varies with the relative solution-surface motion. Thus, in the case of a rotating electrode with a planar disk Levich [10] has shown that V(x) is given by the expression

$$V(\gamma) = -0.51\left(\frac{\omega^3}{\nu}\right)^{1/2} \times x^2 \quad \left[\text{if } x \ll \left(\frac{\nu}{\omega}\right)^{1/2}\right]$$

in which ν is the kinematic viscosity (square centimeters per second) of the liquid and ω is the rate of rotation of the disk in radians per second. The same author has shown that it is not necessary to know V(x) for all values of x; Eq. (1) may be solved in V(x) is shown for $x \cong 0$.

Integration of the equation gives

$$C = C^* + \frac{C_\infty - C^*}{\delta} F(x)$$

in which the function F(x) is given by

$$F(x) = \int_0^x \exp\left[\frac{1}{D}\int_0^\theta V(\alpha)\, d\alpha\right] d\theta,$$

α and θ being integration variables while the constant δ[11] is given by

$$\delta = \int_0^\infty \exp\left[\frac{1}{D}\int_0^\theta V(\alpha)\, d\alpha\right] d\theta.$$

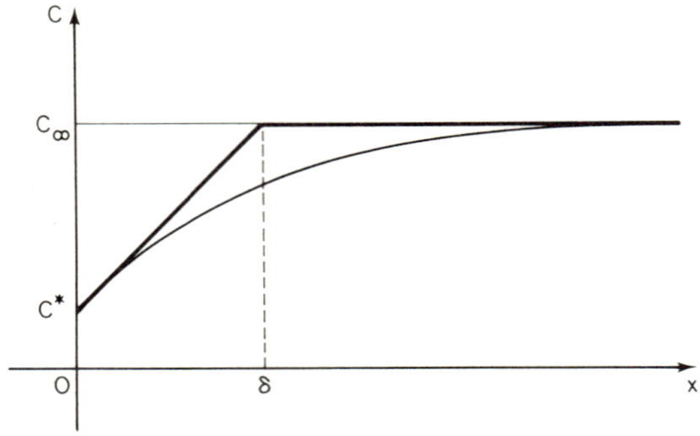

Fig. 12. Variation of the concentration of a species being reduced at a plane electrode as a function of the distance from this electrode. The heavy line shows the variation according to Nernst's empirical approximation.

This constant, which has a mathematical origin, also has a physical significance. It has been shown that in fact it is the abscissa of the point at which the tangent drawn from the origin intersects the straight line representing $C = C_\infty$ (Fig. 12). This then is the basis for the approximation previously given by Nernst on purely empirical grounds, namely, that the change in concentration occurs only in a thin layer called the diffusion layer.[8]

For the rotating disk electrode it has been shown [10] that

$$\delta = 1.61 D^{1/3} \nu^{1/6} \omega^{-1/2},$$

i.e., δ is of the order of 10^{-3} cm for an electrode rotating at 600 rpm (commonly used) in an aqueous solution containing a compound having a diffusion coefficient ca. 10^{-5} cm^2/sec. Under these conditions the flux at the electrode is given by

$$\Phi^{\circ} = \frac{D}{\delta}(C_\infty - C^*).$$

Hereafter, we shall use the ratio

$$m = D/\delta$$

which characterizes the rate of movement of particles and which itself has the dimensions of a rate. In aqueous solutions this ratio is of the order of 10^{-2} to 10^{-4} cm/sec. It can be shown that for simple systems m is practically identical for A and B,

$$A + \bar{e} \rightleftarrows B.$$

The equations of flux then reduce to

$$\Phi^{\circ}_A = m\left([A] - C_A^*\right), \quad \Phi^{\circ}_B = m\left([B] - C_B^*\right)$$

in which [A] and [B] are the bulk concentrations and C_A^* and C_B^* the concentrations at the electrode.

3. Study of Charge Transfer

Each of the two reverse elementary reactions of a simple system

$$A + \bar{e} \underset{k'}{\overset{k}{\rightleftarrows}} B$$

have rates k and k' expressed in centimeters per second. The overall rate of the electrode process can then be written

$$v = kC_A^* - k'C_B^*$$

and the quantity of current involved in unit time and at a unit of electrode surface area, i.e., the current density J, is then given by

$$J = -Fv = F(k'C_B^* - kC_A^*).$$

(In this expression F is the faraday and the minus sign is in accordance with the convention that reduction currents are given negative values.)

The rate constants k and k′ have been subject to theoretical interpretations similar to those involved in absolute rate theory in classical chemical kinetics [13]. In particular, it has been shown that the constants may be expressed as

$$k = A \exp\left(-\frac{\overrightarrow{\Delta G^*}}{RT}\right)$$

$$k' = A' \exp\left(-\frac{\overleftarrow{\Delta G^*}}{RT}\right) \qquad (2)$$

in which A and A′ are constants at a given temperature and $\overrightarrow{\Delta G^*}$ and $\overleftarrow{\Delta G^*}$ are the free energies of activation of the forward and reverse reactions. A detailed analysis of these free energies shows that they have two components, one which is independent of electrical parameters and one which is dependent on the standard potential E_0 of the system A/B and on the potential E of the electrode:

$$\overrightarrow{\Delta G^*} = \overrightarrow{\Delta G_n^*} + \alpha F(E - E_0)$$

$$\overleftarrow{\Delta G^*} = \overleftarrow{\Delta G_n^*} + (\alpha - 1) F(E - E_0)$$

(α is a coefficient which will be discussed later). Further,

$$A \exp\left(-\frac{\overrightarrow{\Delta G_n^*}}{RT}\right) = A' \exp\left(-\frac{\overleftarrow{\Delta G_n^*}}{RT}\right) = k^\circ.$$

Equation (2) can then be written[9]

$$k = k^{o} \exp\left[-\frac{\alpha F}{RT}(E - E_{o})\right]$$

$$k' = k^{o} \exp\left[\frac{(1-\alpha)F}{RT}(E - E_{o})\right].$$

The following comments must be made concerning the various terms involved in these rate expressions.

The standard potential E_o, which has previously been defined, is characteristic of the couple A/B. By definition it is a thermodynamic value

$$E_{o} = -\frac{\Delta G^{o}}{F}. \tag{3}$$

ΔG^{o} is the charge of standard free energy of the reaction [4]; note that it is independent of the nature of the electrode.

The constant k^{o}, called the intrinsic charge transfer rate constant, is on the contrary not uniquely characteristic of the couple A/B. It is independent of electrode potential but varies with the couple under consideration, the electrode metal, the medium, the temperature, etc. [15]. It has been the subject of many theoretical treatments related to statistical mechanics, quantum mechanics [93], and double layer theory (p. 57). The object of these studies is to explain the various values which k^{o} may have. Particularly, on the basis of simple collision theory, it can be seen that k^{o} must have an upper limit since the number of collisions of a particle with a given surface in unit time cannot be infinite. Calculations show that this upper limit is ca. 10^{4} cm/sec at ambient temperature. In fact, many experimental data place the value of k^{o} between 10^{-5} and 10 cm/sec when no extraneous substance is specifically adsorbed (p. 60). Furthermore, it becomes larger as the bulk of A and B increases. This can be explained on the basis that the rate of electron transfer in oxidation or reduction is very high and that the factor which contributes most to the activation energy is the accompanying reorganization of the solvation sphere. Species which, e.g., because of their dimensions are poorly solvated yield the highest values of k^{o}. Organic molecules which are not sterically hindered in general have higher k^{o}'s than small inorganic ions, which are more strongly solvated (Table 1). In this connection it should be mentioned that, when organic compounds show overall slow kinetics in electrochemical reactions, it is generally due to the fact that chemical reactions coupled to the charge transfer intervene and not because the electron transfer is primarily slow. However, adsorption phenomena may also affect the kinetics of charge transfer (p. 69).

The physical significance of the transfer coefficient α is one of the most controversial and least clear issues in the entire development of theoretical electrochemistry.[10] Some authors criticize even the introduction of a nondimensional coefficient which is independent of potential. Since the time when

TABLE 1

Some Examples of the Intrinsic Charge-Transfer Rate Constant

System	Cr^{3+}/Cr^{2+}	Fe^{3+}/Fe^{2+}	$Fe(CN)_6^{3-}/Fe(CN)_6^{4-}$	Anth/Anth[a]
Medium	H_2O KCl 1 M	H_2O $HClO_4$ 1 M	H_2O KCl 1 M	DMF Bu_4NI 0.1 M
Electrode	Hg	Pt	Pt	Hg
k^o (cm/sec)	10^{-5}	5×10^{-3}	10^{-1}	≥ 4.0[b]
Ref.	[16]	[16]	[16]	[17]

[a]Anthracene and its radical anion in DMF.
[b]The exact number is actually too large to measure by the method used.

α was originally introduced on a purely empirical basis [19] only the recent interpretations of electron exchange on the basis of quantum mechanical considerations seem to ascribe to α a satisfactory meaning, even though the matter is not yet settled.

Many experimental determinations [20] seem to show that in a rapid (reversible) simple system, such as the one that has been discussed here, α is generally ca. 0.5 in certain potential domains. Furthermore, this value has received certain support in the theoretical developments involving quantum mechanics [21].

4. Equation of the Current Intensity-Potential Curve

We shall now derive the equation of the i-E curve for the situation in which the number of moles of the species A which arrives at the electrode per unit time and per unit area is equal to the number of moles of A which is consumed per unit time and per unit area in an electrochemical reaction

$$\Phi^o_A = v.$$

The number of moles of A which is consumed equals the number of moles of B formed, then

$$\Phi^o_A + \Phi_B = 0$$

from which (and the preceding expressions) is derived

$$[A] - C_A^* = C_B^* - [B] = \frac{k}{m} C_A^* - \frac{k'}{m} C_B^*.$$

From these two equations C_A^* and C_B^* as functions of k/m, k′/m, [A], and [B] can be calculated, leading to

$$\Phi_A^{\circ} = \frac{k[A] - k'[B]}{1 + k/m + k'/m}.$$

Substituting for Φ_A° the expression for the current density

$$J = F\Phi_A^{\circ},$$

one obtains

$$J = -F\Phi_A^{\circ}$$

$$J = Fm\frac{(k'/m)[B] - (k/m)[A]}{1 + k/m + k'/m}.$$

The general equation for the i-E curve also takes into account the variation of k and k′ as a function of E:

$$J = Fm\frac{[B]\exp\left[\frac{(1-\alpha)F}{RT}(E - E_o)\right] - [A]\exp\left[\frac{-\alpha F}{RT}(E - E_o)\right]}{\frac{m}{k^o} + \exp\left[-\frac{\alpha F}{RT}(E - E_o)\right] + \exp\left[\frac{(1-\alpha)F}{RT}(E - E_o)\right]}. \quad (4)$$

The symbols above have the following meaning: J is the current density in milliamperes per square centimeter; E and E_o are given in volts; m and k^o are in centimeters per second; [A] and [B] are in moles per liter; F is the faraday, 96,484 Cs; R is the (perfect) gas constant taken at 8.3144 joules $mole^{-1}$ $degree^{-1}$; and T is the absolute temperature.

5. Concepts of Limiting Current, Overpotential, and Exchange Current

Reference to Eq. (4) permits us to define certain concepts of current in electrochemistry and to indicate where these concepts may be applied.

a. Limiting Current. When the numerical value of the potential E becomes sufficiently large, the term

$$\exp\left[\frac{(1-\alpha)F}{RT}(E - E_o)\right]$$

becomes very large in comparison to the other exponential term and to m/k^o which, as has been shown, is necessarily limited. The current density therefore tends to reach a limiting value

$$J_{\lim B} = Fm[B]$$

which is called the limiting current density of B.

Likewise, when the potential becomes sufficiently negative, J approaches the limit

$$J_{\lim A} = -Fm[A]$$

which is the limiting current density for species A.

Equation (4) takes the form

$$J = J_A + J_B$$

if the following substitutions are made:

$$J_A = J_{lA} \frac{k}{k + k' + m} = J_{dA} \frac{\exp\left[-\frac{\alpha F}{RT}(E - E_o)\right]}{\frac{m}{k^o} + \exp\left[-\frac{\alpha F}{RT}(E - E_o)\right] + \exp\left[\frac{(1-\alpha)F}{RT}(E - E_o)\right]}$$

$$J_B = J_{lB} \frac{k'}{k + k' + m} = J_{dB} \frac{\exp\left[\frac{(1-\alpha)F}{RT}(E - E_o)\right]}{\frac{m}{k^o} + \exp\left[-\frac{\alpha F}{RT}(E - E_o)\right] + \exp\left[\frac{(1-\alpha)F}{RT}(E - E_o)\right]}$$

The simplified expression conveys the fact that for all values of E the current density of the entire current is the summation of the current density J_A due to the reduction of A and J_B the current density due to the oxidation of B.

In the special case where no current flows, i.e.,

$$J_A + J_B = 0,$$

the system is at the equilibrium potential and the equation reduces to Nernst's classical equation

$$E_{eq} = E_o + \frac{RT}{F} \ln\frac{[A]}{[B]}.$$

b. Overpotential and Exchange Current. Most treatises on electrochemistry introduce these two concepts very early but they are not really essential for an understanding of the kinetic aspects of electrochemical reactions. We shall nevertheless provide the definitions.

The overpotential η is defined by the equation

$$\eta = E - E_q = E - E_o - \frac{RT}{F} \ln \frac{[A]}{[B]}.$$

Then

$$\exp\left[\frac{F}{RT}(E - E_o)\right] = \frac{[A]}{[B]} \exp\left(\frac{F}{RT}\eta\right)$$

and the numerator of Eq. (4) may be written in the form

$$Fm[A]^{1-\alpha}[B]^{\alpha}\left\{\exp\left[-\frac{(1-\alpha)F}{RT}\eta\right]-\exp\left(-\frac{\alpha F}{RT}\eta\right)\right\}.$$

If mass transfer is not limiting — and mathematically this means that the value of m is large — the general equation becomes

$$J = Fk^{\circ}[A]^{1-\alpha}[B]^{\alpha}\left\{\exp\left[\frac{(1-\alpha)F}{RT}\eta\right]-\exp\left(-\frac{\alpha F}{RT}\eta\right)\right\}.$$

The exchange current density is defined by

$$J_o = Fk^{o}[A]^{1-\alpha}[B]^{\alpha}$$

so that the current density can be expressed by

$$J = J\left\{\exp\left[\frac{(1-\alpha)F}{RT}\eta\right]-\exp\left[-\frac{\alpha F}{RT}\eta\right]\right\}. \tag{5}$$

This classical fundamental equation has been the basis on which very many treatments of electrochemical kinetics rest when mass transfer is neglected. If mass transfer is considered, the general equation becomes

$$J = J^{o}\frac{\exp\left[\frac{(1-\alpha)F}{RT}\eta\right]-\exp\left[-\frac{\alpha F}{RT}\eta\right]}{1-\frac{J^{o}}{J_{1A}}\exp\left[-\frac{\alpha F}{RT}\eta\right]+\frac{J^{o}}{J_{1B}}\exp\left[\frac{(1-\alpha)F}{RT}\eta\right]}.$$

It can be seen that the concept of overpotential is linked to that of exchange current while the difference $E - E_o$ is associated with the intrinsic rate constant k_o. It is assuredly easier and more significant from the point of view of kinetics to use the values $E - E_o$ and k^{o} rather than η and J^{o} which latter two have the additional disadvantage of depending on the concentrations [A] and [B]. This is particularly true when analyzing the complex situations which exist when coupled chemical reactions are involved. However, it should be noted that the exchange current J^{o} does correspond to a real physical phenomenon. It is a measure of the current which is associated with the exchange at the electrode between species A and B even though no current is flowing in the circuit, i.e., when, on a macroscale, nothing appears to be occurring.

We shall use the concepts of overpotential and exchange current in explaining what is meant by Tafel lines and Tafel's law. When mass transfer is negligible, i.e., at the foot of an i-E curve, and if the potential is sufficiently different from E_{eq} (in other words, the overpotential in absolute numbers is sufficiently large), one or both exponential terms may be neglected and η becomes a linear function of log J:

$$\eta = a \pm b \ln J \quad (\text{when } \eta \gg \frac{RT}{F}).$$

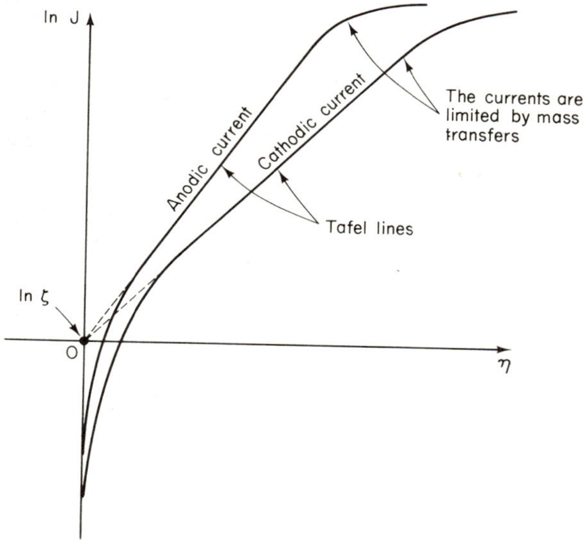

Fig. 13. Relationship between ln J and the overvoltage η for anodic and cathodic reactions of the same A/B system when $\alpha \neq 0.5$.

This relationship was proposed originally by Tafel in 1905 in connection with the system H_2/H^+ (not a simple system) and allows one in principle to determine the exchange current J^o (Fig. 13). The latter is the intersection ($\ln \zeta$) at the ln J axis of the extension of the linear portions of the $\ln J = f(\eta)$ curves. The slope of these linear portions of the curves, called Tafel lines, provides values for α, which is contained in the coefficient b, equal to $RT/\alpha F$ or $RT/(1 - \alpha)F$, depending on whether cathodic or anodic curves are being considered.

The construction of Tafel lines is sometimes used to show that the electron transfer is the critical kinetic step in a complex electrochemical process [22]. Likewise, in principle E_{eq} can also be determined by means of Tafel lines since the curve $\ln J = f(\eta)$ has the axis $\eta = 0$ as an asymptote [23].

6. Rate of a Simple System

"Slow" and "rapid" systems previously defined qualitatively (p. 25) may now be examined more precisely with the aid of general equation (4).

This equation defines J as a function of E but it shows that the inverse is generally impossible, i.e., defining E as a function of J. The equation involves four parameters of which three (E_o, k^o, and α) are characteristic of charge transfer while the fourth ($m = D/\delta$) is characteristic of mass transfer.

In fact, the coefficient k^o does not occur alone but in the form of the nondimensional expression k^o/m. Considering the values which k^o and m may have, the ratio can vary, in a simple system, between 10^{-3} and 10^{-1}.

It can be shown that if $10^2 < k^o/m$ or $k^o/m < 10^{-2}$, E can be expressed as a function of J and that the cathodic and anodic branches of the i-E curves then have the form of Fig. 6, A and B, respectively.

The concept of rapid and slow systems is thus defined quantitatively and in conjunction with parameters which have kinetic meaning. The definitions given previously[11] are only corollaries of the more precise definitions given now. It must be noted, however, that the rate of a system is not defined absolutely by considering k^o only but by comparing k^o and m. This is then a relative concept in which the kinetics of mass transfer and of charge transfer are compared.

a. The Case of a Rapid System ($k^o/m > 10^2$). It can easily be shown that in this case m/k^o is negligible with respect to exponential terms and that the general equation may be written

$$E = E_o + \frac{RT}{F} \ln \frac{J - J_{\lim A}}{J_{\lim B} - J}. \tag{6}$$

This is the classic equation of i-E curves for a redox system which electrochemical and polarographic specialists customarily call "rapid" or, less satisfactorily, "reversible." The appropriate i-E curve is shown in Fig. 6, A.

The half-wave potentials of the total curve $J = f(E)$ and of the individual curves $J_A = f(E)$ and $J_B = f(E)$ obtained when only A or B is in solution are the same. If it appears that the above equation leads to the conclusion that $E_{1/2} = E_o$ it is because ab initio the postulate was made that $m_A = m_B$. The more rigorously correct expression is

$$E_{1/2} = E_o + \frac{RT}{F} \ln \frac{m_B}{m_A}.$$

Consequently, $E_{1/2}$ is also a function of terms applying to the kinetics of mass transfer.

It follows from Eq. (6) that $\ln [(J - J_{\lim A})/(J_{\lim B} - J)]$ varies linearly with potential with a slope equal to F/RT. Verification of this linearity and determination of the slope are often used to demonstrate that a simple rapid system is indeed involved. It should be noted, furthermore, that the parameters k^o/m and α do not appear in Eq. (6). Therefore, obtaining the i-E curve in this case does not lead to a determination of these coefficients.

b. The Case of a Slow System ($k^o/m < 10^{-2}$). An examination of the denominator of the general equation (4) shows that if $E < E_o$ when $m/k^o > 10^2$, the second exponential term is negligible in comparison to the last ratio. It can be shown that if [A] and [B] are of the same order of magnitude, then at small values of potential J_B is negligible compared to J_A so that J differs little from J_A. Then

$$E = E^c + \frac{RT}{\alpha F} \ln \frac{J_{\lim A} - J}{J} \tag{7}$$

with

$$E^c_{1/2} = E_o + \frac{RT}{\alpha F} \ln \frac{k^o}{m}.$$

In the same way, when $E > E_o$,

$$E = E^a_{1/2} + \frac{RT}{(1 - \alpha)F} \ln \frac{J}{J_{\lim B} - J} \tag{8}$$

with

$$E^a_{1/2} = E_o - \frac{RT}{(1 - \alpha)F} \ln \frac{k^o}{m}.$$

Equations (7) and (8) together represent the i-E curve of a slow ("irreversible") redox system, illustrated in Fig. 6, B.

These equations show also that $\ln[(J_{\lim A} - J)/J]$ and $\ln[J/(J_{\lim B} - J)]$ vary linearly with the potential E and have respective slopes of $\alpha F/RT$ and $[(1 - \alpha)F]/RT$. We shall later refer to waves which verify this type of linear relationship (as do also the anodic and cathodic waves of rapid systems) as pure waves.

Equations (7) and (8) show furthermore that the half-wave potentials $E^a_{1/2}$ and $E^c_{1/2}$ are determined not only by m representing terms related to mass transfer but also by k^o and α related to the kinetics of charge transfer. These half-wave potentials have full kinetic significance and in principle permit the determination of the ratio k^o/m. Figure 14 shows the form of i-E curves corresponding to various k^o/m values.

c. The Case of a Moderately Rapid System ($10^{-2} < k^o/m < 10^2$). In an analysis of this case it is necessary to retain general equation (4). It is easy to show that the i-E curve does not have the form of a pure wave, i.e., that there is not a linear relationship between E and expressions like $\ln[(J_{\lim} - J)/J]$.

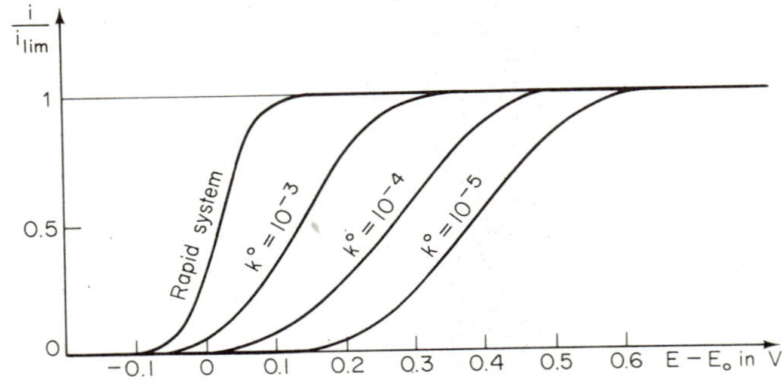

Fig. 14. Form of the current intensity-potential curve for a simple one-electron system with various values of k^o (centimeters per second). The transfer coefficient is taken at 0.5; $D = 10^{-5}$ cm^2/sec, $\delta = 10^{-3}$ cm, and $m = 10^{-2}$ cm/sec.

7. Successive Simple Systems

Phenothiazine was chosen (p. 17) to represent a system in which a compound is capable of exchanging two electrons successively. Such examples occur frequently. Indeed, compounds which appear to undergo two-electron exchanges in reality probably always undergo two successive one-electron exchanges.

The reaction scheme may be represented as follows:

$$A + e \rightleftharpoons B(E_0^{\,1})$$

$$B + e \rightleftharpoons C(E_0^{\,2})$$

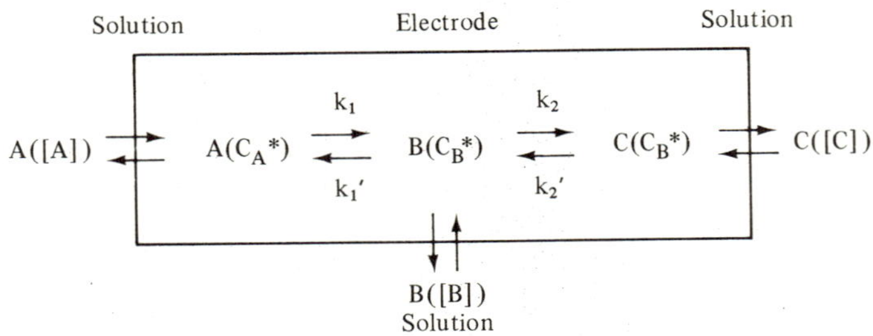

This type of system appears to be a two-electron system only when the diffusion of the intermediate species B into the bulk of the solution is not detectable because of the fact that B undergoes a second electron transfer at the electrode rapidly and is converted to C. Again, one can see that the relative rates of charge transfer and mass transfer will determine the results that are obtained. The values of E_0^1 and E_0^2, of k_1^0 and k_2^0, and of α_1 and α_2 and the various m coefficient will naturally be involved.

As mentioned before, the total current J is the sum of the currents associated with the oxidation and reduction of the several species. The solution of the equations relating mass transfer and charge transfer leads to the equation of the i-E curve. Rigorous treatment must take into account the disproportionation reaction

$$2B \rightleftarrows A + C$$

if its kinetics is important in comparison to the rate of reduction of B.

If E_0^1 and E_0^2 are sufficiently separated, as in the case of phenothiazine, a curve showing two successive waves is obtained (Fig. 7). Conversely, under certain conditions a single wave may be obtained with a limiting current that corresponds to an exchange of two electrons (Fig. 15). A classic

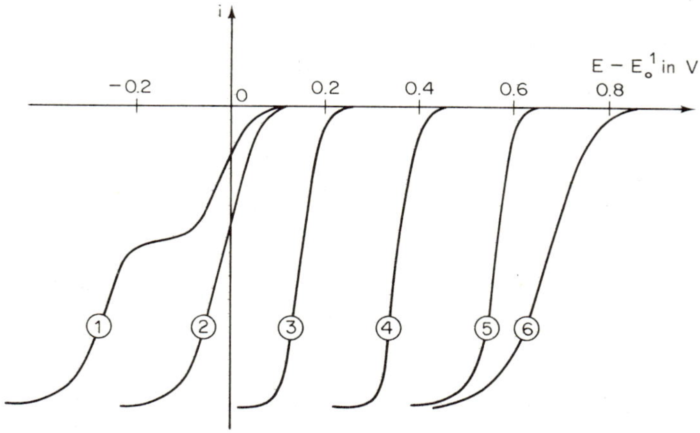

Fig. 15. Current intensity-voltage curve for a two-electron oxidation with $k_1^0/m_1 = k_2^0/m_2 = 10^6$ for various values of $E_0^2 - E_0^1$: −0.240 V (curve 1); 0.000 V (curve 2); 0.360 V (curve 3); 0.84 V (curve 4); 1.2000 V (curve 5); and 1.680 V (curve 6).

example in inorganic electrochemistry is the reduction of Cu^{2+} cations to Cu^0 in a noncomplexing aqueous medium. The i-E curve is not a pure wave. It is impossible to represent it in one equation containing the coefficients α and k^0 as constants and be valid for the entire curve; on the contrary, it is possible to interpret it as the summation of two currents which arise from two one-electron exchanges [24]. Furthermore, it is possible by special

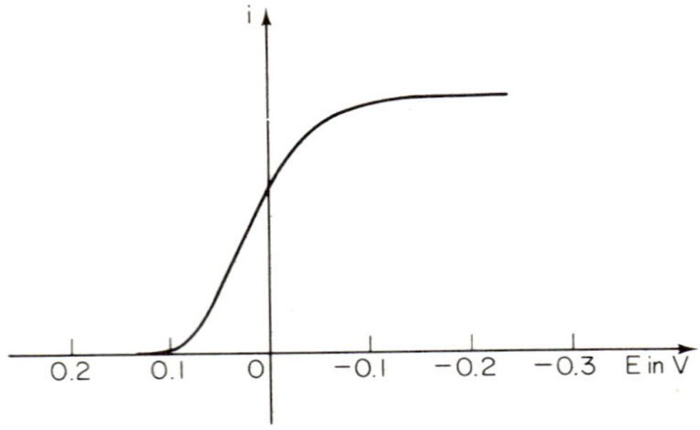

Fig. 16. Current intensity-potential curves corresponding to the reduction of Cu^{2+} (2.83×10^{-4} M) at a mercury electrode. Medium: 10^{-1} M aq. $HClO_4$. Voltages vs. SCE.

techniques such as the ring disk electrode (Chapter II, p. 102) to demonstrate the transitory existence of Cu^{+} [24].

The two-electron wave shown by Cu^{2+} (Fig. 16) is the result of the fact that E_0^1 and E_0^2 are of abnormal order ($E_0^1 = -0.790$ V and $E_0^2 = 0.144$ V referred to a saturated calomel electrode in 0.1 M $HClO_4$). This can evidently come about only because of the respective energies of solvation of Cu^{+} and Cu^{2+}. **Such reversal in the normal order of standard potentials** is less probable in redox reaction of organic molecules because in their case solvation phenomena are of lesser importance. The majority of true[12] two-electron organic systems yields i-E curves having two waves.[13]

III. INTERVENTION OF COUPLED CHEMICAL REACTIONS

A. Introduction

In all the preceding discussions it has been assumed that the only phenomenon accompanying electron transfer itself is an eventual reorganization of the solvation sphere of the species undergoing oxidation or reduction.

A system this simple exists frequently in inorganic chemistry, e.g., in the oxidation of Fe^{2+} to Fe^{3+}. But the fact that the phenothiazine → phenothiazine radical cation oxidation is analogous must not obscure the more complicated nature of the latter due to the making and breaking of covalent bonds in the course of complex reactions.

Because there are chemical reactions coupled to mass transfer and to electron transfer, organic electrochemistry provides a means for complex syntheses. On the other hand, it is the very existence of these reactions which makes the elucidation of the mechanisms of organic electrode reactions so difficult.[14]

The oxidation of p-phenylenediamine (PPD) in aqueous media studied by Adams [26] may be cited to illustrate the degree of complexity which may be involved in the electrochemistry of an organic compound.

In moderately aqueous acidic medium, e.g., in 0.3 M $HClO_4$, PPD exists as a mixture of its mono- $PPDH^+$ and diprotonated $PPDH_2^{2+}$ forms. For many reasons it is believed that in the electroactivity range of this medium only the unprotonated molecule PPD is oxidizable. One or two deprotonation reactions must therefore precede the electron-transfer step proper. This is an example of a chemical reaction preceding charge transfer. The overall oxidation is

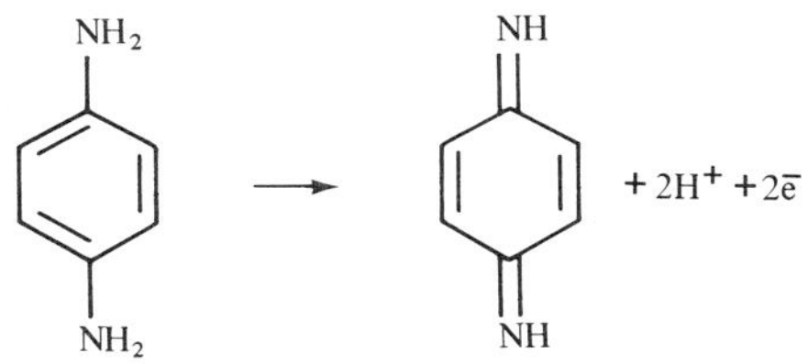

but all the evidence points to the conclusion that there are really four elementary steps involved since two protons and two electrons are lost. Omitting detailed discussion for the moment, it is obvious that chemical reactions are involved which may precede or follow the charge transfers. p-Phenylenediimine (PPI) is not the last of the products that can be formed; it is very reactive and may be partially or completely hydrolyzed according to

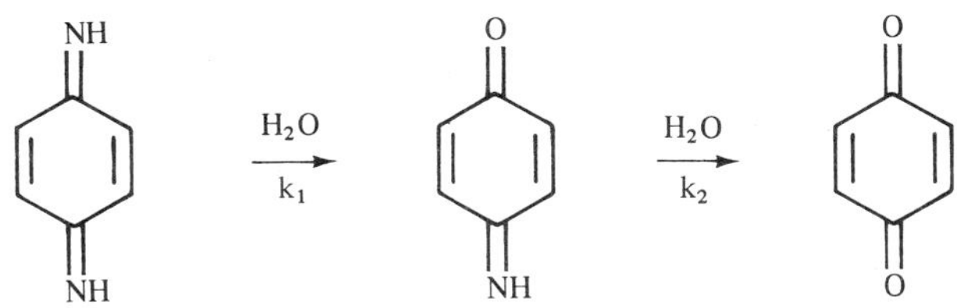

The kinetics, indicated by the values of coefficients k_1 and k_2, are very dependent on the pH of the medium. Consequently, after a certain period of electrolysis the solution contains three oxidants, PPI, the monoimine, and the quinone. The complexity is aggravated even further by the possible occurrence of Michael reactions:

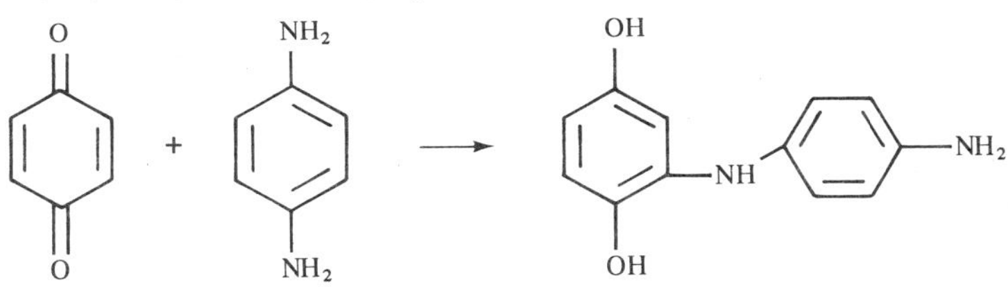

The Michael product can itself be oxidized at the operating potential:

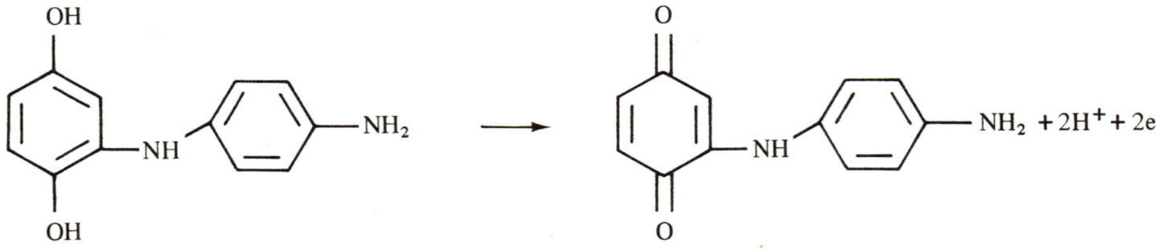

The new quinone product can also undergo the condensation reaction. Thus, the latter stages of the electrochemical process encompass chemical reactions both before and after the electron transfer. Finally, certain of the species formed are easily oxidized by homogeneous electron transfer in solution to the three oxidants formed in the course of the reaction as well as by the electrode.

Fortunately, not all organic electrochemical reactions are so complicated. The rate constants of several theoretically possible chemical reactions in a given system may differ by several orders of magnitude so that some of the reactions are of negligible importance.

In an effort to cope with the complexities, researchers have devised a nomenclature which specifies the sequence of the steps which comprise a reaction mechanism. Thus, a mechanism which involves a chemical step that precedes an electron transfer, e.g., $PPDH^+ \xrightarrow{-H^+} PPDH \xrightarrow{e} \cdots$, is called a CE process (Chemical-Electrochemical). The ensemble of pre- and post-electron-transfer chemical Michael reactions which occur in the course of the oxidation of PPD constitutes an ECE process.

It should be pointed out that there is a certain lack of precision in the literature in using this nomenclature. Certain authors represent in a single E an oxidation like that of PPD to PPI; this really involves four elementary steps because four particles (two protons and two electrons) are lost from each molecule of starting material.

In the following discussion we shall usually use this nomenclature only in a strict sense; that is, each E and C will symbolize one elementary event.

B. The Possibility of Several Reaction Paths for a Given Transformation

Let us now undertake as an example a detailed examination of the relatively simple system in which two electrons and two protons are gained (or lost)

$$AH_2 \rightleftarrows A + 2H^+ + 2e^-.$$

This type of *overall* reaction is very common: oxidation of diamines, hydroquinones, or aminophenols; reduction of quinones to hydroquinones, of aromatic hydrocarbons to their dihydro derivatives, or of a ketone to an alcohol, etc.

It is obvious to organic chemists accustomed to dealing with reaction mechanisms that the above processes can occur only in stages. Electron exchanges occur between structurally similar species (except for solvation spheres) while proton exchanges occur only between species of the same oxidation state and are independent of electron transfer. *In principle*, therefore, there could be seven intermediates involved in the reduction of A to AH_2; pairs of intermediates would be involved in redox or acid-base relationships

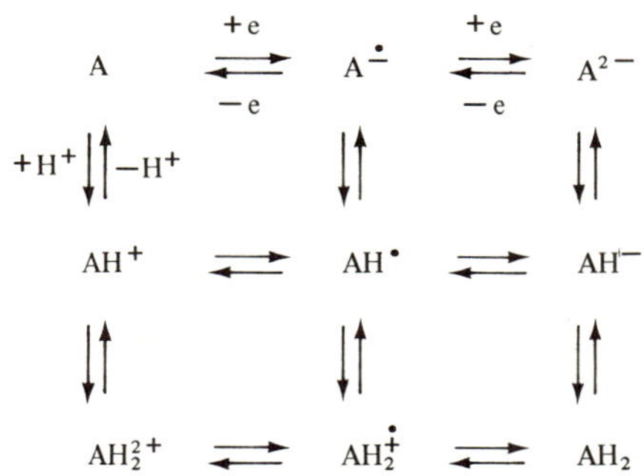

In the scheme diagramed above all electron transfers are given by horizontal lines and all chemical changes are given by the vertical lines.

Six reaction paths or processes are therefore to be considered in the reduction of A to AH_2. The complete mathematical analysis of this problem, of which a brief summary will be given later, leads to the following general conclusions:

(a) The reaction path which is followed depends on the acid-base equilibrium constants which are involved in the general scheme and on the kinetics of electron transfers and chemical reactions.

(b) For a given electrode potential, several reaction paths may be simultaneously followed. The relative importance of each depends on the potential.

(c) Reaction paths may be eventually different in oxidation and in reduction.[15]

The classic case of the reduction of an aromatic hydrocarbon in organic aqueous medium and in aprotic organic medium can be used to illustrate the relationship between the morphology of the i-E curve and the reaction process.

In the latter type of solvent, two one-electron reductions (Fig. 17, A) are observed similar to those described for a simple system (p. 17)

$$R \overset{+\bar{e}}{\rightleftharpoons} R^{\dot{-}} \overset{+\bar{e}}{\rightleftharpoons} R^{=}$$

in which R is the hydrocarbon. However, in an aqueous organic medium, water can protonate the radical anion formed in the first step. The reactions observed then are

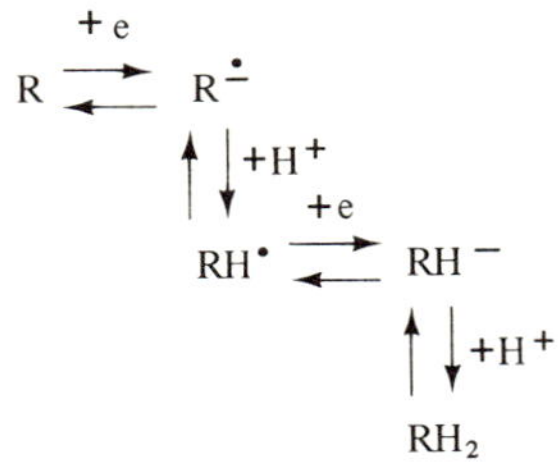

Now it can be shown by means of quantum mechanics [27] that the radical $RH^{\bullet}$ is more easily reduced than the starting material R. Figure 17, B, therefore shows a single wave with a limiting current corresponding to a two-electron transfer. The i-E curves represent an ECEC process in aqueous medium and successively an E and an EE mechanism in aprotic media.

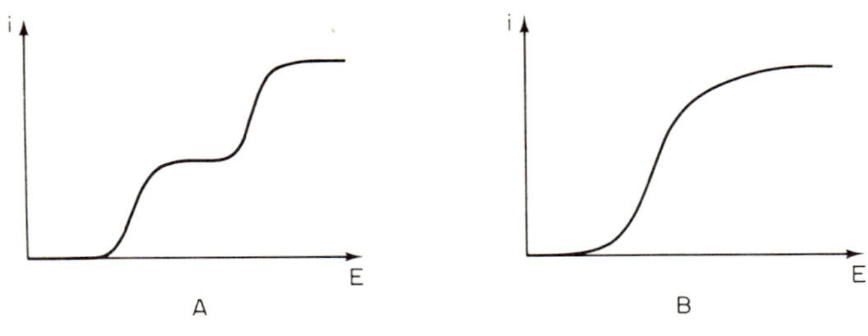

Fig. 17. Current intensity-potential curve corresponding to the reduction of an aromatic hydrocarbon (A) in an anhydrous aprotic solvent and (B) in an aqueous organic medium.

C. Principles in the Determination of the Current-Potential Curves. Results

1. General Considerations

In principle, knowledge of the rate and equilibrium constants for the chemical and electrochemical reactions would permit one to calculate not only which route is taken in each reaction and in each electron transfer but also to derive the equations for the appropriate i-E curves.

Conversely, when a given mechanism is proposed as a hypothesis, comparison of the experimental i-E curves with those calculated using several assumed values for the rate and equilibrium constants should lead to a choice of specific numbers for these constants or to a modification of the original hypothesis.

In fact, however, these calculations and comparisons are very difficult because the reaction paths postulated comprise a very great number of elementary steps. On account of this, many other experimental approaches have been more productive than the establishment of i-E curves (Chapter II). But even with these other means it is often difficult to treat the quantitative aspects of the phenomena. It is therefore not surprising that most publications in organic electrochemistry actually are limited to a qualitative and more or less incomplete discussion of any reaction mechanism proposed.

With these reservations in mind it is, nevertheless, interesting to present the principles of the calculations which can be developed for arriving at the equation for i-E curves for some simplest cases and to give general ideas about the results.

2. Formulation of the Equation in the Case of a Simple System Coupled to a First-Order Chemical Reaction. Results

Consider the processes

$$A_1 + e \underset{k'}{\overset{k}{\rightleftarrows}} B_1$$
$$B_1 \rightleftarrows B_2$$

in which a compound B_1 which is obtained by the reduction of A_1 is chemically transformed into the compound B_2. The electrochemical reaction is characterized by the constants E_0, k^0, and α while the chemical reaction is defined by the rates constants

$$B_1 \underset{k_b}{\overset{k_f}{\rightleftarrows}} B_2$$

and the equilibrium constant

$$\sigma = \frac{[B_2]}{[B_1]} = \frac{k_f}{k_b}.$$

The equation for the i-E curve is obtained in just the same way as in simple systems without coupled chemical reactions. Mass transfer and charge transfer are compared but now the differential equations governing mass transfer include terms which take into account the formation or decomposition of species due to chemical reactions. In the stationary state the equations are

$$D_{A_1} \frac{d^2C_{A_1}}{dx^2} - V(x) \frac{dC_{A_1}}{dx} = 0$$

$$D_{B_1} \frac{d^2C_{B_1}}{dx^2} - V(x) \frac{dC_{B_1}}{dx} = k_f C_{B_1} - k_b C_{B_2}$$

$$D_{B_2} \frac{d^2C_{B_2}}{dx^2} - V(x) \frac{dC_{B_2}}{dx} = k_b C_{B_2} - k_f C_{B_1}.$$

Rigorous integration of these equations is possible in the case of a first-order chemical reaction or of a pseudo-first-order reaction (large excess of the nonelectroactive substance reacting with B_1 and B_2). Under these conditions, as in the case of a simple system (p. 18), it is possible to draw a curve (Fig. 18, curve 2) which represents the variation of C_{B_1} with distance x to the electrode. Because of the chemical reaction the concentration gradient

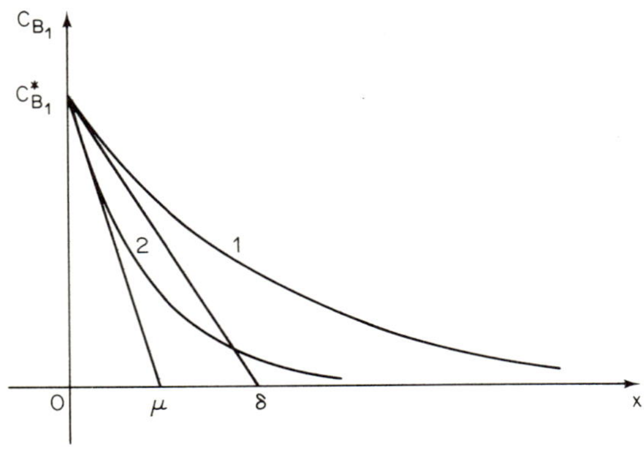

Fig. 18. Variation of the concentration of the species B_1 as a function of the distance from the plane electrode. Curve 1, without chemical reaction; curve 2, with the chemical reaction.

in the vicinity of the electrode is greater than it would have been in the absence of the chemical reaction (Fig. 18, curve 1). The tangent at the origin therefore intersects the x axis at a distance μ which is less than the distance δ previously defined (p. 32). The term *thickness of reaction layer* is used to refer to μ. Like Nernst's diffusion layer the concept of a reaction layer has been used profitably by workers, particularly Brdička and his school [28, 29], who first studied electron transfers coupled to chemical reactions. The same concept appears in mathematical form when the above equations are solved [30].

If μ/δ is called L, then the equation of a reduction i-E curve is

$$\frac{j_l - j}{j} = \frac{k'}{k}L + \frac{m}{k}$$

in which k' and k are the rate constants of the two (inverse) electron transfers (themselves a function of potential [31]). It will be observed that in a simple system L = 1.

With regard to the morphology of the i-E curves the results are as follows: In reduction the limiting current is the same as in the absence of a coupled chemical reaction because it is defined by the rate of transport of species A_1 toward the electrode. However the *form* of the curve is changed; the system appears to be slower than a simple system without coupled chemical reaction. In oxidation the limiting current is defined by the kinetics by which B_1 arrives at the electrode surface. But in this case, now, the kinetics depend not only on mass transfer but also on the rate of the transformation of B_2 to B_1. The observed limiting current may then be lower than that which would be observed if B_1 pre-existed alone in solution at a concentration equal to actual $[B_1] + [B_2]$. This is called a *kinetic current*; its value may be calculated as a function of several constants [32].

The oxidation of aromatic amines in acid medium offers a particularly good example of a kinetic current and of the effect of the medium upon it. These substances cannot be oxidized at an appreciable rate before they are deprotonated. The product of the oxidation is the radical

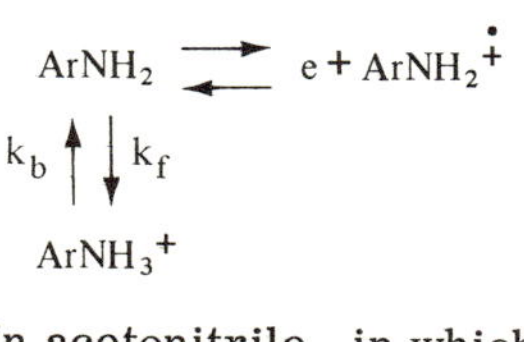

cation of some stability. In acetonitrile, in which deprotonation is relatively slow, the usual curves obtained [33] resemble Fig. 19, A. The deprotonation of $ArNH_3^+$ is so slow that only the free $ArNH_2$ molecules are oxidized and the decrease in the limiting current is proportional to the quantity of acid added. If water is added, the deprotonation rate increases and the limiting current is a true kinetic current: Its value is defined by the constants k_f and k_b and

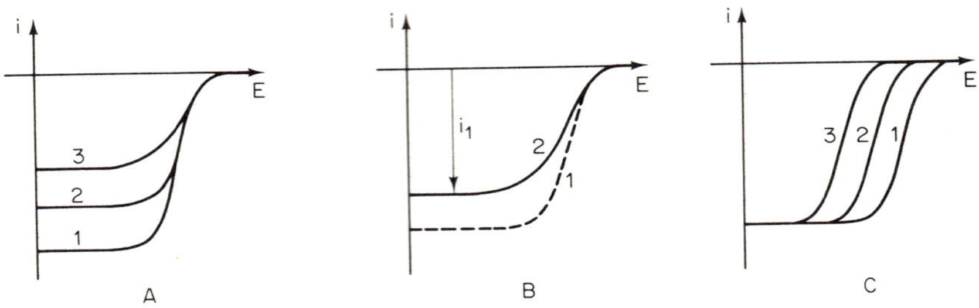

Fig. 19. Current intensity-potential curves corresponding to the oxidation of $ArNH_2$ in the presence of acid. In A, the deprotonation is infinitely slow and the decrease of the limiting current is proportional to the quantity of acid added (curves 1-3). In B, i_1 is a true kinetic current limited by the rate of the deprotonation. In C, the deprotonation is very rapid and the phenomena are controlled only by thermodynamics: the half-wave potential is shifted toward positive potentials when the acid concentration increases (curves 1-3).

the concentration of acid (Fig. 19, B). Finally, in hydroorganic media very rich in water, the deprotonation rate is so high that no appreciable decrease in limiting current may be observed (Fig. 19, C), but the half-wave potential is shifted toward positive potentials when the acid concentration is increased. More precisely, when the deprotonation rate is high enough to allow phenomena controlled only by thermodynamics to occur, the only effect of increasing of acid concentration is to shift the half-wave potential of 60 mV per pH unit (at 25° C).

3. Chemical Reaction Following Charge Transfer and Regeneration of Starting Material (Catalytic Currents)

As mentioned before, a chemical reaction following a charge transfer does not influence the limiting current associated with this transfer. It is evident that this statement is not true if the chemical reaction regenerates the initial electroactive species by a process which, in the case of reduction, may be written as

$$A + e \rightleftarrows B$$

$$B + Z \underset{k_b}{\overset{k_f}{\rightleftarrows}} A + X$$

in which Z is a monoelectroactive reagent. An increase in the reduction current for A is observed. On the plateau of the i-E curve the limiting current is a kinetic current; because there is a reduction of Z and regeneration of A, it is called a catalytic current. The apparent number of electrons involved n_{app} (>1) is the ratio of the limiting current to that which would have been obtained in the absence of chemical reaction. In principle k_f, the rate constant for the regeneration of A [34a], could be determined from n_{app}, the chemical equilibrium constant, and the concentration of the several species present. This type of process is involved whenever there is prepared by oxidation or reduction at an electrode a reagent which oxidizes or reduces a nonelectroactive organic species ([34b]; also Chapter XXIV).

4. The Case of Two Chemical and Two Coupled Electrochemical Reactions (Square Diagram)

The ensemble of two chemical steps and two electron transfers

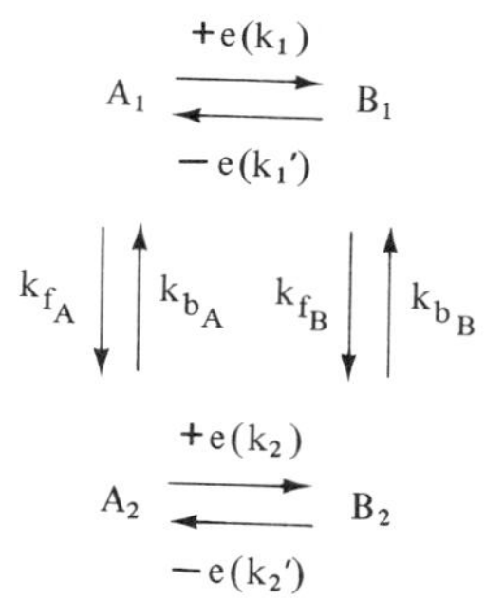

$$\sigma_A = \frac{k_{b_A}}{k_{f_A}}$$

$$\sigma_B = \frac{k_{b_B}}{k_{f_A}}$$

which can be written in the form of a square diagram is the basic model for all complex schemes, such as the one that has already been described in connection with the reduction of A to AH_2.

Once again, formulating expressions for mass transfer and charge transfer leads to the general equation for the i-E curve and shows that the limiting current in oxidations as well as in reductions has a value:

$$J_{lim\ anod} = Fm[B] \quad \text{with} \quad [B] = [B_1] + [B_2]$$

$$J_{lim\ cath} = -Fm[A] \quad \text{with} \quad [A] = [A_1] + [A_2].$$

Consequently, there is here no kinetic current whose value depends on the rate constants of chemical reactions as was the case when a chemical reaction preceded charge transfer [35].

Two limiting cases may be considered following the chemical reactions rates:

(a) The chemical reactions are very slow. The current is then the sum of J_1 and J_2 produced by each of the two simple systems A_1/B_1 and A_2/B_2, the constituents of the two systems being at concentrations defined by the laws of mass action.

(b) The chemical reactions are very rapid. The total system then behaves like a simple redox system, the standard potential of which is a function of the two standard potentials E_0^1 and E_0^2 of the two simple systems A_1/B_1 and A_2/B_2.

When the chemical reactions have intermediate rates, it is again possible to consider two types of cases, depending on the stability of the several species concerned:

(a) Two species which are on the same horizontal line of the square diagram predominate. For example, if $\sigma_A \gg 1$ and $\sigma_B \gg 1$, then A_1 and B_1 will be the predominant species at equilibrium. It can then be shown that only one transfer, corresponding to the A_1/B_1 couple, will be seen and the reaction path going through A_2 and B_2 will be negligible:

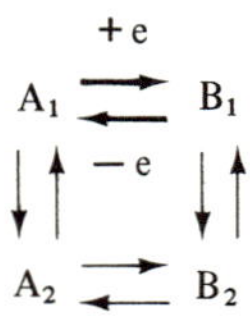

(b) Two species which are situated diagonally opposite on the diagram predominate. Suppose, e.g., $\sigma_A \gg 1$ and $\sigma_B \ll 1$. Two reaction paths may be used to reduce A_1 to B_2 or to oxidize B_2 to A_1. In reduction, a comparison of the kinetics of charge transfer ($A_1 \rightleftarrows B_1$) with the kinetics of the chemical reaction $A_1 \rightleftarrows A_2$ provides a means of deciding which path is preferred. Since the electrochemical kinetics depend on the potential of the electrode, the relative importance of the two paths may vary with potential. The i-E curve therefore may have a form far different from that of a pure wave. It has been shown furthermore that, eventually, the preferred paths may not be the same in oxidation and in reduction.

5. General Case: Grouping of Square Diagrams

If all the equilibrium and rate constants pertaining to a group of square diagrams involving, e.g., the oxidation of AH_2 to A were known (cf. Sec. III.B), it would be possible in principle to establish the reaction path which is followed and to calculate the form of the i-E curve point by point. In reality, however, many of these data are missing.

Nevertheless, in certain simple cases, it is possible to draw at least qualitative conclusions concerning the nature of the reaction path. Certain quinone-hydroquinone couples ($Q-QH_2$) were illustrative. They could be represented by a scheme of nine constituents [36]. When sufficiently basic quinones are present in sufficiently acidic media, the quinone exists in its protonated form QH^+ and only the lower six constituents of the scheme need be considered:

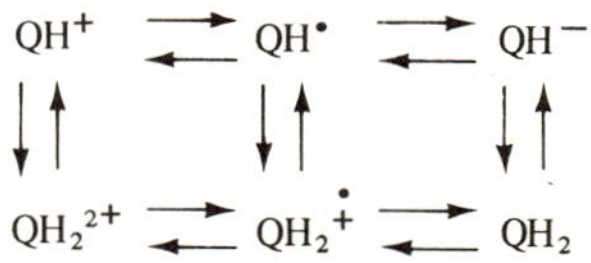

It is then possible to establish by various electrochemical methods [36] that the oxidation of the hydroquinone goes principally by path I or by path II depending on the concentration of acid:

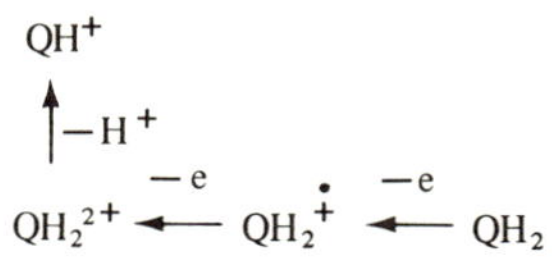

(I)

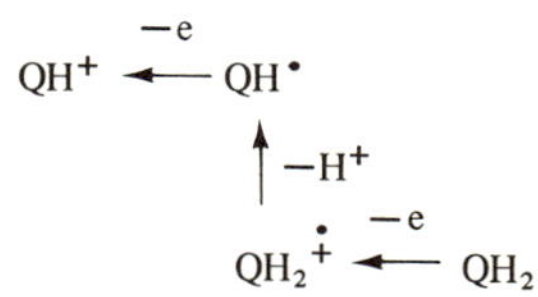

(II)

6. Case of Second-Order Chemical Reactions

The solution of the mass-transfer equations is rendered particularly difficult if second-order chemical reactions are involved. Second-order kinetics may intervene in several ways. Thus, in a simple system coupled to a chemical reaction (cf. Sec. III. C. 2) second-order kinetics may arise from the fact that reagents X and Y reacting with B_1 and B_2

$$B_1 + X \underset{k_b}{\overset{k_f}{\rightleftarrows}} B_2 + Y$$

are not present in sufficient concentration to allow pseudo-first-order kinetics to obtain. The mass-transfer equations are of the form

$$D_{B_1}\frac{d^2C_{B_1}}{dx^2} - V(x)\frac{dC_{B_1}}{dx} = k_f C_{B_1} C_X - k_b C_{B_2} C_Y$$

and two additional equations relative to C_X and C_Y must be considered.

Second-order kinetics may also be the result of dimerization of the product of the electrode reaction

$A + e \rightleftarrows B$

$2B \rightleftarrows B_2$

or, conversely, of the dissociation of a dimer prior to charge transfer.

Finally second-order reactions must also be kept in mind when the products produced at the electrode can undergo mutual redox reactions. Thus, when there are two successive electron transfers (EE)

$A + \bar{e} \rightleftarrows B$

$B + \bar{e} \rightleftarrows C,$

there must of necessity be a dismutation equilibrium

$2B \rightleftarrows A + C.$

Similarly, in an ECE process

$A + \bar{e} \rightleftarrows B$

$B + X \rightleftarrows C + Y$

$C + \bar{e} \rightleftarrows D$

one must, to be rigorous, take into account the redox reaction

$B + C \rightleftarrows A + D.$

A detailed study of many second-order coupled chemical reactions has been reported [37].

IV. ASPECTS OF THE ELECTROCHEMICAL REACTION THAT ARE DUE TO ITS HETEROGENEOUS NATURE

A. Introduction

In all that has preceded, the discussion has been limited to the statements that an oxidizable or reducible species has reached the electrode because of mass-transfer phenomena; that it is capable under certain conditions of undergoing charge transfer with the electrode; and that the kinetics of the charge transfer are governed only by the constants k^o and α, and the magnitude of the potential difference $E - E_0$. This is only an approximate description of the phenomena. It does not take into account the fact that the electrochemical reaction, properly speaking, is a heterogeneous reaction which occurs at the solution-metal interface.

We shall now present more details concerning what occurs at this interface and this will lead to a consideration of two problems:

(a) the specific structure which the solution in contact with the metal possesses, a structure which becomes established as soon as a metal is placed in a liquid containing dissolved electrolyte, even when no electron transfer is occurring between the solute and the metal (an "ideal" electrode under these conditions). Naturally, the nature of this structure and the modifications it undergoes as the electrode potential changes are important considerations affecting charge transfer when the latter occurs. This group of phenomena bear on double layer theory;

(b) certain species present in solution or formed during the electrolysis may adhere to the electrode surface because of specific binding forces; this constitutes adsorption. Adsorption may involve neutral molecules as well as ions, substances capable of undergoing electron exchange with the electrode as well as those which are not electroactive in the potential domain in question. But, in the latter case, substances which are not involved in the electrochemical reaction itself may, nevertheless, affect the electron exchange in which the electroactive species participates; the "inert" substances may retard, inhibit, or in certain cases even accelerate the exchanges.

The organic chemist who is first embarking on a study of electrochemistry finds double layer and adsorption phenomena among the most difficult concepts to master. He should be aware that they comprise a vast field of study which is in the midst of great expansion and this in spite of the fact that much work on the subject has already been done. Like other studies in the chemistry of interfaces, this field seems to possess quite specific obstacles to both experimental and theoretical approaches. Thus, the very concept of double layer, which will be described briefly below, has given rise to descriptions (or models) which differ radically from one another; other authors have taken a fundamentally different approach based on statistical mechanics [38].

The most complete discussions of the present state of our knowledge in this area are found in specialized treatments [39]. It should be noted that most of the experimental results have been obtained in aqueous systems and with dropping mercury electrodes. The latter have many advantages. The electrode material (mercury) can be purified easily and its surface is regularly renewed. Further, the falling drops also avoid the danger of having the electrode slowly contaminated by adsorption of impurities from the medium. Finally, the interfacial tension of this liquid electrode is easily measured; changes in the tension are a convenient indication that adsorption phenomena are involved.

B. Double Layer Theory

1. Existence and Representation of the Double Layer

It has been established that a current of finite value flows through an electrode when its potential or its surface is modified, even if an oxidizable or reducible substance is not present. This nonfaradaic current can be explained only by assuming that in the vicinity of the electrode the solution has a specific structure having the properties of a capacitor which must be of the order of 20 to 50 $\mu F/cm^2$.

Starting with this experimental fact many authors have proposed more or less complicated models to represent this localized structure. The conclusions drawn from these several models are more or less in accord with the experimental facts.

2. Simple Model Neglecting Specific Adsorption

Consider a positively polarized electrode in an aqueous solution of a strong electrolyte such as sodium fluoride. A simple model that may be considered is the following: A first layer of solvent molecules is in contact with the metal. The solvent molecules are oriented because they have a dipole moment. The solvated F^- ions form an orderly layer in front of the electrode in aligning themselves in a plane of closest approach, called the Helmholtz plane (Fig. 20). On the solution side of this plane there is an

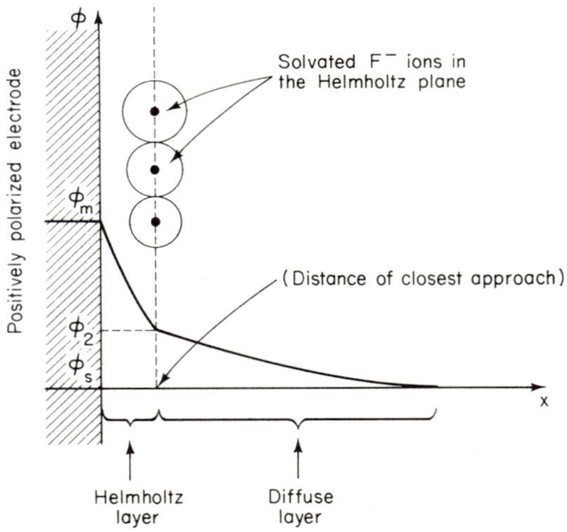

Fig. 20. Model of the double layer and change of potential with distance from the electrode in the absence of specific adsorption. The solvent molecules are not shown.

inhomogeneous distribution of anions and cations — the diffuse layer — in which anions predominate. This distribution is made evident by the existence of a region of negative charge whose absolute magnitude decreases as the distance x from the electrode increases. The potential[16] varies then from

$\varphi_m = E$ of the electrode metal to φ_s, the potential of the solution; a discontinuity in the variation is represented by the value φ_2 [17] which is the potential in the plane of closest approach of the ions.

If the initial positive potential of the electrode is decreased, the attraction for the anions is lowered; it can be seen that there must be a potential E_Z at which there is no charge on the surface of the electrode. This is the potential of zero charge. At this potential $\varphi_2 = \varphi_s$ because there is no local perturbation of the ionic atmosphere.[18] Figure 21 shows that for a given

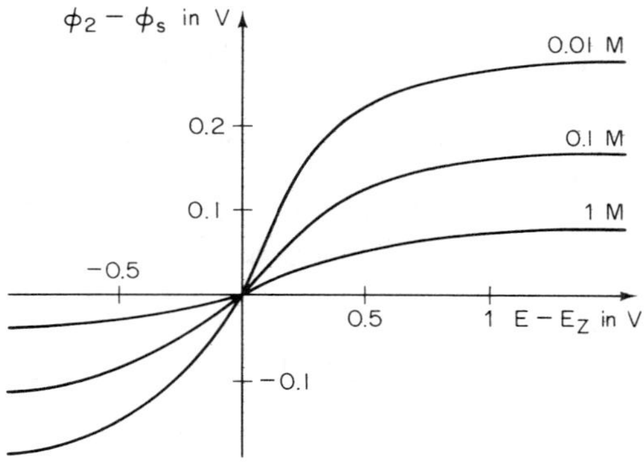

Fig. 21. Change in potential differences between the plane of closest approach and the solution as a function of the difference $E - E_Z$ for several values of the concentration of a 1:1 strong electrolyte.

value of $E - E_Z$ the difference $\varphi_2 - \varphi_s$ is smaller as the concentration of electrolyte is increased. At more negative potentials than E_Z it is the cations which are attracted and $\varphi_2 - \varphi_s$ is negative.

This first simple model can be refined considerably but even it leads to an understanding of certain consequences of the existence of the double layer. Thus, it is obvious that the double layer perturbs the concentration at the electrode of the ionic species which can be oxidized or reduced; the difference in potential which effectively controls the rate of charge transfer is then not exactly E. Most of the double layer models accept the hypothesis that charge transfer occurs at the distance of the plane of closest approach, even when there is little experimental evidence to prove this assumption. Therefore, the potential E must be corrected by the quantity $\varphi_2 - \varphi_s$. It can be shown [40] that the rate constant for the reduction of a charged species, of charge Z_o, is not given completely rigorously by the equation given on p. 34 but by

$$k = k^o \exp\left[-\frac{F Z_o}{RT}(\varphi_2 - \varphi_s)\right] \cdot \exp\left[-\frac{\alpha F}{RT}[E - (\varphi_2 - \varphi_s) - E_o]\right],$$

which can be rewritten

$$k = k^{o} \exp\left[-\frac{F}{RT}(\varphi_2 - \varphi_s)(Z_o - \alpha)\right] \cdot \exp\left[-\frac{\alpha F}{RT}(E - E_o)\right].$$

The intrinsic rate constant defined and used on p. 34 is therefore actually an apparent constant equal to

$$k'^{o} = k^{o} \exp\left[-\frac{F}{RT}(\varphi_2 - \varphi_s)(Z_o - \alpha)\right]$$

and depends on the charge of the electroactive species.

The influence of φ_2 on the kinetics of charge transfer is known as the Frumkin effect.

The difference $\varphi_2 - \varphi_s$ may be considered as constant in the presence of a high concentration of strong electrolyte and at a potential sufficiently separated from E_Z (Fig. 21). However, in the region of E_Z the variation of $\varphi_2 - \varphi_s$ with E may cause a distortion of the i-E curve in the case of slow systems [41].

3. The Double Layer when There is Specific Adsorption

In the model system which is to be described there is no reason for the potential of zero charge E_Z to change with changes in the concentration of electrolyte. In fact, however, this behavior is adhered to only by several electrolytes which have strongly solvated ions (OH^-, F^-, Li^+, Na^+, etc.). Salts composed of ions of weaker solvation energy (I^-, Cl^-, NO_3^-, Cs^+, R_4N^+, etc.) present a more complex behavior in that they exhibit a variation of E_Z with electrolyte concentration.

The latter constitutes a case in which there is specific adsorption of one or both ions of the electrolyte. The model which has been proposed most frequently to represent this situation is based on the hypothesis that certain ions which have sufficiently strong interaction with the metallic surface of the electrode can approach the latter within a distance X, which is less than the distance X_2 within which other ions not having this interaction can come. When the special features allowing an approach to distance X_1 are involved, the solvation shell simultaneously disappears or at least is deformed.

Under these conditions the structure of the double layer is as follows [42] (Fig. 22): At certain sites on the surface of the electrode the monomolecular layer of solvent in contact with the metal is replaced by the specifically adsorbed ions. The imaginary plane passing through the electrical center of these ions is called the inner Helmholtz plane (IHP) or sometimes simply the Helmholtz plane. It is at a distance X_1 from the electrode and the potential there has the value φ_1. Ions which are not specifically adsorbed and which have their normal solvation shells approach

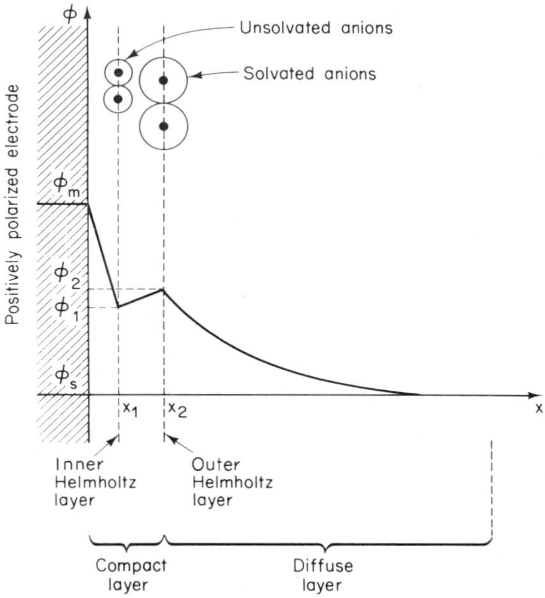

Fig. 22. Model of the double layer and change of potential with the distance from the electrode in the case of specific adsorption of the anion.

to a distance X_2 ($>X_1$). The imaginary plane passing through the center of the latter ions is called the outer Helmholtz plane (OHP) or sometimes the Gouy plane. In this plane the potential has the value φ_2. The curve representing the potential as a function of X, therefore, has two discontinuities in this case.

The region bounded by the metal and the OHP is the inner (compact, rigid, Helmholtz, Stern) layer. On the solution side of the inner layer is the diffuse layer which extends all the way from the OHP into the bulk of the solution. This diffuse layer behaves like an "ionic atmosphere" [42] of the electrode in which the potential drops to φ_S, its value in the bulk of the solution. It should be noted that the region in which there is an effective perturbation of the structure of the solution is very thin. It can be shown [42], e.g., that in a solution 10^{-1} M in a 1:1 electrolyte in water at 25° its thickness is of the order of 10^{-6} cm; this thickness is generally negligible by comparison with δ, the diffusion layer thickness which is of the order of 10^{-3} cm (p. 32). Consequently, it is possible to neglect the existence of the double layer when considering problems of mass transfer. After the latter problems have been solved, it is generally sufficient to apply a correction to account for the double layer.

Like the one previously presented, the model which takes specific adsorption into account is open to many refinements. For example, completely rigorous treatment has to include the variation in dielectric constant in the vicinity of the electrode (a variation which results from the orientation of solvent molecules in the electric field of the double layer) and the discreteness of charges in the double layer [58].

4. When Must Double Layer Phenomena Be Taken into Account?

After this brief discussion of the special structure of the solution in the vicinity of the electrode, the question arises: In the study of a reaction process, must the double layer be considered as a real enough fact but one which, however, has no practical consequences, or on the contrary must its diverse influences be kept in mind?

This question is worth facing because, while certain workers have done important theoretical work on the influence of the double layer on electrochemical kinetics, others have, on the contrary, completely ignored its influence on the mechanisms of electrode reactions.

Actually, the evidence points to the fact that double layer structure plays a very important role; this conclusion is supported by many data from the field of electroorganic synthesis. Thus, it has been noted that in the electrolysis of acrylonitrile in aqueous solutions, hydrodimerization (Chapt. XIX) leading to adiponitrile predominates when the supporting electrolyte is a quaternary ammonium salt, whereas simple saturation leading to propionitrile is favored when alkali metal cations are present [43]. Similar phenomena have been noted in nonaqueous media. Thus, the reduction of ethyl bromide in propylene carbonate at a lead cathode yields tetraethyllead

$$C_2H_5Br + \tfrac{1}{4}Pb + e^- \rightarrow Br^- + \tfrac{1}{4}Pb(C_2H_5)_4$$

when the supporting electrolyte is an onium salt (R_4N^+ or R_3S^+), while at the same potential ethane is formed

$$C_2H_5Br + H^+ + 2e^- \rightarrow C_2H_6 + Br^-$$

when the supporting electrolyte is a salt of Li^+, K^+, Ca^{2+}, or NH_4^+ [44]. It is reasonable to ascribe these phenomena to a preferential adsorption of onium cations at the electrode which then gives rise in this case to a layer at which protonation of the species formed upon charge transfer is disfavored.

In reality a hiatus exists between the empirical facts described above and the state of development of theory, so that at present the data cannot be explained completely nor can a prediction be made concerning the nature and concentration of an electrolyte to be selected for best results in a given synthesis. This matter is discussed further in Chapter III.

The principal conclusions that can be drawn from the theoretical treatments described above may be summarized as follows: where kinetics are concerned, since mass-transfer phenomena are very little influenced by the existence of the double layer (v. s.), it is mainly the influence of the latter upon the kinetics of charge transfer and of coupled chemical reactions which must be analyzed. Such an analysis requires an understanding of the detailed structure of the double layer. But actually only models of this structure are available. To be sure, these models lead to conclusions which are in accord with the results of thermodynamic analysis of various macroscopic consequences, e.g., electrocapillarity (p. 66), of the existence of the double layer. This agreement, however, is not a proof of the validity of these models [43].

Nevertheless, using the model (Fig. 22) which alone is applicable in cases in which specific adsorption is involved, it has been shown that, insofar as the influence of the double layer upon the kinetics of charge transfer is concerned, consideration of the potential φ_2 of the OHP in which the charge transfer supposedly occurs is not sufficient to explain the experimental results obtained when protons are reduced in aqueous media on mercury in the presence of specifically adsorbed halide ions. It is further necessary to postulate that there exists an interaction between the specifically adsorbed ions and the activated state which exists during the transfer. This leads to the conclusion that the transfer occurs not in the OHP but in the inner layer [46].

The influence of the double layer upon the kinetics of coupled chemical reactions has been considered mainly in connection with the reduction of protons released upon dissociation of a weak acid in water

$$AH \underset{k_b}{\overset{k_f}{\rightleftarrows}} A^- + H^+$$

$$H^+ + e^- \rightleftarrows \tfrac{1}{2}H_2 \,.$$

Whether these studies were designed to determine the precise influence of the double layer upon the rate constants for the dissociation reactions [47] or were undertaken in order to evaluate the possibility of determining these constants by electrochemical means [48], they naturally were applied to weak acids AH for which a dissociation step prior to charge transfer is necessarily included in the overall electrochemical process.

Several effects must be kept in mind when considering the fact that in the vicinity of the electrode the dissociation reaction and the reverse recombination do not occur in a homogeneous medium. These effects, which result from the inhomogeneity introduced by the double layer, may include change of the concentration of ions in the double layer compared to the concentration in the bulk, enhancement of the dissociation rate by the field, alteration of the rate and the equilibrium expressions by dielectric saturation of the medium, and acceleration or inhibition of mass transfer of ionic species by the potential and the electric field.

These studies show that the thickness of the reaction layer (p. 51) in which net chemical reaction is assumed to occur and where the chemical system is not in equilibrium may be of the same order of magnitude as the thickness of the diffuse layer. Thus, in the reduction of acetic acid in the presence of ca. 1 M acetate ion, μ is of the order of 4.5 Å; almost all the chemical reaction proceeds in the zone perturbed by the double layer. On the other hand, if the acetate ion concentration is 10^{-2} M, μ is ca. 45 Å and much of the chemical reaction occurs in a field-free region of the solution.

The results above concern the exchange of protons, a reaction having very rapid kinetics. For a slower reaction, almost all the reaction must a fortiori proceed outside the double layer and the influence of the latter is negligible.

The results of the reduction of ethyl bromide cited above show that the overall electrochemical reaction may lead to different products depending on the rate of protonation of the intermediate species produced at the electrode.

The double layer may in certain cases influence the stereochemistry of the product(s) formed in an electrochemical reaction [60]. This phenomenon is discussed in Chapter XXVII.

C. Adsorption Phenomena

1. General Considerations [49]

As was mentioned above, certain ions can be adsorbed at an electrode and thereby impart a special structure to the double layer. The same phenomena may be caused by neutral molecules, particularly if they have a dipole moment. Naturally, solvent molecules are also adsorbed and in fact adsorption of a solute is a displacement in which molecules of solvent are replaced at the electrode surface by molecules of solute. This simple picture makes clear why in organic electrochemistry adsorption phenomena are always more important in aqueous than in nonaqueous media: it can be shown [50] that the energy involved in displacing a molecule of organic solvent previously adsorbed on the electrode by an organic solute molecule is generally much less than that involved in the displacement of a water molecule.

The molecules of a substance adsorbed at an electrode have not usually displaced <u>all</u> the solvent molecules. The surface concentration of the adsorbate, in moles per square centimeter (usually denoted by Γ), is therefore less than the maximum value Γ_m theoretically possible. The two concentrations Γ and Γ_m are related by the <u>adsorption isotherm</u>. The expression formulating this relationship involves the types of interaction between the molecules and the electrode and among the molecules themselves in the adsorbed state [50]. Several isotherms have been proposed [51] corresponding to several approaches to the problem. One of the most frequently used at present is the Langmuir isotherm

$$\Gamma = \Gamma_m \frac{Co}{Co + a}$$

in which Co is the concentration of the dissolved substance and a is a constant characteristic of this compound.

When the electrode potential changes, Γ_m changes. Specifically, at a sufficiently negative or positive potential the attraction of the ions by the electrode becomes strong enough to displace the neutral adsorbed molecules. This is illustrated in Fig. 23 which shows, on the one hand, the variation of

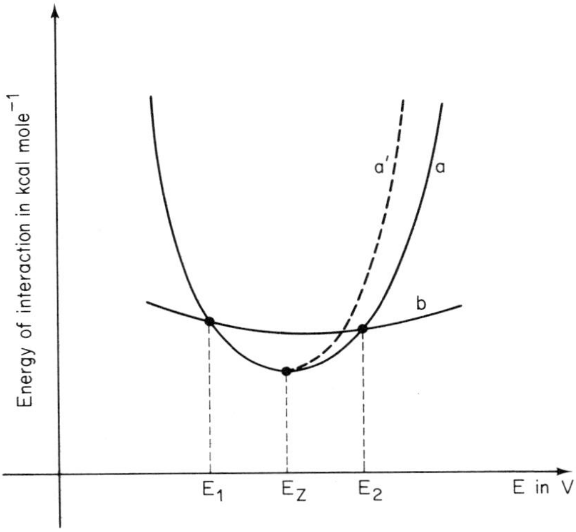

Fig. 23. Variation of the energy of interaction (in absolute values) between the electrode and (a) ions of an electrolyte, e.g., LiF, and (b) a neutral bipolar substance. Curve a′ corresponds to an electrolyte (e.g., LiI) whose anion is specifically adsorbed at the electrode. Outside the potential domain E_1 - E_2 the ions of the electrolyte are adsorbed in preference to the neutral substance.

the energy of interaction between the electrode and a neutral dipolar molecule and, on the other, between the electrode and the ions of an electrolyte. The order of magnitude of the energies involved is in general several kilocalories per mole for aqueous solutions.

2. Electrocapillary Curves

The majority of the studies of adsorption phenomena has been carried out at the mercury electrode, because at this liquid electrode the interfacial tension γ, which is dependent on adsorption phenomena, can easily be

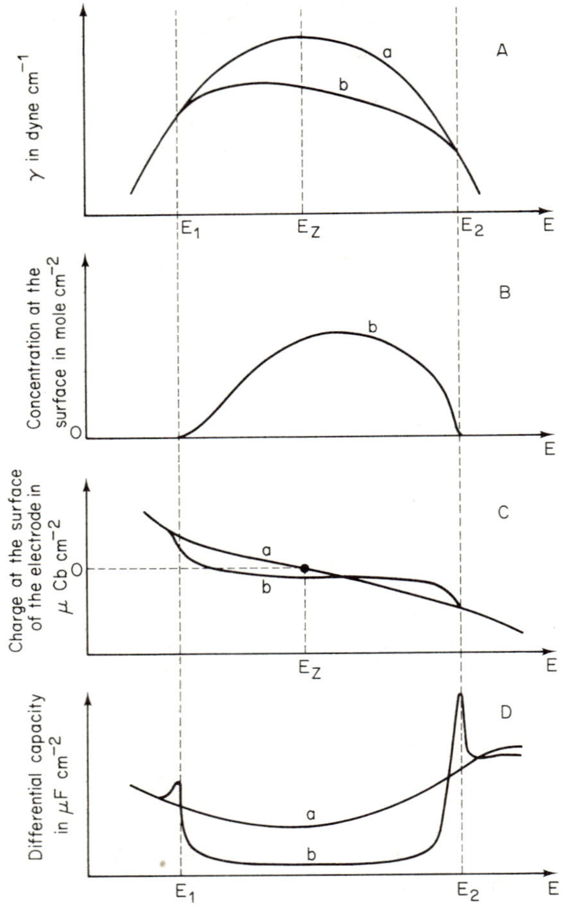

Fig. 24. Variation with potential of: A, interfacial tension; B, the surface concentration of the neutral adsorbed substance; C, the charge density of the electrode; D, the differential capacity of the electrode. For each diagram a relates to the electrolyte solution not manifesting specific adsorption and b to the same solution after an adsorbable substance had been added.

measured, e.g., by Lippmann's capillary electrometer [52] over a range of values.[19] The interfacial tension varies considerably with the potential of the electrode. The curve of Fig. 24, A, showing this variation, is called the electrocapillary curve.

For a solution of an electrolyte neither of whose ions exhibits specific adsorption, the curve is essentially a parabola (Fig. 24, A, a); the peak has the previously defined potential of zero charge E_Z. A very low concentration (e.g., 10^{-4} mole/liter) of an adsorbable substance when added to this solution flattens out the electrocapillary curve between potentials E_1 and E_2 (Fig. 24, A, b). At the latter potentials the substance is desorbed.

In obtaining curves like these it is assumed that adsorption equilibrium has been reached and, importantly, that no charge transfer is occurring at the surface of the electrode (ideal polarized electrode). If equilibrium has been attained, the adsorption phenomena can be studied by thermodynamics [53] from which is derived the following expression called the Gibbs relationship for the surface concentration of a neutral substance:

$$\Gamma = -\frac{1}{RT}\frac{d\gamma}{d(\ln Co)}.$$

From this equation the change of Γ with the electrode potential is obtained (Fig. 24, B).

3. Charge Density and Double Layer Capacity

The presence of adsorbed molecules at the surface of the electrode changes the electric charge per unit surface of the electrode (Fig. 24, C). It can be shown that if adsorption equilibrium has been reached this change is related to the variation of interfacial tension with potential by the Lippmann equation

$$q = -\frac{\partial\gamma}{\partial E}.$$

By definition, the charge on the electrode is zero for $E = E_Z$ when no adsorbable substance is present in the solution, but in aqueous solutions it may reach values of about 10 to 20 $\mu C/cm^2$ if the potential is sufficiently positive or negative. It is the presence of this charge on the electrode and, on the solution side, of an equal but opposite charge which, as already mentioned, confers the properties of a capacitor on the double layer.

Two types of capacity must be considered in this connection. The integral capacity is defined by the relation

$$C_i = \frac{q}{E - E_Z}.$$

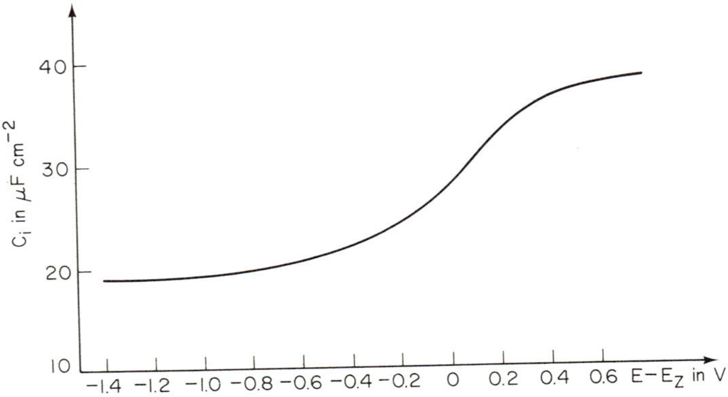

Fig. 25. Variation of integral capacity C_i with potential for a mercury electrode immersed in an aqueous 0.5 M solution of sodium sulfate.

As can be seen, it varies with the potential of the electrode (Fig. 25).[20] For this reason it is advantageous to introduce the concept of differential capacity C, defined by

$$C = \frac{\partial q}{\partial E}$$

and frequently incorrectly called simple double layer capacity. The differential capacity itself varies with potential (Fig. 24, D). It is a quantity relatively easy to measure experimentally by means of a bridge circuit using alternating current [54]. It is important to note that these measurements are possible only because C can be considered to remain unchanged when there is only a very small change in potential, ΔE. Nevertheless, both theory and experiment show that the values of C obtained in this way depend on the frequency used (Fig. 26). It is therefore necessary to indicate the frequency at which a C = f(E) curve was determined, or, better yet, to provide the curve which is obtained when the frequency is extrapolated to zero [55].

In spite of the numerical variations in the values of C, it is clear that the C = f(E) curves have general aspects which are independent of frequency and, specifically, C undergoes large changes at potentials at which neutral substances are adsorbed or desorbed (Fig. 24, D, cf. a and b). Occasionally, in the case of aromatic substances large changes in C occur at potentials situated between the desorption potentials E_1 and E_2; these changes have been attributed to reorientation of adsorbed molecules at the surface of the electrode [56].

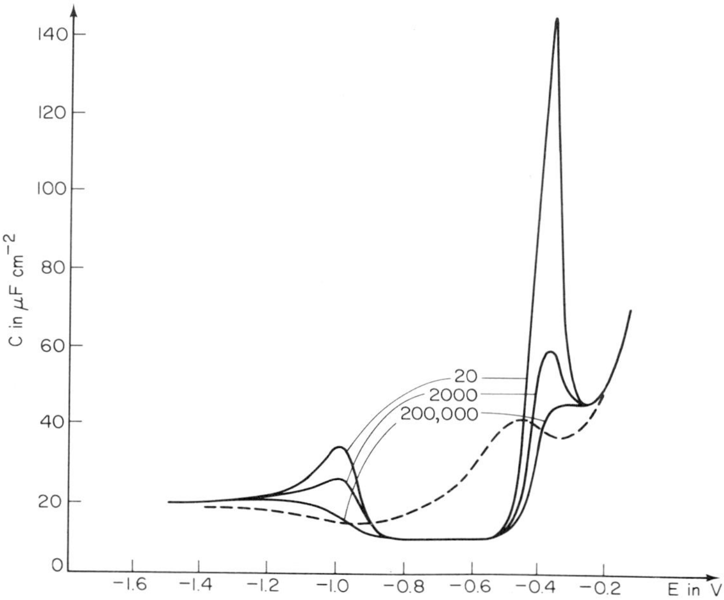

Fig. 26. Variation of the differential capacity at 25° of a hanging mercury drop in a 1 M aqueous solution of KCl containing 10^{-2} M n-amyl alcohol. Measurements were made at 20, 2000, and 200,000 H_Z. The reference electrode is the SCE. The discontinuous curve relates to the solution containing only the electrolyte. Note: in most of the publications dealing with mercury electrodes, potentials increase positively to the left according to the convention adopted by polarographers.

4. Adsorption Kinetics

The several relationships discussed above are the result of thermodynamic considerations which have been applied to an electrode at which adsorption equilibrium has been established and at which no charge transfer is occurring. If these conditions are not satisfied, two kinetic problems must be considered:

(a) the kinetics of the adsorption itself, [21]

(b) the influence of this adsorption on the kinetics of the charge transfer taking place at the electrode.

The first of these two kinetics is really the resultant of the kinetics of several phenomena [57]. The adsorbable species must arrive in the vicinity of the electrode by a mass-transfer process similar to those that have been discussed in connection with i-E curves. This species must then move from solution into the adsorption layer. In certain cases, finally, the adsorbed species must undergo some sort of association or must form a two-dimensional structure comparable to a two-dimensional crystallization.

It is very difficult to study these several phenomena separately. The many experimental studies which have been undertaken to date in general show that the kinetics of the adsorption process proper and the organization of the adsorbed layer are very rapid so that in numerous instances the sequence of phenomena is controlled by mass transfer alone.

5. Effects of Adsorption on the Kinetics of Electrode Phenomena

Adsorption has an important influence on electrode reactions. It is a factor not only in electrochemical syntheses but also in important technological areas, such as the inhibition of corrosion of metals by organic substances [61] and the effect of organic substances on the electrodeposition of metals [62] and the quality of the deposit. Thus, it is not surprising that very much effort has been expended on this question [63]. In spite of all the work that has been done, it seems at present impossible to formulate a general theory which will integrate all the known experimental facts.

The phenomena involved are numerous and complex. First of all, it is necessary to discern that many different situations may obtain depending on whether the adsorbate is the substance actually undergoing the electrochemical change, or is a reaction intermediate, or is the final product, or is even a nonelectroactive species present in the solution. Each of these substances can affect the electrochemical transformation in a different way.

Thus, considering only charge transfer, it must be borne in mind that the surface of the electrode is associated with two regions. In one of these, the double layer is not perturbed by adsorption; here, the intrinsic constant (p. 34) is k'^{o}. In the other, the adsorbate reduces the rate of transfer — even to zero; this corresponds to the existence of another constant k'^{o}_{ads} [64]. Under these conditions the overall transfer occurs with a rate

$$k''^{o} = (1 - \theta)k'^{o} + \theta k'^{o}_{ads}$$

in which θ ($= \Gamma/\Gamma_m$) denotes the coverage of the electrode. But this simple model which pictures the adsorbed molecules as existing in separate little areas on the surface of the electrode does not always account for the results observed. It is often necessary to assume that the molecules may move freely on the surface, much as do molecules of a gas in two dimensions. In this case the presence of an adsorbate is manifested in an increase of the activation energy for charge transfer [65].

The presence of adsorbed species can also modify the potential of the double layer φ_2, particularly if the species are ionic (Frumkin effect). This change may cause an increase just as well as a decrease of k'°, so that often there is an acceleration of the rate of charge transfer in the presence of an adsorbate [66]. However, this rate increase can also arise from another cause: certain specifically adsorbed anions accelerate the rate of charge transfer to an extent that cannot be accounted for by the Frumkin effect alone. In this case charge transfer occurs through the adsorbed species (ligand or bridging effect) [67].

A species adsorbed at an electrode may also modify the rate of mass transfer. This must be taken into account especially when using the model which considers that the electrode surface is only partially covered by the adsorbate [68]. Finally, certain authors have considered that chemical reactions which are coupled to charge transfer may be perturbed by the presence of an adsorbate. Specifically, this phenomenon may be involved in very rapid protonation reactions, e.g., as in the reaction of the radical anion formed by one-electron reduction of nitrobenzene [69].

From the experimental point of view, the i-E polarographic curves obtained with the DME (Chapter II, p. 106) show particularly effectively the influence of adsorption on electrode phenomena. Furthermore, polarography has yielded satisfactory quantitative interpretations [70], particularly in the case of rapid electron transfers. For example, the reduction of methylene blue to its leuco derivative in aqueous medium at pH 7.96 shows a "pre-wave" whose limiting current (Fig. 27) is independent of concentration [71]. The phenomenon has been attributed to adsorption of the leuco form which presumably is the only one of the two forms which is adsorbed under these conditions. It is clear that a rapid and reversible reduction (p. 25) which yields particles that are held on the electrode by adsorption forces should

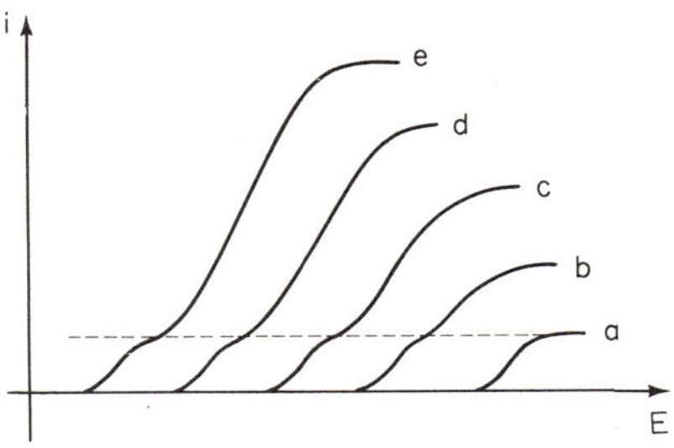

Fig. 27. Current intensity-potential curves for reduction of various concentrations of methylene blue in aqueous medium at pH 7.96 on the DME. The curves have been purposely staggered for clarity. Concentrations: (a) 0.61×10^{-4} M; (b) 1.18×10^{-4} M; (c) 2.79×10^{-4} M; (d) 3.67×10^{-4} M; (e) 4.54×10^{-4} M. (Cf. Ref. [71].)

require less energy than a reduction that leads to free particles. The adsorption of the reduced species, therefore, gives rise to a wave at a more positive potential than the standard potential (Fig. 27, curve a). Since, however, the quantity of reduced product which may be adsorbed at the electrode is limited, the normal reduction of the nonadsorbed species, when present in sufficiently high concentration, is evidenced by a second wave at a more negative potential (Fig. 27, curves b-e). The limiting current of the first wave, the adsorption prewave, is independent of concentration because it is limited only by the amount of substance which can be adsorbed at the electrode. Satisfactory quantitative interpretations have been made [71].

In the above example the electrochemical reaction as well as the adsorption could be considered sufficiently rapid for equilibria to be established instantly. Certain cases in which kinetic steps intervene could also be analyzed [70, 72].

D. Interfacial Phenomena at Solid Electrodes

The mercury electrode then has several great advantages, but unfortunately the oxidation potential of mercury is so low that mercury cannot serve as anode in studies of the oxidation of most organic compounds. Consequently, many such investigations since the 1950s have used solid electrodes, in particular noble metal electrodes of platinum, gold, etc.

Brief mention will now be made of the surface phenomena at solid electrodes; a detailed discussion will be found in Adams' recent book [73] which is devoted entirely to this subject.

As a matter of fact, a platinum electrode, for example, does not possess a surface of bare platinum except within a relatively narrow potential range. This domain is called the "double layer domain" because between these limits the metal is covered by nothing but the solvent in its specific state of organization and electrolyte concentration. At higher potentials, however, the metal is covered with oxides, and at lower potentials by adsorbed hydrogen which has been formed by discharge of the protons of the medium. These facts are particularly well illustrated by Böld and Breiter's [74] cyclic voltammograms (Chapter II, p. 115) taken with a platinum electrode immersed in various aqueous solutions. Figure 28 concerns phenomena in a sulfuric acid solution. The bright platinum electrode immersed in the solution is subjected to a potential which varies linearly with time (upper part of Fig. 28). At time t_1 the direction of the sweep is reversed. In the course of all these potential changes the current passing through the electrode is recorded as a function of the potential (lower part of Fig. 28). Starting at A, there is in the first place a group of two peaks between 0 and 0.4 V corresponding to the oxidation — the current is positive — of the hydrogen adsorbed on the surface of the electrode. Between 0.4 and 0.8 V the metal is bare and sustains only a nonfaradaic current due

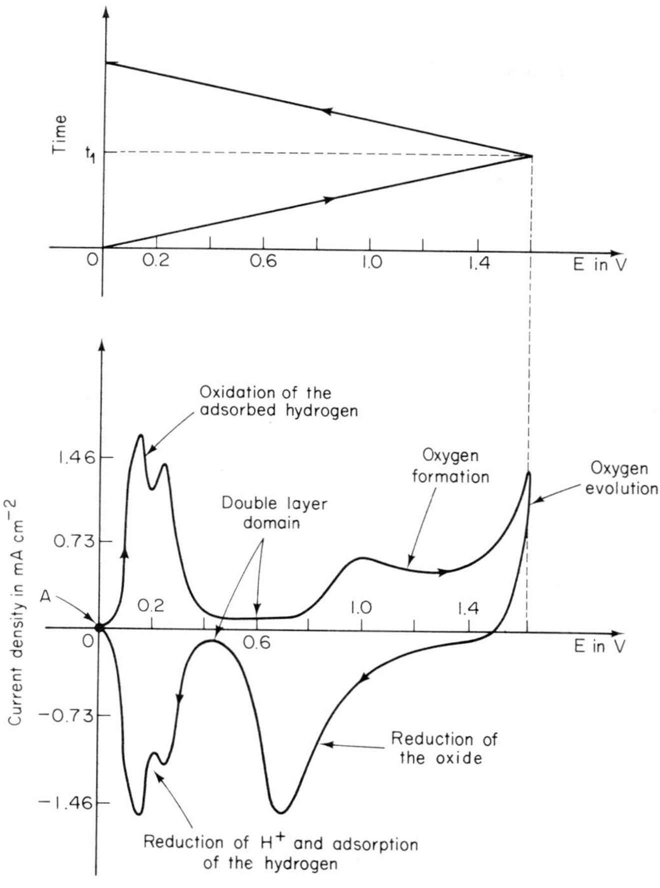

Fig. 28. Cyclic voltammetry curve showing the formation of oxides on a platinum electrode, the reduction of the oxides, and the adsorption of hydrogen; 2.3 M aqueous solution of H_2SO_4, 25°. Reference electrode, H_2/H^+. Sweep rate, 0.5 V/sec.

to the presence of the double layer. Beginning at 0.8 V the current increases again due to the oxidation of water which results in the formation of surface oxides; near 1.6 V massive oxidation of water occurs with evolution of oxygen.

On the return sweep there is a large peak of current centered at 0.7 V. This is produced by the reduction of the oxide which has previously been formed. The surface is not again bare platinum until a voltage of ca. 0.4 V is reached and soon the film of hydrogen covering the metal is reduced. The sweep was not carried sufficiently far in the negative region for massive reduction of protons and liberation of hydrogen to occur.

The first conclusion to be drawn from the analysis of Fig. 28 is that the exact nature of the electrode surface depends on the potential to which it had previously been subjected. Thus, an electrode "cathodized" at 0.0 V and then brought to 0.7 V possesses effectively a platinum surface covered by the double layer. On the contrary, if the electrode had been "anodized" at 1.4 V, change of the voltage to 0.7 would have still left the surface at least partially covered with oxides; on the other hand, if the same anodized electrode had then been brought to 0.4 V, the entire oxide film would have been reduced and the newly regenerated surface of the platinum would have had the same characteristic catalytic properties as has platinized platinum. There are many controversies [75] concerning the exact nature of this regenerated platinum surface.

It should also be noted that the phenomena described are further complicated by the "history" of the electrode [76]. The properties of the electrode are greatly modified if it is used in many successive potential sweeps or if it has previously been used in a variety of electrolyses. "Pretreatment" of the electrodes may, therefore, be very important.

Two supplementary comments should be added: (1) The formation of films of oxides of noble metals and of adsorbed hydrogen is also observed in organic media; the concentration of residual water is enough to cause them. (2) Curves similar to that of Fig. 28 show that the surface of a metal electrode is most definitely the site of complex phenomena which manifest themselves, for example, in the existence of a significant residual current in cyclic voltammetric studies. It is appropriate to point out that the latter method is particularly suitable for detecting change in residual current and that under conditions under which ordinary i-E curves are obtained (Chapter II, p. 106) these changes are generally of small significance. Furthermore, many electrochemical reactions can be carried out at a gold or platinum electrode covered with oxides. It is simply desirable to keep in mind that the nature of the surface may change in the course of a potential sweep and this change is sufficient to change the kinetics of charge transfer considerably and therefore the reaction mechanism.

V. RELATION BETWEEN STRUCTURE AND ELECTROCHEMICAL PROPERTIES

A. General Considerations

The study of the electrochemical properties of organic compounds does not have for its sole purpose the development of new syntheses. Besides the synthetic aspect, which constitutes the major area of this book, organic electrochemistry also has many connotations for physical chemistry. We shall describe here some of the relationships which exist between the structures of compounds and their electrochemical properties.

Right at the start of organic polarography (the reduction of organic compounds at the dropping mercury electrode), J. Heyrovsky, the inventor of the method, stated that the reduction becomes easier (i.e., occurs at more positive potentials), the greater the number of double bonds in the molecule [78]. Soon afterward, Shikata and Tachi's "electronegativity rule" embodied the concept that the reduction potential becomes more positive as the electronegativity of substituents on a molecule becomes greater. Later the rationale for these rules was satisfactorily supplied, thanks to the development of the ideas of inductive and mesomeric effects. More recently, the relationship between structure and reduction (or oxidation) potential has been treated more quantitatively. We shall consider, in turn, the relation between reduction (or oxidation) potential and molecular orbital energies and then the extrathermodynamic types of relations (Hammett or Taft). Detailed discussions of these questions are presented in Refs. [79] and [80], respectively.

B. Relationship between Oxidation and Reduction Potentials and Molecular Orbital Energy

Consider the simple system comprising an aromatic hydrocarbon and its radical anion:

$$R + \bar{e} \rightleftarrows R^{\dot{-}}.$$

As has been mentioned before, the half-wave potential of the i-E curve for the reduction is given by

$$E_{1/2} = \frac{\Delta G^{o}}{F} - \frac{RT}{F} \ln \frac{m_R}{m_{R^{\dot{-}}}}.$$

Now if A stands for the electron affinity of compound R, i.e., the energy liberated when a molecule of R in the gaseous state accepts an electron, and $(G^{o}_{e^-})_{el}$ the free energy of the electron in the electrode metal, and $(G^{o}_{R})_{sol}$ and $(G^{o}_{R^{\dot{-}}})_{sol}$ the free energies of species R and $R^{\dot{-}}$ in solution, then

$$\Delta G^{o} = (G^{o}_{R})_{sol} - (G^{o}_{R^{\dot{-}}})_{sol} + (G^{o}_{e^-})_{el}$$

$$= (G^{o}_{R})_{gas} - (G^{o}_{R^{\dot{-}}})_{gas} + (G^{o}_{e^-})_{el} + \Delta\Delta G^{o}_{sol}.$$

$\Delta\Delta G^{o}$ is the difference in free energy of solvation between R and $R^{\dot{-}}$. In the absence of entropy effects this then becomes

$$\Delta G^{o} = A + (G^{o}_{e^-})_{el} + \Delta\Delta G^{o}_{sol}.$$

In a series of aromatic hydrocarbons the terms m_R and $m_{R^{\dot{-}}}$ differ little from each other and vary little from one hydrocarbon to another because of the small difference between diffusion coefficients D_R and $D_{R^{\dot{-}}}$. Since $(G^o_{e^-})_{el}$ is a constant, it is clear that $E_{1/2}$ varies linearly with A.

This relationship can be correlated with the results of the calculation of the energy of the lowest unoccupied orbital, which is -A. Then,

$$E_{1/2} = -(\alpha_0 + m_{m+1}\beta_0) + c'$$

in which α_0 and β_0 are the coulombic and resonance integrals in terms of the Hückel theory and c' is a constant which is applicable to all the hydrocarbons in the series discussed.

Taking into consideration the changes in $\Delta\Delta G^o_{sol}$, it can be shown [81] finally that there is a relation

$$E_{1/2} = bm_{m+1} + c$$

between the half-wave potential and the lowest unoccupied orbital (Chapter V). Correlations of this type have been verified satisfactorily for a series of aromatic hydrocarbons whose half-wave potentials were determined in various water-dioxane mixtures [82]. The results (cf. Fig. 29) fit the equation

$$E_{1/2} = (2.368 \pm 0.099)m_{m+1} - (0.924 \pm 0.109).$$

It should be noted that when an experimental value for a half-wave potential deviates considerably from the straight line, it is an indication that the hydrocarbon is probably not planar [82].

These studies have stimulated many refinements which have considered possible variations of β_0 [83].

When the same hydrocarbons are oxidized, for example, at a platinum anode in a medium such as acetonitrile, the same types of relationships are obtained between the potential of the anodic half-wave and the energy of the highest occupied molecular orbital m_m [84]. Finally, in the case of alternant hydrocarbons, it is known that the energies m_m and m_{m+1} are symmetrically disposed in reference to a zero level of energies. As a consequence, a linear relationship is obtained between the anodic and cathodic half-waves, as first pointed out by Hoijtink [85]. This constitutes an undeniable success for the Hückel theory; in addition it justifies the distinction between alternant and nonalternant hydrocarbons because compounds of the latter category, e.g., fluoranthene, do not fit this relationship.

These correlations pertain not only to aromatic hydrocarbons. The reduction of ketones is a complex process which involves electrons and protons, but in aprotic media or in aqueous media at very high pH it can be

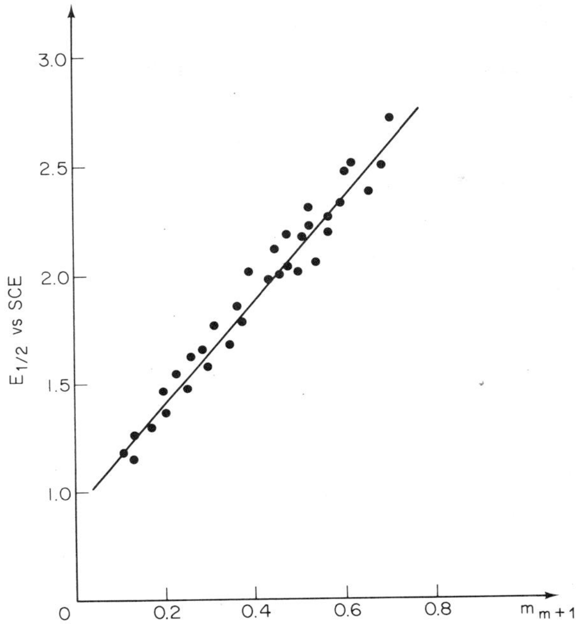

Fig. 29. Relation between the half-wave reduction potential of some aromatic hydrocarbons and the energy of the lowest unoccupied molecular orbital. The medium is a water-dioxane mixture. The i-E curves were obtained at mercury drop electrodes. The potentials are referred to a SCE in aqueous medium.

shown that the electrochemical step recorded in an i-E curve involves the transfer of only one electron

$$RR'C{=}O + \bar{e} \rightarrow RR'C^{\bullet}\text{-}O^{-}.$$

Under these conditions a linear relationship has been found also between $E_{1/2}$ and the energy of the lowest unoccupied orbital m_{m+1} [86].

The following comment must be made about the correlations discussed above. Their very existence presupposes that $E_{1/2}$ is the potential of a simple redox system which is rapid and reversible and that these conditions hold for a series of compounds which exhibit no changes in solvation energies $\Delta\Delta G^{o}_{sol}$. These restrictions are so severe that it is astonishing that such a large number of satisfactory correlations have been published. For example, it is rare indeed to find an aromatic hydrocarbon which is oxidized in a one-electron step to a stable cation radical perfectly stable toward the organic media used in these studies. Much more frequently systems are found to be completely irreversible due to a series of chemical reactions following the electron-transfer step. The correlations that have been obtained between $E_{1/2}$ and the orbital energy in question must be due to the fact that the nature of the follow-up reactions must be identical in the series of compounds studied and certainly also to the canceling of errors.

The simultaneous study of the molecular orbital feature of certain organic compounds and of their electrochemical properties provides information that may be used for more than arriving at the correlations described above. The case of the one-electron oxidation and reduction of nitrosobenzene in an organic medium at a platinum electrode is illustrative [87]. The radical cation and radical anion formed, respectively, are sufficiently stable to permit obtaining their electron paramagnetic resonance spectra (Chapter II, p. 145). From these spectra it has been established that the nitrogen atom a_N has coupling constants of 37.0 and 7.97 G (gauss) in the radical cation and radical anion, respectively. This large difference can be explained only by postulating that in the radical cation the unpaired electron is localized on the NO group (σ radical) while in the radical anion it is delocalized over the entire system of π orbitals of the molecule (π radical).

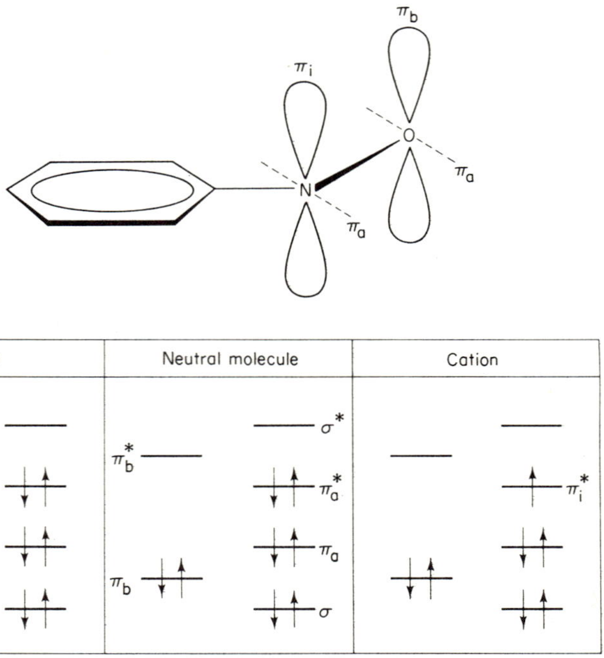

Fig. 30. Electronic structure of the NO group in nitrosobenzene and of the radical cation and radical anion.

These results are clarified by considering the electronic structure of the NO group (Fig. 30). The unpaired electron of the cation radical must be in a π_a^* orbital which, because of the geometry of the NO group, is nonlinear but not coplanar with the benzene nucleus; the above orbital cannot be conjugated with the π system of the ring. Conversely, the unpaired electron in the radical anion is in a π_b^* orbital which is conjugated with the π system of the aromatic ring.

This example illustrates how, by combining an electrochemical method with an appropriate physical method, it is possible to define precisely the electronic system of interest in a given electrochemical reaction: The fact that a σ radical is obtained in the oxidation of nitrosobenzene demonstrates that the electron is taken from the π_a^* orbital of the nitroso group.

C. Extrathermodynamic Relationships

There are many examples in the literature of extrathermodynamic relationships such as the Hammett-Taft between half-wave potentials and coefficients like σ which are characteristic of a variety of substituents in compounds which can undergo electrooxidation or reduction. Actually, it is mainly reductions which have been studied; the work has been published largely by specialists in polarography using the mercury electrode which cannot be used at the positive potentials required to oxidize many organic compounds.

Extensive experimental work and very subtle analyses of these relationships have been published by Zuman [88].

It is not surprising that such relationships have been verified in the case of rapid systems because for a given series of compounds $E_{1/2}$ is proportional to ΔG^o. It can be shown furthermore that for an irreversible system $E_{1/2}$ is proportional to an activation energy ΔG^* which is characteristic of the electrochemical reaction. This is the same sort of situation which leads to Hammett-type relationships, as they apply to both equilibrium and rate constants [89].

As in the case of homogeneous chemical reactions, an attempt has been made to separate polar, mesomeric, and steric effects of substituents upon $E_{1/2}$. Likewise, there have been proposed several σ factors which differ according to the type of interaction between the substituent and the group of atoms undergoing electrochemical reactions [90].

In spite of these refinements it is often found that certain compounds belonging to a group which as a whole obeys the $\rho\sigma$ relationship depart quite considerably from the norm. An example is given in Fig. 31 which shows data obtained in the oxidation of several anilines N-substituted by anthranyl [91]. For R = $N(CH_3)_2$ the $E_{1/2}$ is abnormally low. Zuman [88] has proposed that when there is such a marked deviation the compound in question

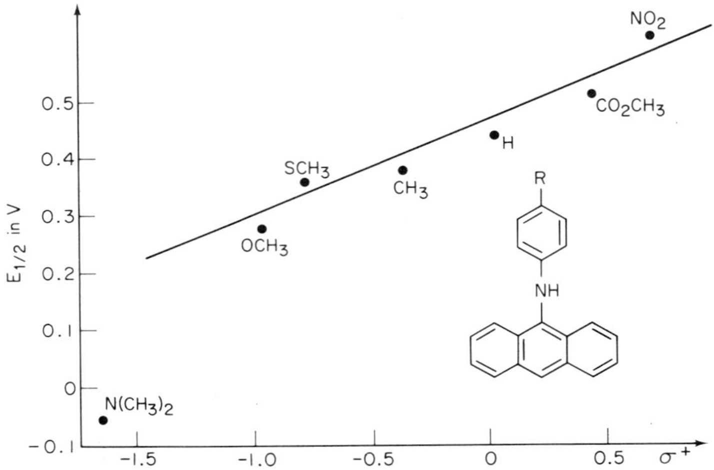

Fig. 31. The relationship $E_{1/2} = f(\sigma^+)$ in the oxidation of several p-substituted N-anthranyl anilines. The i-E curves were obtained at a rotating platinum electrode in CH_3CN containing 10^{-1} M $LiClO_4$. The reference electrode was Ag/Ag^+ (10^{-2} M).

may undergo electrochemical reaction by a mechanism which is different from the others. This does not seem to apply in this case because it is possible to show that for this derivative [R = $N(CH_3)_2$] as well as the others of the series the electrochemical step is a rapid one-electron exchange. However, the coefficient σ^+ of the $N(CH_3)_2$ group does possess a degree of uncertainty.

VI. THE ROLE OF THE REACTION MEDIUM

On several occasions we have referred to the effect which the reaction medium has upon electrochemical phenomena. Since recent publications indicate that the use of a variety of media is becoming more general, it is perhaps appropriate to conclude this introductory chapter by summarizing some reflections on the several factors which may be modified by changing the medium in which an electrochemical reaction is carried out [94].

The medium may influence, firstly, the electrode reaction, and secondly, the chemical reactions which are coupled to the electron transfer.

A. Influence of the Medium on the Electrode Reactions

In the first place, the medium influences mass-transfer phenomena. In this regard it is not only the viscosity which defines the value of D [95], the diffusion coefficient. The diffusion of electroactive species is likewise influenced by the size of their solvation sphere and by the phenomena of exchange of solvating species. To cite an important example: the proton has a mobility in water and in certain other solvents which it does not have in the majority of organic solvents [96].

Likewise, the medium has a very important effect upon double layer phenomena. Three major points must be borne in mind. The double layer capacity is not related in a simple way to the dielectric constant of the bulk solvent. This is not surprising in view of the special structure (oriented molecules) in the vicinity of the electrode [97]. Furthermore, adsorption phenomena, which complicate the interpretation of the results of electrochemical methods (Chapter II), are less frequent in organic solvents than in water for the following reason: the energy involved in replacing a molecule of solvent by a molecule of solute (displacement work) is generally higher in the case of organic solvents than in the case of water [98]. (This is not a minor point of interest in connection with organic solvents.) Finally, the influence of the ions of the supporting electrolyte upon the structure of the double layer should be recalled (p. 60). It emphasizes that the influence of the total medium rather than just of the solvent must be considered.

The medium may also change the characteristics of the charge transfer itself, i.e., the constants E^o, k^o, and α.

Most of the studies related to the variation of the redox potential of a given system as a function of the medium have actually been concerned with the variations of the half-wave potential in a one-electron rapid, reversible system. For the results to be valid, it must be assumed that the i-E curve recorded is truly the result of such an electron exchange without follow-up complications, particularly coupled chemical reactions. Under these conditions the potential of the half-wave, for, e.g., the oxidation of a compound A to a species A^+, is given by the expression

$$E_{1/2} = I + \Delta E_{sol} - \frac{T\Delta S}{F} - \frac{RT}{F} \ln \frac{f^+D}{fD^+} + \text{cte}$$

in which I is the ionization potential of A in the gaseous state, ΔE_{sol} is the difference in solvation energy between A^+ and A, ΔS is the entropy change accompanying the reaction, f and f^+ are the two activity coefficients, and D and D^+ are the diffusion coefficients.

Obviously, changing the solvent can change all the values except that of I, but it is ΔE_{sol} which is subject to the greatest perturbation. Reference [17b] has a detailed discussion of the influence of a variety of characteristics of the medium upon the oxidation and reduction potentials of aromatic compounds.

Much less work has been done on the influence of the medium upon the intrinsic rate constant k^o than on E^o. We have already (p. 59) referred to the effect which the structure of the double layer and adsorption phenomena can have on the kinetics of charge transfer. As a matter of fact, most of the experimental studies which have been published up to now [22] seem to indicate that k^o varies relatively little with the solvent. Thus, in a recent publication relating to the Cd/Cd^{2+} system, even in the most extreme deviation, the ratio of values was only 44 [99]. A comparison of this ratio with the values in Table 1 (Sec. II. C. 3) shows that the nature of the redox system much more than the solvent defines the value of k^o. This is understandable if one remembers that k^o by definition intrinsically takes care of the rearrangement of the solvation sphere during the electron transfer. The activation energies involved in these rearrangements are not changed much by change of solvent but do depend considerably on the species involved.

Very little has been done so far on the influence of the medium on the coefficient α.

B. Influence of the Medium on Coupled Chemical Reactions

As might be expected, the influence of the medium upon chemical reactions coupled to charge transfer is easier to demonstrate than its influence on charge transfer. For example, a change in medium can effect the kinetics of chemical reactions by several powers of 10. Thus, as we have seen, since the rate of charge transfer with organic compounds is high, the overall kinetics of the electrochemical process is most often limited by the kinetics of the coupled chemical reaction. The very great influence of the medium on the chemical kinetics, therefore, explains why it is possible completely to change the reaction pathways by changing the medium.

This fact is well illustrated in an example already cited on p. 47: the reduction of aromatic hydrocarbons, on the one hand, in an aqueous-organic medium and, on the other, in an organic medium which furnishes protons only slowly. All the recent developments in organic electrochemistry using anhydrous solvents have been based on this profound influence of the medium on the coupled chemical reactions; they have made possible the detection of many intermediates and have therefore contributed to the elucidation of reaction mechanisms.

The medium can influence coupled chemical reactions both by its physical and by its chemical properties. The physical properties of the solvent, such as its dielectric constant, dipole moment, and its ability to form hydrogen bonds, may enable the solvent to effect both the rate and the equilibrium constants of a number of coupled chemical reactions. This influence arises from the effect these properties exert upon solvation of intermediates and upon several types of association (ion pairs, charge-transfer complexes, etc.) [100].

The solvent can also participate more directly by virtue of its chemical properties. Formally, two types of intervention may be distinguished.

First, the solvent may be the donor or acceptor of various species. Again, the reduction of aromatic hydrocarbons illustrates the effect of the proton donor property of water. The acid character of this solvent will of necessity be a predominant influence on the course of all reductions carried out in it. The basic character of a solvent may be likewise of importance. Thus, the oxidation of N, N′-diphenylbenzidine [abbreviated $A(NH)_2$] in acetonitrile proceeds in two successive one-electron steps:

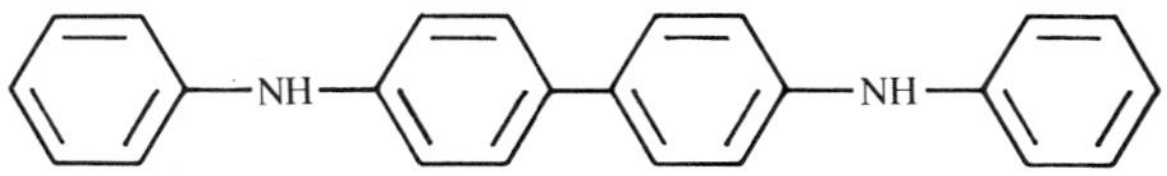

$$A(NH)_2 \xrightarrow{-\bar{e}} A(NH)_2^{\dot{+}}$$

$$A(NH)_2 \xrightarrow{-\bar{e}} A(NH)_2^{2+} .$$

In dimethyl sulfoxide, however, a more basic solvent, oxidation occurs by an ECE mechanism because of the deprotonation of the intermediate radical cation:

$$A(NH)_2 \xrightarrow{-\bar{e}} A(NH)_2^{\dot{+}}$$

$$A(NH)_2^{\dot{+}} \xrightarrow{-H^+} A(NH)N^{\bullet}$$

$$A(NH)N^{\bullet} \xrightarrow{-\bar{e}} A(NH)N^+$$

Secondly, the solvent may also for certain species effect the kinetics of electron transfer. It has already been mentioned (p. 51) that the i-E curves of certain amines $ArNH_2$ undergoing protonation are deformed depending on whether the medium favors rapid or slow deprotonation kinetics. It should be noted that in either type of medium controlled potential electrolysis at an electrode with a large surface leads to the same result, i.e., the oxidation of the free $ArNH_2$ to $ArNH_2^{\dot{+}}$. When acetonitrile is used as the solvent, the kinetics of the deprotonation $ArNH_3^+ \rightarrow ArNH_2$ controls the kinetics of the transformation.

The solvent may also exchange neutral atoms with intermediates involved in an electrochemical process. A recent example [94], among others, involves the reduction of p-bromoacetophenone in acetonitrile. Benzophenone is obtained. Its formation can be explained only by postulating that a hydrogen atom is abstracted from the solvent by a neutral radical:

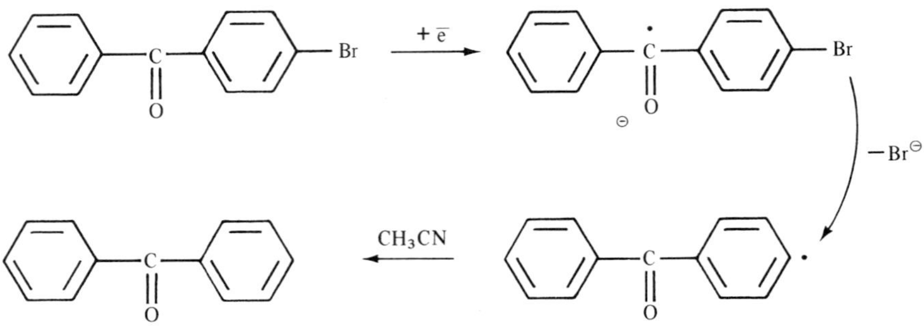

One of the functional groups of a solvent may also involve the solvent in a coupled chemical reaction. There are several examples in Chapter XII in which a positive species formed electrochemically attacks the nucleophilic CN group of CH_3CN. It follows that in planning an electroorganic synthesis it is necessary to choose a solvent which will not react with the reactive intermediates formed at the electrode: unless the reaction were purposely desired, CH_3CN would not be the proper solvent to use when cations are produced.

TABLE 2

Types of Effects Media May Have on Electrochemical Processes

Properties of the medium	Reactions at the electrode	Homogeneous chemical reactions
Physical	Structure of the double layer and its influence on k^o, α, and the potential φ_2 Effect on the adsorption of background electrolyte ions, electroactive substances, and nonelectroactive substances Effects of solvation on E^o and k^o Electroactivity range width	Effects on equilibria and kinetics constants by solvation, ion pairs, CT complexes, etc. (by ϵ, μ, H bonds, etc.)
Chemical	Participation in a possible chemical reaction between adsorbed species	Exchanges of particles (e.g., H^+, OH^-, atoms, etc.) Reactions of functional groups (e.g., CH_3CN, etc.)

Table 2 summarizes the several effects which the medium may have on an electrochemical process. Only an appreciation of these effects will permit the chemist to choose the optimum rational conditions for a given synthesis.

ACKNOWLEDGMENT

I wish to thank Dr. Jacq as well as my co-workers, Messrs. Bonnaterre and Cognard, for very stimulating discussions.

NOTES

[1]Rigorous discussions are given in several classic texts [1].

[2]Sometimes, however, the ions of the "inert" electrolyte can react chemically with the species formed in the electrode process.

[3]The qualitative aspects of this presentation have been greatly influenced by Electrochemical Reactions by G. Charlot, J. Badoz-Lambling, and B. Tremillion, Elsevier, Amsterdam, 1968; and by the chapter by J. Badoz-Lambling in Introduction aux Methodes Electrochimiques (J. Robbin, ed.), Masson, Paris, 1967. The quantitative treatment developed by J. Jacq has been deliberately adopted; see, e.g., J. Jacq, Electrochim. Acta, **12**, 1 (1967), and references cited therein.

[4]This will be referred to hereafter as i-E.

[5]The coordinate axes of an i-E curve may be oriented in eight different ways. We adopt here an I.U.P.A.C. convention recording negative potentials toward the right and negative currents upward.

[6]Also called electroactivity range.

[7]This must not be confused with methods which utilize stationary electrodes (e.g., in cyclic voltammetry, Chapter II) but for which $\delta C/\delta t$ is not necessarily zero.

[8]Strictly speaking, the value of δ which is a function of D is not the same for all species in solution. This is contrary to Nernst's approximation which assumes that *all* concentrations vary similarly at distance δ.

[9]For complete rigor these expressions should contain terms correcting for double layer effects (p. 59) [14].

[10]There are even separate articles devoted to this question alone [18].

[11] Or those which refer to the equality or nonequality of the cathodic and anodic $E_{1/2}$'s.

[12] That is to say, systems in which a chemical step does not intervene between two one-electron transfers.

[13] For some exceptions, see [25].

[14] Complicated further by phenomena of adsorption at electrodes (p. 64).

[15] This apparent contradiction to Hammond's postulate is explained by considering the kinetics involved in the two transformations, which do not occur at the same potential.

[16] Traditionally, authors use φ to denote the various potentials involved in their models.

[17] φ_2 is used here because the symbol φ_1 will be used later.

[18] The potential of zero charge is so important in the discussion of electrode phenomena that the expression rational potential has been coined to designate potentials which are referred to the point of zero charge for an electrolyte with no specific adsorption.

[19] Relative values can also be determined by measuring the mercury drop time, which is approximately proportional to the interfacial tension.

[20] It should be noted that in view of Lippmann's equation the electrocapillary curve would be a true parabola if C_i were independent of E.

[21] The fact that adsorption phenomena may not be infinitely rapid is evident from the change in differential capacity which is observed at high frequency (Fig. 26).

[22] It should be noted that most values of k^o published frequently refer to two-electron transfers (e.g., Cd/Cd^{2+}). They are not, therefore, really the intrinsic constants for one-electron transfer processes as defined on p. 34.

REFERENCES

[1]. See, for example, J. O'M. Bockris and A. K. N. Reddy, Modern Electrochemistry, Plenum, New York, 1970.

[2]. M. Szwarc, in Progress in Physical Organic Chemistry (A. Streitwieser, Jr. and R. W. Taft, eds.), Vol. 6, Wiley-Interscience, New York, 1968, p. 323.

[3]. J.-P. Billon, Ann. Chim., 7, 196 (1962).

[4]. Ref. [1], p. 902.

[5]. Ref. [1], p. 905.

[6]. Ref. [1], p. 876.

[7]. See, for example, G. Briere, G. Cauquis, B. Rose, and D. Serve, J. Chim. Phys., 4928 (1968).

[8]. Ref. [7]; see also G. Cauquis and D. Serve, J. Electroanal. Chem., 27, App. 3 (1970).

[9]. V. G. Levich, Physicochemical Hydrodynamics, Prentice-Hall, Englewood Cliffs, N.J., 1962.

[10]. Ref. [9], p. 60.

[11]. Ref. [9], p. 69.

[12]. Ref. [1], p. 1055.

[13]. (a) P. Delahay, Double Layer and Electrode Kinetics, Wiley-Interscience, New York, 1966, Chap. 7; (b) K. J. Vetter, Electrochemical Kinetics, Academic, New York, 1967.

[14]. Ref. [13a], Chap. 9.

[15]. For a study of the influence of the medium on k^o, see, for example, L. Gierst, E. Nicolas, and L. Tytgat-Vandenberghen, Croat. Chim. Acta, 42, 117 (1970); and T. Biegler, E. R. Gonzalez, and R. Parsons, Coll. Czech. Chem. Commun., 36, 414 (1971).

[16]. N. Tanaka and R. Tamamushi, Electrochim. Acta, 9, 963 (1964).

[17]. (a) A. C. Aten and G. J. Hoijtink, in Advances in Polarography (I. S. Longmuir, ed.), Pergamon, Oxford, 1960, p. 777; (b) M. E. Peover, in Electroanalytical Chemistry (A. J. Bard, ed.), Vol. 2, Dekker, New York, 1967, p. 1.

[18]. See, for example, H. H. Bauer, J. Electroanal. Chem., 16, 419 (1968).

[19]. See, for example, J. A. V. Butler, Trans. Faraday Soc., 19, 729 (1924); and T. Erdey Gruz and M. Volmer, Z. Physik. Chem., 150A, 203 (1930).

[20]. Ref. [18], p. 421.

[21]. R. A. Marcus, J. Chem. Phys., 43, 679 (1965).

[22]. Ref. [13b], Chap. 3.

[23]. Ref. [13b], p. 120.

[24]. J. Jacq, B. Cavalier, and O. Bloch, Electrochim. Acta, 13, 1119 (1968).

[25]. J. M. Fritsch, H. Weingarten, and J. D. Wilson, J. Amer. Chem. Soc., 92, 4038 (1970).

[26]. R. N. Adams, Electrochemistry at Solid Electrodes, Dekker, New York, 1969, p. 245.

[27]. G. J. Hoytink, in Adv. Electrochem. Electrochem. Eng. (P. Delahay, ed.), Vol. 7, Wiley-Interscience, New York, 1970, p. 221.

[28]. R. Brdička, V. Hanus, and J. Koutecky, in Progress in Polarography (P. Zuman and I. M. Kolthoff, eds.), Vol. 1, Wiley-Interscience, New York, 1962, p. 145.

[29]. R. Brdička, in Ref. [17a], p. 655.

[30]. J. Jacq, Electrochim. Acta, 12, 1 (1967).

[31]. Ref. [30], and R. Bonnaterre and G. Cauquis, J. Electroanal. Chem., in press.

[32]. Ref. [30], p. 7.

[33]. G. Cauquis, unpublished.

[34]. (a) Ref. [30], p. 11; Ref. [28], p. 172; (b) see, for example, A. Rashid and R. Kalvoda, J. Electroanal. Chem., 28, 245 (1970); A. Rashid, P. Straka, and R. Kalvoda, ibid., 29, 383 (1971).

[35]. J. Jacq, ibid., 29, 149 (1971).

[36]. J. Bessard, G. Cauquis, and D. Serve, Tetrahedron Lett., 3103 (1970).

[37]. For dimerization, see J. Koutecky and V. Hanus, Coll. Czech. Chem. Commun., 20, 124 (1955); S. G. Mairanovskii, Izv. Akad. Nauk SSSR, 12, 8140 (1961); R. Bonnaterre and G. Cauquis, J. Electroanal. Chem., 32, 194 (1971); and J. Jacq, Rapport Institut Français du Pétrole n° 18198 (1970); for dismutation, see K. Holub, J. Electroanal. Chem., 30, 71 (1971); and R. Bonnaterre and G. Cauquis, ibid., 31, App. 15 (1971).

[38]. D. M. Mohilner, in Electroanalytical Chemistry (A. J. Bard, ed.), Vol. 1, Dekker, New York, 1966, p. 331.

[39]. Ref. [1], p. 623; and Ref. [38].

[40]. Ref. [13a], Chap. 9.

[41]. Ref. [13a], pp. 213 and 230.

[42]. See, for example, Ref. [38], p. 241.

[43]. M. M. Baizer, J. Electrochem. Soc., 111, 215 (1964).

[44]. R. Galli and F. Olivani, J. Electroanal. Chem., 25, 331 (1970).

[45]. Ref. [38], p. 298.

[46]. R. Parsons, J. Electroanal. Chem., 21, 35 (1969).

[47]. See, for example, H. W. Nürnberg and G. Wolff, ibid., 21, 99 (1969), and the references cited therein.

[48]. See, for example, R. R. Schroeder and I. Shain, J. Phys. Chem., 73, 197 (1969), and the references cited therein.

[49]. (a) For a simplified review, see C. N. Reilley and W. Stumm in Ref. [28], p. 81; (b) for an extensive review, see, for example, A. N. Frumkin and B. B. Damaskin, in Modern Aspects of Electrochemistry (J. O'M. Bockris and B. E. Conway, eds.), Vol. 3, Butterworth, London, 1964, p. 149.

[50]. Ref. [13a], p. 82.

[51]. Ref. [13a], Chap. 5.

[52]. G. Lippmann, Ann. Chim. Phys., 5, 486 (1876); 12, 265 (1877); see also Ref. (7), p. 17.

[53]. Ref. [1], p. 688; and Ref. [13a], Chap. 5.

[54]. Ref. [38], p. 355.

[55]. Ref. [38], p. 357.

[56]. Ref. [13a], p. 112; and B. B. Damaskin, Electrochim. Acta, 9, 231 (1964).

[57]. Ref. [38], p. 377.

[58]. C. A. Barlow, Jr. and J. Ross Macdonald, in Adv. Electrochem. Electrochem. Eng. (P. Delahay, ed.), Vol. 6, Wiley-Interscience, New York, 1967, p. 1.

[59]. R. Payne, in idem, Vol. 7, p. 1.

[60]. L. Horner and D. Degner, Tetrahedron Lett., 1968, 5889; see also R. N. Gourley, J. Grimshaw, and P. G. Millar, Chem. Commun., 1278 (1967).

[61]. Ref. [1], p. 1309; and J. M. West, Electrodeposition and Corrosion Processes, Van Nostrand, Princeton, N.J., 1965.

[62]. J. O'M. Bockris and G. A. Razumney, Fundamental Aspects of Electrocrystallization, Plenum, New York, 1967, Chaps. 11 and 12.

[63]. See, for a summary, R. Parson, in Adv. Electrochem. Electrochem. Eng. (P. Delahay, ed.), Vol. 1, Wiley-Interscience, New York, 1961, p. 1.

[64]. P. Delahay and I. Trachtenberg, J. Amer. Chem. Soc., 80, 2094 (1958).

[65]. T. Biegler and H. A. Laitinen, J. Electrochem. Soc., 113, 852 (1966).

[66]. Ref. [63], p. 55.

[67]. See, for example, N. Tanaka, T. Takeuchi, and R. Tamamushi, Bull. Chem. Soc. Japan, 37, 1435 (1964).

[68]. J. Llopis, J. Fernandez-Biarge, and M. Perez-Fernandez, Electrochim. Acta, 1, 130 (1959); Y. F. Frei and J. R. Miller, J. Phys. Chem., 69, 3018 (1965).

[69]. B. Kastening and L. Holleck, J. Electroanal. Chem., 27, 355 (1970).

[70]. J. Heyrovsky and J. Kuta, Principles of Polarography, Academic, New York, 1966, p. 287.

[71]. R. Brdička, Z. Elektrochem., 48, 278 (1942).

[72]. See, for example, E. Laviron, Electrochim. Acta, 16, 409 (1971); and the preceding papers of the same author.

[73]. Ref. [26], Chap. 7; see also S. Gilman, in Electroanalytical Chemistry (A. J. Bard, ed.), Vol. 2, Dekker, New York, 1967, p. 111.

[74]. W. Böld and M. Breiter, Electrochim. Acta, 5, 145 (1961).

[75]. S. D. James, J. Electrochem. Soc., 114, 1113 (1967).

[76]. Ref. [26], p. 205.

[77]. Ref. [26], p. 206.

[78]. P. Zuman, Substituent Effects in Organic Polarography, Plenum, New York, 1967, p. 1.

[79]. A. Streitwieser, Jr., Molecular Orbital Theory for Organic Chemists, Wiley, New York, 1961, Chap. 7.

[80]. Ref. [78], Chap. 3 and subsequent.

[81]. R. M. Hedges and F. A. Matsen, J. Chem. Phys., 28, 950 (1958).

[82]. G. J. Hoijtink, Rec. Trav. Chim. Pays-Bas, 74, 1525 (1955).

[83]. Ref. [79], p. 177.

[84]. Ref. [79], p. 185.

[85]. G. J. Hoijtink, Rec. Trav. Chim. Pays-Bas, 77, 555 (1958).

[86]. Ref. [79], p. 184; and R. W. Schmid and E. Heilbronner, Helv. Chim. Acta, 37, 1453 (1954).

[87]. G. Cauquis, M. Genies, H. Lemaire, A. Rassat, and J.-P. Ravet, J. Chem. Phys., 47, 4642 (1967).

[88]. Ref. [78]; and P. Zuman, in Progress in Physical Organic Chemistry (A. Streitwieser, Jr. and R. W. Taft, eds.), Vol. 5, Wiley-Interscience, New York, 1967, p. 161.

[89]. J. E. Leffler and E. Grunwald, Rates and Equilibria of Organic Reactions, Wiley, New York, 1963.

[90]. Ref. [78], p. 72.

[91]. J.-P. Billon, G. Cauquis, J. Raison, and Y. Thibaud, Bull. Soc. Chim. France, 199 (1967).

[92]. For a general study of this problem, see H. Strehlow, in The Chemistry of Nonaqueous Solvents (J. J. Lagowski, ed.), Vol. 1, Academic, New York, 1966, p. 129.

[93]. Ref. [21]; and V. G. Levich, in Adv. Electrochem. Electrochem. Eng. (P. Delahay, ed.), Vol. 4, Wiley-Interscience, New York, 1966, p. 240.

[94]. For a general discussion of the influence of the medium, see J.-M. Savéant, J. Electroanal. Chem., 29, 87 (1971).

[95]. Ref. [1], p. 296.

[96]. Ref. [1], p. 470.

[97]. Ref. [1], p. 623; and Ref. [59], p. 36.

[98]. Ref. [1], p. 3.

[99]. See R. Parsons in Ref. [15].

[100]. Ref. [17b], p. 15.

[101]. G. Cauquis, unpublished results.

Chapter II

METHODS FOR THE ELUCIDATION OF ORGANIC ELECTROCHEMICAL REACTIONS

Georges Cauquis[1]

Centre d'Études Nucléaires de Grenoble
Grenoble, France

and

Vernon D. Parker

Department of General and Organic Chemistry
University of Copenhagen
Copenhagen, Denmark

I. INTRODUCTION: ELECTROCHEMICAL AND NONELECTROCHEMICAL METHODS

In order to elucidate an electrochemical reaction which an organic compound undergoes, it is necessary in the first instance to establish the sequence of electron transfers (E) and chemical reactions (C) in which it is involved at a given electrode, at a given potential, and in a given medium. Complete understanding of the process requires that the nature of each chemical reaction (C) be known. It is also necessary to determine (a) whether or not several parallel routes exist from the transient intermediates to the final product(s), (b) the role of follow-up homogeneous redox reactions in the bulk of the solution (p. 130), and (c) the influence of adsorption phenomena. Each of these factors may be examined qualitatively or quantitatively, in the latter case, e.g., including the determination of rate and equilibrium constants for the chemical reactions.

The above discussion points up the complexities involved in establishing a complete electrochemical mechanism and makes it clear why the majority of organic electrochemical studies have been restricted to _partial_ descriptions of mechanisms. Actually, in organic electrochemistry as in organic chemistry in general, the term "mechanism of the reaction" is often used rather loosely: it may on the one hand refer to a detailed rigorous experimental study or, on the other, to a hypothesis whose support is analogy with reactions of established mechanism (cf. Chapter XXVI).

For a rigorous experimental investigation of mechanism(s) three types of mutually complementary methods may be used:

(a) _Chemical methods_. The isolation and characterization by the usual techniques of the final products of an electrochemical reaction provide some information concerning the mechanism(s) even though there are often apparent discrepancies between the results of macroelectrolyses and the interpretation of electroanalytical experiments (which provide information about the _initially_ formed products). This is discussed further in Chapter III. It is also possible sometimes to characterize certain intermediate products either by trapping them by an electrochemically inactive reagent present in the medium or by changing the experimental conditions so that the presumed intermediate, itself too unstable to be isolated, is transformed to one more easily isolated. In the former category may be cited the extensive work on the trapping of alkyl radicals (formed by Kolbe anodic oxidation of carboxylates) by means of olefins [1]. An example of the second approach is found in a study of the oxidation of 2, 4, 6-tri-t-butylaniline: in H_2O-MeCN the presumed intermediate (I) is too unstable to be isolated under these conditions; by using a MeOH medium the corresponding ether (II) was formed and could be isolated. The properties of (II) support the hypothesis that (I) had been formed as an unstable intermediate in the first experiment [2]:

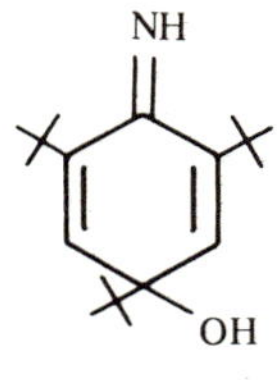

(I)

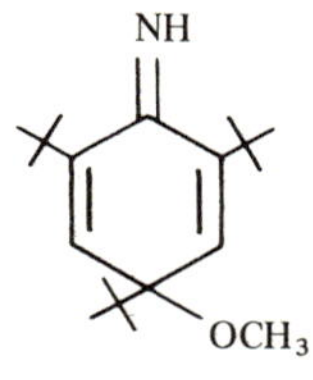

(II)

(b) Electrochemical methods. As has been mentioned, an i-E curve usually reveals the nature of the initial electrochemical step or of the initial electron transfers and the initial coupled chemical steps when the latter are rapid enough to affect the phenomena in the vicinity of the electrode. Many other electrochemical methods, however, provide more useful information than is obtainable from an i-E curve; they often make it possible to choose among two possible mechanisms which lead to the same overall results. A recent study of the mechanism of the two-electron process in which 9, 10-diphenylanthracene is oxidized in the presence of pyridine to yield (III) may be cited as an example [3]:

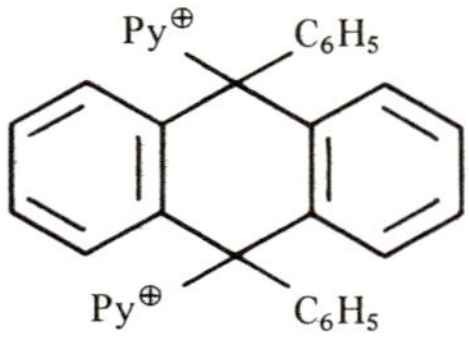

(III)

Frequently, the choice between two possible reaction mechanisms may be made by referring to working curves which are derived from theoretical analyses, often by the aid of a computer [4].

Certain coupled chemical reactions, however, are too slow to be detected by electroanalytical techniques. In these cases, if electron transfers follow the chemical reactions, they cannot be uncovered except by determining the number of faradays of electricity used in a preparative electrolysis. A coulometric determination therefore complements the results of electroanalytical methods.

(c) Physical methods. Physical methods, in addition to permitting the establishment of structures of isolated intermediates and products, may often also be applied in the characterization of nonisolable intermediates. They are the indispensable complement to electrochemical methods: the latter may aid in detecting the existence of intermediates but give no information about their structure.

II. METHODS INVOLVING MASS TRANSFER BY CONVECTION-DIFFUSION

A. General Considerations

The transport of an electroactive substance to the electrode by convection-diffusion may be attained in several different ways.

First, it is possible to have a flow of solution toward a plane or conical electrode or to the interior of a tubular electrode. These various arrangements do not appear to have had important application in organic electrochemistry. Adams' book [5] is recommended for a summary of the principal aspects of this phenomenon.

Second, it is possible to rotate in a motionless solution a metal wire electrode several millimeters in length; the wire is preferably aligned on the axis of rotation or perpendicular to it. The hydrodynamic phenomena involved in this system are relatively complex [6] and the applications are mainly in the realm of analytical chemistry. A vibrating wire electrode also gives interesting results [7].

The rotating disk electrode is much more important in the study of organic electrochemical reactions. Its hydrodynamics has been rigorously studied by Levich [8]; Adams [5] presents a detailed study of its properties and an exhaustive bibliography concerning it. Riddiford's review [9] is likewise comprehensive. Only the results of these studies will be presented here.

B. Rotating Disk Electrode

1. Theoretical Aspects

The rotating disk electrode (RDE), which usually consists of a flat surface of a conducting cylinder (metal, carbon, etc.; see Chapter IV) set into an insulating tube (glass, Teflon, etc.), has been of great service in the study of organic electrochemical reactions, particularly oxidations.

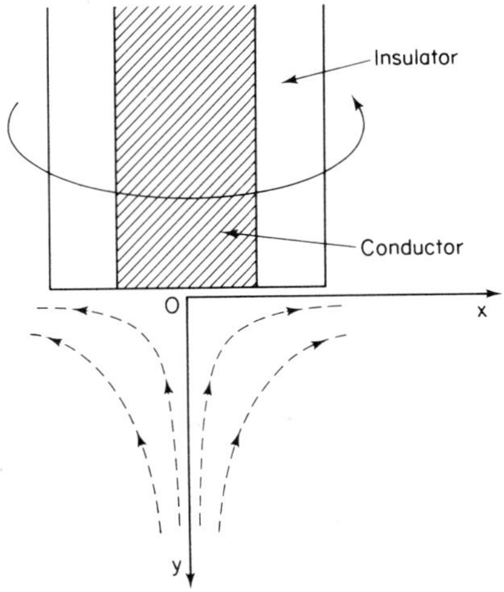

Fig. 1. Rotating disk electrode.

The convection of the solution in the vicinity of the electrode may be described as follows (Fig. 1): the rotation of the disk causes the formation of a layer of liquid which under the effect of the centrifugal force acquires a radial velocity. The horizontal flow causes the liquid which is under the electrode to rise. It has been shown that the thickness δ_0 of the liquid layer which is thrust horizontally by the electrode is approximately equal to

$$\delta_0 \cong 3\left(\frac{\nu}{\omega}\right)^{1/2}$$

in which ν stands for the kinematic viscosity of the medium in square centimeters per second and ω is the angular velocity of the disk. Inside the layer of thickness δ_0 the tangential and radial velocities decrease with the distance from the electrode (measured toward the bottom); at a distance greater than δ_0 the liquid moves solely with a vertical velocity. For an aqueous solution and a rotation speed of ca. 1000 rpm δ_0 is of the order of several tenths of a millimeter.

Furthermore, if laminar flow exists at the surface of the electrode [8], the thickness of the diffusion layer δ which must be used in Nernst's approximation is given by[2]

$$\delta = 1.61 D^{1/3} \nu^{1/6} \omega^{-1/2}$$

in which D is the diffusion coefficient of the electroactive substance in square centimeters per second. The thickness of the diffusion layer is then ca. 10^{-3} cm. when δ_0 is ca. 10^{-2} cm. According to Levich [8] the intensity of the current which flows through the electrode when an electroactive substance undergoes a one-electron reaction controlled simultaneously by mass and charge transfer is given by

$$i = \frac{FADC_o}{1.61D^{1/3}\nu^{1/6}\omega^{-1/2} + D/k} = \frac{FADC_o}{\delta + D/k}$$

in which A is the electrode surface in square centimeters, C_o is the concentration of electroactive substance in moles per liter, and k is the rate constant for charge transfer equal to

$$k = k^o \exp\left[-\frac{\alpha F(E - E^o)}{RT}\right],$$

provided the potential is such that the reverse reaction may be neglected. The relative values of δ and D/k define the two extreme cases in which either mass transfer alone or charge transfer alone is limiting. If, because of the value of E, k is sufficiently large, the term in D/k may be neglected; an expression for the limiting current is obtained which corresponds to mass transfer control alone:

$$i_{lim} = 0.62FAC_oD^{2/3}\nu^{-1/6}\omega^{1/2}.$$

Conversely, if the potential is such that δ becomes very small in comparison with D/k, the expression

$$i = FAkC_o = FAC_ok^o \exp\left[\frac{-\alpha F(E - E^o)}{RT}\right]$$

is obtained which corresponds to control by charge transfer. In the latter case, then, the current is independent of ω.

Figure 2 represents the variation of i with $\omega^{1/2}$. An oxidation is involved and at the lowest potential E_1 only charge transfer controls the value of i. The latter is then independent of the angular velocity ω. At the higher

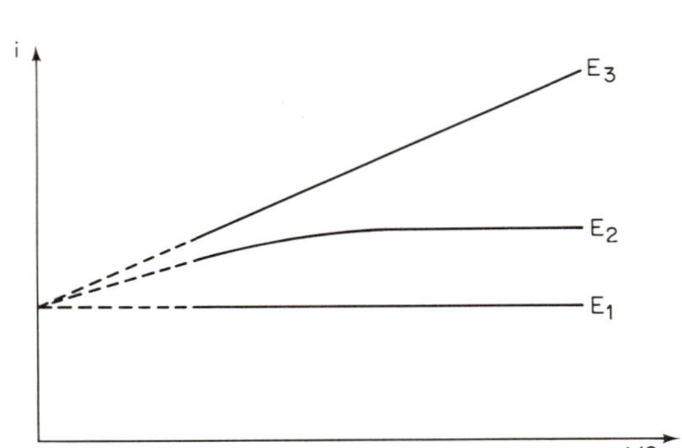

Fig. 2. Variation of current with $\omega^{1/2}$ in the case of an oxidation at potentials $E_1 < E_2 < E_3$.

potential E_3 only mass transfer controls the current which is proportional to $\omega^{1/2}$. Finally, at the intermediate potential E_2, the current is successively controlled by mass and charge transfer. The current is limited by mass transfer for the greater part of the current intensity-potential curve in a rapid system without coupled chemical reactions. A linear $i = f(\omega^{1/2})$ relationship must therefore obtain. Verification of this fact constitutes Levich's criterion.

2. Applications of the Rotating Disk Electrode

The rotating disk electrode has many applications beyond simply providing i-E curves. It makes possible, e.g., the determination of the order of an electrochemical reaction [10], of the kinetics of charge transfer [11], of the coefficients of diffusion [12], and so on.

More important for our purpose, however, is the means it offers for studying the mechanisms by which chemical reactions are coupled to charge transfer. Levich and later authors have studied the behavior of this electrode in the case of several mechanisms. For illustration we shall present here only the conclusions reached by Malachesky et al. [13] and by Feldberg and Auerbach [14] in a quantitative study of an ECE mechanism:

$$A \rightleftarrows B + n_1 e$$

$$Z + B \xrightarrow{k} C$$

$$C \rightleftarrows D + n_2 e.$$

If k, the rate constant for the conversion of B to C, is not too large, it is possible to find a domain of rotation velocity ω beyond which B is removed from the electrode so rapidly that the second electrochemical step becomes undetectable. If the electrode potential is fixed so as to give the limiting current, the latter is then a current which corresponds only to an exchange of n_1 electrons. Conversely, at low rotation speeds, the limiting current corresponds to an exchange of $n_1 + n_2$ electrons. Starting with these fundamentals it is then possible to develop a mathematical treatment which yields a "working curve" (similar to curve A of Fig. 3) which can be used in determining the rate constant k of the chemical reaction. The results have been experimentally verified in the case of the reduction of p-nitrosophenol which has also been studied by other techniques.

Another important type of ECE mechanism is encountered when the chemical step is a dimerization:

$$A \rightleftarrows B + n_1 e$$

$$2B \xrightarrow{k} B_2$$

$$B_2 \rightleftarrows D + n_2 e.$$

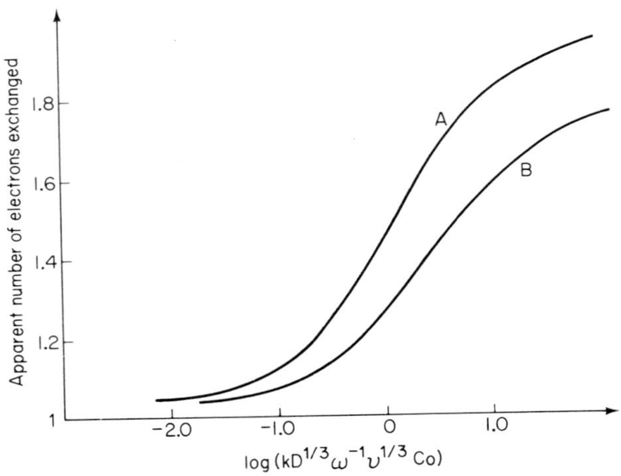

Fig. 3. Working curve for an ECE mechanism involving a first-order (A) or a second-order (B) chemical reaction ($n_1 = n_2 = 1$).

In this case a working curve can again be established (Fig. 3, curve B) [15]. For example, the oxidation of substituted triphenylamines yields tetraphenylbenzidine (TPB) by the following mechanism:

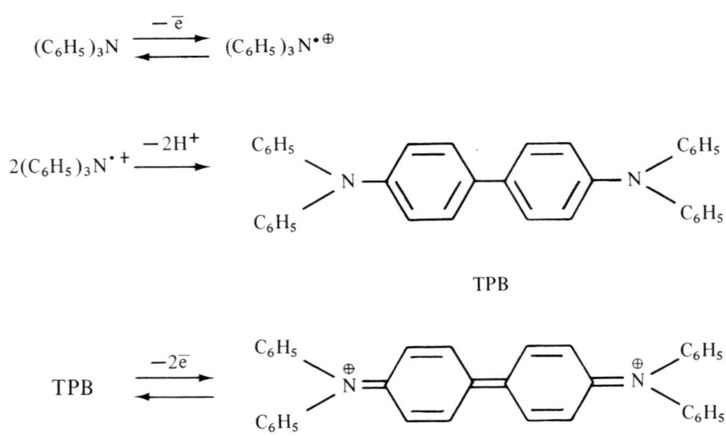

The experimental results were in accord with the theoretical predictions.

The examples which have been cited may have given the impression that the rotating disk is used exclusively for the numerical determination of a variety of constants. It must be added that purely qualitative observations may provide assistance in diagnosing the nature of a mechanism. Thus, simply obtaining an $i_{lim} = f(\omega^{1/2})$, or an $i_{lim}/\omega^{1/2} = f(\omega^{1/2})$, or an $i_{lim}/C_o\omega^{1/2} = f(\omega^{1/2})$ curve (with C_o variable) at the beginning of a new study may provide valuable information [16, 17].

To illustrate, keeping C_o constant, an $i_{lim}/\omega^{1/2} = f(\omega^{1/2})$ curve which is a horizontal line indicates that the system under study is either a rapid reversible system controlled by mass transfer only or that any chemical reactions taking place between or after charge transfer are too rapid to detect in the realm of accessible rotation rates. However, if the curve shows that $i_{lim}/\omega^{1/2}$ decreases as $\omega^{1/2}$ increases, the system is probably a simple but slow system or is following an ECE mechanism.

C. Rotating Ring Disk Electrode

The rotating ring disk electrode (RRDE) is an extension of the technique of the rotating disk electrode. It has great promise in applications involving elucidation of complex organic electrochemical mechanisms. Its utility to date has been limited due to practical considerations, particularly in organic solvents (see Chapter IV).

The principle of this device is the following (Fig. 4): around a disk electrode having a radius r_1 is a ring conductor of inside and outside radii r_2 and r_3, respectively. The assembly is held in two concentric insulating tubes. The ring and the disk may be held potentiostatically at two different voltages E_D and E_R (referred to the same reference electrode). The corresponding currents i_D and i_R may be independently determined.

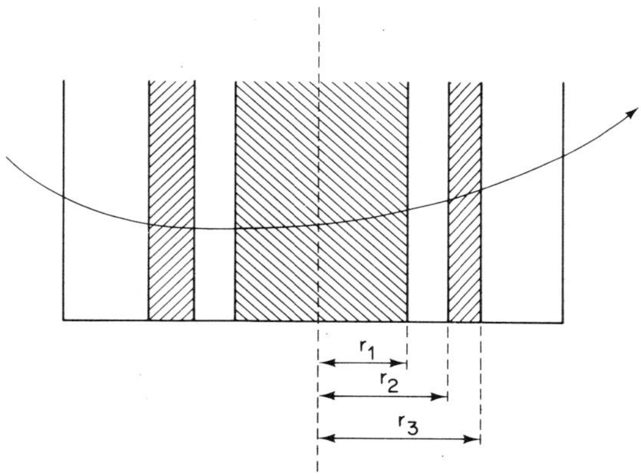

Fig. 4. Schematic of a ring disk electrode.

This type of electrode was first used by Frumkin's school [18]; the theory was developed by Ivanov and Levich [19] and by Albery and Bruckenstein [20, 21].

The general principle of the electrode may be described simply for a system which is free of chemical complications, i.e.,

$$A \rightleftarrows B + \bar{e}.$$

A and B are stable.

Suppose that the disk is brought to a potential E_D such that B is formed on its surface.[3] Because of the centrifugal force to which the solution in contact with the electrode is subjected, species B is transported onto the ring. Let the ring be at a potential at which C_B^* is zero. The complete reduction of B which arrives at the ring is manifested in a current i_R which is usually lower than i_D which corresponds to the formation of B on the disk. The ratio $N = i_R/i_D$, or more precisely $N = -i_R/i_D$ (because the two currents are of opposite sign), is called the <u>collection efficiency</u> of the electrode.

The rotation velocities and the geometry usually employed give N values of ca. 0.30-0.45. These values, which are functions of r_1, r_2, and r_3, give excellent agreement between theory and experiment.

The particular utility of the RRDE is in its employment when B, which is the product of a charge transfer at the disk, is capable of undergoing reactions. Two types of cases need be considered: one in which B undergoes a charge-transfer reaction which is not simply the reverse of the one by which it was formed; and two, in which B undergoes a chemical reaction.

In the first case it is a question of detecting an intermediate involved in a two-electron oxidation or reduction:

$$A \rightleftarrows B + e \quad (E_o^1)$$

$$B \rightleftarrows C + e \quad (E_o^2)$$

with normal potentials E_o^1 and E_o^2 of such values that the i-E curve shows only one wave (cf. the reduction of Cu^{2+} cited previously, p. 44).

From theoretical considerations [22] the conditions can be predicted under which species B may be detected at the ring (by a current i_R), if the ring is held at a constant potential E_R while E_D is slowly changed by sweeping along the i-E curve. Figure 5 shows, on the one hand, the i-E curve obtained from a solution of Cu^{2+} and, on the other, the current observed on the ring which has been brought to a potential at which Cu^+ is reoxidized on its surface.

If the reaction in which B, which has been formed at the disk, is a chemical reaction, the collection efficiency N′ actually obtained is lower

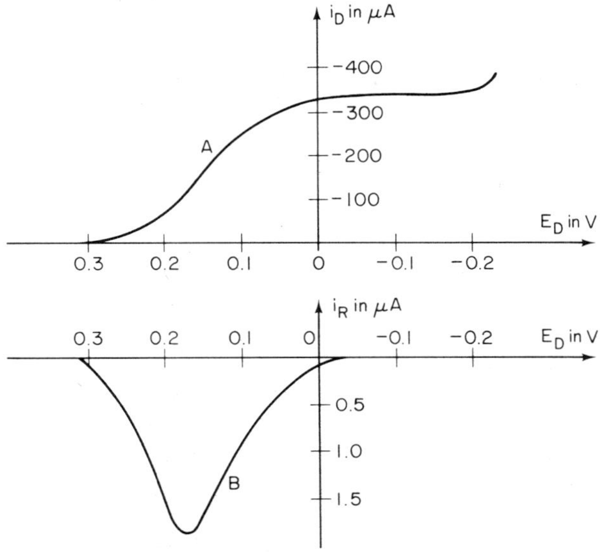

Fig. 5. Current intensity-potential curve (A) of 1.5×10^{-2} M Cu^{2+} solution in 1 M H_2SO_4 and variation of ring current i_R as a function of disk potential (B). The ring potential is held at 0.650 V (vs. normal hydrogen electrode). The disk and ring are of polished Pt, 2 and 7 mm^2, respectively [22].

than the theoretical value N which is obtained for a simple system without chemical complications. Comparison of N and N′ should in principle yield information concerning the kinetics of the several reactions which may be involved. Albery and Bruckenstein [23] have shown in theoretical studies that the solution of this type of problem is far from simple. Nevertheless, the electrode may be "calibrated" using a system which involves a first-order (or pseudo-first-order) reaction of known kinetics and then used for determining the kinetics of an unknown reaction.

Malachesky et al. [24] have provided an example of this type of application of the RRDE; they studied the oxidation of p-aminophenols and the subsequent hydrolysis of the quinone-imine products (EC system)[4]:

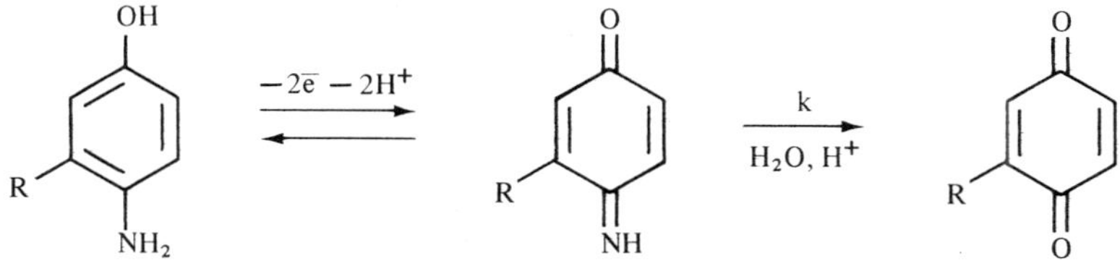

The rate constant k of this hydrolysis was independently determined with the aid of reverse current chronopotentiometry (see p. 132). The authors then showed that very similar values were obtained using the RRDE. This agreement was valid in two cases which differ greatly in k, namely p-aminophenol ($k = 0.052\ sec^{-1}$) and 3-methyl-p-aminophenol ($k = 0.50\ sec^{-1}$).

Besides being useful in quantitative studies, the RRDE can also give qualitative information concerning reaction mechanisms. An intermediate species, e.g., may show characteristic electrochemical behavior on the ring while the precursor is oxidized or reduced on the disk. In this connection it should be noted that this technique approaches the use of cyclic voltammetry (p. 115) for qualitative purposes; the former method, however, has the additional feature of allowing a distinction to be made between adsorbed species which remain on the disk (and are not transported to the ring) and the species which are carried into solution (and are detected on the ring).

The RRDE technique for detecting an intermediate is especially valuable in studies of ECE systems in which the second E occurs at a potential E_o^2 which is such that the intermediate could have undergone a second charge transfer at the same potential at which the first charge transfer had occurred ($E_o^2 < E_o^1$ in the case of two oxidations). At low rotation rates the intermediate is directly converted (on the disk) to its product, while at

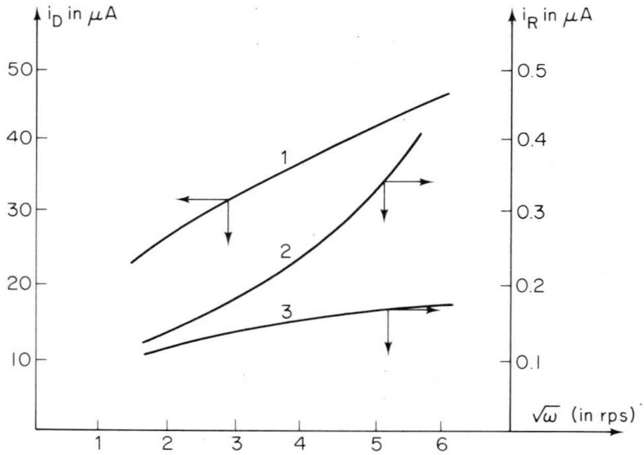

Fig. 6. Detection of the intermediate in the oxidation of 2×10^{-3} M dimethylaniline (DMA) in 0.1 N H_2SO_4 using the ring disk electrode. Curve 1, variation of the disk current i_D with rotation speed, disk at 0.85 V vs. SCE; curve 2, variation of the ring current i_R with ring at 0.60 V while DMA is oxidized at the disk at 0.85 V; curve 3, residual ring current at 0.60 V without reaction on the disk (disk at open circuit).

high rotation rates it is transported at least partially to the ring where it may be detected. Figure 6 illustrates this situation in the oxidation of N, N-dimethylaniline (DMA): the intermediate oxidation product tetramethylbenzidine (TMB) is more easily oxidized than DMA [25]:

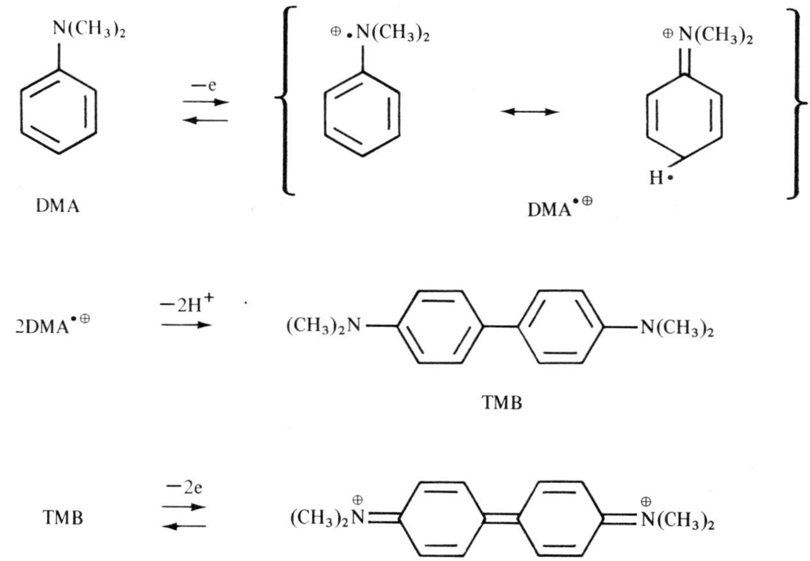

III. METHODS INVOLVING MASS TRANSFER BY DIFFUSION

A. General Considerations

In the methods discussed in the preceding section only the migration of the electroactive substance was suppressed by the device of using a supporting electrolyte. Convection, which is the result of the relative movement of electrode and solution, and diffusion, in the vicinity of the electrode, both cooperated to bring the substance to the surface of the electrode. The present section is concerned with the methods in which diffusion alone obtains, convection having been eliminated by the total absence of any relative electrode-solution movement.

The result of this absence of convection is that when the electrolysis circuit is completed the electroactive substance is progressively depleted in the vicinity of the electrode and the concentration of the products of electrolysis is correspondingly enhanced. In addition to i and E the time t also becomes an important parameter. This situation is common to <u>non-stationary</u> methods.

The relationship between the three parameters i, E, and t may be exploited in several ways. A potential may be impressed upon the electrode and the variation of current with time studied (chronoamperometry). The impressed potential itself may be held constant or may be varied in a programmed manner with time. Specifically, it may vary linearly with time. The resulting method, chronoamperometry with linear variation of potential, is obviously equivalent to the determination of an $i = f(t)$ or $i = f(E)$ curve. The latter is a type of voltammetry which differs from that using a rotating electrode only in the means by which the electroactive substance gets to the electrode. Polarography is such an important case of chronoamperometry with linear variation of potential that it will be discussed in a separate section (III.B). When the potential varies cyclically with time, the method of cyclic voltammetry (III.D) is realized.

It is also possible to control the current passing through an electrode and study the variation of the electrode potential as a function of t (chronopotentiometry); the current may be held constant or may be programmed (Sec. III.E).

The diffusion phenomena involved in each of these methods may be very different. For example, if the electrode is planar and sufficiently large to permit neglect of edge effects (so that it can be considered a planar boundary of the solution), the diffusion is semiinfinite and linear. This approximation of course assumes that the dimensions of the cell are very large compared to those of the electrode. However, spherical or cylindrical diffusion can also be achieved with electrodes of the respective shapes. These several types of diffusion are discussed in detail elsewhere [5, 26, 27].

An additional comment should be made about polarography. It involves a special case of spherical diffusion in which a mercury drop electrode grows at the end of a capillary and in doing so progressively occupies more of the zone of the solution which has been depleted of electroactive substance. Furthermore, the drops fall off periodically. In spite of the complexity of these phenomena, which appear to be very different from simple spherical diffusion, the $i = f(t)$ curve corresponding to the lifetime of one drop as well as the $i = f(E)$ curve corresponding to the average of the current variations in the course of the voltage sweep may be determined and provide useful information [28].

B. Polarography

1. Fundamental Aspects

Polarography is the term applied to voltammetry at a dropping mercury electrode (DME). The method was first described by Heyrovsky in 1922 [29]. The technique involves recording the current at the DME as the applied voltage is gradually increased. The DME consists of a capillary connected to a mercury reservoir. The rate of mercury flow can be regulated by adjusting the height of the reservoir and is usually of the order of about 20 drops/min. The unique feature of the DME is that a fresh electrode surface is generated as each new drop of mercury forms. Polarography is carried out in unstirred solution under conditions that approximate spherical diffusion of substrate to the electrode surface. Normally, conditions are selected which minimize other modes of mass transfer such as convection and migration.

The instrument used for classical polarography is simply a voltage divider which varies the applied voltage between the DME and the reference electrode. Such a system is satisfactory in highly conducting aqueous systems but is totally unsuitable for work in solutions of high resistance due to the error in the potential measurement caused by the iR component of the applied voltage. Thus, a three-electrode polarograph is necessary for nonaqueous work and is desirable for accurate potential measurements in aqueous media. The latter system gives a true current-potential curve since the current is recorded as the potential between the DME and the reference electrode is varied. The corresponding curve obtained with a two-electrode system is an applied voltage-current curve. A thorough description of the DME and polarographic instrumentation can be found elsewhere [31-33].

The polarographic current-potential curve for reduction or oxidation of a substrate shows an increasing current with increasing potential of the DME in the potential region where the substrate is electroactive. The region of increasing current is called a polarographic wave (Fig. 7). Two quantities

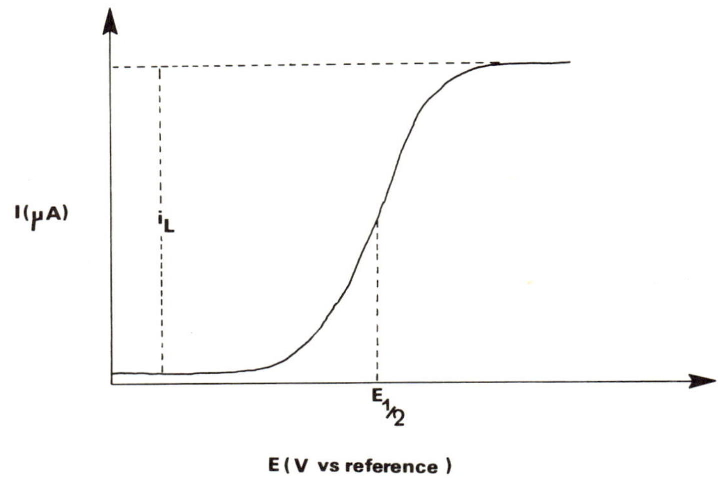

Fig. 7. Polarogram for the reversible reaction: O + ne ⇄ R.

associated with the polarographic wave, the half-wave potential ($E_{1/2}$) and the limiting current (i_L), are of theoretical and practical importance in the study of processes at the DME. The half-wave potential ($E_{1/2}$) is the potential of the DME where the current is equal to one-half of the limiting value ($i_L/2$). A description of the processes which give rise to the particular shape of the polarographic wave can be found in Chapter I.

For a reversible reaction, O + ne = R, where there are no complicating chemical reactions prior to or after charge transfer, the limiting current (i_L) is controlled by diffusion of substrate to the electrode and is thus called a diffusion current (i_d). The diffusion problem has been solved by Ilkovič [34] and the equation giving the diffusion current at the DME is called the Ilkovič equation,

$$i_d = 607nm^{2/3}t^{1/6}D_O^{1/2}O^b,$$

where n is the number of electrons transferred, m is the rate of flow of mercury in milligrams per second, t is the drop time in seconds, D_O is the diffusion coefficient of the substrate in square centimeters per second, and O^b is the bulk concentration of O in millimoles per liter.

The half-wave potential for the reversible process (mass transfer controlled) can be related to E_o (standard electrode potential) for the reaction by a form of the Nernst equation:

$$E_{1/2} = E_o - (RT/nF)\ \ln(D_O/D_R) - (RT/nF)\ \ln\ (f_R/f_O)$$

where R is the gas constant, T is the absolute temperature, n is the number of electrons transferred, F is the faraday, D_O and D_R are the diffusion coefficients of the oxidized and reduced forms, respectively, and f_O and f_R are the activity coefficients. Of course, in the event that the diffusion and activity coefficients of the oxidized and reduced forms are equal, $E_{1/2} = E_o$, and $E_{1/2}$ can then be related to the Gibbs free energy by the equation:

$$-E_{1/2} = \Delta G^{\circ}/nF.$$

This equation forms the basis for the relation of polarography to thermodynamics.

An irreversible process is one which is controlled by the rate of charge transfer rather than by mass transfer of the electroactive species to the electrode. The corresponding equation for the half-wave potential is:

$$E_{1/2} = E^{\circ} - RT/\alpha nF\ \ln\ 0.886k^{\circ}(t_1/D_O)^{1/2}$$

where k° is the heterogeneous rate constant of the electrode process at E_o, α is the transfer coefficient, and t_1 is the drop time. The current for the irreversible process is given by:

$$\ln[i/(i_d - i)] = \ln(0.87k^{\circ}t^{1/2}/D_O^{1/2}) - (\alpha nF/RT)E$$

where i_d refers to the limiting current at very negative potentials in which case the rate of charge transfer is indeed rapid and the process is mass transfer controlled. A plot of (RT/F) ln $i/(i_d - i)$ versus -E is linear with slope αn. The transfer coefficient α is commonly evaluated in this manner. The shape of the polarogram of an irreversible process differs from that of the reversible one in that the rising portion is not as steep. The latter is due to the fact that only a fraction αn of the potential change contributes to the reduction. A more thorough treatment of current-potential curves is presented in Chapter I.

There are four types of limiting currents commonly encountered in organic polarography. They are as follows:

(a) diffusion current (i_d) which is limited by rate of mass transfer of the substrate to the electrode surface (see the description of the reversible process given above),

(b) kinetic current (i_k) which is limited by a chemical step preceding the electrode reaction,

(c) catalytic current (i_c) which is limited by the catalytic effect of a polarographically inactive catalyst,

(d) adsorption current (i_a) which is caused by adsorption of the oxidized or reduced form of the substrate on the surface of the DME.

The general type of the limiting current may be characterized in several ways, the most important of which are: (1) the dependence of i_L on the concentration of the substrate and (2) the dependence of i_L on the height of the mercury column (h). The limiting currents i_d and i_k increase linearly with increasing substrate concentration while i_c and i_a exhibit limiting dependences on concentration. The diffusion current increases linearly with $\sqrt{h}$ while the kinetic current is independent of h. Adsorption currents increase linearly with h. No simple relationship exists between catalytic currents and h.

Anodic voltammetry at rotating solid electrodes is frequently referred to as anodic polarography. Perhaps this is a misnomer and the term polarography should be restricted to voltammetry at the DME. The most commonly used electrode for analogous anodic work is the rotating platinum electrode.

Polarograms are frequently complicated by maxima. The wave for the reduction of a substrate may exceed the height of the true wave and then decrease, giving rise to a sharp peak or a hump on the curve, sometimes making measurements uncertain. Maxima are believed to be due to tangential streaming motion of solution past the surface of the drop. Organic surfactants are commonly used as additives to eliminate maxima.

More detailed discussions of the fundamental aspects of polarography can be found in the literature [28, 29, 32].

2. Applications in the Study of Reaction Processes

The high overvoltage for reduction of the proton on mercury renders the DME ideally suited for the study of cathodic processes. On the other hand, the low anodic dissolution potential of mercury severely limits the DME for study of anodic processes. Thus, the DME can be considered as primarily a tool for studying cathodic reduction.

Almost without exception, the very first experiment that is performed in a study of a cathodic process is to record the polarogram. The half-wave potential ($E_{1/2}$) and the limiting current are measured. The half-wave potential in the electrolyte that is selected for further study is used to determine what potential should be employed for controlled potential electrolysis for product studies. The limiting current provides a means of estimating the number of electrons involved in the reduction. If the polarogram exhibits more than one wave, which is indicative of successive reduction steps,

$$O + n_1e \rightleftarrows R$$

$$R + n_2e \rightleftarrows R',$$

the relative number of electrons n_2/n_1 is easily obtained from measurement of the limiting currents of the two waves. Controlled potential electrolysis near the half-wave potentials of both waves is usually carried out in order to attempt to identify R and R$'$. Since the time span for recording a polarogram is very short in comparison to that of macroscale electrolysis, intermediates exhibiting apparent stability under polarographic conditions are not necessarily isolable from macroscale experiments. In this respect, polarography serves as a very useful guide to synthesis but for the above reason can also be very misleading. A more thorough discussion of this problem can be found in Chapter III.

It is also necessary to be aware of the existence of "latent" or "hidden" limiting currents. At least three types have been described in the literature. Examples of the "third" kind are found in situations in which reduced parent material reacts with parent molecules approaching the cathode by either anionic [110a] or proton-abstracting reactions [110b].

In aqueous or other protic media, $E_{1/2}$ depends not only on E_0 but also on the proton activity. Thus, for the general reversible reaction:

$$O + ne + mH^+ \rightleftarrows R,$$

the dependence of $E_{1/2}$ on pH is given by the equation,

$$E_{1/2} = E^{o}_{1/2} - 0.059m/n\ \text{pH}$$

where $E^{o}_{1/2}$ is the half-wave potential at pH = 0. Therefore, it is common practice to record polarograms over a wide range of pH when studying cathodic processes in protic media. This procedure can be particularly illuminating in cases where more than one wave is observed. For example, if two waves are observed and only the first is pH dependent, it is immediately obvious that protons are involved in the process occurring at the first wave but not in the second process.

The reduction of phthalimide (PM) in buffered aqueous media is a good example of the effect of pH on the two charge-transfer steps [35]

$$PM + H^+ \rightleftarrows PMH^+$$

$$PMH^+ + e \overset{E_1}{\rightleftarrows} \cdot PMH$$

$$\cdot PMH + H^+ \overset{k}{\rightleftarrows} \cdot PMH_2^+$$

$$\cdot PMH_2^+ + e \overset{E_2}{\rightleftarrows} \text{products}$$

$$\cdot PMH + e \overset{E_3}{\rightleftarrows} \text{products.}$$

The rate of the protonation of $\cdot PMH$ is pH dependent. At low pH, this step is fast. The reduction of $\cdot PMH_2^+$ (E_2) occurs more readily than the initial charge transfer (E_1); thus at low pH, the reduction appears as a single 2-e wave. At high values of pH, k is small and protonation does not occur to a measurable extent. Since $\cdot PMH$ is more difficultly reduced (E_3) than substrate, the polarogram consists of two distinct 1-e waves. At intermediate values of pH the height of the first wave is found to be inversely proportional to pH. The polarograms for this system are illustrated in Fig. 8.

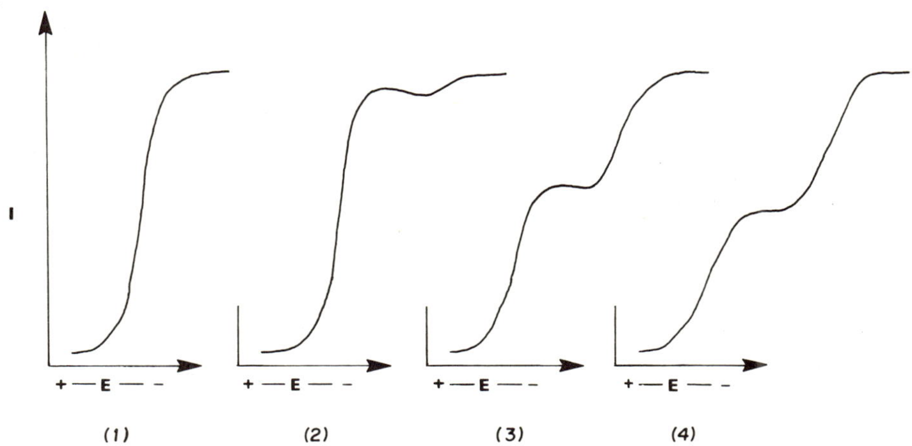

Fig. 8. Polarograms for the reduction of phthalimide at various values of pH: (1) 3.95; (2) 5.65; (3) 6.87; (4) 7.45.

Factors other than pH can give rise to the coalescence of two 1-e waves into a single 2-e wave. An example is the reduction of benzil in dimethylformamide solution. When tetrabutylammonium iodide serves as the supporting electrolyte, two 1-e waves, at -0.58 and -1.28 V (vs. Hg pool), were observed [36]:

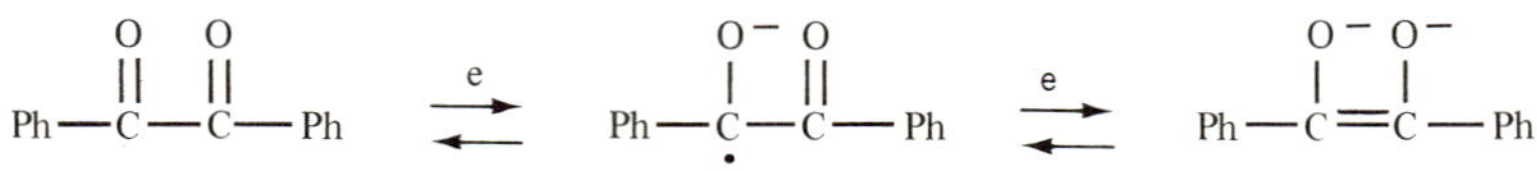

In the presence of lithium chloride, ion pairing between the product of the first electron transfer and lithium ion apparently occurs giving rise to a species reducing less cathodically than benzil and a single 2-e wave is observed at -0.64 V. The polarograms are illustrated in Fig. 9.

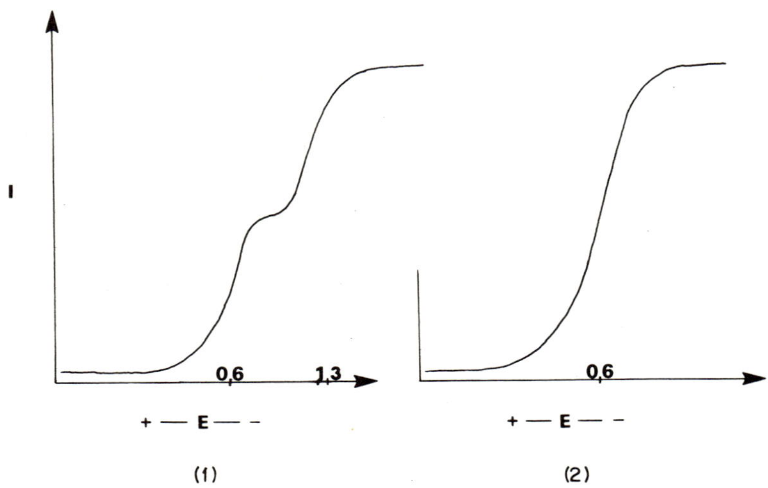

Fig. 9. Polarograms for the reduction of benzil in dimethylformamide solutions: (1) supporting electrolyte is tetrabutylammonium iodide; (2) supporting electrolyte is lithium chloride.

The DME is a convenient tool for following reaction kinetics of homogeneous processes when either the reactants, intermediates, or products are electroactive. Extensive use of this technique has been made and it has been the subject of reviews [37-39].

Polarography can also be used in structure determination of organic molecules. The characteristic reduction waves of certain functional groups may allow the determination of the presence or absence of these groups. Details of such procedures can be found in a book by Brezina and Zuman [40].

Equations which relate $E_{1/2}$ to the standard free energy (ΔG^{o}) of a reversible process or the free energy of activation ($\Delta G^{\ddagger}$) for an irreversible process are the basis for linear free-energy relationships involving polarographic half-wave potentials. This approach has been pioneered by Zuman and the reader is referred to his book on this topic for details [41] as well as to the discussions in Chapters I and III of this volume.

The applications of polarography mentioned here are certainly not exhaustive. Numerous other applications are discussed elsewhere [28-30, 40, 41].

C. Chronoamperometry

1. At Constant Potential

In this method an electrode, generally planar, is suddenly brought to a potential sufficiently negative or sufficiently positive to reduce or oxidize a species (O and R, respectively) present in solution. The i = f(t) curve (Fig. 10, curve 1) is recorded. If the experimental conditions approximate the

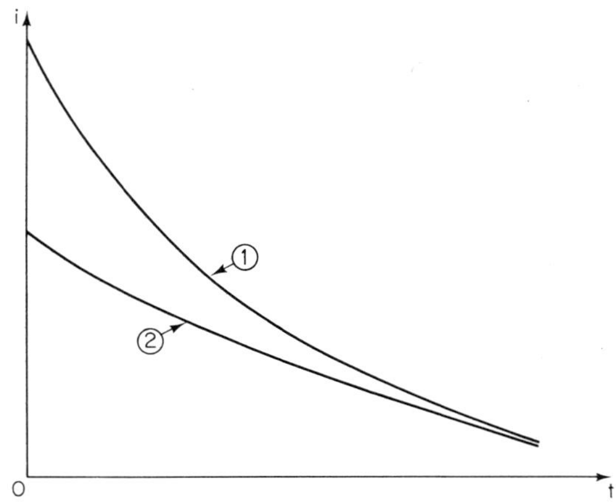

Fig. 10. Variation of current as a function of time in constant potential chronoamperometry. Curve 1, rapid one-electron transfer; curve 2, slow one-electron transfer.

theoretical conditions for semiinfinite planar diffusion, the variation of current with time can be calculated. For a rapid one-electron transfer the Cottrell relationship [26, 42] is obtained

$$i = FAC_{O}(D/\pi t)^{1/2}$$

in which A is the area of the electrode, C_0 is the concentration of the electroactive species, and D its diffusion coefficient. In the case of a slow one-electron transfer, the solution of the diffusion equations shows that the i = f(t) curve is rather different (Fig. 10, curve 2). This sensitivity of the nature of the curve to the kinetics of electron transfer has led to the use of chronoamperometry for studying and determining these kinetics [43]. The $it^{1/2}$ = f(t) curves are used for determining the diffusion coefficient D [44].

2. With Linear Variation of Potential

In this second type of chronoamperometry the potential of the electrode is varied linearly with time; the applied potential is

$$E = E_1 - vt.$$

The i = f(t) or i = f(E) (Fig. 11) curves have a maximum because in the absence of convection the solution is depleted of electroactive substance in

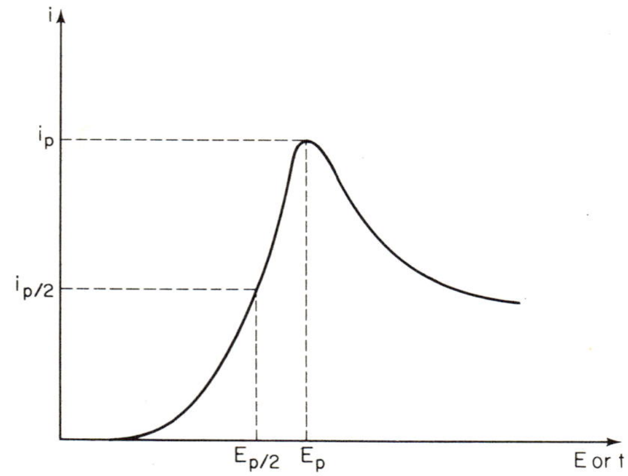

Fig. 11. Variation of current as a function of time or potential in chronoamperometry with linear variation of potential.

the vicinity of the electrode. The presence of a peak in the curve has given the method the names of peak voltammetry or peak polarography; they are used concurrently with linear sweep voltammetry or linear sweep chronoamperometry. Actually the method is generally used as part of a more complex technique in which the direction of the voltage sweep is reversed at a given point (see Cyclic Voltammetry, Sec. III.D). If it is desired to specify that a voltage sweep in only one direction has been used, the method is called single sweep voltammetry. This proliferation of terminology is an unfortunate fact which exists in the nomenclature of electrochemical methods.

Since cyclic voltammetry has played such a prominent role in the study of electrochemical properties of organic compounds, it is discussed in more detail in the following section. In Ref. [42] among others is given a detailed exposition of the scope and possibilities of single sweep voltammetry. In spite of the widespread use of cyclic voltammetry for the purpose, Nicholson and Shain [45, 46] and Savéant [47-49] have shown that single sweep voltammetry often furnishes valuable information and thermodynamic and kinetic data relative to chemical reactions coupled to the electron-transfer steps.

D. Cyclic Voltammetry

1. Theoretical Aspects

Cyclic voltammetry is a technique that allows one to scan the potential of the working electrode in the anodic (cathodic) direction and observe peaks due to the oxidation (reduction) of a substrate. The potential can then be scanned in the cathodic (anodic) direction and peaks due to reduction (oxidation) of intermediates formed during the forward scan may be observed. The electrode system used in cyclic voltammetry is dictated by the nature of the medium as well as by the process being studied. The most common electrodes used are planar platinum disks, platinum wires, hanging mercury drops, and carbon paste electrodes.

Cyclic voltammetry is conducted under conditions of semiinfinite linear diffusion (Sec. III.A) when planar electrodes are used. Conditions are more nearly cylindrical diffusion for wire electrodes and approach spherical diffusion for hanging mercury drops. When short electrolysis periods are employed, the latter two more complicated cases approximate the simpler case, linear diffusion. For a detailed discussion of the diffusion problems encountered during voltammetric studies, the reader is referred to the monograph of Adams [5]. It can be pointed out here that the use of rapid scan techniques minimize diffusional problems and qualitatively the same results can be obtained at various different electrode geometries.

The instrumentation for cyclic voltammetry usually consists of a controlled potential three-electrode system with a triangular wave generator capable of producing voltage sweep rates ranging from about 1 V/min to as high as 1000 V/sec [50]. Voltammograms are recorded on an X-Y recorder (1 sec pen response) for sweep rates less than about 20 V/min and oscilloscopic recording is used for higher sweep rates. Intermediate sweep rates are usually used; extremely high rates have the complication of double-layer charging effects and very slow sweep rates give rise to complications of natural convection.

Ordinarily, an isosceles triangular waveform [Fig. 12, (1)] is used for cyclic voltammetry. The potential of the working electrode is varied linearly with time beginning at point a and continuing to point b (switching potential)

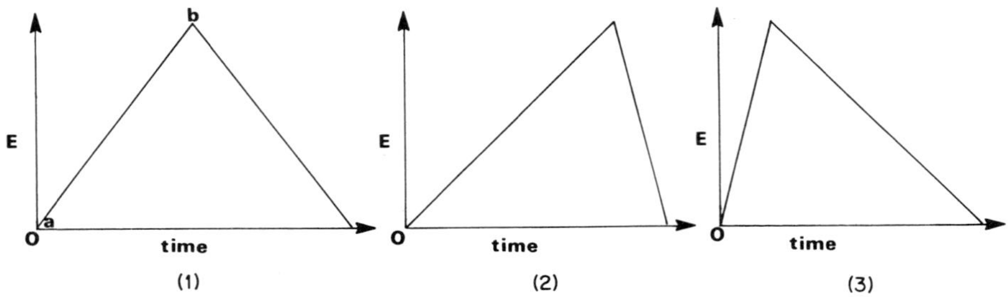

Fig. 12. Waveforms used in cyclic voltammetry: (1) equal forward and reverse sweep rates; (2) slow forward and rapid reverse sweep; (3) rapid forward and slow reverse sweep.

from which the potential is linearly decreased with time to generate the isosceles triangular pattern. Other waveforms can sometimes be used to advantage. For example, a slow initial sweep can be followed by a rapid return sweep [Fig. 12, (2)] or, conversely, a rapid initial sweep can be followed by a slow return [Fig. 12, (3)].

The initial scan up to the switching potential constitutes the recording of a peak voltammogram. Thus, the equations of peak voltammetry apply to cyclic voltammetry. The current at the peak potential is given by the equation

$$i_p = kn^{2/3} A D^{1/2} C^b V^{1/2}$$

where i_p is the peak current in amperes, A is the electrode area, V is the rate of potential change in volts per second, and k is a constant called the Randles-Sevick constant. The other symbols have the same meaning as in the polarographic equations (Sec. II. B). This equation may be rearranged to give the useful expression

$$i_p/V^{1/2} C = k'$$

where k' is comprised of all the constants of the previous equation. Thus, for a reversible charge transfer, in the absence of chemical complications, the product of the peak current divided by the square root of the voltage sweep rate times the concentration of the substrate is a constant independent of the sweep rate. This expression is sometimes called the current function, and when plotted versus $V^{1/2}$ gives a very convenient means of detecting follow-up chemical reactions.

The technique of cyclic voltammetry can be illustrated by the use of the three general hypothetical cases:

(1) reversible (rapid charge transfer) oxidation of R producing O without accompanying chemical reactions;

(2) nonreversible (slower charge transfer) oxidation of R producing O without accompanying chemical reactions;

(3) reversible oxidation of R to O, accompanied by a chemical step which converts O to O$'$.

These three cases were selected only as a means of presenting a brief discussion of the technique and are obviously not exhaustive. Although the examples are discussed as oxidations, they apply equally well for reductions in the opposite sense.

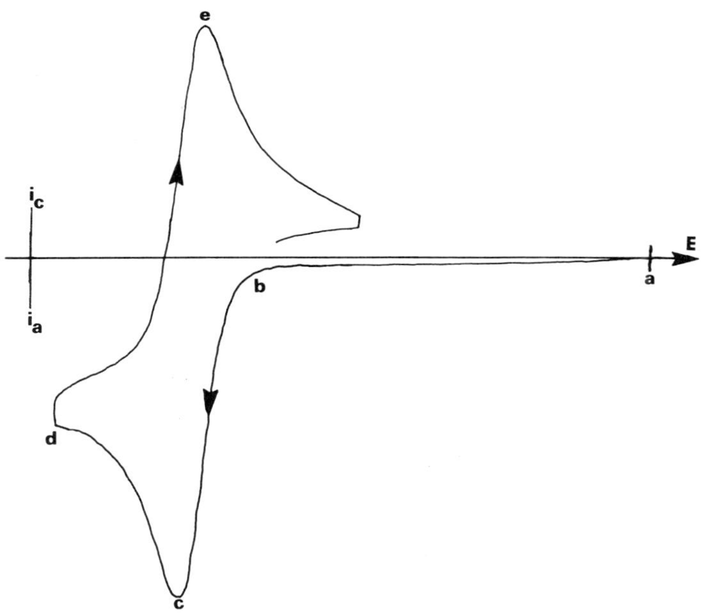

Fig. 13. Cyclic voltammogram for the reversible reaction: R $\rightleftarrows$ O + ne.

The first case, R $\rightarrow$ O + e, is represented by Fig. 13. The starting point a is selected such that no current flows when the cell circuit is closed. From point a, the voltage is scanned in the anodic direction; when point b is reached, oxidation of R begins and the current increases with increasing electrode potential in much the same way as in a polarographic wave (Chapter I) until point c, where depletion of substrate at the electrode brings about a decrease in current and the observation of a peak. From point c, the current decreases in much the same manner as the $it^{1/2}$ at constant potential (Sec. III.A). The direction of the voltage sweep can be changed at point d and a peak is observed at e on the cathodic sweep which corresponds to the reduction of O. Since there are no coupled chemical reactions, no other cathodic peaks are observed and the second cycle differs from the first only slightly due to changes in concentrations of R and O at the electrode. Slight changes continue to occur for a few more cycles until a steady state is achieved. The potential separation ΔE between the anodic and cathodic peaks is given by the equation

$$E_{p_a} - E_{p_c} = 0.58/n \text{ V},$$

and the peak width for the reversible system by

$$E_p - E_{p/2} = 0.59/n \text{ V}.$$

That is, for a one-electron process, both these quantities are equal to about 60 mV for a reversible charge transfer. Furthermore, the ratio of cathodic to anodic peak currents, i_{p_c}/i_{p_a}, is equal to 1.0 for a reversible charge transfer in the absence of coupled chemical reactions. Several methods for the determination of the peak current ratio can be found in the literature [5, 51].

The second general case is illustrated by Fig. 14. The form of the voltammogram is identical to the reversible case with the exception that both the peak width and the peak separation are much greater than the equations for the reversible case predict. The deviation here is a consequence of the fact that a slow charge transfer takes place. Thus, the deviation from

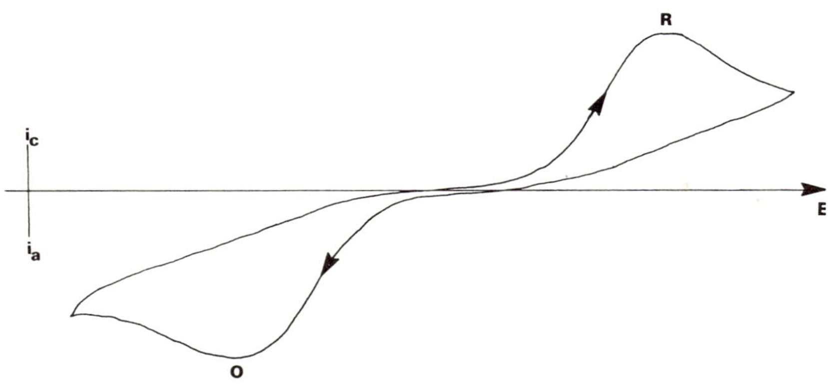

Fig. 14. Cyclic voltammogram for the irreversible (slow charge transfer) reaction: R $\rightleftarrows$ O + ne.

the reversible voltammogram becomes greater as the voltage sweep rate is increased. The dependence of peak separation on rate of voltage scan can be used to determine the rate constant of charge transfer (Sec. III.D.3).

The third general case, reversible oxidation of R to O with a coupled chemical reaction, demonstrates the most useful aspect of cyclic voltammetry in the study of organic electrode processes. In this case (Fig. 15),

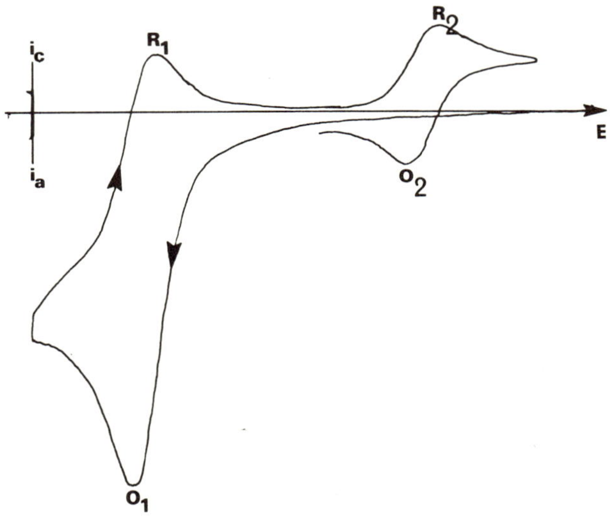

Fig. 15. Cyclic voltammetry for the reversible reaction: R ⇄ O + ne, followed by a chemical step, O → O′, producing an electroactive product.

the second and subsequent cycles are markedly different than the first. The oxidation peak O_1 appears on the first cycle; on the cathodic going scan, a small peak (R_1) is observed along with R_2. The reduction peak (R_1) corresponds to reduction of O generated from oxidation of R. The other cathodic peak (R_2) is due to the product of a chemical follow-up reaction O′. On the second (and subsequent) cycle, a new oxidation peak (O_2) is observed as R′ is being oxidized to O′. The fact that the peak (R_1) due to reduction of O is observed shows that the rate of the chemical step O → O′ is low enough so that the rate of voltage sweep can be great enough to reduce O before it all reacts to form O′. The latter implies, for this hypothetical case, that increasing the voltage sweep rate would increase the height of R_1 at the expense of R_2 and that, at sufficiently low voltage sweep rates, R_1 would essentially disappear, leaving only R_2 on the cathodic scan.

A detailed analysis of the theoretical aspects of cyclic voltammetry is beyond the scope of this chapter. The theory has been worked out elegantly by Nicholson and Shain [45] and interested readers should refer to the original literature.

2. Qualitative Applications

In this section we shall present results of several studies which have used cyclic voltammetry to gain a qualitative picture of the electrode processes taking place during oxidation or reduction of an organic substrate. Cyclic voltammetry allows one to gain a great deal of qualitative information concerning an electrode process. This initial study is generally followed by the use of other electrochemical techniques and methods of product observation.

The voltammetric oxidation of 5, 10-dihydro-5, 10-dimethylphenazine (DMPZ) in acetonitrile very nicely demonstrates the usefulness of cyclic voltammetry in examining a reversible oxidation process [52]:

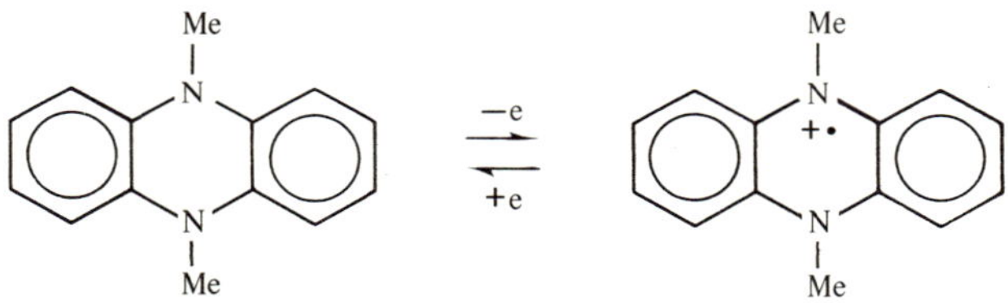

E_1

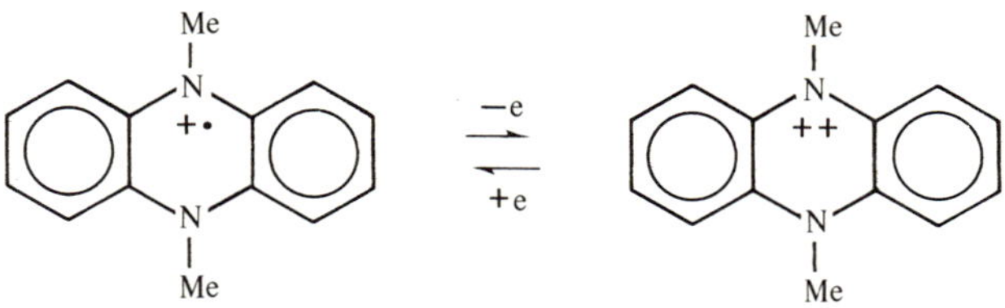

E_2

DMPZ undergoes two successive, reversible 1-e oxidation-reduction steps. Both processes adhere to the criteria of reversibility, i.e., $E_{p_a} - E_{p_c} = 58$ mV, $E_p - E_{p/2} = 59$ mV, and the current function $i_p/v^{1/2}C$ is a constant independent of the voltage sweep rate. This system has been suggested as a model system to which to compare other processes.

Another 1-e reversible system, the oxidation of 9, 10-diphenylanthracene to the cation radical, has been employed to demonstrate a steric influence of the reactivity of anodically generated cation radicals toward nucleophiles [53]. In the absence of an added nucleophile, the oxidation of DPA fulfills the criteria for a reversible process. When the cyclic voltammetric oxidation was carried out in the presence of 3, 5-lutidine (IV),

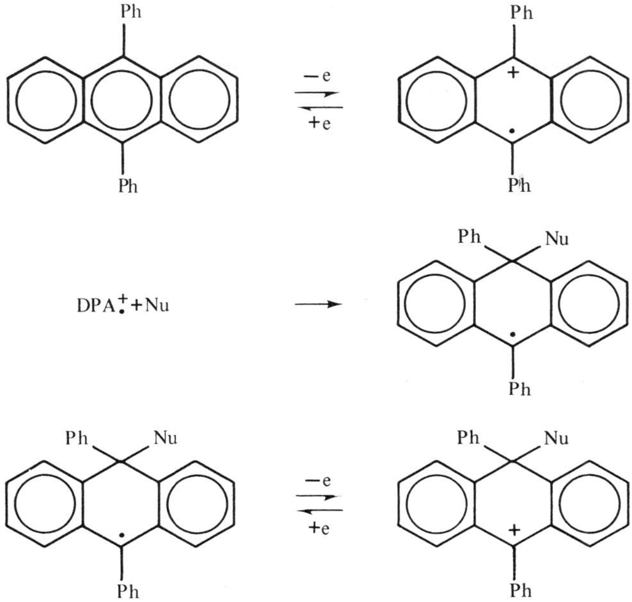

E

C

E

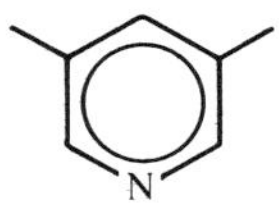

(IV)

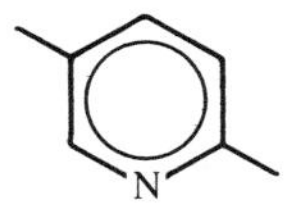

(V)

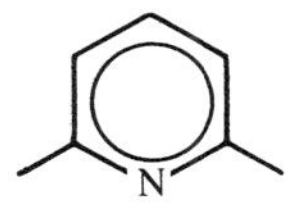

(VI)

a stoichiometric increase in the oxidation peak current (i_{p_a}) was observed at the expense of the reduction peak current (i_{p_c}). In this manner, it was shown that the reaction of DPA cation radical with (IV) is so rapid that it went to completion during the time scale of cyclic voltammetry. The more hindered nucleophile, 2, 5-lutidine (V), was found to be only about one-sixth as reactive as (IV) and about six times as great a concentration of (V) was required to reduce i_{p_c}/i_{p_a} to the same extent as a given concentration of (IV). The severely hindered nucleophile, 2, 6-lutidine (VI), was considerably less reactive.

The voltammetric reduction of DPA provides an example of the use of cyclic voltammetry on the cathodic side [54]. The cyclic voltammogram of DPA in DMF indicates a one-electron reduction peak which gives rise to a corresponding oxidation peak when the potential is switched after the initial charge transfer. When the scan is continued beyond the first peak, a second peak appears at more cathodic potential which has no anodic counterpart and was ascribed to the oxidation of the carbanion (VII). The voltammograms are consistent with the following scheme:

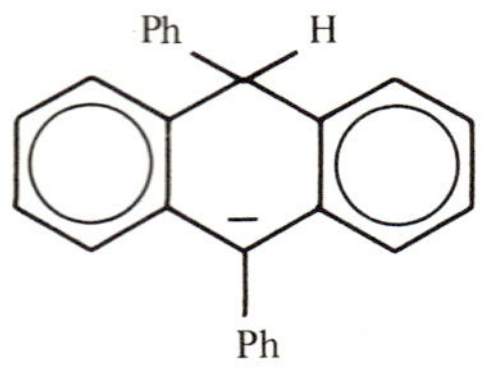

(VII)

$$\mathrm{DPA} + e \underset{}{\overset{E_1}{\rightleftharpoons}} \mathrm{DPA}^{\dot{-}}$$

$$\mathrm{DPA}^{\dot{-}} + e \underset{}{\overset{E_2}{\rightleftharpoons}} \mathrm{DPA}^{-2}$$

$$\mathrm{DPA}^{-2} + \mathrm{HA} \xrightarrow{\text{Fast}} (\mathrm{DPA} - \mathrm{H})^{-} + \mathrm{A}^{-}$$

$$(\mathrm{VII}) - e \longrightarrow (\mathrm{DPA} - \mathrm{H})^{\bullet}$$

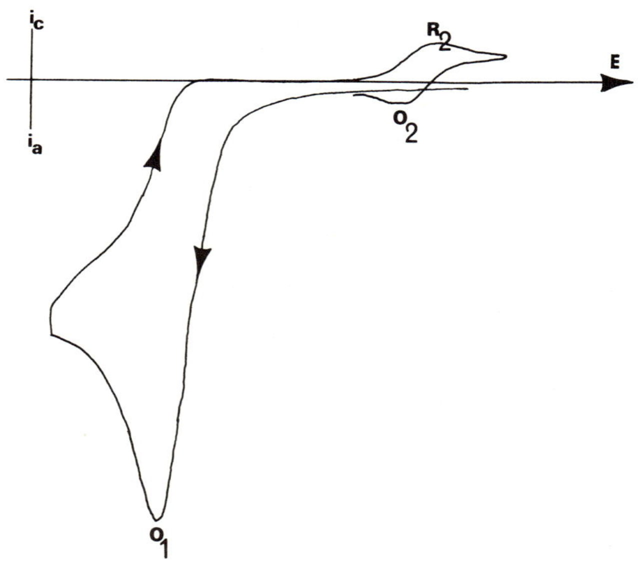

Fig. 16. Cyclic voltammogram of a substituted bibenzyl (VIII) undergoing oxidative cyclization to an oxidized form of a 9, 10-dihydrophenanthrene (IX).

An illustrative example of the application of cyclic voltammetry to the study of a complex process can be found in the anodic oxidation of (VIII) to (IX) [55]. The bibenzyl (VIII) was found to undergo multielectron oxidation in acetonitrile yielding (IX) and further oxidation products. On the initial

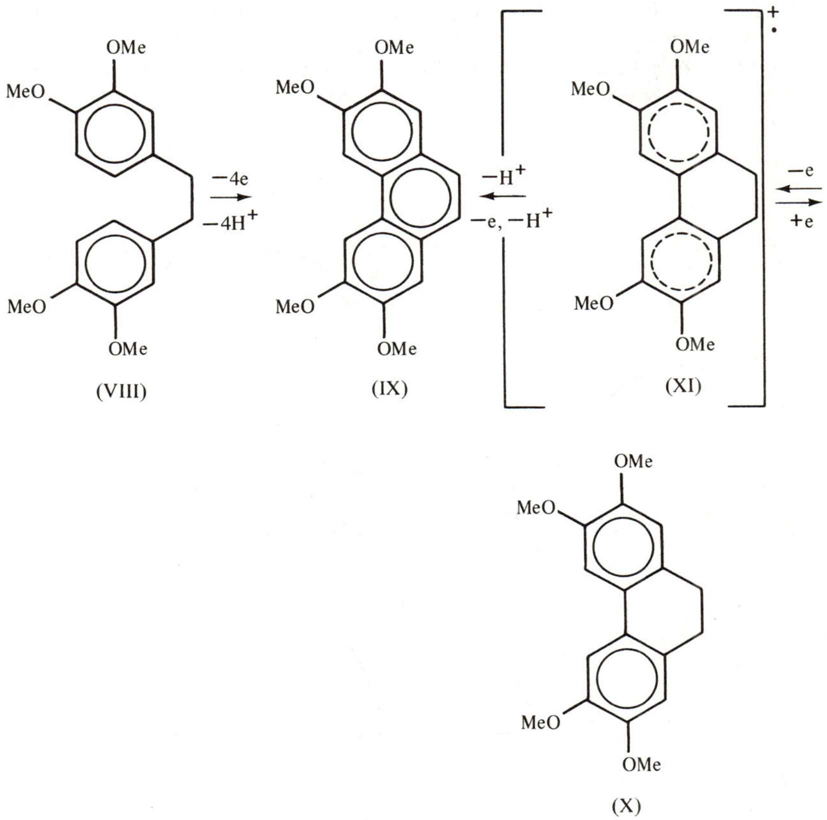

scan of the voltammogram (Fig. 16, a), a single oxidation peak (O_1) was observed. Since all the steps are rapid, the peak current corresponded to greater than 4-e. On the cathodic going scan, a reduction peak (R_2) is observed and on the second and subsequent cycles, a new oxidation peak (O_2) appears. A possible intermediate (X) was synthesized and was indeed found to undergo quasireversible oxidation with redox peaks corresponding to O_2 - R_2 (Fig. 16). Thus, cyclic voltammetry provides strong evidence for the mechanism of the oxidation of (VIII) to (IX). In this case, the voltammetric evidence was supported by the quantitative preparative oxidation of (X) to (IX).

Another very useful aspect of cyclic voltammetry is the variable time scale under which it can be applied. An application which utilizes this variable is the estimation of the stability of intermediates formed during an electrode process. For example, a voltammetric oxidation may appear as a two- or multi-electron process at slow sweep rates due to coupled chemical reactions and further oxidation of initial intermediates. On increasing the voltage sweep rate, it may be possible to "outrun" the chemical steps and observe a reduction peak for the initial product, the cation radical. This technique has been used to observe such reactive species as the anthracene

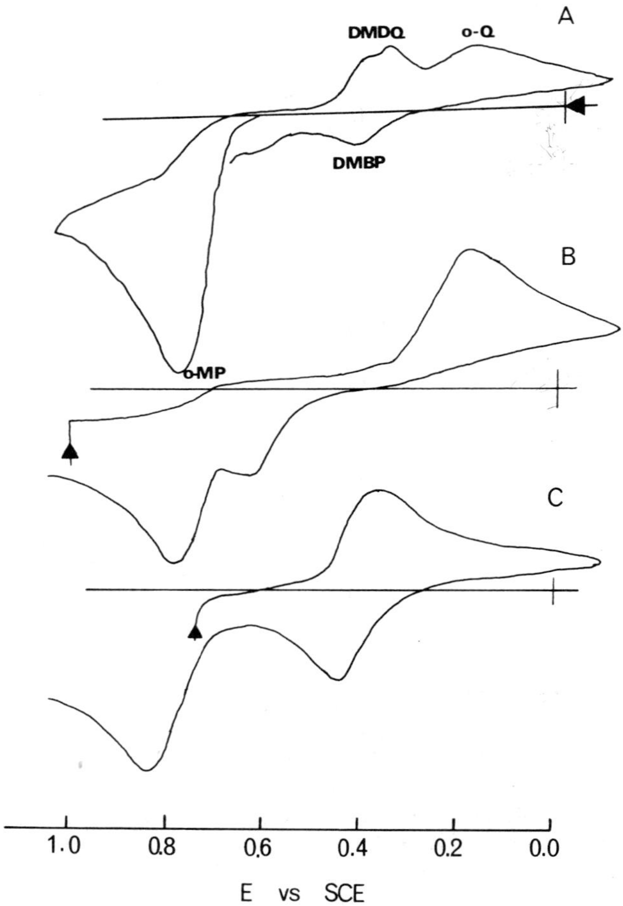

Fig. 17. Cyclic voltammetry and PSCV of o-methoxyphenol in pH 2.3. All solutions ca. 0.5 mM in o-methoxyphenol, scan rate 8 V/min, arrows indicate direction and beginning of each cyclic voltammogram sweep.

cation radical [56]. Semiquantitative estimates of cation-radical stability have been obtained for a variety of substrates using cyclic voltammetry [57].

A variation of the usual use of cyclic voltammetry is potential step cyclic voltammetry (PSCV). This technique is especially useful in cases where two different oxidation states, for example a cation radical and a dication, can be generated at the electrode. If the two species either react at different rates or give different products, PSCV can be useful in establishing the reaction pathways. For example, o-methoxyphenol undergoes two successive 1-e oxidations in aqueous media [58]. The closely spaced electron transfers appear as a single oxidation peak (MP) on the initial anodic sweep of the cyclic voltammogram (Fig. 17, A). On the cathodic going scan, two peaks, DMDQ and o-Q, which correspond to the reduction of the oxidation products, 3, 3′-dimethoxydiphenoquinone and o-benzoquinone. The two processes giving rise to the observed products can be formulated as shown, with DMDQ forming by initial 1-e oxidation and o-Q being the product of a 2-e oxidation of the substrate.

Path A

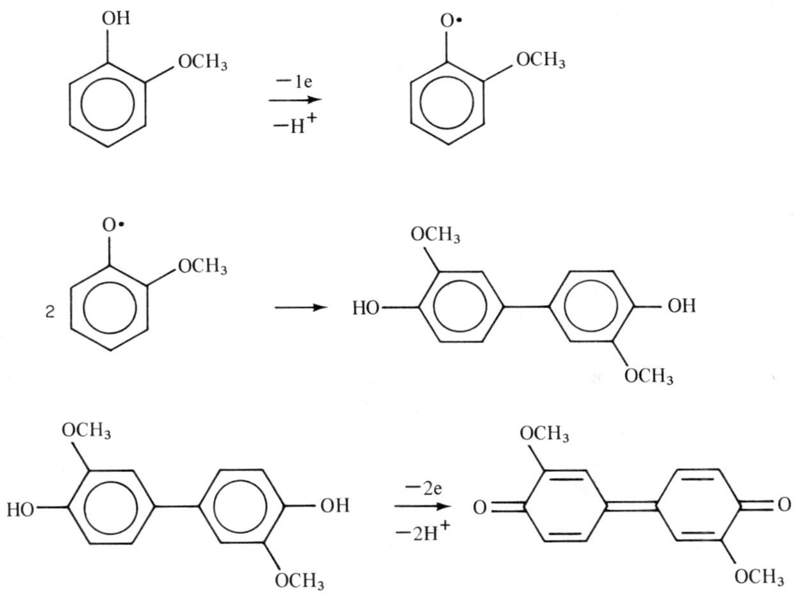

Path B

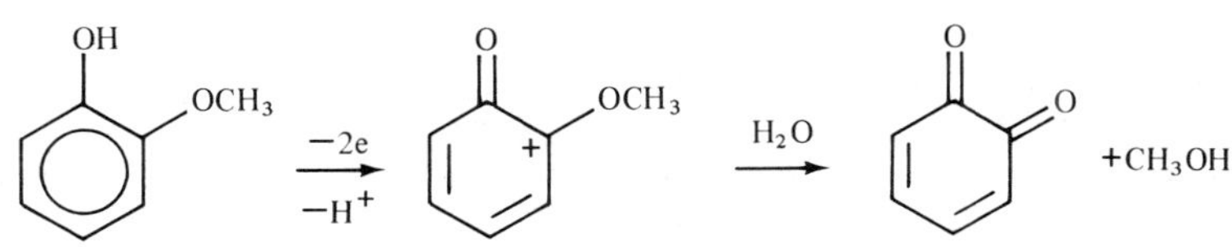

When the potential of a carbon paste electrode is stepped instantaneously from a potential where no current flows to E_1 (near the foot of the oxidation peak) and then a cathodic cycle is initiated, a reduction peak due to the dimer, DMDQ, is observed (Fig. 17, B). On the other hand, if the potential is immediately stepped to E_2 (more anodic than the peak) or more anodic, the initial process consists of a two-electron transfer; o-Q is the principal product and its reduction peak is observed on the cathodic cycle (Fig. 17, C). It follows that stepping the potential to values intermediate between E_1 and E_2 should and does result in the observation of reduction peaks for both products.

A further application of PSCV is the demonstration that the active species in the anodic hydroxylation of thianthrene in acetonitrile is the cation radical rather than the dication as was suggested by a homogeneous kinetic study [59]:

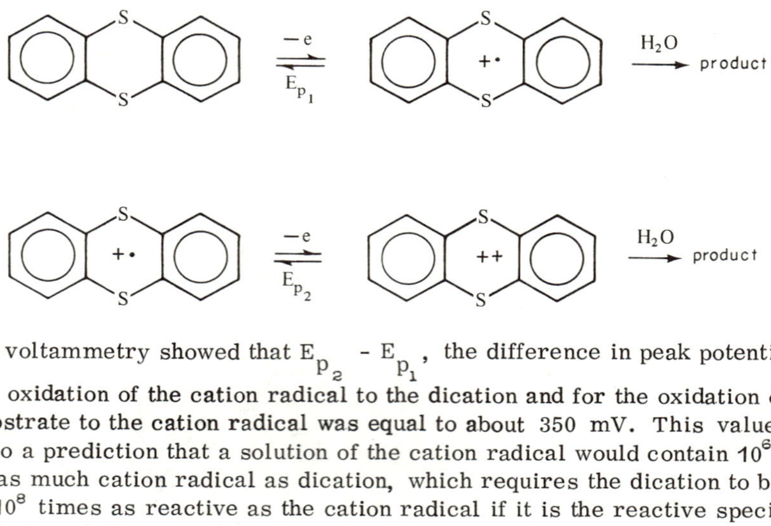

Cyclic voltammetry showed that $E_{p_2} - E_{p_1}$, the difference in peak potentials for the oxidation of the cation radical to the dication and for the oxidation of the substrate to the cation radical was equal to about 350 mV. This value leads to a prediction that a solution of the cation radical would contain 10^6 times as much cation radical as dication, which requires the dication to be about 10^8 times as reactive as the cation radical if it is the reactive species in the hydroxylation reaction. By using the PSCV technique with oxidation potentials ranging from E_1 + 50 mV to E_2, and then measuring the current for the reduction of the cation radical on the cathodic sweep, a measure of the extent of reaction as a function of generating potential was obtained. From these data it was suggested that the dication is indeed about 10^3 times as reactive as the cation radical; however, the reactive species is predominately the cation radical in the potential range E_1 to E_1 + 150 mV.

We can close this section with a note of caution concerning the use of cyclic voltammetry in establishing reaction mechanisms. It is becoming more and more common to see papers in which the only evidence for an intermediate is the observation of a peak on a cyclic voltammogram.

Without having access to the intermediate in order to match the peak with that from the authentic substance, the assignment is on pretty shaky ground. Even if the authentic compound is available, peak matching should be supported by additional evidence in order to produce a convincing case.

3. Quantitative Applications

Cyclic voltammetry is generally considered to be a qualitative rather than a quantitative tool. However, if care is taken, it is possible to determine rate constants of follow-up chemical reactions with results comparing favorably with the other techniques, generally considered to be more suitable for quantitative measurements. For details of the application of cyclic voltammetry to the study of the kinetics of follow-up reactions, the reader is referred to the original literature [45].

Cyclic voltammetry is a useful and convenient tool for the determination of rate constants of electron transfer. A theoretical study has provided a relationship between the peak separation for oxidation to and reduction of an intermediate and the heterogeneous rate constant k_o for the electron transfer [60]. From the results of the latter study it is only necessary to measure E_p at several sweep rates, and, using a working curve, the rate constant can readily be obtained.

E. Chronopotentiometry

1. Theoretical Aspects

Chronopotentiometry is one of the oldest methods for studying electrochemical phenomena. The fundamental equation associated with it was formulated by Sand beginning in 1902 [61]. Nevertheless, in spite of the work of Gierst and Juliard [62] employment of the method appears to have remained limited, no doubt because although the principle is relatively simple its practical application is rather complex. Even so, it is very valuable, particularly in the modern version, reverse current chronopotentiometry.

Chronopotentiometry is concerned with the change in potential of an electrode with time in the course of a short (<60 sec) electrolysis at constant current. Generally, the electrolysis is carried out with an electrode under conditions of linear diffusion; the constant current to be used is chosen so that all the material in the vicinity of the electrode is consumed in a time span which is sufficiently short so that natural convection may not set in.

Under these conditions an $E = f(t)$ curve is obtained similar to that of Fig. 18 which shows a reversible, rapid one-electron oxidation. At a time $t = 0$ a constant current is suddenly impressed on the electrode which now

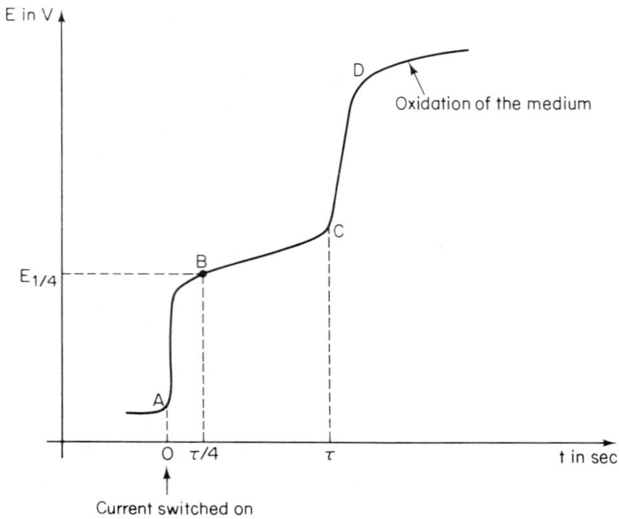

Fig. 18. General form of a chronopotentiometric curve for a one-electron rapid and reversible oxidation.

reaches a potential (point B) at which oxidation starts. Thereafter, the potential varies only slightly until time τ (point C) at which all the electroactive species near the electrode has been consumed. The potential then changes suddenly to a new value (point D) at which the most easily oxidized substances now in the vicinity of the electrode are oxidized (e.g., constituents of the medium, i.e., supporting electrolyte or solvent).

The transition time τ is given by the Sand equation

$$\tau^{1/2} = \frac{\pi^{1/2} F D^{1/2} C_R}{2 i_o}$$

in which C_R is the concentration in moles per cubic centimeter of the oxidizable species, and i_o is the current density in amperes per square centimeter. It can be shown [26] that the equation of the curve is

$$E = E_{1/4} + \frac{RT}{F} \ln \frac{\tau^{1/2} - t^{1/2}}{t^{1/2}} .$$

$E_{1/4}$ is the potential acquired by the electrode at the time $\tau/4$ and is equal to the half-wave of the i-E curve:

$$E_{1/4} = E_{1/2} = E^{\circ} - \frac{RT}{F} \ln \frac{D_O^{1/2}}{D_R^{1/2}} .$$

D_O and D_R are the coefficients of diffusion of the oxidized and reduced species, respectively.

It is obvious that for a given concentration C_o, the product $i_o\tau^{1/2}$ must remain constant.

Actually, many difficulties are involved [63] which arise because the diffusion is not perfectly linear and not all the current i_o is used only in charge transfer; some of it is used in charging the double layer since the potential of the electrode changes rapidly (cf. p. 57). Specifically, these factors result in a deformation of the chronopotentiometric curve which in turn renders the determination of τ difficult and necessitates the use of empirical methods [64]. Bard [65] has proposed an equation for the ratio $i_o\tau^{1/2}/C_R$ which takes into account the several secondary processes.

It should be noted that the Sand equation for τ does not include the rate constant k^o. Therefore, a slow charge transfer gives the same transition time but the form of the curve is changed.

2. Application in the Study of Reaction Processes

It is clear that because of the relationship between $i\tau^{1/2}$ and C_R chronopotentiometry has analytical applications. It is also valuable in the study of reaction processes. Paunovic's review [66] is devoted to this aspect; it considers several possible cases and discusses criteria for diagnosing mechanisms. Two illustrative examples will be presented here.

a. Two Successive Electrochemical Reactions by One Substance.

A sequence of the type

$$O_1 + n_1\bar{e} \rightleftarrows R_1$$

$$R_1 + n_2\bar{e} \rightleftarrows R_2$$

gives two successive waves on the chronopotentiometric curve. Two transition times τ_1 and τ_2 can be determined; it has been shown [26] that the following relationship exists between them:

$$\tau_2 = \tau_1\left[2\frac{n_2}{n_1} + \left(\frac{n_2}{n_1}\right)^2\right].$$

Specifically, in the frequently encountered case in which $n_2 = n_1 = 1$, the relation is $\tau_2 = 3\tau_1$. This result was obtained, for example, in the oxidation of 9-amino-10-phenyl anthracenes which in organic media yield the radical cation and the dication [67]:

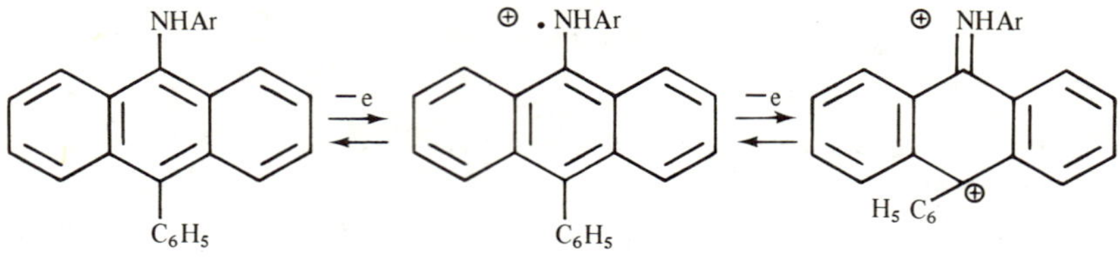

b. Charge Transfer Coupled to a Preceding or Succeeding Chemical Reaction. Consider the two following schemes:

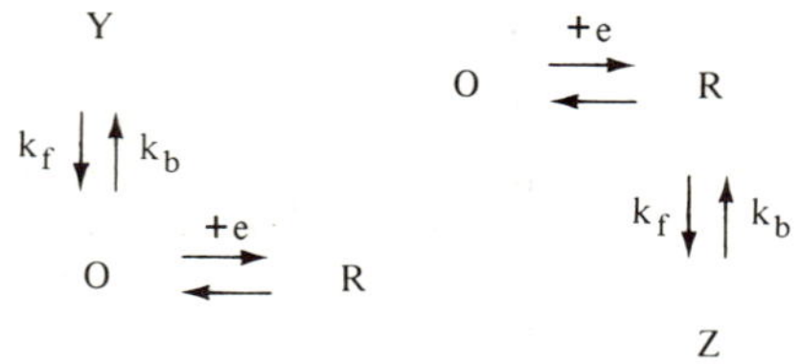

When a succeeding chemical reaction yields a nonelectroactive species Z, the variations in the concentration of species O in the vicinity of the electrode are unaffected by the chemical reaction. The calculations of τ in the absence of chemical complications are still valid here. However, if the chemical reaction precedes the charge transfer, $i_o\tau^{1/2}$ is not a constant independent of i_o [68, 69]; if it is assumed that the charge transfer is reversible and rapid and that all diffusion coefficients are equal, then:

$$i_o\tau^{1/2} = \frac{\pi^{1/2} F D^{1/2} C_o}{2} - i_o \frac{\pi^{1/2}}{2K(k_f + k_b)^{1/2}}$$

In the above equation C_o is the concentration of species O at equilibrium (which is assumed to be established in solution before the chronopotentiometric plot is started) and K is the equilibrium constant $K = C_o/C_Y = k_f/k_b$.

It is clear that in principle the method can be used to differentiate between EC and CE mechanisms because in the former case $i_o\tau^{1/2}$ is independent of i_o and in the latter case it varies with i_o. Mark and Anson [70] have applied these principles in a study of the oxidation of p-phenylenediamines in aqueous acid at low pH. These amines A are capable of yielding two protonated forms, AH^+ and AH_2^{2+}. Previous work had not established which was the species that underwent the first electron transfer. The authors cited demonstrated that the following scheme was valid:

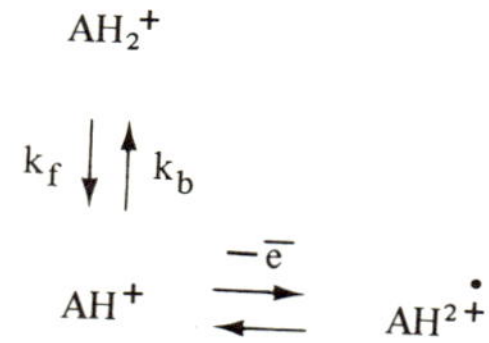

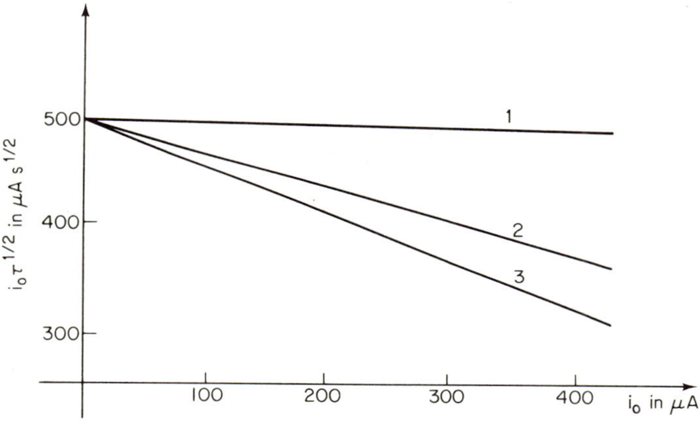

Fig. 19. Variation of $i_0\tau^{1/2}$ as a function of i_0 for an aqueous 5×10^{-3} M solution of p-phenylenediamine at several pH's. Curve 1, pH = 3.4 (buffer H_3PO_4-$H_2PO_4^-$) and pH = 1.9 (buffer HSO_4^--SO_4^{2-}); curve 2, pH = 1.0 (H_2SO_4); curve 3, pH = 0.7 (H_2SO_4). Pt electrode of ca. 0.23 cm^2.

Effectively, a **linear relation** is obtained between $i_0\tau^{1/2}$ and i_0 (Fig. 19), if the pseudo-first-order rate constant for protonation is included in the denominator of the expression for $i_0\tau^{1/2}$

$$k_f = k_b'' [H^+].$$

Here k_b'' is the second-order rate constant related to k_f by $K = k_f/k_b''$. It can be seen from Fig. 19 that deprotonation before charge transfer is evident only at very low pH's. The values of k_f and k_b'', which are of the order of 10^4 sec^{-1} and 10^7 liters $mole^{-1}$ sec^{-1}, respectively, can in addition be obtained.

In Ref. [71] there is a discussion of the difficulties involved in this type of determination when a solid electrode is used. Chronopotentiometry may also be employed to reveal and to study absorption phenomena [72].

3. Chronopotentiometry with Programmed Current

The applications of chronopotentiometry which have been discussed concern the variation of $\tau^{1/2}$ or of $i_o\tau^{1/2}$ as a function of several factors such as i_o or C_R. Reinmuth [73] has demonstrated that a more detailed analysis of the form of the chronopotentiometric curve can provide information concerning the type of mechanism involved at an electrode and criteria for choosing a particular mechanism. But practical applications seem difficult with solid electrodes owing to the lack of reproducibility of the curves.

The several aspects of chronopotentiometry with programmed current are perhaps easier to apply and interpret. Several types of programs have been proposed [74] but the simplest to interpret, at least qualitatively, is reverse current chronopotentiometry. In this technique the current i_o initially imposed is suddenly switched to the current $-i_o$ at a time t_f which is less than or equal to the first transition time τ_f. For a simple one-electron system it has been shown that the transition time τ_r which is observed on the second branch of the recording is given by

$$\tau_r = \frac{t_f}{3}.$$

This is valid if the coefficients of diffusion are or are not equal, whatever the rapidity of the system.

If the product of charge transfer (corresponding to the first part of the chart) undergoes chemical reaction (EC mechanism), the value of τ_r is changed as a consequence and it is often possible to observe in the second branch of the chart the transition time corresponding to the electrochemical reaction of the product which had been formed by the chemical reaction.

A now classic example was worked out by Geske [75] using the oxidation of tropylidene (cycloheptatriene) in acetonitrile. The oxidation leads to the tropylium cation:

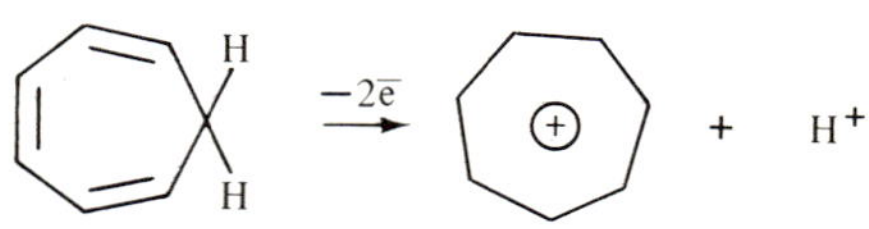

and current inversion gives an E-t curve (Fig. 20) which includes in its second branch the reductions

$$H^+ + \bar{e} \rightleftharpoons 1/2\ H_2$$

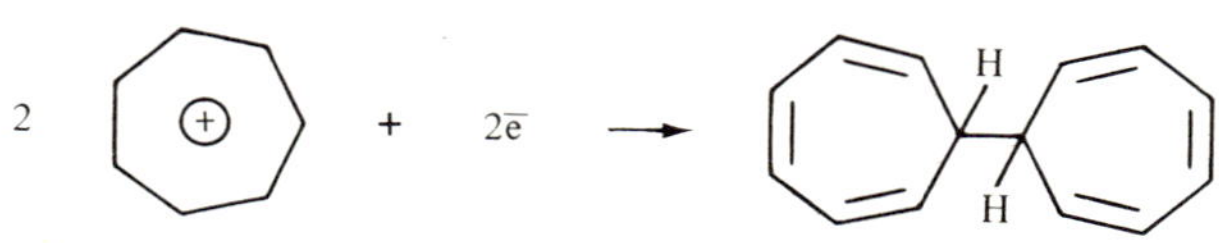

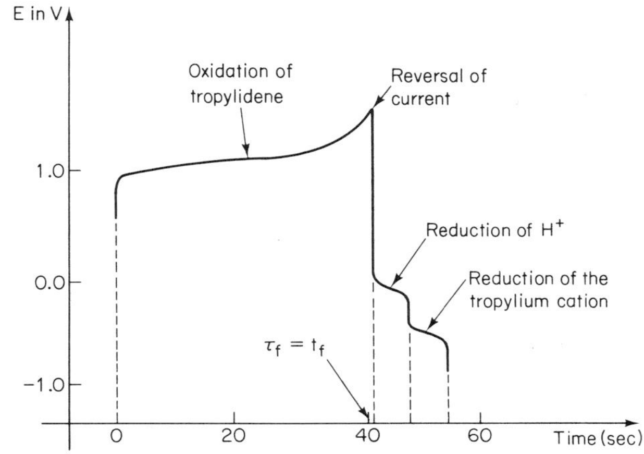

Fig. 20. Chronopotentiometry with current reversal of a 2.89×10^{-3} M solution of tropylidene in acetonitrile. Pt electrode. Supporting electrolyte 0.1 M $LiClO_4$. Reference electrode $Ag/AgNO_3$, 10^{-2} M.

with corresponding transition times τ_{r_1} and τ_{r_2}. Theoretical study of the problem led to the expectation that

$$\frac{\tau_f}{\tau_{r_1}} = 8.0 \qquad \frac{\tau_f}{\tau_{r_1} + \tau_{r_2}} = 3.0 \qquad \frac{\tau_{r_2}}{\tau_{r_1}} = 1.67.$$

These relations were well verified (see Fig. 20).

IV. COULOMETRY

As already indicated, electron transfers which occur after preceding chemical reactions are not detected within the time scale of voltammetric experiments except when the reactions are sufficiently rapid. A voltamperometric study therefore does not always unequivocally lead to the correct conclusion concerning the results to be expected in a macroelectrolysis on an electrode of large area, even if the macroelectrolysis is done carefully at a constant controlled potential so that the electrode phenomena are well defined. Actually, a macroelectrolysis since it occurs over a much larger period than is required for a voltammetric determination, permits slow chemical reactions to intervene.

In such cases the determinations of the i = f(t) curves and of the coulometry, i.e., the number of faradays required for complete electrolysis of 1 mole of substance,[5] are useful supplements to the voltammetric study.

Consider, e.g., a mechanism comprising initially an electron transfer and a very rapid chemical reaction as in

$$A + n\bar{e} + X \rightarrow B,$$

followed by reaction of the starting material with the product B,

$$A + B \xrightarrow{k} Y \text{ (electroinactive)}.$$

The voltamperometric methods give information only about n, the number of electrons involved overall in the first of the above reactions. Coulometry, on the other hand, leads to a value for n_{app} which depends on the kinetics k of the second reaction above. If this reaction is very slow, the effect is to make it appear that only the reaction leading to B had occurred and $n_{app} = n$. Conversely, if the second reaction is rapid enough, n_{app} approaches n/2. For intermediate values of k, n_{app} has values between n and n/2 and is a function of the initial concentration of A.

Let us now consider the change in the current i in the course of the electrolysis. In the first case (when k is very small), A is consumed only in the process involving n electrons; at a given time t the current is given by

$$i = -nF\frac{dN}{dt} = i_o \exp(-p_o t)$$

in which N is the number of moles of A remaining in the cell, i_o the initial current, $p_o = mS/V$ with m being D/δ as previously defined (p. 32). S is the surface of the electrode and V the volume of solution. The log $i = f(t)$ curve is then a straight line (Fig. 21, curve 1). If, however, k is large

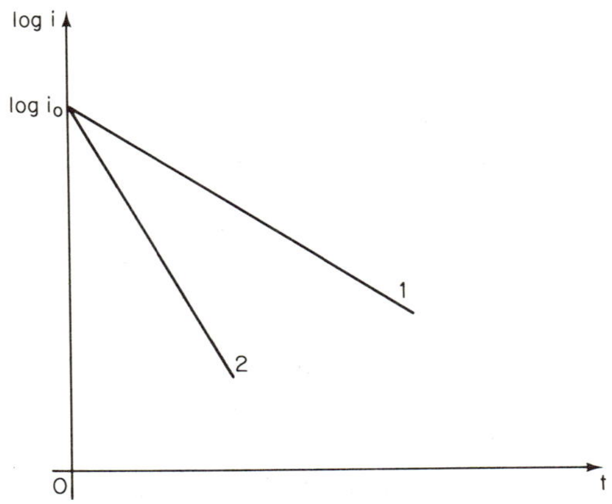

Fig. 21. Log $i = f(t)$ curves illustrating the relationship in the course of a controlled potential electrolysis (see text).

enough, the greater rate of disappearance of B causes a very rapid decrease in current (Fig. 21, curve 2).

A general article dealing with this question has been published by Bard and Santhanam [76] who along with Geske and Meites were among the first to point out the importance for an understanding of electrochemical reaction mechanisms of determining n_{app}. The values of n_{app} for several types of mechanisms and the corresponding shapes of the $i = f(t)$ curves have been determined. The converse procedure is often possible. Criteria have been proposed for choosing a reaction mechanism on the basis of n_{app} values (and their variations under certain conditions) and the forms of the $i = f(t)$ curves. Sometimes it is possible to determine the rate constants of coupled chemical reactions. Meites' review [77] should be consulted for further details.

If the mechanism includes after the electron transfer a chemical reaction which consumes only B,

$$A + n\bar{e} + X \rightarrow B$$

$$B \underset{k_b}{\overset{k_f}{\rightleftarrows}} Y$$

coulometric determinations made during the reduction of A obviously always give the same value ($n_{app} = n$) whatever may be the constants k_f and k_b. However, reversal coulometry provides a means for uncovering this chemical reaction. After a certain period of electrolysis during which B is formed, the electrode potential is brought to a value at which B is reoxidized to A; the number of coulombs Q_f used in the first operation is compared to the number of coulombs Q_b involved in the reoxidation. If the conversion of B to Y is very slow or even reversible, $Q_b = Q_f$. [When the reaction is reversible, the $i = f(t)$ curve obtained in the second step may have an abnormal form; the kinetics of the reoxidation of B can be defined by k_b.] However, if Y is formed in an irreversible reaction, Q_b is less than Q_f. A variety of possible cases has been studied from the theoretical point of view.

It is necessary to keep in mind that the various operations involved in the course of a coulometric experiment take much longer than a voltamperometric determination and that in the former case macroscopic quantities of intermediate products are produced. Precautions which need be taken to avoid the occurrence of undesirable side reactions (reaction with oxygen from the air, with residual water in the medium, etc.) must be observed more stringently in coulometric than in voltammetric experiments [76].

Many other techniques may be used in association with coulometry at controlled potential. Extensive discussions have been published on coulometry at controlled current intensity [78] and its application to the determination of reaction kinetics [79]. A particularly promising form of coulometry is that involved when thin layer cells are used [80]. In these a layer of ca. 10^{-3} cm of solution is electrolyzed while confined between an electrode and an inert membrane or between two electrodes [81].

Finally, it may be interesting in the course of a preparative controlled potential electrolysis (CPE) to employ a physical method which reveals the appearance and disappearance of intermediate species in combination with simultaneous coulometry which reveals how many faradays have passed through the electrolysis cell at any given time. Thus, using a spectrophotometric cell through which is circulated the solution from an operating electrolytic cell, it is possible to determine the variation in optical density corresponding to known absorption maxima of transient species. These optical densities are then examined in the relationship $z = Q/Q_0$ where Q_0 is the number of coulombs corresponding to 1 faraday per mole of electroactive substance and Q is the number of coulombs used for obtaining a given optical density. The theoretical aspects of this method and its application in the study of the reduction of sulfur in dimethyl sulfoxide have been described [82]: in following the optical density at 490 nm (S_8^{2-}) and at 618 nm (S_6^{2-}), it has been shown that the former passes through a maximum at z = 2, i.e.,

$$S_8 + 2\bar{e} \rightarrow S_8^{2-},$$

while the second is at a maximum at z = 8/3, i.e.,

$$3S_8 + 8\bar{e} \rightarrow 4S_6^{2-}.$$

V. POTENTIOMETRY

In potentiometry, which is a stationary method, there is measured the potential (vs. a reference electrode) of an indicating electrode through which no current or a very small constant current is flowing. In the former case, the method is called potentiometry at zero current; in the latter, potentiometry with polarized electrode.

If no current is passing through the electrode it assumes the equilibrium potential (p. 37) given by the Nernst equation. Measurement of the potential then allows the concentration to be determined; as a consequence, potentiometry is a widely used analytical method. Two procedures may be used: either simply the potential is measured when an electrode is placed in a solution of the substance whose concentration is to be determined, or the change of potential is followed as a reagent is added slowly to the solution of the unknown (potentiometric titration) [83].

Potentiometry is not used very often in connection with the elucidation of the mechanism of a complex electrochemical reaction. Sometimes the measurement of the equilibrium potential can serve to characterize the oxidizing or reducing power of a substance provided it is a component of a rapid system. The value of this potential permits a prediction to be made concerning chemical redox reactions when it is considered in conjunction with potential scales (p. 19). Further, potentiometry can serve as one means of following the change of a solution in the course of an electrolysis. Potentiometric analyses have played an important role in the study of rapid redox systems and the complexities involved when a dimer is formed from either the oxidant or the reductant or when a complex is formed from oxidant plus reductant (quinhydrones, e.g.) [84].

It is important to know that the changes in potential which an electrode undergoes in the course of a potentiometric determination can be anticipated by obtaining an i-E curve for each of the two redox systems involved and that the logical interpretation of the analytical results presupposes that these curves have been obtained and utilized [83].

VI. CORRELATIONS AMONG THE SEVERAL ELECTROCHEMICAL METHODS; THE CHOICE OF METHODS

The electrochemical methods most frequently used in organic electrochemistry have been discussed above. It may now be appropriate to attempt to classify them in such a way that the reader may grasp at a glance the individual characteristics of the methods and the correlations among them.

Table 1 classifies these methods [6] into two categories: those in which time t is a variable and those in which it is not. In the case of stationary methods, the three parameters i, E, and C_o (the concentration of electroactive substance) stand in a relationship $f(i, E, C_o) = 0$. In each method one parameter is fixed, the second is varied in some definite fashion, and the resulting variations in the third are studied. In nonstationary methods t must likewise be considered and the fundamental relationship is $f(i, E, C_o, t) = 0$. Each of these latter methods may be similarly described: one parameter is held constant, a second is varied in a defined fashion, and the mutual variation of the third and fourth is studied.

In each of the methods defined as above the variation imposed on a given parameter may follow one of several laws. For example, there may be linear variation with time ($E = E_1 + vt$) or abrupt variation (e.g., potential step from E_1 to E_2), etc. Ref. [27] discusses these various possibilities.

Having classified the methods as above, the problem now remains of choosing the one or more methods which will be of greatest utility in solving a given problem. A complete discussion of this question would exceed the limitations of this introductory chapter; only a few general concepts will therefore be presented.

First of all, it is necessary to distinguish between the aim of arriving at qualitative and at quantitative conclusions concerning a reaction mechanism. (In the latter case obviously electrochemical and chemical rate constants

TABLE 1

A Classification of the Main Electrochemical Methods

General characteristic	Controlled variables	Measured response	Name
Stationary: $f(i, E, C_o) = 0$	$C_o = cte$ $E = E_1 \pm vt$	$i = f(E)$	Potential sweep voltammetry
	$C_o = cte$ $i = i_1 \pm vt$	$E = f(i)$	Current sweep voltammetry
	$i = cte$ (0 or ϵ) C_o variable	$E = f(C_o)$	Potentiometry
Nonstationary: $f(i, E, C_o, t) = 0$	$C_o = cte$ $E = E_1$ (step)	$i = f(t)$	Chronoamperometry
	$C_o = cte$ $E = E_1 + vt$	$i = f(t)$ or $i = f(E)$	Linear sweep chronoamperometry
	$C_o = cte$ $E = E_1 \pm vt$ followed by $E = E_2 \pm v't$	$i = f(E)$	Cyclic voltammetry
	$C_o = cte$ $i = cte$ (step)	$E = f(t)$	Chronopotentiometry
	$C_o = cte$ $i = f(t)$	$E = f(t)$	Programmed current chronopotentiometry
	$C_o = cte$ $E = E_1 - vt$ (or DME)	$i_{av} = f(t)$ or $i_{av} = f(E)$	Polarography

must be determined as well as equilibria constants.) As already mentioned (p. 94), most studies have been concerned with partial qualitative diagnoses of mechanism. The authors' personal laboratory experience and their assessment of the literature indicate that for qualitative conclusions the successive employment of the following methods is likely to be the most fruitful provided in each case all the parameters which can be systematically varied are examined in the exploratory phase of the work:

1. obtention of the i-E curve using the DME and/or the rotating electrode;

2. cyclic voltammetry and the resultant defining of the electroactive intermediates;

3. influence of the medium (pH, presence of reagents, change of solvent, etc.) upon the results of 1 and 2;

4. controlled potential electrolysis with simultaneous determination of the coulometry and the i - f(t) curves;

5. isolation and characterization of the products which are formed by electrolysis under a variety of conditions. Comparison of these results with those obtained by electrochemical methods (Chapter III);

6. study of the reaction intermediates and confirmation by physical methods (Secs. VII, VIII) of the conclusions drawn from the application of the electrochemical methods.

The above remarks are intended only to be of service when an investigation is first launched. As the study develops, the results of one phase usually dictate the choice of the method to be used in the next. Often a conclusion or hypothesis derived at one stage requires that one of the methods already used be tried again slightly differently. Such trial-and-error procedures are not uncommon in many types of research. Examples of this approach are perhaps illustrated in Refs. [85] and [86].

VII. DIVERSE APPLICATIONS OF ULTRAVIOLET, VISIBLE, AND INFRARED SPECTROSCOPY

Ultraviolet absorption spectroscopy may be used in a number of ways for elucidating the mechanism of an electrochemical reactions.

It may contribute in the classic manner to the establishment of the structure of a compound which has been prepared electrochemically, isolated, and purified. To the extent that the structure of the isolated product can be related to the reactions within the cell, this structure determination contributes to the study of the reaction mechanism.

There are two procedures, however, for using this spectroscopy more specifically in elucidating an organic electrochemical mechanism: (a) using a flow cell, and (b) coupling electrochemical and spectroscopic methods.

A. Flow Cells

This technique employs a standard commercial spectrophotometer in which is installed an electrolytic cell. The cell is constructed so that the liquid in one of the compartments may be circulated, as, for example, by the aid of an inert gas. An arrangement similar to that for electron paramagnetic resonance spectroscopy (p. 147) is used with this exception: in place of the tube which is placed in the electron paramagnetic resonance (EPR) cavity, there is used for the uv work a spectroscopic cell with quartz windows. This type of cell can easily be constructed from certain elements of the circulating devices which are available as optional equipment from suppliers of spectrophotometers.

With cells like these it can be determined whether there are formed in the course of an electrochemical reaction species which would, because of their sensitivity to oxygen or the humidity of the air, be difficult to detect by other means. Specifically, only with this type of cell can the existence of isobestic points [7] be determined. Periodic transfer of aliquots of a solution during electrolysis to an ordinary spectrophotometric cell rarely permits the observation of isobestic points; evaporation of the solvent or reaction of the solution with atmospheric constituents in the course of the transfer prevents spectra from being obtained which truly reflect the state of the solution at a given instant of electrolysis.

Furthermore, since use of a flow cell obviates the need for removing samples of solution, the change in the spectrum with the quantity of electricity which has passed through the cell can also be studied. This may have important bearing on the elucidation of a reaction mechanism.

The following example illustrates the capabilities of these cells. The results of investigations by a variety of electrochemical methods lead to the conclusion that the oxidation of phenylhydrazine in acetonitrile containing 0.1 M $LiClO_4$ may be represented by the following equation:

$$3C_6H_5NHNH_2 \xrightarrow{-2\bar{e}} 2C_6H_5NHNH_3^+ + C_6H_5N = NH.$$

One molecule of phenyldiazene is formed concurrently with the protonation of two molecules of the starting material [87]. Since it is known that phenyldiazene is very reactive toward oxygen [88], the spectrum of the solution is periodically recorded in the course of an electrolysis using a flow cell in which circulation is produced by a slow stream of argon. Of particular interest is the spectrum obtained at the time that 0.66 faraday/mole phenylhydrazine has been passed (Fig. 22). If from this spectrum is subtracted the spectrum obtained independently of a solution containing (in protonated form) two-thirds of the initial concentration of phenylhydrazine used here, the resultant curve shows effectively the maxima which might be expected [88] for a solution of phenyldiazene of predicted optical density.

In the several applications of this method it is expedient to interrupt the electrolysis circuit before each spectrum is recorded because in the majority of commercial instruments the recording takes several minutes and in the course of this time the composition of the solution could change significantly if the electrolysis were continuous.

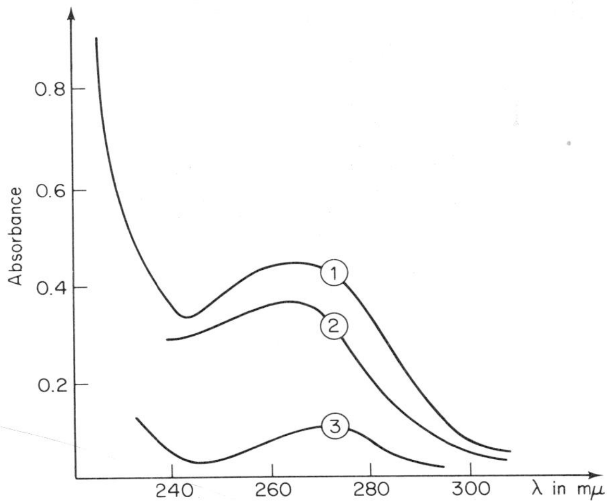

Fig. 22. Ultraviolet and visible spectroscopic evidence for the formation of phenyldiazene during the oxidation of a 2×10^{-3} M solution of phenylhydrazine in acetonitrile containing 0.1 M $LiClO_4$: anode, Pt; reference electrode, Ag/Ag^+ 10^{-2} M. Curve 1, $C_6H_5NHNH_2$ 2×10^{-3} M after partial electrolysis (0.66 faraday/mole); curve 2, $C_6H_5NHNH_3^+$ 0.66×10^{-3} M; curve 3, $C_6H_5N{=}NH$.

B. Combination of Electrochemical and Spectroscopic Methods

The flow cells which have been described make possible the characterization of electrolysis products which are too unstable to be isolated but they do not provide a means for detecting reaction intermediates which have only transitory existence within the medium. This limitation is an important result of the fact that with commercially available spectrophotometers a relatively long time is required for recording a complete spectrum. Therefore, the invention of the rapid scanning spectrophotometer (RSS) has stimulated the development of new approaches which combine spectrophotometry and electrochemical methods.

With the RSS described by Strojek et al. [89], a complete spectrum spanning, e.g., the range between 370 and 700 mμ, can be obtained in ca. 10^{-3} sec. The scanning of the spectrum is brought about by the vibration of a concave mirror. The current which produces this vibration is carried on the X axis of a cathodic oscillograph; the Y axis receives the signal furnished by the photomultiplier after it is logarithmically converted by a suitable circuit. The latter is such as to yield on the screen a curve representing absorbance as a function of wavelength.

This type of spectrophotometer makes possible the use of a number of techniques whereby the light absorption by the intermediate species is directly determined near the working electrode.

1. Transmission Spectrophotometry

The principle is illustrated schematically by Fig. 23. The electrolysis cell contains two parallel surfaces, one of which is a transparent electrode

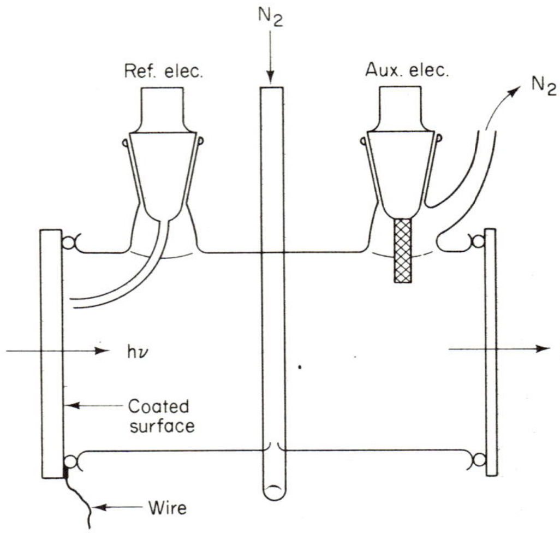

Fig. 23. Schematic of a cell for transmission spectroscopy on a transparent electrode.

at which the absorption spectrum of the substance in question is to be determined. This electrode may consist of a thin plate of quartz or glass covered with a semiconducting layer of tin oxide (transparent between 305 and 700 mμ) or of a thin film of platinum, or even with a thin grid of platinum wires which are placed on the transparent window.

The theory of the method has been expounded by Strojek and Kuwana [90]. Consider the reaction

$$R \rightleftarrows \bar{e} + O$$

in which the stable species O is formed which has a coefficient of absorption ϵ at a wavelength λ at which R does not absorb. The optical density A which is associated with the passage of the light beam is

$$A = \epsilon \int_0^{\delta} (C_O)_x \, dx = \epsilon \int_0^t \frac{1}{S} \cdot \frac{i\, dt}{F}.$$

In this expression δ is the thickness of the layer of the solution containing compound O, $(C_O)_x$ is the concentration of this species at the distance x from the electrode, and S is the surface of the electrode.

With the aid of this relationship it is possible to determine the changes in A which are affected by various electrochemical techniques. For example, if the oxidation is carried out at constant current, A is proportional to time. However, in the course of a chronoamperometric experiment at constant potential, in which it is accepted that the electrochemical processes are controlled by a semiinfinite type of diffusion, the following holds:

$$i = \frac{F D_R^{1/2} C_R^{\circ} S}{\pi^{1/2} t^{1/2}},$$

in which D_R is the coefficient of diffusion of R and C_R° is the concentration of R in solution. It follows then that

$$A = \frac{2\epsilon D_R^{1/2} C_R^{\circ}}{\pi^{1/2}} t^{1/2}.$$

All the relationships obtained in this way are based on the assumption that O is stable. Consequently, if deviations from these equations are noted, it may be concluded that O must be disappearing by some complex process. More precisely, study of several reaction model schemes allows conclusions to be drawn not only concerning the nature of the mechanism by which O is reacting but also concerning the kinetics involved; the latter information is obtained by numerical solution of the equations corresponding to the respective models. Thus, for example, it was shown that in the electrolytic oxidation of o-tolidine (A) at pH 4.0 and at a sufficiently positive potential to convert A to A^{2+}, the latter reacts while diffusing into the solution with A diffusing toward the electrode

$$A + A^{2+} \underset{}{\overset{k_f}{\rightleftharpoons}} 2A^{\dot{+}}.$$

This process is the inverse of a dismutation. The equilibrium constant K of this equilibrium and the rate constant k_f could be evaluated [91]. A more detailed analysis which also takes into account the dimerization of species $A^{\dot{+}}$ and the existence of several protonated forms is also possible [92].

A study of the oxidation of p-aminophenol [90] led to the determination of the rate of hydrolysis of the quinone-imine intermediate to the corresponding quinone and also to the proof that the parent material formed addition products with the quinone and the quinone-imine (Chapter XIV). Such studies could also be carried out in nonaqueous media [93].

Another possible extension of the method is found in the use of thin layer cells. In this type of cell it is possible to electrolyze the starting material completely in the course of a few seconds. It is then possible to record the spectrum of the reaction product undisturbed by the absorption of the starting material.

Employing this relatively simple technique, Murray et al. [94, 95], using this type of cell, have studied the reduction of three Fe^{3+}-acetylacetonate complexes in acetonitrile. They coupled the spectral data with chronoamperometry at triangular potential sweep. As the authors emphasize, this technique, in which changes in spectra of the starting material are obtained during electrolysis of a few minutes' duration, is the only one which is suitable for studying unstable solutions, such as those obtained with Fe^{2+}-acetylacetonates.

This type of cell can also be used for infrared absorption spectroscopy studies. However, the low values of the absorption coefficients in this spectral region have forced investigators to devise a cell which is thick from the point of view of the optical path which must be traversed but thin from the point of view of diffusion phenomena. The desired result is obtained by using several electrodes which are fine grids of platinum (or gold) separated from each other by Teflon spacers. The optical path in such a cell is 300 μ, but the diffusion phenomena associated with the electrolysis are manifest at 50 μ so that the laws governing thin cells still remain valid [96].

2. Spectrophotometry by Internal Reflection

In this second method the light beam does not pass through the solution but only the transparent electrode at whose surface multiple reflections occur (Fig. 24). This type of electrode can be made, for example, of a piece of quartz within which the reflections occur; one of the quartz faces is covered with a semiconducting layer of tin oxide [97, 98] or with a metallic film [99]. Germanium electrodes have also been used [100, 101].

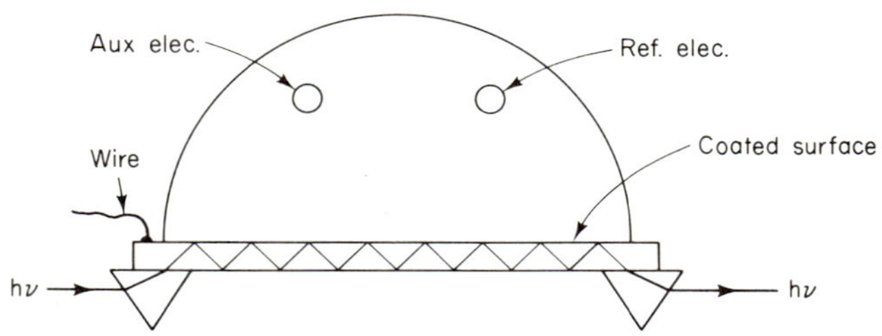

Fig. 24. Schematic for internal reflection spectroscopy at a transparent electrode (top view).

This method is interesting principally because it makes possible the study of species adsorbed at the surface of the electrode. The theory which has been proposed [97, 98] leads to the conclusion that effectively only the first molecular layer adsorbed on the electrode is involved in the change ΔA of the optical density A of the electrode. More precisely, a law similar to Beer's law can be formulated; ΔA is proportional to the concentration of the adsorbed species on the surface. This can be demonstrated experimentally, e.g., by determining the change of ΔA with time in the course of chronopotentiometry with current reversal (p. 132): o-tolidine is first oxidized and then regenerated [97]. Under these conditions, the concentration at the electrode of the oxidized species is proportional to $t^{1/2}$ before current reversal and to $[t^{1/2} - 2(t - t_f)^{1/2}]$ after reversal (t_f is the time at which reversal occurs). If the value of ΔA is substantially proportional to the concentration at the electrode, it too must obey these laws of proportionality. This has been verified for ΔA at 438 mμ which is the wavelength at which the oxidation product of o-tolidine absorbs.

Obviously, other electrochemical methods can be combined with spectroscopic examination [97]. Figure 25 shows a simultaneous recording of ΔA at 438 mμ and of the current in the course of a cyclic voltammetry experiment with an aqueous solution of o-tolidine.

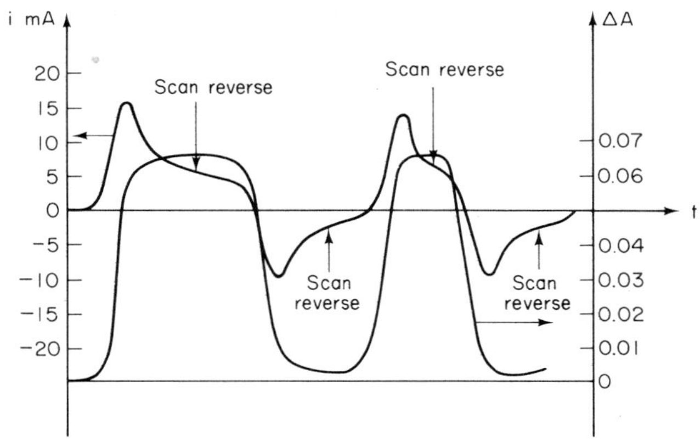

Fig. 25. Chronoamperometry with triangular voltage sweep of a 2×10^{-3} M solution of o-tolidine containing 0.1 M HCl. Absorption at 438 mμ recorded simultaneously. Working electrode covered with tin oxide, reference electrode SCE. Note that since the recorder used was of the X-t type the cyclic voltammetry curve was developed along the axis of the abscissa.

VIII. APPLICATION OF ELECTRON PARAMAGNETIC RESONANCE (EPR OR ESR)

Austen and co-workers [102] were the first to show that EPR could be used to demonstrate that an electrochemical reaction may lead to an intermediate radical species. Since their experiment, however, involved freezing a solution after electrolysis and transferring it to the resonance cavity of a spectrometer, no fine structure could be determined in the spectrum. Geske and Maki [103] first carried out an electrolysis directly in the resonance cavity; this technique permitted them to obtain an easily analyzable spectrum.

Actually, two types of arrangements [104] may be used to observe electrochemically produced radicals: in one the radical is produced within the cavity at an electrode introduced into the cavity (internal generation); in the second (external generation) the radical is produced outside the cavity and then introduced into the latter by appropriate circulation of the solution.

Internal generation allows detection of very unstable radicals, but many technical problems accompany the presence of a metal electrode in a resonance cavity. Furthermore, it is usually very difficult in using this procedure to ensure that complete removal of oxygen has been achieved. This removal is essential in order, first, to avoid the increase of the line width by the paramagnetic oxygen and, second, to eliminate the reaction with oxygen of the radical produced at the electrode.

External generation does not present these inconveniences but on the other hand is of very little value in studying very unstable radicals.

Figure 26 shows an apparatus which may be used with either method of generation of radicals; further, it permits complete deoxygenation of the solution [105]. Using the same principle of a circulating electrolyte, an assembly consisting of two cells operating at two different potentials has been reported [106].

Three types of research activities have evolved from work combining electrochemistry and EPR. In the first — and most common — type, the purpose is to demonstrate the transitory formation of a radical species in order to elucidate a reaction mechanism. An example is found in Ref. [2] which reports, in addition, the change in structure of a radical as a function of the medium. In this application it is assumed that the obtention and analysis of a well-resolved spectrum are sufficient for characterizing a radical intermediate. It must be kept in mind, however, that the analysis of an EPR spectrum does not always lead to a unique conclusion. Furthermore, if a complex process is involved, the radical identified as above is not necessarily the radical initially formed at the electrode: It is often useful to follow the change in the spectrum after the electrolysis has stopped. In this way Hoffman et al. [107] have shown that reduction of nitro-tert-butane in an organic medium yields initially the radical anion $(CH_3)_3CNO_2^{-}\cdot$; the latter is unstable and is converted in a complex process to the more stable radical $[(CH_3)_3C]_2NO\cdot$. In the course of electrolysis both radicals are generated but, if the electrolysis is interrupted, only the nitroxy radical persists.

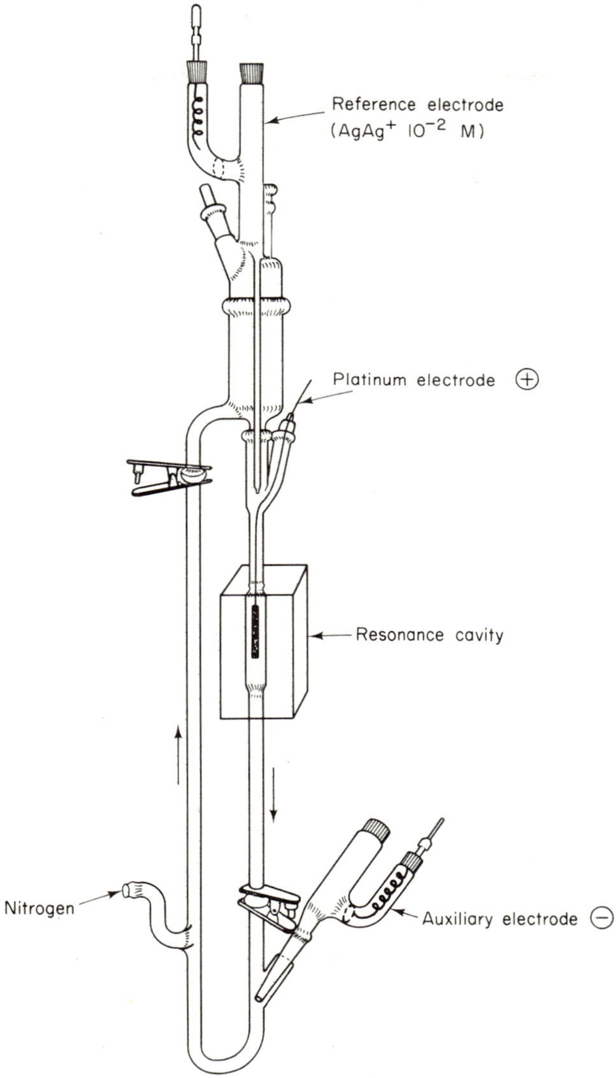

Fig. 26. Apparatus for detection in situ of electrochemically generated radicals. When the assembly is used with internal generation (as shown) the working electrode is a fine metallic grid placed in the resonance cavity. The stream of inert gas used for deoxygenation of the medium and circulating the liquid (see arrows) is interrupted before the electrolysis. When external generation is used, the working electrode is a large rolled-up metallic foil placed in the upper compartment; the gas stream is maintained during the electrolysis.

Failure to observe an EPR signal does not by any means establish that a radical species is not formed in a reaction. Many quantitative studies [104] have shown that EPR is a rather insensitive method for detecting electrochemically generated radicals. Indeed, in many cases radicals were detected by rapid scan cyclic voltammetry which escaped detection by EPR.

In a second line of research workers have taken advantage of the ease of forming radicals electrochemically in order to study their chemical properties. It is possible to elucidate the mechanism of a chemical reaction by forming electrochemically the radicals which are presumed to participate in this reaction. Thus, it was possible to show that the protonation of the radical $R_2NO^{\bullet}$ ($R = p\text{-}CH_3O\text{-}C_6H_4\text{-}$) yielded the radical cation of the corresponding amine ($R_2\dot{N}H^{+}$) and not the protonated form ($R_2\dot{N}OH^{+}$) of the initial radical. The EPR spectrum obtained was identical to that obtained during the oxidation of di-anisylamine at the first wave of its i-E curve. It had been shown that this oxidation involves the loss of one electron [108].

Finally, detection of electrochemically formed radicals by EPR is relevant to studies of quantum chemistry [109].

NOTES

[1] Translated by the Editor.

[2] It should be noted that δ, not to be confused with δ_0, is constant all along the radius of the disk. For this reason it has been called by Levich a uniformly accessible electrode. However, this is only an approximation [8].

[3] In accordance with the statement on p. 97, B is formed uniformly on the entire surface of the disk.

[4] In this case the E part of the EC system is in fact complex and comprises the exchange of $2\bar{e}$ and of $2H^+$.

[5] Or, in other words, the apparent number of electrons, n_{app}, required per molecule.

[6] Table 1 considers only the methods described in this chapter. A great many other methods, used more in electroanalytical than in electroorganic chemistry, have been described; they are also amenable to similar classification [27, 83]. From the discussion that follows the reader can readily visualize the characteristics of these other methods.

[7] Indicating that the electrochemical reaction is simple.

REFERENCES

[1]. See, for example, S. Goldschmidt and E. Stockl, Ber., 85, 630 (1952).

[2]. G. Cauquis, G. Fauvelot, and J. Rigaudy, Bull. Soc. Chim. Fr., 4928 (1968).

[3]. G. Manning, V. D. Parker, and R. N. Adams, J. Amer. Chem. Soc., 91, 4584 (1969).

[4]. S. W. Feldberg, in Electroanalytical Chemistry (A. J. Bard, ed.), Vol. 3, Dekker, New York, 1969, p. 199.

[5]. R. N. Adams, Electrochemistry at Solid Electrodes, Dekker, New York, 1969, p. 45.

[6]. Ref. [5], p. 104.

[7]. Ref. [5], p. 107.

[8]. V. G. Levich, Physicochemical Hydrodynamics, Prentice-Hall, Englewood Cliffs, N.J., 1962.

[9]. A. C. Riddiford, in Adv. Electrochem. Electrochem. Eng. (P. Delahay, ed.), Vol. 4, Wiley-Interscience, New York, 1966, p. 47.

[10]. See, for example, Ref. [5], p. 254.

[11]. Ref. [5], pp. 92 and 240.

[12]. Ref. [5], p. 214.

[13]. P. A. Malachesky, L. V. Marcoux, and R. N. Adams, J. Phys. Chem., 70, 4068 (1966).

[14]. S. W. Feldberg and C. Auerbach, Anal. Chem., 36, 505 (1964).

[15]. E. T. Seo, R. F. Nelson, J. M. Fritsch, L. S. Marcoux, D. W. Marcoux, and R. N. Adams, J. Amer. Chem. Soc., 88, 3498 (1966).

[16]. L. S. Marcoux, J. M. Fritsch, and R. N. Adams, ibid., 89, 5766 (1967).

[17]. Ref. [5], pp. 94 and 253.

[18]. A. N. Frumkin and L. N. Nekrasov, Dokl. Akad. Nauk SSSR, 126, 115 (1959).

[19]. Yu. B. Ivanov and V. G. Levich, ibid., 121, 503 (1958).

[20]. S. Bruckenstein and G. A. Feldman, J. Electroanal. Chem., 9, 395 (1965).

[21]. W. J. Albery and S. Bruckenstein, Trans. Faraday Soc., 62, 1920 (1966).

[22]. J. Jacq, B. Cavalier, and O. Bloch, Electrochim. Acta, 13, 1119 (1968).

[23]. W. J. Albery and S. Bruckenstein, Trans. Faraday Soc., 62, 1946 (1966).

[24]. P. A. Malachesky, K. B. Prater, G. Petrie, and R. N. Adams, J. Electroanal. Chem., 16, 41 (1968).

[25]. Z. Galus and R. N. Adams, J. Amer. Chem. Soc., 84, 2061 (1962).

[26]. P. Delahay, New Instrumental Methods in Electrochemistry, Wiley-Interscience, New York, 1954.

[27]. R. W. Murray and C. N. Reilley, Electroanalytical Principles, Wiley-Interscience, New York, 1963.

[28]. J. Heyrovsky and J. Kuta, Principles of Polarography, Academic, New York, 1966.

[29]. I. M. Kolthoff and J. J. Lingane, Polarography, Wiley-Interscience, New York, 1952.

[30]. R. Annino and K. J. Hagler, Anal. Chem., 35, 1555 (1963).

[31]. G. W. Ewing, Instrumental Methods of Chemical Analysis, McGraw-Hill, New York, 1969.

[32]. L. Meites, Polarographic Techniques, 2nd ed., Wiley-Interscience, New York, 1965.

[33]. J. Heyrovsky and P. Zuman, Practical Polarography, Academic, New York, 1968.

[34]. D. Ilkovič, Coll. Czech. Chem. Commun., 6, 498 (1938).

[35]. A. Ryvolova, ibid., 25, 420 (1960).

[36]. R. H. Philip, Jr., R. L. Flurry, and R. A. Day, Jr., J. Electrochem. Soc., 111, 328 (1964).

[37]. J. E. Page, Quart. Rev., 6, 262 (1952).

[38]. G. Semerano, Proc. First Intern. Congr. Polarogr., Prague, 1, 300 (1951).

[39]. P. Zuman, Proc. Intern. Symp. Microchem., Birmingham (1958), Pergamon, New York, 1960, p. 294.

[40]. M. Brezina and P. Zuman, Polarography in Medicine, Biochemistry and Pharmacy (S. Wawzonek, transl.), Wiley-Interscience, New York, 1958.

[41]. P. Zuman, Substituent Effects in Organic Polarography, Plenum, New York, 1967.

[42]. Ref. [5], p. 45.

[43]. Ref. [5], p. 216.

[44]. Ref. [5], pp. 124 and 143.

[45]. R. S. Nicholson and I. Shain, Anal. Chem., 36, 706 (1964); 37, 178 (1965).

[46]. R. S. Nicholson, ibid., 37, 667 (1965).

[47]. J. M. Savéant and E. Vianello, Compt. Rend., 256, 2597 (1963); Anal. Chim. Acta, 8, 905 (1963); 12, 629 (1967).

[48]. J. M. Savéant, Electrochim. Acta, 12, 753 (1967).

[49]. M. Mastragostino, L. Nadjo, and J. M. Savéant, ibid., 13, 721 and 751 (1968).

[50]. Many commercial instruments are now available and a multitude of circuits appear in the literature. For example, see: Operational Amplifier Symposium, Anal. Chem., 35, 1770 (1963).

[51]. R. S. Nicholson, ibid., 38, 1406 (1966).

[52]. R. Nelson, E. T. Seo, D. Leedy, and R. N. Adams, Z. Anal. Chem., 224, 184 (1967).

[53]. V. D. Parker and L. Eberson, Tetrahedron Lett., 2843 (1969).

[54]. R. Dietz and B. E. Larcombe, J. Chem. Soc. (B), 1369 (1970).

[55]. A. Ronlan and V. D. Parker, Chem. Commun., 1567 (1970).

[56]. M. E. Peover and B. S. White, J. Electroanal. Chem., 13, 93 (1967).

[57]. Ref. [5], Chap. 10.

[58]. L. Papachoudo, J. Bacon, and R. N. Adams, J. Electroanal. Chem., 24, App. 1 (1970).

[59]. V. D. Parker and L. Eberson, J. Amer. Chem. Soc., 92, 7488 (1970).

[60]. R. S. Nicholson, Anal. Chem., 37, 1351 (1965).

[61]. H. J. S. Sand, Phil. Mag., 1, 45 (1901).

[62]. L. Gierst and A. Juliard, J. Phys. Chem., 57, 701 (1953).

[63]. A. J. Bard, Anal. Chem., 33, 11 (1961).

[64]. Ref. [5], p. 183.

[65]. A. J. Bard, Anal. Chem., 35, 340 (1963).

[66]. M. Paunovic, J. Electroanal. Chem., 14, 447 (1967).

[67]. J.-P. Billon, G. Cauquis, J. Raison, and Y. Thibaud, Bull. Soc. Chim. Fr., 199 (1967).

[68]. P. Delahay and T. Berzins, J. Amer. Chem. Soc., 75, 2486 (1953).

[69]. W. H. Reinmuth, Anal. Chem., 33, 322 (1961).

[70]. H. B. Mark, Jr., and F. C. Anson, ibid., 35, 722 (1963).

[71]. Ref. [5], Chap. 7.

[72]. Ref. [66], p. 469.

[73]. R. H. Reinmuth, Anal. Chem., 32, 1514 (1960).

[74]. Ref. [5], p. 178.

[75]. D. H. Geske, J. Amer. Chem. Soc., 81, 4145 (1959).

[76]. A. J. Bard and J. S. V. Santhanam, in Electroanalytical Chemistry (A. J. Bard, ed.), Vol. 4, Dekker, New York, 1970, p. 215.

[77]. L. Meites, in Techniques of Organic Chemistry (A. Weissberger, ed.), Part 4, Vol. 1, Wiley, New York, 1960.

[78]. H. L. Kies, J. Electroanal. Chem., 4, 257 (1962).

[79]. J. Janata and H. B. Mark, Jr., in Electroanalytical Chemistry (A. J. Bard, ed.), Vol. 3, Dekker, New York, 1969, p. 1.

[80]. J. C. McClure and D. L. Maricle, Anal. Chem., 39, 236 (1967).

[81]. A. T. Hubrard and F. C. Anson, in Electroanalytical Chemistry (A. J. Bard, ed.), Vol. 4, Dekker, New York, 1970, p. 129.

[82]. R. Bonnaterre and G. Cauquis, Chem. Commun., 293 (1972).

[83]. G. Charlot, J. Badoz-Lambling, and B. Tremillon, Electrochemical Reactions, Elsevier, Amsterdam, 1962.

[84]. W. M. Clark, Oxidation-Reduction Potentials in Organic Systems, Williams & Wilkins, Baltimore, 1960.

[85]. See, for example, the series of papers devoted to the oxidation of the 2,4,6-tri-t-butylaniline: Ref. [2]; G. Cauquis, J.-L. Cros, and M. Genies, Bull. Soc. Chim. Fr., 3765 (1971).

[86]. See, for example, C. A. Chambers and J. Q. Chambers, J. Amer. Chem. Soc., 88, 2922 (1966); J. Q. Chambers, A. D. Norman, M. R. Bickell, and S. H. Cadle, ibid., 90, 6057 (1968).

[87]. G. Cauquis and M. Genies, Tetrahedron Lett., 3537 (1968).

[88]. E. M. Kosower and P. C. Huang, J. Amer. Chem. Soc., 89, 3910 (1967).

[89]. J. W. Strojek, G. A. Gruver, and T. Kuwana, Anal. Chem., 41, 481 (1969).

[90]. J. W. Strojek and T. Kuwana, J. Electroanal. Chem., 16, 471 (1968).

[91]. J. W. Strojek, T. Kuwana, and S. W. Feldberg, J. Amer. Chem. Soc., 90, 1353 (1968).

[92]. T. Kuwana and J. W. S. Strojek, Discuss. Faraday Soc., 45, 134 (1968).

[93]. T. Osa and T. Kuwana, J. Electroanal. Chem., 22, 389 (1969).

[94]. R. W. Murray, W. R. Heinmann, and G. W. O'Dom, Anal. Chem., 39, 1666 (1967).

[95]. R. W. Heinmann, J. N. Burnett, and R. W. Murray, ibid., 40, 1970 (1968).

[96]. W. R. Heinmann, J. N. Burnett, and R. W. Murray, ibid., 40, 1974 (1968).

[97]. W. N. Hansen, T. Kuwana, and R. A. Osteryoung, ibid., 38, 1810 (1966); see also, J. Amer. Chem. Soc., 88, 1062 (1966).

[98]. V. S. Strinivasan and T. Kuwana, J. Phys. Chem., 72, 1144 (1968).

[99]. B. S. Pons, J. S. Mattson, L. O. Winstrom, and H. B. Mark, Jr., Anal. Chem., 39, 685 (1967).

[100]. A. Reed and E. Yeager, J. Electrochem. Soc., 113, 1044 (1966).

[101]. H. B. Mark, Jr. and B. S. Pons, Anal. Chem., 38, 119 (1966).

[102]. D. E. Austen, P. H. Given, D. J. E. Ingram, and M. E. Peover, Nature, 182, 1784 (1958).

[103]. D. H. Geske and A. H. Maki, J. Amer. Chem. Soc., 82, 2671 (1959).

[104]. R. N. Adams, J. Electroanal. Chem., 8, 151 (1964).

[105]. J.-P. Billon, G. Cauquis, and J. Combrisson, J. Chim. Phys., 374 (1964).

[106]. A. E. J. Forno, Chem. Ind., 1728 (1968).

[107]. A. K. Hoffmann, W. G. Hodgson, D. L. Maricle, and W. H. Jura, J. Amer. Chem. Soc., 86, 631 (1964).

[108]. G. Cauquis and D. Serve, Tetrahedron Lett., 17 (1970).

[109]. See, for example, G. Cauquis, M. Genies, H. Lemaire, and A. Rassat, J. Chem. Phys., 47, 4642 (1967).

[110]. (a) B. I. Khaikin and L. G. Feoktistov, Russ. J. Phys. Chem., 38, 294 (1964) and refs. cited; (b) J. H. Wagenknecht and M. M. Baizer, J. Org. Chem., 31, 3885 (1966); idem, J. Electrochem. Soc., 114, 1095 (1967).

Chapter III

RELATION BETWEEN MICRO AND MACRO PHENOMENA

Petr Zuman

Department of Chemistry
Clarkson College of Technology
Potsdam, New York 13676

I. INTRODUCTION

Beginning with Chapter IV the major emphasis will be on phenomena associated with macroelectrolyses in which relatively large electrodes and relatively concentrated solutions of electrolyte and substrate are involved. Previous discussion, particularly in Chapter II, concerned means for elucidating the mechanisms of electrode reactions. The methods were based on procedures which use microelectrodes, very dilute solutions, and affect infinitesimal chemical changes; they all involve only so much of an ...E...C... sequence as can be revealed by a response of the working electrode.

While it has already been mentioned in passing that great caution must be exercised in proceeding with apparent confidence from a mechanistic investigation of the kind described in Chapter II to a project in electrosynthesis, more emphasis must be placed on this admonition at this point, particularly for the reader whose main interest is synthesis.

II. FACTORS AFFECTING DIFFERENCES BETWEEN MICRO AND MACRO RESULTS

The most important factors which may lead to differences between the results of macroexperiments and those expected from data obtained (and interpreted) on a microscale are [1]:

1. Time. The much longer time involved in preparing an isolable quantity of product allows consecutive chemical reactions to occur. These may involve inter alia the putative end product of a microexperiment.

2. Surface of electrode. On the larger scale surface reactions may intervene, adsorption phenomena are immensely more important — doubly important if the electrode material is different from that used in the "analytical work" — and the rates of the coupled chemical processes may change.

3. Concentration. At higher concentrations of substrate higher-order reactions such as dimerization, oligomerization, and polymerization may be dominant.

4. pH Control. In bulk electrolyses in aqueous media pH changes in the region of the electrode may be rapid and difficult to control [2].

III. POLAROGRAPHIC VS. MACRO RESULTS

The operation of the above factors is well illustrated by comparing the results of a polarographic study with the results of a macroelectrolysis. (This is particularly apropos since very often the first stage of an electroorganic project is the determination of i-E curves, particularly polarograms.) The differences observed can be manifested in shifts of half-wave potentials, in wave shapes, in wave heights, and in the number of waves observed.

The half-wave potential of an irreversible wave at a stirred pool electrode can be related [3] to the rate constant of the heterogeneous electrode process k_e and to the thickness of the diffusion layer (δ) (and hence also to the rate of stirring) by the equation:

$$(E_{1/2})_{\text{Hg pool}} = E^{\circ}_{1/2} + \frac{RT}{\alpha n F} \ln \frac{k_e \delta}{D}$$

where $E^{\circ}_{1/2}$ is the standard potential at which the rate constant k_e is defined. As both k_e and δ and, therefore, their product are considerably smaller than diffusion coefficient D, the half-wave potential of an irreversible process on a mercury pool electrode can be expected to be more negative than the polarographic half-wave potential. For example, for $k_e = 10^{-3}$ cm/sec, $\delta = 10^{-3}$ cm, $D = 10^{-5}$ cm^2/sec, $\alpha = 0.5$, and $n_a = 1$, the half-wave potential at the stirred electrode should be 0.116 V more negative than that found by means of the dropping mercury electrode.

For reversible systems, the half-wave potentials obtained for both types of electrodes should be similar, as

$$(E_{1/2})_{Hg\text{-}pool} = (E_{1/2})_{DME} - \frac{RT}{nF} \ln \frac{\delta_R}{\delta_O}$$

where δ_O is the thickness of the diffusion layer for the oxidized, δ_R for the reduced form.

It is therefore to be expected that the difference between the half-wave potentials at the mercury pool and the dropping mercury electrode will vary from 0 to 0.2 V. Experimentally observed difference of several hundred millivolts between the half-wave potentials obtained with a pool and dropping mercury electrodes indicates a difference in the course of the two electrode processes compared. Similarly, a change in the course of the electrode process can be expected when the half-wave potential at the pool electrode is more positive than that of the dropping electrode. This latter case can, nevertheless, be observed also for $k_e\delta < D$.

The slope of reversible waves at both the dropping and stirred pool mercury electrode should correspond to RT/nF, that for irreversible to $RT/\alpha n_a F$. Assuming completely satisfactory elimination of the iR drop in both systems (particularly for the pool electrode) it is thus possible to assume that both electrodes will give waves of the same shape and any difference in the shape must be caused by factors which affect the dropping mercury and the mercury pool electrode in different ways.

Whereas the absolute values of wave heights at the dropping and pool electrodes are not directly comparable, it is possible to compare ratios of wave heights for compounds which are reduced in more than one step. Any difference in the ratios for the curves obtained with the dropping mercury and mercury pool electrodes indicates a difference in factors governing the limiting current.

If the compound studied gives only one single wave, it has proved useful to compare its wave height to that of a process which is known to be reversible (e.g., cadmium ions or quinone). The ratio of the height of the wave studied and that of the reference compound is then measured by means of the dropping mercury and the mercury pool electrodes. Significant difference in the ratio indicates also a difference in the processes governing the limiting current.

Finally, systems have been found where the number of waves obtained with the two electrodes differed. Sometimes, two waves are observed with the dropping electrode but only one with the pool electrode, and vice versa. In some cases when a well-developed wave was obtained with the dropping electrode, no wave at all has been apparent on the current-voltage curves recorded with the pool electrode. p-Cyanoacetophenone is reduced in 0.1 N sulfuric acid at the dropping mercury electrode in two irreversible reduction waves: The more positive wave i_1 corresponds to a four-electron reduction of the cyano group, and the more negative i_2 which belongs to a one-electron reduction of the protonated form of the substituted acetophone formed in the first reduction step. When a mercury pool electrode was used, one wave at potentials about -0.1 V more negative than wave i_1 was observed [4]. The half-wave potential of this wave has been proved to be independent of concentration of the electroactive substance (Fig. 1): The measured half-

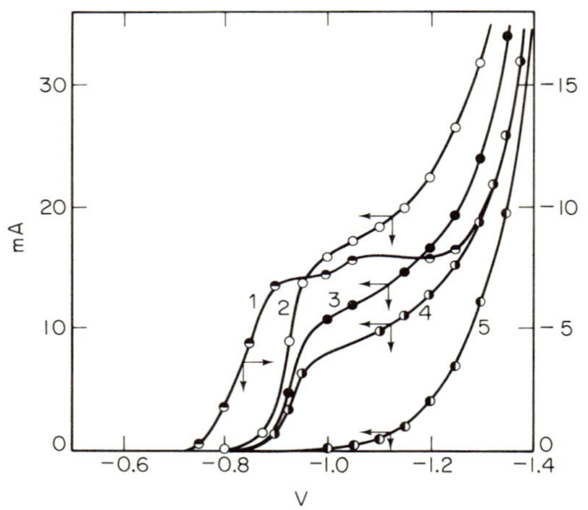

Fig. 1. Comparison of current-voltage curves of 4-cyanoacetophenone using a dropping mercury and stirred mercury pool electrodes. 0.3N-H_2SO_4, 1.10^{-3} M depolarizer. Curves: 1, polarographic wave recorded with a dropping mercury electrode placed directly in the electrolyzer; 2-5, current-voltage curves with a stirred mercury pool electrode; 1 and 2, before electrolysis; 3-5, after controlled potential electrolysis at -1.05 V (SCE) for: 3-5 min, 4-15 min; 5-50 min. The same reference electrode (SCE) was used for the dropping and pool mercury electrodes.

wave potential is thus not affected by the iR drop. This wave seemingly corresponds to the first electron process, shifted to more negative potentials, as predicted by the theory. Absence of the more negative wave i_2 indicates that the reduction of p-aminomethyl acetophenone is preceded by a chemical reaction in which the electroinactive primary product of the first four-electron reduction is transformed into the electroactive form. The yield of this reaction under conditions of electrolysis at the pool electrode is not sufficient to give a measurable wave.

Similarly, two waves at the dropping mercury electrode and one more negative and rather indistinct wave at the pool electrode were reported [5] for some pyridazones. Differences between the half-wave potentials at the two electrodes were in some cases of the order of -0.3 to -0.4 V, considerably more than predicted by the simple theory. Effects of adsorption cannot be excluded.

Examples of different products identified at the dropping mercury and mercury pool electrodes are the behavior of p-diacetyl benzene [6] which in the pH region between 2 and 5 gives a biradical at the dropping electrode and a keto alcohol at the pool electrode where the transformation of the biradical has enough time to be accomplished.

Another example is the reduction of isonicotinic acid amide at pH <1. At the dropping mercury electrode a two-electron process takes place and pyridine-4-aldehyde is formed. At the pool electrode the process proceeds further in a four-electron step to give pyridine-4-carbinol [7]. This was interpreted as resulting from hydration of the aldehyde: At the dropping electrode pyridine-4-aldehyde is formed and immediately hydrated. Since in this pH region the rate of dehydration is relatively slow at the dropping electrode, insufficient dehydrated species is produced to give a measurable faradaic current. This is shown by the small kinetic wave of pyridine-4-aldehyde in this pH region. The time available at the pool electrode is sufficient for the dehydration to take place and the dehydrated aldehydic form can be further reduced to alcohol. Because the reduction potential of pyridine-4-aldehyde is more positive than that of isonicotinamide, the aldehyde reduction wave is not separated and rather a single four-electron step appears. o-Benzoylbenzoic acid diethylamide [8] gives a two-electron reduction wave over the whole pH range when the dropping electrode is used, but controlled potential electrolysis at the mercury pool electrode corresponds to a consumption of only 1.65 electrons per molecule. Dimer was isolated as a product. It can be assumed that the radical formed in the uptake of the first electron does not dimerize during the drop life but has favorable conditions (concentration, time) to do so at the pool electrode where also the adsorption of radicals at the electrode surface can be involved.

Finally, an example on the effect of rate of protonation can be found in the reduction of isonicotinic acid [8]. When the dropping electrode is used, a two-electron wave is observed at pH < 8 corresponding to the reduction of the protonated pyridinium form of isonicotinic acid to pyridine-4-aldehyde. When the electrolysis was carried out at a mercury pool electrode, the formation of pyridine-4-aldehyde was detected only at pH <6. It is possible to deduce that at pH >6 the rate of protonation reaction is not high enough to supply a sufficient amount of the protonated form of the isonicotinic acid at the surface of the pool electrode; whereas, at the dropping electrode the rate of protonation is sufficiently fast to transform all the acid into pyridinium form at pH values two units higher.

IV. SUMMARY AND RECOMMENDATIONS

It is clear that, even if the purpose of a study is exclusively the elucidation of an electrode mechanism, the performance of a bulk electrolysis at controlled potential and the identification of the products are very desirable. The sophistication of the techniques for analyzing the nature of a C reaction in an ...E...C... sequence is improving [9], but the means are not yet available for predicting the structure of the C product. For example, preliminary data may indicate almost certainly that an intermediate radical R• has dimerized to RR, but, if R may exist in isomeric forms, the product may be different from that predicted from simplistic extrapolations [10].

If the purpose of a study is synthesis, the need for bulk electrolyses is obvious.

"Bulk" electrolyses need not be on a large scale. Successful identifications of products have been made by modern organic techniques after electrolyzing only 0.5 to 1.0 ml of solution. Polarograms determined periodically on samples of solution during the course of an electrolysis may provide very useful information on the course of the reaction sequences [1].

It may be that the product it is desired to synthesize is one that is presumably formed during polarography but is lost in "bulk" electrolyses due to subsequent reactions. Syntheses by dropping mercury electrodes, singly or in parallel, or, as has been suggested, by mercury forced through sintered glass, are tedious, time-consuming, and not amenable to up-scaling. It may be that further synthetic work with thin layer [11] and capillary gap cells [12] will reveal that these devices are capable of being used on a macroscale and that the results more closely approach those to be anticipated from mechanistic studies. Nevertheless, for the time being, comparison of i-E curves obtained with the dropping mercury electrode with those with a hanging mercury drop and with mercury pool electrodes has proved useful. Extensive studies of i-E curves with mercury pool and other large-scale electrodes are of essential importance for further progress.

REFERENCES

[1]. P. Zuman, J. Polarogr. Soc., 13, 53 (1967).

[2]. M. Kh. Kadryov et al., Elektrokhim., 7, 94 (1971).

[3]. P. Delahay, New Instrumental Methods in Electrochemistry, Wiley-Interscience, New York, 1954.

[4]. P. Zuman and O. Manoušek, Coll. Czech. Chem. Commun., 34, 1580 (1969).

[5]. P. Pflegel and O. Manoušek, private communication.

[6]. Yu. Kargin, O. Manoušek, and P. Zuman, J. Electroanal. Chem., 12, 443 (1966).

[7]. H. Lund, Paper delivered at the XIXth Congress of IUPAC, London, 1963.

[8]. H. Lund, Abh. Deutsch. Akad. Wiss., Kl. Chem., Geol., Biol., 1, 434 (1964).

[9]. For example, W. V. Childs, J. T. Maloy, C. P. Keszthelyi, and A. Bard, J. Electrochem. Soc., 118, 874 (1971).

[10]. J. Grimshaw and E. J. F. Rea, J. Chem. Soc., C1967, 2628.

[11]. C. R. Christiansen and F. C. Anson, Anal. Chem., 35, 205 (1963); L. B. Anderson and C. N. Reilley, J. Electroanal. Chem., 10, 538 (1965); J. E. McClure and D. L. Maricle, Anal. Chem., 39, 236 (1967).

[12]. F. Beck and H. Gutke, Chem.-Ing.-Tech., 41, 943 (1969).

Section B

PRACTICAL PROBLEMS

Chapter IV

PRACTICAL PROBLEMS IN ELECTROLYSIS

H. Lund and P. Iversen

Department of Organic Chemistry
University of Aarhus
Aarhus, Denmark

I. INTRODUCTION

An attractive feature of the electrolytic method is that it has the possibility of controlling the activity of the reagent, the electron, over a wide range. The success or failure of an electrolysis depends to a great degree on the proper selection of experimental conditions. An intelligent choice of these requires an understanding of the principles of controlling the reaction and of the properties of the different components, such as electrolytic cells, electrodes, solvents, electrolytes, which are part of the electrolytic equipment.

In the classical electrolytic reactions the current density, measured in amperes per square decimeter, was the quantity which was controlled, possibly because it was the easiest factor to measure and keep constant. For a long time nearly all electrolytic reactions were performed with control of the current density, although Haber, as early as 1898, in his famous papers on the stepwise reduction of nitro compounds, realized that the potential of the working electrode was the proper quantity to control.

The differences in the two ways of controlling the electrolytic reaction are illustrated in Fig. 1. In this figure curve I depicts schematically the connection between the current through the cell and the potential of the

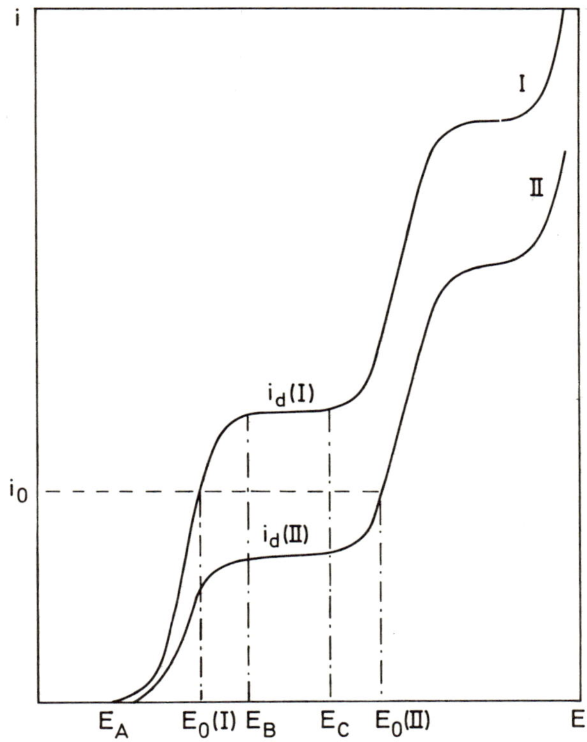

Fig. 1. Schematic representation of the connection between the current and the potential of the working electrode in a solution containing a compound with two groups reducible at different potentials. Curve I, before electrolysis; curve II, after the passage of some current; i_o is an applied current, E_o(I) and E_o(II) the potentials corresponding to i_o; i_d is the limiting current, and E_A, E_B, and E_C are applied potentials.

working electrode. The initial solution contains two reducible compounds or one compound with two groups reducible at different potentials. When the potential at the cathode is between 0 and E_A, no electron transfer across the electrical double layer can take place and thus no current runs through the cell. If the cathode potential is made more negative, the electron transfer becomes possible; that is, the reduction of the most easily reducible compound or group starts. Between E_A and E_B the current rises in dependence on the potential, but when the value E_B has been reached, all the molecules that arrive at the electrode and which can undergo the first reduction are reduced as soon as they reach the electrode. In the potential interval E_B to E_C the current is limited by the transport of the reducible compound to the cathode. This current is called the limiting current i_d, and under fixed conditions it is proportional to the concentration of the electroactive compound.

A further change of the electrode potential in the negative direction results in the occurrence of the second electrode reaction and the current rises; a similar S-shaped curve results from this reduction. At more negative potentials a third reaction or a reduction of the medium takes place.

If a suitable current i_o [$i_o < i_d$ (I)] is sent through the cell, the cathode potential assumes the value E_o(I), and when $i_o < i_d$ (I) this is well below the potential (E_C) at which the second electrode reaction starts; a selective reduction thus occurs at the beginning of the electrolysis. During the electrolysis the concentration of the reducible compound, and thus its limiting current, diminishes and after a while (curve II) the limiting current becomes smaller than the applied current [$i_o > i_d$ (II)]. The cathode potential has then, by necessity, reached the value E_o (II) and at this potential the second electrode reaction also takes place; the electrolysis is no longer selective.

If the second electrode process does not interfere with the first one (e.g. it may be due to the background), the method may still be acceptable in the laboratory (e.g., anodic cyanation [1] or an electrolytic reduction with simultaneous hydrogen evolution [2]). For industrial processes, however, a high current yield[1] is generally desirable and waste of current on the background will be less tolerable. The current yield is the theoretical amount of electricity divided by the amount actually employed for the production of a particular substance (usually expressed in percent).

On the other hand, when the electrode potential is the factor controlled and is kept at a suitable value, e.g., E_B, the second electrode process cannot take place, and the reduction remains selective to the end. The current through the cell is never higher than the limiting current corresponding to the first electrode reaction. This means that the current decreases during the reduction and becomes very small toward the end of the reaction, as the limiting current is proportional to the concentration of the electroactive material.

Another way of keeping the potential constant is to maintain a certain concentration of the reducible compound, e.g., corresponding to the limiting current i_d (I) in Fig. 1. By continuously adding substrate and removing product and by keeping the current at a constant value smaller than i_d (I), e.g., i_o, the potential will stay at E_o (I) and the reaction will remain selective.

In some reactions the required potential lies outside the decomposition potential of the medium and a certain current density is necessary to force the potential to acquire that value; a competition between transfer of electrons to the substrate and to the electrolytic medium takes place. The outcome of this competition, the current yield, depends on factors such as electrode material, current density, specific adsorption, and concentrations of the reacting species.

The information found in Chapters I, II, and III, and the considerations presented above can be translated into practical solutions of the problems of electrolysis in many ways, some of which will be discussed below. While the emphasis in this chapter is on laboratory-scale electrolyses and in Chapter XXVIII on industrial-scale work, it is clear that many of the factors which enter into consideration of cell design, choice of electrode, etc., are common to both.

II. CELLS FOR ELECTROLYSIS

A. General Considerations

Most of the industrial cells discussed in Chapter XXVIII have been developed from laboratory prototypes which are mentioned in this chapter. Polarographic cells and other cells for electroanalytical purposes which are discussed in different monographs [3-6] have been omitted; however, some electron spin resonance (esr) and polarographic cells can be found in Chapter II. The problems connected with the scale-up of organic electrode processes have been treated by MacMullin [7, 8] and the scope of electrochemical engineering by Wagner [9]. Fuel cells have also been omitted here; a review has been given by Eisenberg [10].

A number of problems must be considered when designing a cell for electrolysis, such as potential distribution at the working electrode, position of the reference electrode, ohmic resistance in the circuit, mass and heat transfer, the need for a diaphragm, and the necessity of working in a closed system.

Potential distribution. It is desirable that the potential gradient be the same all over the electrode surface so that a uniform current density distribution is obtained, provided the depolarizer concentration is equal at different parts of the electrode. (Depolarizer was the name given by Heyrovsky [11] for the substance that receives or gives away electrons at the electrode surface.) This can be achieved in two ways by employing concentric, cylindrical electrodes or plane, parallel electrodes of equal size. Generally only the electrode areas facing each other carry the current, so in the former arrangement only one side of the electrode is effective, whereas in the latter arrangement the working electrode may be placed between two auxiliary electrodes; this is of interest when the working electrode is of an expensive material whereas the auxiliary electrodes may be of cheaper material, as, e.g., in the Kolbe reaction [19].

Mercury electrodes present special problems. It is more difficult to make electrodes of the required symmetry and the inequalities in the distance to the anode introduced by stirring the mercury surface result in differences in the potential gradient along the electrode surface and uneven depolarization (see below).

Sometimes, especially in laboratory cells which carry only a small current, the required symmetry of the cell is overruled by other considerations, e.g., convenience of operation, as in the H cells (see p. 176).

Position of reference electrode. Theoretical considerations and actual measurements of the working electrode potential under working conditions show that a uniform potential distribution will be possible only with concentric electrodes of the same area [48, 49]. Thus, none of the intrinsic dissymmetric H cells with mercury electrodes, which have been so widely used, are in fact able to function with a truly constant potential at all parts of the surface. In reduction experiments [49, 50] measurements with a probe electrode have shown that the parts of the electrode near the diaphragm are at more negative potentials than the control potential governed by a reference electrode placed at the mercury surface remote from the diaphragm; as expected the differences are found to be greater with increasing current. Another factor connected with stirred pool electrodes is that the distance between the tip of the reference electrode and the mercury surface will vary due to nonuniform stirring. Thus, the measured potential difference will contain a variable component due to the ohmic voltage (iR) drop in the solution. There is thus a limitation to how exactly the potentiostat should desirably follow small changes of the control potential. In actual practice a controlled potential electrolysis (cpe) will mostly be carried out in a more or less narrow potential range around some average value [51]. As expressed by Harrar and Shain [49]: "In the general case when the potential distribution is known only approximately the best approach is to place the reference electrode as near the working electrode surface as possible on the line of minimum distance between the counter and working electrodes." A controlled potential electrolysis is, thus, an electrolysis in which the potential at the working electrode varies within certain limits. In the authors' experience a selective electrolysis requires a separation of 0.15-0.2 V in polarographic half-wave potential between two electrode reactions.

Ohmic resistance. A low ohmic resistance of the cell is desired from several points of view; this may be achieved by having a small distance between anode and cathode, a large electrode surface, a compact cell, and a low-resistant diaphragm. A distance between anode and cathode as small as 0.1-0.2 mm has been reported [75] although usually larger distances are employed. A large electrode area may involve the use of many electrodes; a further development in this direction is the "fluidized bed" electrode (Chapter XXVIII).

Mass transfer. For small cells stirring, either mechanical or magnetic, may be sufficient to obtain a reasonable mass transfer. Pumping the substrate between the electrodes is a procedure suitable for large-scale electrolysis, as it allows a favorable position of the electrodes and the use of high current densities.

Special forms of mass transfer operate at the wicking electrode [72] and with foaming electrolytes [89]. In the former, the fact that some substrates "creep" on the electrode surface is exploited to transport the substrate to the working electrode. The latter may be useful when electrolyzing gaseous substrates. If a gas evolution process occurs at the working electrode, this will give very efficient stirring. Stirring by gas bubbles may also be achieved with porous electrodes and a gaseous depolarizer in excess.

Heat transfer. For removal of the heat generated during electrolysis the cell may be surrounded by a cooling bath or provided with a water jacket and/or internal cooling spiral; this also permits electrolysis at constant temperature. In larger cells with metallic electrodes the electrodes can be made hollow to allow for a direct water cooling of the electrode itself. However, it is often more advisable to pump the electrolyte through an external heat-exchange system, thus making a flow cell.

Diaphragm. Since the electrode reaction or its products at the auxiliary electrode may interfere with the desired product or the electrode reaction of the working electrode it is very often necessary, especially in reductions, to separate the cell into anode and cathode compartments by some kind of a diaphragm which allows the current to pass, but suppresses mixing of anolyte and catholyte. Cells of this kind are called divided cells.

Because the cell resistance is inevitably increased in divided cells a diaphragm should be used only when unavoidable.

Closed system. It is simplest to use an open cell, but often it is necessary to apply a closed system — either because oxygen or water must be excluded, because high pressure is necessary, or because the reaction involves poisonous reactants.

Often oxygen can be removed to a sufficient degree by bubbling an inert gas through the solution, but when working in systems in which even traces of oxygen must be avoided, all manipulations of substances must occur on a vacuum line and the cell constructed to meet this requirement.

Every practical cell construction represents some sort of a compromise of these demands, and nearly every research group in the field of organic electrosynthesis seems to favor its own approach to the problem of designing the ideal cell. Moreover, it must be remembered that the passage of the current through the cell is always associated with two simultaneous electrode processes: an oxidation at the anode and a reduction at the cathode. From an economic point of view, it is of course desirable that both processes be engaged in the preparation of useful products. This is, however, difficult to realize with purely organic systems, but in principle an organic reduction could easily be combined with an evolution of chlorine or oxygen at the anode and an organic oxidation with production of hydrogen or sodium amalgam at the cathode. In laboratory work one is mostly satisfied by concentrating on only one process at the working electrode and arranging it in such a manner that the product(s) from the counter (auxiliary) electrode will not interfere with the reaction under investigation nor do any harm to the surroundings or to the electrode itself.

The actual design will depend on the type and number of electrodes, operational principle and scale, and on the actual process. Once the relevant parameters of the electrode reactions are known, it is often possible to use a simpler cell with only two electrodes and a minimum of additional facilities for that special reaction. For exploratory work, however, it is more desirable to have a very flexible construction with many additional facilities so that a broad range of experiments can be carried out in the same cell.

It may often be advisable (for cathodic processes especially) to have in the cell the possibility of including an indicator electrode and a pH-control system [74].

In the following some representative examples have been selected for more detailed discussion. References are given to constructions of similar types. Supplementary hints may be found in the book by Allen [12], in a recent Russian book [13] on the electrochemistry of organic compounds, or in the review by Swann [14].

It has been found convenient to organize the presentation according to three criteria: (1) working principle, i.e., batch or continuous operation; (2) absence or presence of a diaphragm; and (3) cell geometry. At the end some special purpose cells will be treated, mainly cells for use in combination with various spectroscopic techniques. The various types are described in order of increasing complexity.

B. Batch Cells

1. Undivided Cells

A simple piece of equipment for the performance of a simple electrolysis may consist of a beaker to contain a conducting solution of the substrate and two working electrodes connected to some dc source. The vessel is usually made of some rigid indifferent material such as glass, but the inner wall of the container can also be made of a conducting material to serve as one of the electrodes. A diaphragm is necessary, unless the product of oxidation at the anode does not interfere with the reduction reaction at the cathode, and vice versa. While this condition is often fulfilled in inorganic electrolytic reactions, it is less often the case in electrolyses of organic compounds.

Solid electrodes may be used in the form of wires, rods, sheets, bars, plates, or gauze, or a liquid such as mercury may be placed at the bottom of the container; electrode materials are treated in a later section. The distance between the electrodes should be kept as small as possible to minimize the ohmic resistance of the cell and thus the heat generated during the passage of the current. To get a larger working area several electrodes [17, 18] may be connected in parallel in the same manner as in a conventional storage battery, thus comprising a "tank cell." In cases (e.g., Kolbe electrolyses) where a film may be formed on the electrode surface during the electrolysis, both electrodes may preferably be made of platinum and the surface cleaned by occasionally changing the direction of the current [20] or using pulse techniques [21, 24].

In more elaborate electrolytic experiments it is often desirable to use a closed cell [25] with accessories for working under an inert atmosphere, at elevated pressure, with reflux, or to allow for the identification of gaseous products; a stirrer and a thermometer are also conveniently added. For the performance of controlled potential electrolyses (cpe) with external control of the working electrode potential by a potentiostat it is necessary to include a reference electrode which should then be placed as near the working electrode surface [26] as possible. However, reference electrodes are not so much used in undivided cells, and it must be stressed that cpe may be realized without the use of a three-electrode system and a potentiostat by having a sufficiently high concentration of the electroactive substance during electrolysis at constant current or by adding another redox system to transfer the electrons to or from the compound under investigation (see Chapters XXV and XXIV). Various kinds of reference electrodes are treated in a separate section.

In Figs. 2 and 3 are sketched two typical examples of undivided cells with additional facilities; the first one (Fig. 2) has been used for electrolyses in ethylenediamine [27], the second (Fig. 3) for electrolysis under a high pressure of carbon monoxide [28]. A similar design has been used in the study of the influence of ultrasound on organic electrode processes [29]. A simple closed undivided cell directly combined with a quartz optical cell [35] has been used for recording uv and visible absorption spectra of electrolytically generated living polymers.

Figures 4 and 5 give examples of a somewhat more complicated design, both with reference electrodes included. The first one has been used for

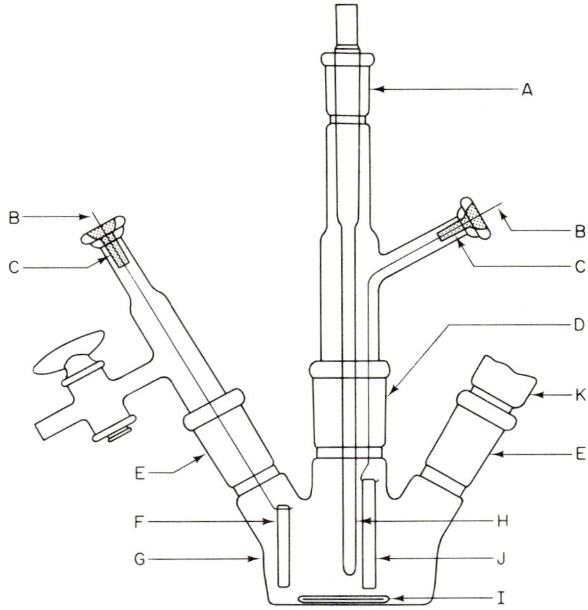

Fig. 2. Electrolysis vessel. A, 12/30 joint; B, platinum wire; C, epoxy resin seal; D, 24/40 joint; E, 19/38 joint; F, carbon anode; G, three-necked 100 ml flask; H, thermocouple well; I, glass enclosed stirring bar; J, carbon cathode; K, condenser. (Reprinted from [27] by courtesy of The Electrochemical Society, Inc.)

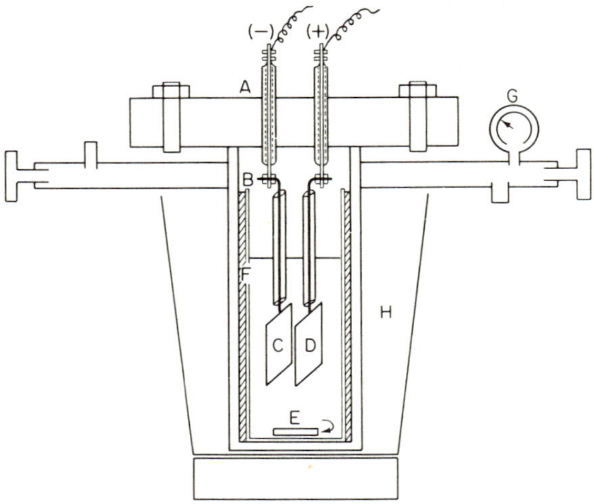

Fig. 3. Undivided cell. A, autoclave (500 cc.); B, cylindrical glass cell; C, cathode, Pt $(2 \times 3\ cm^2)$ or Cu $(2 \times 3\ cm^2)$; D, anode, Pt $(2 \times 3\ cm^2)$; E, magnetic stirrer; F, methanol solution; G, gauge; H, ice salt. (Reprinted from [28] by courtesy of The Chemical Society of Japan.)

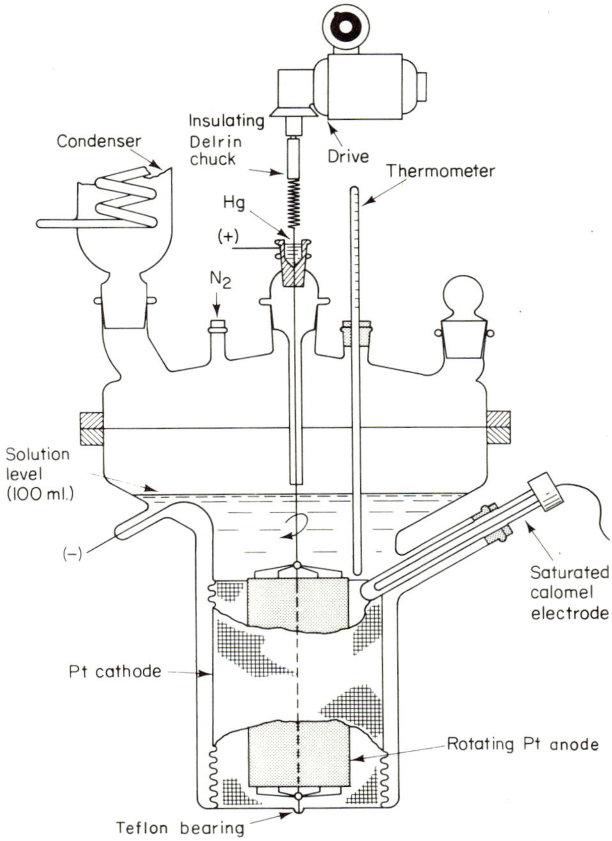

Fig. 4. Preparative scale electrolysis cell. (Reprinted from [31] by courtesy of The American Chemical Society.)

investigation of the hydrodimerization of acrylonitrile [30], though this process is more often studied in divided cells. The construction in Fig. 5 is provided with a rotating platinum anode and has been used for anodic cyanations [31].

Electrolytic fluorination of organic compounds is usually done in undivided cells with several alternating cathodes and anodes [32-34]; the corrosive nature of the HF medium of course presents special technological problems.

2. Divided Cells

a. H Cells. This is the smallest type and can be used only for small-scale

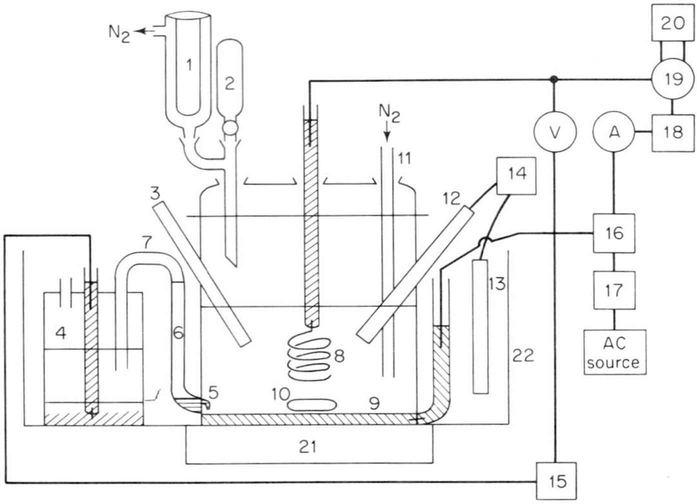

Fig. 5. Apparatus for electrolysis. 1, Dry-Ice acetone cooler; 2, dropping funnel; 3, thermometer; 4, saturated calomel electrode; 5, liquid junction; 6, sat. KCl aq. solution; 7, sat. KCl-agar bridge; 8, platinum cathode; 9, mercury anode; 10, stirrer; 11, N_2 gas inlet; 12, glass electrode; 13, temp. calibrating electrode; 14, pH meter; 15, vacuum voltmeter; 16, constant voltage supply; 17, stabilizer; 18, mercury coulometer; 19, standard resistor (1 Ω); 20, recorder; 21, magnetic stirrer; 22, thermostat. (Reprinted from [30] by courtesy of The Chemical Society of Japan.)

work. In principle an H cell consists of two containers placed beside each other and connected through a diaphragm; the actual design will depend on the types of electrodes and diaphragms used. The prototype for cells with mercury electrodes is the well-known Lingane cell [36] developed for use with a three-electrode system and later modified by Pasternak [37]. Figure 6 shows a modification which is easily constructed from common glassware by a skilled glassworker and has been used for several years in the authors' laboratory [38-39]. It may also be used with a solid working electrode; for nonaqueous work the working electrode compartment may be closed and provided with standard joints. A dropping mercury electrode (dme) is included and allows the concentration of the electroactive substance to be followed during the electrolysis. The electrode surface is renewed by magnetic stirring at the same time as the solution is stirred. An agar plug may be placed in the middle compartment of the cell [18, 37, 38] and in the reference electrode, but this can be done only in aqueous media, and it is often more convenient to use the same solution throughout the cell. A cell of similar type [40] has been reported with two middle parts and a propeller

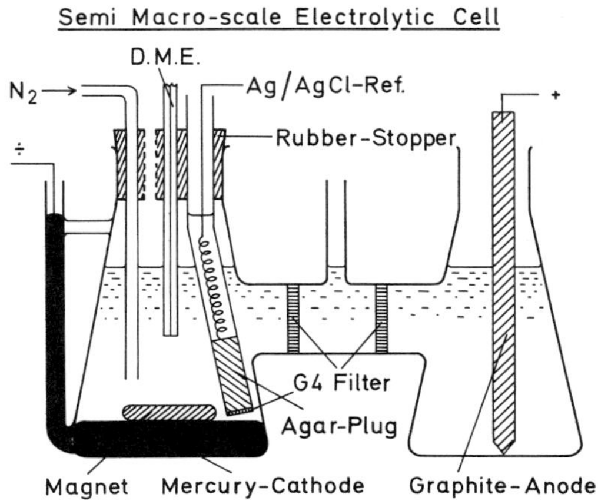

Fig. 6. Semimacroscale electrolytic cell [39].

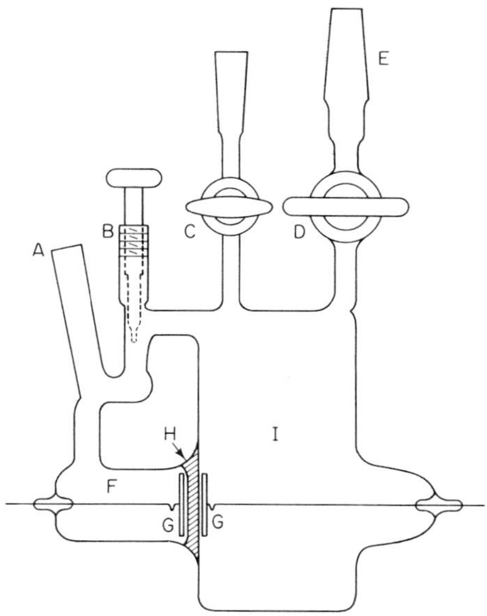

Fig. 7. Bulk electrolysis cell. A, entrance port (fitted with rubber septum) to F, anode chamber; B, Fisher-Porter valve separating F and I, cathode chamber; C, port (with straight-bore stopcock) for removing aliquots of catholyte; D, vacuum stopcock; E, 14/35 inner ground glass joint; G, 30 mm platinum disk electrodes; H, medium porosity glass frit. (Reprinted from [45] by courtesy of the Elsevier Publishing Co.)

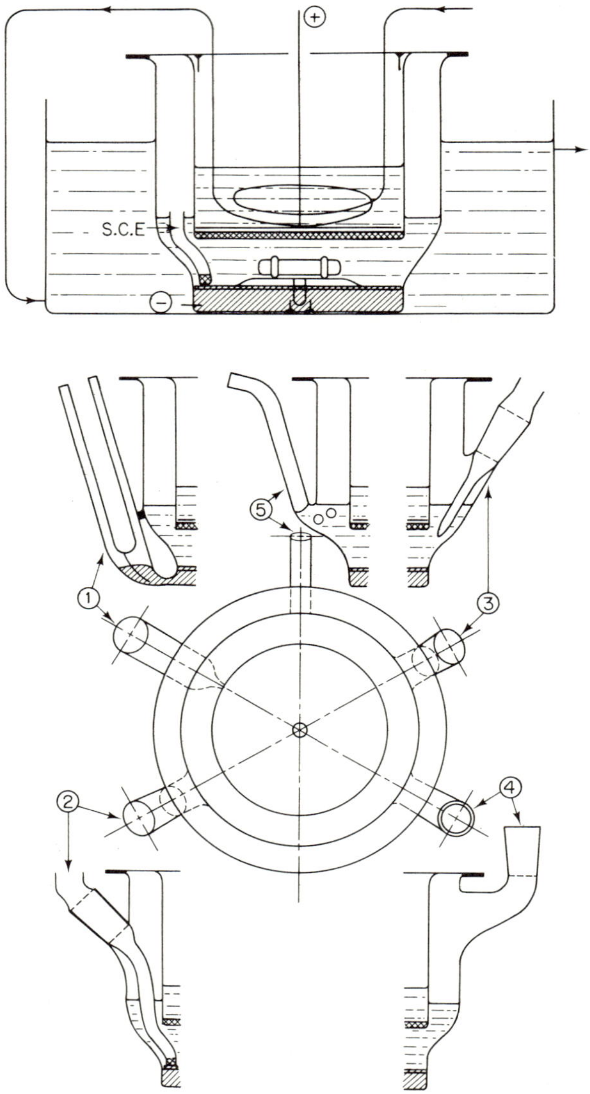

Fig. 8. 1, Cathode connection; 2, SCE reference electrode; 3, thermometer; 4, at disposal, e.g., for reflux condenser; 5, gas inlet. (Reprinted from [51] by courtesy of the Société Chimique de France.)

stirrer as in the original constructions [36, 37] where a stopcock for separation of the mercury was also included. A good description of an all-glass disconnectable H cell construction has been given by Peltier and coworkers [41]. Straightforward H cells without any special features have been designed by Rifi [42] and Metcalf and Whitnack [43]. A detachable H cell for membrane diaphragms has been designed by Allen [12].

H cells with solid plane parallel electrodes of the same area have been used rather often for work in anhydrous media. They give a uniform current distribution. An example [45] of one such cell without a reference electrode is shown in Fig. 7. It has a very small distance between the electrodes which is suitable for work in media of low electrical conductivity. A similar construction including a reference electrode has been used by Sugino et al. [46] and by Dubois et al. [47], and a less compact cell by Funt et al. [35]. A special detachable cell with thermostated hollow brass electrodes and vertical magnetic stirring has been described by Schulz and Strobel [44].

b. Cylindrical Cells. From the above discussion it should be clear that it is difficult to realize a preparative cell with a strictly uniform current and potential distribution, especially with a mercury electrode with which the condition of parallel electrode surfaces of same area is not easily fulfilled. A good approach has been given by Moinet and Peltier [51] and a schematic representation of the construction is shown in Fig. 8. Special attention has been paid to the problem of renewing the mercury electrode surface with minimal variation of the distance to the reference electrode. This has been done by stirring at moderate speed with a magnet mounted on a teflon cross (Fig. 9) turning on a short axis centered at the bottom of the vessel. Good thermal stability is secured by a cooling coil through the anode compartment connected in series with the outer bath. The cells have been proved to work satisfactorily with currents up to 3 A in a well-conducting aqueous medium.

A versatile cell [52] of the same type is shown in Fig. 10.; it could also be constructed from more simple components [53]. A cell for preparative electrolyses on a 0.1-1 mole scale with anolyte flow has been described by Iversen and Lund [54]. A simple cylindrical cell has been used by Horner

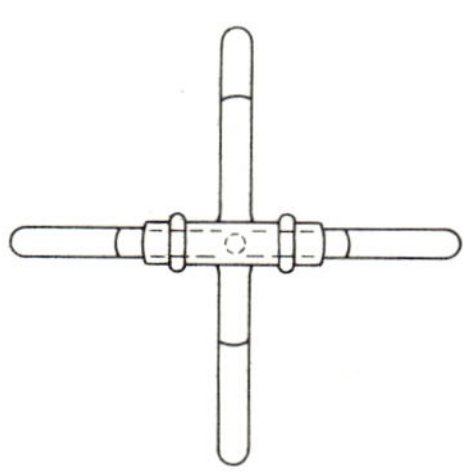

Fig. 9. Cross-type magnetic bar. (Reprinted from [51] by courtesy of the Société Chimique de France.)

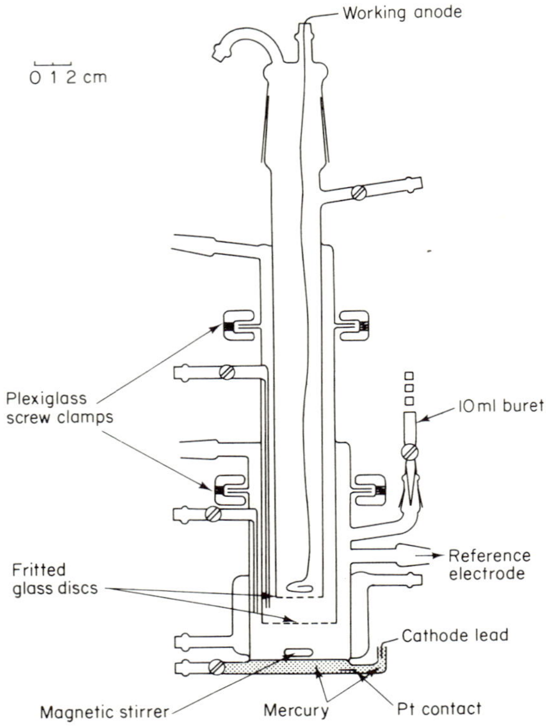

Fig. 10. Versatile three-compartment electrolysis cell. (Reprinted from [52] by courtesy of The Electrochemical Society, Inc.)

and Mentrup [55] for reduction of quaternary salts; in this case a reference electrode is not necessary and the anolyte was simply tap water continuously removed by a water pump. Slightly differing constructions are given in the experimental chapters of Refs. [12] and [13].

When both the working and counter electrodes are solid, concentric cylindrical electrodes separated by a cylindrical diaphragm can be made. This has been done in several published constructions of which two are shown here. A Russian version [56-58] (Figs. 11 and 12) has the shaft of a propeller stirrer protruding through a bearing in the inner compartment; the cell in Fig. 12 includes a reference electrode and a stopcock for easy withdrawal of the electrolyzed solution.

A conventionally designed cell using the same principles has been reported by McMillan [59]; similar constructions (also without a reference electrode) have been used by Izgaryshev and Petrova [60] and by Condit [61]. An all-glass cell designed especially for oxidations in organic media has been described by Barry and Cauquis [62]. Recently, a well-designed closed cell has appeared from the same research group [63].

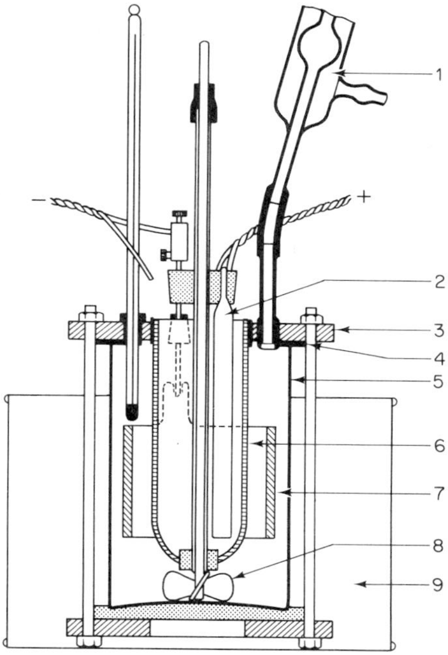

Fig. 11. Electrolyzer: (1) reflux condenser; (2) platinum anode; (3) PTFE lid; (4) rubber gasket; (5) beaker; (6) ceramic diaphragm; (7) cathode; (8) stirrer; (9) water bath. (Reprinted from [56] by courtesy of the Academy of Sciences of the USSR.)

c. Tank Cells. In this section are treated some batch cells which do not fit clearly into Sec. a or b. The simplest way to change an undivided cell into a divided construction is to surround the counter electrode(s) by a porous cup [64, 65]. This principle has been used in the two cells shown in Figs. 13 and 14. The first one (Fig. 13) [66] includes a rotating cathode and the diaphragms consist of asbestos fiber wound directly on the anodes. Cells with rotating electrodes have been intensively used by the Indian workers [66-71] allowing the use of much higher current densities than with stationary electrodes. The cell [72] in Fig. 14 introduces a special concept, the wicking electrode, which makes use of the fact that some substances may have a greater wetting ability than the components of the electrolyte solution on a certain electrode material and thus may form a film covering the available electrode area. This type of electrode was originally developed for the electrolytic reduction of elementary phosphorus to phosphine, but it has been claimed [72] to work well also with organic substances. Though virtually nothing has yet been published, this may very well be of very much interest in future organic processes, because it may overcome the problem of dissolving the organic substance in a cheap well-ionizing solvent, preferably water. As seen from Fig. 14, one end of the electrode dips into a pool of

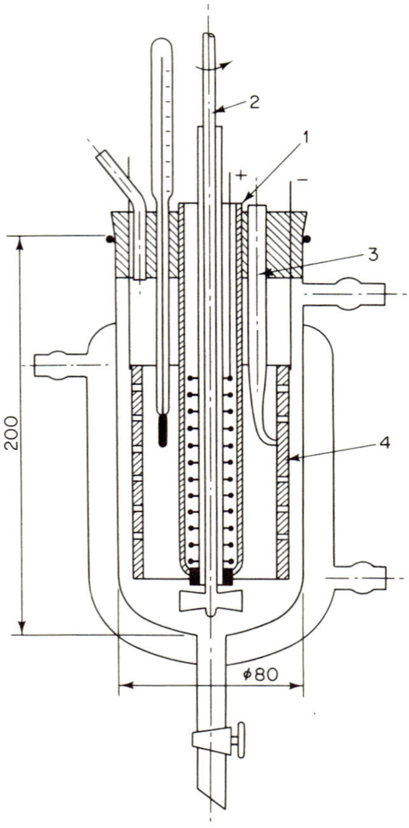

Fig. 12. Electrolyzer with diaphragm: (1) diaphragm; (2) mixer; (3) electrolytic bridge; (4) cathode. (Reprinted from [58] by courtesy of the Academy of Sciences of the USSR.)

molten phosphorus which during the electrolysis continuously will "creep" upwards, renewing the film; these conditions can be maintained at very high current densities with suitable abraded, grooved electrodes [72], as shown in Fig. 15.

The cell sketched in Fig. 16 might also be included under the heading H cells. A rectangular vessel is simply divided into two parts by a crossing ceramic plate. However, the main feature is the unusual arrangement of the mercury cathode with the mercury lying on small polyvinyl chloride (PVC) shelves above each other instead of the conventional pool at the bottom of the vessel.

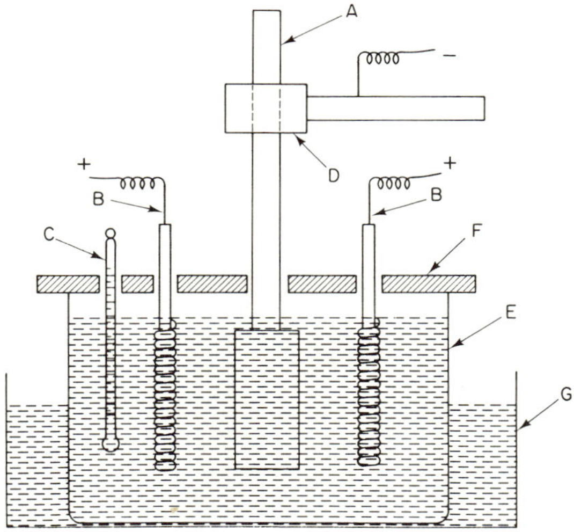

Fig. 13. Cell for the electroreduction of maleic and fumaric acids; A, rotating cathode; B, anodes; C, thermometer; D, contact-cum-guide for the anode; E, cell; F, lid; G, water bath. (Reprinted from [66] by courtesy of the Pergamon Press, Ltd.)

C. Flow Cells

In principle a batch cell can easily be changed to a cell with flow operation by simply adding an external loop with a pump to recirculate the electrolyte [80, 81]. Such a loop will also enable the cell to work with a larger reservoir of electrolyte and depolarizer. An external heat exchanger can also conveniently be included. However, this will not change the basic concept of the cell; only the batch size may be increased.

A cell for continuous flow operation must be designed with a high electrode surface/electrolyte volume ratio, provided with a feeding system, and last but not least, connected with suitable auxiliary equipment for continuous removal of the product(s) of electrolysis and reestablishment of the electrolyte composition. The continuous work-up procedure during electrolysis is somewhat inconvenient in the laboratory and consequently small continuous flow cells have mostly been operated with recycling to a reservoir before scaling up. Large cells and their industrial applications are discussed in Chapter XXVIII.

1. Undivided Flow Cells

In Fig. 17 is shown an example [74] in which a simple batch cell has been provided with a recycle loop which in this case was found necessary in order to make the glass electrode of the pH-stat unit work properly.

Recently there has been a thorough study of cells with a very small electrode distance (of the order 0.1 mm) by the BASF group [75] resulting

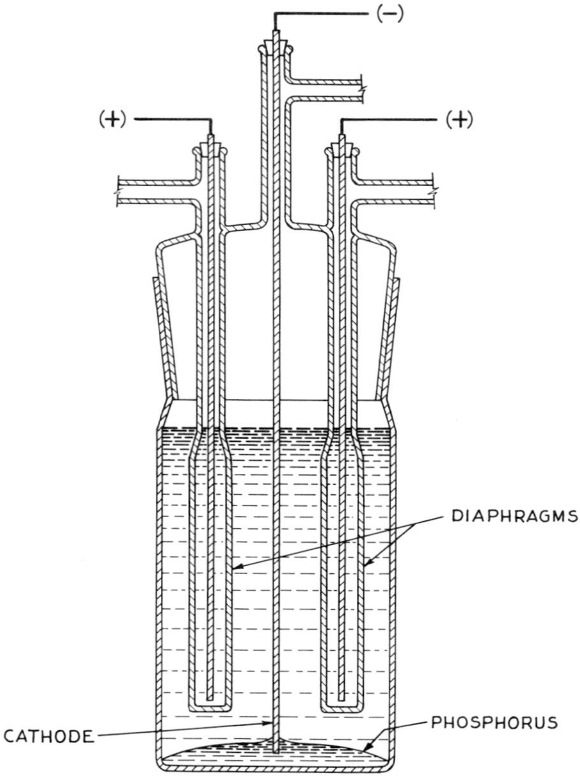

Fig. 14. Wicking electrode cell [72].

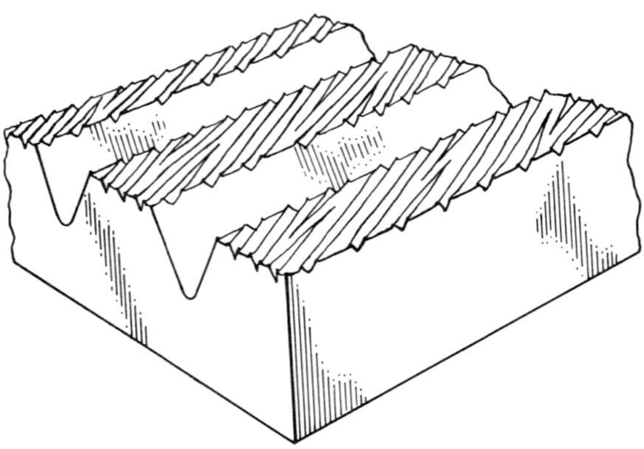

Fig. 15. Detail of machined surface for a wicking electrode [72]. (Reprinted by courtesy of the Hooker Chemical Corporation.)

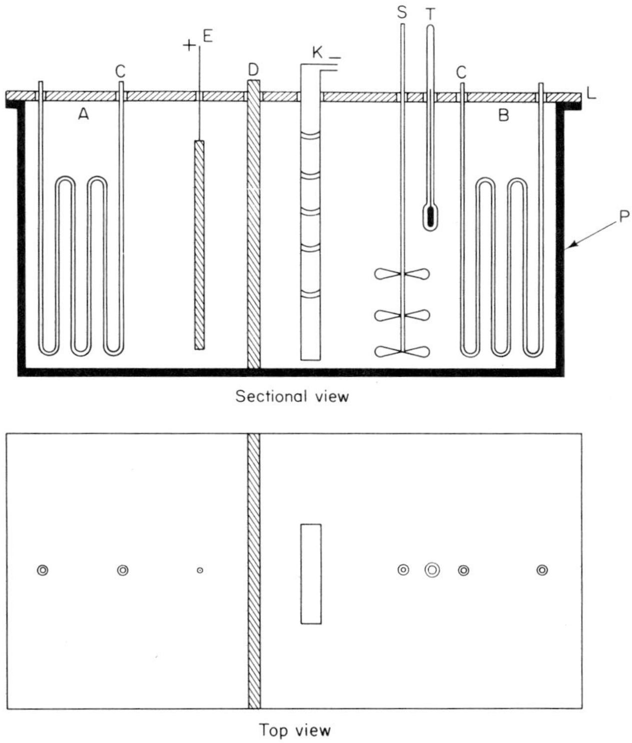

Fig. 16. Electrolyzer for the preparation of aminoguanidine: A, anode compartment; B, cathode compartment; C, cooling coils; D, ceramic diaphragm; E, platinum anode; K, mercury cathode assembly; L, Perspex lid; P, PVC-Perspex container; S, stirrer; T, thermometer. (Reprinted from [73] by courtesy of The Electrochemical Society, Inc.)

in some very elegant constructions which may prove useful in future industrial applications. The "capillary gap" cell [75] with graphite disks is discussed in detail in Chapter XXVIII, but the concept of a capillary electrode gap can also be applied to a cell with a liquid electrode as shown in Fig. 18. The anode which dips into the mercury cathode must be made of some material which is not wetted by mercury. With small anodes the cell is reported to work without short-circuiting even in the absence of artificial electrolyte flow, thus comprising a type of wicking electrode cell. Both constructions have been checked in the acrylonitrile hydrodimerization process, but the cell voltage was lower with the graphite pile (3.9 V, resp. 5.5 V, at 10 A/dm^2) which is also inherently much easier to scale up.

Liquid-permeable gauze electrodes separated by a thin plastic fabric have also been investigated [75]; a single electrode unit is shown in Fig. 19, but several electrode pairs may be placed on a common axis. The electrolyte

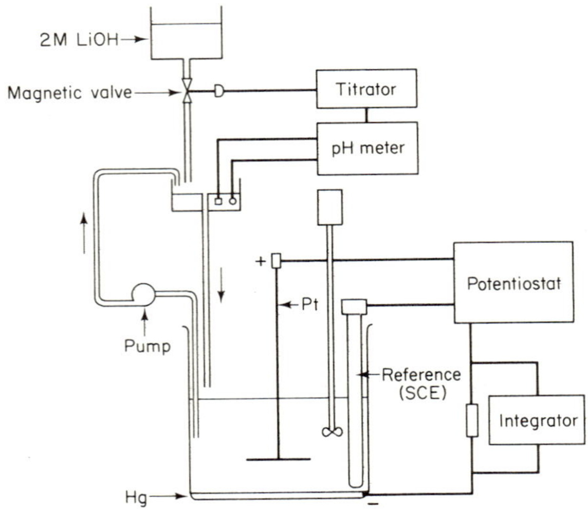

Fig. 17. Electrolysis apparatus. (Reprinted from [74] by courtesy of the Acta Chemica Scandinavica.)

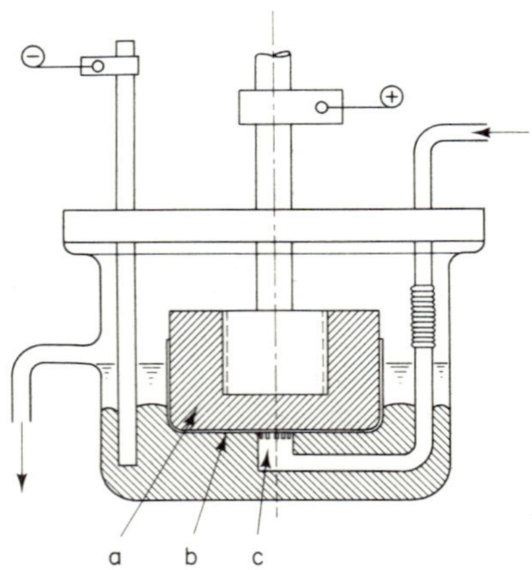

Fig. 18. Cell with an anode dipping into the mercury cathode: (a) graphite cylinder; (b) lead dioxide layer; (c) nozzle for electrolyte inlet. (Reprinted from [75] by courtesy of the Verlag Chemie.)

may either be forced through the electrodes by pumping or the electrode can be incorporated in a vibrator cell; the latter possibility has proved useful especially for Kolbe syntheses employing dicarboxylic acid half-esters.

Filter press (plate and frame) cells are not discussed here as they are treated in detail in Chapter XXVIII; it may be mentioned that the plate

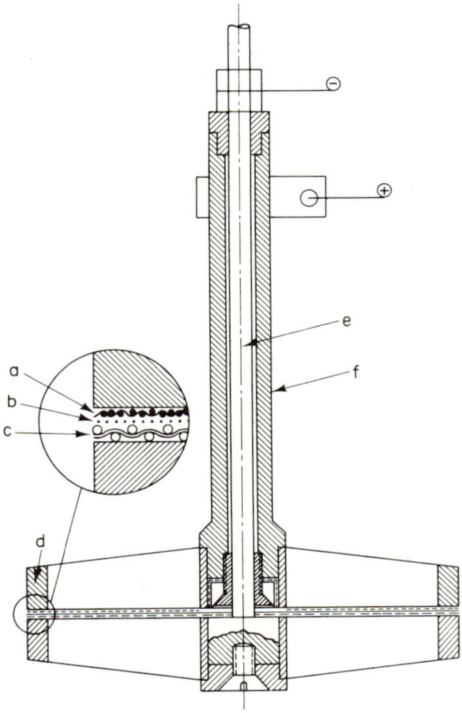

Fig. 19. Liquid-permeable electrode pair for vibrator cell: (a) anode; (b) cathode; (c) polyester fabric; (d) ring-shaped support with spokes; (e) vibrator shaft; (f) metal tubing. (Reprinted from [75] by courtesy of the Verlag Chemie.)

electrodes of such cells can also conveniently be bipolar connected in series using ion-exchange diaphragms [76].

The fluidized bed electrode [77] shown in Fig. 20 is presently being developed actively mainly for the advantages it may have in larger-than-laboratory work. It is discussed in Chapter XXVIII and in Refs. [78], [79], and [79a].

2. Divided Flow Cells

Changing a divided batch cell to flow operation does not need special consideration. Though the cell unit itself (Fig. 21) [82] is in fact a conventional cylindrical batch cell it has been provided with simple facilities for continuous operation of the acrylonitrile hydrodimerization process in the laboratory. The principle is that the substance to be reduced — the

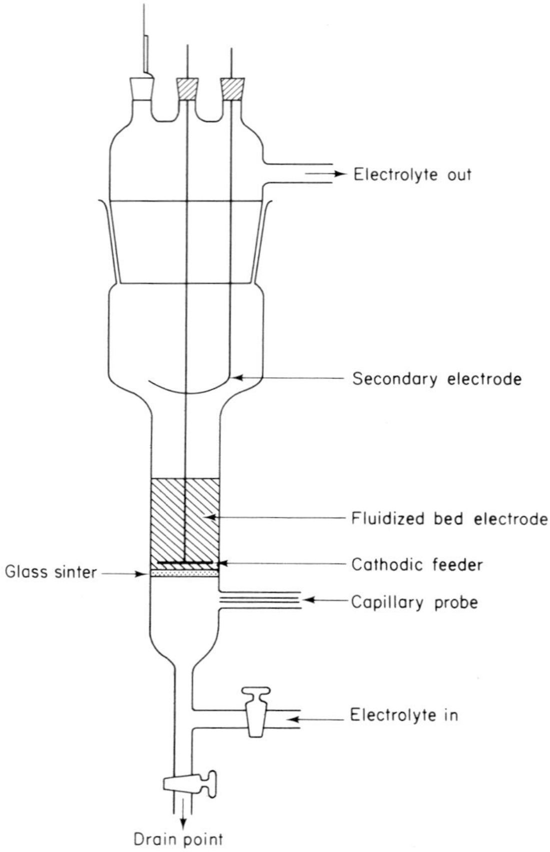

Fig. 20. Details of construction of a cell employing a plane parallel electrode configuration. (Reprinted from [77] by courtesy of The Electrochemical Society, Inc.)

acrylonitrile (AN) – is also used as the extracting solvent.

The cell [83] in Fig. 22 includes the interesting feature of a porous carbon anode; though it works as a divided cell, it has a common outlet for both anode and cathode products.

The construction sketched in Fig. 23 has two liquid-permeable (gauze) electrodes with the space between them fully occupied by a solid polyelectrolyte (cation-exchange membrane in the H^+ form). As practically all the current through the cell is carried by the diaphragm this design permits the electrolysis of very poorly conducting liquids, and in principle the presence

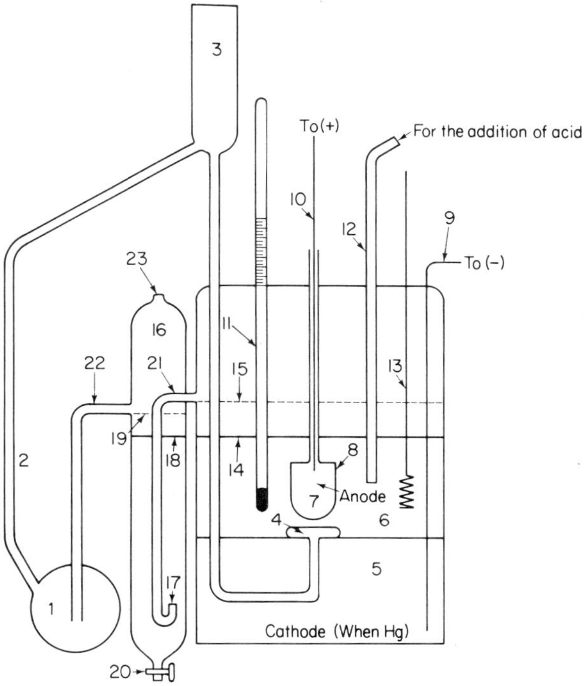

Fig. 21. Schematic drawing of continuous laboratory cell: 1, boiler; 2, riser; 3, condenser; 4, disperser; 5, cathode when Hg; 6, catholyte solution; 7, anode chamber; 8, diaphragm; 9, lead to cathode; 10, lead to anode; 11, thermometer; 12, inlet tube; 13, stirrer; 14, catholyte level; 15, level of supernatant AN; 16, washer; 17, disperser; 18, water level; 19, supernatant level; 20, stopcock; 21, 22, overflow tubes; 23, water inlet. (Reprinted from [82] by courtesy of The Electrochemical Society, Inc.)

of a supporting electrolyte is unnecessary both in the anolyte and catholyte.

The cell [84] in Fig. 24 is of a rather conventional design with the features of a single plate and frame cell unit in the bottom part, but at the same time it includes several additional facilities in one chamber. An all-glass flow cell with ion-exchange membrane has been reported by Cottrell and Farrar [85].

A Russian construction [86] for the electrochemical preparation of perchloric acid anhydride including a cooled anode and a nitrogen bubble lift for electrolyte circulation may also be useful for the preparation of thermally unstable organic intermediates.

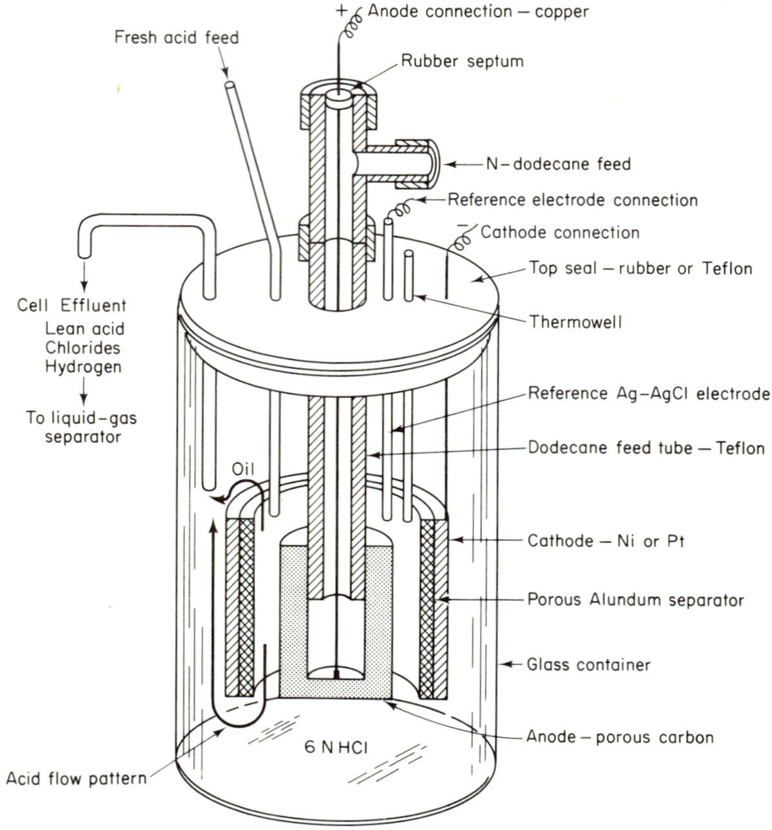

Fig. 22. Chlorination cell with a porous anode. (Reprinted from [83] by courtesy of The Electrochemical Society, Inc.)

A discussion of divided flow cells with fluidized electrodes [77] is omitted here; they are discussed in Chapter XXVIII, and so is a cell with mercury cathode and a flowing film electrolyte described by Condit [61]. A copper tube cell including a cooling jacket has been reported by Schmidt and Neew [87].

A very simple flow cell [88] employing a fixed bed of silver powder has also been reported for the electrolytic preparation of gem-dinitroparaffins. Here the bed electrode was employed mainly to avoid the deterioration of a solid gauze anode during the electrolysis and to facilitate a continuous recovery of the silver compound from chemical reaction in the anode compartment. At the same time the cell could of course be operated easily with higher currents.

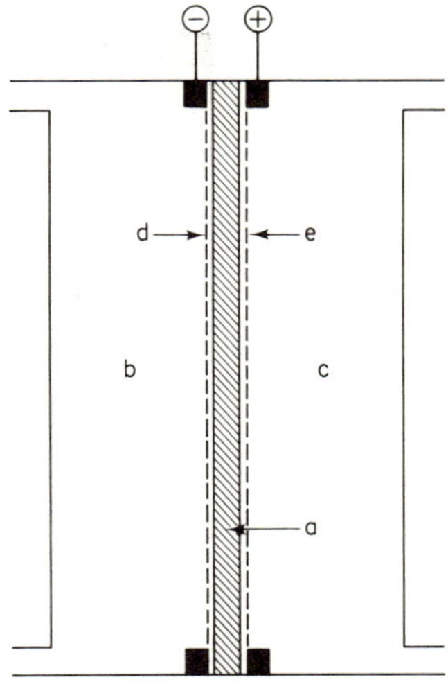

Fig. 23. Cell with gauze electrodes directly on either side of a solid polyelectrolyte: (a) ion-exchange membrane; (b) cathode compartment; (c) anode compartment; (d) cathode; (e) anode. (Reprinted from [75] by courtesy of the Verlag Chemie.)

The possibility exists of using a foaming electrolyte cell for electrolysis of gaseous organic substances. This approach [89] has proved useful in the construction of fuel cells, but results from preparative organic electrolyses have hitherto not been reported.

A continuous flow-cell system for indirect cpe has been proposed by Allen [90]. The design of the cell unit is rather straightforward; the idea is that the potential difference between two probe electrodes, one positioned at the inlet end of the working electrode and the other at the outlet, is used to govern the rate of electrolyte flow through the cell once the working electrode has been brought to the proper potential (measured against the inlet reference electrode) by adjustment of the current.

D. Special Purpose Cells

In this section some designs are described which are not aimed directly at preparative electrolysis of organic substances but merely serve as examples of various techniques which have proved valuable in the elucidation of organic electrode reactions. Thin layer electrolysis may be useful to obtain n values also for organic compounds, but such cells [91-93] will not be discussed here as they are analytical tools.[2] The same applies to a cell [94] with a platinum grid electrode for electrolysis of flowing solutions, though the working characteristics of such cells may also be useful for future preparative applications [95].

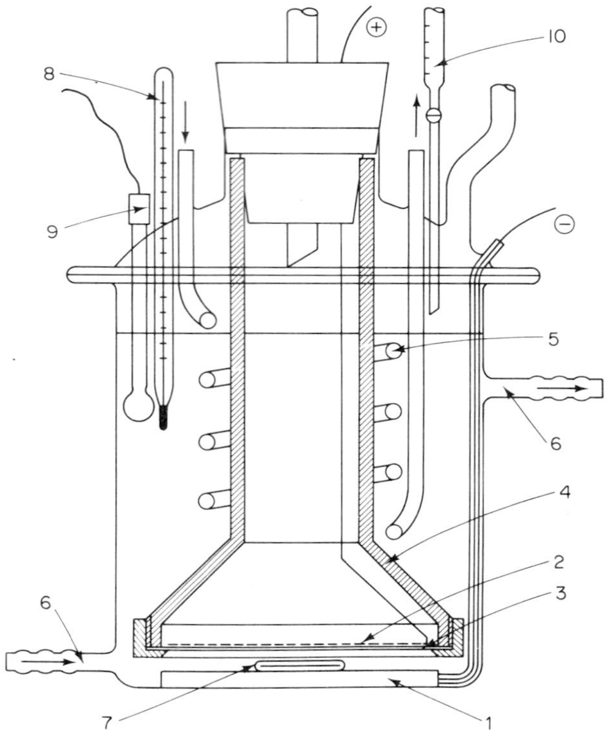

Fig. 24. Electrolytic cell with ion-exchange membrane: (1) cathode (lead, copper, V2A-steel); (2) platinum anode; (3) ion-exchange membrane; (4) polyethylene tube; (5) cooling spiral; (6) electrolyte inlet and outlet; (7) magnetic bar; (8) thermometer; (9) glass electrode; (10) dropping funnel. (Reprinted from [84] by courtesy of the Hüthig & Wepf Verlag.)

Instead, some cells designed for use in combination with various kinds of spectrometers are discussed. One cell [35] having quartz windows for the measurement of uv spectra has already been mentioned earlier in this chapter, but it did not allow in situ investigations of species near the working electrode surface. This requires the presence of optically transparent electrodes (ote). and examples of cells with such electrodes are shown in Figs. 25 and 26. In the first cell (Fig. 25) [96] the ote is made from tin-oxide-coated borosilicate glass or quartz and has the advantage of being able to function both as a cathode and an anode. This also applies to the second cell [97] (Fig. 26) which has a gold minigrid ote consisting of four grids alternated between Teflon tape spacers and pressed between two NaCl plates, thus comprising an optically "thick" but diffusionally "thin" electrode for studies in the infrared region. Gold minigrids [98] and metal films [99] on

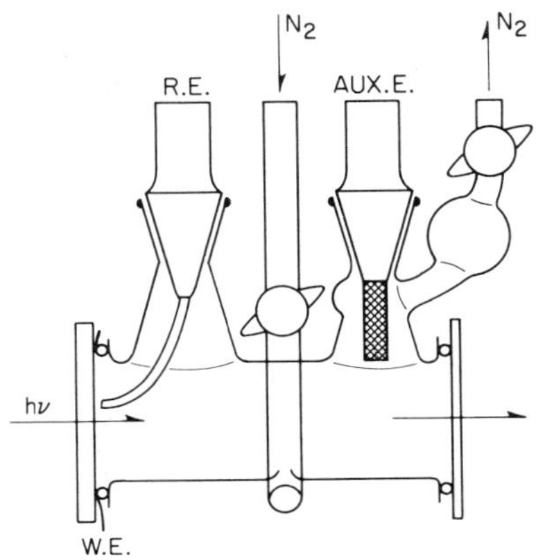

Fig. 25. Ultraviolet and visible region absorption cell: W.E., transparent conducting glass electrode; R.E., reference electrode (SCE); AUX.E., auxiliary electrode. (Reprinted from [96] by courtesy of the Elsevier Publishing Co.)

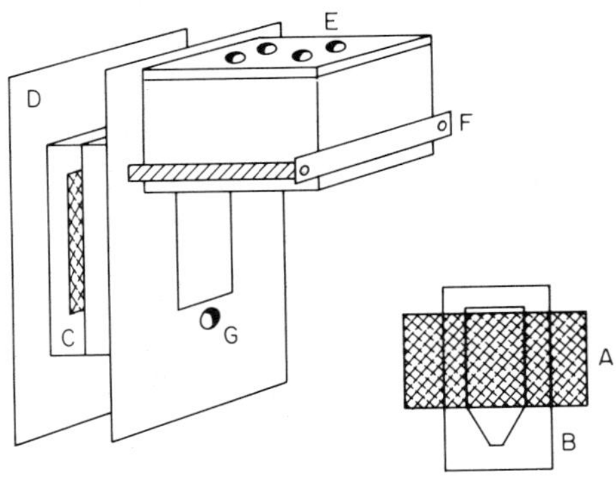

Fig. 26. Infrared transparent thin-layer electrode: A, gold minigrid electrode; B, Teflon spacer; C, NaCl plates; D, steel frame; E, holes for reference and auxiliary electrodes, nitrogen inlet and outlet; F, Teflon cup and support bar; G, outlet to draw solution from cell. (Reprinted from [97] by courtesy of The American Chemical Society.)

glass or quartz have been used in cells for uv and visible spectra. Another method which can be applied in the infrared region potentially most useful for organic compounds is internal reflection spectrometry (irs). An example of such a cell [100] with a gallium-doped germanium working electrode is sketched in Fig. 27; the reference electrode is in this case a platinum wire located in a salt bridge functioning as a quasireference electrode [101].

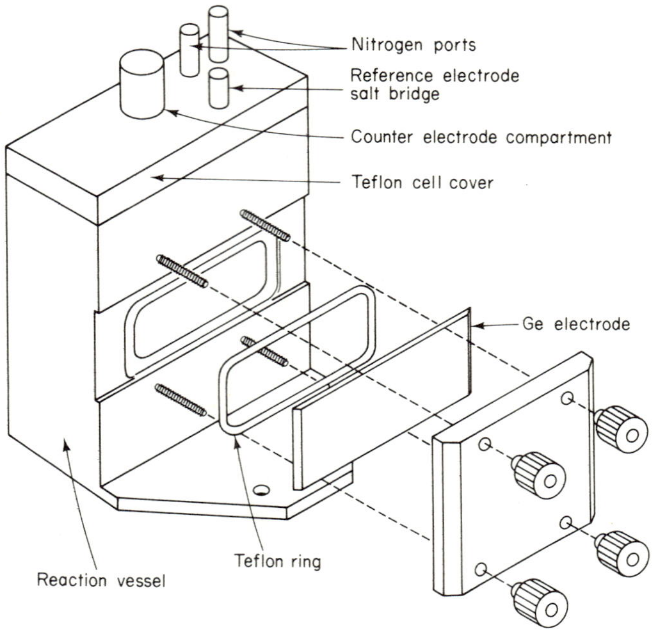

Fig. 27. Electrolysis cell for internal reflection spectrometry. (Reprinted from [100] by courtesy of The American Chemical Society.)

Recently an optical ring disk electrode (orde) system [102] has been reported in which an optically transparent ring surrounds the disk electrode. This arrangement allows recording of unstable species generated at the disk electrode.

Since free radical intermediates are often encountered in organic electrode processes, there has been much interest in studying electrolytically generated radicals by esr technique; a description of the equipment has been included in Chapter II.

For small-scale laboratory use amalgams are usually produced by dissolving the appropriate metal in mercury, but some cells for the reduction of organic compounds with electrolytically generated amalgams have been described [103, 104] and further constructions may be found in the Russian monograph [13]. Recently, a laboratory cell for the continuous generation of sodium amalgam has been reported [105]. As elementary chlorine and chlorinated brine are somewhat inconvenient to handle in the laboratory sodium hydroxide was chosen as the electrolyte for electrolysis with a flowing mercury cathode and a nickel anode to give oxygen as the anodic product.

III. DIAPHRAGM MATERIALS

A. General Considerations

As mentioned earlier, it is often desirable or even necessary to separate the anode chamber from the cathode compartment by some kind of diaphragm. The ideal material should be an electric insulator chemically inert to and preventing the interdiffusion of the anolyte and catholyte components while at the same time presenting only a little resistance to the electric current. For practical reasons the material should also possess reasonable mechanical strength and good dimensional stability. No real material can of course fulfill all these demands simultaneously, and it is generally necessary by trial and error [8] to find a suitable compromise with which the deviations from ideality can be tolerated for the given process. Fortunately there is a rather wide selection of available refractories, membranes, and woven materials, especially for laboratory cells where the engineering demands are less strict than for production cells.

The diaphragm materials can be divided into two categories according to the mechanism by which the electric current is carried through the medium — porous materials and ion-exchange membranes [or nonpermselective and permselective materials (Chapter XXVIII)]. With porous materials a more or less tight barrier is offered to the diffusion of both ions and uncharged species, and the electrical resistivity generally increases drastically with decreasing pore size. Thus there is no change in the transport number of the ions as in the case of ion-exchange membranes which are able to discriminate between ions of opposite charge to a very high degree, according to the type of membrane material. Sometimes a controlled seepage through the diaphragm may even be desirable; this problem has been discussed by MacMullin [8].

B. Porous Materials

From the above discussion of cell constructions it is obvious that a great variety of porous materials have been used in laboratory cells. Most common are sintered glass (especially suitable for all-glass cells) and ceramic refractories such as Alundum, unglazed porcelain, and pottery. These materials are available in many different sizes, shapes, and porosities, which is convenient for laboratory purposes. A serious drawback is that for greater pieces of material the wall thickness must be considerably increased to obtain a satisfactory mechanical strength; this, however, results in a higher voltage loss in the diaphragm with consequent unproductive heat generation and limits the cell size in practice. The ceramic materials are also less suitable for diaphragms for prolonged operation in alkaline media, as the ceramic matrix that holds the grains together is attacked.

Materials of more or less broad utility, reported in the literature, include the following: filter paper [106], asbestos [66, 67, 69, 107, 110], cellophane [108], nylon cloth [70], and porous plastic [77, 109]; other potential candidates are listed in Chapter XXVIII. The porosity of the medium may be decreased by partially filling the voids with another material, e.g., agar-agar or silica gel [51] or magnesium hydroxide.

C. Ion-Exchange Membranes

In most cases the cation-exchange type (H^+ form) of membranes has been used in electrolytic cells, but membranes enabling the anions to pass are also available. It must be remembered that if the membrane is not of completely homogeneous polymeric structure (it may be finely dispersed in some binding matrix), channels will usually exist through the matrix, making it less permselective. In cases where the product of electrolysis or the starting material is an ion with a charge opposite to that of the working electrode (e.g., reduction of trichloroacetic acid in ammonia buffer [111]), it will be especially advantageous to use an ion-exchange membrane as diaphragm to avoid loss of depolarizer or product due to electrical migration out of the working electrode compartment.

The polyelectrolyte membranes are often the material of choice for production cells, and hence a more detailed discussion of their properties is naturally postponed to Chapter XXVIII. Several examples of cells employing ion-exchange membranes have been given in the section on cell constructions.

IV. ELECTRODE MATERIALS

A. General Considerations

The electrode material is an important factor in directing the course of an electrode reaction, but the influence of the electrode material is one of the least understood points in organic electrolysis. It is known that hydrogen or oxygen overvoltage, catalytic effect, adsorption of substrate to the electrode, the nature of impurities and added components, the physical state of — and even the history of — the electrode may influence a branching of an electrode process. Efforts have been made to clarify the situation and although some knowledge has been gained, research in this area is very much needed. Besides the factors mentioned above, the resistance of the electrode toward corrosion must be considered when the electrode material is chosen.

The hydrogen overvoltage of an electrode material is an important factor in reduction in protic solvents, as it determines how negative the potential may be before the reduction of the medium starts to compete seriously with the electrode reaction of the substrate. Metals can be divided into low-, medium-, and high-overvoltage materials. In the first group is platinized platinum; in the second one smooth platinum, nickel, and copper; and in the third group tin, lead, cadmium, and mercury. The values of hydrogen overvoltage of different materials depend on the current density at which they are measured.

Hydrogen overvoltage is caused by a slow step during the electrolytic evolution of hydrogen; the reaction involves the following main steps:

$$H^+_{solv} + e^- \rightleftarrows H_{ads} \quad (1)$$

$$H_{ads} + H^+_{solv} + e^- \rightleftarrows H_{2\,ads} \quad (2)$$

$$H_{2\,ads} \rightleftarrows H_{2\,gas}. \quad (3)$$

The first reaction [112] (1) is believed to be the slow step on mercury and other high-overvoltage metals. This means that no appreciable concentration of adsorbed hydrogen atoms is present on mercury and implies that at mercury cathodes the reduction occurs by a transfer of electrons directly to the substrate and not via adsorbed hydrogen atoms. This implication has been tested [113] and confirmed in the reduction of acrylonitrile in the presence of hydrogen ions.

At electrodes of low hydrogen overvoltage reaction (2) is considered [114-116] to be the slow step, which implies that the electrode surface is covered to a certain extent with adsorbed atomic hydrogen. This is consistent with the finding that reduction at such electrodes in many cases occurs by the electrocatalytic reaction [117] which in many respects is similar to the well-known catalytic hydrogenation reaction.

Generally, electrodes with high hydrogen overvoltage give the most highly reduced product, but there is no firmly established connection, as the catalytic effect of the electrode material may be the factor which determines the product. Phenol, e.g., is not reducible at a lead or mercury cathode, whereas it is reduced to cyclohexanol at a platinized platinum electrode in 2 N sulfuric acid [118, 119].

Oxygen overvoltage plays a similar role in anodic reactions as hydrogen overvoltage in reductions, but the choice of anode materials with different oxygen overvoltages is very restricted as the overriding consideration is the corrosion resistance of the material. Smooth platinum, gold, and lead dioxide are among the resistant anode materials with high oxygen overvoltage.

The catalytic effect of the electrode material on the branching of an electrode reaction is most often apparent at cathodes with low overvoltage, but may also be observed at materials with high hydrogen overvoltage. In the reduction of acetone in 6 N H_2SO_4 at a platinized platinum electrode two independent paths are accessible: One leads to propane and the other to isopropyl alcohol. The rate of formation of these two products depends on the voltage and on the history of the electrode.

Two types of adsorption for hydrogen are known, one of them being "interstitial"; the interstitial hydrogen is important in the electrocatalytic reduction of acetone to isopropyl alcohol [120, 121], but not in the reaction leading to propane. At an anodically treated electrode no interstitial hydrogen is present, but the concentration of it is gradually built up. Besides being of importance in the reduction of substrate, the interstitial hydrogen modifies the adsorption properties of the electrode, and the importance of the adsorption of the substrate at the electrode prior to the reduction is evidenced by the kinetics of the reaction.

The nature of the reaction changes with time. At a freshly anodized electrode acetone is preferentially reduced to propane, whereas later isopropyl alcohol is the main product. Eventually both electrocatalytic reactions are suppressed and hydrogen evolution becomes the main reaction.

The catalytic effect of high-overvoltage electrode materials may be exemplified by the reduction of ketones in hydrochloric acid at mercury and cadmium or amalgamated zinc electrodes; at mercury, alcohols, pinacols, organic mercury compounds, or sometimes hydrocarbons may be formed (Chapter VIII), whereas at cadmium, the reaction is to some degree analogous to the Clemmensen reduction and yields methylene compounds [122].

Some recent examples of the effect of electrode material on the branching of reactions are the reduction of benzyltriethylammonium nitrate in dimethylformamide (DMF) to bibenzyl or toluene at aluminum or platinum electrodes [123] and the reduction of aliphatic aldehydes [124] or alkyl arylketones [125] at different materials.

Adsorption of the substrate at the electrode plays an only partly understood and difficultly predictable role in organic electrochemistry. An example of the influence of adsorption is found in the Kolbe reaction occurring in an aqueous acetate solution at smooth platinum and gold electrodes. Although both materials have high oxygen overvoltage, the Kolbe reaction takes place only at platinum and at a potential higher than that of the oxygen evolution. Addition of indifferent electrolytes which compete with the acetate ions in the adsorption at the electrode lowers the yield of the Kolbe product. In methanol, where adsorption is less important than in water, the Kolbe reaction proceeds both at platinum and gold anodes.

The surface may influence the catalytic activity of the electrode and the adsorption of the substrate. Sometimes, e.g., in the reduction of certain nitro compounds at a tin electrode, the surface changes during the electrolysis as if some of the metal has been dissolved and reduced again [126]. Such a phenomenon is also observed during the reduction of m-nitrobenzenesulfonic acid in a fluidized bed electrode [127]. The role of added stannous salts (Chapter XXIV) to the catholyte in certain reductions has a bearing on this problem [128].

Impurities in the electrode material may modify its properties both in negative and positive direction; the latter may be exploited by using alloys or additions. The desired modification may be either an enhanced mechanical strength, a higher catalytic activity of the electrode material, or a greater resistance toward corrosion. In some cases, however, the amount of side products is increased by the presence of impurities.

B. Cathode Materials

The choice of cathode materials is less restricted by corrosion considerations than the choice of anodes; only a few metals react so readily with the medium that their use is impractical. Some of the most popular electrode materials are mercury, lead, tin, copper, iron, aluminum, platinum, nickel, and carbon.

1. Mercury

Mercury has a very high hydrogen overvoltage. It is obtained commercially sufficiently pure for electrochemical purposes. Stirring of a mercury electrode provides a clean, reproducible surface, which is one of the major advantages of mercury. Another one is that in electrolysis at a mercury cathode results obtained from polarography can be used for selecting the optimal conditions. On the other hand, the liquid state of mercury places certain restrictions on the design of cells employing mercury cathodes. In some cases amalgamated copper or lead electrodes may be used, the amalgamation being done either by treating the solid electrode with a solution of mercuric chloride or simply by rubbing mercury on its surface.

A mercury cathode mostly functions only as an electron donor; numerous examples of such simple reductions have been reported. In some cases, however, the electrode is attacked with the formation of organic mercury compounds. When using mercury on a large scale, suitable precautions must be taken to avoid environmental pollution.

As anode, mercury has a very limited use as it is attacked even at rather low anodic potentials.

2. Lead

Lead has been much used as cathode material ever since the beginning of organic electrochemistry; it has a high hydrogen overvoltage and is easy to work mechanically. The purity of the material is important in some, but not in all cases, and different procedures involving successively anodic and cathodic treatments have been recommended [129] in order to obtain a pure, reproducible surface.

In many reductions a lead cathode yields the same kind of products as a mercury electrode, but in some cases it is either inferior or superior to mercury. The latter is found, e.g., in the reduction in sulfuric acid of aromatic acids to the corresponding alcohols. In such cases specific adsorption of the substrate at the electrode may partly exclude the discharge of hydrogen ions and thus promote transfer of electrons to the substrate, but the nature of this (assumed) adsorption is only partly understood.

3. Tin

Tin electrodes have been used mostly in reductions of nitro compounds. It has been noticed that the electrode surface may change structure during an electrolysis. This has been interpreted [126] as showing that the electrode material reacts "chemically" with the substrate and the stannous (or stannic) ions are discharged subsequently. The negative potential of the cathode would counteract the dissolution of the tin metal, but no investigations using modern electroanalytical techniques have been made to settle the question.

4. Aluminum, Iron, Platinum, and Nickel

Aluminum and iron (steel) have not been so widely used in laboratory experiments, but are preferred in technical installations when possible. Platinum and nickel are used in electrocatalytic hydrogenations: these reactions are discussed in Chapter XXIV.

C. Anode Materials

The choice of anode material is very limited because most metals are attacked in the anodic reaction. In laboratory experiments platinum, gold, and carbon are most widely used; lead containing a small percentage of silver and lead dioxide can be employed in dilute sulfuric acid and iron and nickel in alkaline solution.

1. Platinum

Platinum is used both as microelectrode and in macroscale electrolysis; for larger electrodes platinum plated on titanium may be useful. Platinum is in aqueous solvents probably covered with an oxide at potentials more positive than about 0.75 V (SCE) [130, 131], so the electrode material may, actually, be a platinum oxide.

In some anodic reactions, especially in aprotic media, the products obtained form a layer of tarrish material on the electrode surface which (by covering it) insulates it electrically from the solution. The removal of this layer and the attainment and maintenance of a clean, reproducible surface is often a major problem in the use of noble metal anodes. Heating the platinum in a flame, treatment with nitric acid, or mechanical removal of the insulating layer have all been recommended. Sometimes it is possible to avoid the fouling of the electrode by using pulse electrolysis.

2. Carbon

Graphite electrodes are very useful as anodes; they are, however, not as resistant toward corrosion as platinum electrodes and thus have a limited lifetime. Graphite is available in many forms, one, e.g., resembling cloth, which is convenient for laboratory purposes, another one being glassy carbon.

Oxidation of organic substrates at a graphite anode often results in other products than does oxidation at a platinum electrode; the tendency to further oxidation of a primarily produced radical to a carbonium ion seems higher at carbon than at platinum.

In voltammetry it is important to secure the reproducibility of solid electrodes. In the carbon paste electrode the problem of contamination of the surface is solved by producing a fresh surface for each run.

3. Lead

Lead may be employed pure or alloyed as anode in sulfuric acid; addition of 1% silver, 0.3% tin, and a little cobalt raises its resistance toward corrosion. Other metals may improve the yield of a given electrode process. Thus, addition of antimony and cadmium to a lead anode [132] is advantageous in the oxidation of o-toluenesulfonamide to o-benzoylsulfonimide (saccharin). The same effect may, however, be obtained by using an uncoated, unalloyed lead anode if Sb_2O_3 is added to the anolyte [133]. It seems possible that the dissolution of some antimony from the alloyed lead anode takes place and produces the same effect as the Sb_2O_3 in the anolyte.

In many cases the surface of the anode is oxidized to lead dioxide which thus is the real anode material.

4. "Dimensionally Stable Anodes" (DSA)

These recently introduced materials [306], consisting of precious metal (e.g., ruthenium) oxides coated on baser metals (e.g., titanium), have already had a major impact on the chloralkali industry as replacements for graphite. There is considerable, as yet unpublished work on their utilization in organic electrochemistry. The prospects are good that a number of organic electrosyntheses that were not considered industrially viable because they required platinum will now become candidates for adoption.

D. Semiconductors

Electrodes of semiconductors have been employed mostly in fundamental studies of electron-transfer reactions [134, 135]. Such reactions at semiconductors are a great deal different from those at metals. In the latter case the rate of an electrochemical process is controlled by the high electric field at the surface. On semiconductors the rate is mainly controlled by the concentration of electronic charge carriers in the surface. Therefore, the type of carriers (electrons or holes) in the semiconductor can influence the reaction rate to a very large extent. Of importance for an exchange of electrons between a redox couple and a semiconductor is, inter alia, the redox potential of the former and the energy levels of the conduction and valence bands of the latter.

V. REFERENCE ELECTRODES

A. General Considerations

A reference electrode [136, 137] is a half-cell that defines a potential to which all other measurements are referred. The primary standard electrode is the normal hydrogen electrode (NHE), but as this electrode is inconvenient for practical work, other reference electrodes are used. Such reference electrodes must have a potential which changes very little and is known to within 1 mV or so. In some cases of controlled potential electrolysis it is sufficient to know the potential of the working electrode during the electrolysis within 10-20 mV, because the potential variations between different points of the electrode are of this magnitude (vide supra), and less precise electrodes, the potential of which is known within 10 mV, can be used; such electrodes may be termed comparison electrodes. In principle any electrode at the surface of which an electrochemical reaction with a large exchange current can take place may be used as a reference electrode.

Several problems are encountered when potentials in different solvents are sought compared to the potential scale in water. A variety of approaches [138-145] have been followed to attack this problem; usually the approach has been to introduce some kind of nonthermodynamic assumption, such as the supposition that certain large, monovalent ions (Rb^+, Cs^+) [138, 144] or redox systems [139-141] of the charge type n/n + 1 (preferably 0/ + 1, e.g., ferrocene/ferricinium [139, 140]) have a nearly equal free energy of solvation in the two solvents, so that the free energy of a transfer of the reference ion is small. As the free energy of solvation of an ion is large (and unknown) small differences between two large quantities are difficult to estimate, and the value of the different approaches is difficult to assess. Although this problem is important for certain aspects of electrochemistry it is of minor interest for practical electrolysis, so long as the liquid junction potential between the reference electrode and the test solution can be kept constant and reproducible within acceptable limits.

The ideal reference electrode is indefinitely stable, unpolarizable, easy to handle, temperature independent, has a low resistance, and does not cause contamination of the test solution.

A stable potential of a reference electrode is very important. Instability may be caused by several factors. One of the components may be unstable in the solvent. Thus, mercurous chloride is unstable in acetonitrile and a calomel electrode is unreliable in this solvent. Evaporation of solvent from a standard solution, the concentration of which determines the potential, will change the potential. A liquid junction potential may change; this is observed when an aqueous calomel electrode, connected to the test solution through a potassium chloride-agar bridge, is used for measurements of potentials in acetonitrile for periods longer than about 5 min (vide infra).

An unpolarizable electrode is an electrode the potential of which does not change significantly when a small current passes through it. Most of the potential-measuring devices used draw a small current from the reference electrode, depending on the resistance of the circuit. Polarization is diminished by employing a reference electrode with an electrode reaction with a high exchange current, with a large area, and with a sufficiently high concentration of the potential-determining species, so its concentration is not changed appreciably by the passage of the current.

In some reference electrodes the concentration of the potential-determining species is kept constant by maintaining a saturated solution of a suitable compound. The solubility of a compound is generally temperature dependent, and the potential dependence on temperature of such electrodes is thus higher than for other kinds of reference electrodes.

The contact between the reference half-cell and the test solution requires some consideration. The ohmic resistance of the electric contact should preferably be kept low and contamination of the test solution by the reference electrode and vice versa must be avoided. Usually some kind of salt bridge with or without a diaphragm is employed. The liquid junction potential is lowered if a salt, such as KCl, with nearly equal mobilities of the cation and anion is used. The diffusion between the two half-cells is discouraged by introducing some kind of diaphragm, such as a salt-saturated agar gel, a fritted glass plate, asbestos fibers, or a fritted glass membrane. A considerable ohmic resistance may be introduced simultaneously which puts certain requirements on the internal resistance of the potential-measuring device. In certain cases it has been necessary [146] to employ an electrometer, but generally a pH-meter-type instrument can be used.

The reference electrodes have been divided according to the electrode reaction responsible for maintaining the constant potential: In electrodes of the first kind, the potential of the metal electrode is determined by the concentration of the metal ions; in electrodes of the second kind, the potential is determined by the concentration of an anion which forms an insoluble salt with the metal cation; and in redox electrodes, the potential of an indifferent electrode is determined by the relative concentrations of the two components of a redox system.

B. Electrodes of the First Kind

Metal electrodes of this kind require that neither the metal nor the metal cation react chemically with the solvent; reactive metals may, however, be used as their amalgams. Electrodes of the first kind require a standardized solution in the half-cell which makes such electrodes less convenient to employ than most electrodes of the second kind.

The silver-silver ion electrode is a widely used reference electrode [147] in nonaqueous solvents such as acetonitrile [148]. The electrode consists of a silver wire immersed in a solution of $AgClO_4$ or $AgNO_3$. It is connected to the test solution through a salt bridge containing $NaClO_4$ or a tetraalkylammonium perchlorate and equipped with some kind of device to diminish diffusion.

The Ag/Ag^+ electrode is reversible [149] in many solvents. The main advantage of this and similar reference electrodes is that it allows working in a system containing a single solvent. Silver ions react with certain organic solvents, such as DMF, and other metal electrodes must then be used.

Silver perchlorate is hygroscopic and has a tendency to explode. Handling of $AgClO_4$ may be avoided by electrolytic preparation of it in the reference electrode [150]: current is passed through a silver anode in an anolyte containing a soluble perchlorate, until the desired silver ion concentration is reached. Silver nitrate may be used instead of silver perchlorate, although it is not as fully dissociated in acetonitrile [151]. The influence of the ion pairing on the potential has been discussed [152].

A zinc amalgam electrode has been employed in dimethyl sulfoxide (DMSO) [153] and ammonia [154] as reference electrode. The saturated zinc amalgam was prepared [153] by electrolytic deposition of zinc into a layer of pure mercury from an aqueous $ZnSO_4$ solution. The solution in equilibrium with the Zn(Hg) consisted of a saturated solution of $Zn(ClO_4)_2$, 4DMSO or $ZnCl_2$, $6NH_3$.

The use of a sodium amalgam electrode [$Na(Hg)/NaClO_4(s)$] in DMF has been reported [153], and a similar lithium amalgam electrode has been employed in DMSO [155]. The potential of the cell, $Li(Hg)/Li^+Cl^-$ (DMSO), has been measured for LiCl concentrations from 0.01 to 1.0 M and the Nernst relation was verified within 1 mV; the Li(Hg) electrode obeys the Tafel equation with the transfer coefficient $\alpha = 0.5$ over a current density range from 10^{-7} to 2×10^{-3} A/cm^2 [155].

A special reference electrode is the electron electrode which can operate in solvents capable of solvating electrons, such as ammonia. A sheet of platinum suspended in a solution of sodium metal in ammonia may be used as a reference electrode [156]. The concentration of the reference solution could be varied over relatively wide limits without a change in the potential of the electrode. A sodium concentration of 0.001 M was found suitable [156].

C. Electrodes of the Second Kind

An electrode of the second kind may be used if the metal cation forms an insoluble salt with an anion, but from the point of establishing a satisfactory reference electrode of this kind the most interesting constant is not the solubility constant, but the equilibrium constant for the reaction

$$MeX_{(s)} + X^- \rightleftarrows MeX_2^-.$$

Calomel electrode. The saturated calomel electrode (SCE) is probably the most widely used reference electrode. It consists of a mercury electrode in contact with a solution which is saturated with respect to both Hg_2Cl_2 and KCl. The half-reaction is

$$Hg_2Cl_2 + 2e^- \rightleftarrows 2Hg + 2Cl^-.$$

The activity of the potential-determining mercurous ion is maintained constant through the solubility product of Hg_2Cl_2 by the constant activity of Cl^- in the saturated KCl solution which is in equilibrium with solid KCl. The passage of current thus just changes the quantities of the solid salts, Hg_2Cl_2 and KCl, but not the composition of the solution. The solubilities of Hg_2Cl_2 and KCl change with temperature and the potential of the aqueous SCE vs. NHE is $E = +0.242 - 7.6 \times 10^{-4} \times (t^\circ - 25)$.

The "normal calomel electrode" (NCE), which employs a 1 N KCl solution instead of a saturated solution, has a smaller temperature coefficient [$E = +0.280 - 2.4 \times 10^{-4} \times (t^\circ - 25)$ vs. NHE], but because it requires a standardized solution it is less convenient than the SCE. The potential change of SCE with temperature is less than 1 mV/degree and is of no importance for controlled potential electrolysis.

Many designs of calomel electrodes exist, and several types are commercially available. Some of these are constructed for voltammetric work, whereas others are used in connection with glass electrodes. The latter type generally has a high resistance which must be considered when the choice of voltmeter or potentiostat is made.

The use of SCE presents only few problems in aqueous solution. For measurements in nonaqueous solvents it can be employed in two fashions:

TABLE

Reference Electrodes

	SCE (aqueous)	Hg/Hg_2Cl_2	Ag/AgCl	Ag/Ag^+	$Li(Hg)/Li^+$	$Na(Hg)/Na^+$
CH_3CN	+ [165, 166] [c]	-	(-)[160]	+ [138, 166]		
DMF	+	(+)	(+)[159]			+ [153]
NH_3	–	(+) [165]	-	-		
$(H_2NCH_2)_2$	+	(+) [184]				
C_5H_5N		+		+ [169, 170]		
$((CH_3)_2N)_3PO$			-	+		
DMSO	+	-	(+) [159, 171]		+ [163]	
CH_3OH	+	+	+	+		
CH_3COOH		+ [172]	+ [158]			
$(CH_3CO)_2O$		+ [174]		+ [175]		
H_2SO_4		+				
PC	+					
CH_3NO_2			+ [150]	+ [150]		
THF				+ [179, 180]		
$(CH_3OCH_2)_2$				+ [181, 182]		
CH_2Cl_2	+			-		
THAB	+ [183]					

[a] Symbols: +, stable reference electrode; (+), usable comparison electrode; (-), dubious comparison electrode; -, unsuitable electrode.

[b] DMF = dimethylformamide; DMSO = dimethyl sulfoxide; PC = propylene carbonate, THF = tetrahydrofuran; THAB = tetrahexylammonium benzoate.

[c] Reference numbers.

The aqueous SCE can be connected to the nonaqueous solvent through some device which hinders contamination of the solvent, or an electrode analogous to the aqueous SCE may be constructed in the nonaqueous solvent.

In many nonaqueous solvents a calomel electrode is inapplicable. In acetonitrile, mercurous chloride, although insoluble, undergoes a disproportionation reaction in the presence of excess chloride ion according to

$$Hg_2Cl_2 + Cl^- \rightarrow Hg^0 + HgCl_3^-.$$

1

in Different Solvents [a, b]

	$Tl(Hg)/Tl^{+}$	$Zn(Hg)/Zn^{2+}$	$Cd(Hg)/Cd^{2+}$	$Pb(Hg)/Pb^{2+}$	$H_2(Pt)/H^{+}$	Chloranil	Electrons (solvated)
					- [185]		
	+ [159]		+ [159,163]	(+) [159]			
		+ [167]		+ [162]			+ [156]
		+ [168]					
					- [186]		
	+ [159,161]	+ [153]			- [187]		
					+ [173]	+ [164]	
	(+) [159]		(-) [159]	(-) [159]	+ [178]		
					+ [188]		

It is often possible to employ an aqueous SCE which is connected through a conventional aqueous KCl-agar bridge to a suitable salt solution in the non-aqueous solvent, which in turn dips into the solution under investigation. In such cases gradual deposition of KCl and/or dehydrated agar at the tip of the salt bridge may cause the liquid junction potential to change and thus the potential of the electrode to drift. The drift is often negligible when the measurement is completed within about 5 min [157].

If chloride ions must be avoided, a mercury-mercurous sulfate electrode [$Hg/Hg_2SO_4(s)$, $K_2SO_4(s)$; E = 0.621 V vs. NHE] may be employed. In alkaline solution a mercury-mercuric oxide electrode (E = 0.098 vs. NHE) may be useful.

The saturated silver-silver chloride electrode [Ag/AgCl(s), KCl(s); E = 0.197 V vs. NHE, E = -0.045 V vs. SCE] is especially convenient as it can be constructed by placing a silver wire in a solution of saturated potassium chloride, separated from the test solution by an agar plug and a glass filter (e.g., Fig. 6). An analogous electrode may be employed in certain non-aqueous solvents, e.g., nitromethane [150], acetic acid [158], DMF [159], but it does not work well in others, e.g., ammonia, acetonitrile [160], where complexes are formed which causes the potential to drift.

The thallium amalgam-thallium chloride electrode [Tl(Hg)/TlCl(s), LiCl] is a stable and reversible reference electrode in solvents such as DMSO and DMF [159, 161]. Similarly, the lead amalgam-lead chloride electrode [159, 162] may be employed, but this electrode takes a longer time to reach a stable potential than the Tl (Hg)/TlCl electrode.

In dimethylformamide (DMF) a cadmium amalgam electrode [159, 163] gives a reproducible potential which is stable and well poised.[3] The cadmium amalgam was prepared [163] by heating cadmium with pure mercury for several minutes, and adding, after cooling, sufficient mercury to make a slurry, which was filtered through a pinhole in filter paper. The half-cell was: Cd(Hg)/$CdCl_2$(s), $CdCl_2$-H_2O(s), NaCl(s), DMF. The variation of cell potential with water content has been examined.

D. Redox Electrodes

In a redox electrode both components of the redox system are in solution, and their relative concentrations determine the potential of an inert electrode immersed in the system. Such electrodes have been used as reference electrodes only in special cases, because the potential is dependent on two concentrations, the ratio of which must be kept constant.

Chloranil electrode. The tetrachloroquinone (chloranil)-tetrachlorohydroquinone system has been used as reference electrode in alcohols and acetic acid [164], often in the presence of an excess mineral acid. The half-cell consisted of a platinum electrode in a solution saturated with respect to both chloranil and the hydroquinone. The electrode is well poised and at constant temperature and constant hydrogen ion activity the potential is stable.

Table 1 summarizes the suitability of the various reference electrodes in different solvents.

VI. SOLVENTS FOR ELECTROLYSIS

A. General Considerations

Electrolysis can occur only in a medium which conducts the electric current, and the nature of the solvent is important for the course of electrolytic reactions. Factors such as proton activity, usable potential range, dielectric constant, ability to dissolve electrolytes and substrates, accessible temperature range, vapor pressure, viscosity, toxicity, and price must be taken into consideration when the choice of solvent is made (cf. Chapter I).

1. Proton Activity

The availability of protons is one of the most important properties of a medium by which it can influence an electrolytic reaction, especially a reduction. Anodic reactions are generally less susceptible to the proton activity. The reactivity of some nucleophiles in anodic substitution reactions may, however, depend on proton activity.

At high proton activity a protonation of the substrate may occur prior to the charge-transfer step. A protonated molecule is reduced at a less negative potential than an unprotonated one. At medium proton activity the unprotonated molecule is reduced to a radical anion; this is rapidly protonated to a radical which generally is more easily reduced than the unreduced substrate. In aprotic media the radical anion formed in the charge-transfer step may be so long-lived that it can be regarded as the product. The radical anion is normally more difficult to reduce than the unreduced substrate.

In aqueous solution a change in pH may alter: (a) the sequence of reducibility of different "electrophores"; (b) the degree of reaction of an "electrophore" may be different in acid and alkaline solutions; (c) a branching of a reaction may be influenced by the availability of protons; and (d) the stereochemistry of the product may be pH dependent. As examples can be given: (a) the reduction of isonicotinic amide in which the carboxamide group is reduced in acid solution whereas the pyridine ring suffers reduction at high pH (Chapter XVII); (b) the reduction of aromatic nitro compounds which at low pH at a suitable potential yields an amine and, in alkaline solution, a phenylhydroxylamine (and its condensation products) (Chapter VII); (c) the reduction of acrylonitrile may yield the two-electron reduction product, propionitrile, or the one-electron reaction product, adiponitrile, depending on the availability of protons (Chapter XIX); (d) the stereochemistry of the product mixture of pinacols formed on electrolysis of aromatic ketones is pH dependent (Chapter VIII).

When considering an aprotic solvent, it must be recalled that it is difficult to obtain a strictly anhydrous medium. Especially in voltammetric experiments where the concentration of the substrate is low, the residual water nearly always present in a "dry" solution may influence the results obtained. It must also be considered that the anodic limiting reaction of some anions, e.g., perchlorates [189-191], results in the formation of water. Protons may also be supplied by abstraction by a base, e.g., an anion radical, from the nonaqueous solvent or from tetraalkylammonium ions, in the latter case by a Hofmann elimination.

A determination of water in a "dry" solution is thus often of interest. The Karl Fischer method is unreliable below about 20 ppm (1 mM). A gas chromatographic method has been employed [192] for determination of water in certain cyclic esters, but injection of electrolyte solutions causes

TABLE

Dielectric Constants and Accessible

	Dielectric constant	Working electrode	Reference electrode	Supporting electrolyte
Water	80	Pt	SCE	$HClO_4$
Methanol	33			
H_2SO_4 (96-99%)	>84			
CH_3COOH	6.2	Pt	SCE	NaOAc
CH_3CN	37.5	Pt	Ag/Ag^+	$LiClO_4$
DMF	36.7	Pt	Hg pool	$LiClO_4$
NMP	32	Pt	Hg pool	$LiClO_4$
HMPA	30	Pt	Ag/Ag^+	$LiClO_4$
NH_3	23			
$(H_2NCH_2)_2$	12.5			
Pyridine	13	Graphite	Ag/Ag^+	$LiClO_4$
DMSO	46.7	Pt	SCE	$NaClO_4$
Sulfolane	44	Pt	SCE	$NaClO_4$
PC	69	Pt	SCE	TEAP
CH_3NO_2	36.7	Pt	SCE	$LiClO_4$
THF	7.4	Pt	Ag/Ag^+	$LiClO_4$
$(CH_3OCH_2)_2$	3.5	Hg	SCE	TBAP
CH_2Cl_2	8.9	Pt	SCE	TBAP

[a] The limit for the accessible potential for a solvent depends on many factors such as supporting electrolyte, electrode material, and magnitude of permissible current density, and the numbers given are cited as illustrative values only.

[b] DMF = dimethylformamide; NMP = N-methylpyrrolidone; HMPA = hexamethylphosphotriamide; DMSO = dimethyl sulfoxide; sulfolane = tetramethylene sulfone; PC = propylene carbonate; THF = tetrahydrofuran; TBAP, resp. TBAI = tetrabutylammonium perchlorate, resp. iodide; TEAP, resp. TEAB = tetraethylammonium perchlorate, resp. bromide.

[c] Reference numbers.

2

Potential Range for Some Solvents [a, b]

Anodic limit, V	Working electrode	Reference electrode	Supporting electrolyte	Cathodic limit, V
1.5	Hg	SCE	TBAP	-2.7
	Hg	Hg pool	TEAB	-2.2 [195]
	Hg	Hg pool	None	-0.7 [177]
2.0 [206]	Hg	SCE	TEAP	-1.7 [196]
2.4 [242]	Pt	Ag/Ag^+	$LiClO_4$	-3.5 [242]
1.5 [246]	Hg	Hg pool	TEAP	-3.5 [246]
1.4 [246]	Hg	Hg pool	TEAP	-3.3 [246]
0.8 [242]	Hg	Ag/Ag^+	$LiClO_4$	-3.6 [252]
	Hg, Pt	Hg pool	TBAI	-2.3 [165]
	Hg	NCE	TBAI	-2.35 [184]
1.4 [253]	Hg	Hg pool	$LiClO_4$	-1.7 [253]
0.7 [256]	Hg	SCE	TEAP	-2.8 [256]
2.3 [258]	Pt	SCE	$NaClO_4$	-4 [258]
1.7 [266]	Hg	SCE	TEAP	-2.5 [266]
2.7 [150]	Hg	SCE	$LiClO_4$	-1.2 [150]
1.8 [180]	Pt	Ag/Ag^+	$LiClO_4$	-3.6 [180]
0.65 [181]	Hg	SCE	TBAP	-2.95 [181]
1.8 [197]	Pt	SCE	TBAP	-1.7 [197]

accumulation of salts in the injection chamber and sometimes erratic results due to thermal decomposition of electrolytes. Analytical use of the 1.9 μ water band in the near-infrared [193] may be practical for certain solvents, e.g., propylene carbonate, but not for many of the commonly used solvents. Recently a method has been described [194] in which the water reacts with lead tetraacetate in benzene; the lead dioxide thus formed is determined spectrophotometrically at about 400 nm. As little as 2.5 ppm in a 2 ml sample can be measured. The method was found to be applicable to a wide range of solvents and to certain solutions commonly employed in electrochemistry.

2. Usable Potential Range

The useful potential range which can be obtained in a given system depends on the electrode material, the supporting electrolyte, and the solvent; only the solvent will be discussed here. Approximate values for the anodic and cathodic limits of some commonly used solvents are given in Table 2. The numbers can be taken only as a crude indication of the usable potential range, as the latter depends on the criteria used.

Many solvents are available which are so difficult to reduce that it is the discharge of the supporting electrolyte which sets the cathodic limit for the potential; on the other hand, only few polar solvents are resistant to anodic oxidation. The reason is that most polar groups have loosely bound nonbonding electrons on oxygen or nitrogen which are easily lost at the anode. The solvents useful for anodic reactions at high potentials contain electron-attracting groups such as nitrile, carboxy, sulfone, or nitro groups. For more easily oxidized substrates sulfoxides or dialkylamides may be used, whereas water, alcohols, and aliphatic amines have a very limited useful potential range for anodic reactions.

3. Dielectric Constant

The dielectric constant is of interest primarily because it influences the ohmic resistance of the medium. Solvents with high dielectric constant are preferable because salts dissociate better in such media. The dielectric constants for some solvents commonly used for electrolysis are listed in Table 2. These solvents can be divided roughly into three main groups: those with high dielectric constant ($D > 60$), those with medium dielectric constant ($20 < D < 50$), and those with low dielectric constant ($D < 13$). In the first group are solvents like water, formamide, N-methylamides, and propylene carbonate; the second one comprises compounds like acetonitrile, dimethylformamide, dimethyl sulfoxide, methanol, nitromethane, and ammonia; whereas in the last group belong solvents like acetic acid, ethylenediamine, methylamine, tetrahydrofuran, dioxane, and methylene chloride. The solvents in the last group require a higher concentration of supporting electrolyte in order to acquire a reasonable conductivity.

For acetonitrile, propylene carbonate, 4-butyrolactone, DMSO, DMF, and similar amides the maximum conductance for fully ionized salts occurs at approximately 1 M concentration and is about $10^{-2}\ \Omega^{-1}\ cm^{-1}$, comparable to an aqueous salt solution of 0.5-1 M.

4. Dissolving Power

The choice of solvent is often a result of a series of compromises. Very few solvents have good dissolving power for both organic substrates and inorganic salts. The good electrochemical properties of water are marred by the lack of its ability to dissolve many organic systems. This ability may be raised by using water in a mixture with an organic solvent like ethanol, acetonitrile, dimethylformamide, or dioxane, or by using a "hydrotropic" supporting electrolyte [198, 199] like a tetraalkylammonium p-toluenesulfonate. Tetraalkylammonium salts are soluble in most polar solvents and even in less polar liquids such as chloroform and methylene chloride. Some polar solvents, e.g., acetonitrile and DMF, combine good dissolving power for both organic compounds and a variety of salts.

5. Temperature Range, and Other Factors

It is desirable that the solvent be a liquid in a convenient temperature interval and that the vapor pressure be not too high at the working temperature to avoid excessive loss of solvent. This is not an important point since the electrolysis is often carried out in a closed system anyway. The vapor pressure may also sometimes be lowered considerably by working with a high salt concentration. Too low a vapor pressure of the solvent is inconvenient, as it may be difficult to remove the solvent during the work-up.

The toxicity and odor of the solvent warrant consideration. Proper design of the apparatus and careful handling of the solvent make these properties less objectionable. A low viscosity is preferable both from the point of view of diffusion and from considerations of pumping in flow cells. For some kinds of electroanalytical work, however, a high viscosity is preferable. Diffusion coefficients in some solvents useful for electrolysis have been published [200].

B. Protic Solvents

It is practical to divide solvents into two groups, protic and aprotic solvents. Protic solvents are those which have protons bonded to heteroatoms and include acids, neutral solvents, and some bases. Recently a review of solvents useful for electrochemistry has appeared [201]. Electrochemical reactions in nonaqueous systems [202] and the chemistry of nonaqueous solvents [203] have been treated in recent monographs.

1. Acid Solvents

The electrochemistry of a protonated organic compound differs in many cases from that of the neutral molecule, and the strength of the acid necessary for protonation of the substrate must be considered. Sulfuric acid and acetic acid, often containing some water, have generally been used. The use of "superacids," such as SbF_5-SO_2 or SbF_5-FSO_3H, in electrochemistry has apparently not yet been investigated, but they offer interesting possibilities. A strongly protonating medium which has recently been introduced is fluorosulfonic acid [176].

a. Sulfuric Acid. Concentrated or slightly diluted sulfuric acid is in many respects an excellent medium for electrolysis. It dissolves many organic compounds and it is sufficiently dissociated to make addition of foreign electrolytes unnecessary.

Sulfuric acid can protonate even very weak bases, such as aromatic ethers or some aromatic hydrocarbons; it may promote the formation of carbonium ions either by dehydration of hydroxyl compounds or by addition of a proton to a double bond. These properties are still more strongly developed in the "superacids."

Sulfuric acid has certain disadvantages: It cannot be removed by distillation during the work-up. It is, therefore, often most practical to dilute the medium with water and extract the product. If neutralization is necessary, concentrated ammonia is preferable. Another disadvantage is that sulfuric acid may oxidize or sulfonate the substrate.

Most of the electrolyses in sulfuric acid have been performed without control of the potential. In polarographic work the mercury pool has been used as reference electrode [177]. Sulfuric acid has been used as a medium both for anodic and cathodic reactions: As anode, platinum or lead dioxide has been employed; as cathode, lead, mercury, or platinum.

A 1:1 mixture of sulfuric acid and ethanol was a very popular medium for electrolytic reductions around the turn of the century. The mixture dissolves many organic compounds, but the leveling effect of the alcohol lowers its protonating ability considerably.

b. Acetic Acid. Acetic acid is a good solvent for many organic compounds and some inorganic salts. It has a rather low dielectric constant which makes it necessary to have a considerable electrolyte concentration ($\gtrless$ 0.5 M) to obtain a reasonable conductance.

Acetic acid may be considered as solvent for acetoxylations [204-206] and for reactions which require an acidic medium other than sulfuric acid. The acidity may be enhanced by adding perchloric acid to acetic acid containing a suitable amount of acetic anhydride which reacts with the introduced water.

Usable supporting electrolytes include NaOAc, NH_4OAc, LiCl, HCl, H_2SO_4, $HClO_4$, $NaClO_4$, Bu_4NClO_4, and Bu_4NBF_4. The choice of supporting electrolyte may influence the product distribution. Thus, anodic acetoxylation gives different results in NaOAc/HOAc and in Bu_4NBF_4/HOAc. Several types of reference electrodes work well in acetic acid. The chloranil electrode [164] behaves reversibly in this medium and so does the analog of SCE [207] [$Hg/Hg_2Cl_2(s)$, LiCl(s), HOAc]; the mercury pool [208] and the Ag/AgCl electrode [158] can also be used.

The cathodic limit is about -1.7 V (SCE) and the limiting reaction is evolution of hydrogen. The anodic range ends at about 2.0 V (SCE) when NaOAc is electrolyte [206]. The limiting reaction is probably the discharge of acetate ions.

Acetic acid may be purified by partial freezing [206] or by distilling it from CrO_3 [209] in the presence of acetic anhydride. CrO_3 oxidizes impurities and acts as an acid catalyst for the reaction between water and acetic anhydride. An excess of acetic anhydride is necessary to force the equilibrium $2AcOH \rightleftarrows Ac_2O + H_2O$ sufficiently to the left [210]. Modifications of the method have been described [207, 208].

c. Hydrogen Fluoride. Hydrogen fluoride has been used for electrochemical fluorination of organic compounds [211-214] and is a promising solvent for electrochemical studies. It has a high dielectric constant (80 at 0° C), is less viscous than water, transparent for uv light to 165 nm, and is difficult to oxidize. It is a strongly protonating solvent and unlike sulfuric acid is not an oxidant. It dissolves many metal fluorides, giving highly conducting solutions; it also dissolves many organic substances. The main disadvantages are its poisonous effect, its ability to attack glass, and its relatively low boiling point, 19.5°C.

Hydrogen fluoride can be handled in apparatus of suitable metals (copper, nickel, magnesium, or aluminum, which all form a protective fluoride coating, or platinum) or plastic materials [a.o. polypropylene, Teflon, polyvinylidene fluoride ("Viton")]; polychlorotrifluoroethylene ("Kel-F") can be made into transparent windows. Capillary for a dropping mercury electrode may be made from Teflon [215]. Hydrogen fluoride is obtained commercially in steel cylinders in a purity of 99.5%. The impurities may be removed by distillation [216] or electrolysis [217]. During the electrolytic removal of water the explosive F_2O is formed which must be taken into consideration [214].

The useful potential range in the cathodic direction is rather limited due to evolution of hydrogen, whereas the anodic limit is about 2.6 V (SCE). Sodium or potassium fluoride as supporting electrolyte may be used; addition of these salts lowers the protonating power of the solvent. The solubility of NaF is about 30 g NaF/100 g HF; other fluorides, such as LiF, KF, CsF, NH_4F, AgF, and TlF, are very soluble in liquid HF. The acidity of HF may be enhanced by addition of BF_3 or SbF_5.

The hydrogen electrode has been shown to behave reversibly in HF [218]; as reference electrode for polarographic studies the mercury pool has been used [219].

d. Other Acids. Methanesulfonic acid [220] and formic acid have also been used as solvent for electrolysis; "superacids" offer possibilities for unusual electrode reactions. Also trifluoroacetic acid [221] has been employed as medium for electroanalytical work.

2. Neutral Solvents

This group comprises water, mono- and polyvalent alcohols, and monoethers of polyvalent alcohols. Often mixtures of water and alcohols are used. Such mixed solvents retain to some degree the favorable characteristics of water, but raise its ability to dissolve organic substrates.

a. Water. Water is the solvent of choice for electrolysis, unless some of its properties are undesired or inadequate. An aprotic solvent is sometimes preferable to water since a reaction mechanism of an electrode reaction in an aprotic, organic solvent is often simpler to elucidate than in water. Adsorption phenomena are often a complicating factor in aqueous solution, but much less so in many organic solvents. The usable potential range of water in the anodic direction is quite limited, and, the most important factor, the solubility of many organic compounds in water is quite unsatisfactory.

b. Methanol. The electrochemical behavior of methanol is rather similar to that of water. Methanol is well suited for anodic reactions of the Kolbe type and for methoxylations; for reductions it may be used when the solubility of the substrate in water is too low. The dielectric constant of methanol is fairly high (33); its liquid range (-98° to 64°) is convenient and it is easy to remove during the work-up.

Quite a few electrolytes are soluble in methanol, e.g., NH_4Cl, LiCl, HCl, KOH, KOMe, $NaClO_4$, and tetraalkylammonium salts. For reductions a solution of HCl in MeOH is convenient as it has a high conductance and both components are easy to get rid of during the isolation of the product.

Reference electrodes which can be used in water will work in most cases also in MeOH, thus the Hg/Hg(I) and Ag/Ag(I) electrode may be used. The useful potential range for large-scale electrolysis is about the same as for water, but methanol is less satisfactory for electroanalytical work.

Methanol can in general be used as received; it can be dried by treatment with magnesium [222].

c. Other Alcohols. Ethanol and the higher alcohols have lower dielectric constant and poorer ability to dissolve electrolytes. They have been used mostly in mixture with water or sulfuric acid.

3. Basic Solvents

Basic, protic solvents include ammonia and primary and secondary amines. These solvents are primarily of interest for an organic electrochemist because of their ability to solvate electrons, and solvated electrons have special reducing properties (Chapter XXV). They also permit reductions in a protic medium in the presence of a very strong base, the conjugate base of the solvent.

a. Ammonia. The use of ammonia as solvent for electrochemical reactions present some problems due to its inconvenient liquid range (-77.7° to -33.4°); this difficulty can be coped with in different ways. The electrolysis may be made in a suitable low-temperature thermostat [223], the cell can be equipped with an efficient reflux condenser [224], a high-pressure apparatus can be used [225], or strong solutions (5-10 M) of various salts, such as NH_4SCN, NH_4I, $LiNO_3$, $LiClO_4$, NaSCN, and NaI, can be used [226, 227] as they form a medium in which the vapor pressure of ammonia at ambient temperature is below 1 atm. The easy purification and low price of ammonia compared to the alternative amines favor the choice of ammonia for large-scale preparations.

Ammonia has a medium-high dielectric constant (23.7 at -36°C) and good dissolving power toward inorganic salts. It can act both as an acid and a base; the strongest possible acid in ammonia is ammonium ion and the strongest base the amide ion.

Besides the electrolytes mentioned above, salts such as NH_4Cl, NH_4NO_3, KNO_3, and $NaClO_4$ may be used as supporting electrolyte; the tetraalkylammonium salts are generally rather sparingly soluble in ammonia. Most electrolyses in ammonia, except in voltammetric studies, have been performed without a reference electrode; as such a Zn(Hg)/$ZnCl_2$- [154] or Pb/Pb^{2+}-electrode [165, 223, 228] may be used. In polarography the mercury pool electrode has been applied [165].

Ammonia has been employed mostly for cathodic reactions, but some oxidations [224, 227] have been carried out in this medium, although the potential range in the anodic direction is quite small. The anodically limiting reaction is oxidation to nitrogen and protons [229]; the cathodically limiting reaction is the transfer of electrons to the solvent which occurs about -2.3 V (vs. Hg pool electrode) in a saturated solution of TBAI. In the electrolytic generation of solvated electrons the potential is determined by the surface concentration of electrons and no external reference electrode is needed.

Ammonia is purified by distilling it after treatment with sodium; the distillation from sodium may have to be repeated.

b. Methylamine. Methylamine behaves electrochemically much like ammonia but has a somewhat more convenient liquid range (-93.45° to -6.3°C), although its odor and low boiling point require special considerations. Its dielectric constant is rather low (11.4 at -10°C), but sufficient conductivity for large-scale electrolysis can be obtained using LiCl as supporting electrolyte. Like ammonia, it is both an acid and a base.

Methylamine has mainly been used in large-scale electrolysis [230, 231] and no reference electrodes have been employed so far. Reference electrodes, such as the $Zn(Hg)/Zn^{2+}$ electrode, which work both in ammonia and ethylenediamine, would be assumed to be applicable to methylamine solutions.

Methylamine has been employed only for reductions, especially as solvent for electrolytic generation of solvated electrons. It would be of very limited use for anodic reactions. It may be purified by distillation from sodium.

c. Ethylenediamine. Ethylenediamine (EDA) is a primary amine with a more convenient liquid range (11°-117°) than methylamine. Its dielectric constant ($\epsilon = 12$) is about the same as that of methylamine. It dissolves many inorganic salts and is a better solvent than ammonia for many organic substrates.

LiCl, $NaNO_3$, and tetraalkylammonium salts can be used as supporting electrolyte. For the electrolytic generation of solvated electrons mainly LiCl has been employed [232, 233]. A reversible reference electrode in EDA is the $Zn(Hg)/Zn^{2+}$ electrode [234], but the Hg pool [184] or the aqueous calomel electrode, fitted with a suitable salt bridge, is also applicable.

EDA is oxidized rather easily anodically, but is resistant to reduction; the cathodic limiting current is the discharge of cations or the donation of electrons to the solvent. EDA may be purified by repeated refluxing and distillation from sodium [235].

Other primary or secondary, aliphatic or cyclic amines may be used with similar results.

C. Aprotic Solvents

The use of aprotic solvents is of interest in many cases where aqueous solutions lack desired or have undesired properties. The scarcity of protons makes it generally simpler to elucidate the mechanism of an electrode reaction, as the product of the first electron transfer, a radical ion, is more stable in aprotic than in protic medium, and the same is often the case for other intermediates. Controlled addition of a suitable proton donor or electrophile (reductions) or nucleophile (oxidations) is often useful in determining reaction mechanism. The strength of a proton donor may vary from perchloric acid through acetic acid and a phenol to an alcohol; pyridine, carboxylate ions, alcohols, or alkoxides may be used as nucleophiles. Sometimes it is of interest to determine whether it is the basic or the nucleophilic properties of the compound which are important. The use of two bases with approximately the same pK values but widely differing in nucleophilicity, such as pyridine and 2, 6-dimethylpyridine, might answer the question.

"Dry" aprotic solvents usually contain some water, which is especially of importance in voltammetric studies where the concentrations of water and substrate are about equal. The influence of a certain molar concentration of water depends on the solvent, the supporting electrolyte, and the substrate.

Electrochemical phenomena may also be simplified in organic solvents, compared to aqueous solutions, due to the less important role adsorption plays in organic solvents.

1. Acetonitrile

Acetonitrile is one of the most used polar aprotic solvents both for anodic and cathodic reactions. It is an excellent solvent for many organic substrates and quite a few organic and inorganic salts; it is miscible with water. Salt solutions show a reasonably high conductivity due to its rather high dielectric constant ($\epsilon = 37$).

Besides having good electrochemical properties acetonitrile is an unusually good solvent for uv spectroscopy, being transparent until about 190 nm. It is furthermore a convenient solvent for esr spectroscopy, which is valuable for the investigation of radicals produced during electrolysis. For use in preparative electrochemistry its liquid range (-45° to 82°) is sufficient for most purposes and its relatively low boiling point makes the removal of the solvent easy. On glc, tailing may occur, unless a suitable polar column is used.

Acetonitrile has a wide, usable potential range both in anodic and cathodic direction. The limit is set by the electrode reaction of the supporting electrolyte in both directions. In the presence of sodium or lithium ions their discharge sets the limit; the sodium subsequently reacts with acetonitrile, whereas lithium does not. Quaternary ammonium ions are reduced with dealkylation. Perchlorate ions, a widely used supporting electrolyte, are discharged to perchlorate radicals which probably rapidly form oxygen and chlorodioxide radicals [191]. Transfer of protons in acetonitrile is a rather slow process.

Useful supporting electrolytes in acetonitrile are $NaClO_4$, $LiClO_4$, tetrabutylammonium salts, such as the chloride, bromide, iodide, perchlorate, and tetrafluoborate, as well as other tetraalkylammonium salts. $AgNO_3$ and $AgClO_4$ are very soluble in acetonitrile; AgCl is somewhat soluble, especially in the presence of an excess of chloride ions.

The Ag/Ag^+ electrode [147, 148], which is reversible in acetonitrile, may be used as reference electrode. Quite often an aqueous calomel electrode with a suitable liquid junction is used [165]; a comparison of different reference electrodes has been made [166]. The Hg/Hg(I) is not stable in acetonitrile, as Hg(I) decomposes [236].

Commercial acetonitrile contains usually impurities such as acrylonitrile, acetic acid, aldehydes, amines, and water. Several methods of purification [239] have been proposed: one is to distil it several times from P_2O_5 followed by fractionation from K_2CO_3 [149]; a pure product may be obtained, but the procedure is time-consuming and wasteful. Treatment with CaH_2, distillation from P_2O_5, and fractionation from CaH_2 yield a product useful for cathodic polarographic purposes [238]; traces of aromatic hydrocarbons are not removed, which makes the product less suited for anodic voltammetry and uv spectroscopy. For removal of acrylonitrile sodium hydride has been recommended [239].

Removal of impurities by reaction in turn with benzoyl chloride and $KMnO_4$ followed by careful distillation gives a product suitable both for electrochemical and spectroscopic work [240]. Alternatively, acetonitrile is treated with N_2O_4, then with CaH_2, fractionated under N_2 from CaH_2, and the product further purified by letting it run through a column of activated alumina [236].

Pure acetonitrile is not particularly hygroscopic. The product obtained by these procedures can be stored for several months without deterioration; the water content is approximately 1 mM. It is advisable to use the same glass apparatus repeatedly for the purification and storing of acetonitrile.

Water can be removed from commercial acetonitrile by azeotropic distillation with CH_2Cl_2 [241, 242] or by treatment with molecular sieves. Acrylonitrile, which boils 4° lower than acetonitrile and thus is difficult to separate from it by fractional distillation, may be removed by azeotropic distillation with ethanol [243].

Other nitriles, aliphatic and aromatic, have been investigated as solvents for electrolysis, but they seem not to offer any particular advantages over acetonitrile.

2. Dimethylformamide (DMF)

Dimethylformamide is a polar solvent with a dielectric constant (37) comparable to that of acetonitrile. It is a good solvent for most organic compounds and also for many organic and inorganic perchlorates and organic fluoborates. Owing to these properties DMF has been widely used in electrochemical work. DMF is, however, susceptible to hydrolysis, which yields the easily oxidizable dimethylamine and the reducing agent formic acid.

DMF has an acceptable liquid range (-61° to 153°), although the boiling point is near the upper limit for a convenient removal of it during the work-up by distillation. DMF is inferior to acetonitrile as a solvent for uv spectroscopy.

For reductions DMF has a usable potential range comparable to that of acetonitrile, but it is inferior for oxidations. In the presence of inorganic ions their discharge is the cathodic limiting reaction, whereas it is more uncertain whether it is the solvent or the cation which is reduced in solutions of tetraalkylammonium ions. The anodic limiting reaction at a Pt electrode is an oxidation of DMF, which involves the removal of an electron from the amide nitrogen.

As supporting electrolyte, tetraalkylammonium perchlorates or fluoborates are mostly used, but LiCl or $NaClO_4$ may also be employed. The reference electrode may be a modified Ag/AgCl electrode [159, 163], a $Cd(Hg)/Cd^{2+}$ [159, 163] or $Na(Hg)/Na^+$ electrode [153].

For many purposes the commercial reagent grade DMF, dried with molecular sieves, can be used; DMF decomposes on distillation at atmospheric pressure. The most convenient way to purify DMF is probably to let it pass through a column [243] of alumina (e. g., Woelm Al_2O_3 activity grade one). Other methods involve drying ($CuSO_4$ [244], which also removes amines), azeotropic distillation with benzene, or percolation through molecular sieves [245] followed by fractional distillation at reduced pressure. DMF is difficult to obtain anhydrous, but for most purposes a small water content is not critical.

3. N-Methylpyrrolidone (NMP)

DMF has certain disadvantages which can be traced to its hygroscopic properties and rather easy hydrolysis; N-methylpyrrolidone, which is a N-disubstituted amide just as DMF, is less apt to hydrolyze due to its cyclic structure. Furthermore, the product of hydrolysis is not a reducing agent as is formic acid formed from DMF. Although only a limited number of experiments in NMP have been done [246, 247] yet it seems probable that NMP will replace DMF in a number of cases due to its higher stability [247a].

Like DMF, NMP is of limited value for oxidations, but useful for cathodic reactions. Its dielectric constant is 32, and it is able to dissolve the same types of salts and organic compounds as DMF. NMP has a liquid range of from -24° to 205°; the useful potential range is approximately the same as for DMF. Mercury in contact with its oxidation products has been used as reference electrode; it is more stable than the Ag/Ag^+ electrode [246].

Other organic amides, such as formamide, N-methylformamide, and N-methylacetamide, have been investigated as solvents for electrochemical use primarily due to their high dielectric constant; none of them, however, seem to offer distinct advantages over DMF or NMP and in some respects they are inferior.

4. Hexamethylphosphoramide (HMPA)

HMPA is a polar solvent which is remarkable for its ability to solvate electrons. Even in a mixture with ethanol (67 mole % ethanol, 33 mole % HMPA) reductions via solvated electrons may take place [248, 250]. It has been suggested that HMPA is selectively adsorbed at the electrode, thus providing a layer of low proton activity near the electrode.

The liquid range of HMPA is 7.2° to 235°; its dielectric constant is 30 (20° C). HMPA solvates cations strongly, whereas anions are less solvated.

Unlike carboxamides, e.g., DMF, it is not attacked by aqueous alkali at $t < 80^\circ C$, and it is very resistant toward nucleophilic attack [250] but forms peroxides under the influence of light and oxygen [249].

HMPA is completely miscible with water; it can be extracted from water by chloroform which forms a complex with HMPA [152], $K(CHCl_3/H_2O) = 5.5$; HMPA is somewhat more basic than DMF. HMPA dissolves salts such as LiCl, $LiClO_4$, $NaClO_4$, and R_4NClO_4; the Ag/Ag^+ may be used as reference electrode [252]. Using R_4NClO_4 the accessible potential range is from +0.14 V to -3.35 V (Ag/Ag^+) at a mercury electrode [249]. HMPA can be purified by fractionation under vacuum, bp 88-92° (3 mm), 68-70° (1 mm).

Just like DMF, HMPA is not suitable for anodic reaction, since the amide nitrogen loses an electron easily. For cathodic reactions the useful potential range extends, in the presence of LiCl, to the value at which electrons are given off to the solution. HMPA has not yet been used widely in electrochemical experiments, but its special properties will undoubtedly prove useful in the future.

5. Pyridine

Pyridine is a reasonably strong base and a good nucleophile. Although its dielectric constant is rather low (12) many salts are soluble in pyridine and the solutions have good conductivity. In contradistinction to most bases, pyridine, which is a π-electron deficient heteroaromatic compound, is rather resistant toward oxidation.

Pyridine has a liquid range from -41° to 115°; it is miscible with water and a good solvent for many organic compounds. Salts such as $LiNO_3$, LiCl, $LiClO_4$, NaI, KSCN, and tetraalkylammonium salts are soluble in pyridine. As reference electrode the Ag/Ag^+ [170] or the Hg pool electrode can be employed. Solute-solvent interactions have recently been discussed [305].

The useful potential range is from about -2.3 V [170] (vs. Ag/Ag^+ electrode, DME, Bu_4NI) to +1.4 V (vs. Ag/Ag^+, graphite, $LiClO_4$) [253]. The cathodic limiting reaction might be the discharge of the cation, whereas an electron abstraction from pyridine is the anodic limiting reaction, in which probably 2-pyridylpyridinium ion is formed.

Besides being used as a solvent, pyridine has been employed as a difficultly oxidizable nucleophile [148]. Reagent pyridine is satisfactory for most purposes; it may be dried with molecular sieves, type 4A.

6. Dimethyl sulfoxide (DMSO)

DMSO is an excellent solvent for many inorganic salts and organic compounds. It is difficult to reduce and fairly resistant to electrolytic oxidation. Its dielectric constant is high (47). It thus has many of the qualities desirable for a solvent for electrolysis and it shows promise of being one of the most important electrochemical media [254]. The liquid range is from 18° to 189°, which makes it somewhat inconvenient to get rid of DMSO in the work-up. When used as solvent for electrolysis it must be considered that DMSO is not always inert, but has a fair reactivity in certain reactions. DMSO is unfit for uv spectroscopy.

DMSO has a low toxicity; it is, however, transported through the skin very fast. Thus, minutes after applying a drop of DMSO to the palm of the hand its sweetish taste can be detected in the mouth. It may carry with it dissolved toxic materials which otherwise would be unable to penetrate the skin.

Many salts are soluble in DMSO so the choice of supporting electrolyte is less restricted than in most other nonaqueous solvents. In general, perchlorates, even $KClO_4$, nitrates, and halides are soluble, whereas fluorides, cyanides, sulfates, and carbonates are not; thus not only $NaClO_4$, LiCl, $NaNO_3$, and tetraalkylammonium salts can be used but also salts such as NH_4PF_6 and NH_4SCN. The ability of DMSO to solvate ions is also of importance in the indirect electrolytic hydrodimerization of, e.g., acrylonitrile (Chapter XXV).

The equilibrium constant of the self-ionization of DMSO is 5.10^{-8} (25° C); the conjugate base $(CH_3SOCH_2)^-$, which is formed when DMSO acts as a proton donor during a reduction, is a rather strong base and a fairly good nucleophile.

The potential range of DMSO in the cathodic direction is at a Hg electrode limited by the discharge of the cation. Potassium and sodium react with DMSO, whereas lithium does not; tetraalkylammonium ions are reduced about -2.8 V (SCE) [255, 256]. The anodic limiting reaction at a Pt electrode is the oxidation of DMSO; probably one of the nonbonding electrons is lost, but the limiting reaction has not been investigated yet.

As reference electrode [159] for polarographic work an aqueous or methanolic calomel electrode, connected to the DMSO solution with a suitable salt bridge to avoid contamination, has been used. The most stable reference electrodes in DMSO seem to be the amalgam electrodes such as Tl(Hg)/TlCl [161] or Li(Hg)/LiCl [155].

DMSO is somewhat hygroscopic, and the main impurity in a good commercial grade of DMSO is water; the odor is due to dimethyl sulfide. For many purposes DMSO can be used as received or after drying either by shaking with CaO-MgO or passing through a column of molecular sieves. For further purification fractional distillation or crystallization may be used. It has been recommended [257] that a fractional crystallization be done followed by two fractional distillations in vacuo, one through a column packed with activated carbon and one through a spinning band column. The pure DMSO must be handled without contact with any humidity.

7. Sulfolane

Among sulfones, sulfolane (tetrahydrothiophen-1, 1-dioxide, tetramethylene sulfone) shows promise for electrochemical use. It has a wide liquid range (28.6° to 285°); on dissolving 0.1 M salts in sulfolane its melting point is depressed below 25° C. Sulfolane has a relatively high dielectric constant (43 at 20° C), it is chemically stable toward both reducing and oxidizing agents, and it has a fair dissolving power for inorganic salts. One of its drawbacks is that it is rather hygroscopic.

The accessible potential range is large, both in anodic and cathodic directions. The limiting cathodic reaction seems to be the discharge of the supporting electrolyte, whereas the limiting anodic reaction probably is the oxidation of the solvent. With $NaClO_4$ as supporting electrolyte the anodic limit at a Pt electrode is about 2.3 V and the cathodic limit about -4 V (vs. Ag/Ag^+ electrode) [258].

As supporting electrolyte, $NaClO_4$, $LiClO_4$, NH_4PF_6, and quaternary ammonium compounds can be used. The Ag/Ag^+ [258, 259], Ag/AgCl [260] electrode or an aqueous SCE [261], with a suitable salt bridge, can be used as reference electrode.

Sulfolane may be purified by fractional crystallization followed by fractional distillations at reduced pressure [261]; alternatively, treatment with aqueous hydrogen peroxide, followed by extraction with CH_2Cl_2 and fractional distillation may be employed [258].

Sulfolane is an alternative to acetonitrile or nitromethane as a solvent for anodic reactions; in reductions it generally offers no advantages over solvents such as acetonitrile, DMF, or DMSO, but special uses may be found.

8. Propylene Glycol Sulfite

A series of other sulfur-containing solvents have been investigated for electrolytic use. Of these propylene glycol sulfite [262, 263], a derivative of 1, 3, 2-dioxthiol, was found most suitable. It has a dielectric constant of 33, forms solutions with good conductivity with salts as $LiClO_4$, and does not react with lithium. It has not yet been used for organic electrolysis.

9. Nitromethane

Nitromethane is one of the few solvents useful for anodic reactions, among them anodic coupling of aromatic hydrocarbons. Its application for cathodic reactions is limited, but in some cases where its unreactivity toward certain active halogen compounds is valuable it may be used. The liquid range of nitromethane is -29° to 101°; its dielectric constant is 37, but the dissociation of salts is not as high as could be expected.

Nitromethane dissolves rather few inorganic salts; $LiClO_4$, $Mg(ClO_4)_2$, and tetrabutylammonium salts can be used as supporting electrolyte. The Ag/Ag^+ or Ag/AgCl electrode can be employed as reference electrode [150].

At a platinum electrode potentials about 3 V (vs. Ag/AgCl) can be reached, and the limiting reaction is the discharge of the supporting electrolyte. The useful range in cathodic direction is very dependent on the water content.

Nitromethane has been purified by repeated washings with bicarbonate, sulfuric acid, and water, followed by drying and fractional distillation; the water content of the product is 5-8 mM. Alternatively fractional distillation at reduced pressure followed by repeated fractional freezing may be employed [264].

10. Nitrobenzene

Nitrobenzene has been used for electrolysis [265]; it was found that certain radicals were rather stable in this solvent. Nitrobenzene has a liquid range from 5.7 to 210.9°C; Bu_4NClO_4 may be used as supporting electrolyte. An aqueous SCE separated from the solution by a suitable bridge with porous glass has been used as reference electrode. Nitrobenzene may be purified by passing it through a column of alumina followed by a distillation in vacuo.

11. Propylene Carbonate (PC)

PC is of potential use in organic electrochemistry [307]. It has a high dielectric constant (69) and a wide liquid range (-49° to 242°), but it is more reactive than acetonitrile and the potential limit for anodic reactions is lower than for acetonitrile. It has so far been used mostly for electrochemical generation of cation radicals [266].

The cathodic limiting reaction of propylene carbonate containing $LiClO_4$ has been shown [267] to be a reduction to propylene and carbonate. This reductive elimination is analogous to the reduction of vicinal halogen compounds to alkenes.

As reference electrode a Li/LiCl electrode which, in the presence of excess of tetrabutylammonium chloride is an electrode of the second kind, may be used [268].

12. Ethers

Ethers such as tetrahydrofuran and 1, 2-dimethoxyethane have low dielectric constants (7.4, resp. 3.5) and the choice of supporting electrolyte is very limited. The ethers are difficult to reduce and are inert toward many metal organic reagents [182] which are soluble in the ethers. Ethers are thus suitable as medium for the anodic addition of Grignard reagents to olefins [269, 270]. Dimethoxyethane and polyethylene glycol dimethyl ethers have the ability to solvate electrons [271].

As supporting electrolyte $NaClO_4$ or $LiClO_4$ may be used in both solvents; TBAI in tetrahydrofuran and TBAP in dimethoxyethane have also been employed. As reference electrode the Ag/Ag^+ electrode functions satisfactorily [180, 182].

The ethers form peroxides. These can be removed by distillation from $LiAlH_4$. It must be stored under nitrogen or distilled from $LiAlH_4$ directly into the electrolytic cell.

13. Methylene Chloride

Electrolysis may be carried out in methylene chloride. Its dielectric constant is low (9), and only the larger tetraalkylammonium salts are soluble in methylene chloride.

Methylene chloride has been used both for anodic and cathodic reactions. The major advantages of using methylene chloride seem to be that certain cation radicals are more stable in methylene chloride than in the solvents usually employed in electrochemistry [272], and that methylene chloride is a solvent of low nucleophilicity [273]. A disadvantage is that chloride ions produced at the cathode may diffuse to the anode compartment and interfere with the anodic reactions. This can partially be avoided by addition of small amounts of acetic acid to the catholyte whereby the cathode reaction becomes an evolution of hydrogen rather than formation of chloride ions.

14. Sulfur Dioxide

Sulfur dioxide is a solvent of low nucleophilicity and may be used when nucleophilic trapping by solvent is to be avoided. Its liquid range is somewhat inconvenient (-75° to $-10^\circ C$), but its dielectric constant reasonably high (17.6 at $-20^\circ C$).

As supporting electrolyte Bu_4NClO_4 or Pr_4NPF_6 may be used; a 0.2 M solution of these salts has a conductivity of about $9 \times 10^{-3}\ \Omega^{-1} cm^{-1}$ at $-22^\circ C$. As reference electrode Ag/0.1 M $AgNO_3$ in acetonitrile has been used. The anodic range of perchlorates or hexafluorophosphates in sulfur dioxide is large [274].

D. Salts

In certain electrolytic media the concentration of salts is so high that the medium could be regarded as consisting of a solvated salt. This is thus the case in the strong solutions (10 M) of many salts in ammonia where the vapor pressure of ammonia at ambient temperature is less than 1 atm. It also applies to the strong solutions of tetraalkylammonium p-toluenesulfonates in water, used very often in organic electrochemistry.

A genuine liquid salt is tetrahexylammonium benzoate [183] (THAB), which is a viscous liquid capable of acting both as solvent and supporting electrolyte; it has, however, a rather high resistance. Although it is a fairly polar solvent THAB is miscible with benzene, toluene, and carbon tetrachloride; a mixture of THAB-toluene (75:25) has a rather low specific resistance of 3.8 kΩ and a much lower viscosity than pure THAB, which is about as viscous as glycerol. Relatively little water is miscible with THAB; addition of 1% by volume of water to tetrahexylammonium benzoate results in a two-phase system. THAB is soluble in acetone.

THAB is useful for reductions; the cathodic limit at a Hg electrode is about -2.6 V (vs. Ag/AgCl); at a platinum electrode the anodic limit is at about 0.3 V.

THAB has been prepared by neutralizing tetrahexylammonium hydroxide, obtained from the iodide by treatment with Ag_2O, with benzoic acid and evaporation of the water. The analysis of the dried product corresponds to a hemihydrate. The material is somewhat hygroscopic, but can be dried by brief heating to 90°C.

At higher temperature certain other quaternary ammonium salts (40-150°C) [275] or dimethylamine hydrochloride (170-200°C) [276] may be used as media for electrolysis.

VII. ELECTROLYTES

The passage of electric current through a nonmetallic liquid is dependent on ions. The choice of electrolyte depends on properties such as its solubility, dissociation constant, mobility, discharge potential, and protic activity. It is desirable to employ a salt with high solubility, completely dissociated, high mobility of the ions, and a numerically high discharge potential. The role of adsorption of ions at the electrode is discussed elsewhere.

A. Anions

The choice of anion is of most importance in anodic reactions. Unless the oxidized anion is desired in an indirect electrolytic oxidation (Chapter XXIV) a difficultly oxidizable anion such as perchlorate, tetrafluoborate, hexafluorophosphate, or nitrate, is chosen. In aqueous solution the choice of anion is less critical than in nonaqueous solvents. In cathodic reactions the choice of cation has highest priority and the anion is generally selected on the basis of solubility of the salt or the ability of the anion to assist in dissolving the substrate.

1. Perchlorate

The perchlorate ion has a high anodic discharge potential and only in few solvents, such as acetonitrile or nitromethane, is the perchlorate ion oxidized in preference to the solvent. When oxidized, the primarily formed perchlorate radical decomposes into oxygen and chlordioxide radical [191].

Caution must be exercised when handling perchlorates. Silver perchlorate is explosive; grinding of it in a mortar should be avoided. For use in reference electrodes it may be prepared by anodic dissolution of silver in a perchlorate medium [150]. Evaporation of solutions of organic perchlorates is discouraged but, if necessary, must occur in vacuo and at moderate temperatures. Even sodium perchlorate, which can be dried at 110°C without undue risk, may cause explosions when heated with an organic solvent.

Sodium perchlorate is soluble in many organic solvents. It is hygroscopic, but can be dried at 110°C. Drying of $NaClO_4$, H_2O often results in a sintered mass which must be broken and dried. By recrystallization of $NaClO_4$ at temperatures above 80°C the anhydrous salt can be obtained, which can be dried without difficulties. Lithium perchlorate is very hygroscopic. Tetraalkylammonium perchlorates are less hygroscopic and some of them may be recrystallized from water; they are useful for voltammetric studies in organic solvents.

2. Tetrafluoborate

The discharge potential of the tetrafluoborate ion is slightly more positive [277] than that of the perchlorate ion. Tetrafluoborates do not have a tendency to explode, and they are recommended instead of perchlorates for use in preparative work where the solvent has to be removed during the work-up.

Tetrabutylammonium tetrafluoborate is only slightly soluble in water, but is soluble in organic solvents. It is prepared by mixing solutions of $NaBF_4$ with a soluble tetrabutylammonium salt. Its solubility properties are convenient for its use in preparative electrolysis. During the work-up the organic solvent may be diluted with water whereby the salt precipitates and is recovered for future use.

3. Hexafluorophosphate

Hexafluorophosphates are slightly more resistant toward anodic oxidation than tetrafluoborates and perchlorates [277]; in a 0.1 M solution of these salts in acetonitrile the potentials (vs. Ag/0.01 M Ag^+ electrode) at a platinum disk electrode (0.12 cm^2) at 1 mA were for perchlorate, 2.48 V; for tetrafluoborate, 2.91 V; and for hexafluorophosphate, 3.02 V [277].

4. Nitrate

Nitrate ion is not as resistant to oxidation as perchlorate or tetrafluoborate; in nitromethane [278] solution it is oxidized anodically to NO_2^+ and oxygen, the NO_2^+ reacts either with the water always present in "dry" organic solvents to form nitric acid or with nitrate ion to form N_2O_5. Nitrates, to a higher degree than perchlorates [279], form ion pairs in some solvents. Tetrabutylammonium nitrate can be prepared conveniently by mixing an aqueous solution of $NaNO_3$ with the commercially available tetrabutylammonium hydrogen sulfate. The salt is then extracted into methylene chloride giving an almost quantitative yield of Bu_4NNO_3.

5. Aromatic Sulfonates

The use of aromatic sulfonates stems from their ability to assist the solution of organic substrates in an aqueous medium. This "hydrotropic" effect [198, 199, 280, 281] is accentuated if a quaternary ammonium sulfonate, rather than an alkali metal sulfonate, is used.

The aromatic sulfonates have been employed mostly for reductions, but they would be expected to be reasonably stable toward oxidation. To obtain a reasonable hydrotropic effect rather high concentrations are used; in the hydrodimerization reaction of acrylonitrile the medium may be regarded as either a solution of tetraalkylammonium p-toluenesulfonate in water or the salt containing some water.

6. Carboxylate Ion

In the Kolbe reaction and in anodic carboxylation reactions salts of carboxylic acids function both as substrate and electrolyte. Usually the presence of other anions will diminish the yield of the Kolbe reaction, whereas the presence of difficultly oxidizable anions like bicarbonate or perchlorate ions favors the related Hofer-Moest reaction.

7. Tetraethylaluminate, Tetraphenylborate

The sodium salts of these anions are soluble in tetrahydrofuran and have been found suitable as supporting electrolyte in this solvent [282, 283].

B. Cations

The discharge potential of a cation determines whether it is useful or not, so in practice only alkali and alkali earth metal ions together with ammonium and tetraalkylammonium ions are used. The nondischarge of a cation is a sine qua non, but in addition a choice of cation must take into consideration ion-pair formation, solvation and adsorption of particular species (e.g., see Ref. [304]).

1. Lithium Ions

Many lithium salts, such as lithium perchlorate and the halides, are soluble in nonaqueous solvents. The reduction potential of Li^+ depends on the electrode and the solvent. At a mercury cathode amalgam formation takes place whereas formation of lithium metal occurs at platinum in aprotic media. Lithium metal is less reactive than sodium, and in some solvents sodium will attack the solvent whereas lithium is unreactive.

2. Sodium Ions

In aqueous buffers sodium ion is a common cation, but in nonaqueous solvents practically only $NaClO_4$ is useful. Even $NaBF_4$ is not soluble in most aprotic solvents.

3. Tetraalkylammonium Ions

The most commonly used quaternary ammonium salts are tetrabutylammonium perchlorate (TBAP), the tetrafluoborate (TBAT), the halides (TBAC, TBAB, TBAI), and the corresponding tetraethylammonium salts, e.g., the perchlorate (TEAP), but also the tetramethyl or tetrapropylammonium salts have been employed; the former cannot undergo a base-promoted Hofmann elimination. The tetraalkylammonium ions are soluble in many nonaqueous media, and they may be extracted from an aqueous solution by means of chloroform or methylene chloride [284].

The discharge potential becomes slightly more negative with increasing size of the ion, but in practice there is not much gained by going beyond tetrabutylammonium ions [285]. The reduction route of these ions is discussed elsewhere (Chapter XXI).

In aqueous solution the tetraalkylammonium ions are adsorbed at the electrode where they form a layer with a low proton activity. The adsorption of the ions depends on the potential of the electrode, the electrode material, the composition of the medium, and the size of the cation.

Quaternary ammonium salts are readily accessible; some are commercially available, and their preparation presents no problems [284]. Reducible impurities may be removed by electrolysis at a sufficiently negative potential.

C. Buffers

Control of an electrolytic reaction often requires that the proton activity remain within acceptable limits during the electrolysis. For small-scale electrolytic preparations (less than about 10 g/liter of substrate) a sufficiently high initial concentration of buffer, acid, or base is adequate in aqueous solution; for large-scale electrolysis a controlled addition of protons during a reduction must be provided. This addition may be controlled by a pH-stat or coupled to the current integrator. In aprotic media a proton donor, electrophile, or a nucleophile may play a similar role as buffers in aqueous media.

1. Acids

Hydrochloric acid is a convenient acid for laboratory-scale electrolysis; it is easy to remove during the work-up by evaporation. The oxygen-containing acids, such as sulfuric or perchloric acid, must be neutralized before basic products can be extracted. Sulfuric acid is most conveniently neutralized by concentrated ammonia rather than by sodium hydroxide, whereby the precipitation of sodium sulfate is avoided.

Even in acetic acid perchloric acid is a strong acid. Anhydrous solutions are made by adding acetic anhydride equivalent to the water present in the acetic acid and introduced together with the perchloric acid. Excess of acetic anhydride should be avoided as this solution is a very strong acylating agent and it forms yellow condensation products on standing [286].

In aprotic media the addition of proton donors may be important. If the proton donor is too strong, protons may be reduced in preference to the substrate and a too-weak proton donor may not furnish a sufficient number of protons. Often a phenol is an acceptable compromise.

Extreme acidic conditions may be obtained in "superacids," e.g., SbF_5 or BF_3 in HF, SbF_5 in FSO_3H.

2. Bases

Reductions in aqueous alkaline solution do not present many problems; the choice of cation is discussed above. In nonaqueous solvents bases may be generated electrolytically [287] or formed by reaction of the solvent with an alkali metal. The conjugate base of the solvent is the strongest base obtainable in any medium.

During an oxidation in a nonaqueous solvent the removal of a proton or an attack of a nucleophile on the oxidized substrate may be a chemical step in an ECE reaction. To explore which reaction is operating one can, in parallel voltammetric experiments, use two compounds which are equal in base strength but differ widely in nucleophilicity: Such a pair is pyridine and 2, 6-dimethylpyridine.

3. Buffer Systems

The use of buffers in electrolysis to maintain a desired pH warrants some consideration. During a reduction protons are consumed at the electrode surface and unless they are replenished rapidly, the pH in the reaction layer will be considerably higher than in the bulk of the solution. A high concentration of a buffer system, which exchanges protons fast, is required to maintain a desired pH.

In aqueous solution phosphate, borate, and carboxylate buffers are applicable, but in mixed solvents the solubility of phosphate and borate buffers is low; in such mixtures carboxylate buffers may be used in slightly acid, and amine buffers in slightly alkaline medium. To avoid excessive buffer concentrations it is advisable to maintain pH in the bulk of the solution constant by means of a pH-stat.

VIII. VOLTAGE SOURCES AND COULOMETERS

A. Voltage Sources

To perform a simple electrolysis under constant conditions only some kind of dc source, which may be a battery, a rectifier, or a motor generator in combination with a rheostat for the wanted current range, is needed. Small ripples on the dc voltage usually do no harm in preparative work.

For experiments under controlled potential conditions the electrode voltage may be controlled manually following the readings of a dc voltmeter with high input resistance or automatically by a potentiostat. This is a device that is able to keep the potential of the working electrode within a small range around a fixed value (vs. the reference electrode) independent of the cell current. The apparatus will thus reduce the cell current as the electroactive substance is consumed and will also follow possible changes of the cell resistance.

Since Hickling's first description [288] many variants have been proposed [289-294], and excellent commercial models are now available from several firms. All the devices work on a feedback principle which allows the preset voltage to be maintained within tolerances and time spaces dependent on the particular model. Good reviews on the principles have been given by Booman and Holbrook [295] and by Shain and coworkers [296, 297]. Further diagrams and references can be found in the earlier-mentioned books [3, 6, 12, 13] and in a recent review [303].

B. Coulometers

Very often it is of great interest to determine the n value of an electrode reaction or the current efficiency of an electrosynthesis. Both involve a measurement of the electricity consumption, i.e., an integration of the electric current over the time of electrolysis. In constant current experiments this is of course an easy task. An obvious method is graphical integration of current-time measurements at suitable intervals. In practice it is more convenient to include an integrating device in the circuit.

Coulometers (or integrators) usually belong to one of three types: (a) electrochemical; (b) electromechanical; and (c) electronic, according to their working principle. Electrochemical coulometers may be based on weighing a metal deposit (usually silver or copper), titration of an electrogenerated species like iodine or hydroxide ion, or measuring the volume of an evolved gas mixture. The latter type seems to be the most convenient to work with in practice using either potassium sulfate [6] (hydrogen-oxygen) or hydrazine [298] (hydrogen-nitrogen) solutions. A good description is found in Lingane's book [6] where also some electromechanical integrators are discussed. In the authors' experience the best and simplest (cheapest) instrument of this type is a common household dc counter provided with a suitable shunt and calibrated for the wanted current range [39]. A drawback is that such dc instruments are hardly manufactured any more and that the accuracy is not good at very small currents (less than 20 mA). For the latter purpose a gas coulometer may be used (small-scale electrolyses only), but an electronic integrator with a selection of current ranges is more recommendable. One such instrument has been designed by Harrar and Behrin [293], another by Lamm and Samuelsson [74]; most of the potentiostat makers will also supply electronic integrators.

IX. PRACTICAL HINTS

Besides the major features in electrolysis discussed above there are smaller problems to be solved. It is, of course, easiest to consult an experienced electrochemist on such matters, but such a person might not always turn up when he is wanted, so the following items are briefly mentioned: circuit, removal of oxygen before electrolysis, removal of impurities, detection of intermediates, integration of current, and work-up.

A. Circuit

In Refs. [40] and [74] can be found circuits in which a potentiostat is included. For exploratory work a cheap, manually controlled circuit, as shown in Fig. 28, may be assembled from components available in every organic laboratory. Although many organic electrosyntheses may be performed without control of the potential by means of a reference electrode circuit it is advisable in research work always to include a potential-measuring device. A potentiostat facilitates the control of the electrolysis considerably. Many different potentiostats are now commercially available. When choosing a potentiostat, properties such as output voltage, power,

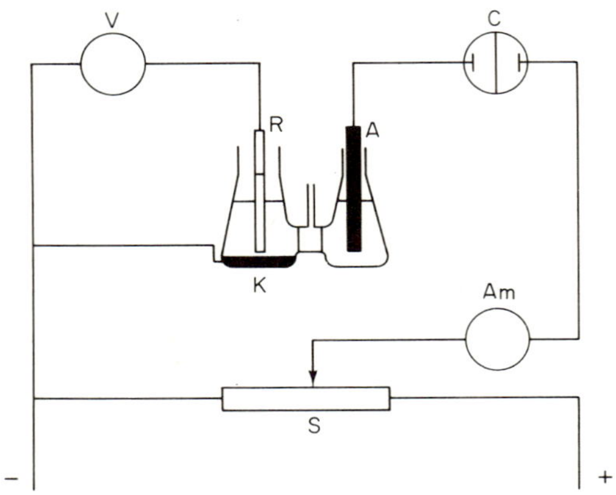

Fig. 28. Circuit for constant potential reduction: K, cathode; A, anode; R, reference electrode; V, potentiometer or pH meter; C, coulometer; Am, ammeter; S, voltage adjuster [38].

response time, and input resistance in reference circuit must be considered. Electrolysis in nonaqueous solvents generally requires a higher voltage across the cell than electrolysis in water. Potentiostats for ordinary preparative electrolysis do not need as fast a response as those intended for pulse electrolysis. The input resistance of the reference circuit must be high compared to that of the reference electrode.

In the circuit there should also be included a current integrator. A silver or gas coulometer may be used in small-scale work, but it is more practical to use an electronic or electromechanical integrator. Old-fashioned electromechanical, digital dc-wattmeters (previously used by the Public Works for measuring electricity consumption) provided with a suitable shunt for the desired current range have been employed in the authors' laboratory for many years.

B. Removal of Oxygen

Before electrolysis most of the dissolved oxygen may be removed by bubbling an inert gas through the solution. It is advisable, especially for small scale in nonaqueous solvents, to saturate the gas with the solvent before it is introduced into the cell. The inert gas, nitrogen or argon, may be obtained commercially sufficiently pure, but if not, nitrogen may be purified by passing it through a tube containing a (commercially available) copper catalyst. Previously, nitrogen has been purified by bubbling it through solutions of Cr(II) or V(II) salts or a Fieser solution [3].

Removal of traces of oxygen from a solution by bubbling an inert gas through it is virtually impossible; if such traces must be avoided, all manipulations must take place on a vacuum line and the cells constructed accordingly.

C. Removal of Impurities

To obtain a meaningful value of electricity consumption it is necessary that electricity not be consumed by impurities. A convenient way to remove impurities is to preelectrolyze the catholyte at a potential somewhat more negative than that to be employed during the electrolysis of the substrate. When the background current has reached an acceptable value, the substrate is added and the electrolysis performed at the required potential.

D. Detection of Intermediates

During the electrolysis it is advisable to obtain some indications of what is going on in the cell. An indicator microelectrode in the electrolytic cell is valuable in order to follow the progress of the electrolysis and to detect intermediates. An intermediate, e.g., more easily reducible than the substrate, may be detected as a small prewave, and its half-wave potential or other voltammetric characteristics used for its identification. Polarographic or cyclic voltammetric curves measured during the electrolysis also present a means of monitoring the process.

E. Determination of n-Values

A direct determination of n values can be made by microcoulometry with a dropping mercury electrode either with inclusion of a coulometer [299] or by comparison of polarographic diffusion currents with a known reaction [300]. A critical review of the methods has been given by Weaver and Whitnack [301]. However, n values derived from electrolyses with macroelectrodes should be considered with some care due to the possibility of further reactions during the time of the experiment. A reliable procedure [302] for compounds which can be dissolved fully is to follow the concentration of the electroactive substance polarographically, plot corresponding values of wave height and coulometer readings, and calculate the n value from the slope of the possibly straight line obtained during the first part of the electrolysis. If a straight line is not obtained, this indicates that secondary reactions take place.

F. Work-Up

Often, from the point of view of electrolysis, the choice of electrolyte or solvent is free within certain limits and the work-up may then be facilitated considerably by a suitable selection of experimental conditions. For example, the reduction of a nitroalkane [54] to a hydroxylamine requires an aqueous acid medium; sulfuric acid is just as good as hydrochloric acid for the electrolysis, but whereas hydrochloric acid may be removed in vacuo at a low temperature, leaving an alkylhydroxylamine hydrochloride, the sulfuric acid must be neutralized and the free, less stable base extracted.

NOTES

[1]Also denoted current efficiency.

[2]For a recent review see Ref. [93a].

[3]In redox systems analogous to buffering in acid-base systems.

REFERENCES

[1]. L. Eberson and S. Nilsson, Discuss. Faraday Soc., 45, 242 (1968).

[2]. A. Dunet, J. Rollet, and A. Willemart, Bull. Soc. Chim. Fr. [5], 17, 877 (1950).

[3]. I. M. Kolthoff and J. J. Lingane, Polarography, Wiley-Interscience, 2nd ed., New York, 1952.

[4]. J. Heyrovsky and J. Kuta, Grundlagen der Polarographie, Akademie Verlag, Berlin, 1965.

[5]. (a) I. M. Kolthoff and Y. Okinaka, (b) Z. P. Zagorski, in Progress in Polarography (P. Zuman and I. M. Kolthoff, eds.), Vol. II, Wiley-Interscience, New York, 1962, pp. 357-381, 549-568.

[6]. J. J. Lingane, Electroanalytical Chemistry, Wiley-Interscience, 2nd ed., New York, 1958.

[7]. R. B. MacMullin, Electrochem. Tech., 1, 5 (1963).

[8]. R. B. MacMullin, ibid., 2, 106 (1964).

[9]. C. Wagner, in Advances in Electrochemistry and Electrochemical Engineering (C. W. Tobias, ed.), Vol. 2, Wiley-Interscience, New York, 1962, pp. 1-14.

[10]. M. Eisenberg, in Advances in Electrochemistry and Electrochemical Engineering (C. W. Tobias, ed.), Vol. 2, Wiley-Interscience, New York, 1962, pp. 235-291.

[11]. J. Heyrovsky, in Polarographie, Springer, Wien, 1941, p. 9.

[12]. M. J. Allen, Organic Electrode Processes, Chapman & Hall, London, 1958.

[13]. A. P. Tomilov, S. G. Mairanovsky, M. Ya. Fioshin, and V. A. Smirnov, Elektrokhimiya Organischeskikh Soedinyenii, Khimiya, Leningrad, 1968.

[14]. S. Swann, Jr., in Technique of Organic Chemistry (A. Weissberger, ed.), Vol. II, Wiley-Interscience, New York, 1956, pp. 385-523.

[15]. N. Clauson-Kaas, F. Limborg, and K. Glens, Acta Chem. Scand., 6, 531 (1952).

[16]. F. Limborg and N. Clauson-Kaas, ibid., 7, 234 (1953).

[17]. W. J. Koehl, Jr., J. Amer. Chem. Soc., 86, 4686 (1964).

[18]. G. Dryhurst and P. J. Elving, J. Electrochem. Soc., 115, 1014 (1968).

[19]. T. Timell, Sv. Kem. Tidsskr., 58, 67 (1946).

[20]. R. P. Linstead, J. C. Lunt, and B. C. L. Weedon, J. Chem. Soc., 1950, 3331.

[21]. T. Dickinson and W. F. K. Wynne-Jones, Trans. Faraday Soc., 58, 400 (1962).

[22]. M. Fleischmann, J. R. Mansfield, and Lord Wynne-Jones, J. Electroanal. Chem., 10, 522 (1965).

[23]. M. Fleischmann, J. R. Mansfield, H. R. Thirsk, H. G. E. Wilson, and Lord Wynne-Jones, Electrochim. Acta, 12, 967 (1967).

[24]. A. Hickling and R. Wilkins, Discuss. Faraday Soc., 45, 261 (1968).

[25]. S. D. Ross, M. Finkelstein, and R. C. Petersen, J. Amer. Chem. Soc., 89, 4088 (1967).

[26]. L. Eberson and K. Nyberg, ibid., 88, 1686 (1966).

[27]. H. W. Sternberg, R. Markby, and I. Wender, J. Electrochem. Soc., 110, 425 (1963).

[28]. T. Inoue and S. Tsutsumi, Bull. Chem. Soc. Japan, 38, 2122 (1965).

[29]. F. I. Kukos and L. I. Antropov, Zh. Fis. Khim., 32, 2294 (1958).

[30]. T. Asahara, M. Seno, and T. Arai, Bull. Chem. Soc. Japan, 42, 1316 (1969).

[31]. S. Andreades and E. W. Zahnow, J. Amer. Chem. Soc., 91, 4181 (1969).

[32]. A. F. Clifford, H. K. El-Shamy, H. J. Emeléus, and R. N. Hazeldine, J. Chem. Soc., 1953, 2372.

[33]. T. Gramstad and R. N. Hazeldine, ibid., 1956, 173.

[34]. M. Schmeisser and P. Sartori, Chem.-Ing.-Tech., 36, 9 (1964).

[35]. B. L. Funt, D. Richardson, and S. N. Bhadani, Can. J. Chem., 44, 711 (1966).

[36]. J. J. Lingane, C. Gardner Swain, and M. Fields, J. Amer. Chem. Soc., 65, 1348 (1943).

[37]. R. Pasternak, Helv. Chim. Acta, 31, 753 (1948).

[38]. H. Lund, Österreich. Chem.-Z., 68, 43 (1967).

[39]. P. E. Iversen, J. Chem. Educ., 48, 136 (1971).

[40]. S. Musha, T. Wasa, and T. Naito, Bull. Chem. Soc. Japan, 39, 1902 (1966).

[41]. D. Peltier, M. Le Guyader, and J. Tacussel, Bull. Soc. Chim. Fr., 1963, 2609.

[42]. M. R. Rifi, J. Amer. Chem. Soc., 89, 4442 (1967).

[43]. H. F. Metcalf and G. C. Whitnack, U.S. Pat. 2,835,631 (1958).

[44]. R. C. Schulz and W. Strobel, Monatsh. Chem., 99, 1742 (1968).

[45]. L. D. McKeever and R. Waack, J. Organometal. Chem., 17, 142 (1969).

[46]. K. Sugino, T. Sekine, and N. Sato, Electrochem. Tech., 1, 112 (1963).

[47]. J. E. Dubois, A. Monvernay, and P. C. Lacaze, Electrochim. Acta, 15, 315 (1970).

[48]. J. E. Harrar and I. Shain, Anal. Chem., 38, 1148 (1966).

[49]. D. Peltier and C. Moinet, Bull. Soc. Chim. Fr., 1968, 2657.

[50]. M. M. Baizer, J. P. Petrovich, and D. A. Tyssee, J. Electrochem. Soc., 117, 173 (1970).

[51]. C. Moinet and D. Peltier, Bull. Soc. Chim. Fr., 1969, 690.

[52]. A. Cisak and P. J. Elving, J. Electrochem. Soc., 110, 160 (1963).

[53]. A. Cisak, Rocz. Chem., 41, 809 (1967).

[54]. P. E. Iversen and H. Lund, Acta Chem. Scand., 19, 2303 (1965).

[55]. L. Horner and A. Mentrup, Lieb. Ann., 646, 49 (1961).

[56]. L. V. Kaabak, A. P. Tomilov, S. L. Varshavskii, and M. I. Kabachnik, Zh. Org. Khim., 3, 3 (1967).

[57]. A. P. Tomilov and B. L. Klyouyev, Elektrokhim., 2, 1405 (1966).

[58]. A. P. Tomilov and V. A. Klimov, ibid., 3, 232 (1967).

[59]. G. W. McMillan, U.S. Pat. 2,485,982 (1949).

[60]. N. A. Izgaryshev and A. A. Petrova, Zh. Fiz. Khim., 24, 745 (1950).

[61]. P. C. Condit, Ind. Eng. Chem., 48, 1252 (1956).

[62]. C. Barry and G. Cauquis, Bull. Soc. Chim. Fr., 1966, 1032.

[63]. G. Cauquis and B. Hammerlé, ibid., 1970, 2000.

[64]. H. Laato and P. Nissi, Suom. Kem., B39, 155 (1966).

[65]. M. W. Leeds and G. B. L. Smith, J. Electrochem. Soc., 98, 129 (1951).

[66]. R. Kanakam, M. S. V. Pathy, and H. V. K. Udupa, Electrochim. Acta, 12, 329 (1967).

[67]. T. D. Balakrishnan, K. S. Udupa, G. S. Subramanian, and H. V. K. Udupa, Chem.-Ing.-Tech., 41, 776 (1969).

[68]. K. Natarajan, K. S. Udupa, G. S. Subramanian, and H. V. K. Udupa, Electrochem. Tech., 2, 151 (1964).

[69]. K. S. Udupa, G. S. Subramanian, and H. V. K. Udupa, J. Electrochem. Soc., 108, 373 (1961).

[70]. R. Ramaswamy, M. S. Venkatachalapathy, and H. V. K. Udupa, ibid., 110, 294 (1963).

[71]. P. N. Anantaraman, G. S. Subramanian, and H. V. K. Udupa, Ind. J. Tech., 4, 271 (1966).

[72]. G. T. Miller, U.S. Pat. 3,361,656 (1958); CA, 68 (1968), 83743, presented for the Detroit Section of E.C.S., November 21, 1968; N. M. Winslow, Trans. Electrochem. Soc., 88, 81 (1945).

[73]. M. S. V. Pathy, Electrochem. Tech., 3, 94 (1965).

[74]. B. Lamm and B. Samuelsson, Acta Chem. Scand., 24, 561 (1970).

[75]. F. Beck and H. Guthke, Chem.-Ing.-Tech., 41, 943 (1969).

[76]. M. J. Allen, J. Electrochem. Soc., 109, 731 (1962).

[77]. J. R. Backhurst, J. M. Coulson, F. Goodridge, R. E. Plimley, and M. Fleischmann, ibid., 116, 1600 (1969).

[78]. P. LeGoff, F. Vergnes, F. Coeuret, and J. Bordet, Ind. Eng. Chem., 61, 8 (1969).

[79]. J. N. Hiddleston and A. F. Douglas, Electrochim. Acta, 15, 431 (1970).

[79a]. M. Fleischmann and J. W. Oldfield, J. Electroanal. Chem., 29, 211, 231 (1971).

[80]. G. B. L. Smith and M. W. Leeds, U.S. Pat. 2,589,635 (1952).

[81]. X. de Hemptinne and K. Schunck, Ann. Soc. Sci. Bruxelles, 80, 289 (1966).

[82]. M. M. Baizer, J. Electrochem. Soc., 111, 215 (1964).

[83]. F. N. Ruehlen, G. B. Wills, and H. M. Fox, ibid., 111, 1107 (1964).

[84]. F. Beck and H. Leitner, Angew. Makromol. Chem., 2, 52 (1968).

[85]. W. R. T. Cottrell and J. Farrar, J. Chem. Soc. A, 1970, 1418.

[86]. E. V. Kasatkin, A. A. Rakev, and V. I. Veselovskii, Elektrokhim., 3, 1034 (1967).

[87]. H.-J. Schmidt and H. Nees, Ger. Pat. 1,078,134 (1960).

[88]. C. M. Wrigth and D. R. Levering, Tetrahedron, 19, Suppl. 1, 3 (1963).

[89]. L. Nanis and F. R. McLarnon, J. Electrochem. Soc., 117, 1527 (1970).

[90]. M. J. Allen, Electrochem. Tech., 1, 315 (1963).

[91]. J. H. Strohl and T. A. Polutanovich, Anal. Lett., 2, 423 (1969).

[92]. J. E. McClure and D. L. Maricle, Anal. Chem., 39, 236 (1967).

[93]. J. C. Sheaffer and D. G. Peters, ibid., 42, 430 (1970).

[93a]. A. T. Hubbard and F. C. Anson, Electroanal. Chem., 1970, 129.

[94]. R. E. Sioda, Electrochim. Acta, 13, 375 (1968).

[95]. R. E. Sioda, ibid., 15, 783 (1970).

[96]. J. W. Strojek and T. Kuwana, J. Electroanal. Chem., 16, 471 (1968).

[97]. W. R. Heineman, J. N. Burnett, and R. W. Murray, Anal. Chem., 40, 1974 (1968).

[98]. R. W. Murray, W. R. Heinemann, and G. W. O'Dom, ibid., 39, 1666 (1967).

[99]. A. Yildiz, P. T. Kissinger, and C. N. Reilley, ibid., 40, 1018 (1968).

[100]. D. R. Tallant and D. H. Evans, ibid., 41, 835 (1969).

[101]. D. J. Fisher, W. L. Belew, and M. T. Kelley, in Polarography 1964 (G. J. Hills, ed.), Vol. 2, MacMillan, London, 1966, p. 1043.

[102]. J. E. McClure, Anal. Chem., 42, 551 (1970).

[103]. S. Takagi and T. Ueda, J. Pharm. Soc. Japan, 58, 44 (1938).

[104]. P. S. Chechel and L. I. Antropov, Zh. Prikl. Khim., 31, 1856 (1958).

[105]. S. Lynn and H. H. Paalman, J. Electrochem. Soc., 116, 1758 (1969).

[106]. T. Asahara, M. Seno, and H. Kaneko, Bull. Chem. Soc. Japan, 41, 2985 (1968).

[107]. R. A. Benkeser, E. M. Kaiser, and R. F. Lambert, J. Amer. Chem. Soc., 86, 5272 (1964).

[108]. J. Hranilovic, D. Koruncev, and E. Gustak, Electrochem. Tech., 6, 62 (1968).

[109]. J. L Weiniger and F. F. Holub, J. Electrochem. Soc., 117, 340 (1970).

[110]. T. D. Balakrishnan, K. S. Udupa, G. S. Subramanian, and H. V. K. Udupa, Chem. Ind. (London), 1970, 1622.

[111]. N. Urabe, T. Seyama, and W. Sakai, J. Electrochem. Soc. Japan (Overseas Suppl. Ed.), 28, E-69 (1960).

[112]. T. Erdey-Gruz and M. Volmer, Z. Phys. Chem., 150, 203 (1930).

[113]. L. G. Feoktistov and M. M. Gol'din, Elektrokhim., 4, 356 (1968).

[114]. J. Heyrovsky, Rec. Trav. Chim., 44, 499 (1925).

[115]. J. Horiuti and G. Okamoto, Sci. Pap. Inst. Phys. Chem. Res. (Tokyo), 28, 231 (1936); CA, 30, 3332 (1936).

[116]. K. J. Vetter and D. Otto, Z. Elektrochem., 60, 1072 (1956).

[117]. X. de Hemptinne and J. C. Jungers, Z. Phys. Chem., 15, 137 (1958).

[118]. F. Fichter and R. Stocker, Ber., 47, 2003 (1914).

[119]. W. D. Bancroft and A. B. George, Trans. Amer. Electrochem. Soc., 57, 399 (1930).

[120]. X. de Hemptinne and K. Schunk, Ann. Soc. Sci. Bruxelles, Ser. I, 80, 289 (1966); CA, 66, 91139 (1967).

[121]. X. de Hemptinne and K. Schunck, Trans. Faraday Soc., 65, 591 (1969).

[122]. C. Schall and W. Kirst, Z. Elektrochem., 29, 537 (1923).

[123]. S. D. Ross, M. Finkelstein, and R. C. Petersen, J. Amer. Chem. Soc., 92, 6003 (1970).

[124]. V. G. Khomyakov, A. P. Tomilov, and B. G. Soldatov, Elektrokhim., 5, 850 (1969).

[125]. I. A. Avrutskaya, M. Ya. Fioshin, S. D. Kabanova, and L. E. Gerasimova, ibid., 5, 951 (1969).

[126]. A. Chilesotti, Z. Elektrochem., 7, 768 (1900).

[127]. F. S. Holland, CITCE Meeting, Prague, 1970, Extended Abstract, p. 414.

[128]. N. A. Izgaryshev and M. Ya. Fioshin, Zh. Obshch. Khim., 24, 766 (1954).

[129]. J. Tafel, Ber., 33, 2209 (1900).

[130]. I. M. Kolthoff and N. Tanaka, Anal. Chem., 26, 632 (1954).

[131]. J. W. Ross and I. Shain, ibid., 28, 548 (1956).

[132]. Ger. Pat. 865,740 (1953); CA, 53, 292 (1959).

[133]. Brit. Pat. 766,578 (1957); CA, 51, 10278 (1957).

[134]. H. Gerischer, in Advances in Electrochemistry and Electrochemical Engineering (P. Delahay, ed.), Vol. 1, Wiley-Interscience, New York, 1961, p. 139.

[135]. H. Gerischer, Chim., 22, 65 (1968).

[136]. D. J. G. Ives and G. J. Janz (eds.), Reference Electrodes, Academic, New York, 1961.

[137]. J. N. Butler, in Advances in Electrochemistry and Electrochemical Engineering (P. Delahay and C. W. Tobias, eds.), Vol. 7, Wiley, 1970.

[138]. V. A. Pleskov, Usp. Khim., 16, 254 (1947).

[139]. H.-M. Koepp, H. Wendt, and H. Strelow, Z. Electrochem., 64, 483 (1960).

[140]. H. Schneider and H. Strelow, J. Electroanal. Chem., 12, 530 (1966).

[141]. L. V. Nelson and R. T. Iwamoto, Anal. Chem., 33, 1795 (1961).

[142]. I. M. Kolthoff and F. G. Thomas, J. Phys. Chem., 69, 3049 (1965).

[143]. O. Popovych, Anal. Chem., 38, 558 (1966).

[144]. J. F. Coetzee and J. J. Campion, J. Amer. Chem. Soc., 89, 2513, 2517 (1967).

[145]. J. Brenet, J. Chim. Phys., 66, 1057 (1969).

[146]. W. H. Tiedemann and D. N. Bennion, J. Electrochem. Soc., 117, 203 (1970).

[147]. V. A. Pleskov, Zh. Fiz. Khim., 22, 351 (1948).

[148]. H. Lund, Acta Chem. Scand., 11, 491, 1323 (1957).

[149]. P. Walden and E. J. Birr, Z. Phys. Chem., 144A, 269 (1929).

[150]. G. Cauquis and D. Serve, Bull. Soc. Chim. Fr., 302 (1966).

[151]. H. Yeager and B. Kratochvil, J. Phys. Chem., 73, 1963 (1969).

[152]. B. Kratochvil, E. Lorah, and C. Garber, Anal. Chem., 41, 1793 (1969).

[153]. D. L. McMasters, R. B. Dunlap, J. R. Kuempel, L. W. Kreider, and T. R. Shearer, ibid., 39, 103 (1967).

[154]. J. Sedlet and T. DeVries, J. Amer. Chem. Soc., 73, 5808 (1951).

[155]. D. R. Cogley and J. N. Butler, J. Electrochem. Soc., 113, 1074 (1966).

[156]. R. N. Hammer and J. J. Lagowski, Anal. Chem., 34, 597 (1962).

[157]. J. F. Coetzee and J. R. Padmanabhan, J. Phys. Chem., 66, 1708 (1962).

[158]. H. W. Salzberg and M. Leung, J. Org. Chem., 30, 2873 (1965).

[159]. J. C. Synnott and J. N. Butler, Anal. Chem., 41, 1890 (1969).

[160]. A. I. Popov and D. H. Geske, J. Amer. Chem. Soc., 79, 2074 (1957).

[161]. W. H. Smyrl and C. W. Tobias, J. Electrochem. Soc., 113, 754 (1966).

[162]. V. A. Pleskov and A. M. Monosson, Acta Physicochim. USSR, 3, 615 (1935).

[163]. L. W. Marple, Anal. Chem., 39, 844 (1967).

[164]. J. B. Conant, L. F. Small, and B. S. Taylor, J. Amer. Chem. Soc., 47, 1959 (1925).

[165]. H. A. Laitinen and C. J. Nyman, ibid., 70, 2241 (1948).

[166]. R. C. Larson, R. T. Iwamoto, and R. N. Adams, Anal. Chim. Acta, 25, 371 (1961).

[167]. J. Sedlet and T. DeVries, J. Amer. Chem. Soc., 73, 5808 (1951).

[168]. H. W. Sternberg, R. E. Markby, and I. Wender, J. Electrochem. Soc., 110, 425 (1958).

[169]. U. Bertocci, Z. Electrochem., 61, 434 (1957).

[170]. A. Čisak and P. J. Elving, J. Electrochem. Soc., 110, 160 (1963).

[171]. M. C. Giordano, J. C. Barzan, and A. J. Arvia, Electrochim. Acta, 11, 741 (1966).

[172]. J. Čihalik and J. Šimek, Coll. Czech. Chem. Commun., 23, 615 (1958).

[173]. J. O'M. Bockris, Discuss. Faraday Soc., 1, 95 (1947).

[174]. W. B. Mather and F. C. Anson, Anal. Chem., 33, 1634 (1961).

[175]. V. Plichon, Bull. Soc. Chim. Fr., 262 (1964).

[176]. G. Adhami and M. Herlem, J. Electroanal. Chem., 26, 363 (1970); J. Bertram, M. Fleischmann, and D. Pletcher, Tetrahedron Lett., 349 (1971)

[177]. J. G. James, Trans. Faraday Soc., 47, 1240 (1951); I. Bergman and J. C. James, ibid., 50, 60 (1954).

[178]. B. E. Conway, N. Marincic, D. Gilroy, and E. Rudd, J. Electrochem. Soc., 113, 1144 (1966).

[179]. J. Badoz-Lambling and M. Sato, Acta Chim. Hung., 32, 191 (1962).

[180]. J. Perichon and R. Buvet, Electrochim. Acta, 9, 567 (1964).

[181]. A. K. Hoffmann, W. G. Hodgson, D. L. Maricle, and W. H. Jura, J. Amer. Chem. Soc., 86, 631 (1964).

[182]. R. E. Dessy, W. Kitching, and T. Chivers, ibid., 88, 453 (1966).

[183]. C. G. Swain, A. Ohno, D. K. Roe, R. Brown, and T. Maugh, ibid., 89, 2648 (1967).

[184]. G. Schober and V. Gutmann, Monatsh., 89, 401, 649 (1958).

[185]. G. J. Hills, in Ref. [136], p. 433.

[186]. A. K. Gupta, J. Chem. Soc., 3473 (1952).

[187]. I. M. Kolthoff and T. B. Reddy, Inorg. Chem., 1, 189 (1962).

[188]. J. Perichon and R. Buvet, Bull. Soc. Chim. Fr., 3697 (1967).

[189]. P. T. Cottrell and C. K. Mann, J. Electrochem. Soc., 116, 1499 (1969).

[190]. R. R. Rao, S. B. Milliken, S. L. Robinson, and C. K. Mann, Anal. Chem., 42, 1076 (1970).

[191]. G. Cauquis and D. Serve, J. Electroanal. Chem., 27, App. 3 (1970).

[192]. R. J. Jasinski and S. Kirkland, Anal. Chem., 39, 1663 (1967).

[193]. R. J. Jasinski and S. Carroll, ibid., 40, 1908 (1968).

[194]. C. D. Thompson, F. D. Bogar, and R. T. Foley, ibid., 42, 1474 (1970).

[195]. W. Rogers and S. M. Kipnes, ibid., 27, 1916 (1955).

[196]. G. B. Bachman and M. J. Astle, J. Amer. Chem. Soc., 64, 1303 (1942).

[197]. Ref. [201], p. 123.

[198]. R. H. McKee and C. J. Brockmann, Trans. Electrochem. Soc., 62, 203 (1932).

[199]. R. H. McKee and B. G. Gerastopolou, ibid., 68, 329 (1935).

[200]. J. M. Sullivan, D. C. Hanson, and R. Killer, J. Electrochem. Soc., 117, 779 (1970).

[201]. C. K. Mann, "Nonaqueous Solvents for Electrochemical Use," in Electroanalytical Chemistry (A. J. Bard, ed.), Dekker, New York, 1969, p. 57 ff.

[202]. C. K. Mann and K. K. Barnes, Electrochemical Reactions in Nonaqueous Systems, Dekker, New York, 1970.

[203]. J. J. Lagowski (ed.), The Chemistry of Nonaqueous Solvents, I, II, Academic, New York, 1966, 1967.

[204]. S. D. Ross, M. Finkelstein, and R. C. Petersen, J. Amer. Chem. Soc., 86, 4139 (1964).

[205]. M. Leung, J. Herz, and H. W. Salzberg, J. Org. Chem., 30, 310 (1965).

[206]. L. Eberson and K. Nyberg, J. Amer. Chem. Soc., 88, 1686 (1966).

[207]. J. Čihalik and J. Šimek, Coll. Czech. Chem. Commun., 23, 615, 1037 (1958).

[208]. I. Bergman and J. C. James, Trans. Faraday Soc., 48, 956 (1952).

[209]. K. J. P. Orton and A. E. Bradfield, J. Chem. Soc., 983 (1927).

[210]. J. A. Knopp, W. S. Linnell, and W. C. Child, J. Phys. Chem., 66, 1513 (1962).

[211]. J. H. Simons, J. Electrochem. Soc., 95, 47 (1949).

[212]. M. Schmeisser and P. Sartori, Chem.-Ing.-Tech., 36, 9 (1964).

[213]. S. Nagase, Fluorine Chem. Rev., 1, 77 (1967).

[214]. M. Kilpatrick and J. G. Jones, in The Chemistry of Nonaqueous Solvents (J. J. Lagowski, ed.), Vol. II, Academic, New York, 1967, p. 43 ff.

[215]. H. P. Raaen, Anal. Chem., 34, 1714 (1962).

[216]. M. E. Runner, G. Baloy, and M. Kilpatrick, J. Amer. Chem. Soc., 85, 1038 (1963).

[217]. J. H. Simons, Chem. Rev., 8, 213 (1931).

[218]. G. I. Kaurova, L. M. Grubina, and Ts. A. Adzhemyan, Elektrokhim., 3, 1222 (1967).

[219]. J. W. Sargent, A. F. Clifford, and W. R. Lemmon, Anal. Chem., 25, 1727 (1953).

[220]. S. Wawzonek, R. Berkey, and D. Thomson, J. Electrochem. Soc., 103, 513 (1956).

[221]. G. Petit and J. Bessière, J. Electroanal. Chem., 25, 317 (1970); 31, 393 (1971); 34, 489 (1972).

[223]. H. A. Laitinen and C. E. Shoemaker, J. Amer. Chem. Soc., 72, 663 (1950).

[224]. A. F. Clifford and M. Giménez-Huguet, ibid., 82, 1024 (1960).

[225]. W. B. Schaab, R. F. Conley, and F. C. Schmidt, Anal. Chem., 33, 498 (1961).

[226]. M. H. Miles and P. M. Kellett, J. Electrochem. Soc., 115, 1225 (1968).

[227]. M. H. Miles and P. M. Kellett, ibid., 117, 60 (1970).

[228]. W. H. Tiedemann and D. N. Bennion, ibid., 117, 203 (1970).

[229]. H. Schmidt and H. Meinert, Z. Anorg. Allgem. Chem., 295, 156 (1958).

[230]. R. A. Benkeser and E. M. Kaiser, J. Amer. Chem. Soc., 85, 2858 (1963).

[231]. R. A. Benkeser and S. J. Mels, J. Org. Chem., 34, 3970 (1969).

[232]. H. W. Sternberg, R. E. Markby, and I. Wender, J. Electrochem. Soc., 110, 425 (1963).

[233]. H. W. Sternberg, R. E. Markby, I. Wender, and D. M. Mohilner, ibid., 113, 1060 (1966).

[234]. W. B. Schaab, R. E. Bayer, J. R. Siefker, J.-Y. Kim, P. W. Brewster, and F. C. Smith, Rec. Chem. Progr., 22, 197 (1961).

[235]. L. M. Mukherjee and S. Bruckenstein, Pure Appl. Chem., 13, 421 (1966).

[236]. E. O. Sherman and D. C. Olson, Anal. Chem., 40, 1174 (1968).

[237]. J. F. Coetzee, Pure Appl. Chem., 13, 429 (1966).

[238]. J. F. Coetzee, G. P. Cunningham, D. K. McGuire, and G. R. Padmanabhan, Anal. Chem., 34, 1139 (1962).

[239]. G. A. Forcier and J. W. Olver, ibid., 37, 1447 (1965).

[240]. J. F. O'Donnell, J. T. Ayres, and C. K. Mann, ibid., 37, 1161 (1965).

[241]. J. W. Teter and W. J. Merwin, U.S. Pat. 2,453,472 (1948); CA, 43, 2630 (1949).

[242]. J. P. Billon, J. Electroanal. Chem., 1, 486 (1960).

[243]. N. S. Moe, Acta Chem. Scand., 21, 1389 (1967).

[244]. R. E. Visco, Meeting, The Electrochemical Society, Dallas, Texas, May 1967.

[245]. S. B. Brummer, J. Chem. Phys., 42, 1636 (1965).

[246]. M. Breant, M. Bazoin, C. Buisson, M. Dupin, and J.-M. Rebattu, Bull. Soc. Chim. Fr., 5065 (1968).

[247]. M. Breant and C. Buisson, J. Electroanal. Chem., 24, 145 (1970).

[247a]. M. Breant, Bull. Soc. Chim. Fr., 1971, 725.

[248]. H. W. Sternberg, R. E. Markby, I. Wender, and D. M. Mohilner, J. Amer. Chem. Soc., 89, 186 (1967); 91, 4191 (1969).

[249]. J.-Y. Gal and T. Yvernault, Bull. Soc. Chim. Fr., 2270 (1971); 839 (1972).

[250]. J.-E. Dubois and G. Dodin, Tetrahedron Lett., 2325 (1969).

[251]. H. Normant, Angew. Chem., 79, 1029 (1967).

[252]. J.-E. Dubois, P.-C. Lacaze, and A. M. de Ficquelmont, Compt. Rend., C262, 181, 249 (1966).

[253]. W. R. Turner and P. J. Elving, Anal. Chem., 37, 467 (1965).

[254]. J. N. Butler, J. Electroanal. Chem., 14, 89 (1967).

[255]. I. M. Kolthoff and T. B. Reddy, Inorg. Chem., 1, 189 (1962).

[256]. I. M. Kolthoff and T. B. Reddy, J. Electrochem. Soc., 108, 980 (1961).

[257]. R. Philippe and J.-C. Merlin, Bull. Soc. Chim. Fr., 4713 (1968).

[258]. J. Desbarres, P. Picket, and R. L. Benoit, Electrochim. Acta, 13, 1899 (1968).

[259]. J. F. Coetzee, J. M. Simon, and R. J. Bertozzi, Anal. Chem., 41, 766 (1969).

[260]. J. B. Headridge, D. Pletcher, and M. Callingham, J. Chem. Soc. A., 684 (1967).

[261]. J. Martinmaa, Suom. Kem. B, 42, 33 (1969).

[262]. R. E. Johnson, NASA Tech. Memo., NASA TMX-1283 (1966).

[263]. R. E. Johnson, NASA Tech. Memo., NASA TMX-1380 (1966).

[264]. A. K. R. Unni, L. Ellias, and H. I. Schiff, J. Phys. Chem., 67, 1216 (1963).

[265]. L. S. Marcoux, J. M. Fritsch, and R. N. Adams, J. Amer. Chem. Soc., 89, 5768 (1967).

[266]. R. F. Nelson and R. N. Adams, J. Electroanal. Chem., 13, 184 (1967).

[267]. A. N. Dey and B. P. Sullivan, Electrochemical Society Meeting, May 1970, Abstract No. 326.

[268]. A. Caiola, G. Guy, and J. C. Sohm, Electrochim. Acta, 15, 555 (1970).

[269]. H. Schäfer, Chem.-Ing.-Tech., 42, 164 (1970).

[270]. H. Schäfer and H. Küntzel, Tetrahedron Lett., 3333 (1970).

[271]. T. Osa, T. Yamagishi, T. Kodama, and A. Misono, "The Synthetic and Mechanistic Aspects of Electroorganic Chemistry," Symposium, Durham, N.C., October 1968, Preprints of papers, p. 157.

[272]. J. Phelps, K. S. V. Santhanam, and A. J. Bard, J. Amer. Chem. Soc., 89, 1752 (1967).

[273]. K. Nyberg, Acta Chem. Scand., 24, 1609 (1970).

[274]. L. L. Miller and E. A. Mayeda, J. Amer. Chem. Soc., 92, 5818 (1970).

[275]. J. E. Gordon, ibid., 87, 4347 (1965).

[276]. A. Kisza, Z. Phys. Chem. (Leipzig), 237, 97 (1968).

[277]. M. Fleischmann and D. Pletcher, Tetrahedron Lett., 6255 (1968).

[278]. G. Cauquis and D. Serve, Compt. Rend. Acad. Sci. Paris, 262, 1516 (1966).

[279]. H. Yeager and B. Kratochvil, J. Phys. Chem., 73, 1963 (1969).

[280]. R. M. McKee, Ind. Eng. Chem., 38, 382 (1946).

[281]. M. M. Baizer, J. Electrochem. Soc., 111, 215 (1964).

[282]. B. L. Funt, D. Richardson, and S. N. Bhadani, Can. J. Chem., 44, 711 (1966).

[283]. N. Yamazaki, S. Nakahama, and S. Kambara, Polym. Lett., 3, 57 (1965).

[284]. A. Brändström and K. Gustavii, Acta Chem. Scand., 23, 1215 (1969); A. Brändström, P. Berntsson, S. Carlsson, A. Djurhuus, K. Gustavii, U. Junggren, B. Lamm, and B. Samuelsson, ibid., 23, 2202 (1969).

[285]. H. O. House, E. Feng, and N. P. Peet, J. Org. Chem., 36, 2371 (1971).

[286]. W. B. Mather and F. C. Anson, Anal. Chem., 33, 1634 (1961).

[287]. P. E. Iversen and H. Lund, Tetrahedron Lett., 3523 (1969).

[288]. A. Hickling, Trans. Faraday Soc., 38, 27 (1942).

[289]. J. J. Lingane and S. L. Jones, Anal. Chem., 22, 1169 (1950).

[290]. R. Schindler, H. Will, and L. Holleck, Z. Electrochem., 63, 596 (1959).

[291]. J. Tacussel, Electrochim. Acta, 11, 381 (1966).

[292]. A. Bewick and O. R. Brown, J. Electroanal. Chem., 15, 129 (1967).

[293]. J. E. Harrar and E. Behrin, Anal. Chem., 39, 1230 (1967).

[294]. H. Matschiner and H. Unger, Wiss. Z. Univ. Halle, Math.-Naturwiss. Reihe, 18, 341 (1969).

[295]. G. L. Booman and W. B. Holbrook, Anal. Chem., 35, 1793 (1963).

[296]. W. M. Schwarz and I. Shain, ibid., 35, 1770 (1963).

[297]. W. L. Underkofler and I. Shain, ibid., 35, 1778 (1963).

[298]. S. J. Reddy and V. R. Krishnan, Indian J. Chem., 8, 626 (1970).

[299]. S. Bogan, L. Meites, E. Peters, and J. M. Sturtevant, J. Amer. Chem. Soc., 73, 584 (1951).

[300]. T. DeVries and J. L. Kroon, ibid., 75, 2484 (1953).

[301]. R. D. Weaver and G. C. Whitnack, Anal. Chim. Acta, 18, 51 (1958).

[302]. H. Lund, "Studier over Elektrodereaktioner i Organisk Polarografi og Voltammetri," Thesis, Aarhus, 1961, p. 93.

[303]. A. J. Bard and K. S. V. Santhanam, Electroanalytical Chemistry (A. J. Bard, ed.), Vol. 5, Dekker, New York, 1970, p. 215.

[304]. T. Fujinaga, K. Izutsu, and T. Nomora, J. Electroanal. Chem., 29, 203 (1971); J. P. Petrovich and M. M. Baizer, J. Electrochem. Soc., 118, 447 (1971); M. M. Baizer and J. P. Petrovich, ibid., 114, 1023 (1967).

[305]. P. Elving, J. Electroanal. Chem., 29, 55 (1971).

[306]. O. DeNora, Chem.-Ing.-Tech., 43, 182 (1971); F. Beck and G. Csizi, Ger. Offen. 1,915,951, Oct. 8, 1970; CA, 74, 37744z (1971).

[307]. H. J. McComsey, Jr. and M. S. Spritzer, Anal. Lett., 3, 427 (1970); J. Courtot-Coupez and M. L'Her, Bull. Soc. Chim. Fr., 1970, 1631; D. R. Cogley, J. C. Synnott, J. N. Butler, and E. Grunwald, Electrochemical Society Meeting, Washington, D.C., May 1971, Extended Abstract 122.

Section C

SYNTHETIC AND MECHANISTIC ASPECTS OF CATHODIC REACTIONS OF ORGANIC COMPOUNDS CLASSIFIED BY ELECTROPHORES

Chapter V

CATHODIC REACTIONS OF HYDROCARBONS

Roy Dietz

Division of Inorganic and Metallic Structure
National Physical Laboratory
Teddington, Middlesex, England

I. INTRODUCTION

The reactions of hydrocarbons treated here are those of hydrocarbon anions, formed by the cathodic addition of one or two electrons to a neutral hydrocarbon molecule. The ease with which electron addition takes place depends on the electron affinity of the hydrocarbon and the solvation energy of the ions formed. The values of these parameters are such that polycyclic aromatic hydrocarbons are reduced cathodically most readily. Hückel molecular orbital (HMO) theory [1] had its greatest success in accounting for the properties of this class of compounds, and the electrochemical behavior is no exception. Hoytink has summarized [2] his classical application of HMO theory in the field. The polarography [3] and physical chemistry of the process have also been reviewed [4, 5]. Here the emphasis is on the chemical reactions of electrogenerated anions.

II. FORMATION OF ANION RADICALS

Current-voltage curves show [6, 7] that many aromatic hydrocarbons are reduced cathodically in two discrete steps in solvents of low proton availability. Those for the first step are of the form discussed in Chapter II for a rapid reversible transfer of one electron to form a product stable enough to diffuse into solution [7, 8]. Controlled potential reduction of anthracene in dimethylformamide (DMF) containing n-Bu_4NClO_4 at a potential just cathodic of the first electron transfer gave an ESR spectrum like that of the product of reduction of anthracene with sodium in tetrahydrofuran [9]. Radical anions of aromatic hydrocarbons were first identified [10, 11] in reactions with alkali metals in ethers. Electrical conductivity of the solutions established [12] the ionic nature of the products. ESR spectroscopy showed [11] the hydrocarbon ion to contain an unpaired electron in the antibonding orbital of lowest energy. The electrode reaction of anthracene is therefore that shown in Fig. 1.

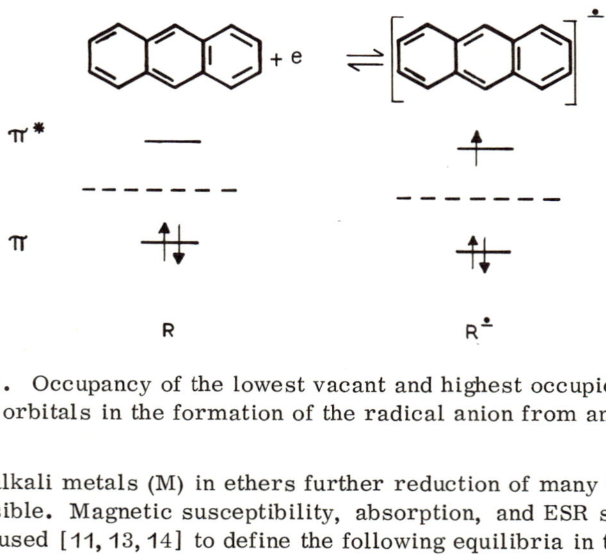

Fig. 1. Occupancy of the lowest vacant and highest occupied Hückel molecular orbitals in the formation of the radical anion from anthracene.

With alkali metals (M) in ethers further reduction of many hydrocarbons (R) is possible. Magnetic susceptibility, absorption, and ESR spectroscopy have been used [11, 13, 14] to define the following equilibria in these systems:

$$R + M \rightleftharpoons R^{\dot{-}},\ M^+ \tag{1}$$

$$R^{\dot{-}},\ M^+ + M \rightleftharpoons R^{2-},\ 2M^+ \tag{2}$$

$$R^{2-},\ 2M^+ + R \rightleftharpoons 2(R^{\dot{-}},\ M^+). \tag{3}$$

The equilibrium constants and degrees of association between ions are very sensitive to solvent and gegenion [15]. A second cathodic reduction step is possible for many hydrocarbons in aprotic conditions, and is identified with dianion formation, the electrochemical analog of Eq. (2) (Sec. IV).

The measured polarographic half-wave potentials for the first reduction step of aromatic hydrocarbons in 75% [16] and 96% aqueous dioxane [17], 2-methoxyethanol [18], and DMF [19] vary linearly with the HMO energies of the lowest vacant orbitals of the hydrocarbons. The importance of this HMO/$E_{1/2}$ correlation in the development of the theory of the chemical steps in cathodic reduction appears below (Sec. III.A). Its success, as that of many applications of HMO theory, is probably [20-22] due to fortuitous compensation of errors. More sophisticated MO methods have been used, but are unlikely to approach the flexibility of the HMO correlation. Differences in hybridization [23] and the effects of substituents [24] and non-planarity [25, 26] are readily allowed for by crude perturbation theory.

Reduction at a potential shown by current-voltage curves to lead to the addition of one electron has been used widely [27-30] in the generation of hydrocarbon anion radicals for ESR work. In this way the range of counterions is broadened to include tetraalkylammonium ions. The control possible in electrogeneration may be exploited in two ways: the electrode potential determines the extent of reduction of a given substrate, and conditions for anion radical generation can be chosen simply. Constant current generation in a flow cell produces [31] the fixed concentration of anion radicals needed to find the rate of homogeneous electron exchange between neutral molecule and anion radical by ESR line broadening.

Hydrocarbon anion radicals produced by reduction with alkali metals in ethers are generally stable indefinitely. Those produced by cathodic reduction in DMF and acetonitrile (CH_3CN) containing R_4N^+ salts are much less so. Current-voltage curves can show only whether an anion radical is stable enough to diffuse from the electrode. Coulometry provides a more searching test. Cathodic reduction to produce a stable anion radical requires only 1 faraday per mole of hydrocarbon; subsequent anodic oxidation should yield an equal amount of charge. By these criteria the anion radicals of anthracene [9, 32] and 9, 10-diphenylanthracene [32] are stable in DMF/n-Bu_4N^+ only at low temperatures under vacuum, or with carefully purified nitrogen passing over the catholyte. Decay may be due to protonation (Sec. III.A) or to electron transfer in solution (Sec. III.C). The ideal voltammetric behavior puts an upper limit on the rates of these reactions. A kinetic study [33] of the decay of electrogenerated anion radicals unfortunately omitted product identification.

III. REACTIONS OF ANION RADICALS

A. Protonation

The classical theory of the cathodic reduction of aromatic hydrocarbons was developed by Hoytink [34-37] who used HMO theory to rationalize the effects of protonation on charge consumed.

Monoprotonation of the anion radical of an aromatic hydrocarbon removes one atom from the conjugated system, producing a π-radical of structure analogous to a Wheland intermediate. The HMO picture of the orbital energies shows the lowest vacant orbital of the π-radical formed from an alternant hydrocarbon to be nonbonding; naphthalene provides an example (Fig. 2). That of a π-radical formed similarly from a nonalternant

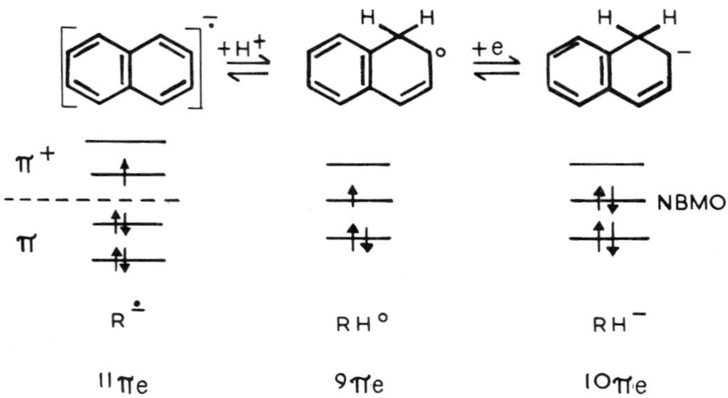

Fig. 2. Occupancy of the lower vacant and higher occupied Hückel molecular orbitals in the protonation of the anion radical of naphthalene.

hydrocarbon would be either nonbonding or bonding. It follows from the HMO/$E_{1/2}$ correlation that the reduction potential of the π-radical is always more anodic than that of the original hydrocarbon, the lowest vacant orbital of which is antibonding. Hence protonation always results in further electron transfer, from either the electrode [Eq. (6)]

$$R + e \rightleftarrows R^{\dot{-}} \tag{4}$$

$$R^{\dot{-}} + \text{Proton source} \rightleftarrows RH^{\bullet} \tag{5}$$

$$RH^{\bullet} + e\ (\text{Electrode}) \rightleftarrows RH^{-} \tag{6}$$

or a second anion radical:

$$RH^{\bullet} + R^{\dot{-}} \rightleftarrows RH^{-} + R. \tag{7}$$

Reduction is followed by further protonation:

$$RH^- + \text{Proton source} \rightleftharpoons RH_2. \tag{8}$$

The reaction is an example of an ECE process; an electron transfer [Eq. (4)] is followed by a chemical step [Eq. (5)] which allows a second electron transfer [Eqs. (6) and (7)] to occur.

The rates of the two competitive paths for the reduction of $RH^{\bullet}$ [Eqs. (6) and (7)] may be estimated theoretically [31]. There emerges a simple prediction based solely on the reduction potential E_R° of the hydrocarbon. Homogeneous electron exchange [Eq. (7)] is fast when $-1 < E_R^{\circ} < -2.3$ V vs. SCE, but becomes slower when $E_R^{\circ} > -2.3$ V. For anthracene and naphthalene the predicted rates are 10^9 and 10^2-10^4 moles^{-1}/sec, respectively. Homogeneous reduction [Eq. (7)] should therefore compete with heterogeneous reduction [Eq. (6)], which is essentially diffusion controlled [38], only in the case of anthracene. Current transients for potential step reduction of the two hydrocarbons in DMF containing phenol as a proton source have the shapes required by this prediction [31].

Equations (6) and (7) ensure that protonation of an anion radical always increases the cathodic current beyond that corresponding to 1 faraday per mole of hydrocarbon; the charge recoverable on current reversal is correspondingly decreased. Formation of a dihydro derivative requires 2 faradays per mole of substrate [Eq. (9)]; more charge implies that RH_2 is itself reduced at the potentials used.

$$R + 2e + \text{Proton source} \rightarrow RH_2. \tag{9}$$

The charge consumed in the first reduction step on a given time scale gives a measure of how the ease of protonation of anion radicals varies with solvent and structure of the parent hydrocarbon. Protonation becomes easier [16-19, 39] in the solvent series DMF, 96% aqueous dioxane, 75% aqueous dioxane, and 2-methoxyethanol. Adding increasing amounts of proton donors (benzoic acid, phenol, and water) to aprotic solvents has a similar effect [40, 41]. With a moderate excess of phenol, reduction of anthracene in DMF consumes [41] 2 faradays/mole. In tetrahydrofuran (THF) a series of buffers is available covering a wide range of pH; protonations of anion radicals are then equilibrium processes [42]. The variation of polarographic wave height and potential with pH generates an equilibrium diagram defining regions of potential and pH in which the reduction products are stable. The pK_a values of the protonated species $RH^{\bullet}$, RH^-, and RH_2 follow from the diagrams.

The ease of protonation of electrogenerated anion radicals is markedly dependent on structure. Several factors may be distinguished:

(i) As a working rule protonation of anion radicals occurs more readily the smaller the electron affinity of the parent polycyclic hydrocarbon, as reflected in the polarographic half-wave potential. In 2-methoxyethanol, for example, polarographic wave heights for hydrocarbons reduced easily correspond to the addition of a single electron; those of the remainder become larger as $E_{1/2}$ becomes more negative [18, 39]. Protonation of such basic anion radicals as those of naphthalene ($E_{1/2}$ = -2.53 V vs. SCE) and stilbene (-2.19 V) is apparent on polarography in 96% dioxane-water but not in DMF [17, 19].

(ii) Anion radicals with terminal methylene groups, such as those of styrene [8] and unsaturated aliphatic hydrocarbons [43], are protonated readily even under nominally aprotic conditions. Phenyl substituents stabilize these anion radicals [44]; those of α,ω-diphenylpolyenes are protonated more readily as the polyene chain becomes longer [45].

(iii) An acidic hydrocarbon, such as tetraphenylpropene [23] or 4, 5-methylenephenanthrene [46], may act as a proton donor toward its own anion radical

$$Ph_2C{=}CH{-}CHPh_2 + \bar{e} \rightleftarrows [Ph_2C{=}CH{-}CHPh_2]^{\dot{-}} \qquad (10)$$

$$[Ph_2C{=}CH{-}CHPh_2]^{\dot{-}} + Ph_2C{=}CH{-}CHPh_2 \rightleftarrows Ph_2\dot{C}CH_2CHPh_2 + Ph_2C{=}CH{-}\overset{\ominus}{C}Ph_2. \qquad (11)$$

Further electron transfer and protonation give tetraphenylpropane.

(iv) An sp-carbon encourages protonation. The anion radical of diphenylacetylene is barely detectable [29] after controlled potential reduction in DMF. Tetraphenylallene provides an extreme example [23]. No deviation from two-electron behavior could be detected in current transients after potential step reduction for 0.2 msec in DMF. The rate of the chemical reaction in the ECE sequence must exceed 10^3 sec^{-1}. In this case the carbanion RH^- is stable enough to be detected by cyclic voltammetry and identified by independent preparation:

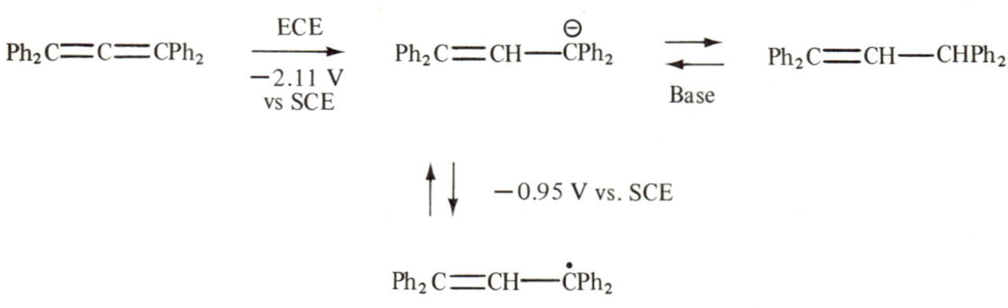

(12)

Several arguments have been advanced to show that protonation of electrogenerated anions takes place by reaction with an un-ionized molecule of the proton donor. The donors commonly added to aprotic solvents (phenol, acetic, and benzoic acids) are weakly ionized, and yet complete protonation occurs [47] rapidly when only a small excess over the hydrocarbon is used. The dependence [40] of the shape, height, and potential of dc and ac polarographic waves of hydrocarbons on proton donor concentration is that expected for reaction with un-ionized donor. And the current transients during potential step reduction [31] of naphthalene and anthracene in DMF with excess of phenol agree with theory for a pseudo-first-order reaction with phenol molecules.

In buffered THF, however, the effect of pH on the polarography of tetracene and anthracene is that expected for coupled electron and proton addition [42]. At high pH values the electrode process corresponds to Eqs. (4) and (13), and at low pH values to Eq. (14). Buffers are not available for the solvents more commonly used for cathodic reduction

$$R^{\dot{-}} + H^+ + e \rightleftarrows RH^- \tag{13}$$

$$R + 2H^+ + 2\bar{\epsilon} \rightleftarrows RH_2. \tag{14}$$

There is evidence (Sec. IV) that π-carbanions, RH^-, are protonated [Eq. (8)] more rapidly than the corresponding anion radicals. If protonation of an anion radical is fast enough to be detected instrumentally, therefore, reduced hydrocarbons, RH_2, are formed in the diffusion layer, and must be responsible for any new reduction processes. In protic solvents the reverse reactions of the protonation equilibria [Eqs. (5) and (8)] may be neglected and we have:

$$R \quad + 2e + 2HX \rightarrow RH_2 + 2X^-, \quad \text{potential } E_1 \tag{15}$$

$$RH_2 + 2e + 2HX \rightarrow RH_4 + 2X^-, \quad \text{potential } E_2. \tag{16}$$

Two cases are distinguishable [48]. When E_2 is sufficiently more cathodic than E_1, a discrete new process appears at the potential E_2. Otherwise, reduction of RH_2 is evident in an increased cathodic current at E_1. Once E_2 is known, reference to the HMO/$E_{1/2}$ plot, or comparison with model compounds, identifies RH_2. One of many examples [37, 48, 49] is provided by the reduction of benz(a)anthracene in 75% aqueous dioxane. The second of the two polarographic waves, each corresponding to addition of two electrons, appears [49] at the same potential as the first wave of naphthalene. For this hydrocarbon, then, RH_2 in Eq. (15) is 7, 12-dihydrobenz(a)-anthracene(I):

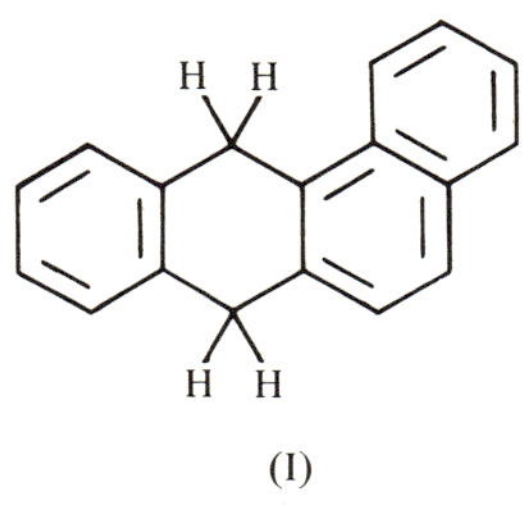

(I)

Products detected in this way are, of course, not necessarily those isolable after macroelectrolysis; isomerization of reduced hydrocarbons may intrude during work-up.

The major products of cathodic reduction of cyclic conjugated hydrocarbons in protic solvents are frequently 1, 4-dihydro derivatives [48-51]. Naphthalene gives 1, 4-dihydronaphthalene in 2-methoxyethanol [50] and cyclooctatetraene gives 1, 3, 6-cyclooctatriene in aqueous ethanol [51] for example. On electronic grounds the conjugated, 1, 2-isomers are more stable, and hence kinetic control of protonation of RH^-, and by analogy $R^{\dot{-}}$, is usually assumed. Steric factors may enter here, however; cyclohexa-1, 4-diene is the more stable isomer [52]. For kinetic control, the course of protonation is determined by the activation energies for the various transition states. Protonations of carbanions in protic solvents are very fast, and the activation energies small. The transition states must therefore be close to the carbanions in structure, and a property of an isolated carbanion may be used as a reactivity index [53, 54]. The one chosen [49, 54] is the charge density q_r; least activation energy is required for protonation at the site r where q_r is largest. HMO charge densities are generally used; allowance must be made for the inductive effect of the methylene group in RH^- if two sites have equally high charge densities. Correct HMO predictions for cathodic reduction of naphthalene and cyclooctatetraene are shown in Fig. 3; the method is successful for the majority of hydrocarbons.

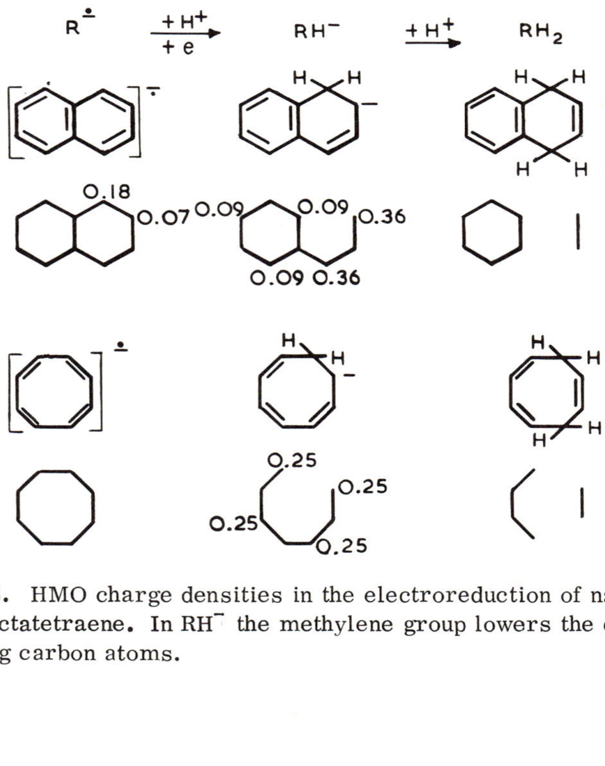

Fig. 3. HMO charge densities in the electroreduction of naphthalene and cyclooctatetraene. In RH^- the methylene group lowers the charge at the neighboring carbon atoms.

Some cases in which the HMO charge density failed as a reactivity index led Hoytink [55] to consider the effect of the electric field on the charge distribution. For the terphenyl anion the q_r values predict protonation in the central ring, and yet polarography shows the product to be isoconjugate with biphenyl. Similar difficulties arise with quaterphenyl and phenanthrene. These three molecules have strongly anisotropic polarizabilities and Hoytink showed semiquantitatively that in a sufficiently strong field the charge distribution in the anions (specifically the dianions) would be perturbed in the direction required by the experimental results. The potential falls rapidly on the solution side of the electrode surface and fields of the strength required ($\sim 10^7$ V/cm) exist there [56]. Whether both protonation steps [Eqs. (5) and (8)] occur within that region is doubtful, however. The electrode charge at the potentials used is negative [57], and anions should be repelled from the surface.

Products of the slow protonation of anion radicals under nominally aprotic conditions may correspond to the addition of protons at sites other than those of the greatest HMO charge density. In contrast to the examples of Fig. 3, reduction of naphthalene in anhydrous acetonitrile containing tetraethylammonium p-toluenesulfonate results [58] in predominant

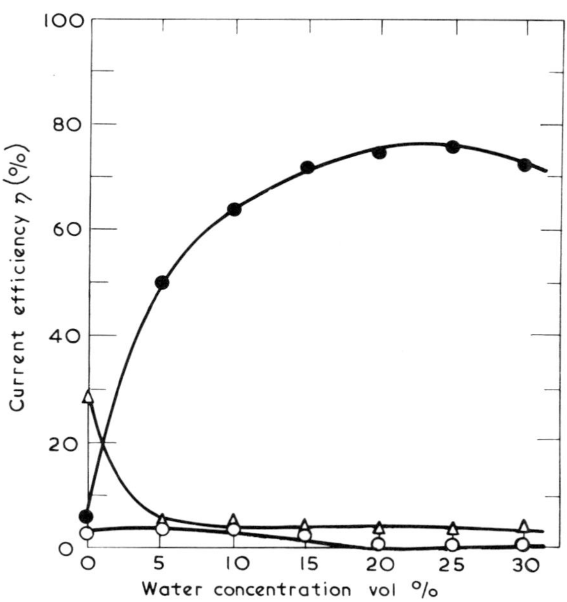

Fig. 4. Effect of added water in the electroreduction of naphthalene in acetonitrile/Et_4N tosylate on the current efficiencies for the production of 1, 4-dihydronaphthalene (●), 1, 2-dihydronaphthalene (o), and tetralin (Δ). Reprinted from Ref. [58] by courtesy of the Chemical Society of Japan.

1, 2-protonation and subsequent reduction to tetralin; when water is added the current efficiency of 1, 4-dihydronaphthalene rapidly overtakes those for tetralin and the 1, 2-isomer (Fig. 4). Similarly reduction of cyclooctatraene in anhydrous DMF gives 1, 3, 5-cyclooctatriene almost exclusively [51]. Isolation of these more stable isomers may reflect a change in the orientation of protonation of the π-carbanions RH^-, or merely base-catalyzed isomerization of the RH_2 species.

There are three possible sources of protons in the reductions under nominally aprotic conditions: protic impurities, the solvent, and NR_4^+ ions. Products attributable to proton loss from DMF and CH_3CN have been found [8] after reduction of hydrocarbons; the cathode potential was not controlled in this work, however. There is good evidence for proton removal from R_4N^+ ions by strongly basic anions (Sec. II.D). The reaction is, of course, a Hofmann elimination, giving a terminal alkene and a trialkylamine. 1-Butene and tri-n-butylamine were found [8] after reduction of stilbene in DMF/n-Bu_4N^+ but the cathode potential was unknown and reduction of n-Bu_4N^+ cannot be ruled out.

The reductions described in this section may, of course, be brought about by such conventional chemical means as dissolving metals [59] or catalytic hydrogenation [60]. And, further, these methods are applicable to hydrocarbons that are not reducible via anion radicals produced bv heterogeneous electron transfer. Cathodic reduction has, however, two advantages: The conditions for a selective reduction may be defined closely by simple preliminary voltammetry; and quaternary ammonium counterions may result in a steric orientation of protonation different from that in a dissolving metal system.

B. Addition of Electrophiles

The stability of anion radicals under aprotic conditions means that reactions with added electrophiles are possible. Surprisingly few are known. The course of reaction follows by analogy with that for protonation [8, 61, 62]. Two faradays per mole are consumed when aromatic hydrocarbons are reduced in the presence of carbon dioxide. The products, dihydrodicarboxylates, are those expected from an ECE process, shown for stilbene [Eqs. 17-21]. Carbon dioxide is reducible at the mercury cathodes commonly used in this work, but at a potential more cathodic than many hydrocarbons [8].

Dihydrodicarboxylates are formed in this way from butadiene [63], phenanthrene, and diphenylacetylene [61] too. In a similar reaction [64] 9, 10-dihydroanthracene bissulfonate is formed when anthracene is reduced in SO_3-saturated DMF. 1, 4-Addition to naphthalene [61, 65] suggests that carboxylation, like protonation, is kinetically controlled. Competition between carboxylation and protonation can occur; some monocarboxylate is formed [66] from naphthalene in DMF and stilbene in acetonitrile [26].

$$\mathrm{PhCH{=}CHPh} + e \rightleftharpoons [\mathrm{PhCh{=}CHPh}]^{\dot{-}} \tag{17}$$

$$[\mathrm{PhCH{=}CHPh}]^{\dot{-}} + \mathrm{CO_2} \longrightarrow \mathrm{Ph\dot{C}H{-}}\underset{\mathrm{COO^-}}{\underset{|}{\mathrm{CHPh}}} \tag{18}$$

$$\mathrm{Ph\dot{C}H{-}}\underset{\mathrm{COO^-}}{\underset{|}{\mathrm{CHPh}}} + e \xrightarrow{\text{electrode}} \mathrm{Ph\bar{C}H{-}}\underset{\mathrm{COO^-}}{\underset{|}{\mathrm{CHPh}}} \tag{19}$$

or

$$\mathrm{Ph\dot{C}H{-}}\underset{\mathrm{COO^-}}{\underset{|}{\mathrm{CHPh}}} + [\mathrm{PhCH{=}CHPh}]^{\dot{-}} \longrightarrow \mathrm{Ph\bar{C}H{-}}\underset{\mathrm{COO^-}}{\underset{|}{\mathrm{CHPh}}} + \mathrm{PhCH{=}CHPh} \tag{20}$$

$$\mathrm{Ph\bar{C}H{-}}\underset{\mathrm{COO^-}}{\underset{|}{\mathrm{CHPh}}} + \mathrm{CO_2} \longrightarrow \underset{\mathrm{COO^-}}{\underset{|}{\mathrm{PhCH}}}{-}\underset{\mathrm{COO^-}}{\underset{|}{\mathrm{CHPh}}} \tag{21}$$

Phenylethylenecarbonate must be formed [66] from stilbene by a different route. The stereochemical questions raised by the addition of a group other than a proton to a hydrocarbon anion radical are discussed in Chapter XXVI.

Cathodic alkylation of hydrocarbon anions is possible [67] in molten tetra-n-butylammonium nitrate (at 150°). Unsubstituted polycyclic aromatics are reduced in a two-electron step, and reduction of anthracene at a potential just cathodic of the single wave gives 9-n-butylanthracene. The potential is close to the cathodic limit, at which tri-n-butylamine and butene are formed. Since naphthalene is neither reduced nor alkylated at that potential, however, alkylation was thought to involve interaction of an anthracene anion with $n\text{-}Bu_4N^+$.

Anion radicals generated by reduction with alkali metals react with hydrogen [68], sulfur dioxide [69], tosylates [70], substituted methyl cyanides [71], carbon monoxide [72], trimethyl silylchloride [73], and alkyl halides [74]. Electrochemical variants of these reactions have received little attention.

C. Homogeneous Oxidation

In this section we consider reactions in which a cathodically generated anion radical of a hydrocarbon loses an electron to another solution species, which is thereby catalytically reduced. The hydrocarbon molecule merely relays an electron from the electrode to a second solute. Electron transfer of this sort will be important when the solution redox potential of the second solute is less negative than that of the original hydrocarbon. Current-voltage curves are not an infallible guide here; heterogeneous reduction may be slow. Tertiary butyl chloride, for example, is reduced in DMF by the phenanthrene anion radical, although the half-wave potential for heterogeneous reduction of t-BuCl is more cathodic [75].

Catalytic reduction of a second species would result in low current efficiency for preparative reduction of a hydrocarbon. Thus, one reason [32] for the instability of electrogenerated solutions of the anthracene radical anion mentioned in Sec. II may be catalytic reduction of traces of oxygen, which has a less negative solution redox potential than most hydrocarbons. Again homogeneous reduction of tetraalkylammonium ions may cause the very low current efficiency for naphthalene reduction in aprotic solvents as follows:

$$\begin{aligned}
&\text{Naphthalene} + e \rightleftharpoons (\text{Naphthalene})^{\dot{-}} \\
&(\text{Naphthalene})^{\dot{-}} + R_4N^{+} \rightleftharpoons \text{Naphthalene} + R_4N^{\bullet} \\
&R_4N^{\bullet} \longrightarrow NR_3 + \text{Alkene} + \text{Alkane}
\end{aligned} \tag{22}$$

Ethylene and ethane were found [58] after reduction of naphthalene in CH_3CN containing tetraethylammonium p-toluenesulfonate. The potential for discharge of R_4N^+ ions at mercury corresponds to amalgam formation and is presumably more cathodic than the solution redox potential.

One variant [76] of the titanium-aluminum process for the fixation of atmospheric nitrogen uses cathodically produced naphthalene anion radicals to reduce Ti^{IV}, and the initially formed Ti^{II}-nitrogen complex. Cathodic reduction of the naphthalene formed makes the process catalytic.

When an anion radical is oxidized at a metallic electrode, formation of the neutral molecule in the ground state is strongly favored over formation of an excited state [77]. No such restriction applies to homogeneous electron transfer which, if sufficiently exothermic, may yield excited states of hydrocarbons. Naively one may suppose that an electron is removed from the bonding MO of highest energy to give either the first excited singlet or triplet; emission of light may then occur as in Eq. (23):

The same species may be produced by homogeneous reduction of cation radicals. Electron-transfer reactions of polycyclic hydrocarbons satisfy the general conditions for efficient electron-transfer luminescence [78], and many examples are known of emission from electrochemical cells (electrochemiluminescence) [78-80].

$$R^{\dot{-}} \xrightarrow[\text{Homogeneous}]{-e} R^* \xrightarrow{-h\nu} R \qquad (23)$$

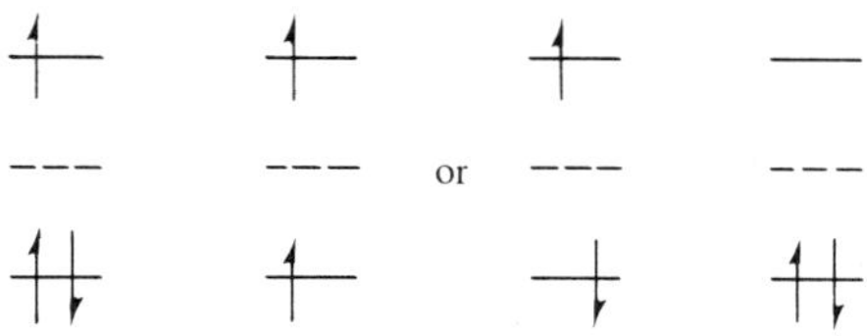

A common experimental technique [81-89] has been to generate a radical cation or anion at an electrode and then to change the electrode potential so that a second species is produced and homogeneous electron transfer may occur. Variants of this use an alternating potential [90-93] so that both species are produced alternately at each electrode, or a ring disk electrode [94]. When the potentials chosen are such that both radical anion and cation are formed, annihilation [Eq. (24)] is possible and the emission is called annihilative:

$$R^{\dot{-}} + R^{\dot{+}} \rightarrow R^* + R. \qquad (24)$$

Preannihilative emission may occur when the potential excursion is too small to produce both radical ions, and is usually two decades less intense [84, 85, 87].

The emission observed, whether annihilative or preannihilative, is usually the fluorescent emission of the hydrocarbon, corresponding to the decay of the first excited singlet. The energy made available by homogeneous electron transfer is given by the difference between the redox potentials of the donor and acceptor plus a small entropy term [95] and in many cases falls short of that needed to populate the singlet state directly. Thus annihilative emission from rubrene in benzonitrile (BN)/n-Bu_4NClO_4 corresponds [87, 88] to the fluorescent spectrum of the hydrocarbon. The potential difference (2.35 V) is close to the singlet excitation energy (2.39 eV). With p-benzoquinone present, however, and the potential excursion shortened so that only the rubrene cation radical and the quinone anion radical are formed, the singlet emission appears again [88]. Here the potential difference is only 1.48 V. For anthracene in DMF, fluorescent emission appears at a potential excursion of 2.8 V; the singlet excitation energy is 3.3 eV [86].

Ample energy is available to populate the lowest triplet states, the excitation energies of which are ~1.3 and 1.8 eV in the last two examples. Singlets could then be produced by diffusion-controlled triplet-triplet annihilation. In the rubrene BN system the annihilative intensity depends [87, 88] on the ratio of the times spent generating the two radical ions in the way required by a theoretical treatment [96], provided that <u>two</u> annihilations (radical anion-cation; triplet-triplet) are involved in singlet production. Spectral evidence of triplets has been obtained too [84, 89, 97].

The nature of the homogeneous electron transfer in preannihilative electrochemiluminescence with no added donor to acceptor is unclear [98]. Impurities which need be present at only 10^{-7} M levels, and ion-radical aggregates [79] have been held to be important. Products of reaction of radical ions with solvents and impurities have been suggested often [84, 87, 89] as electron-transfer agents. The instability of cation radicals is a particular difficulty, not always appreciated in early work.

Low-intensity singlet electrochemiluminescence may be produced without any change in electrode potential when a radical ion is generated in the presence of an appropriate reagent. Oxidation of rubrene in DMF (a nucleophilic solvent) or in acetonitrile containing added nucleophiles [83] and reduction of anthracene with oxygen [99] provide examples.

D. Dimerization and Polymerization

Since hydrocarbon anion radicals are powerful nucleophiles, electron-transfer reduction of a hydrocarbon susceptible to nucleophilic addition can lead to oligomerization. Polymerization of phenyl alkenes was discovered during the classical work on hydrocarbon-alkali metal adducts. Styrene, for example, was polymerized in ethers by alkali metals [100] and by cumyl potassium [101]. In such systems the initially formed [Eq. (25)] anion radical may dimerize [Eq. (26)] or add to a neutral molecule [Eq. (27)]; the electron source may be a metal, a carbanion, or an anion radical [102]:

$$RCH{=}CH_2 + \text{Electron source} \rightleftarrows (RCH{=}CH_2)^{\dot{-}} \qquad (25)$$

$$2(RCH{=}CH_2)^{\dot{-}} \rightleftarrows R\bar{C}HCH_2CH_2\bar{C}HR \qquad (26)$$

$$(RCH_2{=}CH_2)^{\dot{-}} + RCH{=}CH_2 \rightleftarrows R\bar{C}H_2CH_2CH_2\dot{C}HR. \qquad (27)$$

Anion radical species may grow by nucleophilic or radical attack on monomer; the dianions only by nucleophilic attack. Radical centers are quickly destroyed in bimolecular reactions, but anionic centers are removed only if electrophiles are present. With suitable precautions polymeric "living" anions are produced. Electron-transfer polymerization has been reviewed extensively [103]. (See also Chapter XXIX.)

Electrochemical anionic polymerization is possible if protonation is avoided. Hydrocarbons with terminal methylene groups give anion radicals that are typically bases strong enough to remove protons from tetraalkylammonium ions. In THF/n-Bu_4NClO_4 protonation of the anions of 1,1-diphenylethylene and styrene is evident on the time scale of cyclic voltammetry [44]; polymeric anions are destroyed by adding such salts [104]. With alkali or alkaline earth metal salts in ethers, 1,1-diphenylethylene is reduced cathodically [105] to give stable solutions of the dimeric dianions, $Ph_2\bar{C}CH_2CH_2\bar{C}Ph_2$; further growth is inhibited sterically. Monomers having no steric restriction, such as styrene [104, 106-8] and α-methylstyrene [109, 110] give polymeric living anions. α-Methylstyrene is polymerized [110] only in the catholyte of an evacuated cell containing THF and $LiAlH_4$ or $NaAlR_4$; polymers are formed of molecular weight close to that expected for the distribution of available monomer between the anions generated. The number of anionic centers, and hence the stoichiometry of polymerization, are shown by spectral measurements to be controllable coulostatically.

The electrode process in these cathodic polymerizations has not always been well defined [104]. At the platinum cathodes used, the metal ion M^+ and the monomer are reduced at similar potentials. Monomer anion radicals may be produced either directly [Eq. (28)] or catalytically by electrogenerated metal [Eq. (29)]:

$$RCH{=}CH_2 + e\ (\text{Electrode}) \rightleftharpoons (RCH{=}CH_2)^{\dot{-}} \tag{28}$$

$$\begin{aligned} Na^+ + e &\rightleftharpoons Na \\ Na + RCH{=}CH_2 &\rightleftharpoons Na^+(RCH{=}CH_2)^{\dot{-}} \end{aligned} \tag{29}$$

Living anion polymerization of styrene [104] in hexamethylphosphoramide containing NMe_4BPh_4 is an unambiguous case; the cation discharge potential is 40 mV more cathodic than the half-wave potential of styrene.

Many examples of cathodic reductive coupling of alkenes activated by electron-withdrawing groups to give hydrodimers are known, and are treated in detail in Chapter XIX. Evidence is presented there that the essential step is nucleophilic addition of an anion radical to a neutral olefin [cf. Eq. (27)]. Hydrodimers are formed [111] from hydrocarbons such as 9-benzalfluorene, to which nucleophilic addition can occur. More surprising examples are provided by stilbene [8], diphenylacetylene, and phenanthrene [61]. Electrode potentials were not controlled, and the course of phenanthrene reduction was ascribed [61] to insolubility of the dimer. Dimerization of anion radicals of polycyclic aromatic hydrocarbons is energetically unfavorable [112].

IV. FORMATION AND REACTIONS OF DIANIONS

We now turn to the second reduction step that is undergone by many aromatic hydrocarbons at a potential some 500 mV more cathodic than that for anion radical formation. Current-voltage curves frequently show that one electron is transferred [7, 32, 35, 42, 113, 114]. Reduction of

9,10-diphenylanthracene at a suitably cathodic potential consumes 2 faradays per mole [32]. The reduction is attributed to dianion formation.

Two marked differences from the first reduction step are shown by current-voltage curves (Fig. 5); there is little or no reverse peak corres-

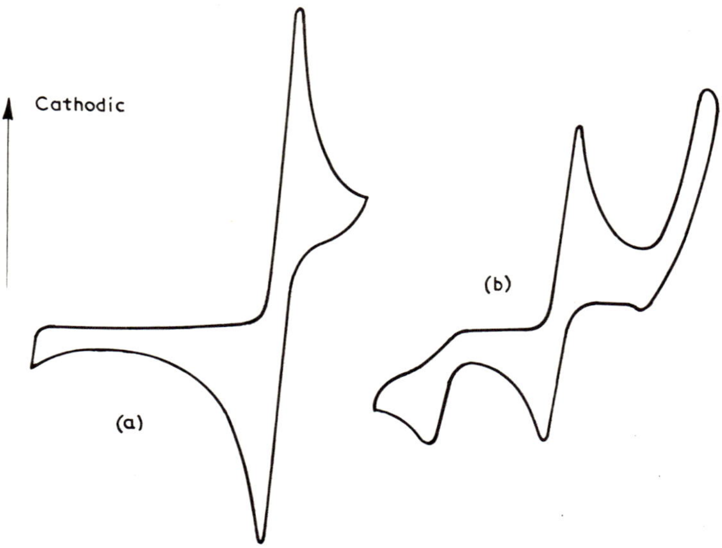

Fig. 5. Current-voltage curves for the reduction of 9,10-diphenylanthracene in DMF/n–Bu_4NClO_4 at a platinum microelectrode. Continuous linear voltage sweep 4.7 V/sec. (a) Anion radical formation: sweep -0.23 V to -2.31 V vs. SCE; (b) dianion formation: sweep -0.69 to -2.72 V vs. SCE. Reprinted from Ref. [121] by courtesy of the Chemical Society.

ponding to reoxidation of the dianion [32, 114, 115], and a new anodic peak often appears at more anodic potentials [115]. Only in such special cases as dibiphenylene ethene, the dianion of which is Hückel aromatic, is the dianion stable [113] in aprotic solvent containing quaternary ammonium salts. Other dianions are consumed in a rapid chemical reaction, which is assumed to be protonation.

In alkali metal-ether systems dianions of polycyclic aromatic hydrocarbons are generally stable [116, 117], and may be formed from anion radicals by electron transfer [Eq. (2)] or disproportionation [Eq. (3)]. In cathodic generation, the equilibrium constant K_d for disproportionation [Eq. (30)] is related [118] to the difference between the potentials for the formation of $R^{\dot{-}}$ and R^{2-}:

$$2R^{\dot{-}} \rightleftarrows R^{2-} + R; \quad K_d. \tag{30}$$

For planar polycyclic hydrocarbons in aprotic solvents containing quaternary ammonium salts the usual difference (500 mV) corresponds to a disproportionation constant of $10^{-8.5}$. Progressively smaller separations between the two reduction steps, and hence progressively larger values of K_d, are found [119] for a series of stilbenes with increasing steric hindrance to planarity. In THF/$n\text{-}Bu_4N^+$, tri- and tetraphenylethylene are reduced [44] in a single two-electron step, consistent with a potential separation of ~100 mV and $K_d \sim 10^2$.

Spontaneous protonation of electrogenerated dianions of polycyclic aromatic hydrocarbons is so rapid that no other reactions are known. Monoprotonation gives a π-carbanion RH^-:

$$R^{2-} + \text{Proton source} \rightleftarrows RH^-, \tag{31}$$

in which one atom is removed from conjugation [120]. Further protonation gives a dihydro compound RH_2. This EEC process [Eq. (32)], in which a chemical reaction follows two charge transfers, should be compared with the ECE reduction [Eq. (33)] under more acidic conditions at less cathodic potentials (Sec. III.A):

$$R \rightarrow R^{\dot{-}} \rightarrow R^{2-} \rightarrow RH^- \rightarrow RH_2 \quad \text{EEC} \tag{32}$$

$$R \rightarrow R^{\dot{-}} \rightarrow RH^{\bullet} \rightarrow RH^- \rightarrow RH_2 \quad \text{ECE} \tag{33}$$

Note that the π-carbanion RH^- is common to both sequences, provided that the dianion is protonated at the same site as the anion radical.

New anodic peaks, like that shown in Fig. 5, appear after formation of dianions of polycyclic aromatic hydrocarbons in several solvent-electrolyte combinations [44, 115, 121]. None is seen, however, when a proton donor is added, or when the scan rate is reduced. Transient carbanion intermediates must therefore be responsible and the EEC mechanism strongly suggests the π-carbanion RH^-. Oxidation of the π-carbanion is probably coupled with proton loss [121], so that the new peak represents the production of neutral hydrocarbon [Eq. (34)] rather than the radical $RH^{\bullet}$,

$$RH^- - 2e \rightarrow R + H^+(\text{base}) \tag{34}$$

at least at slow scan rates. The thermodynamic instability of the radicals $RH^{\bullet}$ derived from tetracene and anthracene in THF/$n\text{-}Bu_4N^+$ has been deduced from pH/potential diagrams [42].

Heterogeneous reduction of the radical $RH^{\bullet}$ begins to compete with proton loss at large scan rates, and a reverse peak appears. Potentials observed [115] for reduction of the radicals $RH^{\bullet}$ agreed well with the HMO/$E_{1/2}$ correlation, assuming protonation of the dianions at sites of highest HMO charge density. Reduction of the radicals $RH^{\bullet}$ formed from an alternant hydrocarbon always involves the nonbonding molecular orbital; the measured potentials for reduction were close to one value (-1.1 V vs. SCE). Gas-phase electron affinities of the radicals $RH^{\bullet}$ could be derived from the measurements. Comparison with calculated self-consistent field MO values suggested that reduction of pyrene and perylene proceeded beyond RH_2 [5].

Detection of π-carbanions in this way has two consequences for the ECE mechanism discussed in Sec. III.A. The radicals $RH^{\bullet}$ are indeed reduced at less cathodic potentials than the hydrocarbons, as is required by Hoytink's mechanism [Eqs. (4)-(8)]. No carbanions could be detected by rapid sweep voltammetry in the ECE reduction of hydrocarbons in the presence of proton donors; π-carbanions are therefore probably protonated more rapidly than the corresponding anion radicals.

No secure foundation exists for the assumption that electrogenerated dianions are protonated at sites of highest HMO charge density. No macroreduction at a potential corresponding to dianion formation has been reported for an unsubstituted polycyclic hydrocarbon. Indeed polarography suggests that in DMF/n-Pr_4N^+ the dianion of cyclooctatetraene gives 1, 3, 5-cyclooctatriene [51] and not the isomer (1, 3, 6-) predicted by the charge density criterion [Fig. 3].

V. REDUCTION OF CARBONIUM IONS

Reduction of stable carbonium ions is a trivial extension of that of hydrocarbons. The triphenylmethyl cation is reduced reversibly [122] in aprotic solvents to the corresponding radical, which is itself reduced at more cathodic potentials. Analogous di- and tri-carbonium ions are reduced stepwise. Radicals formed by the one-electron reduction of the Hückel-aromatic carbonium ions, tropylium [123] and triphenylcyclopropenyl [124], are less stable and dimerize rapidly.

Unstable carbonium ion intermediates in the electrooxidation of aromatic hydrocarbons may be detected by their reduction in rapid potential sweeps [121]. Just as this chapter has been concerned largely with the formation of hydrocarbon anions at electrodes and their nucleophilic reactions, the anodic oxidation of hydrocarbons (Chapter XIII) involves the formation of hydrocarbon cations and their electrophilic reactions. Carbonium ions play the same role in electrooxidation as do carbanions in electroreduction. Under suitable conditions the voltammetric behavior of 9, 10-diphenylanthracene, for example, on reduction (Fig. 5) and oxidation (Fig. 6) is strikingly similar, and reflects the symmetry of the electronic changes involved. Oxidation gives a stable cation radical, an unstable dication, and a new transient species reducible on the reverse sweep. The new cathodic process has been identified with the reduction [Eq. (35)] of a π-carbonium ion, formed by reaction of the dication with a nucleophile, Nu^-:

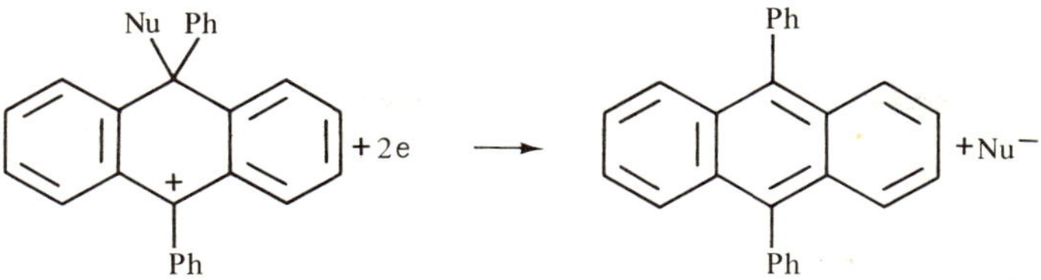

(35)

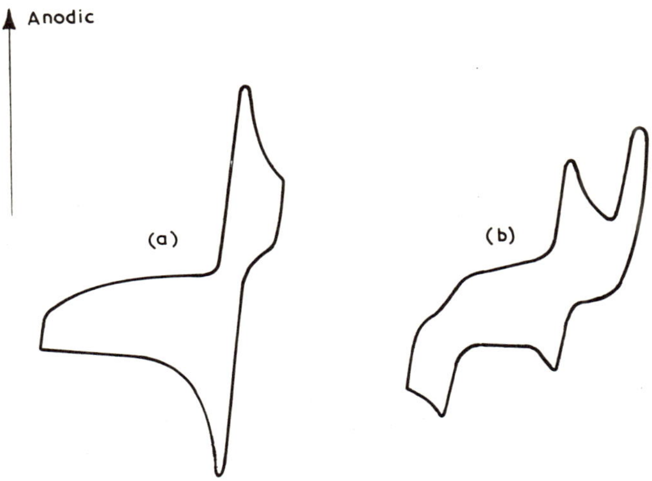

Fig. 6. Current-voltage curves for the oxidation of 9, 10-diphenylanthracene in benzonitrile/n-Bu_4NClO_4 at a platinum microelectrode; continuous linear voltage sweep 15.8 V/sec. (a) Cation radical formation: sweep +0.04 to +1.63 V vs. SCE; (b) dication formation: sweep +0.30 to +1.88 V vs. SCE. Reprinted from Ref. [121] by courtesy of the Chemical Society.

REFERENCES

[1]. A. Streitwieser, *Molecular Orbital Theory of Organic Chemistry*, Wiley, New York, 1961.

[2]. G. J. Hoytink, in *Advances in Electrochemistry and Electrochemical Engineering* (P. Delahay, ed.), Vol. 7, Wiley-Interscience, New York, 1970, p. 221 ff.

[3]. C. L. Perrin, in *Progress in Physical Organic Chemistry* (S. G. Cohen, A. Streitwieser, and R. W. Taft, eds.), Vol. 3, Wiley-Interscience, New York, 1965, pp. 210-219; S. I. Zhdanov, Usp. Khim., *38*, 1390 (1969).

[4]. M. E. Peover, in Electroanalytical Chemistry (A. J. Bard, ed.), Vol. 2, Dekker, New York, 1967, Chap. 1.

[5]. M. E. Peover, in Reactions of Molecules at Electrodes (N. S. Hush, ed.), Wiley, New York, 1972.

[6]. G. J. Hoytink, Rec. Trav. Chim. Pays-Bas, 73, 895 (1954).

[7]. G. J. Hoytink, J. van Schooten, E. de Boer, and W. I. Aalbersberg, ibid., 73, 355 (1954).

[8]. S. Wawzonek, R. Berkey, E. W. Blaha, and M. E. Runner, J. Electrochem. Soc., 102, 235 (1955).

[9]. D. E. G. Austen, D. E. J. Ingram, P. H. Given, and M. E. Peover, Nature, 182, 1784 (1958).

[10]. W. Hückel and H. Bretschneider, Justus Liebigs Ann. Chem., 540, 157 (1939).

[11]. D. Lipkin, D. E. Paul, J. Townsend, and S. I. Weissman, Science, 117, 534 (1953); S. I. Weissman, J. Townsend, D. E. Paul, and G. E. Pake, J. Chem. Phys., 21, 2227 (1953).

[12]. N. D. Scott, J. F. Walker, and V. L. Hansley, J. Amer. Chem. Soc., 58, 2442 (1936).

[13]. D. E. Paul, D. Lipkin, and S. I. Weissman, ibid., 78, 116 (1956); J. Chaudhuri, J. Jagur-Grodzinski, and M. Szwarc, J. Phys. Chem., 71, 3063 (1967).

[14]. E. R. Zabolotny and J. F. Garst, J. Amer. Chem. Soc., 86, 1645 (1964).

[15]. A. I. Shatenshtein and E. S. Petrov, Russ. Chem. Rev., 36, 100 (1967).

[16]. A. Maccoll, Nature, 163, 78 (1949); G. J. Hoytink and J. van Schooten, Rec. Trav. Chim. Pays-Bas, 71, 1089 (1952); 72, 903 (1953).

[17]. G. J. Hoytink, ibid., 74, 1525 (1955).

[18]. I. Bergman, Trans. Faraday Soc., 50, 829 (1954).

[19]. R. Pointeau, Ann. Chim., 7, 669 (1962); P. H. Given, J. Chem. Soc., 2684 (1958).

[20]. R. M. Hedges and F. A. Matson, J. Chem. Phys., 28, 950 (1958).

[21]. M. E. Peover, Electrochim. Acta, 13, 1083 (1968).

[22]. H. B. Mark, Rec. Chem. Progr., 29, 217 (1968).

[23]. R. Dietz, M. E. Peover, and R. Wilson, J. Chem. Soc. B, 3535 (1968).

[24]. A. Streitwieser and I. Schwager, J. Phys. Chem., 66, 2316 (1962).

[25]. L. O. Wheeler, K. S. V. Santhanam, and A. J. Bard, ibid., 71, 223 (1967); A. J. Bard, K. S. V. Santhanam, J. T. Malloy, J. Phelps, and L. O. Wheeler, Discuss. Faraday Soc., No. 45, 167 (1968).

[26]. R. Dietz and M. E. Peover, ibid., No. 45, 154 (1968).

[27]. D. H. Geske and A. Maki, J. Amer. Chem. Soc., 82, 2671 (1960).

[28]. R. Pointeau and J. Favede, Fifth Intern. Sump. Free Radicals, Uppsala, 1961.

[29]. R. E. Sioda, D. O. Cowan, and W. S. Koski, J. Amer. Chem. Soc., 89, 230 (1967).

[30]. A. E. J. Forno, Chem. Ind. (London), 1728 (1968).

[31]. A. E. J. Forno, M. E. Peover, and R. Wilson, Trans. Faraday Soc., 66, 1322 (1970).

[32]. K. S. V. Santhanam and A. J. Bard, J. Amer. Chem. Soc., 88, 2669 (1964).

[33]. R. Pointeau, J. Favede, and P. Delhaes, J. Chim. Phys., 68, 2676 (1964).

[34]. G. J. Hoytink, J. van Schooten, E. de Boer, and W. I. Aalbersberg, Rec. Trav. Chim. Pays-Bas, 73, 355 (1954).

[35]. A. C. Aten and G. J. Hoytink, Z. Phys. Chem. (Frankfurt), 21, 192 (1959).

[36]. G. J. Hoytink, Rec. Trav. Chim. Pays-Bas, 73, 895 (1954).

[37]. G. J. Hoytink, ibid., 76, 885 (1957).

[38]. J. M. Hale, J. Electroanal. Chem., 19, 315 (1968).

[39]. I. Bergman, Trans. Faraday Soc., 52, 690 (1956).

[40]. P. H. Given and M. E. Peover, Coll. Czech. Chem. Commun., 25, 3195 (1960).

[41]. P. H. Given and M. E. Peover, J. Chem. Soc., 385 (1960); Advances in Polarography, Pergamon, New York, 1961, p. 48.

[42]. J. Perichon and R. Buvet, Bull. Soc. Chim. Fr., 3255 (1965); 3697 (1967); 1282 (1968).

[43]. V. G. Mairanovskii, I. E. Valaslek, and G. I. Samokhvalov, Sov. Electrochem., 3, 538 (1967).

[44]. B. L. Funt and D. Gray, Electrochemical Society Meeting, New York, May 1969; J. Electrochem. Soc., 117, 1020 (1970).

[45]. G. J. Hoytink and J. van Schooten, Rec. Trav. Chim. Pays-Bas, 73, 691 (1953).

[46]. J. Janata, J. Gendell, R. G. Lawton, and H. B. Mark, J. Amer. Chem. Soc., 90, 5226 (1968).

[47]. P. H. Given and M. E. Peover, J. Chem. Soc., 385 (1960).

[48]. S. Wawzonek and H. A. Laitinen, J. Amer. Chem. Soc., 64, 2365 (1942).

[49]. G. J. Hoytink and J. van Schooten, Rec. Trav. Chim. Pays-Bas, 71, 1089 (1952).

[50]. I. Bergman, Polarography, 1964, Macmillan, New York, 1966, p. 925.

[51]. R. D. Allendoerfer and P. H. Rieger, J. Amer. Chem. Soc., 87, 2336 (1965).

[52]. R. B. Bates, R. H. Carningham, and C. E. Staples, ibid., 85, 3030 (1963).

[53]. M. J. S. Dewar, Discuss. Faraday Soc., 2, 261 (1947); Adv. Chem. Phys., 8, 102 (1965); Molecular Orbital Theory of Organic Chemistry, McGraw-Hill, New York, 1969, p. 291.

[54]. A. Streitwieser and S. Suzuki, Tetrahedron, 16, 153 (1961).

[54a]. M. J. S. Dewar, J. A. Hashmall, and N. Trinajstic, J. Amer. Chem. Soc., 92, 5555 (1970).

[55]. G. J. Hoytink, Rec. Trav. Chim. Pays-Bas, 76, 775 (1961).

[56]. M. A. Devanathan, Trans. Faraday Soc., 50, 573 (1954).

[57]. D. C. Grahame, Z. Elektrochem., 59, 740 (1955).

[58]. A. Misono, T. Osa, and T. Yamagishi, Bull. Chem. Soc. Japan, 41, 2921 (1968).

[59]. M. Smith, in Reduction (R. L. Augustine, ed.), Dekker, New York, 1968, p. 95.

[60]. R. L. Augustine, Catalytic Hydrogenation, Dekker, New York, 1968.

[61]. S. Wawzonek and D. Wearring, J. Amer. Chem. Soc., 81, 2067 (1959).

[62]. J. Harrison, Tetrahedron Lett., 6457 (1966).

[63]. J. W. Loveland, U.S. Pat. 3,032,489 (1962); through CA, 57, 4470 (1962).

[64]. J. W. Loveland, U.S. Pat. 3,214,356 (1965); through CA, 64, 671d (1966).

[65]. W. C. Neikam, U.S. Pat. 3,344,045 (1967); through CA, 68, 8809 (1968).

[66]. W. C. Neikam, U.S. Pat. 3,344,046 (1967); through CA, 67, 121859 (1967).

[67]. B. J. Woodhall and G. R. Davies, Meeting of Society for Electrochemistry, Chester, England, April 1969.

[68]. S. Bank and T. A. Lois, J. Amer. Chem. Soc., 90, 4505 (1968).

[69]. S. Bank and D. A. Noyd, Tetrahedron Lett., 1413 (1969).

[70]. W. Closson, P. Wriede, and S. Bank, J. Amer. Chem. Soc., 88, 1581 (1966).

[71]. L. Horner and H. Güsten, Justus Liebigs Ann. Chem., 652, 99 (1962).

[72]. W. Büchner, Chem. Ber., 99, 1485 (1966).

[73]. D. Weyenberg and L. H. Toporcer, J. Amer. Chem. Soc., 84, 2843 (1962).

[74]. D. Lipkin, F. R. Galiano, and R. W. Jordan, Chem. Ind. (London), 1657 (1963); J. F. Garst, J. T. Barbas, and F. E. Barton, J. Amer. Chem. Soc., 90, 7159 (1968); G. D. Sargent and G. A. Lux, ibid., 90, 7160 (1968).

[75]. J. W. Sease, R. C. Reed, and A. J. Fry, unpublished work quoted by O. R. Brown and J. A. Harrison, J. Electroanal. Chem. Interfacial Electrochem., 21, 387 (1969).

[76]. E. E. van Tamelen, R. B. Fechter, and S. W. Schneller, J. Amer. Chem. Soc., 91, 7196 (1969).

[77]. R. A. Marcus, J. Chem. Phys., 43, 2654 (1965); Electrochim. Acta, 13, 1217 (1968).

[78]. D. M. Hercules, Accounts Chem. Res., 2, 301 (1969).

[79]. A. Zwieg, in Advances in Photochemistry (W. A. Noyes, G. S. Hammond, and J. N. Pitts, eds.), Vol. 6, Wiley-Interscience, New York, 1968, p. 425.

[80]. D. M. Hercules, in Physical Methods in Organic Chemistry (B. Rossiter, ed.), Wiley, New York, 1970.

[81]. D. M. Hercules, Science, 145, 808 (1964).

[82]. K. S. V. Santhanam and A. J. Bard, J. Amer. Chem. Soc., 87, 139 (1965).

[83]. D. M. Hercules, R. C. Lansbury, and D. K. Roe, ibid., 87, 139 (1965).

[84]. A. Zwieg, D. L. Maricle, J. S. Brinen, and A. H. Maurer, ibid., 89, 473 (1967).

[85]. D. L. Maricle and A. H. Maurer, ibid., 89, 188 (1967).

[86]. L. R. Faulkener and A. J. Bard, ibid., 90, 6284 (1968).

[87]. R. E. Visco and E.A. Chandross, Electrochim. Acta, 13, 1187 (1968).

[88]. J. Chang, D. M. Hercules, and D. K. Roe, ibid., 13, 1197 (1968).

[89]. D. L. Maricle, A. Zwieg, A. H. Maurer, and J. S. Brinen, ibid., 13, 1209 (1968).

[90]. R. E. Visco and E. A. Chandross, J. Amer. Chem. Soc., 86, 5350 (1964).

[91]. E. A. Chandross, J. W. Longworth, and R. E. Visco, ibid., 87, 3259 (1965).

[92]. A. Zwieg, A. H. Maurer, and B. G. Roberts, J. Org. Chem., 32, 1322 (1967).

[93]. A. Zwieg, W. G. Hodgson, W. H. Jura, and D. L. Maricle, Tetrahedron Lett., 182 (1963).

[94]. J. T. Maloy, K. B. Prater, and A. J. Bard, J. Phys. Chem., 72, 4348 (1968).

[95]. R. E. Visco, R. Sonner, and E. A. Chandross, unpublished work quoted in Ref. [87]; G. J. Hoytink, Discuss. Faraday Soc., No. 45, 14 (1968).

[96]. S. W. Feldberg, J. Amer. Chem. Soc., 88, 390 (1966); J. Phys. Chem., 70, 3928 (1966).

[97]. A. Weller and K. Zachariasse, J. Chem. Phys., 70, 3928 (1966).

[98]. See discussion following Refs. [87-89].

[99]. M. D. Malbin and H. B. Mark, J. Phys. Chem., 73, 2992 (1969).

[100]. W. Schlenk, J. Appenrodt, A. Michael, and A. Thal, Chem. Ber., 47, 473 (1914).

[101]. K. Ziegler and K. Bähr, ibid., 61, 253 (1928).

[102]. M. Szwarc, M. Levy, and R. Milkovich, J. Amer. Chem. Soc., 78, 2656 (1956).

[103]. M. Szwarc, Ber. Bunsenges. Phys. Chem., 67, 763 (1963); Carbanions, Living Polymers and Electron Transfer Processes, Wiley, New York, 1968; Progress in Physical Organic Chemistry (A. Streitwieser and R. W. Taft, eds.), Vol. 6, Wiley-Interscience, New York, 1968, p. 323.

[104]. J. D. Anderson, J. Polym. Sci., Part A-1, 6, 3185 (1968).

[105]. L. D. McKeever and R. Waack, J. Organometal. Chem., 17, 142 (1969).

[106]. B. L. Funt and S. W. Laurent, Can. J. Chem., 42, 2728 (1964).

[107]. B. L. Funt and S. N. Bhadini, ibid., 42, 2733 (1964).

[108]. B. L. Funt, D. Richardson, and S. N. Bhadini, ibid., 44, 711 (1966).

[109]. N. Yamazaki, S. Nakahama, and S. Kambara, J. Polym. Sci., Part B, 3, 57 (1965).

[110]. B. L. Funt, S. N. Bhadani, and D. Richardson, ibid., Part A, 4, 2871 (1966).

[111]. M. M. Baizer and J. D. Anderson, J. Org. Chem., 30, 1348 (1965).

[112]. B. J. McClelland, Chem. Rev., 64, 301 (1964).

[113]. A. C. Aten, C. Büthker, and G. J. Hoytink, Trans. Faraday Soc., 55, 324 (1959).

[114]. T. J. Katz, W. H. Reinmuth, and D. E. Smith, J. Amer. Chem. Soc., 84, 802 (1962).

[115]. R. Dietz and M. E. Peover, Trans. Faraday Soc., 62, 3535 (1966).

[116]. T. L. Chu and S. C. Yu, J. Amer. Chem. Soc., 76, 3367 (1954).

[117]. D. Balk, G. J. Hoytink, and J. W. H. Schreurs, Rec. Trav. Chim. Pays-Bas, 76, 813 (1957); G. J. Hoytink and P. H. van der Meij, Z. Phys. Chem. (Frankfurt), 20, 1 (1959).

[118]. G. J. Hoytink, E. de Boer, P. H. van der Meij, and W. P. Weijland, Rec. Trav. Chim. Pays-Bas, 75, 487 (1956).

[119]. M. E. Peover, private communication, 1967.

[120]. N. H. Velthorst and G. J. Hoytink, J. Amer. Chem. Soc., 87, 4529 (1965).

[121]. R. Dietz and B. E. Larcombe, J. Chem. Soc. B, 1369 (1970).

[122]. H. Volz and W. Lotsch, Tetrahedron Lett., 2275 (1969); G. Kothe, W. Summermann, H. Baumgärter, and H. Zimmermann, ibid., 2185 (1969).

[123]. S. I. Zhdanov, Z. Phys. Chem. (Leipzig), 235 (1958).

[124]. R. Breslow, W. Bahary, and W. Reinmuth, J. Amer. Chem. Soc., 83, 1763 (1961).

Chapter VI

ELECTROCHEMICAL REDUCTION OF ORGANIC HALIDES

M. R. Rifi

Union Carbide Corporation
Bound Brook, New Jersey

I. INTRODUCTION

Numerous papers (including some review articles [1, 2, 87]) have appeared in the literature concerning the electrochemical cleavage of carbon-halogen bonds. Nevertheless, few papers have dealt with the isolation and analysis of the products obtained from the cleavage of such bonds. For this reason, the elucidation of the mechanism of this cleavage has been arrived at mainly from polarographic data.

While much valuable information, such as multiple reduction steps, and the stereochemistry of carbon-halogen bonds in molecules [3, 4] can be obtained from polarographic plots, diffusion constants and adsorption phenomena make it difficult to assign potential-determining steps to the proper chemical reactions. Thus, in discussing the mechanism of the cleavage of carbon-halogen bonds, emphasis must be placed on the products of the reaction.

II. REDUCTION PRODUCTS

In general, macroscale electrolysis of monohalides leads to the formation of the corresponding hydrocarbons [5-10]. However, some coupling products have been observed [11, 12]. Vicinal dihalides afford olefins [13] and α,ω-dihalides afford cyclic as well as acyclic hydrocarbons [14-16]. Gem-Halides afford mostly hydrocarbons [17, 18]. In some cases, carbene intermediates have been trapped [17].

III. MECHANISM OF REDUCTION

A. General Considerations

The electrochemical reduction of organic halides is irreversible, follows the Ilkovic equation, and is in general independent [1] of the pH of the medium [13, 20, 21]. Elving and Pullman [73] have advanced the following mechanism:

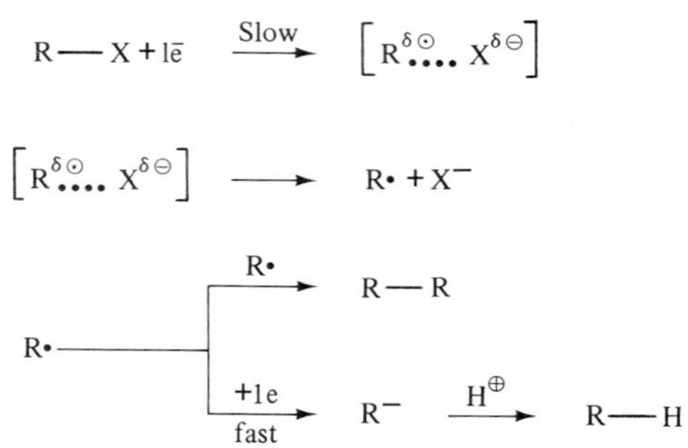

This mechanism suggests that one electron is added in the rate-determining step to produce a transition state in which the carbon atom has radical character. Thus, radicals formed from such a transition state could lead to dimeric products. This hypothesis may be supported by the finding [12] that the electrolysis of p-nitrobenzyl bromide affords p-dinitrobibenzyl.

Klopman also found that p-substituents such as methoxy, methyl, bromo, carboxy, and nitro groups all shift the half-wave potential of benzyl bromide to a more positive value, which is consistent with a radical-like transition state. Similarly, Streitwieser and Perrin [74] compared the half-wave potential of substituted benzyl chlorides. While no satisfactory Hammett σ-ρ plot could be obtained, the deviation was discussed in terms of radical-stabilizing effects of substituents. On the other hand, Sease and coworkers [22] found that plots of half-wave potentials vs. Hammett's σ for substituted benzyl bromides and halobenzenes indicate a negatively charged potential-determining transition state. For the alkyl halide series, Lambert [75] has shown a linear correlation between their half-wave potential and the corresponding Taft polar substituents (σ^*) and has pointed out that the slope of the line indicates a negatively charged transition state. Wawzonek et al. [23] have found that phenylacetic acid was obtained from the macroscale electrolysis of benzyl bromide in the presence of carbon dioxide.

The reduction of alkyl halides in dimethyl formamide (DMF) was considered by Kaabak and coworkers [24] to proceed via the following complex:

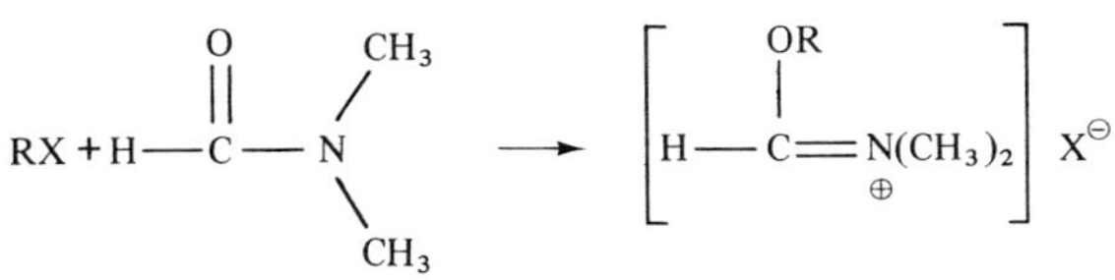

which upon hydrolysis yields esters and ammonium halides. Among the other products isolated were hydrocarbons, amines, dimers, carbon monoxide, and hydrogen. The formation of these products was explained by the authors as follows:

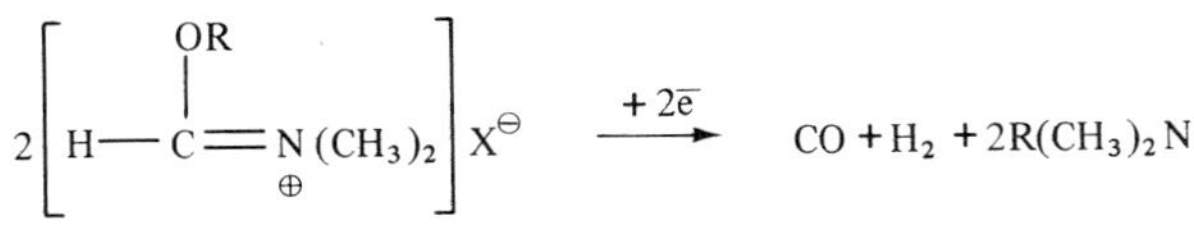

The authors, however, found that alkyl halides in DMF without electrolysis failed to give products. While the above hypothesis is interesting, there is no chemical evidence to substantiate it and other explanations are available for the formation of the products observed.

It can be seen from the above discussion that the mechanism of the electrochemical cleavage of carbon-halogen bond has not yet received unanimous agreement.

For the electrochemical cleavage of carbon-halogen bonds, several factors must be considered, some of which are the following: (a) orientation of the bond with respect to the cathode[2]; (b) effect of the electric field on the polarization of the bond; (c) effect of structure on the ease of reduction of carbon-halogen bonds; and (d) formation of complexes between cathode and organic halides.

TABLE 1

Half-Wave Potential of Organic Halides [a, b]

Organic halide	Solution	$-E_{1/2}$	Reference
Monohalides (aliphatic; group 1)			
Methyl bromide	75% dioxane	2.01	[63]
Methyl chloride	"	2.23	[63]
Methyl iodine	"	1.63	[63]
Ethyl bromide	N, N, -Dimethyl-formamide (DMF)	2.13	[25]
Butyl bromide	"	2.23	[25]
Neopentyl bromide	"	2.37	[25]
Cyclopropyl bromide	"	2.36	[25]
Cyclobutyl bromide	"	2.36	[25]
Cyclohexyl bromide	"	2.29	[25]
Benzyl bromide	"	1.22	[63]
$Br-CH_2-CH_2-CH_2-N^{\oplus}(C_2H_5)_3\,^{\ominus}Br$	"	1.34	[63]
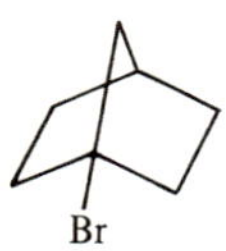		2.17 (Ag/Ag⁺)	[27]
	95% DMF	1.99	[28]
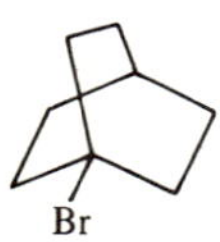	DMF	1.79 (Ag/Ag⁺)	[27]

Table 1 (cont.)

Organic halide	Solution	$-E_{1/2}$	Reference
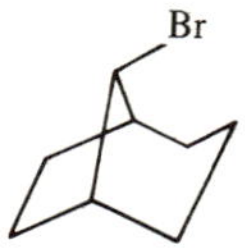	95% DMF	2.15	[28]
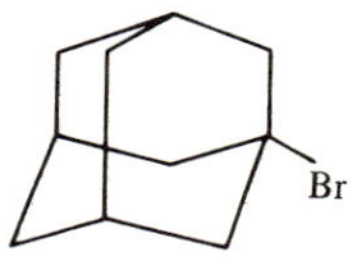	95% DMF	2.02	[28]
1-Bromotripticine	DMF	1.42 (Ag/Ag^+)	[27]
Monohalides (vinyl; group 2a)			
Vinyl bromide	75% Dioxane	2.47	[13]
Vinyl chloride	75% Dioxane	No wave	[13]
Monohalides (aromatic; group 2b)			
Bromobenzene	DMF	2.32	[63]
Chlorobenzene	DMF	2.54	[63]
α-Bromonaphthalene	75% Dioxane	1.96, 2.38	[13]
α-Chloronaphthalene	75% Dioxane	2.10, 2.38	[13]
Monohalide (allylic; group 3)			
Allyl bromide	DMF	1.29	[48]
Allyl chloride	DMF	1.91	[48]
Allyl iodide	DMF	0.23, 1.16	[48]
Dihalides (vicinal-aromatic; (group 4a)			
o-Dibromobenzene	DMF	1.28	[43]
o-Bromochlorobenzene	DMF	1.38	[43]

Table 1 (cont.)

Organic halide	Solution	$-E_{1/2}$	Reference
Dihalides (vicinal-aliphatic; (group 4b)			
1, 2-Dibromoethane	95% DMF	1.23	[28]
1, 2-Dibromoethane	DMF	1.38	[15]
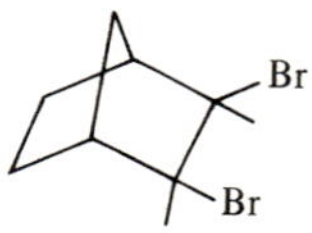	95% DMF	1.53	[28]
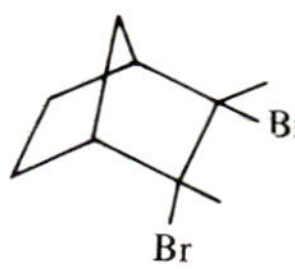	95%	1.21	[28]
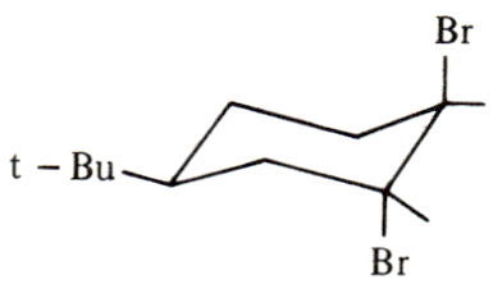	95% DMF	0.86	[28]
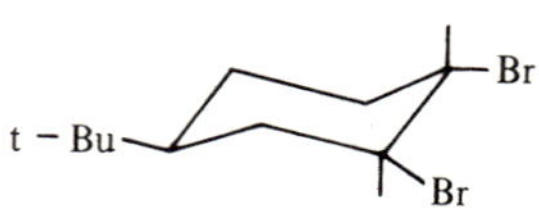	95% DMF	1.67	[28]
Dihalides (nonvicinal aromatic; (group 5a)			
m-Dibromobenzene	DMF	1.40, 1.88	[43]
Dihalides (nonvicinal aliphatic; (group 5b)			
1, 3-Dibromopropane	DMF	1.91	[15]

Table 1 (cont.)

Organic halide	Solution	$-E_{1/2}$	Reference
1, 4-Dibromobutane	DMF	1.99, 1.95 [c]	[15]
1, 5-Dibromopentane	DMF	2.14, 2.20 [c]	[15]
1, 6-Dibromohexane	DMF	2.13, 2.19 [a]	[15]
1, 3-Dibromo-1, 3-dimethyl-cyclobutane (cis, trans)	DMF	2.02	[15]
1-Bromo-3-bromomethyl-cyclobutane	DMF	2.08	[15]
cis-1, 6-Dibromo cyclodecane	95% DMF	1.78	[28]
trans-1, 6-Dibromo cyclodecane	95% DMF	1.73	[28]
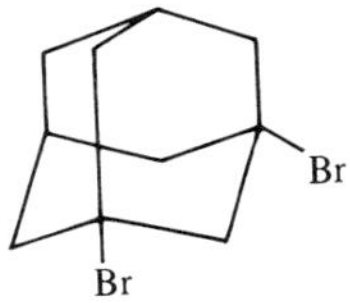	DMF	1.93	[63]
α, α′-Dibromo-m-xylene	DMF	1.32	[16]
α, α′-Dibromo-o-xylene	DMF	0.61, 1.58	[16]
α, α′-Dibromo-p-xylene	DMF	0.80, 1.72	[16]
α, α′-Dichloro-p-xylene	DMF	1.2	[48]
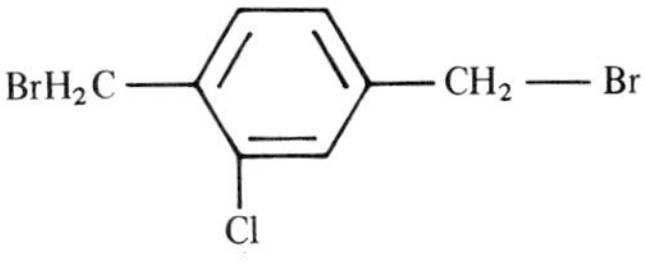	DMF	1.2	[48]
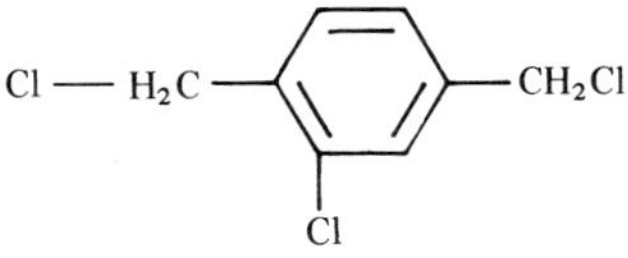	DMF	1.4	[48]
p-Dibromobenzene	DMF	1.53, 1.82	[43]
p-Bromochlorobenzene	DMF	1.60, 2.02	[43]
p-Dichlorobenzene	DMF	2.20	[63]

Table 1 (cont.)

Organic halide	Solution	$-E_{1/2}$	Reference
4, 4′-Dibromodiphenyl	DMF	1.80	[63]
4, 4-Dichlorodiphenyl	DMF	2.01	[63]
gem-Halides (group 6)			
Methylene dibromide	75% Dioxane	1.48	[13]
Methylene dichloride	75% Dioxane	2.23	[13]
Bromoform	75% Dioxane	0.49, 1.09	[13]
Chloroform	75% Dioxane	1.67	[13]
Tetrabromomethane	75% Dioxane	0.3, 0.75	[13]
Tetrachloromethane	75% Dioxane	0.78, 0.71	[13]
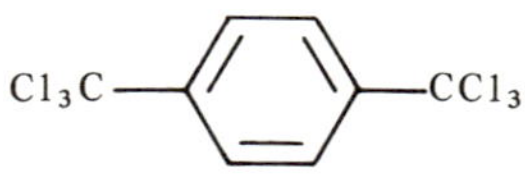	DMF	0.7	[48]
	DMF	1.1	[48]

[a] Unless otherwise specified, mercury was used as the cathode and a calomel electrode as the reference.

[b] For further information the following key references should be consulted: [28, 83].

[c] Ascending part of the curve is deformed by a sharp maximum.

While attempts have been made to describe the above factors mathematically, the organic chemist who uses electrolysis as a tool in organic synthesis has long looked for a unified (hence simple) description. Consequently, the mechanism of the different classes of organic halides advanced below will be discussed in terms of the transition state of the reactions and the products isolated. The reader will be assisted in following the discussion by referring to Table 1 in which the reported half-wave potentials of organic halogen compounds have been grouped to correspond to the sequence followed below.

B. Alkyl Halides (Table 1, Group 1)

The ease of the reduction of alkyl halides follows the orders RI >RBr >RCl >RF. These reductions have been considered by some authors [1, 25] as resembling SN_2 (or SN_1, depending on steric effects [26]) displacement where electrons act as nucleophiles and attack the back side of the molecule, thus displacing a halide anion. This analogy was considered as consistent with the increased difficulty of reduction,[3] with an increase of the size of α-substituents, and with the observation that the following reaction [77] proceeds with inversion of configuration:

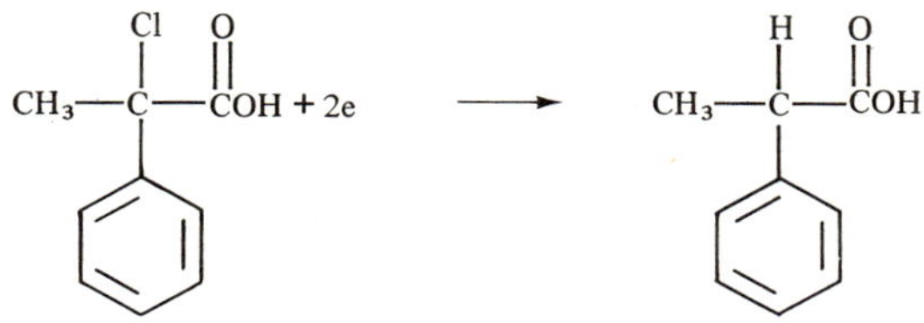

On the other hand, other workers [27] prefer an attack by the electrons on the halogen side and argue that carbon atoms in compounds such as (I), (II), and (III) are too hindered for a back-side displacement:

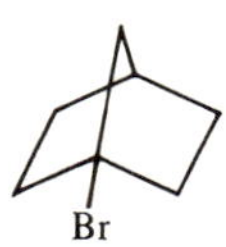

(I)

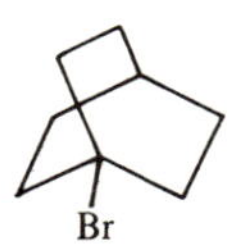

(II)

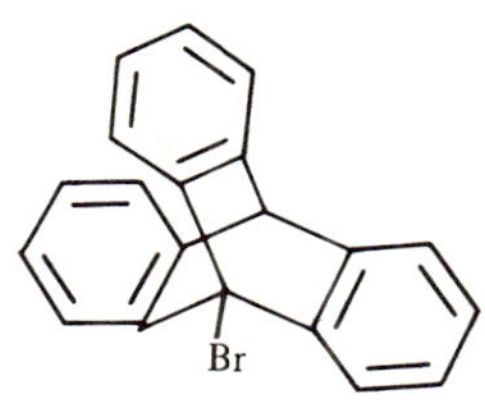

(III)

$E_{1/2}$ (Ag/Ag^+) −2.17 −1.79 −1.42

Whether electrons add to the halogen or carbon atoms has so far provided little, if any, information concerning the cleavage of carbon-halogen bonds. In fact, based on the half-wave potentials of several alkyl bromides and their solvolysis rates, Zavada and coworkers [28] have questioned the value of comparing electroreduction with SN processes.

Further controversy arises from a discussion on the addition of electrons in the rate-determining step for the cleavage of alkyl halides. While the addition of one electron in the slow step is generally assumed, the addition of two electrons [1, 9, 15] should not be neglected, as this results in no violation of any fundamental principles. How many electrons are added in the slow step may depend on the structure of the alkyl halide in question. For example, there is no product arising from the cleavage of simple alkyl halides that requires a radical-like transition state. Any coupling product isolated may be explained easily by the attack of anionic intermediates on the starting material.

In an attempt to explain radical or anionic intervention in the electrolysis of organic halides, Mann and coworkers [29] electrolyzed the optically pure 1-bromo-1-methyl-2, 2-diphenylcyclopropane and obtained the corresponding hydrocarbon with 25% retention of configuration. There was also an indication that some organomercury compound was formed. Since the cyclopropyl radical would have afforded products with complete racemization [30] and the cyclopropyl anion products with complete retention [31], Walborsky (see [29]) proposed the following mechanism:

$$\text{R-Br} + e^- \rightarrow \left[\text{R}^{\delta\ominus}\text{-}\overset{\delta\ominus}{\text{Br}}\right] \qquad (1)$$

$$\left[\text{R}^{\delta\ominus}\text{-}\overset{\delta\ominus}{\text{Br}}\right] \rightarrow \text{R}\cdot + \text{Br}^{\ominus} \qquad (2)$$

$$\text{R}\cdot + e^- \rightarrow \text{R}^{\ominus} \qquad (3)$$

$$\text{R}\cdot + \text{Hg}^{\circ} \rightarrow \text{RHg}\cdot_n \qquad (4)$$

$$\text{RHg}\cdot_n + \bar{e} \rightarrow \text{R}^- + \text{Hg}_n^{\circ} \qquad (5)$$

$$\text{RHg}_n\cdot + \text{RHg}_n\cdot \rightarrow \text{RHg}_n\text{R} \rightarrow \text{RHgR} + \text{Hg}_{n-1}^{\circ} \qquad (6)$$

$$\text{R}_2\text{Hg} + \bar{e} \rightarrow \text{R}^{\ominus} + \text{RHg}\cdot \xrightarrow{\text{Hg}^{\circ}} \text{RHg}\cdot_n \qquad (7)$$

$$\text{R}^- + \text{H}^{\oplus} \rightarrow \text{RH}. \qquad (8)$$

While the above mechanism may be consistent with the observed data, extreme caution should be observed in comparing the retention of configuration of anionic species prepared by different methods. For it seems hardly fair to compare a carbonlithium anion [31], which is somewhat covalent, with the same anion derived from the electrolysis of the corresponding bromide. The organomercuric compound formed in the reduction of the cyclopropyl bromide may have arisen from the reaction of the alkyl halide with mercury [32-34].

Further work on optically active cyclopropyl halides was carried out by Annino and coworkers [5]. These authors concluded that there was no chemical evidence for the intervention of radical or carbonium ion intermediates.

Sease and Reed [10] have recently electrolyzed 1-bromohexane and obtained hexane, hexene, and dihexylmercury. The formation of the last compound was considered to be an indication for the formation of hexyl radical which was trapped by mercury. While mercury is a good trap for radicals, dihexylmercury may have been formed in a manner similar to that reported for the formation of bibenzylmercury [32], namely:

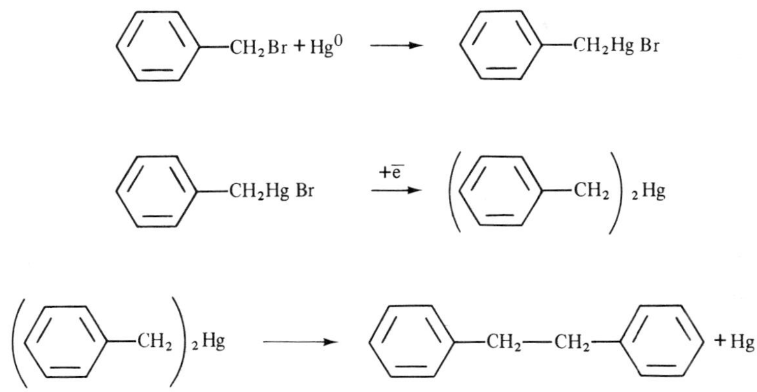

The formation of dihexylmercury may also be explained by the scheme advanced by Kirrman and Kleine-Peter [42] (see discussion of allyl bromide).

Perhaps the best argument for a radical-like transition state can be found in the reduction of activated halides such as compounds (IV)-(VI)

$Br(CH_3)_2C\text{-}CO_2H$ $Br(CH_3)CH\text{-}CO_2H$ $BrCH_2CO_2H$

(IV) (V) (VI).

The ease of reduction of these compounds is (IV) > (V) > (V), which is the reverse of what would be expected from steric effects and anionic stability. However, it is not clear why increasing the length of the alkyl chain in the α-bromoalkyl acids leads to a steady increase in the ease of reduction, although it may be argued that the greater adsorbability of the longer-chain compounds provides an increase in the entropy of activation for the reduction [78].

The reduction of halopropionitrile was studied by Tomilov and Smirnov [89] and Feoktistov and Zhdanov [90]. Thus, the electrolysis of β-iodipropionitrile in 0.5 N sulfuric acid afforded propionitrile (40%) and $(NC\text{-}CH_2CH_2)_2Hg$ (45.9%). These products were explained by the following mechanism:

$$ICH_2CH_2CN + Hg + \bar{e} \rightarrow NC\text{-}CH_2CH_2Hg^{\bullet} + I^{\ominus}$$

$$2NC\text{-}CH_2CH_2Hg^{\bullet} \rightarrow (NC\text{-}CH_2CH_2)_2Hg + Hg.$$

Feoktistov and Zhdanov [19, 91] studied the polarographic behavior of halopropionitriles and found the following interesting observations:

β-Chloropropionitrile did not reduce at the dropping mercury electrode while β-bromopropionitrile exhibited a two-electron wave. β-Iodopropionitrile, on the other hand, exhibited two one-electron waves. These waves, however, could be observed only at high concentration of the nitrile. The authors explained these results as follows:

$$I\text{-}CH_2\text{-}CH_2CN + \bar{e} \rightleftharpoons [I\text{-}CH_2\text{-}CH_2CN]^{\ominus}$$

$$2[I\text{-}CH_2CH_2CN]^{\ominus} \rightarrow NC\text{-}(CH_2)_4CN + 2I^{\ominus}$$

$$[I\text{-}CH_2\text{-}CH_2CN]^{\ominus} + \bar{e} \xrightarrow{H^{\oplus}} CH_3CH_2CN + I^{\ominus}.$$

While the above mechanism may appear to be consistent with the observed data, the nature of the product from the first reduction step is not understood. Furthermore, because of the weakness of the carbon-iodine bond (compared to C-Br and C-Cl), it is possible that a reduction similar to that described for allyl halides [42] is taking place and which would be consistent with the observed data:

$$I\text{-}CH_2CH_2\text{-}CN + Hg \rightarrow I\text{-}Hg\text{-}CH_2CH_2CN$$

$$I\text{-}Hg\ CH_2CH_2CN \rightleftharpoons {}^{+}Hg\text{-}CH_2CH_2CN + I^{\ominus}$$

$${}^{\oplus}Hg\text{-}CH_2CH_2CN + \bar{e} \rightleftharpoons \cdot Hg\text{-}CH_2CH_2CN$$

$$2\cdot Hg\text{-}CH_2CH_2CN \rightarrow NC\text{-}(CH_2)_4\text{-}CN + 2Hg$$

$$\cdot Hg\text{-}CH_2\text{-}CH_2CN + \bar{e} \xrightarrow{H^{\oplus}} CH_3CH_2CN + Hg.$$

Products obtained from the reduction of benzyl halides under different reaction conditions have led to the elucidation of different reaction mechaisms involving radical and anionic intermediates. Since it has been shown from simple molecular orbital (MO) calculations (which do not include solvation energy) that the benzyl radical and anion posess the same resonance energy [35], it is likely, but not necessary, that both intermediates are responsible for the observed products. For example, the formation of 4,4′-dinitrobibenzyl from the reduction of p-nitrobenzyl bromide [12] may have resulted from the dimerization of p-nitrobenzyl radical, or from the attack of p-nitrobenzyl anion on the starting material. The isolation of toluene and phenylacetic acid from the electrolysis of benzyl chloride [23] in the presence of carbon dioxide strongly supports the formation of a benzyl anion.

C. Vinyl and Aromatic Halides (Table 1, Groups 2a and 2b, Resp.)

Aromatic and vinyl halides are more difficult to reduce than alkyl halides and afford the corresponding aromatic and olefinic products. While there is little or no doubt that these products are derived from anionic intermediates, the pathway leading to such intermediates has been the subject of some controversy. Thus, in the vinyl series, an electron [11] (or perhaps two electrons) may add to the carbon-halogen bond to give the radical which undergoes further reduction to the anion. Alternatively, an electron [37] may add to the lowest unoccupied Π molecular orbital to give the anion radical $(\overset{\ominus}{C}-\overset{\odot}{C}X)$ which loses a halide anion to form the vinyl radical. The radical then undergoes further reduction to the anion.

In the reduction of vinyl halides, the configurational stability of the intermediate (radical or anion) becomes important as it will determine the stereochemistry of the product. Unfortunately, this question has not been settled. Thus, while Skell and Allen [64] consider the vinyl radical to be configurationally stable, others [65-67] have suggested that it undergoes rapid cis-trans equilibration.

The electrochemical reduction of monobromomaleic acid, monobromofumaric acid, and their corresponding diethyl esters has been studied by Elving and coworkers [11]. It was found that, with the exception of monobromomaleic acid, the products obtained were fumaric acid and its diethyl ester. The reduction of monobromomaleic acid afforded fumaric acid, maleic acid, and 1, 2, 3, 4-butanetetracarboxylic acid. The relative amounts of these products were pH dependent with the yield of fumaric acid, increasing with an increase of pH at the expense of maleic acid and the dimeric product. Elving and coworkers explained these results by invoking vinyl radicals as intermediates, which couple and accept an electron and a proton to give the observed products. This, however, does not explain why the coupling reaction should be pH sensitive. Furthermore, the dimeric product may have been formed from an anionic attack on the starting material via a Michael-type addition:

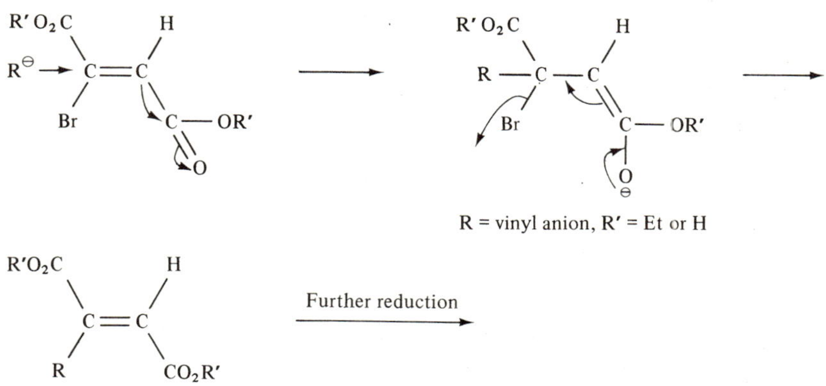

R = vinyl anion, R′ = Et or H

Anionic attack on activated olefins in aqueous solution has been reported by Petrovich and coworkers [36].

Miller and Riekena [37] have suggested that in the reduction of activated vinyl halides, an electron is added to the Π system to give the radical anion:

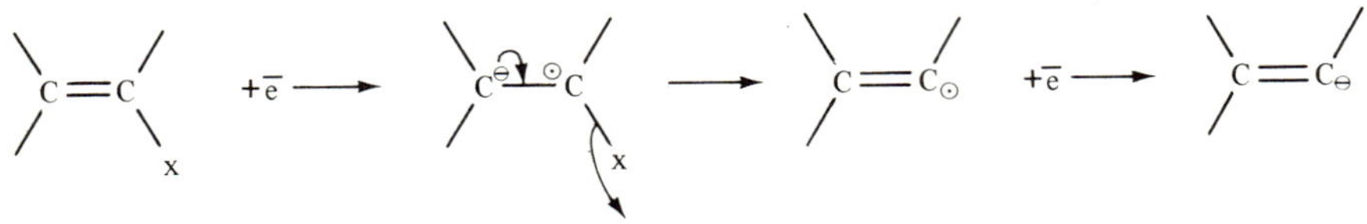

While the elimination of the halide ion may be fast, rotation about the σ-bond may take place prior to elimination to afford the intermediate which would lead to the thermodynamically stable product.

Considerable work has been done on the reduction of nitro-aromatic halides [38-40]. This reduction (with the exception of aromatic iodides) is complicated by the reduction of the nitro group at the same potential required to cleave the carbon-halogen bond. Reduction of nitro-halobenzene may take place at the aromatic ring as was postulated by Lawless and Hawley [41]:

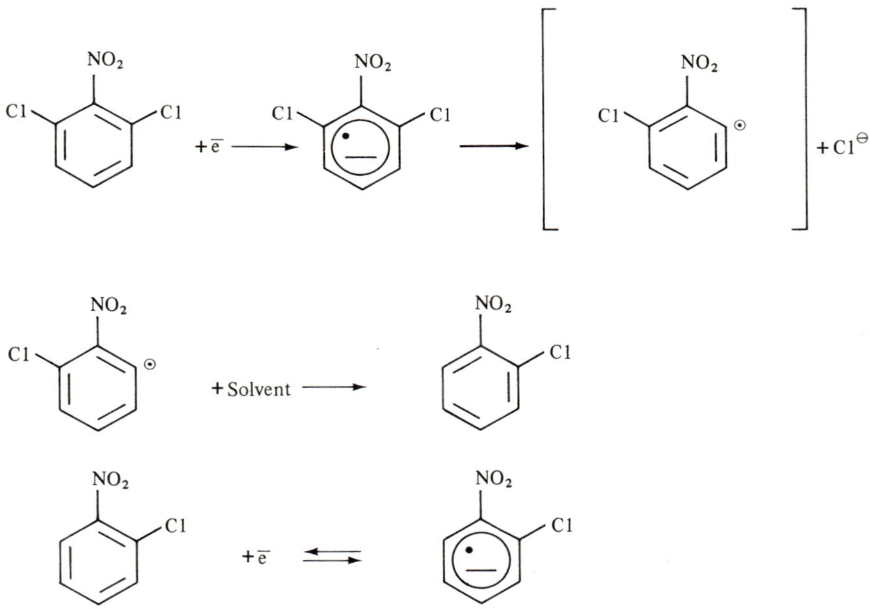

It is of interest that the radical anion of o-chloronitrobenzene does not lose a chloride ion to form nitrobenzene.

The reduction of halopyridines [39, 40, 85] was found (in some cases) to be dependent on the pH of the medium. The reduction occurs at more positive potentials at low pH values. This may be an indication that the pyridinium salt is being reduced in a one-step two-electron reduction via an anionic-like transition state.

D. Allyl Halides (Table 1, Group 3)

The reduction of allyl bromides was studied by Kirrman and Kleine-Peter [42]. From the polarographic behavior of these compounds and their organomercury derivatives isolated, the following mechanism was postulated for their reduction:

$$RBr \xrightarrow{Hg} RHgBr$$

$$RHgBr \rightleftarrows RHg^{\oplus} + Br^{\ominus}$$

$$RHg^{\oplus} + \bar{e} \rightarrow RHg\cdot$$

$$2RHg\cdot \rightarrow R_2Hg_2$$

$$R_2Hg_2 \rightarrow R_2Hg + Hg$$

$$2RHg\cdot \rightarrow R\text{-}R + 2Hg$$

$$RHg\cdot + \bar{e} \xrightarrow{H^{\oplus}} RH + Hg.$$

E. Vicinal Dihalides

1. Aromatic (Table 1, Group 4a)

The reduction of aromatic dihalides was studied by Wawzonek and Wagenknecht [43]. With the exception of o-dibromobenzene, all exhibited two polarographic waves and upon macroscale reduction afforded benzene. o-Dibromobenzene, on the other hand, exhibited one polarographic wave at a potential 0.57 more positive than that of bromobenzene. Its macroscale reduction in the presence of furan afforded benzene and α-naphthol in very low yield. These results were explained by the authors as follows:

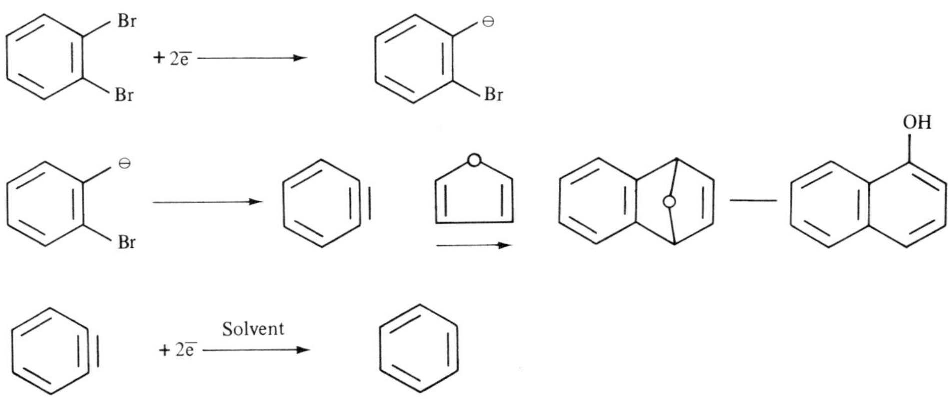

Such a mechanism does not account for a 0.57 V (~13.1 kcal/mole) difference in the reduction o- and m-dibromobenzene. Instead, it is likely that both bromines in o-dibromobenzene are involved in the transition state of a concerted mechanism (a more detailed discussion of such mechanism is described under aliphatic vicinal dihalides below):

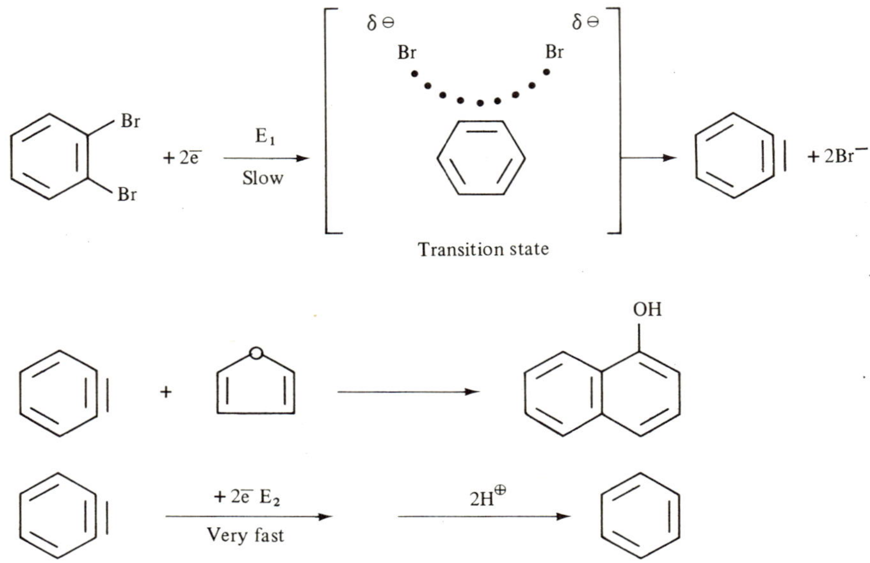

In the above mechanism, $E_2 < E_1$ and the yield of benzyne depends on its rate of diffusion away from the cathode, which will allow its reaction with furan.

2. Vinyl

1, 2-Di-haloolefins undergo two-electron reduction to give acetylenes which are reduced at a more negative potential than vinyl halides. In general, trans-dihalides reduce at a more positive potential than the cis isomers [80].

The reduction of the cis and trans isomers of 2, 3-dichloroacrylonitrile was studied by Jura and Gaul [81]. Based on the polarographic behavior of a mixture of the two isomers together with a knowledge of the reduction potentials of monochloroacrylonitriles they concluded that the two isomers reduce via different mechanisms:

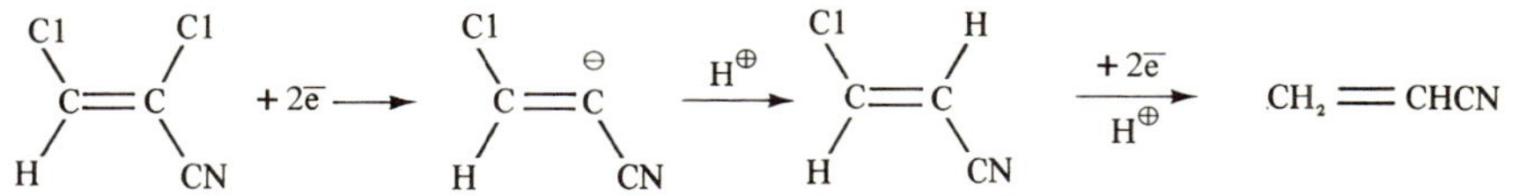

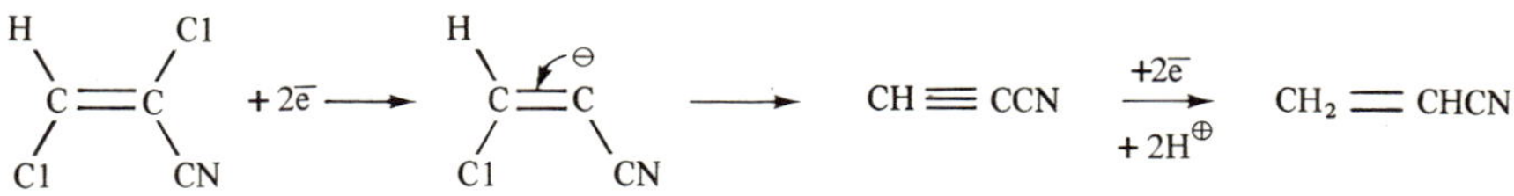

It is questionable whether the anionic intermediate formed from the trans isomer is capable of avoiding protonation to afford cyanoacetylene exclusively. Furthermore, the above mechanism is not consistent with the observation that trans-dihalides reduce at more positive potentials than the cis isomers. Thus, it is likely that the trans isomer reduces in a concerted mechanism which has a transition state that involves the displacement of both halogens.

3. Aliphatic (Table 1, Group 4b)

The polarographic reduction of a number of aliphatic vicinal dihalides was studied by Zavada and coworkers [28]. These authors correlated the ease of reduction of the dihalides with the dihedral angles between the two carbon-halogen bonds. Such angles, however, must be correlated with the transition state of the reaction which is related to the half-wave potentials of the dihalides. Thus, in base-promoted elimination reactions, the lowest energy transition state is the one in which the departing groups are trans to one another [44]. In cases where a trans elimination is not possible, cis elimination does take place [45]. The transition states for both eliminations, however, are of different energies. In all cases it is essential that the products of the reaction be known. Thus, while compounds (VII) and (VIII) reduce at different potentials [28] (products were not analyzed) their reduction may proceed via different transition states to afford different products.

(VII) $E_{1/2} = 0.8$ V

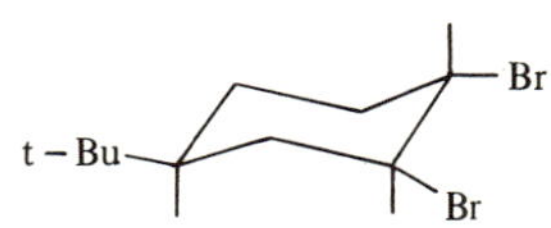

(VIII) $E_{1/2} = 1.67$ V

A mechanism [15] which was scantily described under the reduction of aromatic and olefinic vicinal dihalides above, and which is in general agreement with all the data (including product analysis) observed for the reduction of aliphatic vicinal dihalides, is visualized as follows:

$$\text{X-C-C-X} + 2\bar{e} \rightarrow \left[\underset{\text{Transition state}}{X^{\delta\ominus} \cdots\cdot C \underline{\cdots\cdot} C \cdots\cdot X^{\delta\ominus}} \right] \rightarrow \text{Products.}$$

In this concerted mechanism, the transition state of the reaction has a bond being developed. How well developed this bond is will be revealed in the half-wave potential of the dihalide. Such a mechanism can be extended to 1, 3- and 1, 4-dihalides. Thus, the reduction of 1, 3-dibromopropane [14] afforded cyclopropane as the only product while 1, 4-dibromobutane afforded cyclobutane and n-butane in 25 and 75% yield, respectively. Furthermore, the reduction of 1, 3-dibromopropane in excess water had no effect on the yield of cyclopropane. These observations are quite consistent with a concerted mechanism with a transition state that has resemblance to products.

The reduction of meso and racemic α, α′-dibromosuccinic acids was studied by Elving and coworkers [46] who isolated fumaric acid from both isomers. In this case, the carbonyl group stabilizes the anionic intermediate, which is formed in a stepwise mechanism, and allows the rotation of the σ-bond to afford the thermodynamically favored product:

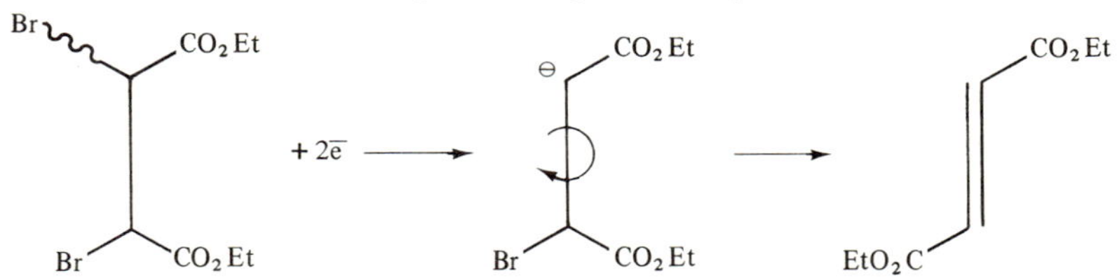

F. α, ω-Dihalides

1. Aromatic (Table 1, Group 5a)

The reduction m- and p-dihalobenzene has been studied by Wawzonek and Wagenknecht [43] and was described under vicinal halides above. The effect of structure on the polarographic reduction of iodobenzene was examined by Colichman and Liu [47]. In this study, a linear relationship was found between the half-wave potentials of the halides and Hammett's α-substituent constants. The reduction of p-fluoroiodobenzene in this series is unusual. It exhibited two polarographic waves at -0.72 and -1.66 V (SCE). However, under similar conditions, fluorobenzene could not be reduced. This was explained by the authors as follows[4]:

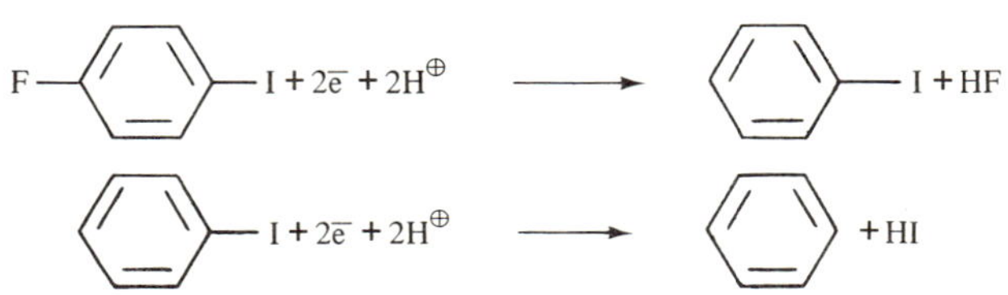

2. Aliphatic (Table 1, Group 5b)

The polarographic reduction of aliphatic α,ω-dihalides has been studied by several investigators [14, 15, 28, 46, 48, 49]. However, very little has been reported on the mechanism for the reduction of these dihalides [15, 16]. In the reduction of α,α'-dibromoxylenes, Covitz [16] found that the ortho isomer exhibited two polarographic waves at -0.61 and -1.58 V (SCE). Its macroscale electrolysis afforded a polymeric product which was identified as poly-o-xylylene. The para isomer reduced at -0.8 and -1.72 V (SCE) and upon macroscale electrolysis afforded poly-p-xylylene and a small amount of the cyclic dimer, p-cyclophane. On the other hand, α,α'-dibromo-m-xylene exhibited one polarographic wave at -1.32 V (SCE) and upon macroscale reduction afforded m-xylene. Under the same condition, benzyl bromide showed a half-wave potential at -1.22 V (SCE). These results were explained by the author as follows[5]:

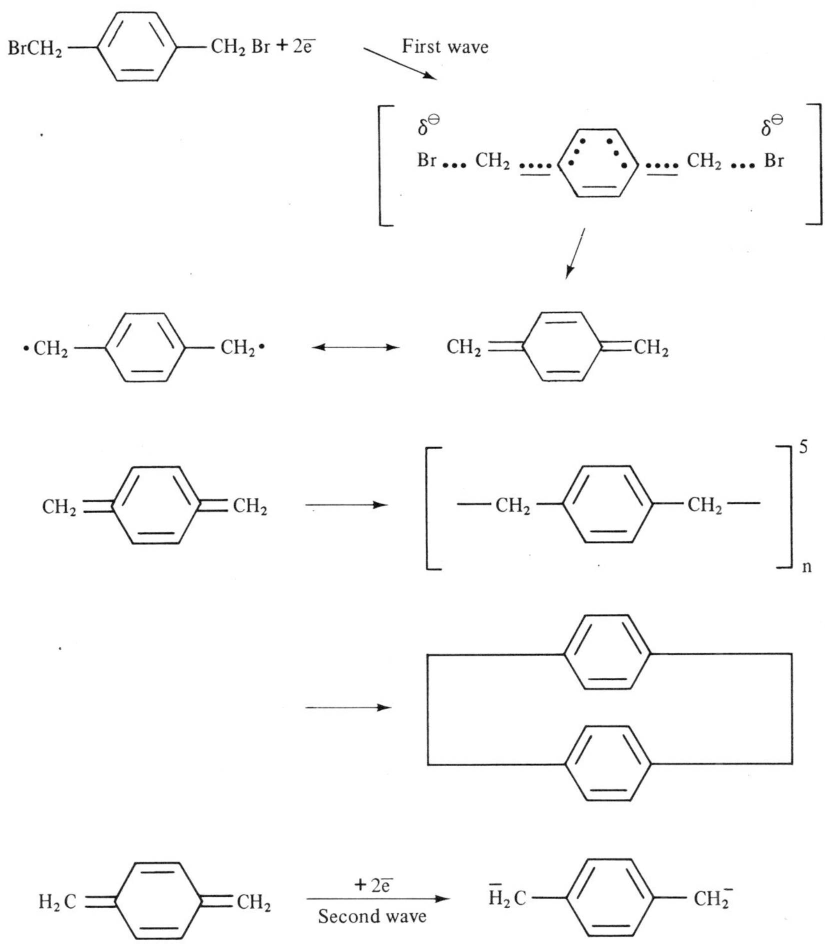

TABLE 2

Reduction of α,ω-Dibromides in 0.01 M Tetra-n-Butylammonium Perchlorate in N, N-Dimethylformamide

Dibromide	Products	$-E_{1/2}$ V (SCE)	αn[a]
$Br-CH_2-CH_2-CH_2-\overset{+}{N}Et_3Br$	 + Et_3N	1.34	0.43
$Br-CH_2-CH_2-CH_2Br$[b]	 only products	1.93	0.263
$Br-CH_2-(CH_2)_2-CH_2Br$	 + $CH_3(-CH_2)_2-CH_3$	1.95, 1.99[c]	0.233
$Br-CH_2-(CH_2)_3-CH_2Br$	$CH_3-(CH_2)_3-CH_3$	2.14, 2.20[c]	—
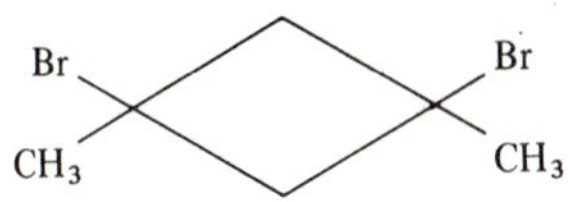 (cis and trans)	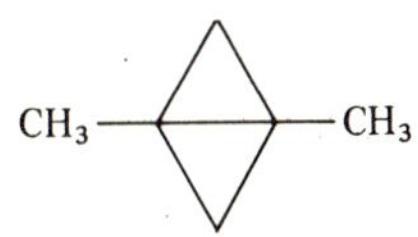	2.02[c]	0.246
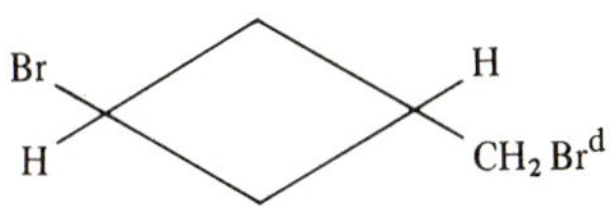 (mixture of isomers)	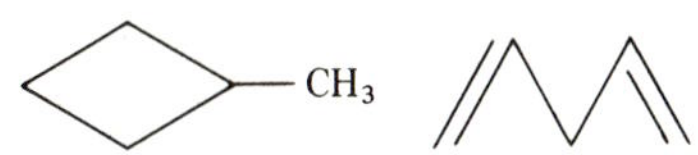	2.1	—
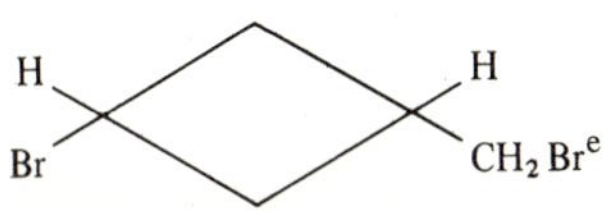		—	—

Footnotes to Table 2

[a]For a detailed discussion on the significance of (α) see Ref. [53].

[b]Reduction in the presence of excess water has no effect on the yield of cyclopropane.

[c]Ascending part of curve is deformed by a sharp maximum.

[d]Kindly supplied by K. B. Wiberg and V. Williams, Yale University.

[e]Reduction carried out in dried hexamethylphosphoramide.

The reduction of α, α'-dibromo-m-xylene occurs in a stepwise fashion under one polarographic wave with diffusion current nearly twice as large as each of the ortho and para isomers. The reduction of several α, ω-dihalides was studied by the author [14, 15] and is summarized in Table 2.

The data in Table 2 can best be explained as follows: The reduction of 1, 3-dibromopropane and 3-bromopropyltriethylammonium bromide proceeds via a concerted mechanism while the rest of the dihalides reduce in a stepwise fashion[6] to form anionic intermediates. The reduction of 1, 3-dibromo-1, 3-dimethylcyclobutane and 1-bromo-3-bromomethylcyclobutane is of interest. The cis and trans isomers of the first compound reduced at the same potential and afforded the same product. This indicates that the anionic intermediate equilibrates prior to cyclization:

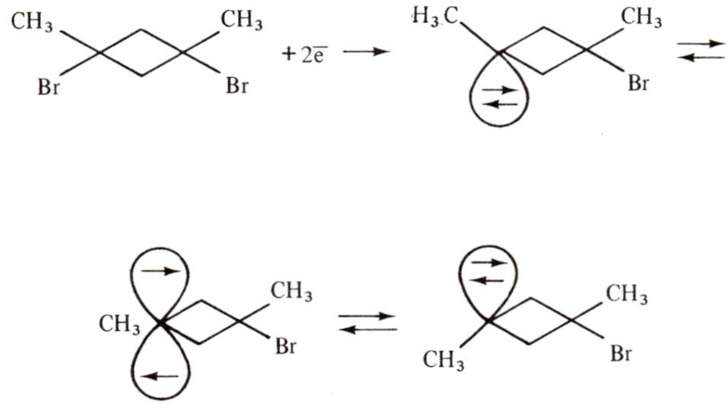

The reduction of 1-bromo-3-bromomethylcyclobutane in DMF and hexamethylphosphoramide (HMPA) demonstrates once again [58-61] that DMF is capable of donating protons to anionic intermediates:

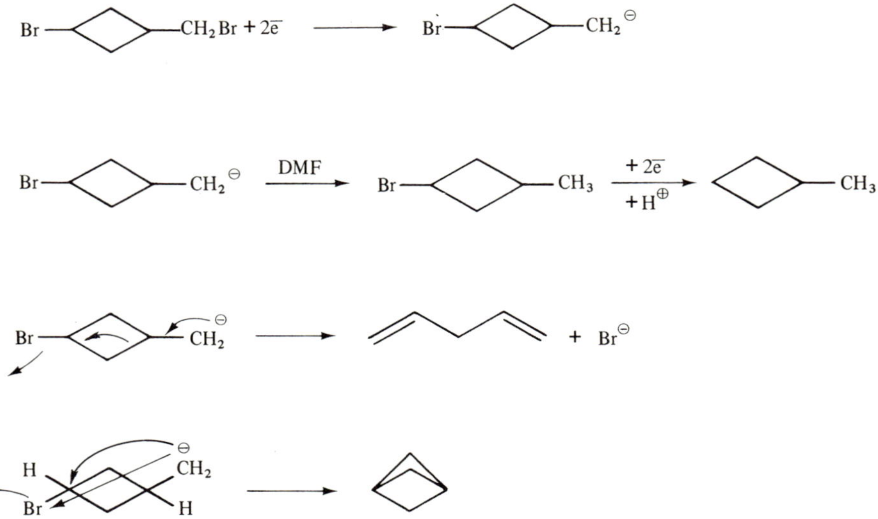

G. gem-Halides (Table 1, Group 6)

The reduction of gem-halides proceeds in a two-electron stepwise mechanism with the replacement of the halogen with hydrogen [17, 18, 54-57]:

$$R\text{-}CF_3 + 6\bar{e} + 3H_2O \rightarrow RCH_3 + 3F^- + 3O\bar{H} \quad [56]$$

$$Br_3\text{-}CO_2H + 6\bar{e} \rightarrow CH_3\text{-}CO_2H.$$

Iversen [92] has described the controlled potential electrolysis of trichloroacetic to dichloroacetic acid.

It is difficult to ascertain from the polarographic behavior of gem-halides whether their reduction proceeds via a transition state with radical or anionic character. However, the products obtained from macroscale electrolysis can best be explained as derived from anionic intermediates.

Wawzonek and Duty [17] investigated the reduction of gem-halides in aprotic solvents and isolated products derived from carbene intermediates:

$$CCl_4 + 2\bar{e} \longrightarrow \overset{\ominus}{C}Cl_3 + \overset{\ominus}{Cl} \longrightarrow :CCl_2 + \overset{\ominus}{Cl}$$

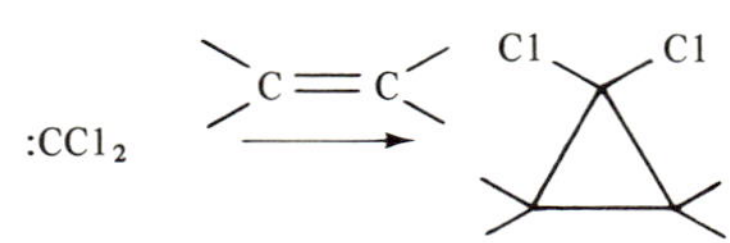

IV. EXPERIMENTAL

A. Choice of Solvent

Due to the insolubility of most organic halides in aqueous solution, their reduction is generally carried out in an organic solvent which may be partially aqueous. Under anhydrous conditions the most common solvents employed are N, N-dimethylformamide (DMF), acetonitrile, hexamethylphosphoramide, and dioxane. Under aqueous solutions, ethanol or dioxane are generally used.

B. Supporting Electrolyte

The choice of supporting electrolyte, in general, depends on the type of reduction being carried out as well as the halide being reduced. For example, activated benzyl halides can be reduced in aqueous dioxane with hydrochloric acid as the supporting electrolyte [48]. When the protonation of anionic intermediates is to be inhibited, alkali halides such as lithium chloride or bromide are employed. These supporting electrolytes, however, do not have sufficient solubility in organic solvents. Other supporting electrolytes used under anhydrous conditions and which are quite soluble in organic solvents are the tetraalkylammonium salts such as tetraethylammonium halides which can be purchased from Eastman Kodak or Matheson Coleman and Bell Companies.

C. Apparatus

The most common cathode used for the reduction of halide is mercury [7]. However, lead, Monel, and platinum have also been used. The choice of the electrolysis cell depends on the type of reaction being carried out. For a variety of cell designs, the reader is referred to Refs. [11], [12], and [62] and to Chapter IV.

V. SUMMARY

The reduction of organic halides is irreversible, follows the Ilkovic equation, and is in general independent of the pH of the medium. The ease of the reduction is in the order R-I > R-Br > R-Cl > R-F. Aromatic and vinyl halides are more difficult to reduce than alkyl halides. Vicinal dihalides and gem-halides reduce at a lower potential than the corresponding monohalides.

Macroscale reduction of monohalides leads to the corresponding hydrocarbon and some coupling products. Vicinal dihalides afford olefins and/or saturated hydrocarbons. α,ω-Dihalides form cyclic and the corresponding hydrocarbon products, depending on the proximity of the halides. gem-Halides generally afford hydrocarbon, although carbene intermediates have been trapped. Synthetic results that have been obtained are summarized in Table 3.

TABLE 3

Reduction Products of Organic Halides

Halide	Reaction Conditions		Products	Yield	Ref.
	Cathode	Solution			
Monohalides (aliphatic)					
B-Halopropionitrile	Hg	Aq solution	Adiponitrile, propionitrile $(NCCH_2-CH_2)_2Hg$	—	[86], [88], [89]
Ethyl bromide	Pb	Propylene carbonate	Tetraethyllead	100	[68]
1-Bromohexane	Hg	DMF	Hexane, hexene	80, 10	[10]
			Dihexylmercury	Trace	[10]
6-Bromo-1-hexene	Hg	DMF	1-Hexene	80	[10]
			Methylcyclopentane	7	[10]
Neopentyl bromide	Hg	DMF	Neopentane	100	[10]
Benzyl chloride	Hg	DMF	Toluene	75	[23]
	Hg	DMF/CO_2	Phenylacetic acid	1	[23]

Table 3 (cont.)

Halide	Reaction Conditions		Products	Yield	Ref.
	Cathode	Solution			
p-Nitrobenzyl chloride	Hg	MeOH		—	[12]
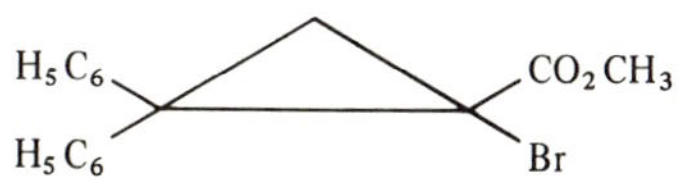	Hg	95% C_2H_5OH	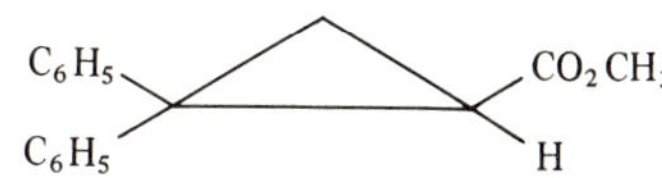	73-95	[5]
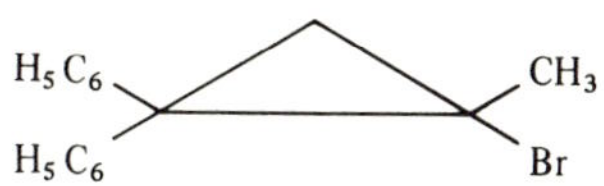	Hg	CH_3CN	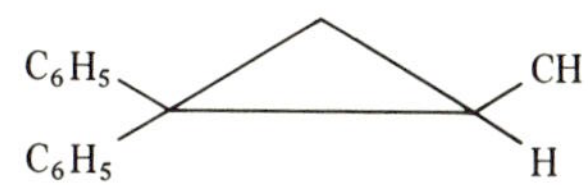 (25% retention)	93	[29]
CH_3-$CBrNO_2$	Hg	Ag, acid	CH_3CH-$N{=}O_2^-$	—	[38]
Br-CH_2-CH_2-$CH_2N^+(C_2H_5)_3Br^-$	Hg	DMF	Cyclopropane, triethylamine	—	[14]
Monohalides (vinyl and aromatic)					
trans-3-Iodo-3-hexene	Hg	DMF	trans-3-Hexene; cis-	90:6	[67]
		0.1 M phenol	Same products	90:6	[67]
cis-3-Iodo-3-hexene	Hg	DMF	trans-3-Hexene; cis-	70:30	[67]
Monobromofumaric acid	Hg	C_2H_5OH	Fumaric acid	89-100	[11]

Table 3 (cont.)

Halide	Reaction Conditions: Cathode	Reaction Conditions: Solution	Products	Yield	Ref.
Monobromodiethyl fumarate	Hg	C_2H_5OH	Diethyl fumarate	89-100	[11]
Monobromodiethyl maleate	Hg	C_2H_5OH	Diethyl maleate	92-97	
Monobromomaleic acid	Hg	C_2H_5OH, pH = 0-12.0	Maleic acid	5-37	[11]
			Fumaric acid	21-100	[11]
			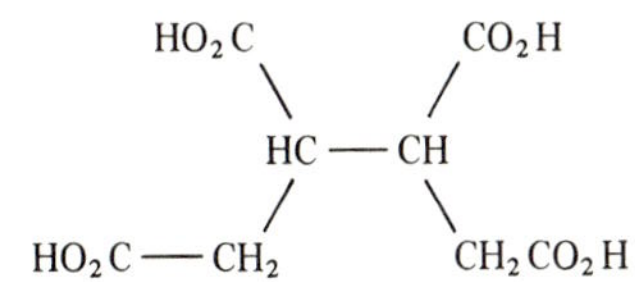	–	[11]
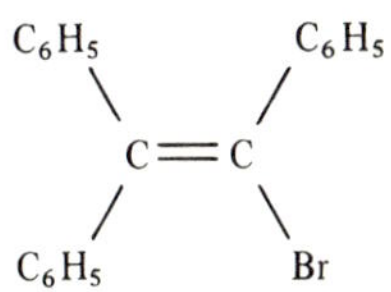			$(C_6H_5)_2CH\text{-}CH_2C_6H_5$	–	[37]
Bromobenzene	Hg	DMF	Benzene	100	[10]
Dihalides (vicinal)					
1, 2-Dibromoethane	Hg	75% Dioxane	Ethylene	–	[13]
1, 2-Diiodoethylene	Hg	75% Dioxane	Acetylene	–	[13]
α, α′-Dibromosuccinic acid	Hg	C_2H_5OH, H_3^+O	Fumaric acid	–	[46]
			Maleic acid	–	[46]

Table 3 (cont.)

Halide	Reaction Conditions		Products	Yield	Ref.
	Cathode	Solution			
$F_3CCHClBr$	Pb	Basic medium	F_2C-CHCl	70	[49]
			Fluoroacetylene	Trace	[49]
o-Dibromobenzene, furan	Hg	DMF (-35°C)	α-Naphthol	~1	[43]
			Benzene	—	
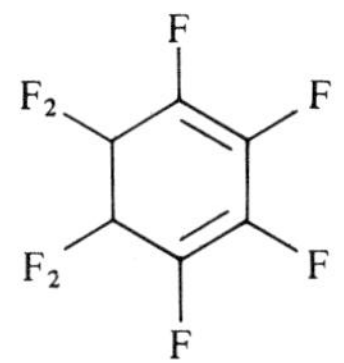	Hg	Aq·C_2H_5OH	Hexafluorobenzene	50	[71]
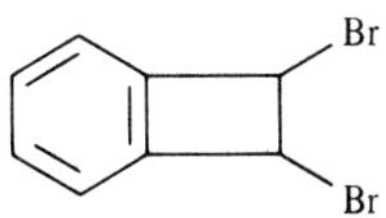	Potassium, glyme		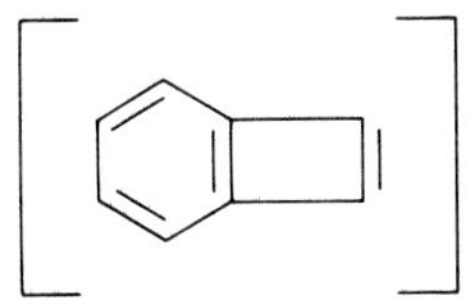 unstable	--	[72]
1, 3-Dibromopropane	Hg	DMF	Cyclopropane	80	[14]
1, 4-Dibromobutane	Hg	DMF	Cyclobutane, n-butane	25, 75	[14]
1, 5-Dibromopentane	Hg	DMF	n-Pentane	80	[14]

Table 3 (cont.)

Halide	Reaction Conditions: Cathode	Reaction Conditions: Solution	Products	Yield	Ref.
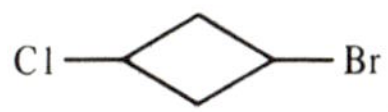	Hg	DMF	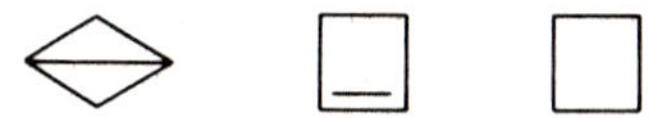	60:20:10	[14]
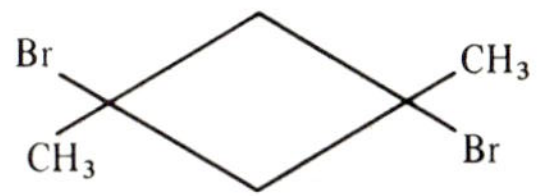	Hg	DMF	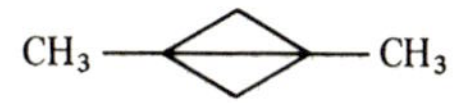	55-94	[14]
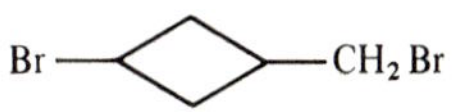	Hg	DMF	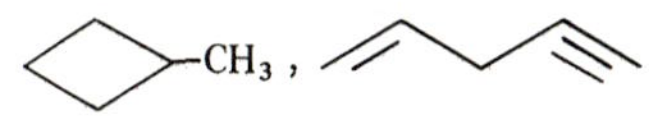	—	[15]
	Hg,	Hexamethylphosphoramide		—	[15]
m-Dibromobenzene	Hg	DMF	Benzene	—	[43]
p-Dibromobenzene	Hg	DMF	Benzene	—	[43]
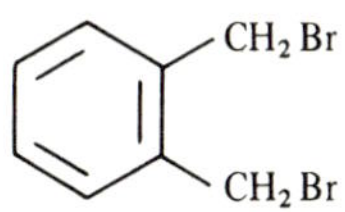	Hg	DMF	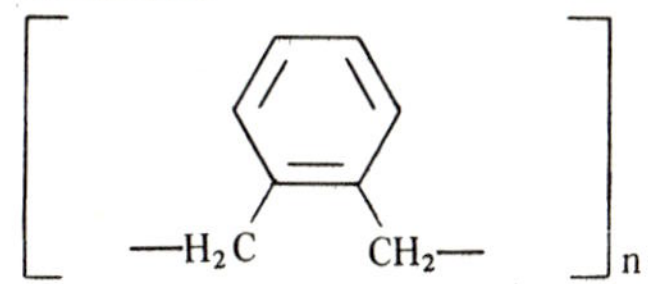	51	[16]
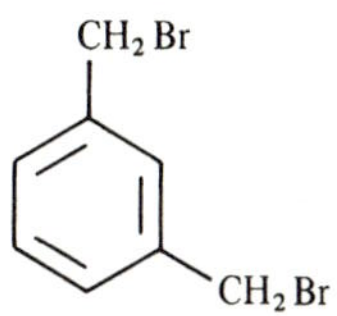	Hg	DMF	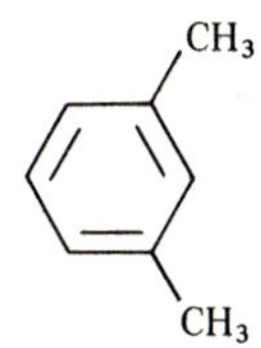	81	[16]

Table 3 (cont.)

Halide	Reaction Conditions		Products	Yield	Ref.
	Cathode	Solution			
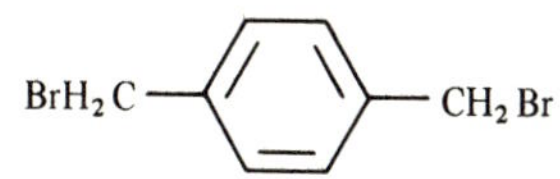	Hg	DMF	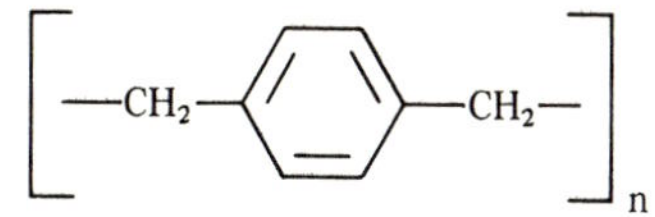	92	[16]
			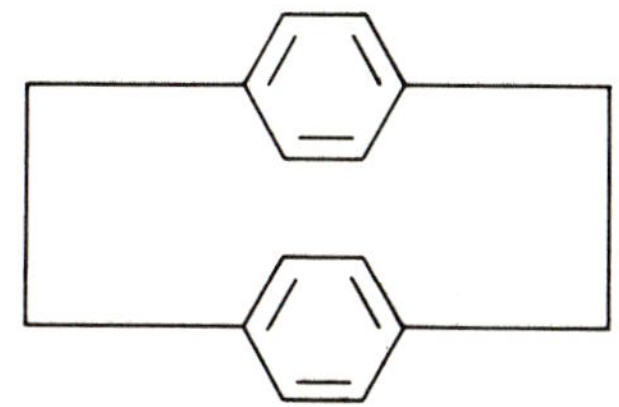	5	[16]
CH_2Cl_2	Hg	DMF	CH_3Cl	–	[17]
$CHCl_3$	Hg	DMF	CH_3Cl	–	[17]
CCl_4	Hg	CH_3CN (-20°C), 2, 3-dimethyl-2-butene	$CHCl_3$ 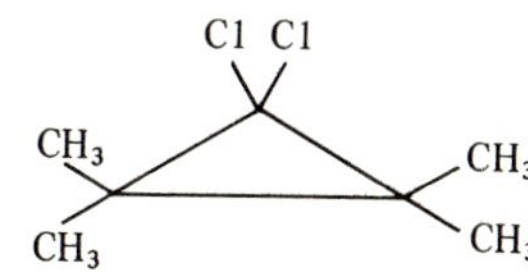	–	[17]
CF_3-CHClBr	Pb	Alkaline medium	CF_2-CHCl	70	[49]
			F-C≡CH	Trace	[49]
$CF_3-C(=O)-C_6H_5$	Hg	Acidic medium	Acetophenone	14	[82]

Table 3 (cont.)

Halide	Reaction Conditions		Products	Yield	Ref.
	Cathode	Solution			
$CCl_3—CO_2H$	Hg	NH_4Cl	$CHCl_2CO_2H$	100	[57]
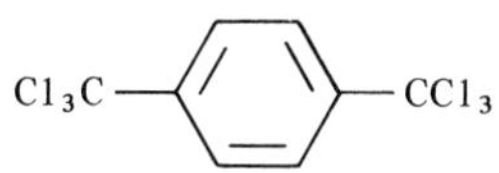	Hg	THF (1-10°C)	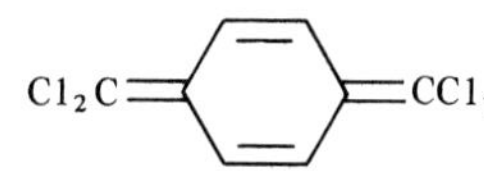	—	[48]
			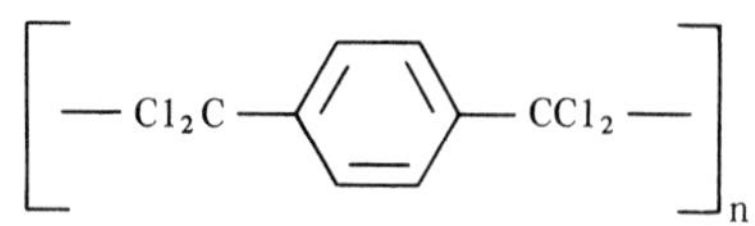		

VI. CONCLUSION

It is quite dangerous for any investigator in the area of electroorganic chemistry to explain all the data concerning the reduction of organic halides by one mechanism. However, for the lack of a simple and unified mechanism which explains all the data on hand, the following one is proposed:

In general, the reduction of monohalides is envisioned to proceed via a one-step, two-electron attack on the halogen, to afford anionic intermediates which undergo a variety of reactions to give products.[8] With the exception of trans-vicinal dihalides and some α, ω-dihalides, which reduce in a concerted mechanism, this concept can be extended to the reduction of all types of organic halides:

$$R - X + 2e^- \xrightarrow{\text{Slow}} R^{\ominus} + X^{\ominus}$$
$$R^- + H^+ \longrightarrow RH$$
$$R^- + RX \longrightarrow RR$$

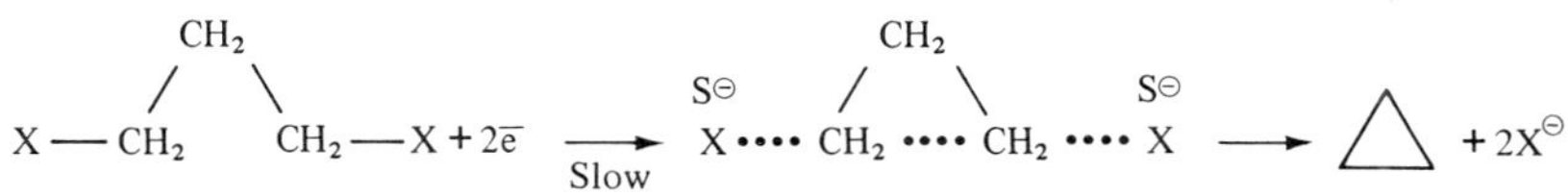

NOTES

[1] For exceptions, see Refs. [81] and [84].

[2] The effect of the electrode on the stereochemical reduction of carbon-halogen bonds has been discussed [5, 76].

[3] The effect of steric hindrance of the $E_{1/2}$ of organic halides is substantial only at high thresholds of structural complexity.

[4] Based on the half-wave potentials of p-dihalobenzenes (see Table 1), it is questionable whether the first reduction affords iodobenzene.

[5] It was shown by Gilch [48] that such polymers can be obtained in acidic solutions, thus suggesting the absence of anionic intermediates.

[6] While inductive effects have some influence on the half-wave potentials of dihalides, especially vicinal dihalides [86, 87], it is questionable whether such effects can satisfactorily explain the difference in the half-wave potentials of 1, 2-dibromoethane and 1, 3-dibromopropane.

[7] It must be borne in mind that allyl and benzyl bromides and iodides may form mercurials before or in the course of the electrolysis.

[8] Proton donor effects have recently been discussed [93].

REFERENCES

[1]. P. J. Elving, Rec. Chem. Progr. (Kresge-Hooker Sci. Libr.), 14, 99 (1953).

[2]. P. J. Elving and B. Pullman, in Advances in Chemical Physics (I. Prigogine, ed.), Vol. 1, Wiley-Interscience, New York, 1961, p. 1.

[3]. A. M. Wilson and N. L. Allinger, J. Amer. Chem. Soc., 83, 1999 (1961).

[4]. P. Kabasakalian and J. McGlozzen, Anal. Chem., 34, 1440 (1962).

[5]. R. Annino, R. E. Erickson, J. Michalovic, and B. McKay, J. Amer. Chem. Soc., 88, 4424 (1966).

[6]. K. Sugino and K. Shirai, J. Chem. Soc. Japan (Pure Chem. Sec.), 70, 111 (1949); 71, 396 (1950).

[7]. R. Galli, J. Electroanal. Chem., 22, 75 (1969).

[8]. P. J. Elving and C. L. Hilton, J. Amer. Chem. Soc., 74, 3368 (1952).

[9]. P. J. Elving and R. E. Van Atta, J. Electrochem. Soc., 103, 676 (1956).

[10]. J. W. Sease and R. C. Reed, Electrochemical Society Meeting, New York, May 4-9, 1969.

[11]. P. J. Elving, I. Rosenthal, J. R. Hayes, and A. J. Martin, Anal. Chem., 33, 330 (1961).

[12]. G. Klopman, Helv. Chim. Acta, 44, 1908 (1964).

[13]. M. v. Stackelberg and W. Stracke, Z. Elektrochem., 53, 118 (1949).

[14]. M. R. Rifi, J. Amer. Chem. Soc., 89, 4442 (1967).

[15]. M. R. Rifi, Tetrahedron Lett., 1043 (1969).

[16]. F. H. Covitz, J. Amer. Chem. Soc., 89, 5403 (1967).

[17]. S. Wawzonek and R. C. Duty, J. Electrochem. Soc., 108, 1135 (1961).

[18]. P. J. Elving, I. Rosenthal, and M. K. Kramer, J. Amer. Chem. Soc., 73, 1717 (1951).

[19]. L. Feoktistov and S. I. Zhdanov, Bull. Acad. Sci. USSR, 2036 (1962).

[20]. P. J. Elving and C. S. Tang, J. Amer. Chem. Soc., 72, 3244 (1950).

[21]. R. Pasternak and H. von Halban, Helv. Chim. Acta, 29, 190, 2595 (1946).

[22]. J. W. Sease, F. G. Burton, and S. L. Nickol, J. Amer. Chem. Soc., 90, 2595 (1968).

[23]. S. Wawzonek, R. C. Duty, and J. H. Wagenknecht, J. Electrochem. Soc., 111, 74 (1964).

[24]. E. V. Kaabak, A. P. Tomilov, S. L. Varshavskii, and M. I. Kabachnik, J. Org. Chem. USSR, 3, 1 (1967).

[25]. F. Lambert and K. Kobayashi, J. Amer. Chem. Soc., 82, 5324 (1929).

[26]. J. Pearson, Trans. Faraday Soc., 44, 683 (1948).

[27]. J. W. Sease, P. Chang, and J. Groth, J. Amer. Chem. Soc., 86, 3154 (1964).

[28]. J. Zavada, J. Krupicka, and J. Sicher, Coll. Czech. Chem. Commun., 28, 1664 (1963).

[29]. C. K. Mann, J. L. Webb, and H. M. Walborsky, Tetrahedron Lett., 2249 (1966).

[30]. H. M. Walborsky, Chen-Jong Chen, and S. L. Webb, ibid., 3551 (1964).

[31]. H. M. Walborsky, F. J. Impastato, and A. E. Young, J. Amer. Chem. Soc., 86, 3283 (1964).

[32]. L. B. Rogers and A. J. Diefenderfer, J. Electrochem. Soc. (Electrochem. Sci.), 942 (1967).

[33]. N. S. Hush and K. B. Oldham, J. Electroanal. Chem., b, 34 (1963).

[34]. I. M. Kolthoff, W. Stricks, and N. Tanaka, J. Amer. Chem. Soc., 77, 4739 (1955).

[35]. A. Streitwieser, Jr., in Molecular Orbital Theory for Organic Chemists, Wiley, New York, 1962, p. 226.

[36]. J. P. Petrovich, M. M. Baizer, and M. R. Ort, J. Electrochem. Soc., 116, 749 (1969).

[37]. L. L. Miller and E. Riekena, J. Org. Chem., 34, 3359 (1969).

[38]. J. Armand, Compt. Rend., 254, 2777 (1962).

[39]. R. F. Evilia and J. Diefenderfer, J. Electroanal. Chem., 22, 407 (1969).

[40]. J. Tirorofled, Adv. Polarogr., 2, 740 (1960).

[41]. J. G. Lawless and M. Hawley, J. Electroanal. Chem. Interfacial Electrochem., 23, Prelim. Note 1 (1969).

[42]. A. Kirrman and E. Kleine-Peter, Bull. Soc. Chem. Fr., 894 (1957).

[43]. S. Wawzonek and J. H. Wagenknecht, J. Electrochem. Soc., 110, 420 (1963).

[44]. S. Winstein, D. Pressman, and W. G. Young, J. Amer. Chem. Soc., 61, 1645 (1939).

[45]. C. H. DePuy, R. D. Thurn, and G. F. Morris, ibid., 84, 1314 (1962).

[46]. P. J. Elving, I. Rosenthal, and A. Martin, ibid., 77, 5218 (1955).

[47]. E. L. Colichman and S. K. Liu, ibid., 76, 913 (1954).

[48]. H. Gilch, J. Polym. Sci., 4, 1351 (1966).

[49]. L. G. Feoktistov, A. P. Tomilov, and M. M. Goldin, Izv. Akad. Nauk USSR Ser. Khim., I, 1352 (1963); through CA, 59, 12624 (1963).

[50]. C. Perrin, Progress in Physical Organic Chemistry (S. G. Cohen, A. Streitwieser, Jr., and R. W. Taft, eds.), Vol. 3, Wiley-Interscience, New York, 1965, p. 165.

[51]. P. Zuman, ibid., Vol. 5, 1967, p. 86.

[52]. R. E. Dessy, W. Kitching, T. Psarras, R. Salinger, A. Chen, and T. Chivers, J. Amer. Chem. Soc., 88, 460 (1966).

[53]. H. H. Bauer, J. Electroanal. Chem. Interfacial Electrochem., 16, 419 (1968).

[54]. A. I. Cohen, R. T. Keeler, N. H. Coy, and H. L. Yale, Anal. Chem., 34, 216 (1962).

[55]. P. J. Elving and E. Bennett, J. Amer. Chem. Soc., 76, 4473 (1954).

[56]. H. Lund, Acta Chem. Scand., 13, 192 (1959).

[57]. T. Meites and L. Meites, Anal. Chem., 27, 1531 (1955).

[58]. S. Wawzonek, E. B. Blaha, R. Berkey, and M. E. Runner, J. Electrochem. Soc., 102, 235 (1955).

[59]. P. H. Given and M. E. Peover, J. Chem. Soc., 385 (1960).

[60]. P. H. Given and M. E. Peover, ibid., 3195 (1960).

[61]. R. H. Philp, Jr., R. L. Flurry, and R. A. Day, Jr., J. Electrochem. Soc., III, 328 (1969).

[62]. M. J. Allen, Organic Electrode Process, Reinhold, New York, 1958.

[63]. M. R. Rifi, unpublished results.

[64]. P. S. Skell and R. G. Allen, J. Amer. Chem. Soc., 80, 5997 (1958).

[65]. A. A. Oswald, K. Griesbaum, B. B. Hudson, Jr., and J. M. Bregman, ibid., 86, 2877 (1964).

[66]. J. A. Kampmeier and G. Chen, ibid., 87, 2608 (1965).

[67]. A. J. Fry and M. A. Mitnick, ibid., 91, 6207 (1969).

[68]. H. E. Ulrey, J. Electrochem. Soc., 116, 1201 (1969).

[69]. S. Swann, Jr. and G. Fuller, J. Amer. Chem. Soc., 75, 5127 (1953).

[70]. J. A. Dougherty and A. J. Diefenderfer, J. Electroanal. Chem. Interfacial Chem., 21, 531 (1969).

[71]. A. M. Doyle, A. E. Pedler, and J. C. Tatlow, J. Chem. Soc. (c), 2740 (1968).

[72]. R. D. Ricke and P. M. Hudnall, J. Amer. Chem. Soc., 91, 3678 (1969).

[73]. P. J. Elving and B. Pullman, Adv. Chem. Phys., 3, 1 (1961); for a later interpretation, see N. S. Hush and G. A. Segal, Discuss. Faraday Soc., 45, 23 (1968).

[74]. A. Streitwieser and C. Perrin, J. Amer. Chem. Soc., 86, 4938 (1964).

[75]. F. L. Lambert, J. Org. Chem., 31, 4184 (1966).

[76]. A. J. Fry and R. H. Moore, ibid., 33, 1283 (1968).

[77]. B. Czochralska, Rocz. Chem., 42, 895 (1968).

[78]. I. Rosenthal, C. Albright, and P. Elving, J. Electrochem. Soc., 99, 227 (1952).

[79]. P. Elving, J. Markowitz, and I. Rosenthal, ibid., 101, 195 (1954).

[80]. S. Mairanovskii and L. Bergelson, Russ. J. Phys. Chem. (English Transl.), 34, 112 (1960); I. G. Markova and L. G. Feoktistov, Zh. Obshch. Khim., 40, 740 (1970); through CA, 73, 41220x (1970).

[81]. W. Jura and R. Gaul, J. Amer. Chem. Soc., 80, 5402 (1958).

[82]. J. Stocker and R. Jenevein, Chem. Commun., 934 (1968).

[83]. P. Zuman, Coll. Czech. Chem. Commun., 15, 1107 (1950).

[84]. P. J. Elving and T. Leone, J. Amer. Chem. Soc., 79, 1546 (1957).

[85]. J. Holubek and J. Volke, Coll. Czech. Chem. Commun., 27, 680 (1962).

[86]. S. Andreades, U.S. Pat. 3,475,298 (1969).

[87]. L. G. Feoktistov, Usp. Electrokhim. Org. Soedin., Akad. Nauk SSSR, Inst. Electrokhim., 135 (1966); through CA, 68, 17126 (1968).

[88]. L. G. Feoktistov, A. P. Tomilov, Yu. D. Smirnov, and M. Goldin, Elektrokhim., 1 (8), 887 (1965); through CA, 63, 16196 (1965).

[89]. A. P. Tomilov and Yu. D. Simirnov, Zh. Vses. Khim. Obshchest. Im. D. I. Mendeleeva, 10 (1), 101 (1965); through CA, 63, 3897 (1965).

[90]. L. G. Feoktistov and S. I. Zhdanov, Electrochem. Acta, 10 (7), 657 (1965); through CA, 63, 9460 (1965).

[91]. L. G. Feoktistov and S. I. Zhdanov, Izv. Akad. Nauk SSSR, Otd. Khim. Nauk, 2127 (1962); through CA, 58, 13449 (1963).

[92]. P. Iversen, J. Chem. Educ., 48, 136 (1971).

[93]. A. J. Fry and R. G. Reed, Electrochemical Society Meeting, Washington, D.C., May 1971, Abstract No. 119.

Chapter VII

CATHODIC REDUCTION OF NITRO COMPOUNDS

H. Lund

Department of Organic Chemistry
University of Aarhus
Aarhus, Denmark

I. INTRODUCTION

The electrolytic reduction of nitro compounds has played an important role in the development of organic electrochemistry in the "classical" period about the turn of the century [1]. It was the investigation of the stepwise reduction of nitrobenzene which led Haber [2, 3] to the idea that the deciding factor in determining the product was the potential of the working electrode, and he was also the first to apply the principle to organic electrolysis.

In this chapter there will be discussed the reduction of aliphatic and aromatic as well as N-nitro compounds and the further reduction of the products obtained such as nitroso, azoxy, azo compounds, and hydroxylamines.

II. ALIPHATIC NITRO COMPOUNDS AND DERIVATIVES

A. Nitroalkanes

1. Primary and Secondary Nitroalkanes

Primary and secondary nitroalkanes are reducible in acid and neutral solution, whereas the formation of the anion of the aci form in alkaline solution prevents their reduction in this medium. Electrolysis of nitroalkanes in acid solution at low temperature yields as main product the corresponding alkylhydroxylamine [4, 7], which is further reducible only in a rather narrow pH region, around pH 7. The electrolytic reduction of nitroalkanes in hydrochloric acid is probably the method of choice for the preparation of alkylhydroxylamines; evaporation of the catholyte after reduction yields the alkylhydroxylamine hydrochloride in a rather pure form [6]. In Table 1 the yields of alkylhydroxylamines and amines from the reduction of some nitroalkanes are given.

TABLE 1

Results from Electrolytic Reduction of Nitroalkanes at 20°C [6]

R	Amount of RNO_2, g	Analysis of catholyte			Current yield, %	% isolated RNHOH
		% RNHOH	% RNH_2	% RNO_2		
Me	28.2	89	3.6	1.2	92.6	77
Et	31.1	82	9.9	0.8	92.9	59
1-Pr	40.1	83	5.9	1.2	91.6	62
2-Pr	39.7	86	4.6	0.8	90.5	81

A small amount of alkylamine and of a carbonyl compound is formed during the reduction. The hydroxylamine is not reduced to amine under these conditions. Reduction of N-methylhydroxylamine [8] has been attempted at -1.15 V (SCE) in hydrochloric acid at 20-30°C for 16 h. Hydrogen was evolved, but no methylamine was detected. The following scheme has been suggested to explain the side products obtained [7, 8]:

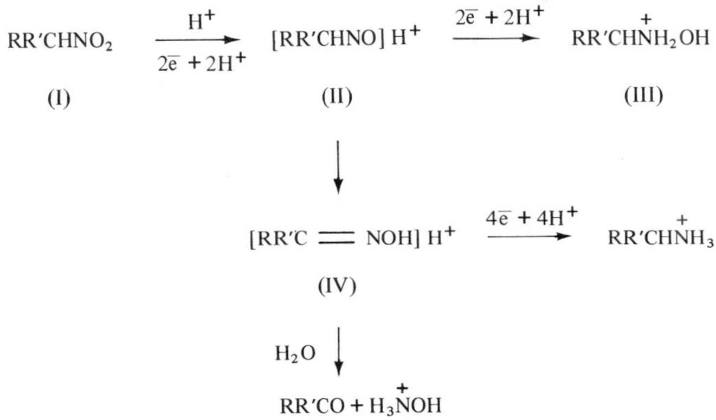

The tautomerization of the nitroso derivative (II) to the oxime (IV) competes with the further reduction, and any parameter which retards the competing chemical reaction raises the yield of the alkylhydroxylamine (III). An increase in temperature would be expected to favor the tautomerization reaction compared to the further reduction, and reduction at a low temperature would thus be essential for the preparation of alkylhydroxylamines.

Accordingly, it has been found that α-hydroxynitroalkanes at 65-85° are reduced to the amino alcohols [9], whereas at 0-35° the hydroxylamino alcohol is the product [10].

In nitroalkanes substituted α to the nitro group with an unreducible electron-attracting group the α-hydrogen dissociates more easily, and the rearrangement (II) → (IV) proceeds at a higher rate; a relatively high proportion of amine must thus be expected. Thus, whereas a single carboxylic ester or sulfamido group α to the nitro group [11, 13] sometimes (but not always) makes the α-hydrogen too labile for the electrolytic preparation of the hydroxylamine, the presence of two carboxylic ester groups always does. It has thus been impossible to prepare 2-hydroxylaminomalonic ester by reduction of the nitro compound. The reduction of 2-nitro-2-alkylmalonic esters, however, produces the expected 2-alkyl-2-hydroxylaminomalonic ester on reduction [14]; here the tautomerization to the oxime is excluded as no α-hydrogen atoms are available.

The form of the polarographic wave of certain nitro compounds, such as 1-(p-bromophenyl)-2-nitroethanol, depends on the buffer concentration; at low buffer concentrations the wave is steeper than at higher concentrations. This has been explained by a catalytic effect of the reduction product on the protonation of the nitro compound, which makes the nitro group more easily reducible. Some compounds of a similar structure, e.g., ephedrine, have a similar catalytic effect on the protonation, whereas others, such as cyclohexylamine, have not [15].

Protonation of the anion of the aci form, the nitronate anion, takes place at the highest rate on the oxygen; the free nitronic acid is unstable and either tautomerizes to the nitroalkane or is hydrolyzed to a carbonyl compound and N_2O. The reduction of a nitronic acid, prepared from the anion, is thus difficult to study, but its polarographic behavior can be investigated during the reduction of α-halonitroalkanes.

It might be mentioned that anodic oxidation of the nitroalkane anions may lead to dimerization or addition reactions; these reactions are treated in Chapter XX.

2. Tertiary Nitroalkanes

These compounds differ from the primary and secondary nitroalkanes in that the intermediate nitroso derivative cannot rearrange into an oxime and this side reaction is avoided. The yield of alkylhydroxylamine is thus high. A small amount of amine (<3%) is, however, usually found [8], but it is not yet clear how this is formed.

The nitroso intermediate can in this case be detected directly in the reduction mixture by its bluish-green color and its polarographic wave; the nitroso compound is, as most other nitroso derivatives, more easily reducible than the corresponding nitro compound.

In nonaqueous, aprotic solvents, such as dimethoxyethane [16] or acetonitrile [17, 18], the primary reduction product from tertiary nitroalkanes is the anion radical. The nitro-t-butyl anion radical is rather unstable (half-life of 0.66 sec) and decomposes into nitrite ion and t-butyl radical. Continued electrolysis results in the formation of some di-t-alkyl nitroxide radical [17, 19]:

$$R_3CNO_2 \xrightarrow{\bar{e}} R_3CNO_2^{\dot{-}} \rightarrow R_3C^{\bullet} + NO_2^-$$

$$R_3C^{\bullet} + R_3CNO_2^{\dot{-}} \rightarrow (R_3C)_2NO_2^- \xrightarrow{+H^+} (R_3C)_2\dot{N}O + {\bullet}OH$$

In acid solution the primary step is probably also formation of a radical anion, but in this medium the latter is protonated and further reduced more rapidly than the decomposition of the radical anion; in more acidic solution a preprotonation may take place.

3. α, β-Unsaturated Nitro Compounds

The reduction of an α, β-unsaturated nitro compound [20, 21], such as β-nitrostyrene (V), occurs in two steps. The first reduction of (V) is a four-electron reaction which yields phenylacetaldoxime (VI); the reaction has been formulated as a reduction of the nitro group followed by a rearrangement of the unsaturated hydroxylamine (VII), but might also be regarded as a 1,4-reduction of the intermediate unsaturated nitroso compound. There is no conclusive evidence for either route, but the former has been chosen here, as the protonation of the heteroatom would be expected to be faster than protonation of the carbon atom:

$$\underset{(V)}{C_6H_5CH{=}CHNO_2} \xrightarrow[-H_2O]{4\bar{e} + 4H^+} \underset{(VII)}{C_6H_5CH{=}CHNHOH} \rightarrow \underset{(VI)}{C_6H_5CH_2CH{=}NOH}$$

The further reduction [21] has been reported to yield a hydroxylamine. A four-electron reduction of the protonated oxime to phenylethylamine would have been the expected reaction.

α-Substituted (and α, β-Unsaturated) Nitroalkanes. β-Bromo-β-nitrostyrene (XVII) [37] gives a six-electron polarographic wave at about the same potentials as β-nitrostyrene (V), but the wave of (XVII) is not followed by a second wave as is the wave of (V). Reduction of (XVII) in 0.6 N H_2SO_4 containing dioxane at E -0.4 V yielded benzyl cyanide and some benzaldehyde; if the reduction was performed in aqueous methanol, some $C_6H_5CH(OCH_3)CN$ was formed. Polarographic data for the nitrile oxide, $C_6H_5CH_2CNO$ [38, 39], show that this compound is not an intermediate in the reduction:

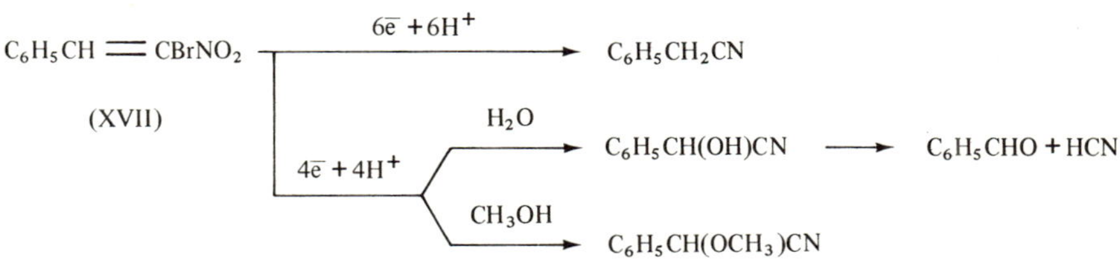

The reduction mechanism is not fully elucidated [40].

A similar reduction route is found for the reduction of β, β-dinitrostyrene [41].

4. α-Substituted Nitroalkanes

Substituents at the α-carbon atom of nitroalkanes may influence their electrode reactions considerably. The influence of electronegative groups on the rate of the nitroso to oxime transformation was mentioned above. If an electronegative group is a good leaving group, it may be reductively removed; substituents of this type are halogens and nitro groups. When a nitro group is reductively lost as a nitrite ion, a further reaction between nitrous acid and a suitable substrate must always be considered as a side reaction.

a. gem-Dinitroalkanes [22-25]. 2, 2-Dinitropropane is in alkaline solution reduced in a two-electron reaction to 2-nitropropane anion; in slightly acid solution the nitronic acid is reduced to acetone oxime. 1, 1-Dinitroethane behaves differently and a wave height corresponding to n = 5-6 is found in acid solution. The following explanation has been suggested [25]:

$$CH_3CH(NO_2)_2 \xrightarrow{2\bar{e}} CH_3CH{=}NO_2^- + NO_2^- \xrightarrow{2H^+} CH_3CH{=}NO_2H + HNO_2 \longrightarrow$$

$$\underset{\text{(VIII)}}{CH_3C(NO_2){=}NOH} \xrightarrow{4\bar{e} + 4H^+} \underset{\text{(IX)}}{CH_3C(NHOH){=}NOH}$$

The ethyl nitrolic acid (VIII) has been shown [23, 26] to be electrolytically reducible to acethydroxyamidoxime (IX) which can be oxidized anodically to ethyl nitrosolic acid. In alkaline solution ethyl nitrolic acid (VIII) forms some azo derivative (XI) [27] in low yield according to

$$\underset{\text{(VIII)}}{CH_3C(NOH)NO_2} \xrightarrow{4\bar{e} + 4H^+} \underset{\text{(XI)}}{\tfrac{1}{2}CH_3C(NOH)N{=}NC(NOH)CH_3}.$$

Tetranitromethane [28] is reduced polarographically to trinitromethane; this is, in turn, reduced at pH 6 to the rather unstable dihydroxyguanidine [24] which in alkaline solution is rearranged and further reduced to hydroxyguanidine.

b. Pseudonitroles. In pseudonitroles, such as 2-nitro-2-nitrosopropane (X) [25], the first reduction in acid solution is a two-electron reaction which results in the loss of the nitro group as a nitrite ion. Protonation of the primarily formed anion on oxygen forms an oxime:

$$\underset{\text{(X)}}{(CH_3)_2C(NO)NO_2} \xrightarrow{2\bar{e}} NO_2^- + [(CH_3)_2CNO]^- \xrightarrow{H^+} (CH_3)_2C{=}NOH.$$

c. α-Halogenated Nitroalkanes. The halogen atom in α-halogenated nitroalkanes is generally removed at a less negative potential than that necessary for the reduction of the nitro group [25, 29]; thus, controlled potential reduction of 2-chloro-2-nitropropane (pH 0.25; E -0.6 V, t = 10°C) gave acetone in 98% yield [25], formed by hydrolysis of the nitrolic acid.

1-Bromo-1-nitroethane (XII) shows three polarographic waves in acid solution. The half-wave potential of the first wave is pH independent, and the third wave corresponds to a reduction of acetaldoxime [25]. A reduction at a potential corresponding to the second polarographic wave yielded 60% acetaldoxime and 30% acetaldehyde:

1. wave:

$$\underset{\text{(XII)}}{CH_3CHBrNO_2} \xrightarrow{2\bar{e}} CH_3CH{=}NO_2^{-} + Br^{-} \xrightarrow[H^+]{\text{Slow}} CH_3CH_2NO_2$$

$$CH_3CH{=}NO_2^{-} \xrightarrow[\text{fast}]{H^+} CH_3CH{=}NO_2H$$

2. wave:

$$\underset{\text{(XIII)}}{CH_3CH{=}NO_2H} \xrightarrow{2\bar{e} + 2H^+} CH_3CH{=}NOH + H_2O$$

3. wave:

$$CH_3CH{=}NOH \xrightarrow[4\bar{e} + 4H^+]{H^+} CH_3CH_2\overset{+}{N}H_3 + H_2O$$

The second wave thus indicates the electrolytic behavior of a nitrolic acid (XIII).

A similar reduction route is found for dihalogenated nitroalkanes such as 1, 1-dichloro-1-nitroethane [25]. This compound shows two polarographic two-electron waves. The second wave does not correspond to the reduction of 1-chloro-1-nitroethane, which is the isolated product from a controlled potential reduction (E -0.2 V, pH = 1) at the potential of the first wave, but rather to a reduction of the chloronitrolic acid, $CH_3CCl{=}NO_2H$. Accordingly, a reduction at the plateau of the second wave (E -1.0 V, pH = 0.2, t = 5°) produced chloroacetaldoxime, which was further characterized by its transformation into acetonitriloxide at pH 6.

The reduction of trichloronitromethane may give methylamine [30] or methylhydroxylamine [31] but, by working with a cooled tin cathode in 35% sulfuric acid at t < 5°, it was possible to obtain dichloroformoxime [32] as a product. By employing a suitable extraction procedure the yield of this compound could be raised [33]. Brintzinger et al. [32] suggest essentially the following reduction path:

$$CCl_3NO_2 \xrightarrow[-H_2O]{2\bar{e} + 2H^+} CCl_3NO \xrightarrow[-Cl^-]{2\bar{e} + H^+} CHCl_2NO \rightarrow CCl_2{=}NOH \xrightarrow{2\bar{e} + 2H^+}$$

$$CHCl_2NHOH \xrightarrow[-Cl^-]{2\bar{e} + H^+} CH_2ClNHOH \xrightarrow[-Cl^-]{2\bar{e} + H^+}$$

$$CH_3NHOH \xrightarrow[-H_2O]{2\bar{e} + 3H^+} CH_3\overset{+}{N}H_3 .$$

A polarographic investigation of trichloronitromethane [25] shows a stepwise removal of chlorine under these conditions before the reduction of the nitro group occurs. Whether the difference in reduction path reflects the difference in the electrode material, electrode potential, or medium is not known. An alternative formulation for the formation of dichloroformoxine could be the following:

$$CCl_3NO_2 \xrightarrow{2\bar{e}} Cl^- + CCl_2{=}NO_2^- \xrightarrow{H^+} CCl_2{=}NO_2H \xrightarrow[-H_2O]{2\bar{e} + 2H^+} CCl_2{=}NOH.$$

5. Nitrohydrazones

The phenylhydrazone of nitroacetaldehyde (XIV) [34] gives at all pH's a single polarographic wave with a height corresponding to n = 2.9-3.5. Electrolysis of the compound in 0.05 N sulfuric acid (E -0.9 V) consumed 3.7 faradays/mole and the hydrazooxime (XV) was detected as product by its anodic wave, which is caused by a two-electron oxidation to the azooxime (XVI):

$$\underset{\text{(XIV)}}{CH_3C(NO_2){=}NNHC_6H_5} \xrightarrow[-H_2O]{2\bar{e} + 2H^+} \underset{\text{(XVI)}}{[CH_3C(NOH)N{=}NC_6H_5]}$$

$$\text{(XVI)} \; \underset{-2\bar{e}\,-2H^+}{\overset{2\bar{e}\,+2H^+}{\rightleftharpoons}} \; \underset{\text{(XV)}}{CH_3C(NOH)NHNHC_6H_5}$$

The corresponding hydrazooxime, obtained by reduction of the nitrohydrazone $O_2NC_6H_4NHN{=}C(NO_2)C_6H_5$ from benzaldehyde and o- or p-nitrophenylhydrazine, is further reducible [35] to the amidrazone; this is in analogy with the finding that benzamidoxime is reducible to benzamidine [36].

6. β-Substituted Nitro Compounds

The polarographic behavior of β-hydroxynitroalkanes is sometimes complicated in acid solution by the dehydration to the more easily reducible nitroalkene [42, 43]. Nitroalkenes are also obtained from α, β-dibromonitroalkanes on reduction [37].

2, 3-Dinitroalkenes-2 like 2, 3-dinitrobutene-2 show [44] a two-electron polarographic wave at all pH's. In acid solution it is followed by another wave. A preparative reduction also indicates n = 2 and a mixture of 2-nitro-2-butene and diacetylmonoxime is the product in acid solution, whereas the former is the major product at higher pH. At pH 6 the reaction is as follows:

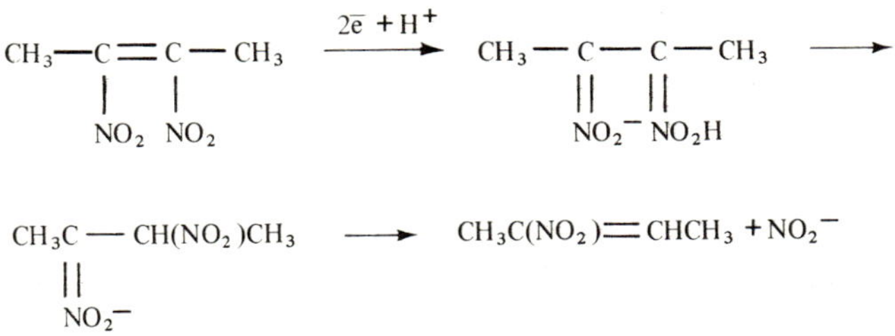

In β-carbonylnitroalkanes the nitro group is more easily reducible than the carbonyl group [45, 46], whereas in 2-bromo-2-nitroacetophenone [45] the bromine is most easily reduced. At pH < 5 2-nitroacetophenone shows a kinetic polarographic wave followed by another wave; the sum of the waves corresponds to n = 4. The kinetic wave is unusual in that the height of it grows with pH. The explanation is that the more easily reducible species is the enolate which is favored by a higher pH; at low pH the nitro ketone is reduced as such

$$C_6H_5COCH_2NO_2 \rightleftharpoons C_6H_5C(OH){=}CHNO_2 \xrightarrow[-H_2O]{4\bar{e} + 4H^+} [C_6H_5COCH_2NHOH].$$

The product from the electrolysis has not been identified; possibly a condensation of a primarily formed ketoalkylhydroxylamine to a pyrazine derivative takes place.

The polarographic behavior of some β-nitrooximes has been investigated [46, 47]; the reduction is not fully elucidated.

B. Nitroso Compounds

Aliphatic primary and secondary nitroso compounds are not stable, but tautomerize to oximes. In nitrosocyclohexane the special steric conditions favor a dimerization rather than a tautomerization. The dimer is reduced polarographically in a six-electron reaction to N, N′-dicyclohexylhydrazine [48].

Tertiary nitroso compounds are stable and are generally reduced similarly to the corresponding nitro compound, but somewhat more easily. The reduction of mono- and dihalogenated nitrosoalkanes also goes analogously to that of the corresponding nitroalkanes; thus 2-chloro-2-nitroso-propane [25] gives acetoxime on reduction (pH 3, E -0.5 V) in 98% yield:

$$(CH_3)_2CClNO \xrightarrow{2\bar{e}} [(CH_3)_2CNO]^- + Cl^- \xrightarrow{H^+} (CH_3)_2C{=}NOH,$$

and 1,1-dichloro-1-nitrosoethane produces chloroacetaldoxime in 76% yield on electrolytic reduction.

C. Oximes

Oximes are tautomeric with nitroso compounds. Oximes are generally reduced to amines in a reaction in which the N-O bond is cleaved before the hydrogenation of the C=N double bond occurs. In acid solution the reaction is [49, 50]:

$$RR'C{=}NOH \overset{H^+}{\rightleftharpoons} [RR'C{=}NOH]H^+ \xrightarrow[-H_2O]{2\bar{e} + 2H^+} RR'C{=}NH_2^+ \xrightarrow{2\bar{e} + 2H^+} RR'CH\overset{+}{N}H_3.$$

A similar route with an initial cleavage of the N-O bond is found in alkaline solution and for acylated [51], N-, or O-alkylated [50, 52, 53] oximes. A reduction of an oxime to a hydroxylamine may occur under certain conditions [54].

Some C-halogenated oximes (hydroxamic acid halides) are reducible in acid solution. The iodides are reducible to the oximes, whereas the reducible bromides or chlorides form amines in a six-electron reduction [38].

D. Alkylhydroxylamines

Aliphatic hydroxylamines are not easily reduced electrolytically. The polarographic wave of such compounds is not well defined and is visible only in a narrow pH region, around pH 7. In acid solution the wave is masked by the hydrogen evolution, and the hydroxylamines are catalytically active in that reaction. It has been suggested [55] that a branching in the reaction takes place after the uptake of the first electron which leads either to a further reduction of the radical or to hydrogen evolution:

$$RNHOH + H^+ \rightleftharpoons R\overset{+}{N}H_2OH \xrightarrow{\bar{e}} R\dot{N}H_2OH$$

$$R\dot{N}H_2OH \xrightarrow{\bar{e}} RNH_2 + OH^-$$

$$R\dot{N}H_2OH \xrightarrow{H^+ + \bar{e}} RNHOH + H_2$$

In acid solution the catalytic reaction is fast and only traces (if any) of amine are formed on reduction. Possibly by lowering the cathode potential and allowing a copious evolution of hydrogen, more amine could be obtained [56].

Alkylhydroxylamines give, however, a well-defined anodic wave in alkaline solution [6, 57]. It might be mentioned that anodic oxidation of hexafluorodimethylhydroxylamine to the stable free radical, hexafluorodimethyl nitroxide, is a convenient way of preparing this radical [58].

E. Azoxy Compounds and Hydrazines

Aliphatic azoxy compounds may be prepared from nitroalkanes by electrolytic reduction to alkylhydroxylamines as described above followed by anodic oxidation in alkaline solution. Azoxymethane and azoxycyclopropane have been prepared by this method [59].

Reduction of aliphatic azoxy compounds gives the symmetrical hydrazines in good yield. This makes it possible to prepare such hydrazines from nitroalkanes in a three-step synthesis by reversing the current twice but without isolating the intermediates [59]:

$$RNO_2 \xrightarrow[-H_2O]{4\bar{e}\ +4H^+} RNHOH \xrightarrow{-2\bar{e}\ -2H^+} [RNO] \xrightarrow[-H_2O]{RNHOH}$$

$$R\overset{\overset{O}{\uparrow}}{N}{=}NR \xrightarrow[-H_2O]{4\bar{e}\ +4H^+} RNHNHR$$

F. Nitric Acid Esters

Such esters are reduced polarographically in a pH-independent two-electron wave to nitrite ion and alcohol [60, 61]:

$$R\text{-}O\text{-}NO_2 \xrightarrow{2\bar{e}} RO^- + NO_2^- \xrightarrow{H^+} ROH.$$

III. AROMATIC NITRO COMPOUNDS AND DERIVATIVES

Numerous investigations [62] have been made on the reduction of aromatic nitro compounds and a number of different reaction products have been obtained. This is caused more by differences in the chemical follow-up reactions rather than by the electrochemical reactions.

The transfer of the first electron to a nitrobenzene in an aprotic solvent produces an anion radical. In the anion radical of an aliphatic nitro compound the charge on the nitro group is virtually unity, whereas in the anion radicals from nitrocyclopropanes [63], nitroolefins [63, 64], and nitrobenzenes [65,

67] a certain delocalization of the charge occurs according to ESR measurements. The degree of localization in the aromatic ring can be varied, particularly by twisting the nitro group out of the plane of the ring by introducing suitable substituents.

A. Mononitrobenzenes

Polarographically nitrobenzene is reduced in aqueous solution at low pH in two waves, and at high pH in one wave. In the presence of a surfactant or higher concentrations of organic compounds, e. g., alcohol, the first wave is a four-electron wave and the second one a two-electron wave [68]. In the absence of surfactants the height of the first wave corresponds to $4 < n < 5$, but the sum of the first and second wave remains constant, $n = 6$ [68]. This

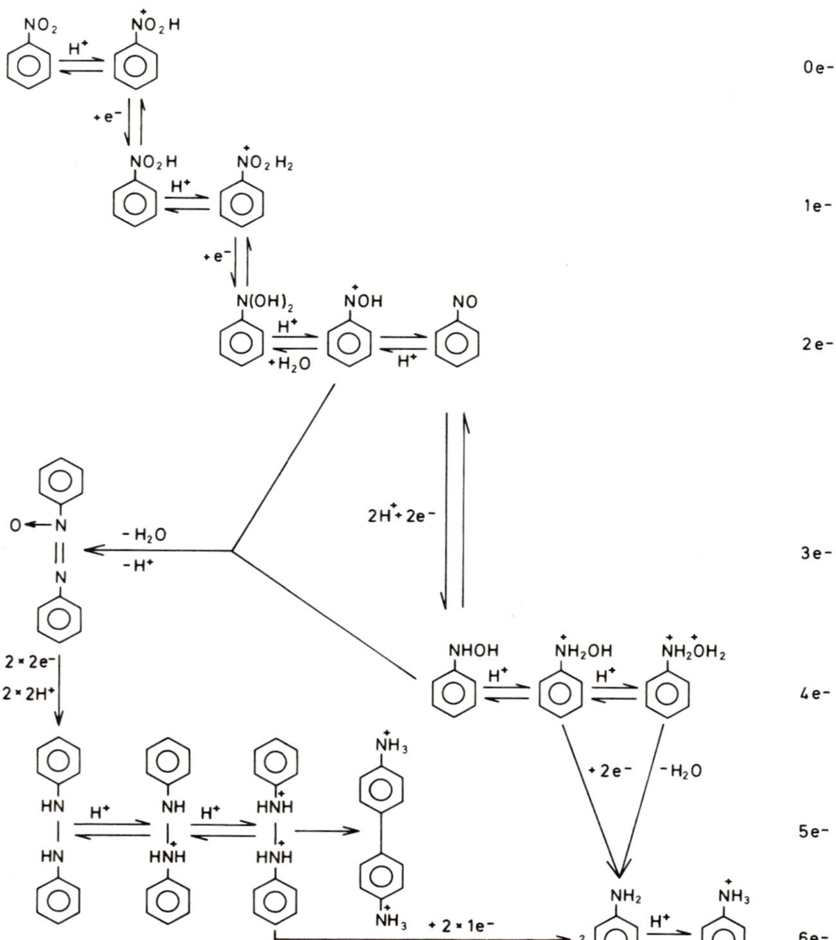

has been interpreted as resulting from two competitive reduction routes, one of them being a four-electron reduction to phenylhydroxylamine and further two-electron reduction at a more negative potential of this compound, and the other one going through a coupling occurring at the surface of the electrode between a protonated nitrosobenzene and phenylhydroxylamine to azoxybenzene and further reduction of this compound. The polarographic reduction may be represented by the scheme [68, 69] on p. 326.

In preparative reductions higher concentrations of organic solvents and/or substrate are present and the surface reaction, the coupling, would not be expected to play a prominent role in acid solution. A reduction at a potential of the first wave would thus result in the formation of a species, here represented by $[C_6H_5\overset{+}{N}H_2OH]$ (XVIII) at the electrode surface, and the diversity of products obtained can be explained by the reactivity of (XVIII). The different reaction paths open to (XVIII) and the products resulting from these processes are discussed below.

1. Desorption

When the electrode potential is such that a further reduction is excluded and no catalyst is present to promote a dehydration reaction, the phenylhydroxylamine is obtained by desorption of the four-electron reduction product. The medium of choice should have pH 3-9 and not contain strong nucleophiles; a mixture of ammonia and ammonium chloride [70] or an acetate buffer [71] is often suitable. The electrode material is of minor importance as long as the cathode potential is controlled in some way; pH must not be too low as rearrangements occur or too high as azoxybenzene then is formed. The temperature must be kept low.

2. Substitution

In strongly acid solution different substitution reactions may occur; the reduction product (XVIII), protonated on nitrogen and/or oxygen, has a tendency to lose water, and this loss may be assisted by the attack of a nucleophile on the aromatic ring. Whether the loss of water occurs before the attack by the nucleophile or in a concerted reaction is not known yet. Any species which can act as a nucleophile in the strongly acidic solution may be employed for this kind of cathodic substitution reaction. The reaction may be represented by

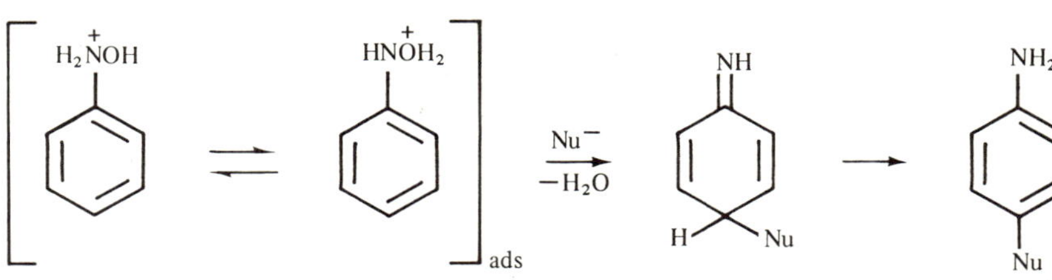

It is well known that phenylhydroxylamine is similarly rearranged in acid solution, but the rate is somewhat slower than that occurring during the electrolytic reduction. This finding has been expressed here by representing the four-electron reduction product as adsorbed phenylhydroxylamine.

In the reduction of nitrobenzene in 4N-hydrochloric acid a mixture of p- and o-chloroaniline and p-aminophenol in the proportion 3:1:1 is found [72, 73]; in sulfuric acid the hydrogen sulfate ion is too weak a nucleophile to compete with the water as nucleophile and the resulting product is p-aminophenol. Substituents capable of delocalizing the positive charge left by the loss of water from the protonated reduction product (in the position ortho to the hydroxylamino group) greatly facilitate the attack para to the hydroxylamino group. The methoxy group is excellent in this respect and o-nitroanisole is easily substituted cathodically.

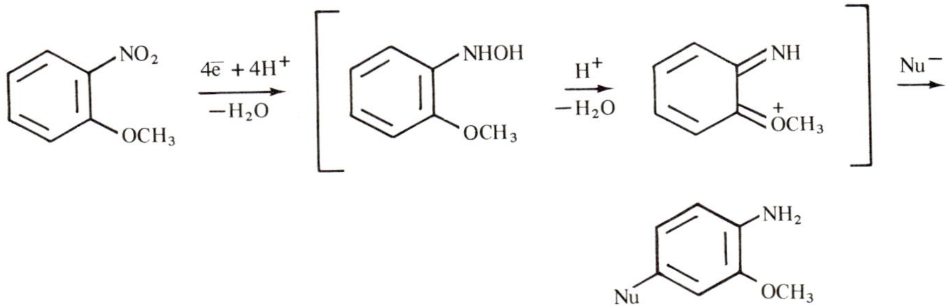

The electrolytic preparation of p-aminophenol was first studied by Gattermann [74]. A wide variety of conditions for the reaction has been described in different patents [75] and the reaction is still under study [76-78]. The reaction is usually run in a fairly strong sulfuric acid at an elevated temperature; if ethanolic sulfuric acid is used, p-phenetidine is the product [79]. The electrochemical preparation of p-aminophenol ought to be competitive with the chemical reaction.

When a substituent para to the hydroxylamino group is present, other reactions may take place as substitution in the ortho position is much less favored. In some cases the substituent is lost according to

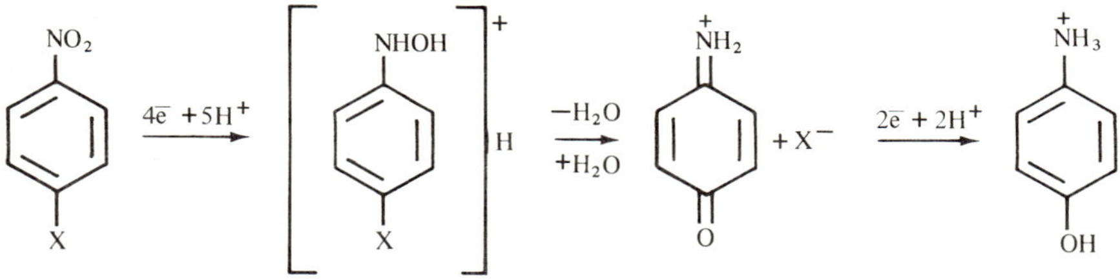

This is found when X = -Cl [80] or $-OCH_3$ [81, 82] and has also been reported to occur during the reduction of p-nitrobenzoic acid in strong sulfuric acid [83].

p-Nitrotoluenes may form different products. In 25% sulfuric acid at 90° p-nitrotoluene is reduced with four electrons and the product is 2,5-dihydroxytoluene (XIX) [80]. The reaction has been formulated as

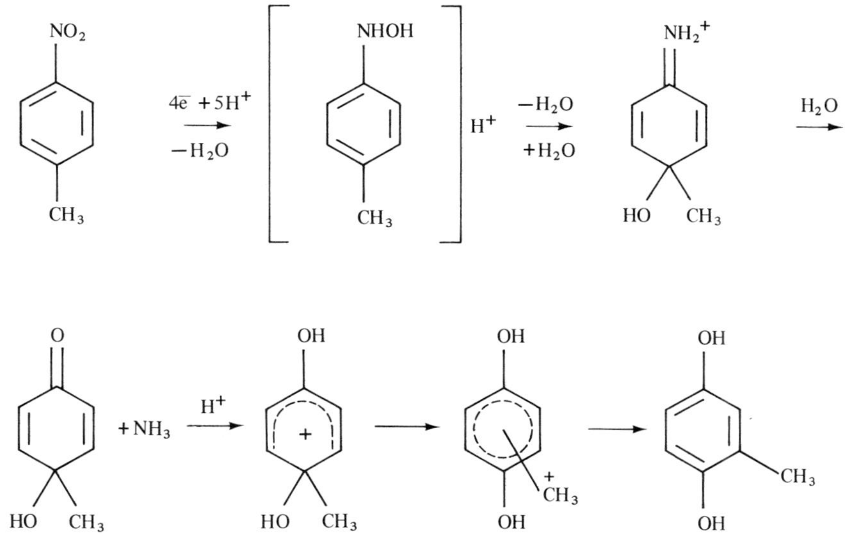

(XIX)

A similar reaction was found [80] in the reduction of 2-methyl-5-nitrobenzoic acid.

p-Methylthionitrobenzene is reduced to the hydroxylamine, which may react in the usual ways; besides those discussed above a rearrangement of the p-methylthiophenylhydroxylamine to p-aminophenylmethylsulfoxide [84, 85] may take place (yield 35-40%)

$$CH_3SC_6H_4NHOH \xrightarrow{[H^+]} CH_3SOC_6H_4NH_2.$$

In concentrated sulfuric acid p-nitrotoluene is reported to give 4-amino-2′-methyl-5′-nitrodiphenylmethane (XX) [74, 86], which has been verified [87]. Gattermann [86] suggested that the primarily formed p-tolylhydroxylamine rearranged into p-aminobenzyl alcohol (XXI) which then attacked unreduced p-nitrotoluene. As (XXI) requires rather drastic conditions for doing so, the following modification may be more likely:

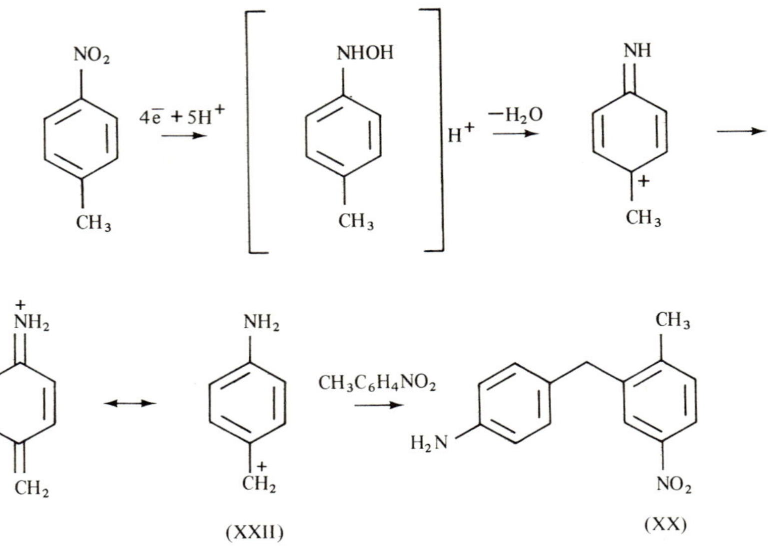

The electrophilic reagent (XXII) may attack any nucleophile present [87]; when the nucleophile is too reactive, it may, however, be attacked by the solvent rather than by (XXII).

3. Condensation Reactions

Phenylhydroxylamine may act as a nucleophile toward any electrophilic center present. As it is the unprotonated phenylhydroxylamine which acts as a nucleophile, a lowering of pH will diminish the concentration of the nucleophile; a higher acidity may, however, make the electrophilic center more reactive, so the overall effect of a change in pH is a function of the properties of both centers.

Nitrosobenzene may act as an electrophilic center and in alkaline solution phenylhydroxylamine reacts with it with formation of azoxybenzene. This reaction occurs generally during the reduction of simple nitrobenzenes in alkaline solution [88-90] and makes azo- and hydrazobenzenes as well as benzidines easily available. Azoxybenzenes may, however, also be formed in acid solution [68, 81]:

$$2C_6H_5NO_2 \xrightarrow[-2H_2O]{6\bar{e} + 6H^+} [C_6H_5NO + C_6H_5NHOH] \longrightarrow C_6H_5\overset{+}{N}(O^-)=NC_6H_5 + H_2O.$$

The preparation of azoxy compounds is possible under a variety of conditions and a number of patents [91] have been issued covering these processes. The yields are generally fair to good (40-90%); substituents ortho to the nitro group retard the formation of azoxybenzenes. A rotating cathode has been employed for this reaction [92, 93] (Chapter IV).

An intramolecular azoxy coupling has been used for the preparation of heterocyclic compounds; this aspect is treated in Chapter XVII.

Carbonyl groups react readily with phenylhydroxylamine and this reaction, in which nitrones are formed, may be used to trap the intermediate phenylhydroxylamine [94]. The reaction may take place intramolecularly with the formation of a heterocyclic N-oxide or hydroxamic acid (Chapter XVII).

A special kind of condensation reaction occurs as a side reaction during the reduction of o-nitroaniline [80, 95]: A phenazine derivative is formed. In the reaction (Chapter XVII) a double Michael addition of o-phenylenediamine to the intermediate o-quinonediimine takes place; other nucleophiles might under suitable conditions attack the diimine.

4. Reduction

In acid solution the protonated phenylhydroxylamine is further reducible at a potential somewhat more negative than that required for the reduction of the nitro compound. Sometimes addition of salts, e.g., Sn^{2+}, facilitates the reduction; this may be regarded as an indirect reduction and is treated in Chapter XXIV.

If an amino or hydroxyl group is situated ortho or para to the nitro group, the intermediate phenylhydroxylamine very rapidly loses water to form the easily reducible quinone mono- or diimine. This loss of water takes place fastest in acid and alkaline solution and somewhat more slowly at pH 5 which is evident from the polarographic behavior of p- and o-nitrophenol [96]. At pH 5 the loss of water is so slow that the four-electron product diffuses away from the electrode before it has lost the water. As mentioned above alkoxy groups para to the nitro group may be lost by substitution and p-aminophenol is the product. A similar reaction may occur in the ortho series [97]. The reduction of p-nitrophenol to p-aminophenol has recently been reinvestigated [98].

A para alkyl substituent may function in the same manner as a hydroxyl group in promoting loss of water from the phenylhydroxylamine and stabilizing the quinoid intermediate. 5-Nitroacenaphthene is thus in acid solution reduced to 5-aminoacenaphthene in a six-electron reduction [99]:

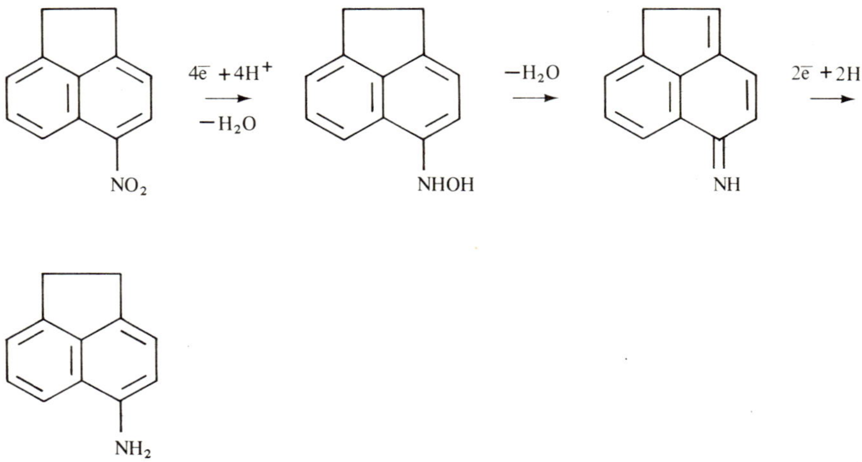

Reduction of a substituted nitrobenzene follows generally one of the reaction paths described above as the nitro group usually is the more easily reducible group. Nitrobenzyl halides [100, 101], however, are reduced in the side chain rather than in the nitro group, and dinitrobibenzyl and nitrotoluene are prominent products (cf. Chapter VI).

Mononitro derivatives of aromatic heterocyclic compounds and condensed aromatic systems behave as described for benzene derivatives. As in the benzene series, any prediction of the most likely reaction route for a given compound must be based upon considerations of the reactivity of the compound and its reduction products in the different types of reactions possible under a given set of parameters.

B. Dinitrobenzenes

Dinitrobenzenes are in principle reduced like the mononitrobenzenes and the same chemical follow-up reactions may occur, but generally not all of the possible reduction products are observed. In substituted dinitrobenzenes also the question arises as to which of the two nitro groups is reduced first.

1. o-Dinitrobenzenes

In dilute mineral acid solution o-dinitrobenzenes may be reduced to an o-nitrophenylhydroxylamine or further to the o-phenylenediamine [102, 103]. The first step in the reduction of 1-X-2, 3-dinitrobenzenes leads to 1-X-2-nitro-3-hydroxylaminobenzene (X = H, COOH, Br, CH_3, OC_2H_5), whereas 1-X-3, 4-dinitrobenzenes in the first reduction form either 1-X-3-nitro-4-hydroxylaminobenzenes (X = COOH) or 1-X-3-hydroxylamino-4-nitrobenzenes (X = Cl, CH_3, OCH_3) [103], e.g.,

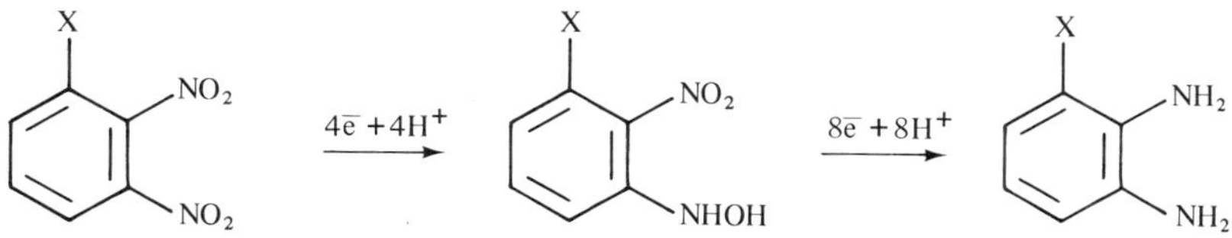

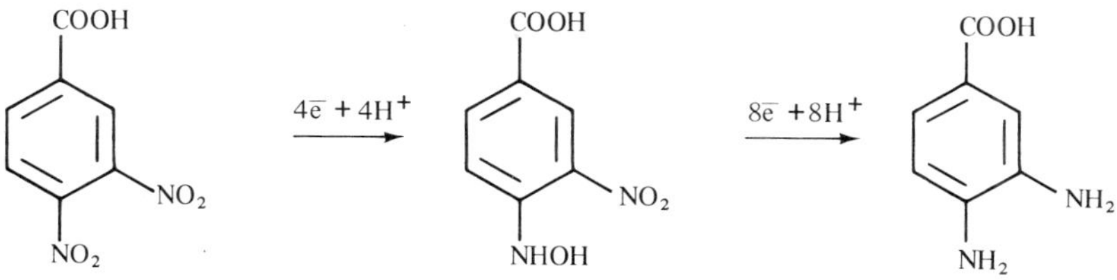

The difference in cathode potential between the first and second reduction is 0.2-0.3 V; a selective reduction thus requires good control of the cathode potential and a suitable cell design.

The hydroxylamino group may react further. A Gattermann reaction has, however, been observed only in the reduction of 2,3-dinitrotoluene. When a carboxyl group is neighbor to the hydroxylamino group, an isoxazolone may be formed; this reaction has been used as an indication of which nitro group is most easily reduced in dinitrobenzoic acids [103].

Reduction of o-dinitrobenzene in alkaline solution at the potential of the first polarographic wave (-0.7 V) also yields o-nitrophenylhydroxylamine [102]. In slightly acid solution some phenazine derivatives may be formed [95].

2. m-Dinitrobenzenes

Reduction in acid solution at room temperature generally may yield a nitrophenylhydroxylamine, a phenylene dihydroxylamine, or a phenylenediamine [82, 102, 104]. In alkaline solution it was found difficult [102] to effect a selective reduction of one of the nitro groups.

For unsymmetrically substituted m-dinitrobenzenes (1-X-2,4-dinitrobenzenes) the first reduction occurs either at the 2-nitro group (X = Cl, Br, I) or at the 4-nitro group (X = COOH, CH_3, C_2H_5) [82]. These differences may be caused by inductive effects and by steric inhibition of coplanarity with the benzene ring of the nitro group or of other substituents. In 2,4-dinitrochlorobenzene which follows the scheme below the inductive effect from the chlorine on the 2-nitro group probably is responsible for its easier reduction:

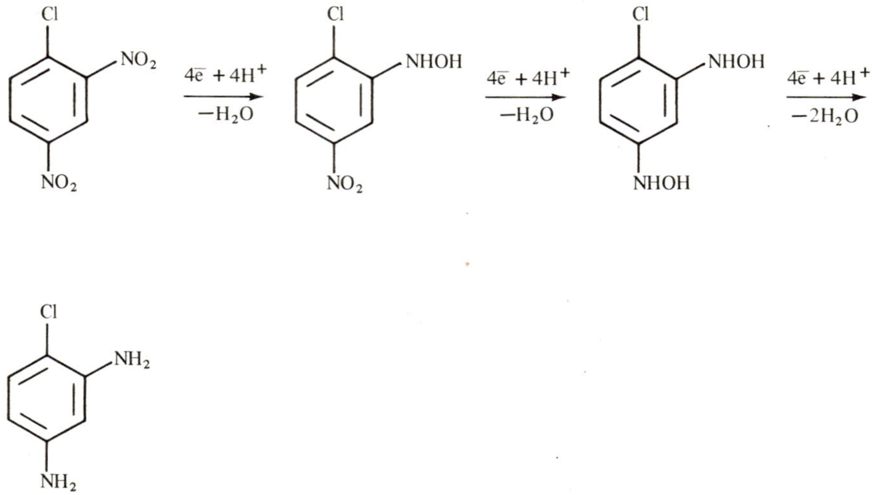

Whereas a dinitroanisole, like 3, 5-dinitroanisole, is reduced similarly, the 2, 4- and 2, 6-dinitroanisoles may or may not lose the alkoxy group, depending on the potential [82]. The reduction of 2, 4-dinitroanisole in a mixture of ethanol and N H_2SO_4 follows this scheme:

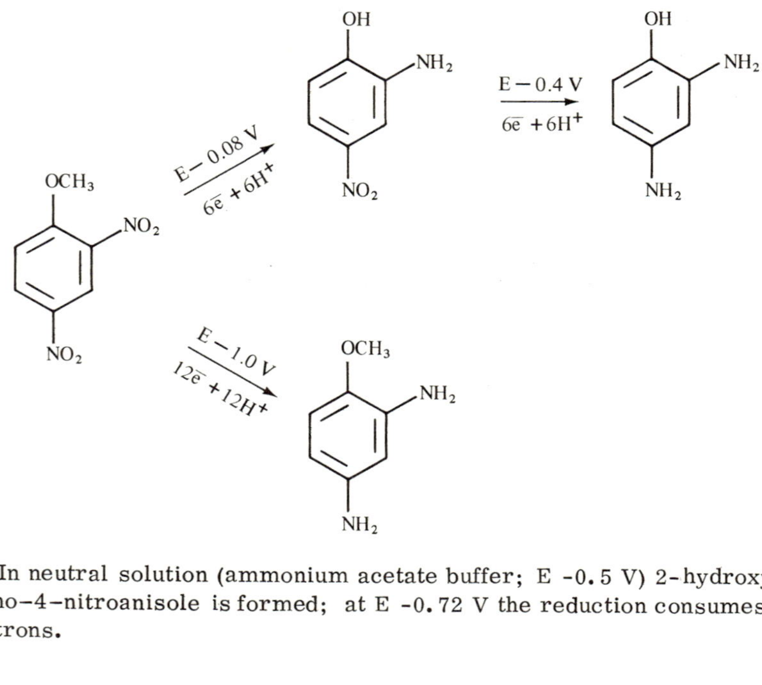

In neutral solution (ammonium acetate buffer; E -0.5 V) 2-hydroxylamino-4-nitroanisole is formed; at E -0.72 V the reduction consumes eight electrons.

In DMSO m-dinitrobenzene is reduced to m-nitrophenylhydroxylamine in the presence of a proton donor, whereas the radical anion, which is reduced further in a second step, is obtained in the absence of protons [105].

3. p-Dinitrobenzenes

Substituted p-dinitrobenzenes may in acid solution be reduced selectively to a nitrophenylhydroxylamine [102, 106, 107]; further reduction leads to the p-phenylenediamine; p, p′-dinitroazoxybenzene has been shown to be an intermediate [69]. 1-X-2, 5-Dinitrobenzenes are reduced first in the 2-nitro group when X = COOH, $CONH_2$, or Br, whereas the 5-nitro group is first attacked when X = CH_3 or OCH_3 [107].

In alkaline solution the polarographic reduction shows [102, 106] a two-electron wave (E -0.35 V) followed by a ten-electron wave (E -1.6 V). The two-electron reduction product $(O_2NC_6H_4NO_2)^{2-}$ is not stable and undergoes secondary chemical reactions; one of the reactions is probably a disproportionation to dinitrobenzene and nitrophenylhydroxylamine. Other intermediates, such as nitrosoaniline or nitroaniline, have been suggested [106], but their presence has not been proved.

Several dinitronaphthalenes have been investigated polarographically [108, 109]. The first wave is usually a four-electron reduction. 1, 2- and 1, 4-Dinitronaphthalenes behave like o- and p-dinitrobenzenes, whereas 2, 3-dinitronaphthalene does not exhibit a conjugation effect similar to that of o-dinitrobenzene. Nitro groups in α-positions are reduced in preference to those in β-positions.

Other dinitro compounds with nitro groups in different rings or one in a side chain may form heterocyclic compounds on reduction; such reactions are treated in Chapter XVII.

C. Nitroso Compounds

Nitroso compounds are reduced like the nitro compounds, but at a less negative potential [110]; nitrosobenzene can thus not be isolated from a reduction of nitrobenzene. The presence of nitrosobenzene as an intermediate in the reduction of nitrobenzene can be shown by trapping it with a suitable nucleophile. The reaction with phenylhydroxylamine to form azoxybenzene has been discussed above; Haber [2] used hydroxylamine and coupled the resulting diazo compound with β-naphthol. Other nucleophiles might be used to trap the nitroso compound and produce products which might have synthetic value. Some applications for the preparation of heterocyclic compounds are discussed in Chapter XVII.

The presence of a nitroso intermediate in nonaqueous solution may also be deduced from results of cyclic voltammetry. Trifluoromethylnitrobenzenes

[111] show two polarographic steps in DMF; the first one is due to a diffusion-controlled one-electron transfer producing a stable anion radical; the second one is a two-electron ECE process resulting in a substituted nitrosobenzene anion radical [112]. Addition of hydroquinone as a proton donor enhances the formation of the nitroso intermediate [111].

The nitrosobenzene cation radical may be formed by anodic oxidation in nitromethane or acetonitrile [113].

D. Phenylhydroxylamines

Polarographic data show that below pH 1 the doubly protonated phenylhydroxylamine is the species reduced at the dropping mercury electrode (DME); between pH 5 and 7 the monoprotonated molecule is reduced. The unprotonated molecule is not reducible [114]. Phenylhydroxylamine is for the main part reduced to aniline, but according to the polarographic data a certain degree of dimerization of an intermediate radical takes place. As discussed above, phenylhydroxylamines may react chemically in several ways.

E. Azoxybenzenes

Azoxybenzenes are reduced in a four-electron reduction to hydrazobenzenes and are generally more difficult to reduce than the azobenzenes; in this respect the system azoxybenzene-azobenzene is similar to that of nitrobenzene-nitrosobenzene. The analogy can be extended: Further reduction of azobenzene yields hydrazobenzene which can be reoxidized to azobenzene and which easily rearranges in acid solution; analogously, phenylhydroxylamine can be reoxidized to nitrosobenzene and can be rearranged to p-aminophenol or similar products in acid solution.

The reduction of azoxybenzene to hydrazobenzene is of preparative interest [92, 115], due to the easy conversion of hydrazobenzenes to benzidines and similar compounds. Several patents [115] describe the conversion of nitrobenzene to azoxybenzene and further to hydrazobenzene, but none of them are continuous processes; a continuous process would stand a good chance of competing economically with the chemical preparation of benzidines.

F. Azobenzenes

The reduction of azobenzenes may follow two routes, either a reversible two-electron reduction or a four-electron reaction with cleavage of the nitrogen-nitrogen bond. The latter reaction is the predominant one for p-hydroxy- and p-aminoazobenzenes [116-118]. In this case the reaction may be formulated as follows:

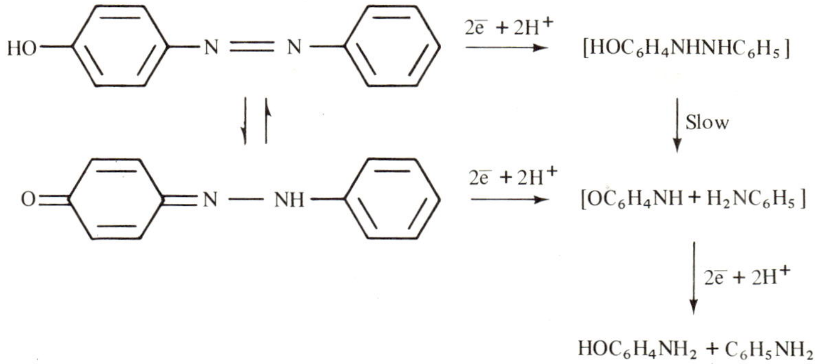

Which of the routes is followed is not yet known; the lower one through a quinone phenylhydrazone is analogous to the reduction of other phenylhydrazones in acid solution [36]. In alkaline solution polarographic data suggest that the upper reaction operates as a two-electron process at high pH, whereas coulometric data showed $n \sim 3$, which means that a slow step occurs after the uptake of the first two electrons.

The system azobenzene-hydrazobenzene is one of the rather few organic redox couples which behaves reversibly or very nearly so at the dropping mercury electrode [119]. In nonaqueous solvents, like DMF, azobenzenes are reduced in two one-electron steps: The first one produces the radical anion, whereas the second one yields a hydrazobenzene dianion. This dianion is easily protonated; it has a tendency to decompose into arylhydrazine [120]:

$$R{-}N{=}N{-}R \underset{}{\overset{\bar{e}}{\rightleftharpoons}} R{-}N{=}N{-}R^{\overset{\bullet}{-}} \overset{\bar{e}}{\rightleftharpoons} R{-}N{-}NR^{2-} \xrightarrow{H^+} RNHNR^-$$

$$R{-}N{-}NR^{2-} \longrightarrow [\text{Intermediates}] \xrightarrow{3H^+} RNHNH_2 + [R^+]$$

The hydrazobenzene dianion is a strong base, the monoanion less so. The reduction of azobenzene has been used for an electrolytic generation of a base in DMF which has been employed in a Wittig reaction [121],

G. Hydrazobenzene

Hydrazobenzene is not polarographically reducible at pH > 4 and in acid solution polarographic curves are difficult to obtain due to the fast rearrangement to benzidine. Polarographic curves of azo- and azoxybenzene show at pH < 4 a second reduction ($n = 2$) corresponding to the formation of aniline. From the pH dependence of the half-wave potentials and a logarithmic analysis of the curves it is proposed that only the diprotonated hydrazobenzene cations undergo reduction, whereas the monoprotonated cation exhibits a catalytic effect on the hydrogen evolution reaction [122].

H. Diazonium Salts

Aryldiazonium salts are reducible at a mercury cathode. Polarograms in acid solution show a pH-independent one-electron wave followed by a larger pH-dependent wave [123, 124]. Reduction on the plateau of the first wave [125] showed n = 1 and only arylmercury salts were isolated, indicating the initial formation of a radical.

During a preparative reduction at a potential of the second wave care must be taken to avoid the reaction between the diazonium salt and the product, phenylhydrazine. By using a flow cell with a small volume/cathode-area ratio and low temperatures the loss by this side reaction was diminished [126].

IV. N-NITRAMINES AND NITROSAMINES

A. N-Nitramines

Protonated N-nitramines are reduced in a single six-electron reaction to the unsymmetric hydrazine [127-130]

$$RR'N\text{-}NO_2 \xrightarrow{6e\ +\ 6H^+} RR'NNH_2 + 2H_2O.$$

In alkaline solution most N-nitro derivatives of primary amines form an unreducible anion $[R\text{-}N\text{-}NO_2]^-$, with nitraminopyridines as exceptions [131], whereas nitramines of secondary amines are reduced in two steps [128]. The first reduction consumes 2 faradays/mol and produces an N-nitrosamine. It is noteworthy that N-nitrosamines in alkaline solution are more difficult to reduce than the corresponding N-nitramines; it seems to be the only class of nitro compounds in which the nitro group is more easily reducible than the nitroso group.

Reduction of nitroguanidine in a 2 M ammonium sulfate solution at a mercury cathode at -1.0 V yields nitrosoguanidine, which precipitates; addition of nickel ions increases the yield [132].

B. N-Nitrosamines

N-Nitrosamines are reducible in both acid and alkaline solutions. Polarographic data show [133-135] that the protonated compound is reduced in a four-electron reaction, whereas the unprotonated nitrosamine consumes only 2 faradays/mole. Controlled potential reduction has [133] the following reactions:

$$\text{Acid solution:}\quad RR'NNO \xrightarrow[4\bar{e} + 4H^+]{H^+} RR'\overset{+}{N}H_3 + H_2O;$$

$$\text{Alkaline solution:}\quad RR'NNO \xrightarrow{2\bar{e} + 2H^+} RR'NH + \tfrac{1}{2}N_2O + \tfrac{1}{2}H_2O.$$

The electrochemical reduction of N-nitrosamines in acid solution is a convenient and general synthetic route to dialkylated, unsymmetric hydrazines [141]. Quite a few nitrosamines were reduced to hydrazines in the "classical" period, but later patents [136] also indicate a continued interest in the reaction.

C. N-Nitrosamides

N-Nitrosamides behave in acid solution like N-nitrosamines; they are reduced to acylated hydrazines. In this way alkylhydrazines may be prepared from primary amines by the following reactions [137]:

$$RNH_2 \xrightarrow[-HCl]{CH_3COCl} RNHCOCH_3 \xrightarrow[-H_2O]{HNO_2} RN(NO)COCH_3 \xrightarrow{4\bar{e} + 4H^+}$$

$$RN(NH_2)COCH_3 \xrightarrow{H_2O} RNHNH_2.$$

Similarly, nitro and nitroso derivatives of urea and guanidine have been reduced to semicarbazides and aminoguanidines in acid solution [130, 138-140].

REFERENCES

[1]. F. Fichter, Organische Elektrochemie, Verlag von Theodor Steinkopff, Dresden und Leipzig, 1942.

[2]. F. Haber, Z. Elektrochem., 4, 506 (1898).

[3]. F. Haber, Z. Phys. Chem., 32, 193 (1900).

[4]. P. Pierron, Bull. Soc. Chim. Fr., 780 (1899).

[5]. M. W. Leeds and G. B. L. Smith, J. Electrochem. Soc., 98, 129 (1951).

[6]. P. E. Iversen and H. Lund, Acta Chem. Scand., 19, 2303 (1965).

[7]. M. Masui, H. Sayo, and K. Kishi, Tetrahedron, 21, 2831 (1965).

[8]. P. E. Iversen and H. Lund, Tetrahedron Lett., 4027 (1967).

[9]. G. W. McMillan, U.S. Pat. 2,485,982 (1949).

[10]. G. B. L. Smith and M. W. Leeds, U.S. Pat. 2,589,635 (1952).

[11]. M. Masui and C. Yijima, J. Chem. Soc., 1101 (1963).

[12]. M. Masui, H. Sayo, and K. Kishi, Chem. Pharm. Bull., 12, 1397 (1964).

[13]. B. Loev and F. Dowalo, Tetrahedron Lett., 781 (1969).

[14]. P. E. Iversen, unpublished observation.

[15]. V. N. Leibzon, V. M. Belikov, and S. G. Mairanovskii, Elektrokhim., 4, 290 (1968).

[16]. H. Schäfer, Chem.-Ing.-Tech., 41, 179 (1969).

[17]. A. K. Hoffmann, W. G. Hodgson, D. L. Maricle, and W. H. Jura, J. Amer. Chem. Soc., 86, 631, 639 (1964).

[18]. H. Sayo, Y. Tsukitani, and M. Masui, Tetrahedron, 24, 1717 (1968).

[19]. H. Sayo and M. Masui, ibid., 24, 5075 (1968).

[20]. M. Masui and H. Sayo, Pharm. Bull., 4, 332 (1956).

[21]. M. Masui, H. Sayo, and Y. Nomura, ibid., 4, 337 (1956).

[22]. M. Masui and H. Sayo, J. Chem. Soc., 4773, 5325 (1961).

[23]. M. Masui and H. Sayo, ibid., 1733 (1962).

[24]. S. Deswarte and J. Armand, Compt. Rend., 258, 3865 (1964).

[25]. J. Armand, Bull. Soc. Chim. Fr., 543 (1966).

[26]. J. Armand and P. Souchay, Chim. Anal., 44, 239 (1962).

[27]. J. Armand, P. Bassinet, and P. Souchay, Compt. Rend., 270, 555 (1970).

[28]. S. G. Mairanovskii and V. A. Petrosyan, Elektrokhim., 2, 115 (1966).

[29]. N. Limosin and E. Laviron, Bull. Soc. Chim. Fr., 4189 (1969).

[30]. Ref. [1], p. 192.

[31]. Ref. [1], p. 151.

[32]. H. Brintzinger, H. W. Ziegler, and E. Schneider, Z. Elektrochem., 53, 109 (1949).

[33]. U.S. Pat. 2,918,418 (19590; CA, 54, 11774 (1960).

[34]. J. Armand, B. Furth, J. Kossanyi, and J.-P. Morizur, Bull. Soc. Chim. Fr., 2499 (1968).

[35]. S. Kwee and H. Lund, Acta Chem. Scand., 23, 2711 (1969).

[36]. H. Lund, ibid., 13, 249 (1959).

[37]. O. Convert and J. Armand, Compt. Rend., 265, 1486 (1967).

[38]. J. Armand, Bull. Soc. Chim. Fr., 882 (1966).

[39]. J. Armand, P. Souchay, and F. Valentini, ibid., 4585 (1968).

[40]. J. Armand and O. Couvert, Coll. Czech. Chem. Commun., 36, 351 (1971).

[41]. J. Armand, to be published.

[42]. I. Ya. Lubyanitskii, P. M. Zaitsev, and Z. V. Zaitseva, Elektrokhim., 1, 990 (1965).

[43]. S. Deswarte, Bull. Soc. Chim. Fr., 522 (1969).

[44]. J. Armand and O. Convert, Compt. Rend., 268, 842 (1969).

[45]. S. Deswarte and J. Armand, ibid., 258, 3865 (1964).

[46]. S. Deswarte, Bull. Soc. Chim. Fr., 534 (1969).

[47]. S. Deswarte, ibid., 545 (1969).

[48]. R. Schindler, W. Lüttke, and L. Holleck, Chem. Ber., 90, 157 (1957).

[49]. P. Souchay and S. Ser, J. Chim. Phys., 49, C172 (1952).

[50]. H. Lund, Acta Chem. Scand., 13, 249 (1959).

[51]. H. Lund, ibid., 18, 563 (1964).

[52]. P. Zuman and O. Exner, Coll. Czech. Chem. Commun., 30, 1832 (1965).

[53]. T. Kubota, H. Miyazaki, and Y. Mori, Bull. Chem. Soc. Japan, 40, 245 (1967).

[54]. H. Lund, Tetrahedron Lett., 3651 (1968).

[55]. M. Heyrovský and S. Vavřička, Coll. Czech. Chem. Commun., 34, 1204 (1969).

[56]. I. A. Avrutskaya and M. Ya. Fioshin, Elektrokhim., 1, 1345 (1965).

[57]. P. E. Iversen and H. Lund, Anal. Chem., 41, 1322 (1969).

[58]. A. P. Tomilov, Yu. D. Smirnov, and A. F. Videiko, Elektrokhim., 2, 603 (1966).

[59]. P. E. Iversen, Chem. Ber., 104, 2195 (1971).

[60]. F. Kaufman, H. J. Cook, and S. M. Davis, J. Amer. Chem. Soc., 74, 4997 (1952).

[61]. M. S. Rusakova, Ya. I. Tur'yan, and R. F. Ustavshchikov, Elektrokhim., 1, 854 (1965).

[62]. Ref. [1], p. 144 ff.

[63]. A. I. Prokof'ev, V. M. Chibrikin, O. A. Yuzhakova, and R. G. Kostyanovskii, Izv. Akad. Nauk SSSR, Ser. Khim., 1105 (1966).

[64]. A. Berndt, Angew. Chem., 79, 240 (1967).

[65]. T. Kitagawa, T. Layloff, and R. N. Adams, Anal. Chem., 36, 925 (1964).

[66]. P. H. Rieger and G. K. Fraenkel, J. Chem. Phys., 39, 609 (1963).

[67]. M. E. Peover and J. S. Powell, J. Electroanal. Chem., 20, 427 (1969).

[68]. M. Heyrovský and S. Vavřička, J. Electroanal. Chem., 28, 409 (1970).

[69]. M. Heyrovský, S. Vavřička, and L. Holleck, Coll. Czech. Chem. Commun., 36, 971 (1971).

[70]. F. Haber and C. Schmidt, Z. Phys. Chem., 32, 272 (1900).

[71]. K. Brand, Ber., 38, 3077 (1905).

[72]. W. Löb, ibid., 29, 1896 (1896).

[73]. K. Elbs and F. Silbermann, Z. Elektrochem., 7, 590 (1902).

[74]. L. Gattermann, Ber., 26, 1844 (1893).

[75]. Ref. [1], p. 156; Ger. Pat. 1,066,589 (1959); CA, 55, 15418 (1961); Fr. Pat. 1,416,966 (1965); CA, 64, 6561 (1966).

[76]. B. B. Dey, T. R. Govindachari, and S. C. Rajagopalan, J. Sci. Ind. Res. (India), 4, 559, 569, 574 (1946); CA, 40, 4965 (1946).

[77]. C. L. Wilson and H. V. K. Udupa, J. Electrochem. Soc., 99, 289 (1952).

[78]. G. S. Krishnamurthy, H. V. K. Udupa, and B. B. Dey, J. Sci. Ind. Res. (India), 15B, 47 (1956); CA, 50, 15289 (1956).

[79]. C. Häussermann, Chem. Ztg., 17, 129 (1893).

[80]. M. Le Guyader, Bull. Soc. Chim. Fr., 1848 (1966).

[81]. M. Le Guyader, ibid., 1858 (1966).

[82]. A. Tallec, Ann. Chim., 3, 347 (1968).

[83]. A. A. Clement, Amer. Chem. J., 16, 511 (1894).

[84]. M. Le Guyader and A. Darchen, Compt. Rend., 267, 1352 (1968).

[85]. A. Darchen, Dissertation, Rennes, 1969, p. 22.

[86]. L. Gattermann and K. Koppert, Ber., 26, 2810 (1893).

[87]. H. Lund, to be published.

[88]. W. Löb, Z. Elektrochem., 7, 335 (1900).

[89]. K. Elbs and C. Wipplinger, ibid., 7, 133, 141 (1900).

[90]. C. Kerns, Trans. Electrochem. Soc., 62, 183 (1932).

[91]. Ref. [1], p. 165.

[92]. K. S. Udupa, G. S. Subramanian, and H. V. K. Udupa, J. Electrochem. Soc., 108, 373 (1961); Indian Pat. 68, 869 (1961); CA, 56, 12670 (1962).

[93]. T. D. Balakrishnan, K. S. Udupa, G.S. Subramanian, and H. V. K. Udupa, Chem.Ing.-Tech., 41, 776 (1969).

[94]. L. Gattermann, Ber., 29, 3040 (1896).

[95]. M. Breant and J. C. Merlin, Bull. Soc. Chim. Fr., 53 (1964).

[96]. D. Stocesova, Coll. Czech. Chem. Commun., 14, 615 (1949).

[97]. B. B. Dey, B. R. Pai, and R. K. Maller, J. Sci. Ind. Res. (India), 10B, 175 (1951).

[98]. H. V. K. Udupa and M. V. Rao, Electrochim. Acta, 12, 353 (1967); A. Darchen and D. Peltier, Bull. Soc. Chim. Fr., 401 (1972).

[99]. S. L. Gupta and N. Kishore, ibid., 15, 1367 (1970).

[100]. J. G. Lawless, D. E. Bartak, and M. D. Hawley, J. Amer. Chem. Soc., 91, 7121 (1969).

[101]. P. Peterson, A. K. Carpenter, and R. F. Nelson, J. Electroanal. Chem., 27, 1 (1970).

[102]. L. Holleck and H. Schmidt, Z. Elektrochem., 59, 56 (1955).

[103]. A. Tallec, Ann. Chim., 3, 164 (1968).

[104]. A. Tallec, ibid., 4, 67 (1969).

[105]. J. S. Dunning and D. N. Bennion, J. Electrochem. Soc., 117, 485 (1970).

[106]. L. Holleck and H. Schmidt, Z. Elektrochem., 59, 1039 (1955).

[107]. A. Tallec, Ann. Chim., 3, 155 (1968).

[108]. G. M. LoPresti, S.-R. Huang, and A. A. Reidlinger, J. Electrochem. Soc., 115, 1135 (1968).

[109]. M. Jubault and E. Raoult, Compt. Rend., 268, 2046 (1969).

[110]. L. Holleck and R. Schindler, Z. Elektrochem., 60, 1138 (1956).

[111]. W. N. Greig and J. W. Rogers, J. Amer. Chem. Soc., 91, 5495 (1969).

[112]. W. Kemula and R. Sioda, Bull. Acad. Pol. Sci., 10, 507 (1962).

[113]. G. Cauquis, M. Genies, H. Lemaire, A. Rassat, and J. P. Ravet, J. Chem. Phys., 47, 4642 (1967).

[114]. M. Heyrovský, S. Vavřička, L. Holleck, and B. Kastening, J. Electroanal. Chem., 26, 399 (1970).

[115]. Ref. [1], p. 174 ff.

[116]. H. A. Laitinen and T. J. Kneip, J. Amer. Chem. Soc., 78, 736 (1956).

[117]. R. Hazard and A. Tallec, Bull. Soc. Chim. Fr., 2917 (1971).

[118]. L. Holleck, D. Jannakoudakis, and A. Wildenau, Electrochim. Acta, 12, 1523 (1967).

[119]. B. Nygård, Ark. Kemi, 26, 167 (1966).

[120]. J. L. Sadler and A. J. Bard, J. Amer. Chem. Soc., 90, 1979 (1968).

[121]. P. E. Iversen and H. Lund, Tetrahedron Lett., 3523 (1969).

[122]. L. Holleck, S. Vavřička, and M. Heyrovský, Electrochim. Acta, 15, 645 (1970).

[123]. E. R. Atkinson, C. E. Garland, and A. F. Butler, J. Amer. Chem. Soc., 75, 983 (1953).

[124]. R. M. Elofson, Can. J. Chem., 36, 1209 (1958).

[125]. J. K. Kochi, J. Amer. Chem. Soc., 77, 3208 (1955).

[126]. P. Rüetschi and G. Trümpler, Helv. Chim. Acta, 36, 1649 (1953).

[127]. H. J. Backer, Rec. Trav. Chim. Pays-Bas, 31, 142 (1912).

[128]. E. Laviron and P. Fournari, Bull. Soc. Chim. Fr., 518 (1966).

[129]. E. Laviron, P. Fournari, and M. Greusard, ibid., 1255 (1967).

[130]. E. Laviron, P. Fournari, and G. Refalo, ibid., 1024 (1969).

[131]. H. Lund, unpublished observation.

[132]. H. F. Metcalf and G. C. Whitnack, U.S. Pat. 2,835,631 (1958); CA, 52, 14394 (1958).

[133]. H. Lund, Acta Chem. Scand., 11, 990 (1957).

[134]. L. Holleck and R. Schindler, Z. Elektrochem., 62, 942 (1958).

[135]. F. Pulidori, G. Borghesani, C. Bighi, and R. Pedriali, J. Electroanal. Chem., 27, 385 (1970).

[136]. U.S. Pat. 2,916,426 (1959); CA, 54, 6370 (1960); Ger. Pat. 1,085,535 (1960); CA, 56, 8562 (1962); Fr. Pat. 1,186,962 (1959); CA., 56, 320 (1962).

[137]. U.S. Pat. 3,267,011; CA, 66, 2188 (1967).

[138]. M. Yamashita and K. Sugino, J. Electrochem. Soc., 104, 100 (1957).

[139]. M. S. V. Pathy, Electrochem. Tech., 3, 94 (1965).

[140]. V. C. Spreter and E. Briner, Helv. Chim. Acta, 32, 215 (1949).

[141]. P. E. Iversen, Acta Chem. Scand., 25, 2337 (1971).

Chapter VIII

SATURATED CARBONYL COMPOUNDS AND DERIVATIVES

L. G. Feoktistov

Institute of Electrochemistry
Academy of Sciences
Moscow, USSR

and

H. Lund
Department of Organic Chemistry
University of Aarhus
Aarhus, Denmark

I. INTRODUCTION

For the purpose of convenience in the discussion all carbonyl compounds except those containing olefinic unsaturation, particularly conjugated olefinic groups, are considered "saturated." Thus, in addition to the paraffinic carbonyls, aromatic and heterocyclic carbonyls are included here, whereas α, β-unsaturated carbonyls are discussed in Chapter IX.

Saturated aldehydes and ketones are usually easily reduced electrochemically. As a rule, the reduction under galvanostatic conditions leads to the derived alcohols, but glycols and pinacols can be obtained as well (Chapter XIX). Potentiostatic conditions essentially enhance the selectivity of the reduction process. Sometimes the electrochemical reduction results in hydrocarbons or organometallic compounds (Chapter XVIII). The choice of electrode material, electrolyte composition (specifically pH), and temperature is of great importance.

Electrochemical reduction of the carbonyl group of acids, esters, amides, and imides is more difficult and sometimes impossible. In some cases, however, aldehydes can be obtained.

II. REDUCTION TO ALCOHOLS

Representative examples of reductions of aldehydes to alcohols are collected in Table 1.

As is clear from the results with paraffinic aldehydes, wide variations in yield and current efficiency have been encountered when conditions were varied. It is particularly noteworthy that hydrocarbon formation, e. g., from propanal, occurs at cadmium and, for the higher aldehydes, increases as temperature and acidity of the medium increase. As the length of the alkyl chain in RCHO increases, glycol formation becomes predominant [45].

Substituted aldehydes (e. g., aldol, glyceraldehyde) form the corresponding alcohols if the substituents themselves do not undergo reduction. However, there may be prior or simultaneous reduction of substituents (chloral, streptomycin). Polyhydroxycarbonyls reduce well: sorbitol and mannitol were prepared from hexoses [1] on an industrial scale (Chapter XXVII).

Aromatic aldehydes readily electroreduce to the corresponding alcohols [36-42], but they are more prone than aliphatic aldehydes to produce glycols, particularly on cathodes of high hydrogen overvoltage under galvanostatic conditions. In 5% sulfuric acid-alcohol on Pt, Ni, and Cu the alcohols are the main reduction products; the yields decrease with increasing temperature.

The carbonyl group of heterocyclic aldehydes is easily reduced to hydroxy. However, resinification (e. g., furfural) in acid may be a serious side

Table 1

Some Reductions of Saturated Aldehydes to Alcohols: $-CHO \rightarrow -CH_2OH$

Aldehyde	Cathodes	Medium; conditions	% Curr. eff.	% Yield	Ref.
Formaldehyde	Cu and Ag			Good	[4-9]
Acetaldehyde	Graphite, Hg	H_2SO_4 and H_3PO_4 solutions			[1-3, 38]
	Cu, Ag	Alkaline			[4-9]
	Pb		91.8	85.9	[10]
	Sn	Phosphate buffer, pH 9, 3-7°	81.5		[45]
Propionaldehyde	Pb	Neutral; weakly alkaline;	96.5		[44]
	Cd, Zn, Hg, C, Cu, Ni, Ag, Al	20°, 4 A/dm^2	11.2 (Al)		[39]
Propionaldehyde	Cd	Acid [a]			
Butyraldehyde	Hg				[11]
Isobutyraldehyde	Hg				[12]
Valeraldehyde	Cd, Pb, Zn	Acid			[139]
Aldol					[13]
Glyceraldehyde	Pb	Na_2SO_4			[16]
Glyceraldehyde		pH 7-8, 10°, 0.2 g/l $FeSO_4$	40.8	98	
Chloral hydrate [b]					[14]

Table 1 (cont.)

Aldehyde	Cathodes	Medium; conditions	% Curr. eff.	% Yield	Ref.
α-Hydroxyisobutyraldehyde [c]	Hg	pH 2.5			[50]
α-Hydroxyisobutyraldehyde [d]	Hg	Phosphate buffer, pH 6			[50]
Streptomycin [e]	Hg				[51]
Streptomycin [f]	Zn	0.01-0.02 A/dm^2, pH 6.8-7	Low	Ca. 100	[53]
Saccharoses	Pb or amalgams	Neutral or alk. or weakly acid		Good	[11, 13, 17-35]
Glucose [g]	Pb, Hg				[33]
Mannose [h]	Usual				
Mannose [i]	Cd				
Glucose [h]	Graphite	Dil. H_2SO_4			[18, 19 46-48]
Mannose [h]	Graphite	Dil. H_2SO_4			
Glucose [h]	Pb/Hg	Alkaline			
Mannose [h]	Pb/Hg	Alkaline			
Lactose	Pb, Zn				[27]
Benzaldehyde	Ni				[52]
2, 4-Dimethylbenzaldehyde				70-75	[42]
3, 5-Dimethylbenzaldehyde				70-75	[42]

Table 1 (cont.)

Aldehyde	Cathodes	Medium; conditions	% Curr. eff.	% Yield	Ref.
p-Methoxybenzaldehyde				70-75	[42]
p-Isopropylbenzaldehyde				70-75	[42]
Salicylaldehyde	Hg	$NaHCO_3$ sol. + CO_2 under press. or OAc^- + H^+		93	[40]
Vanillin	Hg	"		92	[40]
Piperonal	Hg	"		90	[40]
Benzaldehyde	Ni				[52]
Furfural	Zn	pH 8.5		40-72	[43]

[a] Propane also obtained.

[b] Chlorine reductively removed first.

[c] Isobutanol formed at very negative potentials.

[d] Isobutyraldehyde formed at less negative potentials.

[e] Yields dihydrostreptomycin plus dihydrodesoxystreptomycin.

[f] Yields dihydrostreptomycin.

[g] Yields sorbitol and mannitol as major and minor products, resp.

[h] Forms mannitol (as does fructose).

[i] Forms sorbitol.

TABLE 2

Some Reductions of Ketones to Alcohols: $>C{=}O \longrightarrow >CHOH$

Ketone	Cathodes	Medium; conditions	% Curr. eff.	% Yield	Ref.
Acetone	Various	Various			See text
2-Butanone	Pb	$NaHCO_3$		High	[75]
2-Butanone	Pb	Acid [a]			[63]
2-Butanone	Zn	Alkaline			[76, 134]
2-Pentanone	Pb [b]				[65, 77]
3-Pentanone	Pb [b]				[65, 77]
3-Pentanone	Zn	Alkaline		Major	[134]
3-Hexanone	Pb [b]				[65, 77, 133]
Pinacolone	Zn	Alkaline		Major	[134]
Pinacolone	Cd	Acid		Low	[136]
4-Hydroxy-3-methyl-2-pentanone				66	[96]
4-Dimethylamino-2-butanone				62.5	[97]
β-Amino ketones		Acid			[98]
Menthone					[78, 79]
Camphor [c]					[57]
Camphorcarboxylic acids [d]	Hg				[80-82]

Table 2 (cont.)

Ketone	Cathodes	Medium; conditions	% Curr. eff.	% Yield	Ref.
3-Ketocamphononic acid [e]					[82]
Cyclohexanone	Zn, Pb, Tl	Alkaline		Major	[134]
2-Methylcyclohexanone	Hg, Pb [f]				[83]
2-Methylcyclohexanone	Cu [g]				[83]
2-Methylcyclohexanone	Ni [h]				[83]
2, 6-Dimethylcyclohexanone	Hg, Pb [i]				[83]
Acetophenone	Raney Ni			73	[85]
α-Methyldesoxybenzoin [j]	Hg	pH 8		>92	[86]
Benzophenone		Acid			[62]
Benzophenone	Hg	Aq. KOAc, 0.02 A/dm^2, 80-85°		97	[87]
p, p′-Dimethylaminobenzophenone		Acid			[84]
o-Nitrobenzophenone [k]	Pb	Alkaline		86	[88]
3-Methyl-4-phenyl-4-pyrazolinone		NaOAC + $NaHSO_4$		46	[99]
2-Ethyl-1-methyl-3-piperidone				10	[98]

Footnotes to Table 2

[a] Pinacol formation also occurs.

[b] Electrolytically deposited.

[c] Yields borneol and isoborneol.

[d] Mixtures of isomers formed.

[e] Product hydroxy acid lactonizes.

[f] Trans isomer formed.

[g] Cis isomer formed.

[h] Mixed isomers, trans predominant.

[i] Single product formed.

[j] Erythro-1, 2-diphenyl-1-propanol formed.

[k] Product is β-phenyldihydroanthranil.

reaction. Heterocyclic aldehydes exist more in the nonreducible hydrated gem-diol form than do aliphatic and aromatic aldehydes and must, therefore, first be converted to the nonhydrated form. In strongly alkaline media heterocyclic aldehydes are reduced largely to glycols.

Ketones are reduced to secondary alcohols (Table 2) under conditions analogous to those used for reduction of aldehydes.

The reduction of acetone has been studied particularly extensively [54-65]. In acid media at Hg and Pt the current efficiencies for the formation of 2-propanol are 25-50%; pinacol, propane, and organometallics are almost always produced as by-products. In alkaline media at the same electrodes the latter two by-products are absent [54, 56, 66]. Satisfactory yields of 2-propanol on Zn and Cd in sodium acetate solution [62, 67], on Zn in a mixture of sodium acetate and tetraborate [68], and on Hg in sodium chloride-sodium sulfate solutions [56] have been obtained.

Reduction of acetone to the alcohol also occurs on catalytic cathodes. On Pt in 6 N H_2SO_4 the reduction is accompanied by propane formation at low polarizations and on an electrode freshly activated by anodic polarization [69, 70]. On Ni-Al alloys isopropyl alcohol may be obtained in 98% yield at 93% current efficiency using current densities up to 20 A/dm^2 [71].

Ordinary temperatures, high current densities, and, according to some data [72, 73], organic solvents facilitate the reduction of acetone to isopropyl alcohol under galvanostatic conditions. Potentiostatic reduction is apparently not advantageous [74].

Again, it is noted (e.g., 4-hydroxy-3-methyl-2-pentanone and 4-dimethylamino-2-butanone) that nonreducible groups do not interfere with the reduction of the carbonyl.

General consideration of the stereochemistry of electrochemical processes, which would also apply in the reduction of alicyclic ketones, is given in Chapter XXVII.

Alkyl-aromatic and aromatic ketones readily form pinacols on electroreduction (p. 358) but quite often they can be quantitatively reduced to alcohols instead. For the latter purpose cathodes with low hydrogen overvoltage should be employed in dc electrolysis or under potentiostatic conditions.

In the reduction of substituted ketones the substituent itself, depending on its reduction potential, may also be reduced so that the ultimate product obtained may be the result of intramolecular follow-up chemical reactions (cf. p. 331). Thus, o-nitrobenzophenone yields 3-phenyldihydroanthranil, a dehydration ring-closed product of the intermediate α-hydroxylaminobenzhydrol.

Reduction of compounds with adjacent carbonyl groups is complex. Thus, electroreduction at controlled potentials of methylglyoxal on mercury in buffers (pH 7) yields first hydroxyacetone, then at somewhat more negative potentials hydroxypropanal, and at still more negative potentials acetone and 1,2-propanediol [89]. Diacetyl can be reduced to acetoin and in amine-containing buffers the latter can be reduced to 2-butanone and 2,3-butanediol [90]. Benzil is reduced to stilbenediol; the latter changes slowly to benzoin [91]. The ratio of cis- to trans-stilbenediols is greatly dependent on potential, composition of the solution, pH, temperature, and time [92, 94]. Enediol salts were isolated after the reduction of ninhydrin [95].

III. REDUCTION TO GLYCOLS AND PINACOLS

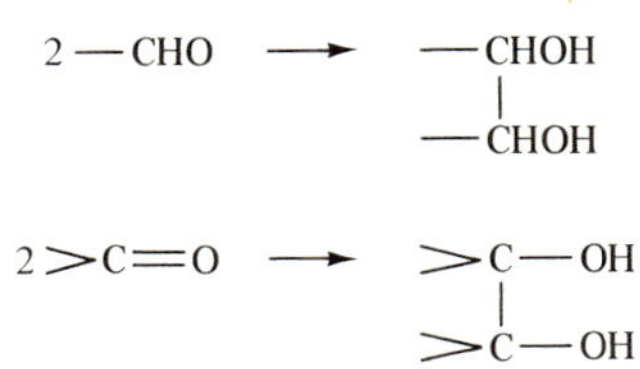

A characteristic feature of the electroreduction of carbonyl compounds, especially aromatic carbonyls, is the ease of obtaining reduced dimeric compounds.[2] The latter types can also be prepared by ordinary chemical reducing agents and sometimes catalytically or photochemically. But, because the reducing ability of a cathode[3] can be varied continuously and controlled at a given value, the electrochemical method is advantageous in many cases. The advantage is most pronounced when controlled potential is employed, although in some cases "hydrodimers" (reduced dimers) can also

TABLE 3

Typical Reductions of Aldehydes to Glycols: $2\,{-}CHO \longrightarrow \begin{matrix} -CHOH \\ | \\ -CHOH \end{matrix}$

Aldehyde	Cathodes	Medium; conditions	% Curr. eff.	% Yield	Ref.
Acetaldehyde	Hg	Acid			[4, 137]
Propionaldehyde	Sn	pH 9, 5-10°, 6A/dm^2	70	90	[44, 45]
Valeraldehyde	Sn	"		Good	[45]
Heptanal	Sn	"		Good	[45]
Heptanal	Hg				[137]
Glyceraldehyde	Zn	pH 4-6, 0.04 A/dm^2	55-70	75-80	[16]
Glyoxylic acid (esters)	C, Ni, Cu, Fe, Ag, Au, Pt			Good	[101-107]
Benzaldehyde [a]	Pt	15% $KHSO_4$, 0.012 A/dm^2		50-65	[36, 37]
Benzaldehyde [b]	Pt	Alkaline			[36]
Benzaldehyde [c]	Pb	Acid			[37, 39]
Benzaldehyde					[146, 148-150]
Benzaldehydes [d]	Ni, Pt, Fe, Cu	Alkaline		70-90	[37]
m-Chloro					[42]
p-Chloro					[42]

Table 3 (cont.)

Aldehyde	Cathodes	Medium; conditions	% Curr. eff.	% Yield	Ref.
0- and p-Methyl-					[42, 145, 149]
2, 4-Dimethyl-					[45]
m-Hydroxy-	—	10% NaOH		20	[41]
p-Hydroxy-				95.2	[144, 145]
4-Hydroxy-3-methoxy-	Pb	Aq. NaOH	—	78	[145, 147]
3, 4-Dimethoxy-					[145]
p-Dimethylamino-	Hg	EtOH + 10% KOH		35 + 58	[141, 142]
p-Acetamino-					[143]
m-Nitro-[e]	Cu	Acid			[9]
Furfural	Pb	Phosphate buffers	Ca. 48	42-63	[43, 151]

[a] Hydrobenzoin and isohydrobenzoin obtained.

[b] One isomer of the product obtained.

[c] Benzyl alcohol and hydrocarbons as by-products.

[d] Similar results were obtained with anisaldehyde, piperonal, salicylaldehyde, and vanillin.

[e] Azoxybenzaldehyde and m, m′-bis-(m-aminobenzylamino) hydrobenzoin formed.

be obtained under galvanostatic conditions. Maintaining a high concentration of substrate favors reductive coupling. Typical hydrodimerizations of aldehydes are given in Table 3.

It is generally true that the highest yields of glycols and pinacols are obtained in alkaline media. In the aliphatic aldehyde series good to excellent yields of glycols have been obtained. The longer the alkyl chain, the higher the yield and the lower the current efficiency [45]. Nonreducible substituents (e.g., as in glyceraldehyde) do not interfere.

Aromatic aldehydes (and ketones) are more prone to hydrodimerize than are their aliphatic analogs. The formation of hydrobenzoins from benzaldehyde and its derivatives has been studied under a great variety of conditions. More recent work [142, 144] shows that under potentiostating 95-97% yields can be obtained.

Typical reductions of ketones to pinacols are shown in Table 4. The conversion of acetone itself to pinacol has been the subject of a formidable number of studies, prompted in part by the fact that this was at one time considered a feasible route to dimethylbutadiene [54, 56-59, 61, 65, 66, 77, 108-126, 128, 129]. The findings have been examined and summarized [127]. Yields are increased as the atomic number of the alkali metal cation is increased and are decreased when a solvent is used. The optimum current density is dependent on the nature of the electrode material [112].

Under the same conditions pinacol yields drop on going from acetone (60%) to 2-butanone (15%) to 3-pentanone to pinacolone (0.6%); 2-octanone forms no pinacol. It has been proposed [134] that substituents with positive inductive effects increase the yields of dimeric products.

Pinacols have also been obtained from alicyclic, aliphatic-aromatic, aromatic, and heterocyclic ketones. Very often the rearranged products, the pinacolones, may be the substances isolated (e.g., [185, 186]). Mixed pinacolization starting with two different ketones has been studied to only a limited extent by reduction of a mixture of two ketones in acid media. The expected three products were obtained [140, 143, 168, 169].

Reductive dimerization of ketones is a versatile reaction and has been used not only for the preparation of the simple pinacols but for the construction of complicated molecules [181], e.g., the methyl ester of (±) cis-bis-dehydrodoisynolic acid (70% at Pb). Attempted pinacol reduction of 1-acetylnaphthalene at mercury in a medium of ethanolic KOH yielded (I) instead of the anticipated product [215]:

TABLE 4

Typical Reduction of Ketones to Pinacols: $2 >C{=}O \longrightarrow \begin{array}{l} >C{-}OH \\ \quad | \\ >C{-}OH \end{array}$

Ketone	Cathodes	Medium; conditions	% Curr. eff.	% Yield	Ref.
Acetone	Zn sponge	Alkaline, 10-16°, 0.005 A/dm^2		50-60	[110-112, 127]
	Mercurated graphite	Alkaline, 10-16°, 0.1-0.3 A/dm^2		50-60	[120, 122, 123]
2-Butanone	Pb	Acid			[63]
2-Butanone	Zn	Alkaline	8	25-28	[76]
3-Hexanone					[130]
Cyclohexanone [g]	Zn	Alkaline	6	15	[134]
Cycloheptanone					[135]
Tetralone					[131]
Dimedone [a]	Hg			30-70	[132]
Androstenedione [b]				High	[187]
Progesterone [b]				High	[187]
Acetophenone				High	[154, 161-162]
"	Pt	Alkaline			[36]
"	Pt	Acetate			[152, 153]
"	Pb	Alkaline			[36]

Table 4 (cont.)

Ketone	Cathodes	Medium; conditions	% Curr. eff.	% Yield	Ref.
Acetophenone	Pb	Acidic			[63]
"	Hg	80% EtOH + KOAc		15.7 meso +44.8 dl	[162]
"	Sn	EtOH + aq. KOAc	54.2	54.2	[160]
Acetophenones					
p-Hydroxy-	Hg	2 N NaOH	—	77	[142]
p-Carboxy-				96	[163]
p-Amino-	Hg, Sn	1.67 N HCl		ca. 60	[164, 165]
o-Amino-	Hg?	20% EtOH + buffers		54,[e] 30, 30	[293]
p-Dimethylamino-	Hg			82	[166]
m-Acetamino-				6.7	
3-Amino-4-Methoxy-		Acid		Low	
"		Alkaline		33	[172]
p-Chloro-[b]	Hg				[149]
p-Methoxy-[b]	Hg				[149]
p-Trifluoromethyl-[b]	Hg				[149]
Propiophenones[b]	Hg, Cu				[149, 150]
p-Methoxy-	Hg			22	[166]
p-Amino-	Hg			57	[166]

Table 4 (cont.)

Ketone	Cathodes	Medium; conditions	% Curr. eff.	% Yield	Ref.
p-Hydroxy-	Sn	Alkaline, 1.5-5 A/dm^2	20	90	[173-177]
p-Hydroxy-	Hg	5-18 A/dm^2; 10% NaOH	48	79.5 [f]	[176-178]
p-Dimethylaminobutyrophenone	Hg			6 + 11	[143]
Benzyl phenyl ketone	Hg [b]	Alkaline			[149]
Cyclohexyl phenyl ketone	Hg	Alkaline			[149]
Mannich bases		Alkaline or acid			[163, 166, 168, 169, 172]
Benzophenone [d]	Pb, Hg	Acid			[62, 63, 87, 153]
Benzophenone	Cu, Fe, Ni	HOAc-H_2SO_4			[155, 157, 158, 170]
Benzophenone		+ Iron salts			[156]
Phenyl p-tolyl ketone					[63]
Phenyl 2, 4-xylyl ketone					[63]
Propiophenones [b, c]	Hg, Cu				[149, 150]
p-Methoxy-	Hg			22	[166]
p-Amino-	Hg			57	[166]

Table 4 (cont.)

Ketone	Cathodes	Medium; conditions	% Curr. eff.	% Yield	Ref.
p-Hydroxy-	Sn	Alkaline, 1.5-5 A/dm^2	20	90	[173-177]
p-Hydroxy-	Hg	5-18 A/dm^2; 10% NaOH	48	79.5 [f]	[176-178]
p-Dimethylaminobutyrophenone	Hg			6 + 11	[143]
Benzyl phenyl ketone	Hg [b]	Alkaline			[140]
Cyclohexyl phenyl ketone	Hg [b]	Alkaline			[149]
Mannich bases		Alkaline or acid			[163, 166, 168, 169, 172]
Benzophenone [d]	Pb, Hg	Acid			[62, 63, 87, 153]
Benzophenone	Cu, Fe, Ni	$HOAc-H_2SO_4$			[153, 157, 158, 170]
Benzophenone		+ Iron salts			[156]
Phenyl p-tolyl ketone					[63]
Phenyl 2, 4-xylyl ketone					[63]
p, p′-Bis-dimethylaminophenyl ketone	Cu				[179]
Benzoin	Pn	Alkaline		10	[36]
Anisoin	Ni, Pt, Fe, Cu	Alkaline			[37]
Phenyl α-naphthyl ketone					[63]
Cumoin					[37]

Table 4 (cont.)

Ketone	Cathodes	Medium; conditions	% Curr. eff.	% Yield	Ref.
Phthaloyl-α-aminobenzophenone					[63]
Acetylpyridines	Hg	Alkaline		68-98	
2-Acetylpyridine	Hg				[149, 182]
3-Acetylpyridine	Hg	Dil. HCl		73.4	[142, 182]
3-Acetylpyridine	Sn	H_2O-alc.-low fatty acid		50	
4-Acetylpyridine					
6-Methyl-3-acetylpyridine					[172]
3-Pyridyl ethyl ketone					[172]
2-Acetylthiophene	Sn	H_2O-DMF-KOAc			[184]
2-Acetylthiophene [d]					[185, 186]

[a] Dehydration product isolated. Yield dependent on concentration and cathode potential.

[b] At controlled potential.

[c] Mixture of isomeric products obtained.

[d] The pinacolone may ultimately be isolated, especially from acid media.

[e] 54% at pH 1.5 (0.5 N HCl), 30% with phosphate buffer or 0.1 M KOH. Mixture of isomers.

[f] Isolated at the pinacolone.

[g] Cyclopentanone yielded a complex unsaturated ketone as well as the expected products [135, 266].

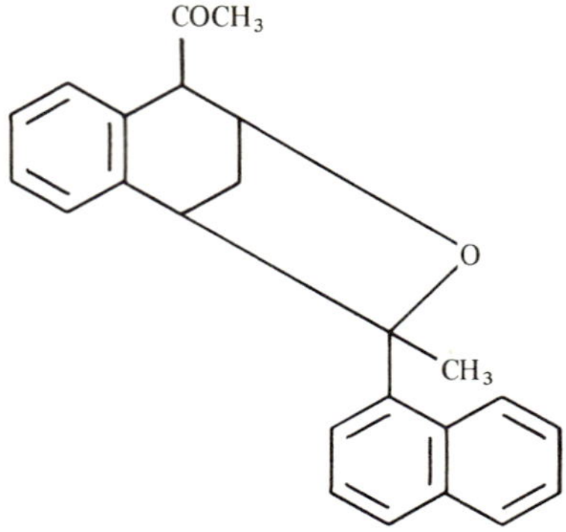

(I)

On the basis of chemical results [194], coupling through the 10, 10′-positions might be anticipated in the reduction of 9-aroyl or 9-alkanoylanthracenes.

Interesting information concerning the effect of molecular geometry upon reaction paths has been obtained from studies of the reduction of diketones which can form inter- and intramolecular products. A cyclopropanediol (II) was isolated (as diacetate) from the electrolysis of (III) in THF containing tributylethylammonium tetrafluoborate and acetic anhydride [216]:

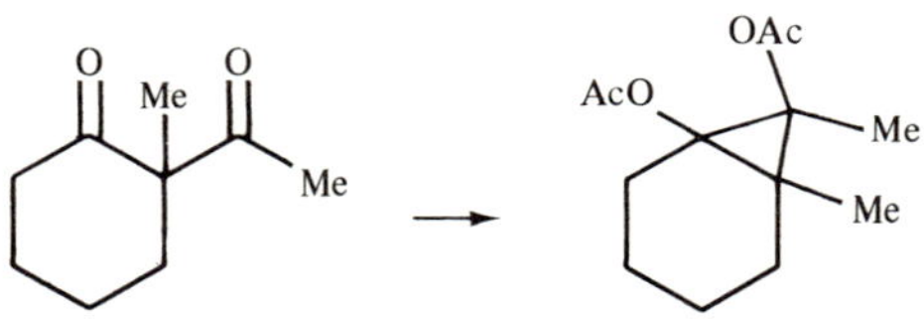

(III) (II)

The trione (IV) behaved similarly:

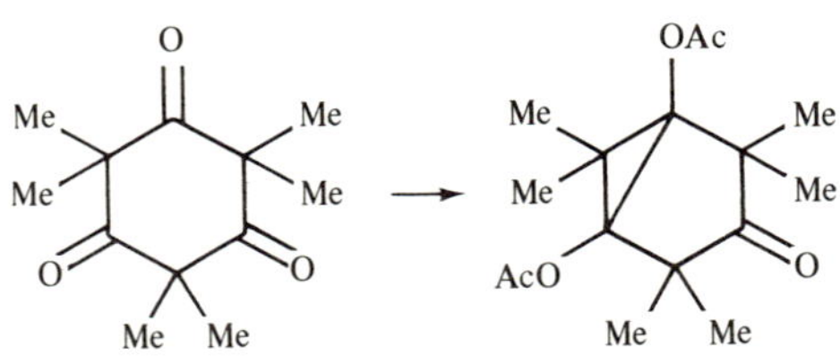

(IV)

Earlier [180] cyclic glycols (V) and (VI) had been obtained from 2, 2′-diaroylbiphenyls and 1, 8-diaroylnaphthalenes, respectively, by reduction in alcoholic or aqueous sodium or potassium hydroxide using a mercury cathode and a lead anode:

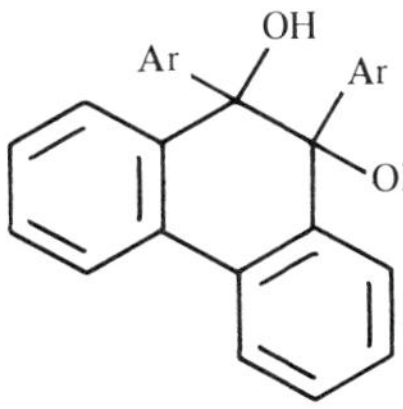

(V)

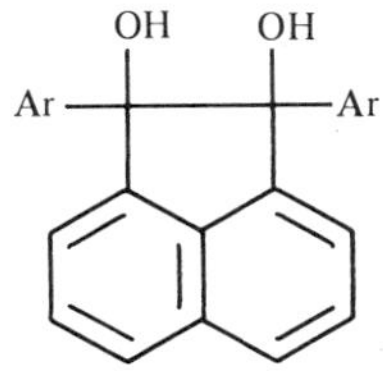

(VI)

$ArCO(CH_2)_3COAr$ yielded cyclic glycols or their derived pinacolones when n = 3; when n was 4 or 5 polymeric materials, presumably glycols, were formed.

Reduction of 2-phenylindan-1, 3-dione in acid medium which might have given an interesting bridged glycol yielded instead the stable diradical (VII) [279]:

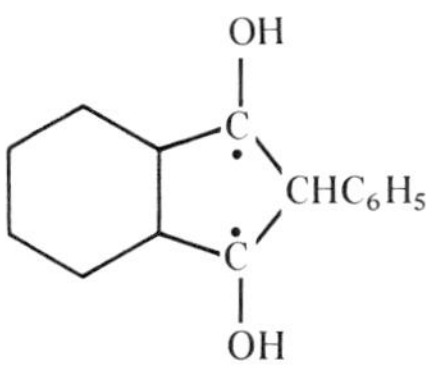

(VII)

Intramolecular pinacolization of acetylacetone yielded a product to which structure (VIII) was assigned [295]; it is more probably (IX) [296]:

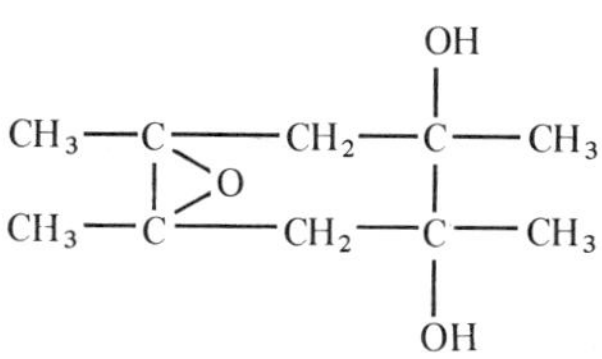

(VIII)

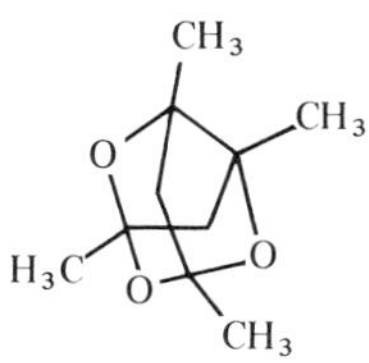

(IX)

Macroelectrolysis of 1, 3-diphenyl-1, 3-propanedione at -1.15 V (pH 4.2) yielded the glycol (X) resulting from the involvement of only one of the two carbonyl groups; at -1.50 V mixtures of isomers of (XI) were probably obtained in which both carbonyl groups had been reduced [297]:

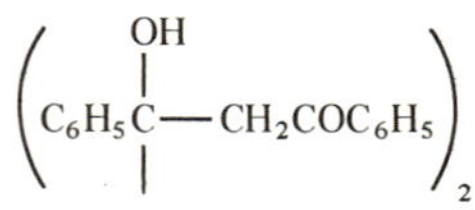

(X)

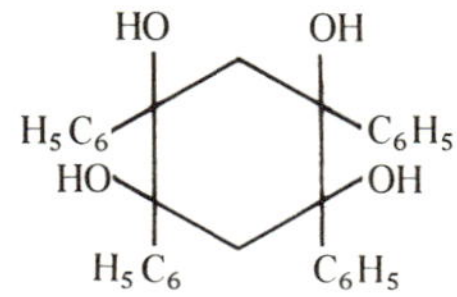

(XI)

A study of the electrochemical reduction of enolized 1, 3-diketones (including 1, 3-diphenyl-1, 3-propanedione) in nonaqueous media [298] showed that the following 0.5 electron process was involved:

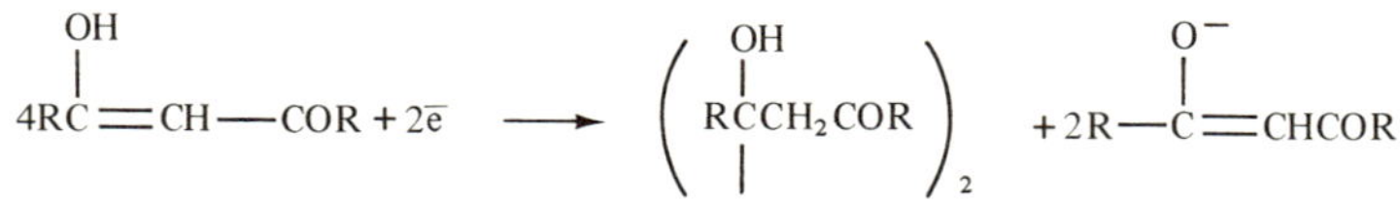

The enolized diketones themselves served as proton donors toward the intermediate anion radicals. A related study involving cyclohexanediones was reported recently [299].

Early attempts [300] to initiate polymerization of vinyl monomers (styrene, acrylonitrile) by the radicals generated by the electrolytic reduction of acetophenone in acid solution were unsuccessful. However, it was recently reported [301] that benzophenone radical anions, formed by reduction of the ketone in DMF with Et_4NClO_4, do cause polymerization of acrylonitrile, methacrylonitrile, methyl acrylate, and methyl methacrylate. A benzophenone moiety did not become incorporated into the polymer; rather, the anion radical disproportionated to ketone and dianion and the latter then transferred an electron to the vinyl monomer.

IV. STEREOCHEMICAL DATA RELATING TO THE PINACOL REDUCTION

Usually, electroreduction of aldehydes and ketones leads to a mixture of diastereoisomeric pinacols, and the stereoselectivity is rather low. But there is no firm evidence that benzaldehyde reduction in alkaline medium [36, 37, 146] resulted only in hydrobenzoin. However, in some cases predominant formation of single (probably meso) isomer was observed in the reduction of p-hydroxybenzaldehyde [144], vanillin, and 2-hydroxy-3-methoxybenzaldehyde [147]. Of the total pinacol yield of 97% in the case of p-dimethylaminobenzaldehyde the fractions of high- and low-melting forms are 39 and 58%, respectively [142]. It was shown that in the reduction of vanillin and p-hydroxybenzaldehyde the pinacols are the meso isomers, while those obtained from veratric aldehyde and p-methylbenzaldehyde are approximately equimolar mixtures of meso and d, l-forms [145]. It was found that on mercury, tin, and copper in alkaline medium the d, l/meso ratio for 2, 3-diphenylbutanediols-2, 3 obtained by reduction of acetophenone (7-C^{14}) is equal to 2.5-3.2, while in acidic medium the ratio becomes 0.9-1.4 [162]. It was also found that the d, l/meso ratio increased somewhat in going from benzaldehyde to acetophenone and further to propiophenone and benzyl phenyl ketone [149, 150]. In this case the stereochemical composition of the products of benzaldehyde hydrodimerization is almost insensitive to a pH change [150]. However, there is strong dependence of stereospecificity in benzaldehyde reduction upon the presence of strongly adsorbing ions in solution: the d, l/meso ratio decreases from 1.0-1.3 to 0.5 in the presence of 5×10^{-4} M tetrabutylammonium perchlorate or 0.8 M sodium iodide [148].

The significance of the distribution of diastereoisomers for the mechanism of the reduction and coupling step [150] is being actively studied (cf. Chapter XXVII).

V. REDUCTION TO HYDROCARBONS

The carbonyl group of aldehydes and especially of ketones can often be reduced electrochemically to methylene [195] in rather high yields. Hydrocarbons usually form in acidic media — sulfuric acid. Representative examples are presented in Table 5.

TABLE 5

Reduction of Carbonyl to Methylene:

Carbonyl	Cathodes	Medium; conditions	% Curr. eff.	% Yield	Ref.
Formaldehyde	Cu, Ag			Low	[1, 4-10]
Acetaldehyde	Cu, Ag			Low	[1, 4-10]
Propionaldehyde	Cd				[39]
Isobutyraldehyde	Pb			To 40	[12]
Valeraldehyde	Cd, Pb, Zn			13	[139]
Enanthal [a]	Cd				[39]
Benzaldehyde					[37, 39]
2, 4-Dimethylbenzaldehyde					[37]
3, 5-Dimethylbenzaldehyde					[37]

Table 5 (cont.)

Carbonyl	Cathodes	Medium; conditions	% Curr. eff.	% Yield	Ref.
o-, m-, and p-Toluylaldehydes					[37]
Salicylaldehyde	Zn				[40]
p-Hydroxybenzaldehyde	Zn, Cd				[40]
Cuminic aldehyde					[39]
Anisaldehyde					[37, 38]
Protocatechualdehyde	Cd				[39]
Vanillin	Cd				[39]
Piperonal	Cd, Zn				[39, 40]
p-Dimethylaminobenzaldehyde					[141]
p-Chlorobenzaldehyde	Cu				[37]
Furfural	Zn			6	[43]
Acetone	Cd			100	[39, 60, 67, 74]
Acetone	Pt[b]				[69, 70, 120]
Acetone	Zn, Zn-Hg, Al			High	[74, 60, 74], resp.
Acetone	Pb, Hg			< 25	[60, 67, 74]
Acetone	Sn, Ag			Very low	[60]
2-Butanone	Cd				[67]
2-Butanone	Zn				[196]
3-Pentanone	Cd				[67]
3-Pentanone	Cd			70	[192]

Table 5 (cont.)

Carbonyl	Cathodes	Medium; conditions	% Curr. eff.	% Yield	Ref.
2-Pentanone	Cd-Bi				[199]
2-Pentanone	Cd	H_2SO_4		To 75	[139, 197, 210]
"	Zn	"		60	[139]
"	Pb	"		17	[139]
"	Hg	"		20	[139]
"	Zr, Pb, Hg	HOAc		0	[198]
Pinacolone	Cd			83.5	[136]
5-Methyl-2-hexanone	Cd	H_2SO_4		83.5	[193]
Acetopropanol	Cd	H_2SO_4		50	[194]
Diacetone alcohol	Pb	H_2SO_4			[200]
Keto esters and derivatives					See text
Acicyclic ketones					See text
Acetophenone	Pt, Pd, Bi			High	[213]
Benzophenone				High	[62, 63, 153, 159]
Benzyl subst.-phenyl ketones	Zn-Hg	H_2SO_4		50-90	[40, 159]
Benzil					
Benzoin					[159]
Desoxybenzoin					[159]
1-Acetylnaphthalene		Acid			[215]
α-Naphthylphenyl ketone	Zn-Hg				[159]
2-Ethyl-1-methyl-3-piperidone	H_2SO_4				[194]
"	Cu[c]	60°			[194]

Table 5 (cont.)

Carbonyl	Cathodes	Medium; conditions	% Curr. eff.	% Yield	Ref.
2-Ethyl-1-methyl-3-piperidone	Pb,[c] Cd[c]				[194]
2-Alkyl-2-hydroxy-1-methyl-5-pyrrolidinones					[214]
3-(N-dimethylaminoethyl)-4-oxoquinolizidine				40	[194]

[a] Yield of hydrocarbon increases with temperature and current density.

[b] Propane is the only product if the electrode has been previously anodically polarized.

[c] Depending on the temperature, various ratios of 1-methyl-2-propylpyrroline, 1-methyl-2-ethyl-3-hydroxy piperidine, and N-methylheptylamine are formed.

The reduction of acetone to propane has been studied very extensively. In addition to the data cited it should be mentioned that when cathodes of different metals are used at the same potential the chemical yields of propane differ substantially, although the total current efficiency for the reduction products varies only by a factor of 2. For example, on mercury the chemical yield and current efficiency, respectively, for propane are 2.2 and 73%; on lead, 25.2 and 86%; on cadmium, 100 and 90%; on zinc, 97 and 86%; and on aluminum, 100 and 45%. The yield of propane increases with increasing cathode potential. The propane yield on cadmium significantly increases with illumination [74].

The keto group of keto esters can be converted also to methylene by simultaneous reduction of the ester group oxygen and elimination of the alkoxy group.

It is not quite clear whether the Tafel rearrangement — a migration of the aceto group from the α-carbon to the carboxyl group carbon — always takes place. Thus, reduction of acetoacetic ester gave butane [203], that of ethylacetoacetic ester n-hexane, instead of methylpentane [204]. Methylbenzylacetoacetic ester gave 1-(1^3-methopentyl)benzene [203-205]. The reduction product of benzylacetoacetic ester was stated to be β-benzylbutane [206], 1-n-butyl-4-methylbenzene [204], or n-amylbenzene [204, 207]. On lead and cadmium in aqueous alcohol-sulfuric acid solution at elevated temperature similar transformations are observed in reduction of propyl- [203, 204], isopropyl- [204, 207], n-butyl- [203, 204], isobutyl [203, 204, 207], sec-butyl- [204, 207], and diethyl- [203, 204, 207] acetoacetic esters and methyl esters of methyl-propyl and methylisopropylacetoacetic acid [204, 207]. The yields approach 50%.

Under the same conditions the reduction of the diethyl ester of α,α'-diacetylsuccinic acid provided the diethyl ester of α,α'-diethylsuccinic acid [203]. From the ethyl ester of oxanilic acid phenylglycine was obtained on mercury in sulfuric acid [208].

The N-methylamide of β-ketocapric acid was reduced on lead in sulfuric acid to capric acid N-methylamide (43.5%) and N-methyl-n-octylamine (12.1%). Moreover, 1-methyl-2-n-propylpiperidone (6.8%) and N-methylconiine [211] were also obtained.

The keto group of alicyclic compounds including steroids can also be reduced to methylene. For example, from menthone menthane is produced in sulfuric acid on lead and tin (50%) and on cadmium and mercury (90%) [79, 159, 201, 202].

The keto group of 5 α-androstanol-17 β-one-3, the methyl ester of $\Delta^{1,3,5,(10)}$-estratrienol-3-one-17, and the acetate of 17α, 20, 20, 21-bismethylenedioxy-Δ^5-pregnenol-3β-one-20 were reduced (85-97%) on lead in an aqueous dioxane-sulfuric acid solution. By this method deuterium may be incorporated at C_{17}, C_{19}, and C_{20}; labeling by ordinary methods is rather difficult [212].

Quite often the reduction of the carbonyl group to methylene results in by-products, particularly unsaturated hydrocarbons. The yields of the latter are negligible and the reaction has no preparative significance. Reduction of isobutyraldehyde led to isobutene [12]; pinacolone to 2, 3-dimethylbutene-2 [136]; benzaldehyde [37, 140] and benzoin [159] to stilbene. Reduction of substituted benzaldehydes also gives the respective substituted stilbenes [37, 38].

The advantages of the electrochemical reduction of carbonyl to methylene must be assessed individually when alternate inexpensive "chemical" methods are available (e.g., Clemmensen, Wolff-Kishner/Huang-Minlon). However, in the reduction of the carbonyl group of amides and lactams, the electrochemical method offers at least a cost benefit.

VI. REDUCTION TO ORGANOMETALLIC COMPOUNDS

Very often the reduction of carbonyl compounds in acid media on Hg, Pb, and some other cathodes results in the formation of organometallic compounds. Yields up to about 40% have been obtained, but conditions have not been found under which the formation of the organometallic is the major reaction.

These reactions are discussed in Chapter XVIII.

VII. REDUCTION MECHANISMS

Electroreduction of carbonyl compounds is a complicated process in which electrodic steps alternate with purely chemical steps and occur partly on an electrode surface and partly in the bulk of the solution. On metals which weakly adsorb hydrogen, the electron, behaving as a reducing agent, is transferred to an adsorbed molecule. On metals which strongly adsorb hydrogen the most advantageous reaction course is electron transfer to the hydroxonium ion or a water molecule; this gives rise to adsorbed atomic hydrogen which adds to a reducing molecule.

Sometimes (mainly in aprotic solvents capable of strong anion solvation) a solvated electron is the reducing agent.

Under conditions of electrochemical reaction a carbonyl compound itself may exist in various forms, and quite often the form predominating in solution is electrochemically inactive or less active than other equilibrium forms. This applies more specifically to the hydrated and nonprotonated forms of carbonyl compounds.

The first reduction step — addition of an electron or hydrogen atom — may be followed either by exactly the same sequence of steps or by a purely chemical step: addition of atoms or atomic groups (most often protonation), dimerization, isomerization, and finally elimination of some atoms or groups. In such cases the nature of the products formed predetermines the possibility of further reduction. Naturally, the rates of the individual steps depend on the nature and concentration of the reacting particles and the electrode potential. The effect of the latter is most specific for electrochemical reactions and it appears either directly in the reaction activation energy change (i.e., through a rate constant of the electrodic stage) or via change of both concentration and (to some extent) properties of the reacting particles.

Thus, generally speaking, electrochemical reduction of the carbonyl compounds involves the same steps as any electrochemical process does. But for carbonyl compounds the most important steps are hydration and dehydration of a gem-diol group (or formation and decay of acetals and ketals), protonation, and dimerization of the intermediate products.

In aqueous media some aldehydes and ketones exist predominantly in a gem-diol form. This is most characteristic of the lowest aliphatic and of some heterocyclic aldehydes, keto acids, and halosubstituted cyclohexanones. This fact is of great significance for the polarography of carbonyl compounds [225-231, 233, 234] in which the observed currents are much lower than the normal diffusion current because gem-diols do not reduce. Base catalysis causes an increase in the current with increasing pH due to enhancement of the dehydration rate. But in strong alkaline media the current again falls, in this case owing to the formation of the anion of the gem-diol. The importance of this step for preparative electrochemical processes has not been elucidated quite sufficiently; however, one may assume that the increasing yield of hydrodimer in going from acetaldehyde to the highest aliphatic aldehydes is above all associated with the lesser degree of hydration of the latter [107]. Formation and decay of acetals and ketals occur considerably more slowly than the hydration-dehydration process [232]. These processes play an important role in the polarography of saccharoses [240-243].

As is true of many other classes of compounds, carbonyl derivatives also may protonate; the protonated forms reduce at less negative potentials than the nonprotonated forms. In polarography this fact is shown in a shift of half-wave potential to more positive values (at a sufficiently high protonation rate); kinetic waves appear also at the less negative potentials and at rather low rates [235-239, 244-253]. A preceding protonation is often decisive for preparative electrochemical processes since one may conduct the first electroreduction step at such low potentials that the intermediate particles are reduced further at only a low rate. Subsequent protonation of the intermediates, specifically that of the anion radicals and dianions, is also very important in view of its preparative application, since neutralization of a negative charge favors a subsequent electrodic stage or simply removes the intermediate particles out of the purely chemical reaction zone. From the above discussion it follows that in terms of synthetic application electroreductions on metals which adsorb hydrogen weakly should be more advantageous than on those with high surface concentration of atomic hydrogen when the steps of charge transfer and proton addition combine in a single step — hydrogen atom addition.

At the present time there is unequivocal evidence that the unstable reduction intermediates of aromatic carbonyl compounds [254-258] are of a radical character, i.e., they have an unpaired electron [259-265]. Polarography played a prominent role in confirming the significance of radical reactions for the reduction of aromatic carbonyl compounds: Polarograms having one-electron waves, change of the number of electrons participating with change in pH, separation of consecutive waves with increasing concentration of the carbonyl compounds, etc.

By means of polarography, cyclic voltamperometry, faradaic impedance, and rectification methods one may elucidate a sequence and determine relative rates of protonation, electron transfer, and radical dimerization. The following reduction steps [244, 249, 256, 257, 267-280] have been proposed for aromatic and related carbonyl compounds in acidic,

$$\mathrm{ArCOR + H^+ \rightleftarrows Ar\overset{+}{C}(OH)R}$$

$$\mathrm{Ar\overset{+}{C}(OH)R + \bar{e} \rightleftarrows Ar\overset{\bullet}{C}(OH)R} \qquad E_1$$

$$\mathrm{2Ar\overset{\bullet}{C}(OH)R \rightarrow ArRC(OH)C(OH)RAr}$$

$$\mathrm{Ar\dot{C}(OH)R + \bar{e} \rightarrow Ar\bar{C}(OH)R} \qquad E_2$$

$$\mathrm{Ar\bar{C}(OH)R + H^+ \rightarrow ArCH(OH)R;}$$

neutral,

$$\mathrm{ArCOR + \bar{e} \rightarrow Ar\dot{\bar{C}}OR} \qquad E_3$$

$$\mathrm{Ar\dot{\bar{C}}OR + H^+ \rightarrow Ar\dot{C}(OH)R}$$

$$\mathrm{2Ar\dot{C}(OH)R \rightarrow ArRC(OH)C(OH)RAr}$$

$$\mathrm{Ar\dot{C}(OH)R + \bar{e} \rightarrow Ar\bar{C}(OH)R} \qquad E_2$$

$$\mathrm{Ar\bar{C}(OH)R + H^+ \rightarrow ArCH(OH)R;}$$

and alkaline media,

$$\mathrm{ArCOR + \bar{e} \rightarrow Ar\dot{\bar{C}}OR} \qquad E_3$$

$$\mathrm{2Ar\dot{\bar{C}}OR \rightleftarrows ArRC\bar{O}C\bar{O}RAr}$$

$$\mathrm{Ar\dot{\bar{C}}OR + \bar{e} \rightarrow Ar\bar{C}\bar{O}R} \qquad E_4$$

$$\mathrm{ArRC\bar{O}C\bar{O}RAr + 2H^+ \rightleftarrows ArRC(OH)C(OH)RAr}$$

$$\mathrm{Ar\bar{C}\bar{O}R + H^+ \rightarrow ArCH\bar{O}R}$$

$$\mathrm{ArCH\bar{O}R + H^+ \rightleftarrows ArCH(OH)R,}$$

where usually $E_1 < E_2$, $E_2 \lesssim E_3$ (often $E_2 > E_3$), and $E_3 < E_4$.

Thus, polarography facilitates importantly the selection of optimal conditions for preparative electrosyntheses of pinacols or alcohols. However, one should bear in mind that the macrocathode conditions are not identical to those in polarography (Chapter III). Moreover, the effect of the nature of the metal is beyond the scope of polarography.

The nature of the metal determines the adsorption conditions of the substance [283-286]. It was shown, for example, that the benzaldehyde and benzophenone reductions occur in an adsorbed state [277, 287]. Probably a radical dimerization also takes place on an electrode surface [148, 292, 277]. In translating polarographic results to other metals one has to take into account the change of adsorption conditions which directly determines the rates of the electrodic and preceding chemical reactions.

Moreover, to a greater or lesser extent, some metals are capable of forming organometallic compounds which may importantly influence a reaction course.

To explain formation of organometallic compounds in the reduction of carbonyl compounds a radical mechanism has been proposed [74, 217, 275, 281], although carbanion reactions with metal cations are not excluded [282]. It is quite probable that hydrocarbon formation in the reduction of carbonyl compounds proceeds via the intermediacy of organometallics [74, 281, 288,

289] during the reduction of protonated molecules [60, 74, 127].

There are some grounds for assuming that in the case of aliphatic carbonyl compounds glycol and pinacol formation occurs via the carbanion rather than radical participation [127]:

$$RR'CO + 2\bar{e} \rightarrow RR'\bar{C}\bar{O}$$

$$RR'\bar{C}\bar{O} + RR'CO \rightarrow RR'C\bar{O}C\bar{O}RR'$$

$$RR'C\bar{O}C\bar{O}RR' = 2H^{+} \rightarrow RR'C(OH)C(OH)RR'$$

or

$$RR'CO + H^{+} \rightleftarrows RR'\overset{+}{C}OH$$

$$RR'\overset{+}{C}OH + 2\bar{e} \rightarrow RR'\bar{C}OH$$

$$RR'\bar{C}OH + RR'\overset{+}{C}OH \rightarrow RR'C(OH)C(OH)RR'.$$

As support (but not proof) of this mechanism, it can be noted that the polarographic data show participation of two electrons in the course of the reduction of aliphatic aldehydes and ketones in both aqueous and dimethylformamide media [219, 248, 290, 291].

VIII. STRUCTURE AND REACTIVITY

Referring to the formula of carbonyl compounds:

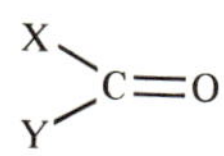

it can be stated that in the case of aliphatic aldehydes (X = Alk, Y = H) increase of the size of the alkyl group increases substantially the tendency of a carbonyl compound to form glycols. On the contrary, for aliphatic ketones (X = Alk, Y = Alk) increase of the size of the alkyl group sharply decreases the yield of dimeric reduction products. It is not clear whether in this case steric hindrance is decisive, which would certainly be the case if an ionic mechanism is involved in hydrodimerization; on this basis there would be a great difference in the bulk of the competing agents (organic molecules and proton donors).

Aromatic aldehydes (X = Ar, Y = H), alkyl-aromatic (X = Alk, Y = Ar), and aromatic (X = Ar, Y = Ar) ketones do not differ in tendencies toward the formation of dimers. This tendency is much higher than in the case of aliphatic compounds. There is also no steric hindrance since in this case the intermediates have a sufficiently long lifetime.

The half-wave potentials for the reduction of all carbonyl compounds were found to decrease when electronegative substituents are introduced. These effects might be qualitatively described by means of Taft and Hammett equations while taking account of steric factors. One should bear in mind that the most electron-attractive α- or p-phenyl ring substituents eliminate or reduce before the carbonyl group is reduced [275].

IX. CARBONYL DERIVATIVES [4]

A. Introduction

Most compounds containing a carbon-nitrogen double bond are reducible electrolytically under suitable conditions; the azomethine group is generally more easily reducible than the corresponding carbonyl group. In several cases only the protonated azomethine compound is reducible in a convenient potential range.

This section discusses briefly the electrochemistry of azomethine compounds of the general formula RR′C=NYR″, where Y may be hydrogen, carbon, or a heteroatom. When Y is carbon or hydrogen, the compound is a derivative of ammonia, when Y is nitrogen or oxygen, the azomethine compound is derived from hydrazine or hydroxylamine, respectively. In the following, the compounds will be treated in the three main groups depending on the nature of Y.

A recent review [302] on the electrochemistry of azomethine compounds gives leading references up to early 1968.

B. Derivatives of Ammonia

Schiff bases and other imines of aldehydes and ketones are polarographically reducible in a wide pH range [303-309], but as the equilibrium between a carbonyl compound and ammonia (or an amine)

$$>C{=}O + HNRR' \rightleftharpoons >C(OH)NRR' \xrightleftharpoons{H^+, -H_2O} >C{=}\overset{+}{N}RR'$$

usually lies rather far to the left, the easy hydrolysis of the imine interferes with an investigation in aqueous solution. A preparative electrolysis may work when the hydrolysis is relatively unimportant or the amine is present in a large excess.

Addition of a large excess of ammonia has been used for the polarographic determination of unconjugated ketones [303, 304]. As the imine is more easily reducible than the ketone, electrolysis of a mixture of a ketone such as cyclohexanone and an amine such as methylamine [305] at a potential where the imine but not the ketone is reducible, produces the substituted amine (in casu methylcyclohexylamine) in high yield. Recently it has been shown [310] that a great variety of ketones and amines (including some secondary amines) produces substituted amines when a mixture of a ketone and an amine is electrolyzed under suitable conditions.

Only a few imines are stable in aqueous solution [311, 312] (Chapter XVII, but sometimes reduction in cold 50% sulfuric acid [313, 314] at a lead cathode gives the amine. In many cases, however, reduction in nonaqueous solvents is preferable.

Some Schiff bases of aromatic aldehydes and ketones give two one-electron waves in aqueous acidic solution [305, 306] or nonaqueous solution [315, 316]. The product of the first one-electron reduction is a radical which may either dimerize or may be reduced further at a more negative potential. Although some dimerized product was formed in a classical electrolysis [312] in 50% sulfuric acid, higher yields may be obtained in nonaqueous solution.

The hydrodimerization reaction has been studied in acetonitrile [318, 319] using immonium salts; these rather than imines were chosen in order to avoid complications with acid-base reactions. Reduction of immonium cations in acetonitrile leads either to a stable free radical or to a hydrodimerization, depending on the substituents at carbon; some of the immonium salts studied exhibited a second wave at a more negative potential [317]. It was possible to show that the reductive coupling is a purely radical one, and the rate constants have been measured in a number of cases using either linear sweep voltammetry or symmetrical cyclic voltammetry. For the electron spin resonance (ESR) investigation of medium stable radicals an indirect electrolytic preparation of the radicals was made by mixing in a flow system a suitable electrolytically produced radical with the immonium cation immediately before the introduction in the cavity [320].

Mixed hydrodimerization is also possible. Reduction of benzalaniline in the presence of acrylonitrile gave a mixture containing 1, 5-diphenyl-2-pyrrolidone(I) [321] and other products:

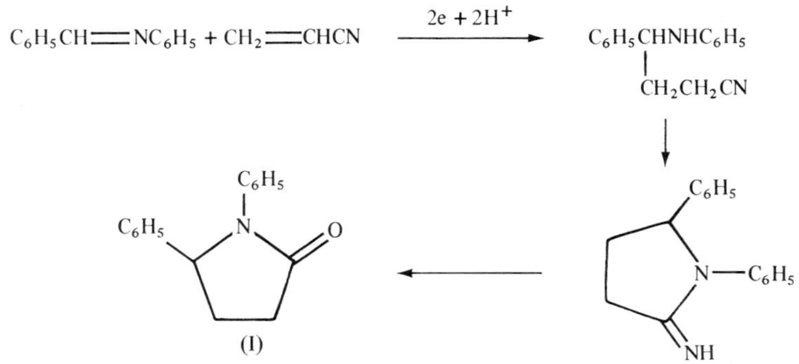

C. Derivatives of Hydrazine

The electrochemistry of hydrazones, azines, and other derivatives of hydrazine is complicated by the possibility of tautomerization. The following forms must be considered [322, 323]:

$$RCH_2\text{-}CH_2\text{-}N{=}NR' \rightleftarrows RCH_2CH{=}NNHR'' \rightleftarrows RCH{=}CHNHNHR'$$

(II) (III) (IV).

The hydrazone (III) is the predominant form, and in acid and neutral solution no indication of either the azo- (II) or the ene-hydrazine form (IV) has been found. In alkaline solution, however, studies with tritium [324] have shown exchange of both the aldehydic hydrogen and of hydrogen atoms at the α-carbon, indicating the presence of both the forms (II) and (IV). The equilibrium constant in 0.1 M ethanolic KOH for the equilibrium ω-benzeneazotoluene ⇄ benzaldehyde phenylhydrazone is about 10^8, but for monoalkylhydrazones of aliphatic carbonyl compounds a higher proportion of the azo form is found. Acetaldehyde propylhydrazone is thus in t-butanol containing 0.02 M potassium t-butylate at 100° in equilibrium with 3.7% of the azo form [325].

The different tautomeric forms may be reduced at different potentials, and the removal from the equilibrium of one of the forms by reduction will lead to a continuous formation of the reducible tautomer in an attempt to restore the equilibrium. The rate of the formation of the reducible tautomer is thus of importance for the electrode reaction.

1. Hydrazones

Phenylhydrazones are generally reducible in acid solution in a four-electron reduction, which leads to the formation of aniline and another amine; in alkaline solutions only activated phenylhydrazones are reducible, and the reduction leads in a two-electron reaction to a substituted phenylhydrazone [306]. Scheme I shows the reduction route in aqueous solution which has been suggested [305, 326].

Polarographic evidence suggests that in acid solution a protonation occurs before the electron transfer; this is supported by the polarographic reduction of some quaternized hydrazones [327]. An argument in favor of this reaction sequence (Scheme I) is that the hydrazine (VIII) is not reducible under the conditions used for the reduction of the phenylhydrazone. Another argument is that the imine (XI) is reduced at about the same potential or more easily than the phenylhydrazone.

After the uptake of a proton and an electron the radical (VI) thus formed may accept either a proton (route A) or an electron (route B). In route A the protonated radical (IX) accepts an electron which results in a cleavage of the nitrogen-nitrogen bond with aniline as leaving group. The imine thus formed (XI) is protonated and reduced further. In route B the radical (VI) forms the carbanion (VII) on electron uptake, but as the aniline anion is a poor leaving group, no cleavage of the nitrogen-nitrogen bond occurs and

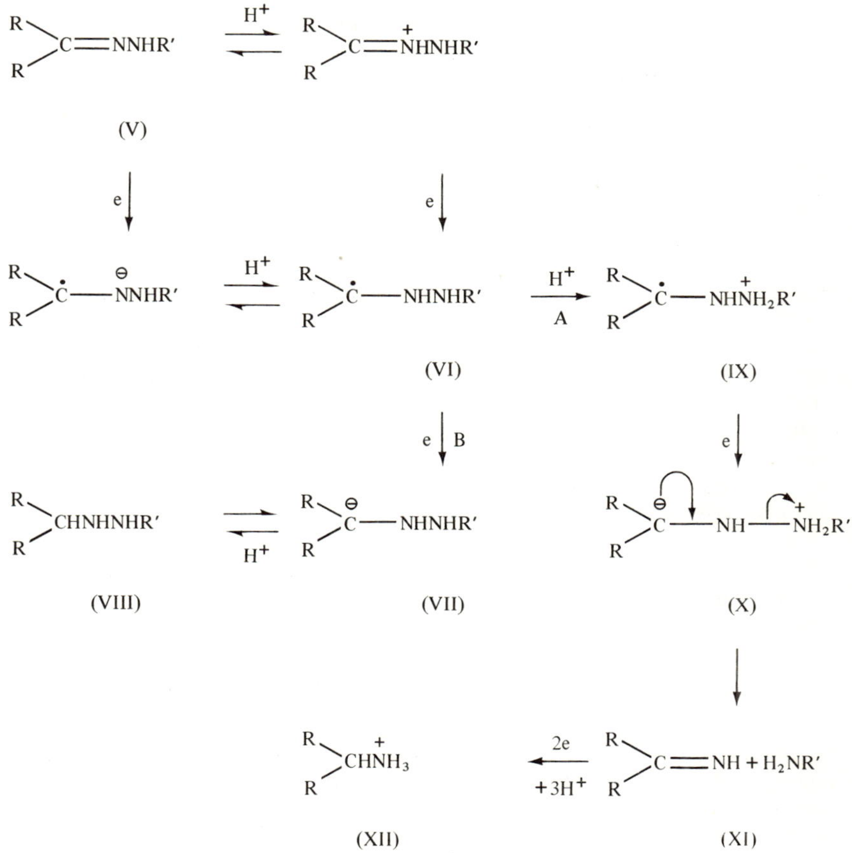

SCHEME I

the hydrazine (VIII) is formed on protonation. (In the scheme H^+ stands for any proton source.)

Nitrophenylhydrazones [328], benzylhydrazones [305], and cyclic hydrazones like 3, 6-diphenyl-2, 3, 4, 5-tetrahydropyridazine [327], 1, 2-dihydrophthalazine [327, 329], or dihydrocinnolines [330] are reduced similarly. The reduction of the cyclic derivatives is discussed in Chapter XVII. Acyclic [305] and cyclic (Chapter XVII] azines also follow a similar reduction scheme.

2. Acylated Hydrazones

In strongly acid solution acylated hydrazones are reduced in a four-electron reduction like the phenylhydrazones. In alkaline solution the acylated hydrazones often are reducible in a two-electron reaction and from reduction of N-benzoyl-N′-benzalhydrazine, N-benzoyl-N′-benzylhydrazine was isolated.

Several polarographic investigations of Girard hydrazones, semi, and thiosemicarbazones have been made [305, 331-340], but relatively few controlled potential reductions have been reported [305, 336-340]. From these it appears that in acid solution a four-electron reduction is found and that the first step is a hydrogenolysis of the nitrogen-nitrogen bond; in some cases a catalytic hydrogen evolution is found [340]. In strongly acid solution there often is found a tendency to a stepwise reduction. Thus, benzaldehyde thiosemicarbazone [336] gives at pH 0 two waves at -0.66 and -0.73 V (SCE); when alcohol is present, the separation of the waves is poor or not visible at all. In alkaline solution some of the compounds are reducible, and then a two-electron reduction is found which results in a saturation of the carbon-nitrogen double bond to the hydrazine. At intermediate pH values both reduction paths may be followed simultaneously and the relative importance of the two routes depends on many factors such as pH, buffer concentration, temperature, and solvent. Thus, benzaldehyde and benzophenone semicarbazone [305], p-acetamidobenzaldehyde thiosemicarbazone [336], p-acetamidobenzaldehyde S-methyl thiosemicarbazone [336], and cyclopentanone and benzophenone Girard hydrazone [337] all yielded the expected amine at low pH; at high pH values benzophenone Girard-D hydrazone [337] and benzaldehyde semicarbazone [339] were reduced in high yield to the acylated hydrazine; and at medium pH several of the compounds gave mixtures of the amine and the hydrazine [305, 336, 337].

Benzaldehyde thiobenzoylhydrazone (XII) is reduced in acid solution according to the general reduction scheme in which the initial step is the hydrogenolysis of the nitrogen-nitrogen bond. At pH <2 three polarographic waves are found [327], and on the basis of controlled potential reductions the following steps were suggested:

1. wave: $$C_6H_5CH{=}NNHCSC_6H_5 \xrightarrow[2e\,+\,2H^+]{H^+} C_6H_5CH{=}\overset{+}{N}H_2 + H_2NCSC_6H_5$$

(XII)

2. wave: $$C_6H_5CH{=}\overset{+}{N}H_2 \xrightarrow{2e\,+\,2H^+} C_6H_5CH_2\overset{+}{N}H_3$$

3. wave: $$C_6H_5CSNH_2 \xrightarrow[2e\,+\,2H^+]{H^+} C_6H_5CH(SH)\overset{+}{N}H_3\,,$$

followed by hydrolysis and reduction of benzaldimine and benzaldehyde.

In alkaline solution a two-electron reduction is found which results primarily in a saturation of the carbon-nitrogen double bond; the reducible species is probably the following azomethine compound, $C_6H_5CH{=}NN{=}C(S^-)C_6H_5$.

The reduction of cyclic acylated hydrazones like pyridazinones [327] and phthalazinones [341] is discussed in Chapter XVII.

D. Derivatives of Hydroxylamine

1. Oximes

Among the azomethine derivatives of hydroxylamine the ordinary oximes are the most important and many polarographic and electrolytic investigations of oximes have been made [305, 342-344]. Most oximes are reducible in acid solution, many fewer in alkaline solution; polarographic data prove that it is the protonated oxime which is the reduible species in acid and neutral solutions. Classical reductions in sulfuric acid at lead cathodes or controlled potential reductions in acid solution at mercury cathodes yield the amine in a four-electron reduction. As the hydroxylamine corresponding to the oxime is not reducible under the conditions employed, it has been suggested [305] that the reduction proceeds as follows:

$$RR'C{=}NOH \rightleftarrows (RR'C{=}NOH)H^+ \xrightarrow{2e\,+\,2H^+}$$

$$(RR'C{=}\overset{+}{N}H_2) + H_2O \xrightarrow{2e\,+\,2H^+}$$

$$RR'CH\overset{+}{N}H_3 .$$

As the site of the protonation at the species to which the transfer of electrons occurs is not known, both a protonation at N and at O may be considered, although N is the most likely basic center.

The postulated intermediate, the imine, can in most cases not be isolated or detected by classical polarography. The reduction of 2,4-dihydroxybenzophenone oxime occurs [345], however, in two steps, and reduction at the potential of the first wave produced the imine which in this case was sufficiently stable to permit its isolation.

If the reaction path shown above is as general as the available evidence suggests, attempts to reduce protonated oximes electrolytically to hydroxylamines are not likely to succeed. Some unprotonated oximes, e.g., benzaldoxime and benzophenone oxime, are reducible in not too strongly alkaline solution; the oxime anion is not reducible. An investigation of the reduction of benzaldoxime in alkaline solution [339] showed that some benzylhydroxylamine is formed under these conditions.

Polarographically there is a difference between the syn- and the anti-oxime in unbuffered solutions having tetraalkylammonium ions as supporting electrolyte [346-348]. syn-Benzaldoxime and other syn-oximes give two waves in this medium; the first wave at -1.84 V (SCE) of syn-benzaldoxime is kinetically controlled, whereas the second one at -2.2 V (SCE) is diffusion controlled. anti-Benzaldoxime gives only one wave [at -1.84 V

(SCE)] which is diffusion controlled. It has been suggested [347] that the two waves of the syn form reflect the tautomeric equilibrium $>C{=}NOH \rightleftharpoons >C{=}\overset{+}{N}HO^-$.

In some cases stereoselective electrochemical reduction of oximes has been found [349]; this is discussed in Chapter XXVII.

2. Alkylated Oximes

Oximes may be alkylated at nitrogen or oxygen; both types are electrolytically reducible [305, 350, 351] in their protonated form. The nitrones are generally also reducible in alkaline solution, whereas the O-methyl ethers of oximes are not reducible in ordinary media containing alkali metal ions. In solutions containing quaternary ammonium p-toluenesulfonates some O-alkylated oximes give a polarographic wave.

N-Alkylsubstituted oximes both of aliphatic and aromatic aldehydes are reduced in a single four-electron wave in slightly acid or alkaline solution [305, 350, 351]. N-Phenylbenzaldoxime is reduced in two or more steps; the first two-electron reduction results in the formation of benzalaniline, which in acid solution is reduced in two one-electron reactions [350].

The reduction of the O-methyl oximes [350] is probably analogous to the reduction of the oximes in that the nitrogen-oxygen bond is cleaved before the saturation of the carbon-nitrogen double bond. The similarity in the reduction in acid solution of the oxime and its alkylated derivatives is understandable when the close resemblance between the electroactive forms is considered:

$$RCH{=}\overset{+}{N}H{-}OH \qquad RCH{=}\overset{+}{N}R'{-}OH \qquad RCH{=}\overset{+}{N}H{-}OR'$$

NOTES

[1]Sometimes electrolytic oxidation of low molecular weight carbohydrates in sulfuric acid [100] and anhydrous alkaline media [49] results in the elimination of one carbon atom at the carbonyl end of the molecule.

[2]Mixed reductive coupling is discussed in Chapter XIX.

[3]A variety of high-overvoltage cathodes is satisfactory, although one particular metal may be optimum for a given reaction.

[4]This section by H. Lund.

REFERENCES

[1]. Brit. Pat. 140, 115 (1918).

[2]. Brit. Pat. 140, 527 (1919).

[3]. P. Pascal, Swed. Pat. 88, 118 (1921).

[4]. Ger. Pat. 274, 201 (1912).

[5]. Ger. Pat. 277, 393 (1913).

[6]. Ger. Pat. 328, 342 (1917).

[7]. Brit. Pat. 335, 638 (1929).

[8]. H. Hibbert and R. R. Read, J. Amer. Chem. Soc., 46, 983 (1924).

[9]. G. Shima, Mem. Coll. Sci. Univ. Kyoto, 9A, 183 (1925).

[10]. M. Y. Kagan and V. G. Kryukov, Chim. Prom. (Russ.), 10, 34 (1933).

[11]. P. Condit, U.S. Pat. 2, 537, 304 (1951); CA, 45, 2341 (1951).

[12]. S. Swann, Jr., E. Onstott, and F. Baastad, Trans. Electrochem. Soc., 102, 113 (1955).

[13]. J. Wolk, U.S. Pat., 2, 419, 515 (1947).

[14]. Tommasi, Traité d'Electrochem., 741 (1889).

[15]. C. Sandonnini and V. Bor̃ghēllo, Atti Acad. Lincei, 21, 30 (1935).

[16]. A. P. Tomilov, A. A. Sergo, and S. L. Varshavskii, Elektrokhim., 1, 1126 (1965).

[17]. Ger. Pat. 140, 318 (1900).

[18]. U.S. Pat. 1, 612, 361; 1, 653, 004; 1, 712, 951; 1, 712, 952 (1926).

[19]. Swed. Pat. 213, 617 (1929).

[20]. N. Belen'kaja and I. Belozerskii, Zh. Obshch. Khim. (Russ.), 19, 1664 (1949).

[21]. H. J. Creighton, Trans. Electrochem. Soc., 75, 289 (1939).

[22]. H. J. Creighton, ibid., 99, 127 (1952).

[23]. J. Bludworth, M. Roulson, and H. Fruby, U.S. Pat. 2, 462, 301 (1949); CA, 43, 3300 (1949).

[24]. M. Wolfrom, M. Konigsberg, F. Moody, and R. Goepp, J. Amer. Chem. Soc., 68, 12 (1946).

[25]. M. Wolfrom, W. Binkley, C. Spencer, and B. Lew, ibid., 73, 3357 (1951).

[26]. P. Regna, ibid., 69, 246 (1947).

[27]. R. Hales, U.S. Pat. 2,289,189 (1943); CA, 37, 42 (1943); U.S. Pat. 2,300,218 (1943); CA, 37, 1660 (1943); U.S. Pat. 2,303,210 (1943); CA, 37, 2277 (1943).

[28]. H. Creighton and R. Hales, U.S. Pat. 2,458,895 (1949); CA, 43, 2104 (1949).

[29]. H. Hepti and W. Kolb, U.S. Pat. 2,507,973 (1950); CA, 45, 230 (1951).

[30]. H. Hirayama, J. Chem. Soc. Japan, 66, 15 (1945); CA, 42, 7169 (1948).

[31]. S. Hayashi, K. Sugino, and I. Mizuguchi, Japan. Pat. 176,226 (1948); CA, 45, 5048 (1951).

[32]. K. Sugino and S. Hayashi, J. Chem. Soc. Japan, 65, 458 (1944).

[33]. E. Parker and S. Swann, Jr., Trans. Electrochem. Soc., 92, 343 (1947).

[34]. K. Brown, Can. Pat. 402,883 (1942); CA, 36, 2481 (1942).

[35]. M. Wolfrom, F. Moody, M. Konigsberg, and R. Joepp, J. Amer. Chem. Soc., 68, 578, 2342 (1946).

[36]. H. Kaufmann, Z. Elektrochem., 2, 365 (1886); 4, 461 (1898).

[37]. H. D. Law, J. Chem. Soc., 89, 1512 (1906); 91, 748 (1907); 99, 1113 (1911).

[38]. J. Tafel and W. Schepps, Ber., 44, 2148 (1911).

[39]. W. Schepps, ibid., 46, 2566 (1913).

[40]. G. Shima, Mem. Coll. Sci. Univ. Kyoto, A11, 1, 407, 419 (1928); A12, 69, 327 (1929).

[41]. W. S. Rapson and R. Robinson, J. Chem. Soc., 1537 (1935).

[42]. H. D. Law, ibid., 91, 748 (1907); 99, 1115 (1911).

[43]. W. O. Reichsfeld and S. K. Kudrna, Tr. LTI Im. Lensoveta (Russ.), 1956, p. 110.

[44]. W. G. Khomyakov, A. P. Tomilov, and B. G. Sildatov, Elektrokhim. 5, 850 (1969).

[45]. W. G. Khomyakov, A. P. Tomilov, and B. G. Soldatov, ibid., 5, 853 (1969).

[46]. A. Findlay and V. H. Williams, Trans. Faraday Soc., 17, 453 (1922).

[47]. H. J. Creighton, Trans. Electrochem. Soc., 75, 289 (1939).

[48]. Fr. Pat. 760, 507 (1934).

[49]. G. W. Hay and F. Smith, Can. J. Chem., 47, 417 (1969).

[50]. K. Tsuji, Agr. Biol. Chem., 25, 915 (1961).

[51]. K. Tsuji, ibid., 25, 432 (1961).

[52]. X. de Hemptinne and J. C. Jungers, Z. Phys. Chem. (Frankfurt), 15, 137 (1958).

[53]. N. E. Khomutov, W. W. Skornyakov, and T. P. Fadeeva, Zh. Fiz. Khim., 38, 102 (1964).

[54]. C. Wilson and K. Wilson, Trans. Electrochem. Soc., 80, 151 (1941).

[55]. A. S. Oganes-jan and L. I. Antropov, Dokl. Akad. Nauk Arm. SSR, 21, 81 (1955).

[56]. G. Hennig and G. Kimball, J. Phys. Chem., 12, 415 (1944).

[57]. J. Tafel and K. Schmitz, Z. Elektrochem., 8, 288 (1902).

[58]. J. Haggerty, Trans. Electrochem. Soc., 56, 421 (1929).

[59]. G. Kirchhof and A. Stepanov, Khim.-Farm. Prom. (Russ.), 1, 21 (1932); CA, 27, 5651 (1933).

[60]. E. Müller, Z. Elektrochem., 33, 253 (1927).

[61]. Ger. Pat. 113, 719 (1899).

[62]. K. Elbs and K. Brand, Z. Elektrochem., 7, 644 (1901).

[63]. K. Elbs and K. Brand, ibid., 8, 783 (1902).

[64]. H. Hibbert and R. Read, J. Amer. Chem. Soc., 46, 983 (1924).

[65]. O. Slotterbeck, Trans. Elektrochem. Soc., 92, 377 (1948).

[66]. M. G. Smirnova, W. A. Smirnov, and L. I. Antropov, Tr. Novochercasse Politech. Inst. Im. Ordzhonikidze, 79, 43 (1959).

[67]. J. Tafel, Z. Elektrochem., 17, 972 (1911).

[68]. W. G. Khomyakov and A. P. Tomilov, Zh. Prikl. Khim., 36, 378 (1963).

[69]. X. de Hemptinne and K. Schunk, Ann. Soc. Sci. Bruxelles, 80, 289 (1966).

[70]. X. de Hemptinne and K. Schunk, Trans. Faraday Soc., 65, 591 (1969).

[71]. I. W. Kirilus, R. P. Silkina, and D. W. Sokol'skii, VINITI, Kat. Deponir. Rukopisei, Estestw. i Tochn. Nauki, Tech., M., 1970, Wyp. 18, no. 631-69.

[72]. W. A. Smirnov, in Uspekhi Elektrokhimiya Organicheskikh Soedinenii, Nauka, M., 1966, p. 238.

[73]. A. P. Tomilov, S. G. Mairanovskii, M. Y. Fioshin, and W. A. Smirnov, Elektrokhimiya Organicheskikh Soedinenii, Khimiya, L., 1968, p. 214.

[74]. T. Sekine, A. Yamura, and K. Sugino, J. Elektrochem. Soc., 112, 439 (1965).

[75]. J. Tafel, Ber., 45, 448 (1912).

[76]. A. P. Tomilov and M. I. Kalitina, Zh. Prikl. Khim., 38, 1574 (1965).

[77]. U.S. Pat. 2,422,468 (1947).

[78]. M. Matsui and S. Shimizu, Mem. Coll. Sci. Kyoto Imp. Univ., 4, 245 (1920); Centralblatt, 93, 764 (1922 IV).

[79]. C. Schall and W. Kirst, Z. Elektrochem., 29, 537 (1923).

[80]. J. Bredt, Ann., 366, 11 (1909).

[81]. J. Bredt and W. Perkin, J. Prakt. Chem., 89, 246 (1914).

[82]. J. Bredt, ibid., 84, 786 (1911).

[83]. P. Anziani, A. Anbry, and R. Cornubert, Compt. Rend., 225, 878 (1947).

[84]. F. Escherich and M. Moest, Z. Elektrochem., 8, 849 (1902).

[85]. B. Sakurai and T. Arai, Bull. Chem. Soc. Japan, 28, 93 (1955).

[86]. L. Mandell, R. M. Powers, and R. A. Day, J. Amer. Chem. Soc., 80, 5284 (1958).

[87]. S. Swann, Jr., S. Briggs, V. Neklutin, and A. Jerome, Trans. Electrochem. Soc., 80, 163 (1941).

[88]. S. Swann, Jr., ibid., 69, 316 (1936).

[89]. M. Fedoronke, J. Königstein, and K. Linek, Coll. Czech. Chem. Commun., 32, 1497 (1967).

[90]. M. Fedoronko, J. Königstein, and K. Linek, ibid., 32, 3998 (1967).

[91]. R. Pasternak, Helv. Chim. Acta, 31, 753 (1948).

[92]. A. Vincenz-Chodkowska and Z. R. Grabowski, Electrochim. Acta, 9, 789 (1964).

[93]. E. Bauer, J. Electroanal. Chem., 14, 351 (1967).

[94]. H. E. Stapelfeldt and S. P. Perone, Anal. Chem., 40, 815 (1968); 41, 623 (1969).

[95]. L. Holleck and O. Lehmann, Monatsh. Chem., 92, 499 (1961).

[96]. L. P. Kyrides, J. Amer. Chem. Soc., 55, 3431 (1933).

[97]. E. A. Steck and W. Boehme, ibid., 74, 4511 (1952).

[98]. N. J. Leonard, S. Swann, Jr., and H. L. Dryden, ibid., 74, 2871 (1952).

[99]. F. Fichter and H. Montmollin, Helv. Chim. Acta, 5, 256 (1922).

[100]. A. Findlay and V. H. Williams, Trans. Faraday Soc., 17, 453 (1932).

[101]. Brit. Pat. 12,467 (1912).

[102]. U.S. Pat. 1,181,555; 1,190,845 (1916).

[103]. Can. Pat. 178,644 (1917).

[104]. Ger. Pat. 292,865 (1912).

[105]. Fr. Pat. 455,972 (1912).

[106]. E. Bauer, Z. Elektrochem., 37, 255 (1931).

[107]. A. P. Tomilov, W. G. Khomyakov, and B. G. Soldatov, Zh. Vses. Khim. Obszhestva Im. Mendeleeva, 14, 230 (1969).

[108]. H. Hoberland, DBR Pat. 890 (1953); CA, 50, 12100 (1956).

[109]. T. Chambers and O. Slotterbeck, U.S. Pat. 2,485,258 (1949); CA, 44, 4807 (1950).

[110]. W. G. Khomyakov and A. P. Tomilov, Avt. Swid. SSSR, 85,564 (1949); Bull. Izobret., 12 (1950); Avt. Swid. SSSR, 114,840 (1958); Bull. Izobret., 9 (1958).

[111]. W. G. Khomyakov and A. P. Tomilov, Zh. Prikl. Khim., 36, 373, 378 (1963).

[112]. E. C. French and R. M. Hurd, U.S. Pat. 3,497,430, Feb. 24, 1970; CA, 72, 117230c (1970).

[113]. Ger. Pat. 252,759 (1912).

[114]. Ger. Pat. 306,523 (1918).

[115]. Ger. Pat. 303,302 (1917).

[116]. Ger. Pat. 310,023 (1920).

[117]. Ger. Pat. 324,919 (1920).

[118]. U.S. Pat. 2,408,036 (1946); CA, 41, 345 (1947).

[119]. J. Herry, P. Drumm, and W. O'Connor, Proc. Roy. Irish Acad., 50B, 219 (1945).

[120]. Ger. Pat. 324,920 (1920).

[121]. W. Calvert, India Rubber Tire Rev., 26, 52 (1926).

[122]. W. A. Smirnov and D. P. Semchenke, in Sbornik Annotacii Nauchnoissled. Rabot Vysshikh Uchebnykh Zavedenii, Vyp. 2, izd. MGU, M., 1962, p. 88.

[123]. W. A. Smirnov, M. G. Smirnova, L. A. Demchuk, and D. P. Semchenko, in Cataliticheskie Reakcii v Zhidkoi Faze, izd. ANKazSSR, Alma-Ata, 1963, p. 314.

[124]. N. A. Izgaryshev and I. I. Aryamova, Dokl. Akad. Nauk SSSR, 84, 313 (1952).

[125]. A. P. Tomilov and B. L. Kluyev, Elektrokhim., 2, 1405 (1966).

[126]. A. P. Tomilov and B. L. Kluyev, ibid., 3, 1168 (1967).

[127]. A. P. Tomilov and B. L. Kluyev, Itogi Nauki. Elektrokhimiya, 1967, VINITI, M., 1969, p. 200.

[128]. V. A. Smirnov, L. I. Antropov, M. G. Smirnova, and L. A. Demchuk, Russ. J. Phys. Chem., 42, 897 (1968).

[129]. V. A. Smirnov, L. A. Demchuk, and L. I. Antropov, Zh. Fiz. Khim., 42, 1716 (1968).

[130]. O. de Coninck, Bull. Soc. Chim. Fr., 2, 25, 10 (1876).

[131]. T. Arai, J. Fac. Liberal Arts and Sci., Shinshu Univ., Natur. Sci., 9, 7; CA, 54, 24028i (1960).

[132]. E. Kariv, J. Hermolin, and E. Gileadi, J. Electrochem. Soc., 117, 342 (1970).

[133]. J. Goldberger and F. Tondler, Monatsh., 26, 1474 (1906).

[134]. A. P. Tomilov and L. A. Ignat'eva, J. Appl. Chem. USSR, 38, 2651 (1965).

[135]. M. Godchot and F. Tabouri, Bull. Soc. Chim. Fr., 4, 13, 12 (1913).

[136]. J. P. Wibaut, H. Hoog, S. L. Langendijk, J. Overhoff, and J. Smittenberg, Rec. Trav. Chim. Pays-Bas et Belg., 58, 329 (1939).

[137]. Ger. Pat. 273,392 (1913).

[138]. P. Pascal, Swiss Pat. 88,188 (1919).

[139]. S. Swann, Jr. and E. W. Field, Trans. Electrochem. Soc., 72, 327 (1937).

[140]. G. Laude and J. Wiemann, Bull. Chim. Soc. Fr., 1946, 256.

[141]. G. R. Clemo and J. M. Smith, J. Chem. Soc., 2423 (1923).

[142]. M. J. Allen, J. Org. Chem., 15, 435 (1950).

[143]. M. J. Allen, J. E. Fearn, and M. Levine, J. Chem. Soc., 2220 (1952).

[144]. M. J. Allen, J. Amer. Chem. Soc., 72, 3797 (1950).

[145]. J. Grimshaw and J. S. Ramsey, J. Chem. Soc., C, 653 (1966).

[146]. J. Dewar and J. Read, J. Soc. Chem. Ind. (Trans.), 55, 347 (1936).

[147]. I. A. Pearl, J. Amer. Chem. Soc., 74, 4260 (1952).

[148]. V. J. Puglisi, G. L. Clapper, and D. H. Evans, Anal. Chem., 41, 279 (1969).

[149]. J. H. Stocker, R. M. Jenevein, and D. H. Kern, J. Org. Chem., 34, 2810 (1969).

[150]. J. H. Stocker and R. M. Jenevein, ibid., 33, 2145 (1968).

[151]. W. C. Albert and A. Lowy, Trans. Electrochem. Soc., 75, 367 (1939).

[152]. F. Kaufler, Z. Elektrochem., 13, 635 (1907).

[153]. E. Müller, ibid., 16, 236 (1910).

[154]. S. Swann, Jr., and G. H. Nelson, Trans. Electrochem. Soc., 67, 201 (1935).

[155]. W. D. Cohen, Rec. Trav. Chim. Pays-Bas et Belg., 38, 83 (1919).

[156]. S. Swann, Jr., Trans. Electrochem. Soc., 63, 239 (1933).

[157]. S. Swann, Jr., ibid., 64, 312 (1933).

[158]. E. Pannain, Atti Accad. Naz. Lincei, 5, 17 II, 499 (1908).

[159]. G. Shima, Mem. Coll. Sci. Kyoto Univ., A13, 315 (1930).

[160]. S. Swann, Jr., P. Autbrose, R. Dale, R. Rowe, H. Ward, H. Kerman, and S. Axelrod, Trans. Electrochem. Soc., 85, 231 (1944).

[161]. Brit. Pat. 6, 290, 420 (1949); CA, 44, 1836 (1950).

[162]. J. H. Stocker and R. M. Jenevein, J. Org. Chem., 33, 294 (1968).

[163]. M. J. Allen, J. Amer. Chem. Soc., 73, 3503 (1951).

[164]. M. J. Allen and A. H. Corwin, ibid., 72, 114 (1950).

[165]. N. J. Leonard, S. Swann, Jr., and G. Fuller, ibid., 75, 5127 (1953).

[166]. M. J. Allen, J. Chem. Soc., 1598 (1951).

[167]. G. I. Hobday and W. E. Short, ibid., 609 (1943).

[168]. M. J. Allen, J. A. Siragusa, and W. Pierson, ibid., 1045 (1960).

[169]. H. A. Levine and M. J. Allen, ibid., 254 (1952).

[170]. S. Swann, Jr., Trans. Electrochem. Soc., 64, 245 (1933).

[171]. S. Swann, Jr. and E. O. Edelmann, ibid., 58, 179 (1930).

[172]. M. J. Allen, M. Cohen, W. G. Pierson, and J. A. Siragusa, J. Chem. Soc., 757 (1961).

[173]. Brit. Pat. 523, 515 (1940).

[174]. R. Juday and W. Sullivan, J. Org. Chem., 20, 617 (1955).

[175]. I. A. Avrutskaya, M. Y. Fioshin, S. D. Kabanova, and L. E. Gerasimova, Sov. Electrochem., 5, 890 (1969).

[176]. M. Y. Fioshin, I.A. Avrutskaja, S. D. Kabanova, and Nguen Van Tchan, ibid., 6, 77 (1970).

[177]. I. A. Avrutskaya, M. Y. Fioshin, L. E. Gerasimova, and Nguen Van Tchan, ibid., 6, 81 (1970).

[178]. M. Y. Fioshin, I. A. Avrutskaja, L. E. Gerasimova, and S. D. Kabanova, Khim.-Farm. Zh., 3, 19 (1969).

[179]. F. Escherich and M. Moest, Z. Elektrochem., 8, 849 (1902).

[180]. R. N. Gourley and J. Grimshaw, J. Chem. Soc., C, 2388 (1968).

[181]. W. R. J. Simpson, D. Babbe, J. A. Edwards, and J. H. Fried, Tetrahedron Lett., No. 33, 3209 (1967).

[182]. M. J. Allen and H. Cohen, J. Electrochem. Soc., 106, 451 (1959).

[183]. W. L. Yost, U.S. Pat. 3, 200, 053 (1965).

[184]. E. V. Kryukova and A. P. Tomilov, Sov. Electrochem., 5, 508 (1969).

[185]. C. Caulett, J. Bessin, and J. Bodard, Compt. Rend., 261, 1848 (1965).

[186]. C. Caulett, M. Salaün, and M. Hebert, ibid., 264C, 228 (1967).

[187]. H. Lund, Acta Chem. Scand., 11, 283 (1957).

[188]. F. Woehler, Ann., 51, 153 (1844).

[189]. F. Haber, Russ. Z. Phys. Chem., 47, 297 (1904).

[190]. K. Kaufler, Z. Elektrochem., 16, 236 (1910).

[191]. H. D. Law, J. Chem. Soc., 101, 1550 (1912).

[192]. I. Houben-Weyl, Methoden der Organischen Chemie, Thieme, Leipzig, 1925.

[193]. J. Tafel, Ber., 42, 3146 (1909).

[194]. S. Ohki and V. Naike, J. Pharm. Soc. Japan, 72, 490 (1952).

[195]. S. Swann, Jr., Bull. Cent. Electrochem. Res. Inst., 2, 6 (1955).

[196]. S. Swann, Jr., R. W. Benoliel, L. R. Lyons, and W. H. Pahl, Trans. Electrochem. Soc., 79, 83 (1941).

[197]. S. Swann, Jr., ibid., 62, 177 (1932).

[198]. S. Swann, Jr., L. F. Deditius, and W. A. Pyrrh, ibid., 68, 321 (1935).

[199]. H. J. Read, ibid., 80, 8 (1941).

[200]. R. R. Read and F. A. Fletcher, ibid., 47, 93 (1925).

[201]. C. Schall, Z. Elektrochem., 38, 31 (1932).

[202]. G. H. Keats, J. Chem. Soc., 2003 (1937).

[203]. J. Tafel and W. Jürgens, Ber., 42, 2548 (1909); 45, 437 (1912).

[204]. J. Tafel, ibid., 45, 451 (1912).

[205]. H. Stenzl and F. Fichter, Helv. Chim. Acta, 20, 846 (1937).

[206]. J. Tafel and H. Hahl, Ber., 40, 3317 (1907).

[207]. H. Stenzl and F. Fichter, Helv. Chim. Acta, 17, 669 (1934).

[208]. Ger. Pat. 210, 693 (1904).

[209]. N. Venkatasubramanian, Sci. Gulf, 29, 361 (1963).

[210]. S. Swann, Jr. and J. Feldman, Trans. Electrochem. Soc., 67, 195 (1935).

[211]. R. Lukeš and V. Dudek, Coll. Czech. Chem. Commun., 27, 467 (1962).

[212]. L. Throop and L. Tökés, J. Amer. Chem. Soc., 89, 4789 (1967).

[213]. J. Mizuguchi and S. Matsumoto, J. Pharm. Soc. Japan, 78, 129 (1958).

[214]. R. Lukeš and J. Gorocholinskii, Coll. Czech. Chem. Commun., 8, 223 (1936).

[215]. J. Grimshaw and E. J. F. Rea, J. Chem. Soc., C, 2628 (1967).

[216]. T. J. Curphey, C. W. Amelotti, T. P. Layloff, R. L. McCartney, and J. H. Williams, J. Amer. Chem. Soc., 91, 2817 (1969).

[217]. T. Arai and T. Ogura, Bull. Chem. Soc. Japan, 33, 1018 (1966).

[218]. J. Tafel, Ber., 39, 3626 (1906).

[219]. A. P. Tomilov and Yu. D. Smirnov, in Elektrokhimicheskie Processy s Uchastiem Organicheskikh Veschestv, Nauka, M., 1970, p. 56.

[220]. J. Tafel, Ber., 44, 323 (1911).

[221]. R. R. Enger, ibid., 44, 337 (1911).

[222]. E. A. Efimov and I. G. Erusalimchik, Zh. Fiz. Khim., 38, 2868 (1965).

[223]. T. Arai, Bull. Chem. Soc. Japan, 32, 184 (1959).

[224]. F. P. Jensen and L. H. Gale, J. Amer. Chem. Soc., 82, 145 (1960).

[225]. M. B. Neiman and M. I. Gerber, Zh. Anal. Khim., 2, 135 (1947).

[226]. K. Vesely and R. Brdička, Coll. Czech. Chem. Commun., 12, 313 (1947).

[227]. S. G. Mairanovskii, D. I. Dzhaparidze, and O. I. Sorokin, Izv. Akad. Nauk SSSR, Otd. Khim. Nauk, 795 (1964).

[228]. J. Volke, Coll. Czech. Chem. Commun., 23, 1486 (1958).

[229]. E. Laviron, Bull. Soc. Chim. Fr., 2325 (1961).

[230]. Dang-Quoc-Quan, Compt. Rend., 251, 2927 (1960).

[231]. L. Holleck, in Osnovnye Woprosy Sovremennoi Teoreticheskoi Elektrokhimiya, Izd. Mir. M., 1965, p. 406.

[232]. V. D. Bezuglyi and T. M. Rapota, Zh. Fiz. Khim., 38, 2182 (1964).

[233]. Ya. I. Tur'jan, M. S. Rusakova, B. F. Ustavschikov, and W. A. Podgornova, Elektrokhim., 4, 721 (1968).

[234]. R. Bieber and G. Trümpler, Helv. Chim. Acta, 30, 706, 971, 1109, 1286, 1534, 2000 (1947).

[235]. M. Tokuoka, Coll. Czech. Chem. Commun., 7, 392 (1935).

[236]. T. G. Jahoda, ibid., 7, 415 (1935).

[237]. G. Semerano and B. Polaczek, Gazz. Chim. Ital., 68, 298 (1936).

[238]. W. P. Davies and D. P. Evans, J. Chem. Soc., 546 (1939).

[239]. H. Scheitzer and E. Laquerir, Rec. Trav. Chim., 55, 959 (1936).

[240]. K. Wiesner, Coll. Czech. Chem. Commun., 12, 64 (1947).

[241]. J. M. Los and K. Wiesner, J. Amer. Chem. Soc., 75, 6346 (1953).

[242]. J. M. Los, L. B. Simpson, and K. Wiesner, ibid., 78, 1564 (1956).

[243]. W. G. Overend, A. R. Peacocke, and J. B. Smith, J. Chem. Soc., 3487 (1961).

[244]. S. G. Mairanovskii and W. N. Pavlov, Zh. Fiz. Khim., 38, 1804 (1964).

[245]. J. W. Baker, W. G. Davies, and M. L. Hemming, J. Chem. Soc., 692 (1940).

[246]. A. Kirrman, J. M. Savéant, and N. Moe, Compt. Rend., 253, 1106 (1961).

[247]. P. H. Given and M. E. Peover, J. Chem. Soc., 385 (1960).

[248]. W. N. Epimakhov, Elektrokhim., 6, 322 (1970).

[249]. R. Pasternak, Helv. Chim. Acta, 31, 763 (1948).

[250]. I. A. Korskunov and L. N. Sazanova, Zh. Fiz. Khim., 23, 202 (1949).

[251]. R. A. Day and J. J. Kirkland, J. Amer. Chem. Soc., 72, 2766 (1950).

[252]. D. M. Coulson and W. R. Crowell, ibid., 74, 1290 (1952).

[253]. P. J. Elving and J. T. Leone, ibid., 80, 1021 (1958).

[254]. P. Valenta, Proc. Intern. Congr. Polarogr., Cambridge, 1959.

[255]. P. Valenta, Coll. Czech. Chem. Commun., 25, 855 (1960).

[256]. W. Kemula, Z. Grabowski, and M. Kalinowski, ibid., 25, 3306 (1960).

[257]. W. Kemula, Z. Grabowski, and M. Kalinowski, Naturwiss., 47, 514 (1960).

[258]. L. N. Nekrasov and A. D. Korsun, Elektrokhim., 4, 489, 539, 996 (1968).

[259]. A. D. Korsun and L. N. Nekrasov, ibid., 5, 212 (1969).

[260]. D. E. Austen, P. H. Given, D. S. Ingram, and M. E. Peover, Nature, 182, 1784 (1958).

[261]. P. L. Nordio, G. Giacometti, and P. Favero, Ric. Sci., a. 33, Ser. 2, Part II-A3, No. 4, 107 (1963).

[262]. P. Favero, G. Giacometti, P. L. Nordio, and G. Rigatti, Nuovo Cimento, 24, 21 (1962).

[263]. P. H. Rieger and G. K. Fraenkel, J. Chem. Phys., 37, 2811 (1962).

[264]. N. Steinberger and G. K. Fraenkel, ibid., 40, 723 (1964).

[265]. G. A. Russel, E. Th. Strom, E. R. Talaty, and St. A. Weiner, J. Amer. Chem. Soc., 88, 1998 (1966).

[266]. S. Goldschmidt, Angew. Chem., 69, 132 (1957).

[267]. I. A. Korshunov, A. S. Kirillova, and Z. B. Kuznetsova, Zh. Fiz. Khim., 24, 251 (1950).

[268]. S. G. Mairanovskii, Izv. Akad. Nauk SSSR, Otd. Khim. Nauk, 2140 (1961).

[269]. P. J. Elving and J. T. Leone, see Ref. [253].

[270]. M. Ashwort, Coll. Czech. Chem. Commun., 13, 229 (1948).

[271]. B. Kastening and L. Holleck, Z. Elektrochem., 63, 166 (1959).

[272]. K. Ito, Rev. Polarogr. (Kyoto), 15, 97 (1968).

[273]. S. G. Mairanovskii, J. Electroanal. Chem., 4, 166 (1962).

[274]. S. G. Mairanovskii and W. N. Pavlov, Elektrokhim., 1, 226 (1967).

[275]. P. Zuman, D. Barnes, and A. Ryvolova-Kejharova, Discuss. Faraday Soc., 45, 203 (1968).

[276]. R. M. Powers and R. A. Day, J. Amer. Chem. Soc., 80, 806 (1958).

[277]. W. N. Pavlov, Ya. M. Zolotovitskii, S. G. Mairanovskii, and G. A. Tedoradze, Elektrokhim., 1, 427 (1965).

[278]. J. P. Stradins and W. W. Teraud, Izv. Akad. Nauk Latv. SSR, Ser. Khim., 169 (1964).

[279]. J. P. Stradins, Electrochim. Acta, 9, 711 (1964).

[280]. A. A. Pozdeeva and S. I. Zhdanov, Proc. III Intern. Congr. Polarogr., Southampton, 1964.

[281]. L. Holleck and D. Marquarding, Naturwiss., 49, 468 (1962).

[282]. A. P. Tomilov and I. N. Brago, in Progress Elektrokhimiya Organicheskikh Soedinenii, Vol. 1, Nauka, M., 1969, p. 208.

[283]. A. N. Frumkin, Tr. Fiz.-Khim. Inst. Im. Karpova, Vyp. 4, 56 (1925); Vyp. 5, 3 (1926).

[284]. A. N. Frumkin, Z. Phys. Chem., 35, 792 (1926).

[285]. L. I. Antropov, Trudy IV Soveschaniya po Elektrokhimiya, Izd. AN SSSR, M., 1959, p. 34.

[286]. A. N. Frumkin, Dokl. Akad. Nauk SSSR, 154, 1432 (1964).

[287]. K. Wettig, Elektrokhim., 3, 269 (1967).

[288]. R. Kalvoda and G. Budnikov, Coll. Czech. Chem. Commun., 28, 838 (1963).

[289]. G. K. Budnikov and R. Kalvoda, Zh. Obshch. Khim., 35, 1131 (1965).

[290]. R. Powers and R. Day, J. Org. Chem., 24, 722 (1959).

[291]. A. Yamura, T. Sekine, and K. Sugino, J. Electrochem. Soc. Japan, 34, 110 (1966); see also A. Albisson, G. Mousset, and J. Simonet, Compt. Rend., 272C, 646 (1971).

[292]. S. G. Mairanovskii, Kineticheskie i Kataliticheskie Toki v Polarografii, Izd. Nauka, M., 1966, p. 208.

[293]. H. Lund and A. D. Thomsen, Acta Chem. Scand, 23, 3567 (1969); A. D. Thomsen and H. Lund, ibid., 23, 3582 (1969).

[294]. J. G. Faugere, R. Calas, and R. LaLande, Compt. Rend., 262, 124 (1966).

[295]. J. Jadot, Bull. Soc. Chim. Belg., 57, 346 (1948).

[296]. P. F. Casals and J. Wiemann, Bull. Soc. Chim. Fr., 1967, 3478.

[297]. D. H. Evans and E. C. Woodbury, J. Org. Chem., 32, 2158 (1967).

[298]. R. C. Buchta and D. H. Evans, J. Electrochem. Soc., 117, 1494 (1970).

[299]. E. Kariv, J. Hermolin, and E. Gileadi, ibid., 117, 342 (1970).

[300]. H. Z. Friedlander, S. Swann, Jr., and C. S. Marvel, ibid., 100, 408 (1953).

[301]. G. Mengoli and G. Vidotto, Makromol. Chem., 129, 73 (1969).

[302]. H. Lund, "Electrochemistry of the Carbon-Nitrogen Double Bond," in The Chemistry of the Carbon-Nitrogen Double Bond (S. Patai, ed.), Wiley-Interscience, New York, 1970, pp. 505-564.

[303]. P. Zuman, Nature, 175, 585 (1950).

[304]. P. Zuman, Coll. Czech. Chem. Commun., 15, 839 (1950).

[305]. H. Lund, Acta Chem. Scand., 13, 249 (1959).

[306]. L. Holleck and B. Kastening, Z. Electrochem., 60, 127 (1956).

[307]. B. Kastening, L. Holleck, and G. A. Melkonian, ibid., 60, 130 (1956).

[308]. V. N. Dmitrieva, V. B. Smelyakova, B. M. Krasovitskii, and V. D. Bezuglyi, Zh. Obshch. Khim., 36, 405 (1966).

[309]. V. N. Dmitrieva, N. I. Mal'tseva, V. D. Bezuglyi, and B. M. Krasovitskii, Zh. Obshch. Khim., 37, 372 (1967).

[310]. P. E. Iversen, Lecture, EUCHEM Conference, Sweden, 1971.

[311]. H. Lund, Acta Chem. Scand., 14, 359 (1960).

[312]. H. Lund, P. Lunde, and F. Kaufmann, ibid., 20, 1631 (1966).

[313]. K. Brand, Ber., 42, 3461 (1909).

[314]. H. D. Law, J. Chem. Soc., 154 (1912).

[315]. J. M. W. Scott and W. H. Jura, Can. J. Chem., 45, 2375 (1967).

[316]. N. F. Levchenko, L. Sh. Afanasiadi, and V. D. Bezuglyi, Zh. Obshch. Khim., 37, 666 (1967).

[317]. C. P. Andrieux and J. M. Saveant, Bull. Soc. Chim. Fr., 4671 (1968).

[318]. C. P. Andrieux, L. Nadjo, and J. M. Saveant, J. Electroanal. Chem., 26, 147 (1970).

[319]. C. P. Andrieux and J. M. Saveant, ibid., 26, 223 (1970).

[320]. C. P. Andrieux and J. M. Saveant, ibid., 28, 446 (1970).

[321]. M. M. Baizer, J. D. Anderson, J. H. Wagenknecht, M. R. Ort, and J. P. Petrovich, Electrochim. Acta, 12, 1377 (1967).

[322]. Yu. P. Kitaev and A. E. Arbuzov, Izv. Akad. Nauk SSSR, Otd. Khim. Nauk, 1037 (1957).

[323]. Yu. P. Kitaev and T. V. Troepol'skaya, ibid., 454, 465 (1963).

[324]. H. Simon and W. Moldenhauer, Chem. Ber., 100, 1949 (1967).

[325]. B. V. Ioffe and V. S. Stopskij, Tetrahedron Lett., 1333 (1968).

[326]. H. Lund and E. T. Jensen, Acta Chem. Scand., 25 (1971).

[327]. H. Lund, Discuss. Faraday Soc., 45, 193 (1968).

[328]. Yu. P. Kitaev and I. M. Skrebkova, Zh. Obshch. Khim., 37, 1204 (1967).

[329]. H. Lund and E. T. Jensen, Acta Chem. Scand., 24, 1867 (1970).

[330]. H. Lund, ibid., 21, 2525 (1967).

[331]. V. Prelog and O. Häfliger, Helv. Chim. Acta, 32, 2088 (1949).

[332]. M. Brezina, V. Volkova, and J. Volke, Coll. Czech. Chem. Commun., 19, 894 (1954).

[333]. P. Souchay and M. Graizon, Chim. Anal., 36, 85 (1954).

[334]. Yu. P. Kitaev, G. K. Budnikov, and A. E. Arbuzov, Izv. Akad. Nauk SSSR, Otd. Khim. Nauk, 824 (1961).

[335]. Yu. P. Kitaev and G. K. Budnikov, Zh. Obshch. Khim., 33, 1396 (1963).

[336]. Y. Asahi, Chem. Pharm. Bull. (Tokyo), 11, 930 (1963).

[337]. M. Masui and H. Ohmori, Chem. Pharm. Bull., 12, 877 (1964).

[338]. B. Fleet and P. Zuman, Coll. Czech. Chem. Commun., 32, 2066 (1967).

[339]. H. Lund, Tetrahedron Lett., 3651 (1968).

[340]. D. Fleury and M. B. Fleury, Coll. Czech. Chem. Commun., 36, 331 (1971).

[341]. H. Lund, ibid., 30, 4237 (1965).

[342]. P. Souchay and S. Ser, J. Chim. Phys., 49, C172 (1952).

[343]. R. M. Elofson and J. G. Atkinson, Can. J. Chem., 34, 4 (1956).

[344]. H. J. Gardner and W. P. Georgans, J. Chem. Soc., 4180 (1956).

[345]. H. Lund, Acta Chem. Scand., 18, 563 (1964).

[346]. N. Tütülkoff and St. Buduroff, Compt. Rend. Acad. Bulg. Sci., 6 (3), 5 (1953).

[347]. N. Tütülkoff and E. Paspaleev, ibid., 14, 159 (1961).

[348]. E. Paspaleev, Monatsh. Chem., 97, 230 (1966).

[349]. A. J. Fry and J. H. Newberg, J. Amer. Chem. Soc., 89, 6374 (1967).

[350]. P. Zuman and O. Exner, Coll. Czech. Chem. Commun., 30, 1832 (1965).

[351]. T. Kubota, H. Miyazaki, and Y. Mori, Bull. Chem. Soc. Japan, 40, 245 (1967).

Chapter IX

α, β-UNSATURATED CARBONYLS

Manuel M. Baizer

Corporate Research Department
Monsanto Company
St. Louis, Missouri

I. INTRODUCTION

The complexities involved in the reduction of carbonyls without conjugated olefinic unsaturation (Chapter VIII) are compounded in the case of α, β-unsaturated carbonyls because the latter are simultaneously aldehydes or ketones and activated olefins. Electrolytic reduction of the system

```
     |  |
>C=C-C=O
 4  3  2  1
```

may cause: (a) saturation of the 3, 4-double bond; (b) reduction of the carbonyl to alcohol; (c) both (a) and (b); (d) coupling of two molecules through C-4 or through C-2, or through C-2 of one molecule and C-4 of another; (e) formation of oligomers and polymers by sequential coupling [16]; and (f) formation of organometallic compounds (Chapter XVIII).

The reduction mechanisms for α, β-unsaturated carbonyls depend critically on acid-base reactions preceding, accompanying, and following electron transfer and importantly on the cathode material, the cation of the supporting electrolyte [1a], and the concentration of substrate.

An outstanding "early" mechanistic study was reported by Pasternak [2]. More recent and more comprehensive investigations have issued from Zuman [3] (aqueous systems) and Simonet [4], and Wawzonek [16] and their respective coworkers (nonaqueous systems). Fortunately polarographic studies have often been accompanied by macroelectrolyses and identification of intermediate and final products. The many researches of Wiemann and his students [5] have placed major emphasis on the synthetic aspects of these reductions and a comparison of electrochemical with nonelectrochemical procedures.

II. ELECTRODE MECHANISMS

A. General Remarks

It has been pointed out [3] that valid generalizations in this area could be arrived at only after many types of representative compounds had been studied and many examples of each type given. However, the following statements may provide a guide through the maze of results that have accumulated in the course of many years of work:

1. The general principle [6] that, when an electroreducible group is conjugated with olefinic groups, the molecule becomes under comparable conditions more easily reducible is well illustrated for α, β-unsaturated carbonyls in Table 1.[1]

2. The double bond is almost certainly reduced before the carbonyl group.

3. As is the case with saturated carbonyls, the preprotonated molecule (I) is reduced anodically of its parent (II); likewise, after the first reduction the (protonated) neutral radical (III) is more easily reduced than the anion radical (IV):

$$\mathrm{H[RCH{=}CHCOR']^+} \qquad \mathrm{RCH{=}CHCOR'} \qquad \mathrm{RCH_2CH_2\dot{C}(O^-)R'} \qquad \mathrm{RCH_2CH_2\dot{C}(OH)R'}$$

(I) (II) (III) (IV)

TABLE 1

Comparison of $E_{1/2}$'s of α, β-Unsaturated Carbonyls with Those of Corresponding Saturated Carbonyls

Compound	$-E_{1/2}$[a]	Conditions	Ref.
CH_3CH_2CHO	1.8	0.1 M LiCl	[7]
$CH_2{=}CHCHO$	1.5	pH 7-11	[8]
$CH_3CH{=}CHCHO$	1.25	0.1 N NH_4Cl	[9]
$CH_3(CH{=}CH)_xCHO$	1.05 to 0.7	pH 7; n = 2 to 5, resp.	[10]
$C_6H_5CH_2CH_2CHO$	1.4	0.1 N NH_4Cl	[11]
$C_6H_5CH{=}CHCHO$	0.8	0.1 N NH_4Cl	[11]
$CH_3CH_2COCH_3$	2.25	0.025 N Me_4NI	[12]
$CH_2{=}CHCOCH_3$	1.42	0.1 N KCl	[13]
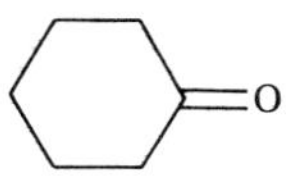	2.45	0.05 N Et_4NI, 75% dioxane	[14]
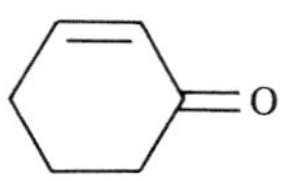	1.55	0.1 N NH_4Cl	[15]

[a]First wave referred to SCE.

4. The hydrated carbonyl group (e.g., in an aldehydrol) is not electroreducible; reversion to the unhydrated form occurs at low or very high pH's.

5. A series of four sequential one-electron reductions exists in the most ideal examples; often the first two or the latter two may be merged into single two-electron steps (see Table 2).

6. In protic systems the electron-transfer steps (often irreversible) are intimately linked with acid-base reactions (reversible) so that CEEC, EECC, etc., reactions are involved (in which C is reaction with a proton and E is the transfer of one electron). A monovalent cation like Li^+ ("hard" acid) may sometimes play the role of a proton in neutralizing an anion radical [1b].

TABLE 2

Polarographic Reduction Phenomena at Various pH Ranges [3]

pH Range[a]	Species reduced	Observations	Mechanisms[b]
<4	(I) and (VII)	Shifts of $E_{1/2}$ with pH; mercurial at first wave; no dimers; the final neutral radical may dimerize, form mercurials, etc.	H^+, ē, ē, H^+, H^+, ē
5-7	(I) and (VIII)	First 2-ē wave (or two 1-ē) followed by 2-ē wave	H^+, ē, ē, H^+, ē, H^+, ē, H^+, ē, H^+, or H^+, ē, ē, H^+, ē, H^+, H^+, ē
8-9	(II) and (VIII)	(II) → (VIII), (VIII) → alcohol	ē, H^+, ē, H^+, ē, H^+, ē, H^+
>11 or 12	(II)	Two one-electron steps	ē, ē, H^+, H^+, ē, ē, H^+, H^+ or ē, ē, H^+, H^+, ē, H^+, ē, H^+

[a]The situation at intermediate ranges has also been analyzed.

[b]Starting with (II).

B. Zuman's Analysis of Aqueous Systems

Zuman and coworkers [3] have found that the most rational interpretation of the polarographic results in protic media can be achieved by examining the behavior of α, β-unsaturated carbonyls within specific pH ranges. The analysis is summarized in Table 2. It will be convenient to refer to the general formulas (I), (II), (III), (IV) above as well as the following:

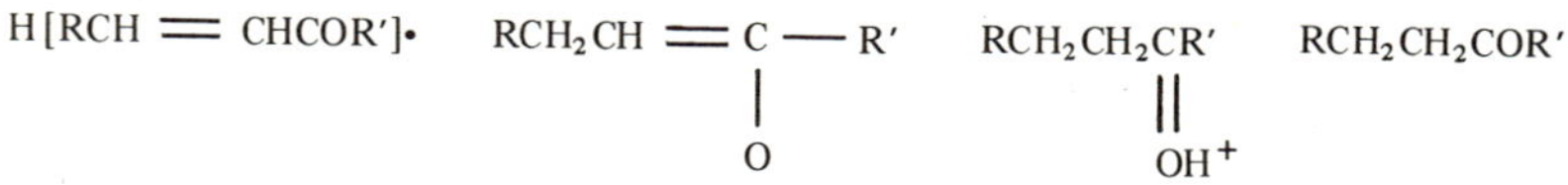

(V) (VI) (VII) (VIII)

The following scheme for reductions at pH 4 exemplifies Zuman's more detailed analyses:

$$\text{(II)} + H^+ \underset{k_{-1}}{\overset{k_1}{\rightleftharpoons}} \text{(I)} \qquad pK_1$$

$$\text{(I)} + \bar{e} \underset{E_1^I}{\rightleftharpoons} \text{(V)}$$

$$\text{(V)} \xrightarrow{k_2} \text{Mercurials}$$

$$\text{(V)} \xrightarrow{k_3} \text{Dimers}$$

$$\text{(V)} \xrightarrow{\text{Solvent}} \text{Products}$$

$$\text{(V)} + \bar{e} \xrightarrow[E_2^I]{} \text{(VI)}$$

$$\text{(VI)} + H^+ \underset{k_{-4}}{\overset{k_4}{\rightleftharpoons}} \text{(VIII)}$$

$$\text{(VIII)} + H^+ \underset{k_{-5}}{\overset{k_5}{\rightleftharpoons}} \text{(VII)}$$

$$\text{(VII)} + \bar{e} \xrightarrow[E_1^{II}]{} RCH_2CH_2\dot{C}(OH)R'$$

C. Aprotic Media[2]

It is clear that since no H^+ reactions intervene before or between the first two 1-ē stages, the anion radical and the dianion intermediates are separately discernible. Since the starting materials are subject to anion attack by anion radicals and dianions (and form oligomers and polymers), the diffusion currents found polarographically are lower than would be expected in the absence of these complications (cf. Chapter II). Two one-electron waves were shown by benzalacetone, benzalacetophenone, and cinnamaldehyde; crotonaldehyde gave only the first one-electron reduction [16]. In another study [4] methyl vinyl ketone showed two one-electron waves but mesityl oxide, isophorone, and ethylidene acetone only one each. Presumably in the latter cases, the second waves were as negative as the discharge potential of the supporting cation.

III. SYNTHETIC ASPECTS

A. Reduction to Saturated Carbonyls and Saturated Alcohols

While this type of electroreduction can be carried out under a variety of conditions and often gives very good yields (e.g., 1-piperonylidene-2-heptanone → 60% 1-(3,4-methylenedioxyphenyl)-3-octanone [17]; benzalacetone → 90% benzylacetone [18]), the same results can usually be achieved by catalytic means. The special synthetic advantages of the electrolytic method are more apparent in the case of reductive coupling reactions.

B. Trapping of Intermediates

Since relatively stable radical anions and anions are formed in the course of the reduction sequence in anhydrous media, it has been possible to trap the intermediates by conducting the electrolysis in the presence of CO_2. Thus, reduction of benzalacetophenone at a Hg cathode using DMF and KI — which limited the reduction to the first wave — while CO_2 was bubbling through the solution [16] yielded two carboxylates, (IX) and (X). It is proposed that (IX) was formed by the sequence of one $\bar{e}$ reduction-

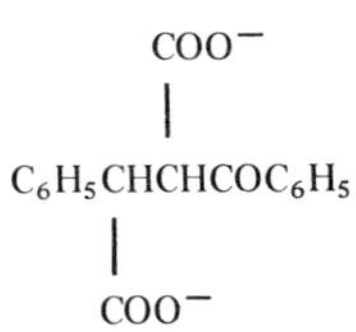

(IX)

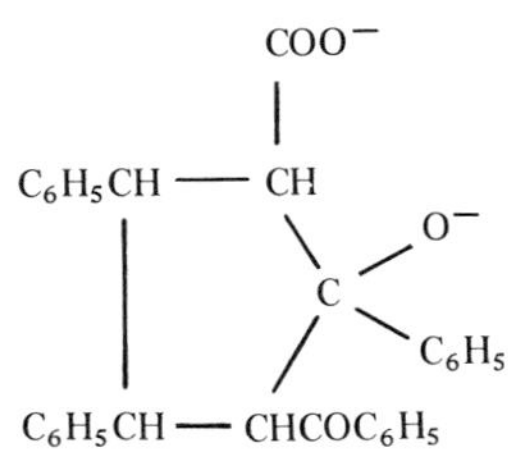

(X)

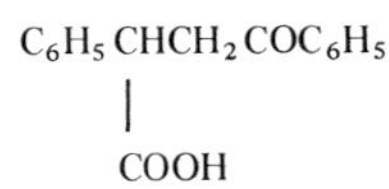

(XI)

carbonation-1-$\bar{e}$ reduction-carbonation (ECEC mechanisms) that obtained in the reductive carbonation of stilbene (Chapter V). Acidification of (IX) caused loss of the carboxyl α to the carbonyl and yielded α-phenyl-β-benzoylpropionic acid (XI) as the isolated product. To explain the formation of (X) it is suggested that the monocarboxylate radical (XII) resulting from

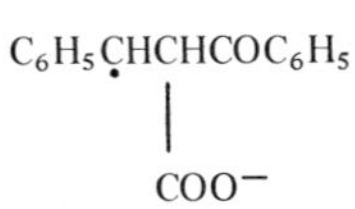

(XII)

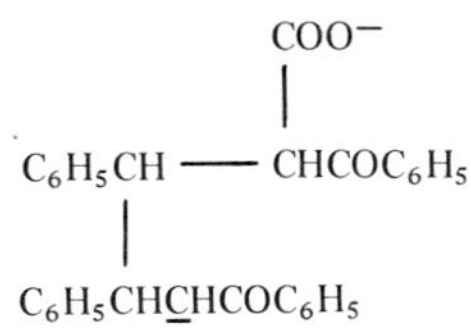

(XIII)

the first EC sequence either before or after reduction adds to a molecule of starting material to yield the intermediate (XIII) which cyclizes to (X).

C. Reductive Dimerizations (Hydrodimerizations)

As has been mentioned (p. 400), reductive coupling may occur via the carbonyl or the β-carbon atoms or both, so that the initial products may be glycols (XIV), diketones (XV), or γ-hydroxycarbonyls (XVI) [5]:

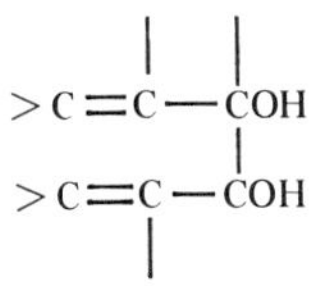
(XIV)

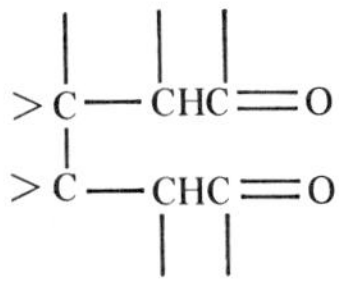
(XV)

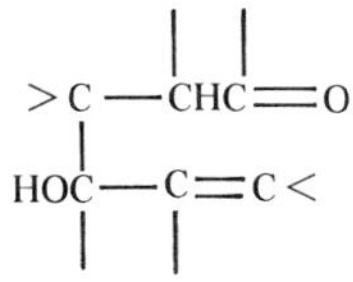
(XVI)

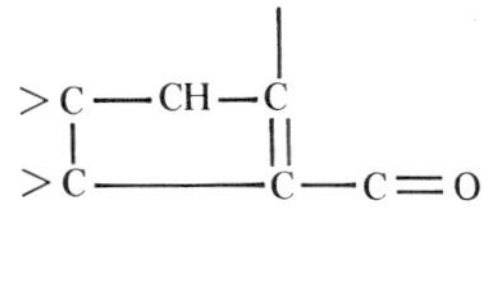
(XVII)

Subsequent well-known chemical reactions may lead to ultimate products which are different from those formed upon electrolysis. For example, the dicarbonyls (XV) readily cyclize and the cyclic tert-alcohols undergo dehydration to α, β-unsaturated carbonyls (XVII). As an example, acrolein on electroreduction yielded cyclopentenaldehyde [19]. The product is of the same type as the starting material, although more substituted at the β-position, so that, without adequate cathode voltage control, it too will be reduced.

Representative reductive dimerizations of α, β-unsaturated carbonyl compounds are presented in Table 3.

Some systematic investigations have been carried out in an attempt to define the optimum conditions for obtaining reduced dimers (XIV), (XV), (XVI), (XVII) (and their further transformation products) by proper choice of electrode, solvent, pH, and cathode potential. Two such studies involve methyl vinyl ketone [21] and mesityl oxide [23].

The role of the supporting cation in influencing selectivity of products has not been sufficiently acknowledged.

It has generally been proposed that in aqueous media at pH <7 the products obtained may be rationalized on the basis of coupling reactions of neutral radicals as indicated below for methyl vinyl ketone:

$$CH_2{=}CHCOCH_3 \xrightarrow[H^+]{\bar{e}} \cdot CH_2CH_2COCH_3 \longleftrightarrow CH_2{=}CH\dot{C}(OH){-}CH_3$$

More complete mechanistic analyses have been made by Zuman et al. [3].

TABLE 3

Representative Reductive Dimerizations of α, β-Unsaturated Carbonyls

Carbonyl compound	Cathode	Medium	Principal products	Ref.
Acrolein	Hg	Aq. Et_4N tosylate	Cyclopentenealdehyde	[19]
Acrolein	Hg	HOAc, KOAc, H_2O, MeCN	Adipaldehyde	[39]
Methacrolein	Hg	Aq. Et_4N tosylate	2-Isopropenyl-4-methyl-2, 3, -dihydrofuran	[20]
Methyl vinyl ketone	Hg	HOAc-NaOAc-EtOH	2, 7, -Octanedione	[21]
Δ^4-3-Ketosteroids	Hg	MeOH-LiCl + aq. buffer	Pinacol	[22]
$\Delta^{1,4}$-3-Ketosteroids	Hg	MeOH-LiCl + aq. buffer	Pinacol	[22]
Mesityl oxide	Hg	$CH_3CN + H_2O$ + KOAc + HOAC + THF	4, 4, 5, 5-Tetramethyl-2, 7-octanedione	[23]
Styryl t-butyl ketone	Hg or Pt	DMF + Et_4NBr	2, 2, 9, 9-Tetramethyl-5, 6-diphenyl-3, 8-decanedione	[24]
2-Cyclohexeneone	Hg	CH_3CN; strongly or weakly acidic; basic	dl- and meso-3, 3′-Dioxo-1, 1′-dicyclohexyl (major) + 3′-oxo-1-cyclohexyl-2-cyclohexene-1-ol	[25]
Benzalacetophenone	Hg	KI in DMF	Hydrotrimer	[16]
Benzalacetophenone	Hg	Bu_4NI in DMF	meso-1, 3, 4, 6-Tetraphenyl-1, 5-hexanedione	[16]
Coumarin	Na-Hg[a]	MeOH, pH 7-8	4, 4-Bihydrocoumarin (2 isomers)	[26]

[a] This is an indirect reduction (Chapter XXIII)

Divinyl ketones have been studied polarographically [10a, 27] but hydrodimerizations in nonacidic media, which present formidable problems, have been largely neglected. As an example, 1, 2-dibenzoylethylene was reduced at mercury, using Et_4N^+ tosylate, DMF, and a controlled amount of water, to dibenzoylethane, a fair yield of dl- and meso-1, 2, 3, 4-dibenzoylbutane, as well as isomers of the hydrotrimer 1, 2, 3, 4, 5, 6-hexabenzoylhexane. Too little water led to oligomers and polymers; too much water yielded dibenzoylethane only [28]. The same hydrodimers were obtained by reductive dimerization with P^{III} esters [31].

An interesting comparison between photochemical and electrochemical behavior of cross-conjugated 4, 4-disubstituted 2, 5-cyclohexadienones (XVIII) has been published [29]. Whereas, calculations had indicated that bond formation between position 3 and 5 in the n, π * state might be expected,

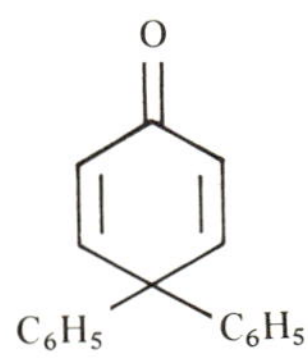

(XVIII)

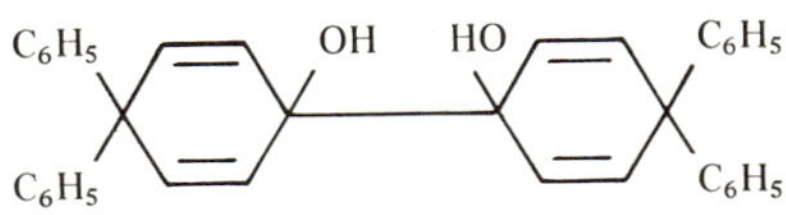

(XIX)

electroreduction of (XVIII) in acid yielded the pinacol (XIX) in ca. 96% yield.

Work on the reductive dimerization of α, β-unsaturated carbonyls has a long history. Metals (Zn, Al, Mg) plus acids (HOAc, HCl) as well as amalgams and cathodes have been used as reducing agents. While a critical comparison of electrochemical vs. "analogous" nonelectrochemical reactions is deferred until Chapter XXVI, it is appropriate to examine this question at this point as it relates to α, β-unsaturated carbonyls (Table 4).

TABLE 4

Comparison of Electrochemical and "Chemical" Reduction of $CH_3COCH=CH_2$

Reducing agent	Reduction products, % yield	
	α,α′-Dimethyl-α,α′-divinylglycol (XIV)	2,7-Octanedione (XV)
Zn + HOAc, 0°	Major diastereoisomers	Minor
Mg + HOAc	80	20
Na-Hg	20	60
Hg cathode, 5°	10	75

Table 4 [30] compares the yields of products obtained by several different reductions of methyl vinyl ketone; Table 5 [25] gives similar types of data for 2-cyclohexeneone.

TABLE 5

Reductive Dimerization of 2-Cyclohexeneone

Reducing agent	Medium	Reduction products, distribution, % 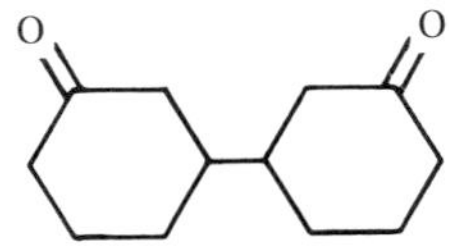 (XV)	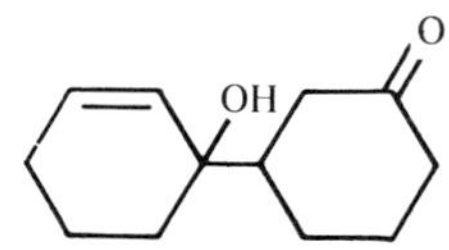 (XVI)
Hg cathode, -1.0 to -1.05 V	CH_3CN, HCl, KCl	74	26
Hg cathode, -1.45 to -1.54 V	CH_3CN, HOAc, NaOAc	78	22
Hg cathode, -1.65 to -1.75 V	CH_3CN, NaOH, KCl	90	10
Na-Hg		97.5	2.5
Mg + HOAc		85	15
Zn + HOAc		57	43

IV. QUINONES

In protic media, the quinone (XX)-hydroquinone (XXI) couple is the classic example of a fast (reversible) system. The $E_{1/2}$ values for quinone, therefore, have unqualified thermodynamic significance and correlate very

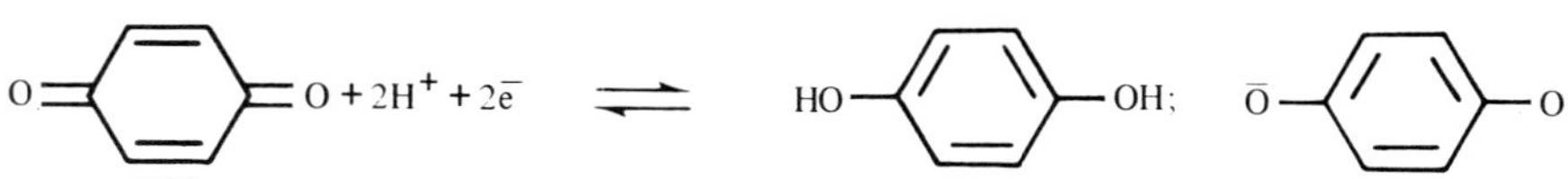

(XX) (XXI) (XXII)

well with oxidation potentials determined by other means [32], e.g., potentiometry.

In alkaline medium the intermediate semiquinone (XXII) formed by one-electron reduction of (XX) has long been known [33]; the radical anion (XXII) may undergo radical dimerization (→ quinhydrone) or may disproportionate to (XX) and (XXI).

Polarography has been used extensively as an analytical method for quinones-hydroquinones [32]; the applications are summarized and compiled periodically in Analytical Chemistry Annual Reviews [34].

The electrochemistry of quinone imines and p-aminophenols is discussed in Chapter XIV.

The electrochemical behavior of quinones in aprotic media has recently been summarized [35]. As might be expected from the earlier discussion in this chapter, the reductions of quinones [e.g., (XX)] to semiquinones [e.g., (XXII)] in a one-electron process are generally separately observable from the second reduction to a dianion. Again, even though the medium is aprotic, a "hard" acid cation of the supporting electrolyte may play the same role as H^+. Thus Eggins [36] found that (XX) is reduced in two 1-ē waves when Et_4NClO_4 in CH_3CN is used as the medium but that the reduction approached a one 2-ē wave when $LiClO_4$ was used as the electrolyte. This is reminiscent of the polarographic behavior of benzil in DMF [1b]. A similar study with naphthoquinones has been reported [40].

A number of α- and β-substituted anthraquinones has been reduced at the dropping mercury electrode in DMF to the radical anion stage. There was good correlation between $E_{1/2}$ and the energy of the lowest vacant molecular orbital (Hückel approximation) [37].

The burgeoning physical chemistry of semiquinones — and more generally of semidione radical ions — has been summarized by Russell [38].

NOTES

[1] On this basis as well as on certain others, the discussion of aromatic carbonyls could optionally have been included here rather than in Chapter VIII provided the restriction "olefinic unsaturation" had been removed.

[2] It must be borne in mind that it has been amply demonstrated that in the absence of stronger acids, N^+R_4 (where $R > CH_3$) is a proton donor toward carbanions; this behavior is a result of a Hofmann-type degradation.

REFERENCES

[1]. (a) M. Ashworth, Coll. Czech. Chem. Commun., 13, 229 (1948); (b) R. H. Philip, Jr., T. Layloff, and R. N. Adams, J. Electrochem. Soc., 111, 1189 (1964).

[2]. R. Pasternak, Helv. Chim. Acta, 31, 753 (1948).

[3]. P. Zuman, D. Barnes, and A. Ryslova-Kejharova, Discuss. Faraday Soc., 45, 202 (1968).

[4]. J. Simonet, Bull. Soc. Chim. Fr., 1533 (1970).

[5]. J. Wiemann, ibid., 2545 (1964).

[6]. J. Heyrovsky, A Polarographic Study of Electrokinetic Phenomena, Actual. Sci. Ind., No. 90, Hermann, Paris, 1934; J. Heyrovsky, Polarographie, theoretische Grundlagen, praktische Ausführung und Anwendungen der Elektrolyse mit der tropfenden Quecksilberelektrode, Springer, Vienna, 1941, p. 185.

[7]. R. Bieber and G. Trümpler, Helv. Chim. Acta, 30, 2000 (1947); 31, 5 (1948).

[8]. R. W. Moshier, Ind. Eng. Chem., Anal. Ed., 15, 107 (1943).

[9]. H. Adkins and F. W. Cox, J. Amer. Chem. Soc., 60, 1151 (1938).

[10]. (a) M. Fields and E. R. Blout, ibid., 70, 930 (1948);
(b) E. Paspaleev and K. Batzalova, Monatsh., 101, 166 (1970).

[11]. A. Winkel and G. Proske, Ber., 69, 1917 (1936).

[12]. M. B. Neiman and Z. V. Markina, Zavod. Lab., 13, 1177 (1947); CA, 43, 4183 (1949).

[13]. E. I. Fulmer, J. J. Kolfeubach, and L. A. Underkofler, Ind. Eng. Chem., Anal. Ed., 16, 469 (1944).

[14]. M. von Stackelberg and W. Stracke, Z. Elektrochem., 53, 118 (1949).

[15]. H. Adkins, R. M. Elofson, A. C. Rossow, and C. C. Robinson, J. Amer. Chem. Soc., 71, 3622 (1949).

[16]. S. Wawzonek and A. Gundersen, J. Electrochem. Soc., 111, 324 (1964).

[17]. S. Ishiwata and T. Nozaki, J. Pharm. Soc. Japan, 71, 1257 (1951).

[18]. G. Shima, Mem. Cocl. Sci. Kyota Imp. Univ., A12, 327 (1929).

[19]. A. Misono, T. Osa, and T. Yamagishi, Kogyo Kagaku Zasshi, 69, 945 (1966); CA, 66, 75482e (1967).

[20]. A. Misono, T. Osa, and T. Ueno, Nippon Kagaku Zasshi, 88, 1184 (1967); CA, 69, 58794c (1968).

[21]. J. Wiemann and M. L. Bouguerra, Ann. Chim. (Paris), 3, 215 (1967).

[22]. H. Lund and A. D. Thomsen, Acta Chem. Scand., 23, 3567 (1969); A. D. Thomsen and H. Lund, ibid., 23, 3582 (1969).

[23]. J. Wiemann and M. L. Bouguerra, Ann. Chim. (Paris), 2, 35 (1967).

[24]. J. Simonet, Compt. Rend., 263C, 1546 (1966).

[25]. E. Touboul, F. Weisbuch, and J. Wiemann, ibid., 268C, 1170 (1969).

[26]. L. N. Lavrishcheva, through CA, 70, 106313q (1969).

[27]. J. Simonet, Compt. Rend., 263C, 685 (1966).

[28]. J. P. Petrovich, M. M. Baizer, and M. R. Ort, J. Electrochem. Soc., 116, 749 (1969).

[29]. A. Mazzenga, D. Lomnitz, J. Villigas, and C. J. Polowcqyk, Tetrahedron Lett., 1665 (1969).

[30]. J. Wiemann, M. R. Monot, and J. Gardan, Compt. Rend., 245, 172 (1957).

[31]. R. G. Harvey and E. V. Jensen, Tetrahedron Lett., 1801 (1963).

[32]. L. I. Smith, I. M. Kolthoff, S. Wawzonek, and P. M. Ruoff, J. Amer. Chem. Soc., 63, 1018 (1941); I. M. Kolthoff and J. J. Lingone, Polarography, 2nd ed., Vol. II, Wiley-Interscience, New York, 1952, p. 699 ff.

[33]. I. M. Kolthoff and J. J. Lingane, Polarography, 2nd ed., Vol. I, Wiley-Interscience, New York, 1952, p. 153 ff.

[34]. E.g., D. J. Pietrzyk, Anal. Chem. Ann. Rev., 38, 278R (1966).

[35]. C. K. Mann and K. K. Barnes, Electrochemical Reactions in Nonaqueous Systems, Dekker, New York, 1970, pp. 190-197.

[36]. B. R. Eggins, Chem. Commun., 1267 (1969).

[37]. A. E. Brodsky, L. L. Gordienko, and L. S. Degtiarev, Electrochim. Acta, 13, 1095 (1968).

[38]. G. A. Russell, "Semidione Radical Ions," in Radical Ions (E. T. Kaiser and L. Kevan, eds.), Wiley-Interscience, New York, 1968, pp. 87-150.

[39]. J. Wiemann, M. Dedieu, and D. Lelandais, Fr. Pat. 1,591,372, June 5, 1970; CA, 74, 70990z (1971).

[40]. T. Fujinaga, K. Izutsu, and T. Nomura, J. Electroanal. Chem., 29, 203 (1971).

Chapter X

CARBOXYLIC ACIDS AND DERIVATIVES

L. Eberson

Division of Organic Chemistry
Chemical Center
University of Lund
Lund, Sweden

I. INTRODUCTION

This chapter is concerned with the cathodic reduction of carboxylic acids and their derivatives, i.e., esters, amides, anhydrides, hydrazides, and nitriles. Cyclic derivatives of substituted carboxylic and polycarboxylic acids, such as lactones, lactams, imides, and anhydrides, will also be included. Only those transformations in which the functional group itself is involved are treated.[1]

As a starting point, it should be mentioned that a number of useful cathodic conversions of the carboxyl group and its derivatives have been known for a long time and that tabulations and discussions of these reactions can be found in several places [1-4].[2] Most of these processes have been performed under constant current conditions using carefully controlled reaction conditions with regard to solvent, temperature, pH, and electrode material. Only recently have mechanistic studies been brought to bear on the nature of these processes. As will be seen later, our knowledge in this area is far from complete.

II. CATHODIC REDUCTION OF CARBOXYLIC ACIDS

The cathodic reduction of a carboxyl group can lead to two types of products, either the corresponding aldehyde in a two-electron process or the hydroxymethyl compound in an over-all four-electron process [Eqs. (1) and (2)]. It would seem to be rather difficult to stop the reduction at the aldehyde

$$RCOOH \xrightarrow[2H^+]{2e^-} RCH(OH)_2 \rightleftarrows RCHO + H_2O \tag{1}$$

$$RCHO \xrightarrow[2H^+]{2e^-} RCH_2OH \tag{2}$$

stage since aldehydes generally are reduced at less negative potentials than carboxylic acids, but as will be discussed below, this transformation is possible in certain cases.

Aliphatic and aromatic acids which are not strongly activated by electron-withdrawing groups and do not contain more easily reducible groups give a well-developed polarographic wave in aqueous [5] or nonaqueous [6, 7] solvents, but this is usually due to reduction of protons to hydrogen. As an example, the hydrogen wave from acetic acid in acetonitrile/0.1 M tetraethylammonium perchlorate appears at -2.3 V (SCE) and in aqueous medium in the intermediate pH region at -1.76 V (SCE). Controlled potential electrolysis (CPE) of acetic acid in acetonitrile at -2.3 V (SCE) demonstrates that the electrode process actually is the reduction of protons. No second wave is found in the accessible potential region of these media, indicating that $RCOO^-$, the species left after reduction of the proton, is very difficult to reduce. Binoxalate ion does not show any polarographic wave in acetonitrile. Hence one would not expect that nonactivated carboxylic acids should be reducible in slightly acidic or alkaline solution or in nonaqueous media. In strongly acidic medium the situation may be different since protonation of the carboxyl group might lead to the formation of a species capable of accepting electrons into the carbon-to-oxygen bonds.

Acids containing certain activating groups can be used as model compounds in polarography without interference from the hydrogen wave. 1-Ethyl-4-carboxypyridinium ion [(I), as the bromide] is such a case [8]. The carboxyl group is activated enough to be the most easily reducible group in the molecule at pH values below 7. The half-wave potential of (I) moves toward more negative potentials with increasing pH in such a way that the $E_{1/2}$-pH curve can be approximated by three lines: the first one in the pH

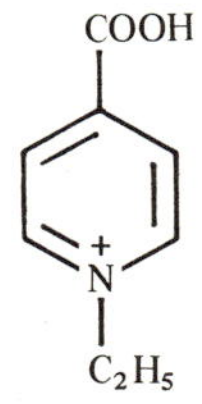

(I)

region of 0.5-1.5 with a slope of 0.029 V/pH; the second one in the pH region of 2-7 with a slope of 0.068 V/pH; and the third one in the pH region of 8-13 with a slope of 0.007 V/pH. The total variation of $E_{1/2}$ is between -0.74 V in strongly acidic and -1.22 V in alkaline solution, an interval which occurs at less negative potentials than the hydrogen wave. From this pH dependence it was concluded that from pH 8 to 13 the reducible species contained the carboxyl group as the carboxylate ion and below 7 existed in its undissociated form. Since the pK of the acid independently was determined to be 1.75, the carboxyl group must be formed in a recombination reaction at the electrode surface in the pH region 2-7, as indicated by the slope of the $E_{1/2}$-pH line in this interval [9].

CPE of (I) in slightly acidic buffers at -1.0 V (SCE) gives the corresponding aldehyde in good yield, provided the reaction is stopped after the consumption of two electrons per molecule (prolonged reaction times led to a further slow reduction of the aldehyde). Table 1 shows the variation of

TABLE 1

Yield of 1-Ethyl-4-Formylpyridinium Bromide from the Cathodic Reduction of 1-Ethyl-4-Carboxypyridinium Bromide as a Function of pH [8] at 25°C

pH	Yield of aldehyde, %
0.15	30
1.35	55
2.55	75
4.45	72
5.15	63
6.29	29
9.15	0

aldehyde yield with pH; the best yields are obtained in the pH region between 2 and 5. In alkaline medium, no aldehyde can be detected, presumably due to the fact that the heterocyclic ring system now is the most easily reducible group of the molecule. At pH values below 0, aldehyde formation also is negligible, which might be explained by further reduction of the aldehyde to the carbinol or even to the completely reduced product, 1-ethyl-4-methyl-pyridinium ion, by analogy with the formation of 2- and 4-picoline by cathodic reduction of pyridine-2- and 4-carboxylic acid, respectively, in strongly acidic solution [10].

At first glance it seems somewhat surprising that the aldehyde can be obtained in such good yields since aldehydes generally are reduced at less negative potentials than carboxylic acids. However, in this case the aldehyde exists predominantly in aqueous solution in its hydrated form [Eq. (1)] which is electroinactive at the potential employed in the CPE runs. Hence, further reduction takes place only slowly. Several other systems have been found to behave similarly [11, 12]: pyridine-2- and 4-carboxylic acid, imidazole-2-carboxylic acid, and thiazole-2-carboxylic acid. In the last case the yield is low due to competing reduction of the ring. Yields are better at low temperatures due to the slowing down of the dehydration reaction.

From the above-mentioned investigations, Iversen and Lund [12] concluded that the following requirements must be fulfilled for the reduction of a carboxyl group to an aldehyde function to take place: (1) The carboxyl group must be activated by a strongly electron-withdrawing group; (2) the carboxyl group must be the most easily reducible group in the molecule (which automatically excludes the nitro group and a number of nitrogen-containing heteroaromatic groups as activating groups); and (3) the carbaldehyde formed must be protected against further reduction by being capable of forming a nonreducible derivative, such as a hydrate or hemi-acetal. The last criterion is probably the most important and stringent one.

Other acids which fulfill these requirements and indeed can be reduced cathodically to aldehydes are oxalic acid (to glyoxylic acid) and lactic acid [4]. Phthalic acid, also an activated acid, cannot be reduced to an aldehyde in neutral or weakly acidic electrolytes at a lead cathode but is converted to phthalide [3], possibly due to the fact that the intermediate phthalaldehydic acid is in equilibrium with hydroxyphthalide [13] which should be easily reducible after protonation and loss of a water molecule [Eq. (3)]:

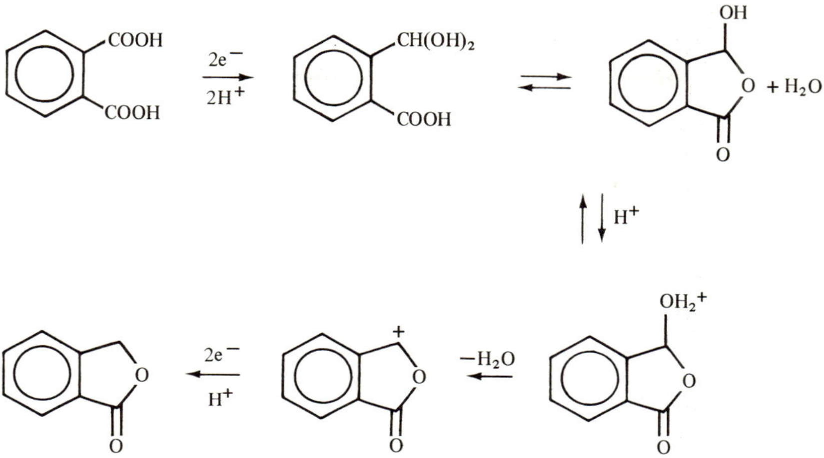

(3)

Reduction of the benzene ring of phthalic acid has also been observed [14, 14a].

In the light of the requirements listed above one would hardly expect that the reduction of benzoic acid would give any useful preparative results since it is not an activated acid and benzaldehyde is not strongly hydrated in aqueous medium. Nevertheless, by using some tricks, it is possible to reduce benzoic acid and certain other acids, such as salicylic acid, to the aldehyde in moderate yield. By adding boric acid to an electrolyte consisting of a solution of sodium benzoate and sodium sulfate in water and by having a layer to extract out free aldehyde as it is formed, Mettler [15] was able to isolate benzaldehyde in 30-50% yield at a mercury cathode; similar yields were obtained from salicylic acid. The function of boric acid is possibly to stabilize the aldehyde hydrate as a borate complex (II):

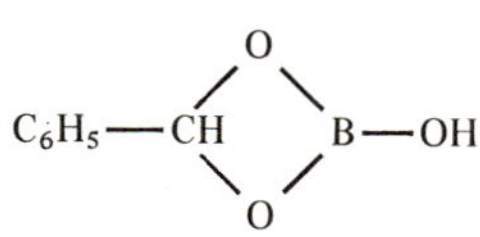

(II)

Also, sodium bisulfite can be added to convert the aldehyde to a nonreducible derivative [16, 17].[3]

In strongly acidic alcoholic media the reduction of nonactivated aromatic carboxylic acids at lead or mercury cathodes gives excellent yields of the corresponding benzyl alcohols [18, 19, 20]. The mechanism of this process is not known; it may be either that the carboxylic acid is reduced in its protonated form, $RCOOH_2^+$, or that the acid first forms an ester with the alcoholic solvent, the ester being more easily reducible than the acid.

Pentafluorobenzoic acid, a strongly activated aromatic acid, is reduced in 20% sulfuric acid at a mercury cathode to give 2, 3, 5, 6-tetrafluorobenzoic acid in 75% yield at -1.20 V (SCE). At more negative potentials ($\leq$ -1.30 V vs. SCE), the carboxyl function in both the pentafluoro and tetrafluoro acid is reducible, a mixture of pentafluoro- and 2, 3, 5, 6-tetrafluorobenzyl alcohol being obtained [Eq. (4)]. Pentafluorobenzyl alcohol, which is not reducible

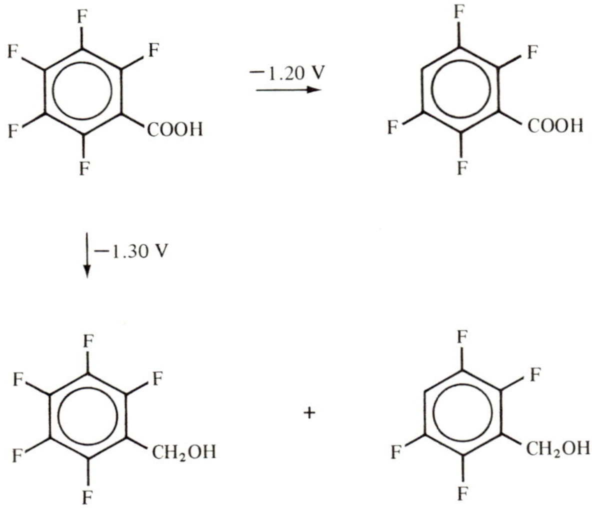

(4)

under the conditions employed, increases in proportion at more negative potentials [21].

Even acids with the carboxyl group attached to an aliphatic carbon atom are reducible to the corresponding alcohol in strongly acidic media. Thus, phenylacetic acid is reduced to β-phenylethanol in 25-35% yield at a lead cathode in 50% sulfuric acid-ethanol [22], and butyric acid to butanol in 80% sulfuric acid in 6.5% yield [23]. Curiously enough, the latter conversion can also be performed in 17% yield in 7% aqueous sodium hydroxide, in spite of the fact that all of the acid must be in the form of the difficultly reducible carboxylate form.

III. CATHODIC REDUCTION OF ESTERS, LACTONES, AND ANHYDRIDES

The first step in the cathodic reduction of an ester in acidic medium leads to an aldehyde, presumably via the hemiacetal [Eq. (5)]. The aldehyde, if not stabilized in the form of an electroinactive addition compound, is then reduced in a second step to form the corresponding carbinol (or sometimes also the pinacol). However, a competing pathway for ester reduction in strongly acidic medium is the formation of ethers [Eq. (6)], sometimes in good yields [18, 24, 25]:

$$RCOOR' \xrightarrow[2H^+]{2e^-} RCH(OH)OR' \rightleftarrows RCHO + HOR' \quad (5)$$

$$RCOOR' \xrightarrow[4H^+]{4e^-} RCH_2OR'. \quad (6)$$

Polarographic data for ester reductions are scarce, but a few investigations on esters of aromatic acids have been reported [26-28]. Half-wave potentials for reduction of Ar-$COOCH_3$ in an unbuffered solution of pH 7.2 are linearly correlated with σ values [26]; $E_{1/2}$ for methyl benzoate was -2.11 V (SCE) and electron-withdrawing substituents in the ring or alkoxy group shift $E_{1/2}$ toward less negative potentials. Generally, esters seem to be more easily reducible than the parent carboxylic acids. Anion radicals of methyl, ethyl, and isopropyl benzoate have been detected by ESR upon cathodic reduction of these esters in acetonitrile/tetrapropylammonium perchlorate [26a].

From the preparative point of view, the cathodic reduction of esters, in spite of its potential synthetic usefulness, is less well known than that of carboxylic acids. Nonactivated esters are not reduced at all or only slowly in strongly acidic medium at a lead cathode [24], whereas methyl or ethyl benzoate is reduced to a mixture of benzyl alcohol and methyl or ethyl benzyl ether, respectively [24, 25]. Phenyl and benzyl benzoate give ethers predominantly [18], as do also ethyl o-chloro- and m-bromobenzoate [25].

In the case of activated esters, the possibility of halting the reduction at the aldehyde stage depends on the same factors as in the reduction of carboxylic acids. Again, the aldehyde must be able to exist in an electroinactive form in the electrolyte in order to be isolable. Thus, cathodic reduction of diethyl terephthalate, diethyl furan- and thiophene-2, 5-dicarboxylate [26, 27] in a buffered ethanolic solution of "pH 4.2" affords the aldehyde ester under CPE conditions (in the interval between -1.50 and -1.65 V). Diethyl oxalate [29] is reduced at a mercury cathode in acidic medium to give the ethyl hemiacetal of ethyl glyoxylate (38-50% yield). On the contrary, diethyl and monoethyl phthalate gave phthalide under similar conditions [14, 30].

Methyl 2-naphthoate, in which the aromatic ring system is also electroactive, is reduced [31] in the presence of boric acid in methanolic solution at a mercury cathode to form 2-naphthalenemethanol in 19% yield, a mixture of diastereomeric pinacols from further reduction of the intermediate aldehyde in 19% yield, and a mixture of methyl 1, 2- and 1, 4-dihydro-2-naphthoates in 42% yield. The tendency for ring attack is still more pronounced in the case of methyl 1-naphthoate which under the same conditions [32] gives methyl 1, 4-dihydro-1-naphthoate (43%) dimer (III) (20%), di-1-naphthylglyoxal (IV) (5%), and 1-naphthalenemethanol (5%):

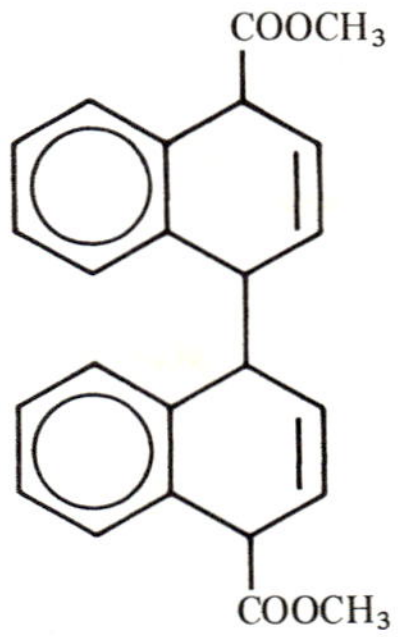

(III)

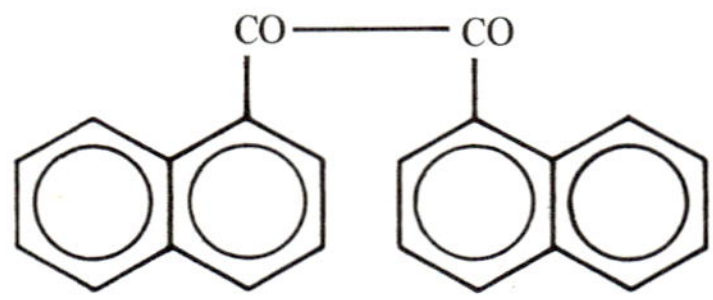

(IV)

Here also it is possible that electrogenerated sodium amalgam is the reducing agent.

2-Alkyl-substituted acetoacetic esters undergo an interesting rearrangement reaction (the Tafel rearrangement) upon reduction at a lead cathode in acidic medium [Eq. (7)] with complete reduction of the functional groups [33-37]:

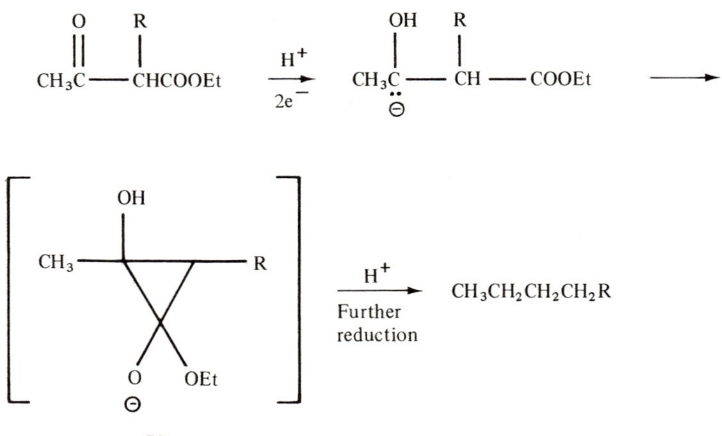

(7)

(V)

In the light of the recent finding that analogous rearrangements during cathodic reduction of β-diketones proceed via cyclopropanediol intermediates [38], it is tempting to suggest that species (V) is an intermediate also in the reduction of these β-keto esters.

As cyclic esters, lactones are reducible to hydroxyaldehydes which in principle should be capable of undergoing the usual aldehyde reduction processes. The reaction has been applied to a number of lactones of sugar acids [4], and good yields of the corresponding sugars have been obtained in these cases because of the stabilization of the carbonyl function by cyclic acetal or ketal formation.

Thiol esters of aromatic acids, Ar-CO-S-R′, are cathodically reduced with fission of the S-CO bond, affording the thiol in good yield [39], e.g., L-cystein in 83% yield from S-benzoyl-L-cystein at a mercury cathode in DMF-water-tetramethylammonium chloride.[4]

Anhydrides have received very little attention as substrates for cathodic reduction. Polarography of cyclic anhydrides [40] in aprotic media (acetonitrile, DMF) displayed two one-electron waves for the most easily reduced anhydrides (pyromellitic, dibromopyromellitic, and tetrachlorophthalic anhydride). Controlled potential electrolyses of phthalic and pyromellitic anhydrides at the potential of the first wave gave highly colored, oxygen-sensitive solutions, which later were shown to give well-resolved ESR spectra of the corresponding radical ions [41]. Phthalic anhydride has been reduced to phthalide on a preparative scale [42], but the reaction conditions employed (ethanol-water-ammonium carbonate) indicate that monoethyl phthalate must have been the actual substrate. The properties of acetic anhydride as solvent for electrochemistry have been studied by chronopotentiometry [43].

IV. CATHODIC REDUCTION OF AMIDES, LACTAMS, IMIDES, AND HYDRAZIDES

The carbonyl group of amides, lactams, and imides is usually reducible to a methylene group in strongly acidic media at lead or mercury electrodes. The carbonyl group can be replaced by a thiocarbonyl group with analogous results. The reaction may be schematically (i.e., neglecting the possible involvement of protonated species) pictured as a two-electron reduction to form an α-amino alcohol [Eq. (8)], followed by elimination [Eq. (9)] of a water molecule in an acid-catalyzed process to form an iminium ion (or an imonium ion in the case of N,N-disubstituted amide groups). A final two-electron reduction gives the protonated form of the amine [Eq. (10)]:

$$\text{—CO—}\overset{|}{\text{N}}\text{—} \xrightarrow[2\text{H}^+]{2e^-} \text{—CH(OH)—}\overset{|}{\text{N}}\text{—} \tag{8}$$

$$\text{—CH(OH)}\overset{|}{\text{N}}\text{—} \xrightarrow{\text{H}^+} \text{—CH}{=}\overset{+}{\text{N}}\langle \tag{9}$$

$$\text{—CH}{=}\overset{+}{\text{N}}\langle \xrightarrow[2\text{H}^+]{2e^-} \text{—CH}_2\underset{|}{\overset{+}{\text{N}}}\text{H} \tag{10}$$

Obviously, side reactions may occur due to loss of amine from the α-amino alcohol instead of water or to hydrolysis of the iminium or imonium ion; in both cases the aldehyde and/or its reduction products would be formed.

Reduction of N, N-diphenyl- and N-methyl-N-phenylbenzamide and N, N, N′, N′-tetraphenylterephthalamide in acetonitrile/tetraethylammonium perchlorate results in solutions which evidence strong ESR signals [44], indicating that the first step is a one-electron transfer. The polarography of pyridine-4-carboxamide and some of its derivatives [45], as well as carboxamides of other heterocyclic systems [11, 12], has been carefully studied in aqueous medium over the whole pH range. As an example, it is instructive to see how structural variations in and at the amide group of pyridine-4-carboxamide itself change the product distribution. Pyridine-4-carboxamide (VI) displays a polarographic wave with $E_{1/2}$ ranging from about -0.6 V at pH 0 to -1.3 V at pH 13, the change being due to the operation of protonation equilibria. CPE of (VI) in 0.8 M hydrochloric acid at -0.80 V (SCE) gave a good yield of pyridine-4-aldehyde, whereas at pH 4.5, at which the aldehyde hydrate does not exist, 4-pyridinemethanol was formed instead. The latter reaction is somewhat difficult to explain since one would expect the aldehyde to be stable as the hydrate at this pH, but one possibility is that the acetate buffer used in this experiment has a strong catalytic effect on the dehydration process. Substitution of one hydrogen of the amide group for a phenyl group changes the product composition considerable, CPE of N-phenyl-4-pyridinecarboxamide in 1 M hydrochloric acid at -0.75 V producing a mixture of 4-pyridinemethanol and 4-anilinomethylpyridine as follows:

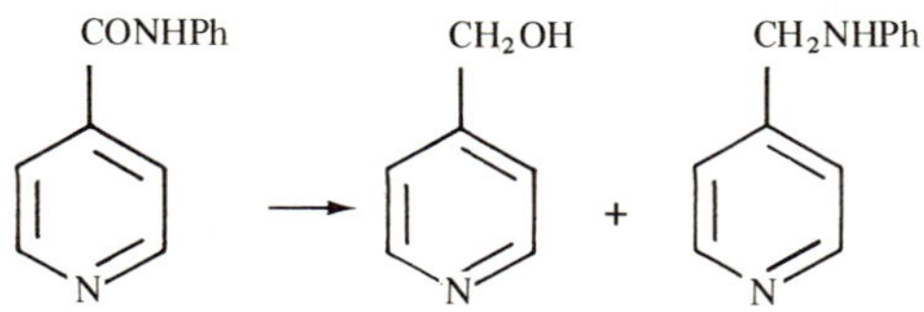

(11)

Finally, 4-pyridinethiocarboxamide is reduced in two two-electron steps in acidic medium; CPE at the potential of the second wave gives 4-aminomethylpyridine. These three cases show that the preparative outcome of an amide reduction is strongly dependent on the substituents at the amide group and that the changes manifest themselves in the chemical step [Eq. (9)].

A large number of preparative applications of amide group reduction have been compiled and discussed by other authors [1-4].[5] Generally, these reactions are performed at lead cathodes in aqueous sulfuric acid or ethanolic aqueous sulfuric acid and result in good yields of the corresponding aminomethyl compounds. Since the media employed are so strongly acidic it is probable that the electroactive species is the protonated amide in these cases. Simple aliphatic amides of the general formula $RCONH_2$ have not been studied; however, a number of N, N-dimethyl- or N, N-diphenyl-substituted aliphatic amides have been successfully reduced. Benzamides give good yields of benzylamines, N, N-disubstitution favoring amine formation also in these cases. Thioamides can be used as substrates with similar results.

In neutral or alkaline solution, N-substituted benzamides undergo cathodic cleavage to form benzyl alcohol and an amine [46].[6] The reaction is performed in an initially neutral solution of tetramethylammonium chloride in methanol, but in a divided cell the catholyte quickly becomes alkaline, so the reaction proceeds under alkaline conditions during most of the electrolysis period.

Phthalimide and substituted phthalimides are reduced in DMF-tetrabutylammonium iodide to give the corresponding anion radicals [41]. In aqueous medium, phthalimide is reduced polarographically in acidic solution in a two-electron process and in alkaline medium in two one-electron steps [47]. The primary product is believed to be hydroxyphthalimidine [Eq. (12)], as has indeed been shown for N-(N′, N′-diethylaminoethyl)tetrachlorophthalimide

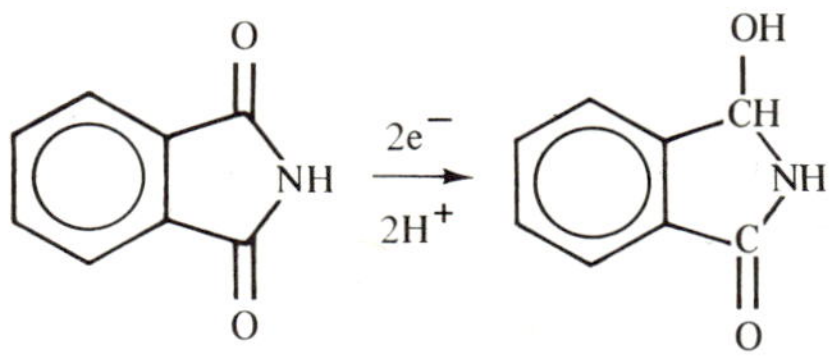
(12)

in preparative experiments [48]. Evidently, the time scale of polarography does not allow observation of further reduction steps, probably because the elimination step [Eq. (9)] is too slow. In macroscale electrolyses further reduction takes place easily, lactams or fully saturated cyclic amines being obtainable in good yields from imides in strongly acidic electrolytes [1-4].

The reduction of acyl hydrazides has been investigated very little. Polarography of 4-pyridinecarboxylic acid hydrazide in acid and neutral solution shows that it is reduced in two two-electron steps [49], the first step being demonstrated by CPE experiments to be cleavage of the N-N bond to form 4-pyridinecarboxamide, which then undergoes reduction to the aldehyde in the second step. Other hydrazides of heterocyclic carboxylic acids behave analogously.

V. CATHODIC REDUCTION OF NITRILES

Both aliphatic and aromatic nitriles are reduced in acidic solution to give amines [Eq. (13)], whereas in neutral or alkaline medium cleavage of the C-CN bond takes place [Eq. (14)]:

$$R\text{-}CN \xrightarrow[4H^+]{4e^-} RCH_2NH_2 \qquad (13)$$

$$\text{R-CN} \xrightarrow[\text{Proton source}]{2e^-} \text{R-H} + \text{CN}^-. \qquad (14)$$

α, β-Unsaturated nitriles, however, undergo hydrodimerization under the last-mentioned conditions (Chapter XIX).

Reactions (13) and (14) have been studied polarographically [50]. In acid solution, aromatic nitriles of the type p-$RCOC_6H_4CN$ (R = H or CH_3) undergo a four-electron reduction at -0.70 V (SCE) with formation of the corresponding benzylamine. The protonated nitrile is the electroactive species. In neutral medium, nitriles of the type XC_6H_4CN (X = p-COO^-, p-SO_2NH_2, and m-SO_2NH_2) display a two-electron wave with formation of cyanide ion and C_6H_5X. This duality in mechanism has been thoroughly studied in the case of 2- and 4-cyanopyridine [51], and it was shown that both pH and electrode material play an important role in determining the mechanism. It was recently found that the electrochemical cleavage of nitriles can also be performed in anhydrous amine media [52], making possible C-CN bond cleavage in aliphatic nitriles such as cycloheptanecarbonitrile and heptanecarbonitrile. This reaction was interpreted as being due to reductive cleavage by solvated electrons (Chapter XXI).

A number of aromatic nitriles including benzonitrile have been reduced electrochemically in DMF-tetrapropylammonium perchlorate to give solutions of relatively stable anion radicals, whose ESR spectra have been recorded [53]. Phthalonitrile, which gives two one-electron waves on polarography in this medium, undergoes decyanation on CPE at the potential of the second wave.

The preparative use of nitrile reduction has been primarily for the synthesis of amines [1, 54-57]. This is done in acidic medium and gives fair to good yields.[7]

Imino ether hydrochlorides can be reduced to amines in good yields at lead cathodes in aqueous sulfuric acid [58], suggesting that cathodic reduction of nitriles in ethanolic sulfuric acid should be a useful procedure, the first step then being the formation of the imino ether salt by acid catalysis.

NOTES

[1] For cathodic coupling of derivatives of α, β-unsaturated acids, see Chapter XIX.

[2] As an example, electroreduction of carboxylic acids to alcohols under mild conditions has been available to the organic chemist long before $LiAlH_4$ was invented.

[3] It should be noted that these reactions may actually be cases of indirect electrochemical processes, i.e., reductions by electrogenerated sodium amalgam.

[4]See also Chapter XXI.

[5]In recent years, significant contributions to this area include selective cathodic reductions of aliphatic amides [45a] to aldehydes or alcohols. These reactions were carried out in an undivided cell at platinum cathodes in methylamine/lithium chloride, alcohols being obtained in the absence of an aldehyde in the presence of a proton source, e.g., ethanol.

[6]See also Chapter XXI.

[7]2-Hexenedinitrile has been reduced to the monoamine [59].

REFERENCES

[1]. F. Fichter, Organische Elektrochemie, Steinkopf, Dresden and Leipzig, 1942.

[2]. S. Swann, Jr., in Technique of Organic Chemistry (A. Weissberger, ed.), 2nd ed., Vol. II, Wiley-Interscience, New York, pp. 385-523.

[3]. M. J. Allen, Organic Electrode Processes, Reinhold, New York, 1958.

[4]. F. D. Popp and H. P. Schultz, Chem. Rev., 62, 19 (1962).

[5]. I. A. Korshunov, Z. B. Kuznetsova, and M. K. Shchennikova, Zh. Fiz. Khim., 23, 1292 (1949); through CA, 44, 2873 (1950).

[6]. J. F. Coetzee and I. M. Kolthoff, J. Amer. Chem. Soc., 79, 6110 (1957).

[7]. I. M. Kolthoff and T. B. Reddy, J. Electrochem. Soc., 108, 980 (1961).

[8]. H. Lund, Acta Chem. Scand., 17, 972 (1963).

[9]. P. Zuman, Fortschr. Chem. Forsch., 12, 1 (1969).

[10]. J. P. Wibaut and H. Boer, Rec. Trav. Chim., 68, 72 (1949).

[11]. P. E. Iversen and H. Lund, Acta Chem. Scand., 21, 279 (1967).

[12]. P. E. Iversen and H. Lund, ibid., 21, 389 (1967).

[13]. P. R. Jones, Chem. Rev., 63, 461 (1963).

[14]. S. Ono, Nippon Kagaku Zasshi, 75, 1195 (1954); through CA, 51, 12704 (1957).

[14a]. P. C. Condit, Ind. Eng. Chem., 48, 1252 (1956); Fr. Pat. 1,516,599 (1968); CA, 70, 92677 (1969); H. Suter, F. Beck, and A. Hrubesch, U.S. Pat. 3,471,381, Oct. 7, 1969.

[15]. C. Mettler, Chem. Ber., 41, 4148 (1908); see also S. Ono and T. Yamauchi, Bull. Chem. Soc. Japan, 25, 404 (1952).

[16]. J. A. May and K. A. Kobe, J. Electrochem. Soc., 97, 183 (1950).

[17]. B. B. Dey and H. V. K. Udupa, Proc. 6th Meeting Intern. Comm. Electrochem. Thermodyn. Kinet., 465 (1955); through CA, 50, 6221 (1956); J. H. Wagenknecht, J. Org. Chem., 37, 1513 (1972).

[18]. C. Mettler, Chem. Ber., 38, 1747 (1905); 39, 2933 (1906).

[19]. F. Mayer and F. A. English, Ann., 417, 69 (1929); F. Mayer, W. Schäfer, and J. Rosenbach, Arch. Pharm., 267, 581 (1929).

[20]. G. H. Coleman and H. L. Johnson, Org. Syn., 21, 10 (1941).

[21]. P. Carrahar and F. G. Drakesmith, Chem. Commun., 1562 (1968).

[22]. S. Ono and T. Hayashi, Bull. Chem. Soc. Japan, 26, 232 (1953).

[23]. M. Masuno, T. Asahara, S. Kuroiwa, K. Shimizu, and J. Nakano, J. Chem. Soc. Japan, Ind. Chem. Sec., 52, 151 (1949); through CA, 45, 1884 (1951).

[24]. J. Tafel and G. Friedrichs, Chem. Ber., 37, 3187 (1904).

[25]. C. Mettler, ibid., 37, 3692 (1904).

[26]. J. Nakaya, H. Kinoshita, and S. Ono, Nippon Kagaku Zasshi, 78, 935 (1957); through CA, 53, 21276 (1959); and P. Zuman, Substituent Effects in Polarography, Plenum, New York, 1967, p. 61.

[26a]. M. Hirayama, Bull. Chem. Soc. Japan, 40, 1822 (1967).

[27]. J. Nakaya and S. Ono, J. Chem. Soc. Japan, Pure Chem. Sec., 74, 907 (1953); through CA, 48, 8082 (1954).

[28]. G. C. Whitnack, J. Reinhart, and E. C. Gantz, Anal. Chem., 27, 359 (1955).

[29]. W. Oroshnik and P. E. Spoerri, J. Amer. Chem. Soc., 63, 3338 (1941).

[30]. W. M. Rodionov and V. C. Zvorukina, Bull. Soc. Chim. Fr., 5, 840 (1938).

[31]. G. T. Mondodoev, N. M. Przhiyalgovskaya, and V. N. Belov, J. Org. Chem. USSR (Engl. Transl.), 1, 1257 (1965).

[32]. G. T. Mondodoev, N. M. Przhiyalgovskaya, and V. N. Belov, ibid., 1, 2048 (1965).

[33]. J. Tafel and H. Hahl, Chem. Ber., 40, 3312 (1907).

[34]. J. Tafel and W. Jürgens, ibid., 42, 2548 (1909).

[35]. J. Tafel, ibid., 45, 452 (1912).

[36]. H. Stenzl and F. Fichter, Helv. Chim. Acta, 17, 674 (1934).

[37]. H. Stenzl, F. Fichter, and H. Arni, ibid., 19, 392 (1936).

[38]. T. J. Curphey, C. W. Ammelotti, T. P. Layloff, R. L. McCartney, and J. H. Williams, J. Amer. Chem. Soc., 91, 2817 (1969).

[39]. Farbwerke Hoechst A.-G., Neth. Pat. Appl. 6,602,261 (1966); through CA, 66, 38237 (1967).

[40]. M. Peover, Trans. Faraday Soc., 58, 2370 (1962).

[41]. R. E. Sioda and W. S. Koski, J. Amer. Chem. Soc., 89, 475 (1967).

[42]. B. Sakurai, Bull. Soc. Chem. Japan, 7, 127 (1932).

[43]. W. B. Mather, Jr. and F. C. Anson, Anal. Chem., 33, 1634 (1961).

[44]. L. Horner and R.-J. Singer, Tetrahedron Lett., 1545 (1969).

[45]. H. Lund, Acta Chem. Scand., 17, 2325 (1963).

[45a]. R. A. Benkeser, H. Watanabe, S. J. Mels, and M. A. Sabol, J. Org. Chem., 35, 1210 (1970).

[46]. L. Horner and H. Neumann, Ger. Pat. Appl. 1,493,665 (1969).

[47]. J. Tirouflet, R. Robin, and M. Guyard, Bull. Soc. Chim. Fr., 571 (1956).

[48]. Ref. [3], p. 75.

[49]. H. Lund, Acta Chem. Scand., 17, 1077 (1963).

[50]. O. Manousek and P. Zuman, Chem. Commun., 158 (1965).

[51]. J. Volke and A. M. Kardos, Coll. Czech. Chem. Commun., 33, 2560 (1968).

[52]. P. G. Arapakos and M. K. Scott, Tetrahedron Lett., 1975 (1968).

[53]. P. H. Rieger, I. Bernal, W. H. Reinmuth, and G. F. Fraenkel, J. Amer. Chem. Soc., 85, 683 (1963).

[54]. K. Ishifuku, H. Sakurai, and H. Okamoto, Japan. Pat. 180,563 (1949); CA, 46, 3432 (1952).

[55]. S. Sawa, Japan. Pat. 5,019 (1951); CA, 47, 2617 (1953).

[56]. A. P. Tomilov, S. L. Varshavskii, and Yu. D. Smirnov, USSR Pat. 132,214 (1960); CA, 55, 6214 (1961).

[57]. T. Nonaka and K. Sugino, J. Electrochem. Soc., 114, 1255 (1967); Yu. D. Smirnov, L. I. Saltykova, and A. P. Tomilov, Zh. Prikl. Khim., 43, 1620 (1970).

[58]. H. Wenker, J. Amer. Chem. Soc., 57, 772 (1935).

[59]. Yu. D. Smirnov and A. P. Tomilov, Zh. Org. Khim., 5, 864 (1969); CA, 71, 38299h (1969).

Chapter XI

ONIUM COMPOUNDS[1]

L. Horner

Organisch-Chemisches Institut
University of Mainz
Mainz, Germany

I. ROLE OF THE SUPPORTING ELECTROLYTE

Although supporting electrolytes play a critical role in preparative electrochemistry, little is yet known about these "inert" adjunctives. In principle, all cations and anions may be used in conjunction with electrochemical reactions. For cathodic reductions alkali metal and onium salts, particularly quaternary ammonium salts, are especially suited. In principle, phosphonium, arsonium, and tulfonium salts are also satisfactory. However, it must be noted that the onium salts may be reductively degraded if their discharge potentials are more positive than that of the substance to be reduced. Chapter XXI discusses the chemical reactions associated with the discharge of these onium cations.

In recent years many electrochemists have been concerned with elucidating the interaction between substrate and electrode surface but have, in general, neglected to consider the influence of both the supporting electrolyte and the solvent. In this connection two recent publications [1, 20], representative of many others, attempt by subtle electrochemical measurements to arrive at conclusions concerning the interaction between the electrode and the substrate.

The composition of the products as well as the stereochemistry may be influenced by the proper choice of the supporting electrolyte. This has been demonstrated most notably in several cases of cathodic hydrodimerization of activated olefins.

The cathodic reduction of acrylonitrile in the presence of alkali metal tosylates leads predominantly to propionitrile. When, however, tetraalkylammonium salts are used, the hydrodimer, adiponitrile, is formed in high yield even in aqueous systems [3]. Similar results were obtained with diethyl maleate [4]. Feoktistov and coworkers [5] as well as Gillet [6, 7] independently have developed a representation (shown in Fig. 1) of the electrical double layer in an attempt to explain the hydrodimerization.

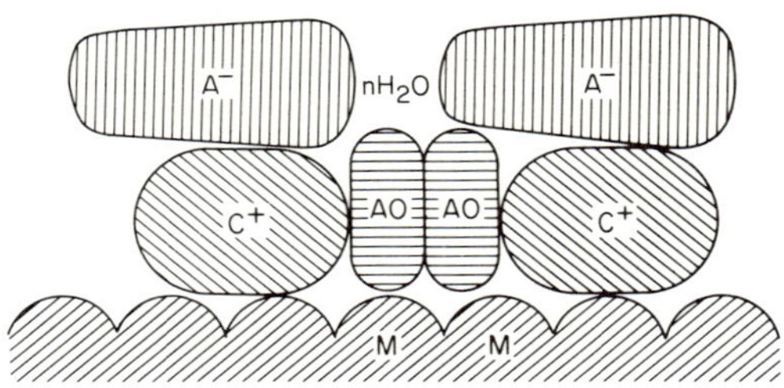

Fig. 1. Model of the electrochemical double layer in the cathodic reduction of acrylonitrile to adiponitrile. M, atoms of the electrode surface; C^+, hydrophobic cation; A^-, hydrophilic anion; AO, activated olefin molecule.

Figure 1 shows that the double layer is built up of two different layers. The first layer contains the molecules of acrylonitrile and the unsolvated tetraalkylammonium ions which are adsorbed in the electrode and thereby displace the water molecules. The radical anions (or dianions) which are formed on electroreduction cannot be protonated immediately and react with a second (unreduced) molecule to form a dimeric anion which, after desorption, is protonated in the second layer which consists of hydrophilic anions and water.

On contrast to the tetraalkylammonium ions, alkali metal cations are hydrated. They therefore bring water into the first adsorption layer and make possible the protonation of the electrogenerated intermediate. In this case the sequence leads to propionitrile.

Whether this simple model is generally serviceable for an explanation of all cathodic dimerizations is not yet certain. The investigations of Tomilov and coworkers [8] indicate that additional factors, e. g., the internaction between adsorbed supporting cation and the substrate molecules, are important in determining product distribution in cathodic hydrodimerizations. (See Table 1.) It is not obvious how Gillet's model can explain the decreasing

TABLE 1

Influence of the Cations of the Supporting Electrolyte upon the Hydrodimerization of Various Compounds [8]

	Acrylonitrile	Methyl acrylate	Mesityl oxide
Cathode material [a]	Pb	Sn	Zn
pH [a]	7-8	7-8	14
% Hydrodimer using K^+	10.8	36	45
% Hydrodimer using Et_4N^+	71	55	41

[a]Cathode material and pH for maximum yields of hydrodimers.

yield of hydrodimer in the presence of tetraethylammonium salts in going from acrylonitrile to methyl acrylate to mesityl oxide.

On the other hand, Smirnov and co-workers [9] reported that in reduction of acetone to pinacol the yields were affected by the counterion in the order $Me_4N^+ \gg Cs^+ > K^+ > Na^+ \gg Li^+$. Nicolas and Pallaud [10] obtained much better yields of the respective hydrodimers of benzophenone, acetophenone, and ethyl acrylate from electrolyses in tetraethylammonium p-toluenesulfonate than in p-toluenesulfonic acid.

The above few examples illustrate that much remains to be done before we can understand the phenomena that occur in the double layer and the role that supporting electrolytes play therein. An attempt to rationalize radical or anionic behavior of electrogenerated radical anions on the basis of the nature of the counterion has been reported [11].

II. REDUCTION POTENTIALS OF AMMONIUM SALTS

Table 2 summarizes the reported [12-17] reduction potentials at a mercury drop of a number of these cations.

TABLE 2

Discharge Potentials of Some Quaternary Ammonium Salts

Cation	-E vs. SCE	-E vs. Ag/AgCl/KCl
NH_4	2.00 [a,b]	—
$C_2H_5NH_3$	2.08 [a,b]	—
$(C_2H_5)_2NH_2$	2.01 [a,b]	—
	2.11 [a,b]	—
	1.90 [a,b]	
$(CH_3)_3NH$	1.82, [a,b] 2.23 [c,n]	
$(C_2H_5)_3NH$	1.83 [a,b]	
$(CH_3)_4N$	2.65, [a,d] 2.7, [e,h] 2.94, [f,d] 2.93 [c,d]	2.41 [g,l]
$(CH_3)_3NC_4H_9$	2.9 [e,h]	
$(C_2H_5)_3NC_6H_{13}$	2.9 [e,h]	
$(CH_3)_3NC_8H_{17}$	2.3 [e,h]	
$(CH_3)_3NC_{10}H_{21}$	2.3 [e,h]	
$(CH_3)_3N$ cetyl	2.0 [e,h]	
$(CH_3)_3NCH_2C_6H_5$	2.08, [a,d] 2.2 [e,h]	
$(CH_3)_3N\text{-}CH{=}CH_2$	1.99 [a,d]	
$(CH_3)_3NC_6H_5$	2.21, [a,d] 2.38 [f,i]	
$(CH_3)_3NCH_2CH_2OH$	2.15, [a,d] 2.72 [c,n]	
$(CH_3)_3NC_2H_4COOCH_3$	2.64 [c,n]	
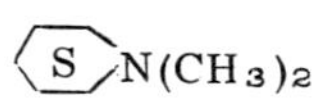	2.67, [a,d] 2.6 [e,h]	

Table 2 (cont.)

Cation	-E vs. SCE	-E vs. Ag/AgCl/KCl
$(C_2H_5)_4N$	2.78,[a,d] 2.500,[o,j] 2.9,[e,h] 2.56,[f,i] 2.67[c,n]	2.45[g,l]
$(C_3H_7)_4N$	2.710,[o,j] 2.52[c,n]	2.54[g,l]
$(C_4H_9)_4N$	2.87,[a,d] 2.750,[o,j] 2.9,[e,h] 2.46,[f,i] 2.57[c,n]	2.70[g,l]
$(C_5H_{11})_4N$	2.890[o,k]	
$(i\text{-}C_5H_{11})_4N$	2.954[o,j]	
$(n\text{-}C_6H_{13})_4N$	2.854[o,j]	2.87[g,l]
$(n\text{-}C_8H_{17})_4N$	3.050[o,j]	2.99[g,l]
$(C_2H_5)_3N\ i\text{-}C_5H_{11}$	2.48[f,i]	
$(n\text{-}C_4H_9)_3NCH_2C_6H_5$	2.5[e,h]	
$(C_2H_4OH)_3NC_2H_5$	2.38[c,n]	
$(CH_3)_2(C_2H_5)(C_6H_5)N$	2.71[c,n]	
$(CH_3)_2(C_6H_5CH_2)(C_6H_5)N$	1.59[e]	
$(CH_3)_2(C_6H_5CH_2)$(cetyl)N	2.0[e,h]	
$(CH_3)_2(i\text{-}C_5H_{11})(C_6H_5)N$	2.44[f,i]	

[a]Ref. [12].

[b]0.1 Mole of amine per liter in 0.1 M aqueous tetramethylammonium halide[m] ≡ 0.01-0.03 M hydroxide.

[c]Ref. [16].

[d]Foot potential of 0.1 M solution of hydroxide in water.

[e]Ref. [14].

[f]Ref. [15].

[g]Ref. [17].

[h]Half-wave potential of the halide ($C = 10^{-4}$ mole/ℓ) in 0.1 M aqueous tetrabutylammonium hydroxide.

[i]Decomposition potential of the hydroxide in water.

[j] Potential of the final rise of the iodides (0.1 mole/ℓ in 90% methanol).

[k] As hydroxide, otherwise as in footnote j.

[l] Foot potential of the halide in 0.1 M methanolic solution.

[m] Tetramethylammonium chloride or bromide.

[n] Half-wave in aqueous tetramethylammonium bromide (conc. not specified).

[o] Ref. [13].

A. Chemical Behavior of Ammonium Salts upon Discharge at a Cathode

Upon electrolysis of alkali metal salts in aqueous solution hydrogen is evolved at the cathode. Upon exclusion of proton donors (e.g., in fused salts) metals are obtained which form amalgams with a mercury cathode. Analogously, electrolysis of ammonium salts at mercury leads to the very unstable ammonium amalgam, first observed in 1808 [18]. At the end of the last century the first alkylated ammonium amalgams were also formed. The preparation of tetramethylammonium amalgam [19], whose structure was determined by Brauer and Düsing [20], is particularly easy.

B. Onium Amalgams

There have been several reports on the preparation and properties of tetramethylammonium amalgam [20, 21]; however, only recently have tetraalkylammonium amalgams with longer side chains become known. Solutions containing 0.1-1.0 mole of $R_4N^+X^-$ in aprotic solvents like acetonitrile or dimethylformamide were electrolyzed at low temperatures and in the absence of oxygen at 10-20 V; R included CH_3, C_2H_5, n-C_4H_9 [23] as well as $C_{16}H_{33}N(CH_3)_3$ [22] and $C_{12}H_{25}N(CH_3)_3$. Dark porous masses were formed on the mercury surface; they were isolated and investigated by the method of Brauer and Düsing [20]. The amalgams are reported to have the composition $1NR_4$: 12-13 Hg [23] or $1C_{16}H_{33}N(CH_3)_3$: 7Hg [22]. The actual formation of amalgams is further supported by the following:

1. the rest potential (measured against a standard electrode) which is established after interruption of the electrolysis;

2. the reduction by the amalgam of aromatics (e.g., naphthalene, anthracene, and diphenyl), alkyl halides, aromatic nitriles, and nitro compounds to anion radicals which were identified by ESR [24].

In order to distinguish between reduction by tetramethylammonium amalgams and by "solvated electrons" of organic substrates, reductions [25] were carried out with the former prepared according to the literature [20].

It was found that benzophenone was reduced to ketyl and stilbene to bibenzyl but neither sulfones nor sulfonamides nor carboxamides were reduced [26, 27].

Using a published procedure [23], Cottrell and Morris [28] have prepared R_4P amalgams including R = CH_3, n-C_4H_9, and C_6H_5 as well as $(C_6H_5)_3PCH_3$, $(C_6H_5)_2P(CH_3)_2$, and $C_6H_5P(CH_3)_3$. They also confirmed the formation of amalgam by measurement of rest potential and by reaction with water and with nitrobenzene.

Thermal decomposition of phosphonium amalgams yields tert-phosphines and hydrocarbons; decomposition of sulfonium amalgams (v.i.) leads to thio ethers and a mixture of saturated and unsaturated hydrocarbons but not to dimeric products. This is further discussed in Chapter XXI.

III. REDUCTION POTENTIALS OF PHOSPHONIUM SALTS [29]

Many polarographic reductions of phosphonium salts have been reported [30-39]. The half-wave data may be compared only cautiously since the polarograms were obtained under a variety of conditions. A further complication is the fact that most of the reductions are irreversible and the half-wave potentials depend on the following factors:

1. concentration of the depolarizer;
2. temperature;
3. drop time;
4. height of the mercury column;
5. nature and concentration of the maximum suppressor.

The interpretation of the polarogram is difficult because there are often several reduction waves. Representative examples are given in Table 3; more complete data are reported in the original literature [29]. Of 56 phosphonium salts examined [29], eight show one wave; 29, two waves; nine, three waves; five, four waves; and five, six waves. The most anodic wave may be ascribed to the discharge of the phosphonium ion. Associated with this is the cleavage of a ligand. The more negative waves may be due to far-reaching reductive cleavage leading to secondary phosphines. This is discussed further in Chapter XXI.

Dependence of the Reduction Potential upon the Substituents

There is no linear relationship between the half-wave potentials of the alkyltriphenylphosphonium salts and Taft's σ^* constant. Only when R in $R\overset{+}{P}Ph_3$ is methyl, benzyl, or benzhydryl is a straight line obtained.

TABLE 3

Polarographic Data for Representative Phosphonium Salts [a]

Phosphonium bromide	Compound No. [b]	Wave	$-E_{1/2}$ V	id (μA)	n_1	n_2
Methyltriphenyl	(I)	1	1.862	2.60	2.7	1.1
Cyclohexyltriphenyl	(II)	1	1.859	2.94	1.6	
		2	2.982	0.28	1.2	
n-Propyltriphenyl	(V)	1	1.822	1.36	2.3	
		2	1.962	0.72	1.4	
		3	2.075	0.14	2.2	
Allyltriphenyl	(X)	1	1.626	2.10	1.4	
		2	1.775	0.20	0.9	
		3	2.065	0.45	2.2	
		4	2.468	1.35	2.1	
Benzyltriphenyl	(XI)	1	1.552	2.80	1.3	
Methylene-bis-triphenyl	(XVIII)	1	1.195	2.50	0.8	
		2	1.817	0.95	1.7	
		3	1.932	1.19	2.7	
Trimethylene-bis-biphenyl	(XX)	1	1.307	0.40	2.7	
		2	1.712	5.40	2.1	
		3	2.054	1.00	2.0	
Hexamethylene-bis-triphenyl	(XXIII)	1	1.822	4.75	1.8	
		2	2.000	1.10	0.8	
Decamethylene-bis-triphenyl [c]	(XXV)	1	1.715	1.25	0.8	
		2	1.916	0.65	2.2	
		3	2.073	0.45	—	
Dodecamethylene-bis-triphenyl [c]	(XXVI)	1	1.664	1.50	2.7	
		2	1.944	2.50	0.9	
p-Anisyltriphenyl	(XXXV)	1	1.835	1.26	1.4	
		2	1.988	0.26	1.8	

Table 3 (cont.)

Phosphonium bromide	Compound No.[b]	Wave	$-E_{1/2}$ V	id (μA)	n_1	n_2
		3	2.128	0.56	1.9	
p-Tolyltriphenyl	(XXXVIII)	1	1.692	2.50	2.0	0.8
		2	2.073	0.26	1.7	
Tetraphenyl	(XL)	1	1.680	2.80	2.1	0.6
		2	2.068	0.15	1.5	
Di-p-tolyldiphenyl	(LII)	1	1.695	2.48	1.5	0.9
		2	2.092	1.18	0.9	
Tri-p-tolylphenyl	(LIII)	1	1.727	2.60	2.1	1.2
		2	2.071	0.32	3.0	
Di-p-anisyldiphenyl	(LIV)	1	1.790	0.98	1.3	
		2+3	1.954	3.12	—	
Tri-p-anisylphenyl	(LV)	1	1.848	1.30	1.2	
		2	2.006	0.22	2.1	
		3	2.128	0.74	2.9	
p-Dimethylaminophenyl triphenyl		1	2.026	7.5	1.05	
Diphenyl-bis-p-dimethylamino-phenyl		1	2.110	11.5	0.55	1.15

[a]10^{-3} M depolarizer, 0.1 M tetra-n-butylammonium bromide; maximum suppressor 0.05% Triton X-100; temp. 25.0 ± 0.2°. Rapid polarography (t = 0.2 sec).

[b]Numbering according to Ref. [29].

[c]Conc. $<10^{-3}$ M.

Replacement of the phenyl groups in tetraphenylphosphonium by one, two, or three methyls or by one or two ethyls shifts the $E_{1/2}$ cathodically by approximately equal amounts (Table 4). Replacement of the phenyl groups in tetraphenylphosphonium bromide [(XL), Table 3] successively by p-tolyl groups causes increasing cathodic shifts [(XXXVIII), (LII), (LIII)]. This

TABLE 4

Dependence of the Half-Wave Potential of Some Phosphonium Salts upon Systematic Changes of Substituents on Phosphorus

Phosphonium bromide	Compound No.[a]	$-E_{1/2}$, V	Change in $E_{1/2}$	id/c (μ/mM)
Tetraphenyl	(XI)	1.680		2.80
			0.182	
Methyltriphenyl	(I)	1.862		2.60
			0.225	
Dimethyldiphenyl	(XIV)	2.087		4.38
			0.184	
Trimethylphenyl	(XV)	2.271		5.10
Tetraphenyl	(XL)	1.680		2.80
			0.156	
Ethyltriphenyl	(III)	1.836		1.79
			0.195	
Diethyldiphenyl	(XVI)	2.031		2.58

[a]From Ref. [29].

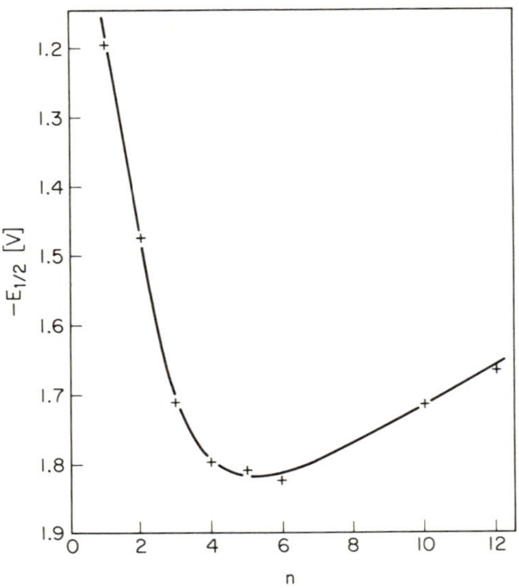

Fig. 2. Dependence of the half-wave potential of bis-phosphonium salts (XVIII)-(XXVI) $(C_6H_5)_3P(CH_2)_nP(C_6H_5)_3Br_2$ upon n.

can be ascribed to the electron-releasing effect of the methyl group. Even more negative shifts are evidenced when the phenyl groups are replaced by p-anisyl or dimethylaminophenyl [cf. (XXXV), (LIV), (LV), Table 3]. Comparisons of the half-wave potentials of the bis-phosphonium salts (XVIII)-(XXVI) (Table 3) reveal a surprisingly strong mutual effect of the two cationic centers. As is shown in Fig. 2, bis-phosphonium salts with five-six methylene groups between the two P atoms have the most negative reduction potentials. It is possible that unfavorable orientation of the molecules at the mercury surface, brought about by conformational requirements, is responsible for the difficulty of the reduction.

IV. REDUCTION POTENTIALS OF QUATERNARY ARSONIUM SALTS

The reduction potentials of 11 arsonium salts were determined (Table 5) under standard conditions (those of Table 3). The arsonium salts are cleaved ca. 0.3 V more anodically than the corresponding phosphonium salts. As a corollary, the bonding energy between ligand and central metal is significantly different in the P and the As salts.

TABLE 5

Polarography of Some Arsonium Salts in Aqueous Solution [a]

Arsonium bromide	Wave	$-E_{1/2}$, V	id (μA)	n_1	n_2
Methyltriphenyl	1	1.498	1.04	0.7	
	2	1.775	0.35	0.8	
	3	2.081	0.48	1.6	
	4	2.378	0.30	1.0	
Ethyltriphenyl	1	1.489	1.12	0.9	
	2	2.054	1.08	1.3	
o-Tolyltriphenyl	1	1.389	1.70	0.9	0.3
	2	2.062	0.19	1.7	
m-Tolyltriphenyl	1	1.384	1.64	0.8	0.3
	2	2.070	1.04	0.9	
p-Tolyltriphenyl	1	1.382	1.82	0.9	0.3
	2	2.041	0.54	1.8	

Table 5 (cont.)

Arsonium bromide	Wave	$-E_{1/2}$ V	id (μA)	n_1	n_2
Tetraphenyl	1	1.377	2.02	1.0	0.3
	2	2.060	0.40	1.8	
p-Cumyltriphenyl	1	1.364	2.20	0.8	0.2
	2	2.038	0.34	2.2	
p-Biphenylyltriphenyl	1	1.224	0.84	1.1	
	2	1.416	0.62	0.5	
	3	2.048	0.42	0.9	
	4	2.330	1.74	1.2	
Benzyltriphenyl	1	1.213	2.83	0.5	0.3
	2	2.057	0.51	2.1	
	3	2.356	0.83	0.7	
	4	2.470	0.24	2.1	
Triphenylallyl		1.20			
Triphenyl-1-propenyl		1.59			

[a] Conditions same as those of Table 3.

V. REDUCTION POTENTIALS OF TERTIARY SULFONIUM SALTS

Tertiary sulfonium salts would be as good supporting electrolytes as quaternary phosphonium and arsonium salts except that they easily decompose at the cathode to thio ethers and other decomposition products. (See Chapter XXI.)

Lüttringhaus and Machatzke [40] carried out a systematic investigation of cytostatic sulfonium salts (Table 6). The authors propose that reductive cleavage occurs in two one-electron stages.

VI. YLID FORMATION IN THE ELECTROLYSIS OF PHOSPHONIUM AND SULFONIUM SALTS

Three waves (-1.48, -2.25, and -2.56 V) were observed [37] in the polarographic reduction of triphenylcyanomethylphosphonium salts. The second was ascribed to triphenylcyanomethylenephosphorane which had been

TABLE 6

Influence of Ligands upon the Half-Wave Potentials of Certain Sulfonium Salts[a]

R	$(C_2H_5)(C_2H_5)\overset{+}{S}R$	$(HOCH_2CH_2)(CH_3CH_2)\overset{+}{S}R$	$(HOCH_2CH_2)(HOCH_2CH_2)\overset{+}{S}R$	$(ClCH_2CH_2)(CH_3CH_2)\overset{+}{S}R$	$(C_6H_5CH_2SCH_2CH_2)(CH_3CH_2)\overset{+}{S}R$	$(ClCH_2CH_2)(ClCH_2CH_2)\overset{+}{S}R$
CH_2CH_3	>2	—	—	—	—	—
CH_2CH_2OH	>2	>2	>1.8	—	—	—
$CH_2CH_2SCH_2C_6H_5$	1.66	—	—	—	1.37	b
$CH{=}CH_2$	1.68	—	1.52	—	—	—
CH_2CH_2Cl	1.60	—	—	1.46	—	1.32
$CH_2CH{=}CH_2$	1.44	1.27	—	—	—	b
$CH_2C_6H_5$	1.30	1.21	1.12	0.87	0.79	b
CH_2CH_2Br	1.10	—	0.92	1.01	—	0.92
$CH_2CH{=}CHCH_2R'$	1.06	0.94	0.88	0.80	—	b
$CH_2-C_6H_4-CH_2R'$	1.03	0.92	0.85	0.81	—	b
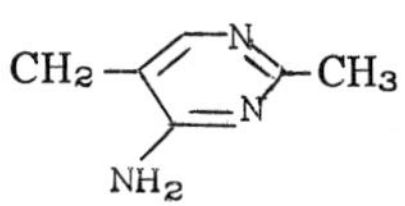	0.94	0.82	0.71	b	b	b

[a]Ref. [40]. Only the first of two waves is indicated.

[b]The compound is unstable.

formed according to the following sequence:

$$[(C_6H_5)_3P\text{-}CH_2CN]^+ \xrightarrow{2e^-} (C_6H_5)_3P + [CH_2CN]^-$$

$$[(C_6H_5)_3PCH_2CN]^+ + [CH_2CN]^- \longrightarrow (C_6H_5)_3P{=}CHCN + CH_3CN.$$

It was recently reported [41] that ylids are formed when phosphonium salts $(C_6H_5)_3\overset{+}{P}R$ (R = CH_3, C_2H_5, $C_6H_5CH_2$, CH_2COCH_3) are electrolyzed in solvents containing carbonyl groups (e.g., C_6H_5CHO, butyraldehyde, cyclohexanone). The ylids react in situ with the carbonyl compound, present in excess, to yield the corresponding olefin.

Sulfonium salts also are reported to form the corresponding ylids upon electrolysis in dimethyl sulfoxide at carbon electrodes [42]; the ylids may then react further with suitable acceptors, e.g., benzaldehyde, benzophenone, and cyclohexanone. The following example is illustrative:

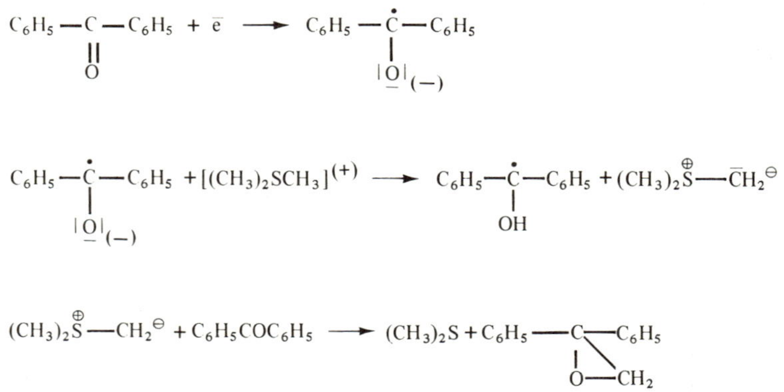

Iversen and Lund [43] were able to provide further insight into the mechanism of ylid formation using a phosphonium salt as an example. Electrolysis of a mixture of azobenzene, phosphonium salt, and a carbonyl compound in anhydrous DMF at a controlled potential at which only the first could be reduced yielded the olefin derived from the carbonyl:

$$C_6H_5\text{-}\overline{N}{=}\overline{N}\text{-}C_6H_5 + 2\bar{e} \rightarrow [C_6H_5\text{-}\underline{\overline{N}}\text{-}\underline{\overline{N}}\text{-}C_6H_5]^{2(-)} \quad \text{I}$$

$$\text{I} + 2[(C_6H_5)_3P\text{-}CH_2R]^{(+)} \rightarrow C_6H_5\text{-}NH\text{-}NH\text{-}C_6H_5 + 2(C_6H_5)_3P{=}CHR \quad \text{II}$$

$$\text{II} + R'R''CO \rightarrow RCH{=}CR'R'' + (C_6H_5)_3P{=}O.$$

VII. ASYMMETRIC INDUCTION BY USE OF OPTICALLY ACTIVE SUPPORTING ELECTROLYTES

This subject is at present in the early stages of exploration. Thus far, it has involved the reduction of ketones in the presence of chiral quaternary ammonium salts. The results obtained are discussed in Chapter XXVII.

NOTE

[1]Translated by the Editor.

REFERENCES

[1]. J. Volke, Chem. Listy, 62, 497 (1968).

[2]. J. Volke and A. M. Kardos, Coll. Czech. Chem. Commun., 33, 2560 (1968).

[3]. M. M. Baizer, J. Electrochem. Soc., 111, 215 (1964).

[4]. M. M. Baizer and J. P. Petrovich, ibid., 114, 1023 (1967).

[5]. L. G. Feoktistov, A. P. Tomilov, and I. G. Sevast'yanova, Sov. Electrochem., 1, 1165 (1965).

[6]. I. E. Gillet, Bull. Soc. Chim. Fr., 1968, 2919.

[7]. I. E. Gillet, Chem. Ing.-Tech., 40, 573 (1968).

[8]. A. P. Tomilov, E. V. Kryukova, V. A. Klimov, and I. N. Brago, Sov. Electrochem., 3, 1352 (1967).

[9]. V. A. Smirnov, L. I. Antropov, M. G. Smirnova, and L. A. Demchuk, Russ. J. Phys. Chem., 42, 897 (1968).

[10]. M. Nicolas and R. Pallaud, Compt. Rend., 267C, 1834 (1968).

[11]. J. P. Petrovich and M. M. Baizer, J. Electrochem. Soc., 118, 447 (1971).

[12]. F. Beck, Ber. Bunsenges., 72, 379 (1968).

[13]. W. Wiesner and K. Schwabe, J. Electroanal. Chem., 15, 73 (1967).

[14]. B. C. Southworth, R. Osteryoung, D. K. Fleischer, and F. C. Nachod, Anal. Chem., 33, 208 (1961).

[15]. V. Gutmann, G. Schöber, and K. Utvary, Monatsh., 88, 887 (1957).

[16]. P. van Rysselberghe and J. M. McGee, J. Amer. Chem. Soc., 67, 1039 (1945).

[17]. O. Spohn, Thesis, University of Mainz, 1967.

[18]. T. J. Seebeck, Ann. Chim., 66, 191 (1808).

[19]. H. N. McCoy and W. C. Moore, J. Amer. Chem. Soc., 33, 273 (1911).

[20]. G. Brauer and G. Düsing, Z. Anorg. Allg. Chem., 328, 154 (1964).

[21]. See also L. Horner and A. Mentrup, Liebigs Ann. Chem., 646, 49 (1961).

[22]. C. Bratu and A. T. Balaban, Rev. Roum. Chem., 13, 625 (1968).

[23]. J. D. Littlehailes and B. J. Woodhall, Chem. Commun., 1967, 665; Discuss. Faraday Soc., 45, 187 (1968).

[24]. J. Myatt and P. F. Todd, Chem. Commun., 1967, 1033.

[25]. O. Spohn, Thesis, University of Mainz, 1965.

[26]. L. Horner and H. Neumann, Chem. Ber., 98, 1715 (1965).

[27]. L. Horner and H. Neumann, ibid., 98, 3462 (1965).

[28]. R. T. Cottrell and R. A. Morris, Chem. Commun., 1968, 409.

[29]. L. Horner and J. Haufe, J. Electroanal. Chem., 20, 245 (1969).

[30]. L. Horner, R. Röttger, and H. Fuchs, Chem. Ber., 96, 3141 (1963).

[31]. E. L. Colichman, Anal. Chem., 26, 1204 (1954).

[32]. M. Shinagawa, H. Matsuo, and H. Nezu, Japan. Anal., 5, 20 (1956).

[33]. H. Matsuo, J. Sci. Hiroshima Univ., Ser. A, 22, 281 (1958).

[34]. S. Wawzonek and J. H. Wagenknecht, Polarography 1964 (G. J. Hills, ed.), Macmillan, New York, 1966.

[35]. H. Matschiner and K. Issleib, Z. Anorg. Allg. Chem., 354, 60 (1967).

[36]. H. Matsuo, J. Sci. Hiroshima Univ., Ser. A, 22, 51 (1958).

[37]. J. H. Wagenknecht and M. M. Baizer, J. Org. Chem., 31, 3885 (1966).

[38]. J. Grimshaw and J. S. Ramsey, J. Chem. Soc. B, 1968, 63.

[39]. J. M. Savéant and M. Veillard-Rayer, Bull. Soc. Chim. Fr., 1967, 2415.

[40]. A. Lüttringhaus and H. Machatzke, Liebigs Ann. Chem., 671, 165 (1964).

[41]. T. Shono and M. Mitani, J. Amer. Chem. Soc., 90, 2728 (1968).

[42]. T. Shono and M. Mitani, Tetrahedron Lett., 1969, 687.

[43]. P. E. Iversen and H. Lund, ibid., 1969, 3523.

Section D

SYNTHETIC AND MECHANISTIC ASPECTS OF ANODIC REACTIONS OF ORGANIC COMPOUNDS CLASSIFIED BY ELECTROPHORES

Chapter XII

HYDROCARBONS

L. Eberson

Division of Organic Chemistry
Chemical Center
University of Lund
Lund, Sweden

I. INTRODUCTION

The purpose of this chapter is to summarize the electrochemical behavior of hydrocarbons, with special reference to the electrode processes involved in the selective oxidation of hydrocarbons to form one or several defined organic products. Thus, the technically important problem of oxidizing hydrocarbons to carbon dioxide and water at catalytic electrodes in fuel cells will not be dealt with here. This subject has been reviewed elsewhere [1].

For many years the study of hydrocarbon electrochemistry at anodes was confined to preparative studies of oxidations of aromatic and alkyl-aromatic hydrocarbons in water-containing electrolytes under constant current conditions [2-4]. Usually complex mixtures were obtained, as exemplified by the isolation of m-methylbenzaldehyde, m-toluic acid, isophthalic acid, 2, 4-dimethylphenol, 2-methylbenzoquinone, 2, 4-dimethyl-4-hydroxy-2, 5-cyclohexadienone, 2, 5-dimethylhydroquinone, 2, 5-dimethylbenzoquinone, and 2,2′-dihydroxy-3, 3′, 5, 5′-tetramethylbiphenyl from the oxidation of m-xylene in acetone-water-sulfuric acid at a platinum or lead dioxide anode [4]. Although this reaction is hardly one that would be chosen for an electrochemical study of mechanistic scope, it nicely shows the diversity of products that can be obtained from the anodic oxidation of a hydrocarbon in the presence of water; it also shows the necessity of finding experimental conditions better suited for quantitative electrochemical measurements.

The use of nonaqueous solvents proved to be the key to such studies of the anodic behavior of hydrocarbons. Following a pioneering study by Lund [5] of the polarography of aromatic hydrocarbons in acetonitrile-sodium perchlorate at the rotating platinum electrode, a large number of electrochemical investigations in nonaqueous solvents have aimed at the elucidation of anodic hydrocarbon oxidation processes. Concurrent preparative studies under CPE conditions, although not always very helpful in clarifying the nature of the initial electrochemical step, have nevertheless been of great help in finding out about the nature of the chemical reactions following electron transfer. The overall mechanistic picture is thus reasonably well established [6, 7], at least for a number of model systems. It remains to be seen whether these findings can be extrapolated to systems of less ideal behavior. In particular, there seems to be a need for a better knowledge of the chemistry of cation radicals in homogeneous media [8, 9] before safe predictions can be made.

II. GENERAL MECHANISTIC CONSIDERATIONS

A. The Direct Oxidation Mechanism

Following the postulate that electrons are always transferred one by one [10], the primary electrode process in the direct oxidation of a hydrocarbon should be the formation of a cation radical [Eq. (1)], analogous to the initial formation of an anion radical in the cathodic reduction of a hydrocarbon [6]. However, due to the high chemical reactivity of cation radicals, a fast reaction will normally occur with a nucleophilic and/or basic species present in the electrolyte, resulting in the formation of either an addition product [(I), between cation radical and nucleophile] or an elimination product [(II), by proton loss from the cation radical to a base] to form neutral radicals (I) and (II) [Eqs. (2) and (3)]:

$$R{-}H \xrightarrow{-e^-} R{-}H^{\dot{+}} \qquad (1)$$

$$R{-}H^{\dot{+}} \xrightarrow{Nu^-} \overset{Nu}{\underset{H}{R\cdot}} \qquad (2)$$

(I)

$$R{-}H^{\dot{+}} + B \longrightarrow R\cdot + BH^+ \qquad (3)$$

(II)

where Nu^- is a nucleophile and B is a base. In either case the radical formed may have an oxidation potential less positive than that of the original substrate, making possible the transfer of a second electron to the anode, as shown below:

$$\overset{Nu}{\underset{H}{R\cdot}} \xrightarrow{-e^-} \overset{Nu}{\underset{H}{R+}} \qquad (4)$$

$$R\cdot \xrightarrow{-e^-} R^+ \qquad (5)$$

Finally, the cations formed are converted to neutral products by a second chemical step with a nucleophile and/or a base in the electrolyte. The processes outlined above correspond to the well-established ECE mechanism. To specify the nature of the chemical step it is convenient to use subscripts N and B, respectively, to denote nucleophile or base action in the chemical step, making, for example, the reaction sequence (1), (3), and (5) an EC_BE mechanism.

Thus, in many cases polarographic studies on hydrocarbon oxidation will give information only about the total number of electrons (two) transferred in the ECE process, not about the number of electrons (one or two) transferred in the first electrochemical step. Preparative studies, which are often run in the presence of nucleophile or base, are even less decisive on this point. Of course, products formed in the ECE process may be electroactive too, so that many-electron processes may appear in polarograms.

It is easy to imagine other processes leading to the same overall result as the ECE mechanism. One obvious possibility is that the cation radical is in fast equilibrium with an extremely reactive dication [Eq. (6)], so that the dication is the kinetically active species

$$R\text{-}H^{\dot{+}} + R\text{-}H^{\dot{+}} \rightleftharpoons R\text{-}H + R\text{-}H^{2+} \tag{6}$$

(as has been assumed to be the case for the reaction between the thianthrene cation radical and water in acetonitrile solution [8]). The mechanism should then properly be denoted an ECC mechanism, but would, of course, be indistinguishable from the ECE process except for the dependence on $[R\text{-}H^{\dot{+}}]^2$ instead of on $[R\text{-}H^{\dot{+}}]$ for the disappearance of the cation radical.

Another possibility is that the cation radical will undergo a second electron transfer to the anode without any intervening chemical step; such an EEC process is likely to occur when the two steps are closely spaced and again the net result could appear to be exactly the same as in the ECE process [8a].

B. The Indirect Oxidation Mechanism

Indirect oxidation mechanisms have been proposed in some cases where voltammetric studies do not give unambiguous information about the nature of the electroactive species. Generally, they are assumed to involve the oxidation of the anion of the supporting electrolyte to form an inorganic radical which subsequently attacks a weak C-H bond in the organic substrate [Eqs. (7) and (8)], e.g., a C-H bond in the benzylic position of an alkylaromatic hydrocarbon,

$$X^- \xrightarrow{-e^-} X\cdot \tag{7}$$

$$R\text{-}H + X\cdot \rightarrow R\cdot + H\text{-}X. \tag{8}$$

It is sometimes difficult to distinguish this type of mechanism from a direct one.

III. POLAROGRAPHIC STUDIES

A. Aromatic Hydrocarbons

The first studies of the anodic polarography of hydrocarbons were performed in an electrolyte consisting of acetonitrile with a perchlorate salt as the supporting electrolyte at a rotating platinum microelectrode [5, 11-14]. This electrolyte is relatively inert chemically and has a high anodic limit, allowing for the determination of half-wave potentials up to around 2.1 V vs. the Ag/0.1 M Ag^+ reference electrode. Generally, aromatic hydrocarbons display a well-developed wave under these conditions, although in some cases waves have been described as slightly deformed or deformed. In

TABLE 1

Half-Wave Potentials of Aromatic Hydrocarbons in Acetonitrile/0.5 M Sodium Perchlorate vs. the Ag/0.1 M Ag^+ Electrode [5, 13]

Compound	$E_{1/2}$, V
Benzene	2.00; 2.08
Toluene	1.98; 1.93
o-Xylene	1.57
m-Xylene	1.58
p-Xylene	1.56
1, 3, 5-Trimethylbenzene	1.55
1, 2, 3-Trimethylbenzene	1.58
1, 2, 4-Trimethylbenzene	1.41
1, 2, 3, 5-Tetramethylbenzene	1.43
1, 2, 4, 5-Tetramethylbenzene	1.29
Pentamethylbenzene	1.28
Hexamethylbenzene	1.16
Naphthalene	1.34; 1.31
Anthracene	0.84; 0.84
Phenanthrene	1.23
Tetracene	0.54 (1.20) [a]
1, 2-Benzanthracene	0.92
1, 2, 5, 6-Dibenzanthracene	1.00 (1.26) [a]
Chrysene	1.13
Pyrene	0.86 (1.12) [a]
Biphenyl	1.48
Indene	1.23
Fluorene	1.25
Acenaphthene	1.11

[a] Value of the second wave.

several cases two anodic waves separated by 0.2-0.7 V appear in the polarograms.

Table 1 lists half-wave potentials for a number of aromatic hydrocarbons, chosen to cover a wide range of structural variations. Excellent correlations have been found between half-wave potentials and Hückel molecular orbital parameters [13-16], ionization potentials [12, 14], charge-transfer transition energies [12], and positions of p-bands in uv absorption spectra of the hydrocarbons [5, 12]. The basis and limitations of such correlations have been examined critically [6].

Following Lund [5], who based his assignment on a preparative CPE experiment with anthracene in the presence of pyridine [Eq. (9)], the anodic waves observed for these hydrocarbons were considered to correspond

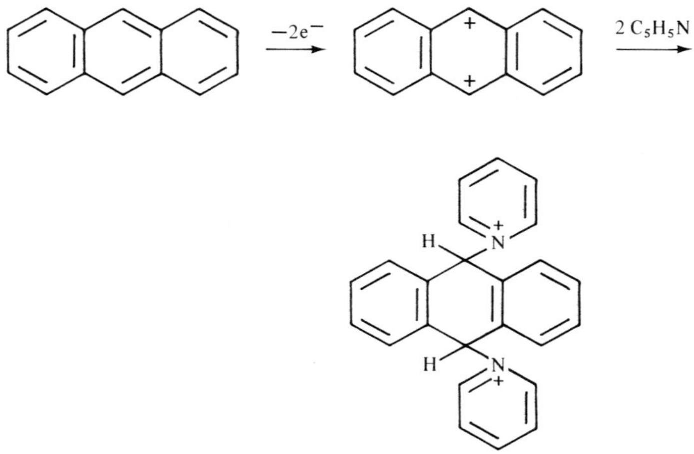

(9)

to direct two-electron transfers with formation of a dicationic intermediate (i.e., an EEC process). Preparative results were generally in agreement with this postulate, although also products which must clearly have been formed via a one-electron process have been isolated in some cases [17]. However, forceful arguments for an initial one-electron transfer as being the common first step for most hydrocarbon oxidations have been brought forward by several groups of investigators, as is discussed below.

Since most preparative studies are performed in the presence of a nucleophile (to give products of substitution, addition, etc.), it is natural that an electrochemical synthesis yields a product which indicates that an overall transfer of two or more electrons has taken place. Also in polarographic investigations, at least one nucleophilic and/or basic species, namely water, cannot be removed completely from the electrolytes. Actually, even with the most rigorous purification procedures, the concentration of water may easily be at the same level as that of the substrate to be studied. Thus, two- or many-electron transfers that are often observed may reflect the occurrence of a series of one-electron transfers with intervening, fast chemical steps. There appears to be one fortunate feature connected with the use of perchlorate salts as supporting electrolytes, in that this anion seems to prevent water molecules from approaching the region near the electrode [18, 19], at least in acetonitrile solution. The fluoroborate ion appears to have the opposite effect.

Several approaches are useful in the study of the primary electrochemical step. One is to employ a model compound whose corresponding cation radical is very stable, 9, 10-diphenylanthracene being the most notable example [20-23]. On cyclic voltammetry in acetonitrile-tetraalkylammonium perchlorate this compound shows a one-electron, reversible anodic process with a peak potential of 1.22 V (SCE), followed by a second one-electron, irreversible process at about 1.6 V. The blue solution obtained by electrolysis of this compound at +1.5 V gives rise to a well-resolved ESR spectrum of the cation radical, identical with that recorded from a solution of 9, 10-diphenylanthracene in concentrated sulfuric acid. Another molecule of the same structural type, 9, 10-di-(α-naphthyl)anthracene, exhibits analogous behavior, but here it is noticeable that the cation radical is less stable than that of 9, 10-diphenylanthracene, presumably due to higher degree of delocalization of unpaired electron density into the naphthalene rings than into the benzene rings [23a]. Similarly, the ESR spectra of the cation radicals of perylene and 9, 10-dimethylanthracene can be observed in the solutions obtained after electrolysis of the parent hydrocarbons [21]. It is, however, appropriate to point out that in the absence of electrochemical evidence regarding the number of electrons transferred in a particular process, the mere observation of an ESR spectrum from an electrolyte during or after electrolysis is not unequivocal evidence for a one-electron transfer: The cation radical might have been formed from the dication through the disproportionation reaction [Eq. (6)], especially when the two electron transfer steps are closely spaced.

Another approach is to use fast sweep techniques in the case of moderately reactive cation radicals [24]. One such case is 1, 2-benzanthracene, which on cyclic voltammetry at low sweep rates shows a complete lack of cathodic current on reverse scan, but at high sweep rates undergoes a quasireversible one-electron process. Thus, at high sweep rates the cation radical does not have time to react chemically before being reduced back to the hydrocarbon. Chrysene, pyrene, and anthracene exhibit similar behavior. Benzo[a]pyrene, the well-known carcinogen, has been very carefully studied using several electroanalytical techniques; its cation radical appears to be intermediate in stability between that of 9, 10-diphenylanthracene and those of the hydrocarbons mentioned above.

A third approach is to reduce the nucleophilicity of the solvent below that of acetonitrile (a nucleophile of about the same strength as acetic acid). The limiting factor then will be the nucleophilicity of the substrate itself. Thus, in nitrobenzene-tetrapropylammonium perchlorate [25] and/or methylene chloride-tetrabutylammonium perchlorate [26], 9, 10-diphenylanthracene, 9, 10-dimethylanthracene, 1, 3, 6, 8-tetraphenylpyrene, rubrene, and tetracene undergo a one-electron oxidation to stable cation radicals. In the former solvent, cation radicals of anthracene, 9-methylanthracene, and naphthacene have been shown to be so unstable that the net observable process appears to involve two or more electrons [25].

A detailed study of the anodic oxidation of hexamethylbenzene [27] shows that in acetonitrile-sodium perchlorate this hydrocarbon undergoes a two-electron oxidation at 1.11 V (Ag/0.1 M Ag^+) followed by a second two-electron process at 1.42 V, both irreversible at low sweep rates in cyclic voltammetry. At high sweep rates, both processes exhibited reversible behavior. The first wave was assumed to correspond to the reversible formation of the dication, which undergoes a rapid proton loss to form the pentamethylbenzyl cation. This finally reacts with acetonitrile and water to give N-pentamethylbenzylacetamide [Eq. (10)]:

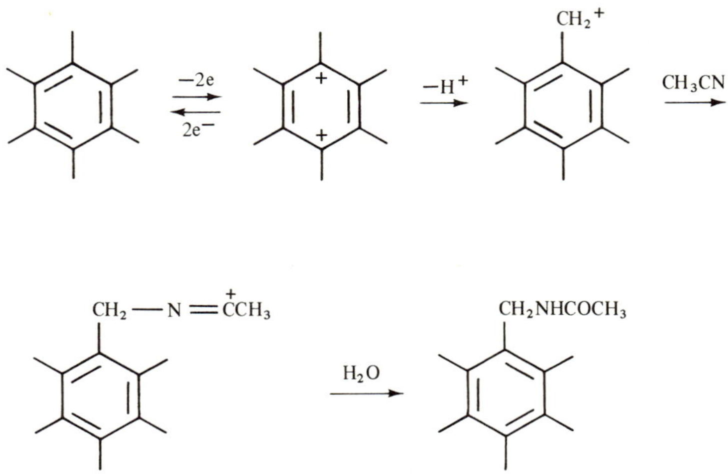

(10)

This mechanism has been criticized because it does not account for the formation of bibenzyls in the oxidation of alkylbenzenes [18, 28], although no bibenzyl derivative is formed in the particular case of hexamethylbenzene. Instead, it is proposed that proton loss occurs from the cation radical to give the pentamethylbenzyl radical, which in a second one-electron oxidation gives the pentamethylbenzyl cation, etc., i.e., the familiar ECE mechanism. Hexaethylbenzene, whose corresponding cation radical would be expected to lose its proton at a lower rate than the hexamethylbenzene cation radical [28], exhibits clear-cut one-electron behavior. However, if the

ECE mechanism is correct for hexamethylbenzene, it is difficult to see why reversibility should appear at high sweep rates. More work is needed to clarify this point.

Other solvent-supporting electrolyte systems employed in the polarographic study of aromatic hydrocarbons at the anode include glacial acetic acid-sodium acetate [17], antimony trichloride [29], and molten tetrabutylammonium nitrate at 150° [30]. Apart from the inherent interest in the preparative aspects of these systems, they do not seem to offer any advantages over the ones discussed previously. For difficultly oxidizable substrates, acetonitrile containing tetrabutylammonium tetrafluoroborate or hexafluorophosphate is an excellent electrolyte, as in the anodic oxidation of benzene [31]. In molten $AlCl_3$-NaCl-KCl at 150° at a pyrolytic graphite anode aromatic hydrocarbons are oxidized at much less positive potentials, e.g., benzene at about +1.0 V vs. NHE (cf. Table 1), than in any of the common solvents, presumably due to π-complex formation between the Lewis acid and the aromatic ring [31a].

B. Aliphatic and Alicyclic Hydrocarbons

The anodic limit of acetonitrile containing a perchlorate salt is not high enough to allow a more general study of nonaromatic hydrocarbons, except in a few favorable cases [12, 32]. With the introduction of tetraethylammonium fluoroborate and hexafluorophosphate [33] as supporting electrolytes, the anodic limit in acetonitrile can be extended considerably, so that even certain saturated aliphatic hydrocarbons can be studied polarographically. Table 2 lists half-wave potentials for a number of aliphatic and alicyclic hydrocarbons; most of these values are well correlated with the ionization potentials of the hydrocarbons. From the diffusion currents it can be estimated that two electrons are involved in the majority of these oxidations [32, 33].

For the simple aliphatic hydrocarbons, the oxidation mechanism probably involves formation of a very reactive carbonium ion:

$$\text{R-H} \xrightarrow[-\text{H}^+]{-2e} \text{R}^+ \xrightarrow[\text{H}_2\text{O}]{\text{CH}_3\text{CN}} \text{RNHCOCH}_3, \qquad (11)$$

which with acetonitrile and water gives an N-substituted acetamide [33]. It is interesting to note that the compounds (1, 4-cyclohexadiene, 1, 3, 5-cycloheptatriene, and cyclooctatetraene) which can be transformed into a 6π-electron system (benzene, tropylium ion, and cyclooctatetraene dication) via a two-electron transfer with or without loss of a proton have exceedingly low half-wave potentials. Somehow the stability of the product must be reflected in the transition state for the potential-determining step in these cases.

TABLE 2

Half-Wave Potentials for Oxidation of Aliphatic and Cycloaliphatic Hydrocarbons, vs. the Ag/0.01 M Ag^+ Electrode

Compound	$E_{1/2}$	Reference
n-Alkanes	>3.4 [a]	[33]
2, 2-Dimethylbutane	3.28 [a]	[33]
2-Methylpentane	3.01 [a]	[33]
2-Methylbutane	3.00 [a]	[33]
Ethylene	2.90 [a]	[33]
Propene	2.84 [a]	[33]
1-Alkenes	2.7-2.8 [a]	[33]
2-Alkenes	2.2-2.3 [a]	[33]
Cyclohexene	2.05;[a] 1.95 [b]	[33], [12]
1, 3-Butadiene	2.09 [b]	[12]
2-Methyl-1-butene	2.03 [b]	[12]
2-Methyl-1, 3-butadiene	1.90 [b]	[12]
2, 3-Dimethyl-1, 3-butadiene	1.89 [b]	[12]
1, 4-Cyclohexadiene	~ 1.3 [c]	[32]
1, 3, 5-Cycloheptatriene	1.13 [c]	[32]
Cyclooctatetraene	~1.2 [d]	[17]

[a] In acetonitrile-tetraethylammonium tetrafluoroborate.

[b] In acetonitrile-sodium perchlorate.

[c] In acetonitrile-lithium perchlorate.

[d] Estimated from the value obtained in acetic acid/sodium acetate.

It has recently been reported [33a] that in anhydrous FSO_3H-acetic acid alkanes $>C_4$ yield carbonium ions on oxidation and α, β-unsaturated ketones as ultimate products.

IV. FOLLOW-UP REACTIONS

A. Different Types of Reactions

We have already mentioned (Sec. II) that the problem of elucidating the nature of the primary electrochemical step requires the application of electrochemical techniques and that product studies generally are of little help in this connection, except that they provide information about the overall number of electrons involved in the formation of a particular product. It has been shown for a variety of model compounds that the primary step is indeed a one-electron transfer to form a cation radical and it is postulated that this is valid for all hydrocarbon oxidations. We also have mentioned the alternative pathways by which the cation radical may be transformed to stable products: the ECE [Eqs. (1)-(3)], the EEC [as exemplified by Eq. (10)], and the ECC mechanism [Eqs. (1) and (6), with the dication as the kinetically active species]. In some cases there is also the possibility that an indirect oxidation mechanism operates [Eqs. (7) and (8), possibly followed by electrochemical and/or chemical steps].

Before we proceed with a mechanistic discussion of the follow-up reactions, it is appropriate to mention briefly the different reaction types possible in hydrocarbon oxidations. These can be divided into substitution, addition, and coupling reactions, the last-mentioned type connected with either elimination or addition, as schematically depicted in Eqs. (12)-(15):

Substitution:

$$\text{R-H} + \text{Nu}^- \rightarrow \text{R-Nu} + \text{H}^+ + 2e^- \tag{12}$$

Addition:

$$>\text{C}=\text{C}< + 2\text{Nu}^- \longrightarrow \text{Nu}-\overset{|}{\underset{|}{\text{C}}}-\overset{|}{\underset{|}{\text{C}}}-\text{Nu} + 2e^- \tag{13}$$

Coupling with elimination:

$$2\text{R-H} \rightarrow \text{R-R} + 2\text{H}^+ + 2e^- \tag{14}$$

Coupling with addition:

$$2\ >\text{C}=\text{C}< + 2\text{Nu} \longrightarrow \text{Nu}-\overset{|}{\underset{|}{\text{C}}}-\overset{|}{\underset{|}{\text{C}}}-\overset{|}{\underset{|}{\text{C}}}-\overset{|}{\underset{|}{\text{C}}}-\text{Nu} + 2e^- \tag{15}$$

In Eqs. (12), (13), and (15) Nu^- denotes a nucleophile, e.g., AcO^-, MeO^-, H_2O, or CH_3CN (all of these do not necessarily work in all processes, though). For alkylaromatic hydrocarbons containing at least one α hydrogen, one can distinguish between nuclear and side-chain substitution, the latter reaction resulting in substitution of one α hydrogen in the alkyl group.

A detailed discussion of these reaction types can be found in Chapters XX and XXII, whereas in this chapter we are concerned mainly with the elucidation of the sequence of reactions leading to each of these reaction types.

B. Addition and Substitution Processes

Convincing evidence has been put forward that the ECE mechanism is the reaction pathway in the oxidative addition of two molecules of a pyridine derivative to 9, 10-diphenylanthracene [analogously to Eq. (9)]. By studying the number of electrons transferred (n_{obsd}) in the oxidation of 9, 10-diphenylanthracene (DPA) in the presence of different nucleophiles at the rotating disk electrode as a function of the rate of rotation (ω), it has been possible to show [34] that the strength of the nucleophile strongly affects the dependence of n_{obsd} on ω. The reaction scheme below summarizes the underlying idea:

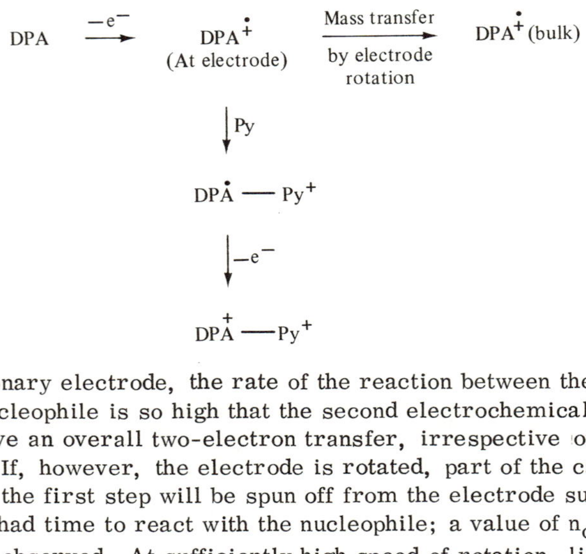

(16)

At a stationary electrode, the rate of the reaction between the cation radical and the nucleophile is so high that the second electrochemical step will proceed to give an overall two-electron transfer, irrespective of nucleophilic strength. If, however, the electrode is rotated, part of the cation radicals formed in the first step will be spun off from the electrode surface before they have had time to react with the nucleophile; a value of n_{obsd} between 2 and 1 is observed. At sufficiently high speed of rotation, limiting one-electron behavior can be observed provided the nucleophile is not too strong. From the n_{obsd}-$\omega^{1/2}$ curves the pseudo-first-order rate constants for the reaction between cation radical and nucleophile can be determined approximately (Table 3). As is seen from Table 3, the rate constants follow the order of nucleophilicity of these pyridines.

TABLE 3

Approximate Pseudo-First-Order Rate Constants for Reaction between 9, 10-Diphenylanthracene Cation Radical and Different Pyridine Derivatives [34]

Nucleophile	Rate constant, sec^{-1}
4-Methylpyridine	1000
Pyridine	300
4-Acetoxypyridine	30
4-Cyanopyridine	3

We have previously referred to the possibility that the species reacting with the cation radical may either react as a nucleophile or a base (Sec. II). In certain cases cyclic voltammetric techniques can be used for differentiating between these types of action [35], in that one adds incremental amounts of the reagent to a system which is reversible on the time scale of cyclic voltammetry in the absence of the added reagent. By determining the dependence of the magnitude of the reduction current for the cation radical on the concentration of the added reagent, one can estimate relative reactivities of different reagents toward the cation radical. In essence, one determines the concentration of the reagent necessary to eliminate the reduction current completely. The scheme below gives relative reactivities for a series of dimethylpyridines of differing steric requirements toward the cation radicals of DPA and 9, 10-dimethylanthracene (DMA):

Cation radical	3, 5-Lutidine	2, 5-Lutidine	2, 6-Lutidine
$DPA^{\dot{+}}$	37	6.5	1.0
$DMA^{\dot{+}}$	1.7	1.3	1.0

$DMA^{\dot{+}}$ shows a small difference only in reactivity toward the three reagents, indicating that the attack of the reagent is on a proton of one of the methyl groups, a relatively unhindered position, i.e., it acts as a base. This is supported by the isolation of a dimer from the CPE runs [Eq. (17)]. On the other hand, in the reaction with $DPA^{\dot{+}}$ only nucleophilic action is possible and hence a fairly large difference in reactivity between the sterically unhindered 3, 5-lutidine and the sterically hindered 2, 6-lutidine for attack on the positive 9-carbon atom is observed.

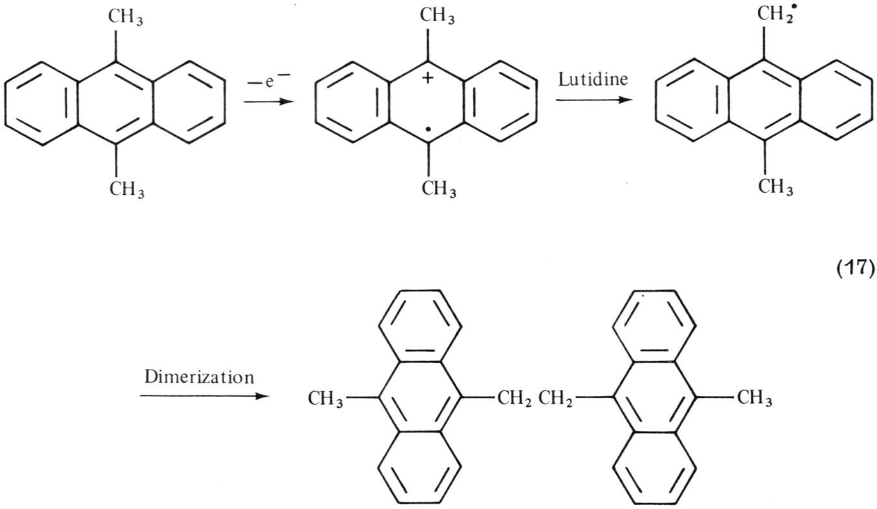

(17)

The situation is, however, not always as simple as in the cases described above. 9,10-Dimethylanthracene undergoes a one-electron transfer oxidation at the potential of the first, reversible peak of the cyclic polarogram in the presence of acetate ion with formation of a dimeric product (III). At the potential of the second, irreversible peak, a two-electron oxidation product (IV) is formed, and it can be shown that in this case the dication must be the product-forming species, e.g., an EEC mechanism [36]:

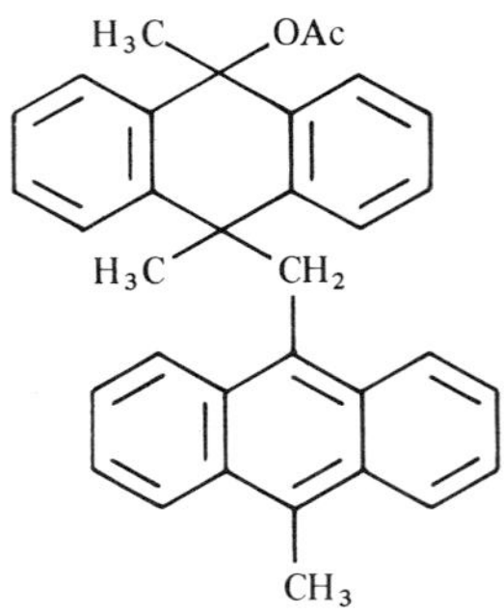

(III)

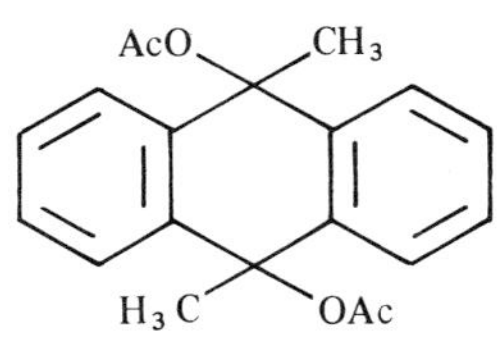

(IV)

Concluding, strong evidence for both the ECE and EEC mechanisms in anodic addition processes of anodic hydrocarbons has been advanced. It has been proposed that the ECE mechanism should be the more general mechanism of the two [34] in both addition and substitution reactions; verifications of this postulate are awaited with great interest.

C. Coupling Reactions

For coupling processes one would imagine that the mechanistic problem would be less complicated since they take place via a one-electron oxidation and therefore radical coupling would be an attractive mechanism. Unfortunately, the situation is not that simple, as is shown by the different reactions taking place in the anodic oxidation of methylsubstituted aromatic hydrocarbons in methylene chloride-tetrabutylammonium tetrafluoroborate [37], a medium of such a low nucleophilicity that the substrate itself is the strongest nucleophile present. In the case of durene the major product is the diphenylmethane derivative (V), presumably via the 2, 4, 5-trimethylbenzyl cation formed in an ECE process [Eq. (18)]. Neither a bibenzyl nor a biphenyl derivative is formed.

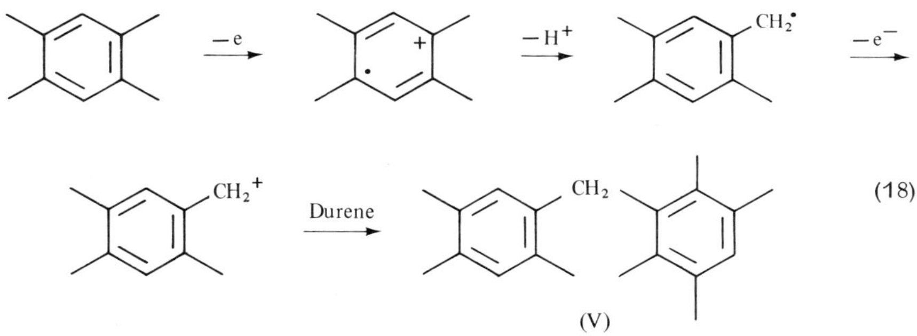

(18)

(V)

The possibility that (V) is formed via an attack by the 2, 4, 5-trimethylbenzyl radical on durene is ruled out by the fact that the yield of (V) is strongly dependent on the nucleophilicity of the medium employed, as shown in the scheme on the next page [Eq. (19)].

A mechanism involving the carbonium ion as a common intermediate is the most likely one for these reactions, since a radical reaction would be expected to be little affected by changes in the nucleophilicity of the medium.

On the other hand, mesitylene on oxidation in methylene chloride-tetrabutylammonium tetrafluoroborate gives the biphenyl derivative (VI) as the major product, no diphenylmethane derivative being formed. In this case the mesitylene cation radical probably attacks mesitylene to form a dimeric cation radical, which undergoes a second electron transfer and subsequently loses two protons [Eq. (20)].

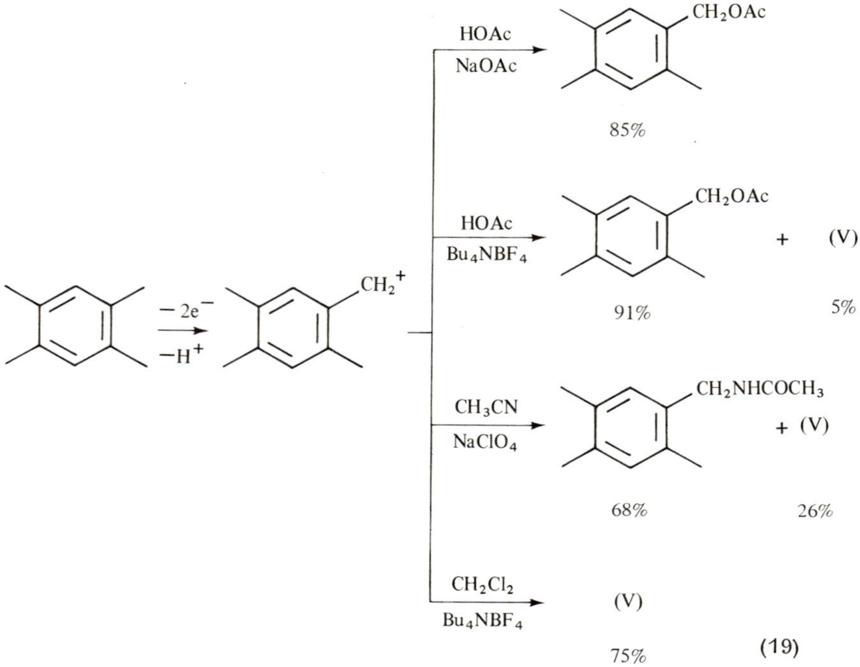

(19)

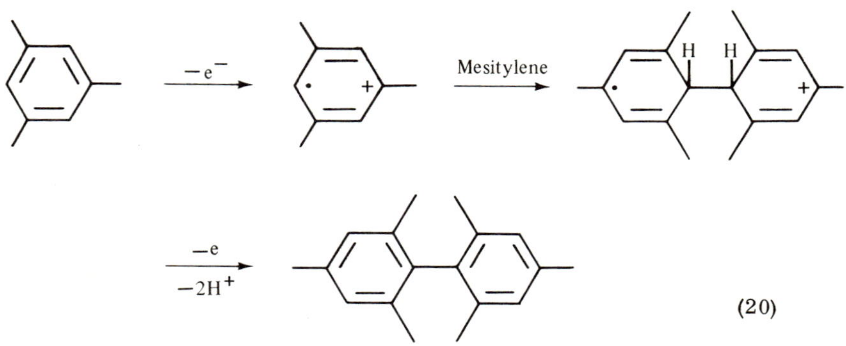

(20)

(VI)

Attack of cation radical on mesitylene rather than dimerization of two cation radicals is preferred since the yield of (VI) increases appreciably with increasing concentration of mesitylene but is not affected by changes in current density. A similar mechanism probably operates in the electro-oxidation of benzene, in which a polymerization process takes place [31], and in the anodic polymerization of aromatic hydrocarbons in the presence of hydrogen chloride-aluminum chloride [38]. However, it should be pointed out that some authors prefer to formulate the formation of dimer as a coupling reaction between two cation radicals, as in the formation of 6, 6′-bibenzo[a]pyrenyl from benzo[a]pyrene [24a].

The anodic oxidation of anthracene in acetonitrile-sodium perchlorate leads to the formation of a dimeric product, bianthrone (VII), in addition to anthraquinone [39]. Bianthrone is the major product when the electrolysis is carried out at low water concentration (<50 mM) and at a low anode potential (1.0 V vs. Ag/0.1 M Ag^+); addition of an excess of an aliphatic

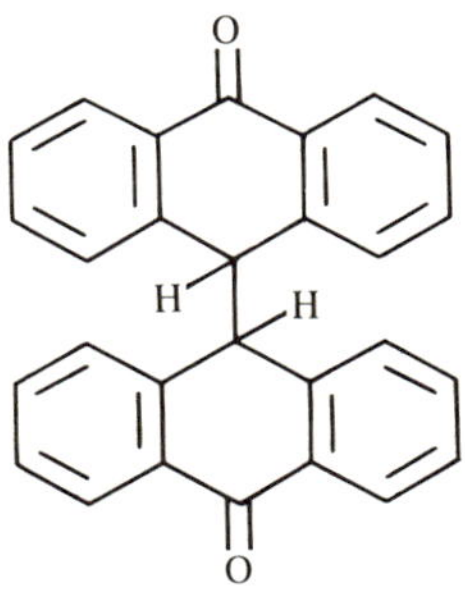

(VII)

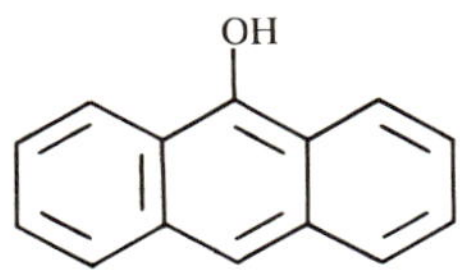

(VIII)

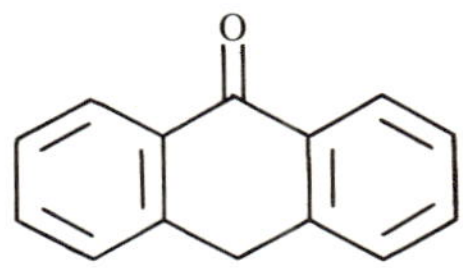

(IX)

alcohol gives identical results. At higher water concentrations and/or at a higher anode potential (1.4 V) the amount of anthraquinone increases. Coulometric studies under conditions favoring the formation of bianthrone have given conflicting results, in that both two- [39] and three-electron [40] oxidations have been observed. The two-electron oxidation was explained by formation of 9-anthranol (VIII) as the primary electrolysis product, which then would undergo air oxidation [as its tautomer, anthrone (IX)] during work-up to form bianthrone. The three-electron oxidation process appears more likely, however, since it was found that anthrone is not converted to bianthrone during work-up and that bianthrone actually is the product directly isolable from the electrolysis mixture [40].

An interesting intramolecular coupling reaction takes place in the anodic oxidation of tetraphenylethylene in acetonitrile-tetraethylammonium perchlorate [41]. This compound displays three oxidation peaks with peak potentials at 1.35, 1.55, and 1.70 V (vs. Ag/0.1 M Ag^+), the first peak corresponding to a two-electron transfer. CPE at the potential of the first peak produces 9, 10-diphenylphenanthrene (X), which in separate experiments was shown to give rise to the second and third wave. Cyclic voltammetry of tetraphenylethylene shows that the first peak can be separated into two peaks at 1.25 and 1.40 V, respectively, the first one associated with a cathodic peak but not the second one. An EEC mechanism [Eq. (21)] is proposed for the formation of the phenanthrene derivative:

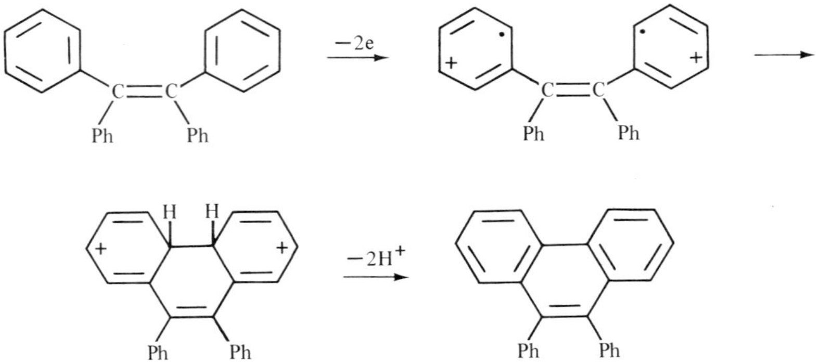

(X)

(21)

Coupling accompanied by addition is a novel and interesting reaction [42, 43], taking place in the electrolysis of suitably substituted alkenes (vinyl ethers, arylalkenes, butadiene) in an electrolyte consisting of methanol-sodium methoxide or methanol-sodium iodide. The reaction is favored by the use of a carbon anode [42]. This reaction involves the rather common situation that the supporting electrolyte is discharged at lower potentials than the half-wave potential of the organic substrate, so that one would immediately suspect an indirect oxidation mechanism. Nevertheless, the mechanism has been postulated to occur via the cation radical of the substrate, formed in a direct process [Eq. (22)]. No electrochemical evidence for this is available yet:

$$\mathrm{RCH{=}CH_2} \xrightarrow{-e^-} \mathrm{R\overset{+}{C}H{-}\dot{C}H_2} \xrightarrow{\text{Dimerization}}$$

$$\mathrm{R\overset{+}{C}H{-}CH_2{-}CH_2{-}\overset{+}{C}HR} \xrightarrow{\text{MeOH}} \mathrm{R\underset{}{\overset{OMe}{\overset{|}{C}}}H{-}CH_2{-}CH_2{-}\overset{OMe}{\overset{|}{C}}HR} \qquad (22)$$

D. Indirect Oxidations

The situation referred to above, that the supporting electrolyte and/or the solvent is oxidized at a lower potential than the organic substrate, is frequently encountered. Anodic methoxylation and cyanation are two typical cases for which homolytic processes involving methoxy and cyano radicals, respectively, have been invoked [44, 45]. However, it can be shown that in cyanation [46-48] and at least some cases of methoxylation [49] one must work at an anode potential around or higher than the half-wave potential of the organic substrate in order to get any substitution product. At lower potentials, current passes through the electrolyte but no substitution takes place. In some cases this method does not give unequivocal results, as for example in the side-chain acetoxylation of alkylaromatic compounds in acetic acid-tetramethylammonium nitrate [50, 51]. Actually both the direct and the indirect mechanisms appear to operate simultaneously (inferred from product studies).

E. Conclusions

In most cases, hydrocarbon oxidations occur via an initial one-electron transfer to the anode to form a cation radical. We have not touched upon the problem of the role of adsorption in this process, but available data suggest that both aliphatic and aromatic hydrocarbons are adsorbed at platinum surfaces in a step preceding electron transfer [1, 52]. In aqueous solutions, the adsorption free energies appear to fall in the range of -(6-9) kcal/mole for aromatic hydrocarbons; it is also suggested that the aromatic hydrocarbon lies flat on the electrode surface, possibly via π-bonding [53, 54].

The further fate of the cation radical depends on its reactivity toward other reagents present. By suitable blocking of reactive sites and/or extensive delocalization of the positive charge [7] the cation radical may be stable for long periods or at least during the time scale of the electrochemical experiment. In the majority of cases, the cation radical is very reactive as an electrophile or proton donor and interacts with nucleophiles or bases present; cases are also known where it can react as a radical, i.e., in a dimerization process. The factors which govern the reactivity of cation radicals are little known; the kinetic study of the reactions of cation radicals, both at the electrode and in homogeneous systems, should be a very fertile area for research.

REFERENCES

[1]. B. J. Piersma and E. Gileadi, in Modern Aspects of Electrochemistry (J. O'M. Bockris, ed.), Vol. 4, Plenum, New York, 1966, Chap. 2.

[2]. F. Fichter, Organische Elektrochemie, Steinkopf, Dresden and Leipzig, 1942.

[3]. M. J. Allen, Organic Electrode Processes, Reinhold, New York, 1958.

[4]. N. L. Weinberg and H. R. Weinberg, Chem. Rev., 68, 449 (1968).

[5]. H. Lund, Acta Chem. Scand., 11, 1323 (1957).

[6]. M. E. Peover, Electroanal. Chem., 2, 1 (1967).

[7]. R. N. Adams, Accounts Chem. Res., 2, 175 (1969); Electrochemistry at Solid Electrodes, Dekker, New York, 1969, Chap. 10.

[8]. H. J. Shine and Y. Murata, J. Amer. Chem. Soc., 91, 1872 (1969); Y. Murata and H. J. Shine, J. Org. Chem., 34, 3368 (1969); however, see V. D. Parker and L. Eberson, J. Amer. Chem. Soc., 92, 7488 (1970).

[8a]. R. Dietz and B. E. Larcombe, J. Chem. Soc. (B), 1970, 1369.

[9]. M. Szwarc, Progr. Phys. Org. Chem., 6, 323 (1968).

[10]. N. N. Semenov, Some Problems in Chemical Kinetics and Reactivity, Vol. 1, Princeton Univ. Press, Princeton, N.J., 1958, p. 198.

[11]. J. W. Loveland and G. R. Dimeler, Anal. Chem., 33, 1196 (1961).

[12]. W. C. Neikam, G. R. Dimeler, and M. M. Desmond, J. Electrochem. Soc., 111, 1190 (1964).

[13]. W. C. Neikam and M. M. Desmond, J. Amer. Chem. Soc., 86, 4811 (1964).

[14]. E. S. Pysh and N. C. Yang, ibid., 85, 2124 (1963).

[15]. G. J. Hoijtink, Rec. Trav. Chim., 77, 555 (1958).

[16]. R. Zahradník and V. Párkányi, Talanta, 12, 1289 (1965).

[17]. For a summarizing discussion, see L. Eberson and K. Nyberg, J. Amer. Chem. Soc., 88, 1686 (1966).

[18]. L. Eberson and B. Olofsson, Acta Chem. Scand., 23, 2355 (1969).

[19]. K. Nyberg, Chem. Commun., 774 (1969).

[20]. R. E. Visco and E. A. Chandross, J. Amer. Chem. Soc., 86, 5350 (1964).

[21]. P. A. Malachesky, L. S. Marcoux, and R. N. Adams, J. Phys. Chem., 70, 2064 (1966).

[22]. R. E. Sioda and W. S. Koski, J. Amer. Chem. Soc., 87, 5573 (1965).

[23]. R. E. Sioda, J. Phys. Chem., 72, 2322 (1968).

[23a]. L. S. Marcoux, A. Lomax, and A. J. Bard, J. Amer. Chem. Soc., 92, 243 (1970).

[24]. M. E. Peover and B. S. White, J. Electroanal. Chem., 13, 93 (1967).

[24a]. L. Jeftic and R. N. Adams, J. Amer. Chem. Soc., 92, 1332 (1970).

[25]. L. S. Marcoux, J. M. Fritsch, and R. N. Adams, ibid., 89, 5766 (1967).

[26]. J. Phelps, K. S. V. Santhanam, and A. J. Bard, ibid., 89, 1752 (1967).

[27]. A. E. Coleman, H. H. Richtol, and D. A. Aikens, J. Electroanal. Chem., 18, 165 (1968).

[28]. V. D. Parker, ibid., 21, App. 1-3 (1969).

[29]. D. Bauer, J.-P. Beck, and P. Texier, Compt. Rend. Acad. Sci. Paris, C, 266, 1335 (1968).

[30]. B. J. Woodhall and G. R. Davies, Abstracts, Meeting of the Society for Electrochemistry, England, April 15-16, 1969, p. 26.

[31]. T. Osa, A. Yildiz, and T. Kuwana, J. Amer. Chem. Soc., 91, 3994 (1969).

[31a]. M. Fleischmann and D. Pletcher, J. Electroanal. Chem., 25, 449 (1970).

[32]. D. H. Geske, J. Amer. Chem. Soc., 81, 4145 (1959).

[33]. M. Fleischmann and D. Pletcher, Tetrahedron Lett., 6255 (1968).

[33a]. J. Bertram, M. Fleischmann, and D. Pletcher, ibid., 549 (1971).

[34]. G. Manning, V. D. Parker, and R. N. Adams, J. Amer. Chem. Soc., 91, 4584 (1969).

[35]. V. D. Parker and L. Eberson, Tetrahedron Lett., 2839 (1969).

[36]. V. D. Parker, Chem. Commun., 848 (1969); cf. also V. D. Parker and L. Eberson, ibid., 340 (1969).

[37]. K. Nyberg, Acta Chem. Scand., 24, 1609 (1970); L. Eberson and K. Nyberg, Preprints, Symposium on Electrochemical Conversion of Hydrocarbon-Based Raw Materials, National Meeting, American Chemical Society, Chicago, September 1970, p. B7.

[38]. N. E. Wisdom, Jr., Extended Abstracts, Meeting of The Electrochemical Society, New York, May 4-9, 1969, p. 338.

[39]. E. J. Majeski, J. D. Stuart, and W. E. Ohnesorge, J. Amer. Chem. Soc., 90, 633 (1968); E. J. Majeski, J. D. Stuart, and W. E. Ohnesorge, Chem. Commun., 353 (1970).

[40]. V. D. Parker, ibid., 1131 (1969); L. Eberson and V. D. Parker, ibid., 1290 (1970); V. D. Parker, Acta Chem. Scand., 24, 2757 (1970).

[41]. J. D. Stuart and W. E. Ohnesorge, Extended Abstracts, Meeting of The Electrochemical Society, New York, May 4-9, 1969, p. 305.

[42]. H. Schäfer and E. Steckhan, Angew. Chem. (Intern. Ed.), 8, 518 (1969); H. Schäfer, Chem. Ing. Tech., 42, 164 (1970).

[43]. B. Belleau and Y. K. Au-Young, Can. J. Chem., 47, 2117 (1969).

[44]. K. Sasaki, H. Urata, K. Uneyama, and S. Nagaura, Electrochim. Acta, 12, 137 (1967).

[45]. K. Koyama, T. Susuki, and S. Tsutsumi, Tetrahedron, 23, 2675 (1966); T. Inoue, K. Koyama, T. Matsuoka, and S. Tsutsumi, Bull. Chem. Soc. Japan, 40, 162 (1967).

[46]. V. D. Parker and B. E. Burgert, Tetrahedron Lett., 4065 (1965).

[47]. L. Eberson and S. Nilsson, Discuss. Faraday Soc., 45, 242 (1968).

[48]. S. Andreades and E. W. Zahnow, J. Amer. Chem. Soc., 91, 4181 (1969).

[49]. N. L. Weinberg and T. B. Reddy, ibid., 90, 91 (1968); N. L. Weinberg, J. Org. Chem., 33, 4326 (1968).

[50]. S. D. Ross, M. Finkelstein, and R. C. Petersen, ibid., 35, 781 (1970), and previous papers of this series.

[51]. K. Nyberg, Acta Chem. Scand., 24, 473 (1970); 25, 3246 (1971).

[52]. B. J. Piersma, in Electrosorption (E. Gileadi, ed.), Plenum, New York, 1967, Chap. 2.

[53]. H. Dahms and M. Green, J. Electrochem. Soc., 110, 1075 (1963).

[54]. J. O'M. Bockris, M. Green, and D. A. J. Swinkels, ibid., 111, 743 (1964).

Chapter XIII

CARBOXYLIC ACIDS

L. Eberson

Division of Organic Chemistry
Chemical Center
University of Lund
Lund, Sweden

I. INTRODUCTION

The anodic oxidation of a carboxylic acid in the form of a solution of an alkali metal salt in a suitable solvent is the oldest and probably most useful electrochemical synthesis procedure in organic chemistry. Commonly denoted the Kolbe electrosynthesis, it makes possible the preparation of the coupling product R-R from $RCOO^-$ [Eq. (1)], as well as the unsymmetrical coupling product in addition to the two possible symmetrical ones from

$$2RCOO^- \rightarrow R\text{-}R + 2CO_2 + 2e^- \qquad (1)$$

a mixture of two acids. As will be seen later, the restrictions imposed on the choice of R are relatively lenient, allowing for a wide variety of functional groups to be present in the starting material without serious interference.

Up until the end of the 1950s, the Kolbe reaction was predominantly used for the preparation of the coupling product [1-6] which was assumed to be formed via a radical coupling process of anodically generated radicals. Without anticipating the later discussion of the detailed reaction mechanism, the mechanism can be pictured as a discharge of the carboxylate ion followed by decarboxylation of the intermediate acyloxy radical:

$$RCOO^- \rightarrow RCOO\cdot + e^- \qquad (2)$$

$$RCOO\cdot \rightarrow R\cdot + CO_2. \qquad (3)$$

The radicals formed can then undergo the usual reactions: coupling to form R-R, disproportionation to form RH and $RH - H_2$, attack on C-H bonds present in substrate or solvent to form RH, and hydrogen atom abstraction to form $RH - H_2$. In many cases RH and $RH - H_2$ were indeed found as by-products, as demanded by this mechanism.

However, it was noted very early that other by-products could be isolated, sometimes as the major components of the reaction mixture. These were alcohols, ROH, and esters, ROOCR, formed on anodic oxidation of $RCOO^-$ in aqueous solution (the Hofer-Moest reaction), and it was also noted that the carbon skeleton of the alcohol and the alcohol portion of the ester had sometimes undergone partial or complete rearrangement during electrolysis. Analogous phenomena were encountered when the electrolysis was performed in other solvents, such as methanol and acetic acid, in which ethers, ROMe, and acetates, ROAc, could sometimes be isolated. These findings led Walling [7] to suggest that the anodically formed radicals could undergo a further oxidation to form carbonium ions, from which the above-mentioned products could be derived [Eqs. (4a) and (4b)]:

$$R\cdot \rightarrow R^+ + e^- \tag{4a}$$

$$R^+ \xrightarrow[B]{Nu} R\text{-}Nu^+ + RH - H_2, \tag{4b}$$

where Nu is a nucleophile and B is a base.

This idea was simultaneously and independently conceived in other laboratories [8, 9] and decisive experimental verification soon followed [9]. It was quickly recognized that many of the puzzling phenomena previously encountered could be rationalized on the basis of this assumption, thus removing a number of less satisfactory radical mechanisms. Much recent work on the anodic oxidation of carboxylates has been centered on the problem of finding preparative uses for the carbonium ion pathway and on studying the properties of anodically generated carbonium ions as compared to those formed by other means. This development has been summarized in a recent review article [10], covering the literature through the first half of 1966. A useful tabulation of "non-Kolbe" oxidations of carboxylic acids can be found in a review article by Weinberg and Weinberg [11]. The denotation "non-Kolbe" here refers to the formation of products other than the coupling product(s).

The term "non-Kolbe" oxidation is rather unfortunate and will not be used any further in this chapter. Instead, a Kolbe reaction will be defined as any anodic process, in which a carboxylate is the electroactive species, irrespective of whether the intermediates formed are predominantly radical or cationic in character. Since these two aspects of the process are usually inseparable, there is no need for a differentiation.

II. EXPERIMENTAL FACTORS

Since the possibility of two competing pathways in the Kolbe reaction exists, the immediate problem to answer is whether one can influence the direction of a given process by proper manipulation of experimental variables so that either the radical or carbonium ion mechanism is favored. As we shall see, unidirectional selectivity can easily be achieved, in that one can always find conditions to favor the cationic mechanism; on the other hand, in many cases, because of structural features in the substrate, it may be impossible to realize the radical pathway even under the most favorable experimental conditions.

A. Anode Potential and Current Density

The Kolbe process exhibits a very characteristic log [current density]-anode potential curve in both aqueous and nonaqueous solvents [10, 12-16]. In an aqueous solution of a carboxylate, starting at a low anode potential, there is first a linear part corresponding to the oxygen evolution process (from oxidation of water) up to a potential of about +2.1 V (NHE). At this value, the anode potential suddenly jumps by some 0.3 V to another, high-potential region of linear log i-E dependency, in which the oxygen evolution reaction is almost completely blocked and the Kolbe reaction takes place almost exclusively (Fig. 1). The same behavior is found in nonaqueous

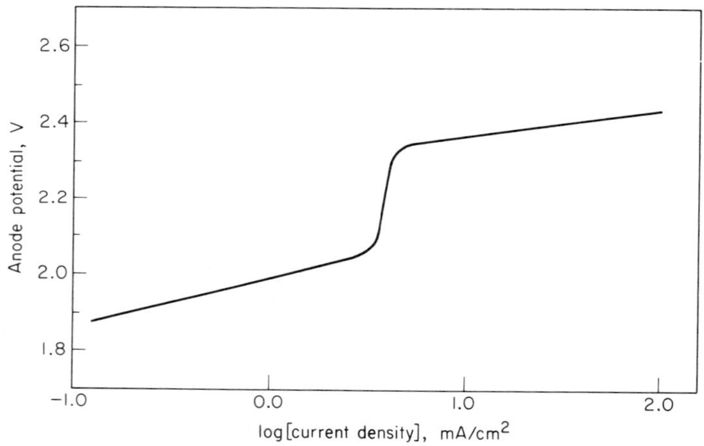

Fig. 1. Anode potential (vs. NHE) as a function of log[current density] for the electrolysis of 0.5 M aqueous sodium acetate [13].

media [14, 16]. The potential at which the Kolbe reaction starts is commonly referred to as the critical potential. Critical potentials for a number of different carboxylates have been compiled [10] and it appears that it generally has a value between 2.1 and 2.8 V (NHE), no correlation between critical potential and structure of the substrate being discernible.

Thus, the most characteristic features of the Kolbe reaction are (1) it takes place at a very high anode potential, independent of the structure of the acid; and (2) it almost completely suppresses a reaction that normally would have taken place at considerably lower potentials, namely the oxygen evolution process. The latter feature is exhibited by some other organic electrode processes, e.g., the anodic oxidation of mercury(II)-olefin complexes in aqueous perchloric acid [17] and of N, N-dimethylbenzylamine in methanol-sodium methoxide [18]. It is generally accepted that this type of behavior indicates that an adsorption process is taking place, which leads to a blocking of the low-potential reaction (see also Sec. III.B).

Looking at the simple mechanistic picture of Eqs. (2) and (3), it appears reasonable that a high current density should favor coupling between radicals since this reaction is second order in [R•]. This is indeed also the rule found empirically. Hence, if one is interested in preparing the coupling product, one should run the Kolbe oxidation at as high a current density as possible, greater than 0.25 and often up to 1.0 A/cm^2. It is obvious from the electrochemical characteristics that no anode potential control (or even check) is necessary since at current densities around 1 mA/cm^2 the critical potential is already exceeded.

Since the discharge of a carboxylate requires a very high anode potential which is largely independent of the structure of the substrate, it can be predicted [19] that the further fate of the radical R• must depend on its resistance toward further oxidation [Eq. (4a)]. Even if all other experimental conditions are chosen to favor the formation of the coupling product, there is nothing we can do about the high anode potential necessary to discharge the carboxylate ion and, therefore, an easily oxidized radical will not survive long enough to couple but instead will be further oxidized. This reasoning is based on the simple mechanism outlined in Eqs. (2), (3) and (4a) but pertains, albeit in different terms, to other possible mechanisms also.

B. Electrode Material

The choice of anode material offers one possibility of at least unidirectional control of the mechanism of the Kolbe reaction. In aqueous solution, only smooth platinum and iridium or carbon anodes will support the discharge of carboxylates. If the structure of the acid is such that a coupling product can be formed, a platinum or iridium anode must be chosen to get an optimum yield of the coupling product; on the other hand, at a carbon anode, many carboxylates give products derived almost exclusively from carbonium ions [20-24]. However, exceptions to this behavior have been described [25, 26]. In nonaqueous solvents, the choice of electrode material is less critical, although carbon anodes still favor the carbonium ion mechanism strongly, and platinum the radical mechanism [20, 24]. Lead dioxide appears to behave similarly to carbon in acetate oxidation [27], but further experimentation is necessary to find out how general this behavior is [28]. The same applies to the claim that glassy carbon gives as good results as does platinum in Kolbe oxidations [27].

The nature of the cathode material is not critical in Kolbe reactions. Since the electrolysis is normally run with a large excess of the acid present, the cathode process is reduction of protons to form hydrogen. In most cases neither starting material nor products are reducible at less negative potentials than that required for hydrogen evolution, so it is not necessary to use divided cells. In doubtful cases, however, check runs in a divided cell are recommended (as for example in the oxidation of α-bromo acids [29]).

C. Medium Effects

Much of the earlier work on the Kolbe reaction centered around the problem of optimizing the yield of coupling product in aqueous solution runs [1-4]. The formation of coupling product is favored in slightly acidic solution, whereas alkaline electrolytes tend to promote the carbonium ion mechanism. In order to keep the electrolyte acidic during electrolysis, the reaction is simply performed in an aqueous solution of the acid, neutralized by an alkali metal hydroxide to an extent of 2-5%. Since hydroxyl ions are formed at the cathode at the same rate as carboxylate ions are consumed at the anode, the concentration of carboxylate ion will remain approximately constant during the whole run. The pH of the electrolyte will be on the slightly acidic side during most of the electrolysis time due to the excess of acid present, but will increase rapidly toward the end of the run, thus signaling the end point of the reaction. This procedure is denoted the salt-deficit method.

In cases where it is necessary to have all of the acid in the carboxylate form but still desirable to prevent the electrolyte from becoming alkaline, a mercury cathode can be used for binding discharged alkali metal ions as an amalgam [30].

However, except in cases of mechanistic studies, water is seldom used as a solvent in Kolbe oxidations nowadays since aqueous-organic or non-aqueous solvents are much simpler and more advantageous to use. In the first place, solubility problems largely vanish so that the salt-deficit method can be used without difficulty; in the second place, nonaqueous solvents in many cases strongly favor the formation of the coupling product (provided the proper anode material is employed). As solvents of choice for the coupling reaction methanol [2-4] and DMF [31, 32] should be mentioned as the best ones, although the latter solvent has the disadvantage of being oxidized according to Eq. (5) at fairly low potentials [33-36]:

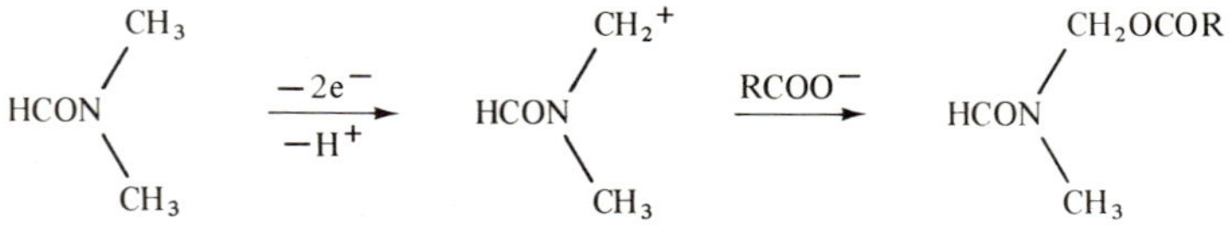

(5)

Acetonitrile can also be used, although the low conductivity and solubility of carboxylates in the anhydrous solvent sometimes necessitate the addition of water to the electrolyte [26]. In acetonitrile, the carbonium ion pathway leads to the formation of N-alkylacetamides [26, 37, 38], an electrochemical analog to the Ritter reaction [39].

In some cases a carboxylate can be electrolyzed in the corresponding anhydrous acid, but this method is of course limited to short-chain fatty acids. Acetic acid should be useful in cases where one wants to prepare acetates via carbonium ions, especially if the acid to be electrolyzed is appreciably stronger than acetic acid. In that case the unsymmetrical coupling reaction with methyl radicals from discharge of acetate ion is suppressed since it is known that the stronger acid is preferentially electrolyzed in mixtures of acids [40]. Water-pyridine (e. g., 10:90) appears to be the solvent of choice for favoring carbonium ion formation [24, 41] even when a platinum anode is used.

D. Temperature Effects

Generally, temperature has only a small effect on the outcome of a Kolbe oxidation. In some cases a slight decrease in the yield of coupling product with increasing temperature has been noted, but it has not been established whether this is accompanied by an increase in the yield of other products. Only in rare cases have drastic temperature effects been observed [42, 43], Eq. (6) representing such an example in which a 15° change in temperature completely alters the product distribution [42]:

$$Br(CH_2)_{10}COO^- \xrightarrow[MeOH]{65^\circ} \underset{71\%}{Br(CH_2)_{10}COOCH_3} + \underset{2.4\%}{MeO(CH_2)_{10}COOCH_3}$$

$$Br(CH_2)_{10}COO^- \xrightarrow[MeOH]{50^\circ} \underset{64\%}{Br(CH_2)_{20}Br} \qquad (6)$$

E. Effect of Additives

Certain anions and cations [4] inhibit the radical path and favor the formation of carbonium ion-derived products. We have already mentioned hydroxide ion as belonging to this category, others include bicarbonate, sulfate, perchlorate, dihydrogen phosphate, and fluoride ion. A variety of cations (such as Pb^{2+}, Mn^{2+}, Cu^{2+}, Fe^{2+}, and Co^{2+}) are detrimental for the coupling process, simply because the anode is quickly covered by a layer of the corresponding metal oxide which does not support the process. (However, it seems that lead dioxide can be used for oxidation of acetate in acetonitrile [27] with formation of carbonium ions.) Alkali and alkaline earth metal ions as well as tetraalkylammonium ions do not affect the Kolbe reaction adversely and therefore are the counterions of choice in preparative applications.

Organic additives can change the course of the Kolbe reaction in several ways. In the simplest case it can undergo an oxidation process at a lower potential than that of carboxylate discharge and thus be the electroactive species in itself (anodic acyloxylation, see Chapter XXII). A more complicated and more common case is that the organic additive can serve as a substrate both for electrochemical oxidation and for attack by radicals or carbonium ions generated by the Kolbe reaction; several cases of this behavior are discussed later (Sec. V.C).

F. Conclusions

Table 1 summarizes the influence of experimental factors upon the mechanism of the Kolbe reaction. In order to obtain optimum yield of

TABLE 1

Influence of Experimental Factors on the Mechanism of the Kolbe Reaction

Formation of R•, favored by	Formation of R^+, favored by
High current density	Low current density
Smooth Pt anode	C or PbO_2 anode
Slightly acidic or neutral solution	Alkaline solution
Methanol or DMF as solvent	Water or water-pyridine as solvent
Low (room) temperature	Effect of temperature not established. Addition of ClO_4^-, HCO_3^-, $H_2PO_4^-$, SO_4^{2-}, F^-

coupling product from a given system, one should always use the conditions specified in the left column of the table; if no coupling product is obtained, the structure of the acid in all probability is such as not to allow the radical to survive further oxidation. It should also be pointed out that even under conditions most favorable for coupling, structural as well as experimental, carbonium ion formation usually competes with radical dimerization. Possible exceptions are simple systems of the type CF_3COO^-/CF_3COOH [14] and CH_3COO^-/CH_3COOH [15] in which the yield of coupling product approaches 100%.

III. MECHANISM OF THE KOLBE REACTION

A. Different Views of the Mechanism

The simplicity of the mechanistic scheme presented in the Introduction [Eqs. (2), (3), and (4a)] should not overshadow the fact that the detailed reaction mechanism of the Kolbe reaction is extremely complex and still subject to lively discussion [44-49], part of which has been summarized in recent reviews [10, 16]. The main problems concern (1) the nature of the first electrochemical step for which a one-electron oxidation of the carboxylate to form the radical and carbon dioxide in a concerted step [Eq. (7)] has been proposed [19] instead of the simple discharge step of Eq. (2); (2) the nature of the follow-up reactions, i.e., whether the initially formed radicals react in the solution or are adsorbed at the electrode surface; and (3) the nature of the process leading to carbonium ions, for which a step involving the formation of $RCOO^+$ has been invoked [Eq. (2) followed by Eq. (8)] by some authors [41, 50]:

$$RCOO^- \rightarrow R\cdot + CO_2 + e^- \tag{7}$$

$$RCOO\cdot \rightarrow RCOO^+ + e^- \rightarrow R^+ + CO_2. \tag{8}$$

Before we examine the pros and cons of the more elaborate mechanisms, it should be pointed out that the simple scheme below [Eq. (9)] provides a reasonably good tool for qualitative predictions regarding the preparative outcome of a particular Kolbe synthesis [19]:

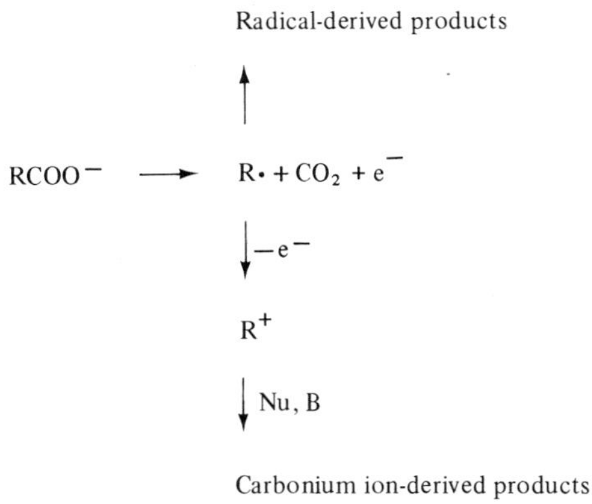

(9)

It can be shown that the dominant parameter in determining the standard potential of the hypothetical electrode process $R\cdot \rightarrow R^+ + e^-$ is the ionization potential of the radical. Making the usual assumption that rate constants are linearly correlated with thermodynamic quantities, the rate of this process must also be strongly dependent on the ionization potential. By relating ionization potentials to a representative selection of systems in which the distribution between radical- and carbonium ion-derived products is known, a useful rule of thumb, valid for Kolbe oxidations at a platinum anode with methanol as solvent, emerged: Radicals with ionization potentials above 8 eV (gas-phase values determined by the electron impact or photoionization method) should be fairly resistant toward further oxidation and give coupling product in fairly good yield; radicals with ionization potentials below 8 eV should undergo further oxidation to a large extent. Since extensive tabulations of ionization potentials are available [51, 52], it is relatively simple to make qualitative predictions on this basis [52a].

B. The Detailed Reaction Mechanism

Vijh and Conway [16] have scrutinized the different views of the detailed mechanism of the Kolbe process and concluded that a mechanism involving adsorbed radical species [Eqs. (10)-(15)] best fits in with available data. Thus, following the rate-determining discharge step, a layer of chemisorbed acyloxy radicals would be formed at the electrode surface which is probably not a "pure" platinum surface but one covered with a platinum oxide at least in aqueous solvents.[1] Decarboxylation of the acyloxy radical, possibly as a dissociative desorption step [45], would then take place, and the resulting radicals would be chemisorbed at the electrode surface, undergoing the product-forming steps — coupling, disproportionation, etc. — as adsorbed species:

$$RCOO^- \qquad RCOO^-(M) \tag{10}$$

$$RCOO^-(M) \xrightarrow{-e^-} RCOO\cdot(M) \tag{11}$$

$$RCOO\cdot(M) \rightarrow R\cdot(M) + CO_2 \tag{12}$$

$$R\cdot(M) \rightarrow \text{Radical-derived products} \tag{13}$$

$$R\cdot(M) \xrightarrow{-e^-} R^+ \tag{14}$$

$$R^+ \xrightarrow[B]{Nu} \text{Carbonium ion-derived products,} \tag{15}$$

where M is a metal site, Nu is a nucleophile, and B is a base.

The transition region observed in the log[current density]-anode potential curve (Fig. 1) would correspond to the buildup of the necessary conditions for the Kolbe process to take place (oxide formation in aqueous medium or formation of an adsorbed layer of acyloxy radicals in nonaqueous solution).

The alternative view that radicals would be formed according to a concerted mechanism [Eq. (7)] and react in the solution [Eq. (9)] has been discussed by Eberson [10]. The very close similarity with regard to product distribution between the Kolbe reaction and analogous homogeneous processes provides the basis for this view; in addition, the stereochemistry observed is more consistent with the behavior of free than adsorbed radicals. Strong evidence, based on stereochemical arguments, that $RCOO^+$ is an intermediate in the formation of carbonium ions has been put forward [41, 50].

Much new experimental material has become available in the period elapsing since the writing of the above review articles. As expected, in view of the complexity of the problem at hand, this new information is not sufficient to rule out any of the mechanisms discussed, but may provide stimuli for the design of further critical experiments. In order to test the mechanism depicted in Eq. (7), a number of aliphatic acids were electrolyzed in aqueous or methanolic alkaline medium, using a competition technique [47]. The results indicate that in water there is no significant difference among the rates of electrolysis of acids with primary, secondary, or tertiary R groups, provided the molecular weight is approximately the same in all cases, and that carboxylates with a higher molecular weight have larger rate constants than those with lower molecular weight. In methanol, no significant rate difference was found for any structural change. This was taken as evidence that the concerted mechanism [Eq. (7)] cannot be valid since this implies that the stability of the radical formed would be reflected in the energy of the transition state, thus making acids with tertiary R groups react considerably faster than those with primary ones. However, the composite nature of the relative rate constants evaluated makes them difficult to interpret [49] and the results do not necessarily rule out the concerted process [Eq. (9)].

Chronopotentiometric oxidations of cesium salts of 2, 2-dimethylpropanoic and 2, 2-dimethylpentanoic acid at a platinum anode in acetonitrile [48] show two steps, the first one being a diffusion-controlled, one-electron oxidation. The second step is not diffusion-controlled and appears to be coupled to a preceding or simultaneously occurring chemical side reaction. The first step was assumed to correspond to the formation of the acyloxy radical and/or its decarboxylation products ($R\cdot + CO_2$) and the second one to oxidation of the radical to the cation ($R\cdot \rightarrow R^+ + e^-$). Acids with less alkyl substitution at the 2-carbon atom displayed similar behavior, although the reproducibility of the second step was poor.

A kinetic study [53] of the oxidation of acetate ion at the platinum electrode in alkaline medium, i. e., under conditions favoring the formation of methanol, has been performed using a square dc pulse technique. The data obtained were best in agreement with the assumption that acetate ion is discharged to form an adsorbed methyl radical which is the common intermediate in the formation of ethane and methanol. The methanol was assumed to be formed by coupling between methyl and hydroxyl radicals (both adsorbed), the latter ones being intermediates in the concurrent oxygen evolution process. A carbonium ion mechanism was ruled out mainly on the basis that no methyl acetate formed by reaction between the methyl cation and acetate ion could be detected. However, the half-life of methyl acetate in the alkaline media employed would be in the range of between 20 and 100 sec [54], so this compound might easily have escaped detection.

Some very interesting observations were recently made [26] in a study of the oxidation of aliphatic carboxylates in acetonitrile containing 15-20% water at a carbon anode. These are optimal conditions for the formation of N-alkylacetamides via cations, as exemplified here for potassium pentanoate [Eq. (16)], in which case up to 60% of the rearranged amide (I) could be

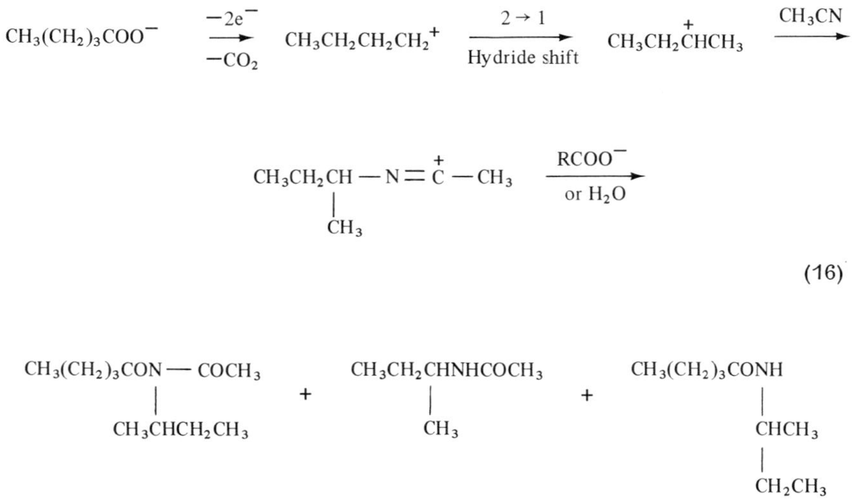

(16)

(II) (I) (III)

isolated in addition to small amounts of amides (II) and (III) as well as elimination products (isomeric butenes and methylcyclopropane). No coupling product was formed in this case. However, when attempts were made to extend this reaction to higher molecular weight fatty acids, the tendency to form coupling products increased with increasing chain length and accordingly the cationic pathway was suppressed. The same behavior is observed in Kolbe oxidations of aliphatic carboxylates at platinum anodes in aqueous solution [4]. This effect of increasing the chain length of the carboxylate is difficult to reconcile with a mechanism involving adsorbed radicals; instead the concerted mechanism [Eq. (7)] was favored.

Another interesting observation made in this study is that alcohols are not formed from oxidation of potassium pentanoate until the water content of the electrolyte reaches 20%. In view of the fact that water is a considerably stronger nucleophile than acetonitrile [55], this is a rather surprising result. Similar observations have been made [55, 56] in the anodic oxidation of aromatic hydrocarbons in mixtures of acetonitrile and water; in these systems the anion of the supporting electrolyte is of decisive importance for the partition of the intermediate carbonium ions between reaction with water or acetonitrile [56]. Hence one must conclude that the composition of the medium in the immediate vicinity of the electrode surface is different from that of the bulk of the solution.

The suggestion that $RCOO^+$ might be an intermediate in the formation of carbonium ion-derived products was recently put forward by Gassman and Zalar [41] on the basis of observation of partial stereospecificity in the electrolysis of cis- and trans-bicyclo[3.1.0]hexane-3-carboxylic acid to form the corresponding 2-alcohols [Eqs. (17) and (18)]:

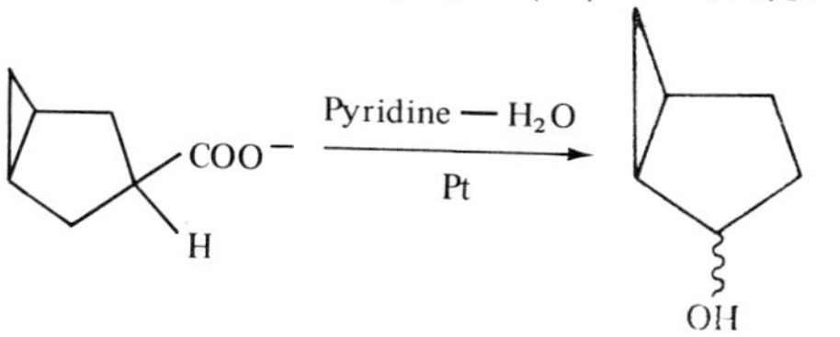

cis-trans ratio = 1.20

(17)

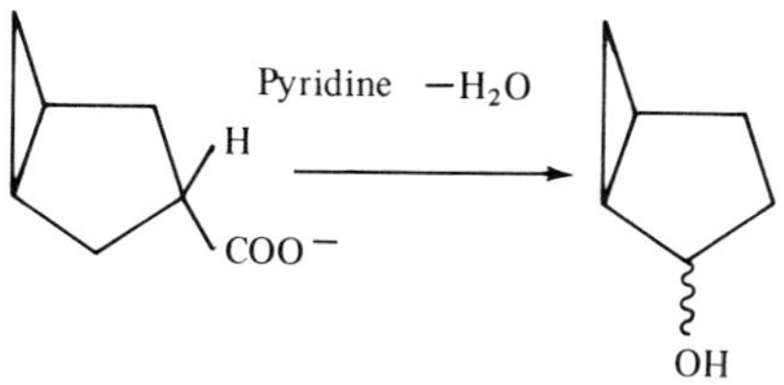

cis-trans ratio = 0.84

(18)

This was interpreted in terms of formation of acyloxonium ions which would undergo a concerted hydrogen-bridging carbon dioxide loss [Eqs. (19) and (20)]:

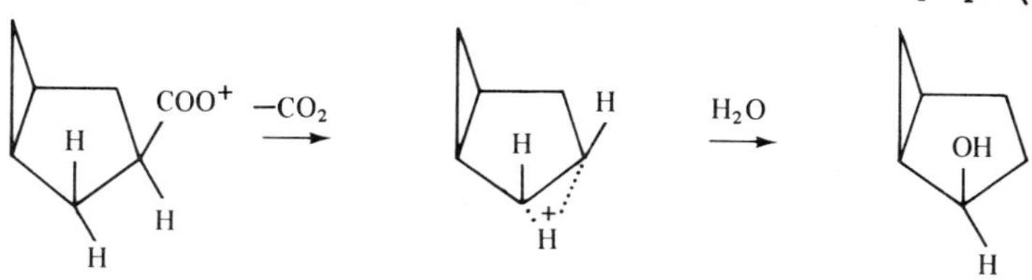

(19)

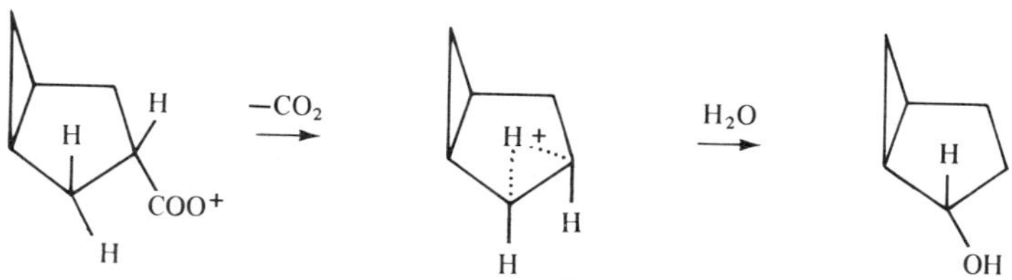

(20)

Although this interpretation is not wholly unambiguous [10], it has received support by some recent results from the anodic oxidation of substituted cyclopropanecarboxylic acids. Anodic oxidation of carboxylates (IV) and (V) produced a complex mixture of products, among which the allylic ethers (VI) and (VII) and (VIII), respectively, were formed in a concerted stereospecific opening of the cyclopropane ring [Eqs. (21) and (22)]:

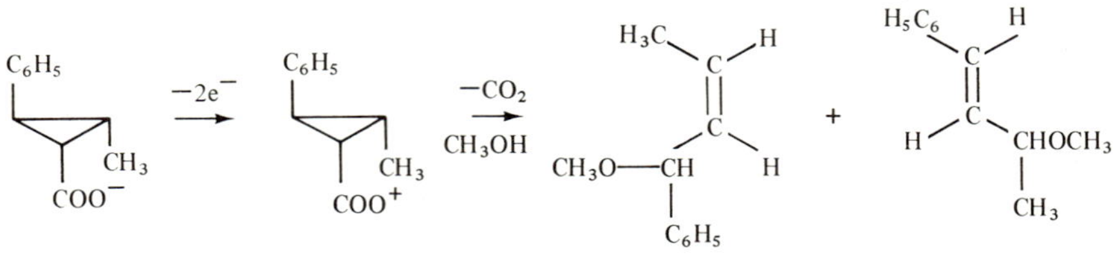

(IV) (VI) (VII) (21)

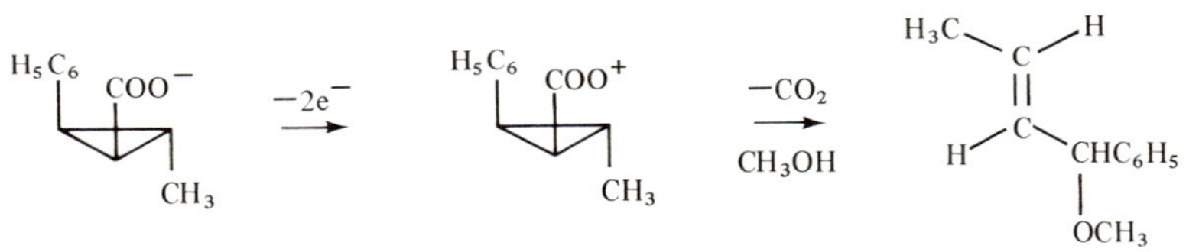

(V) (VIII) (22)

The possibility that the stereochemistry might be explained by retention of the configuration of adsorbed cyclopropyl radicals was rejected on the basis of other observations regarding the stereochemistry of the Kolbe reaction (see Sec. IV.A). Similar results were obtained in the oxidation of cis- and trans-2-phenylcyclopropane-1-carboxylic acid.

Concluding, many problems related to the mechanism of the Kolbe reaction are presently under active discussion, presumably to a certain extent due to the difference in emphasis placed upon the problems by different groups of chemists. For organic chemists, the simpler mechanisms involving reactions between species in solution are sufficient to interpret product distributions, whereas physical chemists require the assumption of adsorbed species as intermediates to interpret electrochemical measurements. Eventually, the two points of view will no doubt merge into a commonly accepted mechanism; this process is likely to take place at an accelerated rate as soon as organic chemists start to find results which cannot be explained in terms of the simple mechanisms. Such results are most likely stereochemical in nature. Another interesting aspect of the Kolbe reaction is that ESR investigations have never given any positive evidence for radical intermediates [16]. This may mean either that adsorbed radicals are involved in the electrode reaction or that the systems hitherto investigated give too short-lived radicals. Here systems designed to give more stable radicals should give valuable information regarding the mechanism.

IV. STEREOCHEMISTRY OF THE KOLBE REACTION

A. Radical Reactions

The stereochemistry of the Kolbe reaction is of interest in connection with the problem of adsorbed vs. free radical intermediates in the product-forming reactions. It appears reasonable to predict that an optically active carboxylate R^*COO^-, in which the optically active center is adjacent to the carboxylate function, should give adsorbed radicals $R^{*\bullet}$ with at least partial retention of configuration [45]. The coupling reaction, if it occurs at the electrode surface, would give a dl-isomer of the coupling product with partial retention of configuration (in addition to the meso isomer). However, the few experiments carried out with optically active acids (ethyl hydrogen methylethylmalonate [57] and 2-methyl-butanoic acid [58]) indicate that racemization takes place, which would be in accordance with a mechanism involving radicals reacting in solution. The same conclusion was reached on the basis of a study of the electrolysis of cis- and trans-methyl hydrogen cyclohexane-1, 2-dicarboxylate [59]. It has, however, been pointed out [16] that the microscopically rough nature of the electrode surface could easily account for these results. Moreover, bearing in mind the structures of these systems, it is possible that some optical activity in the products might have escaped detection. Clearly, more experiments are needed to clarify this problem.

In homogeneous solution reactions, allylic radicals have a defined stereochemistry, the configuration around the double bond in the starting material being largely retained in the product(s). Allylic radicals generated by Kolbe electrolysis are no exceptions, as for example those formed from cis- and trans-hex-3-enoic acid in methanolic solution [Eq. (23)]. The

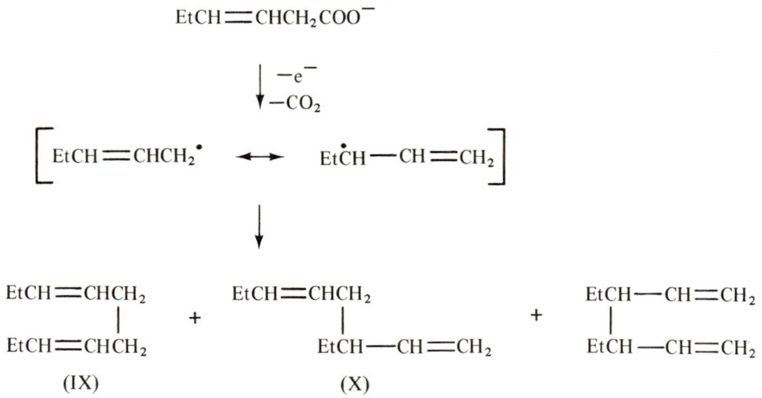

(23)

(IX) (X)

retention of configuration around the double bonds in (IX) and (X) is greater than 90%. When the same allylic radical is generated by electrolysis of $RCH(COO^-)CH{=}CH_2$ all geometrical isomers of (IX) and (X) are formed, as would be expected [60].

B. Carbonium Ion Reactions

We have already touched upon the stereochemistry of two carbonium ion processes [Eqs. (19)-(22)], both of which appear to follow a mechanism involving the formation of $RCOO^+$ or adsorbed radicals which retain their configuration during formation. Other stereochemical investigations of anodically generated carbonium ions are not unambiguous on this point. Thus, electrolysis of (+)-2-methylbutanoic acid in aqueous solution gives inactive 2-butanol and the (±)-2-butyl ester of (+)-2-methyl-butanoic acid [61], showing that racemization must have taken place at some stage of the reaction. Anodic oxidation of etther exo- or endo-norbornane-2-carboxylic acid [(XI) and (XII)] gives exo-2-norbornyl methyl ether (XIV) as the only volatile product [9]; no endo ether was detectable. Electrolysis of optically active endo acid (XII) gives the inactive ether (XIV):

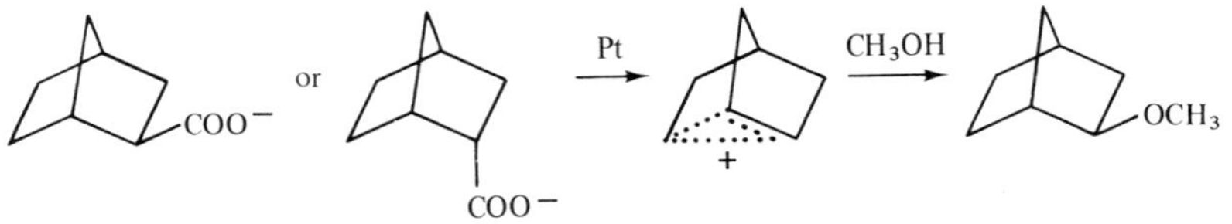

(24)

(XI) (XII) (XIII) (XIV)

Exo- and endo-5-norbornene-2-carboxylic acid [(XV) and (XVI)] give 3-methoxynortricyclene (XVIII) in 56% yield [9]:

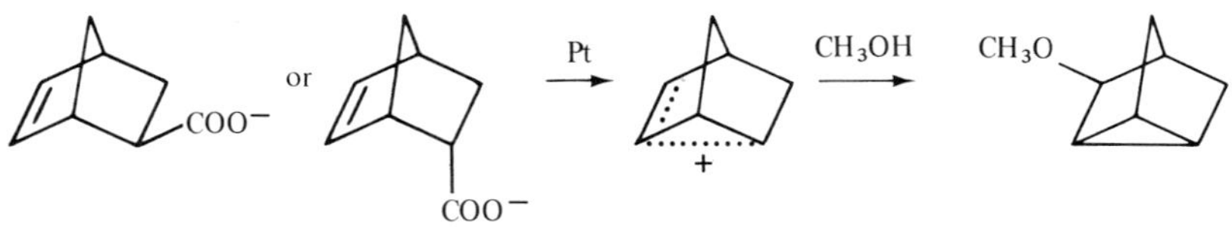

(25)

(XV) (XVI) (XVII) (XVIII)

In these two cases, participation of neighboring carbon or double bond in the formation of carbonium ions (XIII) and (XVII) is indicated, although it cannot be excluded that an initially formed, poorly solvated, classical ion is long-lived enough to collapse into the bridged ion.

In the electrolysis of Δ^5-cholestene-3β-carboxylic acid (XIX) in methanolic solution there is evidence that the cholesteryl radical (XX) initially formed must be free from the electrode, since the products obtained show that oxidation of the radical has taken place either from the α-side (without participation of the Δ^5 double bond) or from the β-side (with participation) to form carbonium ions (XXI) and (XXII), respectively [Eq. (26)]. Oxidation from the β-side is slightly favored [9, 62].

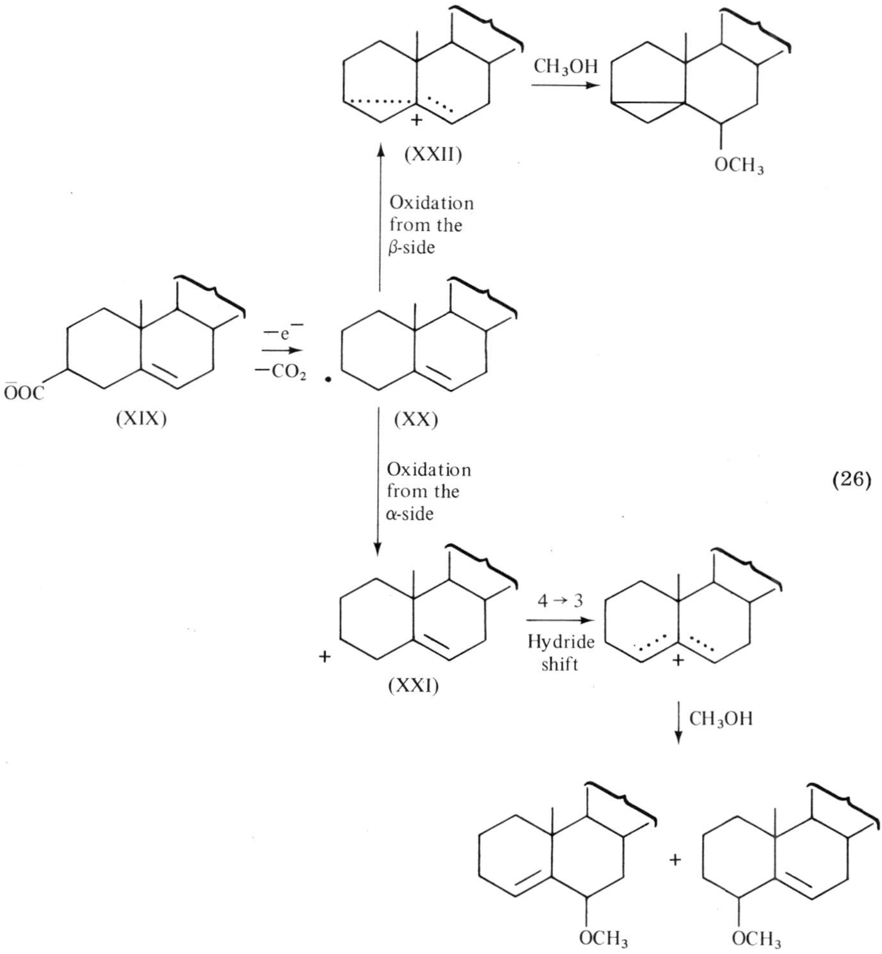

(26)

V. APPLICATIONS OF THE KOLBE REACTION

In this section we shall deal with the applications of the Kolbe reaction in organic chemistry. In general, the classical application of preparing coupling products, be it in a symmetrical or unsymmetrical fashion, is still the most important one. However, the carbonium ion route, besides its inherent interest in connection with studies on carbonium ion chemistry, has given rise to some methods of synthetic value.

A. Coupling between Radicals

The preparative uses of the coupling reaction have been summarized in several excellent treatments [2, 4, 5] and only the most important features are treated here. As a general rule, only substituents at the α-carbon of the carboxylate are of critical importance in determining the yield of the coupling product, since it is only in this position that substituents can exert a strong influence on the ionization potential of the intermediate radical and hence its sensitivity toward further oxidation (see Sec. III.A). Substituents situated further away generally do not affect the coupling reaction adversely. Table 2 is an attempt to summarize the effect of α-substituents for Kolbe couplings; the reactions mentioned have normally been run under conditions favoring coupling (slightly acidic medium, platinum anode, methanol or DMF as solvent in most cases). Inspection of the table shows that substituents that stabilize carbonium ions by their electron-repelling effect (inductively or mesomerically) tend to give low yields in the coupling reaction. Conversely, electron-withdrawing substituents favor high yields of the coupling product.

The Kolbe coupling reaction has found a large number of applications in the synthesis of natural products [4, 5], notably through the elegant work of Weedon and his coworkers, and Ställberg-Stenhagen. The use of simple, optically active intermediates in which the optical activity is associated with a carbon atom β to or further away from the carboxyl group allows stereospecific syntheses to be carried out without difficulty. The geometrical configuration of a double bond situated in the γ, δ-position or further away is preserved during coupling. Thus, syntheses of long-chain branched acids, hydroxy acids (the hydroxy group is preferably protected by acetylation during coupling), unsaturated acids, and cyclic compounds with known configuration are possible.

Other applications include the synthesis of substituted succinic acids and derivatives [74-78], esters of trialkylacetic acids [79], α,ω-dihalides [42, 43, 80, 81], α,ω-dinitriles [82], α,α'-diaminosuberic acid [83], esters of keto acids [84, 85], methyl ketones [86], α,ω-diols [87], α,ω-diamines [87, 88], $\alpha,\alpha,\alpha',\alpha'$-tetramethyladipic acid and derivatives [88a], and perfluorinated compounds [88b]. A large effort has been made to optimize the yield of dimethyl sebacate from methyl hydrogen adipate [89-92], a process of potential interest for the large-scale manufacture of sebacic acid.

What appears to be coupling between organic and inorganic radicals, produced during coelectrolysis of a carboxylate and inorganic species, has been observed in several cases. Thus, Fichter [1, 93] carried out many electrolyses of carboxylates in the presence of sodium nitrate, obtaining

TABLE 2

Yield[a] of Coupling Product from α-Substituted Carboxylates

Substituents in $R^1R^2R^3C$-COO^-			Yield of coupling product, %	Reference
R^1	R^2	R^3		
H	H	H	93	[4, 5]
Alkyl	H	H	30-90	[4, 5]
Alkyl	Alkyl	H	0-8	[4]
Alkyl	Alkyl	Alkyl	0	[4]
$-(CH_2)_n-$		H	0-30	[4, 63, 65]; cf. also [24]
Phenyl	H	H	55; 59	[64]
Ph	Ph	H	0; 8; 36	[66]; [31]; [65]
Ph	Ph	Ph	0	[64, 67]
2, 4, 6-Trialkyl-phenyl	H	H	30-45	[65]
4-$NO_2C_6H_4$	H	H	33	[68]
2-$NO_2C_6H_4$	H	H	0	[68]
4-$CH_3OC_6H_4$	H	H	0; 11	[69]; [70]
$C_2H_5CH{=}CH$	H	H	35-40	[60]
F, Br, Cl, or I	H	H	0	[5, 29]
F	F	F	93	[14]
HO, NH_2	H	H	0	[4, 5]
RCONH	H	H	<15	[71]
ArO	H	H	8-45	[72]
Alkyl-O	H, Ph	H	0	[5]; [73]
Alkyl	EtOOC[b]	H	20-85	[74, 75]
Alkyl	Alkyl	EtOOC[b]	15-35	[74, 76]
Alkylidene		EtOOC[b]	10-25	[74, 76a]
Alkyl	CN	H, Alkyl	30-60	[77]
Alkyl	$CONH_2$	H, Alkyl	5-55	[78]

[a] Based on amount of starting material.

[b] Coupling involving half esters is often called the Crum Brown-Walker reaction.

complex mixtures of products among which there was always a low yield of a nitrate corresponding to coupling between the alkyl radical and a nitrate radical:

$$RCOO^- \xrightarrow[-CO_2]{-e^-} R\cdot + NO_3^{\cdot} \xleftarrow{-e^-} NO_3^- \qquad R\cdot + NO_3^{\cdot} \longrightarrow RONO_2 \tag{27}$$

More recently [94], sodium hexanoate was electrolyzed together with sodium azide at a platinum anode in aqueous solution, 1-azidopentane being obtained in 50-60% yield along with decane under the most favorable conditions (pH 9-12, low azide concentration, high current density). It was established that the reaction occurs at the characteristically high potential of the Kolbe reaction, the products mentioned above being obtained at +2.7 V (SCE) but not at +2.3 and 2.5 V. That the reaction really is radical and not cationic in nature is indicated by the fact that t-butylacetic acid under similar conditions gives an azide with an unrearranged carbon skeleton [95]. However, one cannot entirely rule out the possibility that the strongly nucleophilic azide ion might capture the neopentyl cation before it has had time to rearrange.

Coelectrolysis of difluoramine, HNF_2, and sodium acetate in aqueous solution produces methyldifluoramine in 25-60% yield:

$$CH_3COO^- \xrightarrow[-CO_2]{-e^-} CH_3\cdot + \cdot NF_2 \xleftarrow{-e^-} HNF_2 \qquad CH_3\cdot + \cdot NF_2 \longrightarrow CH_3NF_2 \tag{28}$$

Since difluoramine itself is oxidized in a one-electron step to form the difluoramino radical, a radical coupling mechanism appears probable in this case too. The reaction is claimed to be a good laboratory synthesis method for alkyldifluoramines [96].

B. Disproportionation of Radicals

In competition with the coupling reaction, disproportionation between radicals having at least one hydrogen atom at the β-carbon atom takes place to a certain extent. In cases where one can estimate the disproportionation-to-coupling ratio it is fairly low [4, 5]. Thus, for the 3,3-dimethyl-1-butyl radical it is 0.1 [97], about the same as the gas-phase values for ethyl and 1-propyl radicals. An increasing number of hydrogens on the β-carbon atom favors disproportionation [74, 75, 78].

This reaction has so far not led to any preparative applications.

C. Addition and Additive Dimerization via Radicals

Interesting reactions between anodically generated radicals and 1,3-dienes are addition and additive dimerization [98, 99], as exemplified below [Eq. (29)] with methyl radicals generated by electrolysis of acetate ion in methanol-acetic acid in the presence of butadiene:

$$\begin{array}{l} CH_3\cdot + CH_2{=}CH{-}CH{=}CH_2 \longrightarrow \\ [CH_3CH_2\dot{C}H{-}CH{=}CH_2 \longleftarrow CH_3CH_2CH{=}CHCH_2\cdot] \\ 2\,EtCH{=}CHCH_2\cdot \longrightarrow EtCH{=}CHCH_2CH_2CH{=}CHEt \\ EtCH{=}CHCH_2\cdot \xrightarrow{CH_3\cdot} EtCH{=}CHEt \end{array} \quad (29)$$

The major part of the C_{10}-diene fraction is 3,7-decadiene, in addition to small amounts of the branched isomers. An interesting feature of the addition reaction is that it appears to be stereospecific [99], in that the trans-alkene is formed from butadiene and the cis-alkene from isoprene. Electrolysis of potassium ethyl oxalate in methanol [98] provides a source of ethoxycarbonyl radicals which react with butadiene in an additive dimerization process to form diethyl 3,7-decadiene-1,10-dioate; trifluoromethyl radicals from trifluoroacetate react analogously [98]. The electrolysis of methyl hydrogen adipate in the presence of butadiene has received considerable attention [100], presumably because of the economic potentialities of the process. A total yield of 55% of diesters (XXIII) and (XXIV) is claimed:

$$\underset{\text{(XXIII)}}{MeOOC(CH_2)_5CH{=}CH(CH_2)_5COOMe} \quad \underset{\text{(XXIV).}}{MeOOC(CH_2)_5CH{=}CH(CH_2)_2CH{=}CH(CH_2)_5COOMe} \quad (30)$$

D. Initiation of Polymerization by Radicals [2]

The additive dimerization reaction can of course be extended to a polymerization process with suitable monomers present in the electrolysis mixture. Thus, styrene, acrylonitrile, vinyl acetate, methyl methacrylate, vinyl chloride, and acrylic acid all give polymers upon rapid stirring together with an electrolyzing aqueous solution of potassium acetate [101-103], the polymerization process being initiated by the anodic reaction. [3] Experiments with acetate-2-^{14}C have convincingly demonstrated that methyl radicals generated at the anode indeed serve as polymerization initiators.

E. Radical Substitution Reactions

One of the first cases of a radical substitution process employing anodically generated radicals was the formation of trinitro-m-xylene in 9% yield upon electrolysis of sodium acetate-acetic acid in the presence of trinitrotoluene [104]. Later other substrates have been employed, as in the arylation [105] and alkylation [106] of pyridine, the methylation of nitrobenzene [107] and aromatic hydrocarbons [36, 108], and in one case of an intramolecular alkylation process [109]. Generally, yields of radical substitution products are low and the reaction is not very promising from the preparative point of view. The concurrent acyloxylation reaction [Eq. (31)] is now believed to occur via electrochemical oxidation of the organic additive [10, 36]:

$$RCOO^- + Ar\text{-}H \xrightarrow{-2e^-} Ar\text{-}OCOR + H^+. \qquad (31)$$

F. Generation and Study of Carbonium Ions

Following the experimental verification of carbonium ion intermediacy in the Kolbe reaction {[9], see Eqs. (24)-(26)}, a number of investigations have aimed at a study of the properties of such carbonium ions and at a comparison with carbonium ions generated in homogeneous solution, e.g., by solvolysis, deamination, and deoxidation. It soon became apparent that Kolbe carbonium ions are of the "hot" variety, undergoing substitution and elimination (cyclopropanes are among the elimination products), and rearrangements to form more stable species, which, in turn, give elimination and substitution products. As an example, electrolysis of 2-methylbutyric acid [20] in aqueous solution at a carbon anode gives a mixture of all of the butene isomers, isobutene, methylcyclopropane, 1- and 2-butanol, t-butyl alcohol, and the 1-, 2-, and t-butyl esters of the starting material. Similar product mixtures were obtained from 3-methylbutyric acid and pentanoic acid. Comparisons were made with butyl, 2-butyl, and isobutyl cations generated in homogeneous media by deamination and deoxidation, respectively, and it appears that anodic carbonium ions have properties resembling those of deoxidation cations (thus, cyclopropanes are not formed in deaminations).

The generation of the 3, 3-dimethyl-1-butyl cation (XXV) from the electrolysis of potassium 4, 4-dimethylpentanoate in aqueous solution at a platinum anode has been the subject of a very careful study [110], with regard to the fate of both radicals and cations. The quantity of radical-derived products (from coupling and disproportionation) decreases drastically with increasing pH of the electrolyte, from about 50% at pH 6.6 to about 2%

at pH 14.5, as would be expected from previous experience (Sec. II.C). At pH 10, the cationic pathway accounts for roughly 50% of the products formed, the 3, 3-dimethylbutyl cation partitioning between the different reactions indicated in the following scheme:

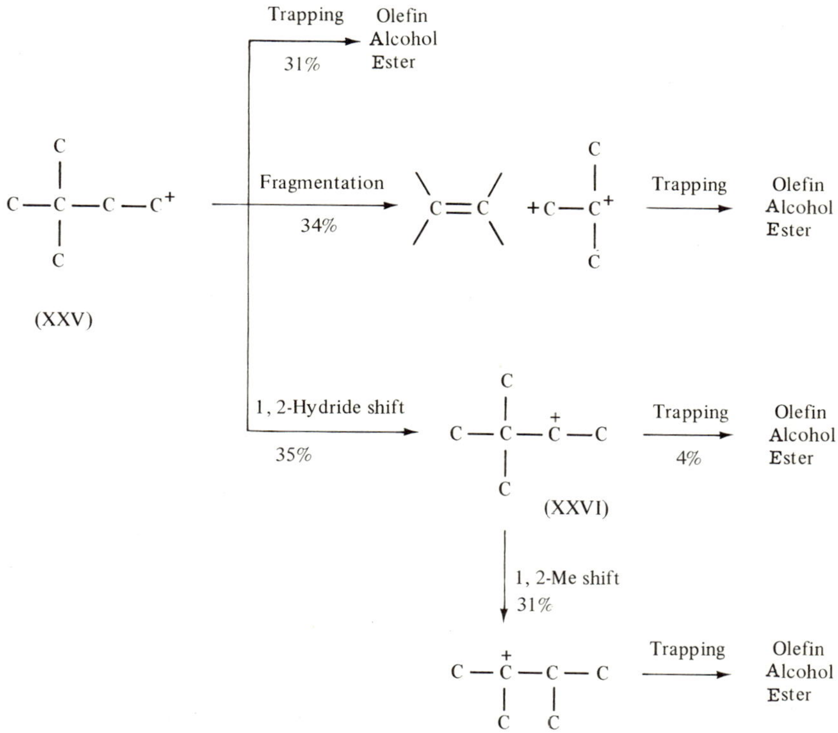

(32)

From the pH dependence of the composition of products derived via the different routes it was concluded that above a pH value around 14, trapping by hydroxide ion instead of water becomes the important substitution process. This is seen as a dramatic decrease in the ratio of rearranged to unrearranged products at this pH. A reasonable explanation is that the unimolecular rearrangement reaction competes favorably with the trapping of the 3, 3-dimethyl-1-butyl cation by water but less effectively with the stronger nucleophile, hydroxide ion.

The anodically generated 3, 3-dimethyl-1-butyl cation undergoes fragmentation [Eq. (32)] to form ethene and the t-butyl cation to a considerable extent (20-55%, depending on pH), which is analogous to the behavior of this cation generated by deoxidation but not by deamination and solvolysis. The quantity of fragmentation products increases with increasing hydroxide ion concentration. This has been explained in terms of an increase in exothermicity of the fragmentation process (initially almost thermoneutral) as the potential gradient near the anode is steepened with increasing hydroxide ion concentration.

The 4, 4-dimethyl-2-pentyl cation shows similar behavior [110]. Here also, deoxidation appears to be the closest analog in homogeneous solution.

The 3, 3-dimethyl-2-butyl cation (XXVI), when generated by electrolysis of potassium 2, 3, 3-trimethylbutanoate [111], reacts mainly by rearrangement to the tertiary cation at pH values below 14 [see lower part of Eq. (32)]. At high concentrations of hydroxide ion (3 M KOH) it can be trapped unrearranged to the extent of 30-60%, in spite of the fact that this cation is considered to be one of the most rearrangement-prone cations.

Medium-ring cycloalkyl cations, formed by anodic oxidation of cycloalkanecarboxylates at a carbon anode in aqueous solution [112], undergo 1, 2-, 1, 3-, and 1, 5-eliminations, as exemplified below [Eq. (33)] for cyclooctanecarboxylic acid:

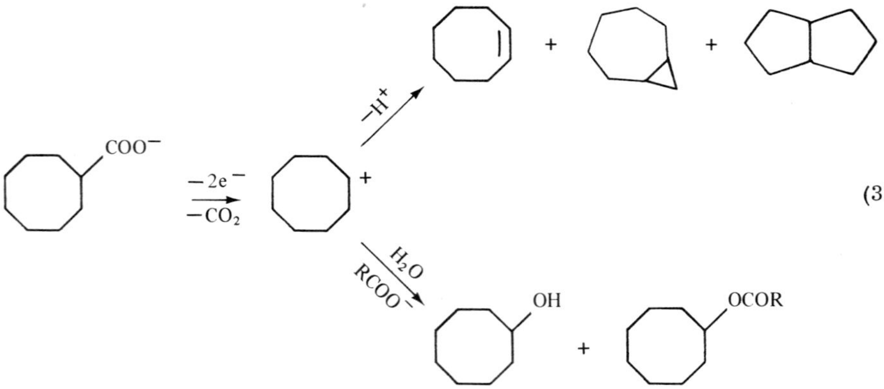

(33)

where R is a cyclooctyl. Anodic oxidation of 1-adamantaneacetic acid in methanol gives 3-methoxyhomoadamantane [Eq. (34)], whereas homoadamantane-3-carboxylic acid gives 1-methoxyhomoadamantane via a

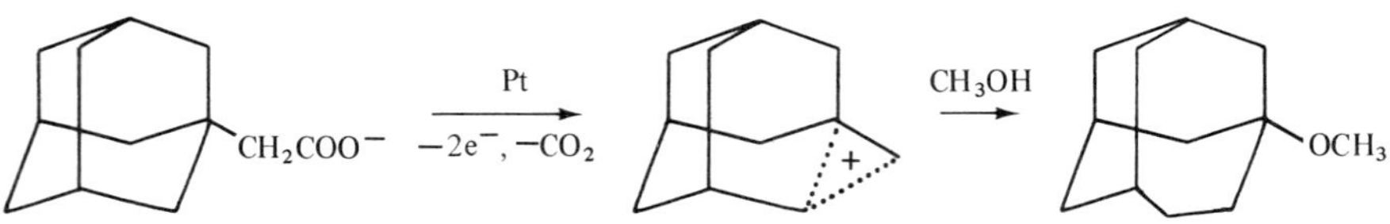

(34)

1, 3-hydride shift [Eq. (35)] assumed to be synchronous with the loss of an electron from the 3-homoadamantyl radical to the anode [52a, 113].

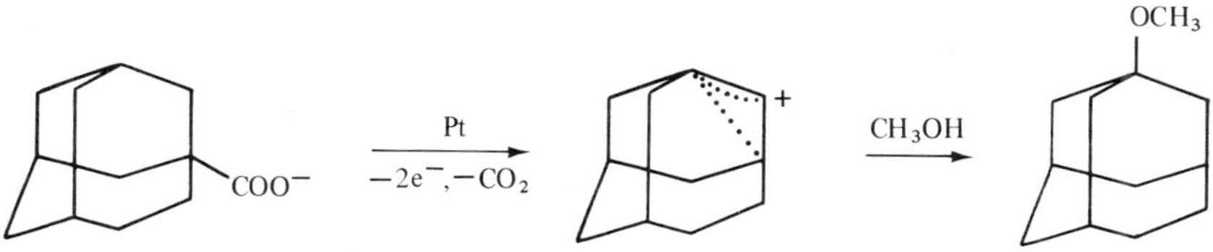

(35)

A careful study of the partitioning between radical and cationic pathways as a function of anode material and solvent in the electrolysis of phenylacetic and 1-methylcyclohexaneacetic acid has been performed [24]. Table 3 shows

TABLE 3

Anodic Oxidation of Phenylacetic Acid [24]

Solvent	Anode material	Ratio radical products/ carbonium ion products
Methanol	Pt	1.3
Methanol	C	0.26
Methanol-pyridine (67:33)	Pt	1.4
Methanol-pyridine (67:33)	C	0.0076
Water-pyridine (67:33)	Pt	0.78
Water-pyridine (67:33)	C	0.00

how the ratio between radical- and carbonium-derived products changes with the anode material and solvent. It is noteworthy that the radical path is completely suppressed only in the case when carbon is the anode material and water-pyridine the solvent (see Sec. II.C). The ability of the carbon anode to promote carbonium ion formation was assumed to be the presence of paramagnetic centers at the carbon surface, which bind the radicals initially formed and thus impede desorption. A second electron transfer is therefore favored over coupling.

Carbonium ions generated from anodic oxidation of cyclopropaneacetic acid, cyclobutanecarboxylic acid, and allylacetic acid in aqueous solution (pH 7) at a platinum anode [114] produce mixtures of substitution products (cyclopropylcarbinol, cyclobutanol, and allylcarbinol, and their esters with the starting material) which are quite similar in composition. That equilibration of cations is not complete, however, is indicated by the nature of the elimination products in the three cases. Cyclopropaneacetic acid gives little hydrocarbon that is not attributable to the radical route, cyclobutanecarboxylic acid gives appreciable amounts of hydrocarbons (cyclobutene and bicyclobutane), whereas allylacetic acid gives predominantly butadiene and ethene, the latter one a fragmentation product.

2, 2, 3, 3-Tetramethylcyclopropanecarboxylate undergoes a very complex set of transformations upon electrolysis in methanol; all of these appear to occur via the carbonium ion pathway [63].

Anodic oxidation of 3, 3-diphenylacrylate in acetic acid at a carbon anode gives both unrearranged (4-phenylcoumarin, 2, 2-diphenylvinyl acetate, and diphenylacetaldehyde) and rearranged products (benzil and related compounds) [115]. The last-mentioned products can also be obtained by anodic oxidation of diphenylacetylene in the same medium. A mechanism involving the formation of the 2, 2-diphenylvinyl cation, followed by rearrangement to the 1, 2-diphenylvinyl cation, was postulated [Eq. (36)]:

$$\begin{array}{ccccc} Ph_2C{=}CHCOO^- & \xrightarrow[-CO_2]{-2e} & Ph_2C{=}CH^+ & \longrightarrow & Ph\overset{+}{C}{=}CHPh \\ & & \Big\downarrow{\scriptstyle CH_3COOH,\ H_2O} & & \Big\downarrow{\scriptstyle \text{Several steps}} \\ & & Ph_2CH_2CHO + Ph_2C{=}CHOAc & & PhCOCOPh \end{array} \tag{36}$$

4-Phenylcoumarin is most likely formed in a reaction analogous to anodic acyloxylation (see Chapter XXII).

G. Preparative Uses of the Carbonium Ion Route

The complex mixtures obtained in the cases described above (Sec. V.F) would seem to preclude the possibility of using the carbonium ion route for preparative purposes. However, some uses have been found, of which the following are the most promising: (1) anodic decarboxylation of monocarboxylic acids; (2) bisdecarboxylation of 1, 2-dicarboxylic acids, and in one case, a 1, 3-diacid; (3) decarboxylative ring expansion of β-hydroxy acids; (4) decarboxylative C_β-C_γ cleavage of γ-hydroxy acids; and (5) decarboxylative acetamidation of monocarboxylic acids. In many of these cases optimum conditions for generation of carbonium ions have not been employed, so there is probably ample room for improvement.

Apart from the reactions already discussed, the anodic decarboxylation of monocarboxylic acids has been used in the syntheses of natural products in several cases (±)-Muscone has been synthesized via decarboxylation of (XXVII) [116], followed by hydrogenation of alkene (XXVIII) [Eq. (37)].

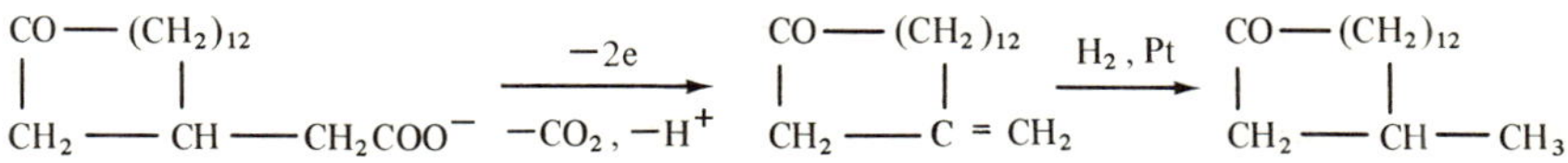

(37)

(XXVII) (XXVIII) (±)-Muscone

Isostevic acid (XXIX) can be decarboxylated anodically to give a good yield of (XXX) and (XXXI) [117];

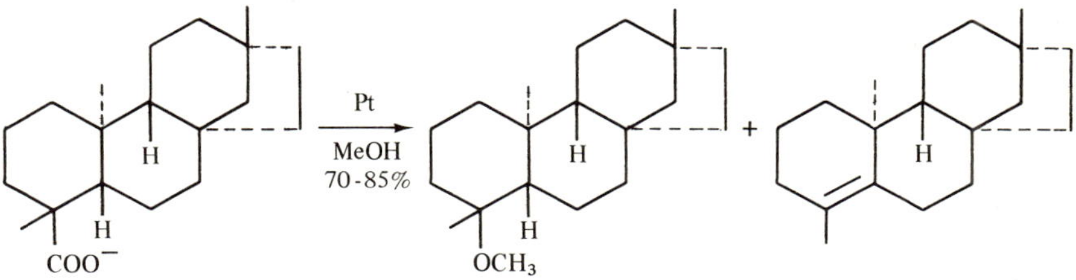

(38)

(XXIX) (XXX) (XXXI)

Decarboxylation of 3β-acetoxybisnorallocholanic acid (XXXII) has also been achieved successfully [118]:

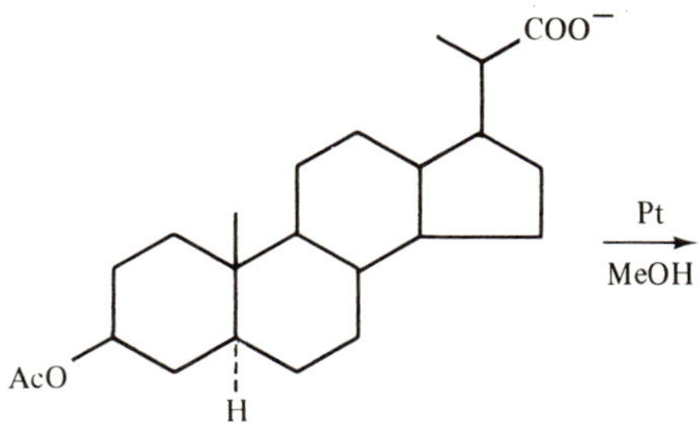

(XXXII)

(39)

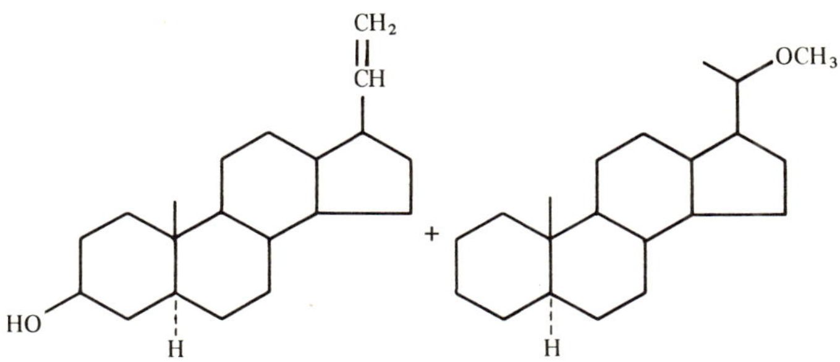

11% 42%

Anodic bisdecarboxylation of 1, 2-diacids, known for many years [4] but not used for synthetic purposes until recently, is sometimes a favorable alternative to the Grob degradation by lead tetraacetate since it appears to be a more lenient procedure. It has been formulated [119] with a monocarboxylic zwitterion (XXXIII) as an intermediate [Eq. (40)]. meso- and dl-2, 3-Diphenylsuccinic acids give a 36 and 40% yield, respectively, of

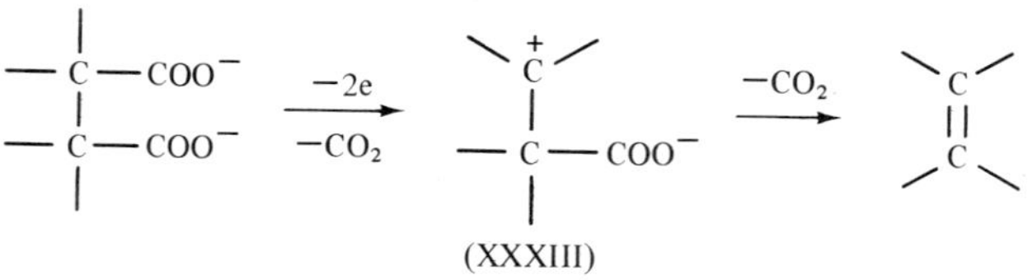

(40)

trans-stilbene with no cis isomer detectable, indicating that the intermediate (XXXIII) is free to rotate to form the thermodynamically most favored isomer in the meso case [119]. The method can be used for diacids containing double bonds [Eq. (41)] and cyclopropane rings [Eq. (42)] without

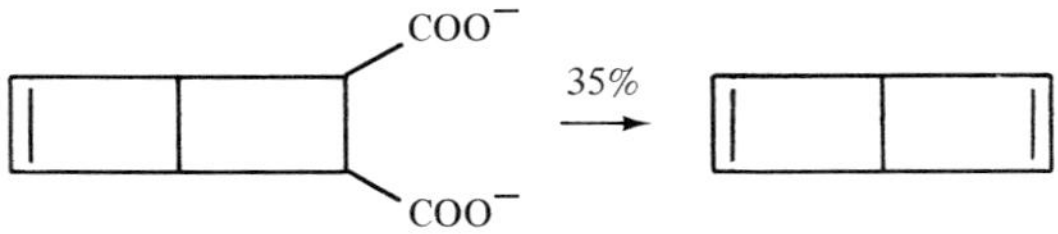

(41)

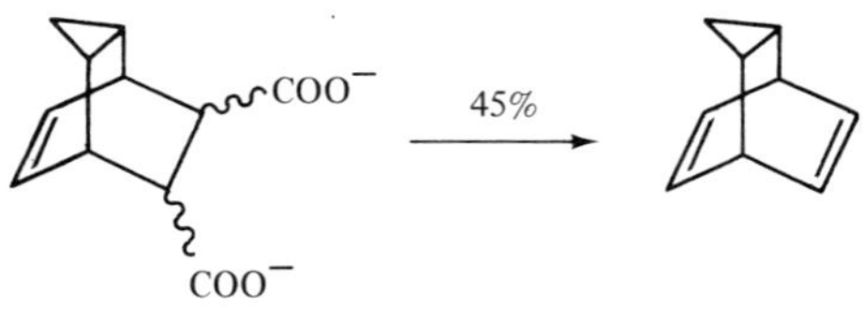

(42)

affecting these sensitive groupings [120, 121]. Here the Grob degradation fails. A number of other bicyclic diacids have successfully been converted to olefins [120-122]. Pyridine-water appears to be the best solvent for this process; moreover, triethylamine is preferably used for converting the acid into the carboxylate form.

The only case of an internal coupling reaction hitherto reported [123] is the formation of a bicyclobutane derivative (XXXIV) by anodic oxidation of trans, trans, trans-1, 3-dicarboxy-2, 4-dimethoxycarbonylcyclobutane (XXXV), presumably via the same type of zwitterionic intermediate as in the case of 1, 2-diacids [Eq. (43)]. Other 1, 3-diacids give lactones [123, 124] or other products [4, 125] on anodic oxidation, so this internal coupling process must be due to the favorable steric arrangement in diacid (XXXV).

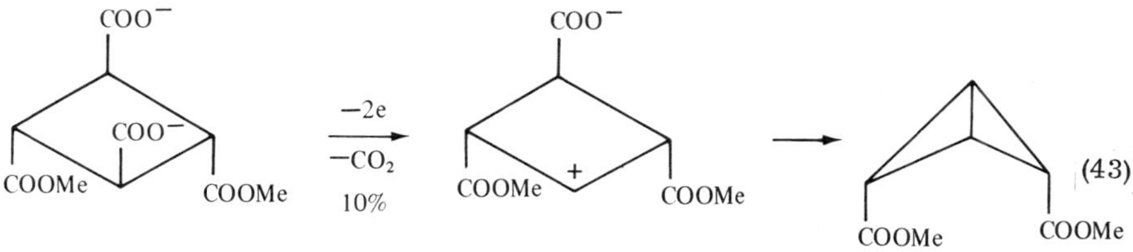

(43)

An intriguing possibility is that anodic oxidation of a 1, 1-diacid might give a carbene-like intermediate [Eq. (44)]. This has been sought in the case of malonic acid [126] but no evidence for a carbene intermediate was obtained.

$$R_2C(COO^-)_2 \xrightarrow{-2e^-} 2CO_2 + R_2C: \quad (44)$$

Electrolysis of a β-hydroxy acid leads to a β-hydroxy carbonium ion, which rearranges to form a protonated ketone [9]. This reaction is useful

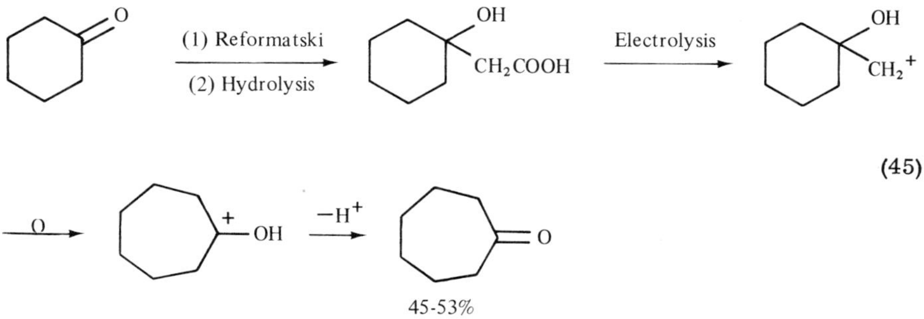

(45)

as a ring expansion method [Eq. (45)]. The epimeric acids (XXXVI) and (XXXVII) give the same mixture of homodecalones (XXXVIII) and (XXXIX) on anodic oxidation [127]:

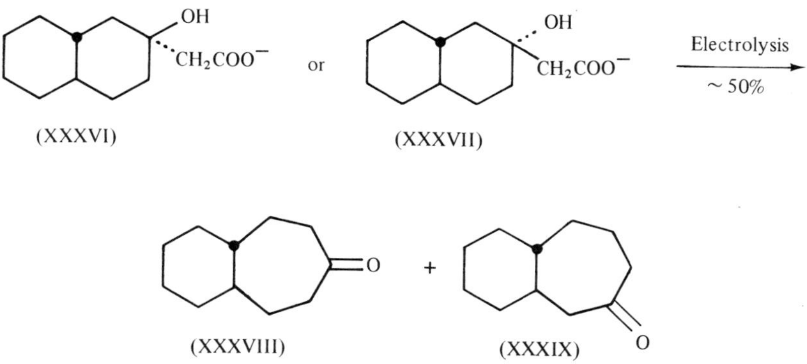

(46)

γ-Hydroxy acids undergo an interesting C_β-C_γ cleavage reaction [9, 128], as exemplified below for γ-benzyl-γ-hydroxyvaleric acid [Eq. (47)].

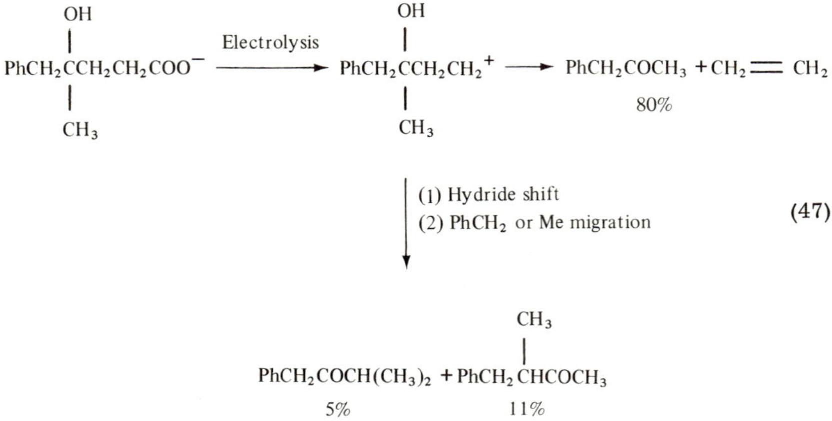

(47)

The reaction is analogous to the fragmentation process discussed above for the 3, 3-dimethyl-1-butyl cation (Sec. V.F). Another case [Eq. (48)] is the ring opening of the γ-hydroxy acid (XL) to form trans-5-cyclodecenone [130]:

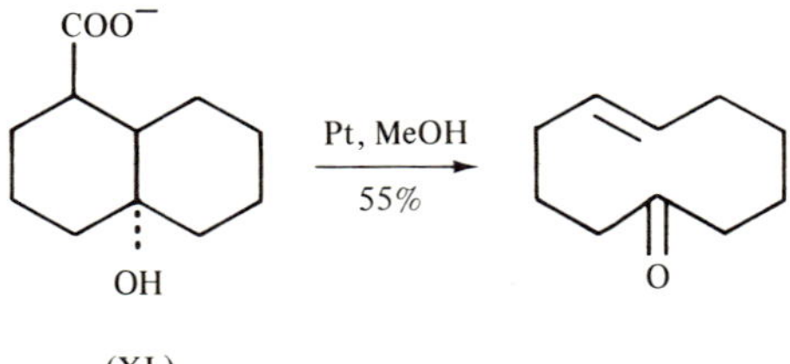

(XL)

(48)

The conversion of acids into the corresponding N-alkylacetamides [26, 37, 38] via the cationic route [Eq. (49)] has already been mentioned:

$$RCOO^- \xrightarrow[-CO_2]{-2e^-} R^+ \xrightarrow{CH_3CN} R\text{-}N{=}\overset{+}{C}\text{-}CH_3 \xrightarrow{H_2O} RNHCOCH_3 \qquad (49)$$

[Secs. II.C and III.B, Eq. (16)]. Although the reaction does not appear to be general [26], it is a mild alternative to the Ritter reaction and may be useful in isolated cases.

VI. MISCELLANEOUS

A. Esterification, Transesterification, and Hydrolysis

A side reaction often encountered during Kolbe electrolysis in methanol is the conversion of a certain portion of the starting material to its methyl ester, or the partial transformation of an ethoxycarbonyl group of a half-ester to a methoxycarbonyl group [10, 24, 29, 63]. At least the transesterification reaction is an anodic process, taking place in the anode compartment of a divided cell during electrolysis (the system studied was a partly neutralized mixture of ethyl hydrogen diethylmalonate and propionic acid in methanol). Under nonelectrolytic conditions, check experiments show that no transesterification takes place [130]. It is also possible to hydrolyze an ester function anodically [130] in a suitable solvent, such as dioxane-water. It seems probable that a common mechanism involving acyl-oxygen cleavage [10, 24] operates in these cases [Eq. (50)]:

$$RCOOR' \xrightarrow{-2e^-} R'O^+ + RCO^+ \xrightarrow{R''OH} RCOOR''. \tag{50}$$

B. Comparison with Other Anodic Processes

The formation of anodic carbonium ions by mechanisms similar to that of the Kolbe reaction has been postulated in other anodic processes, such as oxidation of alkaneboronates [131, 132] and alkyl halides [133, 133a]:

$$R\text{-}B(OH)_3^- \xrightarrow{-e^-} R\text{-}B(OH)_3\cdot \rightarrow B(OH)_3 + R\cdot \xrightarrow{-e^-} R^+ \tag{51}$$

$$R\text{-}X \xrightarrow{-e^-} (R\text{-}X)^{\dot{+}} \rightarrow R^+ + Br\cdot. \tag{52}$$

These processes have been compared with the Kolbe reaction [132] and found to be quite similar with regard to the fate of the carbonium ion at the smooth platinum anode. However, at a carbon anode differences between alkanecarboxylate and alkaneboronate oxidation were noticed.

The anodic oxidation of thio acids does not lead to Kolbe-type reactions [1, 134]. Instead, dimerization takes place via sulfur-to-sulfur coupling.

C. Comparison with Similar Homogeneous Processes

The Kolbe reaction can be partly simulated by certain homogeneous oxidations of carboxylic acids, such as oxidation by persulfate, ion in water {[135], Eq. (53)} or sulfuric acid {[136], Eq. (54)}, cobaltic ion in water or water-acetonitrile {[137], Eq. (55)}, and fluorine {[138], Eq. (56)}:

$$S_2O_8^{2-} + H_2O \rightarrow HSO_4^- + H^+ + SO_4^{\dot{-}} \xrightarrow{RCOO^-} R\cdot + CO_2 \tag{53}$$

$$RCOOH \xrightarrow[H_2SO_4]{K_2S_2O_8} RCOOSO_3H \rightarrow R^+ + CO_2 + HSO_4^- \quad (54)$$

$$RCOOH + Co(H_2O)_6^{3+} \rightarrow H_3O^+ + |RCOOCo(H_2O)_5|^{2+} \rightarrow$$

$$CO_2 + Co(H_2O)_5^{2+} + R\cdot \xrightarrow{Co^{3+}} R^+ \quad (55)$$

$$RCOO^- + F_2 \xrightarrow{H_2O} |RCO\text{-}O\text{-}F| \rightarrow R^+ + CO_2 + F^- \quad (56)$$

Also the decomposition of diacyl peroxides [10, 139] has many features in common with the Kolbe reactions. In all of these cases, quantitative comparisons are difficult to make due to the large differences in concentrations between the species generated in heterogeneous and in homogeneous processes [10].

NOTES

[1]Polaromicrotribometry, a method which measures the changes in coefficient of friction at the anode surface with changes in the anode potential, has given indication that a hydrophobic polymer film may be responsible for the inhibition of water oxidation when the Kolbe potential is reached [140].

[2]Cf. Chapter XXIX.

[3]Propionates have also been used [141].

REFERENCES

[1]. F. Fichter, Organische Elektrochemie, Steinkopf, Dresden and Leipzig, 1942.

[2]. S. Swann, Jr., in Technique of Organic Chemistry (A. Weissberger, ed.), 2nd ed., Vol. II, Wiley-Interscience, New York, 1956, p. 385.

[3]. M. J. Allen, Organic Electrode Processes, Reinhold, New York, 1958.

[4]. B. C. L. Weedon, Quart. Rev. (London), 6, 380 (1952).

[5]. B. C. L. Weedon, Adv. Org. Chem., 1, 1 (1960).

[6]. G. E. Svadkovskaya and S. A. Voitkevich, Russ. Chem. Rev. (English Transl.), 29, 161 (1960); M. Ya. Fioshin, Electrochemistry 1965, (S. S. Krugilov, ed.), VINITI, Moscow, p. 139 ff.

[7]. C. Walling, Free Radicals in Solution, Wiley, New York, 1957, p. 581.

[8]. C. G. Overberger and P. Kabasakalian, J. Amer. Chem. Soc., 79, 3182 (1957).

[9]. E. J. Corey, N. L. Bauld, R. T. LaLonde, J. Casanova, Jr., and E. T. Kaiser, ibid., 82, 2645 (1960).

[10]. L. Eberson, in Chemistry of the Carboxyl Group (S. Patai, ed.), Wiley-Interscience, London, 1969, Chap. 2.

[11]. N. L. Weinberg and H. R. Weinberg, Chem. Rev., 68, 449 (1968), Table XVI.

[12]. M. Ya. Fioshin and Yu. B. Vasil'ev, Proc. Acad. Sci. USSR, Phys. Chem. Sec., 134, 929 (1960).

[13]. T. Dickenson and W. F. K. Wynne-Jones, Trans. Faraday Soc., 58, 382, 388, 400 (1962).

[14]. B. E. Conway and M. Dzieciuch, Can. J. Chem., 41, 21, 38 (1963).

[15]. A. K. Vijh and B. E. Conway, Z. Anal. Chem., 224, 149, 160 (1967).

[16]. A. K. Vijh and B. E. Conway, Chem. Rev., 67, 623 (1967).

[17]. M. Fleischmann, D. Pletcher, and G. M. Race, J. Electroanal. Chem., in press.

[18]. N. L. Weinberg, J. Org. Chem., 33, 4326 (1968).

[19]. L. Eberson, Acta Chem. Scand., 17, 2004 (1963).

[20]. W. J. Koehl, J. Amer. Chem. Soc., 86, 4686 (1964).

[21]. W. J. Koehl, J. Org. Chem., 32, 614 (1967).

[22]. V. D. Parker, Chem. Commun., 1164 (1968).

[23]. J. G. Traynham and J. S. Dehn, J. Amer. Chem. Soc., 89, 2139 (1967).

[24]. S. D. Ross and M. Finkelstein, J. Org. Chem., 34, 2923 (1969).

[25]. L. Eberson and S. Nilsson, Acta Chem. Scand., 22, 2453 (1968).

[26]. D. L. Muck and E. R. Wilson, Electrochem. Soc., 117, 1358 (1970).

[27]. A. Kunugi, H. Urata, and S. Nagaura, Denki Kagaku, 237 (1968); through CA, 69, 82878 (1968).

[28]. N. Sato, T. Sekine, and K. Sugino, J. Electrochem. Soc., 115, 242 (1968).

[29]. R. G. Woolford, J. Soong, and W. S. Lin, Can. J. Chem., 45, 1837 (1967).

[30]. N. Dinh-Nguyen, Acta Chem. Scand., 12, 585 (1958).

[31]. M. Finkelstein and R. C. Petersen, J. Org. Chem., 25, 136 (1960).

[32]. L. Rand and F. Mohar, ibid., 30, 3156, 3885 (1965).

[33]. S. D. Ross, M. Finkelstein, and R. C. Petersen, J. Amer. Chem. Soc., 86, 2745 (1964).

[34]. S. D. Ross, M. Finkelstein, and R. C. Petersen, J. Org. Chem., 31, 128 (1966).

[35]. S. D. Ross, M. Finkelstein, and R. C. Petersen, J. Amer. Chem. Soc., 88, 4657 (1966).

[36]. L. Eberson and K. Nyberg, ibid., 88, 1686 (1966).

[37]. L. Eberson and K. Nyberg, Acta Chem. Scand., 18, 1567 (1968).

[38]. J.-M. Kornprobst, A. Laurent, and E. Laurent-Dieuzeide, Bull. Soc. Chim. Fr., 3657 (1968); ibid., 1490 (1970).

[39]. L. I. Krimen and D. J. Cota, Org. React., 17, 213 (1969).

[40]. R. C. Petersen, M. Finkelstein, and S. D. Ross, J. Org. Chem., 32, 564 (1967).

[41]. P. G. Gassman and F. V. Zalar, J. Amer. Chem. Soc., 88, 2252 (1966).

[42]. R. G. Woolford, Can. J. Chem., 40, 1846 (1962).

[43]. R. G. Woolford, W. Arbic, and A. Rosser, ibid., 42, 1788 (1964).

[44]. B. E. Conway and A. K. Vijh, Electrochim. Acta, 12, 102 (1967).

[45]. L. Eberson, ibid., 12, 1473 (1967).

[46]. M. Fleischmann and F. Goodridge, Discuss. Faraday Soc., 45, 254 (1968).

[47]. P. H. Reichenbacher, M. Y.-C. Liu, and P. S. Skell, J. Amer. Chem. Soc., 90, 1816 (1968).

[48]. P. H. Reichenbacher, M. D. Morris, and P. S. Skell, ibid., 90, 3432 (1968).

[49]. L. Eberson, ibid., 91, 2402 (1969).

[50]. T. Shono, I. Hishiguchi, S. Yamane, and R. Oda, Tetrahedron Lett., 1965 (1969).

[51]. R. I. Reed, Ion Production by Electron Impact, Academic, London, 1962.

[52]. F. Gutmann and L. E. Lyons, Organic Semiconductors, Wiley, New York, 1967, pp. 669-692.

[52a]. J. A. Van Zorge, J. Strating, and H. Wynberg, Rec. Trav. Chim., 89, 781 (1970).

[53]. G. Atherton, M. Fleischmann, and F. Goodridge, Trans. Faraday Soc., 63, 1468 (1967).

[54]. L. Smith and H. Olsson, Z. Phys. Chem., 118, 99 (1925).

[55]. L. Eberson and B. Olofsson, Acta Chem. Scand., 23, 2355 (1969).

[56]. K. Nyberg, Chem. Commun., 774 (1969).

[57]. M. S. Kharasch, J. G. Kuderna, and W. H. Urry, through G. M. Wheland, Advanced Organic Chemistry, 2nd ed., Wiley, New York, 1949, p. 714.

[58]. E. S. Wallis and F. H. Adams, J. Amer. Chem. Soc., 55, 3838 (1933).

[59]. C. G. Overberger and P. Kabasakalian, J. Org. Chem., 21, 1124 (1956).

[60]. R. F. Garwood, C. J. Scott, and B. C. L. Weedon, Chem. Commun., 14 (1965).

[61]. J. Benner and K. Mislow, J. Org. Chem., 21, 1312 (1956).

[62]. N. L. Bauld, Ph. D. Thesis, Univ. Illinois, Urbana, Ill., 1959.

[63]. R. Brettle and B. Cox, J. Chem. Soc. (C), 1227 (1969); T. D. Binns, R. Brettle, and B. Cox, ibid., 584 (1968).

[64]. R. P. Linstead, B. R. Shephard, and B. C. L. Weedon, J. Chem. Soc., 3624 (1952).

[65]. L. Rand and A. F. Mohar, J. Org. Chem., 30, 3885 (1965).

[66]. A. J. van der Hoek and W. T. Nauta, Rec. Trav. Chim., 61, 845 (1942).

[67]. L. Rand and A. F. Mohar, J. Org. Chem., 30, 3156 (1965).

[68]. B. Wladislaw and A. Giora, J. Chem. Soc., 1037 (1964).

[69]. B. Wladislaw and H. Viertler, Chem. Ind. (London), 39 (1965).

[70]. B. Wladislaw and H. Viertler, J. Chem. Soc. (B), 576 (1968).

[71]. R. P. Linstead, B. R. Shephard, and B. C. L. Weedon, J. Chem. Soc., 2854 (1951).

[72]. F. Fichter and K. Kestenholz, Helv. Chim. Acta, 25, 785 (1942); F. Fichter and H. Stenzl, ibid., 22, 971 (1939).

[73]. B. Wladislaw and A. M. J. Ayres, J. Org. Chem., 27, 281 (1962).

[74]. L. Eberson, Acta Chem. Scand., 13, 40 (1959).

[75]. L. Eberson and B. Sandberg, ibid., 20, 739 (1966).

[76]. L. Eberson, ibid., 14, 641 (1960).

[76a]. J. M. Conia and J. M. Denis, Tetrahedron Lett., 3545 (1969).

[77]. L. Eberson, J. Org. Chem., 27, 2329 (1962); L. Eberson and S. Nilsson, Acta Chem. Scand., 22, 2453 (1968).

[78]. L. Eberson, ibid., 17, 1196 (1963).

[79]. L. Eberson, J. Org. Chem., 27, 3706 (1962).

[80]. I. A. Avrutskaya and M. Ya. Fioshin, Zh. Vses. Khim. Obshchest., 12, 359 (1967); through CA, 67, 113194 (1967); L. I. Kazakova, M. Ya. Fioshin, and I. A. Avrutskaya, Zh. Prikl. Khim. (Leningrad), 41, 1326 (1968); through CA, 69, 64097 (1968).

[81]. K. Maruyama and K. Murakami, Nippon Kagaku Zasshi, 89, 196 (1968); through CA, 68, 110809 (1968).

[82]. J. Kamlet, U.S. Pat. 2, 782, 157 (1957); through CA, 51, 10564 (1957).

[83]. K. Mori, Nippon Kagaku Zasshi, 82, 1375 (1961); through CA, 57, 14929 (1962).

[84]. S. Motoki and T. Odaka, Nippon Kagaku Zasshi, 77, 163 (1956); through CA, 52, 268 (1968).

[85]. S. Motoki and Y. Yamada, Nippon Kagaku Zasshi, 81, 665 (1960); through CA, 56, 397 (1962).

[86]. S. Motoki and T. Odaka, Nippon Kagaku Zasshi, 76, 930 (1955); through CA, 51, 17727 (1957).

[87]. J. Haufe and F. Beck, Abstracts, Meeting of Applied Electrochemistry, Bonn, Germany, October 6-8, 1969.

[88]. I. A. Avrutskaya, M. Ya. Fioshin, and E. V. Gromova, Zh. Prikl. Khim., 40, 2857 (1967); through CA, 68, 83686 (1968).

[88a]. G. Cauquis and B. Haemmerlé, Bull. Soc. Chim. Fr., 183 (1970).

[88b]. A. I. Levin, S. V. Sokolov, O. N. Chechina, and K. I. Bil'dinov, J. Gen. Chem. USSR (English Transl.), 39, 413 (1969).

[89]. A. I. Kamneva, M. Ya. Fioshin, S. M. Itenberg, L. I. Kazakova, V. T. Shchugarev, A. I. Efimenkova, L. A. Muzychenko, B. M. Kuindzhi, and M. A. Zepalova, USSR Pat. 131, 751 (1960); through CA, 55, 5200 (1961).

[90]. Yu. M. Tyurin, M. V. Veselova, U. A. Kuratova, and A. P. Belous, Zh. Prikl. Khim., 35, 1082 (1962); through CA, 57, 9575 (1962).

[91]. W. Schwaab and W. Himmele, Ger. Pat. 1, 181, 688 (1964); through CA, 62, 11432 (1965).

[92]. F. Beck, Chem. Ing. Tech., 37, 607 (1965).

[93]. F. Fichter, Trans. Electrochem. Soc., 65, 309 (1939).

[94]. C. M. Wright and G. A. Ward, Extended Abstracts, 127th Meeting of the Electrochemical Society, San Francisco, Calif., May 1965.

[95]. C. M. Wright, private communication.

[96]. G. A. Ward and C. M. Wright, J. Amer. Chem. Soc., 86, 4333 (1964).

[97]. P. S. Skell and P. H. Reichenbacher, ibid., 90, 2309 (1968).

[98]. R. V. Lindsey, Jr. and M. L. Petersen, ibid., 81, 2073 (1959).

[99]. W. B. Smith and H. Gilde, ibid., 81, 5325 (1959); 83, 1355 (1961).

[100]. M. Ya. Fioshin, A. I. Kamneva, L. A. Mirkind, L. A. Salmin, and A. G. Kornienko, Neftekhim., 2, 557 (1962); through CA, 58, 11205 (1963); M. Ya. Fioshin, L. A. Mirkind, and A. G. Kornienko, Zh. Vses. Khim. Obshchest., 9, 600 (1964); through CA, 62, 4906 (1965).

[101]. S. Goldschmidt and E. Stöckl, Chem. Ber., 85, 630 (1952).

[102]. W. B. Smith and H. Gilde, J. Amer. Chem. Soc., 82, 659 (1960).

[103]. W. B. Smith and D. T. Manning, J. Polym. Sci., 59, S45 (1962).

[104]. L. F. Fieser, R. C. Clapp, and W. H. Daudt, J. Amer. Chem. Soc., 64, 2052 (1964).

[105]. D. H. Hey and P. J. Bunyan, J. Chem. Soc., 3787 (1960).

[106]. S. Goldschmidt and M. Minsinger, Chem. Ber., 87, 956 (1954).

[107]. D. R. Harvey and R. O. C. Norman, J. Chem. Soc., 4860 (1964).

[108]. S. D. Ross, M. Finkelstein, and R. C. Petersen, J. Amer. Chem. Soc., 86, 4139 (1964).

[109]. P. J. Bunyan and D. H. Hey, J. Chem. Soc., 1360 (1962).

[110]. P. S. Skell and P. H. Reichenbacher, J. Amer. Chem. Soc., 90, 2309 (1968).

[111]. P. S. Skell and P. H Reichenbacher, ibid., 90, 3436 (1968).

[112]. J. G. Traynham and J. S. Dehn, ibid., 89, 2139 (1967).

[113]. F. N. Stepanov and S. S. Guts, J. Org. Chem. USSR (English Transl.), 4, 1868 (1968).

[114]. J. B. Keating and P. S. Skell, J. Amer. Chem. Soc., 91, 695 (1969).

[115]. W. J. Koehl, Jr., J. Org. Chem., 32, 614 (1967).

[116]. L. Ruzicka and M. Stoll, Helv. Chim. Acta, 17, 1308 (1934).

[117]. J. A. Waters, E. D. Mosettig, and E. D. Becker, J. Org. Chem., 27, 4689 (1962).

[118]. J. A. Waters, ibid., 29, 428 (1964).

[119]. E. J. Corey and J. Casanova, J. Amer. Chem. Soc., 85, 165 (1967).

[120]. P. Radlick, R. Klem, S. Spurlock, J. J. Sims, E. E. van Tamelen, and T. Whitesides, Tetrahedron Lett., 5117 (1968).

[121]. H. H. Westberg and H. J. Dauben, Jr., ibid., 5123 (1968).

[122]. H. Plieninger and W. Lehnert, Chem. Ber., 100, 2427 (1967).

[123]. A. F. Vellturo and G. W. Griffin, J. Org. Chem., 31, 2241 (1966); T. Campbell, A. Vellturo, and G. W. Griffin, Chem. Ind. (London), 1235 (1969).

[124]. G. Smets, X. van der Borght, and G. van Heeren, J. Polym. Sci., 2, 5187 (1964).

[125]. M. Toy, J. Electrochem. Soc., 114, 1042 (1967).

[126]. E. K. Spicer and H.-G. Gilde, Chem. Commun., 373 (1967).

[127]. L. Rand and S. Rao, J. Org. Chem., 33, 2704 (1968).

[128]. E. J. Corey and R. R. Sauers, J. Amer. Chem. Soc., 81, 1743 (1959).

[129]. P. S. Wharton, G. A. Hiegel, and R. V. Coombs, J. Org. Chem., 28, 3217 (1963).

[130]. L. Eberson, unpublished observations.

[131]. A. A. Humffray and L. F. G. Williams, Chem. Commun., 616 (1965).

[132]. J. T. Keating and P. S. Skell, J. Org. Chem., 34, 1479 (1969).

[133]. L. L. Miller and A. K. Hoffman, J. Amer. Chem. Soc., 89, 593 (1967).

[133a]. A. Laurent and R. Tardivel, Compt. Rend. Acad. Sci. Paris, C, 271, 324 (1970).

[134]. Y. Hirabayashi and T. Mazume, Bull. Chem. Soc. Japan, 39, 1971 (1966).

[135]. L. Eberson, S. Gränse, and B. Olofsson, Acta Chem. Scand., 22, 2462 (1968), and references contained therein.

[136]. N. C. Deno, W. E. Billups, J. S. Bingman, R. R. Lastomirsky, and R. G. Whalen, J. Org. Chem., 34, 3207 (1969).

[137]. P. R. Sharan, P. Smith, and W. A. Waters, J. Chem. Soc. (B), 857 (1969), and preceding papers in this series.

[138]. V. Grakauskas, J. Org. Chem., 34, 2446 (1969).

[139]. S. Goldschmidt, W. Leicher, and H. Haas, Ann. Chem., 577, 153 (1952).

[140]. J.-E. Dubois and F. Bruno, Compt. Rend. Acad. Sci. Paris, C, 271, 791 (1970).

[141]. A. Kunugi, T. Shimizu, and S. Nagaura, Bull. Chem. Soc. Japan, 43, 1298 (1970).

Chapter XIV

ANODIC OXIDATION OF AMINES

Vernon D. Parker

Department of General and Organic Chemistry
University of Copenhagen
Copenhagen, Denmark

I. INTRODUCTION

One of the most active areas of research in anodic electrochemistry has been concerned with the oxidation of amines. The ease of removal of electrons from the amino function makes it possible to study the oxidations using quantitative electroanalytical techniques.

Intensive studies have recently been conducted on aliphatic amines. Several reaction pathways have been suggested for a variety of different reactions. Anodic methoxylation results in substitution at carbon alpha to the amino function. The course of the latter reaction is not definitely settled; either direct or indirect mechanisms are advocated by different groups.

Amino alcohols undergo oxidation at the alcohol function during electrolysis in acid solution to give good yields of the amino acid, while in basic solution the amine function is oxidized.

The anodic reaction pathways followed by many aromatic amines are now firmly established. The relation of structure to reaction pathway has been defined clearly for a variety of systems. Stable cation radicals and dications have been observed from anodic oxidation of aromatic amines and have been studied by modern techniques.

The anodic oxidation of amino- and arylaminoethylenes is accompanied by the generation of relatively stable cation radicals and dications. Recent investigations have contributed heavily to the understanding of these systems.

II. ANODIC OXIDATION OF ALIPHATIC AND BENZYL AMINES

Several investigations of the anodic oxidation of aliphatic and benzylamines have been reported recently. The investigators were not always in complete agreement on the mechanisms of all the reactions and, thus, some mechanisms are still open to question.

A. Oxidation of Primary Amines

The products of the anodic oxidation of methylamine, used as the solvent, with lithium chloride as supporting electrolyte were identified as methylamine hydrochloride, 1,3,5-trimethylhexahydro-s-triazine(I), and dimethylformamide [1]. The hydrochloride and (I) were found to be direct products and dimethylformamide was obtained only after hydrolysis of the product mixture. The following scheme [Eqs. (1)-(3)] was suggested to account for the formation of (I):

$$CH_3NH_2 \xrightarrow[\text{oxidation by } Cl_2 \text{ from } Cl^-]{\text{Anodic oxidation or}} CH_2{=}NH + 2CH_3\overset{+}{N}H_3Cl^- \quad (1)$$

$$CH_2{=}NH + CH_3NH_2 \rightleftarrows H_2N{-}CH_2{-}NH{-}CH_3 \rightleftarrows \quad (2)$$

$$NH_3 + CH_2{=}N{-}CH_3 \rightleftarrows$$

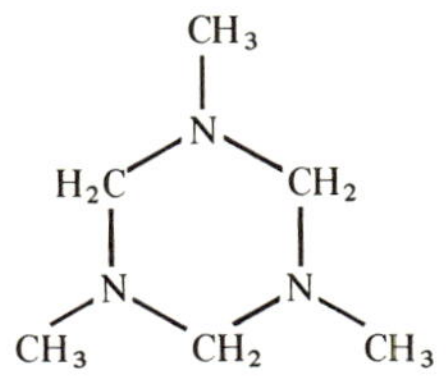

(I) (3)

A complex mechanism was proposed to account for differences in behavior of primary amines at low and high anode potentials [2]. A common first step was suggested for both pathways [Eq. (4)]. The key feature of

$$R{-}CH_2NH_2 \xrightarrow{-\bar{e}} \underset{(II)}{R{-}CH_2\overset{+\bullet}{N}H_2} \overset{-H^+}{\rightleftarrows} \underset{(III)}{RCH{=}\overset{\bullet}{N}H_2} \quad (4)$$

$$\underset{(II)}{RCH_2{-}\overset{+\bullet}{N}H_2} \xrightarrow[\text{Path A}]{} \underset{(IV)}{\overset{\bullet}{N}H_2 + RCH_2^+} \xrightarrow{CH_3CN} \text{Polymer} \quad (5)$$

$$\underset{(III)}{RCH{=}\overset{\bullet}{N}H_2} \xrightarrow[\text{Path B}]{-\bar{e}} RCH{=}\overset{+}{N}H_2 \xrightarrow[-H^+]{H_2O} R\overset{O}{\overset{\|}{C}}H + NH_3 \quad (6)$$

this mechanism, used to explain the differences in electrode potential behavior, is the equilibrium between the cation radical (II) and the radical (III). It was suggested that at low anode potentials, (II) decomposes irreversibly to the carbonium ion (IV), which reacts with acetonitrile to form polymeric products and the amidogen radical. At higher anode potentials (III) is believed to be further oxidized to the aldehyde. t-Butylamine, not having protons on the carbon alpha to the amino function, could not undergo path B. It was pointed out that significant difficulties arise in attempting to determine the relative importance of the mechanisms by quantitative product analysis [2].

The latter mechanism is not completely satisfying for several reasons: (1) it is not apparent that (III) should be more difficult to oxidize than substrate; (2) it seems more likely that the cation (IV), if formed, would react with strongly nucleophilic substrate rather than with solvent; and (3) reaction of (IV) with acetonitrile would be expected to yield stable monomeric amides rather than polymeric material.

B. Oxidation of Triethylamine

The anodic oxidation of triethylamine in dimethyl sulfoxide [3] or acetonitrile [4] results in the consumption of 1.0 faradays per mole of substrate accompanied by the formation of the amine salt. The initial step is believed to consist of generation of the cation radical (V). It was suggested that (V) reacts with dimethyl sulfoxide [3] or acetonitrile [4] to give

$$Et_3N: \rightarrow Et_3N^{\overset{\bullet}{+}} + \bar{e} \quad \text{(V)} \tag{7}$$

$$Et_3N^{\overset{\bullet}{+}} + \text{Solvent} \rightarrow Et_3\overset{+}{N}H + s\cdot \tag{8}$$

$\cdot CH_2\overset{O}{\overset{|}{S}}CH_3$ and $\cdot CH_2CN$ radicals, respectively. However, no solvent-derived products were identified.

C. Anodic Methoxylation of Tertiary Benzylamines

The electrolysis of methanolic solutions of N, N-dimethylbenzylamine (VI) in the presence of KOH was found to produce the methoxymethylamine (VII) and the methoxybenzylamine (VIII) in a ratio of 4:1, respectively [v]. Adsorption of (VI) on the anode was proposed to occur in such a way that

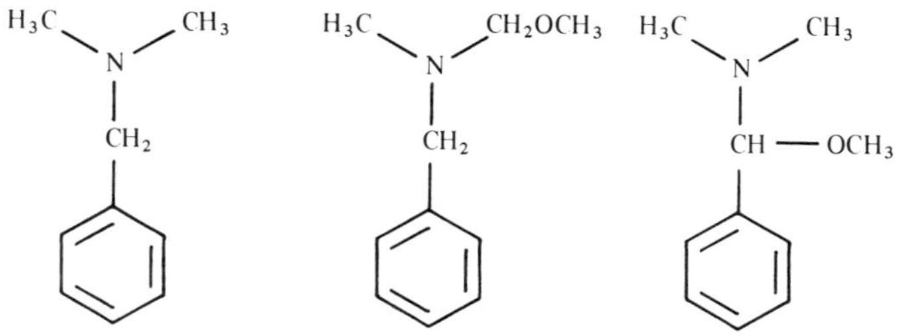

(VI) (VII) (VIII)

the benzyl hydrogens were less available to attack by hydroxide ion than were methyl hydrogens [Eq. (9)].

The latter mechanism has been disputed [6]. Smith and Mann were unable to detect significant amounts of (VII) and (VIII) using 1 M KOH as supporting electrolyte during electrolysis of (VI) at anode potentials of +1.0-1.25 V (vs. Ag/Ag^+) during which 3 to 5 faradays per mole of

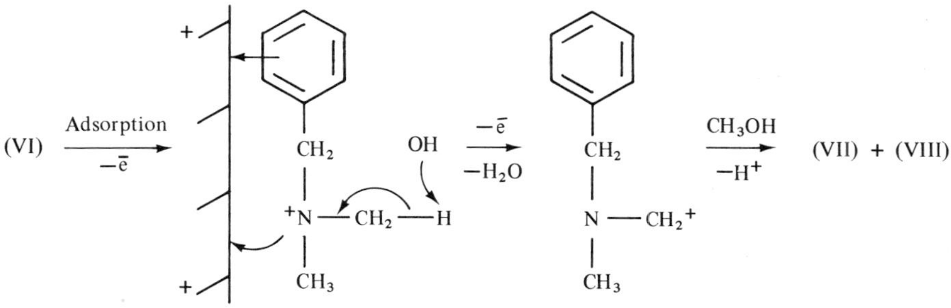

(IX) (X) (9)

substrate were passed. When KOMe was used as the electrolyte, yields of 5-10% of methoxylated compounds were obtained. An indirect mechanism involving generation of methoxyl radicals, which abstract hydrogen from the substrate, was proposed.

Weinberg [7] was able to reaffirm the formation of methoxylated products using KOH as the electrolyte. Complete recovery of starting material was observed at anode potentials lower than +1.0 V (vs. SCE) and at higher anode potentials current efficiencies of the order of 50% were observed for formation of (VII) and (VIII).

III. OXIDATION OF AMINO ALCOHOLS

The anodic oxidation of amino alcohols has been shown to be a good synthetic method for the preparation of amino acids. An indirect mechanism is probable for this conversion since the protonated amine would be expected to be more difficultly oxidized than the aqueous medium. On the other hand, oxidation in neutral or basic solution most likely involves direct discharge of the substrate.

A. Oxidation in Acidic Media

Reasonable yields of several amino acids have been obtained by electrolysis of the corresponding amino alcohols in aqueous sulfuric acid solution at platinum or lead dioxide anodes [8, 9]. For example, electrolysis of ethanolamine resulted in formation of glycine (39-66%), β-amino-n-propanol was converted to α-alanine (39-45%), γ-amino-n-propanol to β-alanine (64-80%), β-amino-n-butanol to α-aminobutyric acid (49%), and ϵ-amino-n-pentanol was converted to δ-aminovaleric acid (31-59%).

B. Oxidation in Basic Media

The anodic oxidation of N-benzyl-N-ethanolamines [5] results in the formation of a mixture of 2-phenyl- and 3-benzyloxazolidines [Eq. (10)]:

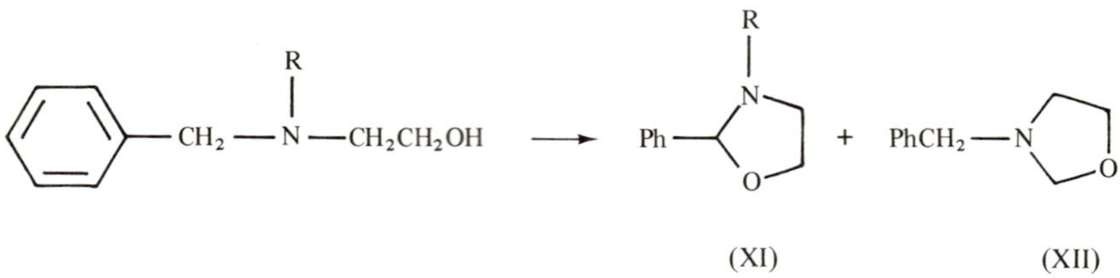

(10)

A competing reaction involves degradation of the hydroxyethyl group to methoxymethyl when the oxidation is carried out in methanol. A mechanism was proposed involving concerted transfer of an electron to the anode from the cation radical (XIII) of the adsorbed N-benzyl-N-alkylethanolamine accompanied by the loss of the elements of formaldehyde and a proton. When R is methyl, the cation (XIV) reacts with methanol to give (XV) while intramolecular cyclization occurs when R is hydroxyethyl to afford (XII):

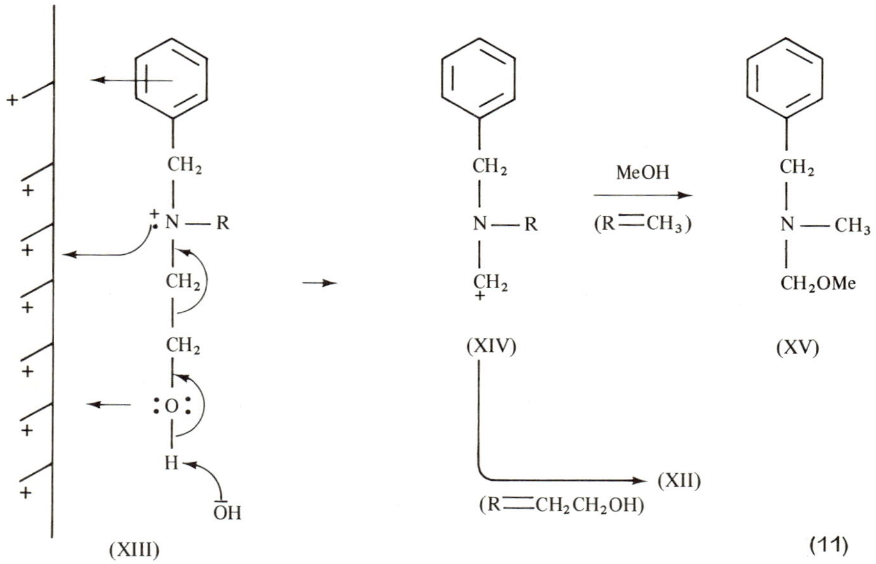

(11)

IV. OXIDATION OF AMINO ACIDS

In the 1930s, Takayama conducted an extensive investigation on the electrolysis of amino acids in aqueous acid solution. A complete tabulation of starting materials and products has recently been compiled [10]. Only a few pertinent examples will be given here.

In an early study of electrolytic oxidation of amino acids [11], it was noted that some aldehyde, along with further oxidation products, was formed.

In an extension of this reaction, it was found that the aldehyde could be distilled off as it was formed if the reaction were carried out at elevated temperatures while bubbling air through the electrolyte. For example, electrolysis of valine gave isobutyraldehyde (32%) and isobutyric acid (27.5%), and leucine gave isovaleraldehyde (53-56%) and isovaleric acid (20-22%). The following scheme was proposed to describe the results:

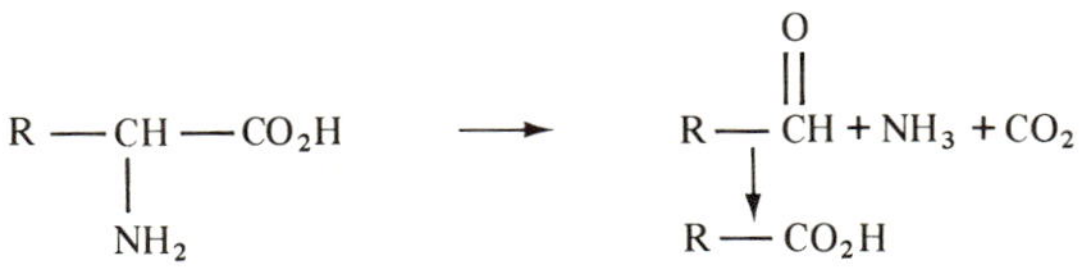

(12)

V. OXIDATION OF AROMATIC AMINES

The aromatic amines, because of their ease of oxidation and high solubility in aqueous solution, have been substrates for numerous anodic studies. Systems such as the aminophenols have served as the testing grounds for many of the methods and theories of modern electroanalytical chemistry.

Due to the large volume of new literature using modern techniques, only recent work will be covered in this survey. For a comprehensive review covering the older work, the reader is referred to the literature [10].

A. Oxidation of Phenylsubstituted Amines

The reactions covered are grouped according to substitution pattern of the amine: primary, secondary, tertiary, with a separate section on p-dimethylaminophenylmethanes.

1. Primary Aromatic Amines (Anilines)

The first modern study of the anodic oxidation of aromatic amines was concerned with the oxidation of aniline at a platinum anode in aqueous sulfuric acid [13]. The product of the reaction was assigned the structure of the octamer, emeraldine. p-Aminodiphenylamine was shown to be a possible intermediate in the oxidation of aniline and it was suggested that ameraldine was formed by successive head-to-tail dimerizations [Eqs. (13), (14), respectively].

A more recent study [14] of the anodic oxidation of para-substituted anilines in aqueous media revealed an interesting coupling reaction involving elimination of the para substituent to give 4′-substituted 4-aminodiphenylamines [Eq. (15)]. It was suggested that the first step, (XVII) → (XVIII), is common for all substrates studied and can be accounted for either by coupling

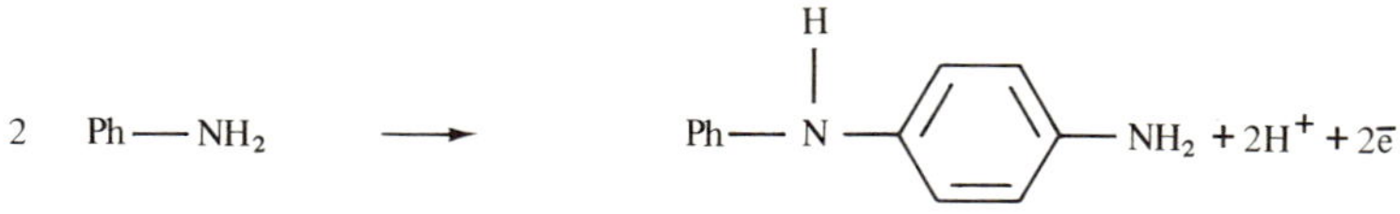

(XVI) (13)

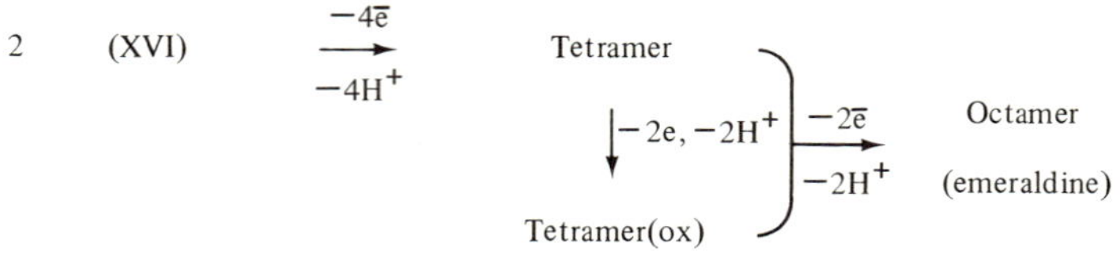

(14)

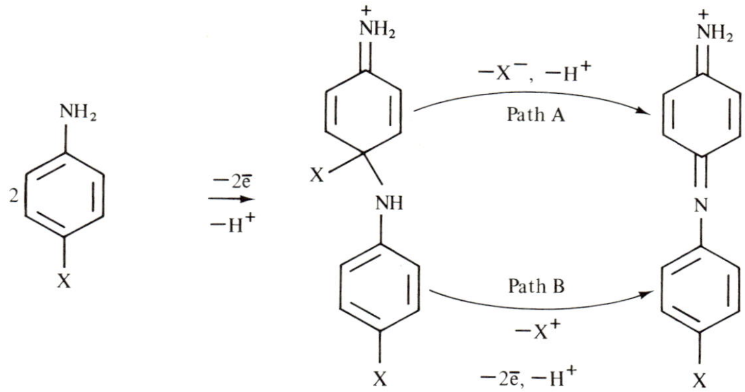

(15)

(XVII) (XVIII) (XIX)

of cation radicals or by nucleophilic attack of substrate on an oxidized intermediate. The mechanism of the second step, (XVIII) → (XIX), was found to be dependent on the nature of X; X can leave either as an anion, resulting in an overall one-electron oxidation, or as a cation (or other oxidized form) accompanied by an overall two-electron oxidation. Quantitative data were presented for p-chloroaniline which fit the 1ē pathway. Others following the 1ē pathway were p-anisidine and p-phenetidine, presumably accompanied by ejection of methoxide and ethoxide ion, respectively. The 2ē pathway was proposed for aniline, o-toluidine, and p-aminobenzoic acid. A competing reaction, benzidine formation, occurs with substrates unsubstituted in the para position.

The anodic oxidation of several para-substituted anilines in the presence of pyridine in acetonitrile was studied at the rotating platinum electrode as well as by product isolation [15]. Two-electron oxidations were observed accompanied by the formation of azobenzenes. The following scheme [Eqs. (16)-(19)] was suggested to account for the results:

$$Ar - \ddot{N}H_2 \quad - \bar{e} \longrightarrow Ar - \overset{+\bullet}{N}H_2 \tag{16}$$

$$Ar - \overset{+\bullet}{N}H_2 + \text{Pyridine} \longrightarrow Ar\dot{N}H + Pyr\bullet H^+ \tag{17}$$

$$2\,Ar - \dot{N}H \longrightarrow Ar - \underset{}{N}(H) - N(H) - Ar \tag{18}$$

$$Ar - N(H) - N(H) - Ar - 2\bar{e} \xrightarrow[Pyr]{} Ar - N{=}N - Ar + 2Pyr\bullet H^+ \tag{19}$$

In some cases, head-to-tail coupling was a competing reaction.

Anodic oxidation of 2, 4, 6-tri, tert-butylaniline in wet acetonitrile [16] is accompanied by the consumption of 4 faradays per mole of substrate and the generation of the quinone (XXII). In dry methanol, (XXIII) is obtained,

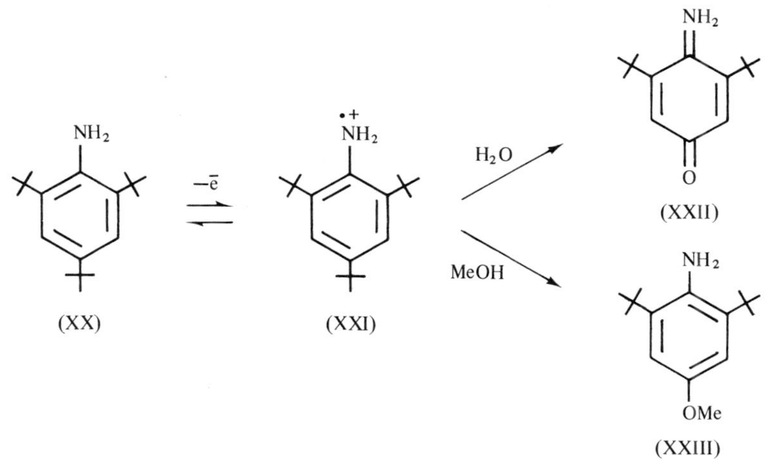

(20)

and in wet solvent a mixture of (XXII) and (XXIII) is formed. Cyclic voltammetry and ESR revealed that the initial step is one-electron oxidation to (XXI).

When the oxidation of (XX) is carried out in acetonitrile in the presence of pyridine, two new compounds, (XXIV) and (XXV), were obtained in yields of 60 and 25%, respectively [18]:

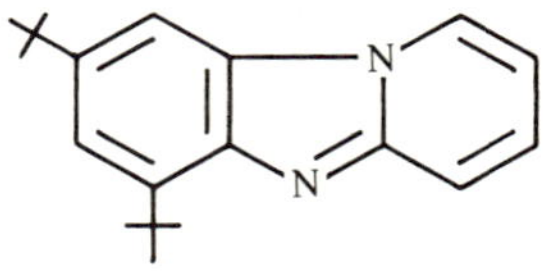

(XXIV)

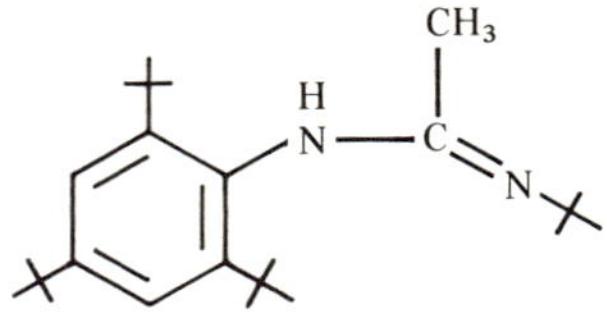

(XXV)

Oxidation of 2, 4, 6-tribromoaniline in acetonitrile containing 2, 6-lutidine gives the corresponding azobenzene as the major product [19] in analogy to the reaction observed with para-substituted anilines in the presence of pyridine [15]. In the absence of lutidine, a competing reaction gives the oxidized diarylamine, (XXVI), and the hydrolysis product, (XXVII):

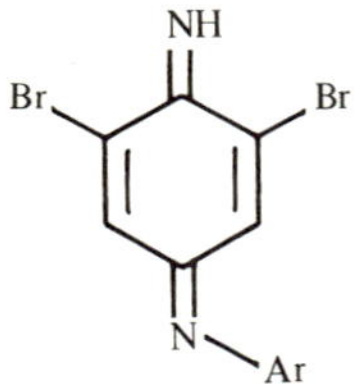

(XXVI)

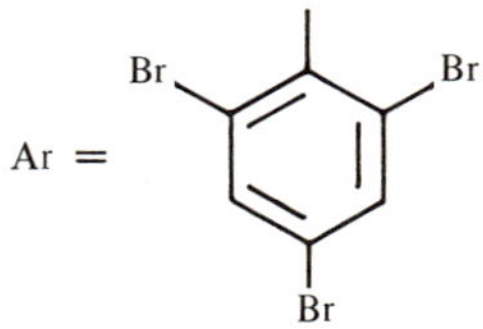

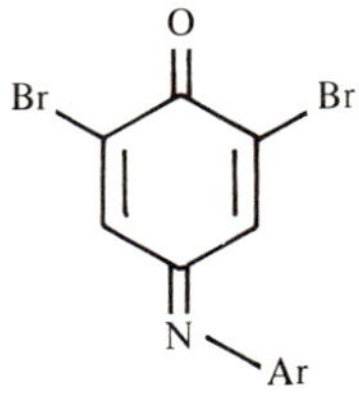

(XXVII)

2. Secondary Aromatic Amines [1]

The anodic oxidation of N-methylaniline in aqueous solution was studied over a wide range of pH [20]. The product of the reaction was suggested to be oxidized dimethylbenzidine (XXXI) on the basis of voltammetric identifica-

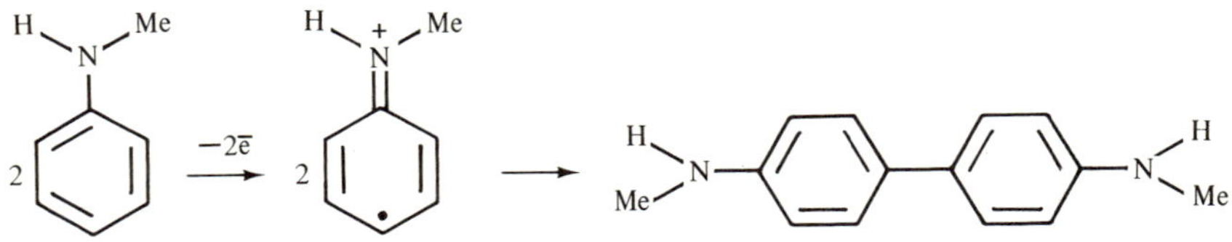

(XXVIII) (XXIX) (XXX)

(21)

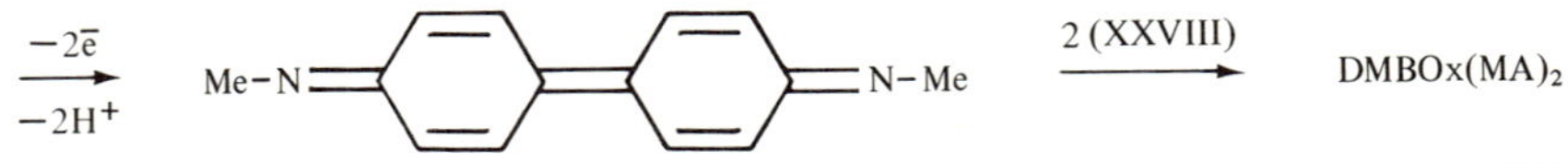

(XXXI)

(22)

tion. The reaction was complicated by further reaction between (XXXI) and substrate [Eq. (22)]. In a similar case [21], diphenylamine was found to undergo oxidation in acetonitrile to form diphenylbenzidine (XXXIII) which was further oxidized to diphenylbenzidine violet (XXXIV):

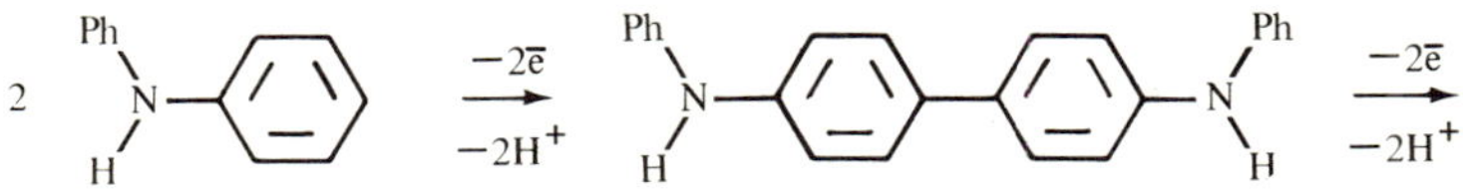

(XXXII) (XXXIII)

(23)

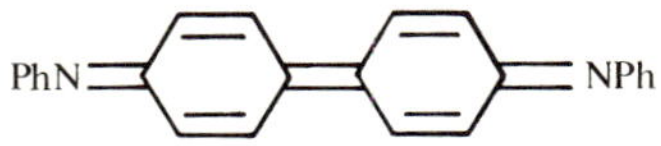

(XXXIV)

The oxidation of di-p-anisylamine to the stable cation radical was recently studied with the aid of electron spin resonance spectrometry [22].

3. Tertiary Aromatic Amines

The anodic oxidation of N, N-dimethylaniline in aqueous solution was studied by a variety of techniques [23-25]. The mechanism of benzidine formation is believed to involve coupling of anodically generated cation radicals [26], as depicted below [Eqs. (24)-(26)]:

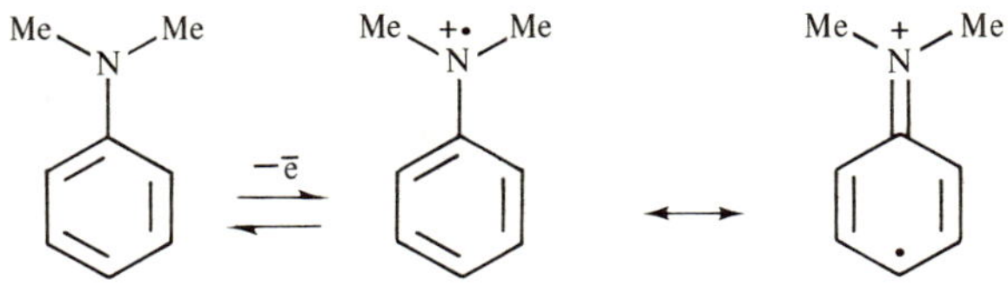

(24)

(XXXV) (XXXVI) (XXXVII)

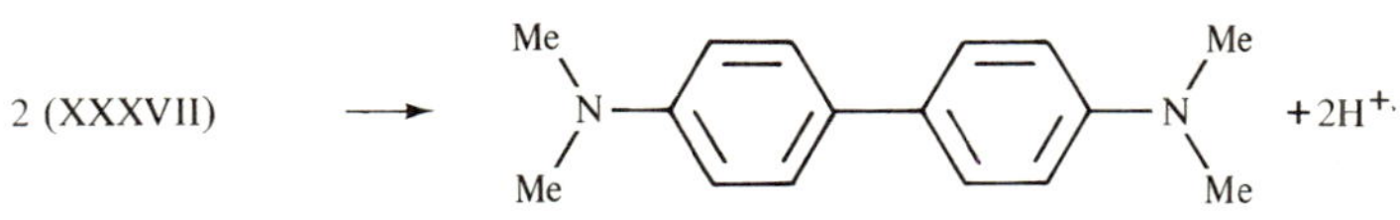

(25)

(XXXVIII)

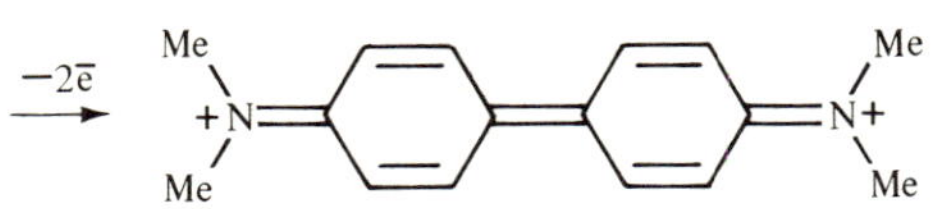

(26)

(XXXIX)

The cation radical (XXXVI) ↔ (XXXVII) has not been observed due to rapid dimerization [Eq. (25)].

The electrolysis of (XXXV) in methanol containing KOH was accompanied by the formation of (XL) and (XLI) [5], while the mononitrate of (XXXVIII) was obtained when the supporting electrolyte was ammonium nitrate [27].

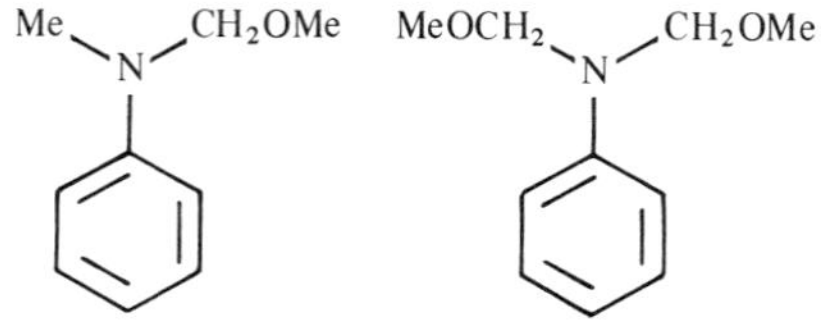

(XL) (XLI)

In a related reaction anodic cyanation of N-methyl-N-ethylaniline (XLII) produced a 2:1 mixture of (XLIII) and (XLIV) [28]:

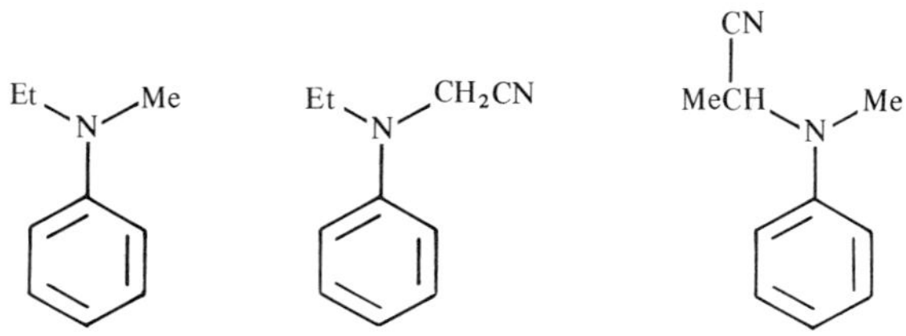

(XLII) (XLIII) (XLIV)

The stability of the cation radicals formed upon anodic oxidation of triphenylamines varies markedly with para substitution [26]. The cation radical of tri-p-anisylamine is extremely stable, not only in nonaqueous solvents but also in aqueous solutions of pH 2-6. The cation-radical stability of a number of triphenylamines [29] and the effect of para substitution on the extent and rate of benzidine formation [30] has been determined.

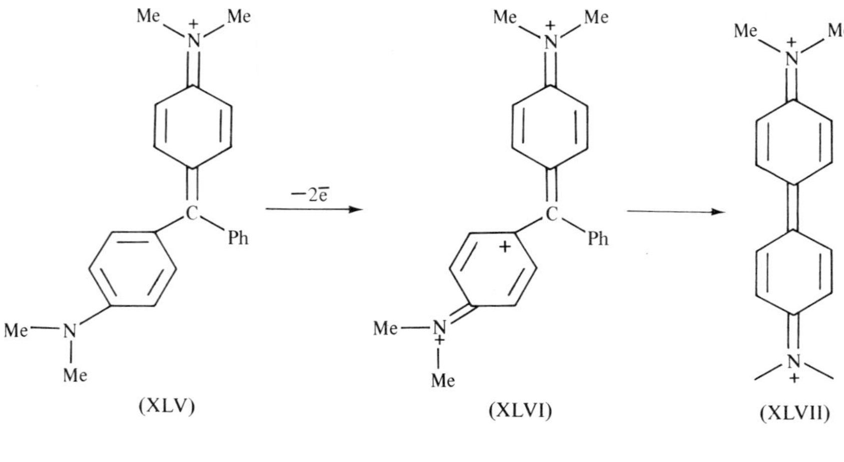

(XLV) (XLVI) (XLVII) (27)

4. p-Dimethylaminophenylmethanes

Several triphenylmethane dyes [31], (XLV) being an example, were found to undergo anodic oxidation accompanied by the ejection of the central carbon fragment and the generation of oxidized benzidines [Eq. (27)]. The initial step was shown to consist of a two-electron transfer. The nature of subsequent steps, (XLVI) → (XLVII), is uncertain; however, it is likely

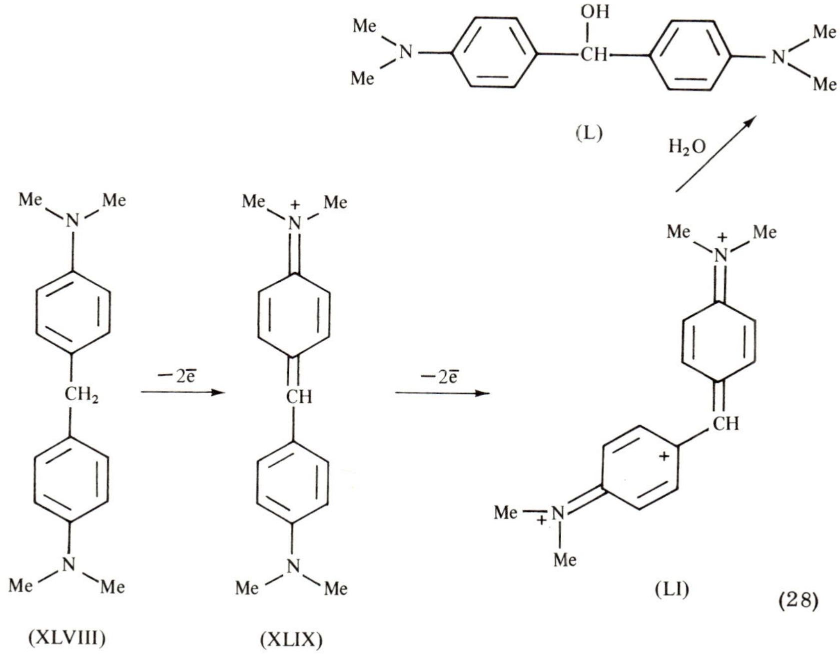

(28)

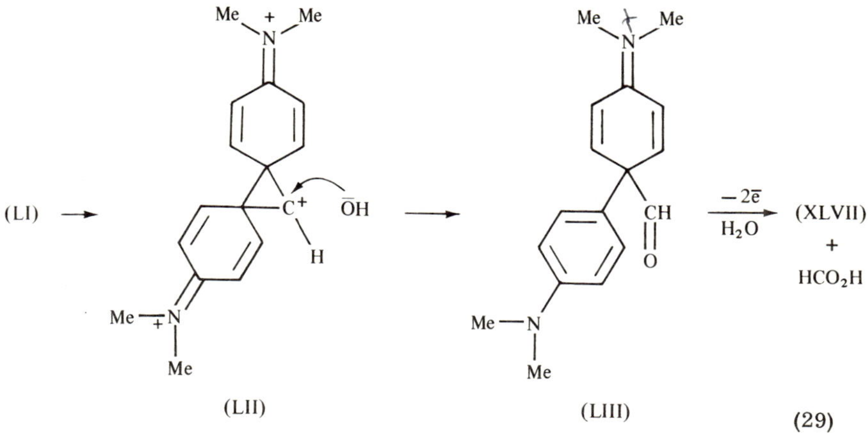

(29)

that a cyclopropane intermediate, as was proposed for the chemical oxidation of (XLV) [32], is involved.

The diarylmethane (XLVIII) forms either (XLVII) [31] or (L) [33] when electrolyzed in aqueous acid solution. The isolation of (L) during the early investigation [33] suggests the reaction sequence [Eqs. (28), (29)].

In the framework of this discussion, it is not surprising that the monoarylmethane, N, N-dimethyl p-toluidine (LIV), undergoes oxidation in the side chain [31] to give the aldehyde (LVIII) and the acid (LX) [Eqs. (30), (31)] rather than dimerization to an orthobenzidine:

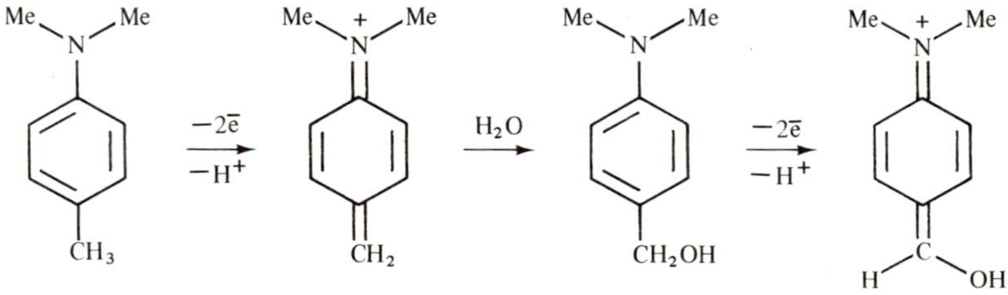

(30)

(LIV) (LV) (LVI) (LVII)

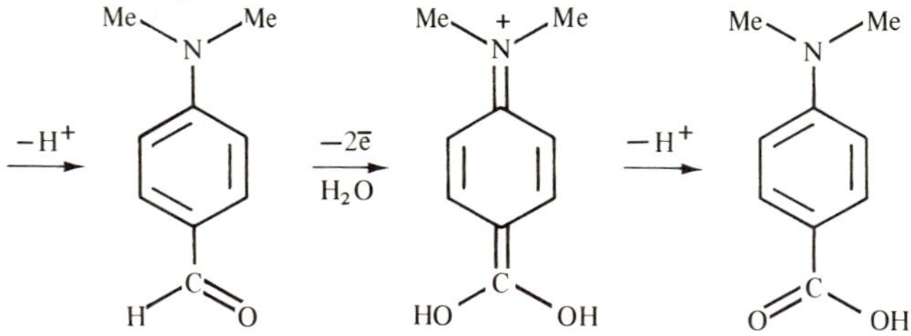

(31)

(LVIII) (LIX) (LX)

In acetonitrile (LIV) undergoes 1-ē oxidation to the cation radical which loses a proton to yield the corresponding benzyl radical [33a]. The benzyl radical dimerizes to the bibenzyl.

B. Oxidation of Phenylenediamines

The anodic electrochemistry of phenylenediamines is complicated by a host of chemical follow-up reactions. Only a brief introduction is presented here; a comprehensive treatment can be found elsewhere [34].

Quantitative studies of the anodic reactions of p-phenylenediamine in aqueous solution have been carried out [35, 36]. The following simplified scheme, ignoring precursor reactions and acid-base relationships, summarizes reactions occurring in aqueous solution [34]. After the initial

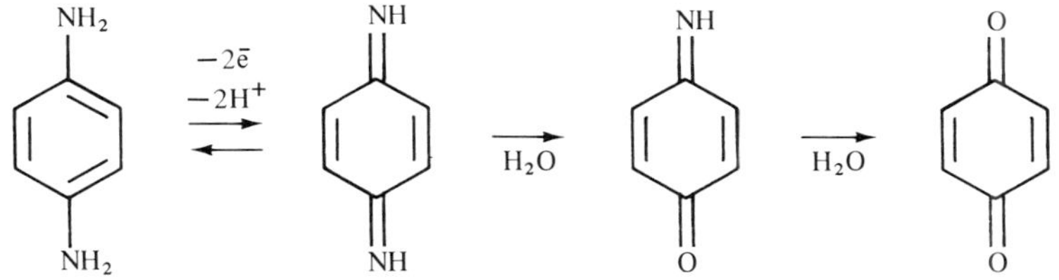

(32)

(LXI) (LXII) (LXIII) (LXIV)

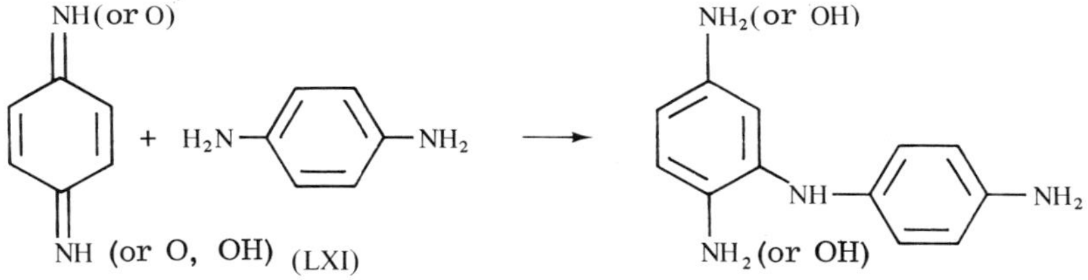

(33)

(LXII) (LXIII) (LXIV) (LXV)

two-electron oxidation to (LXII), rapid hydrolysis and 1, 4-addition reactions take place to give the mixed product, (LXV) [Eqs. (32), (33)]. The electrochemistry is further complicated by further oxidation of (LXV). The reactions of several other phenylenediamines have been summarized [34].

C. Oxidation of Aminophenols

p-Aminophenol undergoes a clear-cut EC process during anodic oxidation in aqueous solution and has become a standard test system for evaluation of the applicability of electroanalytical techniques to the study of EC processes [34].

A two-electron oxidation is observed, followed by hydrolysis of the quinone imine at a rate dependent on the pH [Eq. (34)]. Hydrolysis rates have been measured by reverse-current chronopotentiometry, cyclic voltammetry, potentiostatic, and other methods [37, 38]. These along with the results obtained for other systems have been summarized [34].

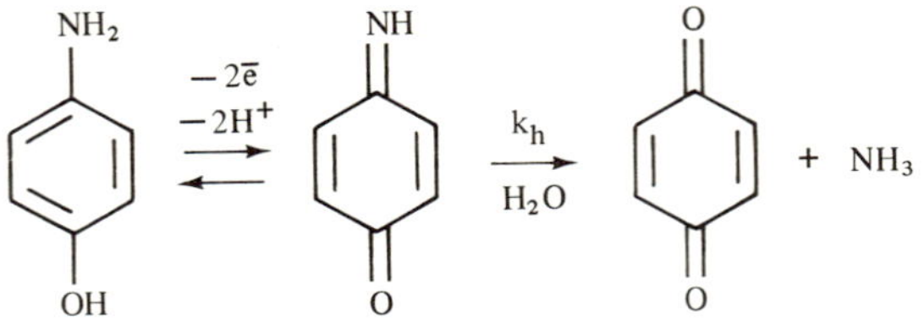

(LXVI) (LXVII) (LXVIII) (34)

D. Oxidation of Aminoanthracenes

The anodic oxidation of 9-aminoanthracenes in acetonitrile has been the object of an intensive investigation [39-42]. The reactions were subjected to analysis by cyclic voltammetry, controlled potential electrolysis, and ESR spectrometry.

9-Amino-10-phenylanthracene (LXIX) shows two anodic polarographic waves, at +0.17 and +1.02 V (vs. Ag/Ag^+), at a platinum electrode. Oxidation at +0.2 V produced the cation radical (LXX), (LXXI) which was observed by ESR [40]:

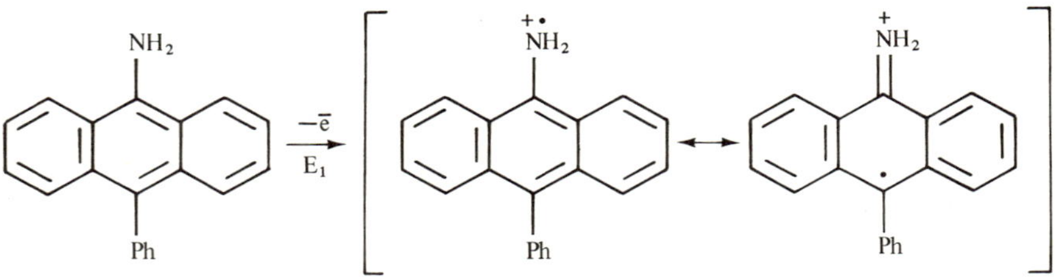

(LXIX) (LXX) (LXXI) (35)

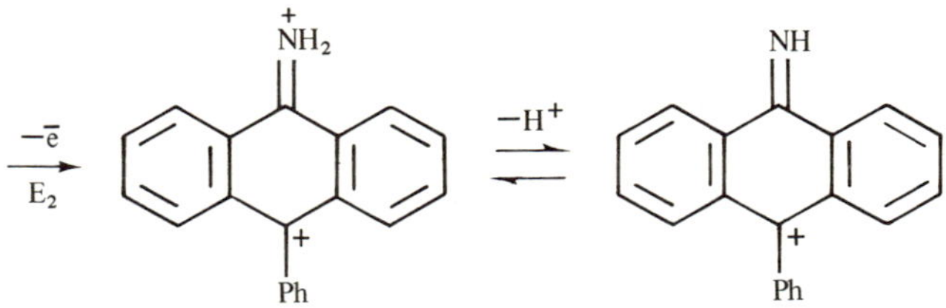

(LXXII) (LXXIII) (36)

The wave at the higher potential was assigned to oxidation of the cation radical to the dication [Eqs. (35), (66)]. Products formed during electrolysis at controlled potential are consistent with this interpretation. Oxidation at a potential between the two waves produced the peroxide (LXXIV), in the presence of oxygen, and a head-to-tail dimer (LXXV), when conducted in the presence of the base, diphenylguanidine. Oxidation at higher potentials produced the dication which reacted with water to give (LXXVI):

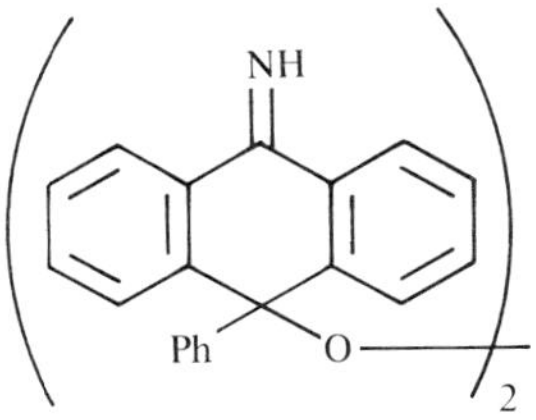

(LXXIV)

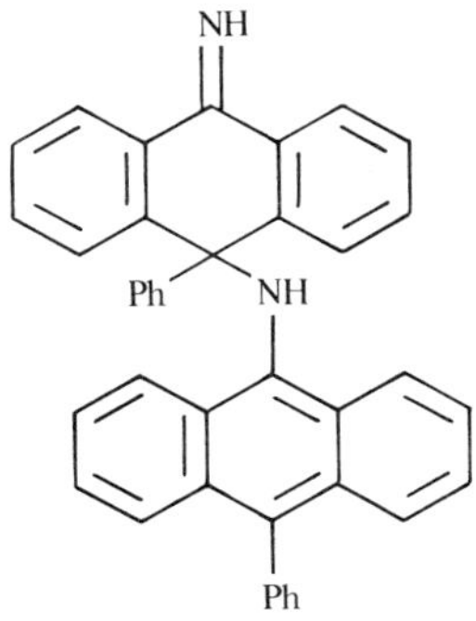

(LXXV)

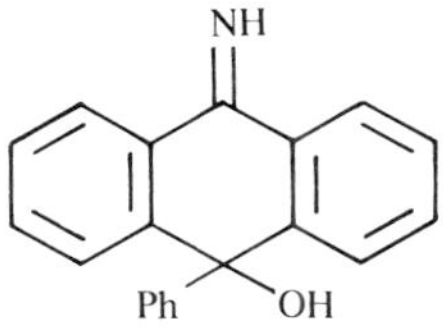

(LXXVI)

The imino function of (LXXIV), (LXXV) and (LXXVI) was observed to undergo hydrolysis to give the corresponding keto compounds.

Several N-substituted 9-aminoanthracenes of general formula (LXXVII) were studied [39, 42] and were found to behave in a manner similar to (LXIX). An additional product (LXXVIII), formed by a tail-to-tail coupling process, was observed when the 10-position was unsubstituted:

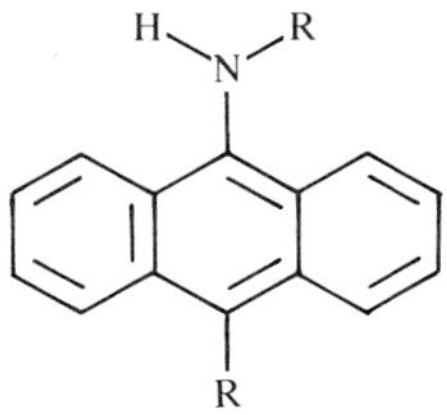

(LXXVII)

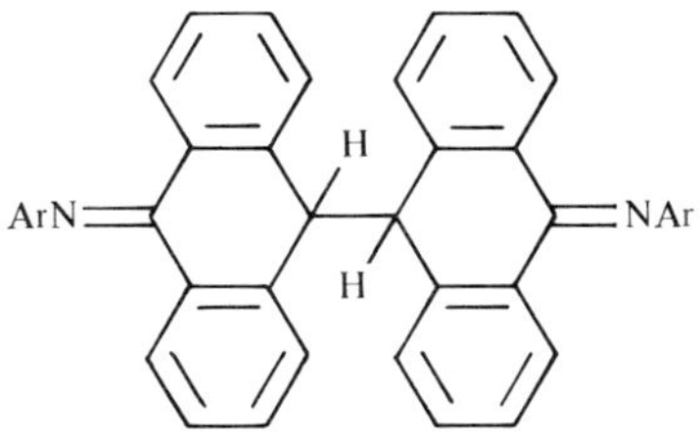

(LXXVIII)

VI. OXIDATION OF AMINO- AND AMINOPHENYLETHYLENES

The anodic oxidation of ethylenes, completely substituted with amino or aminophenyl groups, resulted in the formation of remarkably stable cation radicals and dications. When one of the ethylene carbons is unsubstituted, rapid dimerization reactions occur.

A. Aminoethylenes

Tetrakis(dimethylamino)ethylene (LXXIX) was found to undergo a two-step oxidation at the dropping mercury electrode to produce first the cation radical (LXXX) and then the dication (LXXXI):

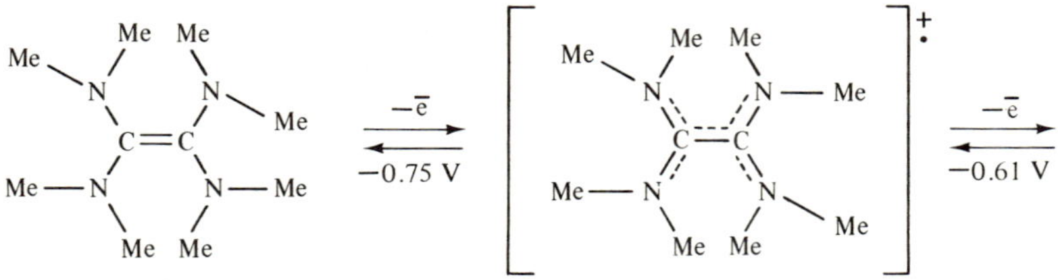

(LXXIX) (LXXX)

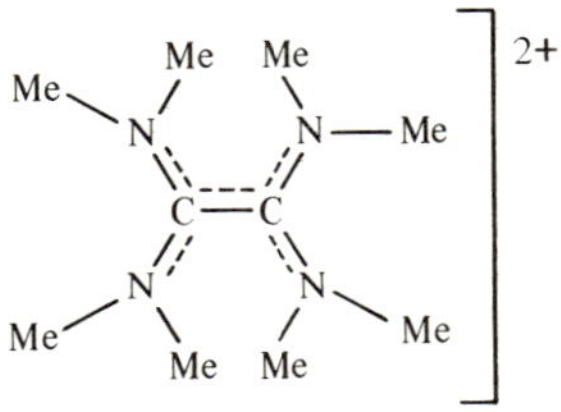

(37)

(LXXXI)

The structure of (LXXX) was studied by ESR and compared with that of N, N, N′, N′-tetramethylphenylenediamine [43]. The spectrum indicates significant restriction to rotation around the ethylene carbon to nitrogen bonds.

Vinylidinebisdimethylamine (LXXXII), having an ubsubstituted position, was oxidized electrolvtically to 1, 1, 4, 4-tetrakis (dimethylamino)butadiene [44]. The butadiene (LXXXIII) is noteworthy in that it undergoes reversible 2-e oxidation at the unusually low anode potential, -0. 8 V vs. SCE:

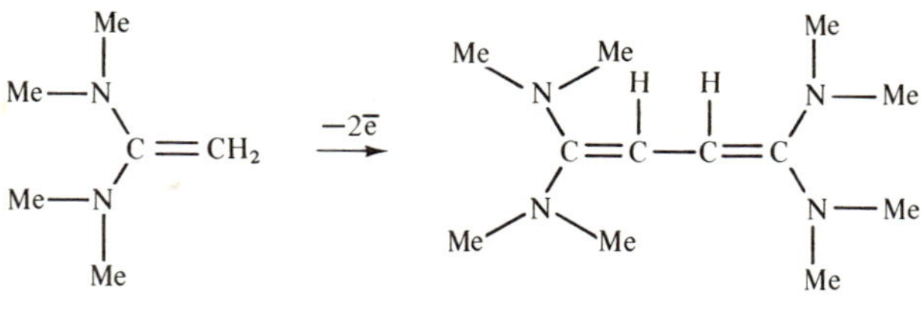

(38)

(LXXXII) (LXXXIII)

In a recent exhaustive study of the anodic oxidation of p-dimethylamino-ethylenes [45] lifetimes of numerous cation radicals were determined by cyclic coulometry and voltammetry. Half-lives in acetonitrile ranging from

3×10^{-3} sec to at least several days were recorded and structural factors contributing to cation-radical stability and follow-up reaction pathway were discussed.

B. Aminophenylethylenes

Cyclic voltammetry, rotating disk electrode voltammetry, and controlled potential coulometry were used to show that tetrakis(p-N, N-dimethylaminophenyl)ethylene (LXXXXIV) undergoes reversible two-electron oxidation to the corresponding dication [46].

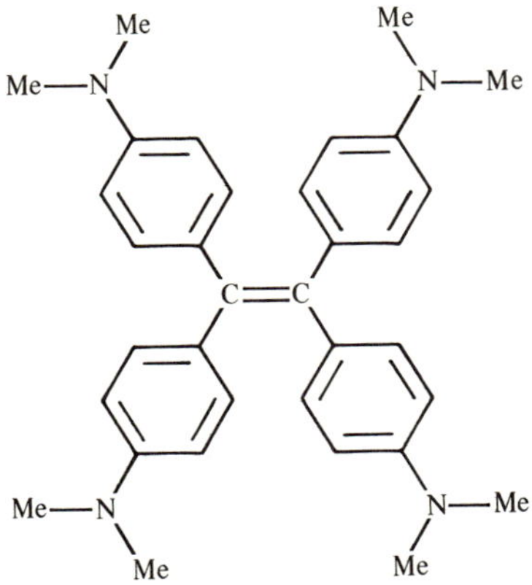

(LXXXIV)

NOTE

[1] For anodic oxidation pathways of diphenylamines see [47].

REFERENCES

[1]. R. A. Benkeser, E. M. Kaiser, and R. F. Lambert, J. Amer. Chem. Soc., 86, 5272 (1964).

[2]. K. K. Barnes and C. K. Mann, J. Org. Chem., 32, 1474 (1967).

[3]. R. F. Dapo and C. K. Mann, Anal. Chem., 35, 677 (1963).

[4]. C. D. Russel, ibid., 35, 1291 (1963).

[5]. N. L. Weinberg and E. A. Brown, J. Org. Chem., 31, 4054 (1966).

[6]. P. J. Smith and C. K. Mann, ibid., 33, 316 (1968).

[7]. N. L. Weinberg, ibid., 33, 4326 (1968).

[8]. S. Mizuno, Y. Hirata, and K. Takei, J. Electrochem. Soc. Japan, Overseas Ed., 29, 31 (1961); CA, 60, 11889 (1964).

[9]. S. Mizuno, Y. Hirata, and K. Takei, Nagoya Kogyo Kaigaju Gakuho, 14, 48 (1962); CA, 60, 11889 (1964).

[10]. N. L. Weinberg and H. R. Weinberg, Chem. Rev., 68, 449 (1968).

[11]. F. Fichter and M. Schmidt, Helv. Chim. Acta, 3, 704 (1920).

[12]. Y. Takayama, T. Harada, and S. Miduno, Bull. Chem. Soc. Japan, 12, 342 (1937).

[13]. D. M. Mohilner, R. N. Adams, and W. J. Argersinger, Jr., J. Amer. Chem. Soc., 84, 3618 (1962).

[14]. J. Bacon and R. N. Adams, ibid., 90, 6596 (1968).

[15]. S. Wawzonek and T. W. McIntyre, J. Electrochem. Soc., 114, 1025 (1967).

[16]. G. Cauquis, G. Fauvelot, and J. Rigaudy, Compt. Rend., 264, 1758 (1967).

[17]. G. Cauquis, G. Fauvelot, and J. Rigaudy, Bull. Soc. Chim. Fr., 4928 (1968).

[18]. J. J. Basselier, G. Cauquis, and J. L. Cros, Chem. Commun., 1171 (1969).

[19]. G. Cauquis, J. P. Coquand, and J. Rigaudy, Compt. Rend., 268, C, 2265 (1969).

[20]. Z. Galus and R. N. Adams, J. Phys. Chem., 67, 862 (1963).

[21]. V. Dvorak, I. Kemec, and J. Zyka, Microchem. J., 12, 99 (1967).

[22]. G. Cauquis and D. Serve, Tetrahedron Lett., 17 (1970).

[23]. T. Mizoguchi and R. N. Adams, J. Amer. Chem. Soc., 84, 2058 (1962).

[24]. Z. Galus and R. N. Adams, ibid., 84, 2061 (1962).

[25]. Z. Galus, R. M. White, F. S. Rowland, and R. N. Adams, ibid., 84, 2065 (1962).

[26]. R. N. Adams, Electrochemistry at Solid Electrodes, Dekker, New York, 1969, p. 351.

[27]. N. L. Weinberg and T. B. Reddy, J. Amer. Chem. Soc., 90, 91 (1968).

[28]. S. Andreades and E. W. Zahnow, Abstracts of Papers, Electrochemical Society Meeting, Electroorganic Section, Dallas, Texas, 1967, p. 15.

[29]. E. T. Seo, R. F. Nelson, J. M. Fritsch, L. S. Marcous, and R. N. Adams, J. Amer. Chem. Soc., 88, 3498 (1966).

[30]. R. F. Nelson and R. N. Adams, ibid., 90, 3025 (1968).

[31]. Z. Galus and R. N. Adams, ibid., 86, 1666 (1964).

[32]. V. Hanousek and M. Matrka, Coll. Czech. Chem. Commun., 24, 16 (1959).

[33]. F. Esherich and M. Moest, Z. Electrochem., 8, 849 (1902).

[33a]. M. Melicharek and R. F. Nelson, J. Electroanal. Chem., 26, 201 (1970).

[34]. Ref. [26], Chaps. 8 and 10.

[35]. H. B. Mark and F. C. Anson, Anal. Chem., 35, 722 (1963).

[36]. M. D. Hawley, Ph. D. Thesis, Univ. Kansas, Lawrence, Kansas, 1965.

[37]. D. Hawley and R. N. Adams, J. Electroanal. Chem., 10, 376 (1965).

[38]. A. C. Testa and W. H. Reinmuth, Anal. Chem., 32, 1512 (1960).

[39]. J. P. Billon, G. Cauquis, J. Raison, and Y. Thibaud, Bull. Soc. Chim. Fr., 199 (1967).

[40]. G. Cauquis, J. Badoz-Lambling, and J. P. Billon, ibid., 1433 (1965).

[41]. G. Cauquis and J. P. Billon, Compt. Rend., 255, 2128 (1962).

[42]. J. Raison, Bull. Soc. Chim. Fr., 959 (1963).

[43]. K. Kuwata and D. H. Geske, J. Amer. Chem. Soc., 86, 2101 (1964).

[44]. J. M. Fritsch and H. Weingarten, ibid., 90, 793 (1968).

[45]. J. M. Fritsch, H. Weingarten, and J. D. Wilson, ibid., 92, 4038 (1970).

[46]. A. J. Bard and J. Phelps, J. Electroanal. Chem., 25, App. 2 (1970).

[47]. P. Berkenkotter and R. F. Nelson, Abstract No. 116, Electrochemical Society Meeting, Washington, D.C., May 1971.

Chapter XV

ANODIC OXIDATION OF OXYGEN-CONTAINING COMPOUNDS

Vernon D. Parker

Department of General and Organic Chemistry
University of Copenhagen
Copenhagen, Denmark

I. INTRODUCTION

The anodic oxidation of oxygen-containing compounds has proven to be a fertile area of research for the organic electrochemist. This chapter is concerned primarily with anodic oxidations in which the oxygen function participates directly in the reaction. Oxidation of carboxylic acids (Chapter XIII) and dimerization of salts of active hydrogen compounds such as β-diketones (Chapter XX) are treated elsewhere as indicated.

Anodic oxidation as a preparative tool is still in its infancy. It is apparent from the material in this chapter that much remains to be done. The optimization of such processes as oxidation of alcohols to aldehydes, aldehydes to acids, and oxidative degradation of sugars, to mention only a few, must be accomplished in order to begin to realize the vast potential of this method.

II. ANODIC OXIDATION OF PHENOLS AND DERIVATIVES

The oxidation of phenols and derivatives is of paramount importance in organic electrochemistry. The importance of phenols and polyols in biological systems provides a link between biochemistry and electrochemistry and cross-fertilization of the two areas is now beginning to take place.

Vermillion and Pearl [1] conducted one of the first studies employing modern techniques to the oxidation of phenolic compounds. They pointed out that two entirely different anodic reactions are possible, depending on the conditions employed. It was postulated that nonionized phenols undergo two-electron oxidation to give the phenoxonium ion while phenolate ions undergo one-electron oxidation to give the phenoxy radical. The fate of phenoxy radicals is generally a coupling reaction. Since anodic coupling is covered in Chapter XX, it will not be covered here. This section is concerned with the other reaction pathway, generation of cationic intermediates.

A. Oxidation of Simple Phenolic Compounds

A comparison of the diffusion currents for a number of phenols with that of hydroquinone in acetonitrile indicated that two-electron transfers are involved [1]. Electrode filming was severe with all compounds except those in which two or more of the reactive ortho and para positions were occupied by blocking groups. Detailed studies were conducted on 2, 6-di-tert-butyl-p-cresol. Based on voltammetric data, it was proposed that the immediate product formed during anodic oxidation in acetonitrile was the quinone methide (II) which undergoes dimerization [Eq. (1)] and further oxidation. When the oxidation was conducted in the presence of methanol, nucleophilic attack on the phenoxonium ion was accompanied by formation of (V) in 65% yield.

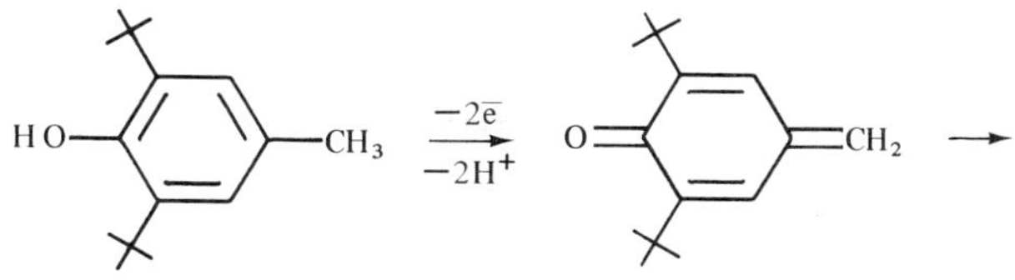

(I) (II)

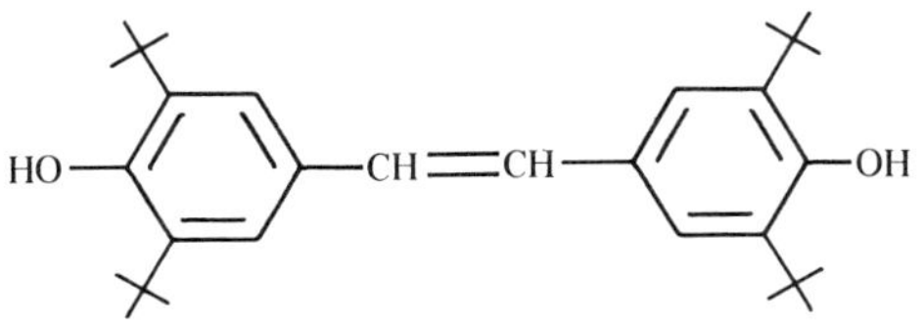

(III)

(1)

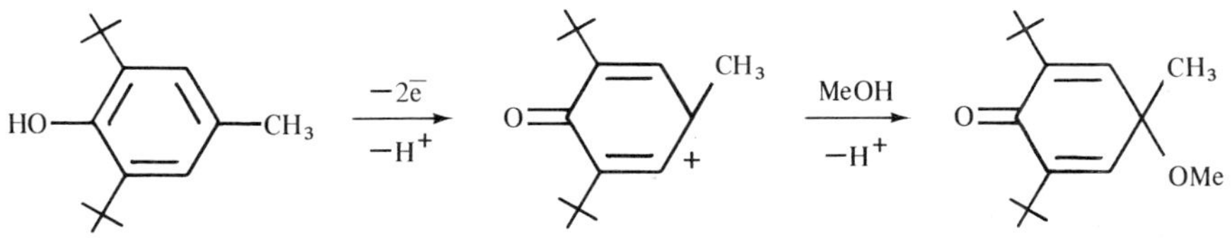

(I) (IV) (V)

(2)

A more recent study of the anodic oxidation of (I) in acetonitrile has shown that (II) is not a primary product [1a]. Both coulometry and rotating disk electrode voltammetry show that 2-e are consumed per mole of (I). Product studies indicate that the initial product is (IV) which reacts with any nucleophile present to give dienones analogous to (V).

Further advantage was recently taken of phenol blocked in the 2, 4 and 6-position with tert-butyl groups [2]. The oxidation of (VI) in acetonitrile containing tetraethylammonium hydroxide produced the blue radical (VII) which on reaction with oxygen generated the peroxide (VIII) [Eq. (3)]. In acetonitrile solution, the phenol (IX) produced the cation (X) which could react with water to give the quinone (XI) or with acetonitrile to form the benzisoxazole (XII), depending on the concentration of water [Eq. (4)].

The oxidation of (IX) is analogous to the behavior of (I) with the exception that the dienones of general structure (XIIa) undergo rapid acid-catalyzed debutylation to give the hydroquinone derivative (XIIb) [2a]. When the oxidation of (IX) is carried out in the presence of stoichiometric amounts of

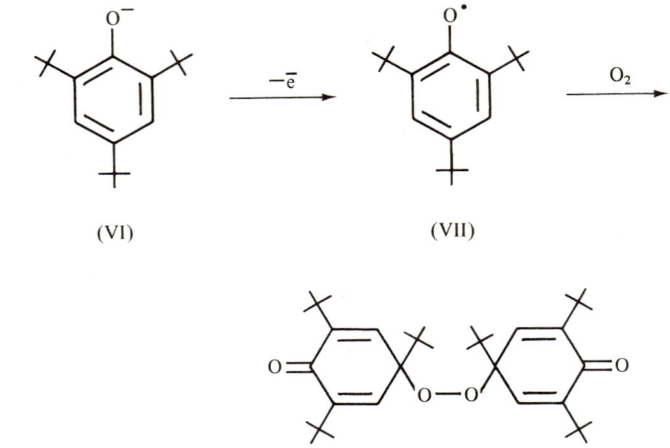

(VI) (VII) (VIII) (3)

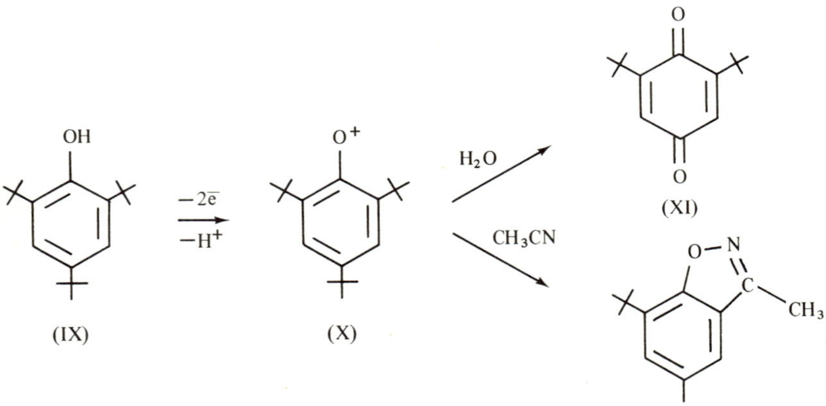

(IX) (X) (XI) (XII) (4)

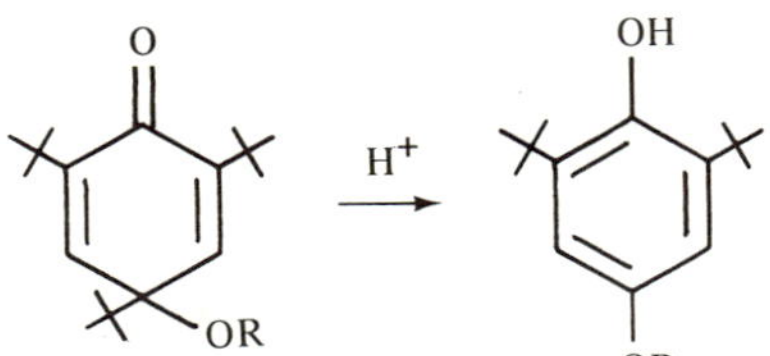

(XIIa) (XIIb)

a base such as 2, 6-lutidine, (XIIa), where R is either H or CH_3, has been isolated in high yield [2a].

Voltammetric oxidation of 2, 4, 6-triphenylphenol produces the radical (XIV) which undergoes further oxidation in acetic acid-sodium acetate to produce a mixture of (XV) and (XVI) [3]:

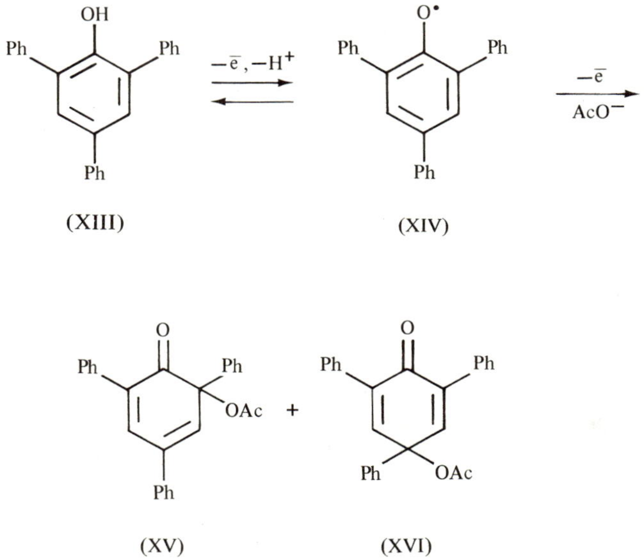

(5)

Anodic oxidation was investigated as a nonenzymatic means of cleavage of tyrosyl-peptide bonds [4, 5]. Phloretylglycine (XVII) was cleaved to (XVIII) and glycine in yields of 30-50%:

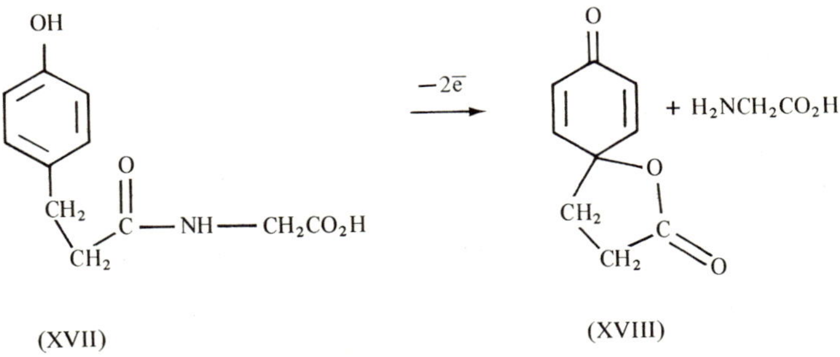

(6)

A competing reaction is the nucleophilic attack of water on the intermediate phenoxonium ion [6].

The older literature contains numerous examples of hydroxylation of phenol and substituted phenols. The reactions were generally carried out under noncontrolled conditions and characterization of the products was not complete. Reference to the latter work can be found in the recent review [7].

B. Anodic Oxidation of Hydroquinones and Catechols

The anodic oxidation of hydroquinone in aqueous acid solution results in the formation of quinhydrone in 75% yield [8, 9]. In pyridine solution, a two-electron oxidation was observed and the product was found to be (XIX) [10]:

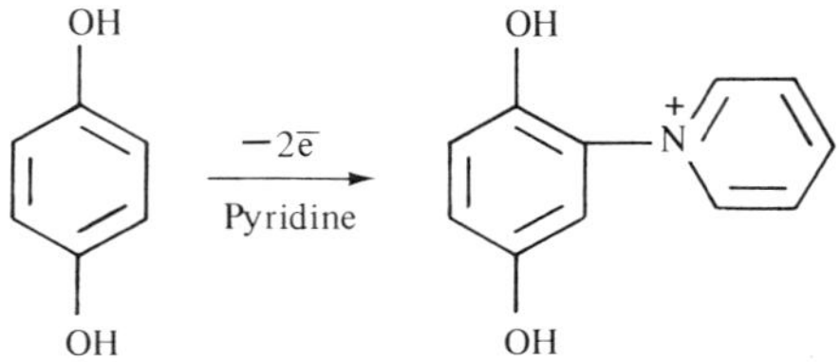

(7)

(XIX)

The course of the anodic oxidation of hydroquinone in acetonitrile is not yet settled. The product of controlled potential electrolysis is benzoquinone. On the basis of rapid scan voltammetry, it was suggested that the initial process is a one-electron oxidation to form a dimer, either quinhydrone [11] or a quinone hemiacetal [12]. An alternative interpretation [13], based on the effect of added bases, i.e., 2, 6-lutidine or water, involves two-electron anodic oxidation accompanied by protonation equilibria, slow during the time scale of rapid sweep voltammetry. The latter interpretation was criticized [12] and it was pointed out that the observed differences could be due to the use of two different supporting electrolytes. However, a rotating disk electrode study [13a] provides convincing evidence that the primary process involves a 2-e oxidation.

Substituted hydroquinones have been observed to undergo 1, 4-addition of water during anodic oxidation in aqueous acid solution [14]. The mechanism of this reaction was shown to consist of anodic generation of the quinone (XX) followed by 1, 4-addition of water, rearrangement to the hydroquinone (XXII) and further two-electron oxidation [Eqs. (8), (9)]. The reaction occurs only when R is a strong electron-withdrawing group.

In aqueous solution, catechol and 4-methylcatechol undergo reversible two-electron oxidation [15]. In the presence of HCl or amines, 1, 4-addition is observed. A modified ECE mechanism was established and kinetic parameters were determined under a variety of conditions [16]. The high reactivity of o-quinones gives rise to an interesting area of anodic electrochemistry. The biologically important compounds, catecholamines,

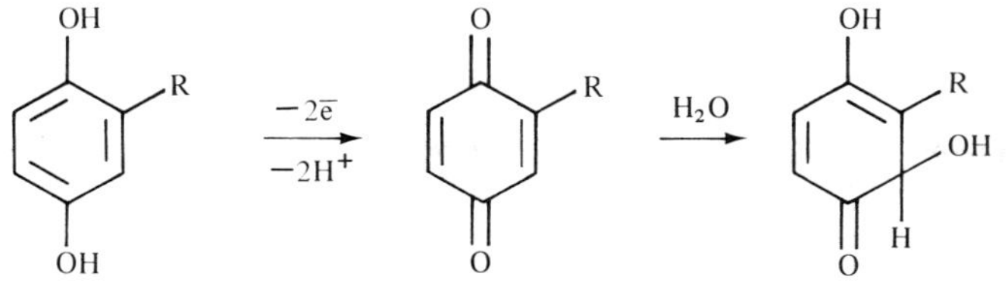

(XX) (XXI)

(8)

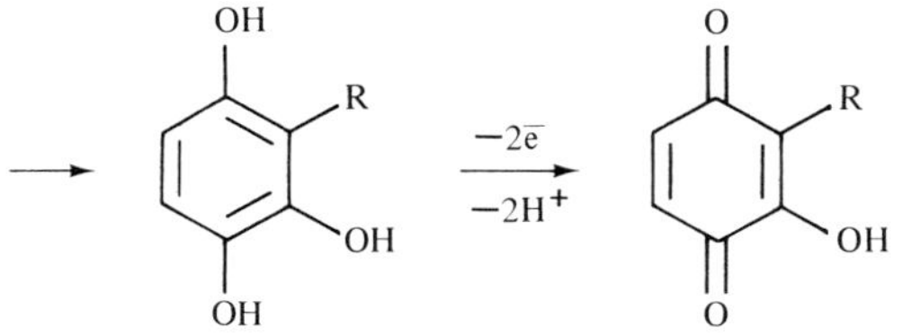

(XXII) (XXIII)

(9)

adrenaline, noradrenaline, dopamine, and related substances, have been studied [17]. The anodic oxidation of adrenaline in acidic solution is a typical example [Eqs. (10)-(12)]:

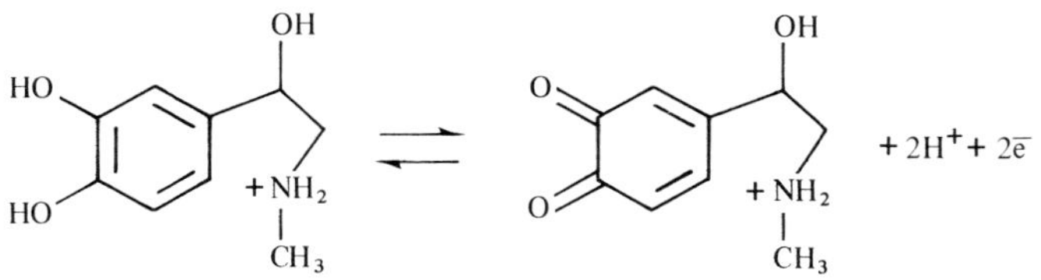

(XXIV) (XXV)

(10)

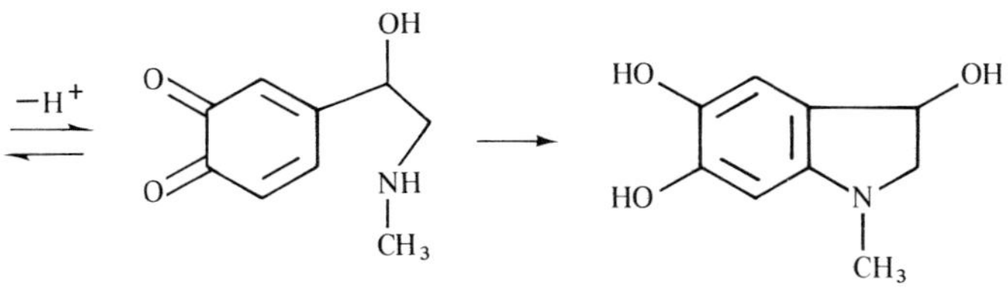

(XXVI) (XXVII)

(11)

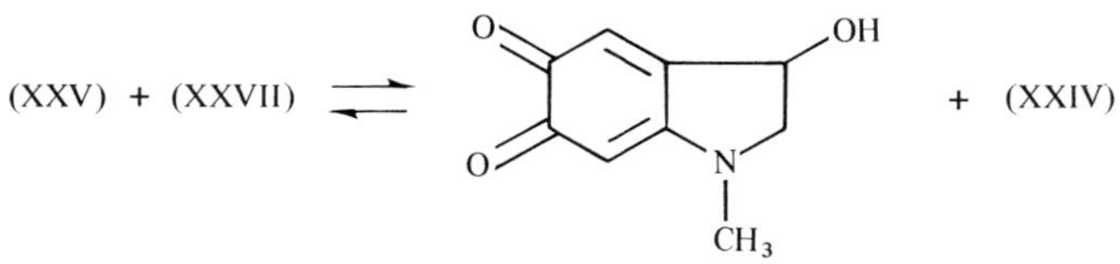

(XXVIII)

(12)

C. Alkyl and Acyl Transfer from Aromatic Ethers and Esters

Electroanalytical techniques have been applied to the study of the anodic oxidation of p-methoxyphenol [18] and 4,4′-oxydiphenol [19]. These reactions were accompanied by the quantitative generation of benzoquinone [Eq. (13)]. The voltammetry is consistent with a two-electron transfer followed by rapid chemical steps.

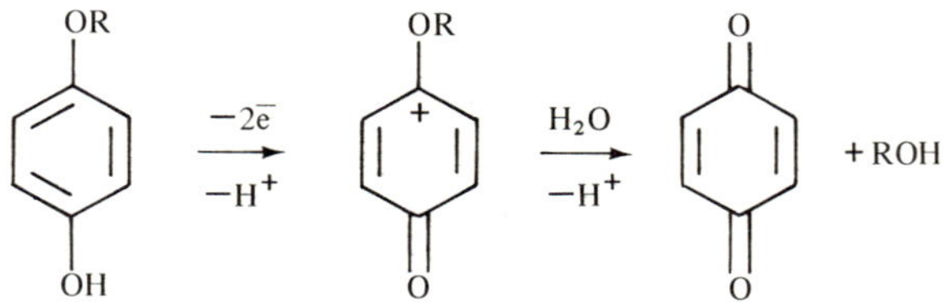

(13)

In an analogous system, a quasi-reversible two-electron oxidation has been observed. Cyclic voltammetry of the α-tocopherol model compound (XXIX) reveals that the dication (XXX) is sufficiently stable in acetonitrile to be observed during the time scale of slow sweep voltammetry [20]. In the presence of sodium acetate and acetic acid (XXX) undergoes ring cleavage to give the quinone (XXXI). In aqueous acetonitrile (XXIX) undergoes

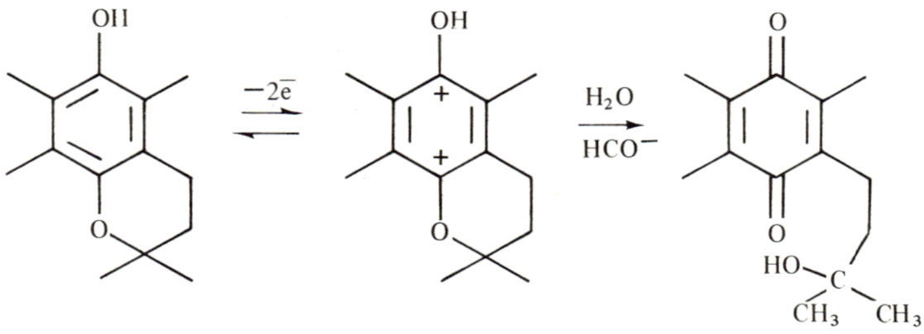

(14)

(XXIX) (XXX) (XXXI)

irreversible 2-e oxidation and the cation (XXXIa) is presumably the primary product [20a]. It was also suggested that the quasi-reversible oxidation in dry acetonitrile also produces (XXXIa). However, the reduction peak originally [20] assigned to (XXX) appears about 0.6 V anodic of that observed for (XXXIb) [which should be a good model for (XXXIa)], and thus it seems more likely that (XXX) [a protonated form (XXXIa)] is the initial product in dry acetonitrile:

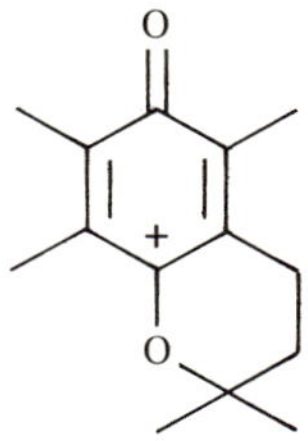

(XXXIa)

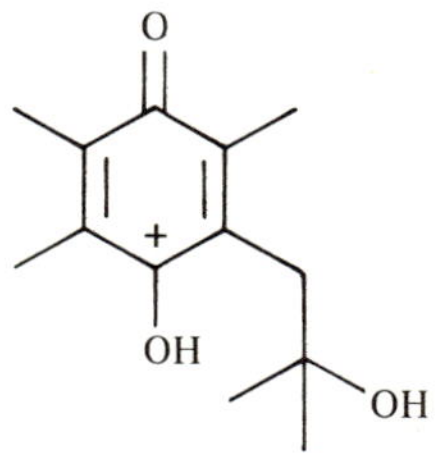

(XXXIb)

Methoxysubstituted aromatic compounds are known to produce cation radicals of some stability [21]. Alkyl transfer from oxygen is a probable decomposition pathway of such intermediates. In a pertinent case, the dimethyl ether of durohydroquinone was found to undergo two-electron anodic oxidation accompanied by the generation of duroquinone [22]. A similar oxidation was observed with the diacetate [22] in contrast to the known stability of hydroquinone diacetates to chemical oxidation [23].

Both methyl and acetyl transfers have been observed to occur during oxidation of derivatives of 9-anthranol [24]. Either 9-methoxy- or 9-acetoxyanthracene was found to undergo one-electron oxidation accompanied by the quantitative generation of bianthrone [Eq. (15)]. The latter results implicate alkoxy and acyloxy derivatives as intermediates in the anodic oxidation of anthracene to bianthrone in the presence of alcohols [25, 26] and acetic acid [27].

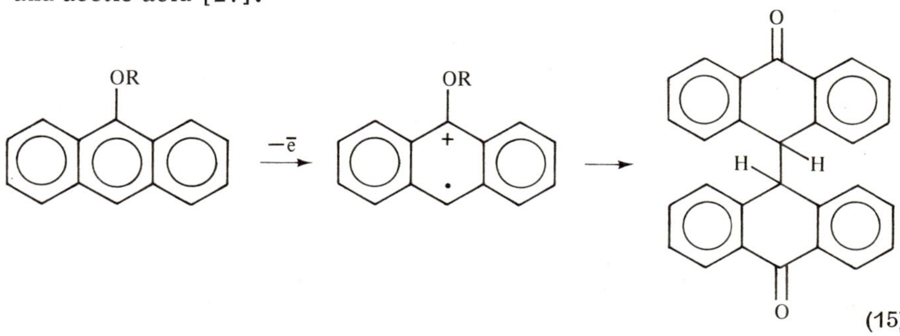

(15)

The oxidation of 9-methoxy-10-methylanthracene (XXXII) in acetonitrile is accompanied by the formation of the dimer (XXXIV) in high yield [28]. The quinone methide (XXXIII) was ruled out as an intermediate when it was found to oxidize about 1 V more anodic than (XXXII). It is interesting to note that (XXXIV) can be prepared in about 50% yield by anodic oxidation of (XXXIII) in acetonitrile in spite of the fact that (XXXIV) oxidizes about 0.30 V less anodic than (XXXIII). This is probably due to the low solubility of (XXXIV) in that medium:

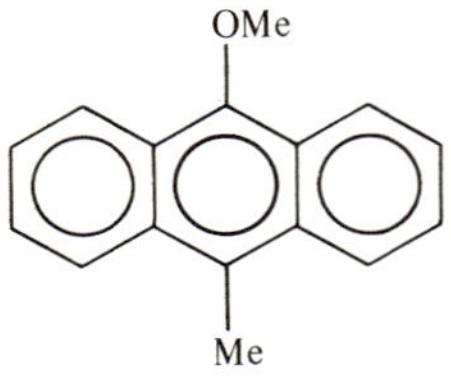

(XXXII)

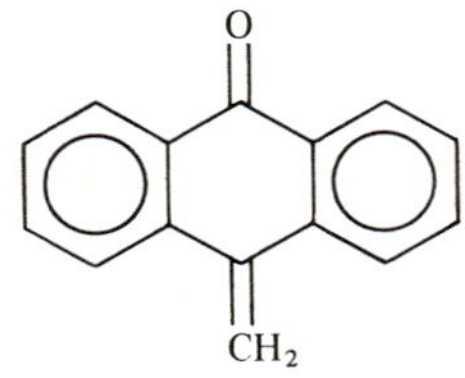

(XXXIII)

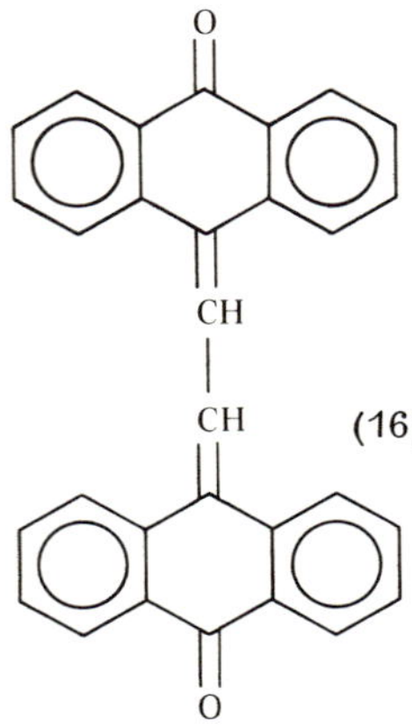

(16)

(XXXIV)

The detailed mechanism of anodic alkyl and acyl transfer has not been established. The reactions could involve either ketal (XXXV) formation or direct nucleophilic displacement by water or solvent on an intermediate dication or cation radical (XXXVI). Mechanistic studies are continuing [28a].

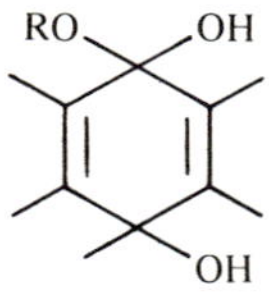

(XXXV)

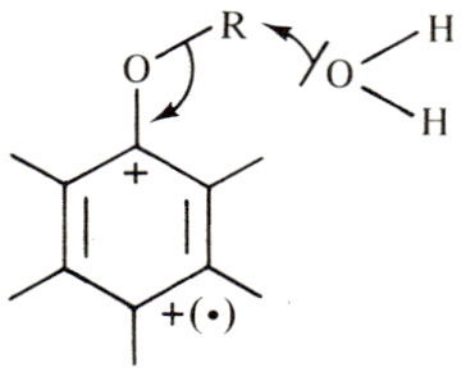

(XXXVI)

(17)

The electrochemical oxidation of 2-methyl-1, 4-naphthaquinol 1-phosphate in aqueous solution was studied over a wide range of pH [29]. The overall reaction is an irreversible two-electron oxidation accompanied by phosphoryl transfer and generation of the quinone. The rates of coupled chemical reactions and the overall oxidation mechanism were shown to be markedly dependent on pH.

III. ANODIC OXIDATION OF ALIPHATIC HYDROXY COMPOUNDS

Primary alcohols may be oxidized either to the aldehyde or to the carboxylic acid. Little work has been done on optimizing the conditions to obtain a particular product and, thus, anodic oxidation has found limited synthetic applications. Secondary alcohols have been observed to give good yields of the corresponding ketone and in other cases fragmentation has been observed, accompanied by the formation of carboxylic acids.

The electrolytic method is potentially an ideal synthetic tool for conversion of alcohols to oxidized products. However, investigations under carefully controlled conditions must be carried out before the potential of the method can be realized fully.

A. Oxidation of Benzyl Alcohols

The anodic polarography of a number of carbinols was investigated by Lund [30]. Many of the compounds gave normal polarographic waves at platinum, lead dioxide, or carbon anodes during oxidation in acetonitrile. Others gave deformed waves due to electrode filming. Preparative electrolysis of p-anisyl alcohol (XXXVII), a typical "normal" compound, resulted in the isolation of (XXVIII) in 72% yield as its 2, 4-dinitrophenylhydrazone:

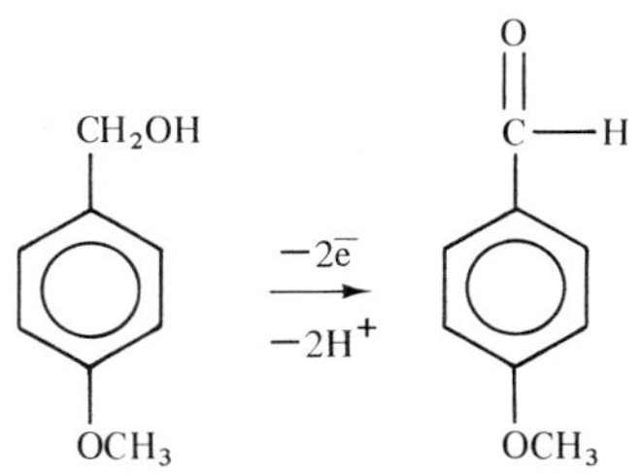

(18)

(XXXVII) (XXXVIII)

Preparative electrolysis of β-naphthylcarbinol, which gave a deformed wave, failed to yield the aldehyde.

B. Oxidation of Primary Aliphatic Alcohols

Propargyl alcohol and butynediol are converted to the corresponding acids in yields of 76 and 74%, respectively, when electrolyzed at a lead dioxide anode in aqueous sulfuric acid. Likewise, isobutyl or isoamyl alcohol are oxidized to the acids in high yield [32]. These transformations most likely involve indirect oxidation, the primary anode process being the discharge of water.

C. Oxidation of Secondary Alcohols

The conversion of 1, 2, 3, 4-tetrahydro-1-naphthol [Eq. (19)] and p-propylanisole to the corresponding ketones in yields of 50 and 13%, respectively, was accomplished by electrolysis in aqueous alkaline tert-butyl alcohol [33]. An indirect mechanism was proposed in which hydroxyl

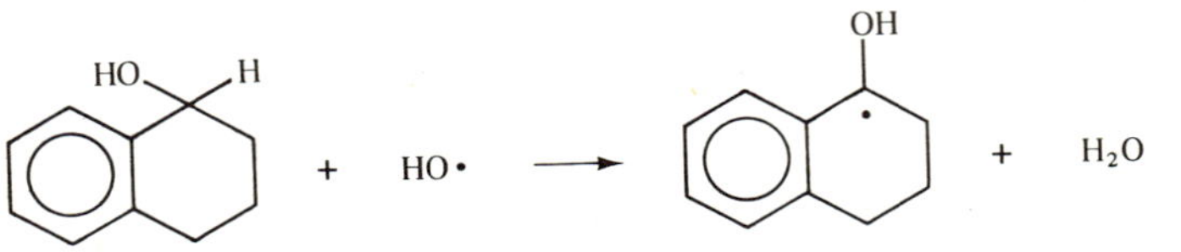

(19)

radicals were generated by electrolysis and reacted with the carbinol [Eq. (19)]. Ketones have been isolated by electrolysis of several other alcohols under noncontrolled conditions. Reference to the latter work can be found in the recent review [7].

D. Oxidation of 1, 2-Diols and Polyols

Ethylene glycol can be oxidized stepwise to glycolic acid or oxalic acid depending on the anode potential employed [34]. Selective oxidation of glycerol could also be accomplished, yielding either dihydroxyacetone, hydroxypyruvic acid, or mesoxalic acid:

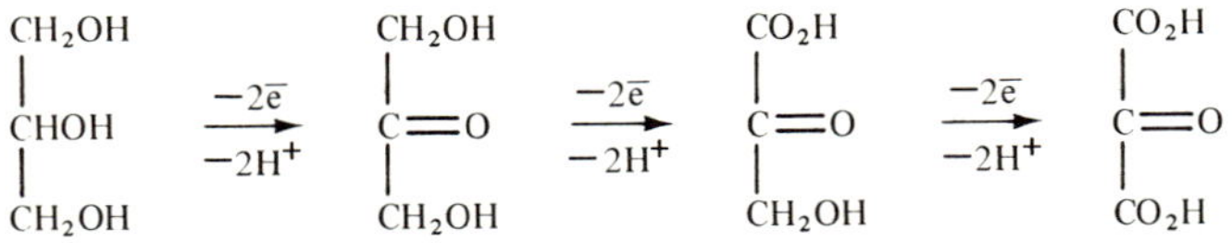

(20)

In alkaline solution, benzopinacol undergoes two-electron anodic oxidation accompanied by the formation of benzophenone [35] [Eq. (21)]. The reaction is distinctly irreversible; the half-wave potential for oxidation of

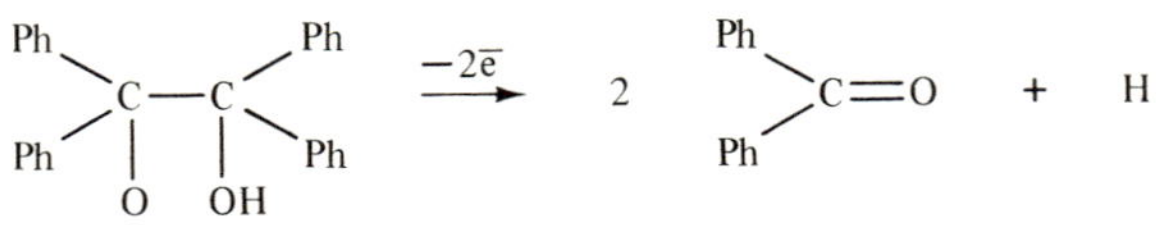

(21)

benzopinacol is -0.58 V and that for benzophenone is -1.52 V in aqueous alkaline ethanol at a dropping mercury electrode.

E. Oxidation of Carbohydrates

Aldoses are successfully oxidized to the corresponding aldonic acid

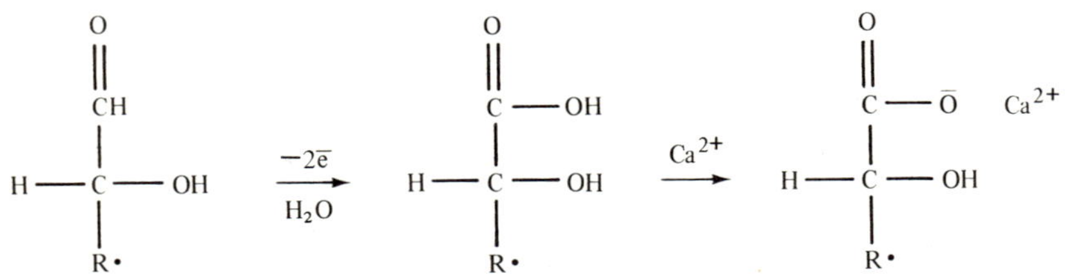

(22)

[Eq. (22)]. When the electrolyte consists of a calcium (or other alkaline earth) salt, the reaction is conveniently stopped at the aldonic acid stage due to insolubility of the andonate salt. For example, glucose is oxidized to calcium gluconate in yields of 85-100% [36-39], d-galactose to calcium galactonate in high yield [36], 3-methylglucose to calcium 3-methylgluconate in a quantitative reaction [40], and d-mannose to calcium mannoate in 92% yield (37%).

A recent study points out the usefulness of the electrolytic method in degradation of carbohydrates [41]. The electrolysis of several monosaccharides was carried out under conditions expected to produce methoxyl radicals. Lower aldoses were detected by chromatographic analysis during electrolysis of d-glucose, d-galactose, and d-mannose as well as other sugars in methanol containing hydroxide or methoxide ions. Although it was not possible to detect aldonic acids chromatographically, a mechanism was proposed involving formation of the aldonic acid as the first, or an early, step in the electrolysis. The α-keto compound may be formed either prior to or after the carboxyl group, thus giving an α-ketoaldonic acid which can decarboxylate to an aldose having one less carbon atom [Eq. (23)]:

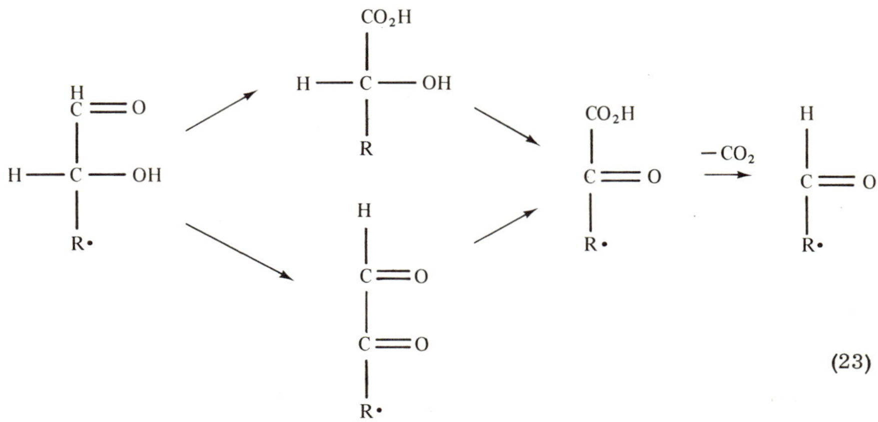

(23)

Optimum conditions are not yet established, but at the present state of development, electrolysis is a practical means of obtaining certain unusual sugars such as d-arabinose and d-threose in semimicro amounts [41].

IV. ANODIC OXIDATION OF ETHERS

Aliphatic ethers are generally electroinactive in the potential range afforded by the experiments which have been carried out. Thus, oxidation products which have been obtained are the result of indirect reactions with intermediates generated from the solvent or supporting electrolyte.

Aryl-methyl ethers have been studied extensively as substrates for anodic addition and substitution reactions. The function of the ether group is to make removal of electrons from the aromatic system occur more readily, and the ether linkage is not generally involved in these reactions. Examples of the latter reactions may be found in Chapter XX.

Enhanced stability of cationic intermediates from anodic oxidation has been established in a number of systems. Since the stability of such intermediates formed by anodic oxidation of aryl ethers is due to the presence of the ether linkage, a brief coverage of relatively stable cationic intermediates is given in this chapter.

Several investigations have been carried out on the electrolytic oxidation of furans. Both direct and indirect oxidation mechanisms have been suggested.

A. Oxidation of Aliphatic Ethers

Electrolysis of tetrahydrofuran in aqueous sulfuric acid at a lead dioxide anode affords succinic acid in 59% yield [42]. The acid could be isolated in high purity as the barium salt.

Electrolysis of aqueous diethyl ether solutions containing perchloric acid results in oxygen evolution at the anode and the ether is not attacked [43]. However, during electrolysis of solutions of perchloric acid in diethyl ether, oxygen is not formed, the platinum anode is not attacked, and acetaldehyde (30%) along with resinous material is obtained.

It has recently been shown that aliphatic ethers can be alkoxylated in the α-position by electrolysis in methanol [44]. For example, electrolysis of a methanolic solution of dioxane using sodium methoxide as the supporting electrolyte resulted in the formation of the monomethoxylated dioxane (28%) as the anodic product [Eq. (24)]. Use of ammonium nitrate as the supporting

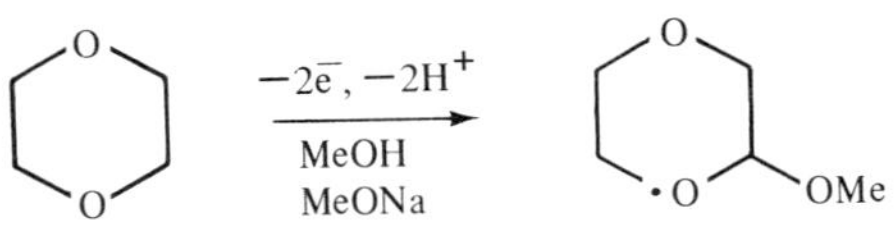

(24)

electrolyte also gave the methoxylated product, whereas no product was obtained when lithium perchlorate was the electrolyte. A mechanism was suggested involving oxidation of the supporting electrolyte [Eq. (25)], followed by hydrogen abstraction [Eq. (26)] and further oxidation at the anode, and, finally, reaction with solvent to give the product [Eq. (27)]. Several other cyclic and open-chained ethers were observed to undergo methoxylation [44].

$$\text{Supporting electrolyte} \xrightarrow{-\bar{e}} \text{S}\cdot \tag{25}$$

$$\text{ROCH}_2\text{R}' \xrightarrow{\text{S}\cdot} \text{RO}\dot{\text{C}}\text{HR}' \xrightarrow{-\bar{e}} \text{RO}\overset{+}{\text{C}}\text{HR}' \tag{26}$$

$$\text{RO}\overset{+}{\text{C}}\text{HR}' + \text{R}''\text{OH} \longrightarrow \text{ROCHR}'(\text{OR}'') \tag{27}$$

B. Relatively Stable Cationic Intermediates from Oxidation of Aromatic Ethers

Zweig and co-workers [21, 45-48] have studied the effect of structure on cation-radical stability for the oxidation of aromatic ethers and related compounds. Successive substitution of methoxy groups for hydrogen on the benzene nucleus [21] has the effect of decreasing the oxidation potential of the aromatic substrate and increasing the stability of the cation radical derived from one-electron anodic oxidation. Still greater cation-radical stability was observed when methoxysubstituted polynuclear aromatic hydrocarbons were examined [46]. The reader is referred to the original literature for more detail.

Tetra-p-anisylethylene undergoes quasi-reversible two-electron anodic oxidation in acetonitrile to form the relatively stable dication [Eq. (28)]. The cyclic voltammogram revealed that $E_p - E_{p/2}$ was greater than that expected for a reversible two-electron transfer and it was suggested that

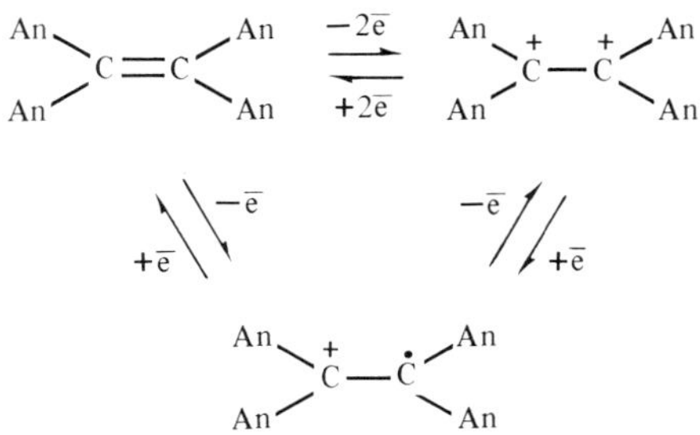

(28)

(29)

this discrepancy was due to deviation from reversibility [49]. Bard and Phelps [50] suggest that the values of $E_{p_a}-E_{p_c}$ and $E_p-E_{p/2}$ are best explained by two closely spaced, reversible one-electron transfers [Eq. (29)], the separation of E° values being only 7 mV. In methylene chloride, a solvent in which cation radicals are more stable [51], cyclic voltammetry clearly shows two successive one-electron transfers with a separation of E°'s greater than 90 mV. Since both the cation radical and the dication are stable species, it should be possible to clarify the electrolytic results in acetonitrile by determining the equilibrium constant for the reaction of the dication with substrate to generate the cation radical.

V. MISCELLANEOUS OXIDATIONS

The salts of carboxylic esters and diketones can be oxidized anodically to dimeric products (Chapter XX). The α-hydroxy ketone, benzoin, depending on the conditions employed, may either be oxidized to the diketone, benzil, in high yield [52] or to benzoic acid [52, 53]. Furoin has been oxidized to furil (30%) and benzofuroin to benzoic acid (32%) at platinum anodes [54]. A polarographic investigation has been carried out on the 2,2′-pyridil system [55]. Several aldehydes have been oxidized to the corresponding carboxylic acids [7], an example being the conversion of anisaldehyde to anisic acid in 95% yield [56].

The oxidation of 1,5-diketones to pyrylium salts was recently reported [57]. The following methylsubstituted compounds were studied:

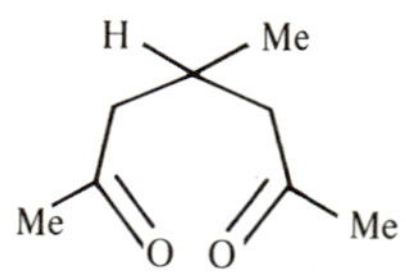

(XXXIX)

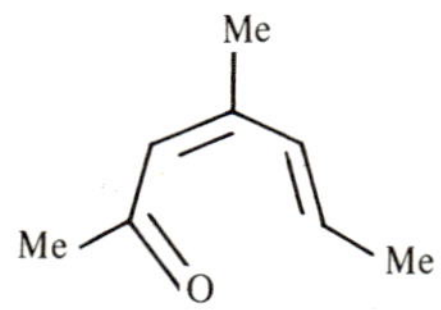

(XL)

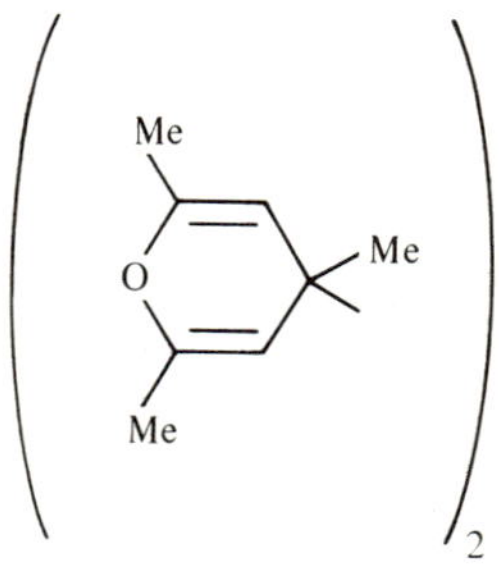

(XLI)

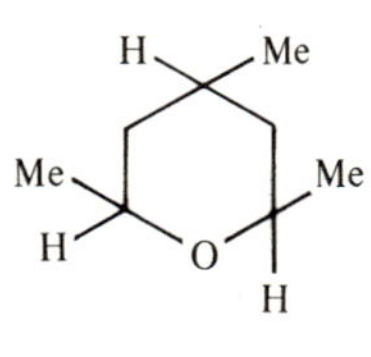

(XLII)

All four compounds gave the same pyrylium perchlorate when oxidized in acetonitrile containing lithium perchlorate. The half-wave potentials were found to be equal to +1.28 (XXXIX), +0.64 (XL), +0.72 (XLI), and +0.92 (XLII) V vs. the Ag/Ag^+ electrode. The corresponding phenyl derivatives were oxidized at considerably more anodic potentials. A mechanism was suggested involving two-electron oxidation to the 1,5-enedione (XLIV) followed by acid-catalyzed cyclization to the pyrylium salt (XLV):

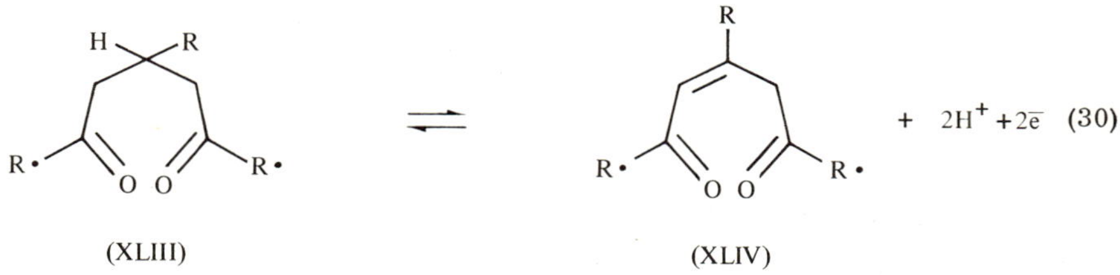

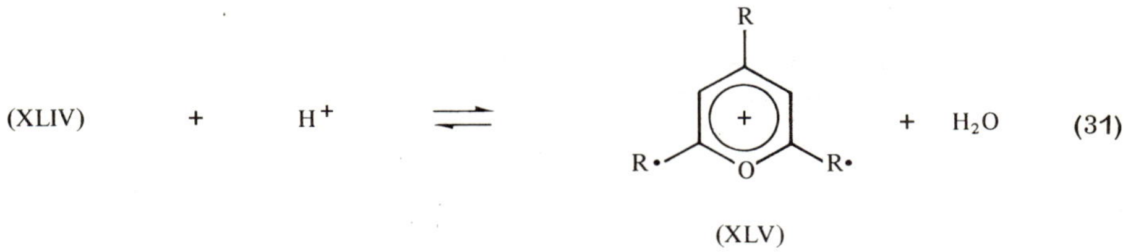

While the proposed mechanism accounts for the formation of pyrylium salts, the ease of oxidation of (XXXIX) compared to that of other saturated ketones, suggests that the reaction does not involve oxidation of saturated carbon-hydrogen bonds [Eq. (30)]. Loss of electrons to the anode may precede dehydrogenation, cyclization providing driving force for the reaction. Thus, the electroactive species could possibly be an enolic form of (XXXIX).

REFERENCES

[1]. F. J. Vermillion, Jr., and I. A. Pearl, J. Electrochem. Soc., 111, 1392 (1964).

[1a]. V. D. Parker and A. Ronlán, J. Electroanal. Chem., 30, 502 (1971).

[2]. A. B. Suttie, Tetrahedron Lett., 953 (1969).

[2a]. A. Ronlán and V. D. Parker, J. Chem. Soc. (C), 3214 (1971).

[3]. F. W. Steuber and K. Dimroth, Chem. Ber., 99, 258 (1966).

[4]. H. Iwasaki, L. A. Cohen, and B. Witkop, J. Amer. Chem. Soc., 85, 3701 (1963).

[5]. M. F. Hawthorne, R. L. Pilling, P. F. Stokely, and P. M. Garrett, ibid., 85, 3702 (1963).

[6]. L. A. Cohen and S. Isoe, Abstracts of Papers, Electrochemical Society Meeting, Dallas, Texas, 1967, p. 17.

[7]. N .L. Weinberg and H. R. Weinberg, Chem. Rev., 68, 449 (1968).

[8]. R. E. Ely, Ind. Eng. Chem., Anal. Ed., 15, 284 (1943).

[9]. G. L. Gupta, Curr. Sci. (India), 33, 680 (1964).

[10]. W. R. Turner and P. J. Elving, J. Electrochem. Soc., 112, 1215 (1965).

[11]. B. R. Eggins and J. Q. Chambers, Chem. Commun., 232 (1969).

[12]. B. R. Eggins, ibid., 1267 (1969).

[13]. V. D. Parker, ibid., 716 (1969).

[13a]. V. D. Parker and L. Eberson, ibid., 1289 (1970).

[14]. L. Papouchado, G. Petrie, J. H. Sharp, and R. N. Adams, J. Amer. Chem. Soc., 90, 5620 (1968).

[15]. J. Doskocil, Coll. Czech. Chem. Commun., 15, 599 (1950).

[16]. R. N. Adams, M. D. Hawley, and S. W. Feldberg, J. Phys. Chem., 71, 851 (1967).

[17]. M. D. Hawley, S. V. Tatawawadi, S. Piekarski, and R. N. Adams, J. Amer. Chem. Soc., 89, 447 (1967).

[18]. M. D. Hawley and R. N. Adams, J. Electroanal. Chem., 8, 163 (1964).

[19]. S. V. Tatawawadi, S. Piekarski, M. D. Hawley, and R. N. Adams, Chem. Listy, 61, 624 (1967).

[20]. V. D. Parker, J. Amer. Chem. Soc., 91, 5380 (1969).

[20a]. M. F. Marcus and M. D. Hawley, Biochem. Biophys. Acta, 201, 1 (1970).

[21]. A. Zwieg, W. G. Hodgson, and W. H. Jura, J. Amer. Chem. Soc., 86, 4124 (1964).

[22]. V. D. Parker, Chem. Commun., 610 (1969).

[23]. V. M. Clark, M. R. Eraut, and D. W. Hutchinson, J. Chem. Soc. (C), 79 (1969).

[24]. V. D. Parker, Acta Chem. Scand., 24, 3151, 3162 (1970).

[25]. K. E. Friend and W. E. Ohnesorge, J. Org. Chem., 28, 2435 (1963).

[26]. E. J. Majeski, J. D. Stuart, and W. E. Ohnesorge, J. Amer. Chem. Soc., 90, 633 (1968).

[27]. V. D. Parker, Chem. Commun., 1131 (1969).

[28]. V. D. Parker and L. Eberson, Tetrahedron Letters, 1762 (1972).

[28a]. V. D. Parker, unpublished work.

[29]. C. A. Chambers and J. Q. Chambers, J. Amer. Chem. Soc., 88, 2922 (1966).

[30]. H. Lund, Acta Chem. Scand., 11, 491 (1957).

[31]. V. Wolf, U.S. Pat. 2,786,022 (1957); CA, 51, 9381b (1957).

[32]. B. P. Ershov and G. Pyatinitskaya, Novosti Tekh., 29 (1940); CA, 34, 4673 (1940).

[33]. R. E. Juday, J. Org. Chem., 22, 532 (1957).

[34]. Siemens-Schuckertverke Aktien Gesellshaft and Varta Petrix Union G.m.b.H., Brit. Pat. 1,051,614 (1966); CA, 66, 81920 (1967); A. S. Vijh, Can. J. Chem., 49, 78 (1971).

[35]. W. Kemula, Z. R. Grabowski, and M. Kalinowski, Coll. Czech. Chem. Commun., 25, 3306 (1960).

[36]. H. J. Isbell and H. L. Frush, J. Res. Natl. Bur. Std., 6, 1195 (1931); 14, 359 (1935).

[37]. C. G. Fink and D. B. Summers, Trans. Amer. Electrochem. Soc., 74, 325 (1938).

[38]. A. N. Kappara and K. M. Joshi, J. Indian Chem. Soc., 29, 69 (1952).

[39]. S. Sathyanarayna and H. V. K. Udupa, Bull. Acad. Pol. Sci. Ser. Sci. Chim., 6, 493 (1958).

[40]. O. T. Schmidt and A. Simon, J. Prakt. Chem., 152, 190 (1939).

[41]. G. W. Hay and F. Smith, Can. J. Chem., 47, 417 (1969).

[42]. British Celanese Ltd., Brit. Pat. 590,310 (1947); CA, 41, 7285 (1947).

[43]. C. Marie and G. Lejeune, An. Soc. Espan. Fis. Quim., 27, 447 (1929); CA, 24, 25 (1930).

[44]. T. Shono and Y. Matsumara, J. Amer. Chem. Soc., 91, 2803 (1969).

[45]. A. Zweig, G. Metzler, A. H. Mauer, and B. G. Roberts, ibid., 88, 2864 (1966).

[46]. A. Zweig, A. H. Mauer, and B. G. Roberts, J. Org. Chem., 32, 1322 (1967).

[47]. A. Zweig and J. E. Lehnsen, J. Amer. Chem. Soc., 87, 2647 (1965).

[48]. A. Zweig, G. Metzler, A. H. Mauer, and B. G. Roberts, ibid., 89, 4091 (1967).

[49]. V. D. Parker, K. Nyberg, and L. Eberson, J. Electroanal. Chem., 22, 150 (1969).

[50]. A. J. Bard and J. Phelps, ibid., 25, App. 2 (1970).

[51]. J. Phelps, K. S. V. Santhanam, and A. J. Hard, J. Amer. Chem. Soc., 89, 1752 (1967).

[52]. J. H. James, ibid., 21, 889 (1899).

[53]. R. H. McKee and J. R. Heard, Trans. Electrochem. Soc., 327, 332 (1934).

[54]. H. D. Law, J. Chem. Soc., 89, 1437 (1906).

[55]. H. Holubek and J. Volke, Coll. Czech. Chem. Commun., 25, 3292 (1960).

[56]. F. Fichter and H. Ris, Helv. Chim. Acta, 7, 803 (1924).

[57]. C. Bratu and A. T. Balaban, Rev. Roum. Chim., 10, 1001 (1965).

Chapter XVI

ANODIC OXIDATION OF SULFUR-CONTAINING COMPOUNDS

Vernon D. Parker

Department of General and Organic Chemistry
University of Copenhagen
Copenhagen, Denmark

I. INTRODUCTION

The anodic reactions of sulfur-containing compounds differ from those of the corresponding oxygen compounds (Chapter XV) primarily in two ways: (a) the former have the tendency to produce compounds resulting from expansion of the valence shell of the heteroatom, a pathway not open to the latter; and (b) the former frequently undergo dimerization reactions accompanied by bond formation between heteroatoms. For example, sulfides can be converted in high yield to the sulfoxides or sulfones and thio acids readily form disulfides. Oxidation of the corresponding oxygen compounds generally result in cleavage of carbon-oxygen, carbon-hydrogen, or carbon-carbon bonds.

A large body of work was reported on the early investigations of the Fichter group. These investigations generally involved the use of aqueous solutions and were conducted under noncontrolled conditions. Applications of modern techniques to the study of the anodic electrochemistry of organic sulfur compounds are very limited. Papers are now beginning to appear describing careful investigations utilizing modern techniques and nonaqueous solvents.

Sufficient data are not now available to justify assigning mechanisms to most of the anodic transformations known to take place. Elucidation of mechanisms of anodic oxidation of sulfur compounds should be a fruitful area for future research.

II. OXIDATION OF SULFIDES AND DISULFIDES

A. Sulfides

In aqueous alcohol or acetic acid solutions, sulfides are oxidized first to the sulfoxide which can either be isolated or further oxidized to the sulfone. For example, anodic oxidation of phenyl sulfide in acetic acid-conc. HCl at a platinum anode gave diphenyl sulfoxide in 82% yield [Eq. (1)]. Under the same conditions, diphenyl sulfoxide was converted to diphenyl

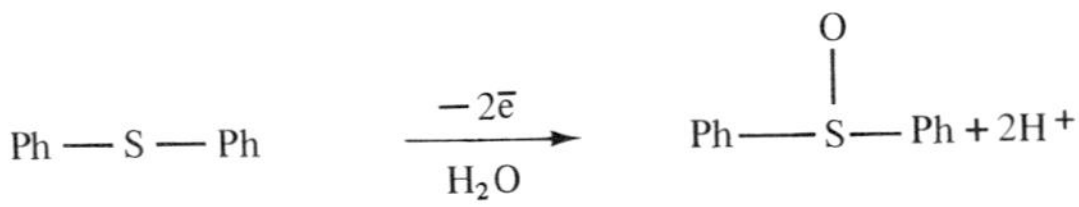

(1)

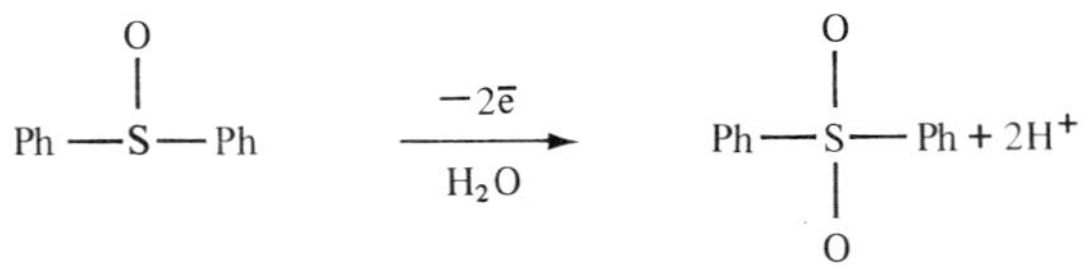

(2)

sulfone [Eq. (2)] in 97% yield [1]. The sulfone could also be prepared directly from the sulfide in 94% yield [2]. This reaction appears to be a generally useful preparative method and has been applied to the preparation of several sulfoxides and sulfones [3, 4]. The electrooxidation of dimethyl sulfide to the industrially important dimethyl sulfoxide has been claimed [4a]. The oxidation of sulfides or sulfoxides assisted by electroregenerated compounds of W, V, Mo, or Se (cf. Chapter XXIV) has also been reported [4b].

A recent investigation of the anodic oxidation of aliphatic sulfides has provided mechanistic evidence for the reaction under both aqueous and non-aqueous conditions [5]. It was suggested that one-electron oxidation of dimethyl sulfide produces the cation radical [Eq. (3)], which undergoes rapid proton loss in anhydrous solution to form the sulfonium ion [Eq. (4)],

$$\mathrm{MeSMe} \xrightarrow{-\bar{e}} \mathrm{Me{-}\overset{+\bullet}{S}{-}Me} \qquad (3)$$

$$\xrightarrow[-\bar{e}]{-\mathrm{H}^+} \mathrm{Me{-}\overset{+}{S}{=}CH_2} \longleftrightarrow \mathrm{Me{-}S{-}\overset{+}{C}H_2} \qquad (4)$$

$$+\mathrm{Me_2S} \rightleftharpoons \mathrm{MeS{-}CH_2{-}\overset{+}{S}(Me)_2} \qquad (5)$$

which on reaction with substrate forms the dimethylmethylthiomethyl sulfonium ion [Eq. (5)] as the immediate major product. A series of slow chemical steps was suggested to result in the ultimate conversion of all the original starting material to the sodium salt of methanesulfonic acid and carbon monoxide. When the oxidation was carried out with 1% water present, nucleophilic attack by water excludes proton transfer and dimethyl sulfide was converted quantitatively to dimethyl sulfone. The products from the anhydrous reaction contained more oxygen than could be accounted for by residual water in the solvent system. Evidence was presented that perchloric acid, produced by reaction of the perchlorate supporting electrolyte with electrochemically generated protons, is a source of water when perchlorate salts are used as supporting electrolytes, the water then being generated by equilibration of the acid with its anhydride [Eq. (6)]. In support of this suggestion, it was reported that no sulfonate was formed when

$$2\mathrm{HClO_4} \rightleftharpoons \mathrm{Cl_2O_7} + \mathrm{H_2O} \qquad (6)$$

the supporting electrolyte was tetramethylammonium fluoborate. However, since the products were not identified in the latter case, the mechanism of this oxidation is not known.

B. Disulfides

Disulfides can be oxidized to the corresponding disulfoxides. For example, benzyl disulfide produced benzyl disulfoxide in 93% yield when oxidized in acetic acid containing sulfuric acid [2].

Several disulfides have been observed to undergo cleavage of the disulfide linkage when subjected to anodic oxidation accompanied by the generation of the sulfonic acid [3, 6, 7]. Thus, oxidation of ethyl disulfide in ethanesulfonic acid is oxidized to the sulfonic acid with a current efficiency of 80% [6].

III. OXIDATION OF THIO ACIDS

Unlike the oxygen analogs, thio acids do not undergo the Kolbe reaction. The major products of the anodic oxidation of thiocarboxylates are the corresponding disulfides [8, 9, 10]:

$$R-\overset{\overset{O}{\|}}{C}-S^- \xrightarrow{-\bar{e}} R-\overset{\overset{O}{\|}}{C}-S\cdot \longrightarrow \left(R\overset{\overset{O}{\|}}{C}-S-\right)_2 \quad (7)$$

The oxidation of potassium thiostearate in aqueous methanol [10] resulted in the formation of distearoyl disulfide (31%) and methyl stearate (37%). It was concluded that only the disulfide resulted from electrolysis, methyl stearate being capable of formation from the disulfide and methanol. It is not clear whether the absence of hydrocarbons is due to the inherent stability of the acyl-thio radical (RCOS•) or whether dimerization of the radical is so fast to preclude formation of R• [Eq. (8)]:

$$R-\overset{\overset{O}{\|}}{C}-S\cdot \longrightarrow R\cdot + COS \quad (8)$$

IV. OXIDATION OF THIOAMIDES AND THIOUREAS

A. Thioamides

Indirect oxidation of thiobenzamide and thiobenzanilide was observed when the substrates were oxidized at a mercury anode [11]. The two-electron oxidation of thiobenzamide is accompanied by the formation of benzonitrile and mercuric sulfide [Eqs. (9)-(11)]. With thiobenzanilide, the decomposition step [Eq. (11)] is not feasible and reaction with a second

$$2Hg \underset{}{\overset{-2\bar{e}}{\rightleftharpoons}} 2Hg^+ \rightleftharpoons Hg_2^{2+} \rightleftharpoons Hg^{2+} + Hg \quad (9)$$

$$Ph-\underset{\underset{S^-}{|}}{C}{=}NH + Hg^{2+} \longrightarrow Ph-\underset{\underset{SHg^+}{|}}{C}{=}NH \quad (10)$$

$$\longrightarrow PhCN + H^+ + HgS \quad (11)$$

molecule of substrate occurs to give a one-electron oxidation product [Eq. (12)]:

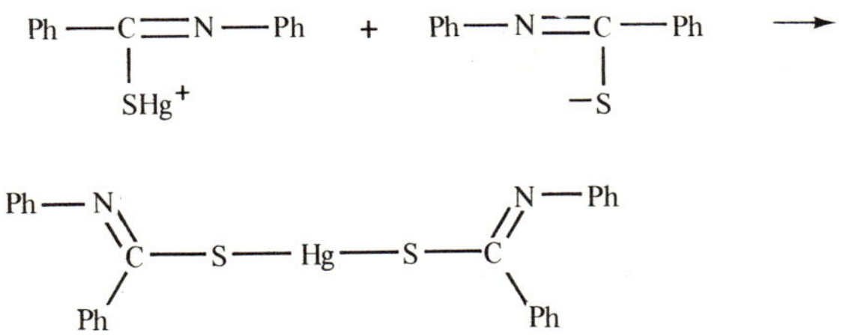

(12)

The anodic oxidation of potassium amidosulfonate [12] in aqueous solution results in a two-electron oxidation accompanied by the formation of potassium azidodisulfonate in 66% yield [Eq. (13)]:

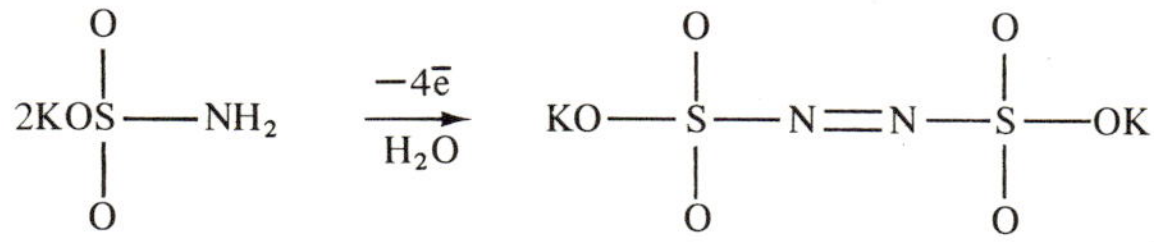

(13)

B. Thioureas

Anodic oxidation of thioureas [Eq. (14)] is a convenient method for the preparation of the corresponding formamidine disulfide salts [1, 3, 13, 14].

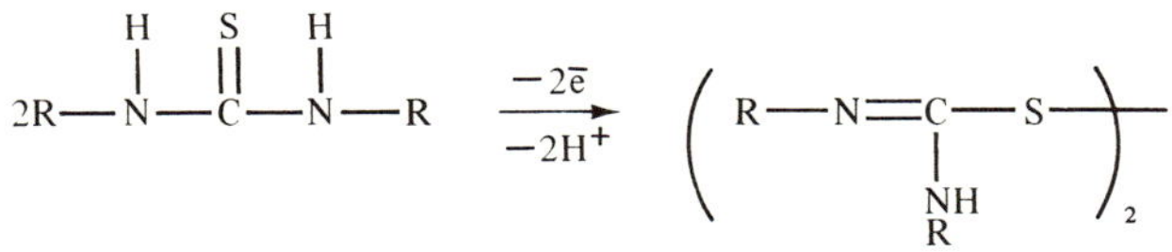

(14)

Well-defined polarographic anodic waves are only obtained in strongly alkaline media [15, 16]. However, a coulometric investigation [14] over a wide range of pH indicates that an acid medium of pH less than 3 is required for the successful coulometric oxidation of thiourea. Under the latter conditions, a quantitative oxidation takes place, the decrease of current with time is exponential, and the final current is negligibly small. In alkaline medium the anode becomes passive after the oxidation is no more than 3% complete.

V. OXIDATION OF MERCAPTIDE, DITHIOCARBAMATE, TRITHIOCARBONATE, AND XANTHOGENATE SALTS

The tendency of sulfur radicals to undergo dimerization is further exemplified by the anodic oxidation of the title salts. No yield data are available. However, the corresponding disulfides from mercaptide [17, 18], and dithiocarbamate [19], trithiocarbonates [19], and dixanthogens [19-21] from xanthogenates have been isolated.

A microcoulometric study of the oxidation of sodium diethyldithiocarbamate at a rotating platinum electrode indicates that the oxidation occurs in two $1\bar{e}$ stages, the first corresponding to the dimerization [Eq. (15) and the second to the oxidation of the disulfide [22]:

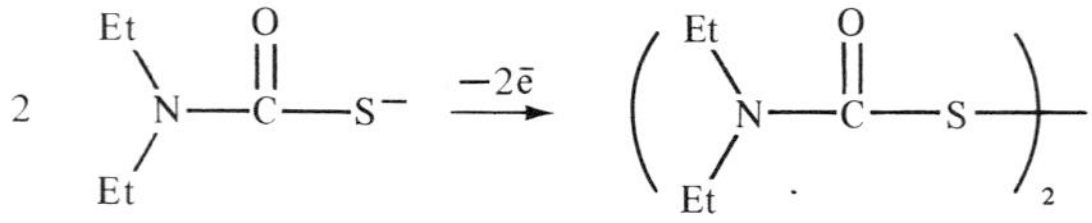

(15)

VI. MISCELLANEOUS OXIDATIONS

A number of aromatic sulfonic acids have been hydroxylated in aqueous solution [6, 23]. These reactions are not of practical interest since the phenolic products are more easily oxidized than substrate and oxidation frequently goes all the way with the production of carbon dioxide.

A novel synthesis of γ- and δ-lactones by anodic oxidation of sodium bisulfite addition products of cycloalkanones has been reported [24]. The reaction is believed to occur by one-electron oxidation [Eq. (16)] followed by ring opening, further one-electron oxidation, and hydride migrations [Eq. (17)] before ring closure [Eq. (18)] to give a 3:2 mixture of the γ- and δ-lactones.

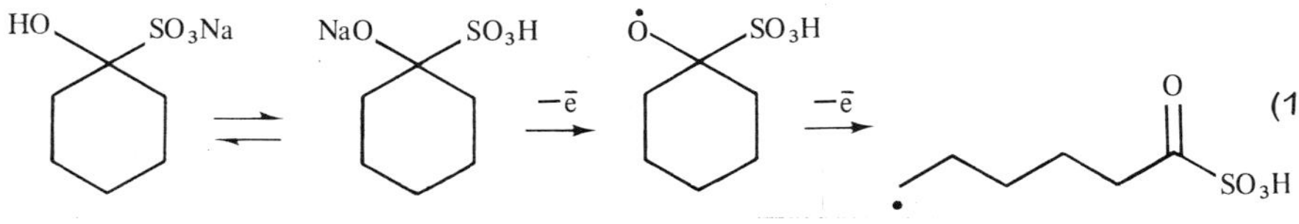

(16)

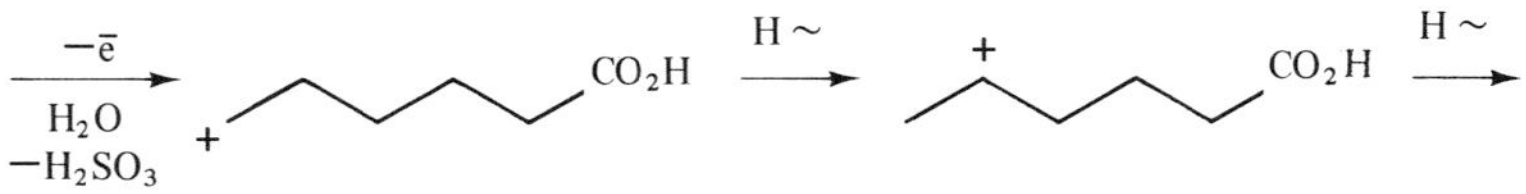

(17)

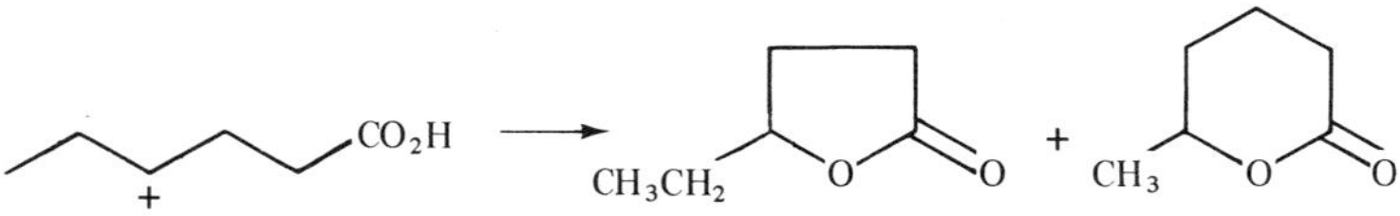

(18)

The oxidation of a multisulfur compound has recently been reported [25]. The orthothiooxalate (I) is oxidized in anhydrous acetonitrile in the region of +1.4 V (vs. SCE). Cyclic voltammetry shows the presence of two reversible one-electron couples. The reversible couples were assigned to the formation of the cation radical (III) and the dication (IV) [Eq. (19)]. The first step in the oxidation is believed to be one-electron oxidation to a cation radical of

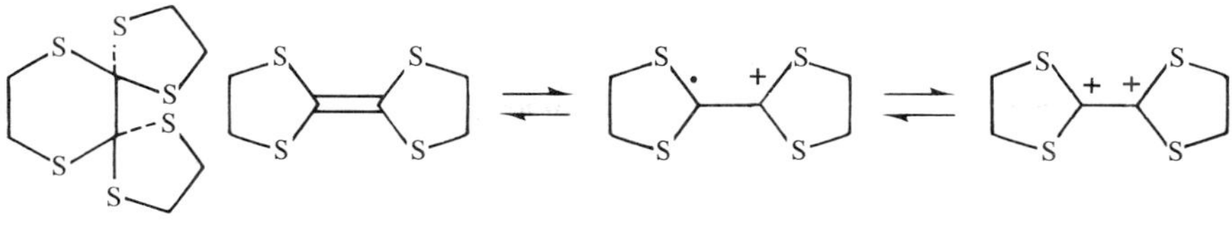

(19)

(I) (II) (III) (IV)

undefined structure which undergoes decomposition to the cation radical of bicyclic tetrathioethylene. The cyclic voltammetry of authentic bicyclic tetrathioethylene (II) supports the proposed oxidation pathway.

VII. DETECTABLE CATION RADICALS

The lifetimes of anodically generated cation radicals from aromatic methylthio compounds appear to be very similar to those of the oxygen-containing compounds and much shorter than those of the corresponding amines [26].

Two anodic waves are observed at a platinum electrode for phenothiazine in acetonitrile [27]. The first corresponds to formation of the cation-radical [Eq. (20)], which was verified by the observation of the ESR spectrum [28], and the second wave is due to further oxidation of the cation radical [Eq. (21)]:

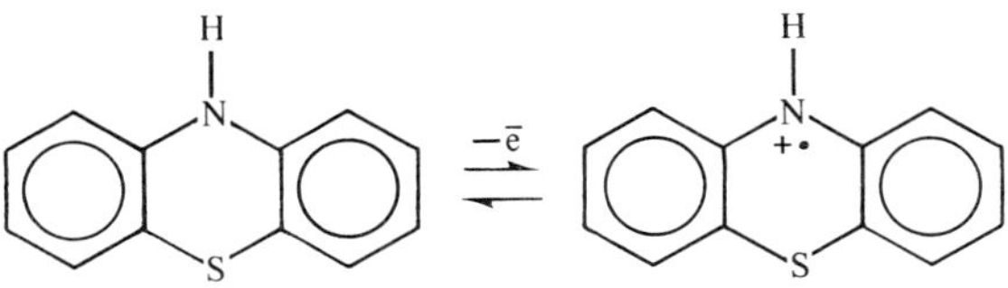

(20)

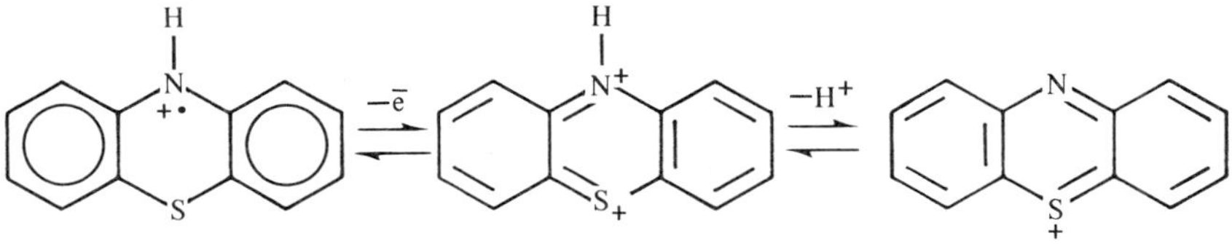

(21)

REFERENCES

[1]. F. Fichter and F. Braun, Chem. Ber., 47, 1526 (1914).

[2]. F. Fichter and P. Sjostedt, ibid., 43, 3422 (1910).

[3]. F. Fichter and W. Wenk, ibid., 45, 1373 (1912).

[4]. J. A. Pursley, U.S. Pat. 3,200,054 (1965); CA, 63, 11023 (1965).

[4a]. E. Worbs, G. Meyer, E. Siggel, and H. D. Rupp, S. Afr. Pat. 6,806,392, March 6, 1969; CA, 72, 3030g (1970).

[4b]. C. F. Bennett and D. W. Goheen, U.S. Pat. 3,418,224 (0000); CA, 70, 43434g (1969).

[5]. P. T. Cottrell and C. K. Mann, J. Electrochem. Soc., 116, 1449 (1969).

[6]. B. K. Brown, U.S. Pat. 2,521,147 (1950); CA, 44, 10556 (1950).

[7]. F. Fichter, R. Brändlin, and E. R. Hallaver, Helv. Chim. Acta, 3, 410 (1920).

[8]. N. Bunge, Chem. Ber., 3, 295 (1870).

[9]. I. M. Pavlova and O. A. Songina, Izv. Vyssh. Ucheb. Zaved., Khim. i Khim. Tekhnol., 77, 378 (1964); CA, 61, 14187 (1964).

[10]. Y. Hirabayashi and T. Mazume, Bull. Chem. Soc. Japan, 39, 1971 (1966).

[11]. H. Lund, Coll. Czech. Chem. Commun., 25, 3313 (1960).

[12]. A. Krettler and W. Teske, Angew. Chem., 71, 69 (1959).

[13]. J. W. Shipley and M. T. Rogers, Can. J. Res., 17B, 147 (1939).

[14]. K. S. V. Santhanam and V. R. Krishnan, Z. Phyz. Chem., 34, 312 (1962).

[15]. M. F. O. Manousek and P. Zuman, Chem. Listy, 49, 1494 (1955).

[16]. L. Jensovsky, ibid., 49, 1267 (1955).

[17]. N. Bunge, Chem. Ber., 3, 911 (1870).

[18]. F. Taboury, Ann. Chim. Phys., 15, 47 (1908).

[19]. C. Schall and S. Kraszler, Z. Electrochem., 5, 225 (1899).

[20]. C. Schall, ibid., 2, 475 (1896).

[21]. C. Schall, ibid., 3, 83 (1897).

[22]. Y. I. Usatenko and F. M. Tulyupa, Tr. Dnepropetrovsk, Khim.-Tekhnol. Inst., 12, 25 (1959); CA, 57, 5709 (1962).

[23]. F. Fichter and E. Stocker, Helv. Chim. Acta, 7, 1064 (1924).

[24]. M. Oyama and M. Ohno, Tetrahedron Lett., 5201 (1966).

[25]. N. D. Canfield, J. Q. Chambers, and D. L. Coffen, J. Electroanal. Chem., 24, App. 7 (1970).

[26]. A. Zweig, A. H. Mauer, and B. G. Roberts, J. Org. Chem., 32, 1322 (1967).

[27]. J. P. Billon, Ann. Chim., 7, 183 (1963).

[28]. J. P. Billon, G. Cauquis, and J. Combrisson, J. Chim. Phys., 61, 374 (1964).

Section E

ELECTROCHEMISTRY OF CERTAIN COMPREHENSIVE CLASSES OF COMPOUNDS

Chapter XVII

ELECTROLYSIS OF HETEROCYCLIC COMPOUNDS

H. Lund

Department of Organic Chemistry
University of Aarhus,
Aarhus, Denmark

I. INTRODUCTION

This chapter discusses some aspects of the electrolysis of heterocyclic compounds. The emphasis is laid on systems in which the electrochemical properties differ from those of the corresponding carbocyclic compounds. The discussion is divided in three main sections: Electrolytic Formation of Heterocyclic Systems, Electrolytic Reactions of Heterocyclic Systems, and Electrolytic Reactions of Substituted Heterocyclic Compounds.

A recent review [1] discusses "Electrolysis of N-Heterocyclic Compounds."

II. ELECTROLYTIC FORMATION OF HETEROCYCLIC SYSTEMS

The electrolytic reactions in which a heterocyclic system is formed are discussed in this section; the reactions are treated under the following headings:

A. Ring Closure Reactions

B. Ring Contractions

C. Ring Expansions

A. Ring Closure Reactions

In the electrolytic formation of heterocyclic systems the role of the current is to bring one or both of the reaction centers to a suitable oxidation state. These reactions may be classified in different ways; below they are treated according to the type of bond formed.

A ring closure will most often involve either the attack of an electrolytically formed nucleophile (alcohol, hydroxylamine, amine, hydrazine) on an electrophilic center (existing or potential carbonyl group, cyano group, nitro or nitroso group) or a reaction between a nucleophile (e.g., hydroxyl or amino group, carboxylate ion) and an electrolytically generated electrophilic center (e.g., nitroso group, carbonium ion). The formation of a nucleophilic center is most often the result of a reduction, whereas the creation of an electrophilic center in many cases is performed by an oxidation.

1. Formation of Carbon-Oxygen Bonds

a. Hydroxyl Group as Nucleophile. An electrolytically formed alcohol group participating in a ring closure is formed in the reduction of some phthalic acid derivatives to phthalides and of o-carboxyphenyl ketones to substituted phthalides; the latter reaction is exemplified in the reduction [2] of o-benzoyl-N, N-diethylbenzamide (I):

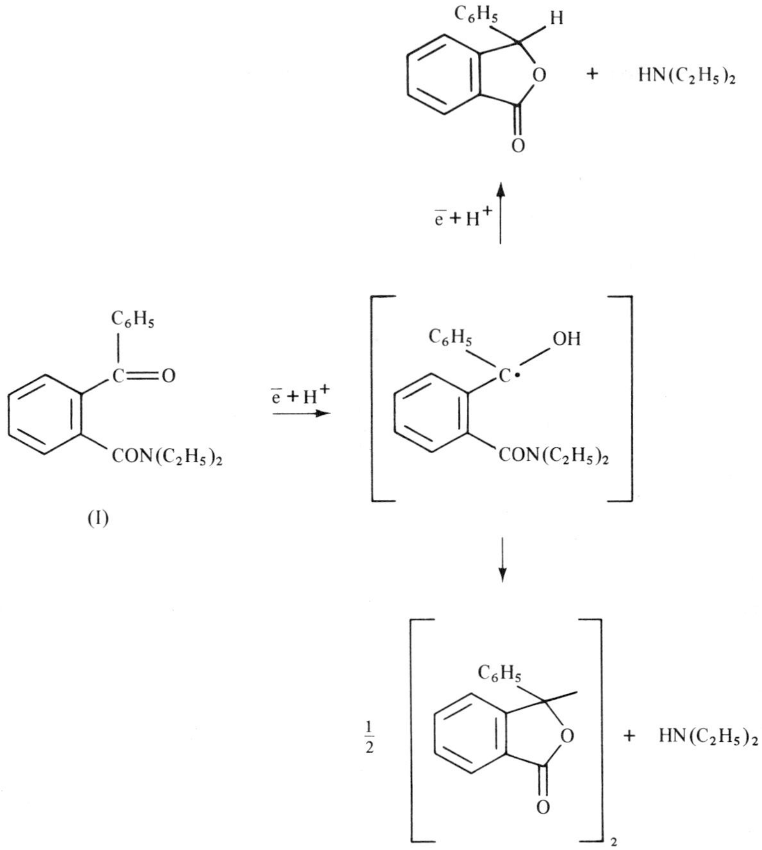

(I)

The reduction of 1-acetylnaphthalene [3] and some α, β-unsaturated carbonyl compounds [4-6] leads to the formation of tetrahydrofuran derivatives as discussed in Chapters VIII and IX.

Another kind of ring formation is found in the anodic oxidation in alkaline methanol of N-benzyl-N-methylethanolamine (II) which produces three main products [7], among these 3-methyl-2-phenyl-oxazolidine (III) and 3-benzyl-oxazolidine (IV); the primary loss of an electron occurs probably from the nitrogen atom; N-benzyldiethanolamine produces similar products on anodic oxidation [7].

The oxidation of some 1, 5-diketones to pyrylium salts has been reported [8]; this reaction is discussed in Chapter XV.

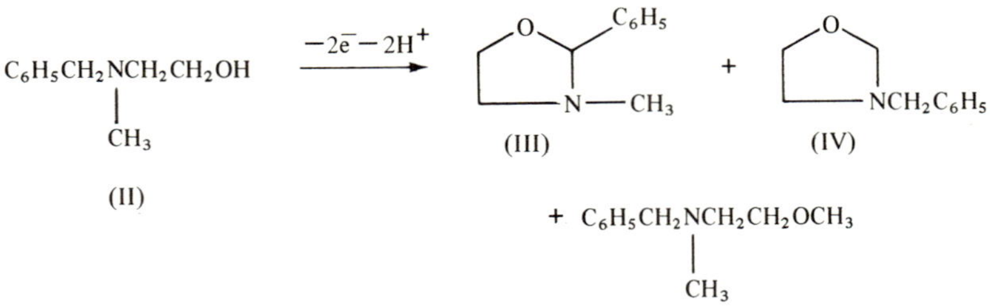

b. Hydroxylamino Group as Nucleophile. A ring closure of this type has been reported only in the reduction of o-nitrocarbonyl compounds to anthranils. Larger rings cannot be obtained by this method as an attack by the nitrogen atom of the intermediate hydroxylamine would be the preferred mode.

o-Nitroacetophenone is reduced to methyl anthranil in sulfuric acid [9] and acetate buffer [10]; in the latter medium the yield is quantitative. o-Nitrobenzaldehyde [9] is reduced similarly and such a reduction might be expected in general from o-nitroaldehydes and -ketones. The reduction of o-nitrobenzophenone [11] has, however, been reported to yield the dihydro-anthranil. This is surprising because a reduction of an intermediate anthranil would be expected to result in a hydrogenation of the nitrogen-oxygen bond rather than in a saturation of the double bond, but the reduction potentials might be quite close.

2. Formation of Carbon-Nitrogen Bonds

a. Hydroxylamino Group as Nucleophile. Aromatic nitro compounds may generally be reduced in a four-electron reduction to hydroxylamines which in acid solution at a more negative potential can be reduced further to an amine. By choosing a suitable cathode potential it is, thus, possible to avoid the further reduction of the hydroxylamine to the amine. When a hydroxyl or amino group is ortho or para to the nitro group, it is not feasible to isolate the hydroxylamino compound, as it is too easily dehydrated to the reducible quinone mono- or diimine. It may, however, be possible to trap the intermediate.

(1). Reaction with Carbonyl Groups. A hydroxylamino group has two nucleophilic centers, the nitrogen and the oxygen atom. Whenever attack of the nitrogen atom on a carbonyl group would result in the formation of a ring having five or more members, this mode is preferred to attack by oxygen. The reaction results in a cyclic nitrone (N-oxide) or hydroxamic acid.

The reduction of o-nitrobenzalacetylacetone (V) in acid solution at the potential of the first polarographic wave gives the cyclic N-oxide (VI) [12, 13]; the isolated yield is 70-75%, but polarographic determination of the product in the reduced solution showed a somewhat higher yield. Compound (VI) may at a more negative potential be reduced further, but the reduction potentials of the N-oxide and the carbonyl group are rather close in acid solution, making a selective reduction of the N-oxide difficult.

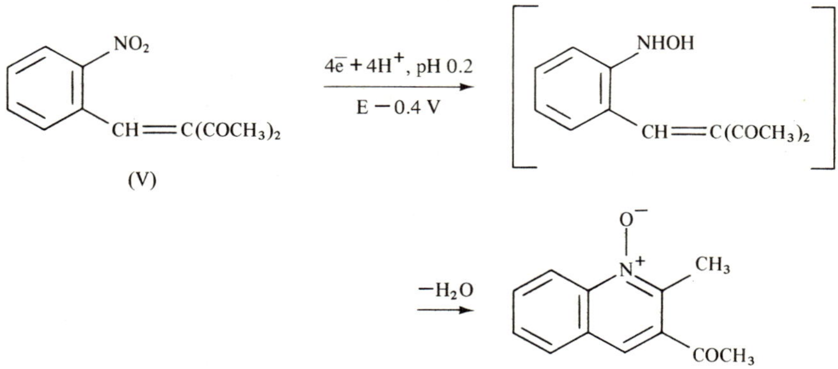

(V)

(VI)

o-Nitrobenzalmalonic acid can be reduced in a four-electron reduction [14], and the resulting cyclic hydroxamic acid is, as many other hydroxamic acids [14], further reducible in acid solution to the amine, the carbostyril.

Reductions at the potential of the first wave of 1-acetamino-2-nitronaphthalene, 2-acetamino-3-nitronaphthalene, and 8-acetamino-1-nitronaphthalene yield cyclic N-oxides, the two former naphthimidazole-N-oxides and the latter 2-methylperimidine-1-oxide [15].

Similarly, o-nitroformanilide gives on reduction in acetate buffer benzimidazole-N-oxide, but the isolated yield is in this case rather low (20%) [13].

Although only few examples have been reported so far, there seems no reason to doubt that a good yield of similar products may be obtained in many cases.

(2) Reaction with Potential Carbonyl Groups. In principle, condensations similar to those described for carbonyl groups would be possible for the corresponding azomethine compounds. An example of this kind is the reduction of 2-nitrophenylisothiocyanate (VII) in ethanolic hydrochloric acid

which in a four-electron reduction forms 2-mercaptobenzimidazole-N-oxide (VIII) [13]:

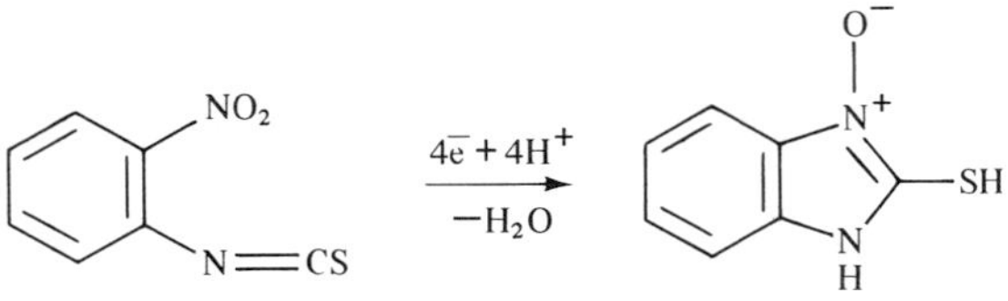

(VII) (VIII)

(3) Reaction with Cyano Groups. These reactions are similar to those described above; the following example, the reduction of o-nitrophenylthiocyanate (IX) [13], may illustrate this:

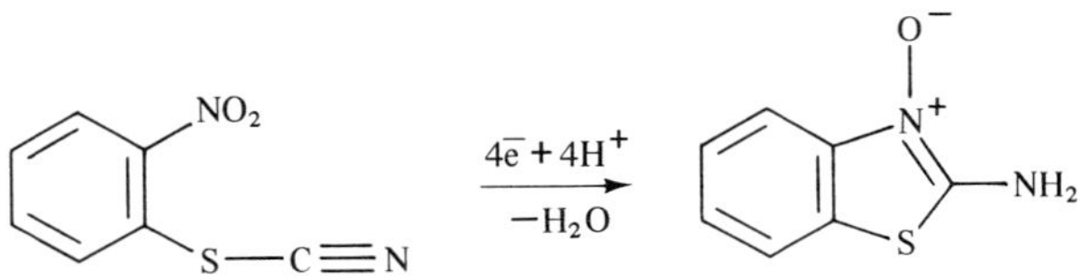

(IX)

b. Amines as Nucleophiles. Amines may be formed by the electrolytic reduction of azomethine derivatives, nitro, nitroso, and hydroxylamino compounds. As the amine is the lowest oxidation state, control of the electrode potential is often not as critical in the preparation of amines as in the synthesis of hydroxylamines. Consequently, most of the "classical" electrolytic reductions of nitro compounds in acid solution resulted in the formation of amines.

(1) Reaction with Carbonyl Groups. Cyclization of an azomethine compound occurs in the reduction of "phenolphthalein oxime" (X) [16] and similar very stable anils [17]:

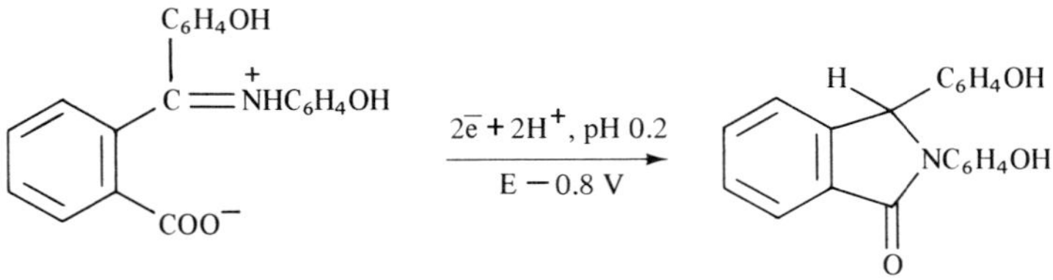

(X)

The reduction of o-nitroacetanilide [18, 19] to 2-methylbenzimidazole, acylated nitronaphthylamines to naphthimidazoles [15], and 2-nitrophenylthiocyanate [18] to 2-aminobenzothiazole has been reported, but, when no other reducible groups are present, the advantage of using electrolytic methods for the reduction is questionable.

Sometimes a "cathodic substitution" reaction takes place during the reduction of o-substituted nitrobenzenes; thus, the reduction of 2-nitrophenoxyacetic acid (XI) in 2 N hydrochloric acid (50% ethanol) yielded 5-chloro-2H-1, 4-benzoxazin-3(4H)-one (XII) (Y = Cl) rather than the expected cyclic hydroxamic acid [13]. In the presence of another nucleophile Y^- (e.g., thiocyanate) a ring substitution with this reagent occurs:

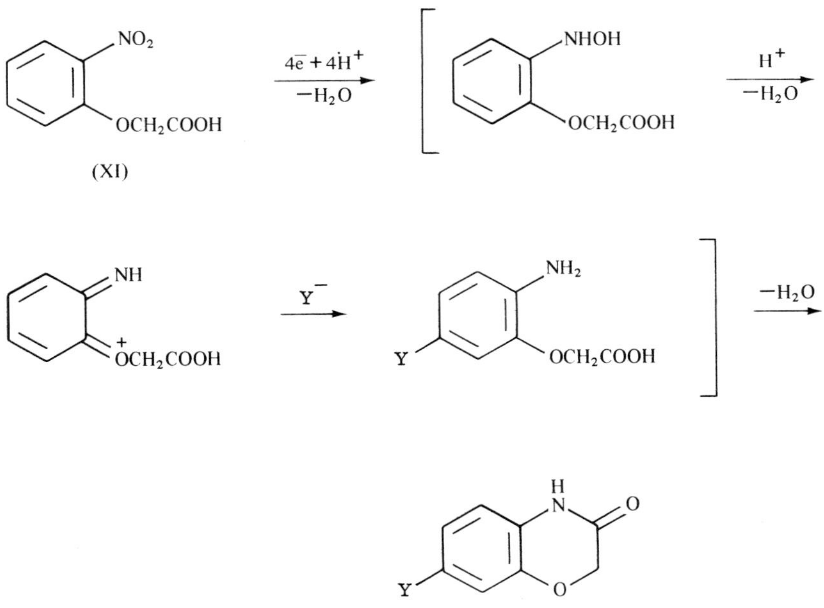

(XII)

(2) Reaction with Azomethine Groups. A useful synthesis of 3-phenylbenzo-1, 2, 4-triazine (XIII) and other benzotriazines is exemplified by the reductive ring closure of o-nitrophenylazophenylnitromethane (XIV) [10, 20]. This compound is converted in an acetate buffer by a four-electron reduction to o-nitrophenylbenzhydrazidoxime (XV) which at a more negative potential is reduced to o-nitrophenylbenzamidrazone (XVI) and further to o-aminophenylbenzamidrazone (XVII). Ring closure through (XVIII) follows to give 1, 4-dihydro-3-phenylbenz-1, 2, 4-triazine (XIX) which may be oxidized to (XIII) by air or anodically. A high yield (85-96%) of this compound may be

obtained in hydrochloric acid solution or acetate buffer containing DMF by reduction at the potential of the second wave.

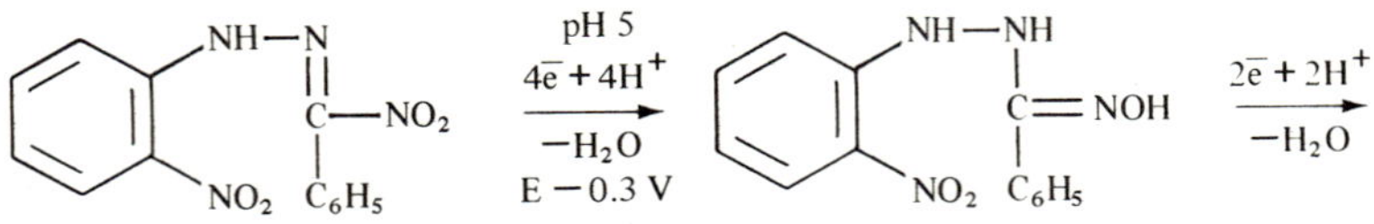

(XIV) (XV)

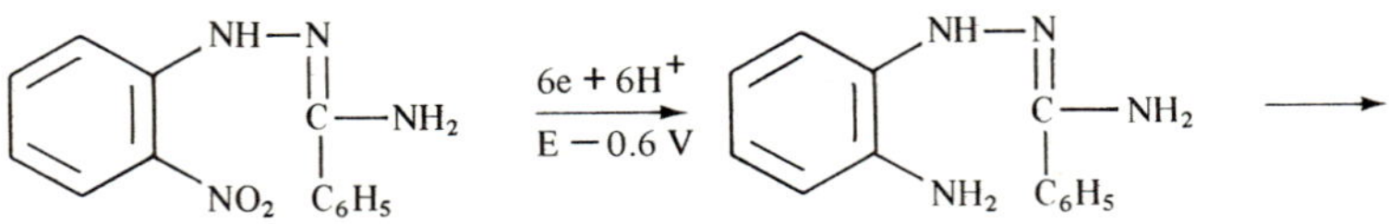

(XVI) (XVII)

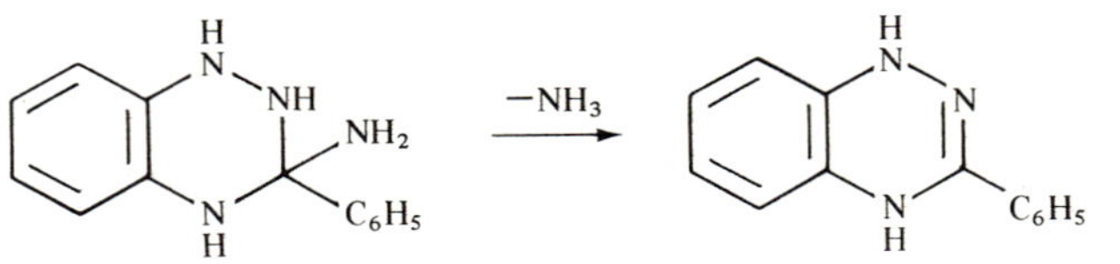

(XVIII) (XIX)

Other leaving groups may be used [20]; thus, thiobenz-(o-nitrophenyl)-hydrazide gives a good yield of (XIX) on reduction, but benz-(o-nitrophenyl)-hydrazide does not. This is probably because the hydroxyl group is a poorer leaving group than the amino or sulfhydryl group. The product analogous to (XVIII) does not lose water but opens to o-hydrazinobenzanilide which forms a benzimidazole on ring closure.

Attempts to reduce benzaldehyde-o-nitrophenylhydrazone to benzaldehyde-o-hydroxylaminophenylhydrazone which on ring closure might form the dihydrobenzotriazine were not successful [20] as it was not possible to avoid the further reduction to amine. A similar situation is also found in the reduction of o-nitroanilines.

It is, thus, essential that the central carbon atom be in the oxidation state of a carboxyl group and carry a good leaving group; if the central carbon atom is in a lower oxidation state, it must be oxidized before reduction of the aromatic nitro group. Benzaldehyde-o-nitrophenylhydrazone may, thus, be nitrosated [20] and further oxidized to o-nitrophenylazophenylnitromethane which, as described, forms (XIX) in two steps with consumption of altogether 12 electrons.

(3) Ring Closure by Michael Addition. Electrolytic reduction of o-aminoacetophenone (XX) or methyl anthranil in alkaline solution yielded [21] some of the expected alcohol (8%), the d, l-pinacol (21%), and meso-pinacol (9%). Besides, there were formed two stereoisomers (23 and 35% yield) of 3-acetyl-4-amino-7-hydroxy-7-methyl-2, 6-methano-2, 5, 6, 7-tetrahydro-1(1H)-benzazonine (XXI). The following reaction mechanism has been suggested [21]:

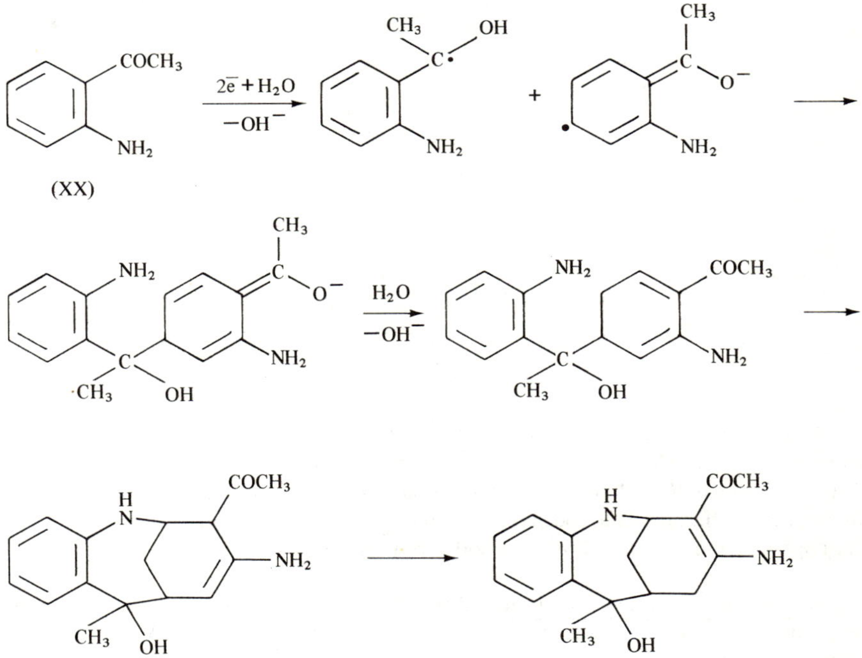

(XX)

(XXI)

An oxidative-induced ring closure occurs during the oxidation of various catecholamines (XXII) [22] at the carbon paste electrode. Whereas the oxidation in M H_2SO_4 yielded the 1, 2-benzoquinone (XXIII), sufficient of the free amine of (XXIII) was present at pH 3 to undergo Michael addition to the adrenalinequinone. As would be expected, the resulting catechol (XXIV) is more easily oxidizable than adrenaline and is converted to the quinone, adrenochrome (XXV), by chemical oxidation by adrenalinequinone:

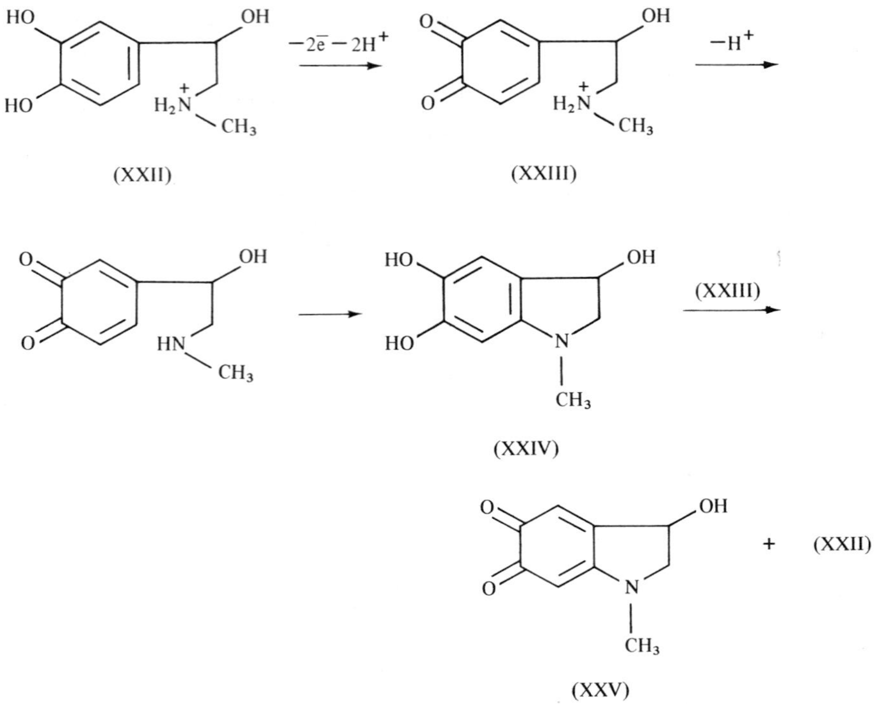

Oxidation of pentafluoroaniline in aqueous acetone containing potassium acetate gives octafluorophenazine and decafluoroazobenzene in 6 and 16% yield, respectively. The octafluorophenazine is believed to be formed through 2-aminononafluorodiphenylamine, as this compound gives octafluorophenazine in 40% yield on anodic oxidation [23].

c. Nitroso Groups as Electrophilic Centers. Nitroso derivatives may be formed as intermediates in the reduction of nitro compounds, but, as the nitroso group generally is more easily reducible than the nitro group, the lifetime of the nitroso stage is rather short. Alternatively, the nitroso compound may be prepared by anodic oxidation of a hydroxylamine at a platinum or carbon anode at a suitable pH.

Activated methyl groups may react with an intermediate nitroso group in acid solution. Thus, o-nitrodimethylaniline yields on reduction in 2 N HCl a mixture of N-methylbenzimidazole and o-aminodimethylaniline [24].

d. Carbonium Ions as Electrophilic Centers. Many aromatic compounds yield carbonium ions on oxidation at a platinum electrode in a nonaqueous solvent as acetonitrile. 2, 4, 6-Tri-t-butylaniline (XXVI) yields on anodic oxidation in acetonitrile containing pyridine the following compounds [25], (XXVII) and (XXVIII):

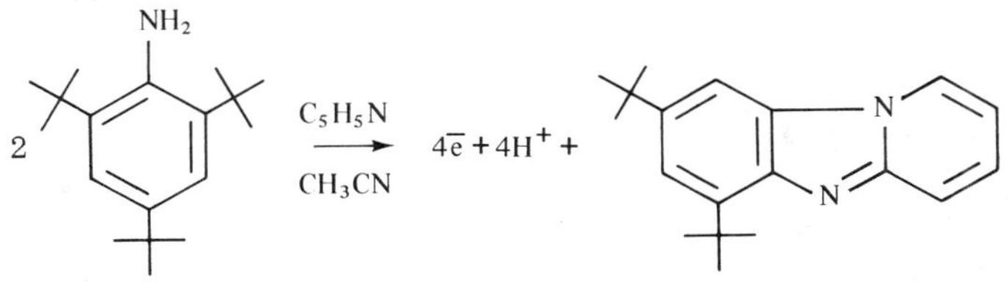

(XXVI) (XXVII)

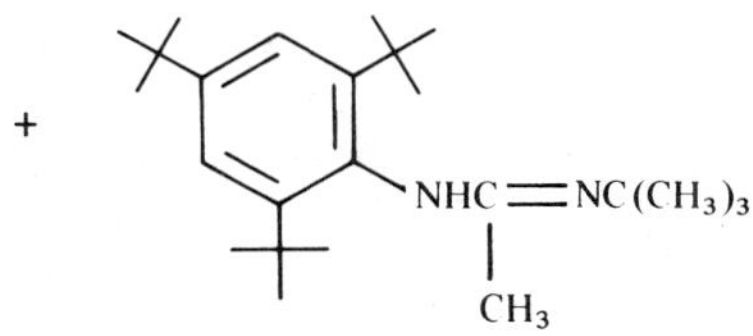

(XXVIII)

The reaction starts with a loss of two electrons from the aromatic system of (XXVI); this is stabilized by loss of a t-butyl carbonium ion, which reacts with acetonitrile in a Ritter-type reaction, and by attack by the nucleophile pyridine. The second loss of two electrons may occur before or after the ring closure. Similar internal substitution reactions are of potential value for the synthesis of heterocyclic compounds.

e. Other Reactions. In the reduction of o-dinitrobenzene (XXIX) at pH 0 there is formed some 2, 3-diamino-9, 10-dihydrophenazine (XXX) [26]:

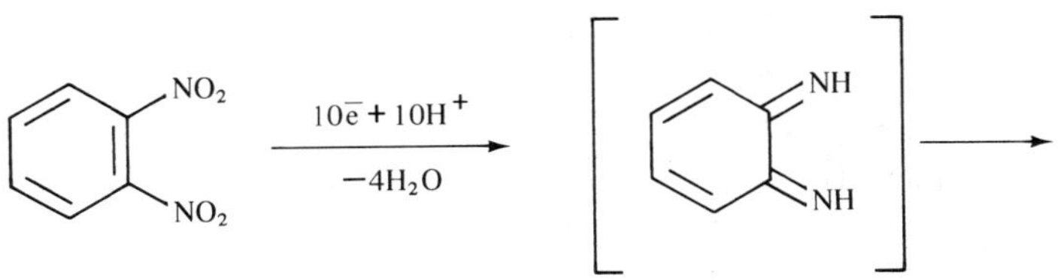

(XXIX) (XXXI)

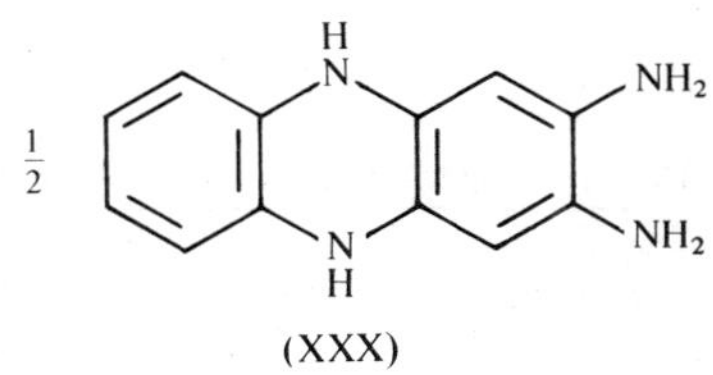

(XXX)

The intermediate responsible for the ring closure is probably the o-quinone diimine (XXXI).

Preparation of sym-triazines by electrolysis of N, N-dimethylcyanamide containing lithium chloride has been reported [27]; the formation of 2, 4, 6-tris(dimethylamino)triazine is, however, not a reduction. The trimerization must be catalyzed by an electrolytically generated base.

A low yield of pyrazine derivatives may be obtained when a carbonyl derivative is electrolyzed in the presence of ammonia; the reduction of thiobenzamide [28] in alkaline solution thus yields some tetraphenylpyrazine, and acetone in ammonia containing potassium iodide yields 7% 2, 5-dimethylpyrazine on electrolysis [29]. The reaction must involve a reoxidation during the work-up.

Anodic oxidation of 1, 1-diphenylhydrazine[1] gives diphenyldiazenium ion $(C_6H_5)_2N{=}\overset{+}{N}H$ (XXXII), which can react with suitable compounds: with styrene (XXXII) forms 1, 4-diphenyl-1, 2, 3, 4-tetrahydrocinnoline, which can be oxidized to the dihydrocinnoline (XXXIII) and further to 1, 4-diphenylcinnolinium ion (XXXIV) [30]:

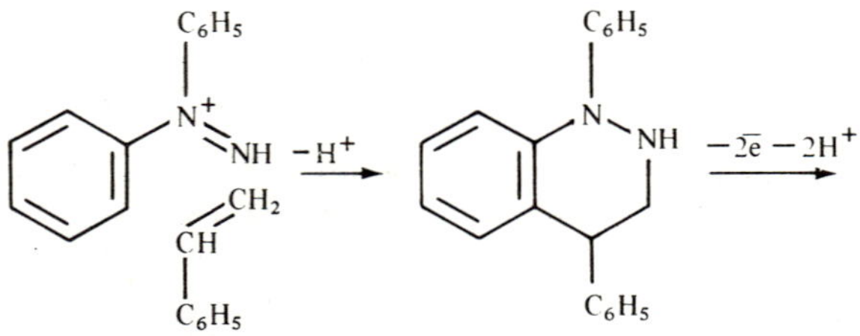

(XXXII)

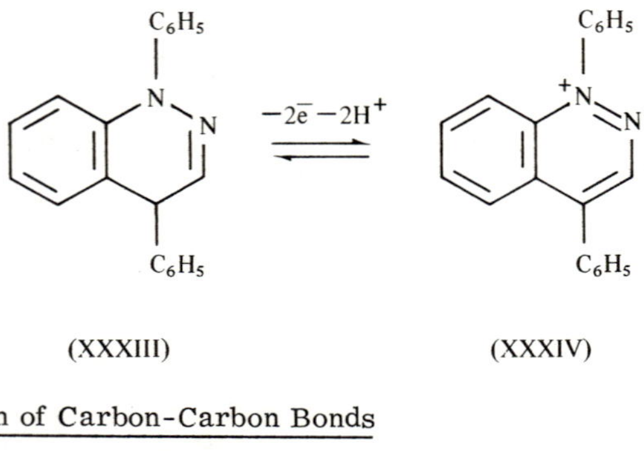

(XXXIII) (XXXIV)

3. Formation of Carbon-Carbon Bonds

Electrolytic formation of carbon-carbon bonds occurs in the reduction of ketones to pinacols, in the Kolbe synthesis, and in the hydrodimerization of activated double bonds [31] or heteroaromatic nuclei [32]. Of these only the last-mentioned reactions have been used in the preparation of heterocyclic compounds. o-Bis(β-bicarbethoxyvinylamino)benzene (XXXV) yields by intramolecular reductive coupling 2, 3-bis(dicarbethoxymethyl)-1, 2, 3, 4-tetrahydroquinoxaline (XXXVI) which on heating gives quinoxaline [31]:

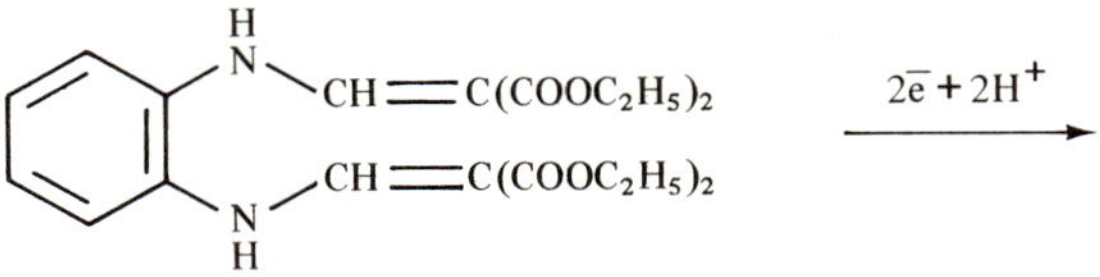

(XXXV)

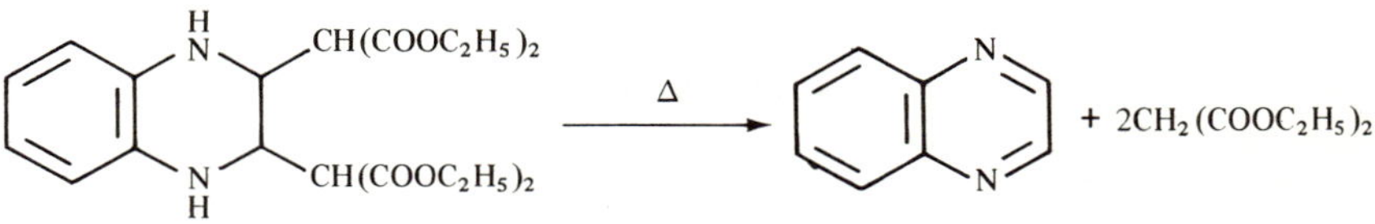

(XXXVI)

4. Formation of Nitrogen-Nitrogen Bonds

This type of ring closure is achieved mostly either by reaction between a hydroxylamino and a nitroso group or between a hydrazine derivative and a nitro or nitroso group.

A reductive coupling is possible whenever two aromatically bonded nitro groups are in such a position that a six- or seven-membered ring may be formed. The coupling occurs probably between a hydroxylamino group and an intermediate nitroso group with the formation of a cyclic azoxy derivative, a heterocyclic N-oxide.

2, 2′-Dinitrobiphenyl [33, 34] is, thus, in alkaline solution at a suitable potential reducible to benzo[c]cinnoline-N-oxide; in acid solution the reduction proceeds to 4, 5-dihydrobenzo[c]cinnoline which is very easily (e.g., by air) oxidized to benzo[c]cinnoline.

1, 8-Dinitronaphthalene could be expected from the coulometric data [35] to undergo ring closure on reduction, but suitable conditions have not been found yet. The diamine [36] and 8-amino-5-hydroxy-1, 4-naphthoquinone-1-imine [37] have been reported as products.

When one of the nitro groups is aliphatically bonded, a coupling may take place, although the yield often is lower. 2-Nitro-1-(2′-nitrophenyl)-ethanol (XXXVII) can in acetate buffer be reduced in two four-electron steps.

The dihydroxylamino compound (XXXVIII) is transformed to cinnoline (XXXIX) by alkali [38]:

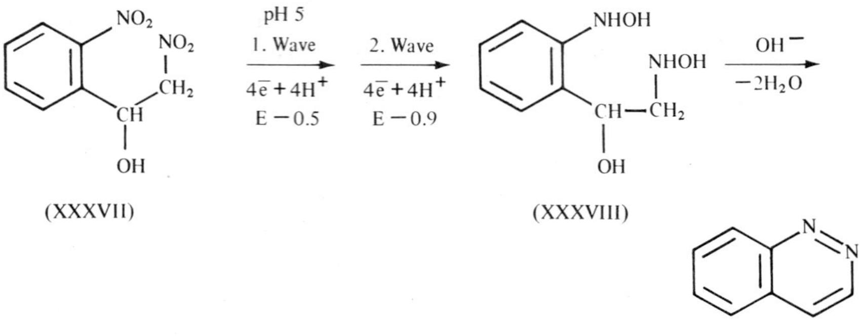

(XXXVII) (XXXVIII) (XXXIX)

Reduction of an azo compound to a hydrazine may lead to a ring closure when a nitro group is suitably situated. Thus, the reduction of 4-hydroxy-2′-nitroazobenzene (XL) in alkaline solution follows the scheme [39, 40]:

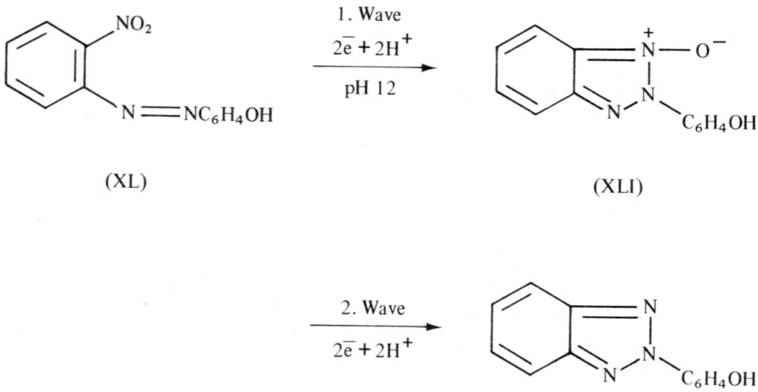

(XL) (XLI) (XLII)

By controlling the electrode potential it is possible to prepare the N-oxide (XLI) in good yield [40], whereas the "classical" procedure yields the benzotriazole (XLII) [39].

A hydrazide may also to some extent condense with an intermediate nitroso group [41], but the yield obtained so far is low. o-Nitrobenzoic hydrazide (XLIII) is in acetic acid buffer reduced to a mixture from which some benzo-1, 2, 3-triazinone-4 (XLIV) has been isolated:

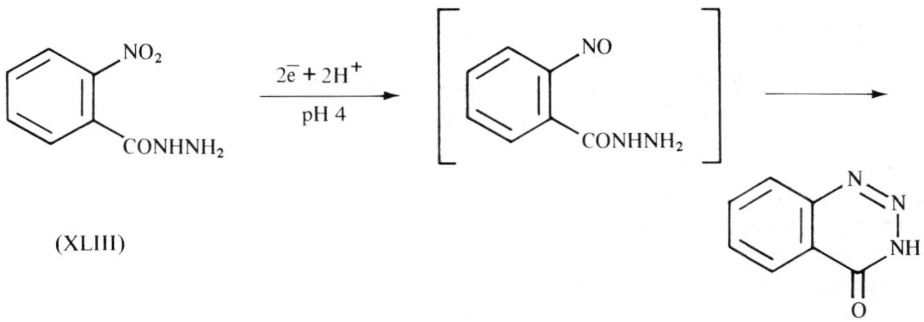

(XLIII) (XLIV)

B. Ring Contractions

In ring contractions occurring during electrolysis of nitrogen-containing heterocyclic compounds a hydrogenolysis of a single bond takes place at some step, which is then followed by a ring closure. These reactions may be classified according to the single bond cleaved in the reduction. Examples of hydrogenolysis of oxygen-nitrogen, nitrogen-nitrogen, and nitrogen-carbon bonds are discussed.

1. Hydrogenolysis of an Oxygen-Nitrogen Bond

The initial step in the reduction of protonated oximes has been shown [42] to be a hydrogenolysis of the oxygen-nitrogen bond; in cyclic derivatives of oximes the same occurs. In 4-(4′-methoxyphenyl)-2, 3-benzoxazin-1-one (XLV) the reduction proceeds in two steps and it is possible to isolate the rather stable intermediate ketimine [43]:

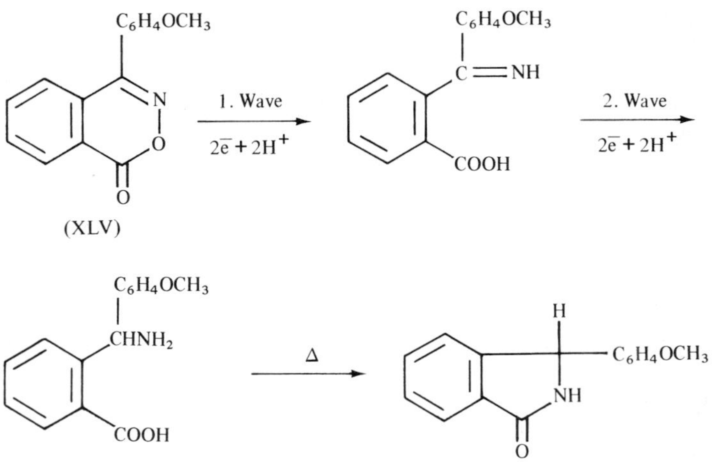

(XLV)

2. Hydrogenolysis of a Nitrogen-Nitrogen Bond

A ring contraction involving this reaction has been found in cyclic hydrazones, hydrazides, and triazines.

The initial step in the reduction in acid solution of azines, hydrazones, and substituted hydrazones has been suggested to be a hydrogenolysis of the nitrogen-nitrogen bond [42]; the ald- and ketimine thus formed are mostly, but not always, reduced further at the potential necessary for the reductive cleavage; this depends on the nature of the substituents. The general reduction scheme for a hydrazone thus is the following:

$$RR'C{=}NNHR'' \xrightleftharpoons{H^+} RR'C{=}N\overset{+}{N}H_2R''$$

$$RR'C{=}N\overset{+}{N}H_2R'' \xrightarrow{2\bar{e}+3H^+} RR'C{=}\overset{+}{N}H_2 + R''NH_3^+$$

$$RR'C{=}\overset{+}{N}H_2 \xrightarrow{2\bar{e}+2H^+} RR'CH\overset{+}{N}H_3 .$$

In the ring closure step the amine $R''NH_2$ attacks the imine with the formation of $RR'C{=}NR''$ and loss of ammonia, and this reaction is to be expected whenever the imine is not reducible at the potential necessary for the reductive ring opening.

Pyridazines to Pyrroles. Pyridazines are generally reduced to dihydropyridazines which can exist in different tautomeric forms; the reduction of these may be different. In the reduction in acid solution of 1-methyl-3, 6-diphenyl-5-t-butylpyridazinium iodide **(XLVI)** the reaction sequence on page 579 has been found [44, 45]. The pyrroles (XLIX) and (L) were formed in the ratio 3:1.

Similar ring contractions in which the nitrogen-nitrogen single bond is cleaved primarily are found in the reductive ring contractions of 1, 4-dihydrocinnolines to indoles [33], phthalazines to isoindoles [46], benzo-1, 2, 4-triazines to benzimidazones [47], benzo-1, 2, 3-triazinone to indazolone [48], and benzo-1, 2, 3-triazine-3-oxides to indazoles [48].

The reduction of phthalazinones is apparently the only observed example of a reduction of a protonated compound containing >C=N-N< in which the C=N double bond has been proved to be more easily reducible than the N-N bond. The differences in reduction potential between the azomethine double bond and the nitrogen-nitrogen bond are not great in noncyclic compounds, and the presence of a phenyl group fused to the heterocyclic ring may alter several parameters. The phenyl group acts as a nonreducible, nondisplaceable unsaturated center, and its presence may influence the orientation of the molecule at and the adsorption of the compound to the electrode.

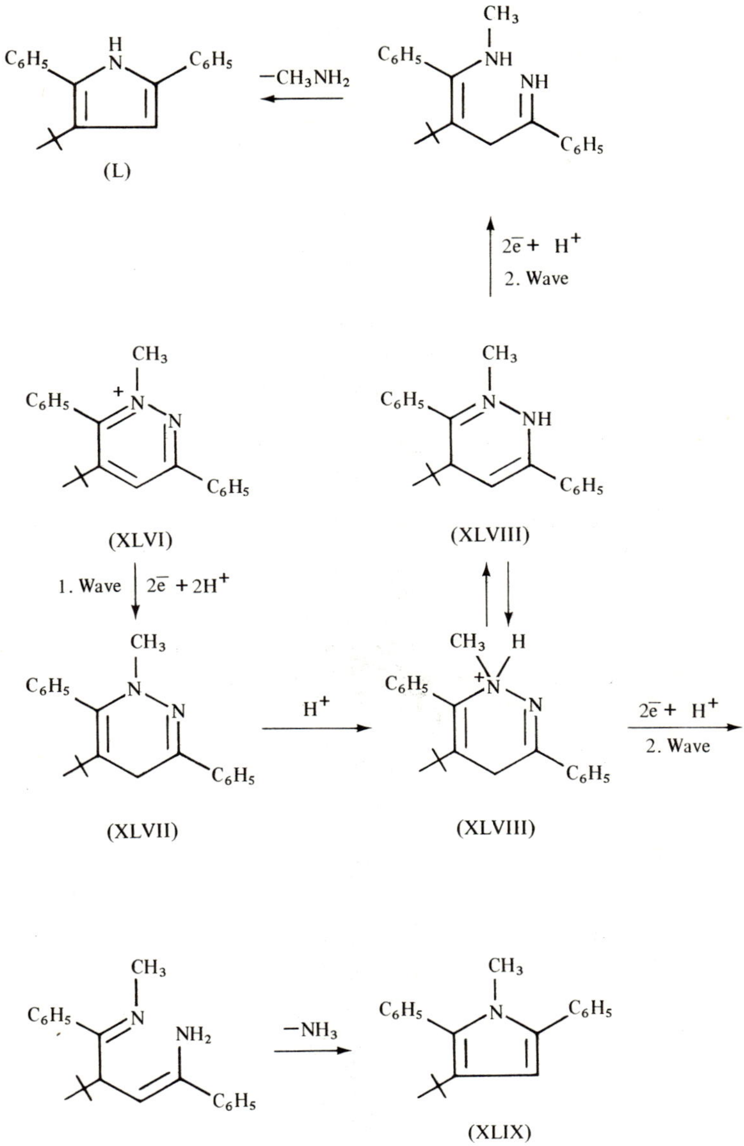

4-Methylphthalazinone(LI) is reduced to 1-methylphthalimidine(LII) in the following steps [49]:

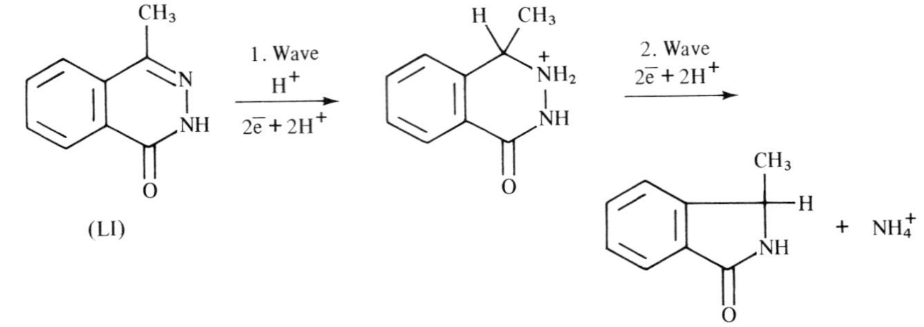

(LI) (LII)

A similar ring contraction has been observed during the reduction in acid solution of 2, 3-dimethyl dihydrophthalazinedione [49].

3. Hydrogenolysis of a Nitrogen-Carbon Bond

An electrolytic cleavage of such a bond in a convenient potential interval occurs only when the bond is activated, e. g., by an adjacent carbonyl or azomethine group.

A ring contraction occurs during the reduction of 7-chloro-2-methylamino-5-phenyl-3H-4, 5-dihydro-1, 4-benzodiazepine (LIII) to 6-chloro-2-methyl-4-phenyl-3, 4-dihydroquinazoline (LIV) [50]. It has been suggested that the reaction takes the following course [12], although another interpretation has also been offered [50]:

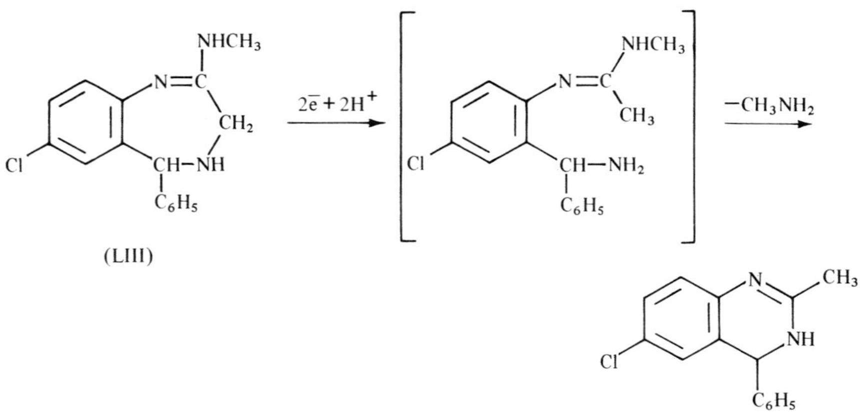

(LIII) (LIV)

C. Ring Expansions

Only a few types of ring expansion involving electrolysis have been reported so far. In the preparation of ψ-phthalazinones (LV) from N-anilinophthalimides (LVI) [51] the phthalimide derivative is reduced to a derivative of phthalaldehydic acid, N-anilinohydroxyphthalimidine (LVII). It has been suggested [12, 52] that this compound suffers a ring opening to a hydrazide of phthalaldehydic acid (LVIII). On ring closure the more nucleophilic nitrogen of the hydrazide attacks the aldehyde group with formation of a six-membered ring, a ψ-phthalazinone. In strongly acid solution the ψ-phthalazinone ring may be opened again and then some of the hydroxyphthalimidine (LVII) forms a phenylhydrazone of phthalaldehydic acid (LIX) on hydrolytic ring opening.

The phenylhydrazone would on ring closure yield 2-phenylphthalazinone (LX); this Rowe rearrangement [52] can thus be explained by assuming that the ψ-phthalazinone is the kinetically controlled product, whereas the 2-arylphthalazinone is the thermodynamically controlled product:

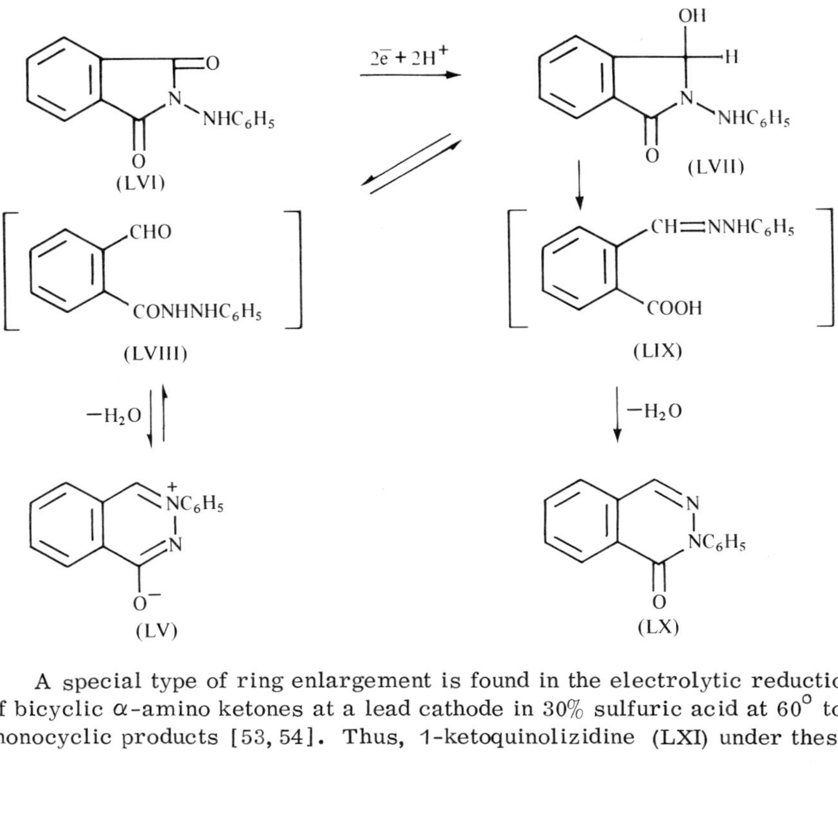

A special type of ring enlargement is found in the electrolytic reduction of bicyclic α-amino ketones at a lead cathode in 30% sulfuric acid at 60° to monocyclic products [53, 54]. Thus, 1-ketoquinolizidine (LXI) under these conditions gave 59% 5-hydroxyazacyclodecane (LXII) and 4.5% 1-azabicyclo-[5, 3, 0]decane (LXIII) in the following way:

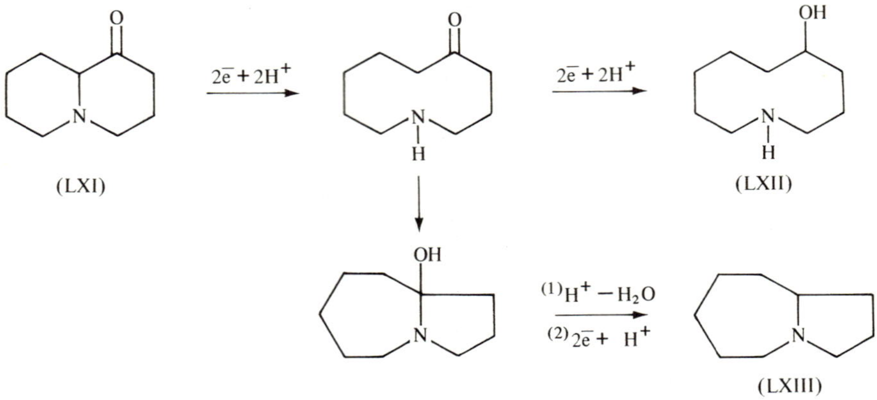

The initial cleavage of the carbon-nitrogen bond is analogous to that found in the reduction of derivatives of 2-aminoacetophenone [55] and to the first step in the ring contraction of (LIII).

In this section the anodic oxidation in nitromethane of tetraphenylpyrrole to tetraphenylpyridine [56] might be mentioned; it is discussed in Sec. III.C.

III. ELECTROLYTIC REACTIONS OF HETEROCYCLIC SYSTEMS

In this section the electrode reactions of heterocyclic compounds in which the nucleus undergoes a reduction or oxidation are discussed. Substituted compounds are included in this section only if the substituent does not become reduced or oxidized.

One advantage of using electrolytic methods in heterocyclic systems lies in the possibility of preparing partly reduced systems, in most cases from π-electron-deficient compounds, and it is probably the most general method for doing so.

Another useful method is the electrolytic substitution of heterocyclic compounds, which are π-electron rich or substituted with electron-donating groups.

A. Compounds with One Oxygen Atom

Furan is, as a π-electron-rich compound, easily oxidized electrolytically. In acetonitrile the oxidation potential of furan is comparable to that of anisole. In methanol the oxidation is an ECEC mechanism in which 2, 5-dihydro-2, 5-dimethoxyfuran (LXIV) is formed [57]:

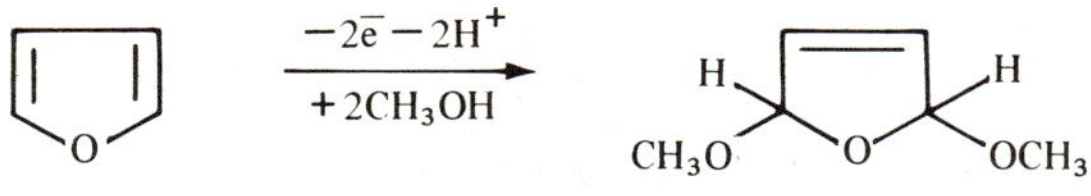

(LXIV)

The reaction may be performed in the presence of an excess of ammonium bromide, where the bromide ions are oxidized to bromine which in turn oxidizes furan. The dimethoxydihydrofuran is a valuable intermediate for the preparation of various heterocyclic compounds such as pyridines [58] and pyridazines [59-61]. The oxidation of a furan to a dimethoxydihydrofuran is thus a key step in a synthesis of pyridoxine [62].

Anodic acetoxylation of furan can be made in acetic acid containing acetate [63, 64] and benzoyloxylation in DMF containing benzoic acid and lithium chloride [65]. A mixture of cis- and trans-benzoyloxylated product is formed. Cyanation of 2, 5-dimethylfuran gives a mixture of cis- and trans-2-cyano-5-methoxy-2, 5-dimethyldihydrofuran [66].

Pyrylium [67] and isobenzopyrylium [68] salts have been shown to be polarographically reducible in a one-electron reduction. The radical thus formed may dimerize, but the actual electrode reaction has not been proved.

Electrochemical reduction of coumarin [69-71] affords the meso and (±) forms of a product hydrodimerized (at C-4) like many other α, β-unsaturated acid derivatives. In the presence of tertiary amines, which are not destroyed during the process but catalyze the evolution of hydrogen, a transfer of hydrogen takes place from the amine radical to the coumarin radical with formation of 3, 4-dihydrocoumarin. Asymmetric amines cause asymmetric induction during the reduction of 4-methylcoumarin [72].

Voltammetric studies in DMF [73] of arylsubstituted isobenzofurans like 1, 3, 4, 7-tetraphenylisobenzofuran (LXV) show that (LXV) gives two one-electron reduction waves and one one-electron oxidation wave. By cyclic voltammetry (CV) it was shown that the first reduction product was stable, whereas the first oxidation product and the second reduction product were unstable.

B. Nuclei Containing One Sulfur Atom

Thiophene. Anodic oxidation in methanol of thiophene and substituted thiophenes at low temperatures (-20 to -30°C) results in methoxylation and ring opening with loss of sulfur as SO_2 [74-77]. From thiophene is isolated butenedialdehyde tetramethylacetal, some methoxysuccinic dialdehyde tetramethylacetal, and a small amount of methyl β-formylpropionate. In general oxidation of thiophenes results in formation of derivatives of α, β-unsaturated γ-dicarbonyl compounds or γ-keto esters. Cyanation of 2, 5-dimethylthiophene yields mainly cis- and trans-2-cyano-5-methoxy-2, 5-dimethyldihydrothiophene and 3-cyano-2, 5-dimethylthiophene [78].

Benzothiophen-1, 1-dioxide(LXV) is polarographically reducible at about -1.3 V at pH >2 [79]. In alkaline solution the polarographic behavior is complicated by addition of nucleophiles to the 2, 3-double bond. Controlled potential reduction in acetate buffer produced 2, 3-dihydrobenzothiophen-1, 1-dioxide (LXVI) [80].

In unbuffered media containing tetraalkylammonium ions as supporting electrolyte (LXVI) gives a wave at about -2.2 V [79]. In view of the results of the reduction of alkylarylsulfones [81] the electrode reaction may be a cleavage of the ring with formation of β-phenylethylsulfinic acid (LXVII):

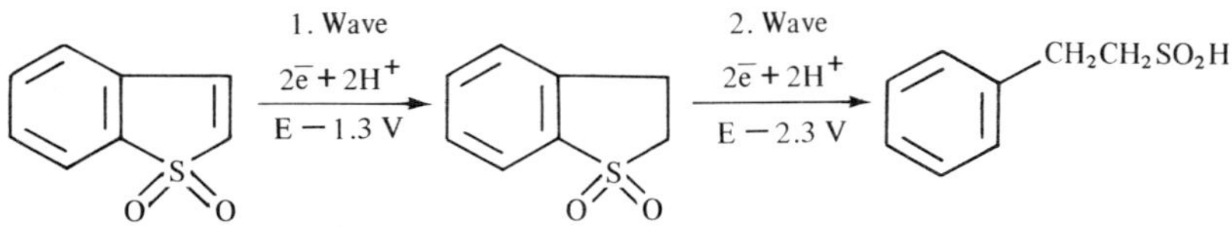

(LXV) (LXVI) (LXVII)

C. Nuclei Containing One Nitrogen Atom

1. Pyrrole Derivatives

Pyrrole is not polarographically reducible, but can be reduced at a lead cathode in dilute sulfuric acid to pyrroline and further to pyrrolidine [82, 83]. Under similar conditions 1, 2-dimethylpyrroline [84] is also reduced to the pyrrolidine, and indoles [85-87] to indolines or dimerized products [88]. Pyrrole may also be oxidized anodically; oxidation of pyrrole at a platinum electrode in acetonitrile yields a tar which covers the electrode with an insulating layer [2]. However, electrolysis of pyrrole in acetonitrile in the presence of benzaldehyde yields tetraphenylporphine besides some tar and other side products [89]. The porphine synthesis is initiated by protons formed in the oxidation of pyrrole. Condensation of protonated pyrrole molecules with benzaldehyde forms tetraphenylporphinogen which is oxidized to the porphine anodically or by a redox reaction with pyrrole.

In acetonitrile pentaphenylpyrrole is oxidized to a rather stable cation radical [90]; in methanolic solution 1-methylpyrrole forms 1-methyl-2, 2, 5, 5-tetramethoxypyrroline [91].

2, 3, 4, 5-Tetraphenylpyrrole (LXVIII) is in nitromethane oxidized in two one-electron steps, the first one being reversible. Oxidation at the potential of the first wave at low temperatures gives n = 2, indicating a disproportionation of the initially formed radical cation (LXIX). The dication thus formed reacts with the residual water to give (LXX) which on heating rearranges to (LXXI). At higher temperatures (LXIX) reacts with nitromethane forming 2, 3, 4, 6-tetraphenylpyridine (LXXII) and nitrite ion together with small quantities of dibenzoylstilbene [56].

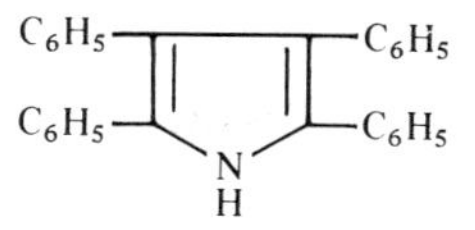

(LXVIII)

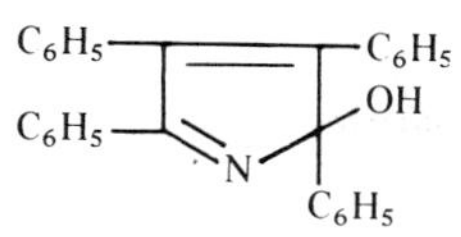

(LXX)

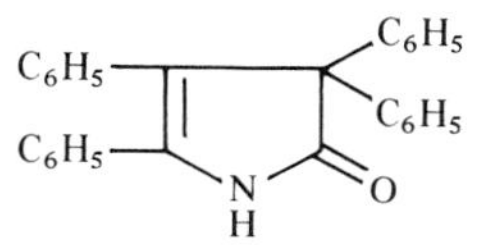

(LXXI)

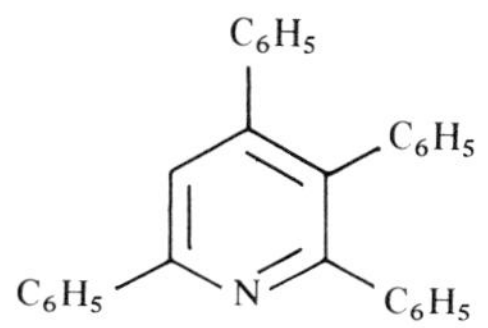

(LXXII)

N-Methyl-1, 3, 4, 7-tetraphenylisoindole(LXXIII) has been investigated by CV in DMF. Contrary to the isobenzofurans discussed above (LXXIII) forms a stable cation radical, but unstable anions [73].

Carbazoles are oxidized at controlled potential at a platinum anode in acetonitrile to cation radicals which are stable when the 3-, 6-, and 9-positions are blocked [92, 93]. The radical cation (LXXIV) from carbazole dimerizes predominantly at the 3-position to 3, 3-dicarbazyl which is further oxidizable to the quinoidal dication at the potential used. In the presence of pyridine, which may cause a rapid deprotonation of the N-H proton of (LXXIV) 9, 9′-dicarbazyl is the isolated product.

2. Pyridine Derivatives

Pyridine is not polarographically reducible and is very difficultly oxidizable at a platinum electrode in acetonitrile. It can, however, be reduced to piperidine at a lead cathode in 10% H_2SO_4 [94, 95]; the yield is very dependent on the purity of the lead and other components. Side products are α, α- and γ, γ-dipiperidyl.

The potential-limiting anodic reaction in pyridine with perchlorate as supporting electrolyte is presumably the oxidation of pyridine to an onium ion which, with excess of pyridine, forms N-(2-pyridyl) pyridinium perchlorate [96].

Substituents may render the nucleus more easily reducible or oxidizable; electron-attracting substituents as, e. g., carbethoxy groups, facilitate the reduction, whereas, e. g., methoxy groups make the oxidation easier. A positive charge on the nitrogen atom also facilitates the reduction of the nucleus, but the charged pyridine ring also makes the electron-attracting substituent more easily reducible. In the cases where the electrode reaction changes with pH, it is generally found that the substituent is reduced in acid and the nucleus in alkaline solution.

The presence of two methoxy groups on the pyridine ring makes it susceptible to anodic oxidation. In alkaline methanol the main products [91] from 2, 6-dimethoxypyridine (LXXXV) are 2, 3, 5, 6-tetramethoxypyridine (LXXVI) and an azaquinone, probably with the structure (LXXVII):

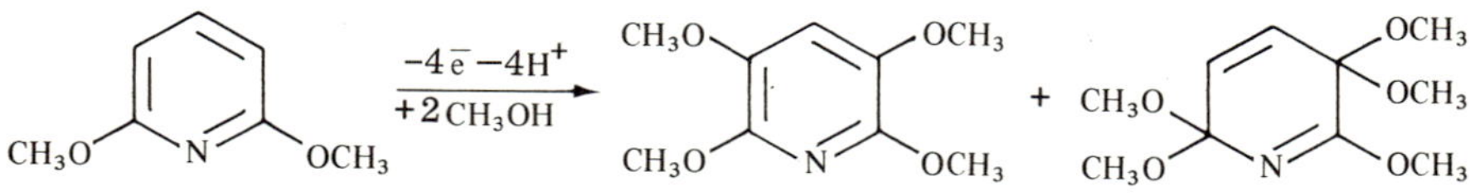

(LXXV) (LXXVI) (LXXVII)

Derivatives of the pyridinecarboxylic acids are polarographically reducible both in acid and alkaline solution; in the latter solvent the nucleus is reduced. In the few cases where a macroscale reduction at a mercury electrode [97, 98] has been made, a 1, 4-dihydropyridine has been isolated.

Pyridinium salts are generally polarographically reducible in alkaline solution and dimerized dihydropyridine derivatives have been postulated as products [99, 100]; further reduction produced the 1, 4-dihydro derivative.

N-Methylpyridinium or N-methylquinolinium salts [101] may be oxidized to the N-methyl-2-pyridone or N-methyl-2-quinolone, respectively, in alkaline medium containing some potassium hexacyanoferrate(III). The reaction is probably an indirect electrolysis where the hexacyanoferrate oxidizes the pseudobase and is, in turn, reoxidized at the anode.

The reduction of quaternary derivatives of nicotinic amide as model compounds for the phosphopyridine nucleotides has been subject to many polarographic [98, 99, 102-106] and electrolytic investigations [98, 102]. At pH 4-7 a one-electron wave is found, whereas in alkaline solution two one-electron waves are seen. Controlled potential reduction in a phosphate buffer at the potential (-1.2 V) of the first wave of 1-methyl-3-carbamido-pyridinium perchlorate (LXXVIII) produced a dimer, suggested to be the 6, 6′-dimer; at a more negative potential (-1.8 V) the dimer was reduced further with two electrons per nucleus. If the reduction of (LXXVIII) was performed at the potential of the second wave (-1.8 V), the 1, 4-dihydro derivative was formed.

Analogous results have been obtained in the electrolysis of the pyridine nucleotides [107-114]. The dimer has, however, been suggested to be a 4, 4′-dimer. The coenzyme activity of the reduction product varies with the experimental conditions. Some reductions yielded totally inactive products, while others produced partly or fully active material. The highest activity was obtained by an indirect electrolytic reduction of TPN [114] (Chapter XXIV), electrochemical reduction of methyl viologen in the presence of ferridoxin-TPN reductase caused a reduction of triphosphopyridine nucleotide to a biochemically active product.

Electrochemical oxidation of analogs of the reduced form (NADH) of nicotinamide adenine dinucleotide has been investigated [115]. The results are consistent with an ECE mechanism yielding the pyridinium salt.

3. Quinoline, Isoquinoline, Acridine

Quinoline and isoquinoline derivatives are polarographically reducible in a way similar to the pyridine compounds, but the reduction potentials required are generally less negative.

The reduction of quinoline at different electrodes in sulfuric acid to dihydro- and tetrahydroquinoline has been studied using constant current [116, 117]. The reaction proceeds with high current efficiency at electrodes with high hydrogen overvoltage (e. g., lead), but not at platinum, which has a low hydrogen overvoltage.

It is of interest that a dihydroisoquinoline like 3, 4-dihydropapaverine is more easily reducible than the parent compound [118]; the electrode reaction is saturation of the azomethine bond.

Anodic oxidation of quinoline at a platinum anode in 75% sulfuric acid yields quinolinic acid (77%) [119]; other quinolines bearing substituents in the pyridine ring in the 3-position may do similarly, whereas some 2- and 4-substituted derivatives form tar [120].

Acridine is polarographically reducible both in acid and alkaline solutions [121]. In strongly acid solution a single one-electron wave is found, and from a preparative reduction compounds dimerized at C-9 [2] were isolated. In less acid and in alkaline solutions two one-electron waves are found.

The acridizinium and phenanthridizinium systems are reducible [122] in a similar manner.

D. Compounds with One Oxygen and One Nitrogen Atom

1. Oxaziridines

Oxaziridines are very easily reducible compounds and their polarographic wave [123] begins at the dissolution wave of the mercury. In the reduction there is formed a gem-amino alcohol which at low pH loses water with the formation of a Schiff base, and at higher pH loses an amine to a carbonyl compound. The Schiff base or the carbonyl compound may be further reducible. The reduction of 2-tert-butyl-3-phenyl-oxaziridine (LXXIX) exemplifies this:

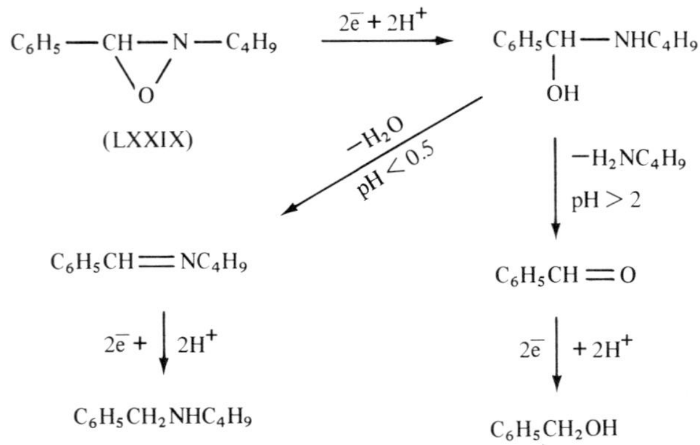

2. Oxazoles

Arylsubstituted oxazoles are reducible [124] at rather negative potentials [-1.9 V to -2.1 V (SCE)] and the electrode reaction consumes in most cases six electrons; 2-(1-naphthyl)-5-phenyloxazole, however, consumes two electrons.

In DMF diphenyloxazole is reduced in two one-electron reductions, the first one resulting in a relatively stable anion radical, the second one presumably in a dihydro derivative [125].

3. Anthranils

These compounds give in acid solution two polarographic waves, and a reduction at a potential corresponding to the foot of the first wave of 3-methylanthranil yields o-aminoacetophenone in good yield [12]. The second wave represents the reduction of this compound. In alkaline medium a single three-electron wave is found, and the reaction mixture contains several dimerized compounds [21].

E. Compounds with One Sulfur and One Nitrogen Atom

1. Thiazole Derivatives

Thiazole and its simple alkyl or aryl derivatives are not polarographically reducible. Substitution in the thiazole ring with electron-attracting groups may render the nucleus reducible. Derivatives of thiazole-2-carboxylic acid are reducible in alkaline solution, and the reduction is assumed to take place in the nucleus [126].

Many thiazolium salts are polarographically reducible [127, 128]; 3-methyl-benzothiazolium iodide [127] is reduced in a two-electron wave, but a preparative reduction gave n between 1 and 2 and a mixture of dimer and monomer 2, 3-dihydro-3-methylbenzothiazole was found. Aryl groups in the 2-position hinder dimerization [128].

2. Isothiazole Derivatives

Substituted benzenesulfonamides bearing strongly electron-attracting substituents can be reduced in slightly alkaline solution at the dropping mercury electrode to ammonia, sulfur dioxide, and a substituted benzene [81]. Saccharin (LXXX) is a compound of this type and the primary electrode reaction is a cleavage of the carbon-sulfur bond; chemical reactions follow the initial cleavage:

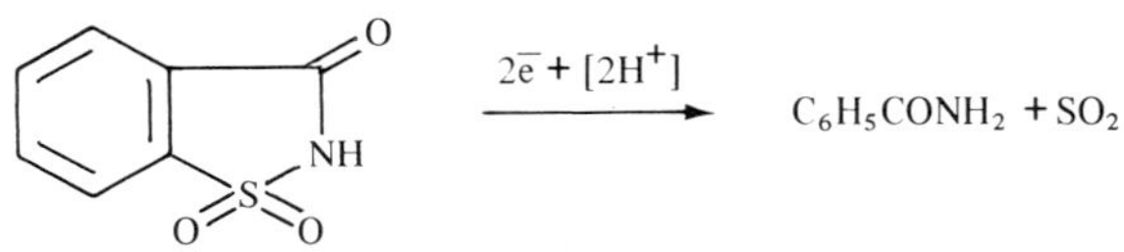

(LXXX)

In acid solution the reduction follows a route similar to that of phthalimide with a primary reduction of the carbonyl group [129, 130].

3. Thiazines

Simple thiazines have not been investigated polarographically. Several derivatives of phenothiazine have been found to be oxidizable at a gold anode in dilute sulfuric acid [131]. The product proved to be the sulfoxide rather than the N-oxide.

In acetonitrile phenothiazine is oxidized in two one-electron waves; the radical cation obtained in the first oxidation loses a proton and is oxidized to the phenazothionium ion, isolated as the perchlorate [132-134].

F. Compounds with Two Oxygen or Sulfur Atoms

1, 4-Dioxins and 1, 4-Dithiins

Voltammetric oxidation of 1, 4-dioxins (LXXXI) and 1, 4-dithiins (LXXXII) in acetonitrile containing perchlorate yields a cation radical in the first one-electron oxidation and a dication in the second one-electron oxidation. Both (LXXXI) and (LXXXII) are typical π-electron-rich heterocycles, whereas the dication contains six π-electrons, and the relatively high stability of the dication is connected with this "aromaticity" [135].

G. Compounds with Two Nitrogen Atoms

1. Diazirines

Diazirines are polarographically reducible in acid solution in a four-electron reduction and in alkaline medium in a two-electron reaction [136]. Diaziridines are reducible only in acid solution [137]; in alkaline solution they can be oxidized anodically to diazirines [136].

The protonated diaziridine is more easily reducible than the diazirine which has too weakly basic properties to be protonated in aqueous solution. In alkaline medium where both are unprotonated, the diazirine is the more easily reducible. This explains why it has been difficult to obtain good yields of diaziridine by chemical and catalytic reduction of a diazirine; if such a reduction is to result in a high yield of diaziridine, it must be performed in an alkaline medium.

2. Pyrazolines

Pyrazolines have been investigated polarographically in a 2:2:1 mixture of DMF, benzene, and methanol with tetraethylammonium iodide as electrolyte. The polarographic data indicate a two-electron reduction of the carbon-nitrogen double bond [138, 139].

3. Pyridazines

Pyridazines are polarographically reducible [140, 141], often stepwise. The first reduction yields generally a dihydropyridazine which can exist in different tautomeric forms, the most stable forms usually being the 1, 4- and 4, 5-dihydropyridazines [140]. The further reduction is complicated by the fact that the different tautomeric forms may be reduced differently and at different potentials and that the rate of transformation of a given tautomer to the most easily reducible one may or may not be fast compared to the further reduction.

The product isolated [140] from the two-electron reduction of 1-methyl-3, 6-diphenylpyridazinium iodide (LXXXIII) in acid solution after extraction from neutral solution was 1, 4-dihydro-1-methyl-3, 6-diphenylpyridazine (LXXXIV); but, as it was found that 1, 4-dihydro-3, 6-diphenylpyridazine in trifluoroacetic acid was transformed to the 4, 5-dihydro derivative [45], this tautomer might be the predominant one in the acid solution.

The reduction in acid solution of di- and tetrahydropyridazines and pyridazinones is often best understood when their resemblance to hydrazones is considered. The reduction of such compounds has been discussed briefly above, and the primary step is a hydrogenolysis of the nitrogen-nitrogen bond. The reduction of 1, 4-dihydro-1-methyl-5-t-butyl-3, 6-diphenylpyridazine to pyrroles was also described above; at more negative potentials pyrrolines and pyrrolidines may be obtained from pyridazines [127].

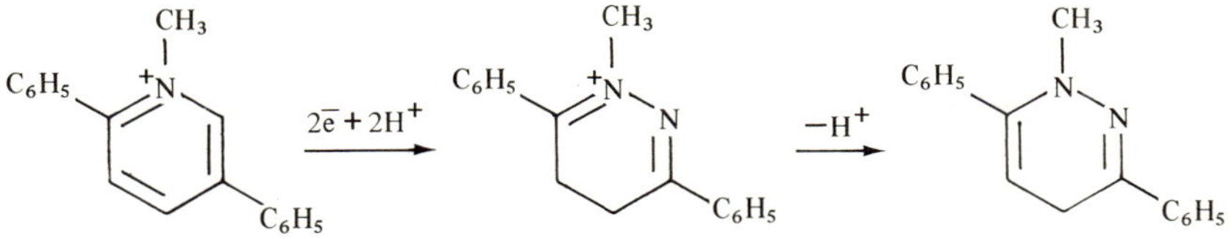

(LXXXIII) (LXXXIV)

3, 6-Diphenyl-2, 3, 4, 5-tetrahydropyridazine (LXXXV) is structurally similar to a benzylhydrazone of propiophenone and, accordingly, a four-electron polarographic wave is observed in acid solution with formation of (LXXXVI) [140]. The electrode reaction is as follows:

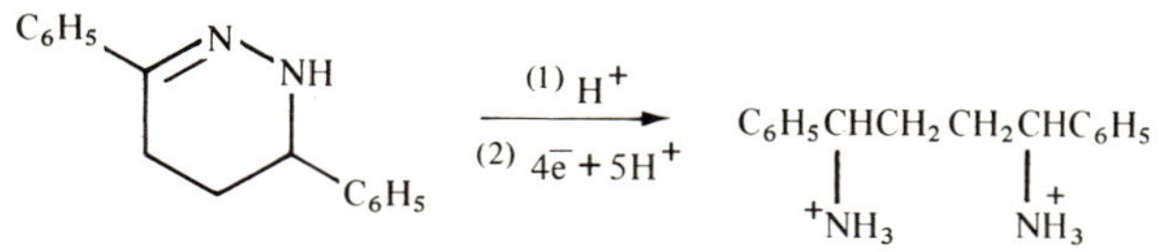

(LXXXV) (LXXXVI)

Reduction in acetate buffer at the potential of the first wave of 3-phenyl-6-dimethylaminopyridazine yields the 4, 5-dihydro derivative in high yield [140].

In a similar way other pyridazines [140], e.g., 3-phenyl-6-methoxypyridazine (LXXXVII) and 3-methyl-6-chloropyridazine, are reduced, but the dihydropyridazines are in these cases rather unstable and may lose methanol or hydrogen chloride, forming the corresponding 4, 5-dihydropyridazinone (LXXXVIII), which is cleaved by further reduction (Chapter XXI):

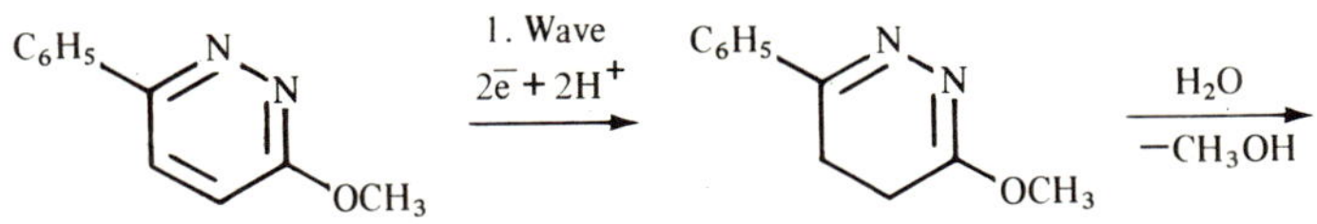

(LXXXVII)

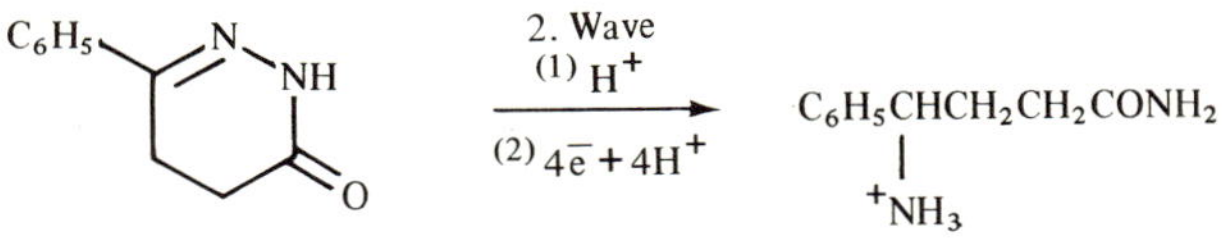

(LXXXVIII)

In alkaline solution 3, 6-diphenylpyridazine gives a two-electron polarographic wave which in strongly alkaline solution is followed closely by a smaller wave. A preparative reduction at the potential of the second wave yields 3, 6-diphenyl-2, 3, 4, 5-tetrahydropyridazine (LXXXV). It has been suggested [45] that the primary reduction product is a 1, 2-dihydropyridazine which tautomerizes to the 4, 5-dihydro derivative; this compound is then reduced further, analogously to the reduction of benzalazine [42].

4. Cinnolines

Cinnolines are generally reducible [33] in two or more steps. The first step in the reduction of aryl- or alkylsubstituted cinnolines is the formation of 1, 4-dihydrocinnolines. This reduction is not reversible. Only in the case of benzo[c]cinnoline is the reduction a saturation of the nitrogen-nitrogen double bond; the reaction resembles the reduction of azobenzene to hydrazobenzene and, like this system, is nearly polarographically reversible [142].

1, 4-Dihydrocinnolines may be regarded as cyclic phenylhydrazones and their reduction is in accordance with the concept that only the protonated compounds are reducible, and the first two-electron step produces an amine and an imine. If the imine is reducible at the potential necessary for the hydrogenolysis of the nitrogen-nitrogen bond (Chapter XXI), the imine is reduced to an amine; if not, a ring closure to an indole takes place, unless the potential is kept so negative that the further reduction of the imine can compete with the ring closure. The ring contraction to indoles has been discussed above. The reduction of 1, 4-dihydro-3-phenylcinnoline (XC) proceeds by a four-electron reaction:

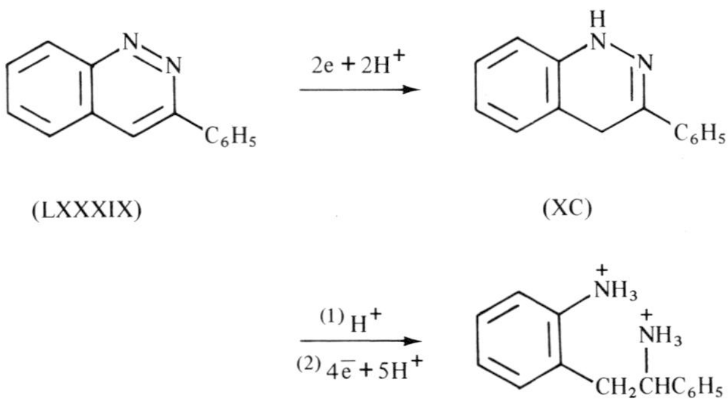

(LXXXIX) (XC)

5. Phthalazines

These compounds are reduced in acid solution somewhat differently from cinnolines. Phthalazine is reduced polarographically in acid solution in a six-electron wave [143] presumably to o-xylene-α, α'-diamine, but 6 N hydrochloric acid [46] is required as medium if the diamine is the desired product. 1-Methylphthalazine(XCI) is reduced as follows:

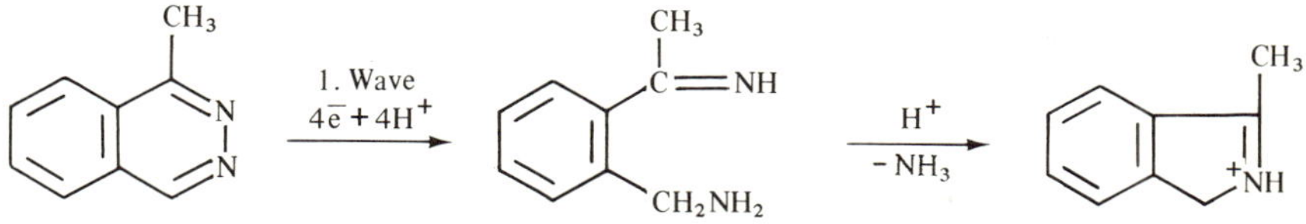

(XCI) (XCII)

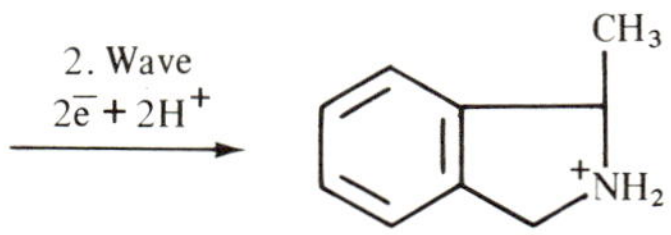

By reducing at -0.65 V (SCE) it is possible to prepare 1-methylisoindole (XCII).

4-Dimethylamino-1-phenylphthalazine can in acid solution be reduced in a four-electron reaction to 2-(1′-amino-1′-phenylmethyl)-N, N-dimethylbenzamidine (XCIII) [144] which at higher pH forms 1-phenyl-3-iminoisoindoline (XCIV); (XCIV) can be reduced to 1-phenylisoindoline (XCV) in a four-electron reduction. From the reduction corresponding to the first wave of 4-methoxy-1-phenylphthalazine the cyclic imidic ester 3-methoxy-1-phenylisoindole (XCVI) can be isolated; (XCVI) may be further reduced to (XCV).

In alkaline solution phthalazine is reduced polarographically in a two-electron reduction, but in a preparative reduction it is possible to isolate mostly dimerized products, resulting from a one-electron reduction, or 1, 2-dihydrophthalazine (XCVII), depending on the conditions; (XCVII) is reducible at a more negative potential to 1, 2, 3, 4-tetrahydrophthalazine, which might be oxidized anodically to (XCVII). Thus, the electrolytic method is advantageous for the preparation of (XCVII) [140, 145]. The dimerized products from the reduction of 2-methylphthalazinium iodide have been shown to have coupled at carbon rather than at nitrogen [46].

The reduction course of substituted phthalazines depends on the substituents; thus 1-methylphthalazine (XCI) is reduced in alkaline solution to 1-methyl-3, 4-dihydrophthalazine, whereas 1-phenyl-4-dimethylamino-phthalazine is reduced to 1, 2-dihydro-1-phenyl-4-dimethylaminophthalazine.

6. Pyrimidines

In acid solution pyrimidine (XCVIII) [146-147] is reduced polarographically in two one-electron waves; two two-electron waves are observed in neutral medium, whereas a four-electron reduction occurs in alkaline solution. These findings have been interpreted in the following way:

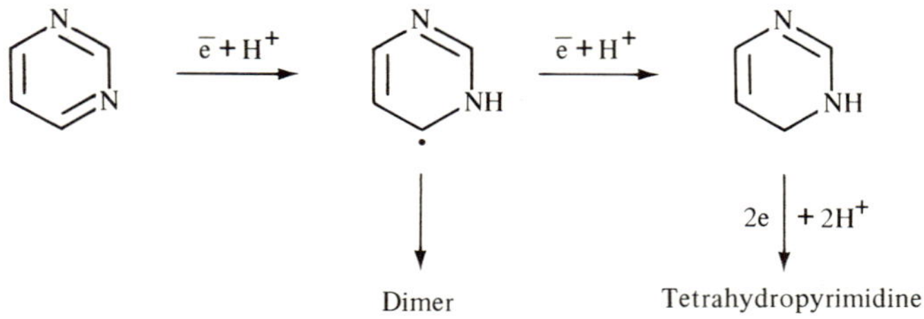

The structure of the tetrahydropyrimidine has not been proved. 2-Aminopyrimidine is reduced similarly, but the dihydropyrimidine is not further reducible.

7. Quinazolines

The polarographic behavior of quinazoline (XCLX) [148] is complicated by the hydration of the quinazolinium ion. The hydrated cation is not reducible, and in the acid region the wave height is partly determined by the rate of its dehydration, and this has been used to find the rate of its dehydration at different pH's [148].

The reduction of (XCIX) resembles that of (XCVIII). In alkaline solution it is reduced stepwise to the 1, 2, 3, 4-tetrahydroquinazoline. 4-Hydroxyquinazoline [140, 149, 150] is reducible in weakly alkaline and neutral solution. The reduction is analogous to the reduction of dihydroquinazoline to tetrahydroquinazoline. At higher concentrations a product dimerized at N-1 was also isolated [149, 150].

8. Pyrazines, Quinoxalines, Phenazines

Pyrazine and its reduction product 1, 4-dihydropyrazine form a system which at a mercury electrode behaves nearly reversibly [151, 152]. In acid solution two one-electron waves are found, indicating a certain stability of the semiquinone-type one-electron reduction product.

The electrochemical reduction of 2, 3-disubstituted 5, 6-dihydropyrazines yields the 1, 4, 5, 6-tetrahydropyrazines [153].

Thus, the diimine-enediamine system is analogous to the dione-enediol system. It is of interest that the 5, 6-dihydropyrazines are somewhat more easily reducible than the pyrazines.

Quinoxaline (C) [153-158] is reduced to 1, 4-dihydroquinoxaline which may be reversibly reoxidized to (C). 2-Methylquinoxaline is reduced to 3, 4-dihydro-2-methylquinoxaline [156], whereas 2-phenylquinoxaline forms 1, 2-dihydro-2-phenylquinoxaline. The 1, 4-dihydroquinoxalines may be formed primarily, but then tautomerize to the more stable dihydroquinoxaline.

2, 3-Diphenylquinoxaline is reduced to cis-1, 2, 3, 4-tetrahydro-2, 3-diphenylquinoxaline; the 1, 4-dihydro compound is transformed to the 1, 2-dihydro derivative, which is further reducible [156]. 1, 2-Dihydro-2, 3-dimethylquinoxaline is reduced in acid solution with ring opening to an imine of 2-butanone, which is hydrolyzed to o-phenylenediamine and 2-butanone [158]. The reaction is similar to that suggested for the ring contraction of (LIII) to (LIV).

5, 10-Dihydro-5, 10-dimethylphenazine (CI) exhibits two successive reversible one-electron oxidation steps in acetonitrile. When (CI) is oxidized in a more nucleophilic solvent or in the presence of a nucleophile, the initially formed dication acts as a methylating agent toward the nucleophile [159] with formation of 5-methylphenazinium cation.

9. Naphthyridines

1, 5-Naphthyridine is polarographically reducible at all pH's. Macroscale reductions consume approximately one electron per molecule and point to a dimerization [160]. The structure of the reduction products has not been established. The diquaternary salt of 1, 5-naphthyridine is polarographically reducible [161], and other naphthyridines would also be expected to be reducible.

10. Benzodiazepines

Both 1, 4- [162-165] and 1, 5-benzodiazepines [160] are polarographically reducible; the electrode reactions of the former have been elucidated. A compound as 7-chloro-2-methylamino-5-phenyl-3H-1, 4-benzodiazepine-4-oxide (CII) is reducible in acid solution in three steps. The first two steps, the reduction of the N-oxide and the saturation of the benzophenone imine,

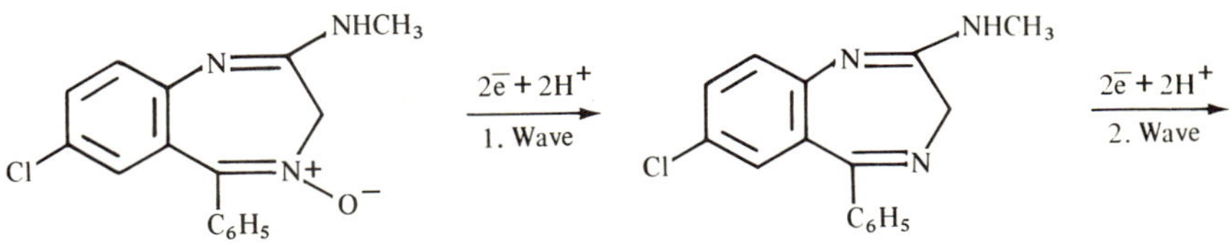

(CII)

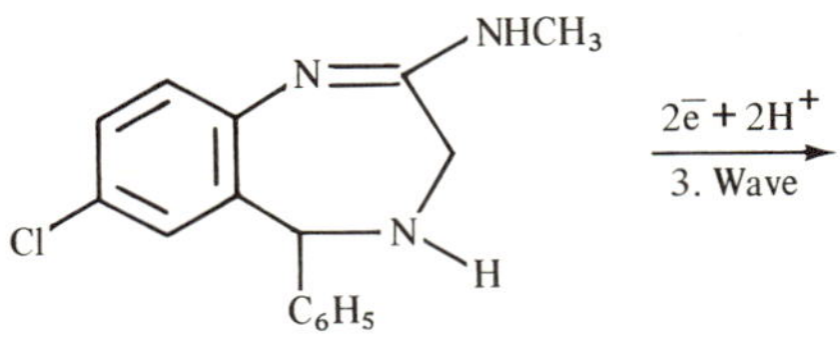

(LIV)

(LIII)

are straightforward. The mechanism of the third step, the ring contraction, is more controversial.

H. Compounds with One Oxygen and Two Nitrogen Atoms

1. Oxadiazoles

Aryloxadiazoles [124] are reducible at rather negative potentials; both four- and six-electron reactions are found, sometimes as two-step reductions. The four-electron reduction of symmetrical diaryloxadiazoles has been suggested to produce the oxadiazolidines.

2. Benzofurazane

This compound is polarographically reducible in the whole pH region in a six-electron reaction [166], and the product is supposedly o-phenylenediamine. At a macroelectrode some condensation reactions might be expected to occur in acid solution as side reactions just as in the case of o-dinitrobenzene reduction [26].

3. Sydnones

3-Phenylsydnones (CIII) are polarographically reducible both in acid and in alkaline medium. In the former case six electrons are consumed per molecule, whereas at high pH, n = 4. The reactions follow the scheme [167]:

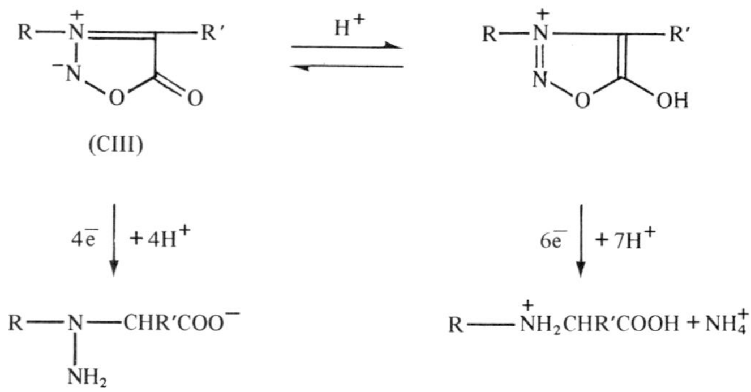

I. Compounds with One Sulfur and Two Nitrogen Atoms

1. 1, 3, 4-Thiadiazoles

Thiadiazoles are often polarographically reducible [45, 168]; not many of them have been investigated by controlled potential electrolysis. The wave height of 2, 5-diphenyl-1, 3, 4-thiadiazole (CIV) corresponds in acid solution to a two-electron reduction to a dihydro derivative, which is probably hydrolyzed to benzaldehyde thiobenzoylhydrazone (CV). This compound is reducible at the reduction potential of (CIV) and will thus be reduced further; (CV) is reduced as follows [45]:

1. wave: $C_6H_5CSNHN{=}CHC_6H_5 \xrightarrow{2\bar{e} + 2H^+} C_6H_5CSNH_2 + HN{=}CHC_6H_5$

2. wave: $HN{=}CHC_6H_5 \xrightarrow{2\bar{e} + 2H^+} H_2NCH_2C_6H_5$

3. wave: Reduction of thiobenzamide [28].

2. Benzo-2, 1, 3-thiadiazoles

These compounds and the corresponding selenodiazoles are reduced [169, 170] as the furazanes in a six-electron reduction; o-phenylenediamine and hydrogen sulfide and hydrogen selenide, respectively, are formed.

J. Compounds with Three Nitrogen Atoms

1. Benzotriazoles

Benzotriazoles (CVI) are reducible in acid solution, but only 2-substituted derivatives have been found to be reducible in alkaline solution. In acid solution the following scheme has been proposed [40]:

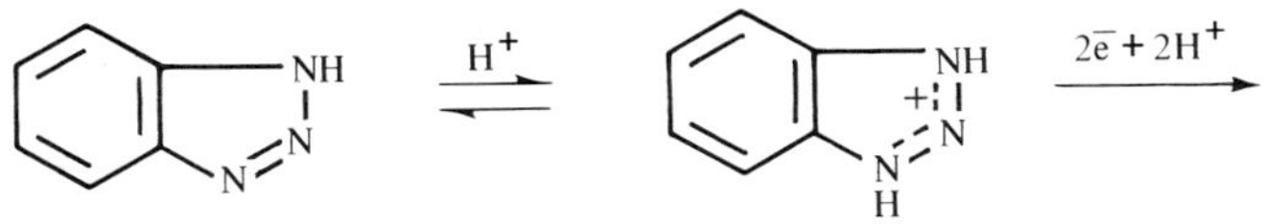

(CVI)

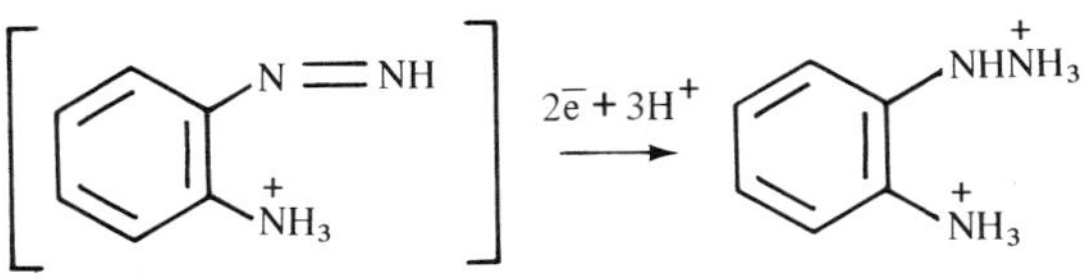

(CVII) (CVIII)

The o-aminophenyldiimide (CVII) cannot be isolated as it is more easily reducible than the benzotriazole, but a small concentration of it can be detected polarographically during the electrolytic reduction. o-Aminophenylhydrazine (CVIII) has three nucleophilic centers and it may be a useful intermediate in the preparation of some heterocyclic compounds.

In alkaline solution 2-methylbenzotriazole is reduced to the 1, 3-dihydro derivative, which may be reoxidized anodically to the starting material [40].

In acid solution the dihydrobenzotriazole forms a phenyldiimide; this compound does not cyclicize in alkaline solution.

2. Benzotriazines

Benzo-1, 2, 4-triazines and dihydrobenzotriazines form a redox system which polarographically behaves nearly reversibly [20, 45]. The further reduction of the dihydrobenzotriazines to benzimidazoles has been discussed above; 3-alkylsubstituted dihydrobenzotriazines are generally not further polarographically reducible, but the unsubstituted and 3-phenylsubstituted derivatives are.

Other triazanaphthalenes would be expected to be reducible and the positions of the nitrogen atoms are the most critical factors for influencing the reduction potential.

K. Compounds with Four Nitrogen Atoms

The more nitrogen atoms inserted in an aromatic nucleus, the more easily reducible the system becomes. When a heteroaromatic system has two (or more) nitrogen atoms in a six-membered ring, the compound will be reducible, at least in acid solution, and, if two nitrogen atoms occupy para positions in such a ring, the redox system will approach reversibility.

1. Purine and Pteridine [2]

The electrolytic reduction of purine (CIX) [171], which has recently been reviewed [147], takes place only in acid solution and, thus, differs from that of pyrimidine and quinazoline. The reduction proceeds in acid solution in two two-electron steps according to [171]:

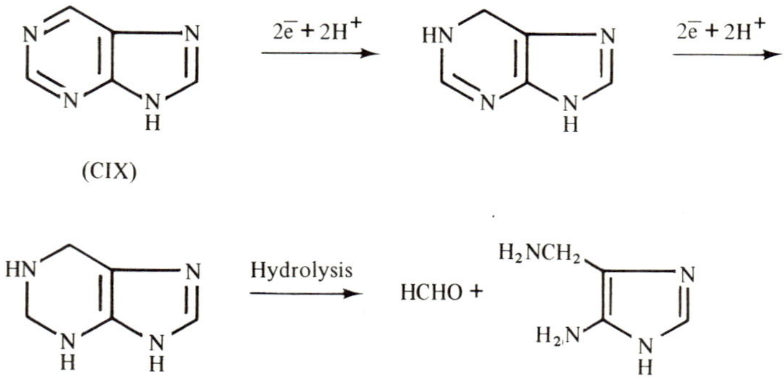

2. Porphines

The polarographic and controlled potential reduction of porphines (CX) has been investigated and it was shown that porphine was reduced polarographically in two two-electron steps [173-175]. Reduction of mesoporphyrinedimethyl ester yielded a product which was easily reoxidized by oxygen. From the uv spectrum of the reduced solution, which has been reduced in deuterium oxide, it could be shown that the reduction yielded a phlorin (CXI) rather than a chlorin (CXII). Reduction of a chlorin yielded a "chlorin-phlorin."

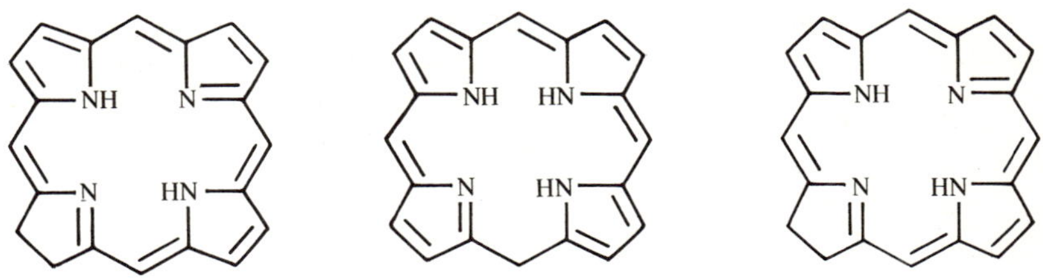

(CX) (CXI) (CXII)

In dimethylformamide porphines are reduced in three or four one-electron steps which results in formation of mono-, di-, tri-, or tetra-negative ions [176].

Anodic oxidation of porphin in alkylnitriles at a platinum electrode takes place in two one-electron steps [177]. The loss of electrons occurs from the lone pairs on the nitrogen atoms.

IV. ELECTROLYTIC REACTIONS OF SUBSTITUTED HETEROCYCLIC COMPOUNDS

This section will treat the electrolysis of substituted heterocyclic compounds, if the electrode reaction involves the substituent directly and if the electrode reaction in some way is different from what would be expected from the carbocyclic series.

A. Alkyl and Alkenyl Substituents

Alkyl derivatives of π-electron-deficient heterocyclic compounds may be oxidized anodically to carboxylic acids; thus, 3-picoline forms nicotinic acid (60%) on oxidation at a platinum anode in 30% sulfuric acid [119] or at a lead anode in 7 N sulfuric acid [178], whereas oxidations in the pyrazole series [179] give only low yields.

Alkenes. Hydrodimerization of activated alkenes is a well-established process. π-Electron-deficient heteroaromatic compounds activate a double bond similarly to a cyano or carbethoxy group, and in accordance with that analogy vinylpyridines can be hydrodimerized. 4-Vinylpyridine [180] forms 1, 4-bis(4-pyridyl)butane in 82% yield on electrolysis in a mildly alkaline solution containing methyltriethylammonium p-toluenesulfonate and some DMF. The mechanism is discussed in Chapter XIX.

If the nucleus contains more than one nitrogen atom, the nucleus would be reduced at a less negative potential than the double bond. No reports of electrolysis of such compounds have been published yet.

Ethynylpyridines are reducible polarographically [181] in a medium containing tetraethylammonium iodide, 2-ethynylpyridine at -1.72 V and 2-methyl-5-ethynylpyridine at -2.06 V (SCE). The reduction potentials of the ethynylpyridines are very close to those of the corresponding vinyl compounds [182]. Similar results have been found for benzene derivatives [183]. The electrode reaction would thus be expected to be a four-electron reduction of the ethynyl group to an ethyl group.

B. Hydroxyl Derivatives

Heterocyclic hydroxyl derivatives can be divided in two groups according to whether the hydroxyl group is exocyclic or not. Heterocyclic compounds having a hydroxyl substituent at the ring have the possibility of existing in different tautomeric forms [184], the hydroxy or the oxo form, and in many cases the oxo form is the predominant one. Nevertheless, regardless of the position of the prototropic equilibria, the compounds will here be treated as "hydroxyl compounds." Phenolic compounds, e.g., the oxidative dimerization of 1, 2, 3, 4-tetrahydroisoquinoline phenols [185], will not be discussed.

1. Compounds with Hydroxyl Groups Attached to a Substituent

Hydroxyl groups are reducible only when they are activated by an electron-attracting group, as, e.g., the phenacyl group [186, 187]. A pyridine ring activates its substituents, but largely in the 2- and 4-positions. This is illustrated by the reduction of pyridoxol (CXIII), which is reducible in an ammoniacal buffer (pH 8.5-10) according to [188]:

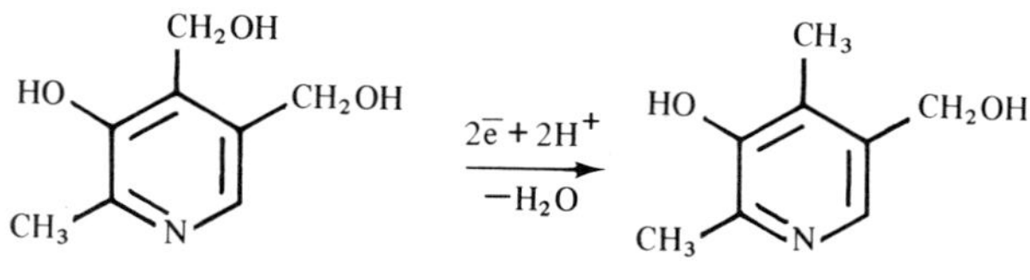

(CXIII)

2. Compounds with Hydroxyl Groups Attached to the Ring

The reduction of such compounds follows generally one of the following routes: (a) Reduction takes place in the nucleus and might involve ring cleavage. This type is exemplified by the reduction of pyridazinones to 4, 5-dihydropyridazinones followed by ring opening (Chapter XXI) and is found for compounds which may be regarded as acid derivatives, such as cyclic esters, hydrazides, amides, and eneamides. (b) The carbonyl group of the oxo form is reduced. This is often followed by loss of the hydroxyl group thus formed and further reduction of the double bond. This type might be found when the oxo group is β to a heteroatom and/or α to a phenyl group and in some dihydroxy compounds.

a. 3-Ethoxy-3-(4′-Methoxyphenyl)phthalides. These compounds have been found to be polarographically reducible in acid solution; reduction in an acetate buffer produced 3-(4′-methoxyphenyl)phthalides [16].

b. Derivatives of Benzothiophene. 3-Hydroxybenzothiophene gives a two-electron reduction wave in acid and a two-electron oxidation wave in alkaline solution [189]. The reduction involves the oxo group and the product is 2, 3-dihydro-3-hydroxybenzothiophene. The oxidation has been proposed to yield thioindigo.

c. Derivatives of Isoindole. Phthalimide in acid solution is reduced polarographically [190, 191] by a two-electron reduction, whereas two one-electron waves are found in slightly alkaline medium. The two-electron reduction product [192], hydroxyphthalimidine (CXIV), a derivative of phthalaldehydic acid, reacts with any nucleophile HY present. In ethanolic solution ethoxyphthalimidine is formed, and in the presence of isoindoline, 1-N-isoindolino-3-oxo-isoindoline [192].

In strongly acid solution (CXIV) dehydrates to the easily reducible 3-oxoisoindole, which is reduced to phthalimidine; a further reduction to isoindoline may take place under conditions similar to those used for the reduction of benzamides [193-195].

The reduction of N-anilinophthalimides [51] has been discussed above.

d. Derivatives of Cinnoline. 3-Hydroxycinnoline [33] is reduced in a two-electron wave from pH 0 to 11. In strongly acid solution the main product is 1-aminoöxindole, which also is obtained on reduction with zinc and sulfuric acid [196]. Reduction of an emulsion in aqueous phosphate buffer (pH 6.5) and n-butanol produced a nearly quantitative yield of 3-keto-1, 2, 3, 4-tetrahydrocinnoline; this compound was easily reoxidized to 3-hydroxycinnoline.

By choosing a pH for the reduction at which the product is stable the preparation of this labile compound could be accomplished.

4-Hydroxycinnoline (CXV) [33] in acid solution is reducible in two two-electron steps. The product from the first two-electron reduction is

4-keto-1, 2, 3, 4-tetrahydrocinnoline (CXVI). This compound is a derivative of phenylhydrazine and can thus be oxidized anodically in alkaline solution to an azo compound (CXVII) which tautomerizes to (CXV). The carbonyl group of (CXVI) is activated by the phenyl ring and — in contrast to 3-keto-1, 2, 3, 4-tetrahydrocinnoline — is thus further reducible; the reduction yields 4-hydroxy-1, 2, 3, 4-tetrahydrocinnoline (CXVIII). On standing, this compound loses water with the formation of 1, 4-dihydrocinnoline. If, however, the compound is oxidized in alkaline solution, the product (CXIX) loses water and forms cinnoline (CXX). As 4-hydroxycinnoline is one of the more easily accessible derivatives of cinnoline, this reaction might be the method of choice for the preparation of cinnoline [33]. The reactions are formulated below:

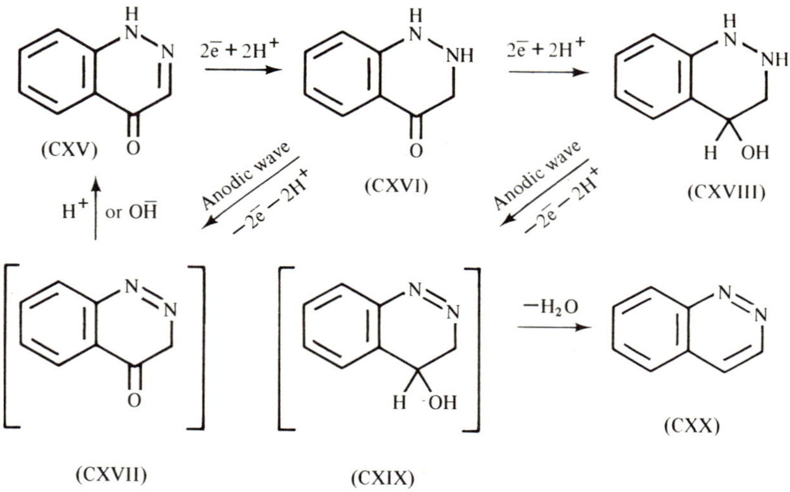

The reduction in alkaline solution of (CXVI) yields dimeric products [33].

e. Derivatives of Phthalazine. Dihydroxyphthalazines [49] as 2, 3-dihydro-2, 3-dimethyl-1, 4-phthalazinedione (CXXI) are at low pH reduced to a phthalimidine (CXXII) in a six-electron reduction. In alkaline solution 4-hydroxy-3, 4-dihydro-2, 3-dimethyl-1-phthalazinone (CXXIII) is formed in a two-electron reaction; in this medium it is not further reducible, but in acid solution it loses water to the reducible phthalazinonium compound (CXXIV). The product from this reaction is 3, 4-dihydro-2, 3-dimethyl-1-phthalazinone (CXXV), which at low pH can be reduced to N-methylphthalimidine (CXXII) and its precursor N-methyl-2-(methylaminomethyl)benzamide. The reduction scheme is given by;

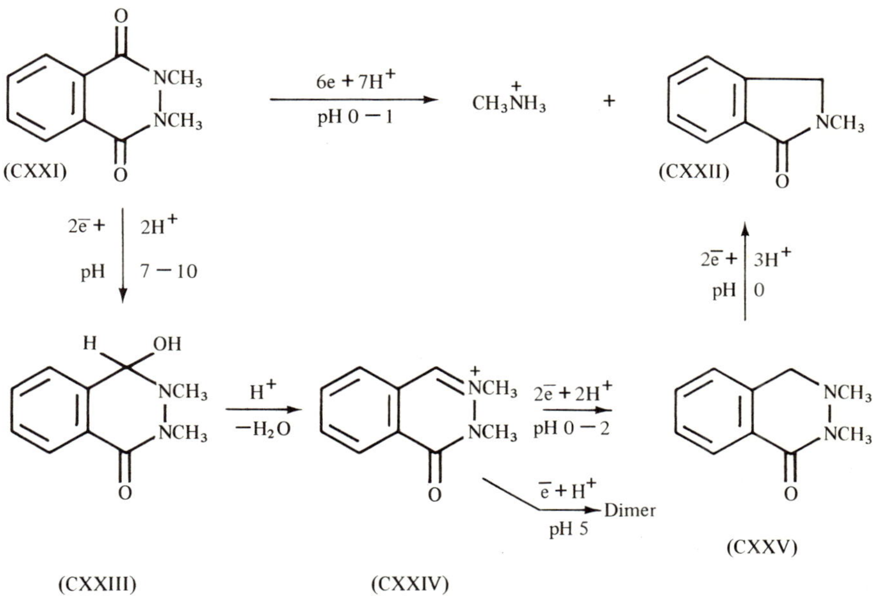

f. Derivatives of Quinazoline. The reduction of 4-quinazolone was discussed in Sec. III.G. 4-Methoxyquinazoline is in acid solution reduced to the 3, 4-dihydro derivative which eliminates methanol; the quinazoline thus formed is reduced to 3, 4-dihydroquinazoline. In alkaline solution is formed, besides quinazoline, some biquinazolyl in a reaction analogous to the benzoin condensation [197].

g. Derivatives of Quinoxaline. Both 2-hydroxy- and 2, 3-dihydroxyquinoxaline are polarographically reducible, the former by a two-electron and the latter in acid solution by a four-electron reduction [198]. The former is more easily reducible than the latter which is understandable since 2-hydroxyquinoxaline (CXXVI) may be regarded as a derivative of glyoxalic acid, whereas 2, 3-dihydroxyquinoxaline (CXXVII) is a derivative of oxalic acid. Derivatives of aldehydes are easier to reduce than the corresponding acid derivatives.

The reduction of (CXXVII) in acid solution may be formulated as:

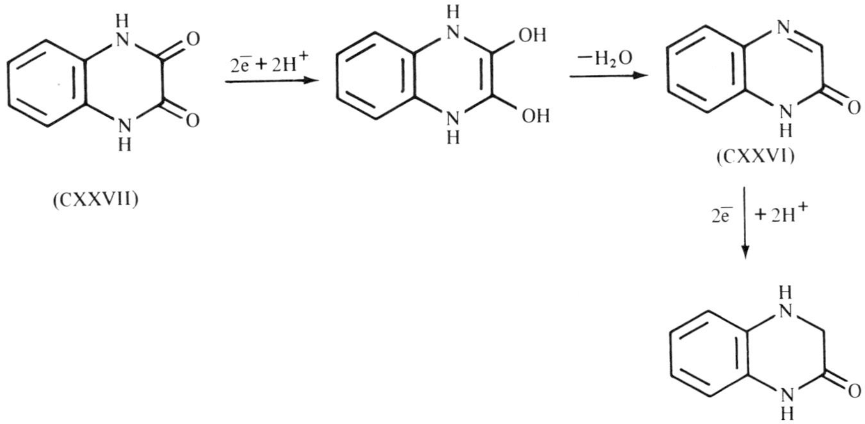

Reduction of 3-alkylated-2-hydroxyquinoxalines produces, besides the 3,4-dihydroquinoxalone, a product dimerized at N-4, whereas 3-arylquinoxalones form only the 3,4-dihydro derivative on reduction [199].

h. Derivatives of Benzodiazepin. A loss of a hydroxyl group is observed in the four-electron reduction of 7-chloro-5-phenyl-4,5-dihydro-2H-benzodiazepin-2,3-(1H)-dione (Oxazepam) which involves both a saturation of the C-N double bond and a removal of the hydroxyl group [200], possibly through a reduction-elimination-reduction reaction.

3. N-Oxides

N-Oxides of heteroaromatic compounds are generally electrolytically reducible in acid solution [201-213]. The most studied compounds of this type are the pyridine-N-oxides [201, 208, 209]. At low pH pyridine-N-oxide gives a two-electron polarographic wave, but at pH >4 the wave height diminishes, as the preprotonation step becomes too slow:

$$C_5H_5\overset{+}{N}\text{-}O^- \underset{}{\overset{+H^+}{\rightleftharpoons}} C_5H_5\overset{+}{N}\text{-}OH \xrightarrow{2\bar{e}+2H^+} C_5H_5\overset{+}{N}H + H_2O.$$

The assumption that it is the O-protonated N-oxide which is reduced is substantiated by the similarity in reduction potential between pyridine-N-oxide and N-methoxy pyridinium ion in acid solution; the latter is also reducible in alkaline solution [45].

A number of different N-oxides have been investigated polarographically, such as N-oxides of quinoline [211], acridine [203], pyrazine [213], quinoxaline [202], phenazine [205], and adenine [207], and the influence of substituents has been discussed.

When the heterocyclic ring contains more than one nitrogen atom, the initial reduction of the N-oxide may take place in the nucleus or the reduction of the nucleus and the N-oxide function occur simultaneously; the reduction of quinazoline-3-oxide [148], 3-methoxy-2, 5-dimethylpyrazine-1-oxide [213], and cinnoline mono- and dioxide [210] is reported to take this route.

At pH <1.5, however, N-oxides like cinnoline-N-oxide [33], 3, 6-diphenylpyridazine-N-oxide [45], or pyrazine-N-oxide [213] give a two-electron polarographic wave followed by the reduction waves of the parent compounds.

4. S-Oxides

Both phenothiazine-S-oxide and -S-dioxide are polarographically [214] reducible at rather negative values in a medium containing tetramethylammonium chloride. The electrode reaction is not known. In acid solution only the S-oxide is reducible; the reduction is probably preceded by a loss of water to the phenothiazonium salt.

C. Aldehydes, Ketones, and Their Derivatives

π-Electron-deficient heterocyclic rings facilitate the addition of nucleophilic reagents to a carbonyl group in conjugation with the ring; protonation of a ring nitrogen enhances this effect. Protonated or quaternized formylpyridines [215-219] are to a high degree hydrated in aqueous solution. This influences their polarographic behavior since the hydrated form, a gem-diol, is not reducible at the dropping mercury electrode. The wave height of such hydrated aldehydes is thus determined at a given pH by the equilibrium constant for the hydration and the rate of the dehydration reaction. Since only the protonated form is appreciably hydrated, the pK_A of the compound is also of importance for interpreting the polarographic data. The dehydration reaction is both acid and base catalyzed, the latter being of importance for the quaternized aldehydes. The acid-catalyzed dehydration reaction has been shown to follow this path:

$$R\overset{+}{N}C_5H_4CH(OH)_2 \underset{}{\overset{H^+}{\rightleftharpoons}} [R\overset{+}{N}C_5H_4CH(OH)\overset{+}{O}H_2] \underset{+H_2O}{\overset{-H_2O}{\rightleftharpoons}} [R\overset{+}{N}C_5H_4\overset{+}{C}H—OH]$$

$$[R\overset{+}{N}C_5H_4\overset{+}{C}H—OH] \underset{+H^+}{\overset{-H^+}{\rightleftharpoons}} R\overset{+}{N}C_5H_4CHO$$

The hydration is most pronounced when the formyl group is in a position strongly activated toward nucleophilic attack; thus, 2- and 4-formylpyridine are hydrated more than the 3-isomer, 2-formylimidazole [220] and 2-formylthiazole [221] more than the 4- and 5-isomers. The quaternary derivatives are hydrated at least as strongly as the protonated parent compounds and in a broader pH region.

The electrode reaction is in the greater part of the pH region a two-electron reduction to the alcohol, but in strongly alkaline solution an appreciable amount of pinacol is formed. The stereochemistry of the pinacol mixture has been investigated in the reduction of 2-acetopyridine [222, 223] but not for the formylpyridines (Chapter XXVII).

The electrochemistry of the azomethine derivatives, such as oximes, Schiff bases, hydrazones, is similar to that of similar carbocyclic compounds and will not be discussed here.

D. Carboxylic Acids and Derivatives

Carboxylic acids and most of their derivatives are difficult to reduce, and a polarographic reduction of a carboxyl group requires activation by a strongly electron-attracting group. The addition of electrons may take place either to the carboxyl group or to the activating group, often depending on pH.

π-Electron-deficient heterocyclic nuclei may activate a carboxyl group sufficiently to render it polarographically reducible [220, 224-227]. An aldehyde is always more easily reducible than the corresponding carboxylic acid, so it is generally not possible to isolate aldehydes from a reduction of carboxylic acids. The electron-attracting groups which activate a carboxyl group toward reduction also activate the aldehyde toward nucleophilic addition reactions, and these aldehydes are in a certain pH region hydrated to a certain degree. It is thus possible in some cases to trap the easily reducible aldehyde formed during the reduction of a carboxylic acid as the nonreducible aldehyde hydrate. As some of the free, reducible aldehyde is always present in equilibrium with the hydrate, a reduction of an acid to an aldehyde must be performed at a suitable pH and at a low temperature, where the rate of the dehydration reaction is low. The reduction must not be allowed to proceed to completion, but must be stopped after the consumption of about two electrons per molecule.

The pyridinecarboxylic acids [226], 2-carboxythiazoles [227], and 2-carboxyimidazoles [220, 228], their esters and the amides [97] are reduced in this way. The corresponding hydrazides [229] and hydroxamic acids [14] are reduced to the amides in the first step and then further reduced.

1. Anodic Reactions

Kolbe reactions of heterocyclic compounds have been studied only in a few cases. Anodic oxidation of 1-azabicyclo[2.2.2]octane-2-carboxylic acid under "Kolbe conditions" produced 2-methoxy-1-azabicyclo[2.2.2]octane [230]. The primary radical, formed by loss of an electron from the carboxylate ion, decarboxylates and is oxidized further to a carbonium ion which is attacked by a methoxide ion.

2. Nitriles

The reduction of 2- and 4-cyanopyridine exhibits special features. In acid solution the protonated species is reduced in a four-electron reaction [231-233] to the pyridylmethylamine, but in alkaline medium the carbon-carbon bond was cleaved with formation of pyridine and cyanide ion [97, 232] (Chapter XXI).

E. Nitrogen-Containing Substituents

The reduction of heterocyclic C- and N-nitro and nitroso compounds exhibits generally no special features and will not be discussed here.

N-Aminopyridinium iodide in acid solution is reduced to pyridine; similar N-amino heteroaromates are probably reduced in the same way [45].

Activated amino groups as found in 4-pyridylmethylamines [234] or 2-pyridyltrimethylammonium iodide [45] are reductively removed in acid solution. The reduction of the latter requires protonation of the ring to obtain sufficient activation.

Elimination of an amino group is sometimes found as a chemical step between the transfer of electrons. The reductions of adenine (CXXVIII) illustrates this [235]:

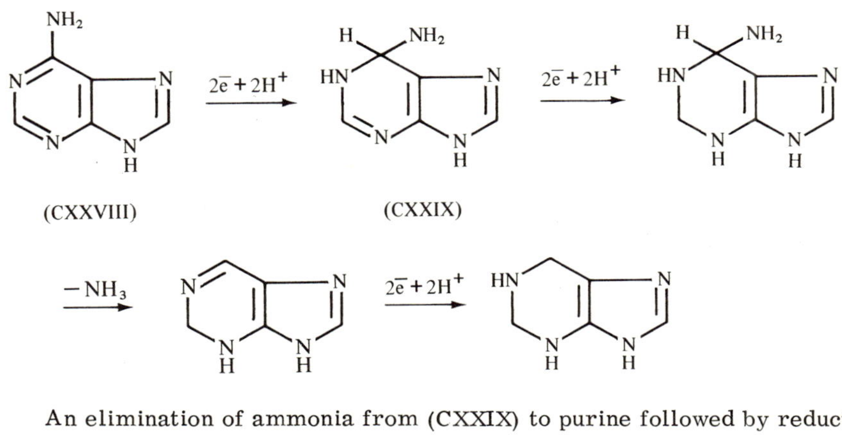

An elimination of ammonia from (CXXIX) to purine followed by reduction of the latter might also be possible; such a reduction route would be analogous to that of 4-aminoquinazoline in acid solution [197] where quinazoline can be detected as an intermediate during the reduction.

Adenine (CXXVIII) gives a single well-defined voltammetric wave at the pyrolytic graphite microelectrode [236]; coulometric data showed a six-electron oxidation. Exhaustive electrolysis at a fixed anode potential in aqueous 1 M acetic acid solution produced a mixture of products. The primary electrochemical oxidation has been proposed to be as follows:

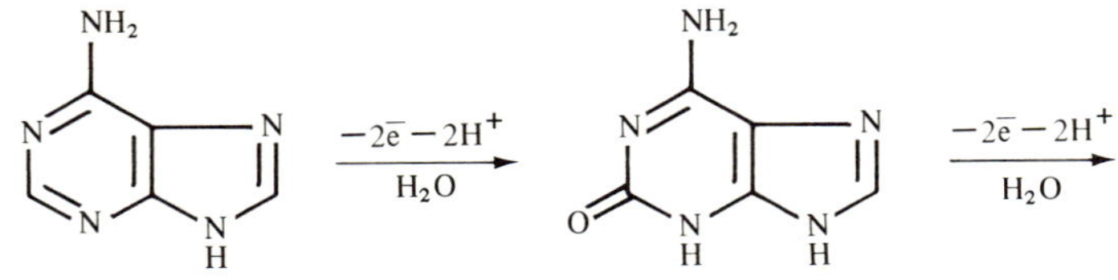

(CXXVIII) (CXXX)

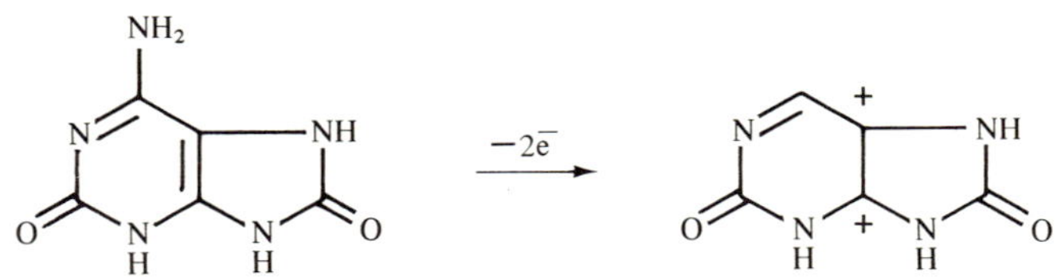

(CXXXI) (CXXXII)

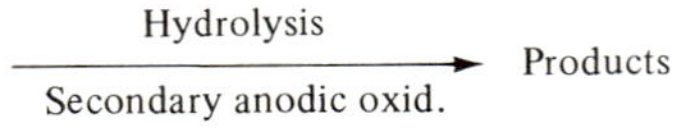

The unstable dicarbonium ion (CXXXII) is hydrolyzed and some products oxidized further. From the electrolyzed solution urea, parabanic acid (CXXXIII), oxaluric acid (CXXXIV), ammonia, 4-aminopurpuric acid (CXXXV), and possibly allantoin (CXXXVI) were isolated:

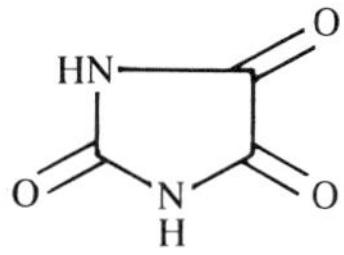

(CXXXIII)

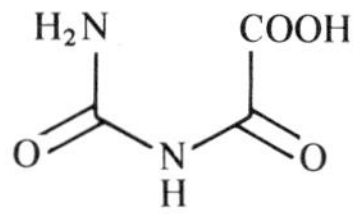

(CXXXIV)

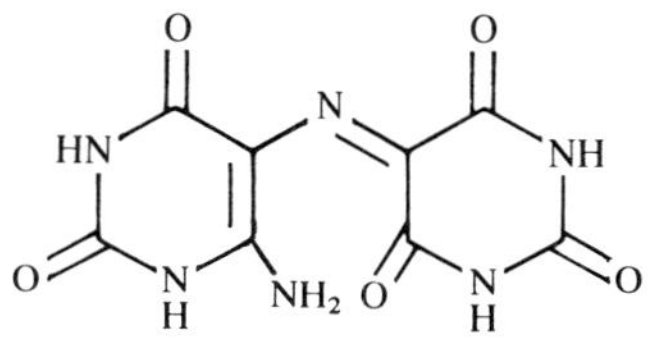

(CXXXV)

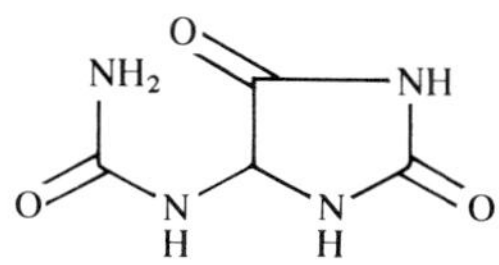

(CXXXVI)

F. Sulfur-Containing Substituents

Reduction of heterocyclic disulfides and anodic reactions of sulfhydryl compounds show no special features. Elimination of a sulfhydryl group may be the slow step in some reactions.

4-Mercaptocinnoline **(CXXXVII)** [33] gives a two-electron polarographic wave, but a preparative reduction consumed four electrons per molecule and produced 1, 4-dihydrocinnoline(CXXXVIII). The reaction must thus involve a slow step after the first two-electron reduction and has been formulated as shown below. The slow chemical step in the reaction is the acid-catalyzed elimination of hydrogen sulfide:

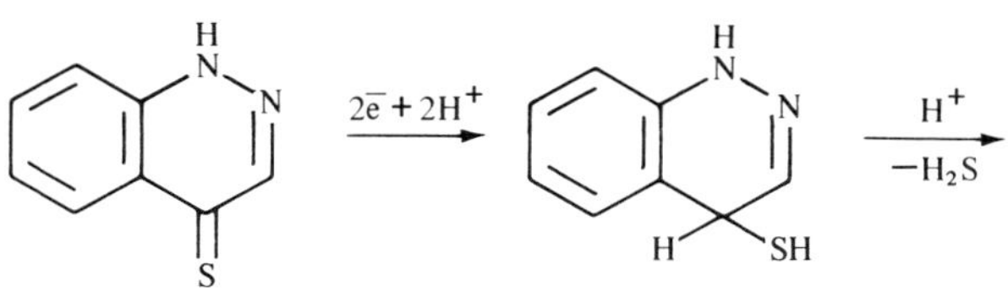

(CXXXVII)

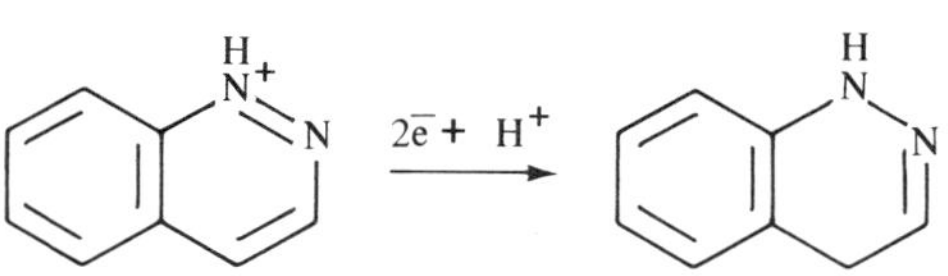

(CXXXVIII)

A similar reaction takes place in 4-mercaptoquinazoline [197] and 2, 6-dimercaptopurine [237].

2-Mercapto-4-hydroxythiazole (CXXXIX) (rhodanine) is reducible in two two-electron steps [238], the second step being found at rather negative potentials. The reduction takes place at the exocyclic sulfur and has been formulated as follows:

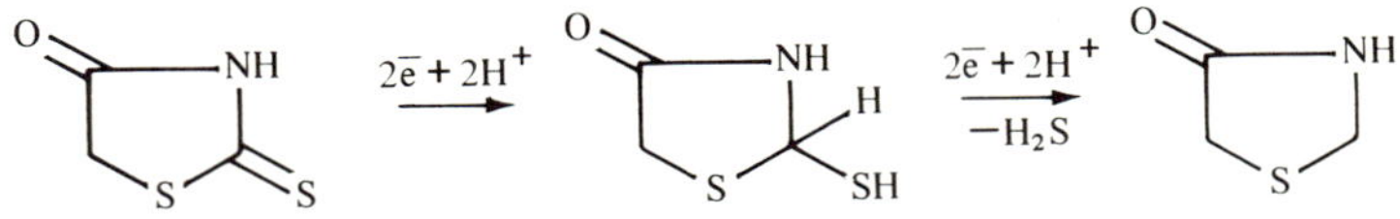

(CXXXIX)

Cyclic voltammetric data at a pyrolytic graphite electrode of 2, 6-dimercaptopurine show three peaks. The first and second one correspond to a disulfide formation from the 6- and 2-mercapto groups, respectively, whereas the third peak is due to an oxidation to purine-2, 6-disulfonic acid. Some disulfonic acid was obtained on preparative oxidation [239].

G. Halogen Derivatives

The electrode reactions of heterocyclic halogen compounds exhibit no special features, but some may be of preparative value. Anodic halogenation may be advantageous in fluorination and iodination reactions, and in halogenations where a low, easily measured concentration of halogen is desirable.

Large-scale reductive removal of the halogens is sometimes of preparative value; thus, it is convenient to remove chlorine from 4-chloroquinazoline [148] and iodine from 4-iodophthalazines [140] by reduction at a mercury cathode, and chlorine from 2-amino-4-chloropyrimidine at a spongy cadmium electrode [240]. Although the polarograms indicate slightly easier reduction of the C-Cl bond than of the phthalazine nucleus, it has been difficult to remove the chlorine from 4-chlorophthalazine without reducing the nucleus.

REFERENCES

[1]. H. Lund, "Electrolysis of N-Heterocyclic Compounds," in Advances in Heterocyclic Chemistry (A. R. Katritzky and A. J. Boulton, eds.), Vol. 12, Academic, New York, 1970, p. 213.

[2]. H. Lund, Elektrodereaktioner i Organisk Polarografi og Voltammetri, Aarhuus Stiftsbogtrykkerie, Arhus, 1961.

[3]. J. Grimshaw and E. J. F. Rea, J. Chem. Soc. (C), 2628 (1967).

[4]. J. Wiemann and P. Maitte, Bull. Soc. Chim. Fr., 430 (1952).

[5]. J. Wiemann and M. Paget, ibid., 285 (1955).

[6]. J. Wiemann, ibid., 2545 (1964).

[7]. N. L. Weinberg and E. A. Brown, J. Org. Chem., 31, 4058 (1966).

[8]. C. Bratu and A. T. Balaban, Rev. Roum. Chim., 10, 1001 (1965); CA, 64, 12489 (1966).

[9]. M. Le Guyader, Compt. Rend. Ser. C, 262, 1383 (1966).

[10]. H. Lund, 13. Nordiske Kemikermøde, Copenhagen, 1968.

[11]. C. Baezner and A. Gardiol, Ber., 39, 2512 (1906).

[12]. H. Lund, Symposium on the Synthetic and Mechanistic Aspects of Electroorganic Chemistry, Durham, N.C., October 1968, Preprints of Papers, p. 197.

[13]. H. Lund and L. G. Feoktistov, Acta Chem. Scand., 23, 3482 (1969).

[14]. H. Lund and J. Hakl, to be published.

[15]. M. Jubault and E. Raoult, Compt. Rend., 270, 1743 (1970); M. Jubault and D. Peltier, Bull. Soc. Chim. Fr., 1561 (1972).

[16]. H. Lund, Acta Chem. Scand., 14, 359 (1960).

[17]. H. Lund, P. Lunde, and F. Kaufmann, ibid., 20, 1631 (1966).

[18]. K. Brand and E. Stohr, Ber., 39, 4058 (1906).

[19]. M. Le Guyader, Bull. Soc. Chim. Fr., 1967 (1966).

[20]. S. Kwee and H. Lund, Acta Chem. Scand., 23, 2711 (1969).

[21]. H. Lund and A. D. Thomsen, ibid., 23, 3567 (1969).

[22]. M. D. Hawley, S. V. Tatawawadi, S. Piekarski, and R. N. Adams, J. Amer. Chem. Soc., 89, 447 (1967).

[23]. A. G. Hudson, A. E. Pedler, and J. C. Tatlow, Tetrahedron Lett., 2143 (1968).

[24]. M. Le Guyader and D. Peltier, Bull. Soc. Chim. Fr., 2695 (1966).

[25]. G. Cauquis and J.-L. Cros, Bull. Soc. Chim. Fr., 3760, 3765 (1971).

[26]. M. Breant and J.-C. Merlin, Bull. Soc. Chim. Fr., 54 (1964).

[27]. S. Takenaka and T. Sekine, Denki Kagaku, 36, 349 (1968); CA, 70, 25097 (1969).

[28]. H. Lund, Coll. Czech. Chem. Commun., 25, 3313 (1960).

[29]. S. H. Wilen and A. W. Levine, Chem. Ind., 237 (1969).

[30]. G. Cauquis and M. Genies, Tetrahedron Lett., 3403 (1970); 3959 (1971).

[31]. J. D. Anderson and M. M. Baizer, Tetrahedron Lett., 511 (1966).

[32]. D. J. McClemens, A. K. Garrison, and A. L. Underwood, J. Org. Chem., 34, 1867 (1969).

[33]. H. Lund, Acta Chem. Scand., 21, 2525 (1967).

[34]. T. Wohlfahrt, J. Prakt. Chem. [2], 65, 295 (1902); E. Laviron and T. Lewandowska, Bull. Soc. Chim. Fr., 3177 (1970); J. W. Rogers and W. M. H. Watson, Anal. Chim. Acta, 54, 41 (1971).

[35]. G. M. LoPresti, S. Huang, and A. A. Reidlinger, J. Electrochem. Soc., 115, 1135 (1968).

[36]. J. Møller, Elektrochem. Z., 10, 201 (1904).

[37]. Bad. Anilin- und Sodafabrik, D.R.P. 79,406 (1894).

[38]. H. Lund, First Intern. Congr. of Heterocyclic Chemistry, Albuquerque, New Mexico, 1967.

[39]. K. Elbs and W. Keiper, J. Prakt. Chem. [2], 67, 580 (1903)

[40]. H. Lund and S. Kwee, Acta Chem. Scand., 22, 2879 (1968); R. Hazard and A. Tallec, Bull. Soc. Chim. Fr., 2917 (1971).

[41]. L. G. Feoktistov and H. Lund, unpublished observation.

[42]. H. Lund, Acta Chem. Scand., 13, 249 (1959).

[43]. H. Lund, ibid., 18, 563 (1964).

[44]. H. Lund and P. Lunde, ibid., 21, 1067 (1967).

[45]. H. Lund, Discuss. Faraday Soc., 45, 193 (1968).

[46]. H. Lund and E. Jensen, Acta Chem. Scand., 24, 1867 (1970).

[47]. S. Kwee and H. Lund, Electrochemical Society Meeting, New York, May 1969, Extended Abstracts, p. 349.

[48]. H. Lund, to be published.

[49]. H. Lund, Coll. Czech. Chem. Commun., 30, 4237 (1965).

[50]. H. Oelschläger and H. Hoffmann, Arch. Pharm., 300, 817 (1967).

[51]. H. Lund, Tetrahedron Lett., 3973 (1965).

[52]. W. R. Vaugham, D. I. McCane, and G. J. Sloan, J. Amer. Chem. Soc., 73, 2298 (1951).

[53]. S. Swann and N. Leonard, ibid., 74, 4620, 6251 (1952).

[54]. L. A. Simonson and C. K. Mann, Tetrahedron Lett., 3303 (1970).

[55]. P. Zuman and V. Horak, Coll. Czech. Chem. Commun., 26, 176 (1961).

[56]. C. Caullet and M. Libert, Journées d'Electrochimie Organique, Clermont-Ferrand, May 1970, Communication No. 3; M. Libert, C. Caullet, and S. Longchamp, Bull. Soc. Chim. Fr., 2367 (1971).

[57]. N. Clauson-Kaas, F. Limborg, and K. Glens, Acta Chem. Scand., 6, 531 (1952); N. Clauson-Kaas and Z. Tyle, ibid., 6, 962 (1952).

[58]. N. Clauson-Kaas, N. Elming, and Z. Tyle, ibid., 9, 1 (1955).

[59]. N. Clauson-Kaas and F. Limborg, ibid., 1, 619 (1947).

[60]. N. Clauson-Kaas, S. Li, and N. Elming, ibid., 4, 1233 (1950).

[61]. N. Elming, ibid., 6, 572 (1952).

[62]. N. Elming and N. Clauson-Kaas, ibid., 9, 23 (1955).

[63]. A. J. Baggaley and R. Brettle, Chem. Commun., 108 (1966).

[64]. K. E. Kolb and C. L. Wilson, ibid., 271 (1966).

[65]. S. Arita, Y. Nagahiro, and K. Takeshita, Kogyo Kagaku Zasshi, 72, 1896 (1969); CA, 73, 10122 (1970).

[66]. K. Yoshida and T. Fueno, J. Org. Chem., 36, 1523 (1971).

[67]. E. Gird and A. T. Balaban, J. Electroanal. Chem., 4, 48 (1962).

[68]. M. Vajda, Adv. in Polarography, Proceed. Second Intern. Congr. of Polarography, 2, 786 (1959).

[69]. A. J. Harle and L. E. Lyons, J. Chem. Soc., 1575 (1950).

[70]. R. Patzak and L. Neugebauer, Monatsh. Chem., 82, 662 (1951); 83, 776 (1952).

[71]. V. S. Griffith and J. B. Westmore, J. Chem. Soc., 1704 (1962).

[72]. R. N. Gourley, J. Grimshaw, and P. G. Miller, J. Chem. Soc. (C), 2318 (1970).

[73]. A. Zweig, G. Metzler, A. H. Maurer, and B. G. Roberts, J. Amer. Chem. Soc., 88, 2864 (1966).

[74]. M Janda, Coll. Czech. Chem. Commun., 28, 2524 (1963).

[75]. M. Janda and J. Radouch, ibid., 32, 2672 (1967).

[76]. M. Janda and L. Paviensky, ibid., 32, 2675 (1967).

[77]. J. Sprogl, M. Janda, and M. Valentova, ibid., 35, 148 (1970).

[78]. K. Yoshida, T. Saeki, and T. Fueno, J. Org. Chem., 36, 3673 (1971).

[79]. H. V. Drushel and J. F. Miller, Anal. Chem., 30, 1271 (1958).

[80]. H. Lund, unpublished observation.

[81]. O. Manousek, O. Exner, and P. Zuman, Coll. Czech. Chem. Commun., 33, 4000 (1968).

[82]. M. Dennstedt, D.R.P. 127, 086 (1901).

[83]. B. Sakurai, Bull. Chem. Soc. Japan, 11, 374 (1936).

[84]. L. C. Craig, J. Amer. Chem. Soc., 55, 2543 (1933).

[85]. O. Carrasco, Gazz. Chim. Ital., 38, 301 (1908).

[86]. J. von Braun and W. Sobecki, Ber., 44, 2158 (1911).

[87]. J. T. Wrobel and K. M. Pazdro, Rocz. Chem., 41, 637 (1967).

[88]. F. Ender, E. Moisar, K. Schäfer, and H.-J. Teuber, Z. Elektrochem., 63, 349 (1959).

[89]. A. Stanienda, Z. Naturforsch., 22b, 1107 (1967).

[90]. G. Cauquis and M. Gewies, Bull. Soc. Chim. Fr., 3220 (1967).

[91]. N. L. Weinberg and E. A. Brown, J. Org. Chem., 31, 4054 (1966).

[92]. J. F. Ambrose and R. F. Nelson, J. Electrochem. Soc., 115, 1159 (1968).

[93]. J. F. Ambrose, L. L. Carpenter, S. C. Creason, and R. F. Nelson, Electrochemical Society Meeting, May 1969, Extended Abstracts, p. 354.

[94]. F. B. Ahrens, Z. Elektrochem., 2, 577 (1895).

[95]. B. Emmert, Ber., 46, 1716 (1913).

[96]. W. R. Turner and P. J. Elving, Anal. Chem., 37, 467 (1965).

[97]. H. Lund, Acta Chem. Scand., 17, 2325 (1963).

[98]. J. N. Burnett and A. L. Underwood, J. Org. Chem., 30, 1154 (1965).

[99]. P. C. Tompkins and C. L. A. Schmidt, Univ. Calif. Publ. Physiol., 8, 237, 247 (1944).

[100]. S. G. Mairanovskii, Dokl. Akad. Nauk SSSR, 110, 593 (1956).

[101]. O. Fischer and K. Neundlinger, Ber., 46, 2544 (1913).

[102]. S. J. Leach, J. H. Baxendale, and M. G. Evans, Aust. J. Chem., 6, 395 (1953).

[103]. J. Nakaya, Nippon Kagaku Zasshi, 81, 1459 (1960); CA, 56, 4514 (1962).

[104]. H. Yasuda and S. Kitagawa, Yakugaku Kenkyu, 27, 779 (1955); CA, 51, 13246 (1957).

[105]. A. J. Cunningham and A. L. Underwood, Biochem., 6, 266 (1967).

[106]. B. Janik and P. J. Elving, J. Amer. Chem. Soc., 92, 235 (1970).

[107]. B. Ke, ibid., 78, 3649 (1956).

[108]. B. Ke, Arch. Biochem. Biophys., 60, 505 (1956).

[109]. R. F. Powning and C. C. Kratzing, ibid., 66, 249 (1957).

[110]. T. Kono and S. Nakamura, Bull. Agr. Chem. Soc. Japan, 22, 399 (1958); CA, 53, 18120 (1959).

[111]. J. N. Burnett and A. L. Underwood, Biochem., 4, 2060 (1965).

[112]. A. J. Cunningham and A. L. Underwood, Arch. Biochem. Biophys. 117, 88 (1966).

[113]. R. W. Burnett and A. L. Underwood, Biochem., 7, 3328 (1968).

[114]. R. J. Day, S. J. Kinsey, E. T. Seo, and H. P. Silverman, The Electrochemical Society Meeting, New York, May 1969, Extended Abstracts, p. 357.

[115]. W. J. Blaedel and R. G. Haas, Anal. Chem., 42, 918 (1970).

[116]. N. E. Khomutov and V. V. Tsodikov, Elektrokhim., 1, 482 (1965).

[117]. N. E. Khomutov and V. V. Tsodikov, ibid., 2, 722 (1966).

[118]. M. Maturova, T. Shun, J. Ctvrtnik, and F. Santavy, Coll. Czech. Chem. Commun., 25, 3321 (1960).

[119]. M. Kulka, J. Amer. Chem. Soc., 68, 2472 (1946).

[120]. J. C. Cochran and W. F. Little, J. Org. Chem., 26, 808 (1961).

[121]. R. C. Kaye and H. I. Stonehill, J. Chem. Soc., 27 (1951).

[122]. J. G. Frost and J. H. Saylor, Rec. Trav. Chim. Pays-Bas, 82, 828 (1963).

[123]. H. Lund, Acta Chem. Scand., 23, 563 (1969).

[124]. V. D. Bezuglyi, N. P. Shimanskaya, and E. M. Peresleni, Zh. Obshch. Khim., 34, 3540 (1964).

[125]. W. N. Greig and J. W. Rogers, J. Electrochem. Soc., 117, 1141 (1970).

[126]. P. E. Iversen and H. Lund, Acta Chem. Scand., 21, 389 (1967).

[127]. H. Lund, to be published.

[128]. Z. N. Timofeeva, M. V. Petrova, M. Z. Girshovich, and A. V. El'tsov, Zh. Obshch. Khim., 39, 54 (1969).

[129]. M. Matsui, T. Sawamura, and T. Adachi, Mem. Coll. Sci. Kyoto Imp. Univ., 15A, 151 (1932); CA, 26, 5264 (1932).

[130]. H. Lund, unpublished observation.

[131]. P. Kabasakalian and J. McGlotten, Anal. Chem., 31, 431 (1959).

[132]. J.-P. Billon, Bull. Soc. Chim. Fr., 1784 (1960).

[133]. J.-P. Billon, ibid., 1923 (1961).

[134]. J.-P. Billon, Ann. Chim., 7, 183 (1962).

[135]. R. Borsdorf, R. Herzschuh, and J. Seidler, Z. Chem., 10, 147 (1970).

[136]. H. Lund, Coll. Czech. Chem. Commun., 31, 4175 (1966).

[137]. J. P. Kitaev and G. K. Budnikov, ibid., 30, 4178 (1965).

[138]. N. P. Shimanskaya, N. A. Lodygin, and V. D. Bezuglyi, Zh. Obshch. Khim., 38, 929 (1968).

[139]. N. P. Shimanskaya, V. D. Bezuglyi, and V. E. Bondarenko, ibid., 39, 2171 (1969).

[140]. H. Lund, Österreich. Chem.-Z., 68, 43 (1967).

[141]. S. Millefiori, Ann. Chim. (Rome), 59, 15 (1969).

[142]. S. D. Ross, G. J. Kahan, and W. A. Leach, J. Amer. Chem. Soc., 74, 4122 (1952).

[143]. C. Furlani, S. Bertola, and G. Murpurgo, Ann. Chim. (Rome), 50, 858 (1960).

[144]. H. Lund and E. T. Jensen, Acta Chem. Scand., 25, 2727 (1971).

[145]. H. Lund, Lecture, 12. Nordiske Kemikermøde, Trondheim, June 1965, Abstract of Papers, p. 33.

[146]. D. L. Smith and P. J. Elving, J. Amer. Chem. Soc., 84, 2741 (1962).

[147]. B. Janik and P. J. Elving, Chem. Rev., 68, 295 (1968).

[148]. H. Lund, Acta Chem. Scand., 18, 1984 (1964).

[149]. P. Pflegel and G. Wagner, Pharm., 22, 60, 643 (1967).

[150]. P. Pflegel and G. Wagner, ibid., 24, 308, 384 (1969).

[151]. L. F. Wiggins and W. S. Wise, J. Chem. Soc., 4780 (1956).

[152]. J. Volke, D. Dumanovic, and V. Volkova, Coll. Czech. Chem. Commun., 30, 246 (1965).

[153]. J. Pinson and J. Armand, Compt. Rend. Ser. C, 266, 1081 (1968).

[154]. G. Sartori and C. Furlani, Ann. Chim. (Rome), 45, 251 (1955).

[155]. M. P. Strier and J. C. Cavagnol, J. Amer. Chem. Soc., 80, 1565 (1958).

[156]. J. Pinson and J. Armand, Compt. Rend. Ser. C, 268, 629 (1969); Bull. Soc. Chim. Fr., 1764 (1971).

[157]. J. Pinson, J.-P. Launay, and J. Armand, Compt. Rend., 270, 1881 (1970).

[158]. M. Fedoronko and I. Jezo, Coll. Czech. Chem. Commun., 37, 1781 (1972).

[159]. R. F. Nelson, D. W. Leedy, E. T. Seo, and R. N. Adams, Z. Anal. Chem., 224, 184 (1967).

[160]. H. Lund, unpublished observation.

[161]. L. A. Summers and J. E. Dickeson, Chem. Commun. (London), 1183 (1967).

[162]. H. Oelschläger, Arch. Pharm., 296, 396 (1963).

[163]. B. Z. Senkowski, M. S. Levin, J. R. Urbigkit, and E. G. Wollish, Anal. Chem., 36, 1991 (1964).

[164]. H. Oelschläger, J. Volke, and H. Hoffmann, Coll. Czech. Chem. Commun., 31, 1264 (1966).

[165]. H. Oelschläger, J. Volke, H. Hoffmann, and E. Kurek, Arch. Pharm., 300, 250 (1967).

[166]. R. Schindler, H. Will, and L. Holleck, Z. Elektrochem., 63, 596 (1959).

[167]. P. Zuman, Coll. Czech. Chem. Commun., 25, 3245 (1960).

[168]. F. F. Medovshchikova and I. Ya. Postovskii, Zh. Obshch. Khim., 24, 1989 (1954).

[169]. L. S. Efros and Z. V. Todres, ibid., 27, 983 (1957).

[170]. V. Sh. Tsveniashvili, S. I. Zdanov, and Z. V. Todres, Z. Anal. Chem., 224, 389 (1967).

[171]. D. L. Smith and P. J. Elving, J. Amer. Chem. Soc., 84, 1412 (1962).

[172]. K. Kretzschmar and W. Jaenicke, Z. Naturforsch., 26b, 225, 999 (1971).

[173]. H. H. Inhoffen and P. Jäger, Tetrahedron Lett., 1317 (1964); ibid., 3387 (1965).

[174]. H. H. Inhoffen and R. Mahlhop, ibid., 4283 (1966).

[175]. H. H. Inhoffen, P. Jäger, R. Mahlhop, and C.-D. Mengler, Ann. 704, 188 (1967).

[176]. D. W. Clack and N. S. Hush, J. Amer. Chem. Soc., 87, 4238 (1965).

[177]. A. Stanienda and G. Biebl, Z. Phys. Chem. (Frankfurt), 52, 254 (1967).

[178]. V. G. Khomyakov, S. S. Kruglikov, and V. M. Berezovskii, Zh. Obshch. Khim., 28, 2898 (1958).

[179]. F. Fichter and H. DeMontmollin, Helv. Chim. Acta, 5, 256 (1922).

[180]. J. D. Anderson, M. M. Baizer, and E. J. Prill, J. Org. Chem., 30, 1645 (1965).

[181]. V. D. Bezuglyi and T. A. Alekseeva, Zh. Anal. Khim., 20, 244 (1965).

[182]. V. D. Bezuglyi, V. D. Dmitrieva, and T. A. Alekseeva, ibid., 16, 477 (1961).

[183]. C. Prevost, P. Souchay, and J. Chauvelier, Bull. Soc. Chim. Fr., 714 (1951).

[184]. A. R. Katritzky and J. M. Lagowski, in Advances in Heterocyclic Chemistry (A. R. Katritzky, ed.), Vol. 1, Academic, New York, 1963, Chaps. 6 and 7; ibid., Vol. 2, Academic, New York, 1963, Chaps. 1 and 2.

[185]. J. M. Bobbitt, H. Yagi, S. Shibuya, and J. T. Stock, J. Org. Chem., 36, 3006 (1971).

[186]. P. Kabasakalian and J. McGlotten, Anal. Chem., 31, 1091 (1959).

[187]. H. Lund, Acta Chem. Scand., 14, 1927 (1960).

[188]. O. Manousek and P. Zuman, Coll. Czech. Chem. Commun., 29, 1432 (1964).

[189]. V. Kucharczyk, M. Adamovsky, V. Horak, and P. Zuman, J. Electroanal. Chem., 10, 503 (1963).

[190]. J. Tirouflet, R. Robin, and M. Guyard, Bull. Soc. Chim. Fr., 571 (1956).

[191]. A. Ryvolova, Coll. Czech. Chem. Commun., 25, 420 (1960).

[192]. A. Dunet and A. Willemart, Bull. Soc. Chim. Fr., 887 (1948).

[193]. E. Späth and F. Brench, Monatsh., 50, 349 (1928).

[194]. B. Sakurai, Bull. Chem. Soc. Japan, 5, 184 (1930).

[195]. F. Fichter and I. Stein, Helv. Chim. Acta, 12, 821 (1929).

[196]. H. E. Baumgarten, P. L. Creger, and R. L. Zey, J. Amer. Chem. Soc., 82, 3977 (1960).

[197]. S. Kwee and H. Lund, Acta Chem. Scand., 25, 1813 (1971); Experientia Suppl., 18, 387 (1971).

[198]. C. Furlani, Gazz. Chim. Ital., 85, 1646 (1955).

[199]. P. Pflegel and G. Wagner, Z. Chem., 8, 179 (1968).

[200]. H. Oelschläger, J. Volke, G. T. Lim, and U. Bremer, Arch. Pharm., 303, 364 (1970).

[201]. E. Ochiai, J. Org. Chem., 18, 534 (1953).

[202]. A. Foffani and F. Fornasari, Gazz. Chim. Ital., 83, 1051, 1059 (1959).

[203]. L. V. Varyukhina and Z. V. Pushkareva, Zh. Obshch. Khim., 26, 1740 (1956); CA, 51, 1960 (1957).

[204]. O. N. Nechaeva and Z. V. Pushkareva, Zh. Obshch. Khim., 28, 2693 (1958); CA, 53, 9229 (1959).

[205]. T. R. Emerson and C. W. Rees, J. Chem. Soc., 1923 (1962).

[206]. N. A. Kudryavtseva, Z. V. Pushkareva, and V. F. Gryazev, Zh. Obshch. Khim., 35, 14 (1965).

[207]. C. R. Warner and P. J. Elving, Coll. Czech. Chem. Commun., 30, 4210 (1965).

[208]. T. Kubota and H. Miyazaki, Bull. Chem. Soc. Japan, 35, 1549 (1962).

[209]. T. Kubota and H. Miyazaki, ibid., 39, 2057 (1966).

[210]. I. Suzuki, M. Nakadate, T. Nakashima, and N. Nagasawa, Tetrahedron Lett., 2899 (1966).

[211]. M. Tachibana, S. Sawaki, and Y. Kawazoe, Chem. Phar. Bull., 15, 1112 (1967).

[212]. J. Hranilovic, D. Koruncev, and E. Gustak, Electrochem. Tech., 6, 62 (1968).

[213]. T. Okano and K. Ohira, Yakugaku Zasshi, 88, 1170 (1968); CA, 70, 16609 (1969).

[214]. N. A. Kudryavtseva, Z. V. Pushkareva, and V. F. Gryazev, Zh. Obshch. Khim., 35, 14 (1965).

[215]. J. Volke, Coll. Czech. Chem. Commun., 23, 1486 (1958).

[216]. J. Volke, Experientia, 13, 274 (1957).

[217]. J. Tirouflet and E. Laviron, Compt. Rend., 247, 217 (1958).

[218]. E. Laviron and J. Tirouflet, ibid., 248, 826 (1959).

[219]. E. Laviron, Bull. Soc. Chim. Fr., 2325 (1961).

[220]. P. E. Iversen and H. Lund, Acta Chem. Scand., 21, 279 (1967).

[221]. J. Tirouflet, E. Laviron, J. Metzger, and J. Boichard, Coll. Czech. Chem. Commun., 25, 3277 (1960).

[222]. J. H. Stocker and R. M. Jenevein, J. Org. Chem., 33, 294 (1968).

[223]. J. H. Stocker and R. M. Jenevein, Symposium on Synthetic and Mechanistic Aspects of Electroorganic Chemistry, Durham, N.C., October 1968, Preprints of Papers, p. 221.

[224]. J. Volke and V. Volkova, Coll. Czech. Chem. Commun., 20, 1332 (1955).

[225]. H. H. G. Jellinek and J. R. Urwin, J. Phys. Chem., 58, 168 (1954).

[226]. H. Lund, Acta Chem. Scand., 17, 972 (1963).

[227]. P. E. Iversen and H. Lund, ibid., 21, 389 (1967).

[228]. P. E. Iversen, ibid., 24, 2459 (1970).

[229]. H. Lund, ibid., 17, 1077 (1963).

[230]. P. G. Gassman and B. L. Fox, J. Org. Chem., 32, 480 (1967).

[231]. J. Volke, R. Kubicek, and F. Santavy, Coll. Czech. Chem. Commun., 25, 1510 (1960).

[232]. J. Volke and J. Holubek, ibid., 28, 1597 (1963).

[233]. E. Laviron, Compt. Rend., 250, 3671 (1960).

[234]. O. Manousek and P. Zuman, Coll. Czech. Chem. Commun., 29, 1432 (1964).

[235]. D. L. Smith and P. J. Elving, J. Amer. Chem. Soc., 84, 1412 (1962).

[236]. G. Dryhurst and P. J. Elving, J. Electrochem. Soc., 115, 1014 (1968).

[237]. G. Dryhurst, ibid., 117, 1118 (1970).

[238]. C. Dreux, M.-L. Girard, and P. Souchay, Compt. Rend., 262, 1565 (1966).

[239]. G. Dryhurst, J. Electrochem. Soc., 117, 1113 (1970).

[240]. K. Sugino, K. Shirai, T. Sekine, and K. Odo, ibid., 104, 667 (1957).

NOTES ADDED IN PROOF

[1] On anodic oxidation of 3, 6-diisobutylpiperazine-2, 5-dione in acetonitrile a compound was obtained which was suggested to be 1, 6-diisopropyl-3, 8-dimethyl-5H, 10H-diimidazo[1, 5-a:1′, 5′-d]pyrazine-5, 10-dione, formed by 1, 3-cycloaddition of the primary oxidation product to solvent [54].

[2] Derivatives of pteridine are generally reduced analogously to quinoxaline; a reversible reduction to a 5, 8-dihydro compound, which may tautomerize into a reducible 7, 8-dihydro derivative, is followed by reduction to the oxidizable 5, 6, 7, 8-tetrahydropteridine derivative [172].

Chapter XVIII

ORGANOMETALLIC SYNTHESES [1]

H. Lehmkuhl

Max Planck Institute
Mülheim-Ruhr, Germany

I. INTRODUCTION

Organometallic compounds have at least one metal-carbon bond [1]. This definition includes coordination compounds in which multiple carbon-carbon bonds form π-bonds with metal but excludes organic compounds in which the metal is bound to heteroatoms such as O, S, N, and P. The last 20 years have seen an enormous growth in our understanding of organometallic chemistry and in the use of these compounds in synthesis and technology. As a consequence, it is difficult, while we are still within this period of development, to present a survey that will still be of value a few years from now. This is especially true if one of the consequences of this book should be an ever increasing use (by electrolysis) of the electron as a simple clean reagent.

This chapter is subdivided as follows:

Part II: Electrochemical Preparation of Organometallic Compounds

Part III: Electrochemical Transformations of Organometallic Compounds.

Experimental examples were provided in the Introductory Survey to illustrate for the experimental organic chemist that the modus operandi in electroorganic work is not particularly obscure or complex. This point cannot be overemphasized. It obtains also in the area of organometallic electrochemistry. Table 8 at the end of this chapter (p. 665) presents in outline form illustrative examples of syntheses which have been well described in the literature.

II. ELECTROCHEMICAL PREPARATION OF ORGANOMETALLIC COMPOUNDS

The electrolytic preparation of organometallic compounds may be carried out by the following principal methods:

1. Hydrocarbon radicals may be produced at a cathode by reduction of organic compounds. The radicals then achieve stabilization by reacting with the cathode material to form organometallics.

2. Metallic compounds may be reduced to a lower valence state of the metal. The latter state is often not stable per se but may be stabilized by suitable organic complexing agents.

3. Carbanions containing organometallics may be oxidized anodically to radicals which under suitable conditions react with the anodic metal to form organometallic compounds. If the original carbanion was also an organometallic compound, this method represents a trans-alkylation

$M'\text{-}R + M'' \rightleftarrows M''\text{-}R + M'$.

A. Organometallics by Reduction of Organic Compounds

Alkyl radicals can be obtained from those organic compounds in which the alkyl group may be regarded as the positive end of a dipole, e.g., $R^{\delta+}\text{-}X^{\delta-}$ and $RR'C^{\delta+}{=}O^{\delta-}$.

1. Metal Alkyls by Cathodic Reduction of

a. Carbonyl Compounds. Electrochemical reduction of ketones or aldehydes in acid solution at cathodes of mercury, lead, or cadmium leads in certain cases to the formation of organometallic compounds which deposit as heavy oils [2, 3]:

$$R_2C{=}O + H^{\oplus} + e^{\ominus} \; [R_2\dot{C}OH] + m \longrightarrow [R_2C(m){-}OH] \xrightarrow{2e^{\ominus},\ 2H^{\oplus}} R_2CH(m) + H_2O$$

where m is the equivalent of Hg, Pb, or Cd. The organomercury compounds are formed from aliphatic or cycloaliphatic ketones in 25-30% yield based on current (Table 1). Ketones in which a phenyl group is bound directly to carbonyls do not form mercurials. Menthone does not [5] form a mercurial in 50% sulfuric acid at 60° but does [6] form di-menthylmercury at 70-77° and higher current densities (17 A/dm^2):

TABLE 1

Organomercurials by Electrochemical Reduction of Ketones

Ketone	Conc. of acid	Temp. (°C)	Organomercurial	Ref.
$C_6H_5CH_2COCH_3$	30% H_2SO_4	20	Bis-[1-phenyl-2-propyl]mercury	[5]
	5% H_2SO_4	55		[5]
	6 N HCl + HOAc (1:1)	18		[5]
CH_3COCH_3	40% H_2SO_4	30	Di-2-propylmercury	[4]
$CH_3COC_2H_5$	20% H_2SO_4	45-50	Di-2-butylmercury	[3]
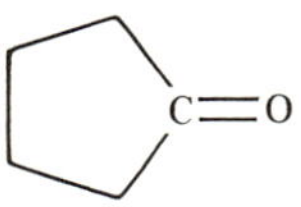	5% H_2SO_4	50	Bis-cyclopentylmercury	[5]
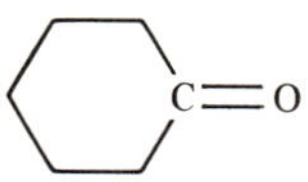	5% H_2SO_4	55	Bis-cyclohexylmercury	[5]
2-, 3-, or 4-Methyl-cyclohexanone	5% H_2SO_4	55	Bis-methylcyclohexyl-mercury	[5]

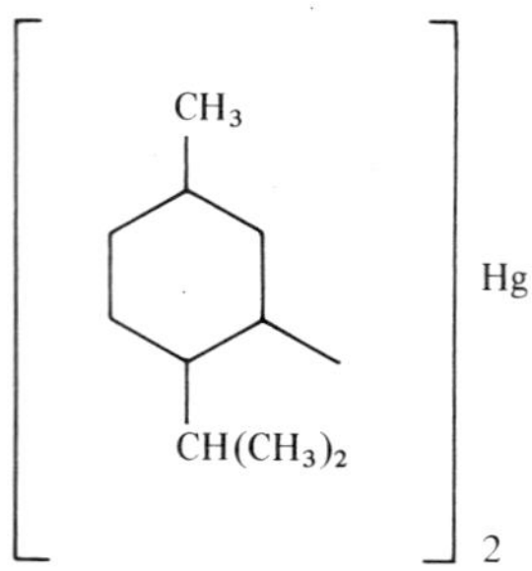

The preparation of di-2-butylmercury is summarized in Table 8, Ex. 1 (p. 665).

When methylvinyl ketone is reduced at mercury at a constant potential of -1.05 V (vs. SCE) and a pH <5, it is not the carbonyl group which is reduced but rather the double bond. The mercury becomes attached to the β-carbon atom [7]:

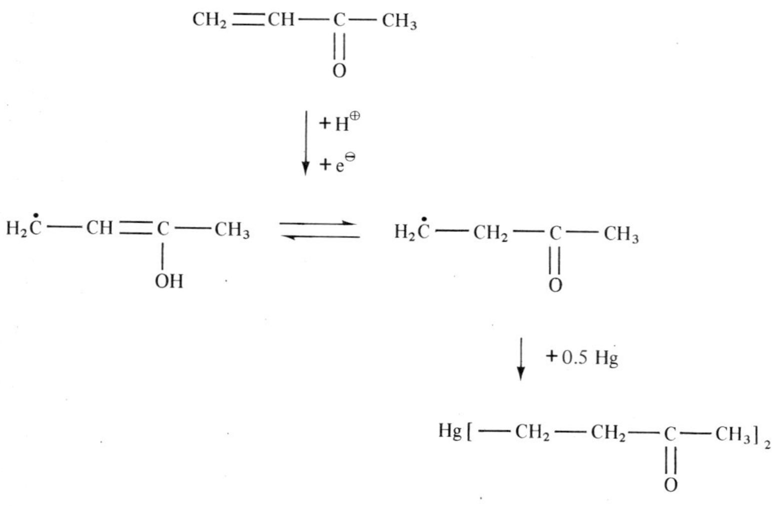

Aliphatic and aromatic aldehydes (except benzaldehyde) do not form organomercurials [8, 9]. Benzaldehyde yields 8.5% of dibenzylmercury [9] when reduced in 50% sulfuric acid at a current density of 19.7 A/dm^2.

Reduction of acetone, methyl ethyl ketone, and diethyl ketone in sulfuric acid at lead yields substantial quantities of tetraalkyllead compounds along with very unstable compounds [10, 11]. These by-products are presumably di- or trialkylleads since on treatment with bromine they form the corresponding dialkyllead dibromides.

α, β-Unsaturated aldehydes also yield organolead compounds on electrolysis [12].

b. Alkyl Halides. In contrast to the anodic formation of alkylleads the cathodic formation [2] always seems to have the disadvantage that in part alkyl derivatives of lower oxidation states of lead are formed and that these intermediates undergo various secondary and disproportionation reactions which involve the separation of lead in the cathode compartment [13-15]. This is also the situation when ethyl iodide is reduced at 55° at lead using alcoholic sodium hydroxide in the electrolyte. The organometallic is formed but hydrogen is also evolved from the cathode. Yields are not specified. It has been proposed that hydrogen evolution can be avoided by using aprotic solvents, e.g., acetonitrile; in this case an electrolyte such as tetraalkylammonium or trialkylsulfonium halide is needed [16]. The yield of tetraethyllead is then reported to be 12%, calculated on the basis of the weight lost by the cathode. The reduction of ethyl bromide in propylene carbonate at lead requires that the overvoltage of ca. 0.5 V be overcome [14a]. At a cathode potential of 1.4 V all the lead that has gone into solution has formed tetraethyllead. At higher potential side reactions diminish the yield [14a]: ethane and ethylene are formed from the ethyl radicals:

$$2C_2H_5\cdot \rightarrow C_2H_4 + C_2H_6, \qquad 2C_2H_5\cdot \rightarrow C_4H_{10}.$$

CPE of alkyl halides in DMF using sodium halides as supporting electrolytes [14d] yields 60-90% (based on current) of tetraalkylleads.

The reduction of methyl bromide in acetonitrile or water gives tetramethyllead in nearly quantitative yield. With ethyl bromide mixtures of tetraethyllead and hexaethyldilead are formed [17]. Mixtures of MeI and EtI in DMF yielded mixtures of methylethylleads [14d]; the composition depended on the potential (-1.40 to -2.1 V).

The cathodic reduction of organic halides may also be used for the preparation of organomercurials. Depending on the nature of the substituents, benzyl halides are reduced to diorganomercurials and to dimeric hydrocarbons in various ratios (Table 2). See also Chapter XIX.

The preparation of bis-(4-methylbenzyl)mercury is summarized in Table 8, Ex. 2 (p. 665).

Electrolysis of 3-iodopropionitrile in 0.5 N sulfuric acid containing sodium sulfate as additional electrolyte yielded the following organometallics using anodes of the appropriate metal [20]:

$(NCCH_2CH_2)_3Sn\text{-}Sn(CH_2CH_2CN)_3$, 12%

$(NCCH_2CH_2)_4Pb$, 13.2%;

$(NCCH_2CH_2)_2Hg$, 45.9%.

TABLE 2

Reduction of Substituted Benzyl Bromides at Mercury [18]

-Cathode V [a]	Substituted benzyl bromide	Yields (%)	
		Dimeric hydrocarbon	Organomercurial
1.45	p-CH_3O-	0	0
1.50	p-$(CH_3)_3C$-	0	50
1.40	H-	0.1	41
1.20	p-CH_3-	0.0	64
1.20	m-Br-	25	0
0.94	3,4-Dichloro-	2.1	39
1.05	3,4-Dichloro-	25	27
1.30	3,4-Dichloro-	21	4.6
0.50	p-NO_2	74	0

[a]Vs. Ag/AgCl/KCl-H_2O (0.1 M).

c. Unsaturated Nitriles. Organotin compounds are formed [21] by electrolysis of methacrylonitrile in 0.7 N sodium hydroxide at 26° and a current density of 1.7 A/dm^2:

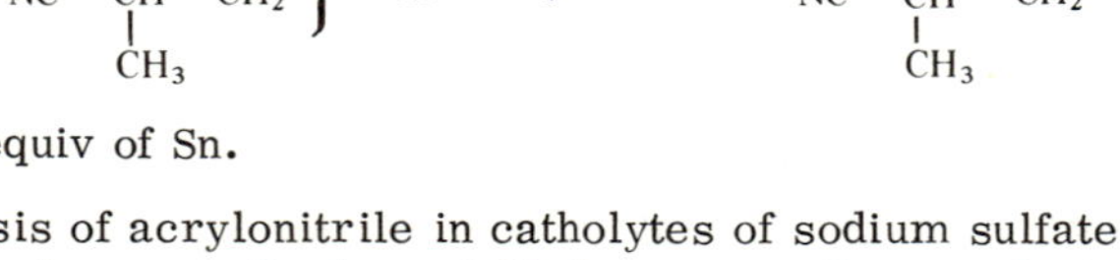

where sn = 1 equiv of Sn.

Electrolysis of acrylonitrile in catholytes of sodium sulfate containing suspended selenium or tellurium yielded, among other products, $(NCCH_2CH_2)_2Se$ and $(NCCH_2CH_2)_2Te$, respectively [19]. In a modified process for formation of metal alkyls the methyl compounds of aluminum, tin, antimony, or mercury are formed in fused halogen compounds of these elements in the presence of an acceptor metal for halogen (Al, Zn, Mg, Sn, or their amalgams, alkali metals) and methyl chloride. The acceptor metal halides formed are electrochemically cleaved into metal and halogen [23]:

$$mCl + CH_3Cl + 2m' \rightarrow CH_3m + 2m'Cl$$

$$2m'Cl \xrightarrow{\text{Electrolysis}} 2m' + Cl_2.$$

d. Other Organometallic Compounds. Electroreduction of organometallics leads to species which can participate in a variety of follow-up reactions. Only those systems will be considered here in which another organometallic is the end product.

Dessy's outline of reaction pathways [24], somewhat enlarged by the present author, is shown below. In this scheme RM denotes an organo-

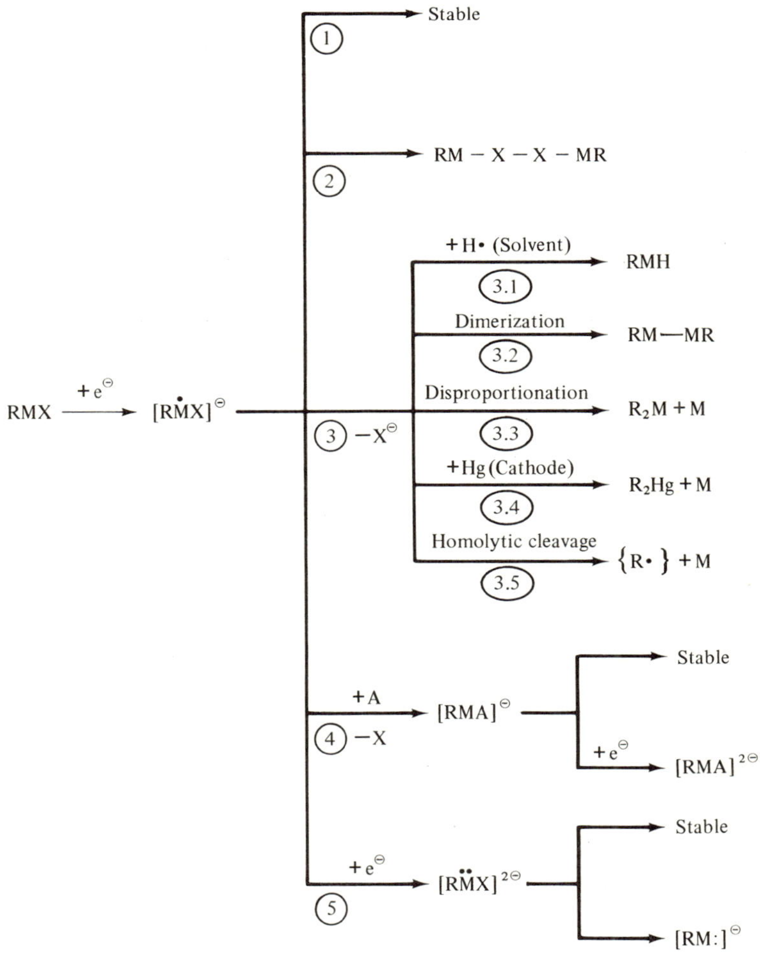

metallic group (neutral or ionic), X is an organic or inorganic moiety or a ligand complexing agent, and A an electrophilic moiety. Although not all the possibilities encompassed in this scheme have been realized preparatively, nevertheless a number of examples within this framework are available.

(1) Pathways ① and ②. Diethylaluminum cations, stabilized by complexation with pyridine, are formed by the following equilibration:

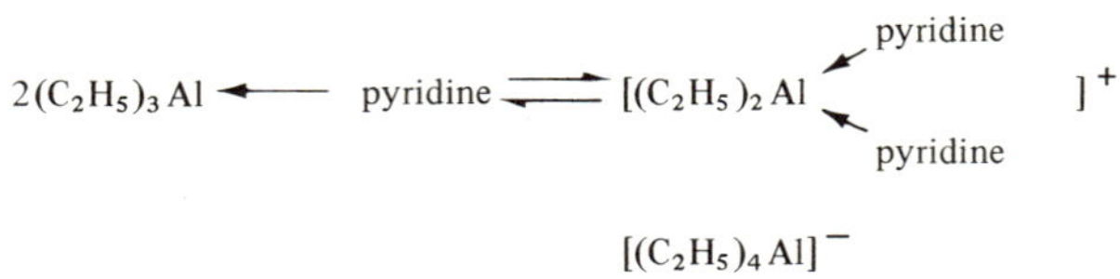

Electrolysis of these cations leads to stable organoaluminum radicals [25]. In diaphragmless cells 30-40% of a blue solid with an analysis corresponding to the composition

$(C_2H_5)_2Al \cdot 2C_5H_5N$

was isolated. An ESR spectrum without hyperfine structure was obtained. Hydrolysis in the presence of air yielded 4, 4′-bipyridyl in addition to unchanged pyridine. One may assume by analogy with the behavior of dichloroaluminum cations in pyridine [26] that the initially formed diethylaluminum-pyridine radical undergoes dimerization:

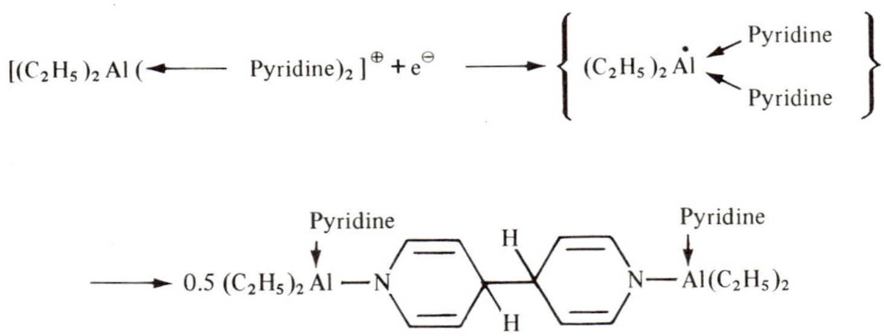

Polarographic investigation of triethylaluminum in pyridine shows that the pyridine-complexed aluminum compound is substantially easier to reduce than pyridine itself [27].

The diethylberyllium radical likewise formed electrochemically and stabilized by complexation with pyridine [28] shows similar relationships and has the composition

$(C_2H_5)Be \cdot C_5H_5N$.

Cathodically formed organometallic radicals may also achieve stabilization in other ways.

(2) Pathway (3.1). Organometallic hydrides [29] may be formed by proton abstraction from the nominally aprotic solvent or from the supporting electrolyte (R_4NX). Thus, in the electrolysis of triphenylchlorosilane and triphenylchlorogermane, triphenylsilane and triphenylgermane, respectively, are formed. The reductions of diphenylmetallo chlorides lead to dihydrides.

$$\{(Ph)_{4-n}M\}Cl_n \xrightarrow[\text{Dimethoxyethane}\ (C_4H_9)_4NClO_4]{+ne^{\ominus}} (Ph)_{4-n}M + nCl^{\ominus} \xrightarrow{+nH(\text{solvent})} (Ph)_{4-n}MH_n$$

	n = 1	n = 2	
M = Si	$E_{1/2}$ = -3.1 V	$E_{1/2}$ = -1.9 V	vs. Ag/0.001 M $AgClO_4$
M = Ge	$E_{1/2}$ = -2.8 V	$E_{1/2}$ = -2.6 V	(0.1 M Bu_4NClO_4)

The corresponding tin and lead compounds behave differently (see (3.2) and (3.4), respectively, below).

(3) Pathway (3.2). The electrochemical dimerization of organic derivatives of metals of the fourth group of the periodic table has been used for a long time for the preparation of hexaethyldilead. For this purpose triethyllead hydroxide in aqueous alcohol is electrolyzed between lead electrodes in a cell with diaphragm [26]. Table 8, Ex. 3 (p. 665) summarizes the preparation of hexaethyldilead. On the other hand, if triphenyllead chloride is reduced at mercury, the phenyl groups are transferred to mercury [29]; diphenyllead dichloride behaves analogously. Both reactions also occur spontaneously without electrolysis ([29, 28] and p. 630):

$$(C_6H_5)_2PbCl_2 + 3Hg \rightarrow (C_6H_5)_2Hg + Pb + HgCl_2$$

$$2(C_6H_5)_3PbCl + 5Mg \rightarrow 3(C_6H_5)_2Hg + 2Pb + HgCl_2.$$

Another example of cathodic dimerization, so far demonstrated only polarographically, is the following:

$$(C_6H_5)_2MX \xrightarrow[-X^{\ominus}]{+e^{\ominus}} 0.5(C_6H_5)_2M\text{-}M(C_6H_5)_2$$

M = As; X = Br	$E_{1/2}$ = 0.9 V [32]	
M = As; X = $OOCCH_3$	$E_{1/2}$ = 1.4 V [32]	vs. Ag/10^{-3} M $AgClO_4$
M = Bi; X = Cl	$E_{1/2}$ = 1.0 V [32]	

Very extensive polarographic investigations have been carried out in the reduction of alkyl- and acyltin halides and hydroxides in protic and aprotic solvents [29, 33-40]. The electrochemical reactions are indicated in the following equations:

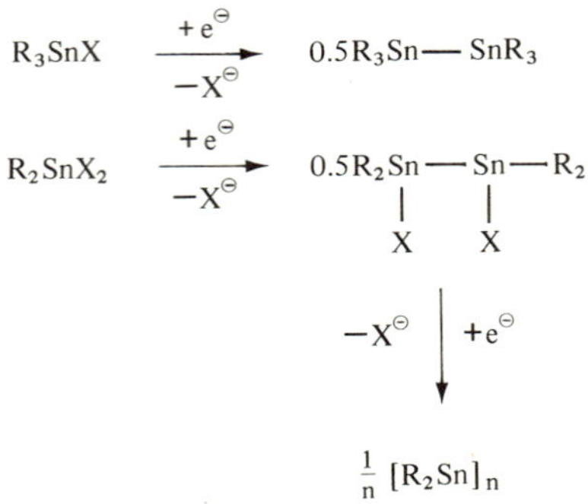

or

$$[R_2Sn(OH_2)_2]^{2\oplus} \xrightarrow[-2H_2O]{+2e^{\ominus}} \frac{1}{n}[R_2Sn]_n.$$

At very negative potentials (-2.3 to -2.7 V) the metal-metal bonds are cleaved to form anions (see pathway (5) below).

In the last few years many polarographic studies have also been made of organic transition metal complexes. These data and a knowledge of the degree of reversibility of the electron-transfer step provide the electrochemist who is synthetically inclined with valuable guides into this large unexploited area.

The cathodic reductions of organometallics which lead to unstable organometallic radicals which disporportionate and deposit metal in the electrode are discussed below (p. 652).

(4) Pathway (3.4). The reduction of certain organometallic compounds at mercury results in the formation of organomercurials. This type of reaction has already been mentioned in the case of phenyllead halides. The electrolysis of diphenyllead diacetate at a constant potential of -2.1 V produces diphenylmercury quantitatively [29].

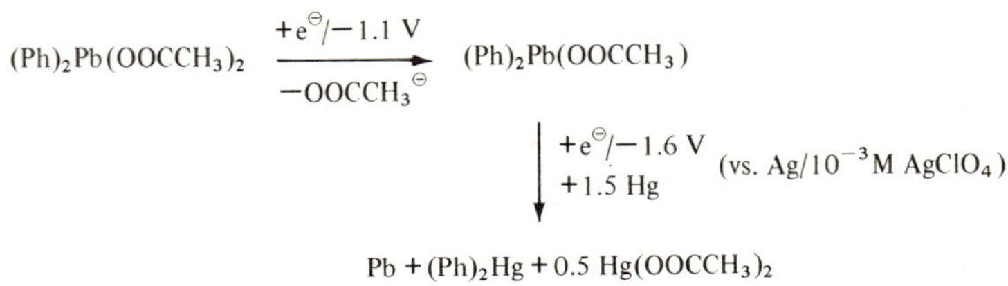

The following are additional examples of electrochemical mercuration:

$[(\pi C_5H_5)Fe(CO)_2]_2Hg$ from $[(\pi C_5H_5)Fe(CO)_2]_2$ or from $[(\pi C_2H_5)Fe(CO)_2]I$;

$[(\pi C_5H_5)Mo(CO)_3]_2Hg$ from $[(\pi C_5H_5)Mo(CO)_3]_2$;

$[Mn(CO)_5]_2Hg$ from $(Mn(CO)_5]_2$; and

$[Re(CO)_5]_2Hg$ from $(Re(CO)_5]_2$ [42].

The reduction of tetraphenylstibonium cation at mercury results in a variety of products, depending on the cathode potential. At the potential corresponding to the first polarographic wave [0.75 to 0.9 V vs. SCE) triphenylantimony and diphenylmercury are formed quantitatively. At the second polarographic wave (-1.4 V) triphenylantimony and benzene are formed [43]:

$$[Ph_4Sb]^{\oplus} + e^{\ominus} + Hg \rightarrow Ph_3Sb + \{PhHg\cdot\}$$

① $$\{PhHg\cdot\} \rightarrow 0.5Ph_2Hg + 0.5Hg$$

② $$\{PhHg\cdot\} + H^{\oplus} + e^{\ominus} \rightarrow PhH + Hg.$$

(5) Pathway ④. This route can be illustrated by an example from the field of organoaluminum chemistry. Molecular compounds formed from trialkylaluminums and electron donors dissociate into dialkylaluminum cations and tetraalkylaluminum anions:

$$2R_3Al \leftarrow D \rightleftarrows [R_2Al(D)_2]^{\oplus} + [R_4Al]^{\ominus}.$$

Electroreduction of the former in ethers (THF, dimethoxyethane, etc.) in the presence of naphthalene using potassium tetraethylaluminate as supporting electrolyte leads to the formation of the dialkyldihydronaphthdiyl anion, presumably by the following sequence [44]:

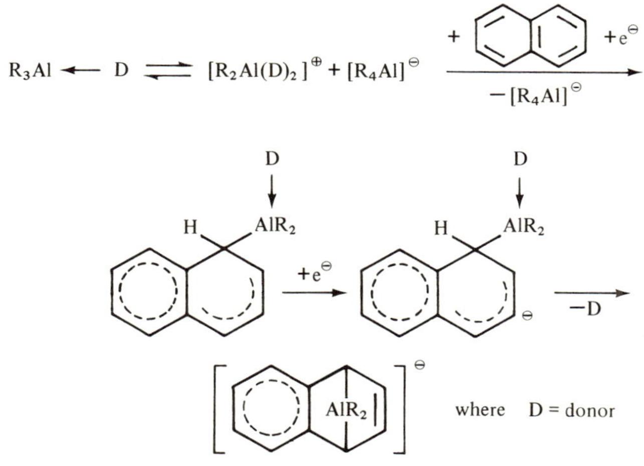

where D = donor

Electrolysis of THF solutions of styrene or isoprene using, e.g., sodium tetraethylaluminate as electrolyte produces "living" polymers of these hydrocarbons [45-47]. It can be shown in divided cells that polymerization occurs via cathodic reduction:

$$\mathrm{PhCH{=}CH_2} \longrightarrow [\mathrm{Ph\dot{C}H - \bar{C}H_2}]^{\ominus}$$

$$2\,[\mathrm{Ph\dot{C}H - \bar{C}H_2}]^{\ominus} \longrightarrow [\mathrm{Ph\bar{C}H - CH_2 - CH_2 - \bar{C}H - Ph}]^{2\ominus}$$

or

$$\mathrm{PhCH{=}CH_2} \xrightarrow{+2e^{\ominus}} [\mathrm{Ph\bar{C}H - \bar{C}H_2}]^{2\ominus} \xrightarrow{+\mathrm{PhCH{=}CH_2}} [\mathrm{Ph\bar{C}H - CH_2 - CH_2 - \bar{C}H - Ph}]^{2\ominus}$$

The typical spectrum of the living anion is obtained from catholyte solutions of α-methylstyrene in THF. The extinction of the solution is strictly proportional to the amount of charge transferred at the electrode [46].

Immediate discharge of the color occurs if the catholyte, colored because of the presence of carbanions, is mixed with anolyte, presumably because of complex formation between the carbanion and the tri- or dialkylaluminum cation. Similar reactions have sometimes been brought about by chemical means.

For example, if a low concentration of styrene is maintained during its reaction with potassium and triethylaluminum in THF, potassium diethyl-1-phenylethdiyl(1, 2)aluminate is formed:

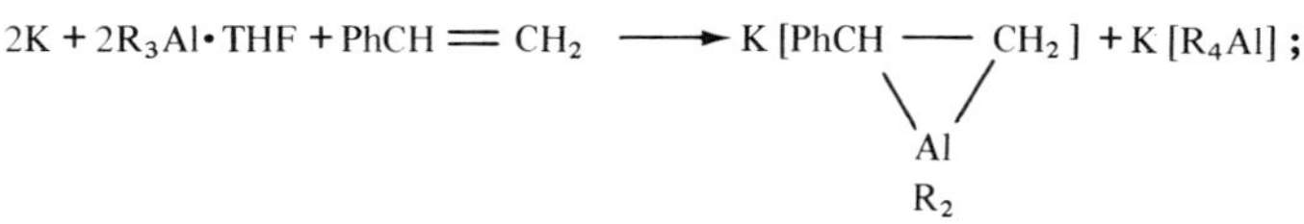

if the concentration of styrene is higher, there is formed in addition potassium diethylaluminum [1, 4-diphenyl-butdiyl]-(1, 4)aluminate:

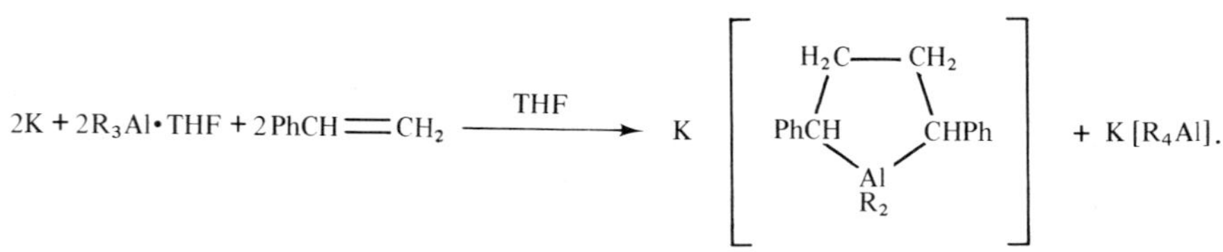

Butadiene polymers are obtained when solutions of $Et_3Al \cdot THF$ and butadiene with sodium tetraethylaluminate as electrolyte are electrolyzed at anodes of iron or copper. If anodes of aluminum are used instead, organoaluminum anions are obtained in place of the polymer. These anions are formed by the addition of the diethylaluminum cation to the butadiene anion and prevent the propagation of the chain via the butadiene anion. The electrolytic reaction is essentially [48] as follows:

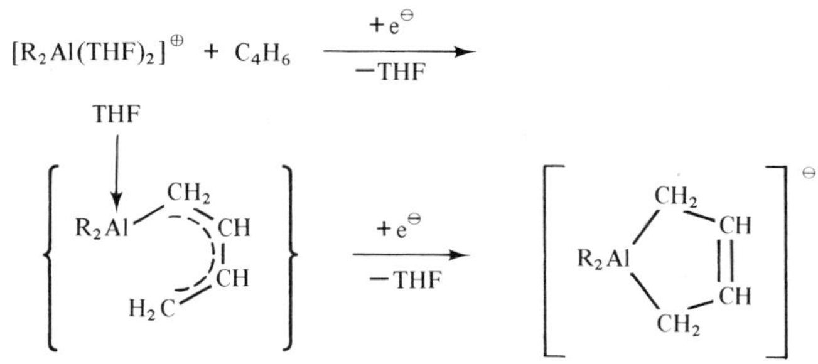

The first exploratory experiments with organoaluminum compounds and electron-accepting olefins show that there are numerous possibilities for preparing new organometallics by electroreduction of organometallics in the presence of the above types of olefins or for preparing known compounds which are now prepared by alternate means. As an example of the latter situation, we may cite the well-studied preparation of trialkylsilyl derivatives of these olefins using alkali metals [49]. Reactions of this type could in general be carried out according to pathway (4).

The phenyl group in phenyltrimethylsilane can be partially reduced in an undivided cell to its cyclohexadiene derivative:

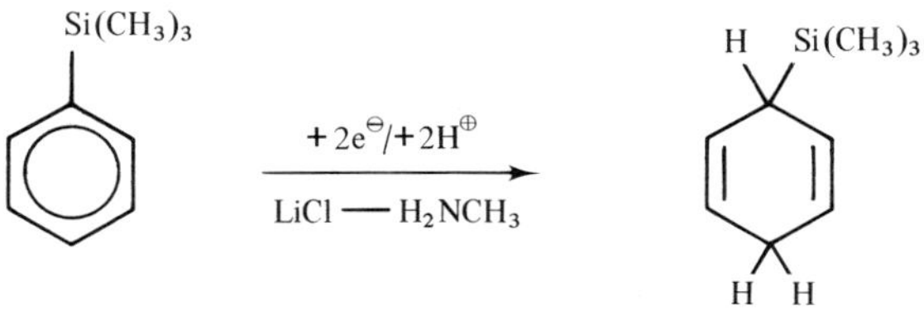

The yield of 3-trimethylsilyl-1, 4-cyclohexadiene is 76% [50].

The dimer organometallics containing metal-metal bonds which arise through pathway (3.2) may often take up one additional electron per metal atom and form anions

$$0.5\,RM\text{-}MR \xrightarrow{e^{\ominus}} [RM]^{\ominus}.$$

The same anions may be formed by a direct two-electron reduction according to pathway (5). Examples are:

$$Ph_3SnCl \xrightarrow[-Cl^{\ominus}]{+2e^{\ominus}/-2.9\ V} [Ph_3Sn]^{\ominus}$$

or when $M \neq M'$:

$$Ph_nM\text{-}M'Ph_n \xrightarrow{+2e^{\ominus}} [Ph_nM]^{\ominus} + [Ph_nM']^{\ominus}$$

$$n = 3 \begin{cases} M = M' = Si & \text{No reduction [29]} \\ M = M' = Ge & E_{1/2} = -3.5\ V\ [29] \\ M = M' = Sn & E_{1/2} = -2.9\ V\ [29] \quad (\text{vs. } Ag/10^{-3}\ M\ AgClO_4) \\ M = M' = Pb & E_{1/2} = -2.0\ V\ [29] \\ M = Sn,\ M' = Si & E_{1/2} = -3.1\ V \end{cases}$$

$$n = 2 \begin{cases} M = M' = As & E_{1/2} = -2.7\ V\ [32] \\ M = M' = Sb & E_{1/2} = -2.45\ V\ [32] \\ M = M' = Bi & E_{1/2} = -2.3\ V\ [32] \end{cases}$$

The diphenylbismuth anion is unstable and decomposes with deposition of metal.

2. Organic Coordination Compounds with Transition Metals by Cathodic Reduction

Organometallic compounds of the transition metals are of great technical and preparative interest. Those compounds in which alkyl or aryl groups are σ-bonded to the transition metal are very unstable and are of interest mainly as reaction intermediates. The transition metal alkyls are often in equilibrium with the corresponding coordination complexes in which multiple C-C bonds are π-bonded to the metal:

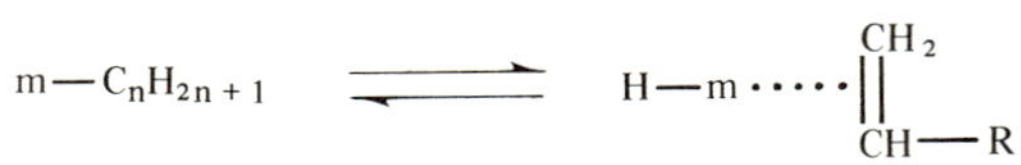

or

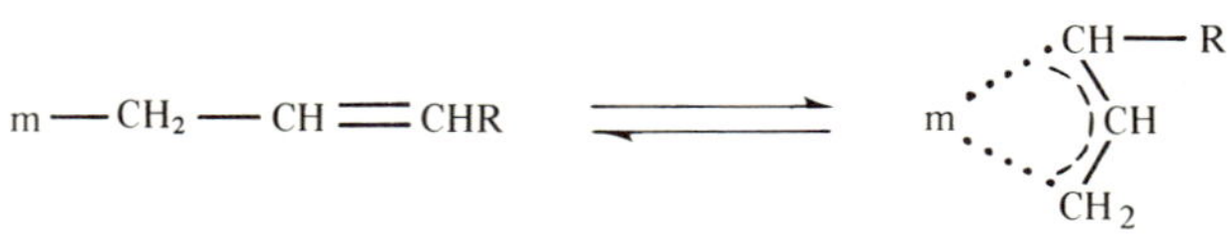

Organic transition metal complexes are valuable as active catalysts for hydrogenation of unsaturated organic molecules; for polymerization, oligomerization, or dimerization of 1, 3-diolefins or alkynes; for isomerization of olefins; for hydroformylation reactions, to mention only a few.

Organic transition metal complexes in which the metal exists in a lower than "normal" valence state may be prepared by the following methods:

(1) by electroreduction of suitable transition metal salts in the presence of organic complexing agents;

(2) by electrochemically reducing the valence state of the metal in a preexisting organometallic complex.

a. Coordination Compounds by Cathodic Reduction in the Presence of a Complexing Agent. Complexes of Cu^{I} halides with cyclic olefins are very unstable, presumably because of the tendency to form complex anions $[CuX_2]^-$:

$$2[(Cyc)CuX] \rightleftharpoons 2Cyc + Cu^{\oplus}[CuX_2]^{\ominus}.$$

In solvating media, which are desirable for electrolysis, the shift of the equilibrium to the right is particularly great because of the solvation of the Cu^{I} cation [51].

Electrolysis of Cu^{II} perchlorate in methanol in the presence of 1, 5-cyclooctadiene and at copper electrodes yields 1, 5-cyclooctadiene-copper^{I} perchlorate:

$$\text{Cathode:}\quad Cu^{2\oplus} + 2COD + e^{\ominus} + [ClO_4]^{\ominus} \rightarrow [(COD)_2Cu(ClO_4)]$$

$$\text{Anode:}\quad Cu + 2(COD)_2 - e^{\ominus} + [ClO_4]^{\ominus} \rightarrow [(COD)_2Cu(ClO_4)].$$

The preparation of bis-(cyclooctadiene-1, 5)-copper^{I} perchlorate is referred to in Table 8, Ex. 4 (p. 665).

Metal carbonyls are obtained in 20-65% yield by electrolysis of metal salts at carbon monoxide pressures of 33-190 atm [57]. The following conditions obtain when tetrabutylammonium bromide is used as electrolyte: 3-150 V, 1-2 A/dm^2 (anode) and 0.1-1 A/dm^2 (cathode).

Nickel complexes, e.g., nickel(II) acetylacetonate, can be reduced to the Ni(0) stage in the presence of suitable complexing agents without depositing metal [52]. Thus, electrolysis of nickel(II) acetylacetonate and tetraalkylammonium halide in THF, Diglyme, dimethyl ether, or pyridine in the presence of the appropriate olefin yields bis-[cyclooctadiene-1, 5]-nickel(0), cyclooctatetraene-nickel(0); and all-trans-cyclodecatriene-1, 5, 9-nickel(0).

Aluminum tris-acetylacetonate is formed as by-product when anodes of aluminum are used; no diaphragm is necessary. The coordination compounds of nickel are obtained in good yield:

$$\mathrm{Ni^{II}(acac)_2} \xrightarrow{+2e^{\ominus}/\text{Al-anode}} \tfrac{2}{3}\,\mathrm{Al(acac)_3} + \left\{\mathrm{Ni^0}\right\}$$

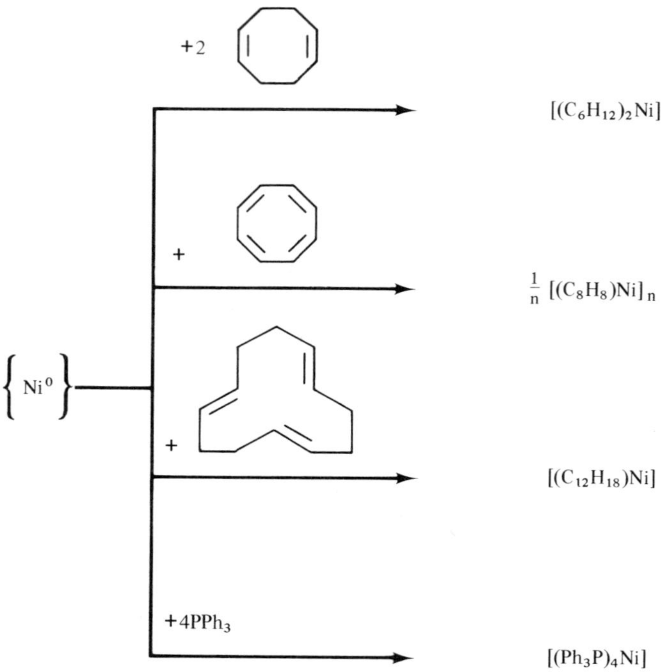

where acac is acetylacetonate. The investigations of Wilke and his co-workers [54] led to the discovery of these compounds and of their utility in oligomerizing diolefins and in codimerizing 1, 3-diolefins with alkenes and alkynes.

Solutions of similar zero-valent nickel complexes can be prepared electrolytically without the need for isolating a definite complex [52]. For example, electrolysis of nickel(II) acetylacetonate in organic solvents produces catalyst solutions which convert butadiene into a mixture of cis-trans isomers of 1, 5, 9-cyclododecatriene. The catalyst solutions prepared in the presence of phosphines (Ni:P = 1:1), just as those made by purely chemical means, convert butadiene into a mixture of 1, 5-cyclooctadiene and 4-vinyl-1-cyclohexene:

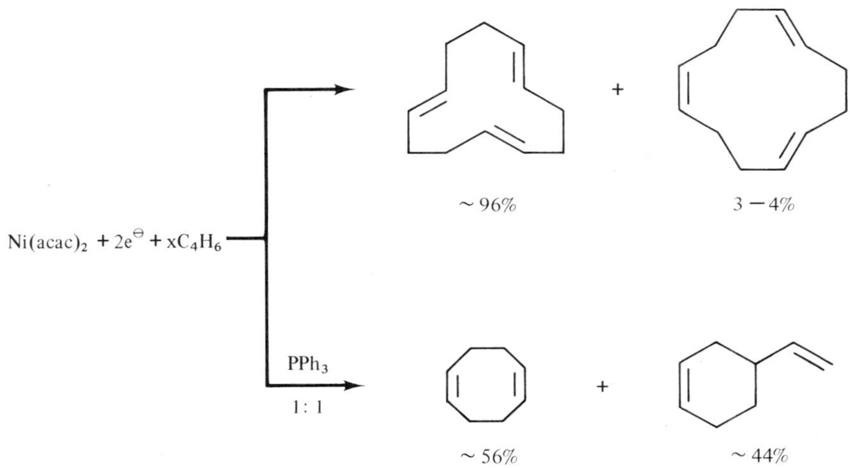

where acac is acetylacetonate.

The preparation of bis-cyclooctadiene-1, 5-nickel(0) and of a nickel complex for the trimerization of butadiene is outlined in Table 8, Ex. 5 and 6, respectively (p. 665).

It is reported that, if butadiene is fed into the electrolyte during electrolysis of tetrakis-pyridine-nickel(II) perchlorate or nickel(II) chloride in ethanol or dimethoxyethane, a complex is formed which yields all-trans-hexadecatetraene upon pyrolysis or hydrolysis with sulfuric acid; the nickel complex which is formed in the electrolysis of bis-(triphenylphosphine) nickel(II) chloride and butadiene is claimed to form 1, 3, 7-octatriene upon decomposition [55].

If a solution of cobalt(II) acetylacetonate or of another cobalt salt soluble, e.g., in THF (say, cobalt bis-triethylaluminate, $Co[(C_2H_5O)_4Al]_2$) is electrolyzed and butadiene is passed into the solution, a mixture of 5-methyl-1, 3, 6-heptatriene and n-octa-1, 3, 7-triene is produced [53]:

$$xC_4H_6 \xrightarrow[\text{Electrolysis}]{Co(acac)_2 + 2e^{\ominus} + Al\ (\text{Anode})} CH_2{=}CH\underset{\displaystyle CH_3}{\underset{|}{C}}H{-}CH{=}CH{-}CH{=}CH_2$$

$$+\ CH_2{=}CH{-}CH{=}CH{-}CH_2{-}CH_2{-}CH{=}CH_2$$

where acac is acetylacetonate.

Other nickel- or cobalt-containing catalysts capable of polymerizing alkynes are obtained by electrochemical reduction of metal bromides in acetonitrile or THF [56].

In this connection it should be mentioned that transition metal carbonyls have been prepared by electrolysis of metal acetylacetonates in pyridine in the presence of carbon monoxide [57]. Similar results are obtained with butadiene and catalysts formed by the electrolysis of cobalt(III) acetylacetonate in MeOH or DMF.

Electrolysis of cobalt(II) acetylacetonate in pyridine or THF in the presence of cyclooctadiene-1, 5 yields π-cyclooctenylcobalt-π-cyclooctadiene-1, 5. In the absence of a proton donor, the hydrogen necessary for the formation of the complex comes from one of the reactants. Alternatively, the electrolysis may be conducted in the presence of a proton donor, such as alcohol. Since the polarographic wave height of cobalt is increased ca. 50% upon the addition of 2:1 cyclooctadiene-1, 5, the following mechanism may be proposed:

$$Co^{2\oplus} + 2e^{\ominus} \longrightarrow Co^0$$

$$Co^0 + 2COD \longrightarrow \{Co(COD)_2\}$$

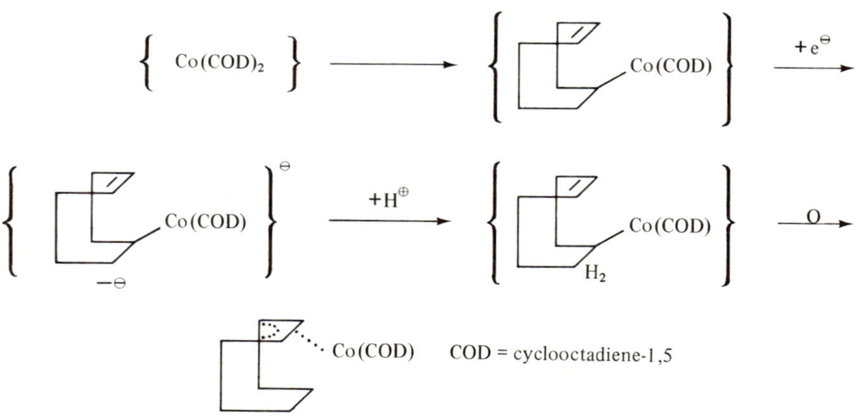

Titanium(IV) compounds and the complexing agent cyclooctatraene form the following Ti-COT complexes upon electrochemical reduction [58]:

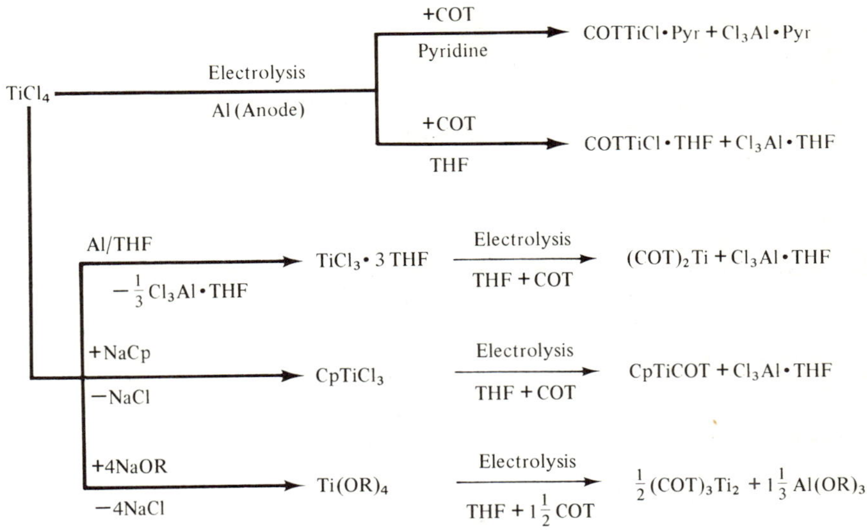

COT = cyclooctatetraene
Cp = cyclopentadienyl
R = alkyl

b. Reduction of Organic Complexes of Transition Metals. Very little preparative use has been made so far of the reduction of organic complexes of transition metals to complexes in which the central metal has a lower valence state. Polarographic data have been obtained for a variety of systems [59] but very often little is known about the character of the products formed by the reduction.

Polarographic reduction of complexes of transition metals often yields information of value in deciding structural questions concerning the π-bonding of ligands to metals and the distribution of the charges (in the complex) between the metal and the ligands. Further, correlations can be made between the oxidation potential of anions of organometallics and their nucleophilicity [60-72].

Electrochemical reduction of coordination compounds which are not organometallics has been used preparatively. Thus, cathodic reduction of cations in which there is complexation with 2, 2′-bipyridyl is a generally useful method for the preparation of bipyridyl complexes in which the metal has a lower valence state, including zero [73, 74]:

$$[M(Bipyr)m]^{n\oplus} + ne^{\ominus} \rightarrow [M(Bipyr)m].$$

An analogous reduction to zero-valent metal occurs when tert-phosphines are the complexing agents, as in the reduction of tris-[triphenylphosphine] rhodium(I) to tetrakis-[triphenylphosphine]rhodium(0) [75]:

$$[(R_3P)_3Rh]^{\oplus} \xrightarrow{e^{\ominus},\, R_3P} [(Ph_3P)_4Rh].$$

See Table 8, Ex. 7 (p. 665).

Even when complex formation does not occur between a given potential ligand and a metal in its higher oxidative state, it is possible to produce the complex having the metal in its lower oxidation step by electrolyzing in the presence of the ligand-to-be [52]. For example,

$$Ni^{II}(acac)_2 + 4Ph_3P \xrightarrow[-2acac^{\ominus}]{2e^{\ominus}} [(Ph_3P)_4Ni].$$

Example 8 of Table 8 (p. 665) summarizes the experimental conditions.

B. Organometallic Compounds by Anodic Oxidation of Systems Containing Carbanions

This method is of wide scope. It has developed from the original observation of Hein et al. [76, 77] that solutions of alkali metal alkyls in diethylzinc or trialkylaluminums are good conductors and yield alkyl radicals upon anodic oxidation. These radicals attack the metal anodes to form alkylmetallics. When metals are used which do not form stable alkylmetallics, the latter intermediates disproportionate or form alkyl dimers:

$$M - C_2H_5 \xrightarrow[-e^{\ominus}]{R_2Zn \text{ or } R_3Al} M^{\oplus} + \{\cdot C_2H_5\}$$

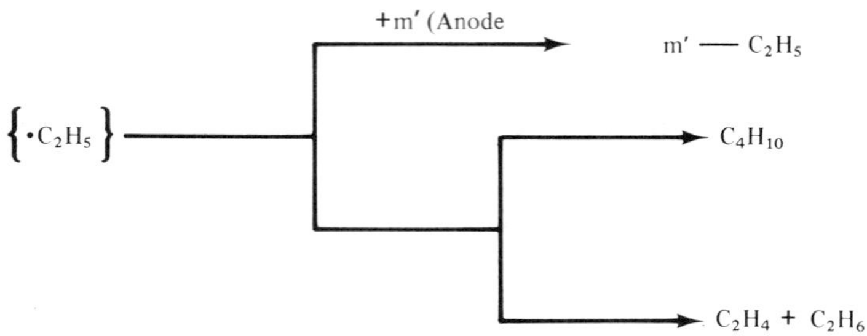

$M = Na, Li, (K), \frac{1}{3}Al, \frac{1}{2}Mg$

If the concomitant cathode process is included in the overall reaction scheme, then the net result is a trans-metallation affected by electrical energy. This type of reaction occurs spontaneously only in the opposite direction [78]:

$$M\text{-}C_2H_5 + m' \rightleftarrows m'\text{-}C_2H_5 + M.$$

These processes achieved technical and preparative interest only after simple means were found for converting the cathodically deposited metal (which serves as the agent for transferring alkyl groups) to a metal alkyl again. When aluminum was the transfer metal, this was accomplished as follows [79]:

$$Al + 1.5H_2 + 3C_nH_{2n} \rightarrow (C_nH_{2n+1})_3Al.$$

For the alkali metals this was achieved by complexation with trialkylaluminums or trialkylborons in a similar scheme:

$$Na + 0.5H_2 \rightarrow NaH$$

$$NaH + R_3Al \text{ (or } R_3B) \rightarrow Na[R_3AlH] \quad [80]$$

or

$$Na[R_3BH] \quad [81]$$

$$Na[R_3AlH] \text{ or } Na[R_3BH] + C_2H_{2n}$$

$$Na[R_4Al] \quad [80]$$

or

$$Na[R_4B] \quad [81].$$

The sodium-aluminum organic system actually corresponds exactly to Hein's systems which contained the complexes $Na[R_4Al]$ or $Na[R_3Zn]$ or $Na_2[R_4Zn]$.

In the multistage process the metal alkyl is made from metal olefin, hydrogen, and electrical energy. The direct reaction of metal, hydrogen, and olefin occurs only with aluminum but can be applied to other metals in the manner indicated if an electrolysis step is included [82].

A second available procedure is the direct alkylation within the cell of the cathodically deposited auxiliary metal by means of alkyl halides. These means are used when alkylmagnesium halides are the electrolytes; the overall process then is [83]:

$$Mg + 2m' \text{ (Anode)} + 2RX \rightarrow MgX_2 + 2\,Rm'.$$

1. Metal Alkyls from Complex Aluminum Alkyls

a. Electrolysis with Simultaneous Cathodic Aluminum Deposition. Larger alkyl groups also participate in metal alkyl formation by electrolysis. However, there are practical limitations because the conductivity of the organoaluminum electrolytes decreases as the length of the alkyl chain is increased. The following discussion is, therefore, limited to a consideration of the ethyl compounds, but in principle it includes the propyl and butyl derivatives; it applies to the methyl derivatives except that because of the high melting points of the pure compounds only suitable alkali metal-methylaluminum complexes may be used.

Organoaluminum electrolytes of the type [84-86] $MX \cdot n(C_2H_5)_3Al$, where M = Na, K, R_4N; X = C_2H_5, F, Cl; $n > 1$, preferably 2, deposit aluminum at the cathode even at high current densities (10-20 A/dm^2). Since alkyl radicals are formed at the anode, the electrolysis of these complexes really involves the decomposition of triethylaluminum which is itself a nonconductor. If the anode is aluminum, the electrolysis simply achieves a transfer of aluminum from the anode to the cathode [84, 85]. Good yields of tetraethyltin (along with a little hexaethylditin) and triethylantimony can be obtained from sodium bis(triethylaluminum) fluoride and anodes of tin and antimony, respectively [86, 87]. Especially with tin, but occasionally also with antimony, passivation phenomena occur. In this case vigorous gas evolution occurs and only 25% of the calculated amount of metal (tin) goes into solution. Satisfactory results are obtained by using freshly cast tin plates of ca. 450 g weight.

In Fig. 1 [88] are shown the elements which are anodically dissolved during electrolysis in organoaluminum electrolytes. The elements within the heavy-bordered area have been shown experimentally to be anodically dissolved; those that are underlined are redeposited on the cathode.

I		II		III	IV	V	VI	VII	VIII
Li		Be		B	C	N	O	F	Ne
Na		Mg		Al	Si	P	S	Cl	Ar
K	Cu	Ca	Zn	Ga	Ge	As	Se	Br	Kr
Rb	Ag	Sr	Cd	In	Sn	Sb	Te	I	Xe
Cs	Au	Ba	Hg	Tl	Pb	Bi	Po	At	Rn

Fig. 1. Elements dissolved anodically in organoaluminum electrolytes [88].

The electrolytic cells [86] used in this work were cylinders which contained two planar electrodes separated by ca. 2 cm. Enough free space is left at the bottom so that from time to time the aluminum could be scraped from the cathode by special means provided. The temperature was kept at ca. 100°. The metal alkyls formed remained dissolved in the electrolyte and were finally recovered by vacuum distillation. Small amounts of triethylaluminum, which were formed by thermal decomposition of the electrolyte, codistilled and were destroyed by careful hydrolysis.

The preparation of tetraethyltin is outlined in Table 8, Ex. 9 (p. 665).

If anodes of lead or mercury are used, the metal alkyls formed are partly destroyed by the aluminum deposited at the cathode; if the anodes are zinc, cadmium, bismuth, gallium, indium, or thallium, the metal alkyls are completely decomposed by the aluminum [86] (cf. Fig. 1). The lead or mercury is codeposited with aluminum in the cathode; the latter series of elements above is deposited in place of aluminum. In most cases [89] this difficulty can be circumvented by electrolyzing in vacuo: Then the metal alkyls distill out in good yield. Their concentration in the electrolyte remains so low that the reaction with aluminum metal becomes negligible. In principle the aluminum deposited in the cathode may be treated with hydrogen and ethylene, thus converted to triethylaluminum, and the latter used to fortify the electrolyte. There are practical difficulties in accomplishing this on a larger scale since even in excellent electrolyses the aluminum is contaminated by the other metal, which is a poison for the aluminum-to-triethylaluminum reaction [90].

Instead of electrolyzing in vacuo, it is possible to use a divided cell with diaphragm in order to prepare those metal alkyls which decompose by reaction with the cathodically deposited aluminum [91]. Even when these means are used, the aluminum is often contaminated (e.g., with ca. 18% of zinc or ca. 9% of cadmium) [86].

b. Electrolysis with Simultaneous Cathodic Alkali Metal Deposition. Sodium is deposited on the cathode when electrolytes such as $Na[R_4Al]$ or $Na[R_3AlOR']$ are used [92]. Since anodic oxidation of the aluminate anion gives rise to uncomplexed aluminum alkyl and the alkyl derivative of the metal used as anode, and since most of the metal alkyls react with sodium metal, it is necessary either to use a divided cell with diaphragm or to carry out the electrolysis in vacuo under conditions which permit the product metal alkyl to distill out.

c. Electrolysis with Cathodic Formation of Alkali Amalgams. When mercury cathodes are used [93-96] it is possible to develop significantly simpler procedures both on a laboratory and on a larger scale for preparing metal alkyls from electrolytes of pure alkali metal tetraethylaluminate or of mixtures of components. Alkali metal amalgams are then formed at the cathode. Sodium amalgam in the presence of triethylaluminum is rather unreactive toward the ethyl derivatives of magnesium [in the form of magnesium-bis-tetraethyl(aluminum)], mercury, aluminum (if there is less than 0.7 wt% sodium in the amalgam) [78], tin, lead, antimony, and bismuth; it is moderately inert to the alkyl derivatives of indium and thallium; it is very reactive toward zinc and cadmium alkyls. Potassium amalgam, however, decomposes uncomplexed triethylaluminum to form potassium tetraethylaluminate and aluminum. Complexed triethylaluminum in the form of potassium triethylaluminum fluoride or potassium bis-triethylaluminum fluoride does not react with potassium amalgam even at 170° [95]. (See Table 3.)

Alkali metal tetraalkylaluminates behave in electrolyses like solutions of alkali metal alkyls in trialkylaluminum. One mole of trialkylaluminum is liberated for each gram-atom of alkali metal and each corresponding equivalent of alkyl group. When these complexes are used exclusively as electrolytes, the desired metal alkyl is always formed in a mixture with n moles of trialkylaluminum (in which n is the valence of the metal forming its alkyl). In the case of the ethyl compounds, two immiscible phases are formed, consisting of the remaining electrolyte and the mixture of metal alkyls. Magnesium forms magnesium-bis-tetraethylaluminate [94].

TABLE 3

Preparation of Ethylmetallic Derivatives by Electrolysis of 1:1 Sodium and Potassium Tetraethylaluminate at Sodium Amalgam [94]

Anode metal	Na in amalgam [a]	Anode loss [a]	Yield of ethylmetallic [a]
Mg	85.5	84	65
Ca	11.9	43	0
Zn	1.4	92	0
Cd	0.9	39	0
Hg	96	[b]	95.5
Al	98.5	90	88
In	55	98	46
Tl	23.5	100	15
Sn	97.5	8	0
Pb	99	100	97.5 [c]
Sb	97.5	13	0
Bi	98.5	75	56

[a] In percent equivalents.

[b] Not determined.

[c] After steam distillation.

Triethylaluminum can also be prepared this way electrochemically; the yield of product is 1.33 times the equivalent of the starting material:

$$3(C_2H_5)_3Al + 3Na + (3/2)H_2 + 3C_2H_4 \rightarrow 3Na[(C_2H_5)_4Al]$$

$$3Na[(C_2H_5)_4Al] + Al_{anode} + Hg_{cath} \rightarrow 3NaHg_x + 4(C_2H_5)_3Al.$$

If a procedure analogous to that used for the preparation of triethylaluminum [95, 97] is applied to the preparation of lead or bismuth alkyls — and in these cases the products are in the lower layer — mixtures of the desired metal alkyls along with triethylaluminum are obtained in a ratio of 1:4 or 1:3. The need for these separations can be avoided by using mixtures of alkali metal tetraalkylaluminates and alkali metal fluoride complexes as electrolytes [95, 96, 98]. A 1:1 mixture of sodium (or potassium) tetraalkylaluminate and triethylaluminum fluoride is converted by the passage of 1 faraday of current into sodium-bis-triethylaluminum fluoride. At the same time there is formed 1 equiv each of alkali metal (as amalgam) and metal alkyl. The liberated triethylaluminum forms the 1:2 alkali metal fluoride complex. When the sodium compound and a lead anode are used, the mixture obtained finally has 23% tetraethyllead and 77% sodium-bis-(triethylaluminum) fluoride. Most of the tetraethyllead settles as a lower layer; the rest can be separated from the electrolysis by vacuum distillation. It is impossible to achieve this without liberating a little triethylaluminum, which is mixed with the tetraethyllead, so that a separation problem still exists, albeit to a smaller degree. Only when potassium is used as the cation is complexation of the liberated triethylaluminum assured [95, 96], as, for example,

$$R_3Al + K[R_3AlF] \rightarrow K[R_6Al_2F].$$

An advantageous electrolyte composition is

$$10K[(C_2H_5)_4Al] + 3K[(C_2H_5)_3AlF].$$

Passage of 3 faradays of current through a lead anode forms the mixture

$$7K[(C_2H_5)_4Al] + 3K[(C_2H_5)_6Al_2F] + \frac{3}{4}(C_2H_5)_4Pb.$$

The use of pure potassium-bis-(triethylaluminum) fluoride is not recommended because of its high melting point (135°); the electrolyte recommended above melts finally at 85°. Of the tetraethyllead formed, 85% separates at 100° in a lower layer and the remainder may be removed in vacuo at 100° without significant contamination by triethylaluminum.

A laboratory procedure which results in ca. 100% yield of tetraethyllead based on the lead that has dissolved and ca. 95% based on current input has been reported [99]. Triethylbismuth and diethylmercury have been prepared similarly. Special cells suitable for pilot plant and larger scale have been described [95, 96]. Cells which have as cathodes rotating disks which are continuously wetted by mercury or amalgam have been constructed with only 2-4 mm spacing between electrodes. Current densities of 35 A/dm^2 at 4 V have been achieved.

Experimental cells have been tested for the preparation of diethylmercury. In these cells alternating anode and cathode disks, mutually electrically insulated, are spaced 2-4 mm apart, rotate on a common axis, and dip into separate circulating mercury streams.

The very elegant synthesis of metal alkyls which uses recycling of the auxiliary metal sodium can be applied only to the preparation of compounds in which the alkyl group has two or more carbon atoms. This restriction is imposed because sodium hydride in the presence of triethylaluminum adds only to ethylene and higher olefins. Since the methyl group in organometallics cannot be formed from an olefin and hydrogen, it must be supplied by a methyl halide, particularly, for practical reasons, by methyl chloride. As a consequence, the liberation of chlorine consumes 1 equiv of the auxiliary metal for each equivalent of the metal alkyl formed. Nevertheless, the electrochemical synthesis according to

$$Na + MeCl + pb \rightarrow NaCl + pbMe$$

has the advantage over the conventional synthesis according to

$$4PbNa + 4MeCl \rightarrow 4NaCl + PbMe_4 + 3Pb$$

that in the former case practically all the lead can be converted to the metal alkyl [100, 101].

A simple adaptation of the ethyl metal process to a methyl metal process is, however, not possible for additional reasons. The potential electrolytes, alkali methylaluminum complexes, melt without exception either above 150° or with decomposition, so that an electrolysis producing the thermally sensitive metal alkyls in a fused electrolyte is impossible. While it is true that the melting point of sodium tetramethylaluminate (238°) can be greatly lowered by the addition of sodium tetraethylaluminate — a 1:1 mixture forms a eutectic melting at 83° — nevertheless, this mixture cannot be used for the preparation of tetramethyllead or mixed ethylmethylleads because electrolysis yields exclusively tetraethyllead. No methyllead compounds are formed. As a consequence, only solutions of alkali metal tetramethylaluminates in THF or Diglyme are suitable for the synthesis of tetramethyllead [100-102]:

$$NaAlMe_4 + xHg\ (Cathode) + Pb\ (Anode) \rightarrow Na(Hg)_x + pbMe + AlMe_3.$$

The trimethylaluminum which is liberated from its complex according to the above equation forms a high-boiling etherate. When Diglyme is the solvent, both it and the trimethylaluminum etherate boil substantially higher than tetramethyllead. The latter can, therefore, easily be distilled out at 50-80°/100 Torr. The sodium amalgam formed at the cathode can easily be reconverted to sodium tetramethylaluminate by means of MeCl and trimethylaluminum in Diglyme or THF [103]:

$$2Na(Hg)_x + CH_3Cl + (CH_3)_3Al \xrightarrow{THF} Na[(CH_3)_4Al] + NaCl + 2xHg.$$

2. Metal Alkyls by Electrolysis of Organoboron Complexes

Many proposals have been made of means for separating the mixtures of triethylaluminum with n equiv of another metal alkyl which arise during electrolysis of alkali metal tetraethylaluminates (p. 656). One of these solutions is to take advantage of the greater spread in boiling points between metal alkyls and triethylboron [104, 105]. Mixtures of sodium tetraethylaluminate and the analogous boron compound are used as the electrolyte. The former serves as an "inert" electrolyte, the latter is decomposed electrochemically. Triethylboron can be reconstituted from sodium hydride and ethylene just as in the case of aluminum compounds (p. 643). Pure boron compounds alone may be used in the preparation of metal alkyls. The procedure then, as in the case of the alkylaluminum system, is to operate in divided cells with simultaneous deposition of sodium [106-108] or to use a mercury cathode [109-110]. In the latter case use of aqueous solutions of sodium tetraethylborate becomes possible.

A cell and procedure for preparing triethylbismuth via sodium tetraethylborate are described in the literature [109]. Tetraethyllead (91%), tetramethyllead (78%), and diethylmercury (81%) were obtained similarly.

3. Metal Alkyls by Electrolysis of Organomagnesium Compounds

These methods date back to the observations of Kondyrev [111] who noted that during electrolysis of alkylmagnesium halides in ether anodes of Zn, Al, and Mg dissolved while Pt, Cu, Fe, Pb (?), Sn (?), Co, Ni, and Ag were unattacked [111, 112]. It is surprising that lead was reported to be inert because a present industrial process produces tetraethyllead just by this means. An occasional passivity of tin anodes has been noted also with organoaluminum and organozinc electrolytes. The quantities of metal which dissolve are in accord with Faraday's law [111]. While the aluminum alkyls remain largely intact, the alkylzinc and alkylcadmium compounds are decomposed at the cathode and deposit their respective metals [112].

The low conductivity of the Grignard solutions in ether militates against the preparative use of these electrolyses, particularly on a large scale [113-115]. Addition of alkyl halides raises the conductivity substantially and also reconverts the magnesium which has deposited at the cathode into alkylmagnesium halide [116]:

$$2RMgX \rightleftharpoons RMg^{\oplus} + [RMgX_2]^{\ominus} \quad [116]$$

$$[RMgX_2]^{\ominus} + m' \xrightarrow[-RMgX]{-e^{\ominus}} m' - X; \quad m' - X + mgR \longrightarrow m'R + mg - X$$

$$RMg^{\oplus} \xrightarrow{+e^{\ominus}} \{RMg\}$$

$$\downarrow$$

$$0.5Mg + 0.5R_2Mg$$

$$0.5Mg + 0.5RX \longrightarrow 0.5RMgX$$

Even after the addition of 0.9:1 of ethyl chloride to a 1.27 N solution of ethylmagnesium in the dibutyl ether of diethylene glycol, 14 V is necessary for a current density of 0.3 A/dm^2 [116]. An electrolyte consisting of 0.95 mole of methylmagnesium chloride, 2 moles of benzene, 0.77 mole of THF, and 1 mole of $C_6H_{13}OCH_2CH_2OC_2H_5$ supports a current density of 2.8 A/dm^2 at 25.5 V after the addition of 0.3 mole of methyl chloride [117-118]. Yields based on current are said to be 156% for tetramethyllead. This indicates that there is a parallel nonelectrochemical route to the product:

$$2RMgX + 2RX + Pb \rightarrow R_4Pb + 2MgX_2 \quad [119].$$

The yields are ca. 98% based on the magnesium compound which enters reaction. Further details of the procedure and cell design are given in the patent literature [120].

Mixtures of tetraethyl- and tetramethyllead are obtained if mixtures of methyl and ethyl chloride (0.3-0.9 mole of alkyl halide) are fed into the electrolyte of methylmagnesium chloride in the course of the electrolysis. Presumably, mixed ethylmethylleads are also formed [117].

Preferred solvents are reported to be diethyl ethers of di- or triethylene glycols to which are added THF and benzene [121, 122]. Periodic current reversal is also proposed [123]. See also [124-128].

4. "Anodic" formation of Metal Alkyls by Electrolysis of Haloaluminum Compounds

In most of the presently known electrolyses of aluminum halides in organic halides using aluminum electrodes only a portion of the products arises by electrochemical means; the remainder, and it is usually the larger portion, results from a direct reaction between the electrode material and the organic halide [129]. As a consequence, the electrolysis of aluminum triiodide in methyl iodide [130] produces mixtures of methylaluminum diiodide and dimethylaluminum iodide [129]:

$$AlI_3 + 3CH_3I + 2Al \xrightarrow{\text{Electrochemical}^3} 3CH_3AlI_2$$

$$3CH_3I + 2Al \xrightarrow[\text{Chemical}]{} (CH_3)_2AlI + CH_3AlI_2 .$$

The ratio of electrochemical to chemical reaction is ca. 1:3 to 1:7.5 and depends, of course, on the electrode surface areas and the time of the contact with the electrolyte [129]. Similar conditions obtain in the electrochemical synthesis [131] of bis-dichloroaluminum methane in dichloromethane. Because of the additional purely chemical reaction [132]

$$2Al + 1.5CH_2Cl_2 \rightarrow 0.5Cl_2AlCH_2AlCl_2 + \frac{1}{n} - [CH_2\underset{\displaystyle Cl}{\underset{|}{Al}}-]_n$$

3.5 to 4 times as much aluminum is lost from the anode than is accounted for by the quantity of current passed.

In any event these examples reveal possibilities of supplementing chemical by electrochemical reactions.

5. Anodic Formation of Transition Metal Coordination Complexes

It has been disclosed that bis-cyclopentadienyl derivatives of metals can be formed by electrolyzing at an anode of the appropriate metal solutions of alkali metal cyclopentadienyls in cyclic or polyfunctional ethers or of other metallo-cyclopentadienyls in DMF [133]. In the presence of cyclopentadiene — or, at a sufficiently high temperature, its dimer — the electrolyte is said to be constantly regenerated:

$$\text{Anode:}\quad Cp^{\ominus} - e^{\ominus} + m \rightarrow mCp$$

$$\text{Cathode:}\quad m'^{\oplus} + e^{\ominus} + CpH \rightarrow m'Cp + \tfrac{1}{2}H_2.$$

The literature describes the following examples:

m' = Na; Cp = methylcyclopentadienyl [133],

m = 0.5Mn; solvent, THF-Diglyme 4:1; curr. eff., 25%;

m' = Tl; Cp = cyclopentadienyl [135],

m = 0.5Fe; solvent, DMF; curr. eff., 91%.

If a carbon monoxide atmosphere is maintained during the electrolysis, cyclopentadienyl-manganese tricarbonyl can be prepared. It is an important technical product used as a gasoline additive. The reported synthesis claims that an electrolyte can be used consisting of a manganese salt, a metal carbonyl, and cyclopentadiene dimer in DMF or a dialkyl ether of diethylene glycol [134]. Since it is necessary to work with CO under pressure, considerable technical difficulties must be involved. Yields of 7% based on current are claimed.

Electrolysis of chlorine-containing organoaluminum compounds at appropriate transition metal anodes (e.g., titanium) leads to formation of Ziegler-type catalysts capable of converting olefins to polyolefins. A patent discloses [131] as examples of these electrolytes $Cl_2AlCH_2AlCl_2$ (p. 649) or the higher molecular weight

$$Cl_2AlCH_2[-\underset{\displaystyle Cl}{\underset{|}{Al}}CH_2]_nAlCl_2$$

III. ELECTROCHEMICAL TRANSFORMATIONS OF ORGANOMETALLIC COMPOUNDS

The second half of this chapter deals with those electrochemical transformations which lead to products which are themselves <u>not</u> organometallic compounds.

Since it can generally be assumed that the metal-carbon bond is polarized in the sense $\overset{\delta+}{M}-\overset{\delta-}{C}$, it is to be expected that organometallics will be electrochemically cleaved, the metal depositing on the cathode and the organic moiety being discharged at the anode. The radicals formed initially at the anode are

$$M\text{-}R \xrightarrow[\text{Anode}]{-e^{\ominus}} M^{\oplus} + || \cdot R$$

$$M\text{-}R \xrightarrow[\text{Cathode}]{e^{\ominus}} M + R^{\ominus}$$

generally unstable and achieve stabilization by follow-up reactions with themselves or solvent molecules or other suitable partners. Of course, before organometallic compounds participate in electrochemical reactions there must be electrical conductivity. The volatile metal alkyls of groups IIa, IIb, IIIa, and IVa of the periodic system have specific conductivities between 10^{-7} and 10^{-13} Ω^{-1} cm^{-1} (Table 4).

The metal-carbon bond in organometallics is polarized by solvents of electron-donating character [136]. The charge density is thereby increased on the organic moiety to which a second molecule of metal alkyl can now add. The ease with which this occurs depends on the electron affinity of the metal. Further, the higher the dielectric constant of the solvent, the greater is the dissociation of the ion pairs into ions [136]:

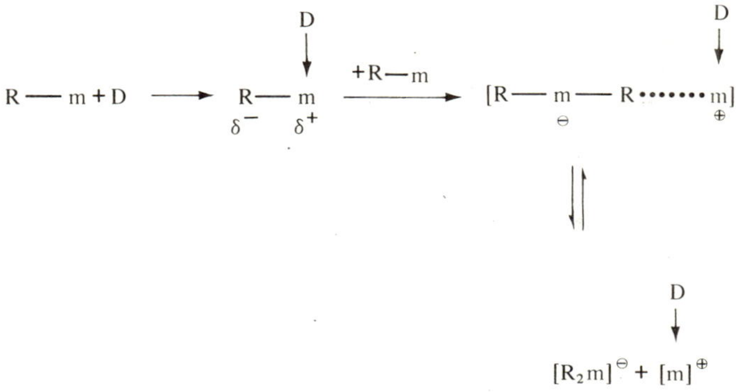

TABLE 4

Specific Conductivities of Certain $(C_2H_5)_nM$: $(\Omega\ cm)^{-1}$ at 20°

M	Spec. conduc.	Ref.
Al	$\sim 10^{-8}$	[137, 138]
Be	10^{-7}	[136]
Zn	3×10^{-11}	[136]
Cd	2×10^{-12}	[136]
Hg	2×10^{-13}	[136]

If Rm, the metal alkyl, is only a weak electron acceptor, the formation of the complex "ate" anion may not occur:

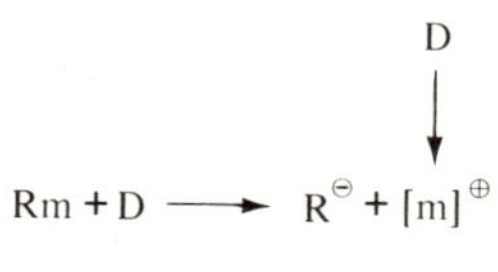

Metal alkyls are 1:1 electrolytes in aprotic solvents (DK: ϵ ca. 2 to 10) regardless of the valence of the metal [140]. Dissociations such as

$$MR_n \rightleftarrows M^{n\oplus} + nR^{\ominus} \quad \text{or} \quad MR_n \rightleftarrows MR^{(n-1)\oplus} + (n-1)R^{\ominus}$$

do not seem to occur. Analogous situations occur with cyclopentadienyl derivatives [141] of the main group metals as well as countless transition metals. At comparable concentrations the conductivity of alkyl metals in solution increases:

(1) with decreasing electron affinity of the metal in the series $R_2Hg < R_2Cd < R_2Zn < R_2Be < R_2Mg$ [136, 139, 142];

(2) with decreasing electron-donating ability of the solvent in the series benzene < dioxane < triethylamine [136];

(3) with the magnitude of the dielectric constant in the series dioxane < triethylamine < diethyl ether < THF [136].

In this way specific conductivities as high as $10^{-5}\ \Omega^{-1}\ cm^{-1}$ can be achieved [143]. Significantly larger enhancements of the conductivity can be attained if the Lewis acid strength of the metal alkyls is sufficiently high to permit the metal alkyls to add alkali metal or tetraalkylammonium alkyls, hydrides, or halides and thus form ionizable complexes [88, 144-152].

A. Cathodic Deposition of Metals

Cathodic deposition of metals from organometallic compounds is especially interesting in the case of very electropositive metals which cannot be deposited from aqueous solutions. If it is so arranged that the organic moieties discharged at the anode dissolve the same metal that is deposited at the cathode, then there becomes available an electrochemical method for metal refining. This is especially true if the organic moieties react selectively at the anode, do not attack contaminant metallic constituents, and simply leave them as an anodic slime. An almost ideal example of this type of system is provided by the refining of aluminum when triethylaluminum electrolytes are used.

1. Metal Deposition for "ate" Complexes [153]

a. Deposition of Aluminum. Triethylaluminum is a volatile metal alkyl which can easily be obtained pure by distillation. This removes the danger that the cathodically deposited metal will be contaminated by derivatives of other metals in the electrolyte.

Trialkylaluminums are practically nonconductors. Their specific conductivity, $10^{-8}\ \Omega^{-1}\ cm^{-1}$, is approximately the same as that of the purest water. Intrinsic dissociation in ions, as in the case of water, is therefore very slight [154]:

Trialkylaluminums are Lewis acids and form stable complex compounds with

Lewis bases. Although the addition of ethers and tert-amines (donors) to trialkylaluminums gives complexes 1000 times as conductive as the starting materials [143, 154-160], the conductivities are still not high enough for electrolytic purposes. The conductivities are raised substantially — to as high as $5 \times 10^{-2}\ \Omega^{-1}\ cm^{-1}$ — if the trialkylaluminums are complexed with sodium or tetraalkylammonium salts [161], e.g., with halides, alkali metal hydrides, alkanoates, and alkyls. Complexes of the type

$$MX \cdot n(C_2H_5)_3Al \quad \text{where} \quad \begin{aligned} M &= Na,\ K,\ R_4N \\ X &= C_2H_5,\ H,\ F,\ Cl,\ Br,\ I \\ n &= 1 \text{ or } 2 \end{aligned}$$

are then formed. If the electrolyte contains triethylaluminum in excess of the 1:1 composition, then except at very high current densities only aluminum is deposited [162, 163, 85]; n = 2 is preferred. The relatively high conductivity ($5.5 \times 10^{-2}\ \Omega^{-1}\ cm^{-1}$ at 100°) of $NaF \cdot 2(C_2H_5)_3Al$ is explained by its special ability to transport charge via the molecular chains formed through association. These chains supposedly align themselves in the electrical field in such a manner that the negative end is turned toward the anode and reacts with the positive aluminum ions. The aluminum then remains in place, the positive charge distributes itself over the chain, and at the end of the chain, near the cathode, a positively charged aluminum ion is discharged [164]. Since it is desired that the cathodically deposited aluminum not be contaminated with sludge that can wander from the region of the anode, the latter is purposely confined within a cotton bag which is sufficiently stable in the electrolytes even at 100° [85, 149].

The anodic reaction is undoubtedly the discharge of alkylaluminate ions. The ethyl radicals which are produced can react with the anode. If the anode contains metals which cannot form stable ethyl derivatives, these metals are not dissolved and go with the anode sludge. This is true of almost all transition metals. The anode sludge contains also other metals such as Mg, Ca, Zn, Sn, Pb, and, surprisingly, in view of its close relationship to aluminum, Ga [149, 165]. The aluminum deposited at the cathode contains Fe, Si, Mg, and Cu; if sodium-containing electrolytes have been used, the Al also contains 10-25 ppm of Na.

In electrolytes which are poor in triethylaluminum ($1 < m < 2$), current densities of only 0.5 A/dm^2 maximum can be attained [166], otherwise, alkali metal will also deposit. The 1:1 complexes, e.g., $Na[R_4Al]$, deposit only the alkali metal (p. 656). The sodium content of the aluminum can be lowered to ca. 0.5 ppm, if electrolytes based on tetraalkylammonium salts are used [149, 167]. Furthermore, ammonium groups with bulky substituents (e.g., benzyl) have the advantage of decreasing the flammability of the electrolyte [149]. To avoid completely the possibility that the deposited aluminum will be contaminated with anodic slime, the electrolysis can be carried out with hydride electrolytes and silver or copper anodes

$$NaH \cdot 2(C_2H_5)_3Al$$

or

$$NaF \cdot 2(C_2H_5)_3Al \cdot n(C_2H_5)_2AlH \quad \text{where} \quad n = 1,\ 2,\ 3.$$

In these systems hydrogen discharge occurs largely or exclusively at the anode. The by-product hydrogen can be used in a separate purely chemical process for converting crude aluminum to diethylaluminum hydride which can be recycled to the electrolyte

$$Al_{(crude)} + 2(C_2H_5)_3Al + \tfrac{3}{2}H_2 \rightarrow 3(C_2H_5)_2AlH. \tag{1}$$

The overall process for refining of aluminum is then a combination of (1) above and (2) below

$$NaF \cdot 2(C_2H_5)_3Al \cdot 3(C_2H_5)_2AlH \xrightarrow[\text{Cu or Ag anode}]{\text{Electrolysis}}$$

$$Al_{\binom{\text{very}}{\text{pure}}} + \tfrac{3}{2}H_2 + NaF \cdot 2(C_2H_5)_3Al + 2(C_2H_5)_3Al. \tag{2}$$

The triethylaluminum which is liberated by the electrolysis in uncomplexed form is an upper immiscible phase in the electrolyte.

The aluminum deposited on the cathode is normally dendritic. A cell has been described [168] in which the aluminum is compressed by rotating rollers.

The form in which the aluminum is deposited from organoaluminum depends on several factors. Aluminum from $NaF \cdot 2(C_2H_5)_3Al$ is more coarsely crystalline than that from ammonium complexes. Crystal size decreases with decreasing temperature of the electrolyte bath and with increasing current density [144]. Addition of toluene causes the formation of more compact and more adherent aluminum deposits [85]. A lower triethylaluminum content has a similar effect [166].

b. Deposition of Gallium, Indium, and Thallium. Extremely pure gallium, indium, or thallium can be deposited by electrolyses of organoaluminum electrolytes at anodes of the appropriate metal [144, 166]. The following complexes were investigated:

$$R_4NX \cdot 2(C_2H_5)_3Al$$

$$R = CH_3,\ C_2H_5$$

$$X = Cl,\ F.$$

For the refining process the fluoro complexes are superior to the chloro. The steady-state concentration of indium in the electrolyte is 0.01 wt% maximum [170]. Cadmium and zinc behave similarly. Organoindium or gallium compounds can be used as electrolytes in place of organoaluminum compounds [171].

c. Deposition of Beryllium. Diethylberyllium forms complexes very similar to those of triethylaluminum [171-174]. For example, considerable refining of beryllium occurs in the electrolysis of $KF\cdot2(C_2H_5)_2Be$ with a beryllium anode [175-177]. The cathodically deposited beryllium contains ca. 20-30% of Be_2C and alkali metal [177, 178]. Diethylberyllium complexes with sodium cyanide or tetraethylammonium chloride decompose easily and are, therefore, not suitable as electrolytes.

d. Deposition of Zinc and Cadmium. The electrolyte $(CH_3)_4NF\cdot2(C_2H_5)_2Zn\cdot2$ xylene, which is liquid at 20°, is suitable for the electrorefining of zinc [149]. Zinc refining should, of course, also be possible using organoaluminum electrolytes. That is true also for the electrodeposition of cadmium if an anode of this metal is employed. Organocadmium electrolytes based on the methyl derivative can also be used [149]. Zinc can be deposited from ether solutions of diethylzinc [179].

e. Concerning the Mechanism of Metal Deposition from "ate" Complexes. In the example discussed above the metal in question was cathodically deposited from its organometallic anionic "ate" complex. Two fundamentally different mechanisms must be considered for the cathodic deposition of metal from anions $[R_nM\text{-}X\text{-}MR_n]^-$:

(a) First a discharge of the cation

$$M'^{\oplus} + e^{\ominus} \rightarrow M'$$

followed by the deposition of M by an associative-displacement reaction involving M' and the uncomplexed organometallic which is easily formed by thermal dissociation of the complex:

$$[R_nM\text{-}X\text{-}MR_n]^{\ominus} \rightleftarrows [R_nMX]^{\ominus} + R_nM$$

$$nM' + (n + 1)R_nM \rightarrow nM'[R_{n+1}M] + M.$$

This associative displacement can be realized chemically and is a possible preparative method for "ate" complexes. The considerably more stable anion $[R_nMX]^-$ is generally unable up to 100° to dissociate into uncomplexed metal alkyl. In agreement with this statement is the fact that electrolysis of a complex such as

$$M'[R_{n+1}M],$$

e.g., sodium tetraalkylaluminate, which has been very thoroughly studied, yields sodium and not aluminum.

To be sure, there is no direct proof for mechanism (a). The fact that in the electrolysis of $NaF \cdot 2(C_2H_5)_3Al$ or $KF \cdot 2(C_2H_5)_3Al$ or $KF \cdot 2(C_2H_5)_2Be$ at a mercury cathode only alkali metal amalgam formation and no aluminum or beryllium deposition is observed is no support for the above argument because in the latter cases the Na^+/Na and K^+/K potentials become more positive by 0.8 V and ca. 1.1 V, respectively, corresponding to the energy of amalgam formation [78]. Another experimental fact which is in agreement with this view is that $NaF \cdot 2(C_2H_5)_3Al$ and its potassium analog do react with alkali metals to form aluminum but do not react with the corresponding amalgams containing 0.1-0.05% of alkali metal [95].

(b) Another possible mechanism ascribes the metal deposition to a prior electrolytic dissociation into metal cations [85, 166]:

$$[R_nM\text{---}X\text{---}MR_n]^{\ominus} \rightleftharpoons [R_nMX]^{\ominus} + R_nM$$

$$2R_nM \rightleftharpoons [R_{n-1}M]^{\oplus} + [R_{n+1}M]^{\ominus}$$

$$[R_{n-1}M]^{\oplus} \xrightarrow{+e^{\ominus}} \left\{R_{n-1}M\cdot\right\} \longrightarrow \frac{n-1}{n} R_nM + \frac{1}{n} M$$

The formation of organometallic cations by electrolytic dissociation and the formation of organometallic radicals are known in certain instances (p. 648).

Concerning the mechanism of the anodic reaction see p. 658.

2. Deposition of Alkali Metals

When organometallic complexes like

$M'[R_{n+1}M]$

are used as electrolytes, sodium can be deposited cathodically when $M' = Na$ and $M = Al$. Nothing is known thus far about the behavior of other metal alkyls R_nM.

a. Deposition of Sodium.

Sodium is deposited cathodically by electrolysis of fused electrolytes (free of uncomplexed trialkylaluminums) of sodium tetraethylaluminate, or mixtures of potassium and sodium tetraethylaluminates containing a maximum of 85% of the potassium complex, or mixtures of sodium tetramethyl- and tetraethylaluminates in a mole ratio of 1:1 [84, 180-183]. If the temperature is maintained above the melting point of sodium, the metal is obtained in liquid coherent form. Since triethylaluminum reacts with sodium triethylaluminum alkoxide [182],

$$Na[(C_2H_5)_3AlOR] + (C_2H_5)_3Al \rightarrow Na[(C_2H_5)_4Al] + (C_2H_5)_2AlOR,$$

and since diethylaluminum alkoxide does not react with molten sodium — with separation of aluminum — before a rather high temperature (>150°) is reached, it is desirable to add a few percent (up to 10%) of sodium triethylaluminum alkoxide. This measure eliminates the possibility that any free triethylaluminum will be present. For the purpose of obtaining cathodically deposited sodium, anodes of crude sodium, or sodium-mercury alloys containing 30-70% Na, or sodium amalgam with ca. 0.05-0.7 wt% sodium are employed. In the last-mentioned case a polarization overvoltage of 0.7-0.8 V must be applied to compensate for the energy of formation of amalgams (ca. 18 kcal/mole) [78]. If inert anodes are used, sodium triethylaluminum alkoxide is, indeed, decomposed electrolytically but 0.5 mole of ethane is lost for every gram-atom of sodium that is formed [114, 115].

$$Na[(C_2H_5)_3AlOR] \xrightarrow{\text{Electrolysis}} Na_{(puriss)} + (C_2H_5)_2AlOR + 0.5C_2H_4 + 0.5C_2H_6$$

$$Na_{(crude)} + 0.5H_2 + C_2H_4 + (C_2H_5)_2AlOR \rightarrow Na[(C_2H_5)_3AlOR] \quad [184, 185].$$

Sodium tetraethylaluminate can be added as electrolyte because the conductivity of the alkoxide-containing complex is very low.

The cathode is a copper plate which reaches to the bottom of the glass cell; the anode is a metal plate shorter at the bottom by 5-10 cm [183, 186]. The sodium runs down the cathode as a film and collects in the bottom free area of the cell. The temperature must, of course, be above the melting point of sodium, suitably 110-120°.

b. Deposition of Potassium and Higher Alkali Metals. Potassium can be obtained in liquid coherent form only within a narrow temperature range (63-70°) from an electrolyte which is a 2:1 mixture of potassium tetraethylaluminate and potassium triethylaluminum alkoxide [187]. Potassium amalgam (e.g., of 0.05-0.1%) can serve as anode. Above this temperature the reaction of potassium with potassium tetraethylaluminate causes decomposition of the electrolyte and deposition of aluminum. The latter then interferes with the formation of a coherent film of liquid potassium and, therefore, with the possibility of devising a continuous process in a simple cell with minimum distance between electrodes. The greater reactivity of the higher alkali metals rubidium and cesium toward potential electrolytes would make the deposition of these metals even more difficult.

3. Deposition of Magnesium from Grignard Solutions

A number of investigators have been concerned since 1912 with the cathodic deposition of magnesium [111, 188-190]. It is reported that electrolysis of a solution of ethylmagnesium iodide in a 35:15 mixture of diethyl ether and dimethylaniline at 1 A/dm^2 and a voltage increasing from 8 to 27 V yields a good deposit of magnesium [190]. The cathodic current efficiency is said to be 80-85%. Magnesium anodes are not attacked by this electrolyte.

B. Anodic Oxidation of Organometallic Compounds

An important segment of this topic, namely, the electrolysis of metal alkyl "ate" complexes at "organophilic" metal anodes M″ which then go into solution as metal alkyls, has already been discussed above (p. 640):

$$[MR_{n+1}]^{\ominus} + \frac{1}{n}M'' \xrightarrow[-e^{\ominus}]{} MR_n + \frac{1}{n}M''R_n.$$

In a particular subclass of this general reaction are the cases in which the metal which is dissolved from the anode is then quantitatively redeposited on the cathode. The metal refining procedures described in subsection c above depend on a certain selectivity in the dissolution of anode materials, whereby the major metallic impurities remain undissolved; if, however, minor metallic impurities do enter into solution, the extraneous metals are deposited at the cathode to a minor extent, if at all. An extreme case of this anodic selectivity has been presented in IIIA2(a) and IIIA2(b) which concerns the refining of alkali metals when a 0.5-0.05% alkali metal amalgam is used as anode. Even with these very low concentrations of alkali metal in mercury, the alkali metal is deposited exclusively; it contains only ca. 40 ppm of Hg in the case of K or 200 ppm in the case of Na [191]. This content of mercury is not due to anodic formation of ethylmercury but rather to the mercury vapor that is dissolved in the electrolyte.

Contrary to the claims in the patent literature [192] lead is not selectively dissolved from anodes of lead amalgam [96] containing up to ca. 10% of lead. Electrolysis using alkali metal tetraethylaluminate electrolytes yields mixtures of tetraethyllead and diethylmercury.

According to Hein et al. [193], ethyl radicals are formed by electrolysis of the "ate" complexes; these radicals react with the anode in a follow-up reaction. It cannot yet be decided which of the following two alternate routes is involved.

(a) The "ate" anion dissociates into an alkyl anion and the latter is then oxidized to an alkyl radical:

$$[MR_{n+1}]^{\ominus} \rightarrow MR_n + R^{\ominus}$$

$$R^{\ominus} \xrightarrow[-e^{\ominus}]{} \{R\cdot\};$$

or

(b) the intact "ate" anion is discharged and then either decomposes before reaction with the anode metal or reacts directly with the anode metal

$$[MR_{n+1}]^{\ominus} \rightarrow \{\cdot MR_{n+1}\}$$

$$(1) \quad \{\cdot MR_{n+1}\} \longrightarrow MR_n + \{\cdot R\}$$

$$(2) \quad \{\cdot MR_{n+1}\} \xrightarrow{(1/n)M''} MR_n + \frac{1}{n}M''R_n .$$

When inert anodes are used, the alkyl radicals dimerize or disproportionate. For example, electrolysis of complex ethylaluminum compound at copper anodes yields a 1:1 mixture of ethane and ethylene [168, 85]

$$2\{\cdot C_2H_5\} \rightarrow C_2H_4 + C_2H_6 .$$

Butane is obtained in addition to ethane and ethylene at anodes of Au, Pt, Ni, Fe, or an alloy of 25% Cr, 20% Ni, and 55% Fe [168, 194]. It follows that the nature of the subsequent reactions of the ethyl radicals depends on the anode metal. It is possible to rationalize these results by postulating that ethyl metal derivatives actually are formed by that group of metals at which ethane and ethylene are the products but that the alkyl derivatives never leave the metal crystal lattice but decompose there by a mechanism which is well-known to apply to metal alkyls.

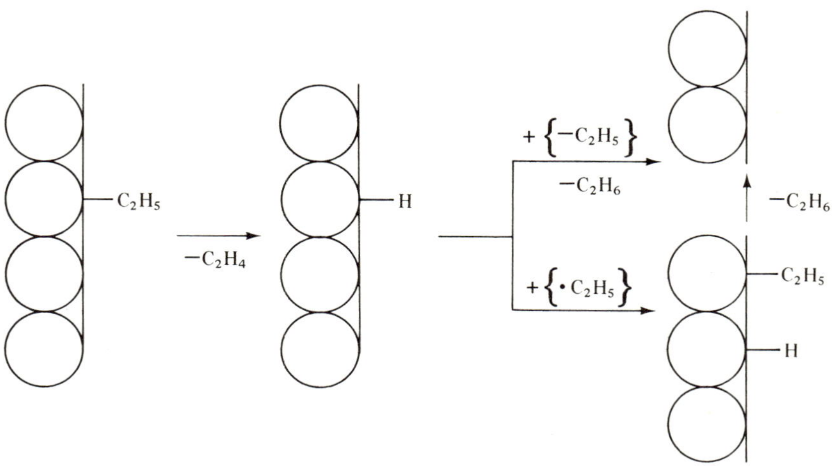

On the other hand, butane formation by combination of two ethyl radicals can occur at the surface of the other metals mentioned. These postulates require experimental testing.

If the organoaluminum electrolyte has aluminum hydride components [85, 168, 194], hydrogen is preferentially evolved at silver and copper anodes. Selectivity toward hydrogen evolution decreases somewhat at higher current densities. Increasing temperature and increasing concentration of hydride hydrogen, likewise, increase preferential hydrogen evolution. See Table 5.

TABLE 5

Influence of Various Anodic Metals [168, 194] on the Composition of the Anodically Evolved Gas [a] in the Electrolysis of $NaF \cdot 2Al(C_2H_5)_3/Al(C_2H_5)_2H$ [b]

Anode metal	H_2, mole %	C_2H_6 mole %	C_4H_{10} mole %	C_2H_4, mole %
Ag	91.2	8.0	0.8	—
Cu	93.5	6.2	0.3	—
Messing	79.0	17.6	3.4	—
Au	44.0	40.2	13.6	2.2
Pd	17.3	68.7	1.0	13.0
Pt	2.8	51.9	29.2	16.1
Ni	4.7	55.2	17.7	21.4
Fe	3.8	39.4	39.5	17.3
C	18.0	55.2	9.0	17.8
Nichrotherm [c]	2.1	51.6	32.8	12.2

[a] Conditions: bath temp. 80; c.d. 10 A/dm^2.

[b] 24.6 Mole % hydride.

[c] 25% Cr, 20% Ni, 55% Fe.

Less preference is shown for hydrogen evolution at anodes of Au, Pd, and graphite; no selectivity for H^- discharge is shown by Pt, Ni, and Fe anodes. In the case of Pt the ratio of H to (C_2H_6 + C_2H_4 + C_4H_{10}) is 4.5:102 which is the $H:C_2H_5$ ratio of 4.5:95.5.

Hydride-containing organoaluminum electrolyte and those having only ethyl groups show marked differences in decomposition potential at copper anodes [194]: 0.2 V (100°) for $NaF \cdot 2(C_2H_5)_3Al \cdot (C_2H_5)_2AlH$ and 0.32 V for

$NaF \cdot 2(C_2H_5)_3Al$. At Pt, which shows no preference for discharge of either of the above two groups, both electrolyte systems have the same decomposition potential (1.1 V). Iron electrodes behave similarly to Pt except that the decomposition potential is raised to 1.8 V. The decomposition potentials of organoaluminum compounds at copper anodes alternate as the alkyl groups are homologated.

TABLE 6

Decomposition Voltage of the Complexes $MF \cdot 2R_3Al$ at Copper and Iron Anodes [194]

Complex R=	Temp., °C	Copper anode, V	Iron anode, V
CH_3	170	0.77	1.9
C_2H_5	100	0.32	1.8
C_3H_7	100	0.62	1.2
C_4H_9	100	0.35	—

Alkylmagnesium bromides behave analogously [195]. The decomposition potential decreases as the branching of the alkyl group increases: $(C_2H_5)(CH_3)CH\text{-}MgBr$, 1.24 V; $(CH_3)_3CMgBr$, 0.97 V. Unsaturated hydrocarbon groups are more easily cleaved than the corresponding saturated groups: $CH_2{=}CHCH_2MgBr$, 0.86 V; $CH_3CH_2CH_2MgBr$, 1.42 V.

It can be seen from Table 6 that methyl groups are considerably more difficult to cleave than ethyl groups. As a consequence, only tetraethyllead is formed at lead anodes from a 20:80 mixture of sodium tetramethylaluminate and sodium tetraethylaluminate [101]. Electrolytes that contain only methyl groups yield mixtures of methane and ethane (17.7:74.9 at 1 A/dm^2 and 170°) at copper anodes. The source of the hydrogen which is needed for the formation of methane is not yet clear. Since a small quantity of propane is also obtained, the following possibility may be considered:

$$2\{CH_3\cdot\} \rightarrow CH_4 + \{:CH_2\}$$

$$\{:CH_2\} + 2\{\cdot CH_3\} \rightarrow CH_3CH_2CH_3.$$

Surprisingly, ether solutions of ethylmagnesium compounds yield predominantly ethane and ethylene along with a little hydrogen and butane even at Pt [196, 189]. Propyl derivatives give 1:1 mixtures of propylene and propane; the presence of hexane could not be established. The ratio of

TABLE 7

Yield of Dimers on Anodic Oxidation of Alkyl- or Arylmetallics R_nM

Hydrocarbyl group on M	Dimer yield with M=				Reaction prod.	Ref.
	Mg	Li	Cd	Zn		
CH_3 (from 2.8 M CH_3MgBr)	100	—	—	—	Ethane	[196]
C_2H_5	Little	—	—	—	Butane	[196]
C_3H_7	Little	—	—	—	Hexane	[199]
C_4H_9	41	44	32	30	Octane	[201]
C_5H_{11}	55-60	49	33	32	Decane	"
C_6H_{13}	45	47	20	37	Dodecane	"
C_7H_{15}	50	42	20	25	Tetradecane	"
C_8H_{17}	47	53	30	35	Hexadecane	"
C_9H_{19}	55-60	52	20	32	Octadecane	"
$C_{10}H_{21}$	35	50-55	25	25-30	Eicosane	"
$C_{11}H_{23}$	5	10	0	0	Docosane	"
$C_{12}H_{25}$	40	38	17	28	Tetracosane	"
$C_{14}H_{29}$	40	40	66	71	Octacosane	"
$C_{16}H_{33}$	52	45	50-55	55	Dotriacontane	"
$C_{18}H_{37}$	54	40	76	75	Hexatriaconane	[214]
$C_6H_5CH_2$	4	—	—	—	1, 2-Diphenylethane	[201, 189]
C_6H_5					Biphenyl	[189]
α-Naphthyl	43	48	34	35	Binaphthyl	[201]
Cyclopentyl	35	44	25-30	28	Bicyclopentyl	"
Cyclohexyl	53	50	20	30	Bicyclohexyl	"
Bornyl	40	40	0	0	Bibornyl	"
Farnesyl	33	30	25	25	Squalene	"
Fenchyl	35	25	25	20	Bifenchyl	"
Geranyl	60	33	10	32	Bigeranyl	"

ethane to methane obtained from methylmagnesium bromide solutions is concentration dependent: ca. 1 M solutions yield methane without ethane; 2.8 M solutions give only ethane; solutions of intermediate concentration give mixtures of these two hydrocarbons [196]. The hydrogen needed for forming methane is derived from the solvent [197]. In this connection dibutyl ether is more readily attacked than diethyl ether [198]. The tendency for the alkyl radicals to dimerize increases as the length of the alkyl chain increases [199]. Isobutyl, sec-butyl, and tert-butyl groups form branched octanes (2,5-dimethylhexane, 3,4-dimethylhexane, and 2,2,3,3-tetramethylbutane, respectively) almost exclusively [200]. The same is true in the electrolysis of n-hexylmagnesium bromide. The yields of dimers obtained from electrolyses of organometallic derivatives of Li, Mg, Cd, and Zn have been investigated as far as the octadecyl compounds [201] (Table 7). The

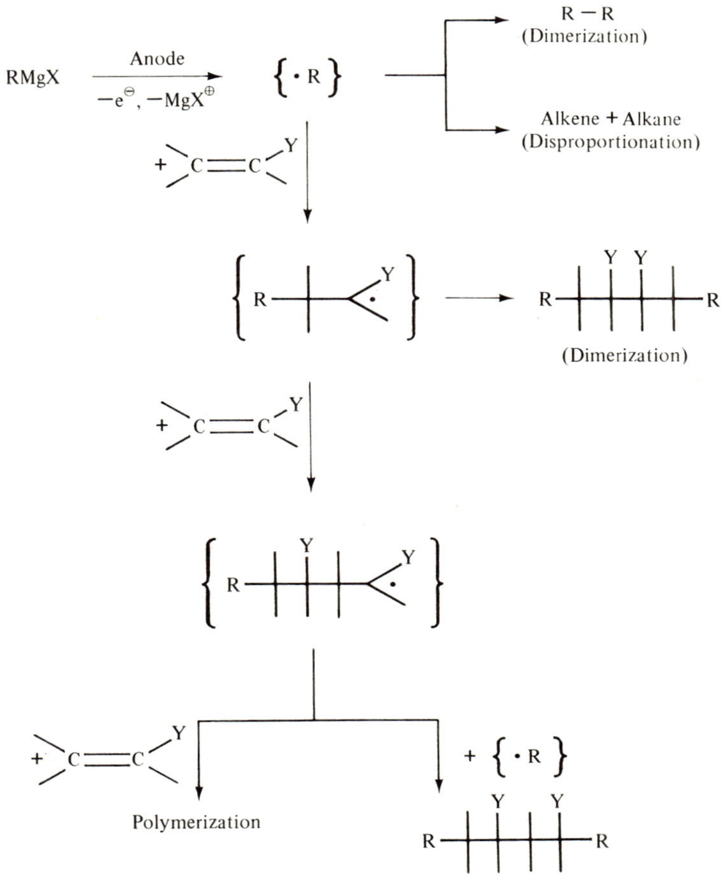

R = butyl, tert-butyl, hexyl, phenyl
Y = phenyl (styrene), vinyl (butadiene), ethoxyl (vinyl ethyl ether)

influence of the anode metal, which was observed in the study of organoaluminum electrolytes, has not yet been considered here.

Radicals formed at the anode during electrolysis of alkylmagnesium halides may add to suitable olefins. The adduct radical either dimerizes or initiates polymerization of the olefin. Chain termination may occur by reaction of the polymer and with an alkyl radical [202] (p. 663).

In the electrolysis of organoaluminum complexes with phenyl substituents even at "organophilic" anodes, e.g., aluminum, only small quantities of the aryl metal compound are formed; the major product is biphenyl [100]:

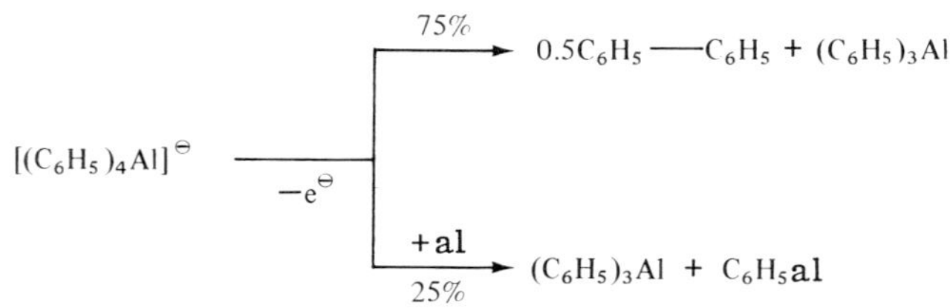

The following conclusions were drawn concerning the mechanism of the oxidation of the tetraphenylborate anion on the basis of voltammetric studies and preparative electrolyses with mixtures of $M^+[(C_6H_5)_4B]^-$ and $M^+[(C_6D_5)_4]B^-$ [203]:

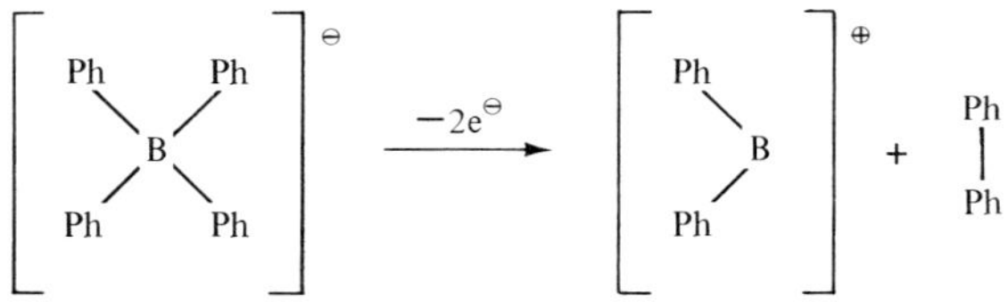

Electrolyses of mixtures of complexes with deuterated and nondeuterated phenyl groups yielded only C_6H_5-C_6H_5 and C_6D_5-C_6D_5. In aqueous solution diphenylboronic acid is formed in addition to diphenyl [218]:

$$[(Ph)_2B]^{\oplus} + H_2O \rightarrow Ph_2BOH + H^{\oplus}.$$

Triphenylgermanate anions are oxidized in liquid ammonia to mixtures of hexaphenyldigermanium and triphenylgermanium hydride [205]:

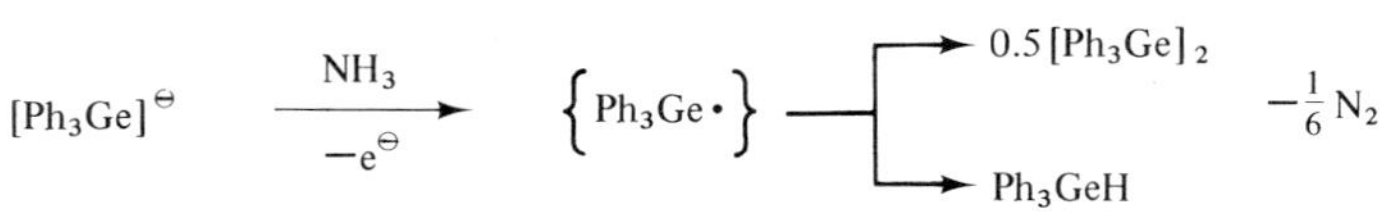

The hydrogen is abstracted from the ammonia. At platinum electrodes a quantity of nitrogen equivalent to that of the Ph_3GeH is formed. The quantity

TABLE 8

Representative Electrochemical Organometallic Preparations

Example No.	Product(s)	Organic reagent charged	Anode	Cathode	Medium; conditions	% Yield	Ref.
1	Di-2-Butylmercury	2-Butanone	Pb	Hg [a]	30% H_2SO_4, 45-50° 7.6-7.8 V, 25 A/dm^2	28	[3]
2	Bis-(4-methylbenzyl) mercury	4-Methylbenzyl bromide	Ag	Hg [a]	2 N LiBr, CH_3OH -1.4 V vs. SCE	64	[18, 19]
3	Hexaethyldilead	Et_3PbCl	Pb	Pb [a]	2 N NaOH or KOH 2.5 V, 0.5 A	97	[31]
4	Bis-[cyclooctadiene-1, 5]-Cu(I) perchlorate	1, 5-Cyclooctadiene	Cu	Cu [a]	$Cu(ClO_4)_2$ 1-2 V, 30 mA	—	[51]
5	Bis-[cyclooctadiene-1, 5]-Ni°	Ni acetylacetonate 1, 5-cyclooctadiene	Al	Al [a]	n-Bu_4NBr; 0° 18 V, 60 mA	70-96	[52]
6	Ni complex [b]	Ni acetylacetonate	Al	Al [a]	n-Bu_4NBr; 20°; THF	—	[52]
7	Tetrakis-(triphenyl-phosphine)-Rh°	$(C_6H_5)_3P$ $[(C_6H_5)_3P]_3RhCl$	—	Pt [a]	CH_3CN-toluene 0.1 M Et_4HClO_4	70	[75]
8	Tetrakis-(triphenyl-phosphine)-nickel°	Ni^{II} acetylacetonate $(C_6H_5)_3P$	Al	Al [a]	n-Bu_4NBr, THF 60 V, 40 mA	83	[52]
9	Tetraethyltin	Sodium bis-triethyl-aluminum) fluoride	Sn [a]	Cu	4.3 V, 1.5 A	60 [c]	[86]

[a] Electrode at which the product is formed.

[b] Butadiene added after electrolysis to cyclic trimers (p. 636).

[c] Some hexaethylditin formed as coproduct.

of nitrogen is considerably smaller at a mercury anode, as shown in the following tabulation:

	Pt anode	Hg anode
m Faraday	0.98	0.98
Equiv. $(Ph_3Ge)_2$	0.35	0.1
Equiv. Ph_3GeH	0.65	0.9
Equiv. N_2	0.72	0.11

The "chemical" formation of olefin-Hg(II) complexes is well-known. They may also be formed by anodically oxidizing Hg in the presence of an olefin [206]. Oxidation of such complexes to acids [207] regenerates Hg(II) so that the process may be considered catalytic in the inorganic reagent (Chapter XXIV).

NOTES

[1]Translated by the Editor.

[2]By cleavage of alkyl halides in propylene carbonate [14a, 14b], acetonitrile [14c], or DMF [14d].

[3]The following steps have been proposed [130]:

$$I^{\ominus} \xrightarrow[-e^{\ominus}]{} I\cdot \xrightarrow{Al} Al^{I}I \xrightarrow{CH_3I} CH_3AlI_2 .$$

This mechanism has not been proved. The following electrode reactions appear to be more plausible:

Cathode: $\frac{1}{3}Al + CH_3I + e^{\ominus} \rightarrow \frac{1}{3}(CH_3)_3Al + I^{\ominus}$

Anode: $\frac{1}{3}Al + I^{\ominus} - e^{\ominus} \rightarrow \frac{1}{3}AlI_3 .$

REFERENCES

[1]. E.g., J. J. Eisch, The Chemistry of Organometallic Compounds, Macmillan, New York, 1967, p. 1.

[2]. T. Sekine, A. Yamura, and K. Sugino, J. Electrochem. Soc., 112, 439 (1965), discuss the reaction mechanism.

[3]. J. Tafel, Ber., 39, 3626 (1906).

[4]. C. J. Haggerty, Trans. Amer. Electrochem. Soc., 56, 421 (1929).

[5]. T. Arai, Bull. Chem. Soc. Japan, 32, 184 (1959).

[6]. C. Schall and W. Kirst, Z. Elektrochem., 29, 537 (1923).

[7]. L. Holleck and D. Marquarding, Naturwiss., 49, 468 (1962).

[8]. W. Schepss, Ber., 46, 2564 (1913).

[9]. T. Arai and T. Oguri, Bull. Chem. Soc. Japan, 33, 1018 (1960).

[10]. J. Tafel, Ber., 44, 323 (1911).

[11]. C. Renger, ibid., 44, 337 (1911).

[12]. H. D. Law, J. Chem. Soc., 1912, 1016, 1544.

[13]. B. Mead (to General Motors), U.S. Pat. 1,567,159 (1923); Chem. Zent , 1926 I, 2052.

[14]. (a) R. Galli, J. Electroanal. Chem., 22, 75 (1969); (b) R. Galli and F. Olivani, ibid., 25, 331 (1970); (c) H. E. Ulery, J. Electrochem. Soc., 116, 1201 (1969); (d) M. Fleischmann, D. Pletcher, and C. J. Vance, J. Electroanal. Chem., 29, 325 (1971).

[15]. G. H. F. Calingaert (to General Motors), U.S. Pat. 1,539,297 (1923); Chem. Zent., 1925 II, 761.

[16]. Du Pont, Brit. Pat. 949,925 (1964); CA, 61, 3935 (1964); E. F. Silversmith and W. J. Sloan (to Du Pont), D.B.P. 1,246,734 (1967); CA, 67, 87255 (1967).

[17]. M. Schuler and M. Kegelman, unpublished results; cf. H. E. Ulery, J. Electrochem. Soc., 116, 1201 (1969).

[18]. J. Grimshaw and J. J. Ramsey, J. Chem. Soc. (B), 1968, 60.

[19]. R. Pasternak, Helv. Chim. Acta, 31, 753, 772 (1948).

[20]. A. P. Tomilov, Yu. D. Smirnov, and S. L. Varshavskii, Zh. Obshch. Khim., 35, 391 (1965); CA, 63, 5238 (1965).

[21]. L. V. Kaabak and A. P. Tomilov, Zh. Obshch. Khim., 33, 2808 (1963); CA, 60, 5534a (1966).

[22]. L. V. Kaabak, A. P. Tomilov, and S. L. Varshavskii, Zh. Vses. Khim. Obshchest. Im. D.I. Mendeleeva, 9 (6), 700-1 (1964); CA, 62, 12751 (1965).

[23]. W. Sundermeyer and W. Verbeek, Angew. Chem., 78, 107 (1966).

[24]. R. E. Dessy et al., e.g., T. Psarras and R. E. Dessy, J. Amer. Chem. Soc., 88, 5132 (1966); or R. E. Dessy, R. King, and M. Waldrop, ibid., 88, 5112 (1966).

[25]. H. Lehmkuhl, G. Fuchs, and R. Koester, Tetrahedron Lett., 1965, 2511.

[26]. E. Ziegler, G. Fuchs, and H. Lehmkuhl, Z. Anorg. Chem., 355, 145 (1957); cf. H. Lehmkuhl, Chim., 24, 182 (1970).

[27]. E. Ladanyi and H. Lehmkuhl, Max-Planck-Institut für Kohlenforschung, Mülheim-Ruhr, unpublished results.

[28]. W. Strohmeier and G. Popp, Z. Naturforsch., 22b, 891 (1967).

[29]. R. E. Dessy, W. Kitching, and T. Chivers, J. Amer. Chem. Soc., 88, 453 (1966).

[30]. T. Midgley, C. A. Hochwalt, and G. Calingaert, ibid., 45, 1821 (1023).

[31]. F. Hein and A. Klein, Ber., 71, 2381 (1938).

[32]. R. E. Dessy, T. Chivers, and W. Kitching, J. Amer. Chem. Soc., 88, 467 (1966).

[33]. M. D. Morris, J. Electroanal. Chem., 20, 263 (1969).

[34]. M. Devaud, Compt. Rend., 263, 1269 (1966).

[35]. G. Costa, Gazz. Chim. Ital., 80, 42 (1950); CA, 44, 9826 (1950).

[36]. M. Mehner, H. Jehring, and H. Kriegsmann, J. Organometal. Chem., 15, 97 (1968).

[37]. V. F. Toropova and M. K. Saikina, Sb. Statei Obshch. Khim., Akad. Nauk SSSR, 1, 210 (1953); CA, 48, 12579 (1954).

[38]. L. Ricciboni and P. Popoff, Atti ist. veneto sci. Venezia, 107; Part II, 123 (1949); CA, 44, 6752 (1956).

[39]. M. K. Saikina, Uch. Zap. Kazan. Gos. Univ. im Lenina Khim., 116, 129 (1956); CA, 51, 7191 (1957).

[40]. G. Tagliavini and L. Doretti, Chem. Commun., 1966, 562.

[41]. E.g., R. E. Dessy, R. B. King, and M. Waldrop, J. Amer. Chem. Soc., 88, 5112 (1966); or S. P. Gubin, et al., Izv. Akad. SSSR (English Transl.), 1969, 258, p. 218.

[42]. R. E. Dessy, F. E. Stary, R. B. King, and M. Waldrop, J. Amer. Chem. Soc., 88, 471 (1966).

[43]. M. D. Morris, P. S. McKinney, and E. C. Woodbury, J. Electroanal. Chem., 10, 85 (1965).

[44]. H. Lehmkuhl, Chim., 24, 182 (1970).

[45]. N. Yamazaki, S. Nakahama, and S. Kambara, J. Polym. Sci., B3, 57 (1965).

[46]. B. L. Funt, D. Richardson, and S. N. Bhadani, Can. J. Chem., 44, 711 (1966); J. Polym. Sci., A4, 2871 (1966).

[47]. B. L. Funt and S. N. Bhadani, ibid., C23, 1 (1968).

[48]. H. Lehmkuhl and J. Čuljković, unpublished results; J. Čuljković, Dissertation, Bochum Univ., Germany, 1969.

[49]. D. Weyenberg and C. H. Toporcer, J. Amer. Chem. Soc., 84, 2843 (1962); J. Org. Chem., 30, 943 (1965); D. Weyenberg, L. H. Toporcer, and A. E. Bey, ibid., 30, 4096 (1965); D. Weyenberg, L. H. Toporcer, and L. E. Nelson, ibid., 33, 1975 (1968).

[50]. R. A. Benkeser and C. A. Tincher, J. Organometal. Chem., 13, 139 (1968).

[51]. St. E. Manahan, Inorg. Chem., 5, 2063 (1966), and citations therein.

[52]. H. Lehmkuhl and W. Leuchte, J. Organometal. Chem., 23. C30 (1970).

[53]. H. Lehmkuhl, W. Leuchte, and E. Janssen, ibid., 30, 407 (1971).

[53a]. H. Matschiner, H. J. Kerrinnes, and K. Issleib, Z. Anorg. Allg. Chem., 380, 1 (1971); Ger. Offen. 1, 962, 758 (1970); CA, 73, 67477 (1970).

[54]. G. Wilke, Angew. Chem., 69, 18 (1963); G. Wilke, et al., Makromol. Chem., 69, 18 (1963); G. Wilke, Angew. Chem., 75, 10 (1963); B. Bogdanović, M. Kroener, and G. Wilke, Justus Liebigs Ann. Chem., 699, 1 (1966); 727, 143 (1969).

[55]. N. Yamazaki and S. Murai, Chem. Commun., 1968, 147.

[56]. W. A. Kornicker (to Monsanto), U.S. Pat. 3, 474, 012 (1964); CA, 72, 22145 (1970).

[57]. R. Ercoli, M. Guainazzi, and G. Silvestri, Chem. Commun., 1967, 927; M. Guainazzi, G. Silvestri, S. Gambino, and G. Filardo, J. Chem. Soc. Dalton, 927 (1972).

[58]. H. Lehmkuhl and K. Mehler, J. Organometal. Chem., 25, C44 (1970).

[59]. General reviews: A. A. Vlcek, Z. Elektrochem., 61, 1014 (1957); Pure Appl. Chem., 10, 61 (1965); Nature, 177, 1043 (1956); Individual studies: C. Furlani and E. O. Fischer, Z. Elektrochem., 61, 481 (1957); S. Valcher, Ric. Sci., 8, 1007 (1965); S. P. Gubin and S. A. Smirnova, J. Organometal. Chem., 20, 241 (1969); R. E. Dessy, et al., J. Amer. Chem. Soc., 88, 471, 5112, 5117, 5124 (1966); 90, 2001 (1968).

[60]. S. P. Gubin, Proceedings of the Fourth Intern. Conf. on Organometal. Chem., The Chemical Society London, Bristol, 1969.

[61]. L. I. Denisovich, S. P. Gubin, Yu. A. Chapovskii, Izv. Akad. Nauk, Ser. Khim., SSSR (English Transl.), 1967, 2738, p. 2271; (a) H. tom Dieck and J. W. Renk, Chem. Ber., 104, 110 (1971); (b) K. R. Bullock, J. Electrochem. Soc., 117, 51C (1970).

[62]. L. I. Denisovich, S. P. Gubin, Yu. A. Chapovskii, and N. A. Ustynok, Izv. Akad. Nauk, Ser. Khim., SSSR (English Transl.), 1968, 891.

[63]. L. I. Denisovich, A. A. Doganson, S. P. Gubin, N. E. Kolobova, and K. N. Anisimov, ibid., 1969, 218.

[64]. R. E. Dessy and L. Wieczorek, J. Amer. Chem. Soc., 91, 4963 (1969).

[65]. R. E. Dessy, F. E. Stary, R. B. King, and M. Waldrop, ibid., 88, 471 (1966).

[66]. R. E. Dessy, R. B. King, and M. Waldrop, ibid., 88, 5112 (1966).

[67]. R. E. Dessy, P. M. Weissmann, and R. L. Pohl, ibid., 88, 5117 (1966).

[68]. R. E. Dessy, R. L. Pohl, and R. B. King, ibid., 88, 5121 (1966).

[69]. R. E. Dessy and R. L. Pohl, ibid., 90, 1995 (1968).

[70]. H. P. Schröer and A. A. Vlček, Z. Anorg. Allg. Chem., 334, 205 (1964).

[71]. A. A. Vlček, ibid., 304, 109 (1960).

[72]. A. Rusina and H. P. Schröer, Coll. Czech. Chem. Commun., 31, 2600 (1966).

[73]. S. Herzog and R. Taube, Z. Chem., 2, 208 (1962).

[74]. A. A. Vlček, Z. Elektrochem., 61, 1014 (1957).

[75]. D. C. Olson and W. Keim, Inorg. Chem., 8, 2028 (1969).

[76]. F. Hein, E. Petzschner, K. Wagler, and F. A. Segitz, Z. Anorg. Allg. Chem., 141, 161, 224 (1924).

[77]. F. Hein and F. A. Segitz, ibid., 158, 153, 159 (1926).

[78]. H. Lehmkuhl and W. Grimme, Justus Liebigs Ann. Chem., 705, 1 (1967).

[79]. K. Ziegler, H.-G. Gellert, H. Lehmkuhl, W. Pfohl, and K. Zosel, ibid., 629, 1 (1969).

[80]. K. Ziegler and H.-G. Gellert (to K. Ziegler), DAS 917, 006 (1951); Chem. Zent., 1955, 5179.

[81]. J. B. Honeycutt and J. M. Riddle, J. Amer. Chem. Soc., 83, 369 (1961).

[82]. K. Ziegler, Chem.-Ing.-Tech., 35, 325 (1963); Angew. Chem., 72, 565 (1960); Brennstoffchem., 40, 204 (1959).

[83]. D. G. Braithwaite (to Nalco), DAS 1,171,919 (1963), CA, 56, 4526 (1964).

[84]. K. Ziegler and H. Lehmkuhl, Z. Anorg. Allg. Chem., 283, 414 (1956); Angew. Chem., 67, 424 (1955).

[85]. H. Lehmkuhl, Dissertation, Technische Hochschule Aachen, 1954.

[86]. E. Huether, Dissertation, Technische Hochschule Aachen, 1957.

[87]. K. Ziegler and H. Lehmkuhl (to K. Ziegler), U.S. Pat. 2,985,568 (1961); see also CA, 53, 17733 (1959).

[88]. R. Doetzer, Chem.-Ing.-Tech., 36, 616 (1964).

[89]. K. Ziegler, ibid., 35, 325, 327 (1963).

[90]. Houben-Weyl, Praeparative Methoden der Organischen Chemie, Band XIII/4, "Aluminium-organische Verbindungen," Thieme, Stuttgart, 1970, p. 32.

[91]. See, e.g., K. Ziegler and H. Lehmkuhl (to K. Ziegler), Belg. Pat. 543,128 (1955); Chem. Zent., 1960, 8016; same inventors and assignee, D.B.P., 1,114,861 (1957); CA, 55, 5199 (1961).

[92]. K. Ziegler and H. Lehmkuhl (to K. Ziegler), DAS, 1,127,900 (1959).

[93]. K. Ziegler and H. Lehmkuhl (to K. Ziegler), D.B.P., 1,161,562 (1959); CA, 60, 11623 (1964).

[94]. W. Grimme, Dissertation, Technische Hochschule Aachen, 1960.

[95]. W. Eisenbach, Dissertation, Technische Hochschule Aachen, 1961.

[96]. K. Ziegler, Chem.-Ing.-Tech., 35, 325 (1963).

[97]. K. Ziegler and H. Lehmkuhl (to K. Ziegler), D.B.P., 1,153,371 (1960); CA, 60, 546 (1964).

[98]. K. Ziegler and H. Lehmkuhl (to K. Ziegler), D.B.P., 1,166,196 (1960); CA, 60, 15910 (1964).

[99]. W. Eisenbach, Thesis, Technische Hochschule Aachen, 1960.

[100]. R. Schaefer, Dissertation, Technische Hochschule Aachen, 1961.

[101]. H. Lehmkuhl, Chem.-Ing.-Tech., 36, 612 (1964).

[102]. K. Ziegler and H. Lehmkuhl (to K. Ziegler), DAS 1,220,855 (1961); CA, 60, 3008 (1966).

[103]. K. Ziegler, H. Lehmkuhl, and R. Schaefer (to K. Ziegler), DAS 1, 174, 779 (1962); CA, 61, 13344 (1964).

[104]. K. Ziegler and H. Lehmkuhl (to K. Ziegler), DAS 1, 149, 005 (1960); CA, 59, 1157 (1963).

[105]. P. Kobetz and R. C. Pinkerton (to Ethyl Corp.), U.S. Pat. 3, 068, 261 (1960); CA, 58, 9135 (1963).

[106]. R. C. Pinkerton (to Ethyl Corp.), U.S. Pat. 3, 028, 323 (1959); CA, 58, 3457 (1963).

[107]. R. C. Pinkerton (to Ethyl Corp.), U.S. Pat. 3, 028, 325 (1959); CA, 57, 4471 (1962).

[108]. K. Ziegler, H. Lehmkuhl, and W. Grimme (to K. Ziegler), DAS 1, 144, 490 (1961); CA, 58, 9880 (1963).

[109]. K. Ziegler and O. W. Steudel, Justus Liebigs Ann. Chem., 652, 1 (1962).

[110]. K. Ziegler and H. Lehmkuhl (to K. Ziegler), DAS 1, 212, 085 (1961); CA, 64, 19675 (1966).

[111]. N. V. Kondyrev, Ber., 58, 459 (1925).

[112]. H. E. French and M. Drane, J. Amer. Chem. Soc., 52, 4904 (1930).

[113]. J. M. Nelson and W. V. Evans, ibid., 39, 82 (1917).

[114]. J. M. Evans and R. Pearson, ibid., 64, 2865 (1942).

[115]. J. Linsk and E. A. Mayerle (to Standard Oil of Indiana), U.S. Pat. 3, 155, 602 (1960), conductivity data given.

[116]. R. E. Dessy and G. S. Handler, J. Amer. Chem. Soc., 80, 5824 (1958); T. Psarras and R. E. Dessy, ibid., 88, 5132 (1966).

[117]. D. G. Braithwaite (to Nalco), DAS 1, 231, 700 (1963); CA, 66, 76155 (1967).

[118]. D. G. Braithwaite (to Nalco), U.S. Pat. 3, 256, 161 (1964); CA, 65, 6751 (1964).

[119]. G. Calingaert and H. Shapiro (to Ethyl Corp.), U.S. Pat. 2, 535, 193 (1950); CA, 45, 3865 (1951).

[120]. D. G. Braithwaite, J. S. D'Amico, P. L. Gross, and W. Hanzel (to Nalco), U.S. Pat. 3, 287, 249 (1962); CA, 66, 51716 (1967); U.S. Pat. 3, 141, 841 (1960); CA, 61, 10323 (1964).

[121]. D. G. Braithwaite (to Nalco), DAS 1, 216, 303 (1963); CA, 65, 8962 (1966).

[122]. D. G. Braithwaite (to Nalco), DAS 1,231,242 (1963); CA, 66, 95196 (1967).

[123]. D. G. Braithwaite (to Nalco), U.S. Pat. 3,234,112 (1961); CA, 59, 7559 (1966).

[124]. Anon., Chem. Week, 95, No. 24, 77 (1964).

[125]. L. L. Bott, Hydrocarbon Proc., 44, No. 1, 115 (1965).

[126]. Anon., Chem. Ind., 18, 272 (1966).

[127]. Anon., Chem. Trade, J. Chem. Eng., 18 Dec. 1964, p. 859.

[128]. Anon., Chem. Eng. News, 42, No. 49, 52 (1964).

([124]–[128]: Concern Nalco process)

[129]. H. Lehmkuhl, Justus Liebigs Ann. Chem., 705, 40 (1967).

[130]. J. R. Chadwick and E. Kinsella, J. Organometal. Chem., 4, 334 (1965).

[131]. E. H. Mottus and M. R. Ort (to Monsanto), Ger. Discl. 1,800,696 (1969); J. Electrochem. Soc., 117, 885 (1970); M. R. Ort, E. H. Mottus, M. M. Baizer, and D. E. Carter, U.S. Pat. 3,546,083 (1970).

[132]. H. Lehmkuhl and R. Schaefer, Tetrahedron Lett., 1966, 2315.

[133]. A. P. Giraitis, T. H. Pearson, and R. C. Pinkerton (to Ethyl Corp.), U.S. Pat. 2,960,450 (1959); CA, 55, 8430 (1961).

[134]. T. H. Pearson (to Ethyl Corp.), U.S. Pat. 2,915,440 (1959); CA, 54, 17415 (1960).

[135]. S. Valcher and E. Alunni, Ric. Sci., 38, 527 (1968).

[136]. W. Strohmeier and F. Seifert, Z. Elektrochem., 63, 683 (1959).

[137]. E. Bonitz, Ber., 88, 742 (1955); Angew. Chem., 67, 525 (1955).

[138]. E. Bonitz and W. Huber, Z. Anal. Chem., 186, 206 (1962).

[139]. W. Strohmeier and K. Huempfner, Z. Elektrochem., 60, 1111 (1956); 61, 1010 (1957).

[140]. W. Strohmeier, F. Seifert, and H. Landsfeld, ibid., 66, 312 (1962).

[141]. W. Strohmeier, H. Landsfeld, and F. Gernert, ibid., 66, 823 (1962).

[142]. W. Strohmeier and K. Nuetzel, ibid., 62, 188 (1958), show that the influence of the alkyl group on the acceptor ability of the metal is slight.

[143]. E.g., H. Lehmkuhl and H.-D. Kobs, Justus Liebigs Ann. Chem., 719, 11 (1968).

[144]. F. Hein, E. Petzschner, K. Wagler, and F. A. Segitz, Z. Anorg. Allg. Chem., 141, 161 (1924); Angew. Chem., 51, 503 (1938).

[145]. F. Hein and F. A. Segitz, Z. Anorg. Allg. Chem., 158, 153 (1926).

[146]. F. Hein and H. Schramm, Z. Phys. Chem., A 151, 234 (1931).

[147]. F. Hein and H. Pauling, Z. Elektrochem., 38, 25 (1932); Z. Phys. Chem., A 165, 338 (1933).

[148]. K. Ziegler, R. Köster, H. Lehmkuhl, and K. Reinert, Justus Liebigs Ann. Chem., 629, 33 (1960).

[149]. R. Doetzer, Chem.-Ing.-Tech., 36, 616 (1964).

[150]. W. Strohmeier, W. Haecker, and G. Popp, Ber., 100, 405 (1967).

[151]. W. Strohmeier and F. Gernert, ibid., 95, 1420 (1962).

[152]. G. Jander and L. Fischer, Z. Elektrochem., 62, 971 (1958).

[153]. R. Doetzer, Chem.-Ing.-Tech., 36, 616 (1964), has a good review of the electrorefining of metals in organometallic electrolytes.

[154]. E. Bonitz, Ber., 88, 742 (1955).

[155]. W. P. Neumann, Habilitationsschrift, Univ. of Giessen, Germany, 1959.

[156]. J. Jakober and C. A. Kraus, J. Amer. Chem. Soc., 71, 2409 (1949).

[157]. Y. Takashi, Bull. Chem. Soc. Japan, 40, 612 (1967).

[158]. R. D. Bushick and R. S. Stearns, J. Polym. Sci., A4, 215 (1966).

[159]. A. K. Tsiomis, Dissertation, Univ. of Tenn., Knoxville, Tenn., 1967.

[160]. D. F. Hoeg, S. Liebman, and L. Schubert, J. Org. Chem., 28, 1554 (1963).

[161]. Ref. [90], p. 110 ff.

[162]. K. Ziegler and H. Lehmkuhl, Z. Anorg. Chem., 283, 414 (1956).

[163]. K. Ziegler and H. Lehmkuhl (to K. Ziegler), D.B.P., 1,047,450 (1954); CA, 54, 19231 (1960).

[164]. H. Cordes and H. Roettger, Z. Elektrochem., 63, 1030 (1959).

[165]. G. Revel, Ann. Chim., 1966, 37.

[166]. A. Boehm, H. Ginsberg, and W. Reuter, Aluminium, 37, 267 (1961); CA, 61, 2719 (1964).

[167]. K. Ziegler and H. Lehmkuhl (to K. Ziegler), DAS 1,056,377 (1955); CA, 52, 19619 (1958).

[168]. K. Ziegler, H. Lehmkuhl, and E. Lindner, Ber., 92, 2320 (1959).

[169]. E. Todt and H. Hauschildt (to Siemens), DAS 1,170,658 (1961); Ger. Discl. 1,483,344 (1965); CA, 59, 3554 (1963).

[170]. Ref. [149], p. 633.

[171]. G. Hans, DAS 1,102,736 (1959); CA, 57, 13800 (1962).

[172]. W. Strohmeier and F. Gernert, Z. Naturforsch., 16b, 760 (1961); 17b, 128 (1962).

[173]. See Ref. [149], p. 625.

[174]. H. Strohmeier, W. Haecker, and G. Popp, Ber., 100, 405 (1967).

[175]. G. Hans, DAS 1,162,576 (1959); CA, 60, 11624 (1964).

[176]. Ref. [149], p. 636.

[177]. W. Strohmeier and F. Gernert, Z. Naturforsch., 20b, 829 (1965).

[178]. W. Strohmeier and G. Popp, ibid., 23b, 870 (1968).

[179]. W. H. Rodebush and J. M. Peterson, J. Amer. Chem. Soc., 51, 638 (1929).

[180]. K. Ziegler, H. Lehmkuhl, and W. Grimme (to K. Ziegler), D.B.P. 1,114,330 (1959); CA, 56, 4524 (1962).

[181]. K. Ziegler, H. Lehmkuhl, and W. Grimme (to K. Ziegler), D.B.P. 1,146,258 (1961); CA, 58, 13458 (1963).

[182]. K. Ziegler, Bl., 1966, 1456.

[183]. E. Huether, unpublished work, Max-Planck-Institut, Mülheim-Ruhr.

[184]. H. Lehmkuhl, Angew. Chem., 75, 1090, 1093 (1963).

[185]. K. Ziegler and H. Lehmkuhl (to K. Ziegler), D.B.P., 1,157,618 (1958); CA, 60, 9311 (1964).

[186]. E. Huether and H. Dislich, unpublished work, Max-Planck-Institut, Mülheim-Ruhr.

[187]. K. Ziegler and H. Lehmkuhl, unpublished results.

[188]. P. Jolibois, Compt. Rend., 155, 353 (1912); 156, 712 (1913).

[189]. L. W. Gaddum and H. E. French, J. Amer. Chem. Soc., 49, 1295 (1927).

[190]. D. M. Overcash and F. C. Mathers, Trans. Electrochem. Soc., 64, 305 (1933).

[191]. W. Eisenbach and E. Blanck, unpublished work, Max-Planck-Institut Mülheim-Ruhr; Ref. [96], p. 325.

[192]. D. K. Wunderlich and L. N. Fussel (to Sinclair Research), U.S. Pat. 3,159,557 (1961); CA, 62, 11432 (1965).

[193]. Refs. [144] and [145].

[194]. E. Lindner, Dissertation, Technische Hochschule Aachen, 1957.

[195]. W. V. Evans, F. H. Lee, and C. Lee, J. Amer. Chem. Soc., 57, 489 (1935).

[196]. W. V. Evans and F. H. Lee, ibid., 56, 654 (1934).

[197]. W. V. Evans and E. Field, ibid., 58, 720 (1936).

[198]. W. V. Evans and E. Field, ibid., 58, 2284 (1936).

[199]. W. V. Evans and D. Braithwaite, ibid., 61, 898 (1939).

[200]. W. V. Evans, D. Braithwaite, and E. Field, ibid., 62, 534 (1940).

[201]. J. L. Morgat and R. Pallaud, Compt. Rend., 260, 5579 (1965); 260, 574 (1965).

[202]. H. Schaefer and H. Kuentzel, Tetrahedron Lett., 1970, 3333.

[203]. D. H. Geske, J. Phys. Chem., 63, 1062 (1959); 66, 1743 (1962).

[204]. W. R. Turner and P. J. Elving, Anal. Chem., 37, 207 (1965).

[205]. L. S. Foster and G. S. Hooper, J. Amer. Chem. Soc., 57, 76 (1935).

[206]. N. L. Weinberg, Tetrahedron Lett., 4823 (1970).

[207]. M. Fleischmann, D. Pletcher, and G. M. Race, J. Chem. Soc. (B), 1746 (1970).

Section F

ELECTRODE REACTIONS CLASSIFIED BY REACTION TYPE

Chapter XIX

ELECTROLYTIC REDUCTIVE COUPLING

Manuel M. Baizer

Corporate Research Department
Monsanto Company
St. Louis, Missouri 63166

I. INTRODUCTION

While a variety of cathodic reactions of organic compounds has been discussed in Section C, the basis of classification there was the nature of the electrophoric group. A separate examination of reductive coupling [1] as a reaction type is warranted because: (a) It is synthetically important in advantageously providing di- and polyfunctional molecules, often in high yield, from simpler starting materials; (b) it has provided great stimulus to mechanistic investigations designed to differentiate between EC and ECE sequences (p. 695); (c) it has been an especially active area in the last decade; and (d) it continues to present many challenging research opportunities.

All of the classes of compounds discussed in Section C — and several others — may under appropriate conditions form reductively coupled products. This formation of new C-C, N-N, C-N, etc., bonds may be intermolecular [Eq. (1)] or intramolecular [Eq. (2)]. In the former case, the product may

$$2(CH_3)_2CO \quad + \quad 2H^+ \xrightarrow{\bar{e}} \begin{array}{l} (CH_3)_2C{-}OH \\ \quad\quad\;\; | \\ (CH_3)_2C{-}OH \end{array} \tag{1}$$

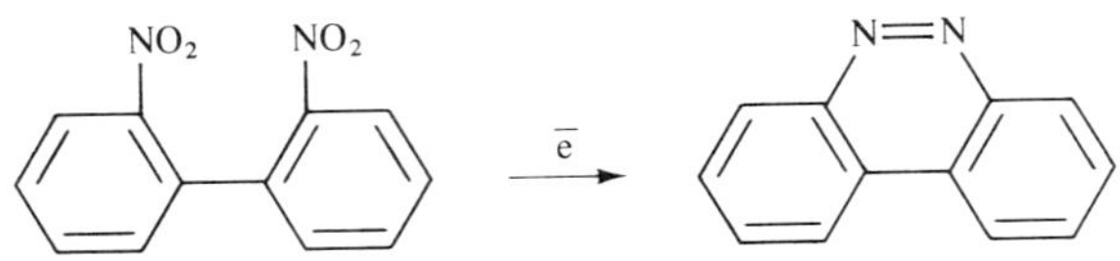

(2)

arise from the joining of two molecules of the same chemical species [Eq. (1)] or two different species [Eq. (3)]. When the starting materials are onium compounds (Chapter XXI), or other substances containing groups

$$\begin{array}{l} CHCOOEt \\ \| \\ CHCOOEt \end{array} + CH_2{=}CHCN \xrightarrow{\bar{e}} \left(\begin{array}{l} CH_2COOEt \\ | \\ CHCOOEt \\ | \end{array} \right)_2 + \begin{array}{l} CH_2COOEt \\ | \\ CHCOOEt \\ | \\ CH_2CH_2CN \end{array} \tag{3}$$

which can "leave" as radicals or anions in the course of the reduction (e.g., halide, tosylate), then coupling occurs only via the organic fragments [Eq. (4)]. Reductive coupling in which more than two molecules are conjoined

$$2RX \xrightarrow{2\bar{e}} RR + 2X^- \tag{4}$$

evolves into the area of oligomerization and polymerization (Chapter XXIX).

The coupled product may be symmetrical [Eqs. (1), (5)] or unsymmetrical [Eq. (3)]. When the coupled product is symmetrical, each of the two

$$2CH_2{=}CHCN \xrightarrow[2H_2O]{2\bar{e}} NC(CH_2)_4CN + 2O\bar{H} \tag{5}$$

halves could have been formed by reduction. In the absence of supporting evidence, however, it must not be taken for granted that two identical moieties have been joined. Unsymmetrical products have been formed from two simultaneously reduced species [Eq. (6)] or by the reaction of one intermediate which has been produced by reduction with a second nonreduced species [Eq. (3)]. For completeness it should be mentioned that sometimes

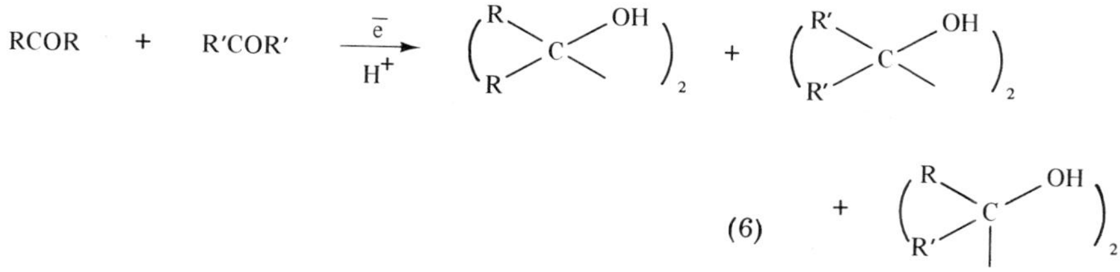

(6)

coupling is merely a consequence of the condensation of two relatively stable, electrically neutral molecules (or sites) which have been simultaneously produced in a reduction [Eq. (7)]:

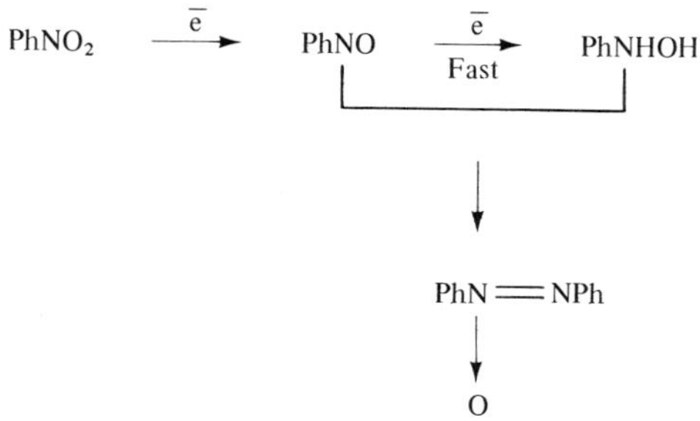

(7)

Except in the cases illustrated by Eq. (7) in which coupling almost certainly occurs as a homogeneous "bulk" reaction, the mechanisms of reductive coupling are complicated because of the heterogeneous nature of electrode reactions. Coupling at the electrode surface or in its vicinity may occur by one route, in the "bulk" by another. Coupling at low pH and with a given electrode material may proceed by a different pathway at high pH and at another electrode. Reaction of an anion radical when it is formed in the presence of a given counterion and in a given degree of association with that counterion may be by one mechanism and when the counterion and/or solvent are changed, by a different mechanism. It is almost always appropriate to keep this pluralism in mind. Allied to this complication is the question of establishing the sequence of E and C steps which lead to the ultimate products. The major sequences are (C)EC, ECE(C), and perhaps EEC, but the rigorous establishment of mechanisms for reductive couplings of a given class of compounds has been achieved only occasionally.

To a great extent incisive mechanistic investigations have become possible only now with the development of sophisticated instrumental tools (Chapter II). Polarography, cyclic voltammetry, chronopotentiometry, ESR spectroscopy, and determination of the structure of products (including their stereochemistry) all contribute to establishment of the final mechanistic conclusions. As Zuman has forcefully pointed out [2], however, the utilization of only one or two of the above techniques in the study of a given problem may lead to completely erroneous rationalizations.

As implied above, the reduced intermediate directly involved in the coupling may be a radical, a radical anion, an anion, or a dianion; the coupling may be: (a) radical dimerization, (b) anionic attack of a radical anion upon substrate, (c) radical-radical anion addition, (d) anionic attack upon substrate. Typical combinations of E and C reactions are summarized in Table 1. It is particularly difficult to distinguish between those EC and ECEC mechanisms which both conceptually lead to the same symmetrical product (e.g., reduction of diethyl maleate).

Success in achieving satisfactory yields in reductive coupling reactions depends obviously on minimizing the probability that the intermediate formed by electroreduction will find more suitable pathways to stable products. Radicals may react with the electrode material to yield organometallics (Chapter XVIII). Radicals formed from olefinic structures are reduced to anions at the same potential as that at which they were formed. Radical anions may disproportionate to dianions and starting materials. Radical anions on protonation are "free" radicals.

II. REDUCTIVE COUPLING REACTIONS

A. Hydrocarbons

Stilbene formed 30% of its reduced dimer ("hydrodimer") upon reduction in anhydrous DMF at a mercury cathode with tetrabutylammonium electrolyte [3]. The intermediate anions or dianions are, however, protonated when acetonitrile is the solvent, so that only saturated monomer is obtained. More sterically hindered vinyl compounds (triphenylethylene, tetraphenylethylene) do not form hydrodimers.

Aromatically substituted ethylenes which are less hindered than stilbene (e.g., styrene, α-methylstyrene) are subject to anionic polymerization (Chapter XXIX), and therefore reduction of these monomers to hydrodimers and other low oligomers would require careful control of proton donor concentration. Whether the dianion $C_6H_5\bar{C}HCH_2CH_2\bar{C}HC_6H_5$ responsible for initiation of "living" anionic polymerization arises by dimerization of anionic radicals (EC) or by anion radical attack upon monomer followed by a second reduction (ECE) is not yet clear [4]. α-Methylstyrene tetramers, whose structure has been the subject of some controversy [5], have been prepared by the reaction of sodium with the monomer. Electrochemical preparation, particularly since it allows a wider choice of counterions, would be interesting.

Reduction of hydrocarbons in the presence of carbon dioxide (p. 262) is sometimes [3] an excellent preparative method for dicarboxylic acids (e.g., stilbene → 92% meso-diphenylsuccinic acid) but gives poor results when the

TABLE 1
Typical Pathways to Reductively Coupled Products

Mechanism type	Example
1. EC	$R_4\overset{+}{N} + \bar{e} \longrightarrow R\cdot + R_3N$ $R\cdot + R\cdot \longrightarrow RR$
2. CEC	$R_2CO + H^+ \longrightarrow R_2\overset{+}{C}-OH$ $2R_2\overset{+}{C}OH \xrightarrow{2\bar{e}} (R_2\underset{\vert}{C}-OH)_2$
3. EEC	$Br(CH_2)_3Br \xrightarrow{2\bar{e}} [BrCH_2CH_2CH_2^-]$ $[BrCH_2CH_2CH_2^-] \longrightarrow$ 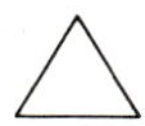
4. ECEC	$\begin{matrix}CHCOOEt\\ \Vert\\ CHCOOEt\end{matrix} \xrightarrow{\bar{e}} \left(\begin{matrix}CHCOOEt\\ \vert\\ CHCOOEt\end{matrix}\right)^{\dot{-}}$ $\left(\begin{matrix}CHCOOEt\\ \vert\\ CHCOOEt\end{matrix}\right)^{\dot{-}} + \begin{matrix}CHX\\ \Vert\\ YCH\end{matrix} \longrightarrow \left(\begin{matrix}CHCOOEt\\ \vert\\ CHCOOEt\\ \vert\\ CHX\\ \vert\\ CHY\end{matrix}\right)^{\dot{-}}$ $\left(\begin{matrix}CHCOOEt\\ \vert\\ CHCOOEt\\ \vert\\ CHX\\ \vert\\ CHY\end{matrix}\right)^{\dot{-}} \xrightarrow[2H_2O]{\bar{e}} \begin{matrix}CH_2COOEt\\ \vert\\ CHCOOEt\\ \vert\\ CHX\\ \vert\\ CH_2Y\end{matrix} + 2O\bar{H}$
5. (CE + E) + CC	$C_6H_5COCH_3 + H^+ \xrightarrow{\bar{e}} C_6H_5\underset{\bullet}{C}(OH)CH_3$ $C_6H_5COCH_3 \xrightarrow{\bar{e}} (C_6H_5COCH_3)^{\dot{-}}$ $C_6H_5\underset{\bullet}{C}(OH)CH_3 + (C_6H_5COCH_3)^{\dot{-}} \longrightarrow \begin{matrix} & & OH & & \\ & & \vert & & \\ C_6H_5 & - & C & - & CH_3\\ & & \vert & & \\ C_6H_5 & - & C & - & CH_3\\ & & & \diagdown & \\ & & & & O^-\end{matrix}$ $\begin{matrix} & & OH & & \\ & & \vert & & \\ C_6H_5 & - & C & - & CH_3\\ & & \vert & & \\ C_6H_5 & - & C & - & CH_3\\ & & \vert & & \\ & & O^- & & \end{matrix} \xrightarrow{H^+} \underset{\vert}{[C_6H_5C(OH)CH_3]_2}$

voltage required (e.g., for butadiene) is close to that for the discharge of the supporting cation [6].[1]

Electrolytic reductive coupling of stilbene with ethyl iodide gave a poor yield of 1, 2-diphenylbutane [3]. Electron transfer from an alkali metal and not from an electrode was used to prepare naphthalene anion radical and dianion, which species underwent interesting coupling reactions with α- and α, ω-dihalides. Evidence has been adduced that the coupling is not due to a nucleophilic attack of the anion radical upon RX but rather an electron transfer followed by radical coupling [7].

Highly conjugated hydrocarbons like 9-benzalfluorene(I), 2-phenyl-1, 3-butadiene, and 8, 8-diphenylbenzofulvene(II) are rather easily electroreducible and form coupled products with nonreduced activated olefins [8]. Thus, (I) and (II) with acrylonitrile formed (III) and (IV), respectively. Hydrodimerization was also achieved with (I) but not with (II).

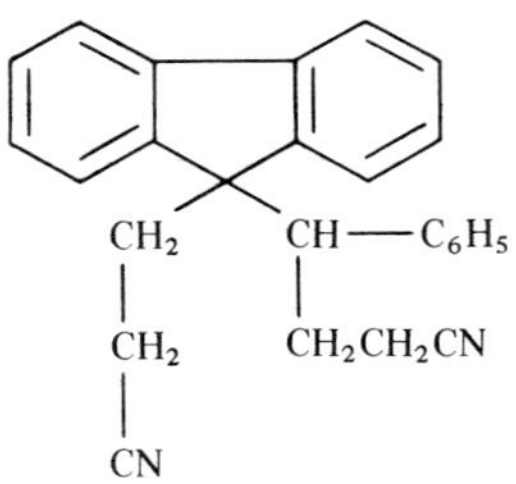

(III)

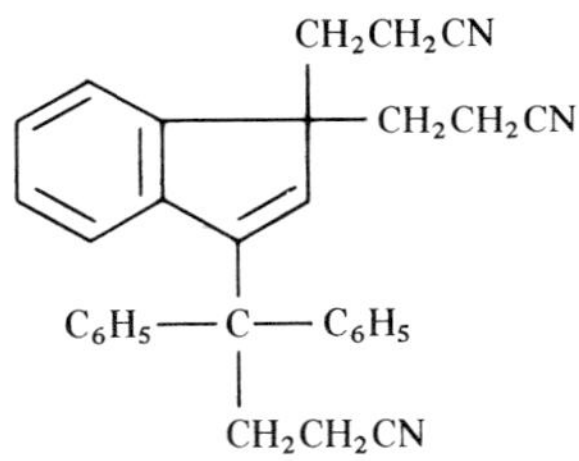

(IV)

B. Halogen Compounds

In the fullest sequence — and very few experimental examples showing all the features have been provided — an organic halide (V) undergoes a one-electron reduction to produce an intact radical ion (VI); the radical ion then

$$\underset{(V)}{RX} + \bar{e} \rightarrow \underset{(VI)}{[RX]^{\overset{\bullet}{-}}} \rightarrow \underset{(VII)}{R^{\bullet}} + X^{-} \xrightarrow{\bar{e}} \underset{(VIII)}{R^{-}} + X^{-}$$

fragments and the radical formed, (VII), may take up another electron and form an anion (VIII). The possible routes to dimers are then: (1) radical recombination of (VII); (2) anionic attack of (VIII) upon (V); (3) interaction of two molecules of (VI) to yield R-R and $2X^-$. The bulk of the experimental evidence has been interpreted on the basis of (1) and/or (2).

Examples of inter- and intramolecular dimerization following cathodic cleavage of organic halides are given in Chapter VI.

C. Nitro and Derived Compounds

The intermolecular reductive coupling of nitro compounds to form azoxy, azo, and hydrazo derivatives is discussed in Chapter VII. The intramolecular reductive coupling of nitro compounds, which leads to heterocycles, is discussed in Chapter XVII.

D. Saturated Carbonyls

1. Hydrodimerizations

Direct and indirect (amalgam) reductive couplings of carbonyls to gem-diols (pinacols) have been the subject of extensive studies ([1, 9] and Chapter

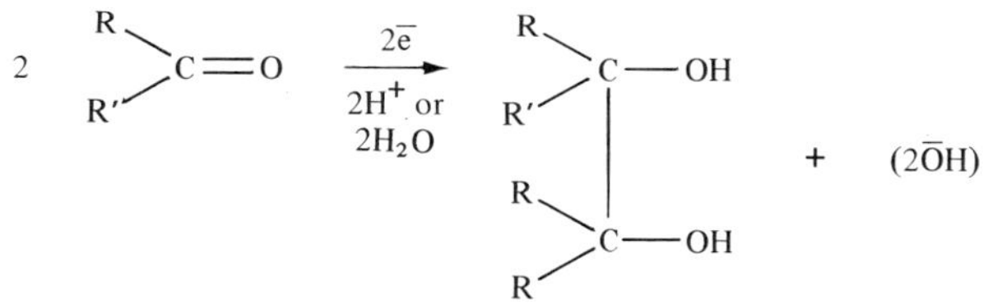

(8)

R = alkyl or aryl
R′ = alkyl or aryl or H

VIII). Because of the great variety of conditions that have been used by a host of workers, no simple statement can be made concerning the optimum method for achieving the dimeric products except that aromatic compounds have given more satisfactory results than aliphatic compounds. The principal side reactions are reduction to alcohol, reduction to hydrocarbon, formation of organometallics, and rearrangement of the pinacol to its pinacolone. When "active methylene" is present or is generated in the presence of alkali additional chemical side reactions are to be expected.

Typical hydrodimerizations have been discussed in Chapter VIII.

2. Mixed Reductive Coupling

Sugino and coworkers [10] have reported reductive couplings between ketones and acrylonitrile, maleic acid, and pyridine. In 20% sulfuric acid, acetone (R = R′ = Me), methyl ethyl ketone (R = Me, R′ = Et), and diethyl ketone (R = R′ = Et) gave the indicated products with 70, 60, and 30% current efficiencies, respectively, at mercury [Eq. (9)]. [Benzophenone (R = R′ = Ph) did not participate in cross coupling.] The cathode voltage required (ca. -1.2 V SCE) indicated reduction of the protonated ketone. When a mixture of acetone and acrylonitrile [Eq. (9)] was electrolyzed in

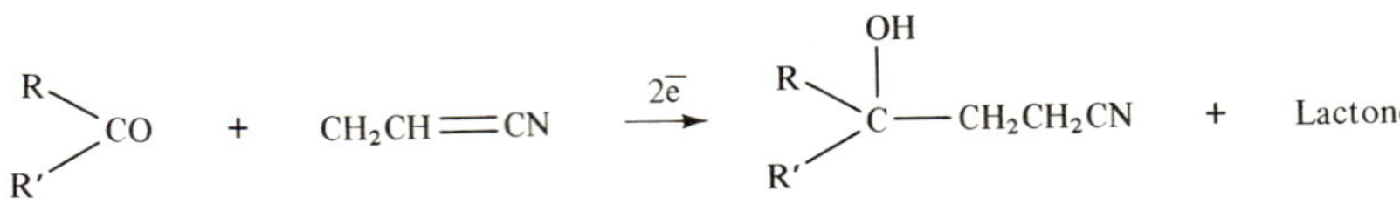 (9)

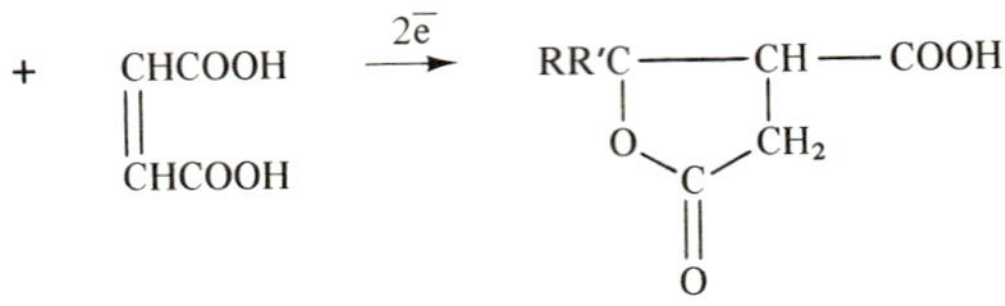 (10)

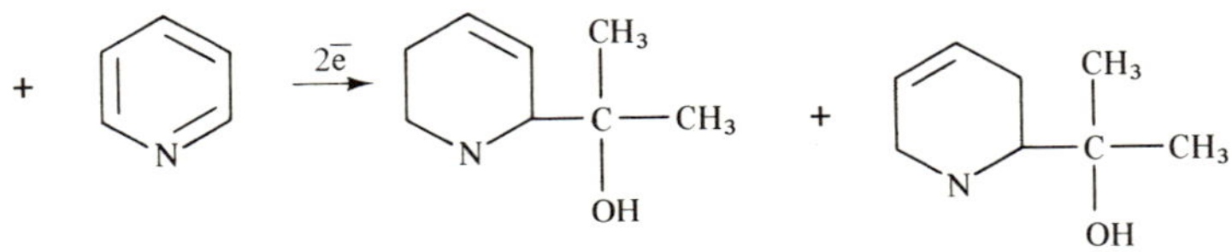 (11)

aqueous tetraethylammonium p-toluenesulfonate, it was the acrylonitrile that was reduced (ca. -1.9 V SCE); the same cross-coupled products were formed with ca. 10% current efficiency together with adiponitrile and propionitrile.

The reactions of Eq. (11) have been further examined by Czech workers [11].

Other syntheses yielding products in which RRCOH has been added to the terminal position of olefins have been carried out. Nicolas and Pallaud [12] obtained current efficiencies of 1-24% using acid-acetone solutions and cyclohexane, 1, 3-butadiene, styrene, α-methylstyrene, and 1-heptene. The products were mono- or di-tert-alcohols. When tetraethylammonium p-toluenesulfonate was used as electrolyte [13], the yield of cross-coupled products increased: acetone + styrene → 2-methyl-4-phenyl-2-butanol (42%); acetone + butadiene → 2-methyl-5-hexen-2-ol (9%) and 2-methyl-4-hexen-2-ol (23%); acetone + trans-stilbene → 2-methyl-3, 4-diphenyl-2-butanol (45%); benzophenone + ethyl acrylate → γ,γ-diphenyl-γ-butyrolactone (>25%); acetone + ethyl acrylate → γ,γ-dimethyl-γ-butyrolactone (51%).

Reduction of a mixture of acetone and allyl alcohol in aqueous alkaline solution at mercury-activated graphite gave 95-98% of 2-methyl-2, 5-pentanediol [14].

Both (C)EEC [46] and (C)ECE [15] mechanisms have been proposed for the process of Eq. (9) (R = R$'$ = Me): the first postulates that 2ē reduction of acetone yields the anion[2] $(CH_3)_2\underline{C}OH$ which nucleophilically attacks acrylonitrile; the second considers that the one-electron reduction product $(CH_3)_2\dot{C}OH$ adds to the β-position of acrylonitrile and that the resulting radical is then rapidly reduced (and protonated) to the final product. The evidence for the anionic mechanism is mainly negative: (a) failure of benzophenone to participate in the reductive coupling; (b) absence of polymerization which might be expected to result from intermediacy of free radicals. The evidence for the (C)ECE mechanism is based on data which demonstrate that a chemical reaction intervenes between the two one-electron reductions.

The older literature has several examples of reductive coupling between carbonyls and alkyl halides. For example, electrolysis of benzaldehyde in the presence of ethyl iodide yielded a small quantity of the ethyl ether of phenylethylcarbinol [17]; electrolysis of benzophenone in the presence of ethyl iodide yielded diphenylethylcarbinol [18].

E. α, β-Unsaturated Carbonyls

Hydrodimerizations have been numerous and are discussed in Chapter IX; cross couplings have been few in number. Typical examples are given in Table 2.

TABLE 2

Representative Cross Couplings with α, β-Unsaturated Carbonyls

Carbonyl compound	Partner	Cathode	Medium	Principal products	Ref.
Mesityl oxide	Acrylonitrile	Hg	DMF + H_2O + $Et_4\overset{+}{N}$ p-T$\bar{S}$	4, 4-Dimethyl-6-cyano-2-hexane	[64]
Mesityl oxide	Ethyl acrylate	Hg	HOAc-NaOAc	4, 4-Dimethyl-6-carbethoxy-2-hexanone	[64]
Methyl vinyl ketone	Diethyl fumarate	Hg	DMF + H_2O + $Et_4\overset{+}{N}$ p-T$\bar{S}$	Diethyl-α-(2-acetyl-methyl) succinate	[33]

F. α, β-Unsaturated Acids

C. L. Wilson [19] reductively dimerized sorbic acid to a mixture of products. In a more comprehensive study, Wilson and co-workers [20] reduced cinnamic acid in aqueous alcoholic sulfuric acid at mercury and obtained a moderately good yield of 3, 4-diphenylcinnamic acid. The method worked well for p-methoxy-, o-chloro- and m-hydroxycinnamic acids and poorly for the o-cyano acid. It failed with o-, m-, and p-nitrocinnamic acids and with the p-amino derivative.

In a recent application of this procedure Wawzonek et al. [21] obtained 2, 3, 4, 5-tetraphenyl-1, 6-hexanedioic acid from α-phenylcinnamic acid.

G. Activated Olefins

The dramatic yield improvement — from 5 to 60% — achieved by Knunyants and co-workers [22] in reductively coupling acrylonitrile (AN) to the important nylon intermediate adiponitrile (ADN) by alkali metal amalgams stimulated research on new aspects of electrolytic reductive coupling.[3] The scope and limitations of hydrodimerization [Eq. (12)], mixed (or

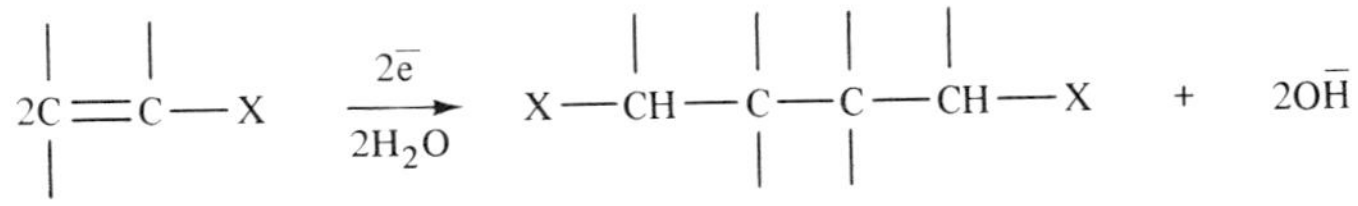

(12)

crossed) reductive coupling [Eq. (13)], and electrohydrocyclization [Eq.

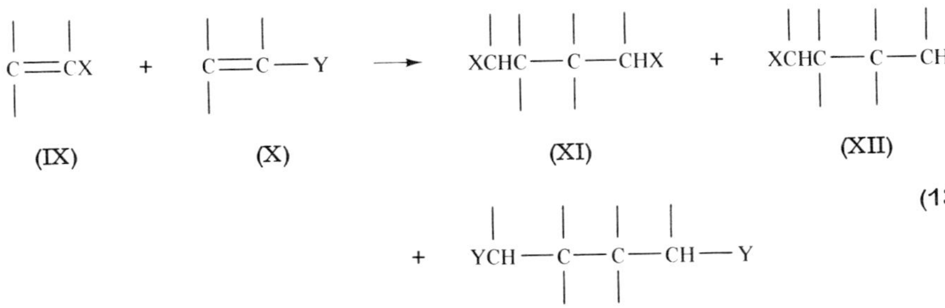

(IX) (X) (XI) (XII) (XIII)

(13)

(XIV)

(14)

(14)] of Michael acceptor olefins (IX), (X), and (XIV), in which X and Y are electron-withdrawing groups (except nitro) and Z is a bridging group like

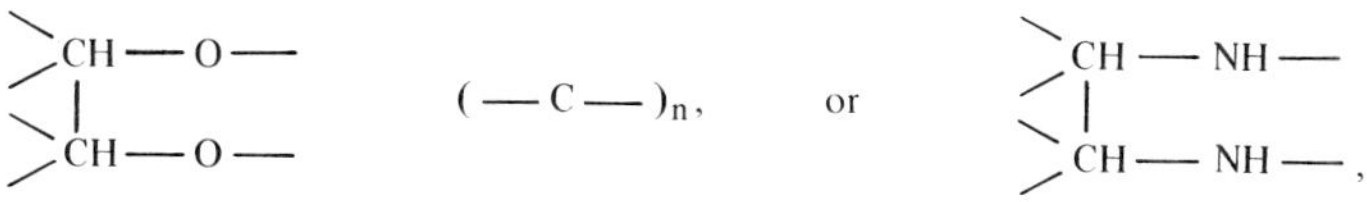

have been discussed in a series of papers and patents [25]. Representative examples of Eq. (12) are given in Table 3 (p. 689), of Eq. (13) in Table 4 (p. 690), and of Eq. (14) in Table 5 (p. 691). Coupling occurs almost exclusively through the β-positions [26]; yields are generally very good. The method has recently [27] been extended to 1, 2-diactivated olefins [Eqs. (15), (16) and Table 6, p. 692].

TABLE 3

Representative Hydrodimerizations of Michael Acceptor Olefins [a]

Olefin	Operating cathode V [b] -E vs. SCE	Current eff. (yield), %	Reference
CH_2=CHCN	1.81-1.91	>90 (>90)	[25b]
CH_2=C(CH_3)CN	2.01-2.05	75.3 (—)	[25]
CH_2=C(CH_2CH_2CN)CN	1.8-1.9	93.5 (—)	[29]
$NCCH_2CH_2$CH=CHCN	—	1.2-66.8 [c]	[32]
CH_2=CHCOOEt	1.85	74-87 (—)	[25]
$(CH_3)_2$C=CHCOOEt	2.10-2.20	66.2 (—)	[25]
CH_3CH=CHCONEt_2	2.03-2.12	61.4 (—)	[25]
CH_2=CH-4-pyridyl	1.6	82 (—)	[30]
CH_2=CHPO(OEt$)_2$	2.0-2.1		[25d]
C_6H_5CH=9-fluorylidene	1.5-1.6	66 (—)	[31]
C_6H_5CH=CHC_6H_5			[3]
CH_2=CHCOCH_3	1.35-1.42	(40-70) [d]	[33]
$(CH_3)_2$C=CHCOCH_3	1.35	(0-56) [d]	[34]
CH_2=CHSO_2CH_3	1.72-1.88	—	[25d]

[a] See Eq. (12), p. 688.

[b] Usually a few tenths of a volt anodic of $E_{1/2}$.

[c] Depending on cathode and electrolyte used.

[d] Strongly dependent on solvent.

TABLE 4

Representative Mixed Reductive Couplings [a]

Olefin	$-E_{1/2}$ SCE	Olefin	$-E_{1/2}$ SCE	Operating V -E SCE	Product	Ref.
CH_2=CHCOOR	1.85	CH_2=CHCN	1.9	1.8-1.9	$ROCO(CH_2)_4COOR$ + $ROCO(CH_2)_4CN$ + $NC(CH_2)_4CN$	[35]
CH_2=CH-CH=CHCN	1.5	CH_2=CHCN	1.9	1.5-1.7	$NC(CH_2)_6CN$ (after hydrogenation) + some isomers	[25c]
CHCOOEt ‖ CHCOOEt	1.3-1.4	CH_2=CHCN	1.9	1.3-1.4	$(EtOCOCH_2CHCOOEt)_2$ + $EtOCOCH_2CH(COOEt)CH_2CH_2CN$	[25]
$(CH_3)_2C$=$CHCOCH_3$	1.59-1.79	CH_2=CHCN	1.9	1.65-1.70	Hydrodimers of mesityl oxide + $(CH_3)_2C(CH_2CH_2CN)CH_2COCH_3$	[25]
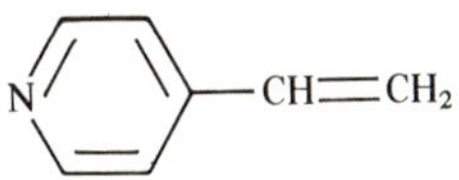	1.5	CH_2=$CHCOCH_3$	1.4	1.4-1.5	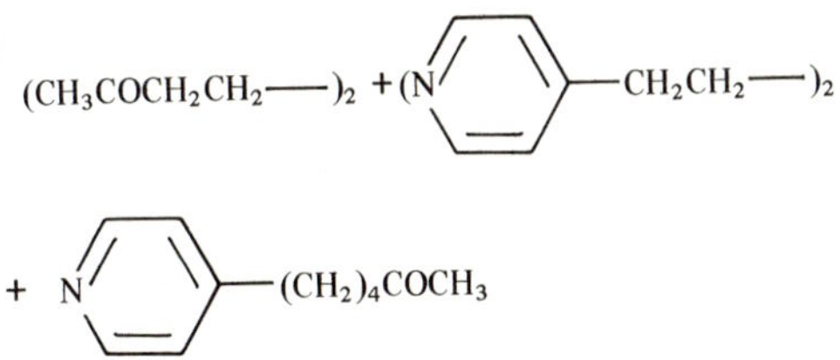	[30]

[a] See Eq. (13), p. 688, and Eq. (3), p. 680.

TABLE 5

Representative Intramolecular Reductive Coupling (Electrohydrocyclization)[a]

Olefin	Product	% Cyclization	Ref.
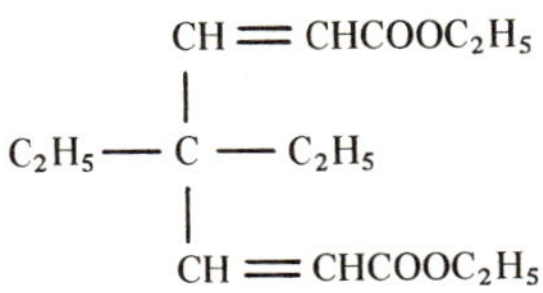	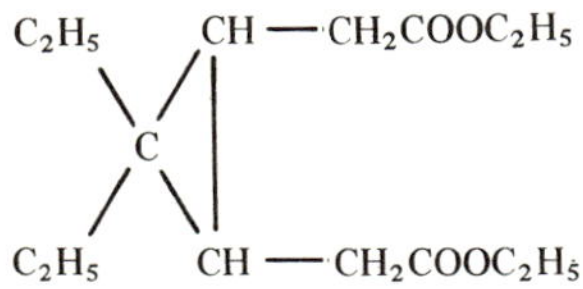	98	[25e]
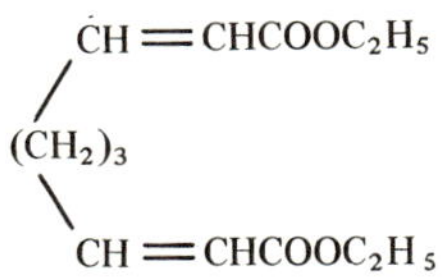	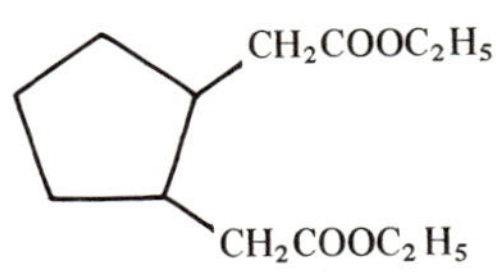	100	[25e]
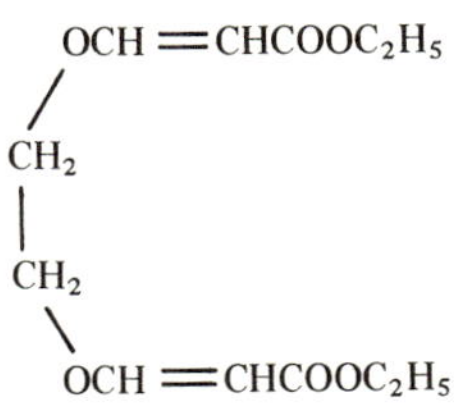	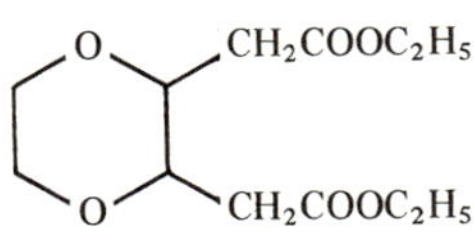	89	[25e]
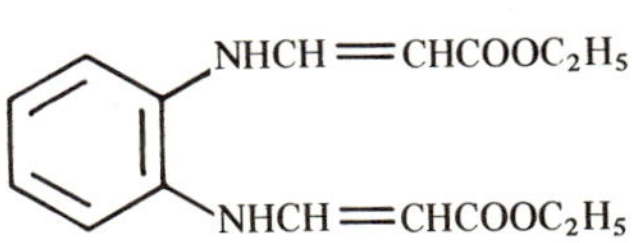	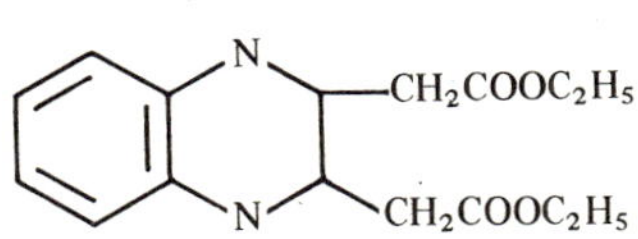	86	[25e]

[a] See Eq. (14), p. 688.

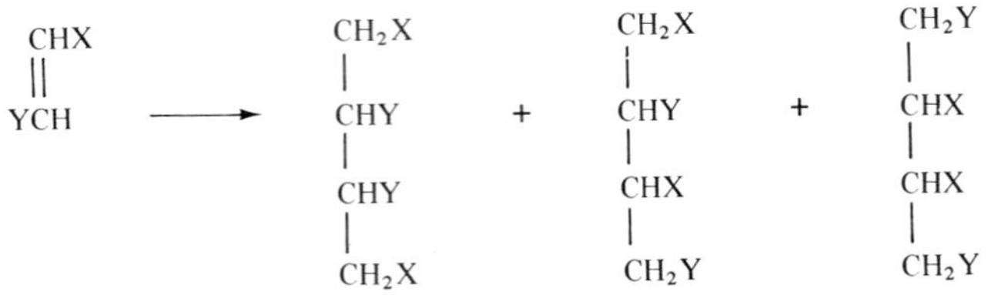

(XV) (XVI) (XVII) (XVIII)

(15)

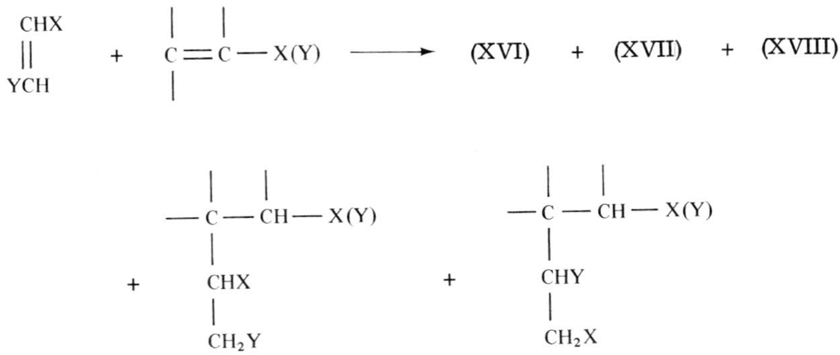

(16)

TABLE 6

Representative Reductive Couplings with trans-1, 2-Diactivated Olefins [a]

Olefin XCH=CHY		Hydrodimers,	Hydrodimer distribution			
X	Y	% of total product	$CHYCH_2X$ / $CHYCH_2X$	$CHYCH_2X$ / $CHXCH_2Y$	$CHXCH_2Y$ / $CHXCH_2Y$	Ref.
C_6H_5CO	C_6H_5CO	44	Symmetrical			
C_6H_5CO	CN	55	100	—	—	[27]
C_6H_5CO	$COOC_2H_5$	96	77	23	—	[27]
$COOC_2H_5$	$CON(CH_3)_2$	96	66	31	2	[27]
CN	$COOC_2H_5$	84	—	5	95	[27]

[a] See Eq. (15) above.

1. Synthetic Aspects

Considering hydrodimerizations [Eq. (12)] first, the effect of the nature of X and of the electronic and steric character of α- and β-substituents upon the reduction potential and the ease of coupling has been summarized [28]. Electron-withdrawing substituents shift the required voltage anodically, electron-donating groups cathodically. Bulky β-substituents may block coupling completely [36] and favor simple saturation of the double bond. Extension of the conjugation, as in 1-cyano-1, 3-butadiene [25c], causes an anodic shift; interestingly, coupling occurs mainly through the ω-position.

The widespread study of the ADN process in particular has had beneficial effects on the development of an understanding of the several procedural factors affecting the success of electrolytic reductive coupling in this series. Side reactions which must be controlled are simple saturation of the double bond, formation of organometallics, oligomerization/polymerization, nonelectrochemical cyanoethylations, saponifications, cyclizations, and electrochemical cleavages. Studies have been concerned with the effect upon these alternate reactions of (a) nature of cathode material, (b) pH of catholyte, (c) supporting electrolyte, particularly its cation, (d) concentration of olefin in catholyte, and (e) type of cell.

a. Cathode Material. High-hydrogen-overvoltage cathodes give best results, but even within this group there are individual variations. Mercury and lead are excellent [25b]; graphite is good [37]; cadmium favors the formation of saturated monomers; and tin forms substantial quantities of tetra-β-cyanoethyltin with AN [38] and lesser quantities of organometallics with other olefins [39]. From the mechanistic point of view it is interesting that, whereas lead and mercury do not form organometallics in the reduction of AN, they and thallium do yield organometallics in the reduction of β-iodopropionitrile [40].

b. pH of Catholyte. The qualification must be borne in mind that in aqueous electrolyses the pH in the vicinity of the cathode may be several units higher than in the bulk of the solution. Too low pH favors oligomerization/polymerization [41], formation of saturated monomer [42], chemical side reactions such as hydrolysis, etc.; too high pH favors cyanoethylation of water (in the case of AN), saponifications, cyclizations [21, 27], etc. Good to excellent results have been reported in ranges of apparent pH of 3.5-9.

c. Supporting Electrolyte. Optimum practical results are obtained with strong aqueous solutions of quaternary ammonium hydrotropic salts (e.g., tetraethylammonium p-toluenesulfonate) which permit one to dispense with or minimize the need for organic cosolvents and to eliminate the possibility that discharge of the cation will occur simultaneously with the reduction of the olefin. There is a further advantage to the quaternary ammonium cations: They are preferentially adsorbed at, e.g., mercury and, unlike the alkali metal cations, are not hydrated so that the intermediate produced by reduction of the olefin has a favorable opportunity to react first with another molecule of parent rather than with water [43]. This effect is manifested even at low concentrations of supporting electrolyte [44], even when quaternary ammonium is added to electrolytes containing alkali metal cations [45], and has been applied to a radical improvement in the results obtained from amalgam reductions (cf. Chapter XXIII).

d. Concentration of Olefin in Catholyte. Optimum practical yields and conversions have been obtained using homogeneous solutions containing more than ca. 5% of olefin. The electrolysis is run to partial conversion with replenishment of the elefin to maintain this level. Lower concentrations and emulsions of the olefin in nonhydrotropic electrolytes are claimed to be operable [46] although there are numerous reports [45, 47] that very low concentrations favor the formation of saturated monomers. Too high a concentration of olefin with concomitant decrease in the concentration of proton donors leads to polymerization [48].

e. Type of Cell. Excellent results are obtained in divided cells [25h] with cation-exchange membranes [25h]. Undivided cells are operable and have the advantage of requiring smaller cell voltages but are subject to the restrictions that (a) the olefin and product not be substantially oxidized at the anode and (b) the oxygen evolved at the anode in aqueous systems not promote undesirable side reactions. For economic reasons the spacing between the electrodes should be minimized. Thin polyolefin spacers [49] as well as various ingenious devices which are intermediate between totally divided and totally undivided cells have been proposed [47].

The synthetic considerations discussed above also apply to mixed reductive coupling [Eq. (13)]. In addition the following special features obtain in this category:

a. When the two participating olefins each reduce individually at about the same cathode voltage, three coupling products are obtained.

b. When the individual reduction potentials are sufficiently far apart ($\leq$0.5 V) that the cathode voltage during the electrolysis of the mixture can be controlled to avoid reduction of the more cathodically reduced olefin [say, (X)] only two products are obtained, (XI) and (XII). Within this voltage range, more cross-coupled product (XII) is obtained toward the negative limit of the range than at more anodic voltages [25g].[4] If the gap between (IX) and (X) is greater than ca. 0.5 V, no cross coupling occurs.[4] This result finds analogy in certain anionic homo- vs. copolymerizations [50].

The reactions of Eq. (14) are an excellent method for producing three- to six-membered cyclic compounds (carbo and hetero) with functional groups in the side chains. In contrast to the acyloin condensation [51] it fails in the construction of larger ring compounds.

The couplings represented in Eqs. (15) and (16) represent the most complex synthetic results obtained in this series. When X = Y, practical yields in one step are obtained of products which could otherwise be prepared only by multiple-step syntheses; when X ≠ Y, the reactions of Eqs. (15) and (16) are less useful for synthesis than for the information they provide concerning the mechanism of coupling.

2. Mechanistic Aspects

The postulate that electron transfers occur one at a time is becoming more and more firmly established as critical experimental examples with suitable models are being provided. If this postulate applies to electrolytic reductive coupling of activated olefins, then the intermediate species responsible for the coupling must be one or more of the following [using the process of Eq. (12) as an example]:

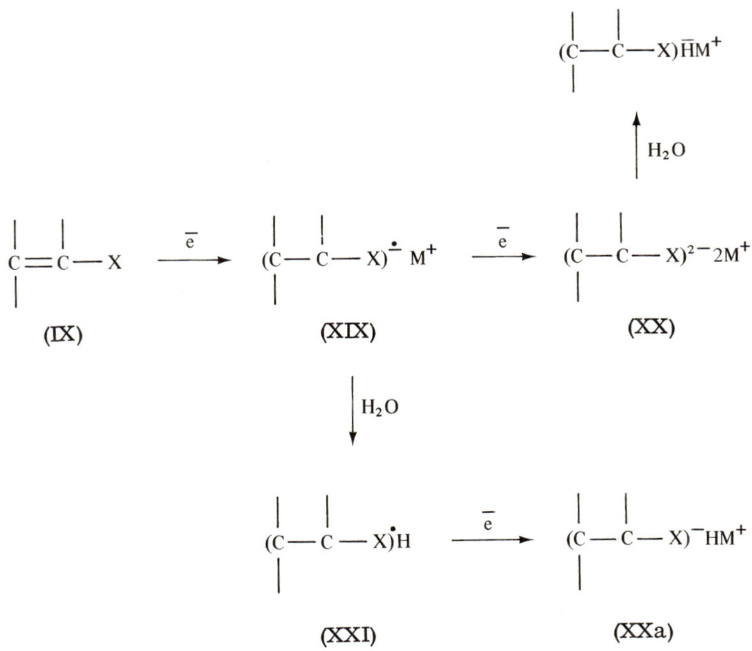

The one or more qualification is important in view of the fact that the pH (and the proton donor availability) and the nature of the counterion M^+ (and solvent) influence the selection of the alternate routes to products [52]. Furthermore, as pointed out above (p. 681), the results of macroelectrolyses are the summation of reactions near the electrode surface and in the "bulk" of the solution [48].

In general, two successive one-electron reductions in suitable cases can be observed by (a) polarography in anhydrous media [(IX) to (XIX), (XX)], (b) reversibility of the first electron transfer as revealed by cyclic voltammetry, and (c) ESR spectrum of the intermediate anion radical. These experimental methods fail for activated olefins (IX) which have only one electron-withdrawing group to help stabilize a potential lone electron and an unshared electron pair (XIX) because (a) they are prone to oligomerize/polymerize during electrolysis in anhydrous media; (b) the electron transfer has been found to be completely irreversible under all conditions; and (c) a corollary of (b), no or only faint ESR signals have been obtained [42]. Only one reduction wave is found. Therefore, (XIX) must disproportionate or be extremely rapidly reduced to (XX) (EE reaction) causing subsequent C steps to be anionic, or it must rapidly undergo a C reaction with water [→ (XXI)] or parent [Eqs. (17), (18)] and then be rapidly reduced:

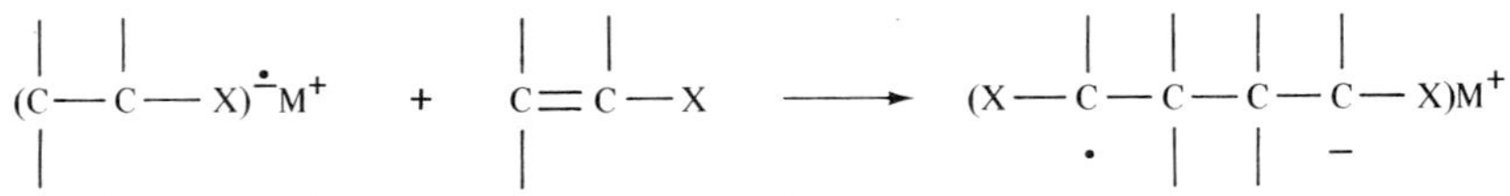
(17)

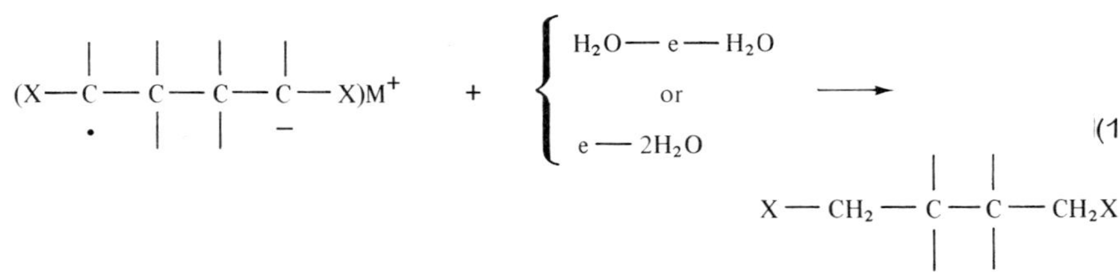
(18)

The most extensively studied system in this series has been AN → ADN, but even here the experimental evidence has not established unequivocally the mechanism(s) of the coupling. The absence of organometallics and polymers and the ability of AN to participate in cross couplings with more easily reduced olefins argue against radical coupling of (XIX) as the major route in catholytes containing quaternary ammonium cations. Beck [42] inferred from his data that one electron and one molecule of water are involved in the rate-determining step [(IX) → (XXI)]. The actual results of the coupling (β to β) require that, if (XX) is the species attacking a neutral AN molecule, the charge in (XXa) be localized at the unstabilized β-position rather than at the α-position. It has been suggested that this dilemma can be resolved by postulating that the β-carbon atom of (XIX) or (XXI) is adsorbed on the mercury surface; the α-position is the one that is protonated before (XXa) is desorbed [43a]. The reaction of (XXa) with AN rather than with water has been explained by the existence of a water-poor zone next to the cathode at which hydrophobic quaternary ammonium ions are adsorbed [43]. Arguments from molecular orbital theory have been used to support both coupling via the dianion [53] and, alternatively [54], the scheme of Eqs. (17) and (18).

The course of events following the reduction of (XV) has been more amenable to experimental elucidation. Anion radicals from (XV) have been observed and their rates of reaction with water, parent olefin, and (XV) estimated [55]. Two reduction waves are manifested.[5] Coupling at the first reduction wave occurs by several pathways: (a) by anionic attack of the anion radical upon parent olefin in such a way that the dimeric species acquires the most stable anionic and (now separated) radical site [Eq. (19)]; (b) by pro-

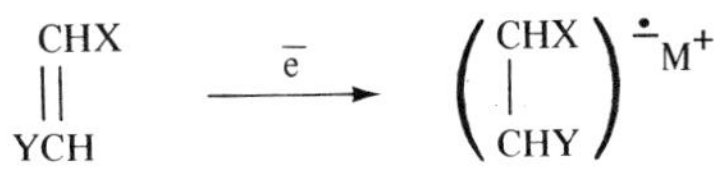

(XV) (XXI)

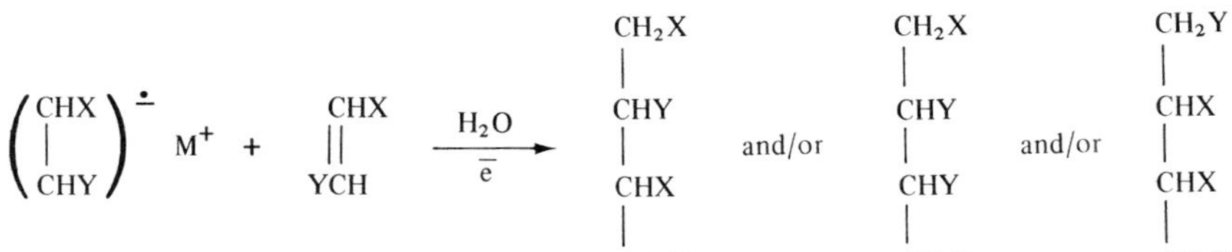

(19)

(XXI)

tonation of (XV) followed by attack of the resulting anion (thermodynamic or kinetic control?) upon starting olefin in the vicinity of the electrode;

(XXII) (XXIII)

(c) by anionic attack of (XXII), (XXIII) upon acceptor olefin(s) in the bulk of the solution; and (d) to a certain extent, depending on the nature of M^+ and solvent, by radical dimerization of (XXI) [56].

Finally, (XIV) shows two reduction waves when, e.g., X = Y = COOR and Z is such that a suitable ring size can form. The relative heights of the two waves correlated well with the degree of cyclization vs. acyclic reduction [25e]. The anodic shift (first wave) was ascribed to <u>concerted</u> one-electron reduction–ring closure [Eq. (20)]; a similar phenomenon was encountered in the reduction of those α,ω-dihalo compounds which formed

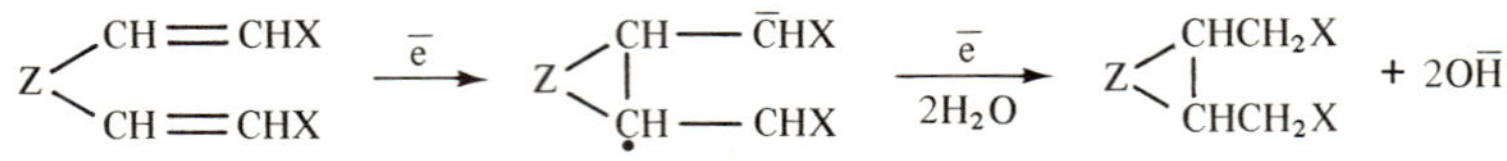

(20)

cyclized products (Chapter VI, p. 295).

Little work has been done on electrolytic coupling of diactivated olefins (XXIV) in which both electron-withdrawing groups are on the same carbon atom. These compounds are very subject to anionic polymerization [57] and

(XXIV) (XV)

have obvious inherent steric problems; both factors contribute to experimental difficulties in evaluating their suitability as coupling partners. It has been observed that (XXIV) reduces at a more anodic voltage than its (XV) isomer [55]. Often only single reduction waves are obtained. It has been suggested [58] that intermediate anion radicals may dimerize.

H. Onium Compounds

Although the reductive cleavage of onium compounds, particularly quaternary ammonium compounds, has been studied very extensively (Chapter XXI), there have been only a few cases reported in which dimeric products were obtained from an R group which was cleaved. Electrolyses in dipolar aprotic solvents yielded 6-35% of R-R when R in $RR^1R^2R^3N^+$ was cinnamyl, p-methoxybenzyl, fluorenyl, α-phenylethyl, or benzyl [59a]. It is obvious that these are the R groups that form relatively stable radicals.

Bibenzyl was formed from benzyltrimethylammonium nitrate in DMF at Al but not at Pt [59b]. An extensive study of the perchlorate salt [59c] shows the effect of four solvents and 17 cathodes on the distribution of the reduction products toluene, bibenzyl, and N, N-dimethyl-N-(2-methyl benzyl) amine.

The establishment of the mechanism(s) for the formation of dimeric products is beset by some of the same problems involved in the reductive dimerization of organic halides (p. 287): dimerization of R • vs. anionic attack of R^- upon $RR^1R^2R^3N^+$. Since electrolysis of d-α-phenylethyltrimethylammonium nitrate gave only dl- and meso-2, 3-diphenylbutane, the former mechanism is favored in this case. It is certainly possible that coupling occurs by a radical mechanism in aprotic solvents and by an anionic mechanism in water.

A recent Dow patent [60] claims a yield of 95% of p, p′-dinitrobibenzyl in the electrolysis of dimethyl-p-nitrobenzylsulfonium chloride at a copper or platinum cathode and carbon anode using KCl as supporting electrolyte.

I. Miscellaneous Classes

1. Immonium Salts

The system below was chosen for study because reductive coupling,

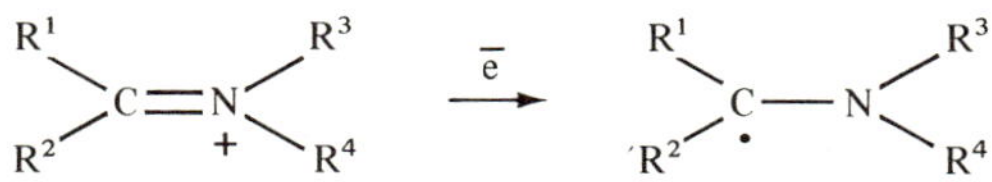

(XXV) (XXVI)

when it occurred, had to proceed unequivocally by free-radical dimerization [61]. One-electron reduction was verified by coulometry and in certain cases by observing the ESR spectra of (XXVI). Cyclic voltammetry was used to estimate the reversibility of the system; in irreversible systems dimer was, indeed, obtained in macroelectrolysis. The dimer could be electro-oxidized to starting material.

2. Schiff Bases and Azo Compounds

The reductive coupling of benzalaniline and azobenzene with AN to yield (XXVII) and (XXVIII), respectively, has been described [62].

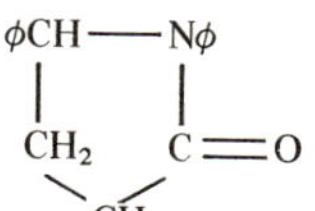

(XXVII)

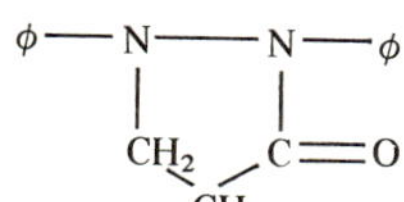

(XXVIII)

The hydrodimerization of Schiff bases has recently [63] been re-investigated. Among the groups R and R^1 employed were phenyl, p-tolyl,

$$2RCH{=}NR^1 + 2\bar{e} + 2H^+ \longrightarrow \begin{array}{l} RCH{-}NHR^1 \\ | \\ RCH{-}NHR^1 \end{array}$$

benzyl, cyclohexyl, and anisyl. The cathode was mercury, and the catholyte a solution of tetra-n-butylammonium bromide in a mixture of ethanol, methyl acetate, and water (5:3:1). Except where the dl product was difficult to isolate from the by-product sec-amine, the yields were 36-67% of ca. a 1:1 mixture of dl and meso hydrodimers.

NOTES

[1] Since CO_2 is electroreducible at ca. -2 V (SCE) at Hg, the mechanism of these trapping reactions is uncertain.

[2] This intermediate has also been proposed to be the precursor to pinacol ([16] cited in [10]) and to organometallics.

[3] Simultaneously, there have been advances in reductive coupling via amalgams (e.g., [23] and p. 806, Chapter XXIII) and via catalytic hydrogenation [24].

[4] Since (XV) in general is reduced more readily than (IX), the reactions of Eq. (16) rather than those of Eq. (13) had to be used in arriving at these generalizations. It is assumed that the mechanisms involved in Eq. (16) are not significantly different from those involved in Eq. (13).

[5] The existence of anionic species even in the presence of bulk water has been noted in a number of cases, e.g., [13].

REFERENCES

[1]. M. M. Baizer and J. P. Petrovich, "Electrolytic Reductive Coupling," in Progress in Physical Organic Chemistry (A. Streitwieser and R. W. Taft, eds.), Vol. 7, Wiley-Interscience, New York, 1970, pp. 189-227.

[2]. P. Zuman, J. Polarogr. Soc., 13, 53 (1967).

[3]. S. Wawzonek, E. W. Blaha, R. Berkey, and M. E. Runner, J. Electrochem. Soc., 102, 235 (1955); S. Wawzonek and J. W. Fan, J. Amer. Chem. Soc., 68, 2541 (1946).

[4]. E.g., M. Levy and M. Szwarc, ibid., 82, 521 (1960).

[5]. A. Vrancken, J. Smid, and M. Szwarc, Trans. Faraday Soc., 58, 2036 (1962); N. M. Berger, J. J. K. Boulton, B. W. Brooks, and M. J. Evans, Chem. Commun., 1967, 8; R. L. Williams and D. H. Richards, ibid., 1967, 414.

[6]. J. W. Loveland, U.S. Pat. 3,032,489 (1962); CA, 57, 4470 (1962).

[7]. J. F. Garst and J. T. Barbas, Tetrahedron Lett., 1969, 3125 and citations; see also, S. Bank and J. F. Bank, ibid., 1969, 4533.

[8]. M. M. Baizer and J. D. Anderson, J. Org. Chem., 30, 1348 (1965).

[9]. (a) S. Swann, Jr. "Electrolytic Reactions," in Technique of Organic Chemistry (A. Weissberger, ed.), Vol. II, Wiley-Interscience, New York, 1956, pp. 495-496; (b) M. J. Allen, Organic Electrode Processes, Reinhold, New York, 1958.

[10]. K. Sugino and T. Nonaka, Electrochim. Acta, 13, 613 (1968); J. Electrochem. Soc., 116, 615 (1969); T. Nonaka and K. Sugino, Denki Kagaku, 38, 105 (1970); M. M. Baizer and J. L. Chruma, J. Electrochem. Soc., 118, 450 (1971); T. Nonaka and T. Sekine, Denki Kagaku, 39, 29 (1971).

[11]. M. Ferles, M. Vanka, and A. Šilhánková, Coll. Czech. Chem. Commun., 34, 2108 (1969).

[12]. M. Nicolas and R. Pallaud, Compt. Rend., 265C, 1044 (1967).

[13]. M. Nicolas and R. Pallaud, ibid., 267C, 1834 (1968).

[14]. A. P. Tomilov and B. L. Klynev, Zh. Obshch. Khim., 39, 470 (1969); CA, 70, 120479e (1969).

[15]. O. R. Brown and K. Lister, Discuss. Faraday Soc., 1968, No. 45, 106.

[16]. A. Yamura, T. Sekine, and K. Sugino, J. Electrochem. Soc. Japan, 34, 110 (1966).

[17]. J. M. Nelson and A. M. Collins, J. Amer. Chem. Soc., 46, 2256 (1924).

[18]. S. Wawzonek and A. Gundersen, J. Electrochem. Soc., 107, 537 (1960).

[19]. C. L. Wilson, Trans. Electrochem. Soc., 75, 353 (1939).

[20]. C. L. Wilson and K. B. Wilson, ibid., 84, 153 (1943); E. P. Goodings and C. L. Wilson, ibid., 88, 77 (1945).

[21]. S. Wawzonek, A. R. Zigman, and G. R. Hansen, J. Electrochem. Soc., 117, 1351 (1970).

[22]. I. L. Knunyants and N. S. Vyazankin, Izv. Akad. Nauk SSSR, Otd. Khim. Nauk, 1957, No. 2, 238-240.

[23]. E.g., Y. Arad, M. Levy, and D. Vofsi, Fr. Pat. 1,553,195, Jan. 10, 1969; CA, 72, 42853x (1970); and very many others; for a review see C. Van Eygen et al., Chim. Ind. (Genie Chim.), 104, 71 (1971).

[24]. E.g., W. J. Linn and A. B. Stiles (Du Pont), Fr. Pat. 1,572,892, June 27, 1969; CA, 72, 66427n (1970); G. Agnes, G. P. Chiusoli, and G. Cometti, Chem. Commun., 1968, 1515A.

[25]. E.g., (a) M. M. Baizer, Tetrahedron Lett., 1963, 973; (b) M. M. Baizer, J. Electrochem. Soc., 111, 215 (1964); (c) M. M. Baizer and J. D. Anderson, ibid., 111, 226 (1964); (d) M. M. Baizer and J. D. Anderson, J. Org. Chem., 30, 3138 (1965); (e) J. P. Petrovich, J. D. Anderson, and M. M. Baizer, ibid., 31, 3897 (1966); (f) M. M. Baizer, J. D. Anderson, J. H. Wagenknecht, M. R. Ort, and J. P. Petrovich, Electrochim. Acta, 12, 1377 (1967); (g) M. M. Baizer, J. P. Petrovich, and D. A. Tyssee, J. Electrochem. Soc., 117, 173 (1970); (h) U.S. Pats. 3,193,475-3,193,510, July 6, 1965, and others.

[26]. G. C. Jones and T. H. Ledford, Tetrahedron Lett., 1967, 615, reported that methacrylonitrile underwent some α- to β-coupling.

[27]. J. P. Petrovich, M. M. Baizer, and M. R. Ort, J. Electrochem. Soc., 116, 749 (1969).

[28]. M. M. Baizer and J. D. Anderson, ibid., 111, 223 (1964).

[29]. M. M. Baizer and J. D. Anderson, J. Org. Chem., 30, 1357 (1965).

[30]. J. D. Anderson, M. M. Baizer, and E. J. Prill, ibid., 30, 1645 (1965).

[31]. M. M. Baizer and J. D. Anderson, ibid., 30, 1348 (1965).

[32]. Yu. D. Smirnov, S. K. Smirnov, and A. P. Tomilov, J. Org. Chem. USSR, 4, 208 (1968).

[33]. M. M. Baizer and J. D. Anderson, J. Org. Chem., 30, 3138 (1965).

[34]. J. Wiemann and M. L. Bouguerra, Ann. Chim. (Paris), 2, 35 (1967).

[35]. M. M. Baizer, J. Org. Chem., 29, 1670 (1964); S. M. Makarochkina and A. P. Tomilov, Zh. Obshch. Khim., 40, 676 (1970); CA, 73, 41210u (1970).

[36]. M. M. Baizer and J. D. Anderson, J. Electrochem. Soc., 111, 223 (1964).

[37]. A. P. Tomilov and L. V. Kaabak, Zh. Obshch. Khim., 33, 731 (1963).

[38]. Ref. [25b] and citation; also I. N. Brago, L. V. Kaabak, and A. P. Tomilov, CA, 67, 104513a (1967).

[39]. D. W. Grisley, Jr. and M. M. Baizer, unpublished work.

[40]. A. P. Tomilov, Yu. D. Smirnov, and S. L. Varshavskii, J. Gen. Chem. USSR, 35, 390 (1965).

[41]. W. Kern and H. Ouast, Makromol. Chem., 10, 202 (1953).

[42]. F. Beck, Chem.-Ing.-Tech., 37, 607 (1965); also T. Asahara, M. Seno, and T. Arai, Bull. Chem. Soc. Japan, 42, 1316 (1969).

[43]. (a) I. E. Gillet, Chem.-Ing.-Tech., 40, 573 (1968); (b) L. G. Feoktistov, A. P. Tomilov, and I. G. Sevast'yanova, Elektrokhim., 1, 1300 (1965); A. P. Tomilov, E. V. Kryukova, V. A. Klimov, and I. N. Brago, Sov. Electrochem., 3, 1352 (1967).

[44]. A. P. Tomilov and V. A. Klimov, ibid., 3, 323 (1967); F. Beck, Ber. Bunsenges., 72, 379 (1968).

[45]. Y. Arad, M. Levy, I. R. Miller, and D. Vofsi, J. Electrochem. Soc., 114, 899 (1967).

[46]. E.g., Asahi Co., Fr. Pat. 1,533,472, July 19, 1968; CA, 71, 21705q (1969); Fr. Pat. 1,531,738, July 5, 1968; CA, 71, 21709u (1969).

[47]. L. G. Feoktistov, A. P. Tomilov, and I. G. Sevast'yanova, Sov. Electrochem., 1, 1165 (1965).

[48]. M. M. Baizer and J. D. Anderson, J. Org. Chem., 30, 1351 (1965), and literature cited therein.

[49]. C. Takagi and K. Ubukata, CA, 69, 66798x (1968), a review of 43 patents.

[50]. K. Höfelmann, R. Sattelmeyer, and K. Hamann, Makromol. Chem., 112, 300 (1968).

[51]. S. M. McElvain, Org. Reac., IV, 256 (1948).

[52]. J. P. Petrovich and M. M. Baizer, unpublished work.

[53]. S. Lazarov, A. Trifonov, and T. Vitanov, Z. Phys. Chem. (Leipzig), 3-4, 221 (1964).

[54]. M. Figeys and H. P. Figeys, Tetrahedron, 24, 1097 (1968).

[55]. J. P. Petrovich, M. M. Baizer, and M. R. Ort, J. Electrochem. Soc., 116, 743, 749 (1969).

[56]. W. V. Childs, J. T. Maloy, C. P. Keszthelyi, and A. J. Bard, Amer. Chem. Soc. Meeting, Chicago, Ill., Sept. 1970, Abstract 20; J. Electrochem. Soc., 118, 874 (1971).

[57]. K. Höfelmann, R. Sattelmeyer, and K. Hamann, Makromol. Chem., 130, 221 (1969).

[58]. M. Bargain, Compt. Rend., 255, 130, 1948 (1962).

[59]. (a) S. D. Ross, M. Finkelstein, and R. C. Petersen, J. Amer. Chem. Soc., 82, 1582 (1960); (b) S. D. Ross, M. Finkelstein, and R. C. Petersen, ibid., 92, 6003 (1970); (c) A. Monvernay and P. C. Lacaze, Bull. Soc. Chim. Fr., 1970, 4206; J. E. Dubois, A. Monvernay, and P. C. Lacaze, Electrochim. Acta, 15, 315 (1970).

[60]. R. A. Wessling and W. J. Settineri (to Dow), U.S. Pat. 3,480,527, Nov. 25, 1969; CA, 72, 66593p (1970).

[61]. C. P. Andrieux and J.-M. Savéant, Bull. Soc. Chim. Fr., 1968, 4671.

[62]. M. M. Baizer, U.S. Pat. 3,410,769, Nov. 12, 1968; U.S. Pat. 3,438,877, April 15, 1969; also Ref. [25f].

[63]. L. Horner and D. H. Skaletz, Tetrahedron Lett., 1970, 1103.

[64]. J. Wiemann and M. L. Bouguerra, Compt. Rend., 265, 751 (1967).

Chapter XX

OXIDATIVE COUPLING

Klas Nyberg

Division of Organic Chemistry
University of Lund
Lund, Sweden

I. INTRODUCTION

Among oxidative coupling reactions the Kolbe reaction is certainly the best known. However, since that reaction is discussed in detail elsewhere (Chapter XIII), no further considerations will be given to it here.

Coupling reactions are generally regarded as the combination of radicals, free or adsorbed, at the electrode surface. The radicals can be either neutral or positively charged, depending on the starting compound and the electrolysis medium. Sometimes it is difficult to tell whether we are actually dealing with a true coupling reaction. As will be seen later, the formation of biphenyls from anodic oxidation of benzene derivatives might occur by combination of two cation radicals, by combination of two neutral radicals (if the cation radicals lose protons), or by attack of cation radicals or neutral radicals on the starting compound. In this chapter we will regard the last cases also as coupling reactions, although they can be regarded as substitution reactions.

II. AROMATIC COMPOUNDS

The transfer of one electron from an aromatic compound to the anode produces a cation radical. The stability of this depends on the degree of delocalization that the substituents will contribute. Usually substituents such as hydroxy, alkoxy, or amino groups will stabilize the positive charge and thus facilitate coupling.

A. Hydrocarbons

The first example of electrolytic oxidative couplings of an alkylaromatic compound is probably the oxidation of 1-methylnaphthalene in acetone-sulfuric acid-water to 4,4′-dimethyl-1-1′-binaphthyl at a lead dioxide anode in 10% yield [1]. Small amounts of coupling products have been detected from oxidation of simple alkylbenzenes, such as toluene and cumene, in acetonitrile-sodium or -lithium perchlorate [2, 3]. The coupling products include biphenyls and bibenzyls. The latter finding has been noted as evidence that anodic side-chain substitution does not arise by oxidation of the aromatic compound but rather by oxidation of the supporting electrolyte (see Chapter XXII).

Anodic oxidation of alkylbenzenes is generally carried out in media that contain nucleophiles of such strength that they react rapidly with the cation radicals produced at the anode. In order to obtain products of the dehydrodimer type (diphenylmethanes, biphenyls, or bibenzyls) it is necessary to decrease the nucleophilicity of the electrolysis medium. In this way the reactive species formed at the anode will find that the starting compound is the most nucleophilic in the medium, and will therefore attack this; coupling will be the predominant reaction.

Anodic oxidation of p-xylene or durene (1, 2, 4, 5-tetramethylbenzene) in methylene chloride containing tetrabutylammonium tetrafluoroborate as supporting electrolyte gives as major reaction products diphenylmethanes [4]. Anodic oxidation of mesitylene in the same reaction medium produces as the major products bimesityl(I) and a m-terphenyl(II):

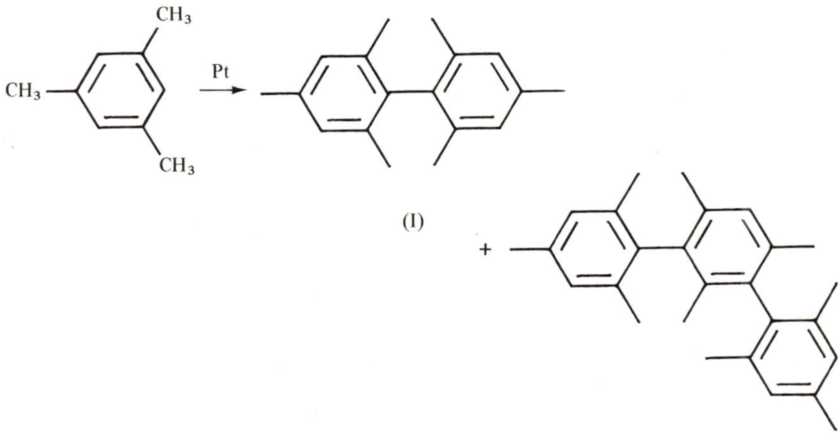

(I)

(II)

The same reaction also takes place in acetonitrile [5]. This is in contrast to other alkylbenzenes that give benzylacetamides in acetonitrile (see Chapter XII). The best yield of (I) and (II) was obtained by running the reaction at a platinum anode (graphite lowered the yield), using rather high concentration of mesitylene (2.0 M) and interrupting the reaction when about 0.4 faraday per mole of mesitylene had been passed. In this way (I) was isolated in 40-50% yield. This is probably the best way of preparing bimesityl. The only comparable chemical method is ferric chloride oxidation of mesitylene [6]. However, that method gives also chlorinated products that are difficult to separate.

The possible reaction paths leading to (I) are outlined on the next page. It is of course difficult to distinguish between these. However, the loss of a proton to form a mesityl radical requires a base and since the basicity of the medium is low, it is perhaps more adequate to suggest that (I) is formed either by coupling of two cation radicals or by attack of a cation radical on mesitylene. Similarly, the formation of (II) might occur by attack of a mesityl cation radical on (I) or by a mixed coupling reaction.

Bimesityl is also formed by anodic oxidation of mesitylene dissolved in a complex composed of mesitylene, hydrogen chloride, and aluminum chloride [7].

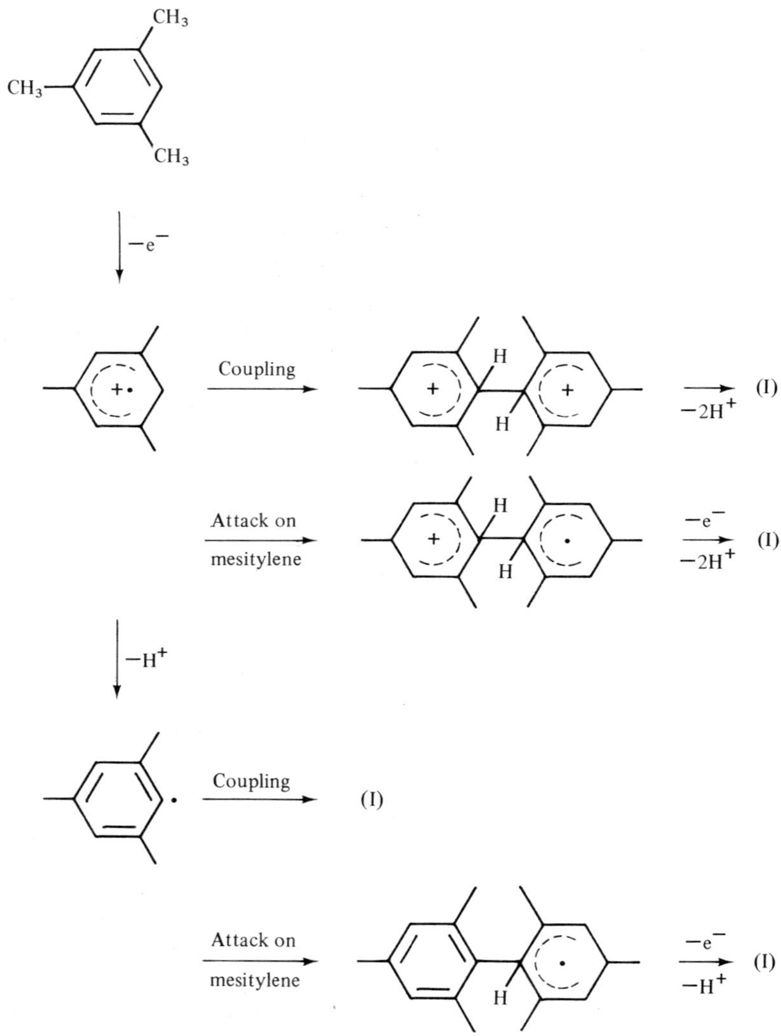

Anodic oxidation of anthracene in acetonitrile-sodium perchlorate containing ethanol has been shown to produce bianthrone(III):

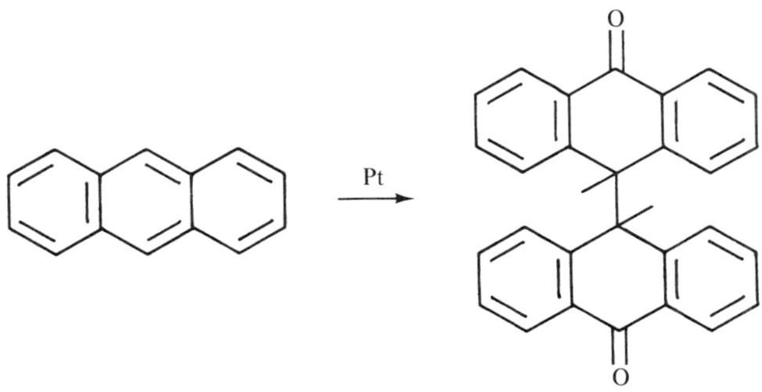

(III)

There are, however, different opinions about the reaction mechanism. Majeski and coworkers [8] believe that anthracene is first oxidized electrochemically to 9-anthrone, which is oxidized by air during work-up to (III). Parker [9] has shown, however, that 9-anthrone is stable and that (III) is a product of the electrolysis formed in a coupling reaction.

A reaction similar to the previous one is anodic oxidation of 1, 5-dichloroanthracene in acetonitrile-sodium perchlorate containing ethanol whereby 1, 1′, 5, 5′-tetrachlorobianthrone is formed in 75% yield [10]. If the electrolysis is conducted in the absence of ethanol, the bianthrone is not formed and the principal products are 1, 5-dichloroanthraquinone and 1, 5-dichloro-9-acetamido-10-anthrone.

Anodic oxidation of 9, 10-dimethylanthracene in acetonitrile-lithium perchlorate containing 2, 6-dimethylpyridine produces the dimer (IV) by coupling of two benzylic radicals [11]:

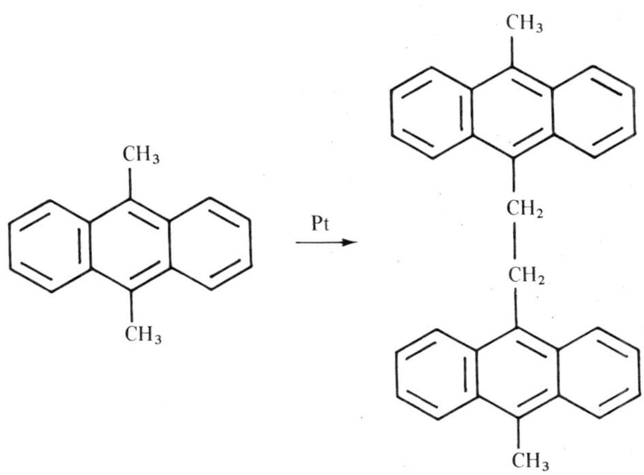

(IV)

Oxidation of the same compound in the presence of acetate ions gave a 9, 10-diacetate when the electrolysis was carried out at 1.8 V (vs. SCE) and the dimeric acetate (V) at 1.1 V [12]:

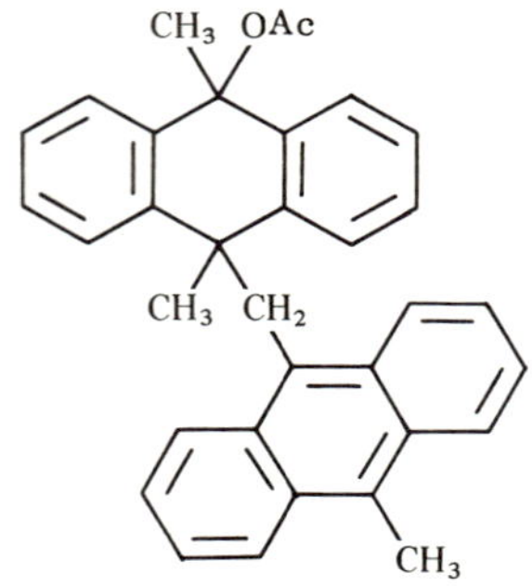

(V)

B. Phenols and Arylalkyl Ethers

The oxidative coupling of phenols has received much attention, and there are excellent reviews of this field [13, 14]. Most of this work has been done by purely chemical means, and, since the early work of Fichter, not much attention has been paid to the electrochemical oxidative coupling of phenols. Only a few examples of the early work will be given; the reader is referred to the review articles for complete details.

The electrochemical coupling of phenols can give several isomeric dihydroxybiphenyls depending on the positions of the substituents. Phenols with the para position unsubstituted usually give para-to-para coupling; with the para position blocked, ortho-to-ortho coupling takes place. However, it should be pointed out that most investigations have not been carried out with modern analytical techniques and therefore the major product is usually the isolated one. Besides carbon-to-carbon coupling, there are also the possibilities of carbon-to-oxygen and oxygen-to-oxygen coupling.

Anodic oxidation of o-cresol is reported to give 4, 4′-dihydroxy-3, 3′-dimethylbiphenyl in 50% yield [15]. Similar oxidation of p-cresol gives ortho-ortho coupling in 36% yield as well as a small amount of carbon-to-oxygen coupling (1%).

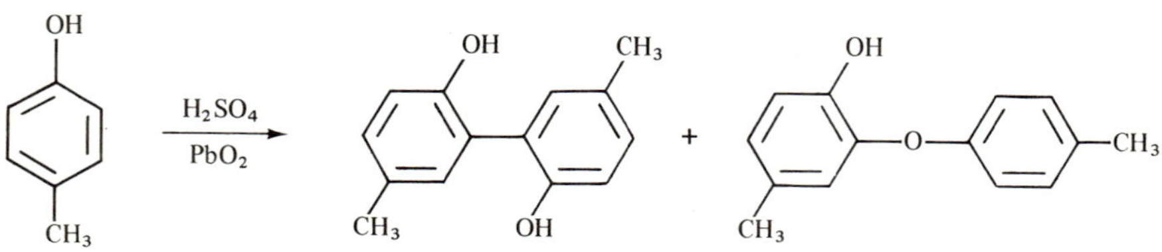

The use of electrolysis for synthesis of isoquinoline alkaloids has recently been investigated by Bobbitt and coworkers [16, 17, 17a]. A series of phenolic tetrahydroisoquinolines (VIa-c) was found to undergo coupling at a platinum anode. The mode of coupling (C-C coupling or C-O coupling) was found to be dependent on the 1-substituent.

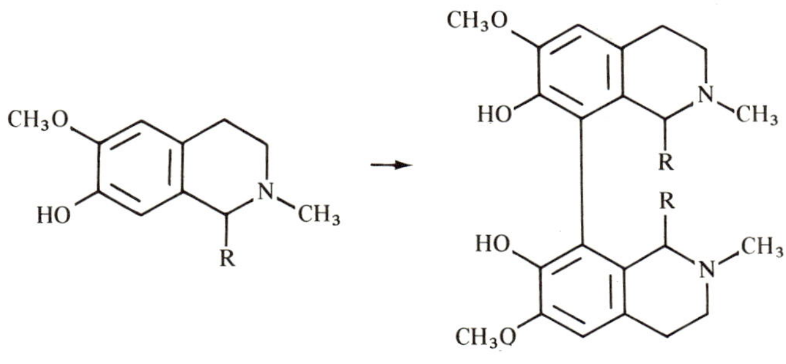

(VI) (VII)

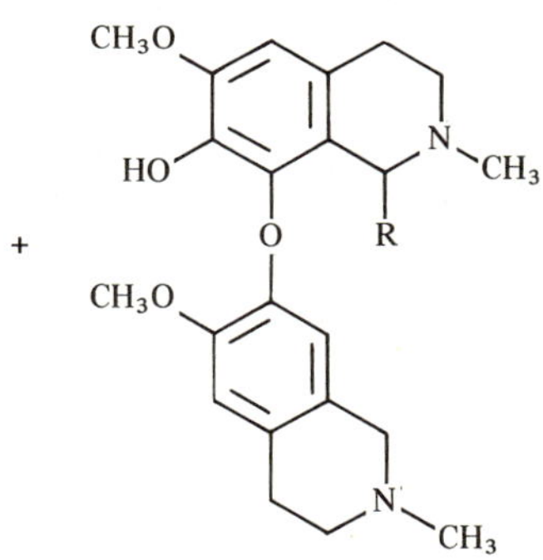

a: R=H
b: R=CH_3
c: R=CH_2CH_3

The yield of (VII) was 34% when R was a hydrogen but decreased readily to less than 1% when R was a methyl or an ethyl group, and, for the latter compounds, the C-O coupling predominated.

In recent work Johnston [18] has examined anodic oxidation of hydroxyacetophenones in potassium hydroxide-methanol solution at a platinum anode. Coupling products were obtained as shown below:

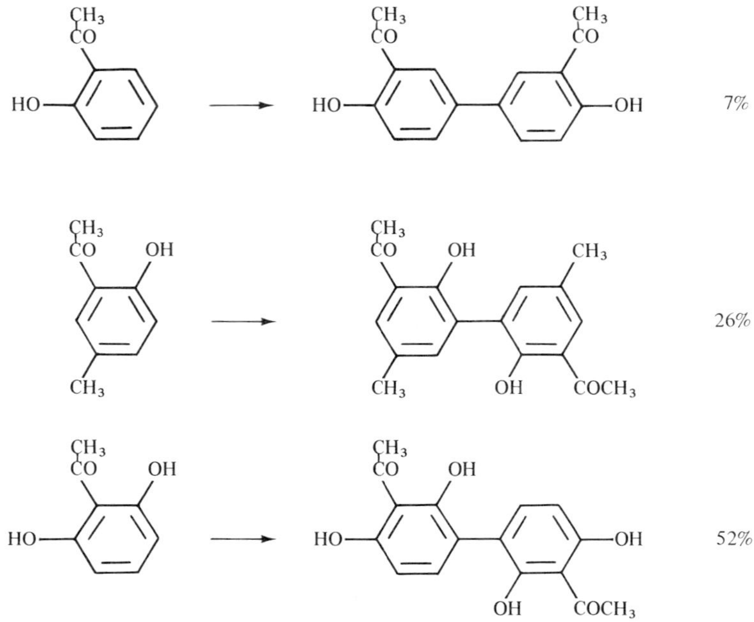

Fichter and coworkers [19-21] investigated anodic oxidation of several methoxybenzenes substituted with hydroxyl or methyl groups.[1] Coupling to biphenyl derivatives generally occurs, but the yields are low. The reactions are generally carried out in dilute sulfuric acid at platinum or lead dioxide anodes. It is also interesting to note that sometimes demethylation takes place as illustrated for anodic oxidation of veratrol:

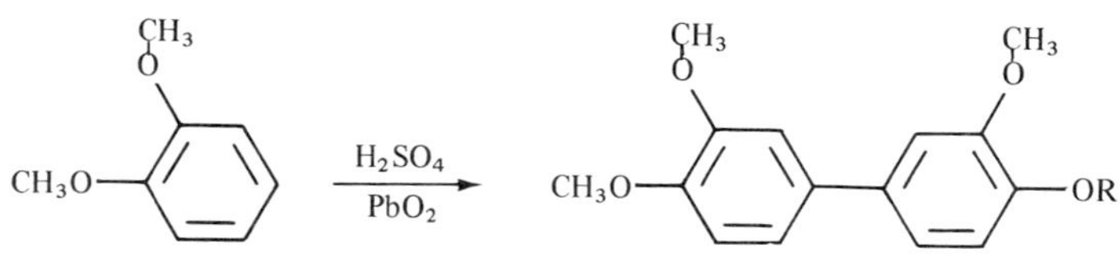

$R{=}CH_3$ or H

An interesting example of preparative anodic coupling is afforded by the work of Erdtman [22]. Electrolysis of 1, 2, 4-trimethoxybenzene in weakly acid solution gave 2, 2′, 4, 4′, 5, 5′-hexamethoxybiphenyl in 85% yield. In strongly acid solution the yield dropped to 15% and in alkaline solution the starting material was recovered unchanged.

2, 5-Dimethoxytoluene has been reported to give the biphenyl (VIII) as the only product from anodic oxidation in acetonitrile-water containing tetraethylammonium perchlorate as supporting electrolyte [23]:

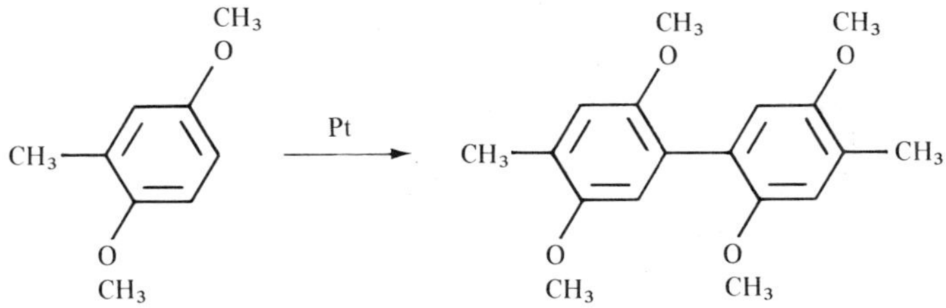

(VIII)

C. Amines

The electrolytic oxidation of aromatic amines has received considerable attention. Since the general aspect of this subject is treated elsewhere (Chapter XIV) we will cover only the principal coupling reactions here. An excellent review has recently been given by Adams [24].

Anodic oxidation of aromatic amines generally produces cation radicals as the primary electrode reaction. These can undergo different chemical reactions. Three principal coupling reactions can be outlined: carbon-to-carbon coupling (benzidines), carbon-to-nitrogen coupling (aminodiphenylamines), and nitrogen-to-nitrogen coupling (hydrazobenzenes). The products are often easily oxidized further.

Triphenylamine is electrolytically oxidized in an ECE reaction as illustrated below [25-27]:

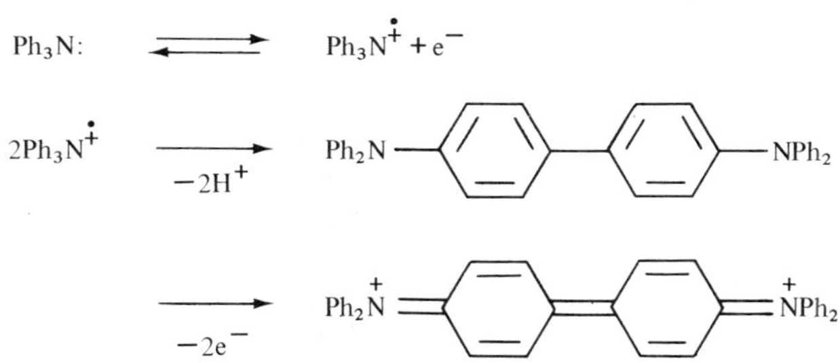

Secondary amines are oxidized in a similar way and according to Adams they undergo coupling not only to benzidines but also to tetraphenylhydrazines by way of nitrogen-to-nitrogen coupling [24].

Anodic oxidation of primary aromatic amines produces coupling products of different types, depending on the pH of the medium. Aniline is oxidized to p-aminodiphenylamine in the pH range 0-6.5. In strongly acid solution a mixture of p-aminodiphenylamine and benzidine is formed [28].

Wawzonek and McIntyre [29] studied oxidation of p-substituted anilines in acetonitrile-pyridine and ran large-scale electrolyses. The main products were azobenzenes formed by oxidation of hydrazobenzenes. The reaction mechanism is given below:

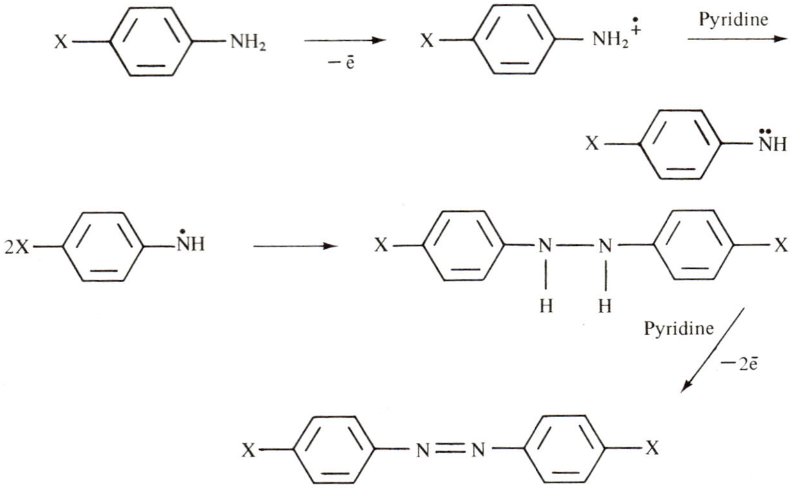

X =		
NO_2	39%	azobenzene
Cl	24%	"
2, 4-di-Cl	30%	"
OCH_3	5%	"

Bacon and Adams [28] studied oxidation of p-substituted anilines in aqueous media and found elimination of one para group in the products that were 4′-substituted 4-aminodiphenylamines.

A fourth type of aniline reaction was recently demonstrated by Melicharek and Nelson [29a] who found that N, N-dimethyl-p-toluidine was converted to 4, 4′-bis-(N, N-dimethylamino)-bibenzyl by anodic oxidation in acetonitrile at a platinum anode.

III. ETHYLENIC COMPOUNDS

Anodic oxidation of ethylenic compounds produces in the first oxidation step cation radicals, as is the case with aromatic compounds. The fate of these reactive species will depend on four principal factors. The substituent at the double bond will determine not only the ease of oxidation of the neutral compound to a cation radical but also the stability of the cation radical by charge delocalization. The electrolysis medium, depending on its nucleophilic strength, may react with the cation radicals. The electrode material can have an influence on the reaction of the cation radicals. The anode potential may determine if the cation radical will be further oxidized to a dication.

Oxidative coupling is generally carried out with ethylenic compounds such as aryl ethylenes, vinyl ethers, or aminoalkenes in which the substituents are well suited for delocalization of the positive charge.

A. Aryl Ethylenes

Anodic oxidation of 3, 4-dimethoxypropenylbenzene in acetonitrile-sodium perchlorate produces the dehydrodimer(IX) formed by dimerization of cation radicals followed by ring closure and loss of protons [30]. If the oxidation is carried out in the presence of pyridine, the reaction takes a different course and no coupling occurs.

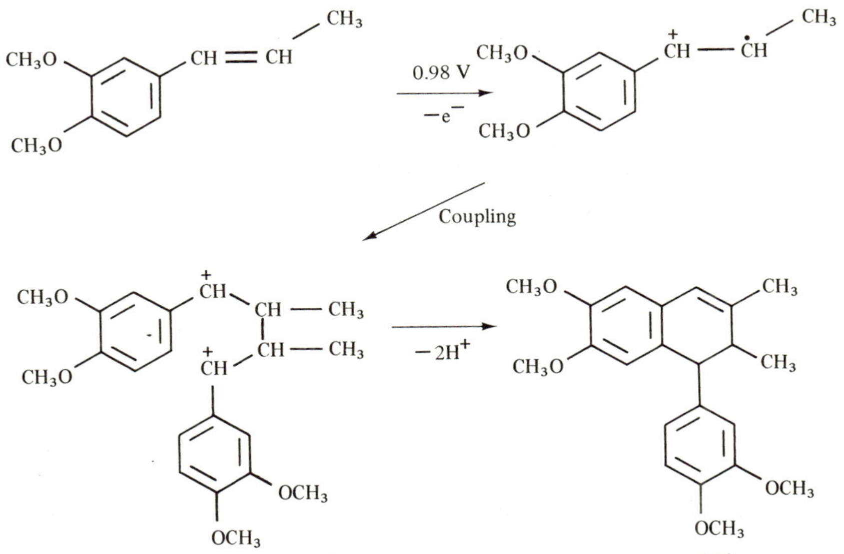

(IX)

Anodic oxidation of 4, 4′-dimethoxystilbene follows a similar route. If the oxidation is carried out in acetonitrile containing acetic acid/sodium acetate, a dimeric acetate (X) is formed at an anode potential of less than 1.05 V (vs. SCE), while at an anode potential greater than 1.20 V a diacetate is formed [31]:

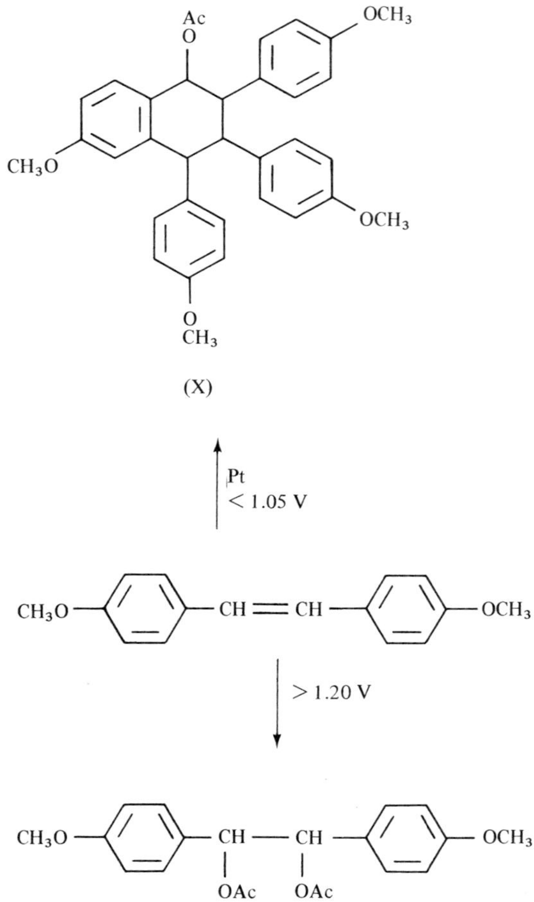

(X)

An internal oxidative coupling takes place when tetraphenylethylene is oxidized in acetonitrile-tetraethylammonium perchlorate. The product is 9, 10-diphenylphenanthrene, which is also obtained on photolysis [32].

The preparative use of oxidative coupling of olefins has recently been investigated by Schäfer and Steckhan [33]. They found that oxidation of aryl ethylenes in methanol using sodium methoxide, sodium perchlorate, or sodium iodide as supporting electrolyte produced dimeric products when the electrolysis was carried out at a graphite anode. Electrolysis at a platinum anode changed the reaction course and the products were the result of addition of two methoxy groups to the double bond. The current yields of the dimeric products are given below:

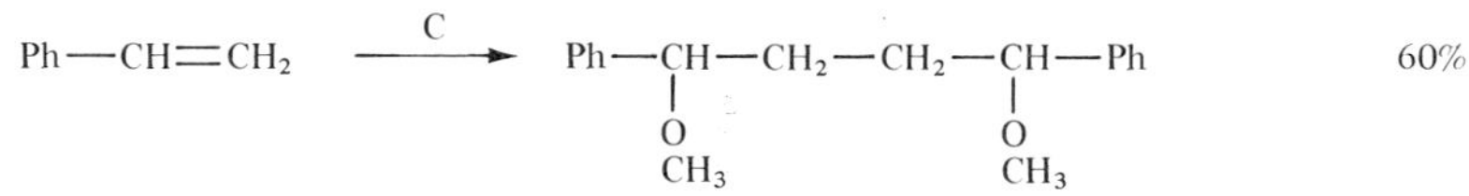

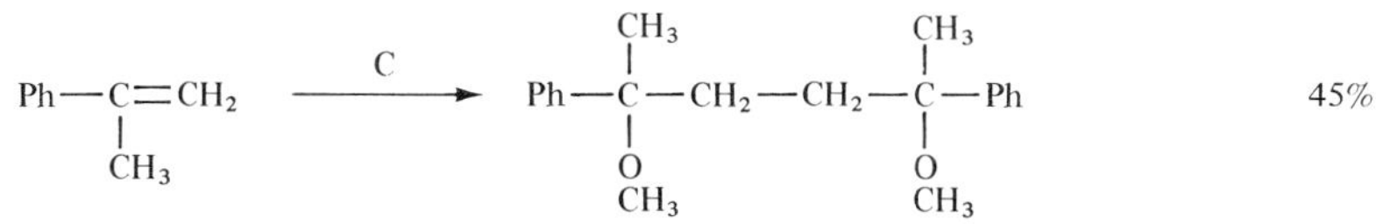

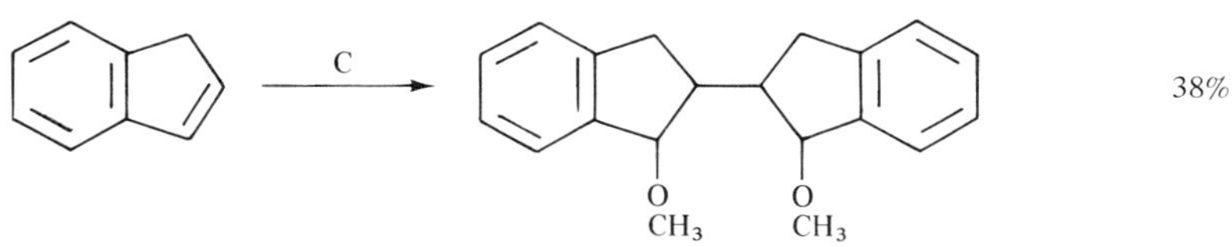

Anodic oxidation of styrene in methanol-sodium methoxide at a platinum anode has been reported to give meso-1, 4-dimethoxy-2, 3-diphenylbutane in addition to 1, 2-dimethoxyphenylethane [34].

The strikingly different results that are obtained at platinum and graphite anodes are probably due to different electron-transfer reactions. At a graphite anode there is no doubt that the aryl olefin is oxidized to a cation radical that undergoes dimerization with another cation radical prior to reaction with methanol or methoxide ions. The transfer step at the platinum anode is, however, more complex. One explanation assumes an initial oxidation of methoxide ions to methoxy radicals that attack the double bond of the olefin. The formation of meso-1, 4-dimethoxy-2, 3-diphenylbutane at a platinum anode supports that explanation. However, since the adsorption phenomena at platinum and graphite are poorly understood one cannot exclude the possibility that oxidation of the olefin takes place at the platinum anode (simultaneous oxidation of olefin and methoxide ions might also occur).

B. Vinyl Ethers

The anodic oxidation of vinyl ethers is a new type of oxidative coupling and is of preparative interest since the products can be hydrolyzed to 1, 4-dicarbonyl compounds. Oxidation of vinyl ethyl ether in methanol-potassium hydroxide produced the coupled dimethoxy derivative (XI) in 22% yield at a platinum anode [35]:

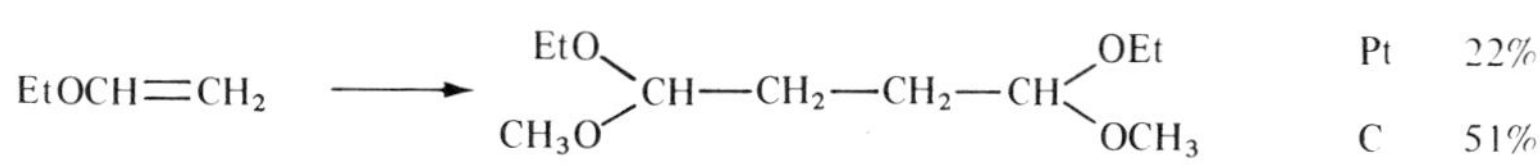

(XI)

A much better yield (51%) was obtained at a graphite anode [33]. The choice of electrode material is also of great importance in this reaction. Similar oxidation of 1-ethoxycyclohexene in methanol gave at a graphite anode the diketone (XII) in 27% yield after hydrolysis [33]:

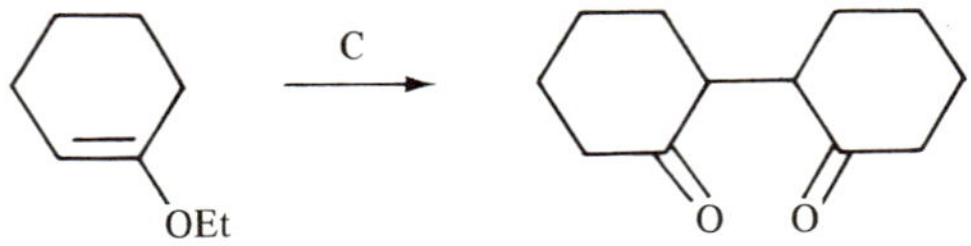

(XII)

Mixed coupling takes place when vinyl ethyl ether and styrene are oxidized in methanol at a graphite anode. The unsymmetrical coupling product (XIII) is obtained in 23% yield together with the two normal coupling products [36]:

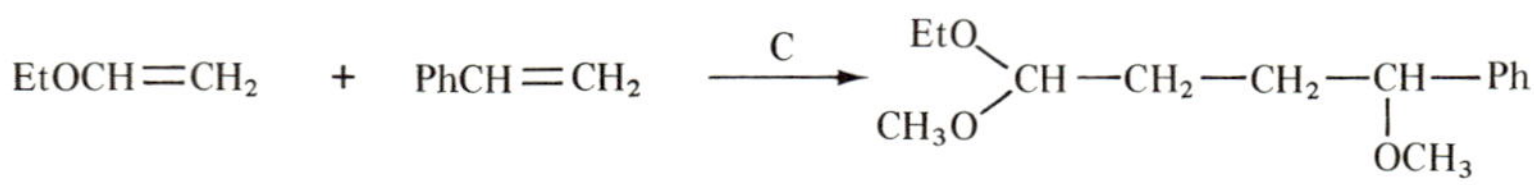

(XIII)

C. Aminoalkenes

Anodic oxidation of dialkylaminoalkenes has recently been studied in great detail [37, 38]. These compounds are very easily oxidized and, depending on the substituents, coupling reaction can take place. Since this work is discussed in Chapter XIV, only one example will be given here. 1, 1-Bis(dimethylamino)ethylene is oxidized to 1, 1, 4, 4-tetrakis(dimethylamino)butadiene according to the reaction outlined:

$$(Me_2N)_2C{=}CH_2 \rightarrow (Me_2N)_2\overset{+}{C}\text{-}\overset{\bullet}{C}H_2 + e^-$$

$$2(Me_2N)_2\overset{+}{C}\text{-}\overset{\bullet}{C}H_2 \rightarrow (Me_2N)_2\overset{+}{C}\text{-}CH_2\text{-}CH_2\text{-}\overset{+}{C}(NMe_2)_2 \xrightarrow[-2H^+]{}$$

$$(Me_2N)_2C{=}CH\text{-}CH{=}C(NMe_2)_2.$$

IV. OXIDATION OF ORGANIC ANIONS

The coupling of radicals formed by anodic oxidation of organic anions is an old electrochemical reaction and although the first such coupling was reported almost 80 years ago, no systematic study of the synthetic applications of the reaction had been carried out up to the last decade.

The most common anions are those derived from nitroalkanes, dialkylmalonates, acetylacetone, and ethyl or methyl acetoacetate. The oxidation of the anions is usually carried out in water, methanol, ethanol, or acetonitrile. Under certain conditions it is possible to use the neutral compound, which is converted to the anion during electrolysis.

The anions are easily oxidized, as can be seen from some polarographic half-wave potentials recorded at a rotating platinum electrode in methanol-benzene using 0.2 M sodium or lithium perchlorate (Table 1) [39].

TABLE 1

Polarographic Half-Wave Potentials of a Few Organic Anions

Sodium salt of	$E_{1/2}$ (vs. Ag/AgCl) V
Dimethyl malonate	0.87
Acetylacetone	0.83
Nitroethane	0.87
2-Nitropropane	0.74
Methyl cyanoacetate	0.65

Polarographic oxidation of nitroalkanes shows a clear dependence on the pH of the electrolysis medium [40]. The nitroalkanes that were investigated gave no oxidation wave at pH 7 and 8, but going to more basic solutions gave oxidation waves of the anions. Increasing the pH showed a corresponding decrease in the oxidation potentials of the anions.

Anodic oxidation of 2-nitrobutane in 25% sodium hydroxide solution at a platinum anode using a divided cell gave the coupling product 3,4-dinitro-3,4-dimethylhexane in 70% yield together with 20% of 2,2-dinitrobutane [41]:

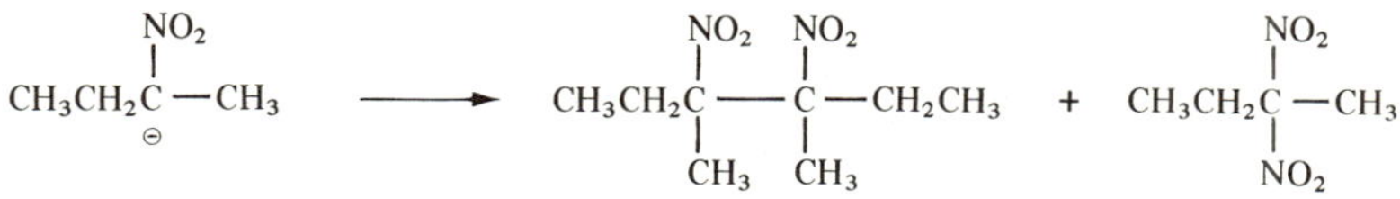

Nitrite ions, formed in some way during the electrolysis, were probably responsible for the formation of the dinitro compound, since it was found that addition of sodium nitrite to the electrolysis mixture increased the yield of the dinitro compound at the expense of the dimer.

Electrolysis of 2-nitropropane in methanol containing trimethylbenzylammonium hydroxide produced 2,3-dinitro-2,3-dimethylbutane in 64% yield [40]. Electrolysis of trinitromethyl anion in toluene yielded $C_6H_5CH_2C(NO_2)_3$ [40a].

As early as 1893 Mulliken [42] showed that electrolytic oxidation of the anions derived from diethyl malonate, triethyl methanetricarboxylate, and acetylacetone produced the coupled products (XIV), (XV), and (XVI):

$$\underset{\ominus}{CH(COOEt)_2} \rightarrow \underset{(XIV)}{(EtOOC)_2CH\text{-}CH(COOEt)_2}$$

$$\underset{\ominus}{C(COOEt)_3} \rightarrow \underset{(XV)}{(EtOOC)_3C\text{-}C(COOEt)_3}$$

$$\underset{\ominus}{CH(COCH_3)_2} \rightarrow \underset{(XVI)}{(CH_3CO)_2CH\text{-}CH(COCH_3)_2}.$$

The electrolytic method for preparing the hexaester (XV) is much better than purely chemical methods (as was shown later). From electrolysis of the triester in the medium sodium hydroxide-water-acetone-ethanol at a platinum anode at 10° a yield of 50% of the hexaester was obtained [43].

A thorough investigation of the anodic oxidation of dialkyl sodiomalonates has been carried out by Brettle and Parkin [44]. They electrolyzed diethyl sodiomalonate in ethanol at a platinum anode and obtained three products: the normal coupling product (XIV), the tetraester (XVII), and the hexaester (XVIII):

$$\underset{\ominus}{CH(COOEt)_2} \xrightarrow[Pt]{EtOH} (XIV) + \underset{(XVII)}{(EtOOC)_2CH{-}CH(CH_3){-}CH(COOEt)_2}$$

$$+ \underset{(XVIII)}{(EtOOC)_2CH{-}C(COOEt)_2{-}CH(COOEt)_2}$$

Oxidation of the malonate anions produces radicals that undergo coupling to form (XIV). The anion of (XIV) can of course also be formed and subsequent oxidation of that will give a radical that can couple with malonate radicals to give (XVIII). The malonate radicals may abstract hydrogens from ethanol and the resulting radicals can disproportionate into acetaldehyde and ethanol; (XVII) is then formed by condensation of the starting material and acetaldehyde. When diethyl sodiomalonate was electrolyzed in an excess of diethyl malonate, (XIV) was isolated in 37% yield.

Electrolysis of dimethyl sodiomalonate in methanol gave the two methyl esters corresponding to (XIV) and (XVIII) and the tetraester (XIX) formed analogously to (XVII):

$$\underset{(XIX)}{(MeOOC)_2CH\text{-}CH_2\text{-}CH(COOMe)_2}.$$

Brettle and Parkin [44] also report that oxidation of diethyl sodiomalonate in dimethylacetamide produced (XIV), (XVIII), and the ethyl ester corresponding to (XIX). They believe that this ester is formed by condensation of formaldehyde and the starting compound, the formaldehyde being produced in the following way:

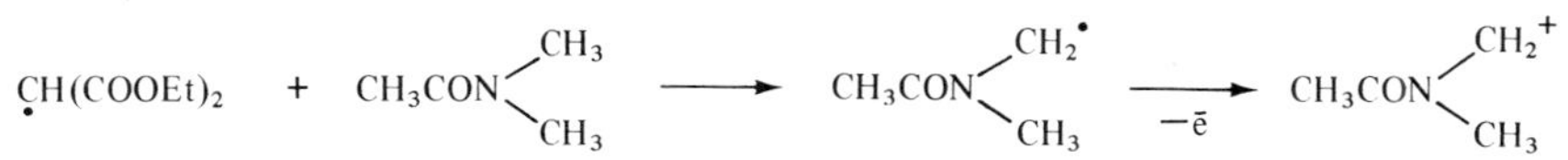

or

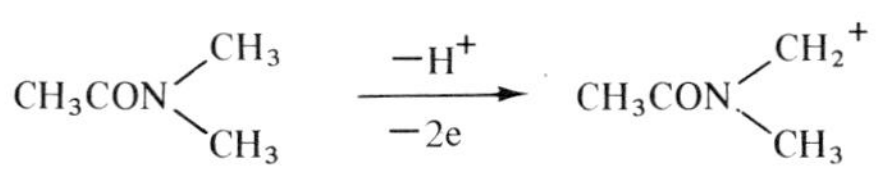

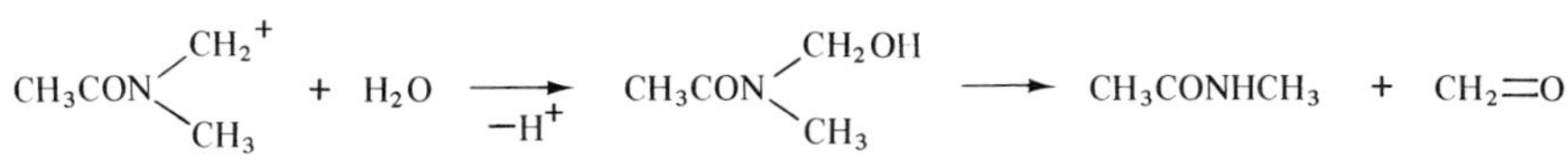

The same types of products were also formed in hexamethylphosphoric triamide [44a].

Okubo and Tsutsumi [45] have investigated the electrolysis of diethyl malonate in acetonitrile containing potassium iodide. By using acetonitrile instead of ethanol they avoid forming the side product (XVII). In the electrolysis iodide ions are oxidized to iodine at the anode and potassium ions are reduced to potassium at the cathode. Potassium reacts with diethyl malonate and forms the anion that can be oxidized to the dimer either electrochemically or chemically (by iodine).

Anodic oxidation of the anion from ethyl acetoacetate will produce the dimer (XX) if a normal oxidative coupling takes place:

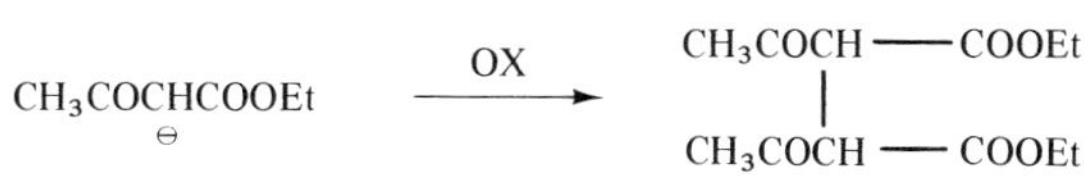

(XX)

There are, however, conflicting reports concerning the oxidation of this anion. Okubo and Tsutsumi showed that oxidation of the anion in ethanol at a platinum anode gave small amounts of (XX). A better yield was obtained by using acetonitrile-potassium iodide as electrolysis medium. They isolated (XX) in about 40% yield [45].

Brettle and coworkers [46, 47] investigated oxidation of ethyl sodioacetoacetate in ethanol or aqueous ethanol but were unable to detect any dimer. Instead they showed that the major product of the electrolysis was (XXI), formed in the same way as the analogous products from the electrolysis of diethyl or dimethyl sodiomalonate in ethanol or methanol [46, 47].

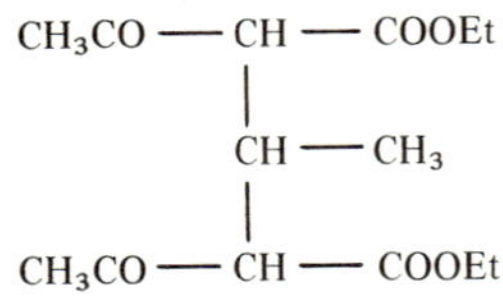

(XXI)

On the other hand, electrolysis of ethyl sodiomethylacetoacetate in N, N-dimethylacetamide did produce the normal coupling product [47a].

If the electrolysis of organic anions is performed in the presence of suitable olefins, the radicals may add to the double bond of the olefin and the radical adduct may undergo several reactions. An extensive investigation of the oxidative addition of radicals to olefins has recently been carried out by Schäfer and coworkers [36, 48, 48a]. In general the reaction can be characterized in the following way:

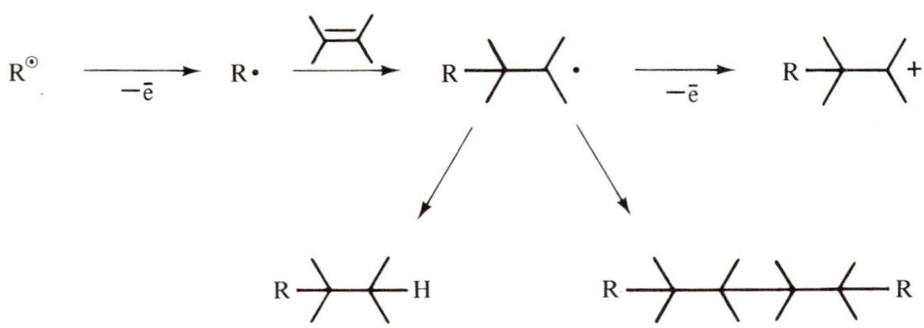

Two types of products are formed, those derived from radical reactions (coupling, hydrogen abstraction) and those derived from carbonium ion reactions. The anode material was platinum (compare Sec. III.A) and the reaction medium was methanol.

Anodic oxidation of sodioacetylacetonate in the presence of butadiene gave a 40% yield of the dimeric product, which after alkali treatment gave trans, trans-5, 9-tetradecadiene-2, 13-dione(XXII):

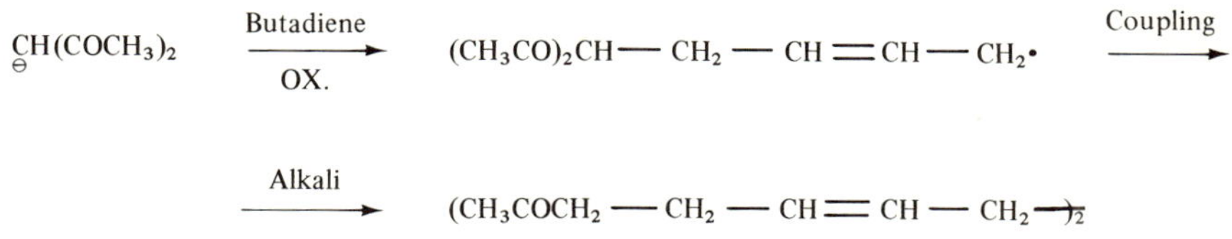

(XXII)

If the electrolysis is carried out in the presence of styrene, the yield of dimeric product is only 10% and the major reaction course is further oxidation of the radical adduct to a carbonium ion.

An interesting reaction from the preparative point of view is the oxidative addition of nitroalkyl radicals to olefins. Oxidation of sodio 2-nitropropane in the presence of styrene afforded two products, depending on the anode potential. At a potential lower than 0.4 V (vs. Ag/AgCl) the main product was the dimer (XXIII), while at potentials higher than 0.7 V the intermediate radical is further oxidized to a cation that reacts with the solvent to give (XXIV):

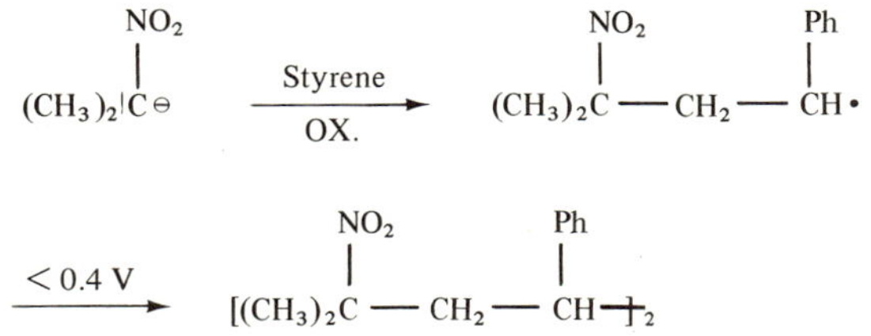

(XXIII)

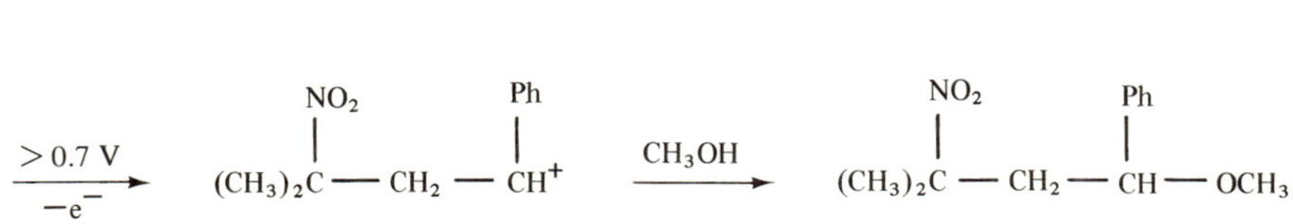

(XXIV)

Oxidation of sodio nitroethane in the presence of styrene did not produce any dimer since the oxidation potential for the anion was high enough to oxidize the radical adduct to a cation.

Anodic oxidative coupling of Grignard reagents is a well-known reaction (see Chapter XVIII). They also undergo oxidative addition to olefins in the same manner as the previous anions. Anodic oxidation of butylmagnesium-bromide in the presence of styrene in ether-lithium perchlorate at a copper anode gave 6, 7-diphenyldodecane in 30% yield. If the concentration of styrene was raised fourfold, the main product was a polymer:

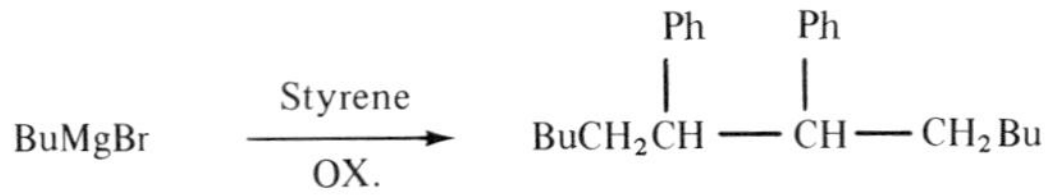

Although azide ions are not organic anions it is of interest to cite here the oxidative addition of azide radicals to olefins. Anodic oxidation of azide ions in acetic acid in the presence of styrene produced the dimeric compound (XXV) in 57% yield, and in the presence of α-methylstyrene the yield of (XXVI) was 45%:

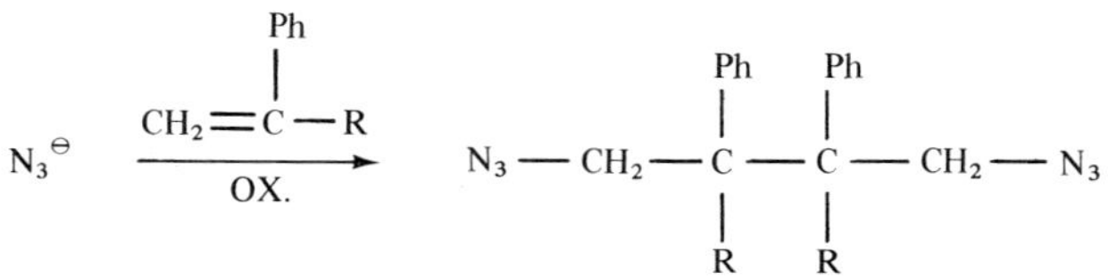

R$=$H (XXV)

R$=CH_3$ (XXVI)

The reaction, however, failed to yield dimeric products in the presence of 1, 1-diphenylethylene or aliphatic olefins.

Finally, attention should be drawn to work that claimed that anodic oxidation of some cyclic 1, 3-diketones in aqueous or aqueous methanolic sodium hydroxide solution produced the coupled products (XXVII) [49]. It has been shown in a subsequent work to be incorrect [50]. The products were shown to be the same that are formed from condensation of the ketones with formaldehyde.

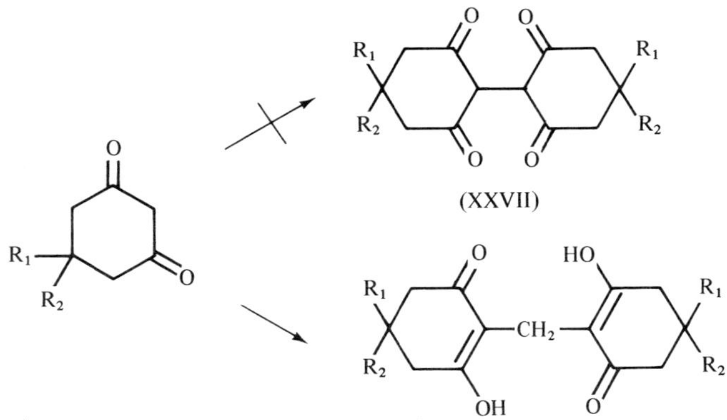

(XXVII)

$R_1 = R_2 = CH_3$
$R_1 = H_1 R_2 = Ph$
$R_1 = H, R_2 = CH_3$

V. SULFUR COMPOUNDS

The electrochemical oxidation of sulfur compounds is treated in detail elsewhere (Chapter XVI) and only some examples of oxidative coupling are given here. It is a well-known fact that mercaptide ions are easily oxidized chemically and it was shown electrochemically a long time ago [51] that the products are disulfides. Thiourea is another sulfur compound that can be oxidized easily electrochemically and the product is formamidine disulfide [52-54]:

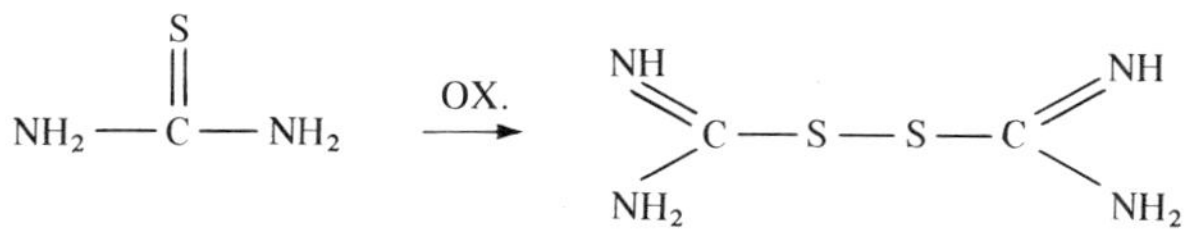

Anodic oxidation of the potassium salt of thiolacetic acid gave a disulfide in 31% yield [51, 55]:

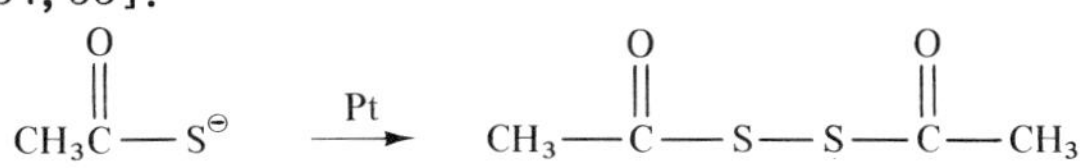

Anodic oxidation of some purine derivatives has been carried out in order to compare the results with those obtained by enzymatic oxidation. It was found that 6-thiopurine was oxidized electrochemically to bis(6-purinyl)-disulfide, which is also obtained by chemical oxidation. These oxidations gave results that were different from the enzymatic oxidation. On the other hand, electrochemical oxidation of adenine and uric acid parallels the enzymatic oxidation [56]:

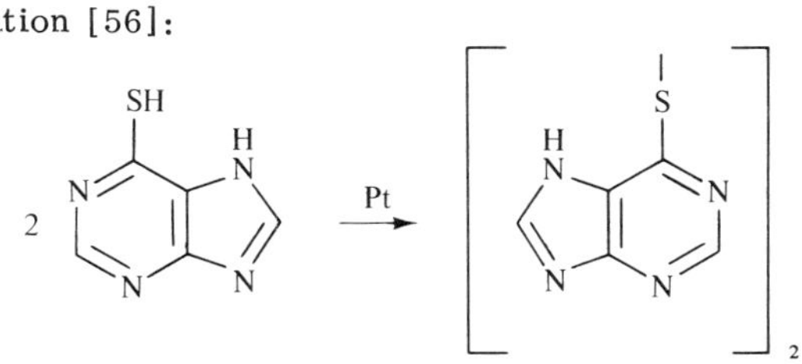

Oxidation of 2, 6-dithiopurine proceeds in two steps. The first step involves the transfer of one electron, and the product is bis(6-purinyl)-disulfide-2, 2′-dithiol. This product can then undergo oxidation to form bis(6-purinyl)-disulfide-2, 2′-disulfide [57].

NOTE

[1] K. Bechgaard et al. [58] have reported excellent results in p, p-coupling of methoxybenzenes using trifluoroacetic acid and methylene chloride.

REFERENCES

[1]. F. Fichter and S. Hersbein, Helv. Chim. Acta, 11, 1264 (1928).

[2]. K. Koyama, T. Susuki, and S. Tsutsumi, Tetrahedron, 23, 2665 (1967).

[3]. V. D. Parker and B. E. Burgert, Tetrahedron Lett., 2411 (1968).

[4]. K. Nyberg, Acta Chem. Scand., 24, 1609 (1970).

[5]. K. Nyberg, ibid., 25, 534 (1970).

[6]. P. Kovacic and C. Wu, J. Org. Chem., 26, 759 (1961).

[7]. N. E. Wisdom, Jr., Abstract No. 138, The Electrochemical Society, Spring Meeting, New York, May 4-9, 1969.

[8]. E. J. Majeski, J. D. Stuart, and W. E. Ohnesorge, J. Amer. Chem. Soc., 90, 633 (1968).

[9]. V. D. Parker, Chem. Commun., 1131 (1969).

[10]. A. E. Coleman, H. H. Richtol, and D. A. Aikens, J. Electroanal. Chem., 18, 165 (1968).

[11]. V. D. Parker and L. Eberson, Tetrahedron Lett., 2839 (1969).

[12]. V. D. Parker, Chem. Commun., 848 (1969).

[13]. A. I. Scott, Quart. Rev., 19, 1 (1965).

[14]. H. Musso, in Oxidative Coupling of Phenols (W. I. Taylor and A. R. Buttersby, eds.), Dekker, New York, 1967, Chap. 1.

[15]. F. Fichter and F. Ackerman, Helv. Chim. Acta, 2, 583 (1919).

[16]. J. M. Bobbitt, J. T. Stock, A. Marchard, and K. W. Weisgraber, Chem. Ind. (London), 2127 (1966).

[17]. G. F. Kirkbright, J. T. Stock, R. D. Pugliese, and J. M. Bobbitt, J. Electrochem. Soc., 116, 219 (1969).

[17a]. J. M. Bobbitt, H. Yagi, S. Shibuya, and J. T. Stock, J. Org. Chem., 36, 3006 (1971).

[18]. K. M. Johnston, Tetrahedron Lett., 837 (1967).

[19]. F. Fichter and W. Dietrich, Helv. Chim. Acta, 7, 131 (1924).

[20]. F. Fichter and H. Ris, ibid., 7, 803 (1924).

[21]. F. Fichter and J. Meyer, ibid., 8, 74 (1925).

[22]. H. G. H. Erdtman, Proc. Roy. Soc. (London), 143 (A), 191 (1934).

[23]. V. D. Parker and R. N. Adams, Tetrahedron Lett., 1721 (1969).

[24]. R. N. Adams, Accounts Chem. Res., 2, 175 (1969).

[25]. E. T. Seo, R. F. Nelson, J. M. Fritsch, L. S. Marcoux, D. W. Leedy, and R. N. Adams, J. Amer. Chem. Soc., 88, 3498 (1966).

[26]. R. F. Nelson and R. N. Adams, ibid., 90, 3925 (1968).

[27]. L. S. Marcoux, R. N. Adams, and S. W. Feldberg, J. Phys. Chem., 73, 2611 (1969).

[28]. J. Bacon and R. N. Adams, J. Amer. Chem. Soc., 90, 6596 (1968).

[29]. S. Wawzonek and T. W. McIntyre, J. Electrochem. Soc., 114, 1025 (1967).

[29a]. M. Melicharek and R. F. Nelson, J. Electroanal. Chem., 26, 201 (1970).

[30]. J. J. O'Connor and J. A. Pearl, J. Electrochem. Soc., 111, 335 (1964).

[31]. V. D. Parker and L. Eberson, Chem. Commun., 340 (1969).

[32]. J. D. Stuart and W. E. Ohnesorge, J. Amer. Chem. Soc., 93, 4531 (1971).

[33]. H. Schäfer and E. Steckhan, Angew. Chem., 81, 532 (1969); Angew. Chem. (Intern. Ed.), 8, 518 (1969).

[34]. T. Inoue and S. Tsutsumi, Bull. Chem. Soc. Japan, 38, 661 (1965).

[35]. B. Belleau and Y. K. Au-Young, Can. J. Chem., 47, 2117 (1969).

[36]. H. Schäfer, Chem.-Ing.-Tech., 42, 164 (1970).

[37]. J. M. Fritsch and H. Weingarten, J. Amer. Chem. Soc., 90, 793 (1968).

[38]. J. M. Fritsch, H. Weingarten, and J. D. Wilson, ibid., 92, 4038 (1970).

[39]. H. Schäfer, Chem.-Ing.-Tech., 41, 179 (1969).

[40]. Tsung-yuan Su, Thesis, Univ. of Iowa, Iowa City, Iowa, 1967.

[40a]. S. A. Shevelev, V. A. Kokorenika, L. G. Feoktistov, and A. A. Fainzil'berg, Soviet Electrochem., 6, 1215 (1970).

[41]. C. T. Bahner, Ind. Eng. Chem., 44, 317 (1952); U.S. Pat. 2,485, 803 (1949).

[42]. S. P. Mulliken, Amer. Chem. J., 15, 523 (1893).

[43]. F. Adickes, W. Brunnert, and O. Lücker, J. Prakt. Chem., 130, 163 (1931).

[44]. R. Brettle and J. G. Parkin, J. Chem. Soc. (C), 1352 (1967).

[44a]. R. Brettle and D. Seddon, ibid., 1153 (1970).

[45]. T. Okubo and S. Tsutsumi, Bull. Chem. Soc. Japan, 37, 1794 (1964).

[46]. T. D. Binns, R. Brettle, and J. G. Parkin, Chem. Commun., 409 (1965).

[47]. T. D. Binns and R. Brettle, J. Chem. Soc. (C), 336 (1966).

[47a]. R. Brettle, J. G. Parkin, and D. Seddon, ibid., 1317 (1970).

[48]. H. Schäfer and A. Alazrak, Angew. Chem., 80, 485 (1968); Angew. Chem. (Intern. Ed.), 7, 474 (1968).

[48a]. H. Schäfer, Angew. Chem., 82, 134 (1970).

[49]. K. M. Johnston and J. D. Stride, Chem. Commun., 325 (1966).

[50]. R. G. Cooks, D. H. Williams, K. M. Johnston, and J. D. Stride, J. Chem. Soc. (C), 2199 (1968).

[51]. N. Bunge, Chem. Ber., 3, 295, 911 (1870).

[52]. F. Fichter and W. Wenk, ibid., 45, 1373 (1912).

[53]. F. Fichter and F. Braun, ibid., 47, 1526 (1914).

[54]. K. S. V. Santhanam and V. R. Krishnan, Z. Phys. Chem., 34, 312 (1962).

[55]. Y. Hirabayashi and T. Mazume, Bull. Chem. Soc. Japan, 39, 1971 (1966).

[56]. G. Dryhurst, J. Electrochem. Soc., 116, 1097 (1969).

[57]. G. Dryhurst, ibid., 117, 1113 (1970).

[58]. K. Bechgaard, O. Hammerich, N. S. Moe, A. Ronlán, U. Svanholm, and V. D. Parker, Tetrahedron Lett., 2271 (1972).

Chapter XXI

CLEAVAGES

L. Horner

Organisch-Chemisches Institut
University of Mainz
Mainz, Germany

and

H. Lund

Department of Organic Chemistry
Aarhus University
Aarhus, Denmark

Part A: L. Horner[1]

Part B: H. Lund

PART A

I. ONIUM COMPOUNDS

A. Introduction

In Chapter XI it was pointed out that when onium compounds are electrolyzed in a protic medium in the absence of a substrate having a more anodic reduction potential hydrogen will be discharged; in aprotic media they are themselves reduced and degraded. There are many onium salts, however, which in protic media are reduced by two electrons to neutral substances:

$$>\overset{+}{Y}—R \quad + \quad 2\bar{e} \quad + \quad H^+ \quad \longrightarrow \quad >Y| \quad + \quad RH$$

B. Quaternary Ammonium

1. Reductive Degradation of Quaternary Ammonium Salts at Lead or Mercury Cathodes

It was demonstrated long ago [1] that anilinium salts with three N-alkyl groups are discharged at a lead cathode. Regardless of the nature of the saturated substituents on nitrogen, benzene and the corresponding tertiary amine were always obtained. These results show that in the discharge of the ammonium ion the bond between N and phenyl is particularly weakened; N-allyl and N-benzyl bonds are weakened even more. Dimethylallyl- (or benzyl-) anilinium ions yield in aqueous solution at a lead cathode propylene (or toluene) and dimethylaniline [2]:

$$[(CH_3)_2RNC_6H_5]^+ \xrightarrow{2\bar{e} + H^+} \underset{95\%}{C_6H_5N(CH_3)_2} + RH \qquad (1)$$

However, in many cases it was shown [2] that reductive cleavage at lead according to Eq. (1) is minor and that Hofmann elimination according to Eq. (2) is dominant:

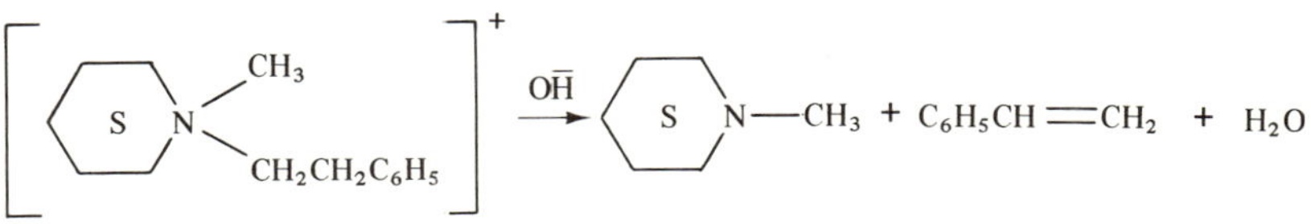

(2)

2. Reductive Cleavage of Quaternary Ammonium Salts at Mercury Cathodes

Cleavage according to Eq. (1) proceeds much more cleanly at mercury (Table 1). The following types resist cleavage and yield hydrogen instead:

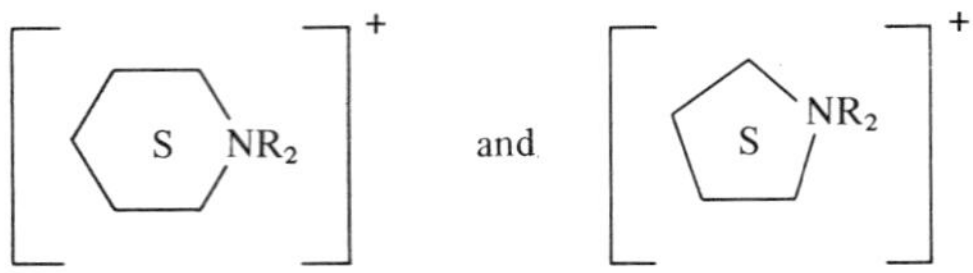

in which R = CH_3, C_2H_5, n-C_4H_9, $HOCH_2CH_2$.

TABLE 1

Reductive Cleavage of Quaternary Aliphatic and Aliphatic-Aromatic Ammonium Salts at Mercury

Quaternary ammonium salt	Tertiary amine hydrocarbon	Yields (% theory)
Trimethylphenylammonium[a]	Trimethylamine	90
	Benzene	83
Dimethylethylphenylammonium[a]	Dimethylethylamine	79
	Benzene	69
Dimethylallylphenylammonium[a]	Dimethylaniline	95
Trimethyl-p-tolylammonium[a]	Trimethylamine	84
	Toluene	86
Trimethyl-β-naphthylammonium[a]	Trimethylamine	80
	Naphthalene	80
Trimethyl-[p-chlorophenyl] ammonium[a]	Trimethylamine	—
	Benzene	88
Trimethyl-[p-bromophenyl] ammonium[a]	Trimethylamine	—
	Benzene	81
Trimethyl-[p-benzoylphenyl] ammonium[b]	Trimethylamine	80
	Benzhydrol	—
N, N, N′, N′-Tetramethyl-4, 4′-diamino-diphenylmethane[c]	Trimethylamine	—
	Diphenylmethane	73
Ethylbenzylpiperidinium[b]	Ethylpiperidine	75
	Toluene	83

Table 1 (cont.)

Quaternary ammonium salt	Tertiary amine hydrocarbon	Yields (% theory)
Allylbenzylpyrrolidinium[a]	Allylpyrrolidine	70
	Toluene	67
	Benzylpyrrolidine	11.5
Benzyl-β-phenethylpiperidinium[b]	β-Phenyethylpiperidine	80
	Toluene	78
Allyl-β-phenethylpiperidinium[a]	β-Phenethylpiperidine	91.5
Methyl-butyl-bis[β-cyanoethyl]-ammonium[a]	Methyl-butyl-[β-cyano-ethyl] amine	61.5
Methylethyldibenzylammonium[a]	Methylethylbenzylamine	83
Methylethyl-n-propylbenzylammonium[a]	Methyl ethyl n-propyl-amine	83
Tetramethylene-bis[N-benzyl-pyrrolidinium][a]	1,4-Bis-[pyrrolidinyl-(1)] butane	87.5 85

[a] As bromide.

[b] As chloride.

[c] As bis-methobromide

As can be seen from Table 1, N-allyl is cleaved at mercury more easily than N-phenyl, and N-benzyl more easily than N-allyl. A simple procedure for the preparation of mixed tertiary amines, as illustrated in the following scheme, is based on this difference in behavior:

$$CH_3N(CH_2C_6H_5)_2 \xrightarrow{+\ C_2H_5Br} [(CH_3)(C_2H_5)(CH_2C_6H_5)_2N]^{(+)}Br^{(-)} \xrightarrow[-C_6H_5CH_3]{2\bar{e}\ +\ H^+}$$

$$(CH_3)(C_2H_5)(CH_2C_6H_5)N \xrightarrow{+n\text{-}C_3H_7Br} [(CH_3)C_2H_5)(n\text{-}C_3H_7)(CH_2C_6H_5)N]^{(+)}Br^{(-)}$$

$$\xrightarrow[-C_6H_5CH_3]{+\ 2\bar{e}\ +\ H^+} (CH_3)(C_2H_5)(n\text{-}C_3H_7)N.$$

The analogous transformations are less satisfactory for $>N-CH_2CH_2CN$ derivatives because the latter are more difficult to prepare in the first place, and secondly, their quaternary salts crystallize poorly.

A useful preparative variation of this "Emmert cleavage" is embodied in the reductive ring opening of anilinium salts in which one of the N-alkyl groups is simultaneously attached to the 2-position of the benzene ring [(Ia)-(Ic)]. The products are of the type (IIa)-(IIc) [3, 4]:

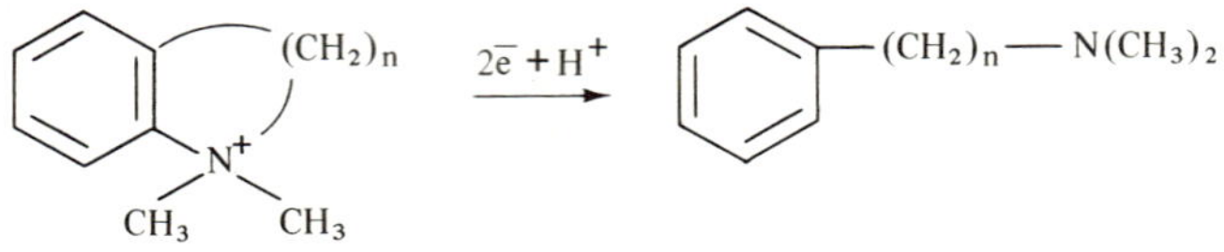

(3)

(Ia) — (Ic) (IIa) — (IIc)

The yield of (IIa) (n = 2) is 90%; of (IIb) (n = 3), 88%; of (IIc) (n = 4), 90%. Tetrahydroisoquinolinium salts are not cleaved at the mercury cathode under ordinary conditions but are cleaved when NaI in liquid NH_3 is used [5].

3. Attempts to Rationalize the Reductive Cleavage of Quaternary Ammonium Salts at a Mercury Cathode

The following observations must be taken into account in arriving at an understanding of the reactions occurring on the mercury surface:

1. Dimethylbenzylanilinium chloride is cleaved to dimethylaniline and toluene both at platinum and at mercury cathodes.

2. In aprotic solvents, e.g., N, N-dimethylacetamide or DMF, 35% of bibenzyl is formed, e.g., from dimethylbenzylanilinium salt. Electrolysis of chiral D-α-phenylethyltrimethylammonium nitrate in DMF yields a mixture of meso- and dl-2, 3-diphenylbutane [6a, 6b]:

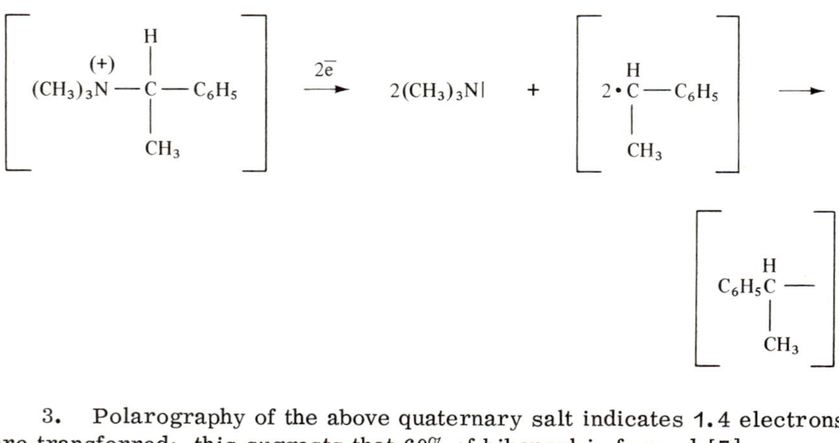

3. Polarography of the above quaternary salt indicates 1.4 electrons are transferred; this suggests that 60% of bibenzyl is formed [7].

4. The bibenzyl fraction of the products increases with increasing concentration of the onium salt and lowering of the temperature [7].

5. ESR spectroscopy reveals the presence of radical fragments [7].

These findings show that:

1. Amalgam formation is not a concomitant of reductive cleavage.

2. The hydrogen overvoltage at the smooth Pt electrode is not sufficient for the cleavage of trimethyl-p-tolylammonium bromide.

3. The cathode cleavage of quaternary ammonium salts proceeds via free radical intermediates.

In view of the fact that tetramethylammonium, stabilized as its amalgam, yields methyl radicals on pyrolysis [8] it is to be expected that the "ammonium radical" discharged at the cathode would likewise expel a ligand as a radical, which, by analogy with the Kolbe reaction, would dimerize. However, no dimeric products have been observed when the reductions are carried out in polar solvents, e.g., water or methanol. This must mean that at very negative overpotentials at mercury the adsorbed $[NR_4]^{\bullet}_{ad}$ decomposes according to Eq. (4) into a tertiary amine and an adsorbed radical $R^{\bullet}_{ad}$; the latter, in turn, takes up a second electron immediately and, now an anion, reacts with a proton to form the end product:

$$\begin{aligned} &NR_4^+ + \bar{e} \rightarrow [NR_4]^{\bullet}_{ad} \rightarrow NR_3 + R_{ad} \\ &R^{\bullet}_{ad} + \bar{e} + H^+ \rightarrow RH. \end{aligned} \qquad (4)$$

This scheme is in agreement with the results found in common in electrochemical [6], sodium amalgam [4], and Na-Pb alloy [9] reductions, namely, that those ligands are preferentially cleaved from nitrogen which form the most stable radicals. The ligands themselves, however, have a noteworthy influence on the reduction potential. This will be illustrated below in the discussion of phosphonium and arsonium reductions.

The mechanism of reduction of $C_6H_5CH_2\overset{+}{N}(C_2H_5)_3\overset{-}{NO_3}$ in DMF at Al and Pt has been studied in detail [6c]. The perchlorate salt has been reduced at several electrodes and in several solvents [6d].

C. Quaternary Phosphonium

1. Reductive Cleavage of Quaternary Phosphonium Salts at Mercury and Lead Cathodes

The reductive degradation of quaternary phosphonium salts to tertiary phosphines does not involve the side reactions incurred with quaternary

TABLE 2

Reductive Cleavage of Quaternary Phosphonium Salts [10]

Phosphonium bromide	Phosphine	Yield, %	
		Mercury [a]	Lead [b]
Methyl-triphenyl-	Methyl-diphenyl-/Triphenyl	83/5.5	40/53
Dimethyl-diphenyl-	Dimethyl-phenyl-/Methyl-diphenyl	69/26	
Trimethyl-phenyl-	Trimethyl-/ Dimethylphenyl-	47/40	
Ethyl-triphenyl-	Ethyl-diphenyl-/Triphenyl-	59/35	—/91
Butyl-triphenyl-	Butyl-diphenyl-/Triphenyl-	50/46	Little /77
Isopropyl-triphenyl-	Isopropyl-diphenyl-/Triphenyl-	25/52	
tert-Butyl-triphenyl- [c]	Triphenyl-	85	
Methyl-ethyl-diphenyl	Methyl-ethylphenyl-	33	
	Methyl-diphenyl-	66	
Benzyl-triphenyl [c]	Triphenyl-	89	85
β-Hydroxyethyl-triphenyl [c]	Triphenyl-	82	85
β-Hydroxyethyl-benzyl-diphenyl [c]	β-Hydroxyethyl-diphenyl	88	
Ethyl-tris(β-cyanethyl)-	Ethyl-bis(β-cyanethyl)	50-70	
Methyl-phenyl-benzyl-benzhydryl	Methyl-phenyl-benzyl	85	
Allyl-triphenyl	Triphenyl		67

Table 2 (cont.)

Phosphonium bromide	Phosphine	Yield, %	
		Mercury [a]	Lead [b]
Acetonyl-triphenyl [c]	Triphenyl		95
p-Dimethylaminobenzyl-triphenyl [d]	Triphenyl		84
Methyl-phenyl-dibenzyl	Methyl-phenyl-benzyl-	86	
Ethyl-phenyl-dibenzyl	Ethyl-phenyl-benzyl-	91	
Propyl-phenyl-dibenzyl	Propyl-phenyl-benzyl	82	
Butyl-phenyl-dibenzyl	Butyl-phenyl-benzyl	81	
β-Hydroxyethyl-phenyl-dibenzyl	β-Hydroxyëthyl-phenyl-benzyl	—	
Methyl-ethyl-phenyl-benzyl	Methyl-ethyl-phenyl-	77	
Ethyl-allyl-phenyl-benzyl	Ethyl-allyl-phenyl	72	
Ethyl-tribenzyl-	Ethyl-dibenzyl-	90	
Methyl-butyl-dibenzyl	Methyl-butyl-benzyl	86	
Ethyl-butyl-benzyl-dodecyl-	Ethyl-butyl-dodecyl	83	
Ethylen-bis-methyl-phenyl-benzyl-	Ethylene-bis-(methyl-phenyl)	83 [11]	
Tetramethylen-bis-methyl-phenyl-benzyl	Tetramethylene-bis-methyl-phenyl	80	

[a] At mercury cathode.
[b] At lead cathode.
[c] As chloride.
[d] As iodide.

ammonium salts, particularly at lead. When phosphonium salts are used, mercury cathodes are not covered with organic materials nor do they powder; as a consequence the electrode can maintain the overvoltage necessary for cleavage. The phosphonium and arsonium salts discharge at a more positive potential than the ammonium salts; this may result from the fact that the abilities of the P and As atoms to expand their octets make other electrode reactions possible. Table 2 illustrates that the composition of the products formed by cleavage of the phosphonium salts depends on the nature of the electrode material. This implies that the cathode material is intimately involved in the reduction process. These differences in the mode of cleavage can be advantageously used for synthetic purposes. In general, mercury is preferred as cathode. It has the advantages of leading to good yield, high purity of the generally easily autooxidizable phosphines, and simple work-up.

2. Method of Alternate Cathodic Cleavage and Quaternization

A convenient synthetic procedure for the preparation [12] of mixed tertiary phosphines and quaternary phosphonium salts has been based on the preferential cleavage of the benzyl group from P (and As). This method also provides a simple means for preparing phosphonium salts with four different ligands. The salts, after resolution into their optical antipodes, can be cathodically cleaved and thus provide optically active tertiary phosphines. The sequence is summarized in Eq. (5):

$$\begin{aligned} &R_3'P \xrightarrow{R^1X} [R_3'R^1P]X \xrightarrow[R'H]{+2\bar{e}+H^+} R_2'R^1P \\ &\xrightarrow{R^2X} [R_2'R^1R^2P]X \xrightarrow[-R'H]{+2\bar{e}+H^+} R'R^1R^2P \xrightarrow{R^3X} \\ &[R'R^1R^2R^3P]X \xrightarrow[-R'H]{+2\bar{e}+H^+} R^1R^2R^3P \xrightarrow{R_4X} [R^1R^2R^3R^4]X \\ &[R^1R^2R^3R^4P]X \xrightarrow[\text{2. Cathodically cleave}]{\text{1. Resolve}} \text{Optically active tert-phosphines.} \end{aligned} \tag{5}$$

The benzyl group is especially suitable as R′. This method can also be applied to the preparation of optically active tertiary arsines [13].

3. Relative Lability of Ligands During Cathodic Cleavage of Phosphonium Salts

In order to be able to plan a synthesis of tertiary phosphines with three different substituents or of phosphonium salts with four different substituents it is necessary to know the lability of the ligands attached to phosphorus [14]. An order of lability can be established by determining quantitatively the cleavage products of triphenylalkyl- or triphenylarylphosphonium salts, Eq. (6):

$$[(C_6H_5)_3PR]^+ \xrightarrow{2\bar{e} + H^+} (C_6H_5)_3P + RH$$
$$[(C_6H_5)_3PR]^+ \xrightarrow{2\bar{e} + H^+} (C_6H_5)_2PR + C_6H_6 \qquad (6)$$

In the orderly arrangement of ligands within this lability series, a statistical factor must be considered since, e.g., in Eq. (6), there are three phenyl groups for one R group.

The experimental results may be summarized as follows:

1. The ligands are reductively cleaved in a statistical manner if they have similar electronic interactions with the positive onium center. In the case of p-anisyl and dimethylaminophenyl groups, however, there is deviation from statistical behavior [13]. The following order of increasing lability is found at least qualitatively:

p-Aminophenyl (-0.660), p-methoxyphenyl (-0.268) > methyl > p-tolyl (-0.170) > p-isopropylphenyl (-0.151) > p-benzhydrylphenyl, m-tolyl (-0.069), o-tolyl > phenyl (±0.000) > ethyl > p-trifluormethyl (±0.551) > p-biphenylyl (+0.009) > naphthyl-(1) > naphthyl-(2) (+0.019) > tert-butyl > allyl > benzyl, fluorenyl-(9), benzhydryl.

There is a parallel increase (numbers in parentheses) in the σ-constants of the ligands. A similar dependence on substituents is found in the cleavage of phosphonium salts to phosphine oxides by alkali hydroxide [15-17].

2. If, however, the series $[R(C_6H_5)_3P]^+$ is arranged in order of decreasing $-E_{1/2}$ values the sequence of Table 3 is obtained (cf. Table 3 of Chapter XI). The ligand R is preferentially lost when the reduction potential of the above phosphonium ion is more positive than that of tetraphenylphosphonium. Salts with more negative $-E_{1/2}$ values than $(C_6H_5)_4P^+$ lose one of the other ligands; the more negative the potential the more favored is the formation of benzene.

This is demonstrated by examining examples 1, 3, and 5-9 of Table 3 (Chapter XI) in conjunction with the preparative results of Table 4 below.

3. Inclusion of alkyl substituents in a systematic arrangement of substituent labilities is complex because, in addition to benzene and alkanes, alkenes and higher-boiling hydrocarbons of unknown structure are obtained.

4. tert-Butyltriphenylphosphonium chloride occupies a special position because it loses practically only the t-butyl substituent. In addition to a small amount of benzene and a yet unidentified compound, 50% of isobutane and 27% of isobutene are formed. The formation of isobutylene is unequivocal evidence for the intermediacy of the isobutyl radical in the electrolysis. Experience to date has shown that alkenes are obtained in significant quantities only from salts with sec- and tert-alkyl substituents.

TABLE 3

Lability of Ligands R in $[R(C_6H_5)_3P]^+$ Arranged in Order of Half-Wave Potentials

No.[a]		$-E_{1/2}$[b]	No.[a]		$-E_{1/2}$[b]
(LI)	p-$C_6H_5SO_2C_6H_4$	1.227	(XXXIX)	m-$CH_3C_6H_4$	1.684
(XIII)	$(C_6H_5)_2CH$	1.271	(XXXVIII)	p-$CH_3C_6H_4$	1.692
(XII)	$NCCH_2$	1.301	(VIII)	i-C_4H_9	1.720
(L)	p-$CF_3C_6H_4$	1.410	(VII)	2-C_4H_9	1.769
(XLV)	α-Naphthyl	1.439	(XXXVII)	p-$(CH_3)_2NC_6H_4$	1.777
(XLVII)	p-$C_6H_5C_6H_4$	1.527	(XXXVI)	p-$H_2NC_6H_4$	1.801
(XI)	$C_6H_5CH_2$	1.552	(VI)	n-C_4H_9	1.802
(X)	Allyl	1.626	(V)	n-C_3H_7	1.822
(XLIV)	p-C_6H_4COOH	1.630	(IV)	i-C_3H_7	1.833
(XLIII)	p-i-$C_3H_7C_6H_4$	1.654	(XXXV)	p-$CH_3OC_6H_4$	1.835
(IX)	tert-C_4H_9	1.657	(III)	C_2H_5	1.837
(XLI)	o-$CH_3C_6H_4$	1.678	(II)	Cyclohexyl	1.859
(XLIX)	C_6H_5	1.680	(I)	CH_3	1.862
			(XXXIV)	p-HOC_6H_4	1.900

[a]Numbering according to Table 3, Chapter XI and citation therein.

[b]Vs. Ag/AgCl/saturated KCl.

The cathodic cleavage of bis-phosphonium salts is even more complicated. Three different P-containing products were isolated and quantitatively determined upon electrolysis of tetramethylene-bis-triphenylphosphonium salts.

$$[(C_6H_5)_3P(CH_2)_4P(C_6H_5)_3]^{2+} \rightarrow \underset{34\%}{(C_6H_5)_3P} + \underset{30\%}{(C_6H_5)_2PC_4H_9} + \underset{68\%}{(C_6H_5)_2P(CH_2)_4P(C_6H_5)_2}$$

TABLE 4

Results of the Cathodic Cleavage of Triphenylalkylphosphonium Salts $[(C_6H_5)_3PR]X$ [14]

R	Benzene, %	Alkane RH, %	Alkene RH, %	C_6H_5R, %	No. of unidentified substances
Methyl	82.3	0.34	—	—	—
Ethyl	42	7	0.04	Trace	1
n-Propyl	44	10	—	—	1
i-Propyl	23	16	4	Trace	2
n-Butyl	41	8	—	—	2
Butyl-(2)	6	43	10	—	2
i-Butyl	19	19	—	—	3
tert-Butyl	1.4	50	27	—	1

TABLE 5

Results of the Cathodic Cleavage of Certain Arsonium Salts $[(C_6H_5)_3AsR]Br$ [14]

	% Hydrocarbons formed	
	C_6H_6	RH
p-Tolyl	92	8
p-Isopropylphenyl	91	9
o-Tolyl	84	16
Methyl	81	19
m-Tolyl	80	20
Phenyl	75	25
p-Biphenylyl	65	35
Ethyl	52	48
Benzyl	0	100

D. Quaternary Arsonium

As already mentioned, optically active tert-arsines can be synthesized by the method of alternate cathodic cleavage followed by quaternization [13]. Here, too, a knowledge of ligand lability is required. The following order of lability of substituents on As is different in several respects from the corresponding order of P substituents [14]: methyl < ethyl < p-tolyl < n-butyl < phenyl < allyl, benzyl, ethoxycarbonylmethyl, phenacyl. The fact that there are these differences leads to the conclusion that the bonding forces between central atom and ligand are significantly different in P compounds from those in As compounds. Support is also found in the finding that on the average arsonium salts are reduced ca. 0.3 V more anodically than the corresponding P salts. Table 5 contains representative examples of the results of cathodic cleavage of arsonium salts at mercury.

E. Quaternary Stibonium

Quaternary stibonium salts are very difficult to prepare since tert-stibines exhibit much more feeble nucleophilicity than do tert-phosphines and tert-arsines. There have, therefore, been few investigations of stibonium salts. Table 6 illustrates the cleavage of a few stibonium salts in aqueous methanol at mercury [18]. McKinney and co-workers [19] found active participation by mercury in the two-stage reduction of tetraphenylstibonium ion. Diphenylmercury is formed quantitatively if the above quaternary is reduced at the more negative potential.

TABLE 6

Cleavage of Some Quaternary Stibonium Salts at a Mercury Cathode

Stibonium salt	Stibine	% Yield
Trimethylbenzyl	Trimethyl	64
Triphenylbenzyl	Triphenyl	90
Tetraphenyl [a]	Triphenyl	89
Methyltriphenyl	Methyldiphenyl	80
	Triphenyl	8

[a] The polarographic behavior on carbon rods uncoated and coated with Hg has been investigated [19b].

In conclusion, the "alternate cathodic cleavage and quaternization" method, useful in the phosphorus and arsenic series, has only limited application for antimony compounds because of the limited success in quaternizing the latter.

F. Iodonium Salts

Diphenyliodonium iodide (and perchlorate) were cleaved in 50% aqueous ethanol at mercury to benzene and iodobenzene [20a]; iodobenzene is further reduced to benzene. The following mechanism, which is in agreement with the polarographic data, was proposed:

$$(R\text{-}I\text{-}R)^+ + \bar{e} + Hg \rightarrow RHg\cdot + RI$$

$$RHg\cdot + \bar{e} + H^+ \rightarrow RH + Hg$$

$$RI + 2\bar{e} + 2H^+ \rightarrow RH + HI.$$

Electrolysis in alkaline medium, however, may lead to organomercurials. For example, cathodic reduction of diphenyliodonium hydroxide in 2 N NaOH produces 51% of diphenylmercury [20b]:

$$2[Ph_2I]^+ + 2\bar{e} + Hg \rightarrow Ph_2Hg + 2PhI.$$

G. Sulfonium Salts

1. Electrochemical Cleavage of Sulfonium Salts

Using triphenylsulfonium bromide as an example, Bonner [21] first demonstrated that sulfonium salts are cathodically degraded to thio ethers. The following year Colichman and Love [22] reported their results with trimethyl as well as o-, m-, and p-cresyldimethyl sulfonium salts. In the opinion of the latter authors, a one-step two-electron cleavage is involved:

$$R_3S^+ \xrightarrow{2\bar{e}} R_2S + R^- \xrightarrow{H^+} R_2S + RH.$$

Polarographic investigations of methylbutylphenacylsulfonium perchlorate were reported in 1963 [63, 24]. About this time Savéant [25] reported on the cathodic cleavage of diethylphenacylsulfonium bromide and methylphenylphenacylsulfonium bromide.

Lüttringhaus and Machatzke [26] postulated two successive one-electron steps in the reductive cleavage of sulfonium salts:

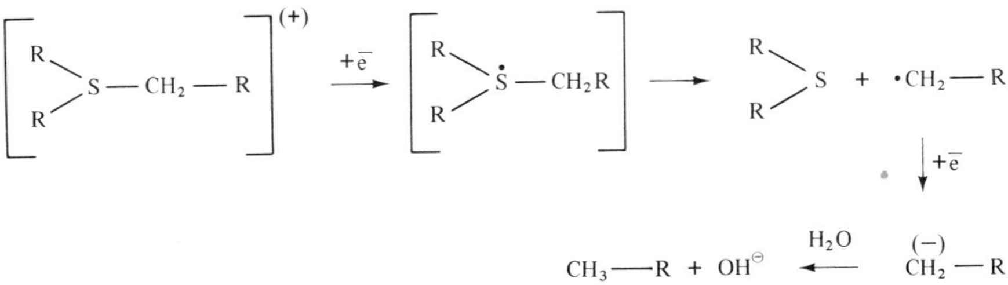

Many sulfonium salts show two reduction waves, e.g., ethylbenzyl-β-chloroethylsulfonium perchlorate at -0.89 and -1.20 V. The proposal of Lüttringhaus was supported and extended by McKinney and Rosenthal [27] in a careful polarographic investigation supplemented by product analysis of triphenylsulfonium bromide. The reduction at mercury, accompanied by strong adsorption phenomena, occurs in two stages: the first (-1.09 V vs. SCE) is pH independent while the second is pH dependent (pH 6, -1.33 V; pH 11, -1.39 V, vs. SCE). The number of electrons transferred, n, is concentration dependent: n = 2 at 0.46 mM; n = 1.6 at 1 mM.

Electrolysis of triphenylsulfonium bromide at the potential of the first wave yields diphenyl sulfide and diphenylmercury quantitatively. No benzene or diphenyl is formed. At potentials more negative than the second reduction wave diphenyl sulfide and benzene are formed quantitatively, provided low substrate concentrations are used. With increasing substrate concentration the yield of diphenylmercury increases at the expense of benzene. These experimental results can be explained by the following mechanism:

1. Stage:

$$\underbrace{[(C_6H_5)_3S]^{+}\cdots\cdots Hg}\quad +\bar{e} \longrightarrow [(C_6H_5)_3S\cdots\cdots Hg\bullet]$$
$$2\,[(C_6H_5)_3S\cdots\cdot Hg\bullet] \longrightarrow 2(C_6H_5)_2S + (C_6H_5)_2Hg + Hg$$

2. Stage:

$$[(C_6H_5)_3S\cdots\cdot Hg\bullet] + \bar{e} + H^{(+)} \longrightarrow (C_6H_5)_2S + C_6H_6 + Hg$$

The formation of diphenylmercury indicates that the mercury electrode participates in the reduction process. This involvement of the electrode surface is indicated by the symbol $(C_6H_5)_3\overset{+}{S}\cdots Hg$.

2. The Suitability of Sulfonium Salts as Supporting Electrolytes

As already mentioned, quaternary ammonium salts are especially well suited as supporting electrolytes. In order to determine whether sulfonium salts were also suited for this role, a series of sulfonium salts was prepared

TABLE 7

Foot Potentials E_F and Reductive Cleavage Products of Certain Sulfonium Salts at Mercury

Sulfonium bromide [a]	$-E_F$, [b] V	Cleavage products (%)
Trimethyl [c]	1.92	Methane (28); dimethyl sulfide (26)
Dimethylethyl	2.10	Methane (33); ethane (1)
Dimethyl n-propyl	2.21	Methane (54); propane (0)
Dimethyl i-propyl	2.11	Methane (16); propane (0.7)
Dimethyl n-butyl	2.15	Methane (50); butane (1)
Dimethylcyclohexyl	2.07	Methane (41); cyclohexane (2)
Dimethylbenzyl	2.11	Methane (55); toluene (5)
Triphenyl	1.07	Benzene (23)
Trimethylsulfoxonium [c]	1.90	Methane (28); dimethyl sulfoxide (25)

[a] 0.1 M solution in water.

[b] Vs. Ag/AgCl/saturated KCl.

[c] As iodide.

and their foot potentials determined [28]. Table 7 summarizes these data and also lists the products formed on discharge of the onium ions. In contrast to the findings with phosphonium and arsonium salts, the methyl group is preferentially cleaved from sulfonium salts. Even in the case of dimethylbenzylsulfonium bromide, ten times as much methane as toluene is formed. The reason for this difference in behavior is still unknown.

Because of their relatively anodic discharge potentials and their relatively facile reductive cleavage sulfonium salts have only limited applicability as supporting electrolytes.

3. Reaction of the Cathodic Cleavage Products of Phosphonium and Sulfonium Salts with Suitable Olefinic Acceptors

In an extension of their work on cathodic hydrodimerization of activated olefins, Baizer and Wagenknecht [29-31] have attempted to couple the fragments arising from the cathodic cleavage of phosphonium and sulfonium salts with activated olefins according to the following scheme:

$$\text{(a)}\quad \overset{+}{A}\,CH_2E \xrightarrow{2\bar{e}} A| + (CH_2E)^- \quad \text{I}$$

$$\text{(b)}\quad \text{I} + CH_2{=}CHE' \rightarrow ECH_2CH_2\overset{(-)}{C}HE' \quad \text{II}$$

$$\text{(c)}\quad \text{II} + H^+ \rightarrow ECH_2CH_2CH_2E'$$

in which A is a leaving group, e.g., $(CH_3)_2S$, $(C_6H_5)_3P$, and E and E′ are electrophilic groups such as CN, COOR, etc.

Electrolysis of dimethyl-α-cyanomethylsulfonium bromide [29] in the presence of acrylonitrile gave glutaronitrile in 20% yield. In contrast, electrolysis of sulfonium salts $(CH_3)_2\overset{+}{S}(CH_2)_n\text{-}CN$ (n = 2-3) in the presence of styrene led to complicated mixtures so that this method is unsatisfactory for preparative purposes.

The cleavage products formed from cyanoalkylphosphonium salts could, however, be captured by styrene:

$$[(C_6H_5)_3\overset{(+)}{P}(CH_2)_nCN] + 2\bar{e} \rightarrow (C_6H_5)_3P| + \overset{(-)}{|}CH_2(CH_2)_{n-1}CN$$

$$C_6H_5CH{=}CH_2 + \overset{(-)}{|}CH_2(CH_2)_{n-1}CN \xrightarrow{+H^{(+)}} C_6H_5(CH_2)_{n+2}CN.$$

In addition, mercurials of composition $Hg[(CH_2)_nCN]_2$ were isolated. When in the above phosphonium salts n = 1, neither coupled product nor mercurial was found.

II. SULFUR COMPOUNDS (EXCLUDING SULFONIUM)

A. Sulfones

Sulfones are generally outstandingly stable compounds. It is therefore necessary to use relatively drastic conditions in order to cleave the C-S bond or to reduce the sulfones to thio ethers. There are numerous publications [32] on their reduction by means of metals, amalgams, bases, organometallic compounds, hydrides, halogens, acid chlorides, etc. Many sulfones may be cleaved under rather mild conditions by Raney nickel [33], sodium amalgam [34], or alkali metals in liquid amines [35-37]. The first indications that electrochemical cleavage might be possible were an outcome of polarographic investigations [38-40]. Later experiments on a preparative scale [41] then showed that many sulfones could be reduced to the corresponding sulfinic acids using a mercury cathode and tetramethylammonium chloride as electrolyte:

$$RSO_2R' + 2\bar{e} + H^+ \rightarrow RSO_2^- + RH.$$

TABLE 8

Electrochemical Cleavage of Some Sulfones

No.	Sulfone	Cleavage products (%)
1.	Diphenyl	Benzenesulfinic acid (87), benzene (62)
2.	Di-p-tolyl	p-Toluenesulfinic acid (74), toluene (71)
3.	Bis-(2, 4, 6-trimethylphenyl)	2, 4, 6-Trimethylbenzenesulfinic acid (74), mesitylene (93)
4.	4, 4′-Dimethoxyphenyl	p-Methoxybenzenesulfinic acid (65), anisole (77)
5.	4, 4′-Dichlorophenyl	Benzenesulfinic acid (77), benzene (74)
6.	Phenyl-[2, 4, 6-trimethyl-phenyl]	Benzenesulfinic acid (80), mesitylene (88)
7.	Phenyl-[2, 4-dimethylphenyl]	Total sulfinic acids (87), m-xylene (58), benzene (20)
8.	Phenyl-p-tolyl	Total sulfinic acids (82), toluene (25), benzene (44)
9.	Phenyl-[p-methoxyphenyl]	Total sulfinic acids (80), anisole (16), benzene (54)
10.	Methyl phenyl	Benzenesulfinic acid (85), methane (84), ethane (1.1)
11.	Ethyl phenyl	Benzenesulfinic acid (75), ethane (70), ethylene (0.5)
12.	Methyl p-tolyl	p-Toluenesulfinic acid (81), methane (84), ethane (1.1)
13.	Allyl p-tolyl	p-Toluenesulfinic acid (80), propylene (75)
14.	Phenylbenzyl	Benzenesulfinic acid (90), toluene (82)
15.	Dibenzyl	Phenylmethanesulfinic acid (83), toluene (78)
16.	Methylbenzyl	Methanesulfinic acid (86), toluene (73)
17.	Ethylbenzyl	Ethanesulfinic acid (49), toluene (74)
18.	Propylbenzyl	Propanesulfinic acid (75), toluene (73)
19.	Butylbenzyl	Butanesulfinic acid (75), toluene (82)
20.	Dipropyl	No reduction
21.	Di-β-phenethyl	No reduction
22.	Di-n-octyl	No reduction

Diaryl, alkaryl, and alkylbenzene sulfones undergo this cleavage; aliphatic sulfones do not react. Table 8 presents illustrative examples.

The following conclusions may be drawn from the data of Table 8:

1. Unsymmetrical diaryl sulfones are cleaved so that the substituent with the greater donor effect is found in the resulting sulfinic acid; the other substituent forms the hydrocarbon.

2. Surprisingly, even sterically hindered diaryl sulfones are electrochemically cleaved in high yield. Contrary to the generalization given in item 1 above, however, phenyl-2,4,6-trimethylphenyl sulfone forms benzenesulfinic acid and mesitylene. This result leads to the conclusion that it is always that C-S bond which is cleaved in which there is the smallest mesomeric interaction with the aryl group.

3. Unsubstituted alkylaryl sulfones retain the aryl-S bond intact. Arylsulfinic acids are formed together with the corresponding hydrocarbons as well as dimerization and disproportionation products.

4. The allyl and benzyl groups are cleaved practically quantitatively from allylaryl and benzylaryl sulfones, respectively.

5. Moderate to good yields of alkanesulfinic acids are obtained by the cathodic cleavage of alkylbenzyl sulfones.

Extensive polarographic and preparative electrochemical investigations of arylsubstituted methyl phenyl sulfones afforded further insight into the cathodic cleavage of sulfones [42]. Electrolysis of sulfones $X\text{-}C_6H_4\text{-}SO_2CH_3$ (where X is an electron-withdrawing group) in aqueous alcohol solution yielded, contrary to item 3 above, alkanesulfinic acids and substituted aromatics:

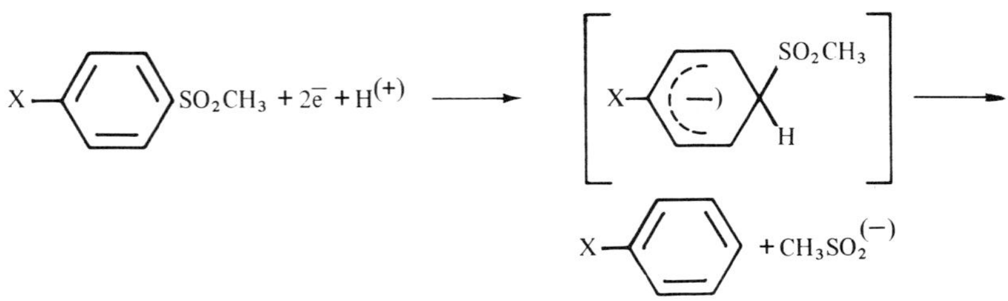

X: 4—COOH, 4—COOMe, 4—COOEt, 4—$CONH_2$, 4—CN, 4—SO_2CH_3

According to Zuman and coworkers [42a], this appears to proceed by an S_N2 mechanism.

Reductive cleavage of β-ketosulfones [123a] and β-ketosulfoxides [123b] in DMF at mercury is a practical method of preparing methyl ketones. In the former case one-electron reduction yielded a radical $RCO\dot{C}HR'$ and a sulfinate anion $R''SO_2^-$. The major fate of the radical was to undergo further reduction to the ketone; in some cases minor amounts of the radical dimer were found. (It is amusing that these dimers, formed here in a reduction process, are also accessible by anodic oxidation of the carbanions $RCO\bar{C}HR'$; see Chapter XX.) Since the anodic reaction was the oxidation of $R''SO_2^-$ to the innocuous $R''SO_3^-$, an undivided cell could be used.

B. Thio Ethers

Thio ethers are reductively cleaved only at very negative potentials [43] which are achievable using tetra-n-butylammonium bromide in DMF.

The following mechanism has been proposed:

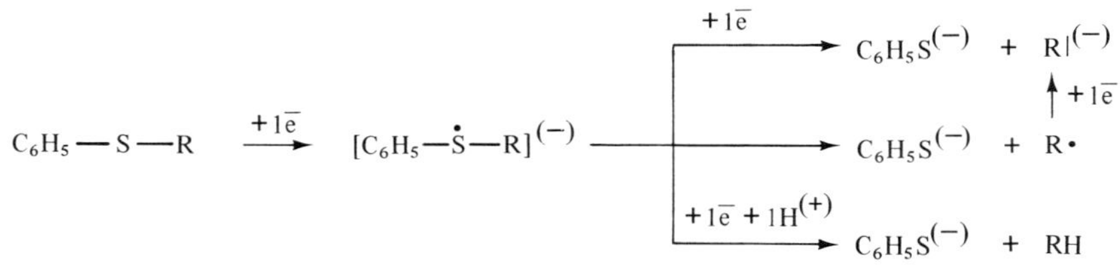

If R is in a sec- or tert-alkyl group, a predissociation of the thio ether on the mercury surface may be the rate-determining step:

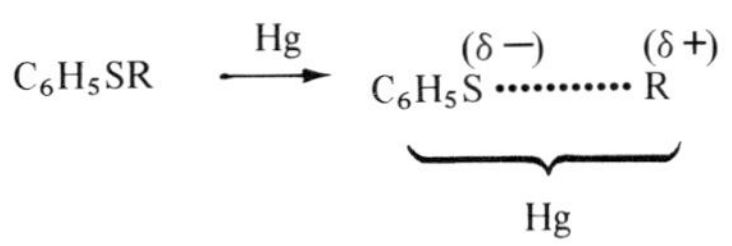

The half-wave potentials of a number of alkylaryl and diaryl sulfides are given in Table 9.

C. Disulfides [2]

A sulfur-sulfur single bond may be cleaved either anodically or cathodically. A sulfur-sulfur bond is generally reducible and compounds such as di-, tri-, and tetrasulfides [44] and thiolsulfonates [45] are polarographically active.

Oxidation of diphenyl disulfide in a mixture of glacial acetic acid and concentrated hydrochloric acid at a platinum anode yielded benzenesulfonic acid [47]; ethyl phenyl sulfide yields benzenesulfonic acid [47].

TABLE 9

Dependence of Half-Wave Potentials of RSR′ on the S Ligands [43]

R	R′	$-E_{1/2}$ [a]
Phenyl	Phenyl	2.549
Phenyl	2-Methylphenyl	2.571
Phenyl	3-Methylphenyl	2.567
Phenyl	4-Methylphenyl	2.595
Phenyl	Ethyl	2.734
Phenyl	i-Propyl	2.703
Phenyl	t-Butyl	2.638
Phenyl	Allyl	2.655
Phenyl	Benzyl	2.569
Phenyl	β-Phenylethyl	2.704
Phenyl	Methoxymethyl	2.704
Phenyl	Alkoxycarbonylmethyl	2.455
Phenyl	Cyanomethyl	2.351
2-Methylphenyl	2-Methylphenyl	2.585
4-Methylphenyl	4-Methylphenyl	2.645
Mesityl	Mesityl	2.738
Phenyl	N, N, N-Trimethylammoniummethyl	2.125

[a] In volts vs. Ag/AgCl/saturated KCl.

Disulfides are easily reducible to mercaptans; a high yield of thioglycolic acid [48] from dithiodiglyolic acid was obtained by reduction in 2 N sulfuric acid at a lead cathode at -0.55 V (SCE). This reduction is one of the few examples of a controlled potential reduction reported prior to the invention of potentiostats.

Thiosulfonic esters which are intermediates in the reduction of aromatic sulfonyl chlorides [49] in ethanolic sulfuric acid at a mercury cathode are further reduced to the disulfides; disulfides have been reported as products in the reduction of substituted benzenesulfonyl chlorides, but the isolation of the disulfide is a consequence of its insolubility. When conditions permitting a reasonable amount of the disulfide to stay in solution are used, e.g., a higher temperature, the mercaptan is the product.

D. Sulfonamides and Carboxamides

The electrochemical reduction of carboxamides at lead in strong sulfuric acid was reported in several early papers [50]. Under these conditions the carbonyl function of the amide group was preferentially reduced to methylene; cleavage to alcohol and amines occurred in low yield only. The latter reaction, however, which requires four electrons per molecule, can be made the principal reaction if mercury is used as cathode and tetraalkylammonium salts as supporting electrolytes [51]:

$$RCONR' + 4H^+ + \bar{e} \rightarrow RCH_2OH + R'NH_2.$$

Table 10 shows the relationship between the structure of the carboxamide and the success of cleavage and indicates the yield obtained.

TABLE 10

Electroreduction of Some Carboxamides Using Tetramethylammonium Chloride as Supporting Electrolyte [51]

Carboxamide	Amine	%	Alcohol	%
N-Hexylbenzamide	Hexyl	66	Benzyl	69
N-Cyclohexylbenzamide	Cyclohexyl	71	Benzyl	67
N-Benzylbenzamide	Benzyl	70	Benzyl	67
Benzanilide	Phenyl	94	Benzyl [a]	53
N-p-Tolylbenzamide	p-Tolyl	62	Benzyl	52
N-Benzoylpiperidine	Piperidine	78	Benzyl	76
N-Benzoylmesidine	Mesityl	81	Benzyl	80
N, N-Diphenylbenzamide	Diphenyl	94	—	
Acetanilide	Phenyl	94	Ethyl	—
N, N-Diphenylacetamide	Diphenyl	93	Ethyl	—
N-Benzylacetamide		No cleavage		
N-Cyclohexylacetamide		No cleavage		
N-Benzylphenylacetamide		No cleavage		
N-Benzylacetanilide	Phenylbenzyl	95	—	—

[a] Along with 4% of benzaldehyde.

As is evident from Table 10, amides derived from aromatic carboxylic acids are cleaved especially easily. Amides derived from aliphatic carboxylic acids are cleaved only when they possess an N-aryl substituent. The successful cleavage of N-benzoylmesidine illustrates that there is no steric hindrance involved in the electron transfer.

It should be noted in passing that aromatic amidines are reduced to primary amines and ammonia in a four-electron process [52]:

$$ArC(=NH)NH_2 + 4\bar{e} + 3H^+ \longrightarrow ArCH_2NH_2 + NH_3$$

Aliphatic and arylallyl amidines are not reduced up to -1.9 V (SCE).

Recently Benkeser [53] reported that primary, secondary, and tertiary aliphatic carboxamides are converted to alcohols in 70-80% yield by electroreduction in liquid methylamine containing LiCl as supporting electrolyte:

$$RCONH_2 \xrightarrow[4\bar{e}]{LiCl/MeNH_2} RCH_2OH + NH_3.$$

In the presence of ethanol the yield of alcohols decreases to ca. 15%; simultaneously, an increased amount of methylimine is formed (by interaction of the intermediate aldehyde and methylamine).

Sulfonamides can be reductively cleaved in alkaline as well as acid media [54]. After polarographic studies had shown that sulfonamides could be reductively cleaved, the cathodic hydrogenolysis of sulfonamides at mercury with tetramethylammonium supporting electrolytes was investigated [51]. It was found that amides of aromatic sulfonic acids are surprisingly cleaved in high yield to amines and sulfinic acids:

$$ArSO_2NHR' + 2\bar{e} + 2H^+ \rightarrow ArSO_2H + R'NH_2.$$

Table 11 summarizes some of the experimental results. In the hydrogenolysis of sulfonamides steric hindrance also plays no role. For example, sulfonamides with spatially bulky groups in the 2, 6-positions are converted to sulfinic acids and amines in high yield. Anilides of aliphatic sulfonic acids are not cleaved electrochemically.

However, sulfonamides $XC_6H_4SO_2NH_2$ (where X is a strong electron-withdrawing group) are cleaved not at the S-N but rather at the C-S bond. Zuman and co-workers [42b] believe that here too, as in the case of sulfones, an S_N2 mechanism is involved.

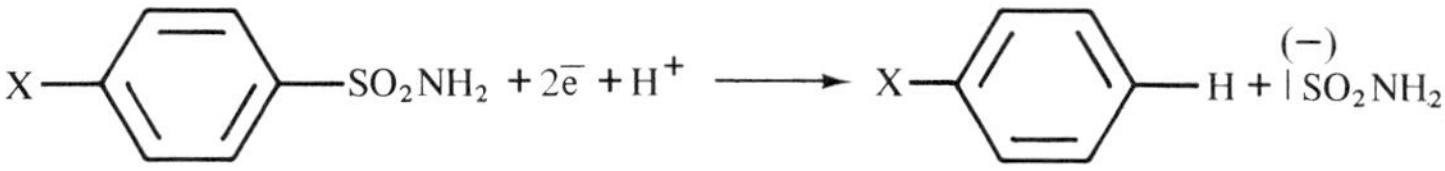

$$\overset{(-)}{|}SO_2NH_2 + H_2O \longrightarrow HSO_3^{(-)} + NH_3$$

$$X = COOH, -COOCH_3, -CN, -SO_2NH_2$$

TABLE 11

Electrochemical Cleavage of Certain Sulfonamides at a Mercury Cathode Using Tetramethylammonium Chloride as Supporting Electrolyte [51]

Sulfonamide	% Amine	% Toluenesulfinic acid
Toluenesulfonamides		
N-Hexyl	94	97
N-Butyl	55	97
N-Cyclohexyl	77	95
N-Benzyl	64	90
N-Phenyl	88	87
N, N-Diphenyl	88	86
N, Phenyl-N-benzyl	95	97
N-[2, 6-Dimethylphenyl]	92	94
N-[2, 6-Diethylphenyl]	94	94
N-Methyl-N-benzyl	98	96
N-[2, 4, 6-Trimethylphenyl]	96	95
N-p-Tolyl	95	98
Toluenesulfonylpiperidine	68	91
Benzenesulfonamides		
Benzenesulfonanilide	87	87
N-Benzylbenzenesulfamide	67	80
Methanesulfonanilide	No reduction	

TABLE 12

Half-Wave Potentials of Acid Amides at the Dropping Mercury Electrode at 25°

No.[a]	Acid amide	$-E_{1/2}$[b, a] A[c]	B[d]
	Sulfonamides		
(I)	Benzenesulfonamide	2.15	2.28
(II)	p-Toluenesulfonamide	2.20	2.36
(III)	p-Aminobenzenesulfonamide	2.33	ca. 2.4
(VII)	N-Tosylglycine	2.18	2.12
(IX)	N, N′-Ditosylethylenediamine	2.10	2.14
(XI)	Benzenesulfanilide	2.07	2.18
(XV)	N-Benzenesulfonyl-N-methylanilide	1.90	1.87
(XVII)	N-Benzenesulfonyldiphenylamide	1.64	1.66
	Carboxamides		
(XIX)	Benzamide	2.15	2.17
(XX)	p-Toluoylamide	2.32	2.30
(XXI)	p-Anisoylamide	2.33	2.32
(XXVIII)	N-α-Benzoyl-dl-lysine	2.28	2.25
(XXXI)	o-Phthalic monoamide	2.06	1.97
(XXXII)	Benzanilide	2.02	2.01
(XXXIII)	p-Toluoylanilide	2.08	2.07
(XXXIV)	p-Anisoylanilide	2.11	2.11
(XXXVI)	p-Chlorobenzanilide	1.90	1.92
(XXXVIII)	m-Trifluoromethylbenzanilide	1.86	1.86
(XL)	N-Benzoyl-p-chloroanilide	1.95	1.94
(XLI)	N-Benzoyl-2, 4-dichloroanilide	1.91	1.90
(XLIII)	N-Benzoyl-N-methylanilide	1.93	1.91
(XLVI)	N-Benzoyldiphenylamide	1.83	1.81
(XLVII)	N-p-Toluoyldiphenylamide	1.90	1.90
(XLVIII)	N-p-Anisoyldiphenylamide	1.93	1.93

Table 12 (cont.)

[a]Numbering according to Ref. [55] which contains additional examples.

[b]Volts vs. Ag/AgCl/KCl.

[c]1 mM depolarizer, 0.1 M tetramethylammonium chloride in ethanol.

[d]1 mM depolarizer, 0.1 M tetramethylammonium hydroxide in ethanol.

[e]Values ±0.01-0.03 V.

1. Reduction Potentials of Sulfonamides and Carboxamides as a Guide to Selective Cleavage [55]

Suitable substituents may so profoundly affect the reduction potentials of sulfonamides and carboxamides as to make possible selective cleavage of one particular acyl group from a diacylsubstituted compound or from one species in a mixture present in the substrate. Table 12 summarizes the reduction potential data for a number of sulfonamides and carboxamides.

The following correlations can be derived from Table 12:

1. Decrease of base strength of the amine components leads to a positive shift of $E_{1/2}$. Cf. (XIX), (XXXII), (XLIII) and (XLVI) on the one hand with (I), (XI), (XV), and (XVII) on the other.

2. There is a significant pH dependence of the $E_{1/2}$'s of N-unsubstituted and N-monosubstituted sulfonamide. The $E_{1/2}$'s are always at a more negative value in the alkaline range since under these circumstances the electron density at the sulfonamide anion is increased.

3. Substituents both in the acid as well as in the amine moieties influence the reduction potentials in a characteristic way as the following comparisons show: (I), (II), (III); (XIX), (XX), (XXI); (XXXII), (XXXIII), (XXXIV); (XLVI), (XLVII), (XLVIII); as well as (XXXII), (XXXVI), (XL), (XLI), and (XXXVIII).

In summary, it may be concluded that electron-donating groups in the acid and amino components shift the $E_{1/2}$'s to more negative values and thereby render the cathodic cleavage more difficult; electron-withdrawing groups have the opposite effect. The Hammett relationship holds for the $E_{1/2}$'s of substituted benzanilides p-R-$C_6H_4CONHC_6H_5$.

2. Selective Cleavage of Acyl Groups of Differing Reduction Potentials

If different molecules, or different positions in the same molecule, possess acyl groups whose reduction potentials are at least ca. 0.2 V apart, the acyl group cleaved at the more positive potential may be selectively

TABLE 13

Competitive Cleavage of Amides: Molar Ratios of Recovered Benzanilide or p-Anisoylanilide To Acid Amide [55]

Amide	$-E_{1/2}$	Benzanilide (-2.02 V)	p-Anisoylanilide (-2.11 V)
Tosylanilide	2.23	1:24 [a]	
N-Methyltosylanilide	1.94	1.2:1	
4-Hydroxybenzanilide	2.23	1:9 [a]	1:2
Benzenesulfonanilide	2.18	1:8	

[a] The reason for the different ratios at the same reduction potential is not yet clear.

removed. The usual electrolysis conditions are used (water, alcohol, or mixture of the two as solvent, tetramethylammonium chloride or hydroxide as supporting electrolyte, and mercury as cathode) but only enough current is passed to cleave one acyl group.

A determination of the ratio of the recovered amides gives an estimate of the selectivity of the cleavage. Typical results are summarized in Table 13.

The cleavage selectivity mentioned above is illustrated on page 757. C and D are formed in ca. 90% yields (calculated on the basis of the quantities of A and B, respectively, which have reacted). By thin layer chromatography only traces of dl-lysine were found in the solution in which B had reacted.

3. Arylacyl and Arylsulfonyl Groups as Protective Groups in Peptide Syntheses [51]

Aroyl and arylsulfonyl groups would not be suitable as protective groups in peptide syntheses because they cannot be cleaved solvolytically with retention of the peptide linkage. However, electrolytic hydrogenolysis permits cleavage of arylsulfonyl groups (as sulfinic acids) and aroyl groups (as alcohols); in this process the peptide bond as well as the configuration at the asymmetric centers is retained. Table 14 illustrates the hydrogenolysis of some acylated α-amino acids as well as di- and tripeptides. The electrochemical procedure for affecting the cleavages has the following advantages:

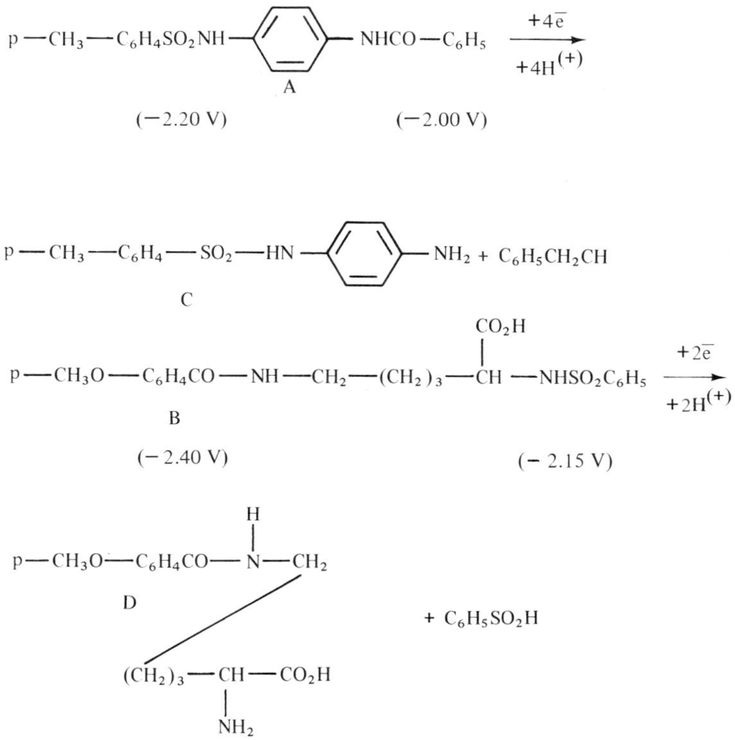

TABLE 14

Electrochemical Cleavage of Certain N-Protected Amino Acids and Peptides [51]

No.	Compound	Yield (%) [a]	No.	Compound	Yield (%) [a]
1.	Tos-L(—)-Tyr	96	9.	Tos-Gly-DL-Try	98
2.	Tos-DL-Meth	87	10.	Benzoyl-Gly-Gly	76
3.	Tos-Gly-Gly	91	11.	Benzoyl-DL-Meth	77
4.	Tos-Gly-DL-Meth	87	12.	Benzoyl-Gly-DL-Met	68
5.	Tos-Gly-DL-Ala	89	13.	Benzoyl-Gly-DL-Phe	63
6.	Tos-Gly-L(—)-Phe	93	14.	Z-DL-Phe	No cleavage
7.	Tos-DL-Ala-Gly	89	15.	Formyl-DL-Phe	No cleavage
8.	Tos-S-BZL-L(—)-Cys	86			

[a] The yields represent isolated chromatographically pure, individual amino acids or peptides.

1. The protective groups may be cleaved at 0-20°.

2. The tosyl (or benzoyl) derivatives of amino acids, dipeptides, and oligopeptides are, in contrast to the underivatized peptides, very soluble in alcohol. The peptides, therefore, precipitate from the alcoholic solution in high purity in the course of the electrolysis.

3. The reduced products (sulfinic acids or benzyl alcohol) formed from the protective groups are soluble in alcohol and, therefore, do not interfere with the subsequent work-up.

4. N-Formyl and N-benzyloxycarbonyl groups are not affected electrochemically. This fact affords further preparative possibilities in designing peptide syntheses.

5. S-Benzyl groups are not attacked electrochemically provided other protective groups which are reducible at a more positive potential are present.

Azlactones can be considered as N-acylated compounds with unsaturation in the side chain. During electrochemical reduction the acyl group is reductively cleaved and simultaneously the olefinic group is saturated [56]:

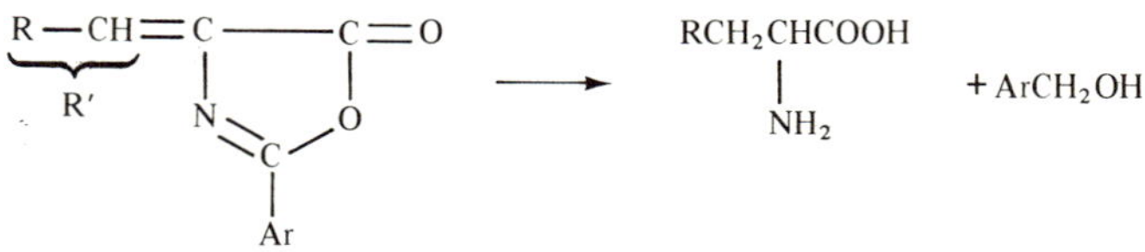

When R = C_6H_5 and R′ = C_6H_5CH (50-60), p-HOC_6H_4CH (50-60), $(CH_3)_2C$ (ca. 30), $(CH_3)_2CH$-CH (75-80), the corresponding amino acids are obtained (yields in parentheses). The aryl substituent is, however, not always removed quantitatively: phenylalanine together with ca. 20% of N-benzylphenylalanine is formed from 2-phenyl-4-benzylidene-5-oxazolone (R′ = C_6H_5CH).

β-Naphthoyl is unsuitable as a protective group. Under the usual electrolysis conditions β-naphthoylanilide E yields 43% of aniline and 29% of the 1,4-dihydronaphthalene derivative F:

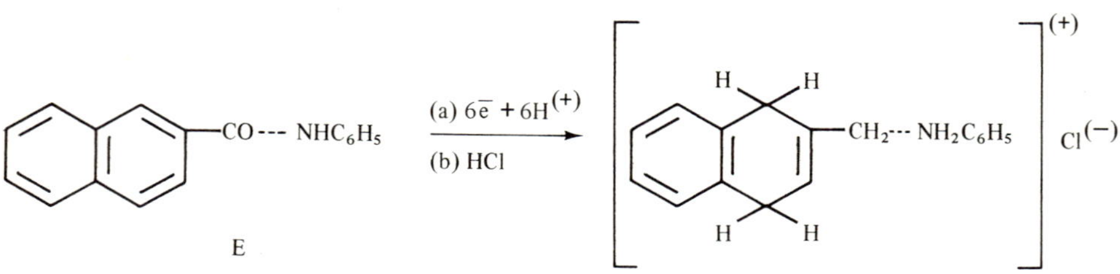

E

F

E. Sulfonic Esters

As Table 15 shows, aromatic sulfonic esters are amenable to electrolytic cleavage [51]. Since esters of alkanesulfonic acids are not cleaved electrochemically and lithium aluminum hydride is not selective in its reduction of aliphatic vs. aromatic sulfonic esters, the electrochemical method affords a welcome supplementary tool.

In order to understand the reaction mechanism it is necessary to take note of the observation [57] that the tosyl esters of optically active alcohols — with the tosyl group at the chiral center — are electrochemically cleaved

TABLE 15

Electrochemical Cleavage of Aromatic Sulfonic Esters at Mercury

Tosylate	Isolated products	% Yield
Methyl	Toluenesulfinic acid	90
Benzyl	Toluenesulfinic acid	99
	Benzyl alcohol	85
Cyclohexyl	Toluenesulfinic acid	94
	Cyclohexanol	81
Phenyl	Toluenesulfinic acid	91

with retention of configuration. The electrochemical method, therefore, supplements the solvolysis procedure (in which racemization or inversion occurs).

The results of cathodic cleavage of several optically active sulfonic esters are summarized in Table 16. Column A shows the rotation of the alcohols before esterification; column B shows the rotation of the alcohols after the cathodic cleavage of the tosylates. It is obvious that the rotations of the charged and recovered alcohols are about the same. The fact that the cleavage has this consequence leads to the conclusion that the electrons are added to the $ArSO_2$ moiety; the C-O bond is unaffected.

TABLE 16

Reductive Cleavage of Optically Active p-Toluenesulfonates at Mercury [57]

Tosyl ester of	% Yield alcohol	A[a]		B[b]	
		$[\alpha]^{21}_{546}$	$[\alpha]^{21}_{578}$	$[\alpha]^{21}_{546}$	$[\alpha]^{21}_{578}$
l-Menthol	73	-57.9	-51.2	-57.2	-50.7
l-Borneol (optically enriched)	95	-33.3	-29.2	-33.9	-29.5
Cholesterine	95	-45.7	-40.0	-45.3	-39.6

[a] Rotation of the alcohol before esterification.

[b] Rotation of the alcohol after cleavage of the ester.

The cathodic cleavage of sulfonic esters takes a different course in aprotic media. A C-S as well as an S-O cleavage occurs. For example, toluene (as well as p-toluenesulfinic acid and the appropriate alcohol) is formed in the electrochemical reduction of alkyl tosylates in acetonitrile at Hg and C electrodes. The following mechanism has been proposed [58]:

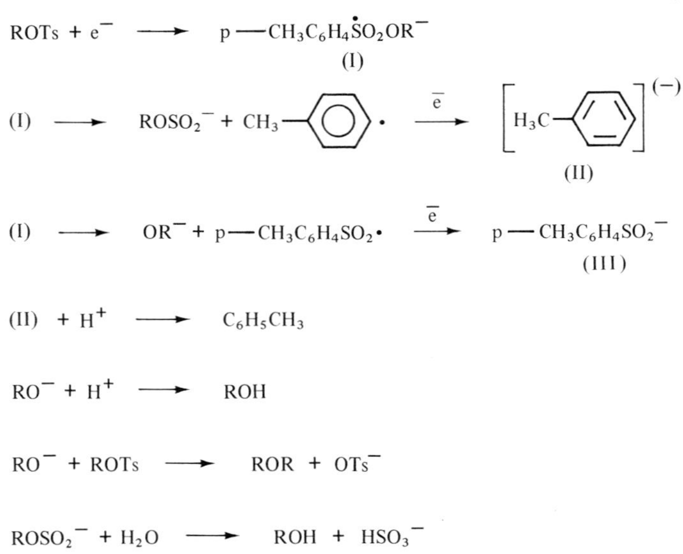

F. Thiol Esters of Aromatic Carboxylic Acids

In contrast to aromatic carboxylic esters, which are reduced to alcohols at mercury in poor yields only [57], the corresponding thiol esters, "active esters," are incomparably more prone to reductive cleavage [57]. As a consequence, aroyl groups assume greater importance as protective groups for SH functions. Some experimental results are collected in Table 17.

TABLE 17

Electrochemical Cleavage of Some Thiol Esters [a] [62]

S-Acyl group	Thiol	% Yield
Benzoyl	Thiophenol	91
"	Benzyl mercaptan	85
"	Z-Cys-OH [b]	87
"	H-Cys-OH	83
"	Z-Cys-Gly-OCH_3	79
"	Z-Leu-Cys-Gly-OH	81
"	BOC-His-Leu-Cys-OH [c]	80

[a] Substrate 10 mM; 30 mM tetramethylammonium chloride in methanol.

[b] Z = carbobenzoxy-.

[c] BOC = tert-butyloxycarbonyl-.

G. Thiocyanates [2]

Two types of electrode reactions are found in the reduction of thiocyanates. 2-Thiocyanatoacetophenones [59] are reduced in a two-electron reduction to the acetophenone and thiocyanate ion, e.g.,

$$C_6H_5COC(CH_3)_2SCN \xrightarrow{2\bar{e}} SCN^- + [C_6H_5COC(CH_3)_2]^- \xrightarrow{H^+} C_6H_5COCH(CH_3)_2.$$

In most other thiocyanates the reduction results in the loss of a cyanide ion; the half-wave potential is nearly independent of pH. Macroscale reductions at a mercury cathode yielded [60] the mercaptan according to

$$R\text{-}SCN \xrightarrow{2\bar{e}} RS^- + CN^-.$$

Microcoulometry [61] at the dropping mercury electrode gave n ~ 1.3, which was interpreted to indicate that the primarily formed radical anion lost a cyanide ion and attacked the mercury with formation of a mercurous salt; this salt either diffused from the electrode or was further reduced to mercaptan.

III. REDUCTIVE ELIMINATION OF SUBSTITUENTS α TO CYANO AND CARBOXYL COMPOUNDS

α-Hydroxyacyl arylacetonitriles can be converted at the mercury cathode to the corresponding arylacetonitriles [63]:

$$\underset{\substack{|\\ OCOR}}{ArCHCN} + 2\bar{e} + 2H^+ \longrightarrow ArCH_2CN + RCOOH$$

The substituents in the aryl group and on the α-hydroxy group have the expected effect on $E_{1/2}$: electron-releasing groups shift the potential cathodically, electron-withdrawing groups anodically. Electrolysis on a preparative scale gives arylacetonitriles in 57-72% yields.

Ketones ArCOCRR′X (R and R′ = H, alkyl, or aryl; X = Cl, OH, OCH_3, OC_6H_5, OCOR, SC_6H_5, SCN, or ONO_2) exhibit two polarographic waves, particularly in acid medium. The more positive wave is associated with the reductive substitution of H for X; the second wave arises from the reduction of the carbonyl group [64]:

$$C_6H_5COCH_2OCOCH_3 + 2\bar{e} + 2H^+ \rightarrow C_6H_5COCH_3 + CH_3COOH.$$

Preparative electrolyses at the more positive potential result in good yields of ketones (X = H).

β-Ketosulfoxides, which are intermediates in the conversion of carboxylic esters by dimethyl sulfoxide into methyl ketones by Corey's method [65], can be cleaved at controlled potential to methyl ketones [66]:

$$RCOCH_2\underset{\substack{\downarrow\\ O}}{S}CH_3 + 4\bar{e} + 4H^+ \rightarrow RCOCH_3 + CH_3SH + H_2O.$$

IV. ELECTROCHEMICAL CLEAVAGE OF TRIPHENYL-PHOSPHINE AND ITS OXIDE

Santhanam and Bard [67] have reinvestigated the electrochemical reduction of triphenylphosphine. The results of polarographic, voltammetric, and coulometric experiments point to the following radical mechanism:

$$(C_6H_5)_3P| \xrightarrow{+1\bar{e}} [(C_6H_5)_3\dot{P}|\,]^{(-)} \longrightarrow [(C_6H_5)_2P|]^{(-)} + C_6H_5\cdot$$

$$[(C_6H_5)_2P|]^{(-)} \xrightarrow{+H^{(+)}} (C_6H_5)_2PH \qquad C_6H_5\cdot \longrightarrow \tfrac{1}{2}\,C_6H_5—C_6H_5$$

This mechanism is further supported by preparative electrolysis of triphenylphosphine: oxidation of the cleavage products leads to diphenylphosphinic acid and diphenyl. On the other hand, the mechanism of the cathodic reduction of triphenylphosphine oxide has not yet been elucidated [67]. Preparative scale electrolysis of the latter leads to various reduction products containing partially reduced phenyl groups.

It can be shown unequivocally that in the cathodic reduction of phenyldibenzylphosphine oxide the phenyl group is partially reduced [68]: 60% of dibenzylcyclohexadienyl phosphine oxide is formed — the position of the double bonds is not yet known — which can be catalytically hydrogenated to cyclohexyl dibenzylphosphine oxide:

$$C_6H_5P(O)(CH_2C_6H_5)_2 \xrightarrow{+2\bar{e}\ +\ 2H^+} C_6H_7P(O)(CH_2C_6H_5)_2 \xrightarrow{4H/Pd} C_6H_{11}P(O)(CH_2C_6H_5)_2.$$

PART B [2]

I. INTRODUCTION

There will be discussed in this section electrode reactions resulting in a cleavage of a single bond, irrespective of whether the splitting occurs as a result of a slow chemical follow-up reaction or is more intimately connected with the electron-transfer step. The reactions are classified according to the single bond cleaved.

II. CARBON-CARBON BOND

An electrolytic cleavage of a carbon-carbon bond may occur in the course of both cathodic and anodic reactions. During an uncontrolled anodic oxidation many organic compounds end up as carbon dioxide and water, which involves splitting of C-C bonds, but there will be discussed here only oxidations in which the reaction does not have the character of an extensive degradation.

A. By Oxidation

One way of avoiding an extensive anodic degradation of organic compounds is to use acetone as a modifying medium, which also is well-known from chemical oxidations, e.g., with $KMnO_4$. Benzoin [69] is thus in sulfuric acid-acetone containing some water cleaved mainly to benzaldehyde (35%) and benzoic acid (45%); a small amount of benzil (5%) and tar (10%) is also formed. Benzil is apparently not an intermediate, as only benzoic acid (67%) was obtained from oxidation of benzil under similar conditions [69]:

$$C_6H_5CHOHCOC_6H_5 \xrightarrow{-2\bar{e}\ -\ H^+} C_6H_5CHO + [C_6H_5CO^+] \xrightarrow[-H^+]{H_2O}$$

$$C_6H_5CHO + C_6H_5COOH.$$

Oxidation of resorcinol dimethyl ether [70] in methanolic KOH produces among other products 5, 5, 6, 6-tetramethoxy-1, 3-cyclohexadiene (I) and hexamethyl cis, cis-orthomuconate (II); (II) could be obtained from (I) in 77% yield on anodic oxidation, conceivably according to

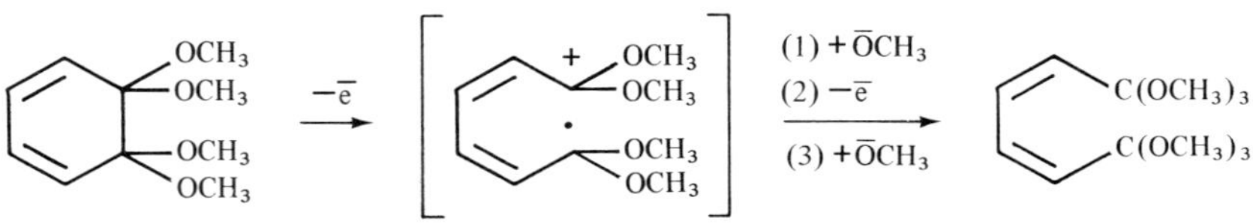

(I) (II)

It is noteworthy that only the least stable cis, cis isomer is formed; this indicates that the intermediates are short-lived and/or adsorbed on the electrode during the fixation of the geometry of the product.

Electrolysis under Kolbe conditions (Chapter XIV) of the γ-hydroxy acid (IIIa) leads to the unsaturated ketone (IV) together with some of the Kolbe coupling products [71, 72]:

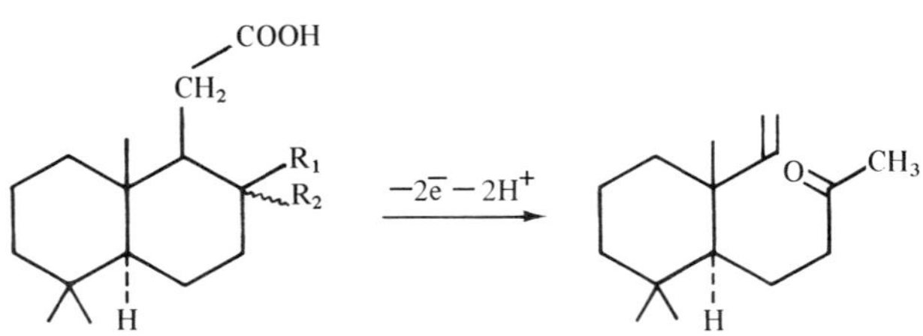

(III) (IV)

a: $R_1 = CH_3, R_2 = OH$
b: $R_1 = OH\ , R_2 = CH_3$
c: $R_1 = CH_3, R_2 = OAc$

The C-8 epimer (IIIb) also yielded the cleavage product (IV), whereas the acetoxy acid (IIIc) on electrolysis gave the corresponding Kolbe product. No cleavage product, neither (IV) nor the enol acetate, was detected. The reaction mechanism has not been proved. The radical, formed on decarboxylation of the primarily formed radical, may either lose a hydrogen atom with carbon-carbon cleavage or be oxidized further to a carbonium ion which then is cleaved with loss of a proton. The former possibility has been considered [72] the more likely as it more easily explains why the acetoxy acid is not cleaved in the same manner.

A ring opening which has also been suggested to occur via the primarily formed radical is found in the anodic oxidation of the sodium bisulfite addition product of cyclohexanones (V) from which the lactones (VI) and (VII) were isolated in a ratio of 2:3. Electrolysis of (V) in methanol rather than in water gives methyl 5-methoxyhexanoate as the major product. The following reaction mechanism has been suggested:

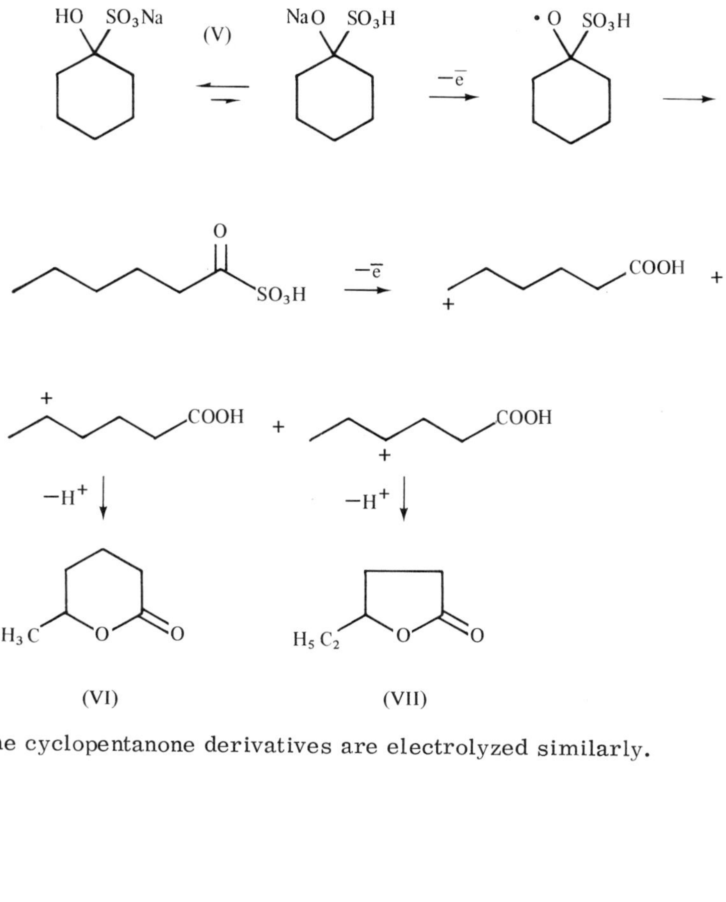

(VI) (VII)

The cyclopentanone derivatives are electrolyzed similarly.

B. By Reduction

Electrolytic cleavage of a carbon-carbon bond is found in certain aromatic nitriles in which the aromatic ring is π-electron deficient, either due to the presence of a heteroatom in the ring or an electron-attracting substituent on the ring. The driving force for the cleavage is the gain in resonance energy on reestablishing the aromaticity of the ring combined with the stability of the cyanide ion. Whereas 2- and 4-cyanopyridine in acid solution are reduced to aminomethylpyridines, the reduction in alkaline solution yields pyridine and cyanide ions [74-77]. Benzonitriles bearing electron-attracting groups like cyano or carbethoxy groups meta or para to the cyano group are also cleaved [78] with the loss of a cyanide ion.

1-Ethyl-4-cyanopyridinium ion is in aqueous solution reduced to a radical which dimerizes with loss of cyanide ion [79], and benzonitrile yields on reduction in DMF [80] a radical ion. In view of the general reduction of pyridines in alkaline solution to 1, 4-dihydropyridines the following reaction path for the reduction of aromatic nitriles is probable [78]:

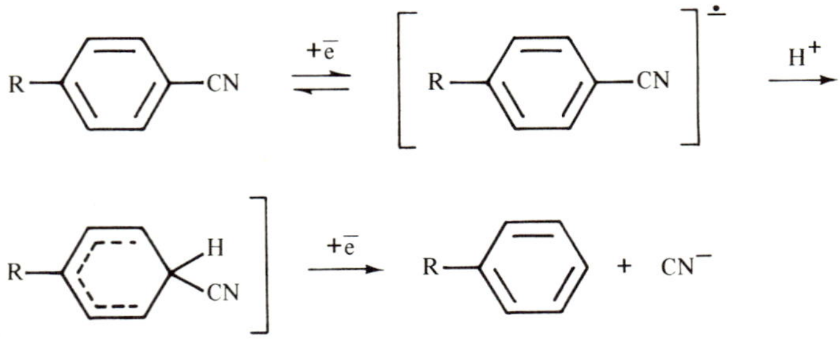

Another type of cleavage during an electrolysis occurs as a slow step in the reduction of certain β-diketones in DMSO [81]; the primarily formed radical dimerizes to a pinacol which is cleaved in a reverse aldol condensation to an α-diketone catalyzed by the base generated during the electrolysis. 1, 3-Diphenyl-1, 3-propanedione is thus reduced in DMSO according to:

$$2C_6H_5COCH_2COC_6H_5 \xrightarrow{2\bar{e} + 2HA} [C_6H_5C(OH)CH_2COC_6H_5]_2 + 2A^-$$

$$\xrightarrow{[A^-]} C_6H_5COCOC_6H_5 + 2CH_3COC_6H_5$$

The benzil is then further reduced to a radical ion.

III. CARBON-OXYGEN BOND

A. By Oxidation

It was an early observation [82] that both hydroquinone mono- and dimethyl ethers on anodic oxidation in dilute sulfuric acid yield quinone, which probably results from a nucleophilic attack by water on the primary oxidation product. Similarly, the oxidation of dimethoxydurene [83] in acetonitrile produced duroquinone rather than a side-chain oxidation product similar to that obtained from p-methoxytoluene.

A special reaction of this type is the oxidation of 2, 2, 5, 7, 8-pentamethyl-6-hydroxychromane (VIII) for which the following reaction was suggested [84] on the basis of polarographic data:

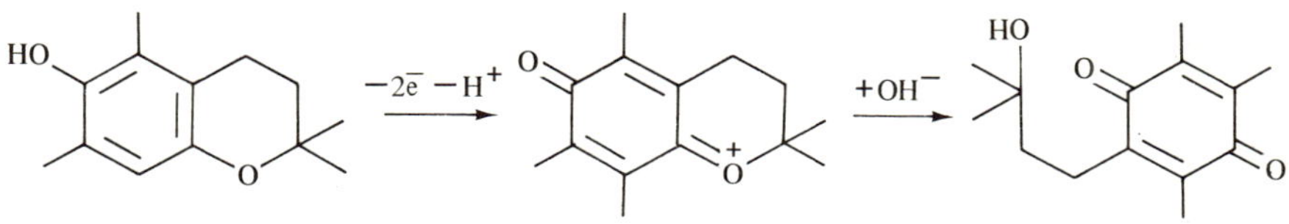

(VIII)

Later the reaction was proved by controlled potential oxidation to be reversible in acetonitrile or acetonitrile-acetic acid containing sodium acetate [85].

B. By Reduction

Reductions involving a splitting of a carbon-oxygen bond are rather rare [85a]; only certain activated compounds are attacked in this manner.

1. α-Hydroxy Carbonyl Compounds

A heteroatom α to a carbonyl or another electron-attracting group may be lost by reduction at a potential less negative than that required for the reduction of the activating group [86-90]. The reaction may be formulated as:

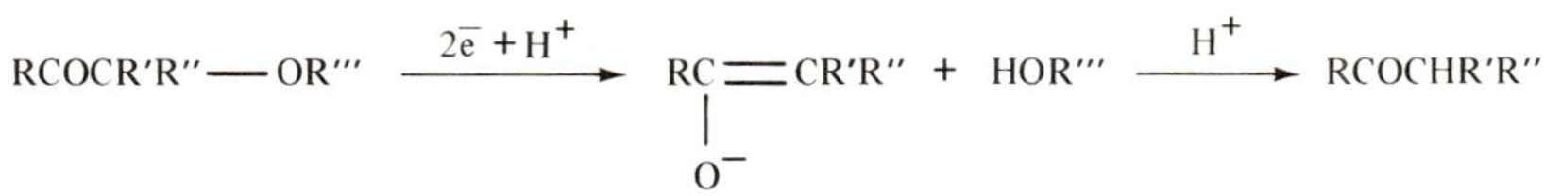

Besides the carbonyl group, a cyano, an amidine group [89], a p-carbethoxyphenyl group [89a], or a π-electron-deficient heteroaromatic ring [90] may act as activating groups.

2. ψ-Esters

o-Benzoylbenzoic acid ψ-ethyl ester (IX), which is an acylated ketal of a benzophenone, may be reduced in an acetate buffer to a phthalide (X) according to

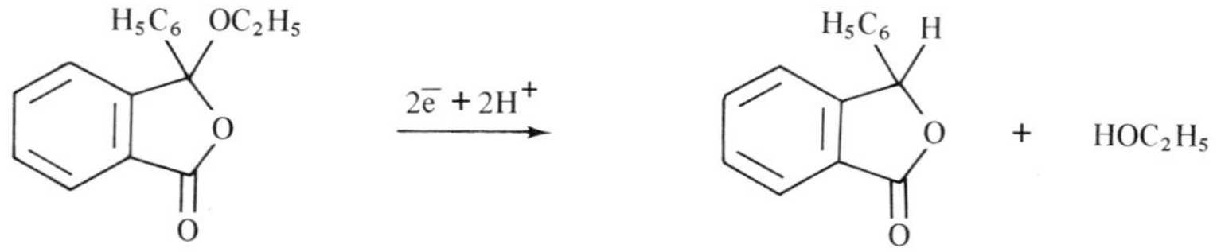

(IX) (X)

The polarographic data do not suggest that the alkoxy group is lost before the reduction as is the hydroxyl group in the case of the reduction of tropyl alcohol [91].

3. Nitroalkoxybenzenes

As described in Chapter VII certain nitroalkoxybenzenes lose the alkyl group during a reduction [93, 94] due to a nucleophilic attack by water. The mechanism of the loss during this reduction and during the oxidation of alkoxybenzenes is similar; in both reactions the essential intermediate has a quinoid structure which is attacked by water.

IV. CARBON-NITROGEN BOND (EXCLUDING QUATERNARY AMMONIUM)

A. By Oxidation

Anodic oxidation of primary [94] and tertiary [95] aliphatic amines in acetonitrile containing a small amount of residual water produced several compounds, including aldehydes. Tripropylamine is oxidized according to

$$(CH_3CH_2CH_2)_3N \xrightarrow{-\bar{e}} (CH_3CH_2CH_2)_3\dot{N}^+ \xrightarrow{-H^+} (CH_3CH_2CH_2)_2N\dot{C}HCH_2CH_3 \xrightarrow{-\bar{e}}$$

$$(CH_3CH_2CH_2)_2N\overset{+}{C}HCH_2CH_3 \leftrightarrow (CH_3CH_2CH_2)_2\overset{+}{N}{=}CHCH_2CH_3 \xrightarrow{H_2O}$$

$$(CH_3CH_2CH_2)_2NH + CH_3CH_2CHO + H^+.$$

B. By Reduction

Several reductions of this type have been discussed elsewhere, e.g., the reduction of gem-dinitroalkanes to nitrite ion and nitroalkane anion [96-98] (Chapter VII), the reduction of pyridylamines to picoline and ammonia [75, 90] (Chapter XVII), and the reduction of quaternary ammonium compounds.

The reduction of α-aminocarbonyl compounds [99, 100] or α-aminonitriles [100a] is similar to the reductive cleavage of the α-hydroxy compounds; the loss of amine occurs at a less negative potential than the reduction of the activating group.

If the amino group is β to the carbonyl group, a reduction takes place at the carbonyl group rather than resulting in a cleavage of the carbon-nitrogen bond; the polarographic reduction of the β-aminoketone and the α, β-unsaturated ketone obtained by elimination of the amine may be used for following the kinetics of the elimination reaction [100b, 100c].

Electrolysis of 4-aminobenzonitrile (XI) in DMF gave a yellow solution which, from the electron paramagnetic resonance (EPR) spectrum, was identified [80] as the radical anion of 4,4′-dicyanobiphenyl (XII); the reaction was formulated as follows:

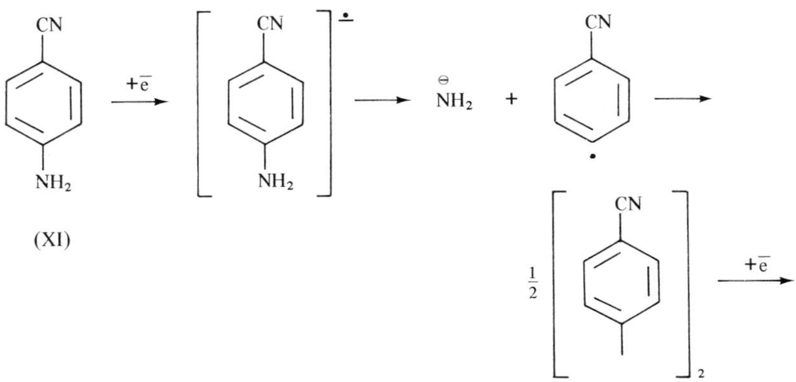

(XI)

(XII)

A carbon-nitrogen bond is broken during a chemical follow-up reaction in many ring closures [101] or ring contractions [102] of heterocyclic compounds in which a nucleophilic attack on an imine followed by an elimination of the amine or ammonia takes place. This is discussed in Chapter XVII, in which is also mentioned the loss of ammonia from the primarily formed dihydro compound in the reduction of certain aminoheterocyclic compounds.

The reduction of amides [104] to alcohols proceeds through a gem-amino alcohol which loses either ammonia (or amine), forming an aldehyde, or water, forming an imine; the aldehyde is then reduced to the alcohol whereas the imine forms an amine on reduction. The intermediate gem-amino alcohol is also formed in the reduction of oxaziridines [105] (Chapter XVII), and a similar gem-diamine during the reduction of amidines to amines [89].

V. CARBON-HALOGEN BOND

Reductions resulting in cleavage of a carbon-halogen bond have been discussed in Chapter VI.

VI. OXYGEN-OXYGEN BOND

Compounds containing an oxygen-oxygen single bond are generally polarographically reducible. Thus, hydroperoxides [107-108a], peracids [109-110], acylperacids [111], ozonoides, aldehyde, and ketone peroxides [112] can be determined polarographically. The electrode reaction, which is a cleavage of the O-O bond, has not been widely used preparatively, although the half-wave potentials of the various types of peroxides are sufficiently different to permit a selective reduction of one type in the presence of others. The presence of many alkyl groups shifts the half-wave potential to more negative values because of the electron-donating power of these and their shielding of the reaction center. In methanol-benzene containing LiCl the half-wave potential [111] of di-tert-butyl peroxide is higher than 2 V; dialkyl peroxides, higher than 1 V, tert-butylperoxyesters, 0.8-1.0 V; hydroperoxides, 0.6-0.9 V; diacyl peroxides, ~0.1 V; and peracids, ~0.05 V. The half-wave potential of an organic peroxide is thus valuable in the elucidation of its structure.

VII. NITROGEN-NITROGEN BOND

Reductive splitting of a bond between a nitrogen atom and a heteroatom occurs in many compounds of the type X=N-Y, where X may be carbon, oxygen, or nitrogen and Y may be oxygen or nitrogen [113]. This rule thus applies to cyclic and acyclic hydrazines, oximes, p-amino and p-hydroxyazobenzenes, triazines in acid solution, and N-nitrosamines in alkaline solution.

A. Hydrazones

Phenylhydrazones may be reduced to aniline and an amine in a four-electron reduction or to a hydrazine in a two-electron reaction, depending on pH. These results have been explained by the scheme on p. 771 [113a].

According to the scheme the branching leading to the formation of different products occurs after one proton and one electron are accepted by the substrate. The radical (XII) may then accept either an electron or a proton; if an electron is transferred to the radical, the carbanion (XIII) may acquire a proton from any weak acid (=[H^+]) present, e.g., water, with the formation of a hydrazine. In acid solution the radical may be protonated and then accept an electron; the zwitterion thus formed may then react as

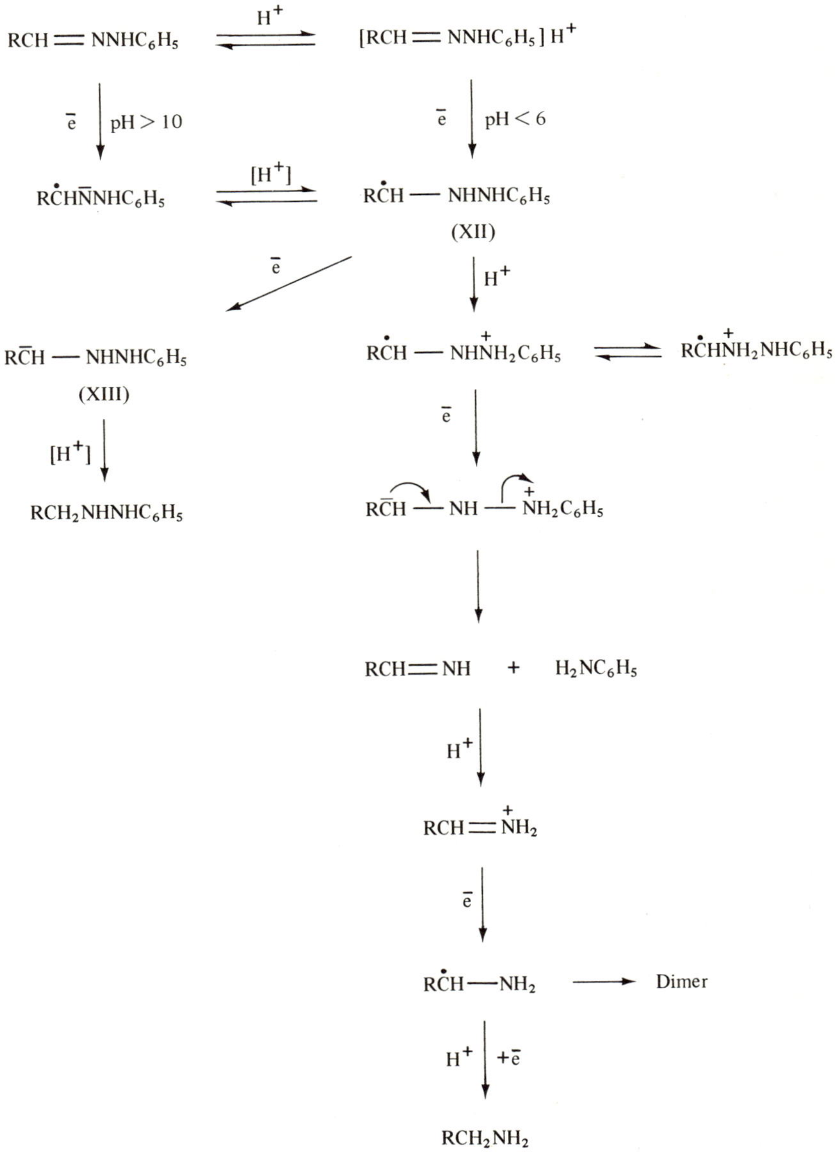

suggested by the arrows, resulting in a cleavage of the nitrogen-nitrogen bond as the protonated nitrogen is part of a good leaving group, aniline. It seems unlikely that the carbanion (XIII) would expel an aniline anion; the carbanion (XIII) could conceivably accept a proton at nitrogen with resulting cleavage, but it might be expected to give at least some hydrazine from protonation at the carbon atom, and this is generally not observed at low pH; at high pH the carbanion (XIII) is generally protonated to a hydrazine.

A similar reaction route seems to be rather general for azomethine derivatives of hydrazine [114] as it is followed by cyclic and acyclic phenylhydrazones, semicarbazones, azines, cyclic hydrazones, and acylated cyclic and acyclic hydrazones. The suggestion that the cleavage of the N-N bond precedes the saturation of the azomethine bond is also an essential part of the interpretation of many of the ring contractions of heterocyclic compounds (Chapter XVII).

The preprotonation step is indicated from polarographic data and from the fact that a quaternized hydrazone like benzaldehyde trimethylhydrazonium iodide, $C_6H_5CH{=}N\overset{+}{N}(CH_3)_3I^-$, is reducible in a four-electron reduction both in alkaline and acid solutions whereas hydrazones generally are either nonreducible or reducible in a two-electron reduction in alkaline solution [113, 114].

The cleavage of an N-N bond in p-hydroxyazobenzenes [115, 116] may be regarded as a variant of the general scheme if the p-hydroxyazobenzene is considered to be reduced in its tautomeric form, the quinone monophenylhydrazone (Chapter VII).

B. Hydrazides, Azides

Certain activated hydrazides [117], such as isonicotinic hydrazide, are reduced to the amides, e. g.,

$$H\overset{+}{N}C_5H_4CONHNH_2 \xrightarrow{2\bar{e} + 3H^+} H\overset{+}{N}C_5H_4CONH_2 + NH_4^+.$$

The reduction of isonicotinoyl azide produces also isonicotinic amide, and most other azides are similarly reduced according to [118]

$$R\text{-}\overset{-}{\bar{N}}\text{-}\overset{+}{N}{\equiv}N \xrightarrow{2\bar{e} + 2H^+} RNH_2 + N_2.$$

Azide groups or activated methylene groups may, however, be reduced to azide ion and methyl compound [118].

C. Triazines

Triazines are reduced to amines and hydrazines. It has not been proved for acyclic triazines whether the cleavage of the N-N bond precedes the saturation of the N=N bond, but in benzotriazoles [119] and benzotriazines [120] (Chapter XII) the first detectable intermediate still contains the N=N bond after the cleavage of the N-N bond.

D. N-Nitrosamines

As discussed in Chapter VII, N-nitrosamines, O=N-NRR′, are reduced in alkaline solution with cleavage of the N-N bond according to [121]

$$O{=}N{-}NRR' \xrightarrow{2\bar{e} + 2H^+} \tfrac{1}{2}N_2O + HNRR'.$$

VIII. NITROGEN-OXYGEN BONDS

The reduction of derivatives of hydroxylamine involves a splitting of the N-O bond. It has been shown [113] that the cleavage of the N-O bond in oximes generally occurs prior to the saturation of the double bond. In alkaline solution at low temperature, however, some reduction of an aldoxime to a hydroxylamine takes place. A similar primary cleavage of an N-O bond is found in the reduction of amide oximes [113] and hydroxamic acids [118]. The reduction of oximes in acid solution has been formulated as

$$RR'C{=}NOH \overset{H^+}{\rightleftharpoons} RR'C{=}\overset{+}{N}H{-}OH \xrightarrow{2\bar{e} + 2H^+} RR'C{=}\overset{+}{N}H_2 + H_2O \xrightarrow{2\bar{e} + 2H^+} RR'CH\overset{+}{N}H_3.$$

Whereas oximes are generally reduced in a single polarographic four-electron wave, a few are reduced in two steps [122] and in such reactions the intermediate imine has been isolated (cf. Chapter XVII).

NOTES

[1]Translated by the Editor.

[2]Contributed by H. Lund.

REFERENCES

[1]. B. Emmert, Ber., 42, 1507 (1909); 45, 430 (1912).

[2]. L. Horner and A. Mentrup, Justus Liebigs Ann. Chem., 646, 49 (1961).

[3]. L. Horner and H. Röder, Chem. Ber., 101, 4179 (1968).

[4]. H. Krauch and W. Kunz, Reaktionen der Organischen Chemie, 3rd ed., Dr. A. Hüthig-Verlag, Heidelberg, 1966, p. 203 for analogous Emde Na-Hg method.

[5]. (a) J. T. Wrobel, K. M. Pazdro, and Z. Bien, Chem. Ind., 1966, 1760; (b) J. T. Wrobel and A. R. Krawczyk, ibid., 1969, 656.

[6]. (a) M. Finkelstein, R. C. Petersen, and S. D. Ross, J. Amer. Chem. Soc., 81, 2361 (1959); (b) S. D. Ross, M. Finkelstein, and R. C. Petersen, ibid., 82, 1582 (1960); (c) idem, ibid., 92, 6003 (1970); (c) A. Monvernay and P. C. Lacaze, Bull Soc. Chim. Fr., 4206 (1970).

[7]. J. S. Mayell and A. J. Bard, J. Amer. Chem. Soc., 85, 421 (1963).

[8]. G. B. Porter, J. Chem. Soc., 1954, 760.

[9]. F. Fichter and H. Stenzl, Helv. Chim. Acta, 16, 571 (1933).

[10]. L. Horner and A. Mentrup, Justus Liebigs Ann. Chem., 646, 65 (1961).

[11]. L. Horner and J. P. Bercz (and coworkers of C. V. Bercz), Tetrahedron Lett., 1966, 5783, report the preparation of optically active biphosphines.

[12]. L. Horner, H. Winkler, A. Rapp, A. Mentrup, H. Hoffmann, and P. Beck, ibid., 1961, 161; for cleavage of optically active phosphonium salts containing benzyl groups by means of $LiAlH_4$ see L. Horner and M. Ernst, Chem. Ber., 103, 318 (1970).

[13]. L. Horner and H. Fuchs, Tetrahedron Lett., 1962, 203.

[14]. L. Horner and J. Haufe, Chem. Ber., 101, 2903 (1968).

[15]. L. Horner, H. Hoffmann, H. G. Wippel, and G. Hassel, ibid., 91, 52 (1958).

[16]. H. Hoffman, Justus Liebigs Ann. Chem., 634, 1 (1968).

[17]. M. Zanger, G. A. Vander Werf, and W. E. McEwen, J. Amer. Chem. Soc., 81, 3806 (1959).

[18]. E. Mayer, Thesis, Univ. of Mainz, 1965.

[19]. (a) M. D. Morris, P. S. McKinney, and E. C. Woodbury, J. Electroanal. Chem., 10, 85 (1965); (b) T. R. Williams and P. S. McKinney, ibid., 30, 131 (1971).

[20]. (a) E. L. Colichman and H. P. Maffei, J. Amer. Chem. Soc., 74, 2744 (1952); (b) J. A. Azoo, F. G. Colli, and J. Grimshaw, J. Chem. Soc. (C), 1969, 2521.

[21]. W. A. Bonner, J. Amer. Chem. Soc., 74, 5078 (1952).

[22]. E. L. Colichman and D. L. Love, J. Org. Chem., 18, 40 (1953).

[23]. P. Zuman and S. Tang, Coll. Czech. Chem. Commun., 28, 829 (1963).

[24]. S. Tang and P. Zuman, ibid., 28, 1524 (1963).

[25]. J. M. Savéant, Compt. Rend., 257, 448 (1963); 258, (1964).

[26]. A. Lüttringhaus and H. Machatzke, Justus Liebigs Ann. Chem., 671, 165 (1964).

[27]. P. S. McKinney and S. Rosenthal, J. Electroanal. Chem., 16, 261 (1968).

[28]. O. Spohn, Thesis, Univ. of Mainz, Mainz, Germany, 1967.

[29]. M. M. Baizer, J. Org. Chem., 31, 3847 (1966).

[30]. J. H. Wagenknecht and M. M. Baizer, ibid., 31, 3885 (1966).

[31]. J. H. Wagenknecht and M. M. Baizer, J. Electrochem. Soc., 114, 1095 (1967).

[32]. Houben-Weyl, Methoden der Organischen Chemie, Vol. IX, Thieme, Stuttgart, 1965, p. 243; C. M. Suter, Organic Chemistry of Sulfur, Wiley, New York, 1948, p. 658.

[33]. R. Mozingo, D. E. Wolf, S. A. Harris, and K. Folkers, J. Amer. Chem. Soc., 65, 1013 (1943).

[34]. R. E. Dabby, J. Kenyon, and R. F. Mason, J. Chem. Soc., 1952, 4881.

[35]. W. E. Truce, D. P. Tate, and D. N. Burdge, J. Amer. Chem. Soc., 82, 2872 (1960).

[36]. Anon., Angew Chem., 66, 654 (1954).

[37]. L. Reggel, R. A. Friedel, and J. Wender, J. Org. Chem., 22, 891 (1957).

[38]. S. G. Mairanovskii and M. B. Neiman, Dokl. Akad. Nauk SSSR, 87, 805 (1952); CA, 47, 6271f (1953).

[39]. E. S. Levin and A. P. Shestov, Dokl. Akad. Nauk SSSR, 96, 999 (1954); CA, 48, 13471e (1954).

[40]. L. Horner and H. Nickel, Chem. Ber., 89, 1681 (1956).

[41]. L. Horner and H. Neumann, ibid., 98, 1715 (1965).

[42]. (a) O. Manousek, O. Exner, and P. Zuman, Coll. Czech. Chem. Commun., 33, 3988; (b) idem, ibid., 4000 (1968); (c) G. Jeminet and J. Simonet, Compt. Rend., 272C, 661 (1971).

[43]. R. Gerdil, J. Chem. Soc. (B), 1966, 1071.

[44]. J. H. Karchmer and M. T. Walker, Anal. Chem., 26, 271 (1954).

[45]. D. Barnard, M. B. Evans, G. McHiggins, and J. F. Smith, Chem. Ind., 20, (1961).

[46]. S. G. Mairanovskii and M. B. Neiman, Dokl. Akad. Nauk SSSR, 87, 805 (1952).

[47]. F. Fichter and W. Wenk, Ber., 45, 1373 (1912).

[48]. E. Larson, ibid., 61, 1439 (1928).

[49]. G. M. Nichols, Electrochemical Society Meeting, New York, May 1969, Extended Abstracts, p. 364.

[50]. T. B. Baillie and J. Tafel, Ber., 32, 68 (1899); K. Kindler, ibid., 57, 773 (1924); 56, 2063 (1923); idem., Arch. Pharm., 265, 389 (1927).

[51]. L. Horner and H. Neumann, Chem. Ber., 98, 3462 (1965); more recent confirmatory work is reported by P. T. Cottrell and C. K. Mann, Abstract No. 118, Electrochemical Society Meeting, Washington, D.C., May 1971.

[52]. P. O. Kane, Z. Anal. Chem., 173, 50 (1960).

[53]. R. A. Benkeser, "The Synthetic and Mechanistic Aspects of Electro-Organic Chemistry," presented at Symposium at Durham, N. C., October 14-16, 1968.

[54]. D. Klamann and G. Hofbauer, Chem. Ber., 86, 1246 (1953); D. Klamann, Monatsh., 84, 651 (1953); L. Field and F. A. Grunwald, J. Org. Chem., 16, 946 (1951).

[55]. L. Horner and R.-J. Singer, Justus Liebigs Ann. Chem., 723, 1 (1969).

[56]. R.-J. Singer, unpublished thesis, University of Mainz, Mainz, Germany, 1965.

[57]. L. Horner and R.-J. Singer, Chem. Ber., 101, 3329 (1968).

[58]. P. Yousefzadeh and C. K. Mann, J. Org. Chem., 33, 2716 (1968).

[59]. H. Lund, Acta Chem. Scand., 14, 1927 (1960).

[60]. K. Schwabe and J. Voigt, Z. Elektrochem., 56, 44 (1952).

[61]. P. Lanza, L. Griggio, and G. Semerano, Ric. Sci., 26, S230 (1956).

[62]. Dr. R. Geiger (Farbwerke Hoechst), patent pending.

[63]. S. Wawzonek and J. D. Fredrickson, J. Electrochem. Soc., 106, 325 (1959).

[64]. H. Lund, Acta Chem. Scand., 14, 1927 (1960).

[65]. E. J. Corey and M. S. Chaykovsky, J. Amer. Chem. Soc., 86, 1639 (1964); 87, 1345 (1965).

[66]. B. Lamm and B. Samuelsson, Acta Chem. Scand., 23, 691 (1969).

[67]. K. S. V. Santhanam and A. J. Bard, J. Amer. Chem. Soc., 90, 1118 (1968).

[68]. L. Horner and H. Neumann, Chem. Ber., 102, 3953 (1969).

[69]. H. D. Law, J. Chem. Soc., 1437 (1906).

[70]. B. Belleau and N. L. Weinberg, J. Amer. Chem. Soc., 85, 2525 (1963).

[71]. E. J. Corey, R. R. Sauers, and S. Swann, ibid., 79, 5826 (1957).

[72]. E. J. Corey and R. R. Sauers, ibid., 89, 1732 (1959).

[73]. M. Oyama and M. Ohno, Tetrahedron Lett., 5201 (1966).

[74]. H. Lund, Acta Chem. Scand., 17, 2325 (1963).

[75]. (a) J. Volke and J. Holubek, Coll. Czech. Chem. Commun., 28, 1597 (1963); (b) J. Volke, R. Kubiček, and F. Šantavý, ibid., 25, 1510 (1960).

[76]. J. Volke and A. M. Kordos, ibid., 33, 2560 (1968).

[77]. A. M. Kardos, P. Valenta, and J. Volke, J. Electroanal. Chem., 12, 84 (1966).

[78]. O. Manousek, P. Zuman, and O. Exner, Coll. Czech. Chem. Commun., 33, 3979 (1968); cf. P. Zuman and O. Manousek, ibid., 34, 1580 (1969).

[79]. W. H. Schwarz, E. M. Kosower, and I. Shain, J. Amer. Chem. Soc., 83, 3164 (1961).

[80]. P. H. Rieger, I. Bernal, W. H. Reinmuth, and G. K. Fraenkel, ibid., 85, 683 (1963).

[81]. R. C. Buchta and D. H. Evans, Anal. Chem., 40, 2181 (1968).

[82]. F. Fichter and W. Dietrich, Helv. Chim. Acta, 7, 131 (1924).

[83]. V. D. Parker, Chem. Commun., 610 (1969).

[84]. L. I. Smith, I. M. Kolthoff, S. Wawzonek, and P. M. Rouff, J. Amer. Chem. Soc., 63, 1018 (1941).

[85]. V. D. Parker, ibid., 91, 5380 (1969).

[85a]. P. E. Given and M. E. Peover, Nature, 184, Suppl. No. 14, 1064 (1959).

[86]. M. Fedoronko, Chem. Anal., 3, 573 (1958).

[87]. P. Kabasakalian and J. McGlotten, Anal. Chem., 31, 1091 (1959).

[88]. H. Lund, Acta Chem. Scand., 14, 1927 (1960).

[89]. P. O. Kane, Z. Anal. Chem., 173, 50 (1960).

[89a]. J. P. Coleman, H. G. Gilde, J. H. P. Utley, and B. C. L. Weedon, Chem. Commun., 738 (1970).

[90]. O. Manousek and P. Zuman, Coll. Czech. Chem. Commun., 29, 1432 (1964).

[91]. P. Zuman, J. Chodkowski, H. Potesilova, and F. Santavy, Nature, 182, 1535 (1958).

[92]. M. Le Guyader, Bull. Soc. Chim. Fr., 1858 (1966).

[93]. A. Tallec, Ann. Chim., 3, 347 (1968).

[94]. K. K. Barnes and C. K. Mann, J. Org. Chem., 32, 1474 (1967).

[95]. P. J. Smith and C. K. Mann, Electrochemical Society Meeting, May 1969, Extended Abstracts, p. 351.

[96]. M. Masui and H. Sayo, J. Chem. Soc,, 4773, 5325 (1961).

[97]. M. Masui and H. Sayo, ibid., 1733 (1962).

[98]. J. Armand, Bull. Soc. Chim. Fr., 543 (1966).

[99]. P. Zuman and V. Horak, Coll. Czech. Chem. Commun., 26, 176 (1961).

[100]. J. Stradins, I. Tutane, A. Arens, and G. Vanags, Zh. Obshch. Khim., 35, 1327 (1965).

[100a]. P. Zuman, O. Manousek, and V. Horak, Coll. Czech. Chem. Commun., 29, 2906 (1964).

[100b]. P. Zuman and V. Horak, ibid., 27, 187 (1962).

[100c]. P. Carsky, P. Zuman, and V. Horak, ibid., 29, 3044 (1964).

[101]. S. Kwee and H. Lund, Acta Chem. Scand., 23, 2711 (1969).

[102]. H. Lund, Discuss. Faraday Soc., 45, 193 (1968).

[103]. D. L. Smith and P. J. Elving, J. Amer. Chem. Soc., 84, 1412 (1962).

[104]. H. Lund, Acta Chem. Scand., 17, 2325 (1963).

[105]. H. Lund, ibid., 23, 563 (1969).

[106]. P. T. Cottrell and C. K. Mann, J. Electrochem. Soc., 116, 1499 (1969).

[107]. D. A. Skoog and A. B. H. Lauwzecha, Anal. Chem., 28, 825 (1956).

[108]. M. L. Whisman and B. H. Eccleston, ibid., 30, 1638 (1958).

[108a]. S. Hayano and N. Shinozuka, Bull. Chem. Soc. Japan, 43, 2039 (1970).

[109]. W. E. Parker, C. Ricciuti, C. L. Ogg, and D. Swern, J. Amer. Chem. Soc., 77, 4037 (1955).

[110]. W. E. Parker, L. P. Witnauer, and D. Swern, ibid., 79, 1929 (1957).

[111]. D. Swern and L. S. Silbert, Anal. Chem., 35, 880 (1963).

[112]. B. N. Moryganov, A. I. Kalinin, and L. N. Mikhotova, Zh. Obshch. Khim., 32, 3476 (1962).

[113]. H. Lund, Acta Chem. Scand., 13, 249 (1959).

[113a]. H. Lund and E. T. Jensen, ibid., 25, 2727 (1971).

[114]. H. Lund, in Chemistry of the Carbon-Nitrogen Double Bond (S. Patai, ed.), Wiley, New York, 1970, p. 505 ff.

[115]. H.A. Laitinen and T. J. Kneip, J. Amer. Chem. Soc., 78, 736 (1956).

[116]. L. Holleck, D. Jannakondakis, and A. Wildman, Electrochim. Acta, 12, 1523 (1967).

[117]. H. Lund, Acta Chem. Scand., 17, 1077 (1963).

[118]. H. Lund, Österreich. Chem.-Z., 68, 43 (1967).

[119]. S. Kwee and H. Lund, Acta Chem. Scand., 22, 2879 (1968).

[120]. S. Kwee and H. Lund, Electrochemical Society Meeting, New York, May 1969, Extended Abstracts, p. 349.

[121]. H. Lund, Acta Chem. Scand., 11, 990 (1957).

[122]. H. Lund, ibid., 18, 563 (1964).

[123]. (a) B. Lamm and B. Samuelsson, ibid., 24, 561 (1970); J. Chem. Soc. (D), 1970, 1010; (b) idem, Acta Chem. Scand., 23, 691 (1969); B. Lamm and B. Samuelsson, Chem. Commun., 1970, 1010.

Chapter XXII

ANODIC SUBSTITUTION

L. Eberson

Division of Organic Chemistry
Chemical Center
University of Lund
Lund, Sweden

I. INTRODUCTION

Anodic substitution reactions are very common in organic electrochemistry. Formally, the overall reaction can usually be represented as in Eq. (1), in which E^+ and Nu^- denote an electrophile and nucleophile, respectively (note that the charges on E^+ and Nu^- are included only as a matter of convenience):

$$R\text{-}E + Nu^- \rightarrow R\text{-}Nu + E^+ + 2e^- . \quad (1)$$

Nu^- can be chosen from the following list: H_2O, ROH, OH^-, RO^-, $RCOO^-$, NO_3^-, SCN^-, $SeCN^-$, CH_3CN, NO_2^-, N_3^-, OCN^-, pyridine, halide ion, and CN^-; E^+ is most commonly H^+, but can also represent a carbonium ion or an alkoxonium ion. The case in which E^+ is equal to CO_2 has already been treated (the Kolbe reaction, Chapter XIII) and will not be dealt with here.

Anodic substitution reactions represent a reaction type that allows for a net substitution by a nucleophile, a reaction not easily achieved by ordinary chemical means. Specifically, it makes possible the direct formation of a C-O or C-N bond by substitution of a C-H bond. This is of considerable synthetic interest.

From the mechanistic point of view, anodic substitution reactions can be divided into indirect and direct processes. In an indirect process, the anode merely serves as a convenient source of an inorganic intermediate, e.g., chlorine by oxidation of chloride ion or nitrate radical by oxidation of nitrate ion, or an organic radical, e.g., methoxy radical from methoxide ion, which then attacks the organic substrate in a reaction which is not different from that taking place when the reagent is generated in any other way. In the direct process, oxidation of the organic substrate takes place at the anode with formation of a positively charged species (cation radical or dication) which undergoes follow-up reactions with nucleophiles and/or bases present in the electrolyte, possibly with further electrochemical steps on the reaction path (Chapter II).

From the outset it should be stressed that not all anodic substitution processes can be classified as belonging strictly to either of these categories at present, simply because available information does not unambiguously rule out either of them. In this chapter, we shall emphasize reactions proceeding according to the direct mechanism, since in many cases the indirect processes do not differ in any important respects from analogous homogeneous processes (cf. Chapter XXIV).

The subject has been discussed in several books and reviews, among which some [1-4] stress the importance of indirect mechanisms, and others the direct ones [5-7].

Depending on the structure of the substrate and the nature of the substitution process, anodic substitution reactions can be divided into three main classes. These are as follows: (1) nuclear substitution of aromatic compounds; (2) substitution at activated carbon atoms, where the activating group is an aryl, vinyl, RCONH-, etc., group; and (3) substitution at nonactivated carbon atoms.

II. FORMATION OF CARBON-TO-OXYGEN BONDS

The possibility of the direct introduction of a carbon-to-oxygen bond in a substitution process according to Eq. (1) is of considerable preparative interest, since in principle the reaction can be run with very cheap starting materials. As an example, one attractive possibility would be to make phenols from aromatic hydrocarbons, using only hydrocarbon, water, and electricity in the process [Eq. (2)]. However, apart

$$\text{Ar-H} + H_2O \rightarrow \text{Ar-OH} + 2H^+ + 2e^- \qquad (2)$$

from the practical problems associated with this type of process, there are, unfortunately, theoretical grounds why this reaction should not be feasible. We here have a situation where the organic starting material is considerably more difficult to oxidize than the product, so that the high anode potential required to oxidize the substrate will be more than sufficient to oxidize the product as it is formed. As an example, under identical experimental conditions benzene and phenol have half-wave potentials of 2.3 and 1.0 V, respectively, so it is easily predicted that the reaction cannot possibly be stopped at the phenol stage. The same type of reasoning pertains to a greater or lesser degree to all nuclear substitution processes involving C-O bond formation in aromatic compounds, since the additional oxygen function in the product will decrease its oxidation potential compared to that of the starting material. Actually, the most surprising aspect of many of these reactions is that any nuclear substitution product can be isolated at all, since one would expect a "cascade" of electrochemical and chemical steps terminating only when a sufficiently oxidation-resistant product has been formed.

A. Hydroxylation

The anodic oxidation of aromatic and alkylaromatic hydrocarbons at platinum or lead dioxide anodes in aqueous media generally leads to a complex mixture of products [1, 2, 7], as has been exemplified previously (Chapter XII). In most cases acidic electrolytes are employed, and acetone, an alcohol, or acetic acid is added as a cosolvent in order to obtain a homogeneous medium. Mechanistically these reactions are not very well understood. It has been suggested that at platinum anodes a layer of a platinum oxide should be the oxidizing agent; however, from the viewpoint of the organic chemist a mechanism involving oxidation of the substrate with initial formation of a cationic species would also be in agreement with experimental facts, at least insofar as products are concerned. In that case the primary product would be a phenol from a nuclear hydroxylation and a benzyl alcohol from a side-chain hydroxylation of an aromatic compound. Both of these are easily oxidized in secondary reactions, so the observed product distributions can easily be accounted for, at least in a qualitative manner.

In nonaqueous solvents, water is an impurity which is impossible to remove completely even after the most rigorous purification procedures [5], thus causing formation of hydroxylation products in addition to the desired ones. An example is the oxidation of alkylaromatic compounds in acetonitrile [8-11], from which the acetamidation product [Eq. (3)] is the major product

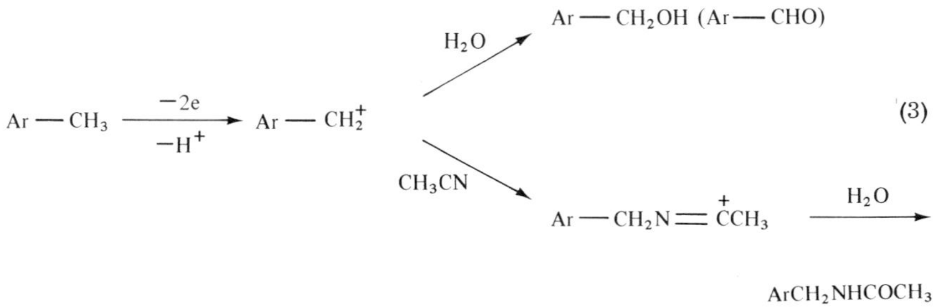

(3)

in a nominally anhydrous solution of sodium perchlorate in acetonitrile [9, 10]. One of the side reactions is formation of the corresponding benzyl alcohol (and its oxidation product, the aldehyde); the alcohol is formed at considerably lower potential than the anodic limit of the solvent-supporting electrolyte, so at least in this case a direct mechanism for hydroxylation appears to operate. A surprising feature is that the intermediate carbonium ion [Eq. (3)] prefers to react with the weakest nucleophile in the system, acetonitrile. This phenomenon is observed in certain systems even at water concentrations as high as 5% [9]. In two cases which were more closely investigated the product distribution was not affected by the water concentration in the region between 2 and 5%. This behavior indicates that the composition of the medium in the vicinity of the anode must be different from that of the electrolyte bulk. The anion of the supporting electrolyte has been found to have a profound influence on the product distribution [12], as can be seen from Eq. (4). Exchanging perchlorate ion for tetrafluoroborate ion changes the product

$$\begin{aligned} &\mathrm{ArCH_3} \xrightarrow[\mathrm{Bu_4NClO_4}]{\mathrm{CH_3CN\text{-}H_2O\ (9{:}1)}} \underset{5\%}{\mathrm{ArCH_2OH}} + \underset{95\%}{\mathrm{ArCH_2NHCOCH_3}} \\ &\mathrm{ArCH_3} \xrightarrow[\mathrm{Bu_4NBF_4}]{\mathrm{CH_3CN\text{-}H_2O\ (9{:}1)}} \underset{95\%}{\mathrm{ArCH_2OH}} + \underset{5\%}{\mathrm{ArCH_2NHCOCH_3}} \end{aligned} \tag{4}$$

distribution by a factor of 400! It appears reasonable to assume that differences in solvation by water between different anions are responsible for this effect, tetrafluoroborate ion being the more strongly solvated ion.

Concluding, anodic hydroxylation is a relatively little studied process, although it is an important (and embarrassing) side reaction in most anodic oxidations due to the presence of trace amounts of water in the electrolytes. The reaction is difficult to study quantitatively by electrochemical methods and appears to be of limited interest from the preparative point of view, unless further oxidation of the hydroxyl derivative gives a single product in good yield. As an example, methyl groups attached to aromatic nuclei can often be oxidized to carboxyl groups in good yield [7].

B. Alkoxylation

Several types of substrates undergo alkoxylation by substitution upon electrolysis in an alcohol containing a suitable electrolyte, such as potassium hydroxide, sodium methoxide, sodium cyanide, or ammonium nitrate. These include aromatic compounds (both nuclear and side-chain alkoxylation is known), alkenes (allylic alkoxylation), N-alkylsubstituted amides, N,N-dialkylamines, and ethers. In many cases alkoxylation by substitution is a side reaction only to concurrent alkoxylation by addition [7].

Conflicting views have been expressed about the mechanism of this process. On the basis of voltammetric measurements, in which it was shown that methanol-supporting electrolyte is oxidized at a much lower potential than that of the substrate, it was suggested that an indirect mechanism operates in the α-methoxylation of toluene, ethylbenzene, and cumene [13]. The nature of the anion present greatly affects the yield of α-methoxylation product, the effectiveness decreasing in the series $F^- \simeq ClO_4^- > CN^- \simeq OH^- > Cl^- \simeq CH_3O^- > Br^-$. The mechanism proposed is depicted in Eqs. (5)-(8):

$$X^- \rightarrow X\cdot + e^- \quad (5)$$

$$RH + X\cdot \rightarrow R\cdot + HX \quad (6)$$

$$CH_3OH \rightarrow CH_3O\cdot + H^+ + e^- \quad (7)$$

$$R\cdot + CH_3O\cdot \rightarrow ROCH_3, \quad (8)$$

where X^- is an anion of supporting electrolyte.

However, this mechanism is unsatisfactory from several points of view. First, it is difficult to see why anions which are as difficult to oxidize as fluoride and perchlorate ion should work much better than easily oxidizable ions, such as methoxide ion. In fact, one would not expect the fluoride or perchlorate ion to be oxidized at all under the conditions used. An alternative would be that methanol is the electroactive species, forming a radical intermediate which could perform the same function as $X\cdot$ in Eq. (6). In that case $HOCH_2\cdot$ is the most likely candidate [instead of $CH_3O\cdot$ of Eq. (7)] since the O-H bond in methanol is much stronger than the C-H bond. In alkaline medium, methoxide ion is most likely the electroactive species, and $CH_3O\cdot$ would then be $X\cdot$.

Secondly, the mechanistic conclusions were based on a study of the substitution products only, which are formed in a very low current yield, of the order of 5%, together with an exceedingly complex mixture of other products [14].

Further support for the indirect mechanism is indicated by the fact that α-methoxylation of tetralin in methanol in methanol-sodium cyanide takes place at potentials as low as 0.5 V vs. SCE, whereas the half-wave potential of tetralin must be around 1.9 V [15]. Similar studies have been performed on a reaction in which the current yield of alkoxylation product is much better, namely alkoxylation of N, N-dimethylamides [Eq. (9)] in alkanol-ammonium nitrate [16]. The use of a nitrate salt results in better yields and

$$R^1CON(CH_3)_2 \quad + \quad R^2OH \xrightarrow{52\text{-}82\%} R^1CON\begin{matrix} CH_3 \\ CH_2OR^2 \end{matrix} \quad + \quad H^+ \quad + \quad 2e^- \tag{9}$$

$R^1 = H, Me, Ph$
$R^2 = Me, Et, n\text{-}Bu$

cleaner reaction than is observed with sodium alkoxide. In methanol-ammonium nitrate current-potential curves at a platinum anode are not affected by the addition of the substrate, indicating that the substrate is more difficult to oxidize. A mechanism involving initial discharge of nitrate ion to form nitrate radical is considered the most likely one in this case [Eqs. (10)-(12)]:

$$NO_3^- \longrightarrow NO_3\bullet \quad + \quad e^- \tag{10}$$

$$RCON\begin{matrix} CH_3 \\ CH_3 \end{matrix} \quad + \quad NO_3\bullet \longrightarrow RCON\begin{matrix} CH_2\bullet \\ CH_3 \end{matrix} \quad + \quad HNO_3 \tag{11}$$

$$RCON\begin{matrix} CH_2\bullet \\ CH_3 \end{matrix} \xrightarrow{-e^-} RCON\begin{matrix} CH_2^+ \\ CH_3 \end{matrix} \xrightarrow{CH_3OH} RCON\begin{matrix} CH_2OCH_3 \\ CH_3 \end{matrix} \tag{12}$$

If water (3.2%) is added to a solution of ammonium nitrate in N, N-dimethylformamide, a good yield of what appears to be a hydroxylation-alkoxylation product (I) is formed [Eq. (13)] from N, N-dimethylformamide. Analogous products are formed in the electrolysis of N-methylsubstituted amides in aqueous sulfuric acid [17].

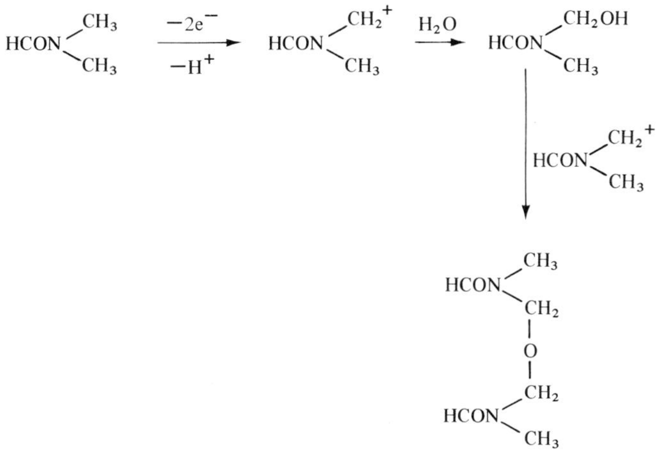

(13)

(I)

Saturated ethers can be α-methoxylated by electrolysis in methanol-sodium methoxide or ammonium nitrate in 10-30% yield [Eq. (14)]; again, an indirect mechanism is considered to be the most likely one [18].

$$RO{-}\overset{|}{\underset{|}{C}}H + MeO^- \longrightarrow RO{-}\overset{|}{\underset{|}{C}}{-}OMe + 2\bar{e} + H^+ \quad (14)$$

Evidence for a mechanism involving electron transfer from the substrate to the anode has been obtained in the methoxylation of N, N-dimethylaniline, which takes place in methanol-potassium hydroxide with formation of (II) as the major product [19, 20]. From current-voltage curves it can be shown that the amine is oxidized at lower potentials than the solvent-supporting electrolyte. Similar behavior was observed in methanol-ammonium nitrate, from which a dimer (III) is isolated as the predominant product. The mechanism in both cases must be an initial one-electron transfer from N, N-dimethylaniline to form the cation radical, which then undergoes different reactions dependent on the basicity of the medium. With strong base present, proton abstraction gives a neutral radical, which is oxidized to a cation [Eq. (15)]; with no strong base present, dimerization of the cation radical takes place [Eq. (16)].

N, N-Dimethylbenzylamine exhibits similar behavior upon electrolysis in methanol-potassium hydroxide [21]. Here, evidence for specific adsorption of substrate is obtained from Tafel plots, and it was concluded that the unusual preference for substitution in the methyl group over that in the methylene group is due to adsorption [Eq. (17)].

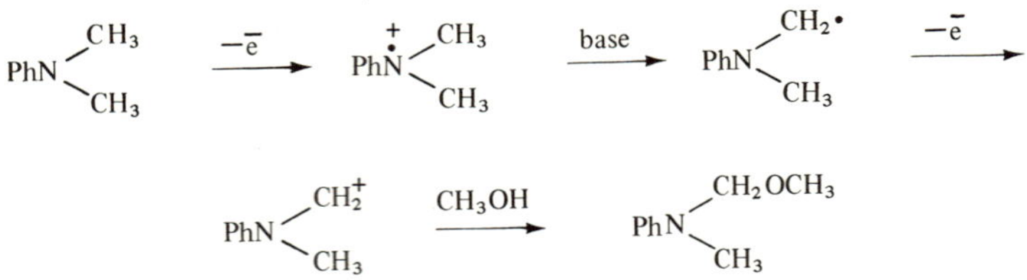

(II)

(15)

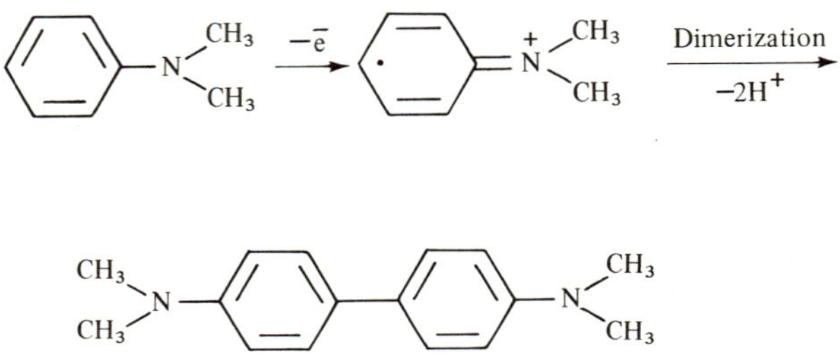

(III)

(16)

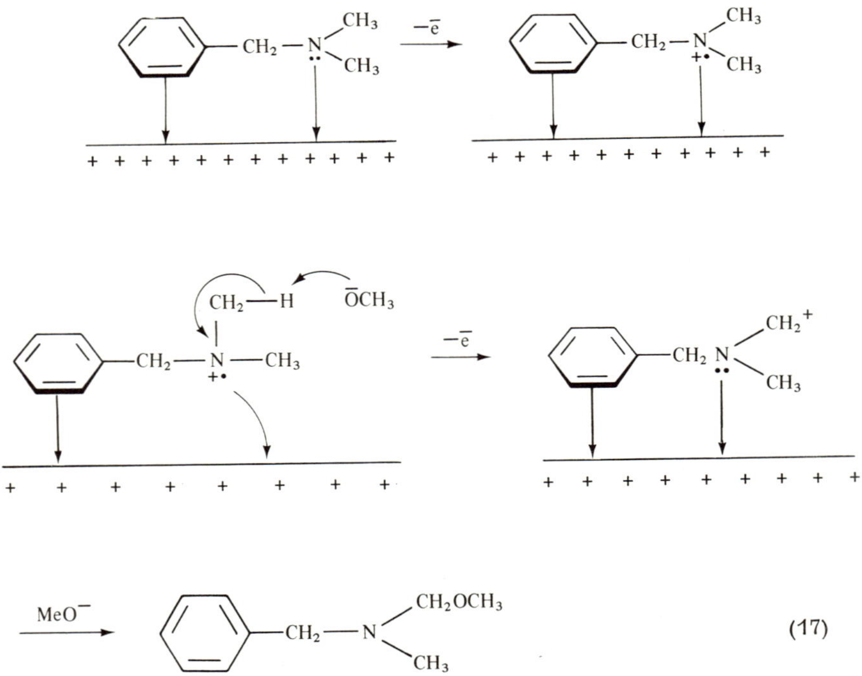

(17)

Methoxylation accompanied by rearrangement has been observed upon anodic oxidation of 1, 1, 3-triphenylindene in SO_2/tetrabutylammonium perchlorate in the presence of a low concentration of methanol. The reaction was interpreted [21a] as being a case of a 1, 5-sigmatropic rearrangement of a cation radical (by analogy with similar rearrangements induced thermally, photochemically, and via radicals), but it is more likely that the mechanism is of the usual ECE type [Eq. (17a)]. Using methanol-lithium perchlorate as supporting electrolyte the normal addition (Chapter XX) of two methoxy groups across the double bond in the indene system takes place.

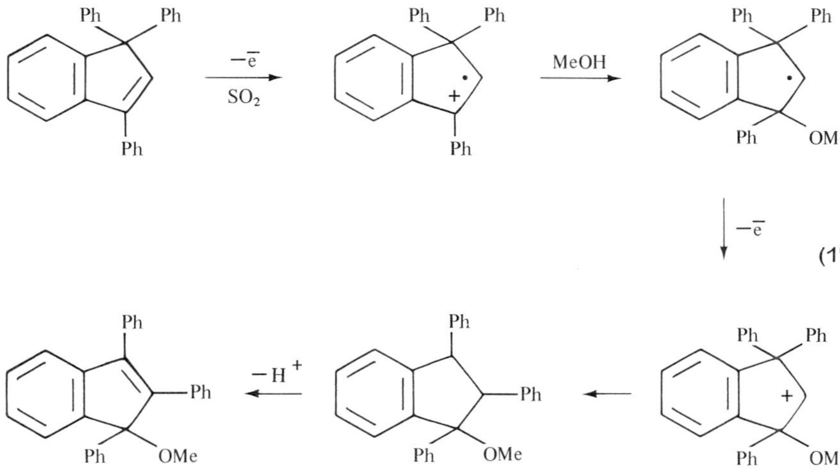

(17a)

Some alkenes have been found to undergo allylic methoxylation during electrolysis in methanol-sodium methoxide, although, besides addition products, allylic ethers are formed in minor amounts only [22, 23].

Concluding, except in the case of easily oxidizable substrates, alkoxylation appears to follow an indirect mechanism involving initial formation of radical species by discharge of solvent and/or supporting electrolyte. The possibilities of preparative applications appear promising.

C. Acyloxylation

Substitution by an acyloxy group can be accomplished by electrolysis of aromatic and alkylaromatic compounds [24-34], alkenes [22, 35, 36], ethers [37], or N-alkylsubstituted amides [16, 38-41] in the presence of a carboxylate in a suitable solvent. The most commonly encountered process is acetoxylation [16, 24-31, 34-36, 39, 40] which takes place in acetic acid at platinum, lead dioxide, or carbon anodes. The supporting electrolyte in this case normally is sodium acetate, although tetramethylammonium nitrate, tetrabutylammonium tetrafluoroborate, tetraethylammonium p-toluenesulfonate, sodium perchlorate, etc., can be employed in certain cases. Other acyloxylations using carboxylates whose corresponding acids cannot be used as solvent have been performed in N, N-dimethylformamide [41] or acetonitrile [32, 33, 42].

The acyloxylation reaction was originally believed to be a case of an indirect oxidation process [24-26] in that an anodically generated acyloxy radical was assumed to attack the organic substrate in a homolytic substitution process. However, since acetic acid-sodium acetate is a rather difficultly oxidizable electrolyte (see Chapter XIII), it can be shown by conventional polarographic technique [27] that compounds which can be acetoxylated successfully also show a polarographic wave at the rotating platinum anode in this medium. CPE at a potential around the half-wave potential of the substrate unambiguously shows that this wave is associated with the acetoxylation reaction. Compounds with estimated half-wave potentials far above the anodic limit of acetic acid-sodium acetate do not undergo acetoxylation [28]. Yields of acetoxylation products vary greatly, depending on the oxidation potential of the product(s) formed. As a general rule, they tend to be low in nuclear acetoxylation of aromatic compounds, and fair to good in acetoxylations at activated carbon atoms.

A characteristic feature of the reaction is that nuclear acetoxylation of aromatic compounds requires the presence of acetate ion [28], whereas acetoxylation of the other types of substrates can be done with either acetate ion or other anions present [16, 30, 36]. This has been interpreted [28] in terms of a concerted two-electron transfer mechanism for nuclear acetoxylation [Eq. (18)], and a mechanism involving carbonium ion formation in other cases [as exemplified by toluene in Eq. (19)].

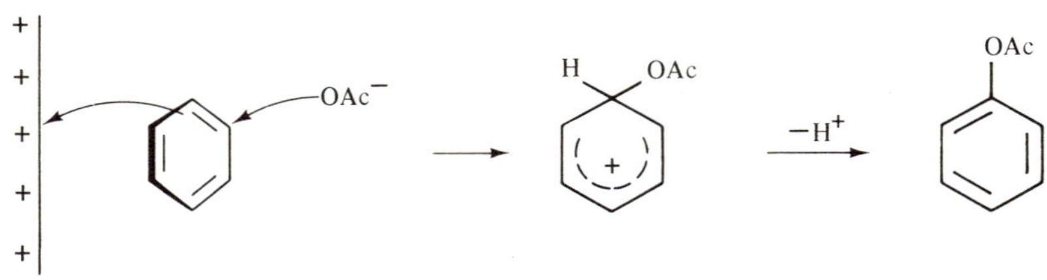

(18)

$$PhCH_3 \xrightarrow{-\bar{e}} PhCH_3^{+\cdot} \xrightarrow{-H^+} PhCH_2^{\cdot} \xrightarrow{-\bar{e}} PhCH_2^+ \xrightarrow{OAc^-} PhCH_2OAc. \quad (19)$$

It is, of course, also possible to assume that the ECE mechanism operates in nuclear acetoxylation too [43], but that in the absence of acetate ion there is no nucleophile present which is strong enough to attack the cation radical in the chemical step of the ECE pathway. If, however, the cation radical has a readily abstractable proton as in the other substrate types, acetic acid is sufficiently strong as a base to remove this proton.

The isomer distribution for anodic acetoxylation of a number of monosubstituted benzenes has been determined [28]. The reaction closely resembles ordinary electrophilic aromatic substitution processes, perhaps a bit on the side of low-selectivity reactions. The isotope effect, K_H/K_D, for nuclear acetoxylation in anisole was found to be 1.0, whereas for α-substitution in ethylbenzene a value of 2.6 was observed.

A number of intramolecular acyloxylations have been reported [27, 44-47], as, for example, in the formation of a spirodienone from β-(p-hydroxyphenyl)-propionic acid [Eq. (20)]. A variation of this reaction has been utilized for specific cleavage of tyrosyl-peptide bonds [48].

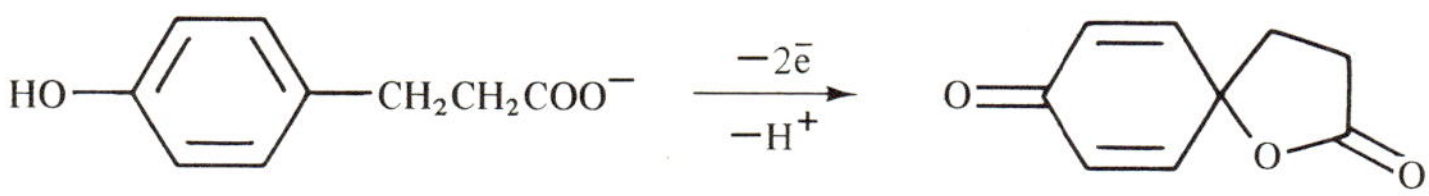

(20)

Acyloxylation of a number of benzyl ethers has been reported [Eq. (21)]. A hemiacetal acetate is presumably formed and subsequently hydrolyzed during the work-up procedure to give the aldehyde [37]:

$$\mathrm{ArCH_2OMe} \xrightarrow[-\mathrm{H}^+]{-2\bar{e}} \mathrm{Ar\overset{+}{C}H{-}OMe} \xrightarrow{\mathrm{OAc}^-} \mathrm{ArCH}\begin{matrix}\diagup \mathrm{OAc} \\ \diagdown \mathrm{OMe}\end{matrix} \xrightarrow{\mathrm{H}^+,\ \mathrm{H_2O}} \mathrm{ArCHO} \qquad (21)$$

30-70% yield

Acetoxylation of alkenes takes place in allylic positions [22, 35, 36] concurrent with addition of acetoxy groups across the double bond or double bond system.

The acyloxylation of N-alkylsubstituted amides has received considerable attention [16, 38-41], both from the mechanistic and the preparative points of view. In electrolytes containing alkali metal carboxylate there seems to be little doubt that a direct oxidation mechanism operates, whereas in cases where a nitrate salt is the supporting electrolyte an indirect mechanism involving hydrogen abstraction from the N-alkyl group by anodically generated nitrate radical [cf. methoxylation, Eqs. (10)-(12)] is indicated but not definitely proved. The same mechanistic problem is encountered in side-chain acetoxylation of alkylaromatic compounds in the presence of nitrate ion [30, 49, 50]. Here, it is believed that both types of mechanisms operate.

D. Substitution by Nitrate Ion

Concurrent with side-chain acetoxylation of alkylaromatic compounds in acetic acid-ammonium nitrate one also observes formation of benzyl nitrates,

$ArCH_2ONO_2$, in about the same yield as the benzyl acetate [30, 49, 50]. It seems likely that nitrate ion and acetic acid compete as nucleophiles for an intermediate carbonium ion [Eq. (22)] although an indirect mechanism cannot be ruled out completely because of the complexity of the supporting electrolyte behavior [50a].

$$ArCH_3 \xrightarrow[-H^+]{-2\bar{e}} ArCH_2^+ \begin{cases} \xrightarrow{NO_3^-} ArCH_2ONO_2 \\ \xrightarrow{CH_3COOH} ArCH_2OAc \end{cases} \tag{22}$$

III. FORMATION OF CARBON-TO-SULFUR BONDS. THIOCYANATION

Phenols and aromatic amines are thiocyanated on electrolysis in an aqueous solution of thiocyanate ion [1-4, 7]. The reaction is a typical example of an indirect process, since it has been shown beyond any doubt that the thiocyanating reagent must be thiocyanogen, formed by anodic oxidation of thiocyanate ion [51]. New thio- and selenocyanations have recently been reported [80].

IV. FORMATION OF CARBON-TO-NITROGEN BONDS

A. Acetamidation

We have already discussed several aspects of the acetamidation reaction (Sec. II.A) which takes place upon electrolysis of an organic compound in acetonitrile containing a perchlorate or tetrafluoroborate salt [Eq. (3)]. The latter supporting electrolyte allows for the oxidation of saturated and unsaturated hydrocarbons [52, 53] to give N-alkylacetamides; this reaction is probably the only established case of a direct, selective oxidation of saturated aliphatics. It should be noted that only methylsubstituted alkylaromatic compounds are good substrates for acetamidation [9]; ethylbenzene and isopropylbenzene give other products predominantly [11]. In order to avoid hydroxylation (from traces of water in the electrolyte) a carbon anode is preferentially employed [9]. An interesting case of acetamidation has recently been described in the anodic oxidation of 3, 6-diisobutylpiperazine-2, 5-dione in acetonitrile-sodium perchlorate [53a]; the complex product isolated is probably formed according to the reaction sequence given in Eq. (23a):

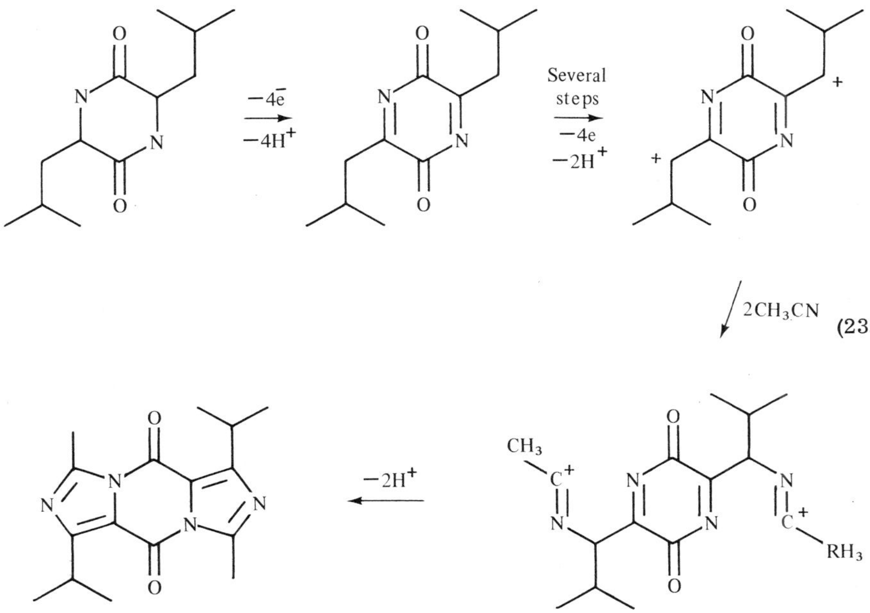

(23a)

Other substrates for anodic acetamidation are alkyl iodides [53b-d] which are cleaved with formation of carbonium ions and iodine [Eq. (23b)] and carboxylates (Chapter XIII):

$$R\text{-}I \xrightarrow{-\bar{e}} R^+ + \tfrac{1}{2}I_2$$

$$R^+ \xrightarrow[H_2O]{CH_3CN} R\text{-}NHCOCH_3. \tag{23b}$$

B. Nitration

The formation of nitro compounds upon electrolysis of a suspension of an aromatic hydrocarbon in dilute aqueous nitric acid has been known for a long time [1-4]. It has been interpreted in terms of formation of a high concentration of nitric acid near the anode surface, in which region an ordinary electrophilic nitration process would take place. Recent studies on the anodic oxidation of nitrate ion in nitromethane [54] would seem to confirm this assumption, since it is shown that dinitrogen pentoxide is the oxidation product, probably formed via reactions (24) and (25):

$$NO_3^- \rightarrow NO_2^+ + \tfrac{1}{2}O_2 + 2e^- \tag{24}$$

$$NO_2^+ + NO_3^- \rightarrow N_2O_5. \tag{25}$$

In an aqueous solution, this would, of course, lead to the formation of nitric acid.

However, studies on the anodic behavior of anthracene [55] in molten tetrabutylammonium nitrate at 150° might indicate another nitration mechanism. Anthracene has a half-wave potential of 0.96 V (Ag/AgCl) for oxidation in this medium; CPE at 1.1 V produces a mixture of 9-nitro, 9,10-dinitro-, and 4,9-dinitroanthracene. That the nitrating species is not formed by oxidation of nitrate ion is shown by the fact that the anodic limit of the molten salt is about 2.4 V; besides, naphthalene ($E_{1/2}$ in this medium around 2.2 V) is not nitrated on CPE at 1.1 V. Thus, a direct nitration mechanism appears to operate in this case. A possibility that cannot be completely excluded, though, is that nitration in the molten salt is a homogeneous process, since it is known [56] that aromatic and even aliphatic hydrocarbons can be nitrated in molten nitrates at elevated temperatures (250-300°).

Aliphatic gem-dinitro compounds can be prepared by anodic oxidation of a nitroparaffin in aqueous alkaline solution [Eq. (26)] in the presence of an excess of nitrite ion [57]. Yields are considerably better when silver ion is

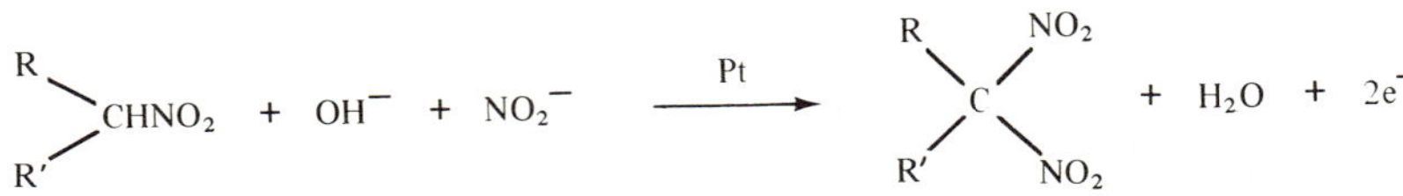
(26)

the oxidizing agent [58], so that an indirect oxidation process involving the generation of Ag^+ at a silver anode was found to give excellent results [59].

C. Substitution by Azide Ion

1-Azido-1-nitroalkanes can be prepared in good yield by anodic oxidation of an aqueous alkaline solution of a 1-nitroalkane in the presence of azide ion at a platinum or nickel anode [60]. The mechanism has been postulated to occur via initially formed azide radicals, which attack the alkylnitronate ion [Eqs. (27)-(29)]:

$$N_3^- \longrightarrow N_3\cdot + e^- \qquad (27)$$

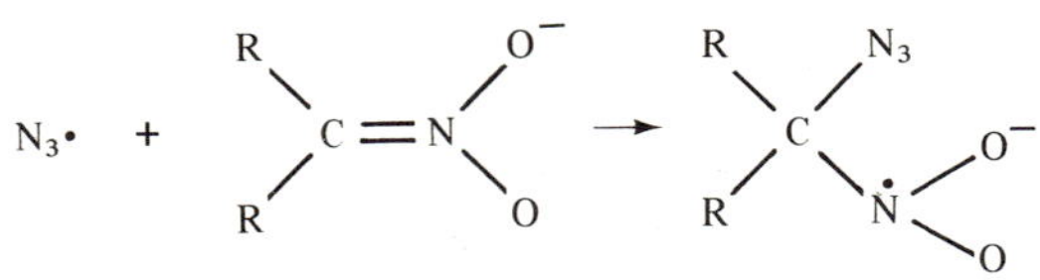
(28)

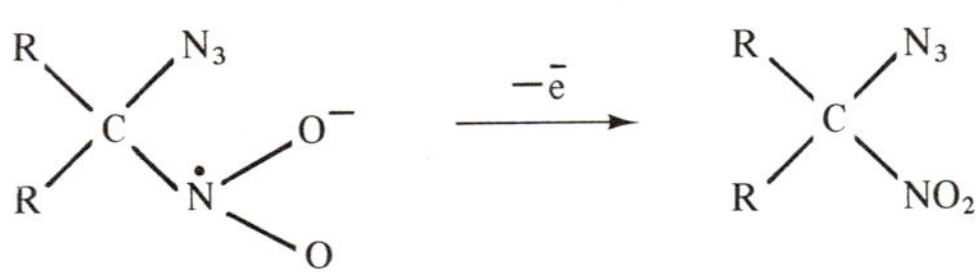
(29)

D. Substitution by Cyanate Ion

Both nuclear and side-chain substitutions of alkylaromatics by cyanate ion have been observed [61]. Thus, electrolysis of anisole in methanol-potassium cyanate results in substitution of hydrogen by a methylcarbamyl group (via the isocyanate); at 1.1 V, no substitution takes place. The reaction has been pictured as a direct oxidation process [Eq. (30)]. Yields are fairly low (<20%):

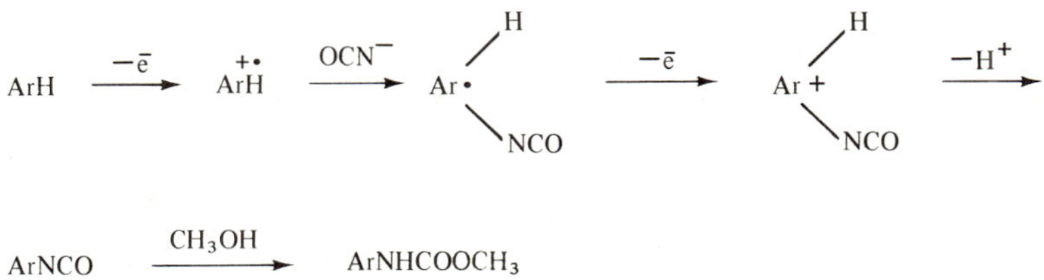

(30)

E. Pyridination

Substitution by pyridine has been observed in one case, namely, the anodic pyridination of tris-(p-methoxyphenyl)-ethylene [62]:

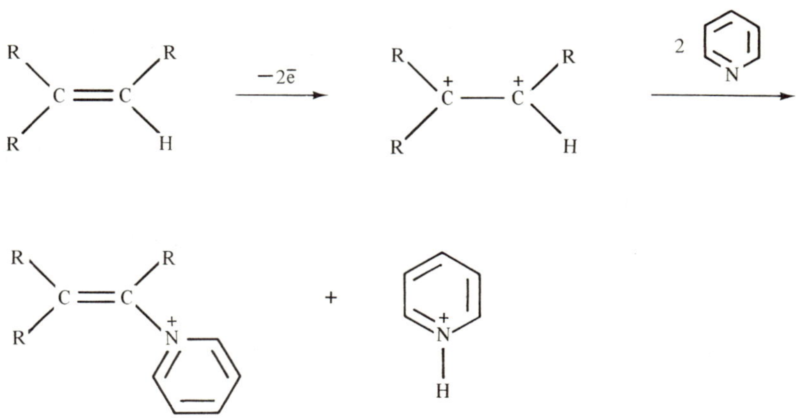

(31)

$R = p-CH_3OC_6H_4-$

V. FORMATION OF CARBON-TO-CARBON BONDS

A. Cyanation

Anodic cyanation is a reaction of considerable preparative interest since it allows for the direct introduction of a cyano group into an aromatic ring [63-69]. Contrary to the situation in acyloxylation and alkoxylation, the product is more resistant toward oxidation than the starting material due to the strong electron-withdrawing effect of the cyano group, so that at least in principle good yields should be obtainable. The reaction is normally run in methanol-sodium cyanide [63-68] but can also be performed in acetonitrile-tetraethylammonium cyanide [69], if one wants to avoid the concurrent methoxylation reaction in methanolic media.

The reaction was initially believed to be a homolytic substitution reaction by anodically generated cyano radicals [63]. However, it was soon found that while significant oxidation of cyanide ion takes place at a potential as low as 0.5 V (SCE), the cyanation process will occur only in the region around or above the half-wave potential of the substrate, so that a direct oxidation mechanism must operate [64, 68, 69]. This has been pictured either as a concerted two-electron transfer assisted by cyanide ion [analogous to the acetoxylation mechanism of Eq. (18)] or as an ordinary ECE mechanism [68, 69]. In both cases adsorption of the substrate is assumed to precede the electron-transfer step.

Isomer distribution for cyanation of aromatic compounds has been determined in a number of cases [67-69]. The results show that the selectivity of the cyanation reaction is higher than that of the anodic acetoxylation process, which would be consistent with a cyanide-ion-assisted electron-transfer step.

In most cases cyanation takes place by substitution of a carbon-hydrogen bond, but in some cases one has observed that also methoxy groups can be displaced, in fact with very good yields of substitution products [69]. Thus, 1, 2- and 1, 4-dimethoxybenzene undergo cyanation by displacement of one of the methoxy groups, presumably via an ECE mechanism [Eq. (32)]. On the other hand, anisole and 1, 3-dimethoxybenzene do not undergo methoxyl displacement but, instead, give nuclear substitution in the usual manner. The presence of an activating group in the o- and p-positions thus seems essential for methoxyl group displacement.

Tertiary amines are also substrates for the cyanation reaction [69], substitution taking place at the carbon atom α to the nitrogen atom. In this case substitution occurs preferentially at a primary carbon atom in mixed amines, as for example N-ethyl-N-methylaniline [Eq. (33)].

Alkylaromatic compounds do not give any side-chain substitution product under conditions which produce nuclear cyanation products [67, 68]. There is no obvious explanation for this failure.

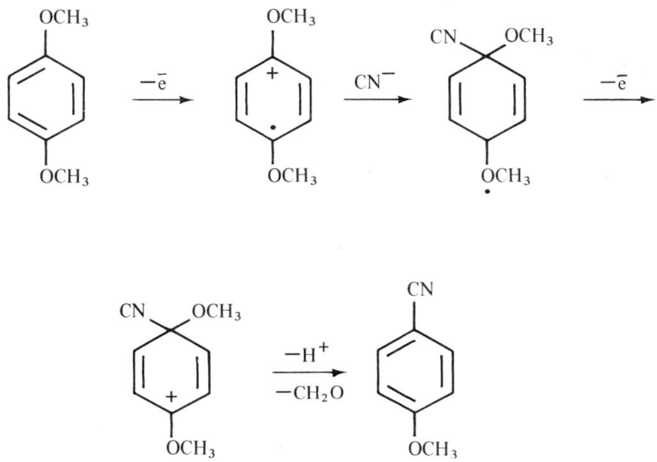

(32)

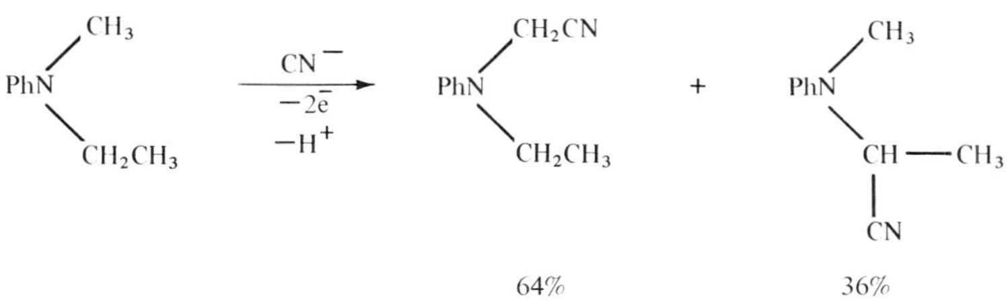

(33)

The fate of the anodically generated cyano radical in the absence of any substrate has been investigated [69] in several solvents. ESR studies convincingly demonstrate that a relatively stable tricyanomethylenimine anion radical is ultimately formed, presumably via the pathway shown in Eq. (34):

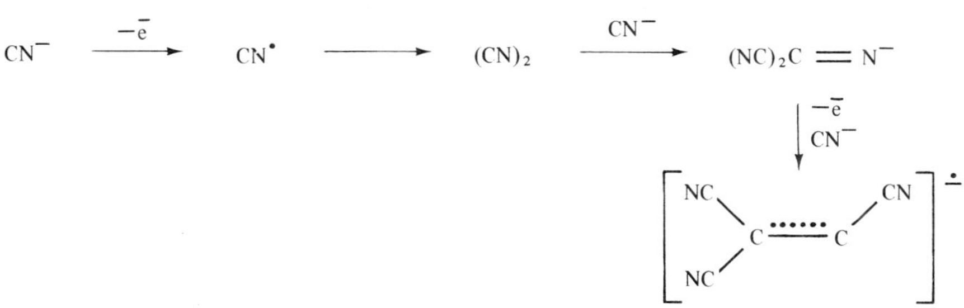

(34)

B. Carbomethoxylation

An unusual and interesting carbomethoxylation reaction occurs [70, 71] during electrolysis of an aryl olefin in methanol-sodium methoxide under carbon monoxide pressure between two platinum electrodes [Eq. (35)]. It has been shown that the platinum cathode dissolves to give a platinum-

$$C_6H_5\overset{R}{\overset{|}{C}}{=}CH_2 + CO + CH_3O^- \xrightarrow{Pt_x(CO)_y} C_6H_5\overset{R}{\overset{|}{C}}{=}CHCOOCH_3 + 2e^- \tag{35}$$

carbonyl complex. This is transformed at the anode to a carbomethoxy-platinum-carbonyl complex, which attacks the aryl olefin to give the carbomethoxylation product. Hence, this reaction is of the indirect type.

Another type of a carbomethoxylation reaction has been reported during electrolysis of anisole in methanol containing 1% of sulfuric acid [72]. Methyl p-anisoate (45%) and o-anisoate (5%) were isolated in addition to products of further oxidative degradation. A mechanism involving an anodically generated hydroxymethyl cation as the attacking species is postulated [Eqs. (36) and (37)]:

$$CH_3OH \rightarrow \overset{+}{C}H_2OH + H^+ + 2\bar{e} \tag{36}$$

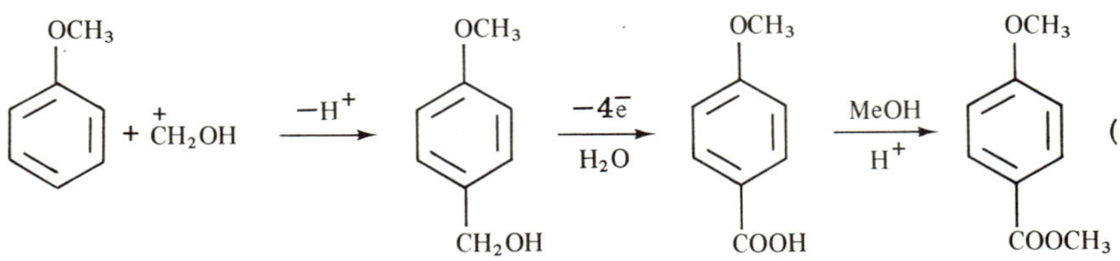

(37)

C. Alkylation and Arylation

Alkylation and arylation products are often formed in small amounts when organic compounds are coelectrolyzed with a carboxylate; the mechanism is most likely a homolytic substitution process by anodically generated radicals (see Chapter XIII). However, in one case, methylation of aromatics by electrolysis in acetic acid-sodium acetate at a graphite anode, it has been suggested that the methyl cation is the methylating species [73].

VI. FORMATION OF CARBON-TO-HALOGEN BONDS

A. Fluorination

Anodic fluorination is a technically very important process for the synthesis of perfluorinated compounds (the Simons reaction). It takes place during electrolysis of virtually any type of organic substrate in liquid hydrogen fluoride at a nickel anode [1] [74, 75]. Normally, no supporting electrolyte is required since protonation of the substrate makes the solution electrically conducting in most cases. If the substrate is a very weak base, however, it may be necessary to add potassium fluoride as supporting electrolyte.

The scope and limitations of this process have been excellently described in review articles [74, 75]. Generally, the yield of the desired perfluorinated compound is not overwhelmingly high due to formation of likewise perfluorinated degradation products, but the simplicity of the procedure nevertheless makes the method an attractive one for fluorination. No mechanistic studies have been reported, no doubt due to the experimental difficulties associated with electroanalytical work in liquid hydrogen fluoride. Several possible mechanisms have been discussed [74], namely fluorination by anodically formed fluorine or nickel fluoride, or a direct mechanism involving radical formation by proton loss from a cation radical [Eq. (38)]. In view of the fact

$$R\text{-}H \xrightarrow{-\bar{e}} R\text{-}H^{+} \xrightarrow{-H^{+}} R\cdot \tag{38}$$

that nonactivated C-H bonds can be attacked anodically in systems with extremely high anodic limit and then give products formed via carbonium ions, and the likelihood that hydrogen fluoride-potassium fluoride has a very high anodic limit, it appears reasonable to add the alternative that the radical [Eq. (38)] may undergo a second electron transfer to form a cation, which would give product by reaction with fluoride ion. The occurrence of a rearrangement in the anodic fluorination of, e.g., 1,1-difluorocyclohexane [76] is an indication of a cationic mechanism [Eq. (39)]:

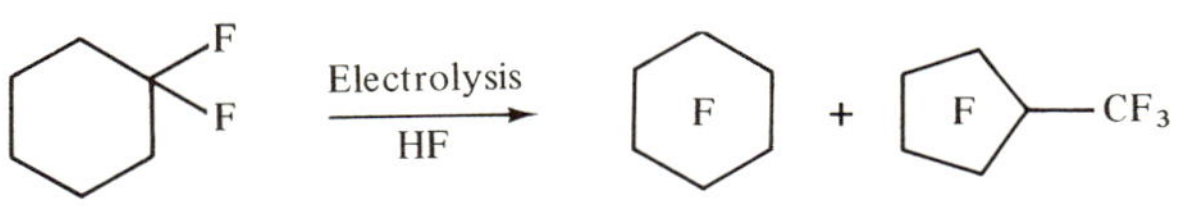

(39)

B. Chlorination, Bromination, and Iodination

Available evidence classifies these substitution reactions as indirect processes [2, 3] occurring via homogeneous reactions of anodically generated halogens. In the iodination of alkylaromatic hydrocarbons, e.g., p-xylene, iodine is converted anodically to a positive iodine species, which attacks the hydrocarbon [53c, 77].

However, recent findings indicate that the anodic bromination of anthracene and naphthalene in acetonitrile-tetraethylammonium bromide at high anode potentials occurs via cation radicals of these substrates [78]. Similarly, allylic chlorination of cyclohexene [79] takes place at high anode potential [Eq. (40)] in acetonitrile-lithium chloride via discharge of substrate, whereas at low potential another process initiated by electro-

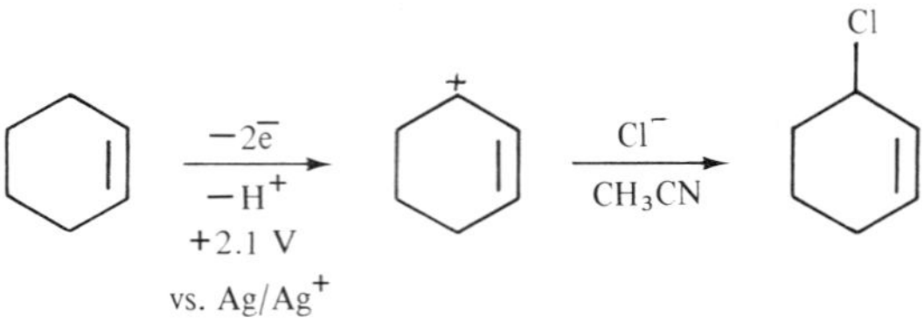

(40)

generated chlorine occurs [Eq. (41)]. Thus, the field of anodic halogenations appears to be open for reinvestigation and, possibly, mechanistic reevaluation.

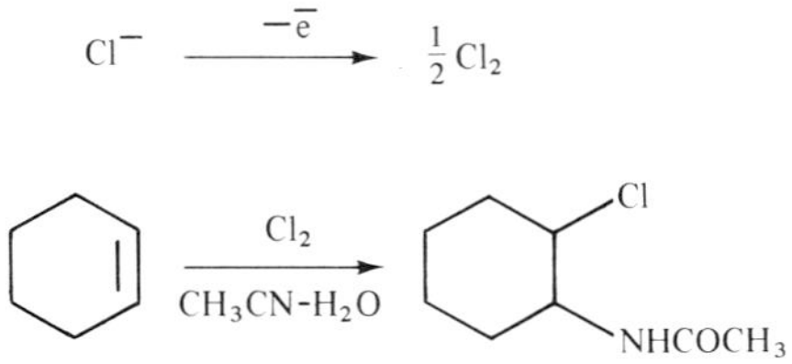

(41)

NOTE

[1] An extremely promising development is the use of porous carbon under certain conditions [81].

REFERENCES

[1]. F. Fichter, *Organische Elektrochemie*, Steinkopf, Dresden and Leipzig, 1942.

[2]. M. J. Allen, *Organic Electrode Processes*, Chapman & Hall, London, 1958.

[3]. A. P. Tomilov, Russ. Chem. Rev. (Engl. Transl.), **30**, 639 (1961).

[4]. A. P. Tomilov and M. Ya. Fioshin, ibid., **32**, 30 (1963).

[5]. R. N. Adams, Accounts Chem. Res., **2**, 175 (1969).

[6]. R. N. Adams, *Electrochemistry at Solid Electrodes*, Dekker, New York, 1969.

[7]. N. L. Weinberg and H. R. Weinberg, Chem. Rev., **68**, 449 (1968).

[8]. L. Eberson and K. Nyberg, Tetrahedron Lett., 2389 (1966).

[9]. L. Eberson and B. Olofsson, Acta Chem. Scand., **23**, 2355 (1969).

[10]. V. D. Parker and B. E. Burgert, Tetrahedron Lett., 2411 (1968).

[11]. K. Koyama, T. Susuki, and S. Tsutsumi, Tetrahedron, 23, 2665 (1967).

[12]. K. Nyberg, Chem. Commun., 774 (1969).

[13]. K. Sasaki, H. Urata, K. Uneyama, and S. Nagaura, Electrochim. Acta, 12, 137 (1967).

[14]. L. Eberson, unpublished observations.

[15]. V. D. Parker and B. E. Burgert, Tetrahedron Lett., 2415 (1968); cf. also T. Inoue, K. Koyama, and S. Tsutsumi, Bull. Chem. Soc. Japan, 37, 1597 (1964).

[16]. S. D. Ross, M. Finkelstein, and R. C. Petersen, J. Amer. Chem. Soc., 88, 4657 (1966); E. J. Rudd, M. Finkelstein, and S. D. Ross, J. Org. Chem., 37, 1763 (1972).

[17]. D. E. Couch, Electrochim. Acta, 9, 327 (1964).

[18]. T. Shono and Y. Matsumura, J. Amer. Chem. Soc., 91, 2803 (1969).

[19]. N. L. Weinberg and E. A. Brown, J. Org. Chem., 31, 4058 (1966).

[20]. N. L. Weinberg and T. B. Reddy, J. Amer. Chem. Soc., 90, 91 (1968).

[21]. N. L. Weinberg, J. Org. Chem., 33, 4326 (1968); cf. also P. J. Smith and C. K. Mann, ibid., 33, 316 (1968).

[21a]. L. L. Miller and E. A. Mayeda, J. Amer. Chem. Soc., 92, 5818 (1970).

[22]. A. J. Baggaley and R. Brettle, J. Chem. Soc. (C), 2055 (1968).

[23]. P. Courbis, Compt. Rend. Acad. Sci. Paris, 266, C, 1703 (1968).

[24]. R. P. Linstead, J. C. Bunt, B. C. L. Weedon, and B. R. Shephard, J. Chem. Soc., 3624 (1952).

[25]. C. L. Wilson and W. T. Lippincott, J. Amer. Chem. Soc., 78, 4291 (1956); J. Electrochem. Soc., 103, 672 (1956).

[26]. D. R. Harvey and R. O. C. Norman, J. Chem. Soc., 4860 (1964).

[27]. L. Eberson and K. Nyberg, J. Amer. Chem. Soc., 88, 1686 (1966); Acta Chem. Scand., 18, 1568 (1964).

[28]. L. Eberson, J. Amer. Chem. Soc., 89, 4669 (1967).

[29]. S. D. Ross, M. Finkelstein, and R. C. Petersen, ibid., 86, 4139 (1964).

[30]. S. D. Ross, M. Finkelstein, and R. C. Petersen, ibid., 89, 4088 (1967).

[31]. W. J. Koehl, Jr., U.S. Pat. 3,347,758 through CA, 67, 63833 (1967).

[32]. K. Koyama, K. Yoshida, and S. Tsutsumi, Bull. Chem. Soc. Japan, 39, 516 (1966).

[33]. K. Koyama, T. Ebara, and S. Tsutsumi, ibid., 41, 2668 (1968).

[33a]. A. Kunugi, T. Shimizu, and S. Nagaura, ibid., 43, 1298 (1970).

[34]. M. Leung, J. Herz, and H. W. Salzberg, J. Org. Chem., 30, 310 (1965); H. W. Salzberg and M. Leung, ibid., 30, 2873 (1965).

[35]. P. Courbis and A. Guillemonat, Compt. Rend. Acad. Sci. Paris, 262, C, 1435 (1966).

[36]. T. Shono and T. Kosaka, Tetrahedron Lett., 6207 (1968).

[37]. R. F. Garwood, Naser-ud-din, and B. C. L. Weedon, Chem. Commun., 923 (1968).

[38]. S. D. Ross, M. Finkelstein, and R. C. Petersen, J. Amer. Chem. Soc., 86, 2745 (1964).

[39]. S. D. Ross, M. Finkelstein, and R. C. Petersen, J. Org. Chem., 31, 128 (1966).

[40]. S. D. Ross, Trans. N.Y. Acad. Sci., Ser. II, 30, 901 (1968).

[41]. L. Rand and F. Mohar, J. Org. Chem., 30, 3156, 3885 (1965).

[42]. J. F. K. Wilshire, Aust. J. Chem., 16, 432 (1963).

[43]. G. Manning, V. D. Parker, and R. N. Adams, J. Amer. Chem. Soc., 91, 4584 (1969).

[44]. H. Iwasaki, L. A. Cohen, and B. Witkop, ibid., 85, 3701 (1963).

[45]. A. I. Scott, P. A. Dodson, F. McCapra, and M. B. Meyers, ibid., 85, 3702 (1963).

[46]. W. A. Bonner and F. D. Mango, J. Org. Chem., 29, 430 (1964).

[47]. W. J. Koehl, Jr., ibid., 32, 614 (1967).

[48]. S. Isoe and L. Cohen, Arch. Biochem. Biophys., 127, 522 (1968).

[49]. S. D. Ross, M. Finkelstein, and R. C. Petersen, J. Org. Chem., 35, 781 (1970).

[50]. K. Nyberg, Acta Chem. Scand., 24, 473 (1970).

[50a]. R. R. Rao, S. B. Milliken, S. L. Robinson, and C. K. Mann, Anal. Chem., 42, 1076 (1970).

[51]. G. Cauquis and G. Pierre, Compt. Rend. Acad. Sci. Paris, 266, C, 883 (1968).

[52]. M. Fleischmann and D. Pletcher, Tetrahedron Lett., 6255 (1968).

[53]. D. Pletcher, Abstracts, Meeting of the Society for Electrochemistry, Thornton, England, April 15-16, 1969.

[53a]. L. A. Simonson and C. K. Mann, Tetrahedron Lett., 3303 (1970).

[53b]. L. L. Miller and A. K. Hoffman, J. Amer. Chem. Soc., 89, 593 (1967).

[53c]. L. L. Miller, E. P. Kujawa, and C. B. Campbell, ibid., 92, 2821 (1970).

[53d]. A. Laurent and R. Tardivel, Compt. Rend. Acad. Sci. Paris, 271, C, 324 (1970).

[54]. G. Cauquis and D. Serve, ibid., 262, C, 1516 (1966).

[55]. B. J. Woodhall and G. R. Davies, Abstracts, Meeting of the Society for Electrochemistry, Thornton, England, April 15-16, 1969.

[56]. R. B. Temple, C. Fay, and J. Williamson, Chem. Commun., 966 (1967).

[57]. C. T. Bahner, Ind. Eng. Chem., 44, 317 (1952).

[58]. R. B. Kaplan and H. Shechter, J. Amer. Chem. Soc., 83, 3535 (1961).

[59]. C. M. Wright and D. R. Levering, Tetrahedron, 19, Suppl. 1, 3 (1963).

[60]. C. M. Wright and G. A. Ward, Extended Abstracts, Meeting of the Electrochemical Society, Electroorganic Division, Vol. 2, 1965.

[61]. V. D. Parker and B. E. Burgert, Tetrahedron Lett., 3341 (1968).

[62]. V. D. Parker and L. Eberson, Chem. Commun., 451 (1969).

[63]. K. Koyama, T. Susuki, and S. Tsutsumi, Tetrahedron Lett., 627 (1965).

[64]. V. D. Parker and B. E. Burgert, ibid., 4065 (1965).

[65]. K. Koyama, T. Susuki, and S. Tsutsumi, Tetrahedron, 23, 2675 (1967).

[66]. T. Susuki, K. Koyama, A. Omori, and S. Tsutsumi, Bull. Chem. Soc. Japan, 41, 2663 (1968).

[67]. S. Tsutsumi and K. Koyama, Discuss. Faraday Soc., 45, 247 (1968).

[67a]. K. Yoshida and T. Fueno, Chem. Commun., 711 (1970).

[68]. L. Eberson and S. Nilsson, Discuss. Faraday Soc., 45, 242 (1968).

[69]. S. Andreades and E. W. Zahnow, J. Amer. Chem. Soc., 91, 4181 (1969).

[70]. T. Inoue and S. Tsutsumi, ibid., 87, 3525 (1965).

[71]. T. Inoue and S. Tsutsumi, Bull. Chem. Soc. Japan, 38, 2122 (1965).

[72]. V. D. Parker, Chem. Ind. (London), 1363 (1968).

[73]. V. D. Parker, Chem. Commun., 1164 (1968).

[74]. J. Burdon and J. C. Tatlow, Adv. Fluorine Chem., 1, 129 (1960).

[75]. S. Nagase, Fluorine Chem. Rev., 1, 77 (1967).

[76]. M. Sander and W. Blöchl, Chem.-Ing.-Tech., 37, 7 (1965).

[77]. L. L. Miller, Tetrahedron Lett., 1831 (1968).

[78]. J. P. Millington, J. Chem. Soc. (B), 982 (1969).

[79]. G. Faita, M. Fleischmann, and D. Pletcher, J. Electroanal. Chem., 25, 455 (1970).

[80]. G. Cauquis and G. Pierre, Compt. Rend. Acad. Sci. Paris, 272, C, 609 (1971).

[81]. H. M. Fox, F. N. Ruehlen, and W. V. Childs, Extended Abstract No. 117, Electrochemical Society Meeting, Washington, D.C., May 1971; J. Electrochem. Soc., 118, 1246 (1971).

Section G

INDIRECT ELECTRO-OXIDATION AND REDUCTION

Chapter XXIII

AMALGAM REDUCTIONS

Henning Lund

Department of Organic Chemistry
Aarhus University
Aarhus, Denmark

I. INTRODUCTION

Alkali metal amalgams function as electron donors in the same manner as a mercury cathode, and in many cases, especially in alkaline solution, a similar product distribution is obtained by direct electrolysis at a mercury cathode and by an amalgam reduction. In some instances, mostly in weakly acidic solution, however, different results have been obtained or electrolysis has not been attempted under strictly similar conditions.

Simultaneously with the transfer of electrons through the double layer a transfer of cations takes place.

Compared to direct electrolysis, reduction via an electrolytically produced amalgam may have some advantages. The preparation of an amalgam is more easily coupled with a useful anodic process, e.g., production of chlorine, than is direct electrolysis. Often a higher current density can be employed. If the production of and reduction by amalgam are performed separately, a higher degree of freedom is possible with respect to choice of solvent, temperature, and pressure in the second step. Amalgam reductions may be run either as batch operations or continuous processes.

It is, however, more difficult to change the reducing power of an amalgam than to change the potential of an electrode, but by working with different metals, different concentrations of metal, and by varying the rate of stirring (which influences the size of the amalgam droplets and thus the effective surface of the amalgam) it is possible to control within certain limits the degree of reduction by an amalgam.

Sodium amalgam is the most widely employed amalgam, but sometimes other amalgams may be used with advantage. Aluminum amalgam is, thus, useful for reduction of alkali-sensitive compounds, but it cannot be prepared in situ by electrolysis of aqueous solutions of aluminum salts.

II. OLEFINIC COMPOUNDS

Reduction of C=C double bonds may result either in saturation of the double bond or in hydrodimerization; the latter reaction is most common in compounds which can act as Michael acceptors.

Reduction of an unpolarized C=C double bond is exemplified in the conversion of biphenylene diphenylethylene to the corresponding ethane [1]. The reduction was performed in boiling amyl alcohol. This procedure illustrates the use of sodium amalgam in a medium which is poorly suited for a direct electrolysis but adequate for dissolving the substrate.

The reduction of cinnamylidene fluorene (I) with aluminum amalgam in moist ether to the dihydride (II) (a 1,4-addition) exemplifies the use of Al(Hg) rather than Na(Hg) for the reduction of alkali-sensitive compounds. In the presence of Na(Hg) (II) is tautomerized to (III), unless special care is taken [1a].

A tetraphenylcumulene like (IV) is reduced by Al(Hg) in tetrahydrofuran containing 4% water to the dihydride (V) in 75% yield [1b].

Of more interest than the saturation of double bonds is the hydrodimerization reaction (Chapter XIX); sodium amalgam hydrodimerizations of α,β-unsaturated ketones were reported early [2], and the product distribution resembled that obtained from direct electrolytic reduction (Chapter IX).

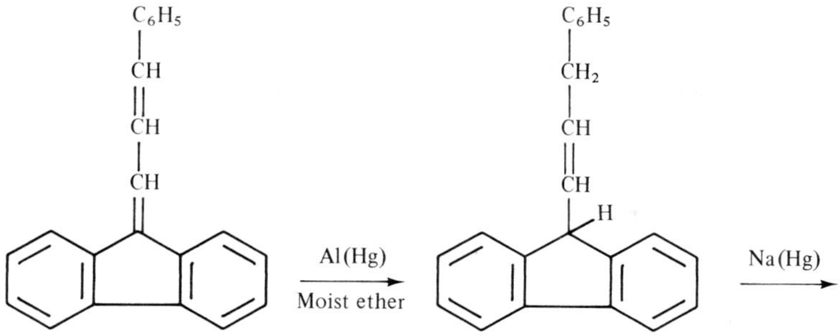

(I) (II)

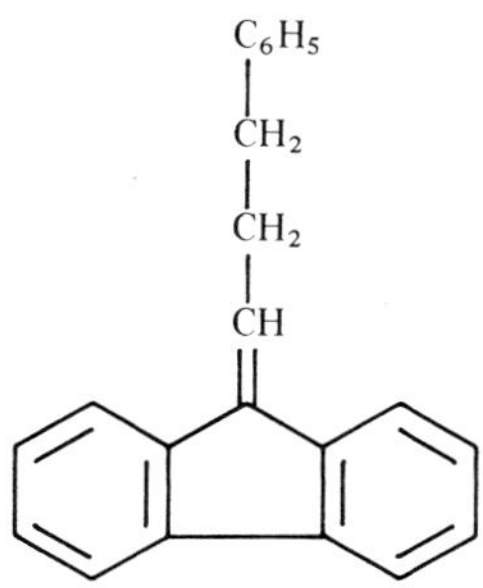

(III)

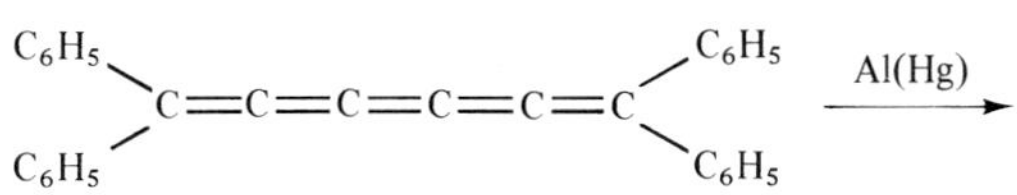

(IV)

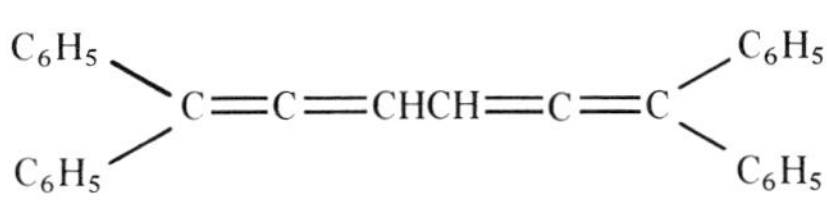

(V)

The first sodium-amalgam-induced dimerization of acrylonitrile to adiponitrile [3] was performed in aqueous solution and the yield was discouragingly low (5%). Mainly, a base-induced addition of water resulting in β, β'-bis(cyanoethyl) ether occurred together with a reduction to propionitrile.

Higher yields (30-60%) of a hydrodimerization with alkali metal amalgams were later reported [4, 5] in the reduction of α, β-unsaturated esters and nitriles. The different alkali metal amalgams often gave different yields of hydrodimerized products; the hydrodimerization of acrylonitrile using potassium amalgam thus yielded 62% adiponitrile, whereas sodium amalgam under similar conditions gave only 37%.

After the development of the electrolytic reductive coupling of acrylonitrile to adiponitrile (Chapter XIX) many investigations have been directed toward developing an alternative, indirect electrolytic process. The electrolytic hydrodimerization reaction in neutral solution [6, 7] is critically dependent on the proton activity at the electrode: If the proton activity is too high, the product is predominantly propionitrile, whereas oligo- or polymerization occurs at a too low proton activity. The establishment of a reaction layer with a suitable proton activity in which conditions are favorable for a hydrodimerization of acrylonitrile to adiponitrile is, thus, of paramount importance. This can be accomplished in different ways.

One possibility is to add quaternary ammonium salts to an aqueous reaction mixture; the positive ions accumulate in the double layer [8-11] around the amalgam and establish here a layer of low water content. The protonation may, thus, be slowed down sufficiently so the dimerization can compete successfully with it.

The unavoidable presence of sodium ions in the double layer during a Na(Hg) reduction introduces differences compared to the electrolytic hydrodimerization with respect to hydration and ion pairing.

The effect of the quaternary ammonium salt depends also on the substrate and on the electrode material [12]. Addition of tetraethylammonium salts thus sharply increases the yield of dimer product from acrylonitrile, has little effect upon the hydrodimerization of methyl acrylate, and does not affect the hydrodimerization of mesityl oxide.

It has been suggested [13] that quaternary ammonium amalgams are intermediates in the hydrodimerizations in an acetonitrile-water-CO_2 medium. In the absence of quaternary ammonium salts the reaction of acrylonitrile with Na(Hg) yields 7% adiponitrile and 90% propionitrile, whereas Na(Hg) in the presence of both quaternary ammonium salts and quaternary ammonium amalgams (prepared by electrolysis in an aprotic medium) yields almost 100% adiponitrile.

Although the product distribution is similar in the two cases, this does not necessarily mean that quaternary ammonium amalgams are intermediates during the Na(Hg) reductions in the presence of R_4N^+ salts; other explanations are possible [10, 11, 14]. A layer of quaternary ammonium salts could first be produced by anodic dissolution of a sufficient amount of the quaternary ammonium cations from the amalgam which afterwards would function as an electron donor.

Another way of producing a low proton activity at the amalgam surface is to add a suitable polar, aprotic compound to the medium. Suitable compounds include urea and its derivatives, amides, lactams, sulfoxides, and sulfones such as 1, 3-dimethylurea, ethyleneurea, biuret, formamide, N-methyl formamide, acetamide, δ-butyrolactam, dimethyl sulfoxide, di-n-butyl sulfoxide, and dimethyl sulfone [15, 16]. DMF or acetonitrile, besides some of the compounds mentioned above, may be used as solvents. One function of the added compound is to establish an aprotic reaction layer at the amalgam surface by selective adsorption of the compound on the amalgam; another one is to solvate the ion pair (consisting of the radical ion and the cation) in a suitable way.

A kinetic investigation [17] of the solvent and salt effect on the reaction of Na(Hg) with acrylonitrile (AN) in Diglyme-DMSO has shown that the slow step is the transfer of an electron to AN and the simultaneous transfer of Na^+ to the solution. In solutions of relatively low water and DMSO content the rate of the formation of propionitrile (PN) and adiponitrile (ADN) is proportional to the concentration of water and DMSO, respectively. This has been interpreted to show that the hydrogenation and the hydrodimerization reactions are independent reactions which can be represented by

$$\text{Na} + \text{AN(H}_2\text{O)} \rightarrow \text{Na}^+(\text{H}_2\text{O})\text{AN}^{\dot{-}} \rightarrow \cdot\text{ANH} + \text{NaOH} \xrightarrow[\text{H}_2\text{O}]{\text{Na}} \text{PN} \tag{1}$$

$$\text{Na} + \text{AN(DMSO)} \rightarrow \text{Na}^+(\text{DMSO})\text{AN}^{\dot{-}} \xrightarrow{\text{AN}}$$

$$\cdot\text{ANAN}^-(\text{DMSO})\text{Na}^+ \xrightarrow[\text{H}_2\text{O}]{\text{Na}} \text{ADN}. \tag{2}$$

As the electron transfer to AN is coupled with a transport of Na^+ to the solution, the rate will be influenced by the energy level of the solvent-separated ion pair $Na^+(solv)AN^{\dot{-}}$, which is higher in Diglyme than in Diglyme/H_2O or Diglyme/DMSO, because H_2O and DMSO are able to solvate both the cation and the anion; the rate of reaction is thus higher in the mixtures than in pure Diglyme.

The water-solvated ion pair reacts rapidly with water forming the easily reducible radical AN• which then is further reduced and protonated to PN. It has been assumed that the DMSO-solvated ion pair $Na^+(DMSO)AN^{\dot{-}}$ reacts more slowly with water, and a reaction with AN takes place. The dimerization under these conditions is thus suggested to be a Michael addition of the radical anion $AN^{\dot{-}}$ to AN rather than a Michael addition of propionitrile monoanion ANH^- or dianion AN^{2-}; the suggestion is supported by calculations based upon the simple HMO approximation [18, 19].

The dimeric anion radical $\cdot ANAN^{-}(DMSO)Na^{+}$ accepts a proton and the resulting radical is reduced to ADN. This interpretation requires that the behavior of the dimeric anion radical be considerably different from that of the monomeric one with respect to the rate constants of the competing reactions.

Hydrodimerization of AN by Na(Hg) may also take place in liquid ammonia [20] in the presence of a suitable proton donor; the dimerization is faster than the amination. If water is used as proton donor, ADN is formed in 80% yield, but when NH_4Cl is used, the yield is raised to nearly 100%. The ratio of PN to ADN depends on the ratio between NH_4Cl and AN; when the ratio is smaller than unity, some polymer is formed because there is not enough proton donor to terminate the anionic polymerization.

In the hydrodimerization of methacrylonitrile (MeAN) by Na(Hg) in liquid NH_3, NH_4Cl is too strong a proton donor; a mixture of isobutyronitrile (IBM) and dimethyl adiponitrile (diMeADN) is formed. Phenol is a suitable weak proton donor in this case, and when the ratio phenol/monomer is unity, only hydrodimerization is observed.

The ratio of the apparent rate constants of the formation of ADN and diMeADN from the monomers is about 2 in NH_3, whereas it is about 100 in Diglyme [21] or Diglyme/DMSO; this means that on cross hydrodimerization of AN and MeAN in NH_3 all the possible combinations are obtained, whereas in Diglyme/DMSO no diMeADN is found. An explanation for this difference might be that the monomer in phenol/NH_3 exists in a protonated form.

Many patents have been issued in the last several years to groups working at I.C.I., U.C.B., and the Weizmann Institute on the amalgam hydrodimerization of AN and related compounds [22].

Reaction between Na(Hg) and acrylic acid (AA) in water, dioxane, or Diglyme produces propionic acid (PA); however, when DMSO (containing about 10% water) is used as solvent, adipic acid (ADA) is the main product [23].

To achieve a high conversion it is necessary to add hydrochloric acid during the reaction. Under such conditions it is possible to obtain 70% ADA and 20-25% PA. Addition of HCl increases the rate of the reaction, but not the ratio of PA to ADA; this is a consequence of the fact that it is the free acid and not the acrylate ion which is reduced. This is in accordance with the less negative reduction potential of the free acid compared to that of the acrylate ion.

The hydrodimerization of acrylic acid to adipic acid is a tail-to-tail addition; hydrodimerization of methacrylic acid and crotonic acid produces mainly the head-to-head and head-to-tail dimers [23].

Another example of this type of reaction is the hydrodimerization of dialkylfulvenes to dicyclopentadienyl alkanes by alkali metal amalgams in an aqueous-alcoholic medium [24].

III. HALOGEN COMPOUNDS

An amalgam reduction of a halogen compound may result in either a cleavage of the carbon-halogen bond, a coupling reaction, a reductive elimination, or in a formation of an organic mercury compound.

Many halogen compounds are reduced to hydrocarbons by Na(Hg). This may be used to obtain reduction of isolated double bonds which are unreactive toward Na(Hg) by addition of hydrogen bromide to the double bond followed by amalgam reduction of the halogen compound to a hydrocarbon; the transformation of 4-pentenoic acid via bromovaleric acid to valeric acid is an example [25].

A reductive coupling of a halogen compound may be exemplified by the reaction of β-chloropropionitrile with Na(Hg) in the presence of some acrylonitrile and a small amount of water to adiponitrile; yields up to 90% have been obtained [26, 27].

Treatment of vicinal halogen compounds with sodium amalgam results in many cases in a reduction-elimination reaction; 4, 5-dibromovaleric acid thus forms 4-pentenoic acid on treatment with Na(Hg) [25]. 1, 1, 2-Trichloro-1, 2, 2-trifluoroethane and 1, 2-dichloro-1, 1, 2, 2-tetrafluoroethane react with Na(Hg) similarly with formation of 1-chloro-1, 2, 2-trifluoroethylene and tetrafluoroethylene, respectively [28].

A special reaction of this type is the formation of benzyne. o-Fluorobromobenzene reacts with lithium amalgam in ether or furan with an intermediate formation of benzyne. In the former medium diphenylene and small amounts of triphenylene and 9, 10-dimercura-9, 10-dihydroanthracene were formed, whereas the major product (76%) in furan was 1, 4-dihydronaphthalene-1, 4-endoxide [29].

A good yield of dialkylmercury compounds [30] may be obtained from alkyl halides and Na(Hg); too high a Na content in Na(Hg) lowers the yield of dialkylmercury due to increased formation of the Wurtz coupling product. The reaction may be formulated as follows:

$$RX + Hg \rightarrow RHgX \xrightarrow{Na} \tfrac{1}{2}R_2Hg + NaX + \tfrac{1}{2}Hg.$$

The ratio of the dialkylmercury to the Wurtz coupling product is dependent on the halogen; thus n-butyl iodide yields 80-81% dibutylmercury and 10% octane, whereas n-butyl chloride gave 11-17% Bu_2Hg and 65-70% octane.

IV. NITRO COMPOUNDS AND DERIVATIVES

Nitro compounds are reduced by amalgams in a manner similar to that found at a mercury electrode. Aliphatic nitro compounds [31] are thus reduced to alkylhydroxylamines by NH_4(Hg), whereas aromatic nitro compounds may form amines, hydrazo, azo, or azoxy compounds and even nitroso compounds [31, 32].

Reduction of nitrobenzene with Al(Hg) in ether at controlled temperature yields phenylhydroxylamine [32a].

Azobenzene has been manufactured by reduction of a suspension of nitrobenzene by Na(Hg). Further reduction to hydrazobenzene is possible if higher temperatures and concentrations of alkali are employed [33]. The corresponding arsenic compounds may be prepared from a phenylarsenoxide by amalgam reduction; thus p-aminophenylarsenoxide yields 80% $H_2NC_6H_4As{=}AsC_6H_4NH_2$ on treatment with 4% Na(Hg) in methanol [34].

It is interesting that treatment of some aromatic nitro compounds with ammonium amalgam yields among other products nitrosobenzenes [32]. 2-Nitrobenzoic acid thus gives about 45% 2-nitrosobenzoic acid together with anthranilic acid (10%), and 2-nitrotoluene (5 g) yields 0.6 g azoxytoluene, 1.5 g o-toluidine, and 2 g 2-nitrosotoluene on treatment with NH_4(Hg).

N-Nitroso and N-nitro compounds may both be reduced to hydrazines. In this case care must be taken to avoid a reductive cleavage of the nitrogen-nitrogen bond [33, 35].

V. CARBONYL COMPOUNDS AND DERIVATIVES

On reaction with Na(Hg) saturated aldehydes or ketones may form either alcohols or pinacols [36, 37]; the reaction is similar to the electrolytic reduction (Chapter VIII).

The use of Al(Hg) in ether for the reduction of alkali-sensitive compounds may give higher yield than Na(Hg); thus, diethyl oxaloacetate is reduced to diethyl malate by Al(Hg) in 70-80% yield, whereas Na(Hg) gives about 50% yield [32a].

α,β-Unsaturated ketones of the structure $RR'C{=}CHCOCH_3$ yield on treatment with sodium amalgam a mixture of products including tetrahydrofurans (VI) and dihydrofuran derivatives (VII), glycols (VIII), ϵ-diketones (IX), and ketols (X) [38-40]. The ratio between the products depends on R

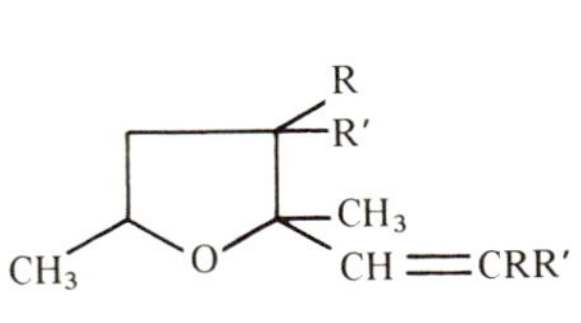

(VI)

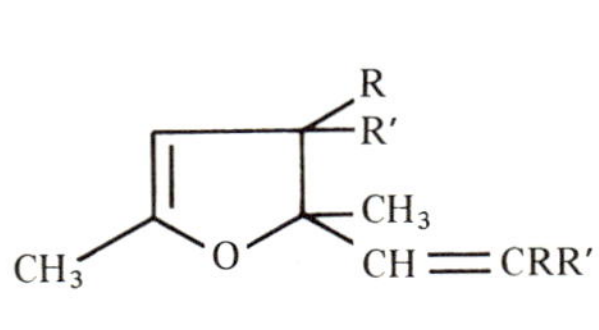

(VII)

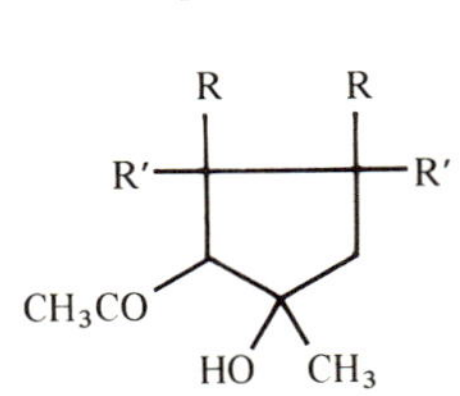

(X)

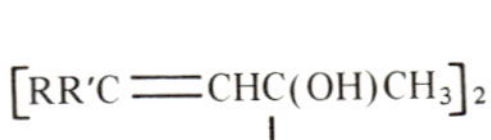

(VIII)

(IX)

and R′; bulky groups as found in steroids favor formation of glycols rather than of ϵ-diketones [41]. The reactions leading to these products are similar to those discussed in Chapter IX.

The reduction of carbonyl derivatives as oximes and hydrazones is similar to the electrolytic reaction. Oximes are reduced to amines [30], whereas hydrazones in alkaline solution may form hydrazines; benzalhydrazine can thus be reduced to benzylhydrazine [42], benzalazine to benzaldehyde benzylhydrazone or further to dibenzylhydrazine [43, 44], and benzaldehyde semicarbazone to benzylsemicarbazide [45].

VI. ACIDS AND DERIVATIVES

Esters of fatty acids are only to some extent reduced to the alcohols by sodium amalgam. Aldonolactones and their esters, however, are reduced to aldoses by Na(Hg) [46]. Optimal conditions are found at pH 3-3.5 and temperatures below 15°C; 2.5-3 atoms of Na [as 2.5% Na(Hg)] are used per mole of lactone. Yields of different aldoses range from 50-84% [47].

The reduction of aromatic carboxylic acids by Na(Hg) may take place in the nucleus or in the carboxyl group. Phthalic acid is reduced to trans-3, 5-cyclohexadiene-1, 2-dicarboxylic acid [48], whereas certain phenolic carboxylic acids may be reduced to aldehydes.

Salicylic acid and some other acids were reduced by Na(Hg) to the aldehyde [49-51] (40-60%) in a slightly acid medium containing p-toluidine and boric acid; the reaction mixture was kept slightly acid by addition of boric acid. A similar electrolytic reduction was not successful unless the mercury cathode was allowed to take up some sodium, thus making the reaction an indirect reduction [52].

The reduction by Na(Hg) of several o-hydroxynaphthalenecarboxylic acids in slightly acid solution containing boric acid has been investigated. 1-Hydroxynaphthalene-2-carboxylic acid [53] is, like salicylic acid, reduced to the aldehyde (62% yield), isolated as the Schiff base, whereas 2-hydroxynaphthalene-3-carboxylic acid [54] under similar conditions yields mainly 2-hydroxy-1, 2, 3, 4-tetrahydronaphthalene-3-carboxylic acid (75-80%) and 5% of the aldehyde. When hydrochloric acid rather than CO_2 was used to neutralize the base formed during the reduction, the 2-oxo-1, 2, 3, 4-tetrahydronaphthalene-3-carboxylic acid [55] was isolated [60%). On heating, the compound, a β-keto acid, decarboxylated.

A loss of a phenolic hydroxyl group occurs during the reduction of 1-hydroxynaphthalene-4-carboxylic acid with Na(Hg) in the boric acid medium from which 1, 4-dihydronaphthoic acid is isolated [56]. The primary reduction probably yields 1, 4-dihydro-2-hydroxy-4-naphthoic acid, which loses water to 1-naphthoic acid; this compound is under these conditions reduced mainly to the 1, 4-dihydro derivative, whereas 2-naphthoic acid predominantly gives the 1, 2-dihydro compound.

Hydrodimerized products are obtained mostly from the carboxylic esters [57] but in some cases also from the free acids [58]. 2-Hydroxy-1-carbomethoxynaphthalene yields 27% 3, 3′-dioxo-4, 4′-dicarbomethoxy-1, 1′, 2, 2′, 3, 3′, 4, 4′-octahydro-1, 1′-dinaphthyl.

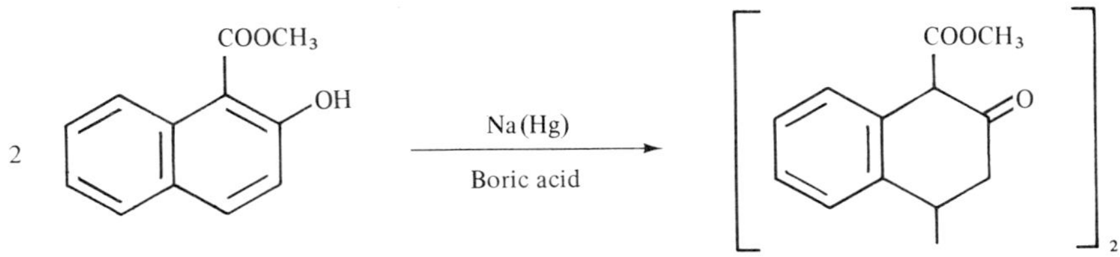

Carbon Dioxide

Carbon dioxide may be reduced by alkali metal amalgams in alcohol to alkali metal formate [59] in a yield of 85-90%; the rate increases with the CO_2 pressure and is dependent on the alkali metal, Na(Hg) < Li(Hg) < K(Hg).

An attempt has been made [60] to hydrodimerize CO_2 to oxalic acid, but only in hexane and with Na(Hg) was sodium oxalate obtained in a reasonable yield (40%).

In other cases a wide variety of products, ranging from simple organic acids such as formic and oxalic acid to totally reduced CO_2 in the form of carbon black, has been obtained by reduction of CO_2 by an appropriate alkali metal amalgam [61].

VII. REDUCTION OF PHENOLIC COMPOUNDS

Reduction of phenolic compounds by Na(Hg) has been performed mostly in slightly acid solutions containing boric acid. The latter acts probably as a complexing agent besides maintaining a suitable pH.

Resorcinol can be reduced in the nucleus by Na(Hg) [62] in slightly acid solution at elevated temperature (90° C); the acidity can be maintained by bubbling CO_2 through the solution:

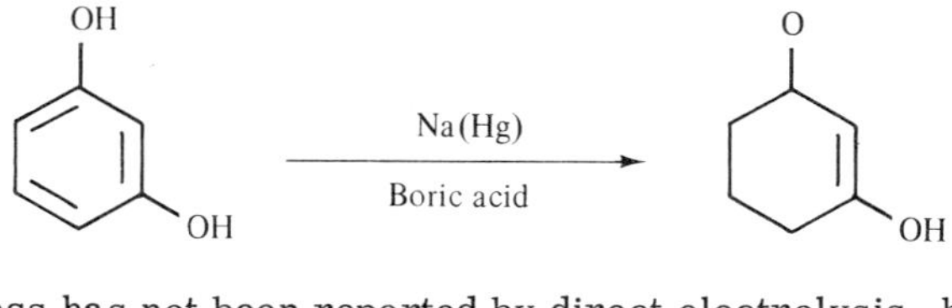

The process has not been reported by direct electrolysis, but it might be possible to do it by applying a suitable potential somewhat more negative than the decomposition potential of the medium at that pH in the absence of the substrate.

Similarly, phloroglucinol is reduced to the saturated trihydroxy compound [63].

VIII. CLEAVAGE REACTIONS (OTHER THAN CARBON-HALOGEN)

A reaction which involves both a carbon-carbon and a nitrogen-nitrogen cleavage is found in the reduction of 2-phenyl-4-carboxytriazole by Na(Hg) in slightly alkaline solution [64]. The reaction probably starts with a reduction in the ring which then opens up and loses cyanide ion according to

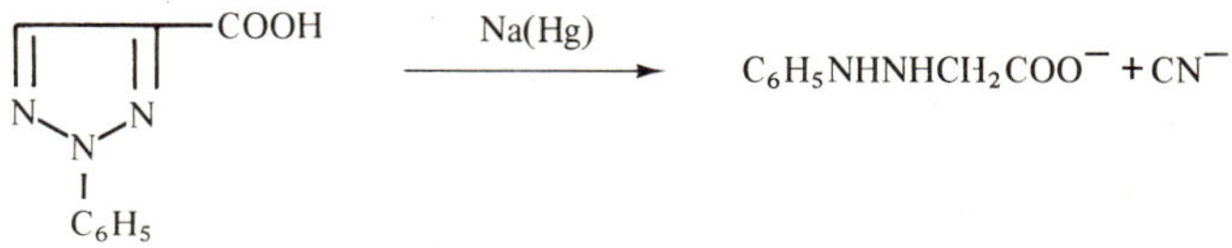

Treatment of benzopinacol with Na(Hg) in a mixture of anhydrous ether and benzene gives a dark-blue solution which fades. Benzhydrol can be isolated in a quantitative yield. It has been suggested that the disodium salt of benzopinacol disproportionates with C-C cleavage into benzhydrol and benzophenone; the latter is reduced by Na(Hg) to benzopinacol, which then decomposes, etc. [65].

A cleavage of carbon-oxygen bonds caused by amalgam reduction is found, e.g., in the reduction of dihydroxyacetoxime to isopropylamine [66], and 2-hydroxy-4-phenyl-3-butenoic acid to 4-phenyl-3-butenoic acid [67]. These reactions are analogous to those discussed in Chapter XXI.

Whereas sodium amalgam is unreactive toward phenyltrimethylammonium chloride, benzyltrimethylammonium chloride is cleaved to toluene and trimethylamine. A similar reaction is responsible for the reductive ring opening of N, N-dimethyl-tetrahydroquinolinium chloride to 2-propyl-N, N-dimethylaniline [68].

Sodium amalgam has been used to cleave sulfur-sulfur bonds. Solutions or suspensions of organic di- or polysulfides can be reduced with Na(Hg) in the presence of an alkali sulfide to the mercaptide [69].

Certain sulfonic acids may be cleaved by Na(Hg); thus 1-aminonaphthalene-3, 8-disulfonic acid can be reductively cleaved to 1-aminonaphthalene-3-sulfonic acid. The sulfonic group in the 8-position is more easily removed reductively than sulfonic groups in the 2-, 4-, or 6-positions [33].

β-Ketosulfoxides react rapidly and cleanly with aluminum amalgam in a mixture of 90% tetrahydrofuran and 10% water to the ketones [70]:

$$RCOCH_2SOCH_3 \xrightarrow[(H_2O)]{Al(Hg)} RCOCH_3 + CH_3SH.$$

For preparation on a larger scale the direct electrolytic cleavage of β-ketosulfoxides [71] or β-ketosulfones [72] has been found preferable.

IX. POLYMERIZATION REACTIONS

Finally, it might be mentioned that Na(Hg) can act as a catalyst in a polymerization of acetaldehyde to a polyvinyl alcohol-type polymer [73, 74].

The polymerization may be performed in a cooled vessel at high pressure under nitrogen with vigorous stirring. The product was suggested to be a linear polymer of the type $CH_3(CHOHCH_2)_nCHO$, but later investigations suggested a branched structure [75].

An oligomerization of ethylene in the presence of cyclopentadienyltitanium dichloride, $C_5H_5TiCl_2$, is possible using Na(Hg). When the ratio of ethylene to $C_5H_5TiCl_2$ was 12, a mixture of 80% n-butane and 20% n-hexane was obtained [76].

REFERENCES

[1]. V. Kaufmann, Ber., 29, 73 (1896).

[1a]. J. Thiele and F. Henk, Ann., 347, 290 (1906).

[1b]. R. Kuhn and H. Fischer, Chem. Ber., 94, 3060 (1961).

[2]. C. Harries and F. Kaiser, Ber., 31, 1806 (1898).

[3]. O. Bayer, Angew. Chem., 61, 229 (1949).

[4]. I. L. Knunyants and N. S. Vyazankin, Dokl. Akad. Nauk SSSR, 113, 112 (1957); CA, 51, 14637 (1957).

[5]. I. L. Knunyants and N. S. Vyazankin, Izv. Akad. Nauk SSSR, Otd. Khim. Nauk, 238 (1957); CA, 51, 11235 (1957).

[6]. M. M. Baizer, Tetrahedron Lett., 973 (1963).

[7]. F. Beck, Chem.-Ing.-Tech., 37, 607 (1965).

[8]. L. G. Feoktistov, A. P. Tomilov, and I. G. Sevast'yanova, Elektrokhim., 1, 1300 (1965).

[9]. M. M. Baizer and J. P. Petrovich, J. Electrochem. Soc., 114, 1023 (1967).

[10]. F. Beck, Z. Elektrochem., 72, 379 (1968).

[11]. I. E. Gillet, Bull. Soc. Chim. Fr., 2919 (1968).

[12]. A. P. Tomilov, E. V. Kryukova, V. A. Klimov, and I. N. Brago, Elektrokhim., 3, 1501 (1967).

[13]. J. D. Littlehailes and B. J. Woodhall, Discuss. Faraday Soc., 45, 187 (1968).

[14]. F. Beck and I. E. Gillet, ibid., 45, 234 (1968).

[15]. F. Matsuda, Tetrahedron Lett., 6193 (1966).

[16]. A. Katchalsky, D. Vofsi, and J. I. Padova, Isr. Pat. 13, 842 (1961); CA, 56, 9977 (1962). A. Katchalsky, D. Vofsi, and J. I. Padova, Isr. Pat. 13, 844 (1961); CA, 56, 11450 (1962). I.C.I., Neth. Appl. 6, 504, 863 (1965); CA, 64, 12557 (1966). I.C.I., Neth. Appl. 6, 609, 103 (1967); CA, 66, 115334 (1967). U.C.B., Neth. Appl. 6, 609, 241 (1967); CA, 67, 53724 (1967). A. Misonou and T. Naga, Japan. Pat. 6, 907, 925 (1969); CA, 71, 35536 (1969).

[17]. Y. Arad, M. Levy, H. Rosen, and D. Vofsi, J. Polym. Sci. A-1, 7, 2159 (1969).

[18]. M. Figeys, Tetrahedron Lett., 2237 (1967).

[19]. M. Figeys and H. P. Figeys, Tetrahedron, 24, 1097 (1968).

[20]. Y. Arad, M. Levy, H. Rosen, and D. Vofsi, J. Polym. Sci. B, Polym. Lett., 7, 197 (1969).

[21]. H. Rosen, Y. Arad, M. Levy, and D. Vofsi, J. Amer. Chem. Soc., 91, 1425 (1969).

[22]. E.g., U.C.B., Fr. Pat. 1, 505, 954 (1967); CA, 69, 95999 (1968). U.C.B., S. Afr. Pat. 6, 704, 827 (1968); CA, 70, 37239 (1969). Toyo Koatsu Ind., Fr. Pat. 1, 535, 374 (1968); CA, 71, 38378 (1969). I.C.I. of Australia and New Zealand, Fr. Pat. 1, 543, 322 (1968); CA, 72, 12163 (1970). Y. Arad, M. Levy, and D. Vofsi, Fr. Pat. 1, 553, 195 (1969); CA, 72, 42853 (1970). I.C.I., US Pat. 3, 458, 559 (1969); CA, 71, 123625 (1969). I.C.I. of Australia and New Zealand, US Pat. 3, 523, 967 (1971); CA, 74, 12646 (1971).

[23]. Y. Arad, M. Levy, and D. Vofsi, J. Org. Chem., 34, 3709 (1969).

[24]. V. A. Smirnov, V. G. Kul'nevide, I. V. Chemikov, G. N. Soltovets, and S. V. Babicheva, U.S.S.R. Pat. 271, 500 (1970); CA, 73, 120180 (1970).

[25]. A. Messerschmidt, Ann., 208, 92 (1881).

[26]. Knapsack-Griesheim A.-G., Ger. Pat. 1, 151, 791 (1963); CA, 60, 1604 (1964).

[27]. A. Misono, T. Osa, T. Asano, and T. Suzuki, Bull. Chem. Soc. Japan, 41, 2434 (1968).

[28]. Farbenfabriken Bayer, Ger. Pat. 847, 442 (1950); 857, 956 (1950); 875, 808 (1950).

[29]. G. Wittig and L. Pohmer, Chem. Ber., 89, 1334 (1956).

[30]. N. N. Mel'nikov, Zh. Obshch. Khim., 16, 2065 (1946).

[31]. S. Takaki and T. Ueda, J. Pharm. Soc. Japan, 58, 276, 533 (1938); CA, 32, 5376, 8384 (1938).

[32]. S. Takaki, T. Ueda, and T. Sakaguti, J. Pharm. Soc. Japan, 58, 938 (1938); CA, 33, 2113 (1939).

[32a]. H. Wislicenus and L. Kaufmann, Ber., 28, 1323 (1895).

[33]. W. Funke, Chem.-Ing.-Tech., 35, 336 (1963).

[34]. P. Ehrlich and A. Bertheim, Ber., 44, 1260 (1911).

[35]. Ger. Pat. 1,130,821 (1960); CA, 57, 12320 (1962); Belg. Pat. 612,838 (1961); CA, 58, 4586 (1963).

[36]. S. Takaki and T. Ueda, J. Pharm. Soc. Japan, 58, 44, 141, 152, 156, 353, 422, 648 (1938); CA, 32, 4149, 5777, 6635, 8384 (1938).

[37]. S. Takaki and T. Tamai, J. Pharm. Soc. Japan, 58, 701 (1938); CA, 33, 536 (1939).

[38]. N. V. Thoai and J. Wiemann, Compt. Rend., 247, 212 (1958).

[39]. J. Wiemann and R. Nahum, ibid., 238, 2091 (1954).

[40]. R. Nahum, ibid., 243, 849 (1956).

[41]. R. Dulou, J. Chopin, and Y. Raoul, Bull. Soc. Chim. Fr., 616 (1951).

[42]. T. Curtius, Ber., 33, 2459 (1900).

[43]. T. Curtius, J. Prakt. Chem., 62, 83 (1900).

[44]. T. Curtius, ibid., 85, 393 (1912).

[45]. S. Kessler and H. Rupe, Ber., 45, 26 (1912).

[46]. E. Fischer and O. Piloty, ibid., 24, 4214 (1891).

[47]. N. Sperber, H. E. Zaugg, and W. M. Sandstrom, J. Amer. Chem. Soc., 69, 915 (1947).

[48]. A. Baeyer, Ann., 269, 145 (1892).

[49]. H. Weil, Ber., 41, 4147 (1908).

[50]. H. Weil, M. Traun, and S. Marcel, ibid., 55, 2664 (1922).

[51]. R. R. Davies and H. H. Hodgson, J. Soc. Chem. Ind., 62, 128 (1943).

[52]. C. Mettler, Ber., 41, 4148 (1908).

[53]. L. N. Lavrishcheva, N. M. Przhiyalgovskaya, S. A. Voitkevich, and V. N. Belov, Zh. Obshch. Khim., 27, 1264 (1957).

[54]. N. M. Przhiyalgovskaya, L. N. Lavrishcheva, G. T. Mondodoev, and V. N. Belov, ibid., 31, 2321 (1961).

[55]. N. M. Przhiyalgovskaya, L. N. Lavrishcheva, and V. N. Belov, ibid., 27, 1266 (1957).

[56]. N. M. Przhiyalgovskaya, G. T. Mondodoev, and V. N. Belov, ibid., 31, 3375 (1961).

[57]. N. M. Przhiyalgovskaya, L. N. Lavrishcheva, G. T. Mondodoev, and V. N. Belov, ibid., 33, 632 (1963).

[58]. L. N. Lavrishcheva, N. M. Przhiyalgovskaya, and V. N. Belov, ibid., 31, 2762 (1961).

[59]. V. A. Smirnov, V. D. Semchenko, D. P. Semchenko, and L. I. Antropov, Zh. Prikl. Khim., 40, 2345 (1967); CA, 68, 63071 (1968).

[60]. V. A. Tushnalobova, V. A. Smirnov, and D. P. Semchenko, Zh. Prikl. Khim., 41, 275 (1968); CA, 68, 110961 (1968).

[61]. R. W. Treharne and C. M. Cox, U.S. Clearinghouse Fed. Sci. Tech. Inform., A.D. 1968, A.D. 679597; CA, 70, 99413 (1969).

[62]. G. Merling, Ann., 278, 20 (1894).

[63]. W. Wislicenus, Ber., 27, 357 (1894).

[64]. A. Jonas and H. von Pechmann, Ann., 262, 277 (1891).

[65]. W. E. Bachmann, J. Amer. Chem. Soc., 55, 355, 770 (1933).

[66]. O. Piloty and O. Ruff, Ber., 30, 1656 (1897).

[67]. G. H. Lahib and P. Cordier, Compt. Rend., 241, 1051 (1955).

[68]. H. Emde, Ann., 391, 88 (1912).

[69]. Ger. Pat. 737,334 (1940); CA, 38, 3666 (1944).

[70]. E. J. Corey and M. Chaykovsky, J. Amer. Chem. Soc., 86, 1639 (1964).

[71]. B. Lamm and B. Samuelsson, Acta Chem. Scand., 23, 691 (1969).

[72]. B. Lamm and B. Samuelsson, ibid., 24, 561 (1970).

[73]. T. Imoto and T. Matsubara, J. Polym. Sci., 56, S-5 (1962).

[74]. T. Imoto and T. Matsubara, J. Polym. Sci. A, 2, 4573 (1964).

[75]. M. Modena, S. Carraro, and G. Corri, J. Polym. Sci. B, Polym. Lett., 4, 613 (1966).

[76]. K. Shikata, Y. Miura, S. Nakao, and K. Azuma, Kogyo Kagaka Zasshi, 68, 2266 (1965); CA, 64, 15721 (1966).

Chapter XXIV

ELECTROGENERATED REAGENTS

Roy Dietz

Division of Inorganic and Metallic Structure
National Physical Laboratory
Teddington, Middlesex

and

Henning Lund

Department of Organic Chemistry
Aarhus University
Aarhus, Denmark

I. INTRODUCTION

For the most part the reagents dealt with in this chapter are generated and used in aqueous solution. They must, therefore, be at least moderately stable under these conditions and are, hence, generally familiar reagents. The recent use of aprotic solvents has made available reagents unfamiliar from conventional organic chemistry, and the chapter closes with some examples of these.

The reactions treated here fall into two broad classes. The first employs the added reagent in a catalytic role; the primary electrode process is followed by a reaction with the organic substrate that regenerates the original reagent. We will call reagents of this sort mediators. In the simplest case an electron is relayed from electrode to organic substrate. Since most of the examples of this kind of catalysis involve transition metal ions, reaction with the substrate probably involves complex formation. The second, and more heterogeneous, class of reactions comprises those in which the electrogenerated reagent is consumed in reaction with the organic substrate.

II. HALOGEN

Anodic oxidation of a solution containing halide ion may result in addition or substitution. Substitutions are dealt with in Chapter XXII. Addition of halogen may be followed by further reactions.

A. Addition of Halogen

When a carefully controlled rate of halogen addition is needed, anodic generation is the method of choice; an indicator electrode can monitor the halogen concentration continuously. Efficient and safe addition of radiohalogen is possible in this way. In a kinetic application [1] a small concentration of bromine was maintained in the presence of a series of alkenes. The rate of bromine generation is then a measure of the rate of addition; rate constants spanning two decades were determined.

Oxidation of a fluoride melt at a carbon anode at 700-1000°C gives fluorocarbons [2]. However, it is doubtful whether free fluorine is involved.

B. Epoxidation of Alkenes

An anodic addition that has gained recent importance is the formation of halohydrins from alkenes. Aqueous halogen is formed by oxidation of alkali or alkaline earth metal chlorides at an inert anode. Addition to alkenes can then take place. Thus, ethylene chlorohydrin is formed [3] with a current yield of 90% when ethylene is bubbled through a cell in which chloride is oxidized at graphite [Eqs. (1) and (2)]:

$$2Cl^- - 2\bar{e} \xrightarrow[\text{anode}]{\text{Graphite}} Cl_2 \tag{1}$$

$$RCH{=}CH_2 + Cl_2 + H_2O \rightarrow RCH(OH)CH_2Cl + Cl^- + H^+. \tag{2}$$

At low chloride concentrations hydrolysis of chlorine is rapid; much of the hypochlorite formed is then oxidized at the anode to chlorate [4]. The yield of chlorohydrin is correspondingly reduced. Symmetrical adducts, alkene dihalide and glycol, are by-products. Dihalide is minimized [5] by small halide concentrations and by using chloride rather than bromide. Glycol is minimized by working at low temperatures [6].

Formation of halohydrins by anodic addition is significant because of their ready base-catalyzed conversion to epoxides. The cathodic reaction of the cells described above in which halohydrins are formed is proton discharge [Eq. (3)]; the catholyte develops a pH value high enough to bring about [7] the elimination [Eq. (4)]:

$$2H^+ + 2\bar{e} \rightarrow H_2\uparrow \tag{3}$$

$$RCH(OH)CH_2Cl + OH^- \longrightarrow \underset{\diagdown O \diagup}{RCH - CH_2} + Cl^- + H_2O \tag{4}$$

Addition of Eqs. (1) to (4) gives the net cell reaction [Eq. (5)]:

$$RCH{=}CH_2 + H_2O \longrightarrow \underset{\diagdown O \diagup}{RCH - CH_2} + H_2\uparrow \tag{5}$$

Ideally, chlorine and chloride play a catalytic role. The competitive industrial process involves the treatment of alkene chlorohydrin with lime; chlorine is lost as the useless $CaCl_2$. Technological work [8][1] in the electrochemical process has concerned the fabrication of a cell separator and the choice of a noncorroding anode. With a porous titanium anode coated with Pt/Ir and a polypropylene diaphragm in 5% aqueous KCl, an 89% conversion of propene to propylene oxide at a current efficiency of 90% is possible [9]. Propene is introduced through the anode.

Bifunctionalization of olefins by oxidation in the presence of a halide and a nucleophile has recently been reported [68].

C. Alkoxylation of Furans

In methanol the initial adduct of anodically generated bromine and furan is trapped by methoxide, and further substitution of bromide gives 2, 5-dihydro-2, 5-dimethoxyfuran [10]:

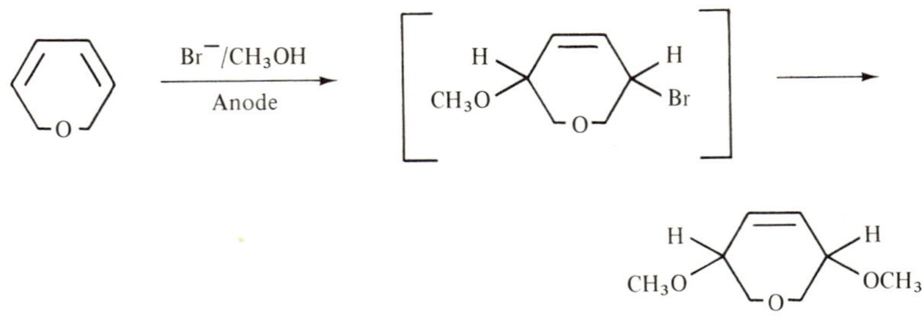

cis and trans

Compounds like this are precursors of several heterocyclic systems [11] (p. 583). The electrolysis is thought to be indirect since a film of bromine appears on the anode, and the anode material is unimportant. Direct anodic oxidation of furans may occur, however, when other anions replace bromide in the electrolyte [12], or, under alkaline conditions, when the product is sensitive to anode material [13].

In ethanol dihydrodiethoxyfurans are formed [14]. Extension of the reaction to include other nucleophiles (e.g., acetate from acetic acid) or other π-electron-rich heterocycles should be possible.

D. Oxidation of Sugars

Glucose [15] can be oxidized to gluconic acid, and other sugars to the corresponding acids, by electrolytically generated bromine according to

$$HOCH_2(CHOH)_4CHO \xrightarrow[\text{NaBr, CaCO}_3]{\text{Graphite anode}} [HOCH_2(CHOH)_4COO]_2Ca.$$

An alkaline solution of the sugar is electrolyzed in a diaphragm-free cell using sodium bromide as electrolyte at 40°C. In order to minimize the loss of bromine the current may be reversed every 15 min. Under these conditions the yield of gluconic acid is nearly quantitative. Other sugars can be oxidized similarly.

III. INORGANIC MEDIATORS

The literature of classical electroorganic chemistry contains many examples of inorganic "accelerators" — salts that were found to increase the current efficiency for formation of a desired product. Those effective in oxidation and reduction processes were called oxygen and hydrogen "carriers," respectively. Thus, $KMnO_4$ facilitates [16] the anodic oxidation of o-toluenesulfonamide to saccharin, and yields of amines in cathodic reductions of nitro compounds are increased [17] by added tin salts.

Many of these empirical observations may be rationalized by supposing that regenerative processes occur. To take a cathodic example, homogeneous reduction of an organic molecule R may be maintained by heterogeneous regeneration of an inorganic reagent:

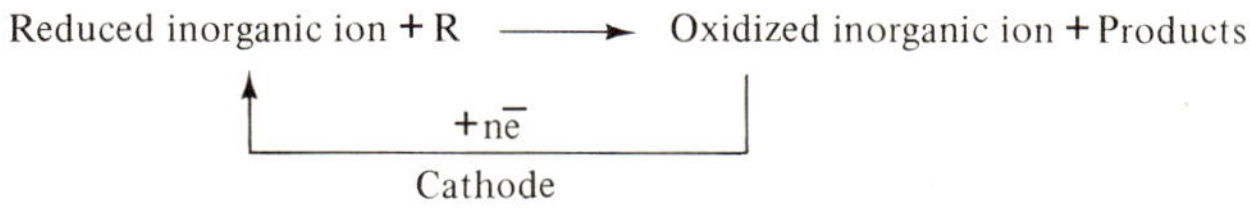

(6)

No real preparative reaction will be as simple as this. Side reactions, particularly at the anode, will intervene sooner or later. The added salt may modify the electrode surface, by either deposition or etching, and thus change the overpotential; and complexes may be formed between the organic and inorganic species. In this early work, of course, the electrode reactions were ill defined; at best, currents were controlled.

More recent work, still largely empirical, has concentrated on two-stage processes. Sufficient inorganic reagent is added to complete the desired chemical reaction in one reactor. The spent reagent is removed, regenerated electrochemically in a separate cell, and recycled. Two-phase systems are particularly convenient [18]. The importance of this work is more economic than scientific. Clean and efficient electrochemical regeneration of an expensive inorganic reagent has obvious attractions.

Two effects of an added redox system (Sn^{II}/Sn^{IV}) are distinguishable [19] in the electroreduction of nitrobenzene in aqueous acid. p-Aminophenol, the desired product, is formed by rearrangement of phenylhydroxylamine. Further reduction of phenylhydroxylamine to aniline becomes important at

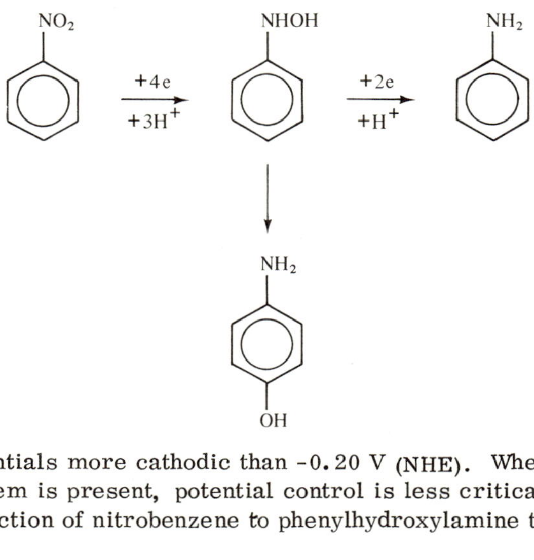

(7)

potentials more cathodic than -0.20 V (NHE). When the stannous/stannic system is present, potential control is less critical because homogeneous reduction of nitrobenzene to phenylhydroxylamine takes place even when the potential is cathodic enough to reduce the hydroxylamine. The cell current and the yield of p-aminophenol are also increased at -0.10 to -0.20 V. Here, both stannic ion and nitrobenzene are reduced at the cathode, and more nitrobenzene is reduced by the stannous ion formed; this effect is equivalent to increasing the rate of mass transfer.

Catalysis [20] of the formation of ethylene from acetylene by chromium salts:

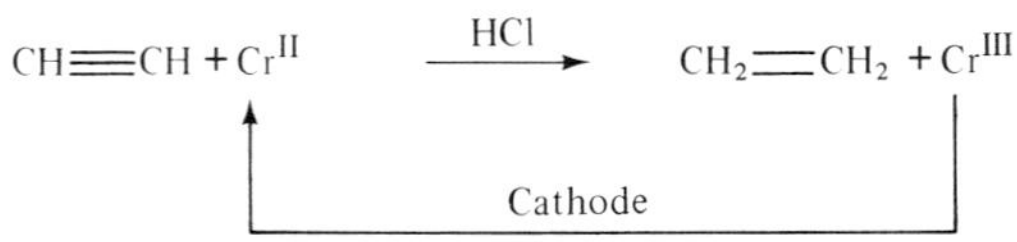

(8)

and of methylphenols from toluene [21] by copper salts:

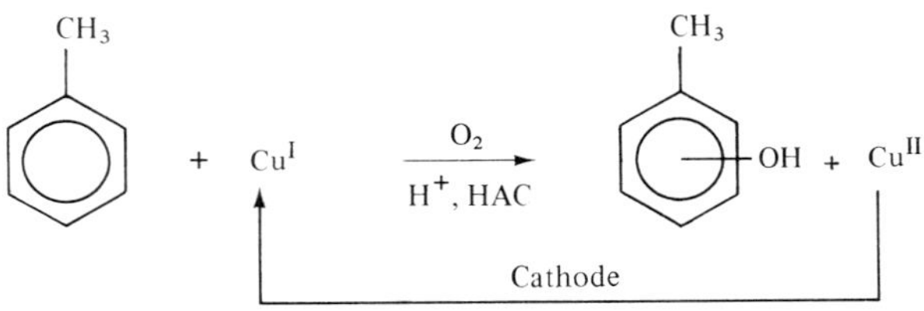

(9)

is presumably effective for similar reasons.

The anodic analog of reaction (6) would involve heterogeneous oxidation of an inorganic ion, which could then oxidize an organic species in solution. Metals with ions of two stable solution oxidation states are obvious candidates. Small amounts of cerium or vanadium salts lead [22] to high yields of anthraquinone in the anodic oxidation of anthracene in 40% H_2SO_4. Electrogenerated manganic salts (in stoichiometric quantities) oxidize [23] substituted toluenes to benzaldehydes in aqueous H_2SO_4. An in situ reoxidation of the manganous ion formed led to a rapid decrease in current efficiency; separation and external regeneration [24] were preferred. Kinetic and product analyses [25] of reactions of Mn^{III} and Ce^{IV} salts with aromatic hydrocarbons lend plausibility to the view that electron-transfer oxidation of aromatic hydrocarbons takes place in these systems [Eq. (10)]:

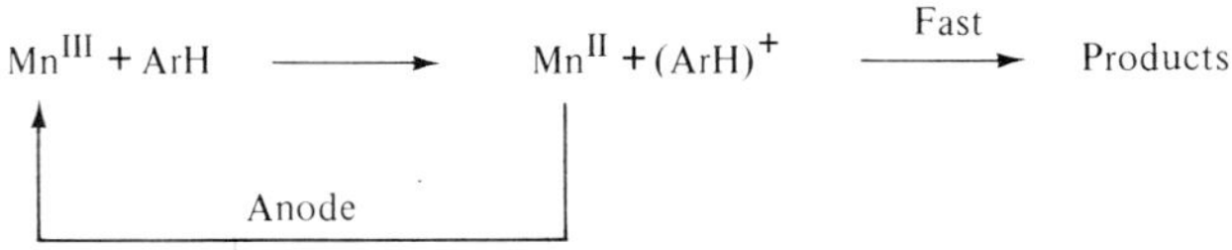

(10)

The need to conserve an expensive reagent has led to much industrial interest in oxidations of this sort. In the first three of the following examples the most recent work favors external regeneration.

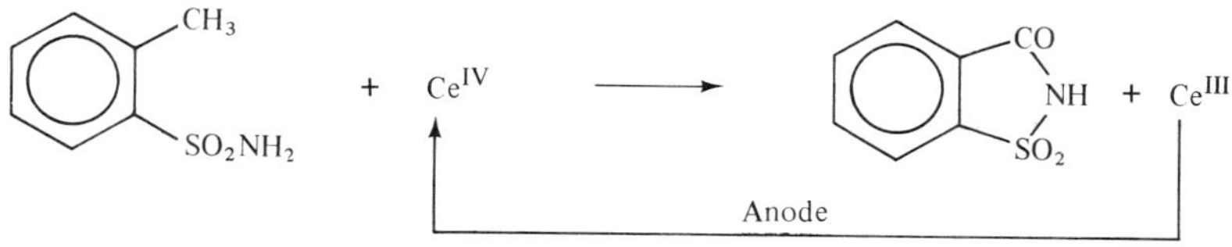

Ref. [26]

$$CH_2{=}CH_2 + Cl^- + Ce^{IV} \longrightarrow Cl(CH_2CH_2)_nCl + Ce^{III}$$

(Ce^{III} regenerated to Ce^{IV} at the anode)

Ref. [27]

$$\text{Starch} + HIO_4 \longrightarrow \text{Dialdehyde starch} + HIO_3$$

(HIO$_3$ regenerated to HIO$_4$ at the anode)

Ref. [28]

$$CH_2{=}CH_2 + Pd^{II} \longrightarrow CH_3CHO + Pd^{0}$$

(Pd^{0} regenerated to Pd^{II} at the anode)

Ref. [29]

Regenerative oxidation has also found application in fuel cell technology, where complete oxidation of small organic molecules is important. High oxidation states of molybdenum [30], rhenium [31], and osmium [32] adsorbed on platinum or platinum black oxidize alcohols, formaldehyde, and formic acid. In the example shown [Eq. (11)] no bulk oxidation by Mo^{VI} occurs [30]:

$$Mo^{VI}_{ads} + HCHO \xrightarrow[H_2SO_4]{\text{Pt black}} Mo^{V}_{ads} + CO_2 + H_2O \tag{11}$$

(Mo$^{V}_{ads}$ regenerated to Mo$^{VI}_{ads}$ on Pt black, $-\bar{e}$)

Other recent examples of the use of inorganic mediators include: oxidation of acetylene to oxalic acid by OsO_4 [63]; reduction of maleic and fumaric acid by V^{II} [64]; oxidations by ferricyanide [65]; and oxidations of amines on Pt via Cu^{III} [66].

IV. ORGANIC MEDIATORS

There are few examples of the use of an organic redox system to relay an electron between electrode and substrate. The technique may be useful in redox reactions of large, biologically active molecules for three reasons: (i) Their diffusion coefficients are small, so that mass transfer in direct electrolyses is inefficient; (ii) the adsorption that often accompanies electrode reactions of organic molecules in aqueous systems may lead to an undesired product; and (iii) the heterogeneous electron transfer may be slow.

An illustrative example of this kind [33] is the reduction of triphosphopyridine nucleotide (TPN^+), which is reduced enzymatically to the 1,4-dihydropyridine form (TPNH) in a two-electron reduction. Direct electrochemical reduction at a mercury cathode produces an enzymatically inactive product, probably the 4,4′-dimer [34], which is formed by coupling two one-electron reduction products; the $E_{1/2}$ of this reaction is -0.93 V (SCE) at pH 7.5.

The thermodynamic redox potential of the system TPN^+-TPNH at pH 7.5 is -0.59 V (SCE), so that methyl viologen dication (1,1′-dimethyl-4,4′-dipyridinium chloride), for example, which is reduced to the cation radical at -0.68 V, should energetically be able to transfer electrons to TPN^+ at a potential where a direct electrolytic reduction of TPN^+ is excluded.

The cation radical from methyl viologen was found to react too slowly with TPN^+ both in the presence or absence of the enzyme diaphorase; however, in the presence [33] of ferredoxin-TPN-reductase (FTR) the cation radical rapidly reduced TPN^+ to a product which was biologically active. It was found that TPNH could be produced electrochemically by the methyl viologen-FTR reduction of TPN^+ and that the electrochemical process did not interfere with the biochemical processes.

V. HYDROGEN

No electrochemical reaction is more familiar than the discharge of protons to give hydrogen gas. The electrogenerated reagent we consider here is the adsorbed hydrogen atom — the intermediate in this process.

Cathodic reduction of an organic molecule R in protic media may take place by electroreduction or electrohydrogenation:

Electroreduction: $$R + \bar{e} \rightleftarrows R^{\dot{-}} \quad (12)$$

$$R^{\dot{-}} + H_2O \text{ (or } H_3O^+) \rightarrow \text{Products} \quad (13)$$

Electrohydrogenation: $$H_3O^+ + \bar{e} \rightleftharpoons H^o_{ads} + H_2O \tag{14}$$

$$H^o_{ads} + H_3O^+ + \bar{e} \rightarrow H_2 + H_2O \tag{15}$$

$$H^o_{ads} + R \rightarrow \text{Products.} \tag{16}$$

The two mechanisms are not always easy to distinguish experimentally. If anion radicals $R^{\bullet -}$ are known to be formed and protons are not reduced at the potential used, electroreduction is clearly indicated. Chapter V described the electroreduction of hydrocarbons. Conversely, if reduction of R occurs only under conditions in which hydrogen is evolved, electrohydrogenation is likely. Very rapid addition of H^o_{ads} to R may make electrohydrogenation occur at potentials more anodic than the hydrogen discharge potential, however.

The pH dependence of the cathode potential during cathodic reduction is complex, but may aid the distinction. Two cases may be isolated. If proton discharge [Eq. (14)] is much faster than both hydrogen desorption [Eq. (15)] and addition to R [Eq. (16)], the potential dependence [35] for electrohydrogenation is 59 mV/pH unit. The potential should be independent of pH for electroreduction if the initial electron addition is irreversible and if there are no protonation reactions preceding the electron transfer [36]. A sharp change of dependence with reduction conditions for a given compound is strong evidence of a change of mechanism. 1-Hexene, cyclohexene, styrene oxide, and dibenzoyl peroxide undergo cathodic reduction [37] at platinum and palladium in aqueous solution. The pH dependence of the cathode potential is ~59 mV/pH unit for all four compounds at platinum, but for only the first two at palladium; pH is without effect on the potentials at palladium for styrene oxide and dibenzoyl peroxide, which have larger electron affinities than the hexenes and are probably electroreduced.

Electrohydrogenation is closely related to conventional catalytic hydrogenation. The key intermediates [35, 38] in both processes are adsorbed hydrogen atoms. They are produced by dissociation of hydrogen molecules in catalytic hydrogenation and by discharge of protons in electrochemical hydrogenation [Eq. (17)]:

$$\begin{array}{ccccc} H_2 & \rightleftharpoons & 2H^o_{ads} & \xrightarrow{R} & RH_2 \\ & & \updownarrow & & \\ & & 2H_3O^+ + 2\bar{e} & & \end{array} \tag{17}$$

The equilibrium between adsorbed hydrogen atoms and protons defines the redox potential for proton reduction at a given cathode. Cathodes of low overvoltage (Pt, Pd, Ni) are those at which that potential is close to the standard homogeneous value; cathodes of high overvoltage (Hg, Sn, Pb) are those at which proton reduction is inhibited.

Electrohydrogenation is more likely to occur at cathodes of low overvoltage and with substrates of low electron affinity. Cathodic reduction of protons in the presence of an unsaturated electroinactive substrate is not a sufficient condition for electrohydrogenation, however. No hydrogenation of acrylonitrile and β-bromopropionitrile occurs [39] when protons are reduced at mercury. Here, addition of hydrogen atoms to the substrate cannot compete with their rapid desorption [Eq. (15)]. Metals from which desorption of hydrogen atoms is slow are necessary for electrohydrogenation; the same metals are effective catalysts for conventional catalytic hydrogenation.

Like catalytic hydrogenation, electrohydrogenation is inhibited by poisons. Arsenic, for example, suppresses the cathodic reduction of aldehydes in aqueous solution at platinum, but not that of acetone, which may be electroreduced at the poisoned surface [40]. When increasing amounts of mercury are plated on a platinized platinum cathode, $[H_{ads}]$ falls and electrohydrogenation of acrylic acid becomes slower; with 35% mercury, $[H_{ads}] \sim 0$ and the only reduction possible is electroreduction, at a potential about a volt more cathodic [41]. Substrates whose reduction is only partly suppressed by poisons, or which give a pH/potential dependence intermediate between those mentioned above, are assumed [42] to be reduced by a mixed mechanism, electrohydrogenation and electroreduction taking place at different electrode sites. The two processes may even occur at opposite sides of a single palladium membrane separating two solutions [43].

Electrical measurements may be used [35, 38] to follow catalytic hydrogenation. A catalyst in an aqueous solution containing hydrogen will adopt a potential determined by the H^{o}_{ads}/H_3O^+ ratio. When a reducible substrate R is added, (H^{o}_{ads}) is decreased by addition to R, and a new equilibrium potential is adopted while R remains. Added species that are adsorbed, but not reduced, do not [38] affect the potential. The course of hydrogenation is reflected in the potential-time curve; inflections correspond to the end points of successive reduction steps. Loss of adsorbed hydrogen from palladium may be followed [44] by the decrease in electrical resistance too. The electrical measurements correlate well with rates of product formation and hydrogen uptake.

The close analogy with catalytic hydrogenation suggests that the extent of electrohydrogenation should be very sensitive to the nature of the interface. Many isolated empirical observations have been made, but few generalizations seem possible [45]. Alkynes may be selectively reduced at a suitable electrode since the product alkenes have slightly different rates of hydrogenation. For example, vinylacetylene gives [46] butadiene quantitatively at Cu/Pt, and $HOCH_2C{\equiv}CCH_2OH$ gives [47] $HOCH_2CH{=}CHCH_2OH$ (90%) at Cu/Ag. In aqueous alkali, adiponitrile gives [48] 1, 6-hexaminediamine at Cu/Fe and aminocapronitrile at Ni/Fe. Electroreduction of diphenylacetylene is known to occur at mercury in aqueous dioxane; no aliphatic unsaturation remains in the product [49] because stilbene is reduced at the potential used. Cathodic reduction at nickel in aqueous acid gives [50] cis-stilbene — the product of catalytic hydrogenation. At palladized platinum in acids and bases ethane is produced cathodically from ethylene at low current densities; ethane and hydrogen are formed when the rate of ethane production reaches a limit determined by the rate of diffusion of ethylene to the electrode [44].

The large surface area of electrodes developed for fuel cells is likely to stimulate interest in electrohydrogenation as a synthetic tool. Propene is reduced essentially quantitatively [51] at a platinum black/Teflon electrode of geometric area 20 cm^2 in acid or base with a throughput of 0.02 mole/h. Since hydrogen must be produced before reduction occurs, electrohydrogenation is likely. The fuel cell principle has been applied [52] in "electrogenerative hydrogenation." Power is generated by a cell that has hydrogen oxidation at the anode, alkene hydrogenation at the cathode, and a separator through which protons, but neither reactant, can pass. Its use as an analytical method for alkenes is suggested.

VI. NOVEL ELECTROGENERATED REAGENTS

Electrode reactions of inorganic molecules may involve species that are not readily accessible by conventional chemical means. In dipolar aprotic solvents they may be stable enough to be used in controlled chemical reactions; under protic conditions coupled chemical reactions are usually so fast that no reaction other than that with the solvent is possible.

The superoxide ion O_2^- provides an excellent example. In such aprotic solvents as dimethylformamide (DMF), dimethyl sulfoxide (DMSO), and acetonitrile (MeCN) containing tetraalkylammonium salts, oxygen is reduced [53] at -0.85 V vs. SCE at mercury, gold, and carbon cathodes. Current-voltage curves show that one electron is added to yield a product stable enough to diffuse from the electrode [Fig. 1(a)]. Controlled potential reduction produces [54] the ESR signal expected from the superoxide ion at low temperatures. The lack of an ESR signal at room temperature is due to relaxation and not to instability of the superoxide ion, the half-life of which is about 40 min at 20° for a 0.1 M solution. The slow decay [55] probably involves attack on the solvent. Solutions of superoxide at this concentration may be made by controlled potential reduction of a solution continuously saturated with oxygen with no more difficulty than a Grignard reagent.

Here, then, we have a novel electrogenerated inorganic reagent. Alkali metal superoxides are well known but are too insoluble to be of preparative use [56]. Some nucleophilic reactions of tetraalkylammonium superoxides are fast enough to be detected [55] by voltammetry. Figure 1(b) shows the effect of added n-butyl bromide during oxygen reduction at mercury in DMF/n-Bu_4NClO_4. Note that the reverse peak, corresponding to the oxidation

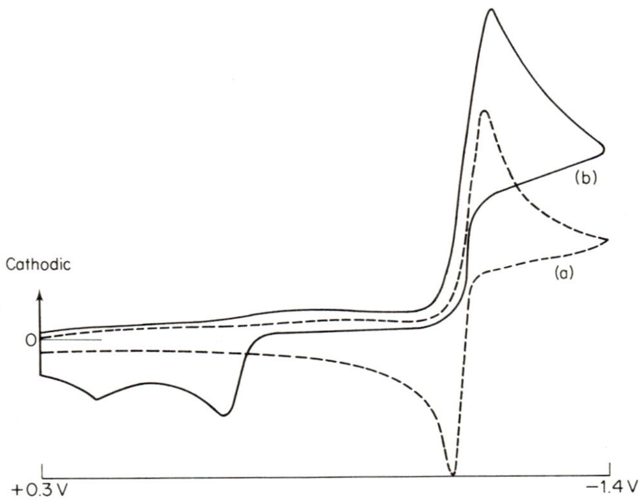

Fig. 1. Current-voltage curves for (a) reduction of oxygen (---) and (b) oxygen + n-BuBr (——) in DMF/n-Bu_4NClO_4 at a mercury drop. Linear voltage scan +0.3 to -1.4 V vs. SCE at 0.1 Hz. Reprinted from Ref. [55] by courtesy of the Chemical Society.

of O_2^- to O_2, disappears and that the charge used in the primary reduction step is increased. There are also two new anodic peaks at potentials corresponding to the oxidation of bromide at mercury. These changes are consistent with a nucleophilic substitution [Eq. (19)]:

$$O_2 + \bar{e} \rightleftarrows O_2^- \tag{18}$$

$$O_2^- + \text{n-BuBr} \rightarrow Bu^nO_2^o + Br^-. \tag{19}$$

Further reduction of n-BuO_2^o should occur at the electrode potential used to give the net electrode reaction [Eq. (20)]:

$$O_2 + \text{n-BuBr} + 2\bar{e} \rightarrow \text{n-}BuO_2^- + Br^-. \tag{20}$$

As required by Eq. (20), the oxygen wave height in polarography under aprotic conditions is doubled [55] by adding excess of alkyl halides and related compounds. Relative wave heights for solutions containing a range of reactants at one concentration give a rough measure of the relative rates of substitution. No change in the oxygen wave height would imply that substitution [Eq. (19)] is negligible during drop life; a doubled wave height shows substitution to be complete. The rates of substitution fell in the series n-BuBr > sec-BuBr ~ i-BuBr > t-BuBr and n-BuBr > n-BuOTs > n-BuCl. Both suggest an S_N2 mechanism provided that O_2^- is a powerful nucleophile.

With excess of alkyl halide a further nonelectrochemical reaction is likely [Eq. (21)]:

$$n\text{-BuO}_2^- + n\text{-BuBr} \rightarrow n\text{-Bu}_2\text{O}_2 + \text{Br}^- . \qquad (21)$$

Since dialkyl peroxides are reduced at much more cathodic potentials than oxygen, controlled potential reduction of oxygen-saturated solutions of alkyl halides in aprotic solvents is a useful synthetic method [57]. Di-n-butyl peroxide, for example, is produced in 80% yield and may be extracted in hexane. Since one halide ion per faraday is formed in the cathode reaction [Eqs. (20) and (21)] and consumed in the anode reaction [Eq. (22)], no depletion of supporting electrolyte occurs:

$$2\text{Br}^- - 2\bar{e} \rightarrow \text{Br}_2 . \qquad (22)$$

Nucleophilic addition [55] of superoxide to Michael acceptors is indicated by the epoxidation (30%) of cyclohexen-3-one but not cyclohexene, and by the formation of fluorenone (70%) and benzoate from benzylidene fluorene.

Dipolar aprotic solvents are particularly useful for base-catalyzed reactions. Superoxide acts as a base in elimination and autooxidation. Reaction [55] of tertiary butyl- and 2-phenylethyl halides with superoxide gives but-2-ene and styrene, respectively. Base-catalyzed elimination must occur [Eq. (23)]:

$$\begin{aligned} &(\text{CH}_3)_3\text{CCl} + \text{O}_2^- \rightarrow \text{HO}_2^\circ + (\text{CH}_3)_2\text{C{=}CH}_2 + \text{Cl}^- \\ &\qquad \text{HO}_2^\circ + \text{O}_2^- \rightarrow \text{HO}_2^- + \text{O}_2 . \end{aligned} \qquad (23)$$

Proton removal [58] from moderately acidic hydrocarbons can also take place. When superoxide is formed in DMF containing fluorene with continuous passage of oxygen, fluorenone is formed rapidly with a current efficiency of over 5000%. Following Russell's work [59] on base-catalyzed autooxidations, the mechanism must be that shown [Eqs. (18), (24), and (25)]:

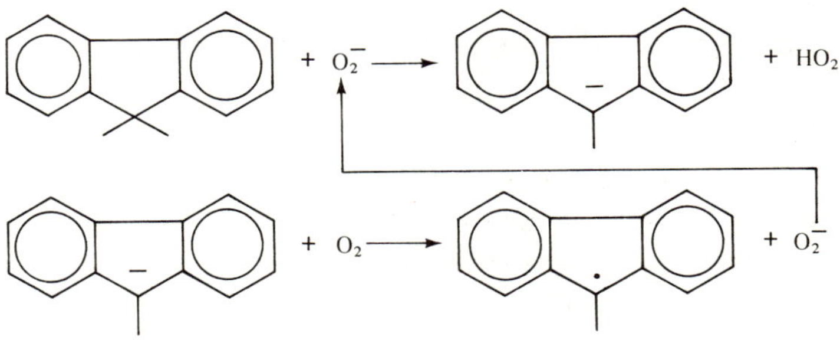

(24)

(25)

Fluorenone is formed very rapidly from the fluorenyl radical and oxygen [59].

The correlation (Chapter V, Sec. II.A) between half-wave potentials and Hückel molecular orbital energy coefficients predicts that the fluorenyl radical should be reduced (and hence the fluorenyl anion oxidized) at about -0.8 V vs. SCE. Since oxygen is reduced at close to that potential, the electron transfer [Eq. (25)] may take place. In similar reactions diphenylmethane and 9,10-dihydroanthracene are autooxidized with high current efficiencies [58].

Anions capable of the same classes of reactions as $O_2^{\dot{-}}$ should be produced by reduction of many inorganic molecules. In aprotic solvents the reduction [60] of phosphorus [Eq. (26)]

$$P_4 + \bar{e} \rightleftharpoons P_4^- \tag{26}$$

is complicated by disproportionation [Eq. (27)]

$$2P_4^- \rightleftharpoons P_4^{2-} + P. \tag{27}$$

It seems likely, however, that the preparative formation [61] of organophosphorous compounds by reduction of phosphorous and butyl bromide in DMF involves nucleophilic substitution [cf. Eq. (19)]. Again, the production of sulfur anions by reduction in aprotic solvents means that their reactions with organic substrates can be studied [62]. Reduction of S in the presence of acrylonitrile has been shown [67] to yield $S(CH_2CH_2CN)_2$.

NOTE

[1] It was also suggested that, if cathodic H_2 were combined with O_2 in a fuel cell arrangement, significant power savings could be achieved.

REFERENCES

[1]. W. Walisch and J. E. Dubois, Chem. Ber., 92, 1028 (1959); J. E. Dubois, Z. Elektrochem., 64, 143 (1960).

[2]. F. Olstowski and J. J. Newport, U.S. Pat. 3,033,767 (1962); W. R. Wolfe, U.S. Pat. 3,017,335 (1961).

[3]. R. Haddon, Brit. Pat. 140,831 (1917); K. P. McElroy, U.S. Pat. 1,253,615 (1918).

[4]. N. Ibl and A. Selvig, Chem.-Ing.-Tech., 42, 180 (1970).

[5]. J. W. Cornforth, F. Jones, K. H. Clifford, and D. T. Green, Society for Electrochemistry Meeting, Chester, England, April 1969.

[6]. S. K. Battacharyya, M. S. Muthanna, and A. D. Patankar, J. Sci. Ind. Res. India, 11B, 365 (1952); through CA, 47, 2064e (1953).

[7]. J. A. M. Le Duc, U.S. Pat. 3,288,692 (1966); through CA, 66, 43288 (1967).

[8]. J. A. M. Le Duc, U.S. Pat. 3,379,627 (1968); through CA, 69, 73488 (1968); W. Kroenig and P. Konrad, Ger. Pat. 1,258,856 (1968); 1,258,857 (1968); through CA, 68, 59423 (1968); J. A. M. Le Duc, P. Konrad and G. Troemel, Ger. Offen., 2,020,590, Nov. 12, 1970; CA, 74, 64199g (1971).

[9]. W. Kroenig and P. Konrad, Ger. Pat. 1,290,926 (1969); through CA, 70, 10636 (1969).

[10]. N. Clauson-Kaas, F. Limborg, and K. Glens, Acta Chem. Scand., 6, 531 (1952). N. Clauson-Kaas, F. Limborg, and P. Dietrich, ibid., 6, 545 (1952). N. Clauson-Kaas, ibid., 6, 556 (1952). J. T. Nielsen, N. Elming, and N. Clauson-Kaas, ibid., 9, 9 (1955). P. Nedenskov, N. Elming, J. T. Nielsen, and N. Clauson-Kaas, ibid., 9, 17 (1955).

[11]. N. Clauson-Kaas, N. Elming, and Z. Tyle, ibid., 9, 1 (1955). N. Clauson-Kaas and P. Nedenskov, ibid., 9, 14 (1955). N. Elming and N. Clauson-Kaas, ibid., 9, 23 (1955). N. Clauson-Kaas and P. Nedenskov, ibid., 9, 27 (1955).

[12]. N. Clauson-Kaas and F. Limborg, ibid., 6, 551 (1952). N. Clauson-Kaas and Z. Tyle, ibid., 6, 962 (1952).

[13]. K. E. Kolb and C. L. Wilson, Chem. Commun., 271 (1966).

[14]. N. Clauson-Kaas, Acta Chem. Scand., 6, 569 (1952).

[15]. H. S. Isbell, H.L . Frush, and J. Bates, Ind. Eng. Chem., 24, 375 (1932); C. G. Fink and D. B. Summers, Trans. Electrochem. Soc., 74, 625 (1938); I. A. Aurutskaya and M. Y. Fioshin, Zh. Prikl. Khim. (Leningrad), 42, 2294 (1969); through CA, 72, 27697f (1970).

[16]. Chem. Fabr. von Hayden, Ger. Pat. 85, 491 (1895).

[17]. M. Y. Fioshin, K. K. Babievskii, and N. A. Izgaryshev, Dokl. Akad. Nauk USSR, 104, 744 (1955).

[18]. R. A. C. Rennie, Ger. Pat. 1,804,726 (1969); through CA, 71, 35535 (1969).

[19]. H. C. Rance and J. M. Coulson, Electrochim. Acta, 14, 283 (1969).

[20]. W. Barker and D. Kennedy, Trans. Electrochem. Soc., 92, 391 (1947); G. Noyori, Japan. Pat. 4356 (1950); through CA, 47, 3327c (1953).

[21]. C. M. Wright, R. F. Heck, and W. Yokoyama, Electrochemical Society Meeting, New York, May 1969.

[22]. K. Shirai and K. Sugino, Denki Kagaku, 25, 284 (1957); through CA, 52, 8794f (1958).

[23]. R. Ramaswamy, M. S. V. Pathy, and H. V. K. Udupa, J. Electrochem. Soc., 110, 202 (1963).

[24]. S. Chidambaram, M. S. V. Pathy, and H. V. K. Udupa, J. Electrochem. Soc. India, 17, 95 (1968); through CA, 70, 25168a (1969).

[25]. P. J. Andrulis, M. J. S. Dewar, R. Dietz, and R. L. Hunt, J. Amer. Chem. Soc., 88, 5473 (1966); E. I. Heiba, R. M. Dessau, and W. J. Koehl, ibid., 91, 6830 (1969), and earlier papers in this series.

[26]. H. Löwe, Brit. Pat. 174, 913 (1922); through CA, 16, 1710 (1922); W. Klages, Swiss Pat. 78, 277 (1918); through CA, 13, 2374 (1919); K. Dürkes, Ger. Pat. 920, 186 (1954); through CA, 52, 19624 (1958); Ger. Pat. 1, 065, 834 (1959); through CA, 55, 6214 (1961).

[27]. A. F. MacLean, U.S. Pat. 3, 413, 203 (1968); through CA, 70, 43437 (1969).

[28]. C. L. Mehltretter and C. S. Wise, Ind. Eng. Chem., 51, 511 (1959); H. F. Conway and V. E. Sohns, ibid., 51, 637 (1959); V. F. Pfeifer, V. E. Sohns, H. F. Conway, E. B. Lancaster, S. Dabic, and E. L. Griffin, ibid., 52, 201 (1960).

[29]. D. L. Klass, U.S. Pat. 3, 147, 203 (1964); through CA, 61, 13195 (1964); F. Goodridge and C. J. H. King, Trans. Faraday Soc., 66, 2889 (1970).

[30]. J. A. Shropshire, J. Electrochem. Soc., 112, 465 (1965).

[31]. J. A. Shropshire, ibid., 114, 773 (1967).

[32]. R. Kalvoda, J. Electroanal. Chem. Interfacial Electrochem., 24, 53 (1970); A. Rashid, P. Straka, and R. Kalvoda, ibid., 29, 383 (1971).

[33]. R. J. Day, S. K. Kinsey, E. T. Seo, and H. P. Silverman, Electrochemical Society Meeting, New York, May 1969, Extended Abstracts, p. 357.

[34]. R. W. Burnett and A. L. Underwood, Biochem. 7, 3328 (1968).

[35]. F. Beck and H. Gerischer, Z. Elektrochem., 65, 505 (1961).

[36]. P. Zuman, in Progress in Physical Organic Chemistry (A. Streitwieser and R. W. Taft, eds.), Vol. 5, Wiley-Interscience, 1967, pp. 118-119.

[37]. A. M. Sokolskaya, S. M. Reshetnikov, and S. A. Ryanibina, Elektrokhim., 3, 1487 (1967).

[38]. F. Beck, Ber. Bunsenges. Phys. Chem., 69, 199 (1965).

[39]. L. G. Feoktistov and M. M. Gol'din, Elektrokhim., 4, 356 (1968).

[40]. T. M. Grishina, G. P. Khomchenko, and G. D. Vovchenko, Vestn. Mosk. Univ. Ser. II, 18, 52 (1963); through CA, 60, 14123c (1964).

[41]. M. E. Manzhelei and L. V. Voitenko, Russ. J. Phys. Chem., 34, 12 (1960).

[42]. I. I. Kulakova and A. V. Shashkina, Vestn. Mosk. Univ. Ser. II, 17, 36 (1962); through CA, 57, 7012e (1962); C. A. Martinyuk and A. I. Shlugin, Zh. Fiz. Khim., 32, 2073 (1958); through CA, 53, 6831f (1959).

[43]. M. von Stackelberg and P. Weber, Z. Elektrochem., 56, 806 (1952).

[44]. L. D. Burke, C. Kemball, and F. A. Lewis, Trans. Faraday Soc., 60, 913 (1964).

[45]. A. L. Klebanskii, L. D. Popov, and N. Y. Tsukerman, Zh. Obshch. Khim., 16, 2083 (1946); through CA, 42, 857h (1948); D. V. Sokolskii, Hydrogenation in Solution, Israel Program for Scientific Translation, Jerusalem, 1964, Chap. 2; G. A. Bogdanovskii, Sov. Electrochem., 5, 432 (1969).

[46]. J. Kato and M. Sakuma, Denki Kagaku, 25, 126, 331 (1957).

[47]. J. Kato and M. Sakuma, Japan. Pat. 8378 (1956); through CA, 52, 15311i (1958); J. Kato, M. Sakuma, and T. Yamada, J. Electrochem. Soc. Japan, 26, 43 (1958); through CA, 55, 16219i (1961).

[48]. A. P. Tomilov, S. L. Varshavskii, and Y. D. Smirnov, Russ. Pat. 132, 214 (1960); through CA, 55, 6214e (1961).

[49]. S. Wawzonek and D. Wearring, J. Amer. Chem. Soc., 81, 2067 (1959).

[50]. K. Campbell and E. Young, ibid., 65, 965 (1943).

[51]. H. J. Barber, J. Org. Chem., 35, 1489 (1969).

[52]. S. H. Langer and H. P. Landi, J. Amer. Chem. Soc., 85, 3043 (1963); 86, 4694 (1964); S. Yurchak, I. Feiz, and S. H. Langer, Abstracts, A.C.S. Meeting, Chicago, Ill., Sept. 1970, PETR 31.

[53]. D. L. Maricle and W. G. Hodgson, Anal. Chem., 31, 1562 (1965); M. E. Peover and B. S. White, Electrochim. Acta, 11, 1061 (1966); D. T. Sawyer and J. L. Roberts, J. Electroanal. Chem. Interfacial Electrochem., 12, 90 (1966).

[54]. W. Slough, Chem. Commun., 420 (1965).

[55]. R. Dietz, A. E. J. Forno, B. E. Larcombe, and M. E. Peover, J. Chem. Soc. (B), 816 (1970).

[56]. A. Le Berre and Y. Berguer, Bull. Soc. Chim. Fr., 2363, 2368 (1966).

[57]. R. Dietz and M. E. Peover, Brit. Pat. 1, 217, 509, Dec. 31, 1970; CA, 74, 70987d (1971).

[58]. R. Dietz, M. E. Peover, and P. Rothbaum, Chem.-Ing.-Tech., 42, 185 (1970).

[59]. G. A. Russell, E. G. Jantzen, A. G. Bemis, E. J. Geels, A. J. Moys, S. Mak, and E. T. Strom, Adv. Chem., 51, 112 (1965).

[60]. I. N. Brago and A. P. Tomilov, Sov. Electrochem., 4, 623 (1968).

[61]. M. I. Kabachnik, A. P. Tomilov, and S. L. Varshovskii, J. Gen. Chem. USSR, 36, 2052 (1966).

[62]. M. V. Merritt and D. T. Sawyer, Inorg. Chem., 9, 211 (1970).

[63]. J. S. Mayell, U.S. Pat. 3,555,084, Jan. 12, 1971; CA, 74, 60256a (1971).

[64]. E. Vrachnou-Astra, P. Sakellaridis, and D. Katakis, J. Amer. Chem. Soc., 92, 811 (1970).

[65]. A. Rashid and R. Kalvoda, J. Electroanal. Chem., 28, 245 (1970).

[66]. I. Fried and D. Meyerstein, ibid., 29, 429 (1971).

[67]. A. P. Tomilov, L. V. Kaabak, and S. L. Varshavskii, CA, 60, 11618a (1964).

[68]. N. L. Weinberg and A. K. Hoffman, Can. J. Chem., 49, 740 (1971).

Chapter XXV

USE OF THE SOLVATED ELECTRON

H. Lund

Department of Organic Chemistry
Aarhus University
Aarhus, Denmark

I. INTRODUCTION

In the electrolytic reductions discussed so far the electron transfer occurs between the electrode and an electron acceptor; the presence of the latter facilitates the release of the electron from the electrode. At sufficiently negative potentials, however, an electron may be released from the electrode although an acceptor is not present; if thus not caught by an acceptor, the electron may end up in a potential energy "trap" induced by its own polarization field. Such electrons are called "solvated electrons" [1, 2].

Reductions by means of electrolytically generated solvated electrons stand between clearly indirect electrolytic reactions, in which the electrons are transferred from the electrode to a species which then chemically reduces the substrate, and electrochemical reductions, in which the charge-transfer step involves a transfer of electrons from the electrode to the substrate without any intermediate electron carriers.

The distinction between the latter type of reactions and the reductions by means of solvated electrons may not be clear-cut since in some cases a layer of solvent molecules may be found between the electrode and the reducible molecule during the electron transfer. In this chapter there will be included only reductions in which the presence of solvated electrons directly has been shown by their UV spectrum [3] or EPR signal or where there are other indications that a direct electron transfer to the substrate is very unlikely.

A generally accepted explanation of the stability of solvated electrons in polar liquids is not at hand [4-6]. Two models for the solvated electrons have been used to interpret the properties of the solutions containing these loosely defined entities. In the "cavity model" the electron is trapped in a cavity created in the bulk solvent; orientation of solvent dipoles appears to be an important factor in such a stabilization. In the "expanded metal" model the electron occupies an "expanded orbital" defined by the hydrogen atoms of the first solvent sphere around the cation.

The mobility of the solvated electron exceeds that of ordinary ions in the same solvent, but not by a large factor. In liquid ammonia, for example, the mobility of the solvated electron [7] is about eight times that of the sodium ion, while in water [8] it is less than four times that of a sodium ion. The observed mobility is much smaller than would be expected for a free electron, but high enough to require an electron-jump mechanism rather than migration of the entire cavity.

Solutions containing solvated electrons may be produced in different ways, inter alia by dissolving metals in suitable polar solvents, e.g., ammonia, amines, amides, or by electrolysis in such solvents containing certain salts. Absorption and EPR spectroscopy have shown, at least for dilute solutions, that the properties of the solutions are strictly similar [9] whether these have been prepared by electrolysis or by dissolving metals.

The solvated electron may form an ion pair with the solvated cation, especially in solvents of low dielectric constants, in which each species retains its essential characteristics. At concentrations higher than about 10^{-2} M another kind of aggregate is formed which may be of the type $e^-M^+e^-$ (M^+) found as an ion pair. This difference in the nature of the solvated electron at high and low concentrations may sometimes result in different types of products obtained in such solutions. In solutions of low concentrations a Birch type of reduction is found, whereas a metallation may occur at higher concentrations [9].

The electrolytic generation of solvated electrons can be represented by

$$e^-_{cath} + HS \rightarrow e^-_{HS} \quad (1)$$

where HS is a polar solvent. The solvated electron can then react in different ways; it may react with the solvent and abstract a proton [Eq. (2)] or with the protonated solvent [Eq. (3)]. Two solvated electrons may react according to Eq. (4) or with the substrate A [Eq. (5)]. To obtain a high current efficiency in the reduction of the substrate conditions must be chosen so that (5) competes successfully with (2), (3), and (4) for the solvated electrons:

$$e^-_{HS} + HS \rightarrow \tfrac{1}{2}H_2 + HS + S^- \quad (2)$$

$$e^-_{HS} + H_2{}^+S \rightarrow \tfrac{1}{2}H_2 + 2HS \quad (3)$$

$$e^-_{HS} + e^-_{HS} \rightarrow H_2 + 2S^- \quad (4)$$

$$e^-_{HS} + A \rightarrow A^{\overset{\bullet}{-}} + HS. \quad (5)$$

The reaction between the solvated electron and a solvent molecule is generally slow in the absence of a proton donor. The stability of solvated electrons in ammonia is well known, and even their reaction with water in ammonia is slow [10]. The mechanism of the disappearance of solvated electrons in this medium has been found to be as follows:

$$H_2O + NH_3 \rightleftarrows NH_4{}^+ + OH^- \quad (6)$$

$$e^-_{NH_3} + NH_4{}^+ \rightarrow 2NH_3 + \tfrac{1}{2}H_2 \quad (7)$$

and, accordingly, the overall reaction rate can be decreased by reduction of the ammonium ion concentration. The lack of solvation energy of hydroxyl ions in ammonia makes (6) unfavorable; in pure water the higher concentrations of protons make the disappearance of solvated electrons much faster than in ammonia. Reductions via solvated electrons can thus not be important in the presence of an appreciable concentration of protonated solvent molecules.

Reaction (4) is most important at the electrode surface; this reaction may set the limit for the useful current density.

The radical $A^{\overset{\bullet}{-}}$ formed in (5) will abstract a proton from the solvent [Eq. (8)]; the resulting radical AH• is in general more easily attacked by the solvated electrons than the substrate A and the further reduction [Eq. (9)] will generally be fast. Proton abstraction [Eq. (10)] completes the sequence:

$$A^{\overset{\bullet}{-}} + HS \rightarrow AH\bullet + S^- \quad (8)$$

$$e^-_{HS} + AH\bullet \rightarrow AH^- \quad (9)$$

$$AH^- + HS \rightarrow AH_2 + S^-. \quad (10)$$

Reductions by means of solvated electrons would thus be expected to occur in polar solvents of low proton activity when a direct electron transfer to the substrate is not possible for energetic reasons. The standard potential of the hydrated electron [11, 12] has been calculated to be -2.68 V vs. NHE from the estimated free energy change of the reaction

$$\bar{e}_{aq} + H^+ \rightarrow \tfrac{1}{2}H_2 \qquad (11)$$

so that in the presence of a substrate which requires a less negative potential for its reduction solvated electrons are not formed to any appreciable extent; their concentration may be calculated by means of the Nernst equation which shows that the concentration of solvated electrons at potentials (in water) more anodic than about -2.4 V (SCE) is negligible.

Solvated electrons have been generated electrolytically in solvents such as ammonia [13, 14], methylamine [15-19], ethylamine [20], ethylenediamine [21-23], diethylene glycol dimethyl ether (Diglyme) [24], and hexamethylphosphotriamide (HMPA, hexametapol) [9] containing ethanol [25-27] as proton donor. The HMPA-ethanol medium is different from the others in that it is a proton-rich solution; reductions in this medium will be discussed separately.

As electrolyte lithium chloride has mostly been used. This is partly due to its higher solubility, compared to other alkali halides, in the solvents used, but to a certain extent is also due to the ease with which the lithium ion and the solvated radical ion from the substrate form an ion pair. Tetrabutylammonium salts have been used, whereas in some cases tetramethyl- and tetraethylammonium salts have been found [23, 28] unsuitable as supporting electrolytes for such reactions.

The anion may have a large effect on the current efficiency and product distribution [23]. Iodides are, at least in amine media, inferior to chlorides; this is probably caused by the catalyzing effect of the iodide ion on the reaction of the amine with lithium to give amide and hydrogen:

$$Li^+ \cdots\cdots e^- + RNH_2 \xrightarrow{(I^-)} RNHLi + \tfrac{1}{2}H_2.$$

The electrode material also affects both the current yield and the product distribution of reductions run in amine media. However, the cathode material is not critical as far as electron release is concerned since cathodes made from such different materials as carbon, platinum, aluminum, copper, and lead have served as electron-releasing electrodes.

II. REDUCTION IN PROTON-RICH MEDIUM

Hexamethylphosphotriamide is a polar, aprotic medium in which the solvated electrons have properties similar to those in amine media [29-32].

Addition of ethanol to HMPA makes the medium proton rich without destroying its ability to form solvated electrons of reasonable stability.

It has been suggested [26] that the hydrogen evolution found in ethanol on electrolysis is suppressed in a HMPA-ethanol (67 mole % ethanol) mixture due to a selective adsorption of HMPA on the electrode surface. In this layer of polar, aprotic HMPA molecules the reactions (2), (3), and (4) are slow. For (3) this is due to the low proton activity, and, as HMPA is a highly polar compound, (1) is favored.

The content of HMPA in the ethanol is of importance. The current efficiency in the reduction of benzene in ethanol at an aluminum cathode is about 0.01% [26] and 95% in HMPA-ethanol (67 mole % ethanol); in HMPA-ethanol (90 mole % ethanol) the current efficiency is 5% [33].

A. Reduction of Aromatic Compounds

Benzene has been reduced in ethanol-HMPA (66.6 mole % ethanol) at -2.4 ± 0.1 V (vs. Ag wire) at an aluminum cathode to give a mixture of 22.8% cyclohexadiene, 10.0% cyclohexene, and 67.2% cyclohexane with an overall current efficiency of 95% [26]. The reduction was carried out in the presence of a large excess of benzene.

In amine media (vide infra) the electrochemical reduction of benzene [15] produces only cyclohexadiene and cyclohexene, but no cyclohexane, showing that the reduction of an isolated double bond under these conditions is much slower than the reduction of the aromatic system. In ethanol-HMPA, however, cyclohexane is the major product even in the presence of an excess of benzene, suggesting that different factors are controlling the reaction in the two types of media.

The rate of reduction of benzene with lithium and ethanol in ammonia has been shown [34] to be proportional to the concentrations of benzene, metal (electrons), and proton donor. Assuming that the same rate law holds for the reduction of cyclohexene, a competitive reduction in which benzene and cyclohexene are exposed to the same electron and proton donor concentration gives, as a first approximation, the ratio of specific rate constants K_b/K_c for the reduction of benzene (K_b) and cyclohexene (K_c).

In ethylenediamine (EDA) the ratio K_b/K_c has been found to be 200, which readily explains why in amine solvents the reduction of benzene stops when isolated double bonds have been formed. In ethanol-HMPA the ratio K_b/K_c was only 1.4. It has been suggested [26] that this ratio is not really the ratio of the reaction rate constants but rather essentially a ratio of the corresponding mass-transfer coefficients. The high concentration of proton donor increases the rate of reaction to such an extent that it essentially becomes diffusion controlled in the reaction layer near the electrode surface, where the concentration of solvated electrons is high.

The suggestion of diffusion control has been substantiated by the findings that a diminution of the rate of reaction by decreasing the concentration of electrons (lowering of the current density), the concentration of protons (lower ethanol concentration), or the temperature results in an increase in the ratio K_b/K_c (Table 1) and a change in the product distribution.

TABLE 1

Competitive Reduction of Benzene and Cyclohexene in EtOH-HMPA [26]

Cathode material	Mole % EtOH	Temp, °C	Current density, A/cm^2	K_b/K_c	Benzene product distribution, %		
					Cyclo-hexadiene	Cyclo-hexene	Cyclo-hexane
Al	66.6	28	0.091	1.4	20.2	8.5	71.3
Al	5	28	0.091	8.2	81.3	16.4	2.3
Al	66.6	28	0.0036	2.5	53.1	10.1	36.8
Al	66.6	-2	0.0036	36.0	93.0	5.8	1.2
Pt	5	28	0.091	7.1	85.7	12.9	1.4

Reduction of tetralin in ethanol-HMPA [25] at -2.5 V (vs. Ag wire) gave results similar to those in the reduction of benzene: The major product is the fully reduced species decalin. Reduction of naphthalene [16, 27] took place at -2.4 ± 0.1 V (vs. Ag wire) under these conditions. It is noteworthy that the electrochemical reduction of naphthalene takes place about 1 V less negative [26] than that of benzene, and the fact that in ethanol-HMPA the reduction occurs at essentially the same potential strongly suggests that the potential-determining step is the release of electrons from the electrode to the solvent. It seems that HMPA adsorbed at the electrode prevents naphthalene from approaching the electrode surface closely enough for a direct electron transfer to occur between electrode and naphthalene.

B. Olefins

Olefins like cyclohexene are easily reduced by solvated electrons in ethanol-HMPA. Even sterically hindered alkenes such as 2, 3-dimethyl-2-butene, $(CH_3)_2C=C(CH_3)_2$ [26], are reduced at only a slightly lower rate than cyclohexene, and this kind of reduction is thus a very promising tool for such reactions.

Only a few types of compounds have up to now been subjected to electrolytic reductions in ethanol-HMPA, but quite a few difficultly reducible compounds are candidates for this kind of experiment. The usefulness of the method can, undoubtedly, be expanded.

III. REDUCTIONS IN AMINE MEDIA

Reductions of different substrates by means of solvated electrons have been investigated mostly in ethylenediamine [21-23] and methylamine [15-19, 36]. The electrolyses have been carried out both in divided and undivided cells and often the product distribution is different in the two cases. It has been suggested [36] that in the undivided cell the protons liberated during the anodic electrode process form amine hydrochloride which then acts as a proton donor in the cathodic process.

A. Aromatic Hydrocarbons

Reduction of benzene [21, 36] and alkylbenzenes [36] at platinum or carbon cathodes in an undivided cell in ethylenediamine (EDA) or methylamine (MA) gave 1, 4-dihydrobenzenes as the main product. In Table 2 the product distribution from reductions in MA is given.

TABLE 2

Electrolytic Reduction of Various Aromatic Hydrocarbons in Methylamine-Lithium Chloride Solutions without Cell Divider [36]

Substrate	Dihydro-benzenes, %	Cyclo-hexenes, %	Recovered substrate, %
Benzene	95	4	1
Toluene	95	5	—
Ethylbenzene	96	4	—
Cumene	81	6	13
t-Butylbenzene	75	5	20

More than 90% of the dihydroalkylbenzenes possessed the 2, 5-dihydro structure; no isomerization takes place under these conditions, and a compound such as 2, 5-dihydrocumene remains virtually unchanged after refluxing with sodium ethoxide in ethanol for 4 h [37]. The addition of a proton to the initially formed radical anion must thus take place ortho or meta to the alkyl group. In this connection it is interesting that the hyperfine splitting in the EPR spectrum of toluene radical anion [38] indicates about equal density of the unpaired electron at the ortho and meta positions, but little or none at the para position. Steric shielding of the ortho positions may favor an attack at the meta positions.

Bulky groups retard the reduction; this has also been found in other solvents [28].

The major product of the reduction of benzene [22, 36] and alkylbenzenes [36] in a divided cell or in a catholyte to which was added lithium alkylamide was cyclohexenes. Some results are given in Table 3. The cyclohexenes were

TABLE 3

Electrolytic Reduction of Various Hydrocarbons in Methylamine-Lithium Chloride Solutions with a Cell Divider [36]

Substrate	Alkyl-cyclohexene, %	Recovered substrate, %
Toluene	83	17
Ethylbenzene	89	11
Cumene	81	19
t-Butylbenzene	65	35

mixtures, with 1-alkylcyclohexene as the major product.

The cyclohexenes are probably formed via the 1,4-dihydrobenzenes, which are isomerized in the presence of lithium alkylamide to the conjugated dienes; these are reduced faster than the starting material by solvated electrons. Alkenes with isolated double bonds are not reduced under these conditions.

B. Acetylenes and Alkenes

Electrolytic reduction of dialkylacetylenes [16] in methylamine in an undivided cell at a platinum cathode yields predominantly trans-alkenes; cis-olefins which are stable under the reaction conditions are formed only as a minor component (2-8%); the reduction by solvated electrons thus differs in the stereochemistry of the product from the electrocatalytic hydrogenation of acetylenes [39].

Nonconjugated aromatic acetylenes are reduced similarly, whereas phenylacetylene and its derivatives are further reduced to the alkylbenzene

[16]. This is consistent with the finding [17, 18] that aromatic alkenes in which the double bond is conjugated with the benzene ring or which may be isomerized under the reaction conditions to such compounds are reduced to the alkylbenzenes.

Addition of ethanol to the methylamine retards the amide-ion-catalyzed isomerization and under such conditions solvated electrons attack the benzene ring of nonconjugated aromatic olefins rather than the alkene. Allylbenzene is thus reduced according to

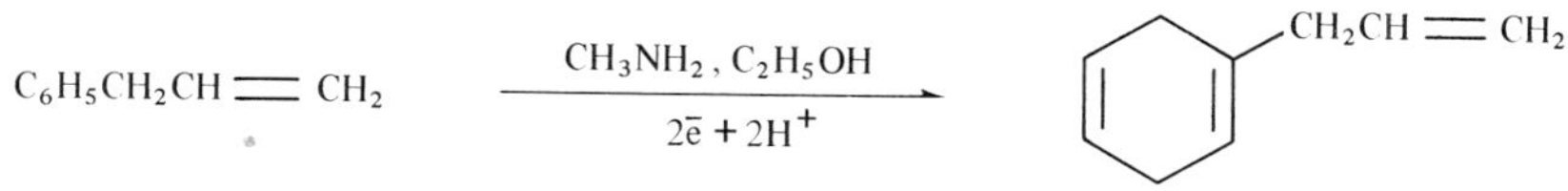

whereas n-propylbenzene is formed in the absence of ethanol, under which conditions propenylbenzene is an intermediate.

C. Amides

Amides of aliphatic acids are reduced to alcohols in methylamine [17, 19]. Primary, secondary, and tertiary amides are reduced in fair to good yields (Table 4). It seems to be the only general reaction for reduction of all aliphatic amides to alcohols.

In the presence of a hydrogen donor stronger than the amine, e.g., alcohol, a high yield of aldehyde or its reaction product with methylamine is obtained; no reduction product from an N-methylimine has been isolated. From these observations the following scheme was proposed [17, 19]:

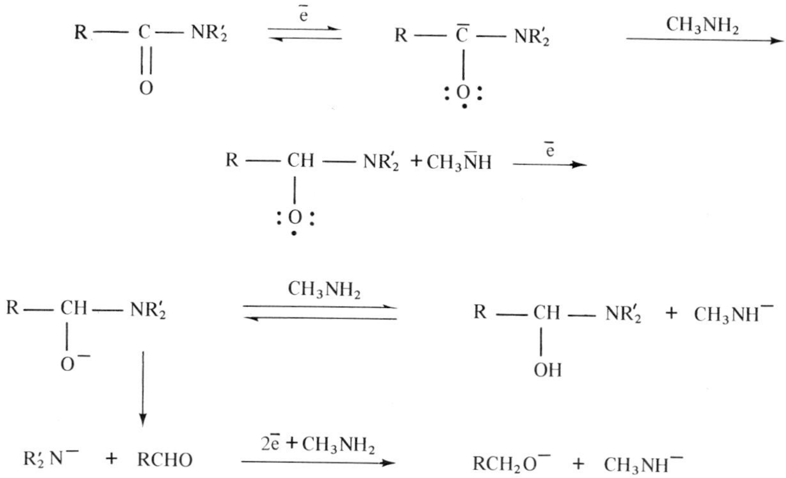

In the presence of a proton donor stronger than methylamine the gem-amino alcohol survives until the reaction mixture is worked up.

TABLE 4

Yields of Alcohol and Aldehyde in the Electrolytic Reduction of Aliphatic Amides at a Smooth Platinum Cathode in Methylamine Containing Lithium Chloride [17, 19]

No.	Amide	Moles	Methylamine, ml	Ethanol, g	Alcohol yield, %	Aldehyde yield, %
1	$CH_3(CH_2)_4CONH_2$	(0.05)	700	—	58	—
2	$CH_3(CH_2)_6CONH_2$	(0.01)	350	7	—	22
3	$CH_3(CH_2)_8CONH_2$	(0.05)	700	—	59	—
4	$CH_3(CH_2)_8CONH_2$	(0.005)	350	5	—	28
5	$CH_3(CH_2)_{12}CONH_2$	(0.02)	450	—	92	—
6	$CH_3(CH_2)_{14}CONH_2$	(0.01)	450	—	86	—
7	$CH_3(CH_2)_{16}CONH_2$	(0.01)	450	—	79	—
8	$CH_3(CH_2)_4CONHCH_3$	(0.05)	600	5	4	50
9	$CH_3(CH_2)_4CONHCH_3$	(0.05)	600	—	51	—
10	$CH_3(CH_2)_8CONHCH_3$	(0.02)	300	—	81	—
11	$CH_3(CH_2)_8CONHCH_3$	(0.01)	300	5	24	58
12	$CH_3(CH_2)_8CON(CH_3)_2$	(0.008)	350	—	93	—
13	$CH_3(CH_2)_8CON(CH_3)_2$	(0.01)	350	5	—	45
14	$CH_3(CH_2)_{14}CON(CH_3)_2$	(0.02)	450	—	97	—
15	$CH_3(CH_2)_8CONH_2$	(0.005)	350	5	50	4

Solvated electrons generated in nonelectrolytic ways have been shown to react with many organic substrates; sometimes the rate of reaction of the solvated electron with a series of compounds differs from what would be expected from considering the electron as a nucleophilic reagent [40]. Reduction of some of these substrates by electrolytically generated solvated electrons may not be able to compete with the direct electron-transfer reaction.

REFERENCES

[1]. E. V. Hart and M. Anbar, The Hydrated Electron, Wiley-Interscience, New York, 1970; E. J. Hart, Accounts Chem. Res., 2, 161 (1969).

[2]. U. Schindewolf, Angew. Chem., 80, 165 (1968); J. L. Dye, Accounts Chem. Res., 1, 306 (1968).

[3]. B. D. Michael, E. J. Hart, and K. H. Schmidt, J. Phys. Chem., 75, 2798 (1971).

[4]. R. A. Ogg, Phys. Rev., 69, 668 (1946); J. Jortner, J. Chem. Phys., 30, 839 (1959).

[5]. E. Becker, R. H. Lindquist, and B. J. Alder, J. Chem. Phys., 25, 971 (1956); E. Arnold and A. Patterson, ibid., 41, 3089, 3098 (1964).

[6]. K. Fucki, D.-F. Feng, and L. Kevan, J. Phys. Chem., 74, 1976 (1970); K. Fucki, D.-F. Feng, L. Kevan, and R. E. Christoffersen, ibid., 75, 2297 (1971); R. Catterall, Nature, 229, 10 (1971).

[7]. C. A. Kraus, J. Amer. Chem. Soc., 36, 864 (1914); ibid., 43, 749 (1921).

[8]. K. H. Schmidt and W. L. Buck, Science, 151, 70 (1966).

[9]. J. E. Dubois and G. Dodin, Tetrahedron Lett., 2325 (1969).

[10]. R. R. Dewald and R. V. Tsina, Chem. Commun., 647 (1967).

[11]. J. H. Baxendale, Radiat. Res. Suppl., 4, 139 (1964).

[12]. E. J. Hart, S. Gordon, and E. M. Frielden, J. Phys. Chem., 70, 150 (1966).

[13]. A. J. Birch, Nature, 158, 60 (1946).

[14]. R. K. Quinn and J. J. Lagowski, J. Phys. Chem., 72, 1374 (1968).

[15]. R. A. Benkeser and E. M. Kaiser, J. Amer. Chem. Soc., 85, 2858 (1963).

[16]. R. A. Benkeser and C. A. Tincher, J. Org. Chem., 33, 2727 (1968).

[17]. R. A. Benkeser, The Synthetic and Mechanistic Aspects of Electro-organic Chemistry, Symposium, Durham, N.C., October 1968, Preprints of Papers, p. 189.

[18]. R. A. Benkeser and S. J. Mels, J. Org. Chem., 34, 3970 (1969).

[19]. R. A. Benkeser, H. Watanabe, S. J. Mels, and M. A. Sabol, ibid., 35, 1210 (1970).

[20]. P. G. Arapagos and M. K. Scott, Tetrahedron Lett., 1975 (1968).

[21]. H. W. Sternberg, R. E. Markby, and I. Wender, J. Electrochem. Soc., 110, 425 (1963).

[22]. H. W. Sternberg, R. E. Markby, I. Wender, and D. M. Mohilner, ibid., 113, 1060 (1966).

[23]. H. W. Sternberg, C. L. Delle Donne, R. E. Markby, and I. Wender, Advances in Chemistry Series No. 55, American Chemical Society, Washington, D.C., p. 516 (1964).

[24]. A. Misono, T. Osa, T. Yamagishi, and T. Kodoma, J. Electrochem. Soc., 115, 266 (1968).

[25]. H. W. Sternberg, R. E. Markby, I. Wender, and D. M. Mohilner, J. Amer. Chem. Soc., 89, 186 (1967).

[26]. H. W. Sternberg, R. E. Markby, I. Wender, and D. M. Mohilner, ibid., 91, 4191 (1969).

[27]. T. Asahara, M. Seno, and H. Kaneko, Bull. Chem. Soc. Japan, 41, 2985 (1968).

[28]. T. Osa, T. Yamagishi, T. Kodama, and A. Misono, The Synthetic and Mechanistic Aspects of Electroorganic Chemistry, Symposium, Durham, N.C., October 1968, Preprints of Papers, p. 157.

[29]. T. Cuvigny, J. Normant, and H. Normant, Compt. Rend., 258, 3502 (1964).

[30]. G. Fraenkel, S. H. Ellis, and D. T. Dix, J. Amer. Chem. Soc., 87, 1406 (1965).

[31]. R. Catterall, L. P. Stodulski, and M. C. R. Symons, J. Chem. Soc. (A), 437 (1968).

[32]. J. M. Brooks and R. R. Dewald, J. Phys. Chem., 72, 2655 (1968).

[33]. H. W. Sternberg, private communication, 1970.

[34]. A. P. Krapcho and A. A. Bothner-By, J. Amer. Chem. Soc., 81, 3658 (1959).

[35]. A. Pullman, B. Pullman, and G. Berthier, Bull. Soc. Chim. Fr., 591 (1950).

[36]. R. A. Benkeser, E. M. Kaiser, and R. F. Lambert, J. Amer. Chem. Soc., 86, 5272 (1964).

[37]. R. A. Benkeser, M. L. Burrous, J. J. Hazdra, and E. M. Kaizer, J. Org. Chem., 28, 1094 (1963).

[38]. T. R. Tuttle and S. I. Weisman, J. Amer. Chem. Soc., 80, 5342 (1958).

[39]. N. N. Campbell and E. E. Young, ibid., 65, 965 (1943).

[40]. E. J. Hart, E. M. Fielden, and M. Anbar, J. Phys. Chem., 71, 3993 (1967).

Section H

(HETEROGENEOUS) ORGANIC ELECTROCHEMISTRY VIS-À-VIS HOMOGENEOUS ORGANIC CHEMISTRY

Chapter XXVI

COMPARISON BETWEEN ELECTROCHEMICAL AND ANALOGOUS CHEMICAL REACTIONS

L. Eberson

Division of Organic Chemistry
Chemical Center
University of Lund
Lund, Sweden

I. INTRODUCTION

A frequently used approach to the problem of elucidating mechanisms of organic electrode processes is to try to find a homogeneous process which as closely as possible duplicates the product distribution of the electrochemical reaction, if not quantitatively, at least qualitatively. If the results are reasonably similar, the wealth of mechanistic information available for reactions in homogeneous media is carefully extrapolated to the electrochemical process under study. It is the purpose of this chapter to examine the validity of this approach and to define its limitations.

A. What Is an "Analogous" Chemical Reaction?

From the structure of and distribution between the products obtained in an organic electrode reaction it is usually easy to make a reasonable estimate as to the nature of the intermediate(s) immediately preceding the products. This constitutes a natural starting point for the selection of one or several analogous chemical reactions for purpose of comparison. At a higher level of sophistication, one is interested in finding an analogous chemical reaction in which the intermediate(s) immediately preceding the products are formed in a sequence of reactions corresponding to the electrochemical reaction sequence.

It is immediately obvious that the last-mentioned alternative is difficult to realize in most cases, at least when direct electrochemical reactions are studied. Direct electrochemical processes proceed via an initial one-electron transfer between the substrate and the electrode, followed by a sequence of chemical and possibly electrochemical steps. This means that the homogeneous analog should also be of the electron-transfer type, and this is a relatively rarely encountered reaction type in organic chemistry [1]. In addition, the homogeneous solution chemistry of radical ions [2], in most cases the first intermediate in a one-electron transfer reaction, is not as well developed as the chemistry of other intermediates, viz., carbonium ions, neutral radicals, carbanions, and carbenes.

Thus, even if one should look for an analogous process which as closely as possible duplicates the electrochemical reaction sequence, one most often has to restrict oneself to a comparison with a reaction in which the intermediate immediately preceding the product(s) is the same as in the electrochemical reaction. Since many electrochemical reactions proceed via a blend of radical and ionic mechanisms one often has to find two reactions for comparison, one for the radical and one for the ionic component.

B. Experimental Problems in Comparing Chemical and Electrochemical Processes

Having selected one or several homogeneous reactions for comparison with a given electrode process, precautions must be taken that primary products in any of the reactions under study do not remain undetected due to further conversion to secondary products or that secondary products are not mistakenly considered to be primary ones. This problem is especially serious in electrochemical reactions since primary products are formed very close to the electrode and hence can react further immediately. One can guard oneself against this complication — at least to a certain extent — by analyzing for products at very low conversion [3]. Preferably, one should follow the product yield and distribution at several stages in the beginning of the reaction to see if there are any trends pointing toward a preferential formation or consumption of one or several primary products. This precaution should be taken with homogeneous process too since cases are known in which the product distribution changes considerably over the reaction period [3].

Even if this precaution is taken, it may not be possible to detect a primarily formed product in an electrochemical process. As an example, if the anodic oxidation of indan [4] in acetic acid-sodium acetate proceeds via the 1-indanyl cation [Eq. (1)] one would expect that the elimination product,

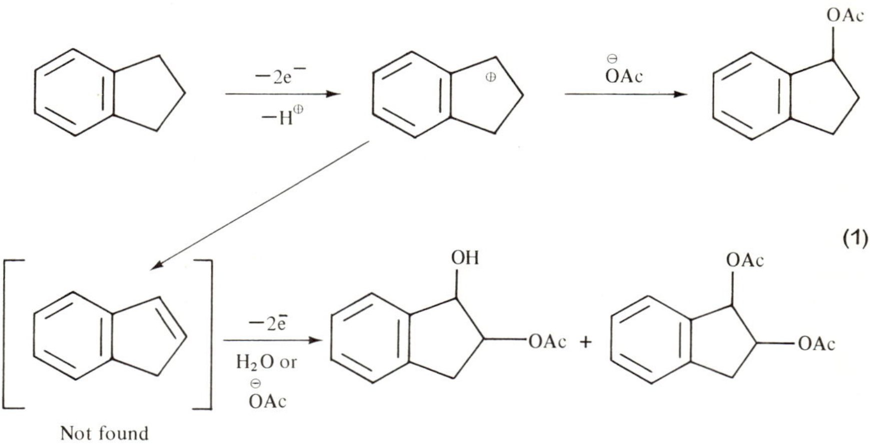

(1)

indene, should be formed in small amounts, analogously to the formation of 3-4% indene from the 1-indanyl cation generated by a solvolysis process in the same medium. However, analysis at different stages of the electrolytic process does not reveal any trace of indene. Either indene is not formed — thus making an important difference between electrochemically and solvolytically generated 1-indanyl cation — or it is formed but consumed in a secondary process without any of it leaking into the bulk of the solution. The latter alternative was found to be the correct one, since a small yield (9%) of the oxidation products of indene [Eq. (1)] was detected at all conversions in the product mixture formed in the oxidation of indan.

C. Concentration and Adsorption Effects

There are several reasons why one need not be overly concerned about finding a homogeneous process which exactly duplicates the electrochemical reaction sequence. Neglecting for a moment the possibility that the electrode reaction may take place as a surface reaction (i.e., via adsorbed intermediates), the concentrations of the generated intermediates will usually be vastly different from those prevailing during an analogous chemical reaction (provided that we are dealing with the normal set of very short-lived intermediates). Outside the electrode there will be a concentration gradient of intermediate(s) with a fairly high concentration next to the electrode; on the contrary, in a homogeneous process a uniform, very low steady-state concentration of intermediate(s) will be built up in the solution. This means that reaction between intermediates, such as coupling reactions between radicals, will have a higher probability in an electrochemical process, whereas in its homogeneous counterpart reactions between intermediates and components of the medium have the higher probability of occurring. It is apparent that this difference will be smaller as the lifetime of the intermediate(s) is increased; at the other extreme, we may and do use electrode reaction for generating fairly concentrated solutions of stable radical ions, radicals, etc.

Apart from this effect, purely due to the difference in conditions of concentration, adsorption may play an important role in the electrochemical process. The substrate molecules may be adsorbed at the electrode surface, making reactions between an unreacted substrate molecule and a radical ion or any other adsorbed species highly probable. Moreover, there are indications that the composition of the double layer may be different from that of the electrolyte bulk, making the order of reactivity between different species appear very peculiar indeed (cf. Chapter XXI, p. 784).

The role of adsorption in organic electrode processes and especially its possible influence on their stereochemistry are discussed to some extent in Chapter XXVII.

The Kolbe reaction provides a reasonably clear-cut example of the concentration effect referred to above, even if it cannot be excluded that adsorption effects are also involved. The anodic oxidation of propionate ion in anhydrous propionic acid at 100° has been compared with the decomposition of propionyl peroxide [Eq. (2)] under as closely similar conditions as possible with the results shown in Table 1 [5]. Here one wants to compare the fate of electrochemically and chemically generated radicals ($C_2H_5COO\cdot$ and $C_2H_5\cdot$). Only products formed via the radical pathway are included in the table (remember that the Kolbe process has a cationic pathway too; see Chapter XIII).

Examination of the yields shown in Table 1 [see also Eq. (2)] shows that reaction between radicals is favored in the electrolytic process. Thus, the

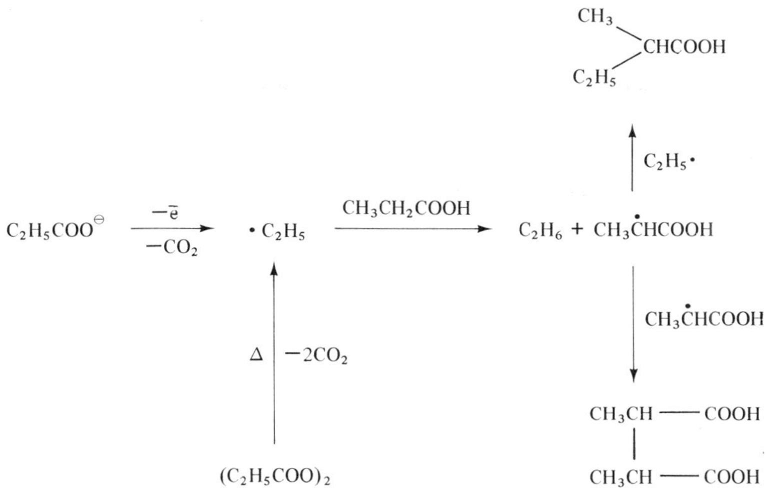

(2)

TABLE 1

Products Formed by Decomposition of Propionyl Peroxide or by Electrolysis of Propionate in Anhydrous Propionic Acid [5]

Product	Composition of product from the decomposition of 1 mole of $(C_2H_5COO)_2$	Composition of product from electrolysis of 2 moles of $C_2H_5COO^-$
CO_2	61.6 [a]	63.4 [b]
C_2H_4	3.0 [a]	5.6 [b]
C_2H_6 (from disproportionation [c])	3.0 [a]	5.6 [b]
C_2H_6 (from attack on solvent [d])	25.1 [a]	8.4 [b]
C_4H_{10}	6.9	15.8
Methylethylacetic acid	0.038 mole	0.002-0.014 mole
2, 3-Dimethylsuccinic acid	0.106 mole	0.00015-0.002 mole

[a] Percentage of total gaseous products.

[b] Percentage of anode gases.

[c] Estimated from the amount of ethylene formed.

[d] Residual ethane after subtraction of the yield of ethane formed via disproportionation.

coupling product, butane, is formed in higher yield in the electrode process than in the homogeneous one. Disproportionation (formation of ethane and ethylene) also occurs to a higher degree in the electrochemical process; on the other hand the proportion of ethane that can be traced to the reaction between ethyl radicals and solvent is higher in the propionyl peroxide decomposition. Moreover, the fate of the $CH_3\dot{C}HCOOH$ radical formed by attack of an ethyl radical on propionic acid is different in the two cases. In the electrochemical process, this radical is formed near the anode where there is a high concentration of ethyl radicals; hence, methylethylacetic acid is the predominant product from this intermediate. On the other hand, in propionyl peroxide decomposition the steady-state concentration of ethyl radicals is extremely low, thus allowing the $CH_3\dot{C}HCOOH$ enough lifetime to undergo a self-coupling reaction forming 2, 3-dimethylsuccinic acid.

D. Comparison with Analogous Heterogeneous Processes

Before we proceed with the discussion it is pertinent to consider briefly comparisons between electrode reactions and analogous heterogeneous reactions. Such comparisons have frequently been made, at least for cathodic reactions for which a number of analogs are available in the various dissolving metal reductions. However, dissolving metal reductions are really nothing but electrochemical processes, proceeding in very much the same way as cathodic reductions at a metal electrode, e.g., via direct electron transfer to substrate at the metal surface, via solvated electrons, or via addition of hydrogen (at a catalytic electrode). Hence, the mechanistic problems encountered in the elucidation of a dissolving metal reduction are the same as in cathodic reductions, and comparisons, although very interesting from other points of view, cannot be of much help unless one of the mechanisms is completely known.

For anodic processes, comparisons with related heterogeneous processes are difficult due even to the lack of suitable reactions with which to compare them. The best candidate, oxidation by certain metal oxides [6], e.g., manganese dioxide [7] or nickel peroxide [8], does not usually bear any close resemblance to anodic reactions with respect to products.

II. ANODIC PROCESSES

An anodic process takes place via successive electron transfers from the substrate to the electrode, in most cases with intervening and/or subsequent reactions with nucleophiles and/or bases in the medium. In order to find suitable homogeneous reactions for comparison one might consider the following reaction types:

(1) oxidation reactions by electron-transfer oxidants, e.g., metal ions or ion radicals;

(2) other oxidation processes, which produce similar product mixtures as in the anodic reaction for comparison; and

(3) reactions which are not oxidation processes, but nevertheless generate the same type of intermediate(s) as in the anodic process, e.g., neutral radicals by homolytic decomposition of suitable starting materials and carbonium ions by solvolysis, deamination, and deoxidation.

A. Oxidation by Electron Transfer

Metal ion oxidations are the most promising candidates for comparison with anodic processes. As an example, it has been shown that oxidation of alkylaromatic compounds with manganese(III) [9] and cobalt(III) acetate [10] most likely proceeds via a mechanism analogous to the electrochemical ECE mechanism [Eqs. (3)-(5)]. The possibility of changing the oxidation potential of a metal ion by changing the nature of the ligands around it (either in the

$$ArCH_3 + Co(III) \rightarrow ArCH_3^{\dot{+}} + Co(II) \quad (3)$$

$$ArCH_3^{\dot{+}} \xrightarrow{Base} ArCH_2^{\bullet} \quad (4)$$

$$ArCH_2^{\bullet} + Co(III) \rightarrow ArCH_2^{+} + Co(II) \quad (5)$$

oxidized or reduced form) gives a further similarity to the electrode process, since this is in effect analogous to a change of anode potential. As an example of this principle, the rate constant for oxidation of triethylamine with a series of ferric ion complexes (a one-electron transfer reaction) as well as some other oxidants has been found to be well correlated with the standard potential of the corresponding redox couple [11].

Lead tetraacetate oxidation has sometimes been used for simulating the cationic pathway of the Kolbe reaction (Chapter XIII). Comparisons between these processes show many similarities [11a], although it has been noted that the stereospecific ring opening of the cyclopropane ring in the anodic decarboxylation of 2-phenylcyclopropanecarboxylic acid (Chapter XIII) does not take place upon lead tetraacetate decarboxylation [11b]. Here the reaction is nonstereospecific.

Oxidation by ion radicals, most commonly cation radicals, has been relatively little studied up till now, but holds great promise for consideration as analogs of anodic processes [2a-2c]. As an example, oxidation of 1,1-bis-(p-dimethylaminophenyl)ethylene [2a] with the stable cation radical, tris-(p-bromophenyl)aminium perchlorate, gives a new cation radical which dimerizes to a dication. Treatment of this stable dication with sodium hydroxide gives 1,1,4,4-tetrakis-(p-dimethylaminophenyl)-1,3-butadiene [Eqs. (6)-(8)]. This interesting reaction sequence is the same as that proposed to account for the additive and eliminative dimerization reactions which occur upon the anodic oxidation of certain olefins at the carbon anode in the presence of nucleophiles [12]. Another example of oxidation by ion radicals is the

$$R_2C{=}CH_2 + R'_3N^{\dot{+}} \rightleftharpoons R_2\overset{\oplus}{C}{-}\dot{C}H_2 + R'_3N: \tag{6}$$

$$2R_2\overset{\oplus}{C}{=}\dot{C}H_2 \longrightarrow R_2\overset{\oplus}{C}{-}CH_2CH_2{-}\overset{\oplus}{C}R_2 \tag{7}$$

$$R_2\overset{\oplus}{C}{-}CH_2CH_2{-}\overset{\oplus}{C}R_2 \xrightarrow{OH^{\ominus}} R_2C{=}CH{-}CH{=}CR_2 \tag{8}$$

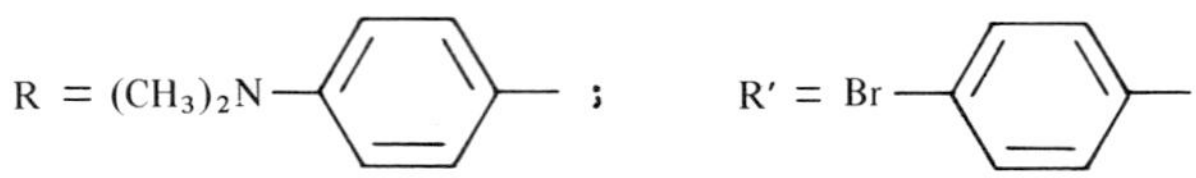

oxidation of carboxylate ions by sulfate anion radicals [13], produced by thermal decomposition of persulfate ion [Eqs. (9) and (10)]. This reaction

$$H_2O + S_2O_8^{2-} \xrightarrow{\Delta} SO_4^{-\cdot} + HSO_4^{-} + \dot{O}H \tag{9}$$

$$RCOO^{\ominus} + SO_4^{\dot{-}} \rightarrow R^{\cdot} + CO_2 + SO_4^{2-} \tag{10}$$

very closely duplicates the radical pathway of the corresponding Kolbe electrolyses (but not the cationic pathway).

Oxidation by quinones, commonly assumed to proceed via a hydride-transfer mechanism [14], may in some cases involve oxidation by electron transfer. Thus, 2, 3-dichloro-5, 6-dicyanobenzoquinone oxidizes 1, 1-diphenylethylene in methanolic solution to produce an adduct [15], most likely via an electron-transfer mechanism through an intermediate charge-transfer complex (a deep-brown color develops immediately after mixing the reagents and eventually fades away):

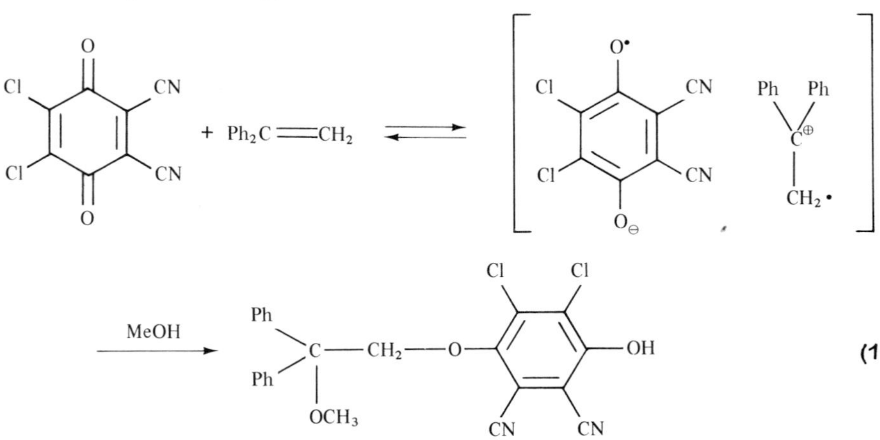

(11)

A similar mechanism has been proposed for the nucleophilic substitution that occurs upon acidification of charge-transfer complexes of pentacyclic aromatic hydrocarbons and tetrachloro-o-benzoquinone with hydrochloric or hydrobromic acid [15a]. The formation of a charge-transfer complex is in a certain sense analogous to the formation of an electrode-substrate complex by adsorption (Chapter XXVII) and, hence, is a partial duplication of a feature of the electrode process which in other reaction types is absent.

B. Other Oxidation Processes

Many metal ion oxidations are not or are only partly of the electron-transfer type but nevertheless produce results very similar to those obtained in anodic oxidations. Lead tetraacetate oxidation, although commonly assumed to react via a radical chain mechanism, can simulate most anodic acetoxylation processes, at least in a qualitative sense [16]. The oxygenation reaction for aromatic compounds based on reaction with diisopropyl peroxydicarbonate-cupric chloride (producing aryl isopropyl carbonates [17]) also gives results closely similar to those obtained by anodic acetoxylation, as for example can be seen from the product distributions in Table 2. Since in this particular case the mechanisms of the two types of processes have independently been established beyond reasonable doubt to be dissimilar, there is reason to warn against too far-reaching conclusions based on

TABLE 2

Comparison between Product Distribution from Anodic Acetoxylation and Oxygenation by Diisopropyl Peroxydicarbonate-Cupric Chloride of Aromatic Compounds

Substrate	Anodic acetoxylation [3]			Oxygenation by diisopropyl peroxydicarbonate-$CuCl_2$ [17]		
	o	m	p	o	m	p
Toluene	43	11	46	57	15	28
Cumene	44	17	39	48	16	36
Biphenyl	31	1	68	47	4	49
Naphthalene	96(α)	4(β)		92(α)	8(β)	
Fluorobenzene	34	8	58	33	22	45
Chlorobenzene	37	6	58	54	13	33
Benzotrifluoride		No reaction		47	51	2

comparison with analogous homogeneous processes. The anodic acetoxylation reaction is an ECE process (Chapter XXII), whereas oxygenation by diisopropyl peroxydicarbonate-cupric chloride takes place via a mechanism which in electrochemical terminology would be called a CCE process [Eqs. (12)-(14)]:

$$(i-PrOCO_2)_2 \longrightarrow 2i-PrOCO_2\cdot \tag{12}$$

$$i-PrOCO_2\cdot + ArH \longrightarrow \dot{Ar}\begin{matrix} \diagup H \\ \diagdown OCOOPr-i \end{matrix} \tag{13}$$

$$\dot{Ar}\begin{matrix} \diagup H \\ \diagdown OCOOPr-i \end{matrix} + CuCl_2 \longrightarrow \overset{\oplus}{Ar}\begin{matrix} \diagup H \\ \diagdown OCOOPr-i \end{matrix} \tag{14}$$

Another type of oxidation process is photooxidation, which in some cases has been compared with electrolytic processes. An example is cyanation of aromatic compounds which occurs electrochemically [18] or photochemically [19] in methanol-sodium cyanide solution. These reactions exhibit strikingly similar isomer distributions in aromatic systems, distinctly different from other types of direct aromatic cyanations [20].

C. Nonoxidative Processes

Such reactions are generally used for the generation of intermediates of the same type as that immediately preceding products in the electrochemical reaction; hence, all reactions producing neutral radicals or carbonium ions are of potential interest in this connection, provided they take place under as nearly identical conditions (solvent, added compounds, temperature) as possible. One should guard against possible complications due to the occurrence of cage processes (in certain types of radical-generating reactions) and ion-pairing phenomena (in certain types of solvolysis reactions).

III. CATHODIC PROCESSES

A cathodic process takes place via successive electron transfers from the electrode to the substrate, in most cases with intervening and/or subsequent reactions with electrophiles in the medium. The following types of homogeneous reactions should be considered for comparison:

(1) reduction processes by electron-transfer reductants, e.g., ion radicals and solvated electrons (for the use of dissolving metal reductions, see Chapter XXIII);

(2) other reduction processes which produce product mixtures similar to those of the cathodic reduction under comparison with, e.g., photoreductions; and

(3) reactions which are not reduction processes, but which generate the same type of intermediates as in the anodic process, e.g., neutral radicals, carbanions, and carbenes.

A. Reduction by Electron Transfer

Homogeneous electron-transfer reductions generally involve reduction by an anion radical [2h] or a dianion [21]. Thus, sodium naphthalenide reacts with alkyl halides [22] according to a mechanism which is the same as that proposed for cathodic halide reduction, i.e., the first step is an electron transfer to the halide with formation of an anion radical, which subsequently undergoes a fast cleavage reaction to form a neutral radical and halide ion. In a second step, the neutral radical is reduced to a carbanion, which is protonated by solvent.

Sodium naphthalenide [23] and cathodic [24] reduction of alkyl toluenesulfonates are likewise closely similar, the result in both cases being a cleavage of the sulfur-oxygen bond with regeneration of the corresponding alcohols.

Heterocyclic anion radicals might also be useful for simulating electrode processes [25].

B. Other Reduction Processes

Homogeneous reductions proceeding according to mechanisms other than direct electron transfer generally do not resemble cathodic processes. Certain photoreductions, however, have been used for comparison with electrochemical processes; a particularly striking and revealing case is the photochemical and electrochemical pinacolization of aromatic ketones, where the stereochemistry is almost identical in the two cases. In combination with other experimental evidence, this similarity strongly suggests that in the cathodic process the coupling between radicals must take place in the solution and not at the electrode surface (Chapter XXVII).

C. Nonreductive Processes

Homogeneous reactions producing neutral radicals or carbanions are of course of great interest for purposes of comparison with cathodic reactions. Again, one has to be careful of possible complications, e.g., with respect to the problem of the nature of the carbanion-counterion interaction (ion pairing or covalent bonding).

IV. CONCLUSIONS

A. Mechanistic Comparisons between Electrochemical and Chemical Reactions

Comparisons with homogeneous processes can help considerably in the elucidation of the mechanism of an electrolytic reaction. If possible, one should select several homogeneous reactions for comparison, preferably of different mechanistic types (say, one of the electron-transfer type and one of the nonoxidative or nonreductive type). If product mixtures are qualitatively similar, there is good reason to believe that mechanisms are similar. A quantitative agreement between product distributions may indicate that at least some intermediates are fairly long-lived so that the follow-up reactions of the electrochemical reaction take part in the bulk of the electrolyte. For reactions taking place through very short-lived intermediates such an agreement is not to be expected for reasons outlined above.

B. Synthetic Uses of Electrochemical vs. Chemical Reactions

In this chapter very little has been said about comparisons between electrochemical and chemical synthesis methods, simply because one of the main aims of the whole book is to show that electrochemical methods of synthesis are in many cases superior to their chemical counterparts and in many cases lead to products not easily accessible via other routes. In his classical text, Fichter wrote two sentences that do not seem to be in accord with at least the latter statement: "Man kann den allgemeinen Grundsatz aufstellen, dass alle elektrochemischen Oxydationen und Reduktionen auch auf rein chemischem Wege durchführbar sein müssen, und umgekehrt," and "Trotzdem muss an dem Grundsatz festgehalten werden, dass die elektrochemischen Reaktionen nicht wesensverschieden sind von den rein chemischen Reaktionen." However, this was written at a time when very little was known about mechanisms of electrode processes, and other processes too for that matter, and we now have many examples which show that electrochemical methods provide a host of synthetically useful methods that cannot at all or at least not easily be duplicated by chemical oxidation or reduction methods. This appears to be especially true for coupling reactions which are strongly favored by the very nature of the electrode process. Also, selective reductions, or oxidations, of certain functional groups in the presence of others became possible electrochemically through the use of controlled-potential conditions [26].

REFERENCES

[1]. K. A. Bilevich and O. Yu. Okhlobystin, Russ. Chem. Rev., 37, 954 (1968).

[2]. For reviews and articles containing leading references, see: (a) C. E. H. Bawn, F. A. Bell, and A. Ledwith, Chem. Commun., 599 (1968); cf. also F. Effenberger and O. Gerlach, Tetrahedron Lett., 1669 (1970). (b) F. A. Bell, A. Ledwith, and D. Sherrington, J. Chem. Soc. (C), 2719 (1969). (c) P. Beresford and A. Ledwith, Chem. Commun., 15 (1970). (d) Y. Murata and H. J. Shine, J. Org. Chem., 34, 3368 (1969). (e) H. J. Shine and Y. Murata, J. Amer. Chem. Soc., 91, 1872 (1969). (f) E. T. Kaiser and L. Kevan (eds.), Radical Ions, Wiley-Interscience, New York, 1969. (g) J. F. Garst and J. T. Barbas, Tetrahedron Lett., 3125 (1969). (h) M. Szwarc, Progr. Phys. Org. Chem., 6, 323 (1968).

[3]. L. Eberson, J. Amer. Chem. Soc., 89, 4669 (1967).

[4]. E. Bernhardsson, L. Eberson, K. Nyberg, and B. Rietz, Acta Chem. Scand., 25, 1224 (1971).

[5]. S. Goldschmidt, W. Leicher, and H. Haas, Ann. Chem., 577, 153 (1952); see also Ref. [16].

[6]. W. F. Pickering, Rev. Pure Appl. Chem., 16, 185 (1966).

[7]. For leading references, see I. M. Goldman, J. Org. Chem., 34, 3289 (1969).

[8]. For leading references, see R. Konaka, S. Terabe, and K. Kuruma, ibid., 34, 1334 (1969).

[9]. P. Andrulis, Jr., M. J. S. Dewar, R. Dietz, and R. L. Hunt, J. Amer. Chem. Soc., 88, 5473 (1966).

[10]. E. I. Heiba, R. M. Dessau, and W. J. Koehl, Jr., ibid., 91, 6830 (1969).

[11]. L. A. Hull, G. T. Davis, and D. H. Rosenblatt, ibid., 91, 6247 (1969).

[11a]. K. Torssell, Ark. Kemi, 31, 401 (1969).

[11b]. T. Shono, I. Nishiguchi, and R. Oda, Tetrahedron Lett., 373 (1970).

[12]. H. Schäfer and E. Steckhan, Angew. Chem., 81, 532 (1969).

[13]. L. Eberson, S. Gränse, and B. Olofsson, Acta Chem. Scand., 22, 2462 (1968).

[14]. D. Walker and J. D. Hiebert, Chem. Rev., 67, 153 (1967).

[15]. H.-D. Becker, J. Org. Chem., 34, 1203 (1969).

[15a]. M. Wilk and U. Hoppe, Liebigs Ann. Chem., 727, 81 (1969).

[16]. L. Eberson, in The Chemistry of Carboxylic Acids and Esters (S. Patai, ed.), Wiley-Interscience, New York, 1969, Chap. 2.

[17]. P. Kovacic, C. G. Reid, and M. E. Kurz, J. Org. Chem., 34, 3303 (1969), and earlier papers of this series.

[18]. S. Tsutsumi and K. Koyama, Discuss. Faraday Soc., 45, 247 (1968); L. Eberson and S. Nilsson, ibid., 45, 242 (1968).

[19]. R. L. Letsinger and J. H. McCain, J. Amer. Chem. Soc., 88, 2884 (1966).

[20]. L. Eberson and S. Nilsson, unpublished results.

[21]. J. Jagur-Grodzinski and M. Szwarc, J. Amer. Chem. Soc., 91, 7594 (1969).

[22]. For leading references, see Ref. [2g].

[23]. W. D. Closson, P. Wriede, and S. Bank, J. Amer. Chem. Soc., 88, 1581 (1969).

[24]. P. Yousefzadeh and C. K. Mann, J. Org. Chem., 33, 2716 (1968).

[25]. B. Angelo, Bull. Soc. Chim. Fr., 1710 (1969).

[26]. A. J. Fry, M. Mitnick, and R. G. Reed, J. Org. Chem., 35, 1233 (1970).

Chapter XXVII

STEREOCHEMISTRY OF ORGANIC ELECTRODE PROCESSES

L. Eberson

Division of Organic Chemistry
Chemical Center
University of Lund
Lund, Sweden

and

L. Horner

Organisch-Chemisches Institut
University of Mainz
Mainz, Germany

Part A: L. Eberson

PART A

I. INTRODUCTION

The study of the stereochemistry of organic reactions still remains one of the most powerful and widely used tools available for the study of reaction mechanisms in homogeneous media. In most cases, a postulated reaction mechanism can be subjected to crucial tests by stereochemical investigations, something which may be difficult or more ambiguous by other means.

Stereochemical studies on organic electrode processes have been much less frequently reported, however. This does not mean that such studies are experimentally more difficult to perform for electrochemical processes than for reactions in solution, but rather indicates that the interpretation of the stereochemistry observed for an electrode reaction is a difficult task, beset with a number of theoretical problems. As a consequence, an electrode reaction mechanism cannot normally be ruled out by the results of a stereochemical study.

Even if studies on the stereochemistry of organic electrode reactions are not very helpful in mechanistic discussions at our present state of knowledge, there are several practical reasons for considerably extending our efforts in this area. First, an electrochemical reaction may give a significantly different ratio between stereoisomers from that obtained in similar homogeneous reactions and hence be of preparative value. Secondly, a suitably manipulated electrode may be a dissymmetric factor in itself or be converted into one by adsorption of an optically active, electroinactive species (solvent molecules or ions from the supporting electrolyte) onto the surface, providing the possibility of inducing optical activity in the product(s) formed. Reactions of this type are of considerable interest for performing asymmetric syntheses, since — at least in principle — a limited amount of optically active material should be sufficient for converting an unlimited amount of optically inactive substrate to optically active product(s).

II. ROLE OF ADSORPTION FOR ORIENTATION OF SUBSTRATE MOLECULES

The first problem at hand is to try to make some educated guesses about the possible role of the electrode surface in the orientation of the substrate molecule during the reaction; i.e., does the molecule have to be adsorbed at the electrode surface in a preferential orientation for the reaction to occur [1]? Here it is useful to try to distinguish between the possible effect of physical adsorption, i.e., adsorption due to van der Waals interactions between the substrate and the electrode surface, and chemical adsorption, in which there is significant overlap between substrate and electrode orbitals.

A. Physical Adsorption

Purely electrostatic considerations would lead to the prediction that the substrate molecule should approach the electrode surface in such a way that electrostatic interactions are as small as possible, e.g., a dipole would try to orient itself with respect to a negatively charged surface with its positive end pointing toward the electrode and the negative end as far away as possible. From the point of view of steric interactions, it also appears reasonable to predict that the molecule will approach the electrode so as to minimize van der Waals repulsions with the surface (which, in the common fashion and in the absence of detailed knowledge, will be approximated as a planar surface).

This type of reasoning is of course an extrapolation of the concepts of homogeneous solution chemistry to a heterogeneous reaction, and it is immediately obvious that this approach has some serious limitations. In the first place, we may ask whether it is at all necessary for the molecule to be so close to the electrode during the electron-transfer step that these interactions can exert a decisive influence on its orientation. One might envisage a situation where the molecule can undergo electron transfer while tumbling in a random fashion at a certain distance from the electrode, say, 10-15 Å from the metal-solution interface. As an example, it should be mentioned that for the cathodic reduction of aromatic substrates Hale [1a] has adopted the position that the critical distance separating the reactant from the electrode during electron transfer might well be of the order of 15 Å or more. In such a case no orientational effect from the electrode surface can be exerted.

There is also the possibility that the electrode is covered by an adsorbed layer of electroinactive species from the solvent-supporting electrolyte, and that the substrate molecule will assume a preferential orientation with respect to the adsorption layer and electron transfer will take place through this layer.

B. Chemical Adsorption

The role of chemical adsorption [1] in the orientation of a substrate molecule is somewhat easier to assess, since for chemical adsorption to occur there has to be significant overlap between substrate and electrode orbitals, i.e., a weak chemical bond has to be established. This implies that there must be an orientational effect, depending on the symmetry of the orbitals involved, and that the substrate molecule must be fairly close to the electrode surface. Thus, it remains only to find out what kind of bonding occurs.

As an example, polycyclic aromatic hydrocarbons are adsorbed on platinum and other metal anodes with adsorption enthalpies in the range between -6 and -10 kcal/mole [2,3], and it has been calculated that adsorption must occur with the plane of the hydrocarbon parallel to the surface. It has also been suggested that the type of bonding involved should be of the same kind as in π-complexes. If this assumption is correct, the distance between the hydrocarbon molecule and the surface would probably be of the same magnitude as in π-donor/π-acceptor complexes, or around 3.5 Å. This situation, although based on a number of uncorroborated assumptions, should have experimentally detectable consequences, even if a fully unambiguous interpretation may not be possible.

The discussion above has been concerned with the orientation possibly required for the first electron-transfer step to occur. Considering the fact that most organic electrode processes are of the multistep type, composed of a sequence of chemical and electrochemical steps, it is evident that the same type of reasoning should be applied to all intermediates involved, a very difficult if not impossible task. One then faces questions about the adsorption of neutral radicals, ion radicals, and other charged species, questions which require much experimentation before one can even attempt to give answers. Consequently, it is common practice to assume that the orientation required for the initial step is fully or partially preserved during the following steps, which amounts to saying that these steps are fast relative to any intermediate desorption step(s).

C. Criteria for Assessing the Role of Adsorption

One very important problem in connection with assessing the role of the electrode for determining the stereochemistry of a given electrode reaction is whether an observed effect actually is due to the heterogeneity of the process. The most obvious way of doing this is to study a closely related homogeneous reaction to see if there is any significant difference. Now we face the problem of defining what a "closely related process" should be: Do we mean that the homogeneous reaction should have exactly the same sequence of steps as the electrode reaction or that the intermediate leading to products should be identical in both cases?

A simple example (not involving stereochemistry) may clarify the nature of the problem: It is generally agreed (Chapter XXII) that methyl-substituted aromatic hydrocarbons are oxidized according to an ECE mechanism with the formation of a carbonium ion as the intermediate forming the ultimate product [Eqs. (1)-(4)]. Now, if we want an as exact duplication as possible of this

$$ArCH_3 \rightarrow ArCH_3^{\dot{+}} + e^- \quad (1)$$

$$ArCH_3^{\dot{+}} + B \rightarrow ArCH_2\cdot + BH^+ \quad (2)$$

$$ArCH_2\cdot \rightarrow ArCH_2^+ + e^- \quad (3)$$

$$ArCH_2^+ + Nu^- \rightarrow ArCH_2Nu \quad (4)$$

$$B = \text{base}, \quad Nu^- = \text{nucleophile}$$

mechanism in homogeneous medium we might choose cobaltic ion oxidation of $ArCH_3$ as a model reaction [4]; on the other hand, if we are interested only in a homogeneous reaction leading to the formation of benzyl cations, we might instead choose a simple solvolysis process of the corresponding benzyl halide.

Of course, the former alternative is the more attractive one but has severely limited applications, since our knowledge of electron-transfer processes of organic compounds in homogeneous solution is only in its infancy at present [5]. The second alternative is less satisfactory but still justifiable: It does tell us if an electrochemically generated intermediate differs from its counterpart in homogeneous medium, but for interpretation it depends strongly on the assumption that the orientation of the initial step is preserved at least partially during the reaction sequence.

Other ways of establishing the role of adsorption would be to change the nature of the solvent and the electrode material. Organic molecules are more strongly adsorbed on an electrode from water or aqueous organic solvents than from nonaqueous solvents, so that, if there is a trend toward a higher degree of stereospecificity with increasing water concentration, this may be taken as an indication that adsorption is a factor involved. Also, different electrode materials have different affinities for organic molecules (e.g., neutral molecules are more strongly adsorbed on Pt than on C), so that a definite trend in this respect would constitute further evidence for adsorption-controlled stereochemistry.

Finally, there is the possibility of changing the electronic properties of the substrate, although this approach must necessarily rest on more or less unsafe assumptions. One might for example assume that in an anodic process the π-donor properties of the substrate should play an important role in the adsorption process, provided this takes place via π-bonding to the electrode. A stronger π-donor would then be expected to be more strongly adsorbed than a weaker one, implying that any electrode control of stereochemistry should be more pronounced in the former case.

The discussion above is merely an attempt to suggest possible modes of attacking the stereochemical problem in electrode mechanisms. As will be evident from the cases to be discussed below, our knowledge is too fragmentary to allow for a detailed description of the stereochemistry of any electrode process. What can be done at present is to try to find promising trends in the available experimental material, hoping that these might generate new ideas to be tested experimentally.

III. ADDITIONS TO MULTIPLE BONDS

A. Anodic Addition to Carbon-Carbon Double Bonds

The stereochemistry of several cases of anodic addition to double bonds has been reported and illustrates the difficulties involved in the interpretation of results.

Anodic oxidation of trans-stilbene [6] in glacial acetic acid-sodium acetate gives predominantly meso-diacetate (I) (R = CH_3) together with a small amount of the threo-hydroxyacetate (II) (R = CH_3). In wet acetic acid-sodium acetate (II) (R = CH_3) is the major product. None of the erythro form of (II) (R = CH_3) was detected in either case. These results were interpreted in terms of an acetoxonium ion intermediate [Eq. (5)], formed by oxidation of an adsorbed trans-stilbene molecule followed or assisted by reaction with acetate ion. Adsorption would thus be responsible for the stereoselective formation of the trans-acetoxonium ion (III) (R = CH_3) which then reacts with water with retention of configuration to form (II) (R = CH_3) or acetate ion with inversion to form (I) (R = CH_3) in accordance with the known behavior of acetoxonium ions generated in homogeneous processes. The same stereochemistry is found for anodic oxidation of trans-p, p$'$-dimethoxystilbene in acetic acid-sodium acetate [7].

However, later investigations [8] have shown that one cannot exclude the possibility that the stereoselective formation of (III) (R = CH_3) is due to the fact that the trans-acetoxonium ion is the thermodynamically favored stereoisomer. Anodic oxidation of stilbene in acetonitrile containing benzoate ion produces a mixture of (I) (R = C_6H_5) and (II) (R = C_6H_5) irrespective of whether stilbene has cis or trans configuration. The obvious conclusion from these experiments is that benzoxonium ion (III) (R = C_6H_5) must be involved as the product-forming species, but that the formation of (III) must occur in a stepwise fashion, allowing for rotation around the central carbon-carbon bond [e.g., as in Eq. (6)] at some stage of the process to form the thermo-

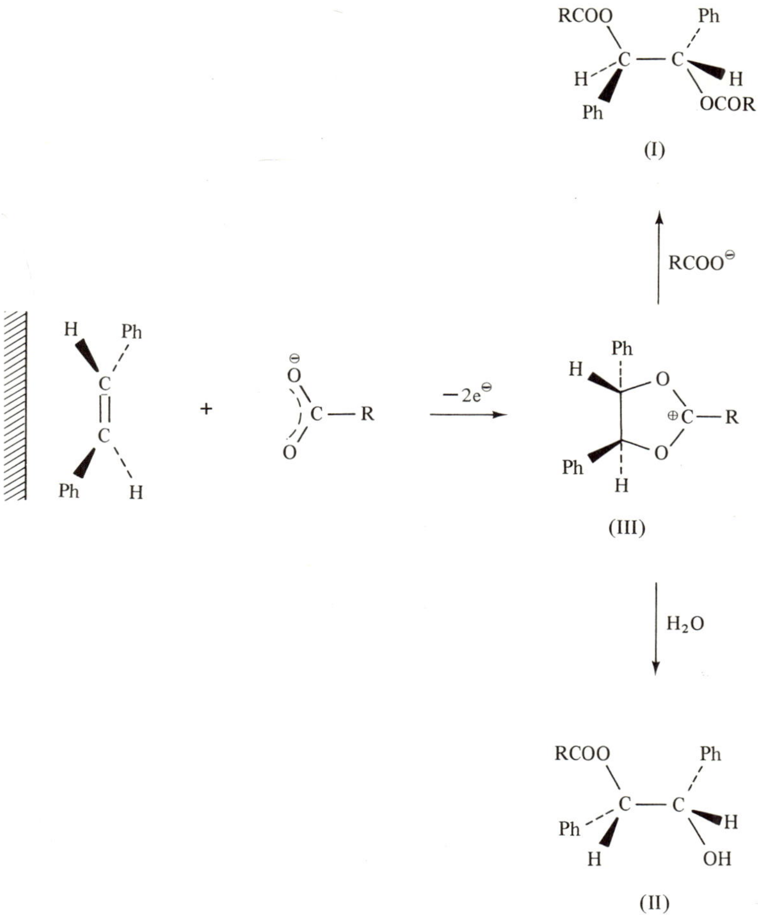

(5)

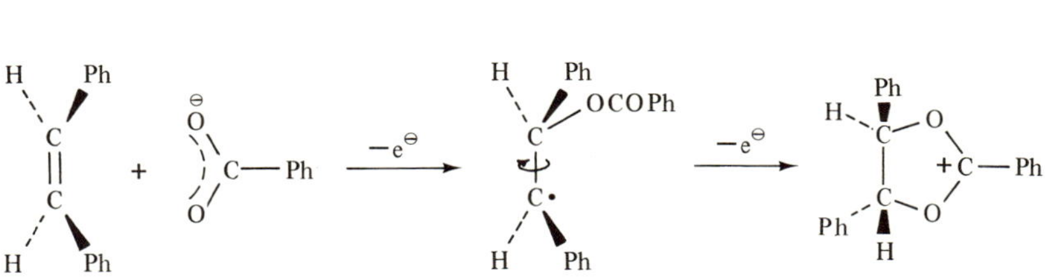

(6)

dynamically stable trans-benzoxonium ion from cis-stilbene. It is likely that the same applies to the reactions in acetic acid.

The anodic addition of two acetoxy groups to cyclooctatetraene at a platinum or carbon anode in sodium acetate-glacial acetic acid [Eq. (7)]

exhibits striking stereochemical differences from analogous chemical oxidations in homogeneous media [9]. Both lead tetraacetate-boron trifluoride and mercuric acetate oxidation in glacial acetic acid lead exclusively to the trans isomer (IV), whereas the anodic reaction gives a mixture of cis and trans isomer, with the thermodynamically less stable cis isomer (V) slightly predominant.

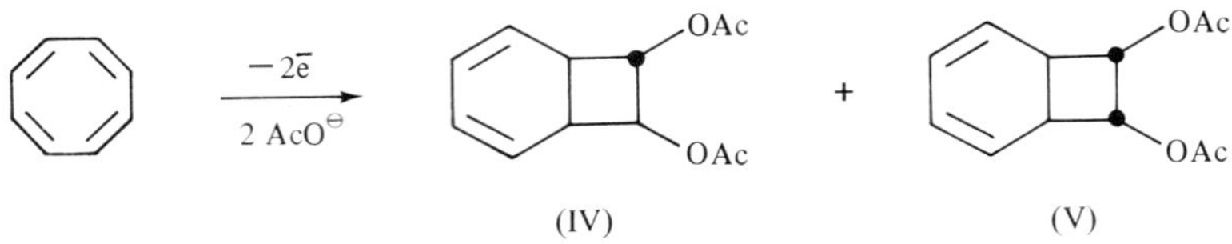

(7)

This result was interpreted in terms of a series of homogeneous reaction steps following the electron-transfer step with no electrode control of stereochemistry. It is obvious, however, that electron transfer from an adsorbed cyclooctatetraene molecule followed by attack of acetate ion from the inside of the tub conformation [Eq. (8)] would lead to the formation of an endohomotropylium ion (VI), analogously to the mechanism of electrophilic addition to cyclooctatetraene [10, 11]. Attack by acetate ion on (VI) followed by ring contraction would then give (V), whereas inversion of (VI) to exohomotropylium ion (VII) would similarly lead to (IV). However, the inherent

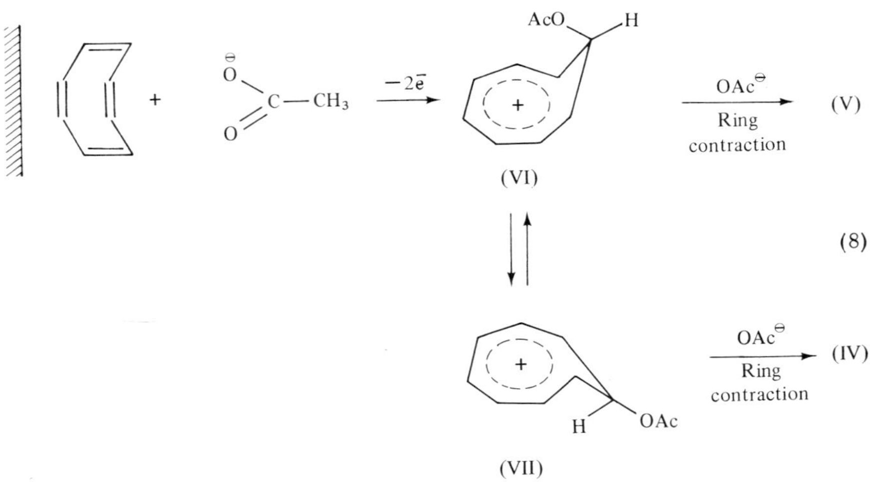

(8)

complexities of the reactions of the cyclooctatetraene system preclude an easy distinction between the possible mechanisms. The practical result, that the cis-diol (VIII) is easily available by the anodic route, is perhaps the most important one [12].

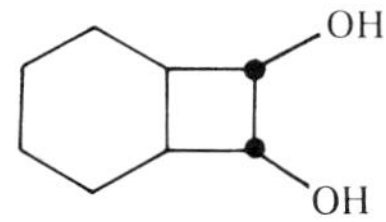

(VIII)

Anodic addition of two acetoxy groups to 1, 3-cyclohexadiene [13] gives, among other products, cis and trans isomers of the 1, 2- and 1, 4-addition products [Eq. (9)]. Lead tetraacetate oxidation proceeds according to the same stereochemistry, so there is no indication of electrode-controlled stereochemistry.

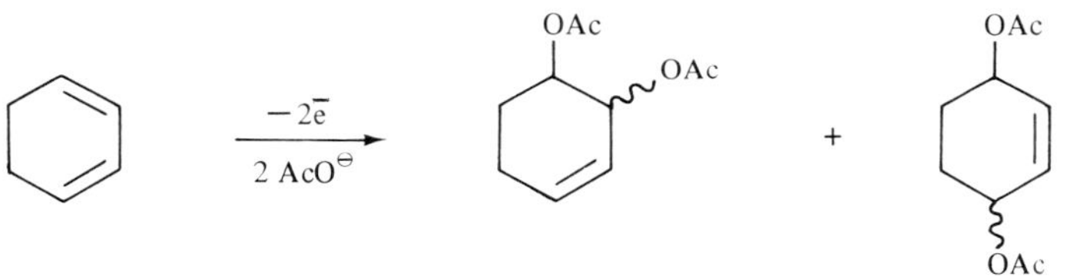

(9)

cis/trans ratio ≈ 1:3   cis/trans ratio ≈ 1:1

Anodic addition of two methoxy groups to cis- and trans-stilbene [14] displays a slight preference for cis addition [Eqs. (10) and (11)]. This result

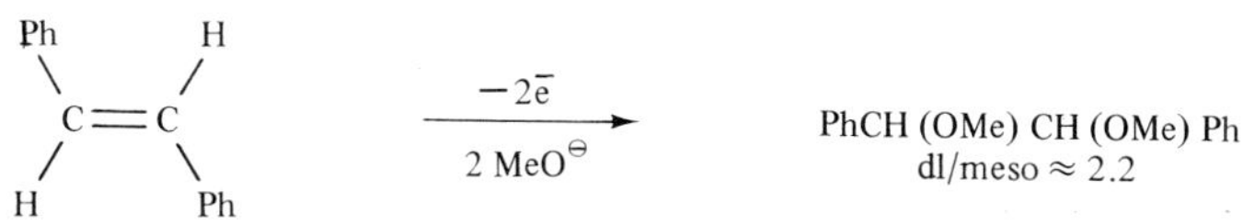

(10)

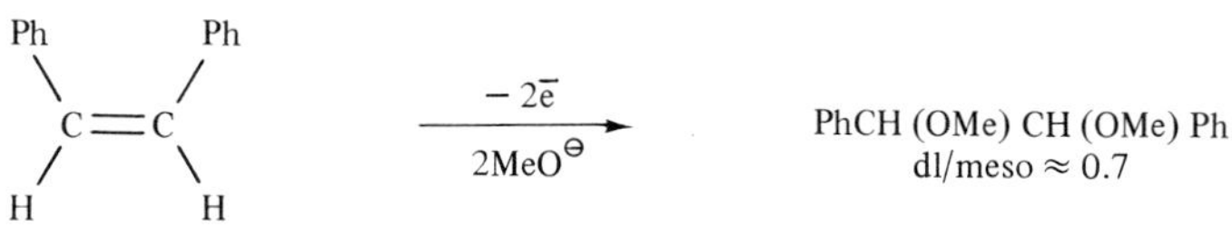

(11)

was interpreted in terms of adsorption control. For anodic 1, 4-addition of methoxy groups to 2, 5-dimethylfuran it was found [15] that cis and trans-2, 5-dimethoxy-2, 5-dimethyldihydrofuran are formed in equal amounts, exactly the same result as in chemical preparations in homogeneous medium. A mixed anodic addition of one methoxy and one cyano group to 2, 5-dimethylfuran [Eq. (12)] has been found to give a 2.4:1 mixture of cis- and trans-1, 4-addition products [16].

Interesting observations have been made on the anodic addition of methyl radicals to 1, 3-dienes [17, 18]. The methyl radicals are produced by anodic

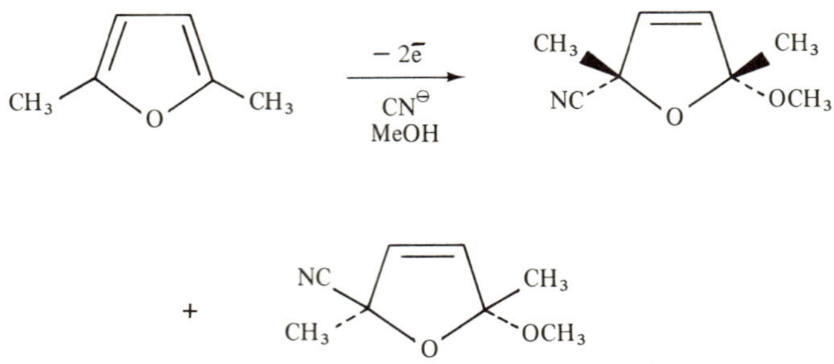

(12)

oxidation of acetate ion (see Chapter XIV) in methanol. The 1, 4-addition product from butadiene was exclusively trans-3-hexene, whereas isoprene gave the cis isomer only of 3-methyl-3-hexene. Similarly, ethyl radicals generated by oxidation of propionate ion add to butadiene to give trans-4-octene only. To explain these results it was assumed that the diene is adsorbed at the electrode in its most stable conformation (the s-trans form for butadiene and the s-cis form for isoprene) at the time for radical attack [Eqs. (13) and (14)]:

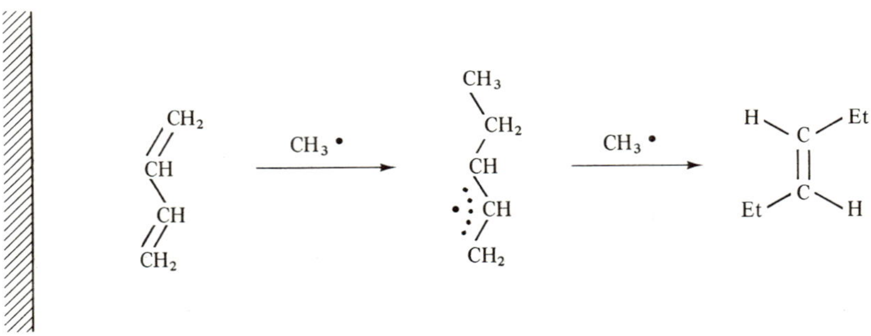

(13)

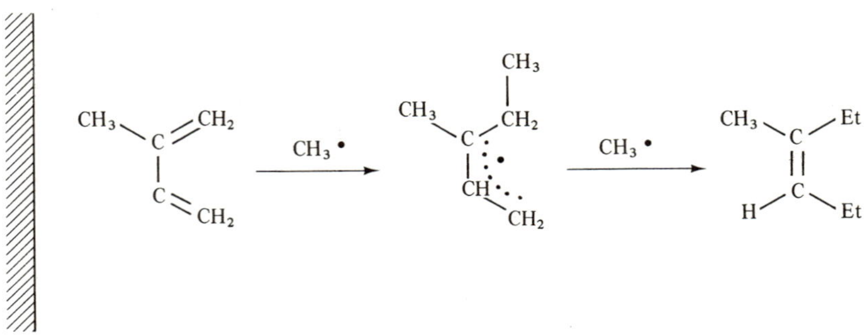

(14)

This would create an allylic radical with defined stereochemistry (cf. [19, 20]), which would be retained long enough for the addition of the second radical to be stereospecific. From 1, 3-cyclohexadiene, in which the configuration of the double bonds necessarily must be s-cis, both cis and trans isomers (with regard to the ring) are formed, indicating that the addition of the second radical may take place in solution.

B. Cathodic Addition to Carbon-Carbon Triple and Double Bonds

Cathodic addition of hydrogen to alkynes with formation of olefins takes place readily and the stereochemical course has been investigated in a number of cases. Cathodic reduction of dialkylacetylenes and diphenylacetylene on a spongy nickel cathode in ethanol/5% sulfuric acid gives exclusively the cis-olefin [21], which was taken as evidence for the reduction taking place at the surface of the electrode (cf. catalytic reduction which gives cis-olefin too). Using the same solvent-supporting electrolyte no reduction of dialkylacetylenes was observed at cadmium, lead, or amalgamated lead electrodes. Another mechanism was invoked [22] for the cathodic reduction of dialkylacetylenes in methylamine-lithium chloride on a mercury electrode. In a divided cell (i.e., under conditions which do not allow lithium methylamide to accumulate in the electrolyte) the trans-olefin is formed predominantly (> 92% in all cases studied). This result is analogous to that obtained when lithium dissolved in methylamine is used as the reducing agent, suggesting that the reaction does not take place via direct electron transfer to the substrate at the electrode surface but instead by lithium formed in situ and dissolved in the amine solvent.

Reduction of butyn-2-diol-1, 4 at a mercury cathode in methanol-tetramethylammonium chloride gives trans-butene-2-diol-1, 4 exclusively, whereas the corresponding diacetate gives a 60:40 mixture of trans- and cis-butene-2-diol-1, 4 diacetate [23]. In both cases it was suggested that the alkyne reacts in the adsorbed state, forming an adsorbed dianion; for the diol the exclusive formation of the trans isomer might be explained by an intramolecular proton-transfer [Eq. (15)] mechanism. For the diacetate it was assumed that the adsorbed transoid and cisoid conformations would be of similar stability [Eq. (16)].

Cathodic 1, 4-addition of hydrogen to 1, 3-dienes [23] at a mercury electrode in methanol-tetramethylammonium chloride displays the stereochemistry shown in Eq. (17). Also in these cases adsorbed intermediates were invoked to explain the stereochemistry observed. Since the effects are fairly small it is difficult to judge the validity of these assumptions.

Some alkenes were also reduced under the same conditions as in the last-mentioned cases. Both for the dianion of dimethylmaleic acid and dimethyl dimethylmaleate exclusive cis addition was observed. For

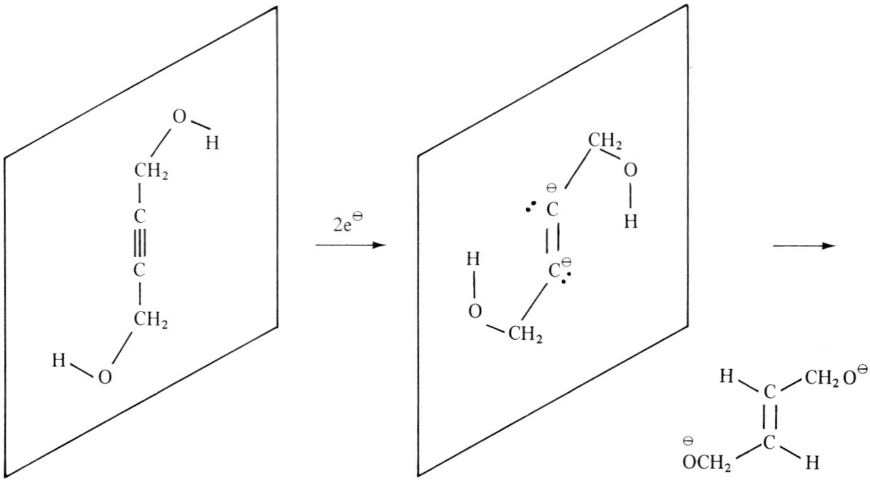

(15)

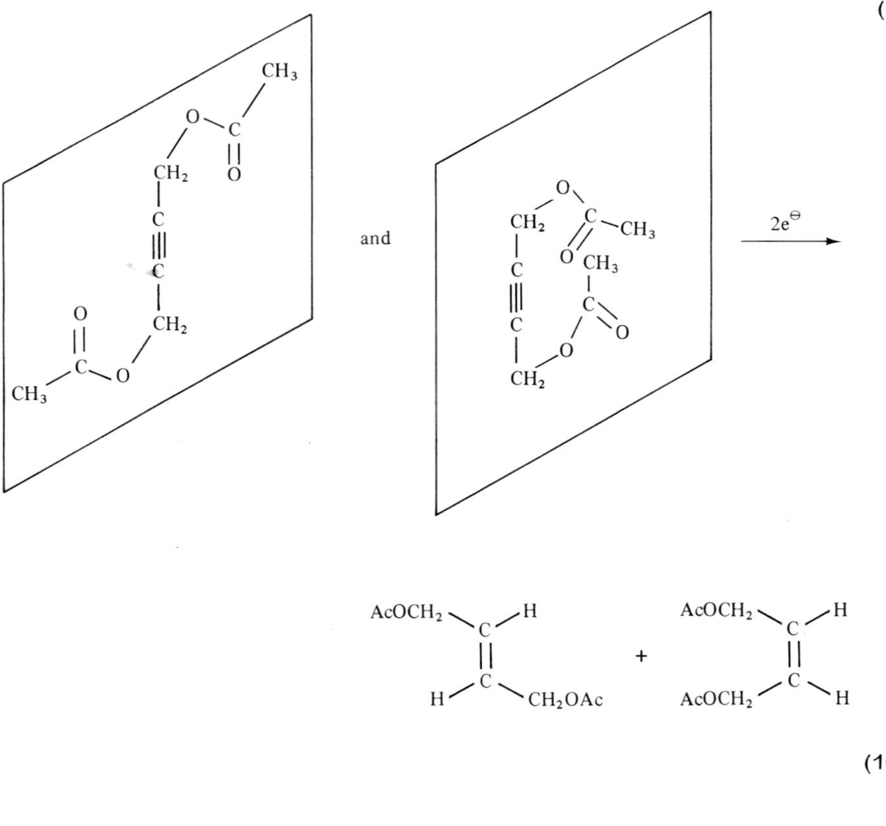

(16)

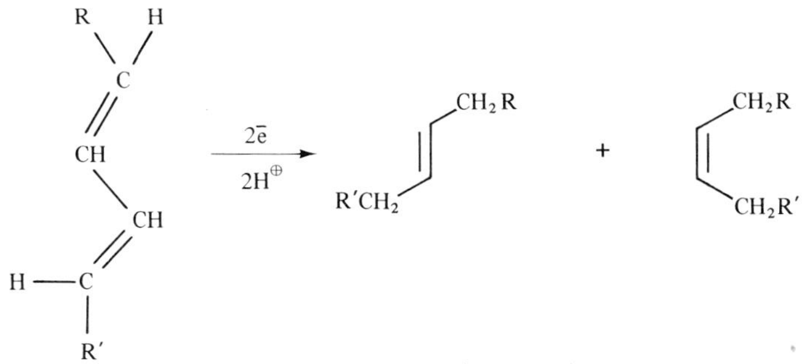

(17)

R	R′	trans	cis
H	Ph	74	26
Ph	Ph	50	50
CH_3	COOH [a]	74	26
CH_3	CH_2OH	45	55

[a]In this case, the carboxyl group was reduced to a CH_2OH group.

cis- dimethylstilbene the ratio between cis and trans addition was 53:47 and for trans-dimethylstilbene, 70:30.

A very interesting case of asymmetric synthesis involving the cathodic addition of hydrogen has been reported [24]. 4-Methylcoumarin is reduced at a mercury cathode at -1.8 V (SCE) in 40% aqueous methanol containing citrate buffer (pH 5-6) in the presence of an optically active alkaloid to form optically active 3,4-dihydro-4-methylcoumarin [Eq. (18)]. The mechanism was assumed to involve formation of an aminium radical from the alkaloid cation, which would donate a hydrogen atom to the radical formed in a 1 e^-/1 H^+ reaction with the coumarin [Eqs. (19) and (20)]. Another possibility is that the double layer is composed of adsorbed electroinactive ammonium ions, which create a dissymmetric milieu for the reaction of the substrate.

Cathodic addition may also involve electrophiles other than protons. A well-studied case is the cathodic addition of two carbon dioxide molecules to stilbene which was first reported [25] to give meso-2,3-diphenylsuccinic acid only (this is the product of trans addition). Later experiments performed at a potential controlled to give only the anion radical show that under these conditions a mixture of cis and trans addition product is obtained [26]. In DMF-tetrabutylammonium iodide cis-stilbene gives a 1.4:1 mixture of dl-/meso-diacid, whereas trans-stilbene gives a 2.7:1 mixture of dl-/meso-

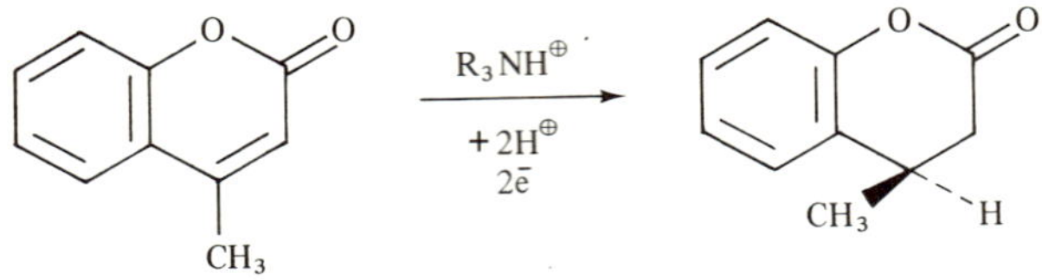

Maximum optical yield attained, 19%

(18)

$$R_3NH^{\oplus} + \bar{e} \longrightarrow R_3\dot{N}H \qquad (19)$$

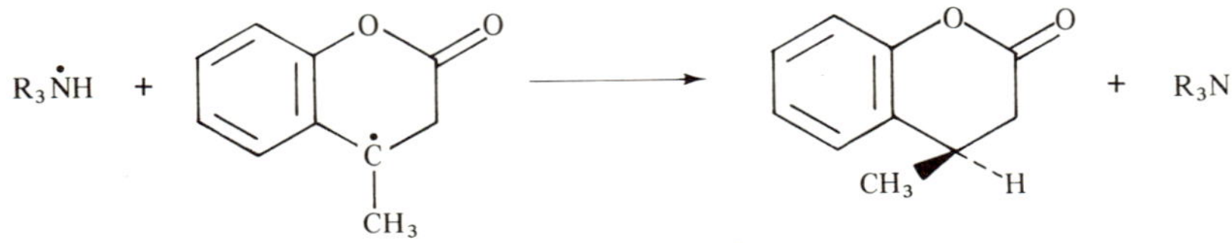

(20)

diacid. It could be shown that the isomerization of anion radicals can be neglected since the anion radical from cis-stilbene has a half-life for isomerization of at least 15 sec. The results were explained in terms of incomplete equilibration between the different conformations of carboxylate ion radicals (IX), (X), and (XI). Transfer of a second electron to any of these species should take place from the side furthest from the carboxylate group due to electrostatic repulsion forces with regard to the electrode. This would give dianion (XII) from conformation (IX) and (X), and dianion (XIII) from (XI). Since conformation (XI) must be considered less stable than (X), both cis- and trans-stilbene should give the dl-diacid preferentially [see Eq. (21)].

C. Cathodic Addition to the Carbon-Oxygen Double Bond

Also in this type of reaction asymmetric induction by optically active supporting electrolytes has been observed (cathodic reduction of acetophenone in the presence of ephedrine and derivatives thereof; see Sec. VI). Since also quaternary ammonium salts have this effect it must be caused by a dissymmetry in the double layer.

D. Cathodic Addition to the Carbon-Nitrogen Double Bond

Electrochemical reduction of camphor oxime (XIV) and norcamphor oxime (XV) at a mercury cathode in 80% methanol-water containing lithium chloride as supporting electrolyte proceeds with a high degree of stereoselectivity [Eqs. (22) and (23)].

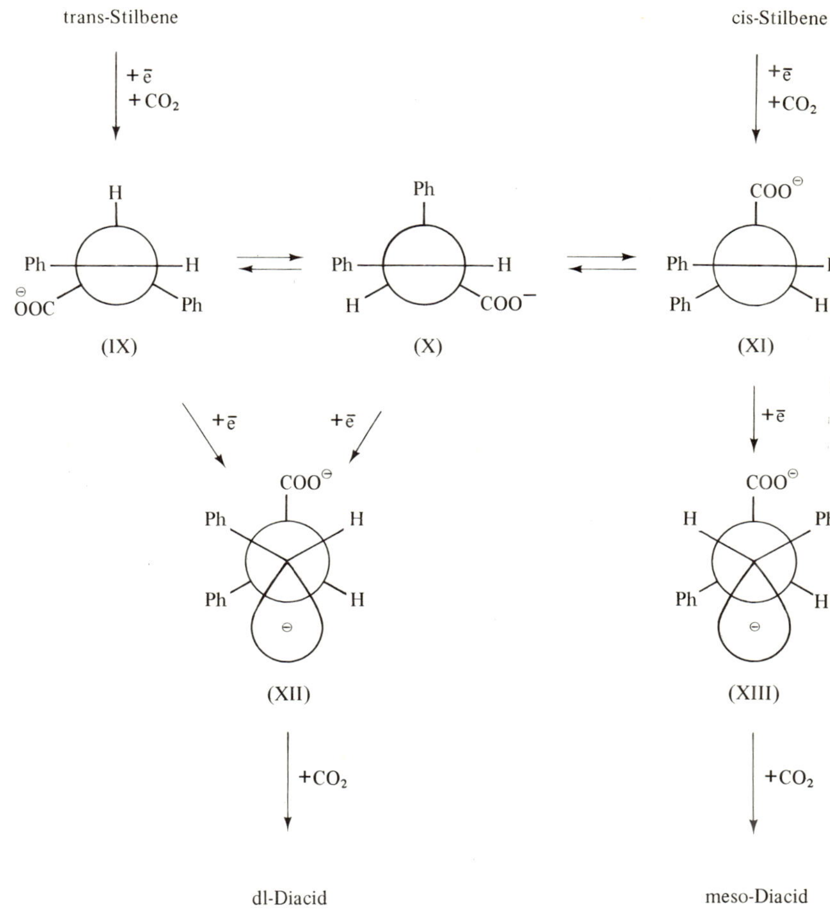

(21)

The stereochemistry of these two processes is exactly the same as that observed for lithium aluminum hydride reduction but opposite to that observed for sodium-alcohol reduction of (XIV) and (XV). These reactions were interpreted in terms of a sterically controlled approach of the oxime to the electrode surface. The oxime would be attacked from the less hindered side [endo attack on (XIV) and exo attack on (XV)], possibly with formation of an alkylmercuric species at the electrode surface. Protonation would then occur with retention of configuration [27]. Asymmetric induction by chiral supporting electrolytes has been observed in the cathodic reduction of acetophenone N-benzylimine [27a]. The optical purity of the dihydro product is 5-8%. See Sec. VI.

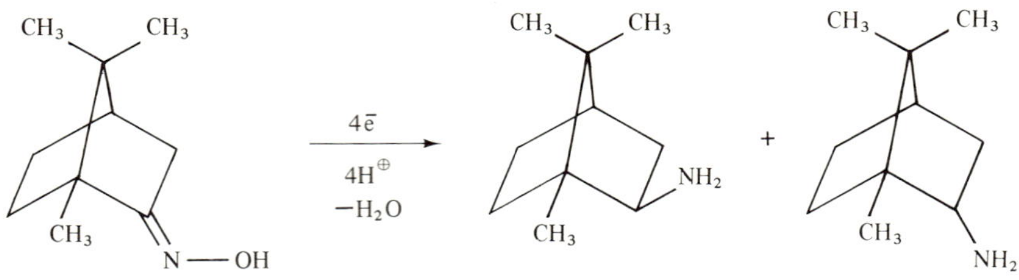

(XIV) 99% 1% (22)

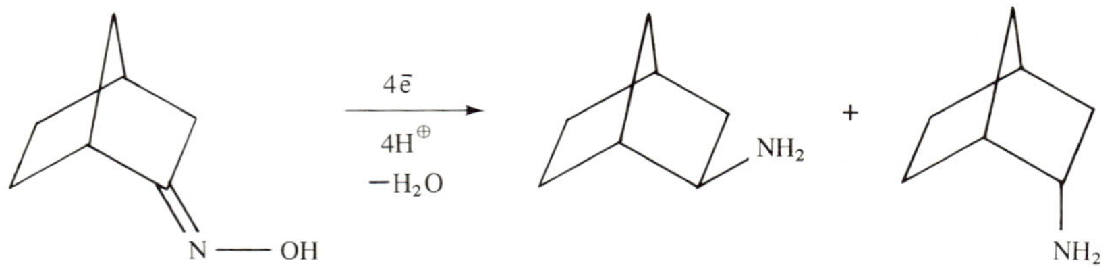

(XV) 0% 100% (23)

IV. SUBSTITUTION REACTIONS

A. Anodic Substitution Reactions

The stereochemistry of the anodic side-chain acetoxylation of 2-t-butylindan (XVI) has been investigated in some detail [28]. If adsorption plays any role for the stereochemistry of the reaction, one would expect that the hydrocarbon would be adsorbed with the less hindered side toward the anode surface. If this electrode-substrate configuration is kept during the initial ECE process which gives rise to carbonium ion (XVII) [Eq. (24)], the electrode would shield the molecule from attack from the less hindered side. Consequently, the thermodynamically less stable cis isomer (XVIII) would be formed predominantly. The actual result is that a 16:84 mixture of cis and trans isomer is formed, far from the expected predominance of the cis compound. However, comparison with two homogeneous processes, lead tetraacetate oxidation of 2-t-butylindan and solvolysis of cis- or trans-1-(p-nitrobenzoyloxy)-2-t-butylindan, which both result in the formation of the 1-acetoxylation product via carbonium ion (XVII), shows that the cis/trans ratio is abnormally high in the anodic process (in the homogeneous reactions, the cis/trans ratio is about 0.02). Moreover, using a less strongly adsorbing electrode material, graphite, the cis/trans ratio decreases to less than 0.07.

These results lend some support to the view that the electrode exerts some control on the stereochemistry of side-chain acetoxylation of (XVI),

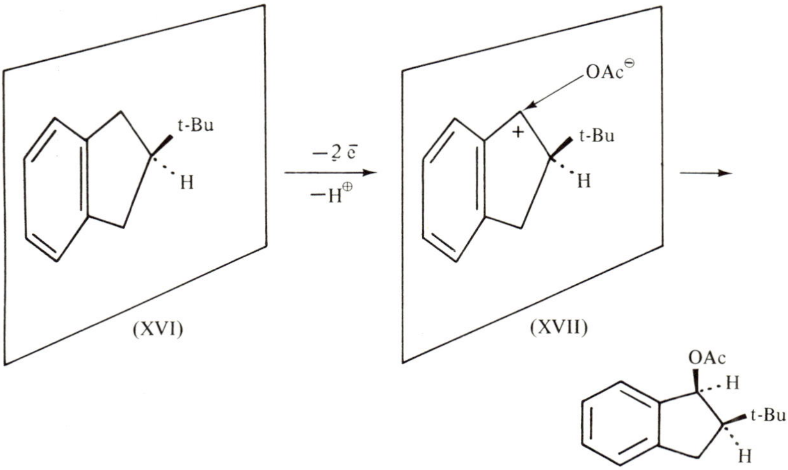

(XVI) (XVII) (XVIII) (24)

although the effect is not overwhelmingly large. It is, however, apparent that 2-t-butylindan is not an ideal substrate for the purpose at hand since the t-butyl group forces the molecule into a conformation in which the difference in steric interactions between each side of the molecule, respectively, and the electrode is not as large as originally expected.

B. Cathodic Substitution Reactions

One of the most widely studied electroorganic reactions from the stereochemical point of view is the cathodic reduction of halides, not surprising in view of the wealth of studies concerned with the homogeneous reactions of optically active halides. Stimulus for this work has been provided by a thought-provoking article by Elving and Pullman [29], in which it was suggested that the mechanism of cathodic halide reduction should be either S_N1- or S_N2-like. In the former mechanism, the strong electric field near the electrode would cause a dissociation of the carbon-halogen bond to form a carbonium ion and a halide ion [Eq. (25)]; two electrons would then be transferred stepwise to the carbonium ion to give a carbanion, which would finally undergo protonation:

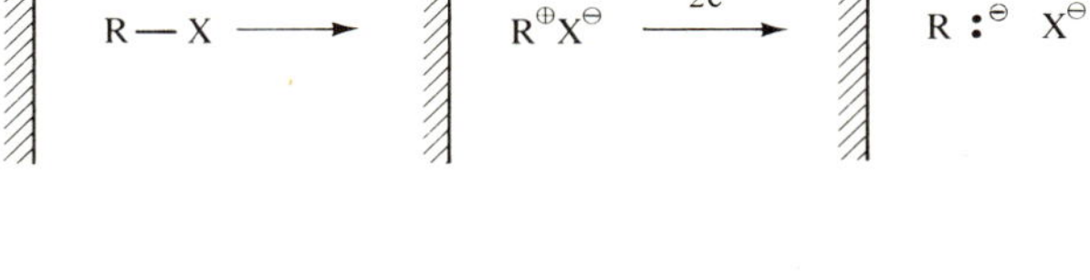

(25)

In the latter mechanism, the halide molecule would approach the cathode with the carbon-halogen dipole oriented perpendicular to the surface, the carbon atom being closest to the electrode due to its partial positive charge [Eq. (26)]. The transition state for the reaction would then resemble an ordinary S_N2 transition state, except that the electrode surface has taken the role of the nucleophile.

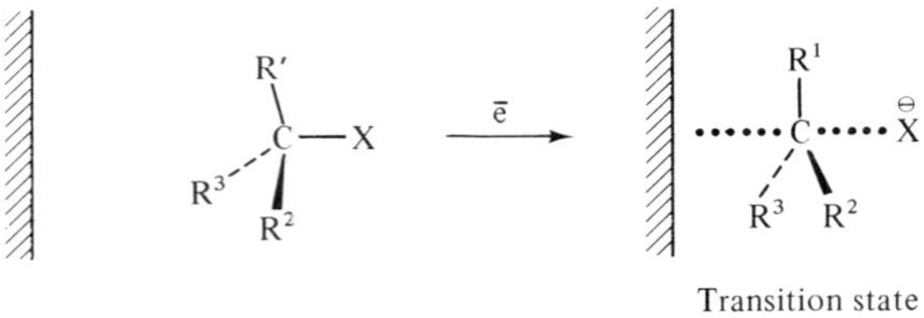

(26)

However, these mechanisms — or at least the proposal common to both of them that the electrode reaction should occur with the halogen atom as far as possible from the electrode surface — were soon challenged. On the basis of measurements of half-wave potentials for a number of bridgehead-brominated bicyclic compounds [39], which cannot conceivably undergo back-side attack on carbon, it was proposed that front-side attack on bromine must occur in these compounds. Of course this does not settle the problem of the mode of attack in acyclic halides, but later studies [31] on the quantitative correlations between half-wave potentials of acyclic bromides and Taft polar and steric parameters indicate that the steric effects operative in the cathodic process are far smaller than in ordinary S_N2 displacements. On the basis of these results a mechanism involving attack by an electron from the electrode surface in a direction perpendicular to the carbon-halogen bond was proposed as a general mechanism for halide reduction. Such a mode of attack would be stereochemically feasible for donation of an electron to the σ-antibonding orbital of the carbon-halogen bond. The last-mentioned mechanism requires that the potential-determining transition state be negatively charged. Indeed, $E_{1/2}$-σ plots for a number of substituted chloro-, bromo- and iodobenzenes and benzyl bromides [32-34] indicate that the transition state must have a negative charge. Incidentally, this rules out the S_N1-like mechanism [Eq. (25)].

The first study on the cathodic reduction of an optically active halide, (+)-S-1-bromo-1-methyl-2, 2-diphenylcyclopropane, established some important features of the process [35]. This reaction was performed in acetonitrile-tetraethylammonium bromide at a controlled potential of -2.7 V (SCE) and gave a 93% yield of 1-methyl-2, 2-diphenylcyclopropane with 63% retention of configuration. It was noted that the variation of current with time during controlled-potential electrolysis was not exponential (as was the disappearance of substrate), indicating the formation of an intermediate prior to hydrocarbon formation. The nature of this intermediate was carefully considered, bis(1-methyl-2, 2-diphenylcyclopropyl)mercury (XX) emerging as the most likely candidate. In support of this, analyses of the electrolyte during electrolysis showed the presence of a soluble mercury compound; furthermore, (XX) could actually be isolated from the reduction of 1-iodo-1-methyl-2, 2-diphenylcyclopropane under the same conditions but at a much lower potential, -1.7 V (SCE). At this potential, (XX) is not so easily reduced to the hydrocarbon.

To account for these observations, the following mechanism was proposed:

$$R\text{-}Br + e^- \rightarrow R\dot{-}Br^- \quad (27)$$

$$R\dot{-}Br^- \rightarrow R\cdot + Br^- \quad (28)$$

$$R\cdot + e^- \rightarrow R^- \quad (29)$$

$$R\cdot + Hg_n^o \rightarrow RHg_n{}^{\bullet} \rightarrow R^- + Hg_n^o \quad (30)$$

$$RHg_n\cdot + RHg_n\cdot \rightarrow RHg_{2n}R \rightarrow RHgR + Hg^o_{2n-1} \quad (31)$$

$$R_2Hg + e^- \rightarrow R^- + RHg\cdot \quad (32)$$

$$R^- + Et_4N^+ \rightarrow RH + CH_2{=}CH_2 + Et_3N. \quad (33)$$

Here the surface nature of the reaction is stressed; radicals are formed at the surface and can be converted to carbanions in a number of ways. The surface nature of the reaction, coupled with the configurational stability of cyclopropyl carbanions, accounts for the stereochemistry observed. It should be noted that lithium or magnesium reduction of the same optically active bromide proceeds with retention of configuration also.

Another investigation [36] of the cathodic reduction of optically active cyclopropyl bromides established that the substituent located at the carbon atom involved in the carbon-halogen bond has a decisive influence on the stereochemistry of the process [Eq. (34)]:

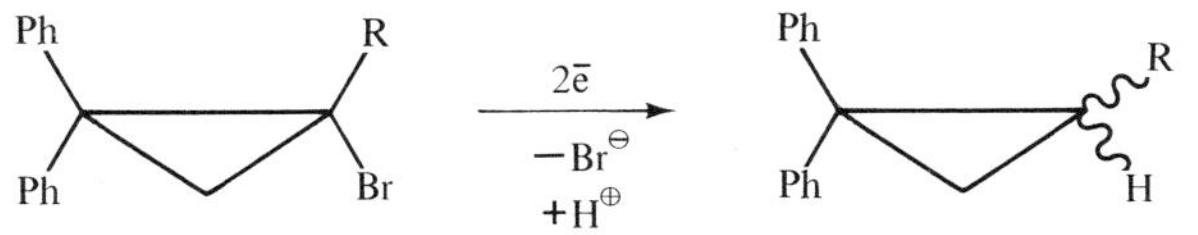

(34)

R	Stereochemistry
COOH	26-35% inversion
COO^-	31-38% retention
$COOCH_3$	30-56% inversion
CH_3	21% retention (cf. above)

These reactions were performed at controlled potential in 95% ethanol with different supporting electrolytes, using a stirred mercury cathode. The mechanism was proposed to involve an initial attack on bromine, leading to an electrode-carbanion complex of retained configuration [Eq. (35)]:

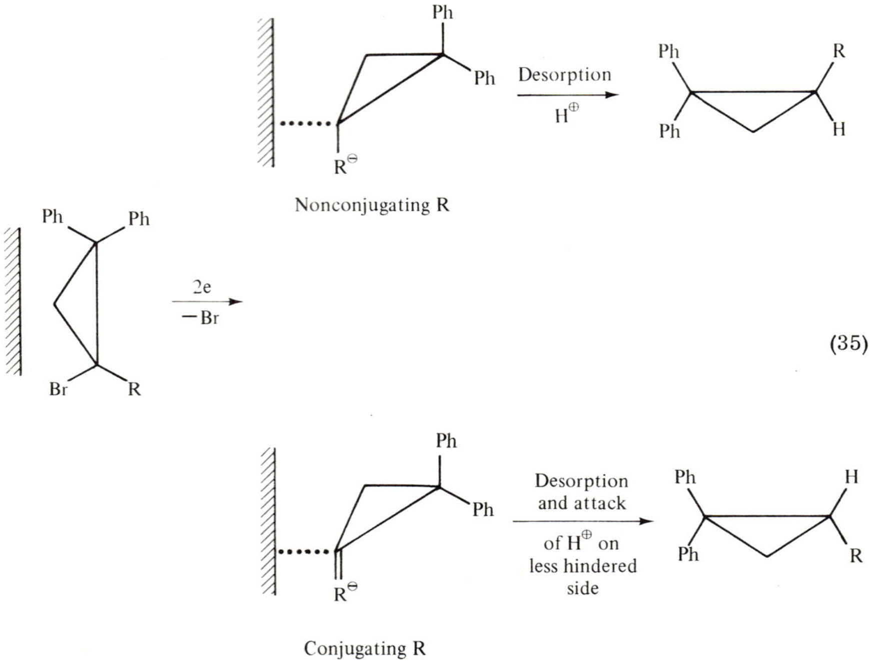

(35)

If the carbanionic center carries a nonconjugating substituent (COO^- or CH_3), the pyramidal carbanion would be desorbed with retention of configuration and be protonated in the solution immediately outside the electrode. For a planar carbanion carrying a conjugating substituent (COOH or $COOCH_3$) protonation from the side not shielded by the electrode would lead to a product of inverted configuration. The fact that partial racemization always occurs in these processes indicates that cyclopropyl radicals might be intermediates at some stage of the reaction.

A very high degree of inversion of configuration (80-90%) was observed in the cathodic reduction of optically active 2-chloro-2-phenylpropionic acid [Eq. (36)] (at a mercury cathode in ethanol-tetraethylammonium chloride) [37]. The mechanism shown in Eq. (36) is the same as that proposed for reduction of cyclopropyl bromides carrying a conjugating substituent. The unusually high degree of inversion of configuration may be explained by the fact that protonation may occur intramolecularly. It would be interesting to see how the corresponding ester behaves in this respect. For a simpler halide,

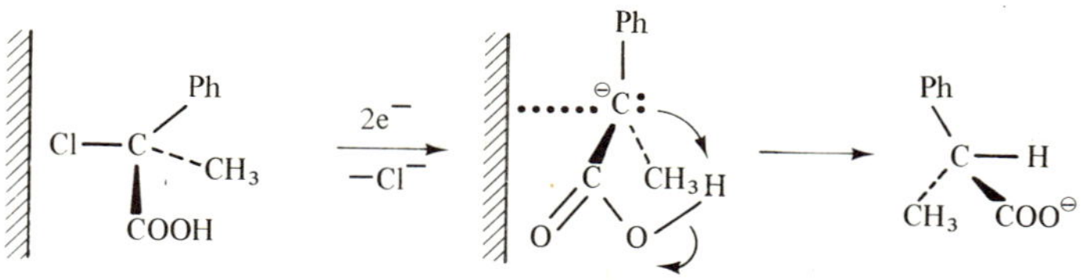

(36)

optically active 1-phenylethyl chloride, cathodic reduction in DMF-D_2O-tetrabutylammonium tosylate at a mercury cathode leads to almost complete racemization [Eq. (37)], indicating that the electrode does not exert any

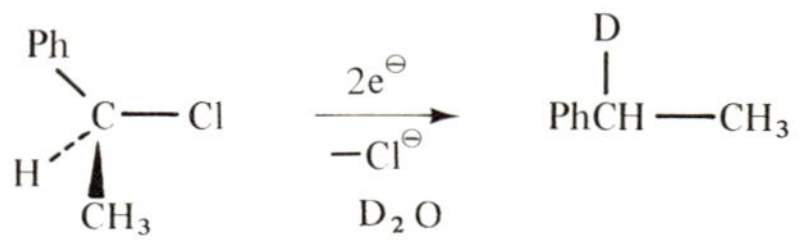

(37)

shielding effect in this case, or that racemization takes place at the radical stage [38]. Almost complete racemization is observed in some cases of carbon-oxygen single-bond cleavage [Eq. (38)] also [38a]. This is in sharp contrast to the result discussed above [Eq. (36)].

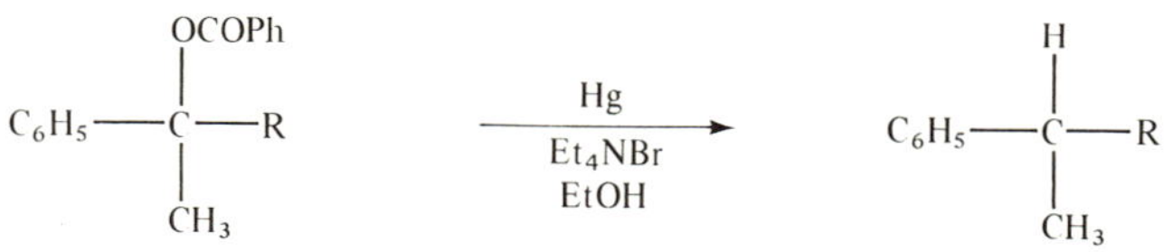

(38)

R = COOH, $COOCH_3$

Reduction of geminal dihalocyclopropanes has been studied in some detail by two groups [39, 40]. Cathodic reduction of (XXI) [39] results in the formation of cis isomer (XXII) predominantly, the proportion of this isomer increasing with increasing water content in the electrolyte. A mechanism

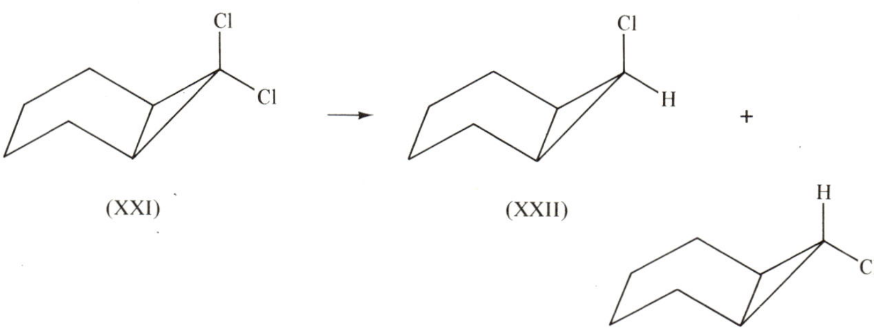

(XXI) (XXII) (XXIII)

(39)

involving reduction of the endo-halogen with retention of configuration is favored by one group [39], whereas the other group [40] concludes that the partial stereospecificity observed might be caused by two competitive stereospecific processes, one on each halogen, occurring in unequal amounts. Support for this view is obtained from the reduction of the two isomers (XXIV) and (XXV), in which exclusive removal of bromine occurs with predominant retention of configuration. Thus, the steric interactions between

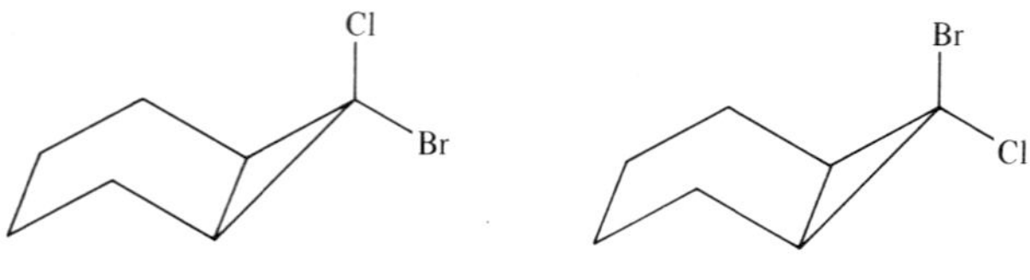

(XXIV) (XXV)

electrode and substrate in the transition state for reduction of the exo-halogen atom, invoked to explain why attack on the endo-halogen is preferred [39], do not seem to be of any critical importance.

Cathodic reduction of the stereoisomeric 3-iodo-3-hexenes gives evidence for the existence of vinyl radicals as intermediates in the process [41]. These exist on a time scale sufficiently long for partial equilibration to occur before reduction to the vinyl carbanion [Eq. (40)]; the possibility that stereochemistry of the starting materials is lost through isomerization of a vinyl carbanion is rendered unlikely by the fact that the very good carbanion-trapping agent, phenol, does not influence the cis/trans ratio of the product when present in even a tenfold excess over substrate.

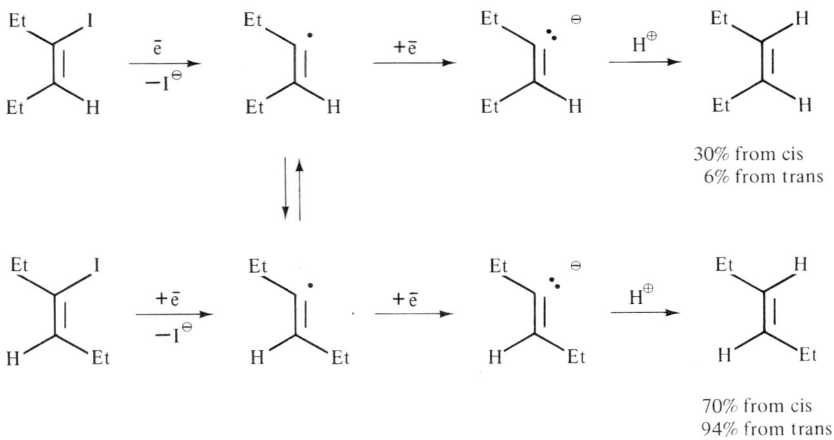

(40)

Concluding, much remains to be done before the detailed mechanism of cathodic halide reduction is established, if it is at all possible to encompass all phenomena within a single mechanism. It appears to be of particular interest to find out the role of organomercury intermediates at mercury electrodes in determining stereochemistry, and to study the stereochemistry at other cathode materials. It may also be fruitful to test the prediction based on MO calculations that the stability of the initially formed anion radical $R\dot{-}X^-$ should increase with increasing atomic number of the halogen atom. This may have stereochemical consequences [42].

V. COUPLING, CLEAVAGE, AND ELIMINATION REACTIONS

A. Coupling Reactions

The stereochemistry of the Kolbe process (anodic coupling of carboxylates) has already been discussed (Chapter XIII).

Intramolecular reductive coupling of bisactivated olefins [43-45], proposed to occur via a concerted electron-transfer-cyclization mechanism [Eq. (41)], generally gives a mixture of cis and trans isomers of the cyclic product. In the cases where the cis/trans ratio is known, it appears that the

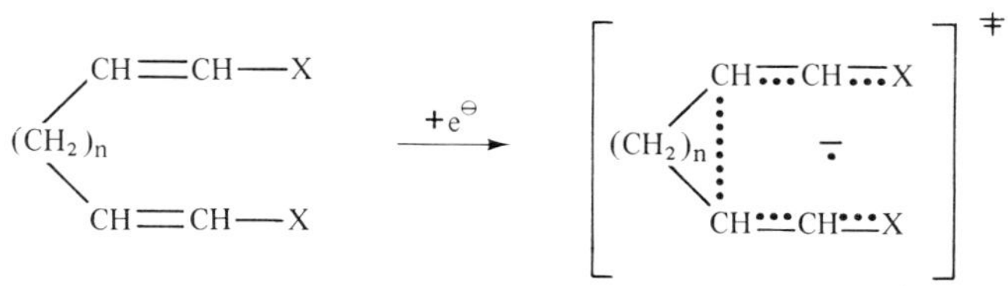

(41)

thermodynamically favored trans isomer is the predominant product. However, in view of the two reactions shown in Eqs. (42) and (43), opinion must be reserved on this point until further data accumulate.

The cathodic reductive coupling of aromatic ketones has been studied in great detail [46-51] with respect to the distribution between dl- and meso-pinacol under different conditions [Eq. (44)]. This reaction shows some stereoselectivity, in that the dl/meso ratio lies in the region 0.9-1.4 in acidic aqueous ethanol and 2.5-3.2 in alkaline aqueous methanol for ketones which do not allow an intramolecular hydrogen bond in the intermediate. These ratios are not significantly influenced by the choice of electrode material (mercury, tin, or copper), electrolysis time, potential, or current. In aprotic media, the dl/meso ratio increases strongly (to about 7 for acetophenone).

These findings, combined with the fact that photochemical bimolecular reduction of the same substrates in homogeneous medium gives the same stereochemical result, indicate that the coupling between radicals in the

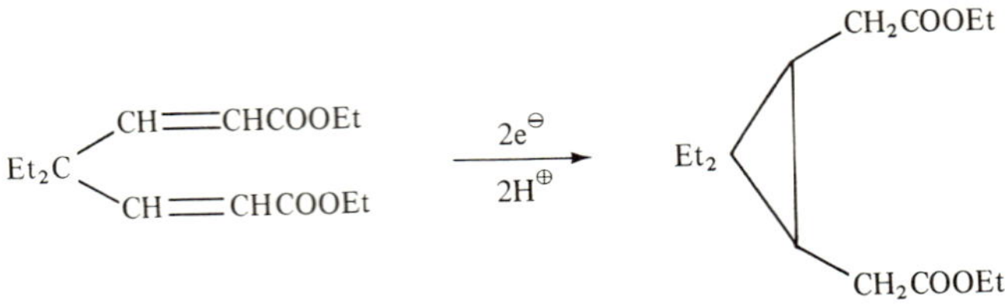

(42)

cis/trans ratio ≈ 0.9

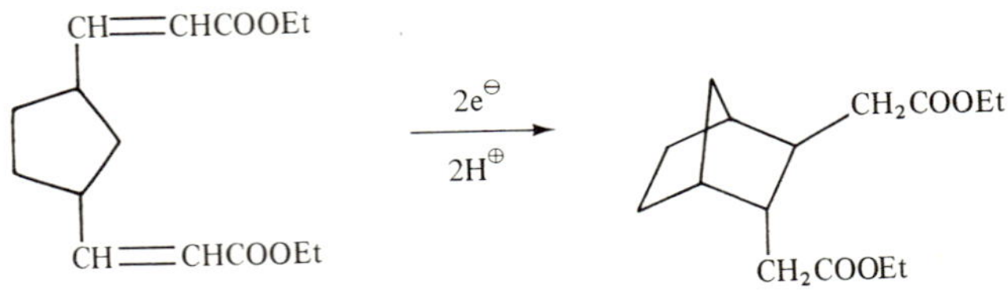

(43)

Only trans isomer formed

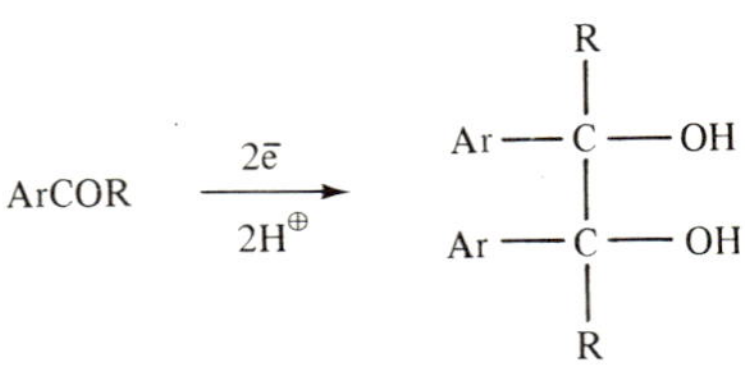

(44)

electrochemical reaction is not electrode controlled. Instead, it was proposed that interspecies hydrogen bonding between dimerizing radicals in acid solution [Eq. (45)] and between a radical and a radical anion in basic solution [Eq. (46)] would lead to the predominant formation of the dl isomer (in the hydrogen-bonded transition states, repulsions between nonbonded groups are smaller in the conformations leading to the dl isomer). The stereoselectivity would be more pronounced in basic medium since a hydrogen bond involving a negatively charged atom as an acceptor is the stronger one.

This mechanism could be tested by reducing a ketone in which intramolecular hydrogen bonding in the intermediate radical would compete with intermolecular hydrogen bonding between radicals. Indeed, it was found that 2-acetylpyridine gives predominantly the meso-pinacol in both acidic and basic media, as the above mechanism requires.

Another indication that the pinacol coupling process takes place in the solution and not at the electrode surface stems from the fact that the dl isomer formed from the reduction of acetophenone in the presence of chiral

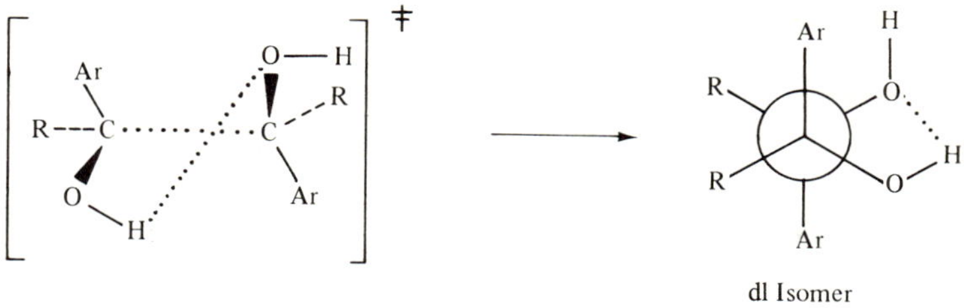

dl Isomer

(45)

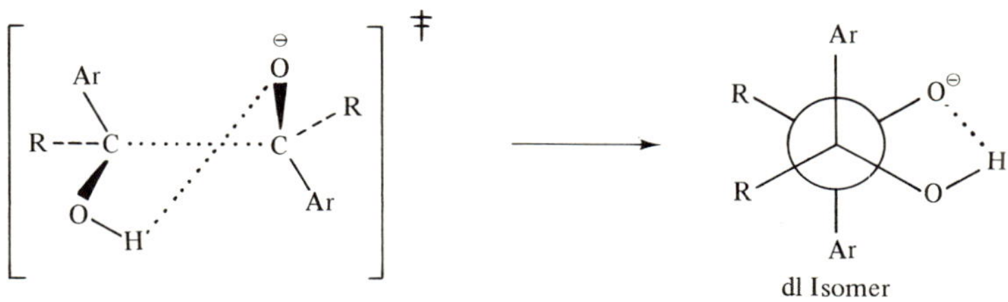

dl Isomer

(46)

supporting electrolytes is optically inactive, whereas the simultaneously formed methylphenylcarbinol displays optical activity (see Sec. VI).

Intramolecular pinacol formation from diketones of the general formula (XXVI) in alkaline ethanol at a mercury cathode gives the cis-pinacol as the major isomer [52], somewhat surprising in view of the results discussed

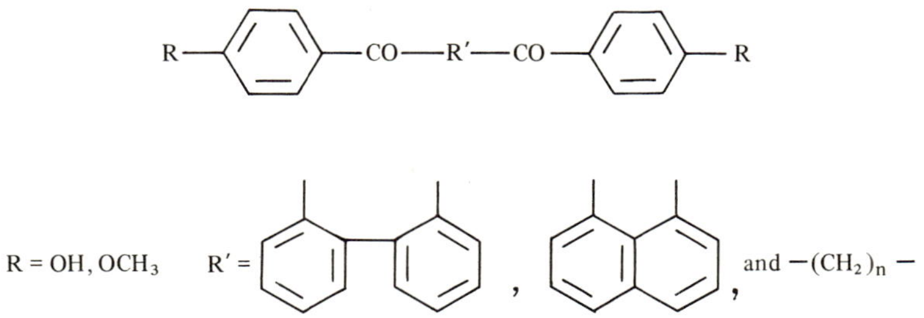

(47)

above (the cis-pinacol corresponds to the meso-pinacol in an intermolecular coupling). It has also been shown that the presence of adsorbed ions (iodide and tetraethylammonium ions) may influence the reductive coupling of benzaldehyde, in that the dl/meso ratio is lowered appreciably [53].

B. Cleavage Reactions

Optically active phosphines and arsines are readily available through the cathodic cleavage of optically active benzyltrialkylphosphonium and -arsonium salts. In both cases the reaction proceeds with retention of configuration [54-57]. See Chapter XXI.

C. Elimination Reactions

Vicinal dihalogenides eliminate the halogen atoms on cathodic reduction to form alkenes. The ease of reduction of vicinal dibromides has been found to be strongly dependent on the dihedral angle between the C-Br bonds, the reduction being easiest for dibromides with dihedral angles of 0 and 180°, respectively [58]. At a dihedral angle around 90° the half-wave potential assumes its most negative value. The compounds below demonstrate the unusually large effects of changing the dihedral angle:

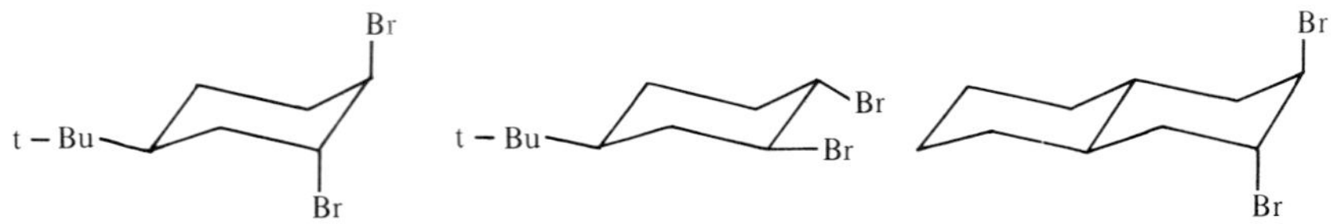

$E_{1/2}$,V −0.86 −1.67 −0.82

−1.21 −1.56

The meso and dl forms of 2, 3-dibromosuccinic acid and the corresponding diethyl esters have been studied polarographically at different pH values [59]. The meso-diacid, which is the more easily reducible isomer, gives fumaric acid over the whole pH range, consistent with a mechanism in which the thermodynamically favored conformation with all-trans substituents is the electroactive species:

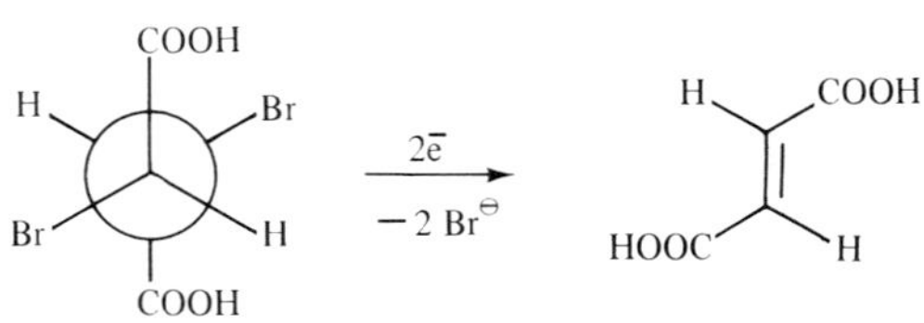

(48)

The dl-diacid was converted to fumaric acid at low pH values (below 0.4) and at high values (above 6.9), whereas in the intermediate pH range also maleic acid was formed with a maximum yield of 70% at pH 4. At this pH, the dl form exists as the monoanion predominantly, the only species capable of assuming a thermodynamically favored conformation with an intramolecular hydrogen bond between the carboxyl and carboxylate group and the bromine atoms trans:

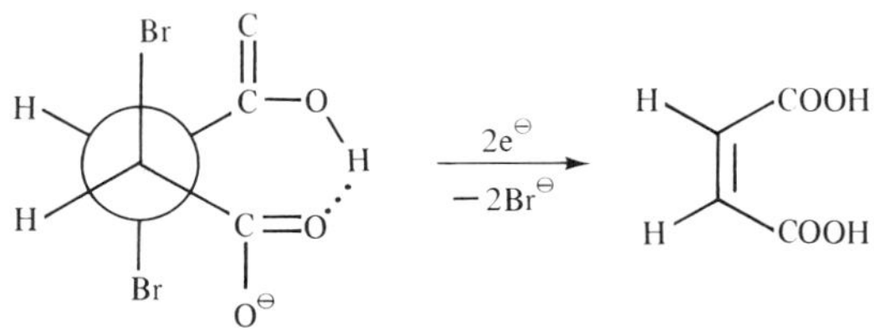

(49)

PART B [1]

VI. ASYMMETRIC INDUCTION BY USE OF OPTICALLY ACTIVE SUPPORTING ELECTROLYTES [60]

The double layer (Chapter I) has been visualized as consisting of an inner and an outer (diffuse) zone; the constitution of the double layer(s) depends on the physical properties of the electrode, supporting electrolyte, substrate, and solvent. Electrostatic considerations suggest that the concentration of the cation of the supporting electrolyte should be increased significantly in the inner zone. In order to gain further insight into the function of the supporting electrolyte, acetophenone was cathodically reduced in the presence of chiral salts. When (-)- or (+)-ephedrine hydrochloride [i.e., (-)-I and (+)-I below] is used, the methylphenylcarbinol formed (II) is optically active. The pinacol (III) which is produced simultaneously (41% yield) is optically inactive and consists of a mixture of dl and meso forms. When (-)-I•HCl is used, 44% of

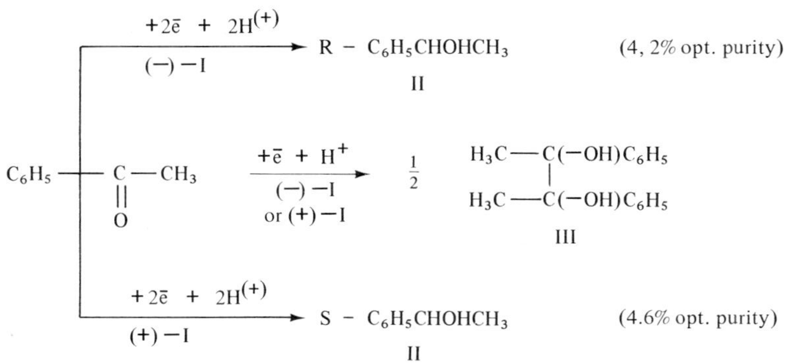

R(+)-methylphenylcarbinol is obtained in 4.2% optical purity; (+)-I•HCl leads to 38% of S-(-)-methylphenylcarbinol in 4.6% optical purity. This is the first demonstration that supporting electrolytes are critically involved in the construction of the double layer; they influence the mode of adsorption of the substrate and therefore the stereochemical course of the reaction.

TABLE 1

Dependence of the Configuration of Methylphenylcarbinol Upon N- and α-Substituents

Expt. No.	Ephedrine derivative: Supporting electrolyte		Ephedrine derivative: $C_6H_5CH(OH)-CH_3$ % optical purity	Expt. No.	Deoxyephedrine derivative: Supporting electrolyte		Deoxyephedrine derivative: $C_6H_5CH(OH)-CH_3$ % optical purity
1	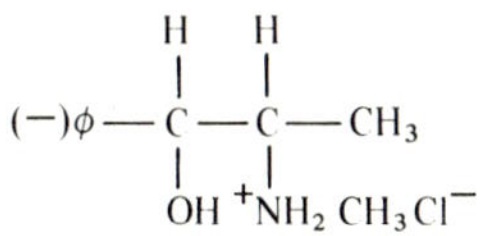	R, S	R: 4–5	7	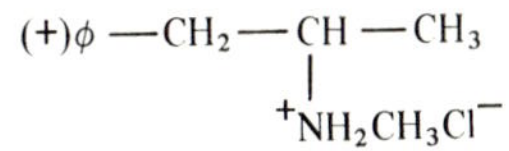	S	R: 1–2
2	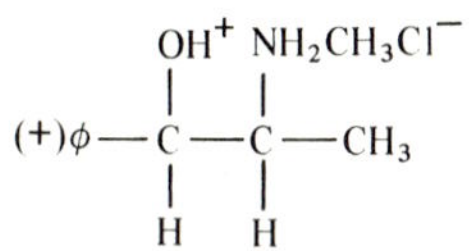	S, R	S: 4–5	8	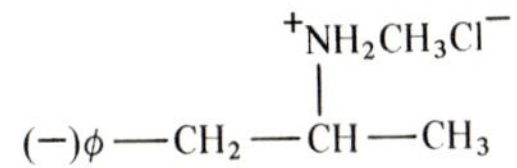	R	S: 1–2
3	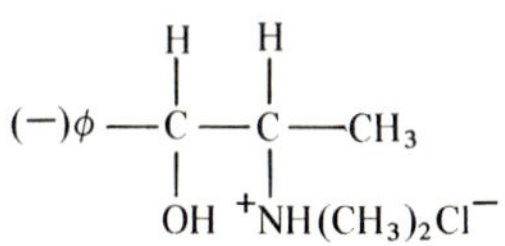	R, S	S: 2–3	9	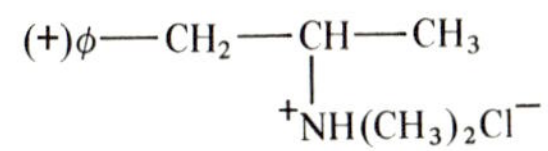	S	S: 1–2
4	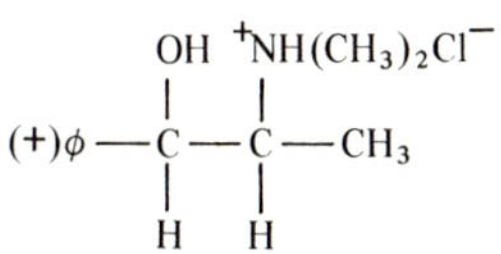	S, R	R: 2–3		—		—

Table 1 (cont.)

Ephedrine derivative			Deoxyephedrine derivative		
Expt. No.	Supporting electrolyte	$C_6H_5CH(OH)-CH_3$ % optical purity	Expt. No.	Supporting electrolyte	$C_6H_5CH(OH)-CH_3$ % optical purity
5	$(-)\phi-CH(OH)-CH(\overset{+}{N}(CH_3)_3Cl^-)-CH_3$ R, S	S: 7–10	10	$(-)\phi-CH_2-CH(\overset{+}{N}(CH_3)_3Cl^-)-CH_3$ S	S: 2
6	$(+)\phi-CH(OH)-CH(\overset{+}{N}(CH_3)_3Cl^-)-CH_3$ S, R	R: 7–10		—	—

The difficulty of arriving at a detailed "conformational analysis" of the interaction between substrate and supporting electrolyte in the double layer is illustrated in Table 1. In using ephedrine salts of like configuration of different degrees of N-methylation the configuration of the carbinol formed changes from R (Table 1, Expt. 1) toward S (Expts. 3 and 5). As expected, when the optical antipodes of the supporting electrolyte are employed, the optical antipodes, in the same optical purity, of the carbinols are obtained (Expts. 2, 4 and 6).

The hydroxy groups α to the phenyl ring also exert a significant influence. The deoxyephedrine salts of varying degrees of N-alkylation yield methylphenylcarbinols of the same configuration as is obtained from ephedrine salts but in lower optical purity.

Attempted Rationalization

Upon acceptance of one electron the acetophenone is transformed into a planar anion radical which leaves the double layer and subsequently dimerizes to dl- and meso-pinacol III.

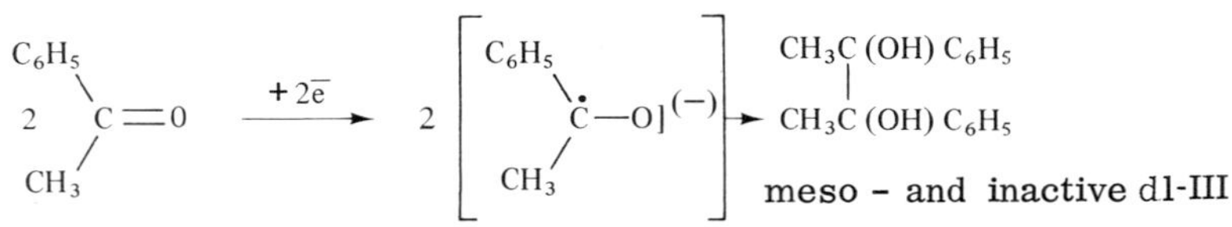

One-electron reduction of the anion radical produces a tetrahedral carbanion which, depending on the environment (in this case the chiral double layer), may retain configurational stability [34], accept a proton from "below," i.e., via the OH or NH group of the supporting cation, and thereby be partially converted to the chiral carbinol.

As a working hypothesis, it is possible to explain the asymmetric electrochemical reduction of acetophenone in the presence of dimethylephedrinium ion (Expts. 5 and 6 of Table 1) on the basis of the following assumptions:

1. The onium center $\overset{+}{N}(CH_3)_3$, the hydrogen atoms at the α- and β-carbon atoms, as well as the benzene ring lie unsolvated on the mercury surface; OH and CH_3 are oriented toward the solution.

2. A hydrogen bridge is formed between the carbonyl group of the substrate and the OH group of the supporting cation.

3. The phenyl group of the acetophenone, according to Cram's rule, prefers a position above the phenyl group of the supporting electrolyte; the CH_3 group of the acetophenone is oriented preferentially near the largest group, the $\overset{+}{N}(CH_3)_3$, of the electrolyte.

4. The electrons are transferred from the mercury surface to the carbonyl group of the preoriented acetophenone and thereby generate a chiral carbanion. The representation of the transition state, for which further experimental support will be sought, is shown in Fig. 1.

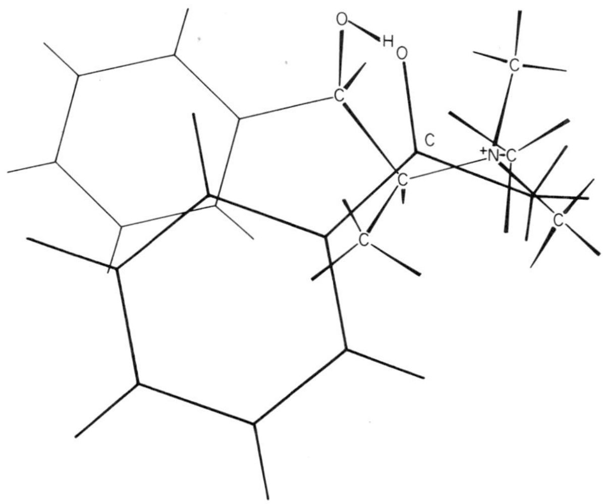

Fig. 1. Possible transition state in reduction of acetophenone.

A second representation of the transition state is proposed to account for the difference in results between Expt. 1, Table 1 (in which using ephedrine hydrochloride as chiral electrolyte the R configuration of the carbinol predominates), and Expt. 3, Table 1. In the former case too the benzene ring, the two H atoms of the side chain, and the onium center probably lie on the mercury; the OH and CH_3 groups are then away from the mercury. The acetophenone molecule approaches the supporting electrolyte in such a manner that, in conformity with Cram's rule, the carbonyl oxygen is close to the OH group and the carbonyl C is opposite the onium center (Fig. 2). In this transition state the electrons are transferred from "below" to the carbonyl C; the chiral carbanion accepts a proton from the same side, perhaps from the protonated nitrogen.

Comparison of the results with deoxyephedrine derivatives and with the various ephedrine salts leaves no doubt that the OH group of the latter is involved. While it is true that the deoxy derivatives lead to carbinols of the same configuration as those from the ephedrine derivatives, the extent of asymmetric induction is smaller in the former case. This shows that in accordance with Cram's rule the size of the groups involved in the transition state is responsible for the chirality of the transition state.

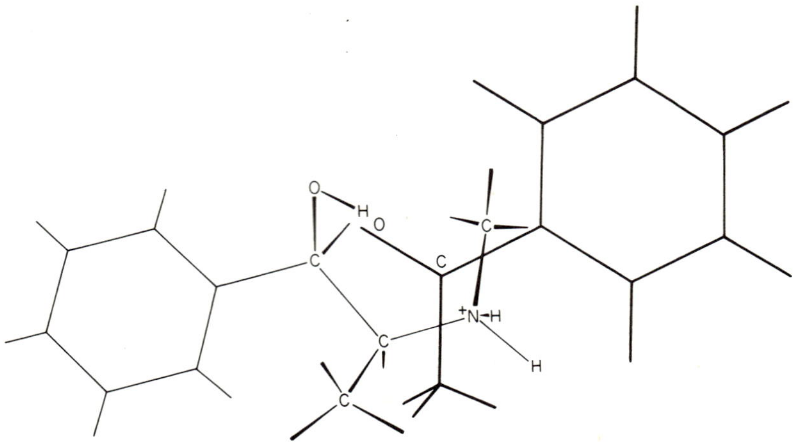

Fig. 2. Possible transition state in reduction of acetophenone.

Cathodic reduction of propiophenone in the presence of some of the above ephedrine derivatives yields carbinols of the same configuration as those derived from acetophenone, but the extent of asymmetric induction is reduced about 20-30%.

Finally it must be emphasized that the above presentation of the possible mode of interaction between the substrate and the chiral supporting electrolyte is only a working hypothesis which must be tested by further experiments. The asymmetric inductions described are unambiguous and, within experimental limits, reproducible; other spatial models might be proposed to account for the results.

NOTE

[1]Translated by the Editor.

REFERENCES

[1]. B. J. Piersma, in Electrosorption (E. Gileadi, ed.), Plenum, New York, 1967, Chap. 2.

[1a]. J. M. Hale, in Reactions of Molecules at Electrodes (N. S. Hush, ed.), (Wiley-Interscience, New York, in press.

[2]. H. Dahms and M. Green, J. Electrochem. Soc., 110, 1075 (1963); J. O'M. Bockris, M. Green, and D. A. J. Swinkels, ibid., 111, 743 (1964).

[3]. X. de Hemptinne, Ann. Soc. Sci. Bruxelles, 80, 140 (1966).

[4]. E. I. Heiba, R. M. Dessau, and W. J. Koehl, Jr., J. Amer. Chem. Soc., 91, 6830 (1969).

[5]. K. A. Bilevich and O. Yu. Okhlobystin, Russ. Chem. Rev. (English Transl.), 37, 954 (1968).

[6]. F. D. Mango and W. A. Bonner, J. Org. Chem., 29, 1367 (1964).

[7]. V. D. Parker and L. Eberson, Chem. Commun., 340 (1969).

[8]. K. Koyama, T. Ebara, T. Tani, and S. Tsutsumi, Can. J. Chem., 47, 2484 (1969).

[9]. L. Eberson, K. Nyberg, M. Finkelstein, R. C. Petersen, S. D. Ross, and J. J. Uebel, J. Org. Chem., 32, 16 (1967).

[10]. S. Winstein, C. G. Kreiter, and J. I. Braumann, J. Amer. Chem. Soc., 88, 2047 (1966).

[11]. R. Huisgen, G. Boche, and H. Huber, ibid., 89, 3345 (1967).

[12]. M. Finkelstein, R. C. Petersen, and S. D. Ross, Tetrahedron, 23, 3875 (1967).

[13]. A. J. Baggaley and R. Brettle, J. Chem. Soc. (C), 2055 (1968).

[14]. T. Inoue, K. Koyama, T. Matsucka, and S. Tsutsumi, Bull. Chem. Soc. Japan, 40, 162 (1967).

[15]. S. D. Ross, M. Finkelstein, and J. J. Uebel, J. Org. Chem., 34, 1018 (1969).

[16]. K. Yoshida and T. Fueno, Bull. Chem. Soc. Japan, 42, 2411 (1968).

[17]. W. B. Smith and H. Gilde, J. Amer. Chem. Soc., 81, 5325 (1959).

[18]. W. B. Smith and H. Gilde, ibid., 83, 1355 (1961).

[19]. R. F. Garwood, C. J. Scott, and B. C. L. Weedon, Chem. Commun., 14 (1965).

[20]. C. Walling and W. Thaler, J. Amer. Chem. Soc., 83, 3877 (1961).

[21]. K. N. Campbell and E. E. Young, ibid., 65, 965 (1943).

[22]. R. A. Benkeser and C. A. Tincher, J. Org. Chem., 33, 2727 (1968).

[23]. L. Horner and H. Röder, Liebigs Ann. Chem., 723, 11 (1969).

[24]. R. N. Gourley, J. Grimshaw, and P. G. Millar, Chem. Commun., 1278 (1967).

[25]. S. Wawzonek, E. W. Blaha, R. Berkey, and M. E. Runner, J. Electrochem. Soc., 102, 235 (1955).

[26]. R. Dietz, M. E. Peover, and B. E. Larcombe, Discuss. Faraday Soc., 45, 161 (1968).

[27]. A. J. Fry and J. H. Newberg, J. Amer. Chem. Soc., 89, 6374 (1967).

[27a]. L. Horner and D. H. Skaletz, Tetrahedron Lett., 3679 (1970).

[28]. L. Eberson and H. Sternerup, Acta Chem. Scand., 26, 1431 (1972); cf. also J. Dirlam, L. Eberson, and H. Sternerup, Chem. Ing. Tech., 44, 178 (1972); and J. Dirlam and L. Eberson, Acta Chem. Scand., 26, 1454 (1972).

[29]. P. J. Elving and B. Pullman, Adv. Chem. Phys., 3, 1 (1961).

[30]. J. W. Sease, P. Chang, and J. L. Groth, J. Amer. Chem. Soc., 86, 3154 (1964).

[31]. F. L. Lambert, J. Org. Chem., 31, 4184 (1966).

[32]. J. W. Sease, F. G. Burton, and S. L. Nickol, J. Amer. Chem. Soc., 90, 2595 (1968).

[33]. W. W. Hussey and A. J. Diefenderfer, ibid., 89, 5359 (1967).

[34]. J. Grimshaw and J. S. Ramsey, J. Chem. Soc. (B), 60 (1968).

[35]. C. K. Mann, J. L. Webb, and H. M. Walborsky, Tetrahedron Lett., 2249 (1966).

[36]. R. Annino, R. E. Erickson, J. Michalovic, and B. McKay, J. Amer. Chem. Soc., 88, 4424 (1966).

[37]. B. Czochralska, Chem. Phys. Lett., 1, 239 (1967); Z. R. Grabowski, B. Czochralska, A. Vincenz-Chodkowska, and M. S. Balasiewicz, Discuss. Faraday Soc., 45, 145 (1968).

[38]. L. Eberson, unpublished observations.

[38a]. R. E. Erickson and C. M. Fischer, J. Org. Chem., 35, 1604 (1970).

[39]. A. J. Fry and R. H. Moore, ibid., 33, 1283 (1968).

[40]. R. E. Erickson, R. Annino, M. D. Scanlon, and G. Zon, J. Amer. Chem. Soc., 91, 1767 (1969).

[41]. A. J. Fry and M. Mitnick, ibid., 91, 6208 (1969).

[42]. N. S. Hush and G. A. Segal, Discuss. Faraday Soc., 45, 23 (1968).

[43]. J. D. Anderson and M. M. Baizer, Tetrahedron Lett., 511 (1966).

[44]. J. D. Anderson, M. M. Baizer, and J. P. Petrovich, J. Org. Chem., 31, 3890 (1966).

[45]. J. P. Petrovich, J. D. Anderson, and M. M. Baizer, ibid., 31, 3897 (1966).

[46]. J. H. Stocker and R. M. Jenevein, ibid., 33, 294 (1968).

[47]. J. H. Stocker and R. M. Jenevein, ibid., 33, 2145 (1968).

[48]. J. H. Stocker and R. M. Jenevein, ibid., 34, 2807 (1969).

[49]. J. H. Stocker, R. M. Jenevein, and D. H. Kern, ibid., 34, 2810 (1969).

[50]. J. H. Stocker, and R. M. Jenevein, Symposium Abstracts, The Synthetic and Mechanistic Aspects of Electroorganic Chemistry, U.S. Army Research Office, Durham, N.C., Oct. 14-16, 1968, p. 221.

[51]. J. H. Stocker and R. M. Jenevein, Extended Abstracts, Meeting of the Electrochemical Society, New York, May 4-9, 1969, p. 359.

[52]. R. N. Gourley and J. Grimshaw, J. Chem. Soc. (C), 2388 (1968); cf. also J. Grimshaw and J. S. Ramsey, ibid., 653 (1966).

[53]. V. J. Puglisi, G. L. Clapper, and D. H. Evans, Anal. Chem., 41, 279 (1969).

[54]. L. Horner, H. Winkler, A. Rapp, A. Mentrup, H. Hoffman, and P. Beck, Tetrahedron Lett., 161 (1961).

[55]. L. Horner and A. Fuchs, ibid., 203 (1962).

[56]. L. Horner, H. Fuchs, H. Winkler, and A. Rapp, ibid., 965 (1963).

[57]. L. Horner and H. Fuchs, ibid., 1573 (1963).

[58]. J. Závada, J. Krupicka, and J. Sicher, Coll. Czech. Chem. Commun., 28, 1664 (1963).

[59]. P. J. Elving, I. Rosenthal, and A. J. Martin, J. Amer. Chem. Soc., 77, 5218 (1955).

[60]. L. Horner and D. Degner, Tetrahedron Lett., 5889 (1968).

[61]. D. J. Cram, Fundamentals of Carbanion Chemistry, Academic, New York, 1965.

Section I

PRESENT AND FUTURE APPLICATIONS OF ORGANIC ELECTROCHEMISTRY

Chapter XXVIII

INDUSTRIAL ELECTROORGANIC CHEMISTRY

D. Danly

Textiles Division
Monsanto Company
Pensacola, Florida 32502

I. INTRODUCTION

Although the history of the electrochemical industry can be traced back to the early 1800s, few processes for electrolytic production of organic chemicals have been commercialized. The lack of interest in producing organic chemicals by electrolysis cannot, in general, be attributed to a deficiency in the basic chemical research since considerable work was carried on in this area during the nineteenth and early part of the twentieth centuries. The failure to become a common mode of large-scale chemical processing is probably due to several factors: (1) scale-up limitations imposed by the dimensionality of an electrode process (two dimensional); (2) relatively poor selectivities and current efficiencies stemming from polarization effects; and (3) inadequacies in diaphragms in separating anolyte and catholyte constituents. In recently commercialized electroorganic processes, the inherent problem of dimensionality has been offset, at least to a degree, either by simplicity of design of plate-and-frame cells or through the use of beds of small electrode particles. Research and development studies in the field of electrodialysis have contributed some novel approaches to the solution of polarization problems and have been both an effect and cause of major advances in the technology of ion permselective membranes. Also, with ever-mounting concern over environmental pollution, the high-selectivity, low-by-product electrochemical process may offer a distinct advantage over a less efficient chemical route, even though the processing costs are comparable.

The electrochemical engineering aspects of the commercialization of an electrolytic process are thoroughly covered in Mantell's monograph [1], while problems peculiar to the scale-up of organic processes are discussed by MacMullin [2]. The technology of electrochemical cells and techniques in Europe are extensively covered in texts by Billiter [3] and Regner [4]. The book by Wilson [5] on electrodialysis is a helpful reference with regard to turbulence-promoting spacers, cell configurations, and current leakage analysis. A review of the literature of industrial electrochemistry has been published by Haglind [6a]; an assessment of recent trends has been made by Kuhn [6b].

While the research electroorganic chemist is accustomed to considering such variables as electrode material, current density, electrolyte composition, pH, and temperature, certain other information is important in the study of a potentially commercializable process. Some of these concerns which are discussed further in this chapter are: (1) the physical and chemical stability of the electrodes, (2) the susceptibility of the electrolyte to deterioration and the electrodes to fouling, (3) the diaphragm characteristics, (4) the optimum balance between the cost of power and the cost of cells and auxiliaries, and (5) means of providing adequate mass-transfer rates at the electrode and diaphragm interfaces. Questions that will likely face the research chemist during the scale-up are as follows: How much temperature rise can be tolerated? What is the limiting current density? What is the life of the electrodes and of the diaphragm? What other materials are suitable? How much agitation is needed? How is the product recovered and what are the problems in recycle of electrolytes? Although many of these questions will not be fully answered without an extensive process development program, knowledge of the kind of information that will be needed can guide the researcher in conducting an effective laboratory study.

A number of general articles reviewing recent technology for production of organic chemicals by electrolysis have appeared [7-13], the most comprehensive of which is that by Fioshin and Tomilov [7]. The patent literature on electroorganic processes has blossomed in the past decade, giving testimony to an awakening of industry to the potentialities of this relatively old area of organic chemistry.

This chapter is divided into two main sections: the first covers the considerations and techniques involved in scale-up of an electrolytic process for production of organic chemicals, while the second part describes the technology of several processes practiced commercially or developed to the point of commercialization.

II. CELL DESIGN

A. General Considerations

1. Continuous vs. Batch Operation

While laboratory studies in electroorganic synthesis are generally conducted using batchwise cell operation, the processes are usually adapted to continuous cell operation for commercial installations. The advantages of batch operation during the exploratory work are similar to those in conventional chemical reactions. One of the foremost advantages is the ability to follow the rates of competing reactions as a function of the changing electrolyte composition. Curves relating product selectivity and current efficiency to the total current passed are extremely helpful in subsequent scale-up of the research setup.

Continuous operation of large-scale electrolysis systems is preferred, however, since it provides for better control of electrolyte temperature and composition, which in turn generally leads to higher product selectivity and better current efficiency. Also in continuous systems the current and heat removal loads are constant with time, allowing more efficient design of the electrical supply equipment and heat exchangers.

In continuous operation, catholyte and anolyte of fixed composition are added at a steady rate to the cell and equal quantities are continuously withdrawn to maintain a constant inventory of reactants and products within the cell. The electrolytes may make only a single pass through the cell or they may be constantly recirculated through an external flow loop with continuous makeup and withdrawal (Fig. 1). This forced circulation through an external loop provides a convenient means of removing the heat generated in the cell. The degree of conversion of reactant is determined by the ratio of the cell current to the feed rate of reactant to the cell (or circulation loop, if one is employed). For reaction systems in which selectivity to the desired product and/or current efficiency fall off with the conversion level, the yields of a

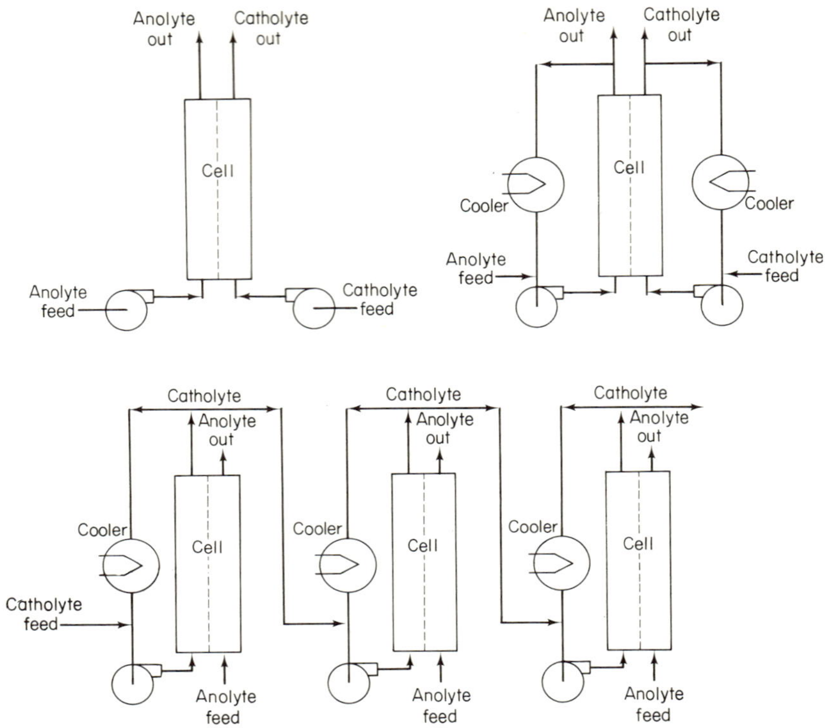

Fig. 1. Arrangements of continuous cell units.

batch system can be approaching by staging of several continuous cells as shown in Fig. 1. Overall conversion of reactant might, for example, be 20% leaving the first cell loop, 40% leaving the second, and 60% from the third. Any number of stages might be employed, but three to five are generally sufficient to approximate batch performance. The selectivity of the staged system would be higher than if all three cells were operated at 60% conversion with parallel feed. These considerations are seen to be analogous to those encountered in the analysis of the kinetics of a conventional organic reaction. From the appropriate batch experiment one can predict the performance of a continuous "once-through" cell, a circulating loop cell, or a series of either of these types.

2. Voltage-Current Relationships

While the overall voltage required to give the desired cell current is usually of only minor concern to the research chemist in his exploratory studies, the success or failure of the commercial process may rest heavily on the ability to design a cell with a favorable voltage-current relationship. Although electrical power consumption in electroorganic processes generally represents less of an economic factor than in inorganic electrolytic processes, minimization of power usage is nevertheless important in most instances.

The overall cell voltage is composed of several components, as listed below:

(1) thermodynamic cell potential

(2) anode overvoltage

(3) anolyte ir drop

(4) diaphragm ir drop (if diaphragm used)

(5) catholyte ir drop

(6) cathode overvoltage.

Methods of calculating these voltages have been described by MacMullin [2]. In addition to the above, there may be electrical power losses external to the cell itself in the busbars and electrical leads. Current leakage through electrolyte conduits is also a concern in cell design.

In cells employing membranes, a concentration potential may be developed across the membrane. In the case of completely permselective membranes this potential is described by the Nernst equation; for partially permselective diaphragms, evaluation of the potential can be calculated by other means [14].

While design for minimum electrode overvoltages through the selection of low-overvoltage electrode materials may be feasible, the choice of suitable candidates is generally limited by other factors, such as electrode stability and reaction selectivity. The voltage drops through the electrolytes and diaphragm are usually more susceptible to optimization in the cell design than are the other components. The electrolyte ir drops are minimized by the cell geometry, that is, positioning the electrode to minimize the electrical flow path through the anolyte and catholyte. Diaphragm voltage drops are varied through changes in thickness, porosity, and direction of permeation, in the case of nonpermselective diaphragms. If ion permselective membranes are employed, the voltage drop will also depend on such factors as ion-exchange capacity and solvent content of the membrane. Development of a satisfactory membrane can be a major undertaking in itself.

B. Cell Geometry

A wide variety of geometrical configurations have been described for use in commercial electroorganic synthesis. The choice of cell geometry is guided predominantly by two factors, both of which can ultimately be translated into process economics. These are (1) cell cost per unit of productivity and (2) cost of electrical power per unit of product.

Since multiple cells are normally required for the commercial production rates, the total cost of the cells is reduced as their current-carrying capacity increases. This can be accomplished by raising the current density and/or the active electrode area of the cells. Other factors affecting cell cost are materials of construction and complexity of design.

Power usage is minimized by designing for the lowest practical cell voltage for a given current density, requiring minimum spacing between electrodes. High current density, which favors minimum cell investment, also increases cell voltage and power cost per unit of product. The opposing economic factors lead to an economic optimization problem with regard to cell current density, which is discussed in more detail in a later section.

The ideal electrolytic cell has both a low cost per ampere passed and a low voltage drop. The relative importance of cell cost versus power usage, together with a number of special process requirements related to polarization effects and electrode and diaphragm type, gives rise to numerous cell designs.

The general types of cell designs can be classified as follows:

(1) filter press or plate-and-frame cell

(2) plate electrodes in tanks

(3) particulate electrode cells

  (a) fixed beds
  (b) suspended beds.

1. Filter Press or Plate-and-Frame Cells

Probably the most common type of cell for large-scale electroorganic preparations is the filter press or plate-and-frame type. This design consists of plate electrodes separated by insulating gaskets or frames, which form compartments for the electrolyte. The electrodes may be unsupported or they may be mounted on nonconducting plates. A diaphragm may be employed between a pair of gaskets to provide hydraulically separate anode and cathode compartments. A number of individual cells are combined in a single closure, which may consist of a filter press or simply end plates drawn together by tie bolts. Electrolyte flow to the cells may be in series or in parallel, as may also be the flow of electrical current.

A design that is claimed to be particularly suited to electrolytic preparation of organic compounds is due to Foreman and Veatch [15]. As seen in Fig. 2, the design is very simple, consisting of a stack of alternating

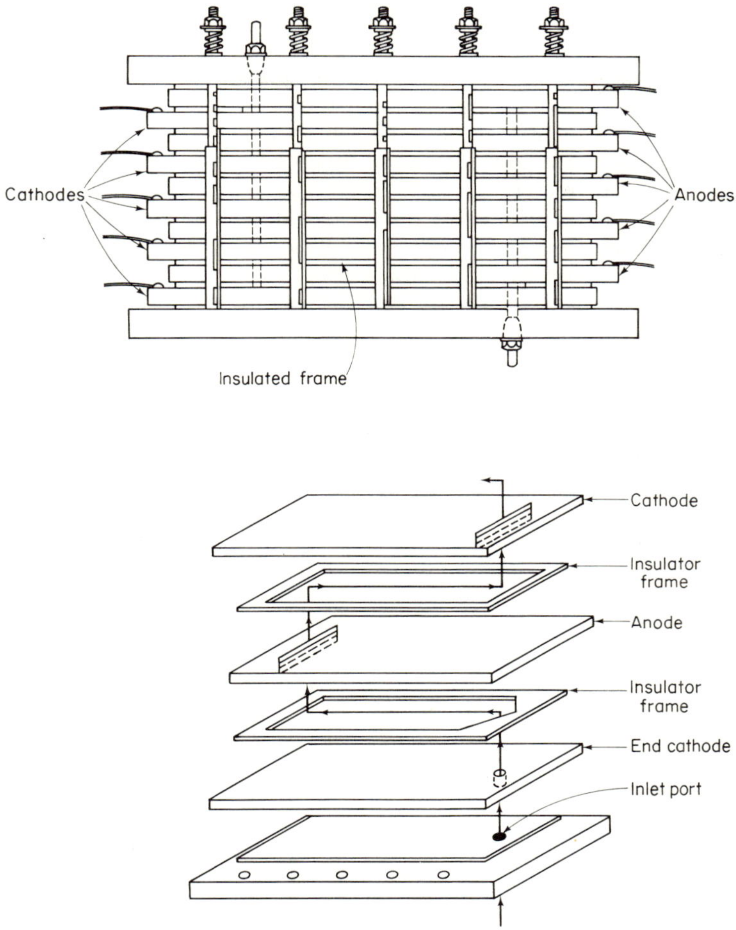

Fig. 2. Foreman/Veatch cell.

electrode plates and insulating frames. Electrolyte flows serially from compartment to compartment through slots in the electrodes, while the supply of electrical current is by parallel connections to alternate electrodes. Cooling may be provided, if required, by inserting a heat-exchanger plate, consisting of a hollow metallic box containing internal cooling coils.

The Foreman/Veatch design is suitable for either oxidation or reduction reactions, but is limited to systems in which a diaphragm is not needed. Spacings of the anode and cathode at distances as close as $\frac{1}{8}$ in. have been demonstrated. The preferred range of current densities is reported to be

30-500 A/ft^2, used in services such as the manufacture of bromohydrin from acrolein.

A 2000 A cell of the plate-and-frame type has been described by Mantell [10]. This cell, designed for commercial production of periodic acid, employs 34 in. × 38 in. electrodes. Cation permselective membranes are used to separate anode and cathode compartments, which are 2 in. deep. These compartments, formed by 3 ft × 3 ft frames of PVC, are fed by parallel flow of electrolyte. Electrical flow is also in parallel. Although not specifically designed for electroorganic reactions, this cell could undoubtedly be adapted to organic processes with only minor modifications. One change that would probably be required for organic systems would be the drastic reduction in electrode compartment depth in order to keep electrolyte ir drops within an acceptable range.

A plate-and-frame design incorporating diaphragms is used by Monsanto in its commercial facilities for production of adiponitrile via electroreductive coupling of acrylonitrile [16]. In the Monsanto design (Fig. 3) a lead cathode and a lead alloy anode are attached to opposite sides of a polypropylene plate, with the electrodes connected electrically by lead lugs passing through the plate. The internal connection provides an efficient and low-cost means of

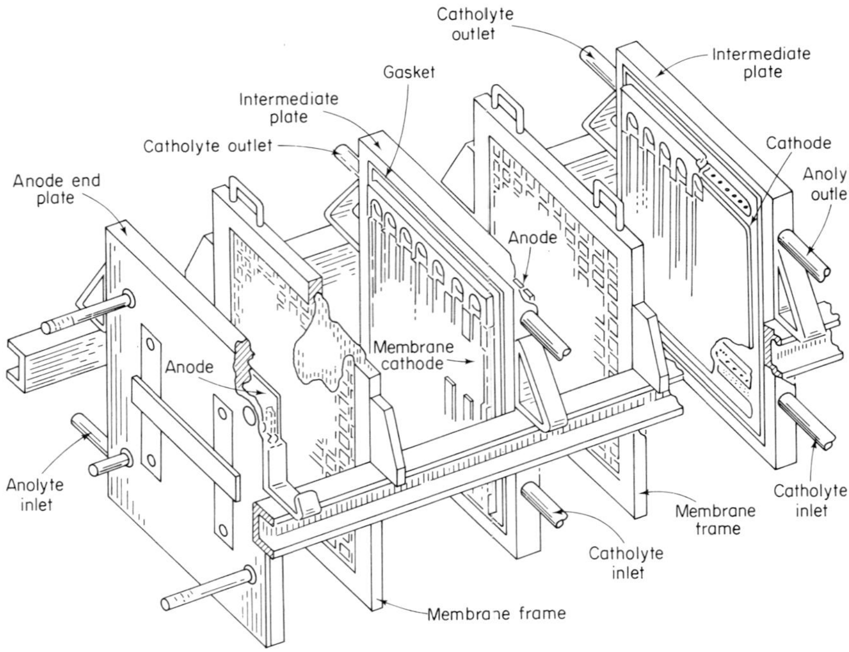

Fig. 3. Monsanto cell.

series electrical connection between the anode of one cell and the cathode of the adjacent one. An ion-exchange membrane, mounted between a pair of gasketed polypropylene frames, is positioned between two electrode plates to form the anode and cathode compartments of a single cell.

Plastic spacers are used to keep the membranes from touching the electrodes. An open-mesh polypropylene screen, fused to the anode side of the membrane frame, serves the dual function of providing reinforcement against outward distortion from hydraulic forces and preventing the membrane from touching the anode. Thin plastic strips are fastened to the cathode on about 1 in. centers to support the membrane at the desired distance from the electrode; a strip thickness from 1/16 to 3/16 in. may be employed. By regulating the internal pressures such that the anolyte compartment is at a slightly higher pressure than the catholyte compartment, the membrane is held against the cathode strips, maintaining precisely the desired cathode-to-membrane distance over the entire electrode.

For maximum yields in the electroreductive coupling of acrylonitrile, high rates of diffusion of both reactants and products at the cathode surface are necessary. The Monsanto design minimizes diffusional effects by operating in the turbulent flow regime, employing flow velocities of 4-8 ft/sec through the cathode compartment. Anolyte and catholyte are introduced and distributed across the bottom of the electrode compartments through cavities formed in the plastic plates behind the bottom edge of the vertically mounted electrodes. Similar cavities near the top of the frame provide for withdrawal of anolyte and catholyte. Thus, electrolyte flow through the cells is in parallel, while electrical flow through the bank of cells is in series.

Current densities as high as 1.0 A/cm^2 are possible with this design, by virtue of the high electrolyte recirculation rates which can be accommodated. The high rates not only ameliorate polarization effects but also provide for economical removal of heat via external exchangers.

A variation of the plate-and-frame design, having many similarities to electrodialysis cells, has been proposed by Murakoshi et al. [17]. The cell, shown in Fig. 4, consists of a stack of individual cells, each of which consists of a sequence of anode sheet, anolyte gasket, ion-exchange membrane, catholyte gasket, and cathode sheet. Anolyte and catholyte are distributed to the respective electrolyte compartments by means of internal channels formed by holes in the sheets. A flow diffuser zone in the vicinity of the peripheral channels serves to distribute the liquid evenly across the breadth of the electrode compartment and to lengthen the electrical path through the internal electrolyte headers, reducing loss in current efficiency due to current bypassing within the stack of cells. A turbulence-promoting plastic screen is positioned within the gasket frames to maintain the membrane at the desired distance from the electrode and to improve diffusion rates at the membrane and electrode surfaces.

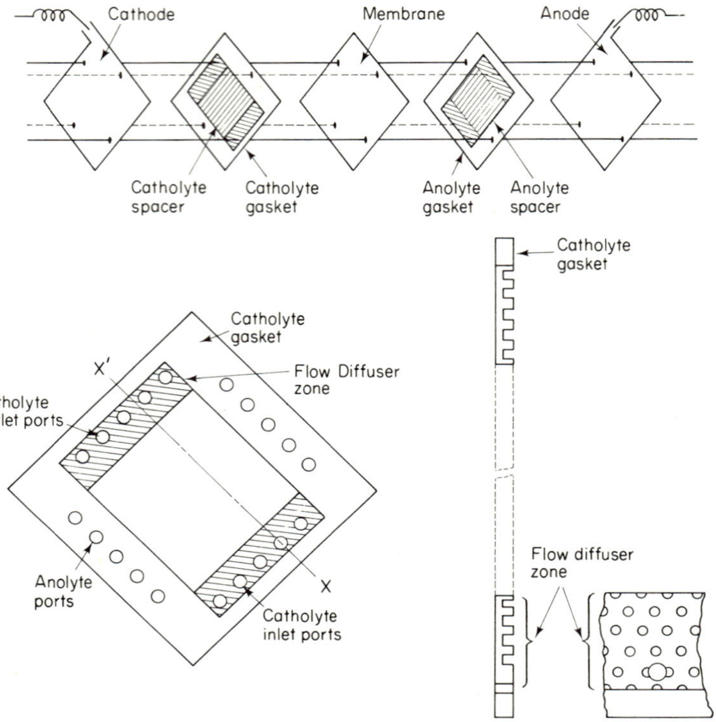

Fig. 4. Asahi cell.

2. Plate Electrodes in Tanks

The use of plate electrodes immersed into a lined, rectangular tank represents a direct extension of laboratory beaker cells. The tank may be fitted with a cover if collection of gaseous products is desired. One example of this type of cell proposed by Holmes and Gardiner [18] is 64 cm by 70 cm by 137 cm high. It is fitted with 20 platinum-coated titanium anodes 50 cm by 75 cm by 2.5 mm on 19 mm centers. Cathode sheets 6 mm thick are positioned between each anode to give a cathode-to-anode spacing of 5 mm. Since both sides of the electrodes are active, 186 dm^2 of anode surface is provided.

A 1000 A tank-type cell has been described for semicommercial production of isobutyric acid [19]. The body of the 70 × 40 × 40 cm cell is constructed from sheet vinyl using welded joints (Fig. 5). Several rows of anodes of lead dioxide electroplated on a nickel grid are suspended from the tank cover. A vinyl frame encases each anode to protect the nickel base metal from attack by the dilute sulfuric acid electrolyte. These frames may be

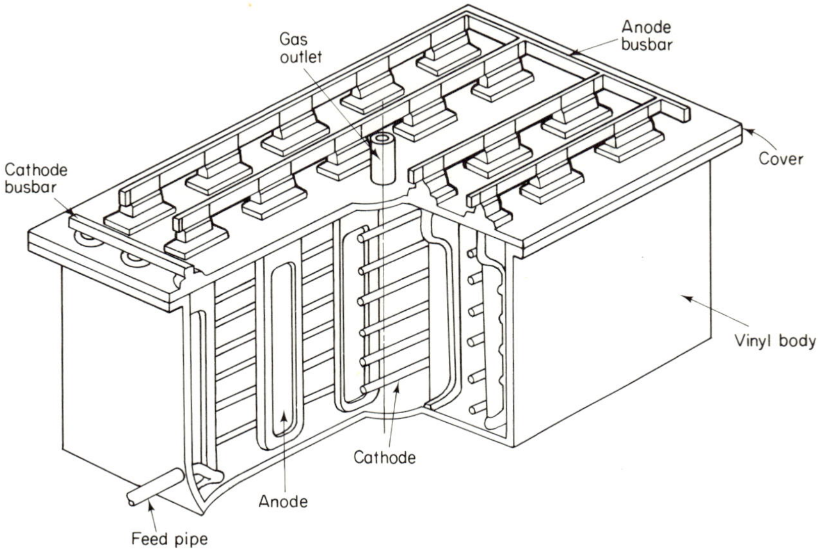

Fig. 5. Kryschenko cell for synthesis of isobutyric acid.

eliminated if passivating base metals such as tantalum or titanium are used. Planar coils of copper are positioned between the rows of anodes to serve the dual function of cathode and cooling means.

A direct scale-up of the rotating electrode concept, frequently employed in bench-scale studies, has been proposed for the electrolytic reduction of salicylic acid to salicylaldehyde [20]. Three cylindrical copper cathodes, 10 cm in diameter, are rotated vertically at 1800 rpm in an 80 × 35 × 28 cm deep tank. Perforated lead alloy anodes are positioned in anode compartments, formed by wrapping microporous rubber diaphragms around wooden supports. Current densities of 0.12-0.15 A/cm^2 are employed, giving the cell a total current-carrying capacity of about 300 A.

An elaborate variation of the tank-type cell has been proposed by Le Duc [21] for electrochemical production of olefin oxides. Special features of this cell include (1) hollow porous anode plates of graphite through which olefin gas is introduced; (2) flattened tubular cathodes formed from steel wire screen; and (3) a diaphragm of asbestos fiber deposited on the outside surfaces of the cathode tubes (Fig. 6). The cathodes and anodes are spaced alternately in the tank, providing an anode-to-cathode gap of about 10 mm. Chambers in the base and sides of the tank provide means of distributing the incoming olefin feed and withdrawing cathode products. A dilute brine solution is fed into the tank, which serves as the anode compartment, and passes through the diaphragm into the cathode chamber, formed within the asbestos-

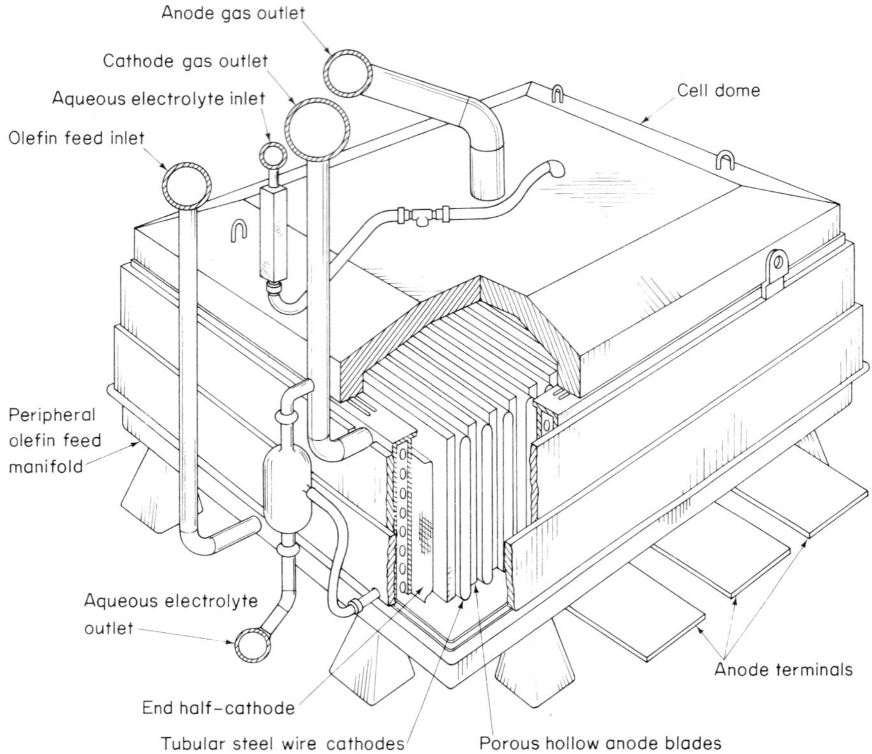

Fig. 6. Le Duc cell for olefin oxidation.

covered steel tubes. An assembly containing 16 anode plates, 15 intermediate cathodes, and two half-cathodes at the ends is capable of passing 31, 000 A, equivalent to a current density of about 110 A/dm^2 of apparent electrode area.

3. Particulate Electrode Cells

A number of cell designs have been proposed employing electrodes in the form of small particles rather than flat or cylindrical plates. The primary advantage of this design feature is to provide a high ratio of electrode surface to cell volume. The variations of particulate-electrode cells closely parallel the designs of catalytic reactors and may be divided into similar categories — fixed bed and suspended bed.

a. Fixed Bed. Fixed bed electrolytic cells are typically cylindrical vessels filled with metal shot. The wall of the vessel frequently serves as one electrode, while the shot bed acts as the other. An example of such a cell, used by Braithwaite of Nalco Chemical [22] to produce tetraalkyllead, consists of a 50 mm diameter by 75 cm long steel pipe filled with lead pellets. A lining of woven polypropylene screen is placed inside the pipe to insulate the lead from the steel. The pipe is made cathodic and a lead rod is inserted in the center of the shot bed to serve as electrical connector to the anodic lead particles. Electrolyte consisting of an ether solution of Grignard reagent and excess alkyl halide is circulated through the cell, combining electrolytically with the lead anode to produce tetraalkyllead.

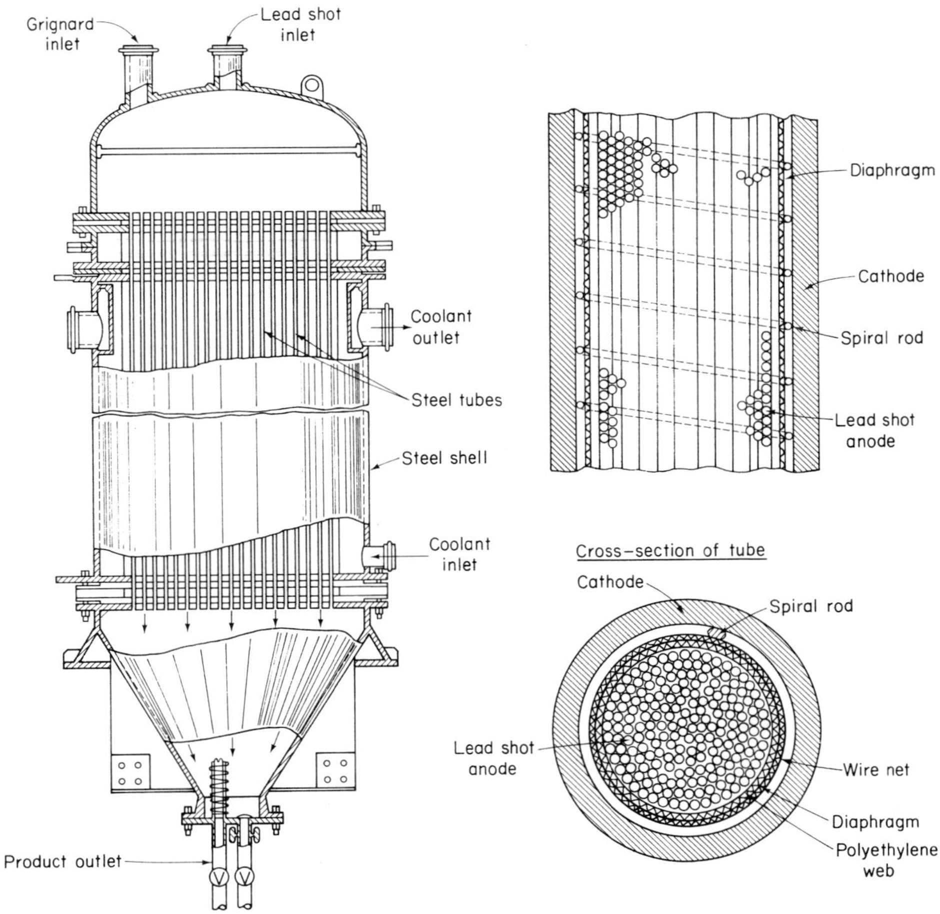

Fig. 7. Nalco cell for TML/TEL.

The commercial cells employed by Nalco at Freeport, Texas have a capacity of 3-4 million lb/yr tetraalkyllead per unit. These cells contain a large number of cylindrical tubes, arranged in an outer vessel as in a shell-and-tube heat exchanger [23]. The tubes are made of steel and serve as cathodes as well as external walls of the individual cell elements (Fig. 7). A cylindrical diaphragm assembly made of polyethylene webbing is positioned within the tube. A metallic wire net surrounds the diaphragm and a spirally wound wire separates the net from the tube wall. Lead shot is placed inside the cylindrical diaphragm, acting as a sacrificial anode. Means are provided at the top of the cell element for introduction of additional lead shot as it is consumed in the electrode reaction. The ether solution of alkylmagnesium halide enters the central anode compartment and flows outward through the diaphragm and down the spiral catholyte chamber to the collection sump at the base of the multitubed cell unit. An improved version of the Nalco cell has been described [24] which employs a porous ceramic lining on the inside walls of the tubular steel cathode in place of the plastic fabric diaphragm assembly. The ceramic lining, which is formed in situ, greatly simplifies the design.

b. Suspended Bed. Several cell designs have been proposed in which a slurry or fluidized suspension of electrically conductive particles is employed in one or both electrode compartments, acting as a discontinuous extension of the electrode surface. One such electrode system is described [25] for electrolysis of sodium chloride, wherein oxygen or air is introduced

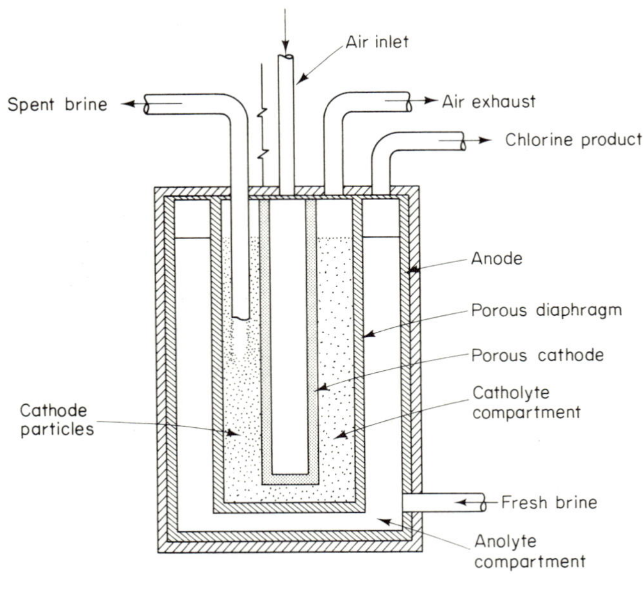

Fig. 8. Butler/Fawcett cell.

in contact with the cathode in order to prevent the evolution of hydrogen and thereby lower the cell voltage requirements. The cell components are contained in a glass or polypropylene cylindrical vessel (Fig. 8), which is lined with an anode of carbon or platinum. A fritted glass or asbestos diaphragm forms an inner chamber for catholyte. Within the catholyte chamber is concentrically positioned a hollow, porous, graphite cathode through which the oxygen-containing gas is introduced. In the annular space between the cathode and the diaphragm are placed small (74-1650 μ) particles of graphite, which are kept in suspension by the oxygen or air passing into the catholyte compartment. Addition of the particles was shown to be effective in raising the cell current at a fixed cell voltage by a factor of 2.

An extension of the concept of suspended electrode particles for commercial electroorganic processing has been proposed by Backhurst and co-workers [26]. In this design one or both electrodes consists of a bed of small (75-1000 μ) metal or metal-coated spheres, fluidized to provide 10-20% bed expansion by upward flow of electrolyte (Fig. 9). Fluidized

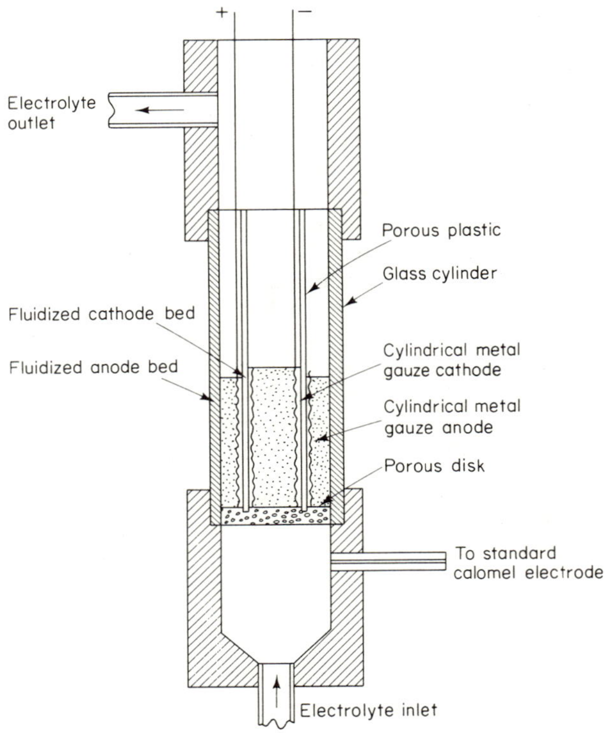

Fig. 9. Backhurst fluidized-electrode cell.

particles studied as cathodes include solid copper, copper-coated glass, and copper-coated polystyrene. Flow of current to the fluidized bed is provided through gauze conductors. The cell is conveniently cylindrical in shape, in which case a cylindrical diaphragm may be employed to provide an annular outer electrode compartment and an inner compartment. Electrolyte flow to both compartments is introduced through a sintered-glass distributor, forming the base of the cell. Particles of metal-coated glass or polystyrene beads are preferable to solid metal beads because they are in general cheaper and may be used over a wide range of electrolyte flow rates through variation in the substrate density and particle size.

The fluidized-bed design provides extremely high electrode area per unit cell volume. For example, one configuration places 11,500 cm^2 of electrode area in a cell volume of only 153 cm^3. However, more relevant to the process economics is the current-carrying capability of the cell and voltage-current relationship. The fluidized electrode is reported to be able to sustain a current of 1 A/cm^3 of bed. Development of large-scale cells employing the fluidized-electrode concept has been undertaken by Constructors John Brown in England [26a].

4. Other Designs

A variety of unusual cells are described in the literature, but most are specifically designed for laboratory studies or preparations and do not readily lend themselves to scale-up. One unique design that might have commercialization potential has been proposed by Beck and Guthke [27]. The distinctive feature of this cell is the extremely narrow anode-to-cathode spacing (100-125 μm), which may be advantageously used with very low conductivity electrolytes in cases where a diaphragm is not required. The cell (Fig. 10) consists of a series of cylindrical, bipolar graphite plates, 1 dm^2 in area and 10 mm thick. The plates are precisely spaced by means of 0.125 mm thick strips of polyester foil, positioned radically. The anode side of the plates is covered with a 200 μm thick layer of electrolytically applied lead dioxide. Electrolyte enters through a central channel and flows radially to the periphery of the cell, where it is collected and withdrawn.

Another unusual cell design proposed by Beck and Guthke and co-workers [28] features a solid polyelectrolyte coating on one or both electrodes. The polyelectrolyte possesses ion-exchange properties and also must be permeable to any gases that may be released at the electrodes. The polyelectrolyte coating is typically 0.5-1.0 mm thick and an anode-to-cathode spacing of 1-3 mm is generally preferred.

As an example of such a cell, an anode consisting of a porous graphite tube is surrounded by a layer of sulfonated styrene/divinylbenzene copolymer. An amalgamated silver gauze, concentrically positioned around the anode, serves as the cathode. Pure water is placed inside the graphite tube and

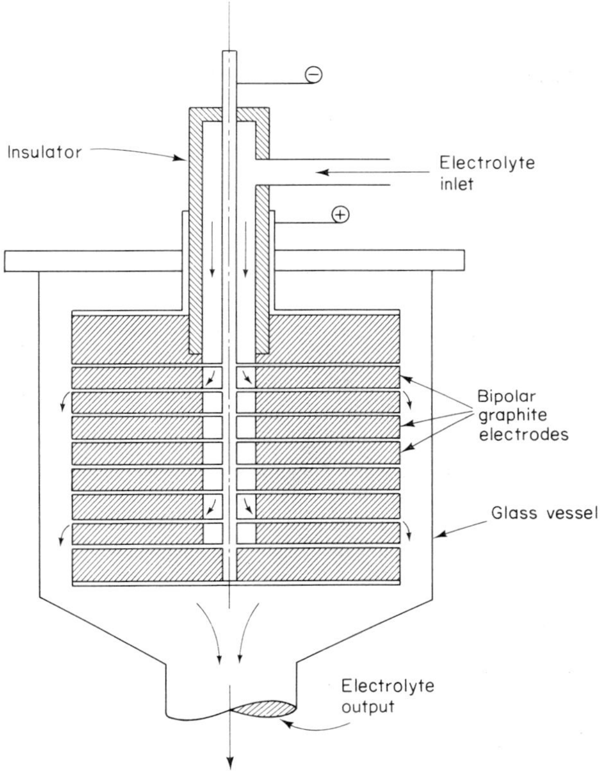

Fig. 10. Beck/Guthke cell.

catholyte solution fills the space outside the encapsulated anode. Since the polyelectrolyte layer serves as the medium for current transfer across a large portion of the anode-cathode gap, liquids with very low conductivity may be electrolyzed at relatively low voltage drops. The cell described above requires 6 V at 0.06 A/cm^2 with a catholyte solution having a resistivity of about 40 Ω-cm.

For some applications an annular cell design offers the advantages of simplicity of construction and low cost since often standard pipe or tubing may be used. The outer tube usually serves as one electrode, while a concentrically positioned rod or tube is used as the counter electrode. A cylindrical diaphragm may be placed between the electrodes. An example of this type of design has been proposed for the oxidation of iodic acid to periodic acid in the indirect dialdehyde starch process [28a]. One design consisted of a 9 cm i.d. by 45 cm long cylindrical vessel of 1% silver-lead, housing a 6 cm diameter thimble-shaped diaphragm of Alundum. A section of 6 mm o.d. steel pipe was placed inside the thimble as the cathode. Agitation at the anode surface, which is important to achieving good current efficiencies in this electrolysis system, was provided by circulation through the annular anode compartment at linear velocities of about 1 m/min. Current densities in the range of 2-8 A/dm^2 were satisfactorily employed.

A power recovery technique has been described by Le Duc [29], which might be adapted to a number of the cell designs previously discussed. In many anodic oxidations hydrogen is evolved at the cathode and is of significant by-product value. By introducing an oxygen-containing gas through a porous cathode an oxygen half-cell is established with water being formed rather than molecular hydrogen. If peroxide ion formation is prevented by means of cathode activators such as silver, manganese, cobalt, nickel, or iron, the half-cell potential is 0.40 V. Porous carbon or graphite is a suitable cathode base material. Thus, with such a cathode system the cell voltage can be lowered by 0.2-0.4 V, which may provide a 5-10% reduction in power consumption. Another variation of this concept has been studied for power recovery from caustic-chlorine cells [30]. In this case, a sodium amalgam-oxygen fuel cell is employed in place of the conventional amalgam decomposer. The fuel cell is physically separated from the amalgam-producing cell and the amalgam is transferred to the fuel cell. Mild steel serves as the anode and silver-activated carbon as the cathode. Oxygen or air is fed to the back of the cathode and the amalgam flows downward over the surface of the grooved anode. An electrolyte of 50% caustic is used and the cell operated at about 150°F. Fuel cell voltages up to 1.3 V were measured at a current density of 0.108 A/cm^2, equivalent to a power recovery of up to 35%. The economics of such a scheme demand a low-cost fuel cell. For example, 1.3 V at 0.108 A/cm^2 will support an investment of about \$60/dm^2 for the fuel cell system, based on capital charges of 10% per year and power at 0.5¢/kWh.

C. Electrodes

The scale-up of an electroorganic reaction from the laboratory cell to the commercial system poses two types of problems with regard to electrodes. The first is the selection of the lowest-cost material which will give the desired electrode reactions and structural stability. The second is maintenance of the active life of electrodes.

1. Electrode Materials

While the choice of electrode materials may be found in the research studies to be limited to one or two materials, there is usually some latitude in the electrode design for the large-scale cells. For example, an electroreduction reaction may require a high hydrogen overvoltage cathode, such as lead, mercury, or tin. From the standpoint of the geometry and cost of the commercial cell, this choice may indeed be a significant one. Even when a single metal is found to produce the desired electrode reaction, several different forms may be possible. In many electrolytic reductions lead is the preferred cathode material and since pure lead is soft and readily deformed, its use presents problems in large cells where maintenance of close tolerances is necessary. One means of overcoming this problem is the electroplating of lead on copper, steel, or other rigid base metal. Alternatively, it may be possible to introduce a few percent of an alloying element such as copper, antimony, or calcium into lead, improving the dimensional stability without affecting electrode performance. This technique must, however, be used with caution since frequently minor quantities of alloying elements can markedly affect by-product formation. In electrolytic processes in which the cathode material is not critical, steel is usually employed because of its low cost and high mechanical strength.

Selection of the anode material for a commercial cell is usually even more difficult than choosing the cathode because of the inherent problem of corrosivity. For an anodic process such as the Kolbe synthesis, which may give satisfactory results only on smooth platinum, the choice is probably narrowed to platinum-clad or platinum-plated titanium or tantalum. Use of unsupported platinum is generally impractical because of the high cost of a sheet having the necessary structural strength and current-carrying capability.

Anodes for use in dilute sulfuric acid have been studied by a number of investigators [31, 32]. A four-component alloy (98.7% Pb, 1% Ag, 0.3% Sn, and 0.02% Co) has shown significantly superior stability compared to the 1% silver/99% lead alloy previously accepted as standard for electrolysis of sulfate solutions. Another anode material suitable for use in dilute sulfuric acid is lead dioxide. Methods of supporting disks of compressed lead dioxide in a lead alloy grid have been developed for the industrial storage-battery industry. A second method of producing lead dioxide anodes involves the electrodeposition of lead dioxide on a base substrate, such as graphite, titanium, or tantalum [33, 34].

Anodes for oxygen-producing reactions in an alkaline environment are generally made of nickel, mild steel, or stainless steel. For anodic halogenations of organics, graphite is frequently employed as the anode; the ready availability of various grades, shapes, and sizes of graphite anodes from the suppliers to the caustic-chlorine industry makes this material a convenient choice. In recent years, there has been considerable activity on development of "dimensionally stable anodes" (DSA) for use in halogen systems [35]. The early efforts on noncorroding anodes for electrolysis of halides involved covering nonnoble metals with a noble metal — the platinized titanium anode is probably the most widely studied example of this type. Coatings of nonnoble metal oxides such as magnetite and lead dioxide have also been employed. More recently, permanent anodes have been described which employ precious metal oxides on titanium substrate. Ruthenium oxide, mixed with oxides of the "valve metals" (titanium and tantalum), has proven to be particularly successful [36]. Dimensionally stable anodes which are said to be suitable for electroorganic redox processes are currently produced in commercial sizes by Oronzio de Nora in Milan.

2. Electrode Activity

In the scale-up of an electroorganic process, specification of the electrode materials and geometry is necessary, but may not be sufficient to ensure an operable process. The nature of the active electrode surface may change chemically or physically with time, altering the selectivity and/or current efficiency of the reaction. Possible chemical changes include the dissolution of the active electrode species, such as the loss of mercury from an amalgamated lead cathode.

Another possible chemical change involves the electrodeposition of extraneous metals on the cathode surface. The detrimental effect of trace quantities of silver and copper in the electroreductive coupling of acrylonitrile to adiponitrile has been reported. While electrode surface contamination is frequently not recognized as a problem in the laboratory experiments with short runs and reagent grade chemicals, it may prove to be a major consideration in the development of a commercial process. Means of counteracting electrode surface contamination include (1) periodic mechanical or chemical cleaning of the electrode, (2) current reversal with purging, (3) stringent raw material specification with regard to offending impurities, (4) chemical or electrolytic treatment of entering streams, (5) replacement of electrodes. In one process, the offending metallic ions were removed from the electrolysis system feed by passing the feed through a bed of lead shot, onto which the metal was electroplated [36a].

Mechanical deterioration of the electrode surface may occur due to either corrosion or erosion. Selection of electrode material is of course the primary means of minimizing such problems. Erosion may be due to high rates of circulation of electrolyte containing solid particles of reaction products or electrode corrosion products, in which case removal of the solids external to the cell may be possible. If the source of the problem cannot be eliminated, renewal of the electrode surface is sometimes possible by electropolishing or replating in place. In commercial cells removal of the electrodes for refurbishing is generally considered to be a last-resort measure.

D. Diaphragms

In many electroorganic systems the use of a diaphragm to separate the anolyte from the catholyte is necessary in order to cut down on side reactions, which lower product yield and current efficiency, and to facilitate the separation of cell products. In cells in which hydrogen and oxygen (or chlorine) are produced, a diaphragm may be needed to avoid the handling of explosive gas mixtures. Selection of a suitable diaphragm for large-scale cells often presents a serious obstacle to the development of a commercially feasible process. The preferred electrolysis diaphragm would possess the following properties: (1) low electrical resistance, (2) limited interdiffusion of anolyte and catholyte constituents other than the current-carrying ions, (3) resistance to plugging (with consequent resistance increase), (4) good dimensional stability, (5) long operating life, and (6) low cost. As might be expected, the chances of finding all of these qualities in a single material are diminishingly small.

Two basic types of diaphragm materials are available — nonpermselective and permselective. The nonpermselective material merely offers a barrier to the diffusion of the ionic and molecular species, with little or no change in electrolyte transport numbers. Permselective materials, or ion-exchange membranes, have the ability to allow ready passage of ions of one charge with a high degree of exclusion of ions of the opposite charge. Thus, cation permselective membranes (frequently misnamed cation membranes) can be prepared in which the mobility of the cation may be many times that of the anion, so that the bulk of the current transfer is the movement of cations from anolyte to catholyte. Similarly, anion permselective membranes may be employed if movement of the cations is to be suppressed.

1. Nonpermselective diaphragms

In some cases a small amount of flow between the anolyte and catholyte can be tolerated or may even be desired, in which case nonpermselective diaphragms may be entirely satisfactory. In the laboratory electrolysis studies, an Alundum cup is often employed. Alundum and other forms of unglazed porcelain are not, however, very practical for use in production cells because of their high susceptibility to fracture. As possible alternatives, a variety of woven and nonwoven fabrics and porous materials are available in a wide range of porosities. A partial list of potential candidates is given below:

(1) Filter cloth

- (a) glass
- (b) polypropylene
- (c) polyvinyl chloride
- (d) Teflon

(2) Nonwoven mats

- (a) asbestos
- (b) plastic felts

(3) Porous plastics

(a) polyethylene
(b) polyvinyl chloride
(c) Teflon
(d) rubber.

Both electrical and diffusional resistance of porous materials are related to their permeability, porosity, pore size, and thickness. Methods of evaluating and correlating the electrical conductivity of different porous media have been described by MacMullin and Muccini [14].

2. Permselective Diaphragms

There has been considerable activity in recent years on development of ion permselective diaphragms (commonly termed ion-exchange membranes) for use in electrodialysis processes. Several general reviews of available commercial membranes have been published [37-39]. One approach to the preparation of thin membranes possessing ion-exchange characteristics involves the casting of a finely divided powder of ion-exchange resin in a plastic binder. These membranes have several disadvantages resulting from their heterogeneous character — relatively large channels usually exist through the matrix between resin particles leading to high rates of electro-osmotic water transfer and molecular diffusion.

Homogeneous ion-exchange membranes, on the other hand, can be produced which are essentially a continuous polymeric structure. One matrix used by several membrane manufacturers is a copolymer of styrene and divinylbenzene, which is sulfonated for cation exchange or aminated for anion exchange. The polymer is frequently reinforced with a screen of Saran, glass, or PVC to improve the mechanical properties.

Typical properties of commercially available membranes are given in Table 1.

Properties of ion-exchange membranes can be varied by altering (1) polymer composition, (2) polymer porosity, (3) exchange capacity, (4) backing material, and (5) thickness. Thus the possibility of tailoring a membrane to the requirements of a specific electroorganic application is good.

E. Mass and Heat Transfer

In the design of a commercial electroorganic cell system there are two important concerns in addition to selecting suitable materials and geometry of cell components — these are mass and heat transfer.

TABLE 1

Properties of Commercial Ion-Exchange Membranes

Thickness, mm	0.1-1.0
Width, cm	90-110
Length, cm	90-any
Resistance,[a] Ω-cm^2	1.5-20
Transport number [a]	0.80-0.99
Water content, %	10-60
Mullen burst strength, psi	25-300
Cost, \$/ft^2	0.50-8.0

[a]Measured in 0.5 N NaCl.

1. Mass Transfer

In an operating cell the concentration of the reactant at the electrode surface will be less than in the bulk electrolyte solution and this concentration difference provides the driving force for transfer of matter across the boundary layer at the electrode. If the rate of transfer is too slow, a deficiency of reactant species will exist at the electrode surface, resulting in increased voltage, side reactions, and reduced current efficiency. This condition of concentration polarization becomes more severe as current density is raised and reactant concentration lowered.

In many electrode reactions it is as important to transfer the products away from the vicinity of the electrode as it is to move the reactants to the electrode. For example, cathodic reductions in aqueous solution frequently produce hydroxyl ions which may lead to undesired reactions with either the reactants or products. Even though the pH of bulk catholyte may be well below 7.0, a highly alkaline layer at the electrode surface can significantly lower product selectivity. Frequently, ensuring adequate rates of transfer is not a serious problem in the laboratory studies because of the geometry of the lab cell, agitation means employed, or the range of current densities and concentrations investigated. However, the tendency in scale-up is to raise conversions and current densities which will increase the importance of mass-transfer considerations.

Methods that have been proposed for improving mass-transfer rates in large-scale cells are the following:

(a) rotation of cylindrical or disk electrodes

(b) flow of electrolyte by external pumps

 (1) open channels
 (2) with turbulence-promoting spacers

(c) turbine or propeller agitators

(d) fluidizing a bed of electrode particles

(e) fluidizing a bed of nonconducting particles

(f) vibration of the electrode

(g) use of a foamed electrolyte [65].

While rotating electrodes have been preferred by many laboratory researchers, they are an expensive solution to the mass-transfer problem in commercial cells. Probably the most common approach is the circulation of electrolyte by external means. The effects of free and forced convection on mass-transfer rates at horizontal planar electrodes have been studied by Tobias and Hickman [40]; other studies in the area of convective mass transport have been described by Ibl [41].

The Monsanto cell for electroreductive coupling reactions exemplifies the use of high fluid velocities in open channels to provide the turbulence necessary to overcome polarization effects [16]. For current densities in the range 0.5 to 1.0 A/cm^2, linear velocities on the order of 2 m/sec through a 3 mm deep catholyte compartment are employed, corresponding to a Reynolds number of about 5700. (The Reynolds number is the dimensionless group $4\,LV\rho/\mu$, where L is the cross-sectional area of the catholyte channel divided by the perimeter of the cross section, V is the fluid velocity, ρ is the density, and μ is the viscosity.) Using this cell in the reductive coupling of acrylonitrile it has been found that a Reynolds number greater than 2000 was required to give adequate depolarization of the cathode as indicated by the product selectivity.

In an alternate approach to solution of the polarization problem in electroreductive coupling of acrylonitrile, a cell design by Asahi [41a] employs a turbulence-promoting spacer in the catholyte compartment, similar to those used in electrodialysis cells. With these spacers, velocities on the order of 30 cm/sec in 1 mm deep channels are claimed to provide adequate mass transport rates at 0.2 A/cm^2.

Rotation of cylindrical electrodes as a means of avoiding concentration polarization has been mentioned previously in this chapter with regard to a tank-type cell for reduction of salicylic acid to salicylaldehyde. A correlation of mass-transfer rates with rotating cylinders for three solid dissolution systems and two electrolytic redox reactions has been developed by Eisenberg and co-workers [42]. These investigators found that the mass-transfer coefficient (and hence limiting current density) is proportional to the peripheral velocity of the cylinder to the 0.7 power and the cylinder diameter to the -0.3 power.

Vibration of the electrode pair at 10-100 Hz and 0.5-1.5 mm amplitude has been shown by Beck and Guthke [27] to effectively reduce concentration polarization in the Kolbe synthesis of dimethyl sebacate from monomethyl adipate. There does not, however, appear to have been any extensions of this concept to large-scale cells. Also in the category of potential, but as yet uncommercialized, means of enhancing mass transfer in electrolysis is the fluidization of nonconducting particles in the electrolyte in the vicinity of the electrode. Such a technique has been found by Le Goff and co-workers [43] to allow a substantial increase in the current density under conditions in which concentration polarization is limiting, and application to the industrial scale-up of electrode processes is forecast.

The fluidized-electrode cell design of Backhurst described previously has the advantage of high rates of mass transfer as well as a large ratio of surface area to cell volume. Increases in current density by a factor of 10-1000 through addition of fluidized electrode particles have been demonstrated.

2. Heat Transfer

In nearly every electroorganic process heat removal is a problem that must be faced in the process scale-up. In the laboratory, temperature control may simply be handled by immersing the cell in a cooling bath or circulating coolant through an internal coil. Common methods of heat removal in large-scale cells are

(1) circulation of electrolyte through external heat exchangers;

(2) staging of cells with interstage cooling;

(3) internal cooling coils and/or cell cooling jackets;

(4) external or internal evaporative cooling.

The high current densities sought for industrial processes coupled with the inherently poor electrical conductivities of electroorganic systems make economical heat removal an important consideration in scale-up. Conventional engineering design techniques are applicable to determination of the heat-transfer requirements.

F. Economic Considerations

In commercialization of an electrolytic process for production of organic chemicals, the major economic considerations are (1) reaction selectivity,

(2) electrolysis system investment, and (3) power usage. Compared with an inorganic electrochemical process, such as caustic chlorine manufacture, the reaction selectivity of the electroorganic process is likely to be relatively more important and the cost of power a less important factor. In general, the capital cost of the electrolysis system is greater for organics than for inorganics per unit of production capacity because of the greater complexity of the cells needed to achieve high selectivities and the smaller scale of production.

The major items of equipment in an electrolysis system for large-scale electroorganic processing are (1) electrolytic cells, (2) rectifiers and buses, and (3) heat exchangers (may be part of cells). If the selectivity and current efficiency are relatively constant over a range of current densities, selection of the most economical cell current involves a balance between the cost of power per unit of product and the cost of capital for the electrolysis system. Since cell voltage, and hence power usage per unit mass of product, increases with current density, savings resulting from reduction in the number of cells at high current densities are offset to a varying degree by increased cost for electricity and higher investment in the electrical system and in heat exchangers, whether internal or external to the cell. Selection of the economic optimum will depend on such factors as (1) the unit cost of power, (2) required rate of return or interest rate on capital (3) shape of the voltage-current density curve, and (4) the relative costs of the three major equipment

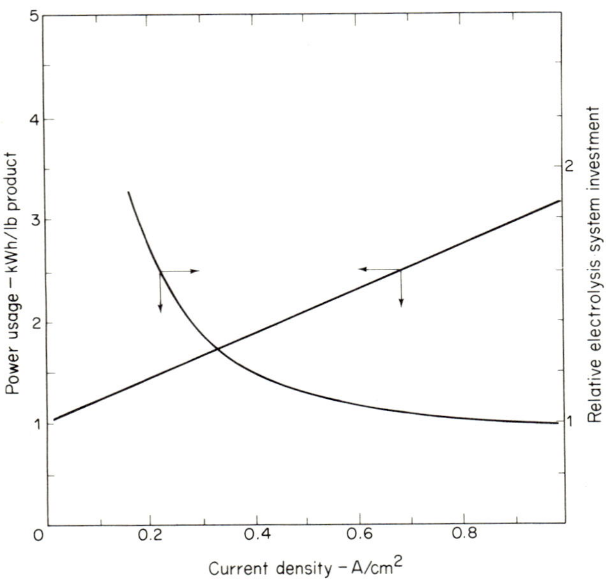

Fig. 11. Variation of electrolysis power usage and investment with current density for a typical system using plate-and-frame cells.

types. The interrelationship between power usage and electrolysis investment is shown in Fig. 11 as a function of current density for a typical system employing plate-and-frame cells. It is seen that a fivefold increase in the current density from 0.2 to 1.0 A/cm^2 reduces the investment by one-half (not four-fifths, as might be expected by consideration of cell cost alone). Current densities employed in most commercial electroorganic processes are in the range of 0.1-1.0 A/cm^2.

III. INDUSTRIAL ELECTROORGANIC PROCESSES

A. General

Electrochemical preparation of ingredients or precursors of certain pharmaceuticals and perfumes has been of commercial interest for a number of years. However, production of organic chemicals at multithousand ton per year rates is only relatively recent. One of the first large plants was that of Atlas Powder Company, which was built in 1937 for the manufacture of 1400 ton/yr of sorbitol and mannitol by cathodic reduction of glucose [44]. This early success was rather short-lived and the process was replaced within a few years by a high-pressure catalytic hydrogenation route. It was not until 1964 that a truly large-scale electroorganic manufacturing facility was put on-stream with the start at Freeport, Texas, of Nalco's tetraalkyllead plant. The capacity of the original installation was 15,000 ton/yr tetramethyllead or 18,000 ton/yr tetraethyllead. The following year Monsanto began production of adiponitrile electrolytically at Decatur, Alabama, at rates of the same order of magnitude as Nalco's [46]. As contrasted to the Atlas plant, both the Nalco and Monsanto facilities have been expanded significantly since their start.

Electroorganic processes that have been investigated on a pilot-plant or semicommercial scale cover a wide range of reaction types. In addition to the reductive coupling of acylonitrile and the preparation of sorbitol/mannitol, examples of other cathodic processes considered for commercialization include the reduction of (1) salicylic acid to salicylaldehyde [47-48], (2) phthalic acids to the corresponding dihydrophthalic acids [49, 50], (3) N,N-dimethylaminoethyltetrachlorophthalimide to N,N-dimethylaminoethyltetrachloroisoindoline [51], and (4) benzoic acid to benzyl alcohol [52]. Anodic processes claimed to have favorable economics are (1) the oxidation of propylene to propylene oxide [21, 53], (2) the Kolbe synthesis of dimethyl sebacate from monomethyl adipate [54], (3) the bromination of hydrogen cyanide to give cyanogen bromide [55], and (4) the oxidation of starch to the dialdehyde [56].

A summary of cell operating conditions and performance for a number of commercial or near-commercial electroorganic processes is given in Table 2.[1] In general, these are characterized by high product selectivity (> 90%), relatively high current efficiency (> 50%), and low power usage (1.5-5 kWh/kg).

TABLE

Cell Operating Conditions and Performance of Some Commercial

Product	Ref.	Reactants	Cell type	Anode material	Cathode material
Adiponitrile					
Monsanto	[46]	Acrylonitrile	P&F [a]	Pb alloy	Pb
Asahi	[41a]	Acrylonitrile	P&F	Pb-Sb	Pb
B.A.S.F.	[27]	Acrylonitrile	P&F	PbO_2/C	C
U.C.B.	[58]	Acrylonitrile	--	Fe oxides	C
Tetraalkyllead					
Nalco	[45]	RCl, Pb, & RMgCl	Bed	Pb	Steel
Ziegler	[63]	Pb & $KAlR_4$	Tank	Pb	Hg
Propylene oxide					
Kellogg	[21]	Propylene	Tank	C	Steel
Bayer	[53]	Propylene	Tank	C	Steel
Sebacic acid					
USSR	[54]	Monomethyl adipate		Pt/Ti	
Salicylaldehyde					
	[52]	Salicylic acid	Tank	Pb-Ag	Hg/Cu
Cyanogen bromide					
Sohio	[55]	HCN & NH_4Br	P&F	C	Steel
Tetraethylammonium hydroxide					
Monsanto	[64]	$(C_2H_5)_4NCl$	P&F	C	Steel
Sorbitol					
Atlas	[44]	Glucose	Tank	Pb	Hg/Pb
Isoindoles					
CIBA	[41]	Phthalimides	Tank	Pb	Hg
Cyclohexadiene-dicarboxylic acid					
Cal. Research	[49]	Phthalic acid	Tank	Pb	Hg

[a] Plate-and-frame

2

and Near-Commercial Electroorganic Processes

Diaphragm material	Current density A/dm^2	Cell voltage	Prod. selec., %	Current effic., %	Power usage, kWh/kg	Status
Ion-exch.	20-100	6-12	90-93	90-92	3-6	Commer.
Ion-exch.	10-20	5-6	85-95	80-95	3-4	Planned
None	7-10	4-5	82-92	80-83	3	Pilot plant
None	4-15	5-6	72-92	84-93	1.5-5	Pilot plant
Polyethylene fab.	1.5-3	15-30	96-99	175	4-8	Commer.
None	46	2.5	100	100	1	Pilot plant
Asbestos	80-110	3.5-4.0	82-86	70-80	5	Pilot plant
Acrylic fabric	2-5	3-5	90	88	4-5	
	40-60		82-85	75		
Porous rubber	12-15	13-15	45	35-45	12-15	Pilot plant
Ion-exch.	10-30	3-4	98	98	1.5-2	Pilot plant
Ion-exch.	20-60	6-13		30-60		Pilot plant
Ceramic	4	20				Excommer.
Ceramic	3.5	15-25	88-92	50-55		Semicommer.
Ceramic	3-30	4-11	98	90-95	1.6-4	Pilot plant

Current densities are typically in the range of 10-100 A/dm^2, contrary to a commonly encountered misconception that electrolysis of organics is usually conducted at very low current densities. It is generally more economical to pay the penalty for low conductivity of organic-containing electrolytes in the form of higher power usage than by the increased number of cells required at low current densities.

B. Commercial Process for Adiponitrile

In excess of a billion pounds per year of adiponitrile is produced worldwide as a precursor to hexamethylenediamine, one of the monomers used in the manufacture of nylon-6, 6 polymer. The importance of the production cost of adiponitrile on the economics of making nylon is well recognized and has been responsible for considerable research activity on the preparation of this intermediate. The possibility of electroreductive coupling of acrylonitrile to adiponitrile was reported in the 1940s, but it was not until the discovery in 1963 by Monsanto's M. M. Baizer [56a] of the importance of the supporting electrolyte that commercially attractive selectivities and current efficiencies were realized. Baizer found that through the use of concentrated solutions of certain quaternary ammonium salts, such as tetraethylammonium p-toluenesulfonate, nearly quantitative yields of adiponitrile could be achieved at lead or mercury cathodes. Following an intensive process development program, Monsanto designed and built a multimillion pound per year facility in Decatur, Alabama, for production of adiponitrile via this technology [46].

In the Monsanto process a catholyte consisting of an aqueous solution of acrylonitrile, quaternary ammonium salt (QAS), and reaction products is circulated through the cathode compartments of a bank of divided plate-and-frame cells of the type shown in Fig. 3. The QAS serves the joint function of increasing the solubility of acrylonitrile in water and improving the electrical conductivity of the solution. The anolyte, a dilute solution of sulfuric acid, is similarly circulated through the opposing anolyte compartments. The cathode is lead; a lead alloy, stable in sulfate electrolysis, is used as the anode material. The electrodes are mounted on polypropylene plates with the anode of one cell joined electrically to the cathode of the next via internal connectors, providing series flow of current through the bank of cells. A cation permselective membrane separates the anode and cathode compartments. Electrolyte is pumped through each cell in parallel and both anolyte and catholyte are recirculated through external exchangers for removal of the heat generated in the cells (see Figs. 12A and 12B). A side stream is withdrawn from the circulating catholyte and sent to an extraction system for removal of the product ADN.

The withdrawn catholyte stream is cooled and countercurrently contacted with excess acrylonitrile in a multistage extraction column. The acrylonitrile/adiponitrile extract stream is then back-extracted with water in a

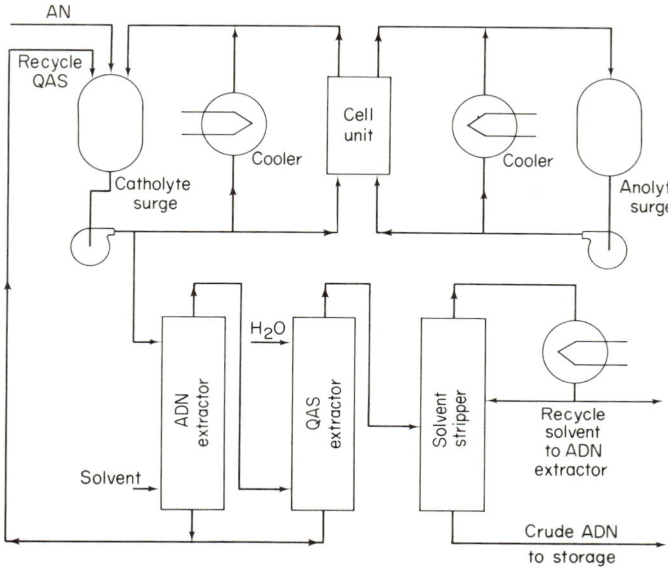

Fig. 12A. Monsanto adiponitrile process.

FIG. 12B. Monsanto adiponitrile synthesis cell house at Decatur, Alabama.

similar column of dissolved QAS. The organic stream leaving the second extraction column is sent to a stripping column, where the acrylonitrile is distilled overhead and recycled to the process, following removal of by-product propionitrile. The bottoms from the acrylonitrile stripper are further refined by distillation to produce a 99+% pure product, suitable for hydrogenation into hexamethylenediamine.

The principal cathode reaction is:

$$2CH_2{=}CHCN + 2H_2O + 2e^- \rightarrow NC(CH_2)_4CN + 2\ OH^-.$$

Water is decomposed at the anode:

$$H_2O \rightarrow 2H^+ + \tfrac{1}{2}O_2 + 2e^-.$$

Major by-products are formed according to the following reactions:

Electrolytic reactions:

(1) $$\underset{\text{AN}}{CH_2{=}CHCN} + 2H_2O + 2e^- \longrightarrow \underset{\text{Propionitrile}}{CH_3CH_2CN} + 2OH^-$$

(2) $$3\underset{\text{AN}}{CH_2{=}CHCN} + 2H_2O + 2e^- \longrightarrow NCCH(CH_2)_3CN\ (\text{with } (CH_2)_2CN \text{ on CH}) + 2OH^-$$

Trimer

Cyanoethylation reactions:

(3) $$\underset{\text{AN}}{CH_2{=}CHCN} + H_2O \xrightarrow{OH^-} \underset{\text{Hydroxypropionitrile}}{HOCH_2CH_2CN}$$

(4) $$CH_2{=}CHCN + HOCH_2CH_2CN \xrightarrow{OH^-} \underset{\text{bis-Cyanoethylether}}{CN(CH_2)_2O(CH_2)_2CN}$$

The Monsanto cells can be operated at current densities ranging from 25-100 A/dm^2 with cell voltages ranging from 6-15 V, corresponding to power consumptions of 3-8 kWh/kg adiponitrile produced.

Several variations of the electroreductive coupling of acrylonitrile have been proposed by other investigators. A process developed by Asahi Chemical Industry [57] in Japan employs a concentration of reactant acrylonitrile below 7% by weight, compared with a value over 10% used in the Monsanto catholyte stream. Special means are employed for suppressing formation of propionitrile, which tends to form at low acrylonitrile concentrations. The Asahi process is similar to Monsanto's in that it employs sulfuric acid anolyte, a cation-exchange membrane, and a quaternary ammonium salt in the catholyte.

Another version developed by Union Chimique-Chemische Bedrijven (U.C.B.) uses diaphragmless cells and an electrolyte consisting of a concentrated solution of alkali perchlorates or thiocyanates, together with a polar solvent such as dimethyl sulfoxide, dimethylacetamide, or dimethylformamide [58]. The acrylonitrile concentration in the U.C.B. electrolyte is about 10-15%, while the water content is kept low (6%). Yields of adiponitrile above 85% are obtained at a graphite cathode. By operating at low current densities the power usage per unit of adiponitrile production[2] is maintained in the range 1.8-3.6 kWh/kg.

Several cell designs have been described by Beck and co-workers [27-28] at B.A.S.F. for electrolytic production of adiponitrile (see Sec. II.B above). Cells both with and without diaphragms have been studied, employing a catholyte consisting of 0.5-3% QAS (for example, tetraethylammonium p-toluenesulfonate or tetraethylammonium ethylsulfate), 20-80% acrylonitrile, 10-20% water, and one or more solvents such as acetonitrile, isopropanol, or dimethylformamide. Product selectivities ranging from 74-91% are reported at adiponitrile current efficiencies from 74-87% of theoretical.

Indirect electrochemical processes for hydrodimerization of acrylonitrile employing alkali amalgams have been described by a number of investigators, Knunyants and Gambaryan [59] being the first to report practically significant yields of adiponitrile. Development of an amalgam process for production of adiponitrile was subsequently undertaken by Imperial Chemical Industries (I.C.I.) and U.C.B. in conjunction with the Weizmann Institute of Science in Israel.

In the I.C.I. process [60], a homogeneous solution consisting of about 10% (by weight) acrylonitrile, 3-5% water, 5-10% QAS, and the balance a polar organic solvent, such as acetonitrile, is contacted with 0.3% sodium amalgam. The quaternary ammonium salts employed in the amalgam process are similar to those reported for the direct electrolysis, i.e., the tetraethyl- and triethylmethylammonium halides, bicarbonates, and p-toluenesulfonates. The reaction is conducted at 30-45°C and the pH is maintained in the range 8-10 by addition of CO_2. Vigorous agitation of the amalgam-solution mixture is necessary to avoid local regions of high alkalinity. Yields of adiponitrile above 95% of theoretical are realized on both the alkali metal and acrylonitrile consumed. The I.C.I. process has been operated on a pilot-plant scale but has not, as yet, been commercialized. In the U.C.B. amalgam process for adiponitrile production, 0.5-2% QAS (trimethylbenzylammonium p-toluenesulfonate, perchlorate, or phosphate) is used in polar solvents such as tetrahydrofuran, dioxane, or diethylene glycol dimethyl ether. Pilot-plant studies have been conducted by U.C.B. and the process has been offered for licensing [61].

C. Commercial Process for Tetraalkyllead

Prior to the introduction by Nalco Chemical in 1964 of its electrolytic process for tetraalkyllead, over a billion pounds per year of this important

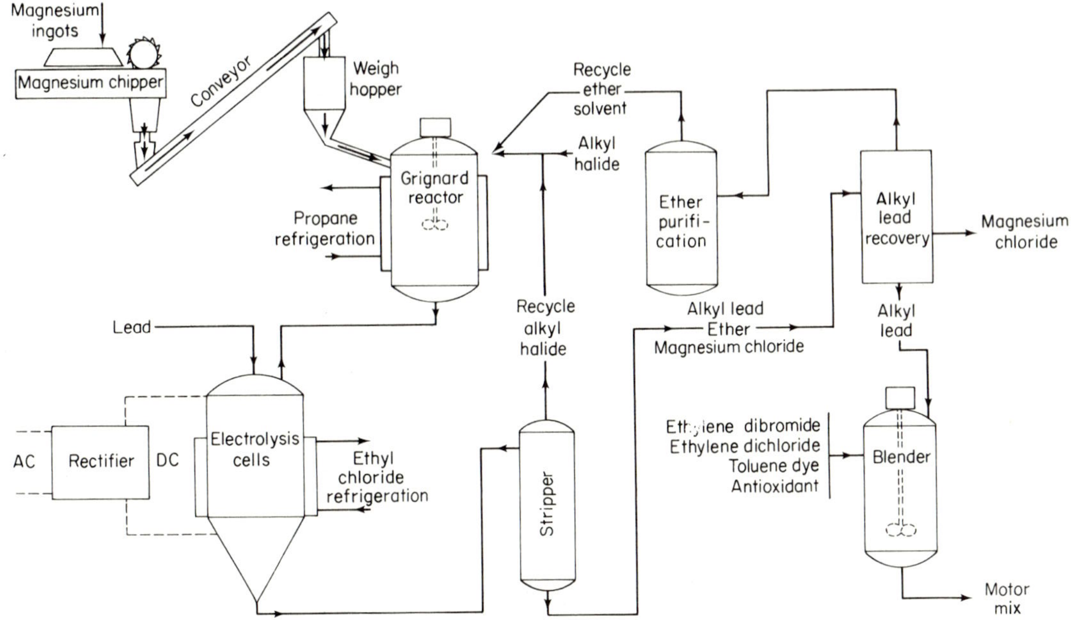

Fig. 13. Nalco tetramethyllead process. (Source: Hydrocarbon Processing, January 1965.)

gasoline additive was produced via a 40-year old process involving the reaction of methyl or ethyl halide with sodium-lead alloy. Nalco combined the large-scale production of Grignard reagent with a novel electrolysis system to pioneer this new process and installed facilities to produce tetramethyllead (TML) or tetraethyllead (TEL) at its plant in Freeport, Texas [46]. A simplified flowsheet of the Nalco TML process is shown in Fig. 13.

The first step is the production of the Grignard reagent from magnesium metal turnings and alkyl halide in a mixture of anhydrous ethers, such as tetrahydrofuran and the diethyl ether of tetraethylene glycol. The reaction is carried out in several 8000-gal, propane-cooled reactors, operating at 10-20 psig and 35-40°C. Yields above 98% are realized.

The reactor product, containing excess alkyl chloride, is fed to ten 8000-gal electrolysis vessels containing a bank of steel-tube cathodes filled with lead shot. The lead constitutes a sacrificial anode and is replaced through a feed arrangement in the top head of the vessel. The bed of shot is insulated from the retaining steel cathode tubes by porous diaphragms of plastic or ceramic (see Sec. II.B.3). The electrolysis is conducted at 40-50°C, cooling being provided by an ethyl chloride refrigeration system. Current densities are typically 1.5-3.0 A/dm^2 at 15-30 V. The electrode reactions may be written as follows, assuming dissociation of the Grignard reagent into two ions, R^- and $MgCl^+$:

Anode reaction:

$$4R^- + Pb \rightarrow R_4Pb + 4e^-$$

Cathode reaction:

$$4MgCl^+ + 4\,e^- \rightarrow 2Mg + 2MgCl_2.$$

Although the theoretical cathode reaction shows formation of a mole of free magnesium for each mole of magnesium chloride, the excess alkyl halide added to the feed combines with the magnesium to re-form Grignard reagent equivalent to half of that consumed. Thus, the overall reaction is

$$2RMgCl + 2RCl + Pb \rightarrow R_4Pb + 2MgCl_2.$$

If the concentration of alkyl halide in the electrolyte is too high, there is a loss of reactants through Wurtz-Fittig-type side reactions such as:

$$RCl + R'MgCl \rightarrow RR' + MgCl_2.$$

On the other hand, insufficient levels of alkyl halide allow magnesium to be plated out on the cathode, ultimately bridging the cathode-anode and causing short-circuits. By proper control of concentrations, temperature, and flow rate, yields up to 96% on the Grignard reagent can be achieved at conversion levels as high as 99%, with yields based on the magnesium consumed of 95% [62].

The effluent from the cells is stripped with natural gas to remove unreacted methyl chloride, which is subsequently reabsorbed in ether for recycle to the Grignard reactors. The liquid phase from the stripper is contacted with water in a series of extraction stages for separation of the magnesium chloride, TML, and mixed ethers. The ethers are purified by azeotropic distillation and gas-drying for recycle to the process. By using different alkyl groups on the Grignard reagent and the alkyl chloride fed to the cells, the process may be employed to prepare mixtures of tetraalkylleads such as Me_4Pb, Me_3EtPb, Me_2Et_2Pb, and Et_4Pb.

An alternate process for tetraethyllead has been developed by Karl Ziegler in Germany [63], involving the electrolysis of molten potassium-aluminum ethyl complex ($KAlEt_4$) at a lead anode and mercury cathode according to the following equation:

$$4KAlEt_4 + Pb \rightarrow PbEt_4 + 4K + 4AlEt_3.$$

If good contact between the resulting potassium amalgam and the triethylaluminum is provided, three-fourths of the potassium-aluminum complex is reconstituted as follows:

$$4AlEt_3 + 3K \rightarrow 3KAlEt_4 + Al.$$

Thus, the overall cell reaction is

$$KAlEt_4 + Pb \rightarrow PbEt_4 + K + Al.$$

The aluminum formed at the cathode is suspended in the mercury in finely divided form and is readily separated together with the dissolved potassium from the product TEL. The conductivity of the electrolyte is very high at the operating temperature of 100-120°C, allowing current densities as high as 45 A/dm^2 at 2.5 V with a 2.5 mm anode-to-cathode spacing. Essentially quantitative yields of TEL are realized.

The potassium amalgam containing suspended free aluminum is processed in four auxiliary steps:

(1) The potassium in the amalgam is exchanged with sodium by countercurrent contacting with molten $NaAlEt_4$ at 130°C.

(2) The resulting mixture of equimolar quantities of aluminum and sodium in mercury is reacted at 200-210°C with 3 moles of $NaAlEt_4$ to yield 4 moles of $AlEt_3$ and 4 moles of sodium amalgam.

(3) The sodium amalgam is freed of sodium by a secondary electrolysis in which liquid sodium is liberated at the cathode, with the mercury being returned to the primary cell.

(4) The liquid sodium is hydrogenated to form NaH which is then reacted with $AlEt_3$ and ethylene at 200°C to give $NaAlEt_4$ for recycle to the exchange reactor (1) and displacement reactor (2).

The overall stoichiometry of the primary and auxiliary reactions is represented by $Pb + 4CH_2CH_2 + 2H_2 \rightarrow Pb(C_2H_5)_4$. In an analogous manner, the Ziegler process is also suitable for preparation of tetraethyltin, triethylantimony, or diethylmercury. (See Chapter XVIII).

NOTES

[1]Few details are available on the reported pilot-plant production of $Pb(OAc)_4$ [66].

[2]UCB, in cooperation with the Soviet Ministry of the Chemical Industry, has developed another adiponitrile process involving electrolysis of an emulsion of acrylonitrile in an aqueous alkali phosphate solution using a graphite cathode and magnetite anode [67].

REFERENCES

[1]. C. L. Mantell, Electrochemical Engineering, McGraw-Hill, New York, 1960.

[2]. R. B. MacMullin, Electrochem. Tech., 1, 5 (1963); 2, 106 (1964).

[3]. J. Billiter, Die Technische Electrolyse der Nichtmetalle, Springer, Berlin, 1954.

[4]. A. Regner, Electrochemical Processes in Chemical Industry, Constable, London, 1957.

[5]. J. R. Wilson, Demineralization by Electrodialysis, Butterworth, London, 1960.

[6]. (a) J. B. Haglind, Adv. Chem. Ser., 78, 24 (1948);
(b) A. Kuhn, Brit. Chem. Eng., 16, 149 (1971).

[7]. M. Ya. Fioshin and A. P. Tomilov, Intern. Chem. Eng., 9, 375 (1969); Brit. Chem. Eng., 16, 154 (1971).

[8]. F. Goodridge, The Chemical Engineer, July-Aug. 1965, CE 166.

[9]. D. E. Danly, Hydrocarbon Proc., June 1969, 159.

[10]. C. L. Mantell, Chem. Eng., June 5, 1967, 128.

[11]. S. Wawzonek, Science, 155, 3758, 39 (1967).

[12]. C. L. Wilson, Intern. Sci. Technol., March 1969, 82.

[13]. H. V. K. Udupa, G. S. Subramanian, and M. S. Venkatachalapathy, Chem. Age India, 15, 14, 501 (1964).

[14]. R. B. MacMullin and G. A. Muccini, Amer. Inst. Chem. Eng. J., 2, 393 (1956).

[15]. R. W. Foreman and F. Veatch (to Standard Oil of Ohio), U.S. Pat. 3,119,760, Jan. 28, 1964.

[16]. D. E. Danly and R. W. McWhorter (to Monsanto), Belg. Pat. 699,284, May 31, 1967.

[17]. M. Murakoshi, Z. Uchio, S. Nozaki, and T. Hatakeyama, (to Asahi Kasei), Japan. Pat. 69 30,253 (1969).

[18]. A. M. Holmes and W. C. Gardiner (to Olin Mathieson), U.S. Pat. 3,055,821, Sept. 25, 1962.

[19]. K. I. Kryshchenko, M. Ya. Fioshin, N.G. Bakhchisaraytsʼyan, and G. A. Kokarev, Sov. Chem. Ind., No. 7, July 1969, 12.

[20]. S. Ganesan, K. S. Udupa, G. S. Submaranian, and H. V. K. Udupa, Chem. Age India, 13, 346 (1964).

[21]. J. A. M. Le Duc (to Pullman, Inc.), U.S. Pat. 3,342,717, Sept. 19, 1967.

[22]. D. G. Braithwaite (to Nalco Chemical), U.S. Pat. 3,391,067, July 2, 1968.

[23]. D. G. Braithwaite, J. S. d'Amico, P. L. Gross, and W. Hanzel (to Nalco Chemical), W. Ger. Pat. 1,221,223, July 21, 1966.

[24]. L. L. Bott (to Nalco Chemical), U.S. Pat. 3,479,274, Nov. 18, 1969.

[25]. R. M. Butler and W. R. Fawcett, Esso Research & Engineering, Can. Pat. 700,933, Dec. 29, 1964.

[26]. J. R. Backhurst, J. M. Coulson, F. Goodridge, R. E. Plimley, and M. Fleischmann, J. Electrochem. Soc., 116, 11, 1600 (1969); M. Fleischmann and J. W. Oldfield, J. Electroanal. Chem., 29, 211, 231 (1971); M. Fleischmann, J. W. Oldfield, and D. F. Porter, ibid., 29, 241 (1971); J. R. Backhurst, M. Fleischmann, F. Goodridge, and R. E. Plimley, Brit. Pat. 1,194,181, June 10, 1970.

[26a]. D. S. Salter and D. H. Smith, Ger. Offen. 1,950,379, Sept. 10, 1970; CA, 74, 9057j (1971).

[27]. F. Beck and H. Guthke, Chem.-Ing.-Tech., 41, 943 (1969).

[28]. H. Guthke, F. Beck, J. G. Floss, and W. Habermann (to Badische Anilin- & Soda-Fabrik), Fr. Pat. 1,476,162, April 14, 1966.

[28a]. H. F. Conway and E. B. Lancaster, Electrochemical Technology, 2, No. 1-2, 46 (1964).

[29]. J. A. Le Duc (to Pullman, Inc.), U.S. Pat. 3,427,235, Feb. 11, 1969.

[30]. J. A. Le Duc, J. C. Kuorilo, and C. Lurie, J. Electrochem. Soc., 116, 546 (1969).

[31]. G. Z. Kiryakov and V. V. Stender, J. Appl. Chem. XXIV, 12, 1263 (1951); XXV, 1, 23 (1952).

[32]. G. Z. Kiryakov, N. F. Razina, and Y. D. Dunaev, Acad. Sci., Kazakh SSR, 6, 4 (1960).

[33]. F. D. Gibson, Jr., U.S. Pat. 2,945,791, July 19, 1960.

[34]. P. Ruetschi, R. T. Angstadt, and B. D. Cahan, J. Electrochem. Soc., 106, 547 (1959).

[35]. Eur. Chem. News, Oct. 31, 1969, 34.

[36]. H. Beer, Neth. Pat. 6,606,302, Nov. 14, 1966.

[36a]. J. N. Maloney, C. R. Campbell, and R. Johnson (to Monsanto), U.S. Pat. 3,457,152, July 22, 1969.

[37]. H. Shaffer, J. Electrochem. Soc., 113, 1 (1966).

[38]. H. P. Gregor, Chem. Eng., Dec. 25, 1961, 73.

[39]. J. B. Farrell and R. N. Smith, Ind. Eng. Chem., 54, No. 6, 29 (1962).

[40]. C. W. Tobias and R. G. Hickman, Z. Phys. Chem., 229, 145 (1965).

[41]. N. Ibl, Chem.-Ing.-Tech., 33, 69 (1961); 35, 353 (1963); 36, 601 (1964).

[41a]. Asahi Kasei Kogyo K. K., Fr. Pat. 1,503,244, Oct. 16, 1967.

[42]. M. Eisenberg, C. W. Tobias, and C. R. Wilke, J. Electrochem. Soc., 101, 307 (1954).

[43]. P. Le Goff, P. Vergnes, F. Coevret, and J. Bordet, Ind. Eng. Chem., 61, No. 10, 8 (1969).

[44]. R. L. Taylor, Chem. Met. Eng., 44, 588 (1937).

[45]. L. L. Bott, Hydrocarbon Proc., 44, No. 1, 115 (1965).

[46]. J. H. Prescott, Chem. Eng., Nov. 8, 1965, 238.

[47]. J. A. May and K. A. Kobe, J. Electrochem. Soc., 97, No. 7, 183 (1950).

[48]. K. S. Udupa, G. S. Subramanian, and H. V. K. Udupa, Ind. Chem., May 1963, 238.

[49]. P. C. Condit, Ind. Eng. Chem., 48, 1252 (1956).

[50]. H. Suter, H. Nohe, F. Beck, and A. Hrubesch (to Badische Anilin- & Soda-Fabrik), U.S. Pat. 3,471,381, Oct. 7, 1969.

[51]. J. W. Drew and G. J. Moll, Ind. Eng. Chem., 53, No. 9, 48 A (1961).

[52]. K. Natarajan, K. S. Udupa, G. S. Subramanian, and H. V. K. Udupa, Electrochem. Tech., 2, No. 5-6, 151 (1964); T. D. Balakrishnan, et al., Chem. Ind. (London), 1622 (1970); V. D. Bezuglyi, V. A. Ekel, M. Ya. Fioshin, E. K. Nechiporenko, and R. F. Ramakaeva, Khim. Prom., 46, 822 (1970); CA, 74, 70879v (1971).

[53]. W. Kronig and J. Grolig (to Farbenfabriken Bayer), Brit. Pat. 1, 090, 006, Nov. 8, 1967.

[54]. A. I. Kamneva, M. Ya. Fioshin, L. I. Kazakova, and Sh. M. Itenberg, Neftekhim., 2, 550-6 (1962).

[55]. R. W. Foreman and J. W. Sprague, Ind. Eng. Chem. Prod. Res. Develop., 2, 303 (1963).

[56]. H. F. Conway and V. E. Sohns, Ind. Eng. Chem., 51, 637 (1959).

[56a]. M. M. Baizer, J. Electrochem. Soc., 111, 215 (1964).

[57]. Anon., Chem. Eng. News, Oct. 2, 1967, 58.

[58]. Union Chimique-Chemische Bedrijven, Brit. Pat. 1, 117, 677, June 19, 1968.

[59]. I. L. Knunyants and N. P. Gambaryan, Usp. Khim., 23, 781-820 (1954).

[60]. R. A. Dewar, V. E. Maier, and M. A. Riddolls (to Imperial Chemical Industries), Can. Pat. 775, 138, Nov. 15, 1966.

[61]. Anon., Eur. Chem. News, 51, Dec. 19/26, 1969; C. Van Eygen, A. Hendrickx, J. Ramioulle, J. Walravens, and A. Verheyden, Chim. Ind. (Genie Chim.), 104, 71 (1971).

[62]. D. G. Braithwaite and L. L. Bott (to Nalco Chemical), U.S. Pat. 3, 380, 900, April 30, 1968.

[63]. K. Ziegler and H. Lehmkuhl, U.S. Pat. 3, 372, 097, March 5, 1968.

[64]. Monsanto Co., Brit. Pat. 1, 066, 930, April 26, 1967.

[65]. L. Nanis and F. McLarnon, J. Electrochem. Soc., 117, 1527 (1970).

[66]. M. S. Sataev, N. G. Bakhchisaraits'yan, M. Ya. Fioshin, G. Kh. Fursman, and K. I. Kryschenko, Khim. Prom., 46, 892 (1970); CA, 74, 49005x (1971).

[67]. C. Van Eygen, A. Hendrickx, J. Ramioulle, J. Walravens, and A. Verheyden, Chimie et Industrie, 104, No. 1 (1971).

Chapter XXIX

ELECTROCHEMICAL POLYMERIZATION

G. Parravano

Department of Chemical and Metallurgical Engineering
University of Michigan
Ann Arbor, Michigan 48104

I. INTRODUCTION

In spite of early observations of electrochemical polymerization (ECP) it is only within the last 20 years that investigations aimed at understanding the chemical phenomena connected with the formation of polymers in an electrolytic cell have been undertaken. From this point of view, ECP is a young branch of electrochemistry that treats phenomena typical of polymerization reactions. The understanding of its nature, the assessment of its possibilities, and the recognition of its aspects are presently emerging. The upsurge in interest in ECP within the last few years has already produced reviews and discussions in the literature [1-8].

There is at present a twofold interest in ECP processes. First, ECP reactions provide a new method of polymerization with a fine control of the initiation and termination steps, and offer new possibilities as a research tool for fundamental studies by permitting the application of fast, transient electrochemical methods to the study of the rate of elementary polymerization steps. These studies are today possible only for photopolymerizable monomers and, in a few cases, for chemically polymerizable monomers through the application of the stopped flow method [9], which, however, permits the study of relatively long reaction times only ($1\text{-}10^{-3}$ sec).

Secondly, ECP could have technological potential. If commercial applications of electrode reactions in general are some guide, the outlook for the development of commercial ECP processes is promising. Preliminary considerations suggest that several existing ECP systems could be competitive with modern commercial polymerization processes and future studies may discover ECP conditions for the synthesis of unusual polymers which cannot be obtained economically by any other method. Some recently studied ECP systems are summarized in Table 1.

In this chapter we discuss the characteristic chemistry and the electrode processes relevant to ECP and point out the interesting and novel aspects in relation to electroorganic reactions in general and to the more classical chemically initiated polymerizations.

II. REACTION CHARACTERISTICS

In ECP the electrically important steps occur at the electrode surface, but the complete polymerization sequence includes reactions in the surrounding liquid phase also. Thus, ECP reactions are essentially heterogeneous polymerizations, in which interphase boundaries play a critical role, chemically and electrically. For this reason ECP systems possess physical and chemical characteristics of electrochemical and electrode processes on the one end and of polymerization reactions on the other. It is from the combination of principles of electrochemistry with those of polymer science that the typical features of ECP emerge.

TABLE 1

Representative Electrochemical Polymerizations [a]

Initiation	Monomer	Electrolyte	Medium	Inter-mediate	Ref.
Cathodic reduction of H^+	MMA, AA, EMA	H_2SO_4	H_2O	FR	[1]
Cathodic reduction of inorganic redox couple	AN	Fe^{+3}, $S_2O_8^{2-}$	H_2O	FR	[10]
Anodic oxidation of carboxylate anion	ST, MMA, AN	$CH_3CO_2^-K^+$	CH_3COOH	FR	[11]
Cathodic electron transfer to monomer	ST	NaI	Anhydrous pyridine	Anionic	[19]
Cathodic reduction of N_4R^+	ST, AN, ST-AN copolymer	$R_4N^+I^-$	DMF bulk	FR and/or anionic	[4]
Anodic oxidation of inorganic anion	ST, αMST, VE	BF_4^-, ClO_4^-	Bulk nitro-benzene	FR and/or cationic	[4]
Cathodic electron transfer to monomer	MMA, AN, ST	$NaNO_3$	DMF	Anion radical	[7]
Cathodic electron transfer to monomer	αMST	$NaAlEt_4$ $LiAlH_4$	THF	"Living" anionic	[63]
Cathodic electron transfer to monomer	Sulfonium compounds	H_2O, THF DMF, MeOH EtOH, CH_3COOH	←	?	[67]

[a] MMA, methyl methacrylate
AA, acrylic acid
EMA, ethyl methacrylate
AN, acrylonitrile
ST, styrene
αMST, α-methylstyrene
VE, vinyl ether
DMF, dimethylformamide
FR, free radical
AR, anion radical.

Electrode processes which take place in ECP (adsorption, desorption, and electron transfer) can be analyzed and discussed within the conceptual schemes of classical electrochemistry. Electrode variables (chemical and physical nature of electrodes and their surface, electrode overpotential, half-cell reaction potential, and current density) are in general expected to influence the structure, size, size distribution, and yield of the polymer obtained.

In a similar fashion, chain growth may include homogeneous and/or heterogeneous reactions and be sustained by free radical and/or ionic intermediates. The processes of chain transfer and termination may be similar to the corresponding steps in homogeneous and heterogeneous chemical polymerizations, but new processes involving electrode reactions may appear.

In other aspects, ECP reactions depart from the behavior of classical electrochemical and polymerization reactions. In most electrochemical processes, the products are formed at the electrode or in a small region in its immediate neighborhood where the sequence of reaction steps is practically localized.

A typical ECP system consists of an electrolytive cell with electrodes, electrolyte, monomer, and solvent. The solvent is generally needed whenever the monomer is not soluble in the electrolyte. Where an electrical current flows through the system, electrode reactions take place. Electron transfer may occur at the cathode with an adsorbed vinyl monomer M, namely:

$$M + e^- \rightarrow M^-$$

followed by desorption of a radical anion

$$M^- \rightarrow \cdot M^-.$$

The radical anion may dimerize to a dianion $^-M_2^-$ and grow, according to the schemes of anionic addition polymerization. Alternatively, the radical anion may be neutralized, and free radical addition polymerization results.

Diffusion phenomena are important since they tend to influence the rate of reaction and the product distribution. In ECP, mixing sometimes inhibits the formation of polymer [12-17] and it is conceivable that mixing may influence the concentration of intermediates, their chemical nature, the polymerization mechanism, and the molecular weight distribution of the polymer. In ECP, chain growth and termination may occur in an environment very different, with respect to solvent medium, reagent concentrations, electric field strength, physical phase, and temperature, from that prevailing during the initiation step.

In some instances chemical activation of certain monomers cannot be achieved for lack of a sufficiently energetic or specific initiation reaction (ketones, aldehydes, thioketones, and thioaldehydes). ECP systems may be activated by means of electrodes with a wide energy spectrum for electron transfer with monomer or initiator precursor; the control of electrode potential permits variations in thermodynamic activity of the initiators in a more powerful manner than by variation in the concentrations of heterogeneous or homogeneous chemical initiators. Since the type of electrode process occurring depends on the electrode potential, the initiation reaction in a given system may be controlled by proper adjustment of the electrode potential. Also, the nature of the electrode material (metals and alloys, inorganic and organic semiconductors, thin coatings on metals, etc.) may provide varying degrees of specificity in the electron-transfer and polymerization reactions yet unknown to polymer chemistry.

In principle, it is possible to produce by ECP methods ions with different electron affinities, thus giving a fine control in the tailoring of copolymers. As a result of these effects, polymers differing from those obtained by chemical or photochemical initiations may be produced by ECP. These effects include the configuration and conformation of the chain (linear, branched, and/or stereoblock structures) and the composition of copolymers. If the propagation step is restricted to the electrode surface, the formation of stereoregular chain structures is conceivable. In favorable cases, modifications of the surface concentration of activators, via manipulations of the current density, could produce polymers with a preponderant type of tacticity.

Finally, the rate of the initiation step may be modified through the cell current. This provides a simple method for the control of the molecular weight distribution, and for the production of polymers with complex molecular weight distributions to fit particular end uses.

Several types of electrode reactions are available for ECP processes: direct electron transfer between electrode and monomer (cathodic production of monomer anions and anodic formation of monomer cations); cathodic deposition and anodic dissolution of metals; discharge of cations and anions; organic redox reactions; and formation by electrode reaction of polymerization activators. Because of the variety of electrode reactions available, ECP processes can occur over a wide temperature range, specifically -100 to 200°C. This fact points to additional control through temperature of the polymer structure and polymerization character.

A comparison of the reaction efficiency between chemical and ECP polymerizations is difficult for lack of accurate results. However, the electrochemical efficiency, relating the number of coulombs passed through the cell to the average number of polymer molecules produced, may be equal to unity for polymerizations not involving chain transfer.

III. CHEMICAL AND ELECTROCHEMICAL STEPS IN ECP

The production of macromolecules in an operating electrochemical cell is the result of chemical and electrochemical steps, linked together in a complex kinetic sequence. The steps include the exchange and transfer of electrons, ions, neutral atoms or molecules, and molecular fragments (radicals, carbanions, carbonium ions), and the transfer may include the crossing of interphase boundaries. As a result, physical transport phenomena (diffusion and convection) must be considered not only in the formulation of the expression for the reaction rate but also in the derivation of the molecular weight and molecular weight distribution. In this section we shall review the chemical and electrochemical principles for the most important of these reaction steps. This information constitutes the basis for deriving criteria for the prediction of reaction rates and of molecular weight of polymers in terms of macroscopic parameters (concentration, voltage, current, temperature).

A. Vinyl Addition Reactions

Consider a vinyl molecule, $H_2C{=}CHR$ (R an electrophilic or electrophobic group). The mobility of π-electrons in such a molecule leads to the following principal electron distributions:

$$\underset{\text{(I)}}{H_2C{=}CHR,}\quad \underset{\text{(II)}}{H_2\dot{C}{-}\dot{C}HR,}\quad \underset{\text{(III)}}{H_2\overset{\oplus}{C}{-}\overset{\ominus}{C}HR,}\quad \underset{\text{(IV)}}{H_2\overset{\ominus}{C}{-}\overset{\oplus}{C}HR.}$$

The stability of ionic forms (III) and (IV) depends primarily on the electron-withdrawing ability of R. The large charge separation in the ionic forms is further stabilized by the electric field established near an electrode. The diradical form (II) is more probable in the absence of an electric field and for monomers with weakly electrophilic or electrophobic substituents.

From quantum mechanical calculations [18] and measurements on rates of adsorption and associated reactions [19] (hydrogenation, polymerization), there is evidence for the formation of π-bonded complexes during the adsorption of vinyl monomers on a neutral metal surface. This mode of adsorption, however, may be significantly altered in the presence of an electrically charged electrode surface. The probable structures of the adsorption complex at the cathodic surface (without electron transfer) may be schematically illustrated as follows:

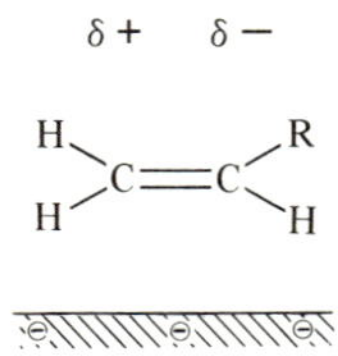

(Va)

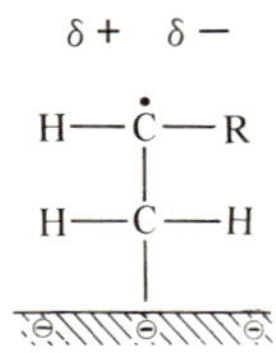

(Vb)

(VI) (VII)

In form (Va), π-bonding occurs with the vinyl group acting as the acceptor, and the electrode as the electron donor. In forms (V) through (VII), the electrode retains its electrons, but one electron from the disrupted double bond is shared with electrons from the electrode surface. Resonance with the R-group electrons produces types (VI) and (VII) in which the unpaired electron undergoes repulsion from the electrode by the electric field of the cathode and the electronegativity of the R group. The adsorption of a monomer containing an electron-releasing group R$'$ at an anode but with electrons shifted toward the electrode may be described in a similar fashion.

The electron transfer between adsorbed monomer and electrode occurs by further rearrangement of the electron distribution in structures (V) through (VII). For cathodic adsorption the result is as follows:

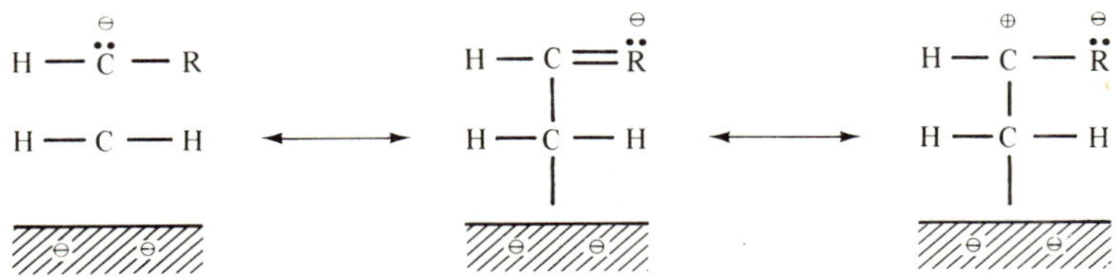

(VIII) (IX) (X)

This electron transfer may be accompanied by simultaneous desorption of the adsorbate. This is visualized as a concerted shift of bonding electrons to give a radical anion [18, 20] whose electron density is highest in the vicinity of the electronegative R group. Structures (XI) and (XII) result:

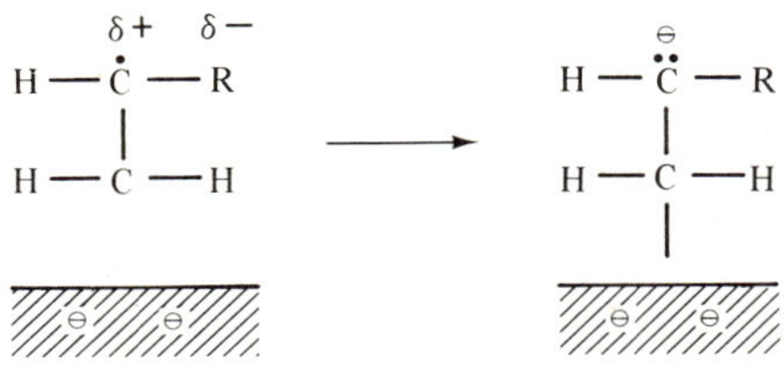

(XI) (XII)

Interesting attempts have been made to correlate the polarographic reduction potentials of vinyl monomers with their ease of anionic polymerization [68] and copolymerization [69]. Fueno and co-workers [68] examined a variety of olefins including acrylonitrile, acrolein, methyl methacrylate, α-methylstyrene, vinyl acetate, isobutylene, and others and concluded that anionic polymerizability roughly paralleled ease of reduction and low electron density at the β-carbon atom (calculated by molecular orbital theory). Höfelmann et al. [69] concluded that monomers which were separately capable of being anionically polymerized would form copolymers only if their polarographic reduction potentials were no more than ca. 0.2 V apart.

Vinyl ECP may be initiated also via nonpolymerizable electroactive substances, which act as electrochemical activators. They may operate by direct chemical combination, transfer, decomposition, and catalytically. With each type of activator, the processes of adsorption and desorption, and of polymer growth and termination at the electrode surface and/or in the bulk of the solution occur.

The behavior of combining-type activators parallels closely that of directly electroactivated vinyl monomers. Catalytic activators act as agents to carry electrons from the electrode to monomers:

$$C^+ + e^- \text{ (Cathode)} \rightarrow C\cdot$$

$$C\cdot + M \rightarrow C^+ + M^{\dot{-}}$$

where C^+ may be an alkali metal [19, 20] or a tetraalkylammonium cation. A converse but analogous process may occur at anodes, leading to cationic polymerization [4]. Transfer activators undergo a multistep change in oxidation state. A simple case may be illustrated for anodic initiation by the corrosion product of the anode:

$$A \text{ (Anode metal)} \rightarrow A^+ + e^- \text{ (Anode)}$$

$$A^+ + M \rightarrow A^{2+} + M^{\dot{-}}.$$

This peculiarly results in the possibility of initiating anionic polymerization at an anode. However, the lifetime of the anion radical may be quite short unless the initiation occurs at some distance away from the anode or unless there is a special interaction between A^{2+} and $M^{\dot{-}}$ which makes oxidation of the complex more difficult than that of "free" $M^{\dot{-}}$. Otherwise the anion radical may be reoxidized back to the original monomer by electron transfer with the anode. This type of ECP initiation was attempted with Mg anodes in pyridine and styrene monomer, but polymer was formed only at the cathode [21] (apparently by direct electron transfer from cathode to monomer).

Electrode reduction or oxidation of a decomposing activator leads to breakdown of the activator into fragments which are sufficiently reactive to initiate chain growth by reaction with monomers. Consider the electrochemical Kolbe reaction which produces free radicals by anodic oxidation of carboxylate anions [23]:

$$RCO_2^- \xrightarrow{-e^-} RCO_2\cdot \longrightarrow R\cdot + CO_2$$

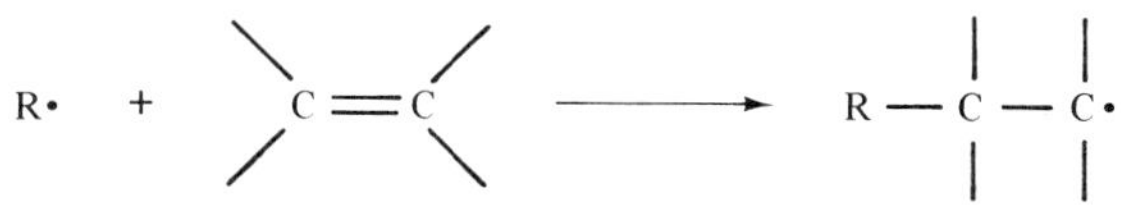

The growth of the polymer may occur on the electrode surface or in the bulk of the electrolytic solution. The latter propagation reaction is similar to that taking place in chemically initiated polymerizations, whereas the former offers some unique growth characteristics. For example, any one of the surface-bound monomer forms (V)-(VIII) may add other monomer molecules to form polymers. Such growth involves monomers in the adsorbed (XIII), or in the liquid phase (XIV) for the case of anionic growth:

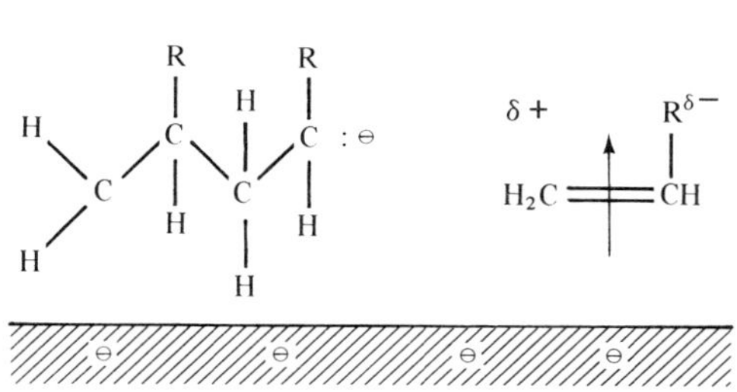

(XIII)

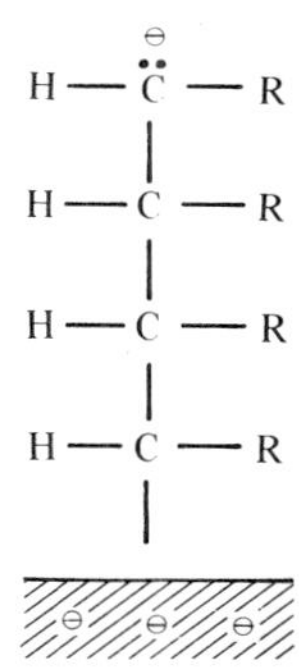

(XIV)

In (XIII), there is competition between the adsorbed carbanion and the electrode electrons for addition to the neighboring adsorbed monomer, hence chain length should be dependent on the electrode potential. Furthermore, it is feasible that in case (XIII) the monomer adsorption centers have characteristics which favor a stereoselective addition of each monomer molecule to the growing chain, thereby resulting in stereoregular chain structure. This is less likely to happen in case (XIV).

Desorption of the reduced monomer as a radical anion (XII) may lead to surface growth by adding adsorbed monomer, or to growth in solution by adding dissolved monomer. It is conceivable that monomers, and radical ions, may be sufficiently reactive only when polarized by the strong electric field of the electrode double layer, hence restricting growth to regions on and/or very close to the electrode surface. Furthermore, the radical ions may have a strong propensity for dimerization to dianions [18, 22, 24-26] (XV). Dimerization is competitive with monomer addition [26, 27] (XVI):

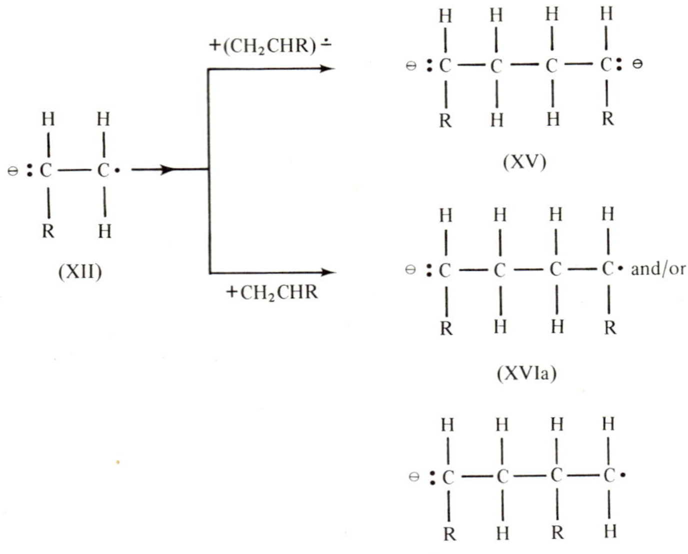

(XII) (XV) (XVIa) (XVIb)

As the chain grows, the negatively charged chain end is repelled electrostatically away from the other (adsorbed) chain end, and the transfer of a second electron to the latter becomes increasingly favorable, due to the lower electron density in the environment of the adsorbed group. This second electron transfer leads to desorption of a polymeric dianion (XVII):

(XVII)

The second electron transfer requires a more negative electrode potential than the first one and the new carbanion is much more reactive than the first one. Various monomers could provide the stabilizing factors (strongly electronegative or π-resonating groups attached to each end of the monomer) required to make this second electron transfer more probable. Without such stabilizing substituents, the very reactive carbanion chain end may either quickly add tail to tail to monomer or be terminated by a proton donor, such as monomer, solvent, electrolyte, transfer agents, or impurities, which otherwise are unreactive toward the carbanion adjacent to R.

Termination of ECP reactions may take place by several mechanisms. The active chain ends, either ionic or radical, adsorbed or dissolved, may react with a terminating agent HX (XVIII) or with a transfer agent HT (XIX):

$$M_n^* + HX \rightarrow M_nH + X^* \quad \text{(Unreactive)} \qquad \text{(XVIII)}$$

$$\left.\begin{array}{ll} M_n^* + HT \rightarrow M_nH + T^* & \text{(Reactive)} \\ T^* + M \rightarrow TM^* \text{ or } M^* + T & \text{(Unreactive)} \\ TM^* \text{ or } M^* \rightarrow TM_{n+1}^* & \end{array}\right\} \quad \text{(XIX)}$$

Here M and M_n are monomer and polymer of length n, respectively.

Free radical chain ends, adsorbed or not, may become terminated by the combination or disproportionation reactions between themselves.

A termination mechanism peculiar to ECP takes place when the active polymer chain transfers an electron with the electrode. Such a transfer is facilitated by a change in electrode potential toward the opposite sign required to initiate the chain, e.g., anionic ECP at a cathode may be terminated by making the cathode potential more positive. Thus on oxidation, carbanion ends would become free radicals, and free radical ends would become carbonium ions (or yield protons). These processes may be visualized as a continuous sequence between carbanions and carbonium ions depending on the electrode potential:

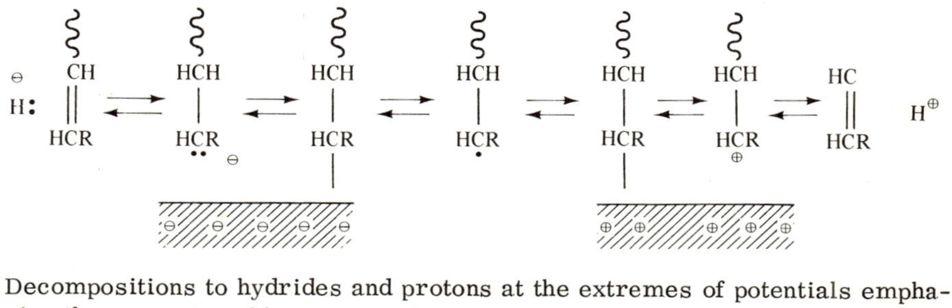

Decompositions to hydrides and protons at the extremes of potentials emphasize the necessity of having stabilizing groups in the monomer molecule to generate the positive and/or negative chain ends. Furthermore, formation of "stable" carbanions and carbonium ions of the types of polymer end groups in a single solvent imposes nearly incompatible restrictions on the properties of the solvent. Since many monomers undergo both cationic and anionic polymerizations, their ECP would allow the control of the polymerization mechanism simply by controlling the electrode potential.

Further possibilities are the control of average polymer chain length and copolymer composition by using monomers of differing electron affinities and reactivities and by oscillating the electrode potential between the various electron-transfer potentials of the system.

Side reactions may accompany the main processes of chain initiation, growth, and termination. Some examples of side reactions which may influence the outcome of vinyl ECP are the following:

Hydrogenation [25]:

$$HX + e^- \text{ (Cathode)} \rightarrow \dot{H} + X^-$$

$$2\dot{H} + CH_2{=}CHR \rightarrow CH_3{-}CH_2R$$

Oxidation [26]:

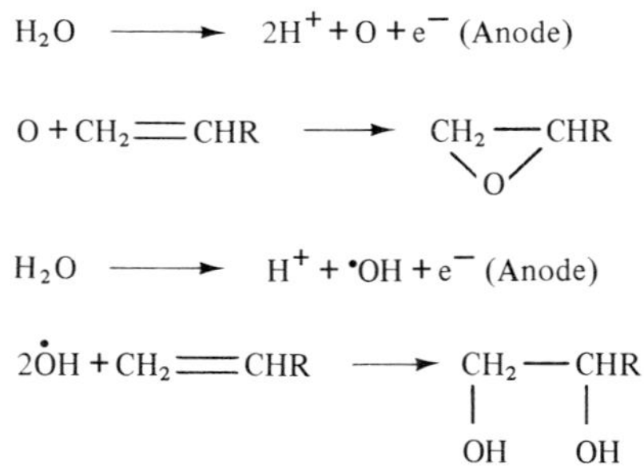

Alkylation:

$$R{-}CO_2^- \xrightarrow{-e^-} R\bullet + CO_2$$

$$R{-}X \xrightarrow{+e^-} R\bullet + X^-$$

Initiator formation

Dialkylation:

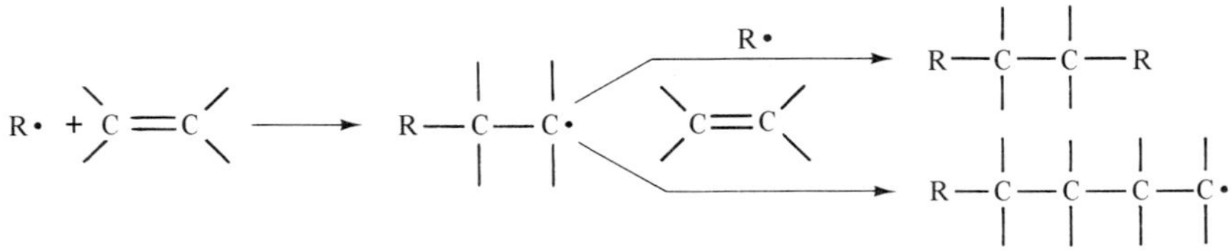

(Monomer or unsaturated polymer)

Grafting, branching, or polymerizing alkylation

Cyclization [29-31]:

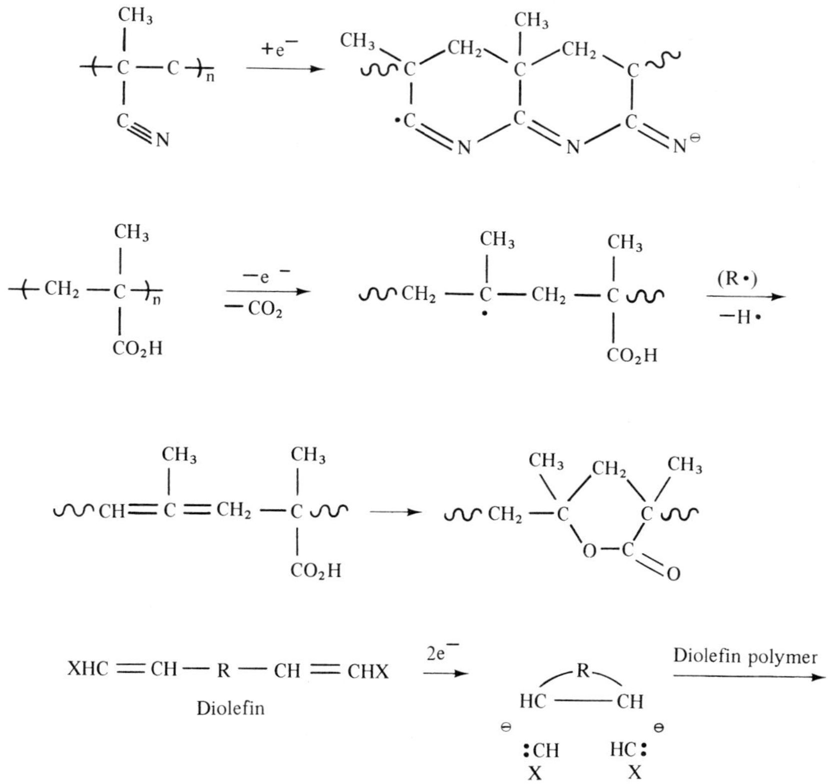

Cross-linking [14]:

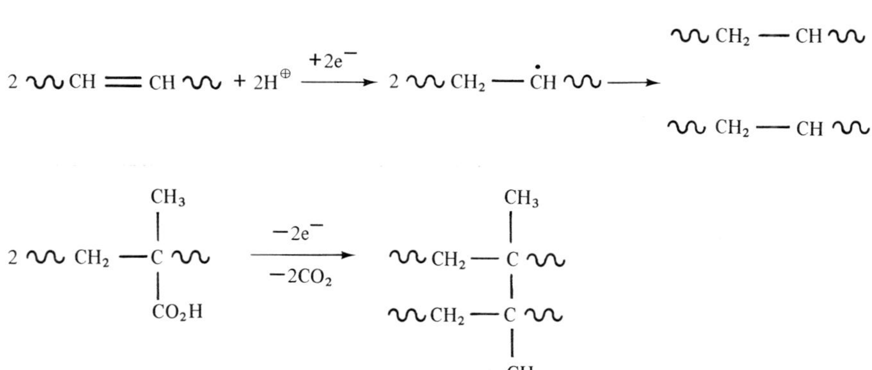

Grafting [32]:

$$\text{Cellulose} + H^{\oplus} + n(CH_2CHR) \xrightarrow{+e^-} \text{Cellulose-poly}(CH_2CHR).$$

Alkylation, isomerization, cyclization, cross-linking, and grafting reactions may involve the dead or growing polymer. These reactions result directly from the intervention of an electrode process or of species generated at the electrodes but reacting in the liquid phase.

B. Condensation Reactions

Electrochemical condensation polymerization requires one electron-electrode transfer for each monomer molecule rather than one electron transfer for each polymer molecule as in the case of vinyl addition ECP. Similarly to the latter, however, electrons may be transferred either directly to the polymer-forming species or indirectly by way of reactive intermediates. Consider the formation of polyalkyls by the Kolbe coupling of dicarboxylates at an anode [31-33]:

$$^{\ominus}O_2C - R - CO_2^{\ominus} \xrightarrow{-e^-} {}^{\ominus}O_2C - R\cdot + CO_2 \xrightarrow{-e^-} \cdot R\cdot + 2CO_2$$

$$2\,^{\ominus}O_2C - R\cdot \longrightarrow {}^{\ominus}O_2C - R - R - CO_2^{\ominus} \xrightarrow{-e} -{}^{\ominus}O_2C - R - R\cdot + CO_2$$

$$\cdot R_n\cdot \longrightarrow \widehat{R_n} \text{ (Cyclization side reaction)}$$

The occurrence of significant cyclization depends mostly on the nature of the R group.

Several compounds have been polymerized by electrochemical condensation: disubstituted xylenes [36], phenols [37, 38], and pyridine [39]. Since artificial resins were first reported to be made by ECP in 1884 [40], other aromatic condensation products [41, 42] and linear polyesters [43] have been made.

C. Ring Opening Addition Reactions

Kinetically similar to vinyl polymerizations, ring opening polymerization tions usually involve ionic intermediates. Molten caprolactam containing sodium benzoate as electrolyte and an acetyl caprolactam (isocyanate) as initiator has been polymerized in this fashion [44, 45]:

Activation:

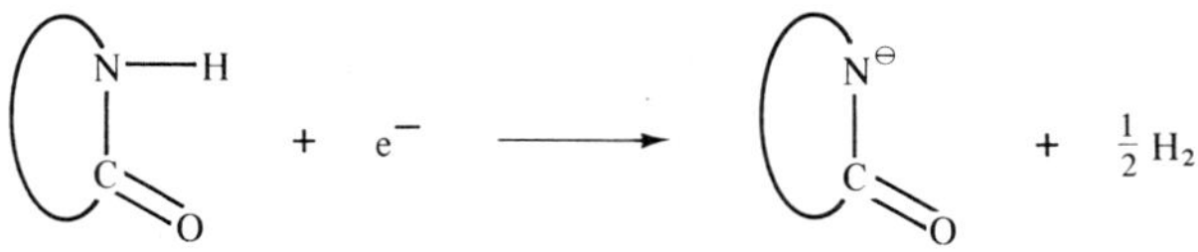

Initiation:

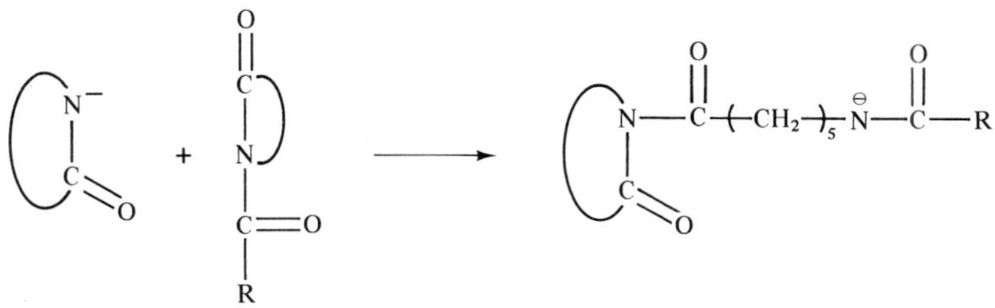

Activation:

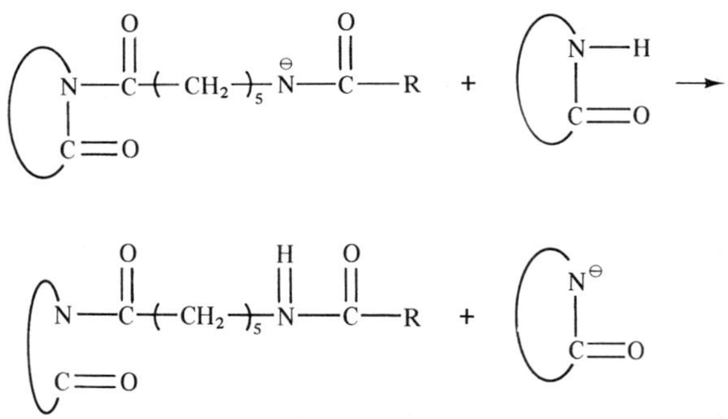

Addition:

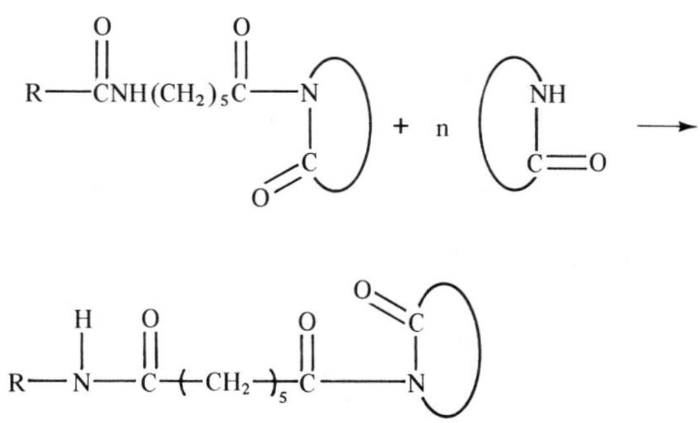

Since ring opening polymerizations require less electric current per polymer molecule than comparable polycondensations, and since these polymers are much different from those obtainable from addition polymerizations of unsaturated compounds, there is great interest in applying ECP to known ring opening monomers: butyrolactams, ethylene oxide, propylene oxide, 1, 3-dioxolane, ethylene imines, halogenated propio- and butyrolactones, oxazolones.

Cyclic ethers may be polymerized cationically by anodic electrolysis [46, 47], either directly:

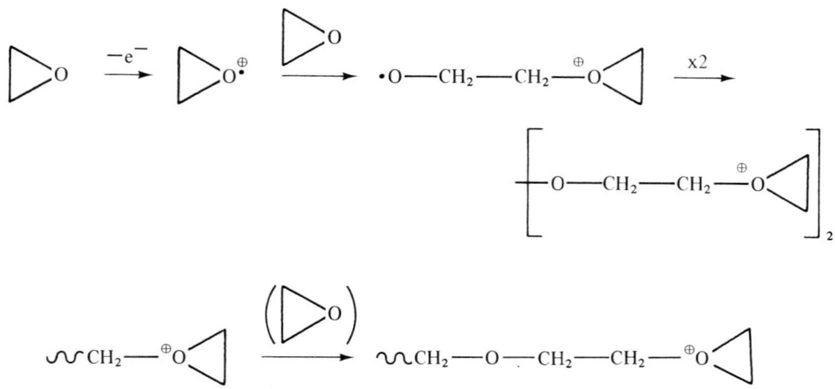

or indirectly:

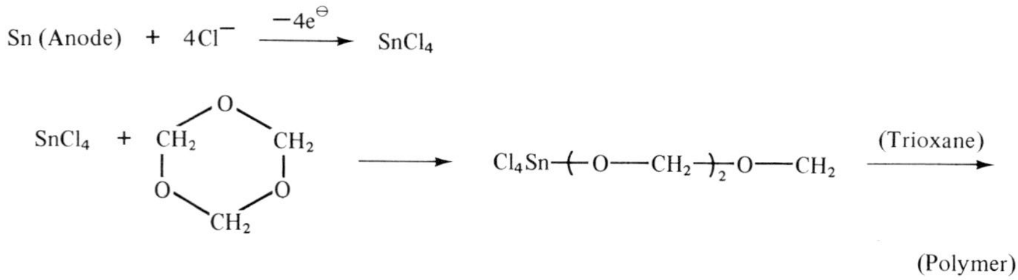

Suitable solvents for these ECP's include acetonitrile, 1, 1-dichloroethane, nitromethane, nitrobenzene, and liquid sulfur dioxide.

D. Physical Effects

Phase changes may take place during the course of ECP reactions. Gaseous monomers must be dissolved under pressure in appropriate solvents. Ethylene and halogenated ethylenes have been polymerized by ECP in an aqueous dispersion [48] or halogenated solvents [49-51]. The polymer often precipitates as a new phase (liquid or solid) which remains attached to the electrode surface [52-54]. This requires the transport of ions, electrons, or neutral species through the deposit. This occurrence has important kinetic implications, depending on the permeability of the polymer coating. Physical effects in the electrode double layer, in the adsorption of the electrolyte, and as a result of concentration gradients may be expected. Interfacial ECP of benzene in hydrogen fluoride has been reported [55].

IV. ELECTRODE PROCESSES

The considerations of the previous section led to the conclusion that electrode processes have a profound effect upon the polymerization reaction and upon the structure and characteristics of the polymer formed. In this section we shall discuss this role in connection with two effects, one dependent on thermodynamic considerations and one derived from kinetic conditions.

A. Electrode Potentials and Overpotentials

In complex systems such as those encountered in ECP, several chemical constituents of the system may be electroactive, depending on the electrode potential and on the overpotential values of the constituents at the current density employed. The predominant electrode reaction takes place among the species with the lowest total overpotential. Due to the different chemical and physical properties of the ECP constituents, their individual overpotentials vary differently with the current density. Hence, by judicious choice of the experimental variables and chemical constituents, a preferred type of electrode reaction may be favored.

The diffusion overpotential is related to the current in a manner which depends on the nature of the particular electrode process. Consider the case of reduction followed by dimerization of the reduced species:

$$2M + 2e^- \rightleftarrows 2\dot{R}^- \xrightarrow{k} {}^-C^-$$

where M may be a vinyl monomer, R its radical anion, and $^-C^-$ the dimer dianion. The half-wave potential of this process at a rotating disk (planar) electrode with very slow and negligible dimerization is given by [56, 57]:

$$E_{1/2} = E^\circ + \frac{RT}{F} \ln\left(\frac{D_R \delta_M f_M}{D_M \delta_R f_R}\right)$$

where the diffusion-layer thicknesses are given by:

$$\delta_{M,R} = \frac{1.62 D_M^{1/3} \mu^{1/6}}{\omega^{1/2}}$$

D's are the diffusion coefficients, f's the activity coefficients, μ the kinematic viscosity, and ω the angular velocity. For the case of more rapid dimerization of R, when concentration of R near the electrode is effectively decreased by reaction [58]:

$$E_{1/2} = E^o + \frac{2RT}{3F} \ln \frac{D_R}{D_M} + \frac{RT}{F} \ln \frac{\mu^{1/9} k^{1/3} \left(C_M{}^b\right)^{1/3}}{0.91 \omega^{1/3} D_R{}^{1/9}} \tag{1}$$

where $C_M{}^b$ is the bulk concentration of monomer. Equation (1) includes kinetic as well as diffusional contributions to the overpotential, and points out the fact that electrode potential may depend on the relative rates of diffusion and chemical reactions associated with the electron-transfer step.

Reaction overpotentials can be very large (especially with phase transformations), depending on the nature of the electrode surface, electrode material, and chemical species involved.

Aside from electrode reactions and mode of initiation, the success of a particular ECP system depends on the effects of the constituents and conditions on the polymer growth and termination reactions. For example, cathodically initiated anionic ECP of styrene in dimethylformamide gives more reproducible results and a faster rate with K salts as supporting electrolytes than with salts of other alkali or alkaline earth metals. Li salts are less effective even though they permit the attainment of more negative potentials than the other salts. Apparently, the ineffectiveness of Li^+ ions results from their greater tendency to form less reactive ion pairs with the growing polymer ends. Such effects limit the usefulness of polarographic data in predicting the success of proposed ECP systems.

It is conceivable that the electrode potential directly influences thermodynamic quantities of the polymerization reaction. We shall discuss the case of the polymerization ceiling temperature T_c, namely, the temperature at which the polymerization growth step is at equilibrium with the depolymerization step, or, in other words, where the rate of addition of monomer to polymer is equal to the rate of the reverse process. This concept, similar to vapor-liquid equilibrium, has the same significance in both condensation and addition polymerizations. The quantity T_c is expressed by:

$$T_c = \frac{\Delta H_p}{\Delta S_p}$$

where ΔH_p and ΔS_p are the enthalpy and entropy changes, respectively, which accompany the polymer growth. It should be emphasized that T_c is determined only by the growth step, and is not affected by the preceding (initiation) or following (termination) processes. For a given polymerization system, T_c may be influenced by the concentration of reactants, the pressure, the electric field, the adsorption on solid or liquid surfaces, and the nature of the polymer growth step.

A unique feature of ECP derives from the influence of electric field strength on the spatial orientation, extent of adsorption and dissociation, and degree of polarization of the reactants and reaction medium. These effects have been discussed and demonstrated for nonpolymerizing systems [60-62]. It is then conceivable that the ceiling temperature of polymerization reactions carried out by electrolysis be dependent on, and controlled by, the electrode potential and/or the current density, particularly for polymerizations in which growth occurs predominantly within the high electric field of the electrode ionic double layer or on the electrode surface. This dependence has not yet been demonstrated experimentally, and even meaningful approximate calculations of the effect are difficult.

If the monomer is more polar or polarizable than the active chain end, the former species will be more oriented by an applied electric field than the latter. Consequently, the entropy change for monomer addition to the active polymer end will become smaller and T_c higher as the electric field strength is increased. This situation is anticipated for vinyl polymerizations generally because of the greater number of π-electrons and resonance possibilities for the monomer [(XX) and (XXII)] compared to those of the live polymer end group [(XXI) and (XXIII)]:

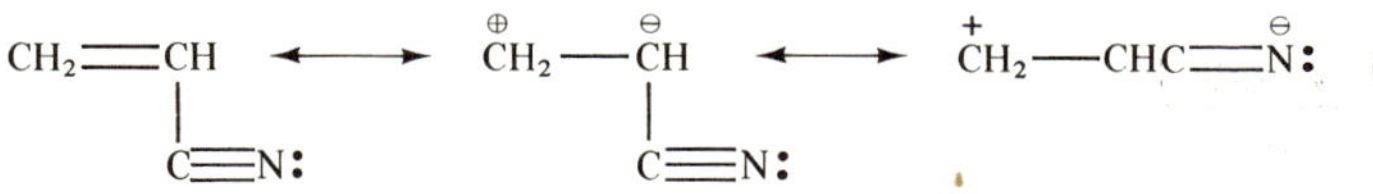
(XX)

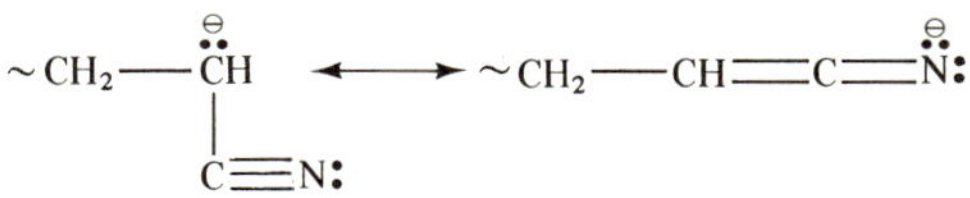
(XXI)

Similarly, for free radical vinyl polymerizations:

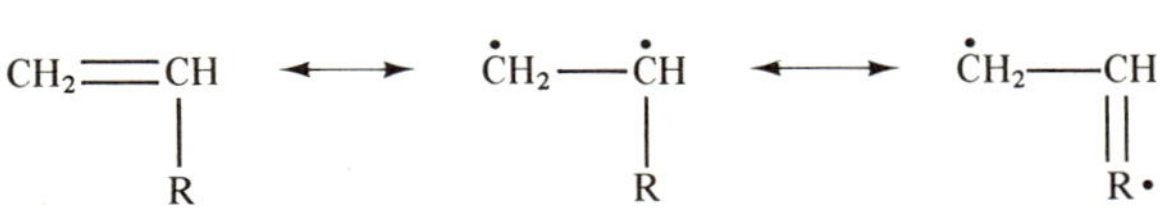
(XXII)

(XXIII)

Recent observations [62] of increased polymerization rates and polymer molecular weights upon application of electric fields to vinyl polymerizations tend to confirm the above reasoning, although this conclusion must be tempered by the fact that these observations are the result of kinetic phenomena whereas T_c is a thermodynamic quantity.

B. Transient ECP

In ECP, the electrical current passed through the electrodes is a reagent whose addition to the reacting system can be programmed easily in time and local density. Particularly attractive is the possibility of varying polymerization conditions to influence the molecular weight of the polymer by using alternating electrolysis current. Intuitively this approach should permit the tailoring of molecular size and of molecular size distribution by the correct current time program. To demonstrate this control we shall discuss the case of a "living" polymerization system, i.e., without chemical termination, the polymerization being controlled by electrochemical initiation at one electrode and termination at another electrode [63]. The initiation and termination reactions may be either by direct or indirect electron transfer with the electrodes.

The reaction scheme to be considered is the following:

Activation:

$$I_2 + 2e^- \xrightarrow[i_c]{\text{Cathode}} 2I^\theta$$

Initiation:

$$I^\theta + M \xrightarrow{k_i} P_1^\theta$$

Growth:

$$P_n^\theta + M \xrightarrow{k_p} P_{n+1}^\theta, \qquad n = 1, 2, 3, \ldots$$

Termination:

$$P_n^\theta - e^- \xrightarrow[\text{(Anode)}]{i_a} P_n^\bullet$$

$$P_n^\bullet + P_m^\bullet \xrightarrow{k_t} M_n + M_m \quad \text{or} \quad M_{n+m}, \qquad m = 1, 2, 3, \ldots .$$

The goal is to program the cathodic and anodic currents, i_c and i_a, so that the kinetic chain length ν is constant, or:

$$\nu = R_p/R_i + 1 = \text{Constant} \tag{2}$$

where

$$R_p = k_p[M][P^\theta] \tag{3}$$

and (for rapid initiation)

$$R_i = k_i[M][I^-] = i_c'. \tag{4}$$

Using Eqs. (2)-(4)

$$-\frac{d[M]}{dt} = R_i + R_p = \nu i_c' \tag{5}$$

which integrates, for constant i_c', to give

$$[M] = [M]_o - \nu \int_o^t i_c' \, dt = [M]_o - \nu i_c' t. \tag{6}$$

From Eqs. (6) and (2)-(4):

$$[P^\theta] = \frac{R_p}{k_p[M]} = \frac{(\nu - 1)i_c'}{k_p([M]_o - \nu i_c' t)} \tag{7}$$

and since

$$\frac{d[P^\theta]}{dt} = i_c' - i_a'. \tag{8}$$

The manner in which i_a' should be controlled, to maintain constant ν, is obtained by differentiating Eq. (7) and combining with Eq. (8) to get:

$$i_a' = i_c' - \frac{(\nu - 1)}{\nu k_p\{([M]_o/\nu i_c') - t\}^2} \tag{9}$$

since $i_a' > 0$ for all times, Eq. (9) requires that:

$$i_c' > \frac{(\nu - 1)}{\nu k_p\{([M]_o/\nu i_c') - t\}^2} \tag{10}$$

from which

$$t < \frac{[M]_o}{\nu i_c'} - \left(\frac{\nu - 1}{\nu k_p i_c'}\right)^{1/2} \tag{11}$$

and

$$i_c' < \frac{k_p[M]_o^2}{\nu(\nu - 1)} \equiv (i_c')_{max} \tag{12}$$

Equation (9) shows that the controlled current i_a' decreases with time starting initially at:

$$i_a'(0) = i_c'\left[1 - \frac{\nu(\nu - 1)}{k_p[M]_o^2} i_c\right] = i_c'\left[1 - \frac{i_c'}{(i_c)_{max}}\right].$$

Alternating square-wave current at one active electrode may be used to control the initiation and termination of this type of electrochemical polymerization. For example, with the cathodic part of the current cycle being held constant at the level i_c', the amplitude of the anodic current pulses should be varied with time according to the preceding Eq. (9). Alternately, the same average molecular weight may be obtained with an ac function whose anodic and cathodic current intensities are equal but of different durations.

V. EXPERIMENTAL RESULTS

A. Oxidation of Carboxylate Anions

One of the best known methods of generating free radicals electrochemically is the Kolbe electrolytic oxidation of carboxylic acids and their salts. The accepted reaction steps are

$$RCOO^- \rightarrow RCOO\cdot + e^- \quad \text{(Anode)}$$

$$RCOO\cdot \rightarrow R\cdot + CO_2$$

$$2R\cdot \rightarrow R_2.$$

Generally, low current densities favor formation of olefins, paraffins, and oxygen resulting from reactions of first order in $RCOO\cdot$ or $R\cdot$; whereas, coupled products and CO_2 appear at higher current densities. In the presence of vinyl monomers M, polymerization may be initiated by either or both of the free radical species, depending on the particular system and conditions:

$$RCOO\cdot + M \rightarrow RCOOM\cdot \xrightarrow{M} RCOOMM\cdot \xrightarrow{M} \text{Etc.}$$

$$R\cdot + M \rightarrow RM\cdot \xrightarrow{M} RMM\cdot \rightarrow \text{Etc.}$$

Systems containing MMA in DMSO and zinc acetate have been extensively studied [21]. The following reaction scheme is consistent with the results:

$$RCO_2^{\ominus} \underset{k_r}{\overset{k_c}{\rightleftharpoons}} RCO_2\cdot + e^- \quad \text{(Anode)} \qquad \text{(Rapid equilibrium)}$$

$$RCO_2\cdot \xrightarrow{k_d} R\cdot + CO_2$$

$$R\cdot + M \xrightarrow{k_p} M_{n+1}\cdot$$

$$M_n\cdot + R\cdot \xrightarrow{k_t} P.$$

This indicates that initiation in these systems is primarily by the secondary alkyl free radical, very little being caused by the primary carboxylate radical; the preponderant termination is via the secondary alkyl radical rather than via a second-order process involving the growing chains.

It was further established that polymer yield was linear with time and current density and that polymer molecular weight varied inversely to polymerization rate, and hence to current density.

B. Anionic and Cationic Intermediates

A thoroughly studied anionic ECP system is that of AN in DMF with $NaNO_3$ [4, 63, 64]. The following reaction scheme is consistent with the experimental result:

$M + \bar{e} \rightarrow M^{\dot{-}}$	Initiation	k_1
$M^{\dot{-}} + M^{\dot{-}} \rightarrow M_2^{2-}$	Dimerization	k_2
$M_n^{2-} + M \rightarrow M_{n+1}^{2-}$	Propagation of dianion	k_3
$M_n^{2-} + M \rightarrow M_n^- + M^-$	Chain transfer of dianion	k_4
$M_n^- + M \rightarrow M_{n+1}^-$	Propagation of anion	k_5
$M_n^- + M \rightarrow P_n$	Chain transfer of anion	k_6

From which is deduced:

$$R_p = k_3\,[\text{Dianion}]\,[M] + k_5(\text{Monoanion}]\,[M]$$

$$= k_3\,[\text{Ion site}]\,[M]$$

and the termination rate is

$$R_t = k_6\,[\text{Ion site}]\,[M]$$

and the degree of polymerization is

$$\overline{\text{D.P.}} = R_e/R_t = k_3/k_6 = \text{Constant.}$$

The experimental molecular weights were found to be independent of monomer concentration, polymerization rate, and current density, as predicted by the above analysis.

The rate of polymerization depended on the first power of monomer concentration, and the polymerization efficiency, expressed as moles of polymer per faraday of electrolysis current, was shown to decrease with increasing current density and monomer concentration.

A most interesting cationic ECP is that of vinyl ethers in nitrobenzene [4, 35] with various salts of perchlorate (ClO_4^-) or tetrafluoroborate (BF_4^-). In nitrobenzene the ECP of isobutylvinyl ether was rapid and exothermic.

C. Condensation Reactions

Condensation of dicarboxylic acids has been attempted only recently [33]. Related to this work is the ECP grafting of vinyl monomers onto polymers which contain carboxylic acid groups, such as polymethacrylic acid in methanol [30]. These condensation processes are based on the Kolbe anode reaction, as discussed previously, in which there are successive Kolbe couplings of free radicals and evolution of CO_2. These condensations are different from the common type in that high overall conversion of monomer is not required in order to form high molecular weight polymers. This is due to the absence of an equilibrium between the polymers and monomers, since the Kolbe coupling is essentially irreversible. Furthermore, since diffusion of the coupled products away from the electrode is slower the larger their molecular weights, each polymer will tend to grow many units before it is displaced by other smaller molecules by diffusion through the laminar sublayer next to the electrode surface; therefore, the molecular weight distribution of ECP condensed polymers may tend to be broader than homogeneous chemical condensation polymerizations.

There are some limitations on the chemical structure of the diacids used in condensation ECP's of this type, since no polymers would form from malonic, succinic, glutaric, or β,β-dimethyl glutaric acids, whereas polymers would form with adipic, azelaic, sebacic, and hexadecane-1,16-dicarboxylic acids [66].

There is some question of the ability of the condensation to proceed to indefinitely high degrees of polymerization due to: (1) insolubilization of the polymer in the medium because of its decreased polarity from loss of CO_2 and of branching and cross-linking; (2) coating of the anode by insulating by polymer which decreases the electrode efficiency and current-carrying capability (possibly alleviated by use of liquid electrodes such as the DME or other moving liquid metal electrodes).

VI. CONCLUSION

The numerous possible combinations of sequences of the elementary steps, considered in earlier sections, determine various ECP mechanisms, where complexity grows quite rapidly since alternate reaction paths may occur simultaneously or consecutively. These may be entangled even further by combination with purely physical transport effects through the liquid electrolyte, the electrical double layer, or at the electrode surfaces. Thus, in order to advance the understanding of ECP reactions, relatively simple systems must be studied and analyzed at first. This approach provides evidence for or against the occurrence or predominance of the numerous reactions which have been discussed qualitatively in this section.

ACKNOWLEDGMENT

I wish to thank Dr. D. G. Laurin for collecting the references cited in this chapter.

REFERENCES

[1]. C. L. Wilson, Rec. Chem. Progr., 10, 25 (1949).

[2]. A. P. Tomilov and M. Ya. Fioshin, Plast. Mass., 10, 2 (1960).

[3]. A. P. Tomilov and M. Ya. Fioshin, Russ. Chem. Rev., 32 (1), 30 (1963).

[4]. J. W. Breitenbach and Ch. Srna, Pure Appl. Chem., 4, 245 (1962).

[5]. C. L. Wilson, in Encyclopedia of Electrochemistry (C. A. Hampel, Ed.), Reinhold, New York, 1964, pp. 963-964.

[6]. H. Z. Friedlander, in Encyclopedia of Polymer Science and Technology (H. F. Mark, N. G. Gaylord, and N. M. Bikales, Eds.), Wiley-Interscience, New York, 1966, pp. 629-641.

[7]. B. L. Funt, in Macromolecular Reviews, (A. Peterlin, M. Goodman, S. Okamura, B. H. Zimm, and H. F. Mark, eds.), Vol. 1, Wiley-Interscience, New York, 1967, pp. 35-56; N. Yamazaki, Adv. Polym. Sci., 6, 377 (1969).

[8]. F. Laborie, Double Liaison, 148, 1489 (1967).

[9]. R. V. Figini, H. Hostalka, K. Hurm, G. Löhr, and G. V. Schulz, Z. Phys. Chem., 45, 269 (1965); H. Hostalka and G. V. Schulz, ibid., 286 (1965).

[10]. J. M. Kolthoff and I. L. Ferstandig, J. Polym. Sci., 6, 563 (1951).

[11]. S. Goldschmidt and E. Stocke, Ber., 85, 630 (1952).

[12]. N. S. Tsvetkov, Vysokomol. Soyed., 3, 549-54 (1961); Polym. Sci. USSR, 3, 570 (1962).

[13]. F. D. Williams, Ph. D. Thesis, Univ. of Manitoba, Winnipeg, Can., 1962; B. L. Funt and F. D. Williams, J. Polym. Sci., A2, 865 (1964).

[14]. R. Schwing, unpublished results, Univ. of Michigan, Ann Arbor, Mich., 1959-1960.

[15]. N. S. Tsvetkov and Z. F. Glotova, Vysokomol. Soyed., 5, 997 (1963); Polym. Sci. USSR, 5, 49 (1964).

[16]. A. I. Fedorva, Li Kno-Tung, and I. V. Shelepin, Zh. Fiz. Khim., 38, 920 (1964).

[17]. I. V. Shelepin and A. I. Fedorova, ibid., 2676 (1964).

[18]. K. Higasi, A. Rembaum, and H. Baba, Quantum Organic Chemistry, Wiley-Interscience, New York, 1965.

[19]. G. C. Bond, Discuss. Faraday Soc., No. 41, 20 (1966).

[20]. E. T. Kaiser and L. Keran (eds.), Radical Ions, Wiley-Interscience, New York, 1968.

[21]. J. Y. Yang, W. E. McEwens, and J. Kleinberg, J. Amer. Chem. Soc., 79, 5833 (1957).

[22]. D. G. Laurin, Ph. D. Thesis, Univ. of Michigan, Ann Arbor, Mich., 1967; Diss. Abstr., B28 (7), 2823 (1968); D. G. Laurin and G. Parravano, J. Polym. Sci., B4, 797 (1966); Part C, No. 22, 103 (1968).

[23]. B. L. Funt and K. C. Yu, ibid., 62, 359 (1962).

[24]. J. Smid and M. Szwarc, J. Amer. Chem. Soc., 8, 1253 (1961).

[25]. Kh. S. Bagdasar'yan and A. I. Nepomnyaschchii, Kinet. Katal., 4 (1), 60 (1963).

[26]. M. M. Baizer, J. D. Anderson, J. H. Wagenknecht, M. R. Ort, and J. P. Petrovich, Electrochim. Acta, 12, 1377 (1967); M. M. Baizer and J. D. Anderson, J. Org. Chem., 1351 (1965).

[27]. M. Szwarc, Z. Elektrochem., 67, 763 (1963); M. Szwarc, "Chemistry of Radical Ions," in Progress in Physical Organic Chemistry (A. Streitweiser and R. W. Taft, eds.), Vol. 6, Wiley-Interscience, New York, 1968, pp. 323-438.

[28]. N. L. Weinberg and H. R. Weinberg, Chem. Rev., 68, 449-532 (1968).

[29]. F. Beck and H. Leitner, Angew. Makromol. Chem., 2, 51 (1968).

[30]. G. Smets, X. van der Borght, and G. van der Haeren, J. Polym. Sci., A2, 5187 (1964).

[31]. M. M. Baizer, U.S. Pat. 3,413,202 (1968); J. D. Anderson and M. M. Baizer, Tetrahedron Lett., No. 5, 511 (1966); J. D. Anderson, M. M. Baizer, J. P. Petrovich, J. Org. Chem., 31, 3890, 3897 (1966).

[32]. J. A. Epstein and A. Bar-Nun, J. Polym. Sci., B2, 27 (1964).

[33]. W. E. Garrison, Ph. D. Thesis, Univ. of Illinois, Urbana, Ill., 1959; Diss. Abstr., 20, 77 (1959).

[34]. M. S. Toy, J. Electrochem. Soc., 114, 1042 (1967).

[35]. B. L. Funt, Ref. [7], p. 54.

[36]. F. H. Covitz, J. Amer. Chem. Soc., 89, 5403 (1967).

[37]. E. A. Rembold, Ph. D. Dissertation, Ohio State Univ., Columbus, Ohio, 1947, and references therein; W. F. H. Borman, U.S. Pat. 3, 335, 075, Aug. 1967.

[38]. D. S. McKinney and J. P. Fugassi, U.S. Pat. 2, 961, 384, Nov. 1960.

[39]. S. Ya. Vasina, G. F. Astakhova, R. F. Karysheva, and A. I. Fedorova, Elektrokhim., 4, 468 (1968).

[40]. A. Bartoli and G. Papasogli, Gazz. Chim. Ital., 14, 103 (1884); Ber., 17, 472R (1884).

[41]. Brit. Pat. 130, 608, July 1969.

[42]. A. F. Shepard and B. Dannels, U.S. Pat. 3, 386, 899, June 1968.

[43]. H. O. vom Orde, U.S. Pat. 3, 341, 436, Sept. 1967.

[44]. H. Gilch and D. Michael, Makromol. Chem., 99, 103 (1966); H. Gilch, U.S. Pat. 3, 399, 124, Aug. 1968.

[45]. Neth. Pat. 6, 603, 440, Sept. 1966.

[46]. R. C. Schultz and W. Strobel, Monatsh., 99, 1742 (1968).

[47]. R. Chen, D. G. Laurin, and G. Parravano, unpublished results, Univ. of Michigan, Ann Arbor, Mich., 1967.

[48]. H. Gehrke, Ger. Pat. 1, 014, 744 (1957).

[49]. D. Goerrig, H. Jonas, and W. Moschel, Ger. Pat. 935, 867, Dec. 1955; CA, 53, 3949 (1959).

[50]. D. Goerrig and H. Jonas, Ger. Pat. 937, 919, Jan. 1956; CA, 53, 3950 (1959).

[51]. J. W. Loveland, Can. Pat. 566, 274, Nov. 1958.

[52]. S. D. Ross and D. J. Keiley, J. Appl. Polym. Sci., 11, 1209 (1967).

[53]. G. S. Shapoval, E. M. Siobets, and N. P. Markova, Dokl. Akad. Nauk SSSR, 173 (2), 392 (1967).

[54]. S. D. Ross, R. C. Petersen, and M. Finkelstein, U.S. Pat. 3, 417, 003, Dec. 1968.

[55]. A. F. Shepard and B. F. Dannels, J. Polym. Sci., Part A-1, 4, 511 (1966).

[56]. P. Delahay, New Instrumental Methods in Electrochemistry, Wiley-Interscience, New York, 1954.

[57]. K. J. Vetter, Electrochemical Kinetics (trans. by Scripta Technica, Ind., S. Bruckenstein and B. Howard, eds.), Academic, New York, 1967.

[58]. Z. Galus and R. N. Adams, J. Electroanal. Chem., 4, 248-250 (1962).

[59]. W. L. Jolly and C. J. Hallada, in Nonaqueous Solvent Systems (T. C. Waddington, ed.), Academic, New York, 1965, pp. 1-41.

[60]. P. Delahay, Double Layer and Electrode Kinetics, Wiley-Interscience, New York, 1965, pp. 81-114.

[61]. B. E. Conway, Electrode Processes, Ronald, New York, 1965, pp. 75, 89, 91.

[62]. I. Sakurada and N. Ise, Chem. Eng. News, 44 (9), 39-40 (1966).

[63]. D. G. Laurin and G. Parravano, Paper presented at the 65th National A.I.Ch.E. Meeting, Cleveland, Ohio, 1969.

[64]. B. L. Funt, D. Richardson, and S. N. Bhadani, Can. J. Chem., 44, 711 (1966).

[65]. N. Yawazaki, I. Tanaka, and S. Nakahama, J. Macromol. Sci., Chem., A2, 1121 (1968).

[66]. W. E. Garrison, Ph. D. Thesis, Univ. of Illinois, Urbana, Ill., 1958, Diss. Abstr., 20, 77 (1959).

[67]. R. A. Wessling and W. J. Settineri, U.S. Pat. 3,480,525, Nov. 25, 1969.

[68]. T. Fueno, K. Asada, K. Morokuma, and J. Furukawa, J. Polym. Sci., 40, 511 (1959).

[69]. K. Höfelmann, R. Sattelmeyer, and K. Hamann, Makromol. Chem., 112, 300 (1968).

Chapter XXX

ORGANIC ELECTROCATALYSIS

Manuel M. Baizer

Corporate Research Department
Monsanto Company
St. Louis, Missouri 63166

Despite the wide variety of organic electrochemical phenomena and reactions which have been summarized in this book there remains a scarcely diminished opportunity for new research. The main guidelines have perhaps been established; fundamental new insights will come only from more basic understanding of organic electrode mechanisms. To be sure, at any given state of knowledge, the unexpected will occur with sufficient frequency to enchant the academic and industrial chemist working in this field. But for the synthetic chemist the greatest challenges, it would appear, exist at the borders where the field of organic electrochemistry abuts upon territories that are waiting to be annexed or at least allied: e.g., polymerization (Chapter XXVIII) and catalysis.

The term electrocatalysis may be a redundancy since in most electrochemical reactions [1] the electrode is merely an agent for electron transfer and itself remains chemically unchanged.

Organic electrocatalysis has come to be a polyvalent term. It may refer to phenomena associated with fuel cells wherein electrodes catalyze the combustion of organic fuels and facilitate the recovery of the energy as electrical energy. This area is outside the scope of this book. However, the suggestion has often been made (e.g., [1]) that "poor" fuel cell electrodes may be excellent catalysts for electrosyntheses starting with low molecular weight hydrocarbons.

Electrocatalysis may also refer to the electrochemical preparation of a species which in or ex situ catalyzes a given process such as polymerization [2] [2a] or carbonylation [2b] or oxidation [2c]. The term is also used in connection with electroregeneration of inorganic reagents which are capable of oxidizing or reducing an organic species in a controlled manner (Chapter

XXIV, p. 825). As has been pointed out, when the reagent is exotic (expensive), electrochemical maintenance of it in the proper valence state may have economic if not scientifically glamorous advantages.

Electrochemical hydrogenation of organic substrates using as cathodes metals (Pt, Pd, Ni, etc.) which in a proper state of subdivision are also heterogeneous hydrogenation catalysts has been studied in some detail [3], particularly by Russian workers. With the anticipated continuous decrease in the cost of electric power and with improvements in cell design, it is possible that many catalytic hydrogenations can be advantageously replaced by electrochemical hydrogenation. Storage of water at a plant site is more attractive than storage of hydrogen.

The last 15 years have seen a tremendous upsurge of activity in the field of transition metal chemistry: the preparation of new compounds; the sophisticated elucidation of structure of new and old compounds; and to a lesser extent the establishment of relationships between structure and reactivity. Many new homogeneous catalysts have been discovered [4] which are effective under mild conditions, e.g., hydrogenations with pentacyanocobaltate, carbonylations with Wilkinson's catalyst. Many of the complexes which are catalytic toward a given process become so only after a (chemical) oxidation or, more often, reduction step. Obviously, an electrochemical operation which can produce and maintain the desired oxidation state will often be advantageous (cf. p. 825). Many of the complexes as prepared have no known catalytic activity. Yet studies of the electrochemical behavior of transition metal complexes have been restricted largely to polarographic investigations [5] and to determinations of the formal valence state of the central metal, i.e., the relative affinity of the metal and the ligands for electrode-supplied electrons [6].

Unlimited opportunities await the investigator who is willing to work at the organic chemistry-electrochemistry, catalysis interface. He has at his disposal "clean" means for supplying or withdrawing electrons, techniques for incorporating potential catalysts into electrode matrices [15], the advantages of electric-field-dependent adsorption on electrodes, a choice of solvents and, when needed, supporting electrolytes with fairly well-defined effects upon the electrode processes and the intermediates in solution. The following literature citations may be suggestive: the partially catalytic electroreduction of a titanium-nitrogen complex to ammonia [7]; the fixation of nitrogen by an electrochemically produced Ru^{II} species [8]; the electrolytic formation of $C_5H_5Fe(CO)_2H$ [9]; the oxidation of Hg^{II}-olefin complexes [10]; the nucleophilic character of (nonelectrochemically produced) Co^{I} species [11] as well as the electrochemical formation of Co^{I} species [12]; the elegant studies by Kochi and co-workers [13] on organic reactions mediated by metal ions of "unusual" valence.

This brief discussion has purposely omitted a consideration of the utilization of electrochemical techniques to help elucidate the mechanisms of nonelectrochemical catalytic processes. Work of this nature has been of increasing assistance to catalyst chemists [1, 14].

NOTES

[1]Except, e.g., when the anode or cathode is purposely sacrificed to synthesize an organometallic compound.

[2]See Chapter XVIII, p. 636, for preparation in situ of a Ni^{o} species for trimerization of butadiene.

REFERENCES

[1]. S. Trasatti, Chim. Ind., 51, 1063 (1969), and literature cited therein.

[2]. (a) Monsanto Co., Belg. Pat. 696, 593, Oct. 4, 1967; M. R. Ort, E. H. Mottus, M. M. Baizer, and D. Carter, U.S. Pat. 3, 546, 083, Dec. 8, 1970; (b) T. Inoue and S. Tsutsumi, Bull. Chem. Soc. Japan, 38, 2122 (1965); (c) M. Savy, Compt. Rend., C1968, 863.

[3]. E.g., A. M. Sokolskaya and S. M. Reshnetikov, Vestn. Akad. Nauk Kazan, SSR, 20, 42 (1964); V. L. Kochetkov and I. V. Kudryashov, CA, 68, 95090v, 95091w (1968); M. E. Mazhelei and L. V. Voitenko, Zh. Fiz. Khim., 34, 27 (1960); I. V. Kudryashov and V. L. Kochetkov, Russ. J. Phys. Chem., 43, 728 (1969); G. A. Bogdanovskii, Elektrokhim., 5, 472 (1969); CA, 73, 65784k (1970); X. D. Hemptinne and K. Schunk, Trans. Faraday Soc., 65, 591 (1969); H. J. Davitt, Jr., Diss. Abstr., 30, 1653 (Int. B 1969); V. I. Gants, E. N. Smirnova, V. V. Sysoeva, and N. N. Storchak, J. Appl. Chem. USSR, 42, 2345 (1969).

[4]. J. P. Collman, Accounts Chem. Res., 1, 136 (1968).

[5]. R. E. Dessy and co-workers in a series of papers: I: J. Amer. Chem. Soc., 88, 453 (1966); XV: ibid., 91, 6800 (1969); A. A. Vlcek, Pure Appl. Chem., 10, 61 (1965); J. Masek, Talanta, 12, 1173 (1965); V. Gutmann and M. Michlmayr, Monatsh., 99, 310, 326 (1968); S. P. Gubin and S. A. Smirnova, J. Organometal. Chem., 20, 229 (1968); L. Denisovich, S. P. Gubin, and Yu. A. Chapovskii, Bull. Acad. Sci. USSR, 1967, 2271.

[6]. J. Manassen and A. Bar-Ilan, J. Catal., 17, 86 (1970); A. Wolbert and J. Manassen, J. Amer. Chem. Soc., 92, 2982 (1970); N. E. Tokel, V. Katovič, K. Farmery, L. B. Anderson, and D. H. Busch, ibid., 92, 400 (1970); R. J. Wilson, L. F. Warren, Jr., and M. F. Hawthorne, ibid., 91, 758 (1969); L. D. Rollmann and R. T. Iwamoto, ibid., 90, 1455 (1968); B. M. Higson and E. D. McKenzie, Inorg. Nucl. Chem. Lett., 6, 209 (1970); C. Elschenbroich and M. Cais, J. Organometal. Chem., 18, 135 (1969); P. K. Aggarwal, W. R. W. Bradman, and D. R. Crow, J. Polarogr. Soc., 14, 93 (1968).

[7]. E. E. van Tamelen and D. A. Seeley, J. Amer. Chem. Soc., 91, 5194 (1969); E. E. van Tamelen, Accounts Chem. Res., 3, 361 (1970).

[8]. I. J. Itzkovitch and J. A. Page, Can. J. Chem., 46, 2743 (1968).

[9]. D. Gresova and A. A. Vlcek, Inorg. Chim. Acta, 1, 482 (1967).

[10]. M. Fleischmann, D. Pletcher, and G. M. Race, J. Electroanal. Interfacial Electrochem., 23, 369 (1969).

[11]. G. N. Schrauzer, Accounts Chem. Res., 1, 97 (1968); F. R. Jensen, V. Madan, and D. H. Buchanan, J. Amer. Chem. Soc., 92, 1414 (1970).

[12]. J. Hanzlik and A. A. Vlček, Chem. Commun., 1969, 47.

[13]. E.g., J. K. Kochi and D. M. Singleton, J. Amer. Chem. Soc., 90, 1582 (1968).

[14]. T. Freund and W. P. Gomes, Catal. Rev., 3, 1 (1969); V. I. Gants, E. N. Smirnova, V. V. Sysoeva, and N. N. Storchak, J. Appl. Chem. USSR, 42, 2345 (1969); K. Le Tran and J. Brenet, Compt. Rend., C1967, 1517.

[15]. Siemens A. G., Fr. Pat. 1, 591, 825, June 12, 1970; CA, 74, 37773k (1971).

Appendix I

CLASSIFIED BIBLIOGRAPHY

G. Cauquis and D. Serve

Centre d'Études Nucléaires de Grenoble
Grenoble, France

I. BOOKS

A. Theoretical Electrochemistry and Various Electrochemical Techniques

[1]. K. J. Vetter, Electrochemical Kinetics, Academic, New York, 1967.

[2]. B. E. Conway, Theory and Principles of Electrode Processes, Ronald, New York, 1965.

[3]. P. Delahay, Double Layer and Electrode Kinetics, Wiley-Interscience, New York, 1965.

[4]. R. N. Adams, Electrochemistry at Solid Electrodes, Dekker, New York, 1969.

[5]. G. Charlot, J. Badoz-Lambling, and B. Tremillon, Electrochemical Reactions, Elsevier, Amsterdam, 1962.

[6]. I. M. Kolthoff and J. J. Lingane, Polarography, Wiley-Interscience, New York, 1952. (Two volumes.)

[7]. L. Meites, Polarographic Techniques, Wiley-Interscience, New York, 1965.

[8]. J. Heyrovsky and J. Kuta, Principles of Polarography, Academic, New York, 1966.

[9]. S. G. Mairanovskii, Catalytic and Kinetic Waves in Polarography, Plenum, New York, 1968.

[10]. J. O'M. Bockris and A. K. N. Reddy, Modern Electrochemistry, Plenum, New York, 1970. (Two volumes.)

[11]. A. T. Kuhn (ed.), Industrial Electrochemical Processes, Elsevier, Amsterdam, 1971.

B. Organic Electrochemistry

[12]. C. J. Brockman, Electroorganic Chemistry, Wiley, New York, 1926.

[13]. F. Fichter, Organische Elektrochemie, Steinkopf, Dresden and Leipzig, 1942.

[14]. M. J. Allen, Organic Electrode Processes, Chapman & Hall, London, 1958.

[15]. P. Zuman, Organic Polarographic Analysis, Pergamon, New York, 1964.

[16]. P. Zuman, Substituent Effects in Organic Polarography, Plenum, New York, 1967.

[17]. P. Zuman, The Elucidation of Organic Electrode Processes, Academic, New York, 1969.

[18]. A. D. Tomilov, S. G. Mairanovski, M. Ya. Fioshin, and V. A. Smirnov, Electrochemistry of Organic Compounds (published in Russian), Khimia, Leningrad, 1968.

[19]. C. K. Mann and K. K. Barnes, Electrochemical Reactions in Non-aqueous Systems, Dekker, New York, 1970.

II. ADVANCES AND ANNUAL SURVEYS

[20]. J. O'M. Bockris (ed.), Modern Aspects of Electrochemistry, Vols. 1-5, Academic, New York, 1954 on.

[21]. P. Delahay and C. W. Tobias (eds.), Advances in Electrochemistry and Electrochemical Engineering, Vols. 1-7, Wiley-Interscience, New York, 1961 on.

[22]. A. J. Bard (ed.), Electroanalytical Chemistry, Vols. 1-4, Dekker, New York, 1966 on.

[23]. P. Zuman and I. M. Kolthoff (eds.), Progress in Polarography, Wiley-Interscience, New York, 1962. (Two volumes.)

[24]. Annual Reviews of Physical Chemistry, Annual Reviews, Palo Alto, Calif., 1950 on.

[25]. Annual Reports of the Progress of Chemistry, The Chemical Society, London, 1903 on.

[26]. Annual Reviews, Fundamental Analytical Chemistry, The American Chemical Society, Washington, D.C., section on "Organic Polarography since 1949." This is part of Analytical Chemistry [39].

[27]. A. N. Frumkin and A. B. Ershler (eds.), Progress in the Electrochemistry of Organic Compounds (in Russian), Nauka Moscow, 1969; Specialist Periodical Report, Vol. 1, Electrochemistry, The Chemical Society, London, 1970.

Comments

Modern Aspects of Electrochemistry and Advances in Electrochemistry and Electrochemical Engineering are two series of interest to electrochemists. These volumes include authoritative reviews on four or five topics per volume with the main emphasis on theoretical electrochemistry. Electroanalytical Chemistry is a series of advances designed to provide reviews in the field of modern electroanalytical chemistry defined in its broadest sense.

Progress in Polarography is a two-volume book describing the progress made in polarography mainly in the years 1950-1960. The entire field is covered in chapters by numerous authors. For a broad survey of recent work, also of value are the sections on electrode processes appearing irregularly in the Annual Reviews of Physical Chemistry and the section on electrochemistry in the Annuals Reports of the Progress of Chemistry. In the latter a section devoted to electroorganic chemistry appeared for the first time in 1969.

Extensive biennial bibliographies of organic polarography (with coverage of cathodic and anodic processes) initiated by Wawzonek and continued by Pietrzyk are excellent compendia of research in the field. The 1968 issue contains no less than 1475 references. In the same volume of Analytical Chemistry some other reviews are of special interest for the chemist interested in electroorganic reactions: "Electroanalysis and Coulometric Analysis"; "Polarographic Theory, Instrumentation, and Methodology"; "Titrations in Nonaqueous Solvents."

In the Specialist Periodical Report (item [27] above) A. Bewick and D. Pletcher, pp. 98-165, present a review of the synthetic aspects of organic electrochemistry 1968-1969.

III. JOURNALS

[30]. Journal of the Electrochemical Society, Inc., New York.

[31]. Journal of the Electrochemical Society of Japan (Overseas Ed.), The Electrochemical Society, Tokyo, Japan.

[32]. Electrochimica Acta, Pergamon, Oxford, England. (Monthly.)

[33]. Journal of Electroanalytical Chemistry, Elsevier, Amsterdam. (Monthly since 1960.)

[34]. Elektrokhimiya (Soviet Electrochemistry), Consultants Bureau, New York. (Monthly since 1965.)

[35]. Journal of the Polarographic Society, The Polarographic Society, London.

[36]. Journal of the Electrochemical Society of India, Indian Institute of Science, Bangalore 12, India.

[37]. Review of Polarography, The Polarographic Society of Japan, Kyoto, Japan. (Published since 1953.)

[38]. Journal of Physical Chemistry, The American Chemical Society, Washington, D.C.

[39]. Analytical Chemistry, The American Chemical Society, Washington, D.C.

[40]. Analytica Chimica Acta, Elsevier, Amsterdam. (Published monthly.)

[41]. Transactions of the Faraday Society, Aberdeen University, Aberdeen, Scotland.

[42]. Collection of Czechoslovak Chemical Communications, Prague. (Monthly.)

[43]. Journal of Applied Electrochemistry, Chapman and Hall, London, England (Quarterly), 1971-.

Comments

The number of journals published which are primarily devoted to electrochemistry are few. Since the organic chemist is supposedly not familiar with them, this list is included. His common journals contain papers dealing with electroorganic studies leading to isolated products more than with extensive mechanistic studies.

The Journal of the Electrochemical Society of Japan (Overseas ed.) has included original manuscripts in English. Electrochimica Acta was first issued in 1959 as an outgrowth of the publication of papers presented at the annual conference of the International Committee for Electrochemical Thermodynamics and Kinetics (C.I.T.C.E.). It is a journal with articles appearing in English, French, or German.

Journal of Electroanalytical Chemistry and Interfacial Electrochemistry is an international journal devoted to all aspects of electroanalytical chemistry, double layer studies, electrokinetics, colloid stability, and electrode kinetics. Published monthly in English it is a valuable source with original papers, short communications, and reviews. A new Russian journal, Elektrokhimiya, began publication in 1965 and contains the greatest part of work devoted to electrochemistry in this country. Its English translation, Soviet Electrochemistry, is available from the Consultants Bureau. The Journal of the Polarographic Society has been published by the Polarographic Society in Great Britain since 1955. Other journals which publish articles of value are listed equally ([39]-[42]).

IV. PROCEEDINGS OF CONGRESS AND SPECIAL ISSUES OF JOURNALS

[50]. Advances in Polarography (I. S. Longmuir, ed.), Vols. 1-3, Pergamon, New York, 1960.

[51]. Polarography 1964 (G. J. Hills, ed.), Vols. 1-2, Macmillan, New York, 1966.

[52]. Symposium on the Synthetic and Mechanistic Aspects of Electroorganic Chemistry (G. R. Ghirardelli, ed.), U.S. Army Research Office, Durham, N.C., Oct. 1968.

[53]. Talanta, Dec. 1965, Vol. 12, No. 12, Heyrovsky Honor Issue.

[54]. Modern Aspects of Polarography (a tribute to Isamu Tachi, J. Kambara, ed.), Plenum, New York, 1966. (Reprint of Vol. 11, No. 1-6, of Review of Polarography, 32 papers on various topics.)

[55]. Fresenius' Zeitschrift für Analytische Chemie, 224, No. 1 (1967).

[56]. Discussions of the Faraday Society, No. 45 (1968), "Electrode Reactions of Organic Compounds," The Faraday Society, London.

[57]. Chemie Ingenieur Technik, 42 (1970). (No. 4 contains all the papers presented in Bonn, Oct. 1969, at a congress on electroorganic chemistry.)

Comments

Advances in Polarography is the proceedings of the Second International Congress on Polarography held at Cambridge (1959). The proceedings of the third congress have appeared under the title Polarography 1964. No proceedings have been published for the fourth congress held in Prague (1966).

We have included a list of special issues of journals which have been devoted to electroorganic or electrochemistry in a broader coverage and contain sometimes papers or reviews of value.

Reference [53] contains 18 reviews by representative polarographers. Applications of polarography to various groups of inorganic and organic species are discussed; special attention is paid to the application of organic solvents.

Reference [55] is a special issue for the 70th birthday of Professor Von Stackelberg. It contains 39 original papers with about a dozen devoted to electroorganic chemistry.

Reference [56]: reports with discussions of papers presented at a meeting held in Newcastle, April 1968. (One can find the progress made recently in the following areas: primarily theoretical, adsorption, cathodic reactions, stereochemistry in electrochemistry, anodic reactions.)

In order to be acquainted with very recent work, one can look to papers presented at meetings of the Electrochemical Society of U.S.A. and of the C.I.T.C.E.[1] These two organizations have divisions devoted to electroorganic chemistry. The abstracts of papers presented at national meetings of the Electrochemical Society are available in two forms: as published in the "journal" (about 75 words) and as extended abstracts (500-1000 words) available from the Society's divisions. Papers presented at C.I.T.C.E. meetings are available as extended abstracts.

V. REVIEWS

In this classification by subjects we are concerned with the main reviews, already mentioned for the most part in various chapters of this book. It is hoped that this selected compilation will provide the reader with a large body of information.

A. Media Available for Electroorganic Reactions

[60]. C. K. Mann, in Ref. [22], Vol. 3 (1969), pp. 57-134. (Numerous references on all organic solvents used as media for electrochemical studies.)

[61]. J. J. Lagowski (ed.), The Chemistry of Nonaqueous Solvents, Academic, New York: Vol. I, Principles and Techniques, 1966; Vol. II, Acidic and Basic Solvents, 1967; Vol. III, Inert, Aprotic, and Acidic Solvents, 1970.

[62]. J. F. Coetzee and C. D. Ritchie, (eds.), Solute-Solvent Interactions, Dekker, New York, 1969. (Eight reviews selected in order to acquaint the reader with the general advances that have been made in understanding solvent effects.)

[63]. D. S. Reid and C. A. Vincent, J. Electroanal. Chem., 18, 427 (1968), "Electrochemistry in the Amides."

[64]. J. N. Butler, J. Electroanal. Chem., 13, 89 (1967), "Electrochemistry in Dimethyl Sulfoxide" (a review with 101 refs.).

[65]. J. Badoz-Lambling and G. Cauquis, in Advances in Analytical Chemistry and Instrumentation (H. Nürnberg, ed.), Wiley-Interscience, New York, in press.

[66]. J. N. Butler, in Ref. [21], Vol. 7, 1970, p. 77, "Reference Electrodes in Aprotic Organic Solvents" (a review with 458 refs.).

B. Mechanistic Studies

[70]. P. J. Elving and B. Pullman, Advances in Chemical Physics (I. Prigogine, ed.), Vol. III, Wiley-Interscience, New York, 1961, p. 1.

[71]. G. Le Guillanton, Bull. Soc. Chim. Fr., 2359 (1963). (Review with 248 refs.)

[72]. C. L. Perrin, Progress in Physical Organic Chemistry (S. Cohen, A. Streitwieser, and R. W. Taft, eds.), Vol. 3, Wiley-Interscience, New York, 1965, pp. 165-316. (Review with 465 refs.)

[73]. P. Zuman, J. Polarogr. Soc., 13, 53 (1967). (Preparative methods used in the elucidation of organic processes.)

[74]. P. Zuman, Advances in Physical Organic Chemistry (V. Gold, ed.), Vol. 5, Academic, New York, 1967, p. 1. (Polarography and reaction kinetics.)

[75]. L. Meites, Pure Appl. Chem., 18, 35 (1969).

C. Reactions Intermediates — Radicals

[80]. A. P. Tomilov and M. Ya. Fioshin, Russ. Chem. Rev., 32, 30 (1963), "Free Radical Reactions in the Electrolysis of Organic Compounds" (review with 222 refs.).

[81]. R. N. Adams, J. Electroanal. Chem., 8, 151 (1964), "E.P.R. in Electrochemistry" (43 refs.).

[82]. O. R. Brown and J. A. Harrison, ibid., 21, 387 (1969), "Reactions of Cathodically Generated Radicals and Anions" (review with 146 refs.).

[83]. B. Kastening, in Advances in Analytical Chemistry and Instrumentation (H. W. Nurnberg, ed.), Wiley-Interscience, New York, in press.

[84]. M. Fleischmann and D. Pletcher, Royal Inst. Chem. Rev., 2, 87 (1969). (135 refs.)

D. Cathodic Reactions

[90]. F. D. Popp and H. P. Schultz, Chem. Rev., 62, 19 (1962). (Review concerned mainly with preparative electrolysis, 386 refs.)

[91]. J. Volke, in Physical Methods in Heterocyclic Chemistry (A. R. Katritzky, ed.), Vol. I, Academic, New York, 1963, p. 217. (307 refs.)

[92]. M. E. Peover, in Ref. [22], Vol. 2, 1967, p. 1. (Review on aromatic hydrocarbons and related compounds, 112 refs.)

[93]. M. M. Baizer and J. P. Petrovich, "Electrolytic Reductive Coupling," in Progress in Physical Organic Chemistry (A. Streitwieser and R. W. Taft, eds.), Vol. 7, Wiley-Interscience, New York, 1970, pp. 189-227.

[94]. See also the reviews by P. Zuman, H. Lund, J. Volke, J. Tirouflet and E. Laviron, S. Wawzonek, S. G. Mairanovskii, in Ref. [54].

[95]. G. H. Joytink, in Ref. [21], Vol. 7, 1970, p. 221, "Electrochemical Reduction of Aromatic Hydrocarbons."

E. Anodic Reactions

[100]. A. P. Tomilov, Russ. Chem. Rev., 30, 639 (1961). "Anodic Substitution Reactions" (203 refs., including many patents).

[101]. K. Sasaki and W. J. Newby, J. Electroanal. Chem., 20, 137 (1969) "Anodic Reactions of Aromatic Compounds" (critical review with 128 refs.).

[102]. N. L. Weinberg and H. R. Weinberg, Chem. Rev., 68, 449 (1968), "Electrochemical Oxidation of Organic Compounds" (excellent review with 896 refs. which succeeds in bringing Fichter's book up to date).

[103]. B. C. L. Weedon, in Advances in Organic Chemistry (R. A. Raphael, E. C. Taylor, and A. Wynberg, eds.), Vol. 1, Wiley-Interscience, New York, 1960, p. 1; and Quart. Rev. (London), 6, 300 (1952).

[104]. L. Eberson, in Chemistry of the Carboxyl Group (S. Patai, ed.), Wiley-Interscience, New York, 1969, p. 53. (132 refs.)

[105]. A. K. Vijh and B. E. Conway, Chem. Rev., 67, 623 (1967), "Electrode Kinetic Aspects of the Kolbe Reaction" (177 refs.).

[106]. B. E. Conway, Rev. Pure Appl. Chem. (Aust.), 18, 105 (1968), "Mechanistic Aspects of Anodic Oxidation" (80 refs.); see also Chap. 10 cf. [4].

F. Electrochemical Synthesis

[110]. S. Wawzonek, Science, 155, 39 (1967) "Synthetic Electroorganic Chemistry (95 refs.).

[111]. H. Hoelemann, Chem. Inorg. Tech., 35, 340 (1963).

[112]. J. Newman, Ind. Eng. Chem., 60, 12 (1968).

[113]. A. P. Tomilov and M. Ya. Fioshin, Khim. Prom., 44, 84 (1968), "Industrial Electrolytic Cells for Synthesizing Organic Compounds."

[114]. C. L. Mantell, Electroorganic Chemical Processing, Noyes Development Corp., New Jersey, 1968.

[115]. A. P. Tomilov and M. Ya. Fioshin, Khim. Prom., 43, 243 (1967).

[116]. J. D. Anderson, J. P. Petrovich, and M. M. Baizer, in Advances in Organic Chemistry, Vol. 6, Wiley-Interscience, New York, 1969, p. 257, "Electrochemical Preparation of Cyclic Compounds" (64 refs.).

[117]. M. M. Baizer, Naturwiss., 56, 405 (1969).

[118]. L. Eberson and H. Schäfer, "Organic Electrochemistry," in Fortschritte der Chemischen Forschung, No. 21, Springer Verlag, Berlin, Heidelberg, New York, 1971.

VI. ABSTRACTING SERVICES AND BIBLIOGRAPHIES

Chemical Abstracts includes articles and patents on electrochemistry in its Sec. 77. Some papers devoted to reactions of organic compounds which are abstracted in other sections are indexed at the end of Sec. 77. The Russian abstract journal Referativnyi Zhurnal has an electrochemistry section for articles and an electrochemical technology section where patents are cited.

Current Chemical Papers, while not an abstract journal, contains a section on electrochemistry which lists current journal articles. The Journal of the Electrochemical Society of Japan (Overseas Ed.), Ref. [31], contains extensive abstracts in English of papers published in Japanese in Denki Kagaku.

Electroanalytical Abstracts (G. Milazzo, ed.), Birkhaüser, Basel, Switzerland, has appeared bimonthly since 1963. It is an international journal dealing with the documentation of all aspects of electroanalytical chemistry, including fundamental electrochemistry. The abstracted papers are divided into classes following the methods of investigation (e.g., voltammetry, polarography, coulometry, electrolyses, etc.)

Swann has published under the title "Bibliography of Electroorganic Reactions" in Chemical Abstracts a compilation of the work related to this topic between 1961, CA Vol. 55, and 1964, Vol. 60.

S. Swann, Electrochem. Tech., 1, 308 (1963); 5, 53, 101, 393, 549 (1967); 6, 59 (1968).

The same author has also listed the "British Patents on Electroorganic Chemistry to 1960" in Electrochem. Tech., 5, 467 (1967).

For biennial reviews on organic polarography see also Ref. [26].

Bibliografica Polarografica is an alphabetical list of the polarographic papers (polarographic is taken in a broad sense, oxidations on platinum electrode are included) published each year. Since 1963 (No. 16), the list of papers, the author, and subject indexes in English have appeared in a single issue. The compilation, performed by the Research Center for Polarography in Padua, is published by the National Research Council of Italy, Roma.

A Bibliography of Polarographic Literature 1922-1967 has been published by the Sargent-Welch Scientific Co., Skokie, Illinois, 1969.

A. Compilations of Half-Wave Potentials

[130]. K. Schwabe, Polarographie und Chemische Konstitution Organischer Verbindungen, Akademie Verlag, Berlin, 1957. (List of $E_{1/2}$'s of organic compounds on Hg electrode in aqueous media.)

[131]. W. M. Clark, Oxidation Reduction Potentials of Organic Systems, Williams & Wilkins, Baltimore, 1960. (Mainly compounds of biological interest, E^{o} determined by potentiometry.)

[132]. See Also Appendix 3 in Ref. [91]. (Numerous polarographic half-wave potentials of heterocyclic compounds.)

[133]. The first compilation of anodic half-wave potentials of organic compounds in the solvents acetonitrile and acetic acid is found in [102].

B. Polarographic Literature-Data Cards

The Polarographic Society of Japan publishes Polarographic Literature-Data Cards with the plan to compile abstracts of all papers dealing with polarography and voltammetry. Each paper is represented by one punched card. Each card has holes marked to be punched by the user. If these holes are cut, a survey of the polarographic literature can be made very conveniently according to a scheme of classification with many entries.

NOTE

[1] Now the International Society for Electrochemistry.

Appendix II

SPECIALIZED EQUIPMENT

P. Iversen

Department of Organic Chemistry
University of Aarhus
Aarhus, Denmark

The following compilation has been made possible only through the kind cooperation of the firms involved. For space reasons, only a limited (but hopefully representative) selection of the current production programs could be included. The approximate prices (in US$) apply at the manufacturing location. By means of the addresses listed after the table, one should be able to get information on the nearest representative and the price of the apparatus delivered. Some of the models have a built-in current integrator; if not, most of the firms can offer that separately. For each instrument there is given the maximum output voltage, current, and power, the approximate response (or rise) time. It must, however, be remembered that the latter parameter normally will depend strongly on the cell attached to the potentiostat, so that the stated values are only meant as a rough guide.

Commercial Potentiostats

Name	Address [a]	Type	Max. output (V)	Max. output (A)	Max. output (W)	Sensitivity (mV)	Response time	Approx. price ex works (US$)
Amel	1	550	30	3	90	<5	—	774
		555	30	10	300	<5	5×10^{-6} sec	1325
		557	100	0.3	30	<5	$<10^{-5}$ sec	1160
Anotrol	2	4700M	20	10	100	0.5	$<2 \times 10^{-6}$ sec	2400
Chemical Electronics	3	TR40/3A	40	3	120	<1	$<10^{-5}$ sec	960
		100A	20	100	2000	<1	$<10^{-4}$ sec	3300
		150V/1.6A	150	1.6	240	<1	$<10^{-5}$ sec	1900

Name	Address [a]	Type	Max. output			Sensitivity	Response time	Approx. price ex works
			(V)	(A)	(W)	(mV)		(US$)
Duffers	4	600	4	2	8	<1	$<4 \times 10^{-6}$ sec	495
Electrolab	5	4-3000	4	1	4	<4	10^{-4} sec	3990
Elscint	6	CHP-1	30	0.5	15	<5	$<10^{-4}$ sec	585
		CHP-2	30	2	60	<5	$<10^{-4}$ sec	1600
Heinzinger	7	HNs	600	1	600	—	$<2 \times 10^{-1}$ sec	2450
		TNs	10	50	500	—	$<5 \times 10^{-2}$ sec	2165
Jaissle	8	1000 T-A	20	1.5	30	—	$5V/10^{-5}$ sec	770
		100 T-B	100	0.1	10	—	$5V/10^{-6}$ sec	1025
		10000 T	10	10	100	—	$1V/10^{-6}$ sec	1110
Juul	9	734	100	3	300	<10	$<10^{-2}$ sec	840
		735	320	1	320	<10	$<10^{-2}$ sec	945
		834	100	10	1000	<10	$<10^{-2}$ sec	1200
MPI	10	MP-1026	10	5	50	<0.5	10^{-6} sec/V	470
PAR	11	173	100	1	100	<1	$<2 \times 10^{-6}$ sec	2310
Tacussel	12	PRT 20-2A	20	2	25	<0.5	$<3 \times 10^{-6}$ sec	905
		PIT 20-6X	20	6	90	<0.5	$<1.5 \times 10^{-5}$ sec	3870
		ASA 4 SHT	150	0.6	45	<5	$<5 \times 10^{-3}$ sec	1035
		ASA 50-10	100	10	500	<5	$<5 \times 10^{-3}$ sec	3510
TRW	13	200 A	20	5	100	—	2×10^{-6} sec/V	2530
Wenking	14	70HC3	40	3	75	<1	$3 < 10^{-6}$ sec	1195
		70HP10	30	10	200	<1	10^{-5} sec	1495
		68FR0.5	100	0.1	25	<1	10^{-6} sec	1040
Witton	15	Mark 4S	100	0.5	50	5	$<10^{-6}$ sec	1275
		T4	15	10	150	2	$<10^{-5}$ sec	1925
		T7	20	0.75	10	<1	$<5 \times 10^{-6}$ sec	575
Yanagimoto	16	V8-2010	20	1	20	0.5	$<5 \times 10^{-5}$ sec	2260
		V8-2030	20	3	60	0.5	$<5 \times 10^{-5}$ sec	2580
		V8-2010L	10	20	200	0.5	$<5 \times 10^{-5}$ sec	2580

1 APPARECHIATURE DI MISURA ELETTRONICHE
Via Bolzano, 30
I-20127 MILANO
Italy

2 MAGNA CORPORATION
11808 S. Bloomfield Ave.
SANTA FE SPRINGS,
California 90670
USA

3 CHEMICAL ELECTRONICS CO.
Durham Road, Birtley
Co. DURHAM
England

4 DUFFERS ASSOCIATES INC.
R.D.5, Box 85
TROY, New York 12180
USA

5 NATIONAL INSTRUMENT LABORATORIES, INC.
12300 Parklawn Drive
ROCKVILLE, Maryland 20852
USA

6 ELSCINT INC.
Building 812,
Raritan Center
P.O. Box 978
EDISON, New Jersey 08817
USA

7 HEINZINGER REGEL-UND MESSTECHNIK
D-8200 ROSENHEIM
Happinger Strasse 71
West-Germany

8 JAISSLE ELEKTRONIK-LABOR
D-7051 NEUSTADT
Spitalhaldenweg 4
West-Germany

9 JUUL ELECTRONIC
Tvedvangen 297
DK-2730 HERLEV
Denmark

10 McKEE-PEDERSEN INSTRUMENTS
P.O. Box 322
DANVILLE, California 94526
USA

11 PRINCETON APPLIED RESEARCH CORP.
P.O. Box 565
PRINCETON, New Jersey 08540
USA

12 TACUSSEL ELECTRONIQUE SOLEA
72 a 78, rue d'Alsace
69-VILLEURBANNE
France

13 TRW INSTRUMENTS
139 Illinois Street
EL SEGUNDO, California 92245
USA

14 GERHARD BANK ELEKTRONIK
D-34 GÖTTINGEN
Werner von Siemensstr. 3
West-Germany

15 WITTON ELECTRONICS LTD.
Tame Road, Witton
BIRMINGHAM 6 7HT
England

16 YANAGIMOTO Mfg. CO. LTD.
28 Joshungamae-cho
SHIMOTOBA, Fushimi-Ku
KYOTO
Japan

AUTHOR INDEX

Numbers in brackets are reference numbers and indicate that an author's work is referred to although his name is not cited in the text. Underlined numbers give the page on which the complete reference is listed.

A

C

G

I

M

O

P

Q

R

T

SUBJECT INDEX

A

B

C

E

N

P

Q